# Unit Operations and Unit Processes

## Including Computer Programs

**VOLUME 2**

Simultaneous Heat and Mass Transfer

Extraction and Adsorption

Membrane-Based Processes

Reaction Kinetics and Reactor Design

# Unit Operations and Unit Processes

**Volume 1**
**Volume 2**
and

This consolidated but versatile textbook provides detailed coverage of all UNIT OPERATIONS and UNIT PROCESSES employed in process industries in two volumes. An exhaustive coverage of each operation/process has been presented, starting from the fundamentals to the most recent advances. The treatment is fully objective and application-oriented with emphasis on simulation, design and analysis of each operation/process. Lengthy description of equipment has been avoided. The book blends UNIT OPERATIONS with UNIT PROCESSES since apart from physical processes (unit operations), chemical processes (reaction kinetics and reactor design) are also covered in full.

This volume covers **SIMULTANEOUS HEAT AND MASS TRANSFER, EXTRACTION AND ADSORPTION, MEMBRANE-BASED PROCESSES,** and **REACTION KINETICS AND REACTOR DESIGN.**

An excellent number of solved, illustrative examples included in each chapter. An objective summary of each chapter is given at the end under the heading "THINGS TO REMEMBER". Objective questions are also listed for the reader to ponder on and to verify himself how far he has assimilated the principles discussed in the preceding text. Sample COMPUTER PROGRAMS prepared on a user-friendly format presented at the end of each chapter.

The book provides complete study material to students of **chemical, mechanical, metallurgical** and **biochemical engineering** and related areas. It would also be a veritable text and reference for teachers, practising engineers, professional consultants and research scientists.

***Contents***

# Unit Operations and Unit Processes

## Including Computer Programs

**VOLUME 2**

**C.M. Narayanan**
Professor, Department of Chemical Engineering
National Institute of Technology
Durgapur-713209, India

**B.C. Bhattacharya**
Ex-Professor, Department of Chemical Engineering
Indian Institute of Technology
Kharagpur-721302, India

CBS

**CBS PUBLISHERS & DISTRIBUTORS**

New Delhi • Bangalore

**Unit Operations and Unit Processes**

**Including Computer Programs** **Vol 2**



First Edition 2007

ISBN Cardcover: 81-239-1487-3
Hardcover: 978-81-239-1505-0



*Production Director :* Vinod K. Jain

*Published by :*
Satish Kumar Jain for CBS Publishers & Distributors,
4596/1-A, 11 Darya Ganj, New Delhi - 110 002 (India)
E-mail : cbspubs@vsnl.com
Website : www.cbspd.com

*Branch Office :*
Seema House, 2975, 17th Cross, K.R. Road,
Bansankari 2nd Stage, Bangalore - 560070
Fax : 080-6771680 • E-mail : cbsbng@vsnl.net

***Typeset at CBS P&D Typesetting unit and printed at***
India Binding House, Noida, UP (India)

*to*

**Young Students of Today**

**World Citizens of Tomorrow**

*The Authors*

**C.M. Narayanan** received his BSc Engg from Banaras Hindu University, ME from Indian Institute of Sciences, Bangalore, and PhD from Indian Institute of Technology, Kharagpur. He has published over 57 research papers in journals of international repute and is the author of two well-known books — *Computer Aided Design of Chemical Process Equipment* and *Mechanical Operations for Chemical Engineers.* His fields of interest are transport phenomenon, computer-aided design, energy engineering, mechanical operations, biotechnology and biochemical engineering. He is an honorary software consultant to a large number of chemical industries and is regarded as an established expert on computer aided design and software development.

Prof. Narayanan has coordinated several national workshops on CAD and has convened international/national seminars and symposia. He has been to many countries abroad as a visiting scientist and is a life member of many professional societies such as Institution of Engineers (India), Indian Institute of Chemical Engineers, Indian Society for Technical Education and Indian Society for Heat and Mass Transfer. He has been conferred upon *Bharat Jyoti Award* for his outstanding contribution to research and development, particularly in the areas of computer aided design and software development. His name has been included in the international *Who Is Who* published by the American Biographical Society, USA.

**B.C. Bhattacharya**, ex-Head of Department of Chemical Engineering, Indian Institute of Technology, Kharagpur, is known as "father of biotechnology" at IIT–Kharagpur for establishing Biotechnology Unit for teaching and research, and also a Regional Centre for Biogas Development and Training. He has published over 100 research and design papers in chemical engineering and biotechnology, four well-known textbooks in chemical engineering and a monograph on biogas technology.

Prof. Bhattacharya visited Germany as DAAD (German Academic Exchange Service) Fellow and France, UK and USA as a Government of India delegate. He guided a large number of PhD scholars in chemical engineering and biotechnology and possesses a number of patents on the new products and equipment developed, and transferred a few of such technologies to the industries. He is on the board of directors of a few chemical and biochemical industries, and has also established a few SSI units. He is a Fellow of Institution of Engineers (India) and a visiting professor at two engineering colleges.

# Preface

An engineer is said to be one who builds. Is he only a builder? Or, is he a designer, a manager and/or a system integrator? Our answer to this question is that he is all of these. An engineer knows how to build, design and manage, and, therefore, he also knows to integrate the system with all its characteristics and diversifications.

However, for building, designing or integrating a system, one must posses a thorough knowledge of all the unit operations and unit processes taking place in the system, and must be able to analyse and simulate all these operations and processes using veritable mathematical/analytical tools. It is, therefore, beyond doubt that for every successful engineer, the study of unit operations and unit processes is practically essential. We are, therefore, happy to present a book that deals with almost all unit operations and unit processes encountered in the process industries.

We have tried out best to provide an exhaustive coverage of each operation/process, starting from the fundamentals to the most recent advances. The treatment is fully objective and application-oriented with more emphasis on simulation, design and analysis of each process/operation. Lengthy descriptions of equipment used are avoided.

The book blends unit operations with unit processes since apart from physical processes (unit operations), chemical processes (reaction kinetics and reactor design) are also covered in full.

For the efficient analysis and simulation of engineering processes, the aid of computers has become practically invariable. Accordingly, we have discussed a number of sample computer programs (prepared on a user-friendly format) at the end of each chapter.

An excellent number of illustrative examples have been included in all chapters so that the reader can grasp the subject matter more objectively, simultaneously maintaining his interest in the topic. We have been purposefully selective regarding these numerical examples. What we have kept in mind is that once the reader has fully familiarized himself with the solved numerical examples of each chapter and has worked out all the analogous unsolved problems given at the end of the chapter, he can be fairly confident of having mastered the basics of the process/operation discussed in that chapter.

An objective summary of each chapter is given at the end under the heading "Things to Remember". Objective questions are also listed for the reader to ponder on and to verify himself how far he has assimilated the principles discussed in preceding text.

The book has been prepared on a very wide canvas so that students and engineers of diversified disciplines such as chemical, mechanical, electrical, metallurgical, biochemical and energy engineering will find this book informative and useful.

Readers shall, however, notice that this book does not contain detailed treatment of mechanical operations such as operations involving particulate matter (size reduction, screening, classification, storage and transport of solids) and mechanical phase separations (sedimentation, filtration, flotation, centrifugation). One of our earlier publications, *Mechanical Operations for Chemical Engineers,* is a full-fledged textbook on these operations and therefore, we have excluded the same from this book.

Our vote of thanks to all our young students, friends and fellow academicians, who assisted us enormously through valuable suggestions and recommendations during the course of the preparation of this book.

We invite all our eminent readers to send us their detailed comments and reviews at liberty. It is such interaction that keeps our spirits high and motivates us to enrich the engineering literature by preparing newer and improved learning materials.

**Prof. (Dr.) C.M. Narayanan**
**Prof. (Dr.) B.C. Bhattacharya**

# Contents

## 6 EXTRACTION AND ADSORPTION 1177–1363

## 7 MEMBRANE-BASED PROCESSES 1365–1476

# Volume 1

## 1 INTRODUCTION 1–146



## 2 FLUID DYNAMICS 147–317

## 3 HEAT TRANSFER 319–688



## 4 MASS TRANSFER 689–868



## APPENDICES 869–895

## *Index* 899

# Unit Operations and Unit Processes

Including Computer Programs

VOLUME 2

# Chapter 5

# Simultaneous Heat and Mass Transfer

## 5.1 DRYING AND DEHUMIDIFICATION

The term *drying*, broadly speaking, could be used to imply any process that involves removal of moisture from a substance. In that sense, it can encompass an extensively large number of operations or processes employed in process industries. For example, small amounts of water can be removed from gases and liquids by scrubbing them countercurrently with a drying agent (also called, dehumidification agent) in a packed or plate tower. An excellent example in this connection is the drying of air by scrubbing it with ethylene glycol in what is called an *absorption–dehumidification plant*. Organic gases and liquids can also be dried by passing them over a solid adsorbent such as silica gel or activated alumina. In the former case, the process is that of absorption (discussed in Chapter 4) since water is being absorbed off into the dehumidification agent, while the latter is an adsorption process (discussed in Chapter 6). Such processes, though may be classified under drying, can be therefore, analysed using the principles and methodologies discussed in Chapter 4 and Chapter 6. It is also possible to separate water from fluids by using membrane-based processes such as ultrafiltration and reverse osmosis. These are discussed in Chapter 7. In this chapter, we shall confine our discussion to the removal of relatively small quantities of water or other liquid from a solid or nearly solid material.

Removal of moisture from solids can be achieved either by mechanical means (by mechanical pressing or by centrifugation) or by thermal vapourisation. Mechanical processes are relatively cheaper and should be preferred if they serve the purpose. They may also be employed prior to drying by evaporation since this will reduce the overall heat load of the thermal process.

Before proceeding further, let us understand the significance of some of the important terminologies associated with the processes of drying and dehumidification.

The *moisture content* of the solid may be expressed on a dry basis or on a wet basis, though the former is more popular. For example, the moisture content of the solid expressed on dry basis ($X$) is

$$X = \frac{\text{Mass of moisture}}{\text{Mass of dry solid}} \quad \text{... (5.1.1)}$$

If expressed on the wet basis, the moisture content will be

$$x = \frac{\text{Mass of moisture}}{\text{Mass of wet solid}} = \frac{X}{(1+X)} \quad \text{... (5.1.2)}$$

The moisture content of a gas (say, air) is usually expressed in terms of what is called the *humidity.* The absolute humidity of air (or, any gas) can be expressed on a molal basis or on a mass basis. The *molal absolute humidity* ($Y$) is given by

$$Y = \frac{\text{Moles of vapour}}{\text{Moles of dry gas}} \quad \text{... (5.1.3)}$$

For air–water system,

$$Y = \frac{\text{Moles of moisture or water vapour}}{\text{Moles of dry air}} \quad \text{... (5.1.4)}$$

If ideal gas behaviour is assumed, then

$$Y = (p_A/p_B) = p_A/(P - p_A) \quad \text{... (5.1.5)}$$

where $p_A$ = partial pressure of vapour (say, water vapour)

$p_B$ = partial pressure of gas (say, air)

$P$ = total pressure

On the mass basis, the *absolute humidity* can be expressed as

$$Y' = \frac{(\text{Mass of vapour})}{(\text{Mass of dry gas})} \quad \text{... (5.1.6)}$$

$$= Y\,(M_A/M_B) \quad \text{... (5.1.7)}$$

where,

$M_A$, $M_B$ = molecular weight of $A$ (that is, vapour) and that of $B$ (that is, the gas) respectively.

For air–water system, $M_A$ = 18.0 and $M_B$ = 28.97. Therefore

$$Y' = (18/28.97)\ Y = 0.622\ Y \quad \text{... (5.1.8)}$$

When the partial pressure of $A$ ($p_A$) in the vapour–gas mixture at a given temperature becomes equal to the vapour pressure of $A$ ($P'_A$) at the same temperature, then the gas becomes saturated with vapour and its absolute humidity is then designated the *saturation humidity* ($Y_S$). Thus

$$Y_S = P'_A\ (P - P'_A) \qquad \text{... (5.1.9)}$$

$$Y'_S = Y_S\ (M_A/M_B) \qquad \text{... (5.1.10)}$$

For air–water system, $Y'_S = 0.622\, Y_S$. The ratio of $Y$ to $Y_S$ (expressed as percentage) is called the *percentage saturation* or the *percentage absolute humidity.* The ratio of the partial pressure of $A$ (say, water) in the gas to its vapour pressure at the given temperature is termed as the *relative humidity* (also called *relative saturation*). Thus, relative humidity = $(p_A/P'_A)$.

In many commercial drying processes, moisture is vapoursied from the solid surface into a purge gas. Air is the most commonly used purge gas. What we mean by the *equilibrium moisture content* ($X^*$) is the limiting moisture content to which a given material can be dried under specific conditions of air temperature and humidity. The equilibrium moisture content can be, therefore, considered to be an operating parameter since its value depends on the operating conditions such as the temperature and humidity of the air or purge gas used. For a given temperature and humidity of air, the moisture content of the material can be reduced only up to $X^*$. However, it must also not be forgotten that the value of $X^*$ depends very much on the nature of the solid as well. For nonporous materials, the equilibrium moisture content is essentially zero at all temperatures and humidities. For organic materials such as wood, paper, soap, etc. the value of $X^*$ varies over wide ranges as temperature and humidity change. Equilibrium moisture content can be measured by placing a sample of the material in a $U$ tube through which a stream of air of controlled-humidity is passed continuously. The sample is weighed periodically until a constant weight is reached. Moisture content at this final weight represents the equilibrium moisture content for the particular conditions.

What we mean by the *free moisture content* is the moisture content of the material in excess of the equilibrium moisture content. If $X$ is the total moisture content of the material and $X^*$ its equilibrium moisture content under the prescribed operating conditions, then $(X - X^*)$ is its free moisture content. It is the free moisture content that can be evaporated in a particular drying operation. The free moisture may include the bound moisture as well as the unbound moisture.

The total moisture in a solid may be bound or unbound or may be both. *Bound moisture* in a solid is that moisture which exerts vapour pressure less than that of the pure liquid at the given temperature. Liquid may get bound by retention in small capillaries, by chemical or physical adsorption on solid surfaces, by dissolution in cell or fiber walls or by homogeneous solution throughout the solid*. Substances that contain bound moisture are called *hygroscopic* substances. *Unbound moisture,* on the other hand, is the moisture contained by the material that exerts an equilibrium vapour pressure equal to that of pure liquid at the same temperature. It is, in fact, the moisture content of the material in excess of that corresponding to 100 percent relative humidity. All moisture in a non-hygroscopic material is unbound moisture.

These terms are illustrated in Fig. 5.1, which is a typical equilibrium moisture curve. For this material shown in Fig. 5.1, the moisture content corresponding to 100 percent relative humidity is $X_{100}$. Let $X_{100}$

---

* Since this moisture thus could contain a high concentration of dissolved solids or it is held in fine capillary pores that have highly curved surfaces or it is in physical combination in a natural organic structure, it tends to exert a vapur pressure lower than that of pure liquid.

= 0.25. This means that any sample of this material can contain bound moisture up to 25 percent If the total moisture content of the material is 30 percent, then it contains 5 percent unbound moisture and 25 percent bound moisture. If the total moisture content of the substance is 20 percent, then it means that the substance contains only bound moisture. Now, let this material is being dried by passing air of relative humidity $A$ over it. The value of equilibrium moisture content $X^*$ can be directly read from the figure as shown. Let $X^* = 0.07$. It means that the free moisture content of the material under these drying conditions is 30 − 7 = 23 percent. With air of relative humidity $A$, the moisture content of the material can be reduced only to 7 percent.

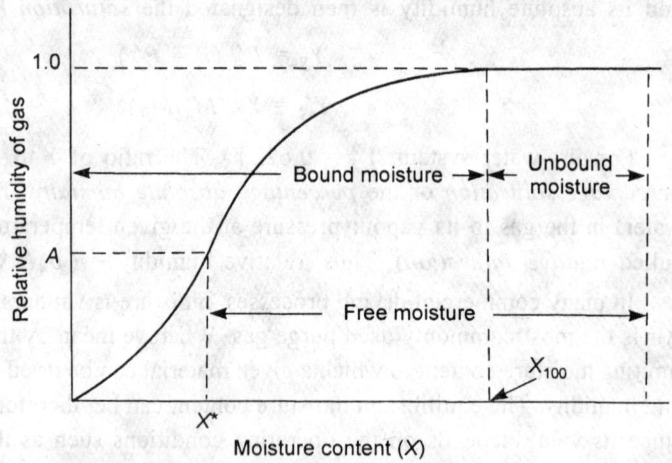


*Figure 5.1: Typical Equilibrium moisture curve showing types of moisture involved in drying of solids*

The temperature of the vapour–gas mixture is usually designated the *dry-bulb temperature* ($t_G$). This will be the reading of a thermometer when immersed in the vapour-gas mixture. The notation *dry-bulb* is purposely used to distinguish it from what is called the *wet-bulb temperature* which stands for the temperature attained by a liquid surface when brought into dynamic equilibrium with the surrounding gas. We shall be elaborating on the significance of wet-bulb temperature subsequently in this section.

The *humid volume* ($v_H$) of a vapour–gas mixture is the volume (in $m^3$) of 1 kg of dry gas and its accompanying vapour at the specified temperature and pressure. If we assume that ideal gas law is valid, then

$$v_H = (22.4)\left(\frac{1}{M_B} + \frac{Y'}{M_A}\right)\left(\frac{t_G + 273}{273}\right)\left(\frac{1}{P}\right) \quad \text{... (5.1.11)}$$

$$= (0.082)\left(\frac{1}{M_B} + \frac{Y'}{M_A}\right)\left(\frac{t_G + 273}{P}\right) \quad \text{... (5.1.12)}$$

where $P$ = total pressure in *atmospheres*

$t_G$ = dry-bulb temperature, °C

For air–water system, since $M_B = 28.97$ and $M_A = 18.0$,

$$v_H = (0.00283 + 0.00456\,Y')\,(t_G + 273)/P \quad \text{... (5.1.13)}$$

It may be noted that when $Y' = 0$, the above equation gives the specific volume of dry gas.

The *humid heat* ($C_S$) is the heat required to raise the temperature of 1 kg of dry gas and its accompanying vapour by 1° C at constant pressure. If $C_B$ is the specific heat at constant pressure of dry gas and $C_A$ that of vapour, then

$$C_S = C_B + Y'C_A \quad \text{... (5.1.14)}$$

Note that when $Y' = 0$, $C_S$ becomes equal to the specific heat of dry gas. For the air–water system, since $C_B$ = specific heat of dry air = 1.0 kJ/(kg K) and $C_A$ = specific heat of water vapour = 1.87 kJ/(kg K), the humid heat [in kJ/(kg K)] is given by

$$C_S = 1.0 + 1.87\,Y' \qquad \text{... (5.1.15)}$$

The *enthalpy* (more precisely, relative enthalpy) *of a vapour–gas mixture* is not difficult to compute. It is the sum of the sensible heat of dry gas, sensible heat of vapour associated with it and the latent heat of vapourisation of the vapour at the reference temperature. Thus

$$H = C_B\,(t_G - t_{ref}) + Y'C_A\,(t_G - t_{ref}) + Y'\lambda_0 \qquad \text{... (5.1.16)}$$

$$= C_S\,(t_G - t_{ref}) + Y'\lambda_0 \qquad \text{... (5.1.16a)}$$

where $H$ = enthalpy of mixture, J/kg of dry gas
$t_{ref}$ = reference temperature, °C
$t_G$ = dry-bulb temperature, °C
$\lambda_0$ = latent heat of vapourisation at $t_{ref}$.

It is conventional to choose 0° C as the reference temperature. Then, the above equation reduces to

$$H = C_S t_G + Y'\lambda_0 \qquad \text{... (5.1.17)}$$

The enthalpy of saturated mixture can be obtained by substituting $Y' = Y'_S$ in the above equation. *For air–water system,* since $C_S$ can be substituted from Eq. (5.1.15) and since $\lambda_0$ = latent heat of vapourisation of water at 0° C = 2502.3 kJ/kg,

$$H = (1.0 + 1.87\,Y')\,t_G + Y'\,(2502.3) \qquad \text{... (5.1.18)}$$

where $H$ is kJ/kg of dry gas.

The *dew point* of a vapour–gas mixture is nothing but its saturation temperature. In other words, it is the temperature at which the vapour–gas mixture becomes saturated. If the temperature of the mixture is decreased below the dew point, the vapour will start condensing.

The *adiabatic saturation temperature* ($t_{as}$) is a term that is widely used in drying and humidification problems. Its significance, however, is not difficult to understand. Suppose a gas (originally at temperature $t_{G1}$ and humidity $Y'_1$) is intimately mixed with a liquid at temperature $t_{as}$ in an adiabatic system (a system is said to be adiabatic when no heat is lost from it to the surroundings neither any heat supplied to it from the surroundings). The liquid will absorb heat from the gas and thereby, will get evaporated into the gas. As a result, the temperature of the gas will drop and its humidity will increase. If the temperature $t_{as}$ is such that the gas leaving the system is saturated with vapour, then $t_{as}$ is called the *adiabatic saturation temperature.* The mathematical relationship for the process can be easily written down since we understand that the sensible heat given up by the gas in cooling to $t_{as}$ equals the latent heat required to evaporate the added vapour, and thereby increase the humidity of gas to $Y'_{as}$. Thus

$$C_{S1}\,(t_{G1} - t_{as}) = (Y'_{as} - Y'_1)\,\lambda_{as} \qquad \text{... (5.1.19)}$$

where $\lambda_{as}$ = latent heat of vapourisation at $t_{as}$
$Y'_{as}$ = *adiabatic saturation humidity* of gas
$C_{S1}$ = *humid heat* corresponding to humidity $Y'_1$

Equation (5.1.9) may also be written (after dropping the suffix 1) as

$$(t_G - t_{as}) = \lambda_{as}(Y'_{as} - Y')/C_S \quad \text{... (5.1.19a)}$$

The wet-bulb temperature, $t_W$ (mentioned earlier) is similar to adiabatic saturation temperature by definition and often, confusion arises in distinguishing one from the other. The wet-bulb temperature is the steady state temperature attained by a tiny mass of liquid, such as that on the wet-bulb of a thermometer, when made to evaporate into a very large mass of gas. Since the gas forms virtually an infinite medium, it is assumed that the temperature and humidity of the gas do not undergo any change. The steady state temperature is reached when a dynamic equilibrium is established such that the rate of heat transfer to the liquid surface by convection equals the rate of mass transfer (by evaporation) away from the surface. Mathematically,

$$h_c(t_G - t_W) = \lambda_W k_g M_A (P'_{AW} - p_A) \quad \text{... (5.1.20)}$$

where $h_c$ = convective heat transfer coefficient, W/(m$^2$ K)

$k_g$ = mass transfer coefficient, kmoles/(m$^2_S$ atm)

$\lambda_W$ = latest heat of vapourisation at $t_W$,

$M_A$ = molecular weight of liquid, kg/kmole

$P'_{AW}$ = vapour pressure of liquid at $t_W$, atm

$p_A$ = partial pressure of vapour in the gas, atm

Since we know that $k_g = k_y/P$ [see Eq. (4.1.19) of Chapter 4], Eq. (5.1.20) may also be written as

$$h_c(t_G - t_W) = \lambda_W k_y M_A \left(\frac{P_{AW}'}{P} - \frac{p_A}{P}\right) \quad \text{... (5.1.21)}$$

where $k_y$ = mass transfer coefficient, kmoles/(m$^2\cdot$s$\cdot\Delta y$).

We know from Eq. (5.1.9) that

$$(P'_{AW}/P) = \frac{Y_W}{(1+Y_W)} = \frac{Y_W'(M_B/M_A)}{1+Y_W'(M_B/M_A)} \quad \text{... (5.1.22)}$$

where $Y_W$ = *molal saturation humidity* at $t_W$.

Similarly, from Eq. (5.1.5),

$$(p_A/P) = \frac{Y}{(1+Y)} = \frac{Y'(M_B/M_A)}{1+Y'(M_B/M_A)} \quad \text{... (5.1.23)}$$

Equation (5.1.21), therefore, becomes

$$h_c(t_G - t_W) = \lambda_W k_y \left(\frac{Y_W'}{\dfrac{1}{M_B} + \dfrac{Y_W'}{M_A}} - \frac{Y'}{\dfrac{1}{M_B} + \dfrac{Y'}{M_A}}\right) \quad \text{... (5.1.24)}$$

If we assume that $(Y'_W/M_A)$ and $(Y'/M_A)$ are small as compared to $(1/M_B)$, then these can be conveniently dropped from the denominator of the above equation. Equation (5.1.24), thus, reduces to

$$h_c(t_G - t_W) = \lambda_W k_y M_B (Y'_W - Y') \quad \text{... (5.1.25)}$$

or

$$(t_G - t_W) = \frac{\lambda_W (Y_W' - Y')}{(h_c/k_y M_B)} \quad \text{... (5.1.26)}$$

It can be seen that Eq. (5.1.26) is analogous to Eq. (5.1.19a). The chief difference between the mechanism of wet-bulb process and that governing adiabatic saturation is that in the former case, it has been assumed that the gas forms an infinite medium and therefore, the changes in gas temperature and gas humidity are neglected.

Strictly speaking, the heat transfer to the liquid surface can be by convection as well as by radiation. However, in wet-bulb thermometry, the effect of radiation can be minimised by using radiation shields or by maintaining a high velocity of gas to keep $h_c$ relatively high. Gas velocity past the bulb can be increased either by swinging the thermometer through the gas as is done with the sling psychrometer or by inserting the wet-bulb thermometer in a constriction in the flow path of the gas. If it is essential to include the effect of radiation in the analysis, then the term $h_c$ in Eq. (5.1.26) must be replaced by $(h_c + h_r)$, where $h_r$ is the equivalent radiation heat transfer coefficient.

For the air–water system, it has been found that $(h_c/k_y\ M_B)$ is approximately equal to the humid heat, $C_S$. Thus

$$\left(\frac{h_c}{k_y M_B}\right) = C_S$$

or

$$h_c/(k_y M_B C_S) = 1.0 \qquad \text{... (5.1.27)}$$

The above relation is called the *Lewis relation*[1]. The term on the left hand side of the above equation is called the *psychrometric ratio.* Since the psychrometric ratio is very nearly equal to unity for air–water system, the wet-bulb temperatures and adiabatic saturation temperatures are substantially equal and can be used interchangeably. For systems other than air–water however, the adiabatic saturation temperature is lower than the wet-bulb temperature. For such systems, the psychrometric ratio may be obtained from the empirical or semi–empirical correlations reported in literature. The correlations due to Bedingfield and Drew[2] are given below:

$$\left(\frac{h_c}{M_B k_y C_S}\right) = Le^{0.56} \qquad \text{... (5.1.28)}$$

where $Le$ = Lewis number

$= (Sc/Pr) = k/(C_S \rho D_{AB})$ ... (5.1.29)

$\rho$, $k$ = density and thermal conductivity respectively of vapour–gas mixture

It the gas is air, then the above relation reduces to

$$\left(\frac{h_c}{k_y M_B}\right) = 0.294\ Sc^{0.56} \qquad \text{... (5.1.30)}$$

The plot of $t_G$ versus $Y'$ at any given value of $t_W$ [with reference to Eq. (5.1.26)] is called the *psychrometric line.* Similarly, adiabatic saturation lines are plots of $t_G$ versus $Y'$ at specific values of $t_{as}$ [with reference to Eq. (5.1.19a)]. For air–water system, these two are identical and are nearly straight lines. For other systems, these plots are curved concave upward, the psychrometric lines being steeper than adiabatic saturation lines (see Figs 5.2 and 5.3).

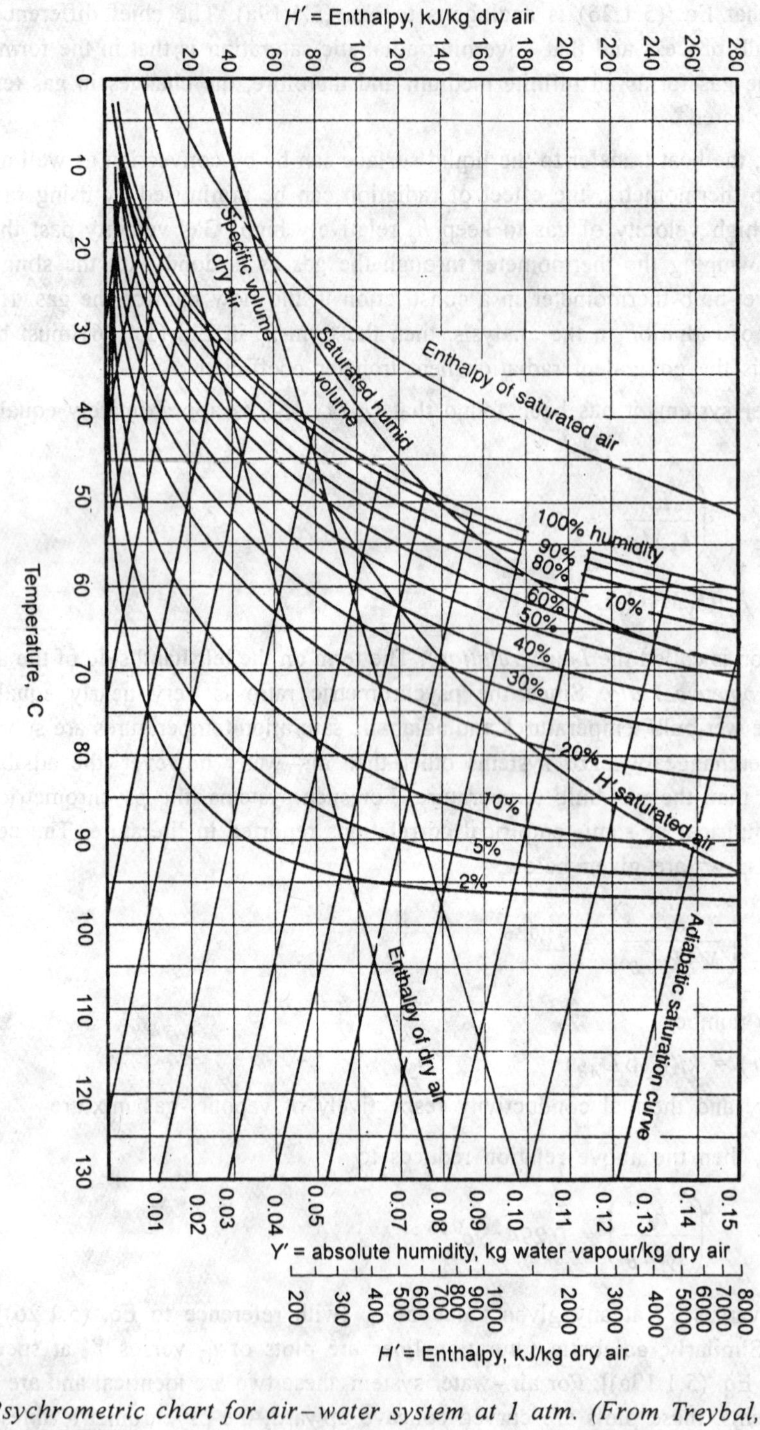


*Figure 5.2: Psychrometric chart for air–water system at 1 atm. (From Treybal, RE, Mass Transfer Operations, McGraw Hill, New York, 1981, by permission)*

Almost all of the above-discussed parameters can be determined from what are called the *psychrometric charts* (also called *humidity charts*). Such charts are quite useful in the analysis of drying problems and in the design of drying equipment. The psychrometric chart for air–water system at a total pressure of 1 atm is shown in Fig. 5.2 and that for air–carbon tetrachloride system in Fig. 5.3. The use of these charts is not difficult. We shall illustrate this with a numerical example.

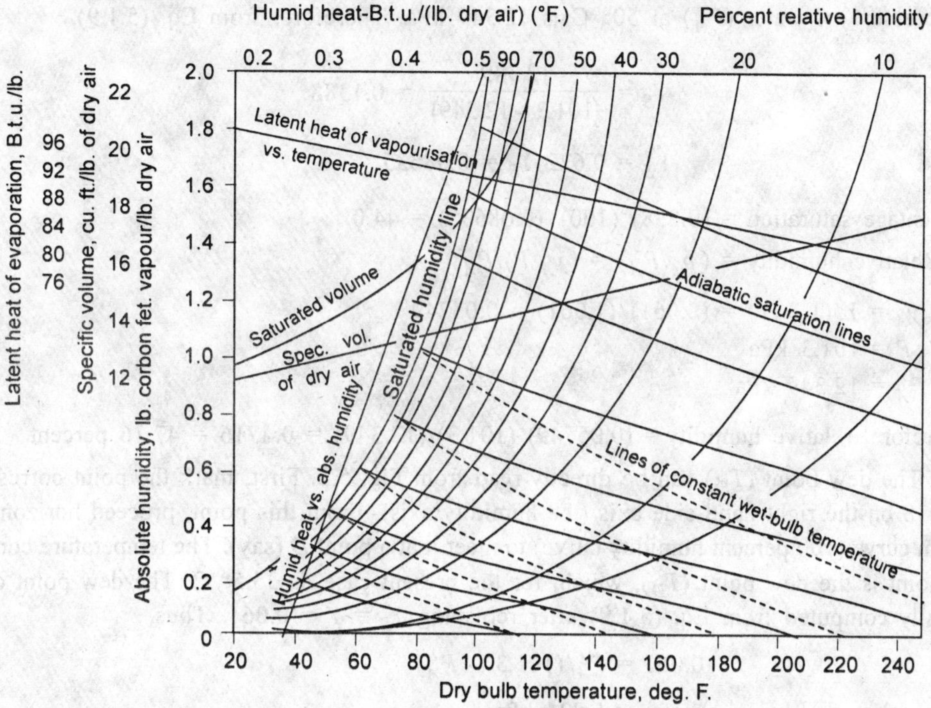


*Figure 5.3: Psychrometric chart for air–carbon tetrachloride system at 1 atm. (From Perry, RH, Chemical Engineer's Handbook, sixth edition, McGraw Hill, New York, 1984, by permission)*

**Example 5.1:** A stream of air that is admitted to a rotary dryer is at 50° C (dry-bulb) and 1 atm pressure and has an absolute molal humidity of 0.061. Compute the following characteristics of this air sample:

(*i*) the percentage saturation,
(*ii*) the relative humidity,
(*iii*) the dew point,
(*iv*) the wet-bulb temperature,
(*v*) humid heat,
(*vi*) humid volume,
(*vii*) the (relative) enthalpy.

**Solution:** It is given that the absolute molal humidity ($Y$) of air is 0.061. Therefore,

$$Y' = 0.622Y = (0.622)(0.061) = 0.038$$

$P$ = total pressure = 1 atm = 101.3 kPa

(*i*) From the psychrometric chart Fig. 5.2, the saturation humidity ($Y'_s$) corresponding to 50° C is 0.086. Therefore,

Percentage saturation = $(Y'/Y'_S)$ (100) = (0.038) (100)/(0.086) = 44.18

The saturation humidity may also be computed analytically from Eq. (5.1.9). Thus, from steam tables, vapour pressure of water ($P'_A$) at 50° C is 12.349 kPa. Therefore, from Eq. (5.1.9),

$$Y_S = \frac{12.349}{(101.3 - 12.349)} = 0.1388$$

$$Y'_S = 0.622\ Y_S = 0.08635$$

Percentage saturation = (0.038) (100) /(0.08635) = 44.0

(*ii*) Relative humidity = $(p_A/P'_A) = (y_A\ P)/P'_A$

where $y_A = Y/(1 + Y) = (0.061)/(1.061) = 0.05749$

$P$ = 101.3 kPa

$P'_A$ = 12.349 kPa

Therefore, relative humidity = (0.05749) (101.3)/(12.349) = 0.4716 = 47.16 percent

(*iii*) The dew point ($T_D$) can be directly read from Fig. 5.2. First, mark the point corresponding to $Y' = 0.038$ on the right hand side axis (the humidity axis). From this point, proceed horizontally to the saturation curve (100 percent humidity curve) to meet it at a point $D$ (say). The temperature corresponding to this point is the dew point ($T_D$), which, for the present case, is 35.5° C. The dew point can also be analytically computed from Eq. (5.1.9) after replacing $Y_S = Y = 0.061$. Thus,

$$0.061 = P'_A/(101.3 - P'_A)$$

or

$$P'_A = 5.824 \text{ kPa}$$

Now, from steam tables, the saturation temperature at 5.824 kPa is 35.55° C. Therefore, $T_D$ = 35.55° C.

(*iv*) The wet-bulb temperature ($t_W$) can also be read directly from the psychrometric chart. First, locate the point that corresponds to $Y' = 0.038$ and $t_G$ = 50° C. For the present case, this point lies exactly on the psychrometric line for 37.5° C (which is also the adiabatic saturation line for the air–water system). Therefore, $t_W$ = 37.5° C. If the point lies between two of the psychrometric lines, then the value of $t_W$ may be deduced by interpolation with reasonable accuracy. For computing $t_W$ analytically, we have to solve Eq. (5.1.26).

$$(t_G - t_W) = \lambda_W\ (Y'_W - Y')/C_S$$

or

$$(50 - t_W) = \lambda_W\ (Y'_W - 0.038)/C_S \quad \text{... (i)}$$

The humid heat $C_S$ is obtained from Eq. (5.1.5) as

$$C_S = 1.0 + (1.87)\ (0.038) = 1.071 \text{ kJ/(kg K)}$$

Equation (i), therefore, becomes

$$50 - t_W = \lambda_W\ (Y'_W - 0.038)/(1.071) \quad \text{... (ii)}$$

The above equation can be solved only by trial. Thus, let $t_W = 36° \text{ C}$ . Now, from steam tables, the latent heat of vapourisation at $t_W$ (that is $\lambda_W$) is 2416.22 kJ/kg. From Eq. (5.1.9),

$$Y'_W = \frac{0.622\, P_{AW}{}'}{(P - P_{AW}{}')} = \frac{0.622\, P_{AW}{}'}{(101.3 - P_{AW}{}')}$$

Since, from steam tables, the saturation pressure corresponding to $t_W = 36°$ C is 5.979 kPa,

$$Y'_W = (0.622)\,(5.979)/(101.3 - 5.979) = 0.039$$

Substituting in Eq. (ii), we get

$$50 - t_W = (2416.22)\,(0.039 - 0.038)/(1.071)$$

or

$$t_W = 47.74° \text{ C}$$

Since the computed value of $t_W$ differs significantly from that assumed at the outset, we have to make another trial. Now, let $t_W = 38°$ C.

$$\lambda_W \text{ (from steam tables)} = 2411.46 \text{ kJ/kg}$$

$$P'_{AW} \text{ (from steam tables)} = 6.6816 \text{ kPa}$$

$$Y'_W = (0.622)\,(6.6816)/(101.3 - 6.6816) = 0.0439$$

$$t_W \text{ [from Eq. (ii)]} = 36.66° \text{ C}$$

Since the agreement between the assumed value and the computed value is still not satisfactory, the trial is to be continued. Thus, let $t_W = 37.8°$ C.

$$\lambda_W = 2411.936 \text{ kJ/kg}$$

$$P'_{AW} = 6.61136 \text{ kPa}$$

$$Y'_W = 0.04343$$

$$t_W = 37.77° \text{ C.}$$

We can terminate the trials here since the computed value of $t_W$ is very close to that assumed. Therefore, $t_W = 37.8°$ C.

(*v*) The humid heat ($C_S$) has already been computed earlier. Thus

$$C_S = 1.071 \text{ kJ/(kg K)}$$

(*vi*) From the psychrometric chart, the specific volume of dry air at 50° C is 0.915 m$^3$/kg and the humid volume of saturated air at 50° C is 1.04 m$^3$/kg of dry air. The humid volume of the given sample of air whose percentage saturation is 44.0 can be, therefore, obtained through interpolation between these two values. Thus

$$v_H = 0.915 + (1.04 - 0.915)\ 0.44 = 0.97 \text{ m}^3\text{/kg of dry air.}$$

The value of $v_H$ can also be computed analytically from Eq. (5.1.13).

$$v_H = (0.00283 + 0.00456\,Y')\,(t_G + 273)/P$$

where $P = 1$ atm

$t_G = 50°$ C

$Y' = 0.038$

Therefore, $v_H$ = 0.97 m$^3$/kg

(*vii*) From the psychrometric chart, enthalpy of dry air at 50° C is 50 kJ/kg of dry air and that of saturated air is 274 kJ/kg of dry air. Interpolating for 44 percent humidity,

$$H = 50 + (274 - 50)\ (0.44) = 148.56 \text{ kJ/kg dry air}$$

Analytically, the relative enthalpy ($H$) may be computed from Eq. (5.1.18).

$$H = C_S t_G + 2502.3\, Y'$$

$$= (1.071)\ (50) + (2502.3)\ (0.038) = 148.637 \text{ kJ/kg dry air}$$

Thus, it can be seen that all the required information can be computed analytically or conveniently read from the psychrometric chart. The slight discrepancy between the two is due to personal error involved in reading the chart.

It must be kept in mind that humidity charts are precise only at the pressure for which they have been developed. If the system pressure is different from this, then the values obtained from the chart must be corrected for pressure

### 5.1.1 Mechanism of Drying

The process of drying involves simultaneous heat and mass transfer. Heat is transferred to the solid to evaporate the liquid or the moisture. Mass is transferred as liquid or as vapour within the solid and as vapour from the surface. The rate of drying is, therefore, determined by the factors governing the rates of these two processes.

Commercial dryers may be broadly classified into direct dryers and indirect dryers according to the method of heat transfer employed. In direct dryers, there is direct contact between the wet solid and the drying medium (say, hot air or hot gases). The vapourised liquid is carried away by the drying medium which may be circulated across the solid or through the material. Such dryers may also be called *convection dryers.* In indirect dryers, the wet solid is dried by contacting it with a hot surface. The surface is heated indirectly by steam, hot water or by electricity. The vapourised liquid is removed independently of the heating medium. Such dryers may be termed as *conduction dryers.*

Heat is thus transferred in industrial dryers by convection, conduction, radiation or by a combination of these. The rate of mass transfer depends on the mechanism of liquid flow within the solid (by diffusion or by capillary flow) and the external conditions such as the temperature, humidity, air flow, state of subdivision of the solid, agitation of the solid and the contact between the hot surface and the wet solid. Due to the complex interdependency of these variables, it is common practice to perform a batch drying test on a sample of the feedstock. A sample of the material is dried experimentally, its loss in weight is continuously recorded and from the results, a plot of moisture content (normally expressed on dry basis) versus time is prepared. A sample plot of this kind is shown in Fig. 5.4. It is important to keep in mind that any standard drying test must be performed under *constant drying conditions,* in the sense that the air velocity, temperature, humidity and pressure must be kept constant during the test.

From Fig. 5.4, it can be seen that the moisture content ($X$) decreases with time, but the rate of decrease (that is, the rate of drying) varies with time. It will be, therefore, more explanatory if we plot the rate of drying ($dX/d\theta$) versus time or versus $X$ as shown in Figs 5.5 and 5.6 respectively. These are often called the *drying curves.*

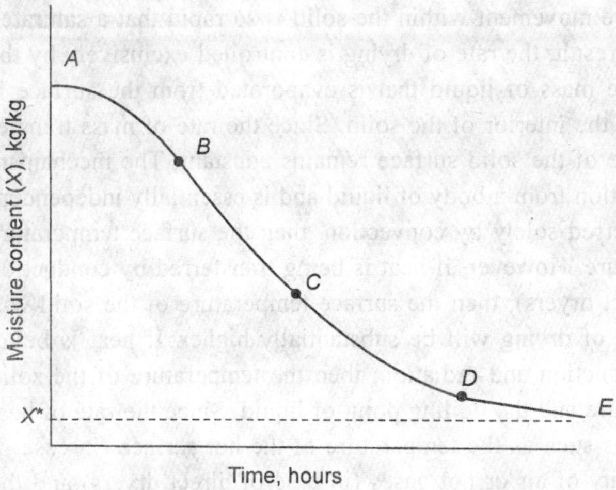


*Figure 5.4: Typical plot of moisture content versus time*

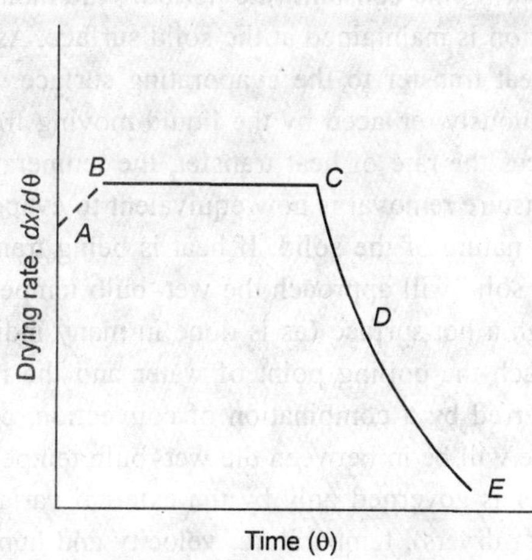


*Figure 5.5: Typical drying rate versus time plot*

Let us consider Fig. 5.6 as for example. The section *AB* of the curve represents the *warming up period* of the solids. Very little drying takes place during this period. Section *BC* represents the *constant rate period* during which the rate of drying remains essentially constant with time. The moisture content corresponding to point *C* at which the constant rate period ends is called the *critical moisture content* ($X_C$). The value of critical moisture content depends on the type of material as well as on the drying conditions. It increases with increased drying rate and with increased thickness of the material being dried. We shall also see subsequently that for many organic solids that are amorphous, fibrous or gel–like in nature such as starch, glue, cereals, eggs, soluble coffee extract, etc. the critical moisture content is very close to the initial moisture content and the constant rate drying period is very short.

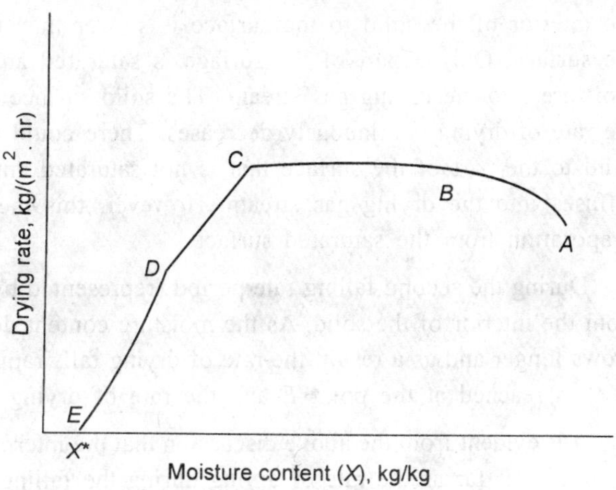


*Figure 5.6: Typical drying curve (drying rate as a function of moisture content)*

Section *CE* on the curve in Fig. 5.6 represents the *falling rate period* typified by the continuous decrease in the rate of drying. The portion *CD* is defined as the first falling rate period and the portion *DE* the second falling rate period. The point *D* at which the first falling rate period ends is often called the second critical point. At *E*, $X = X^*$ (that is, the equilibrium moisture content) and it marks the end of the drying process.

During the constant rate period*, the moisture movement within the solid is so rapid that a saturated condition is maintained at the solid surface. As a result, the rate of drying is controlled exclusively by the rate heat transfer to the evaporating surface. The mass of liquid that is evaporated from the surface is continuously replaced by the liquid moving from the interior of the solid. Since the rate of mass transfer balances the rate of heat transfer, the temperature of the solid surface remains constant. The mechanism of moisture removal is now equivalent to evaporation from a body of liquid and is essentially independent of the nature of the solid. If heat is being transferred solely by convection, then the surface temperature of the solid will approach the wet-bulb temperature. However, if heat is being transferred by conduction through a hot surface (as is done in many indirect dryers), then the surface temperature of the solid will approach the boiling point of water and the rate of drying will be substantially higher. If heat is being transferred by a combination of convection, conduction and radiation, then the temperature of the solid surface will be in between the wet-bulb temperature and the boiling point of liquid. Since the rate of heat transfer is governed only by the external variables such as the temperature of the hot surface (in case of indirect dryers), temperature, velocity and humidity of air or hot gases (in case of direct dryers) and the area exposed to the drying medium, the rate of drying during the constant rate period depends solely on these external variables and is independent of the mechanism of liquid movement within the solid.

During the first falling rate period (represented by section *CD*), unsaturated surface drying occurs. The entire solid surface can no longer be maintained saturated since the rate of liquid movement from the interior of the solid to the surface is slower than the rate of mass transfer (by evaporation) from the surface. Only a part of the surface is saturated and this part dries fast due to the evaporation of moisture into the drying gas stream. The solid surface becomes more and more depleted in liquid and the rate of drying continuously decreases. There could be movement of vapour from the interior of the solid to the part of the surface that is not saturated and after reaching the surface, this vapour further diffuses into the drying gas stream. However, this mechanism is very slow compared to the rate of evaporation from the saturated surface.

During the second falling rate period (represented by section *DE*), the entire evaporation takes place from the interior of the solid. As the moisture content decreases, the path for diffusion of heat and mass grows longer and as a result, the rate of drying falls rapidly. Eventually, the equilibrium moisture content ($X^*$) is reached at the point *E* and the rate of drying reduces to zero.

It is evident from the above discussion that the internal mechanism of liquid movement is a controlling parameter as far as the rate of drying during the falling rate period is concerned. The liquid movement within the solid may be by capillary flow or by diffusion through the solid. Accordingly, it is possible to divide the materials into two broad categories. Materials of the first category are granular or crystalline solids which hold moisture in the interstices between particles or in shallow, open surface pores. Many inorganic substances such as sand, minerals, paint pigments (titanium dioxide, chrome yellow, zinc sulfate monohydrate, etc.) catalysts and sodium phosphates belong to this class. In these materials, moisture movement is relatively unhindered and occurs as a result of gravitational and surface tension or capillary

*As the moisture is vapourised from the surface of the solid, more liquid moves from the interior of solid to the surface. If the rate at which the moisture is removed from the surface by evaporation balances the rate at which moisture moves from interior to surface, then the rate of drying remains constant. This is precisely what happens during constant rate period.

forces. The critical moisture content of these materials is relatively low and the constant rate period continues for a longer duration of time. The equilibrium moisture contents for these materials are usually very close to zero. In the falling rate period, the drying rate decreases essentially linearly with decrease in moisture content (that is, the section *CE* of the curve shown in Fig. 5.6 will be a straight line extending to $X = 0$, since $X^* \approx 0$). The rate decreases to zero at $X = 0$ since $X^*$ for these materials is very close to zero. The solid does not get deteriorated during the drying process. In other words, these materials are not affected by the drying conditions over wide ranges of temperature and humidity. This is an advantage in industrial practices since drying conditions can be chosen on the basis of convenience and economic advantage with little concern over the effect of the conditions on the properties of the dried product. In materials like textiles, paper and leather also, all moisture above the fibre saturation point move by capillary flow. Since the rate drying varies essentially linearly with $X$, the drying time of falling rate period ($\theta_f$) may be computed from Eq. (5.1.51) for these materials.

The second class of materials are those in which the internal movement of liquid takes place by diffusion through the solid structure. This is what happens during the last stages in the drying of clays, starches, cereals, textiles, paper and wood. This is also true to systems in which moisture and solid are mutually soluble. Examples are soap, gelatin and glue. In materials of this kind, the moisture is held as an integral part of the solid structure either in solution or trapped within fibres or fine interior pores. Since liquid movement by diffusion is much slower than that by gravity or capillarity, materials of this type exhibit very short constant rate periods or sometimes no measurable constant rate period at all. The critical moisture content for these substances is fairly high and could often be very close to the initial moisture content. For the same reasons, the first falling rate period also is much reduced and most of the drying takes place in the second falling rate period (namely, the last stage). The values of equilibrium moisture content are generally high for these materials and as a result, a significant amount of moisture remains behind in the solid as residual moisture that cannot be removed further in commercial dryers. Since moisture is present as an intimate part of the solid, these solids are sensitive to the drying conditions. The surface layers of the solid tend to dry more rapidly than the interior. This causes large differences in moisture content through the solid and as a result, warping or cracking occurs. The material also tends to shrink during drying. The drying conditions are to be, therefore, carefully chosen.

If the *moisture movement is exclusively* by molecular diffusion, then the rate of drying can be expressed by Fick's second law of diffusion given in Eq. (4.2.24) of Chapter 4. Thus

$$\frac{\partial X}{\partial \theta} = D_L \frac{\partial^2 X}{\partial x^2} \qquad \text{... (5.1.31)}$$

where $X$ is the moisture content (on dry basis) at any time $\theta$ and $D_L$ is the liquid diffusivity.

The above equation can be solved for the falling rate drying period of a slab of thickness $d$ and which is being dried from the top surface only (all the sides and the bottom surface are insulated), if we make the following assumptions:

(*i*) The diffusivity $D_L$ is constant (this is far from true since $D_L$ varies with moisture content, temperature and humidity).

(*ii*) The initial moisture content (namely $X_c$, since we are considering the falling rate period only) is uniformly distributed through the slab.

Based on the above assumptions, Eq. (5.1.31) can be integrated to give

$$\left(\frac{X - X^*}{X_c - X^*}\right) = \left(\frac{8}{\pi^2}\right) \sum_{n=0}^{n=\infty} \left(\frac{1}{2n+1}\right) \exp\left[-(2n+1)^2 \eta\right] \qquad \text{... (5.1.32)}$$

where $\eta = (D_L \theta \pi^2)/(4d^2)$*. For long drying times, Eq. (5.1.32) may be simplified to

$$\left(\frac{X - X^*}{X_c - X^*}\right) = \left(\frac{8}{\pi^2}\right) \exp\left(\frac{-D_L \theta \pi^2}{4d^2}\right) \qquad \text{... (5.1.33)}$$

By taking logarithm on both sides and rearranging, we get

$$\theta = \theta_f = \left(\frac{4d^2}{\pi^2 D_L}\right) \ln\left(\frac{8}{\pi^2} \frac{X_c - X^*}{X - X^*}\right) \qquad \text{... (5.1.34)}$$

where $\theta_f$ = drying time in the falling rate period.

Differentiating Eq. (5.1.33) with respect to $\theta$, we get the expression for rate of drying as

$$\frac{dX}{d\theta} = \frac{-\pi^2 D_L}{4d^2} (X - X^*) \qquad \text{... (5.1.35)}$$

Since it is difficult to ascertain accurately which of the mechanisms is true for a particular solid and also, since a lot of simplifying assumptions are often involved in the analysis as the one described above, for a reliable estimate of the drying time, we have to mostly depend on the experimental drying test data available. For example, let an experimental drying curve as the one shown in Fig. 5.6 is available to us. The rate of drying may be expressed as $(-dX/d\theta)$ or $(-dm_W/d\theta)$ where $m_W$ is the mass of the wet solid, since these two are closely inter-related. For example, if $m_S$ is the mass of dry solid (which remains constant during the entire drying process), then

$$m_W = m_S (1 + X) \qquad \text{... (5.1.36)}$$

Therefore

$$-\frac{dm_W}{d\theta} = -m_S \frac{dX}{d\theta} \qquad \text{... (5.1.37)}$$

The minus sign is used to indicate that $m_W$ or $X$ decreases with time. It is also common practice to represent the rate of drying as $\left(-\frac{1}{A}\frac{dm_W}{d\theta}\right)$ where $A$ is the exposed area of wet solid. Thus, the rate of drying ($R$) is given by

$$R = -\frac{1}{A} \frac{dm_W}{d\theta} = -\frac{m_S}{A} \frac{dX}{d\theta} \qquad \text{... (5.1.38)}$$

$$= -\frac{m_S}{A} \frac{d\bar{X}}{d\theta} \qquad \text{... (5.1.39)}$$

---

*The above equation holds good even if drying takes place from both top and bottom surfaces of the slab (all the sides, however, remaining insulated), but in that case, $d$ will be one-half of the thickness of the slab.

where $\bar{X}$ = free moisture content = $(X - X^*)$. Rearranging Eq. (5.1.38) and integrating, we get an expression for the time of drying as

$$\theta = -\frac{m_S}{A}\int_{X_1}^{X_2}\frac{dX}{R} = \left(\frac{m_S}{A}\right)\int_{X_2}^{X_1}\frac{dX}{R} \quad \text{... (5.1.40)}$$

where $X_1$ = initial moisture content of the solid
$X_2$ = final moisture content of the solid.

During the constant rate period, the rate of drying is constant. That is, $R = R_c$ = constant. Therefore, Eq. (5.1.40) can be directly integrated to give

$$\theta_c = \left[\frac{m_S\,(X_1 - X_2)}{A R_c}\right] \quad \text{... (5.1.41)}$$

The above equation is applicable only if $X_2 \geq X_c$ so that the entire drying process takes place within the constant rate period and $\theta$ (total drying time) = $\theta_c$. If $X_2 < X_c$, then

$$\theta = \theta_c + \theta_f \quad \text{... (5.1.42)}$$

where
$$\theta_c = \left[\frac{m_S\,(X_1 - X_c)}{(A R_c)}\right] \quad \text{... (5.1.43)}$$

The value of $\theta_f$ (that is, the drying time in the falling rate period) is to be determined from Eq. (5.1.40) by graphical or numerical integration from $X = X_c$ to $X = X_2$. Thus

$$\theta_f = \left(\frac{m_S}{A}\right)\int_{X_2}^{X_c}\frac{dX}{R} \quad \text{... (5.1.44)}$$

The integral on the right hand side may be evaluated graphically (by plotting $1/R$ versus $X$ and finding the area under the curve between $X = X_c$ and $X = X_2$) or numerically (say, by using trapezoidal rule or Simpson's rule).

If $X_1 < X_c$, then the entire drying process takes place in the falling rate period and the total drying time $\theta$ (which will be equal to $\theta_f$) is to be obtained from Eq. (5.1.40) by graphical/numerical entegration.

There can be some special cases for which the graphical or numerical integration could be avoided. For example, the drying curve for the first falling rate period could be linear as shown in Fig. 5.6. In that case, the rate (for the first falling rate period only) can be expressed as

$$R = aX + b \quad \text{... (5.1.45)}$$

where $a$ and $b$ are constants. Differentiating with respect to $X$, we get

$$dR = a\,dX \quad \text{... (5.1.46)}$$

Substituting in Eq. (5.1.40), we get

$$\theta_{f1} = \left(\frac{m_S}{A}\right)\int_{R_2}^{R_1}\frac{dR}{a\,R} = (m_S/Aa)\ \ln\ (R_1/R_2) \quad \text{... (5.1.47)}$$

Since Eq. (5.1.45) is valid for $X = X_c$ to $X'$ (where $X'$ is the moisture content of the solid at the end of the first falling rate period), we can write it separately for $X = X_c$ and $X = X'$ and solve for $a$. Thus,

$$R_c = aX_c + b$$

$$R' = aX' + b$$

Therefore,
$$a = (R_c - R')/(X_c - X') \quad \text{... (5.1.48)}$$

Equation (5.1.47), therefore, becomes

$$\theta_{f1} = \frac{m_S\,(X_c - X')}{(R_c - R')}\ \ln\,(R_1/R_2) \quad \text{... (5.1.49)}$$

It may sometimes happen that the drying curve for the entire falling rate period is linear. In that case, Eq. (5.1.45) will be applicable for $X = X_c$ to $X^*$. Since $R = 0$ at $X = X^*$,

$$R_c = aX_c + b$$

$$0 = aX^* + b$$

Therefore,
$$a = R_c/(X_c - X^*) \quad \text{... (5.1.50)}$$

The drying time in the falling rate period ($\theta_f$) will be then [from Eq. (5.1.47)],

$$\theta_f = \left[\frac{m_S\,(X_c - X^*)}{A\,R_c}\right] \ln\,(R_1/R_2) \quad \text{... (5.1.51)}$$

where $R_1$ is the rate of drying at $X = X_1$ and $R_2$ is the rate of drying at $X = X_2$.

The value of $R_c$ (that is, the rate of drying during the constant rate period) can be computed analytically once the heat transfer coefficient or the mass transfer coefficient is known. We have already stated earlier that during the constant rate period, the surface temperature of the solid remains constant at $t_S$. The solid surface also remains saturated with moisture and the process of moisture removal is analogous to evaporation from a body of liquid to the surrounding gas. Therefore, if $h_t$ is the *total* heat transfer coefficient, then the rate of heat transfer to the solid surface will be $h_t\,(t_G - t_S)\,A$, where $A$ is the drying area. The rate of drying will be then

$$R_c = h_t\,(t_G - t_S)/\lambda_S \quad \text{... (5.1.52)}$$

where $\lambda_S$ = latent heat of vapourisation at $t_S$

The rate of drying can also be expressed in terms of the *mass transfer coefficient* ($k_y$) using an equation similar to (5.1.25).

$$R_c = k_y M_B\,(Y'_S - Y') \quad \text{... (5.1.53)}$$

where $Y'_S$ = saturation humidity at $t_S$.

The surface temperature of the solid can be estimated by equating the above two equations. Thus,

$$(t_G - t_S) = \frac{(Y_S{}' - Y')\,\lambda_S}{\left(h_t / k_y M_B\right)} \quad \text{... (5.1.54)}$$

The above equation is only implicit in $t_s$ since both $\lambda_S$ and $Y'_S$ are functions of $t_S$. It is to be, therefore, solved for $t_S$ by trial with the help of the humidity chart or Eq. (5.1.9).

Several modifications of Eq. (5.1.54) are possible depending on the manner in which the total heat transfer coefficient ($h_t$) is expressed. It is to be noted that $h_t$ takes care of all the available modes of heat transfer such as convection, conduction and radiation. Some of the special cases are discussed subsequently in this section.

**Example 5.2:** A rectangular slab of top surface area 0.5 $m^2$ is being dried by passing hot air across its surface. The sides of the slab are sealed and drying takes place only from the top and bottom surfaces. The sample when tested in the laboratory gave the results tabulated below.

(*a*) Plot the drying rate curve.

(*b*) Determine the critical moisture content and the equilibrium moisture content.

(*c*) Determine the time required to dry the material under the same operating conditions from an initial moisture content of 20 percent (on wet basis) to 2.0 percent (on wet basis).

| *Time, hr* | *Mass of wet solid, kg* | *Time, hr* | *Mass of wet solid, kg* |
|---|---|---|---|
| 0.0 | 5.30 | 5.0 | 4.426 |
| 0.1 | 5.29 | 5.4 | 4.386 |
| 0.2 | 5.27 | 5.8 | 4.350 |
| 0.4 | 5.23 | 6.0 | 4.335 |
| 0.8 | 5.15 | 6.5 | 4.30 |
| 1.0 | 5.11 | 7.0 | 4.2825 |
| 1.4 | 5.03 | 7.5 | 4.2675 |
| 1.8 | 4.95 | 8.0 | 4.2535 |
| 2.2 | 4.87 | 9.0 | 4.2345 |
| 2.6 | 4.79 | 10.0 | 4.2240 |
| 3.0 | 4.71 | 11.0 | 4.2160 |
| 3.4 | 4.6365 | 12.0 | 4.2115 |
| 3.8 | 4.575 | 14.0 | 4.210 |
| 4.2 | 4.522 | 16.0 | 4.210 |
| 4.6 | 4.472 | | |

Mass of dry solid = 4.15 kg.

**Solution:** Since drying takes place only from the top and bottom surfaces of the slab, the total drying area ($A$) is (0.5 + 0.5) = 1.0 $m^2$. The rate of drying can be determined by plotting the moisture content of the solid ($X$) against time ($\theta$) and finding the slope of the tangent to the curve at any particular value of $\theta$. Alternately, the rate may be computed as $R = \left(\frac{1}{A}\right)(dm_W/d\theta) = \left(\frac{1}{A}\right)(\Delta m_W/\Delta\theta)$, provided $\Delta m_W$ and $\Delta\theta$ are quite small. We shall use the second method here. For example, from the given data, at $\theta$ = 0.1 hr, $m_W$ = 5.29 kg and at $\theta$ = 0.2 hr, $m_W$ = 5.27 kg. Therefore,

$$R = \left(\frac{1}{A}\right)\left(\frac{5.29 - 5.27}{0.2 - 0.1}\right) = \frac{(0.02)}{(1.0)(0.1)} = 0.2 \text{ kg}/(\text{m}^2\text{ hr})$$

The corresponding average moisture content will be,

$$X = \left[\left(\frac{m_{W1} - m_S}{m_S}\right) + \left(\frac{m_{W2} - m_S}{m_S}\right)\right]/2$$

$$= \left(\frac{m_{W1} - m_{W2}}{2m_S}\right) - 1 = \frac{(5.29 + 5.27)}{(2.0)(4.15)} - 1 = 0.2723$$

The values of $R$ and $X$, so computed, are listed below:

| *Drying rate, kg/(m² hr)* | *Average moisture content (X)* | *Drying rate, kg/(m² hr)* | *Average moisture content (X)* |
|---|---|---|---|
| 0.1 | 0.2759 | 0.1150 | 0.072 |
| 0.2 | 0.2723 | 0.10 | 0.0617 |
| 0.2 | 0.2650 | 0.09 | 0.0525 |
| 0.2 | 0.2506 | 0.075 | 0.04638 |
| 0.2 | 0.2361 | 0.070 | 0.04 |
| 0.2 | 0.220 | 0.035 | 0.034 |
| 0.2 | 0.2024 | 0.03 | 0.0301 |
| 0.2 | 0.183 | 0.028 | 0.0266 |
| 0.2 | 0.1638 | 0.019 | 0.02265 |
| 0.2 | 0.1445 | 0.0105 | 0.0191 |
| 0.18375 | 0.12608 | 0.008 | 0.01687 |
| 0.15375 | 0.1098 | 0.0045 | 0.01536 |
| 0.1325 | 0.096 | 0.00075 | 0.01464 |
| 0.1250 | 0.0836 | 0.0 | 0.01445 |

The drying–rate curve is plotted in Fig. 5.2.1. From the figure,

$X_c$ = critical moisture content = 0.144

$R_c$ = 0.2 kg/(m² hr)

$X'$ = moisture content at the second critical point = 0.04

$R'$ = 0.07 kg/(m² hr)

$X^*$ = equilibrium moisture content = 0.01445

It is given that the initial moisture content of the solid is 20 percent (on wet basis). Therefore,

$$X_1 = \frac{(0.20)}{(1.0 - 0.20)} = 0.25$$

Similarly,

$$X_2 = \frac{(0.02)}{(1.0 - 0.02)} = 0.0204$$

The value of $X_2$ is, thus, in the second falling rate period. Therefore, the total drying time ($\theta$) will be

$$\theta = \theta_c + \theta_{f1} + \theta_{f2}$$

The value of $\theta_c$ can be obtained from Eq. (5.1.43).

$$\theta_c = m_S\,(X_1 - X_c)/(A\,R_c) = \frac{(4.15)\,(0.25 - 0.144)}{(1.0)\,(0.2)} = 2.2 \text{ hr}$$

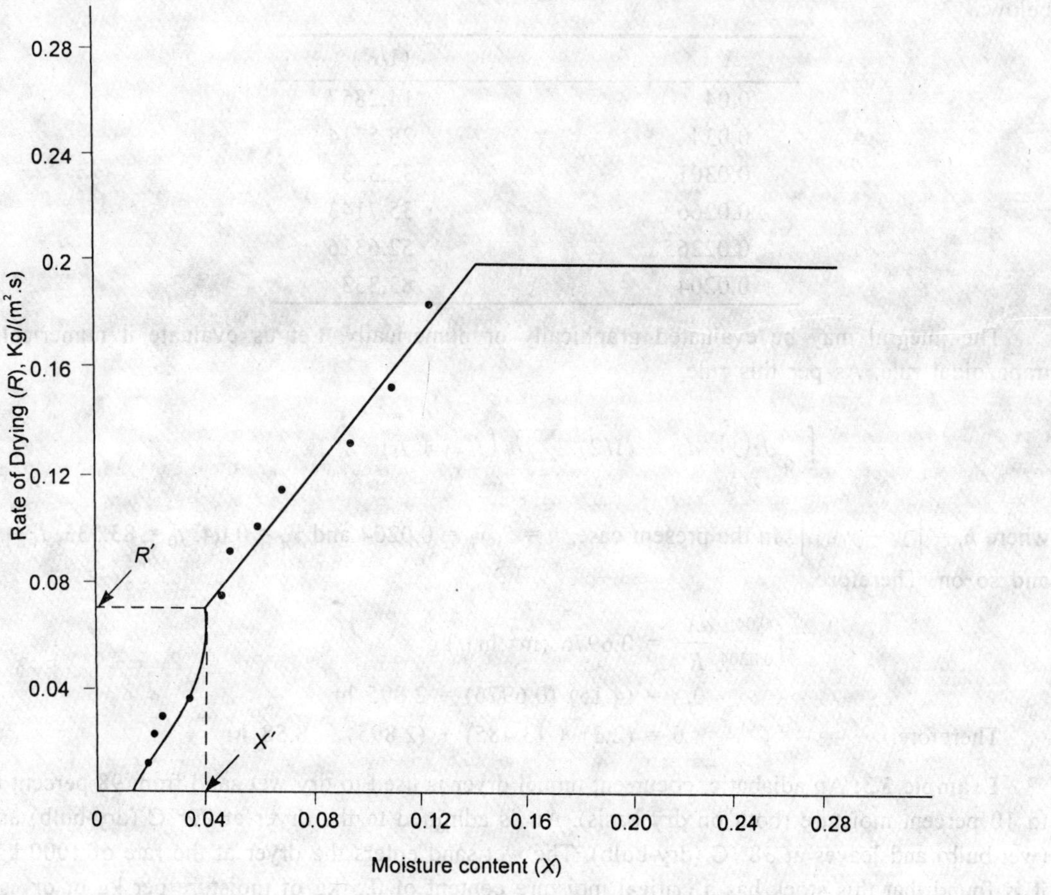


*Figure 5.2.1*

The first falling rate period extends from $X = X_c$ to $X = X'$ and the curve is linear as shown in figure. Therefore, $\theta_{f1}$ can be computed from Eq. (5.1.49) after rewriting it as

$$\theta_{f1} = \frac{m_S\,(X_c - X')}{A\,(R_c - R')}\ \ln\,(R_c/R')$$

$$= \frac{(4.15)\,(0.144 - 0.04)}{(1.0)\,(0.2 - 0.07)}\ \ln\,(0.2/0.07) = 3.485 \text{ hr}$$

The value of $\theta_{f2}$ (that is, the time of drying from $X = X'$ to $X = X_2$) is to be computed from Eq. (5.1.40).

$$\theta_{f2} = (m_S/A) \int_{X_2}^{X'} \frac{dX}{R} = \frac{(4.15)}{(1.0)} \int_{0.0204}^{0.04} \frac{dX}{R}$$

The values of $(1/R)$ at different values of $X$ (ranging from $X = 0.0204$ to $X = 0.04$) are tabulated below:

| $X$ | $(1/R)$ |
|---|---|
| 0.04 | 14.2857 |
| 0.034 | 28.5714 |
| 0.0301 | 33.333 |
| 0.0266 | 35.7143 |
| 0.02265 | 52.6316 |
| 0.0204 | 83.333 |

The integral may be evaluated graphically or numerically. Let us evaluate it numerically using trapezoidal rule. As per this rule,

$$\int_{y0}^{yn} f(y)\ dy = (1/2) \sum_{i=1}^{i=n} h_i \left(f_{i-1} + f_i\right)$$

where $h_i = \left| y_i - y_{i-1} \right|$. In the present case, $n = 5$, $y_0 = 0.0204$ and $y_n = 0.04$. $f_0 = 83.333$, $f_1 = 52.6316$ and so on. Therefore,

$$\int_{0.0204}^{0.04} \frac{dX}{R} = 0.6976\ (\text{m}^2\ \text{hr})/\text{kg}$$

$$\theta_{f2} = (4.15)\ (0.6976) = 2.895\ \text{hr}$$

Therefore,

$$\theta = (2.2) + (3.485) + (2.895) = \mathbf{8.58\ hr}$$

**Example 5.3:** An adiabatic, cocurrent tunnel dryer is used to dry wet sand from 98 percent moisture to 10 percent moisture (both on dry basis). Air is admitted to the dryer at 90° C (dry-bulb) and 35° C (wet-bulb) and leaves at 38° C (dry-bulb). The wet sand enters the dryer at the rate of 1000 kg/hr and it is found that this stock has a critical moisture content of 0.5 kg of moisture per kg of dry solid and its equilibrium moisture content is negligible. Above the critical moisture content, the rate of drying was 5.0 kg/(m$^2$ hr) and below the critical moisture content, the rate fell to zero at zero moisture content along a straight line. The dryer has an effective drying area of 0.0557 m$^2$/kg of dry solid and it can hold 50 kg of dry material per metre of length. Neglecting shrinkage, estimate the required length of the dryer.

**Solution:** It is given that

$$X_1 = 0.98$$

$$X_2 = 0.10$$

$$X_c = 0.50$$

$$R_c = 5.0\ \text{kg}/(\text{m}^2\ \text{hr})$$

$$X^* = 0.0 \text{ (since the rate falls to zero at zero moisture content)}$$
$$(A/m_S) = 0.0557 \text{ m}^2/\text{kg}$$

Since $X_2$ is less than $X_c$, the total drying time ($\theta$) is given by

$$\theta = \theta_c + \theta_f$$

The value of $\theta_c$ can be computed from Eq. (5.1.43).

$$\theta_c = \frac{(X_1 - X_c)}{(A/m_S)\,R_c} = \frac{(0.98 - 0.50)}{(0.0557)\,(5.0)} = 1.7235 \text{ hr}$$

Since the rate falls to zero along a straight line in the falling rate period, we can use Eq. (5.1.51) to compute $\theta_f$. Incidentally, since at $R = 0$, $X = X^* = 0$, the expression for $R$ can be written as, $R = aX$. Now, since at $X = X_1$, $R = R_1 = R_c = a\,X_c$ and at $X = X_2$, $R = R_2 = a\,X_2$, $(R_1/R_2) = (R_c/R_2) = (X_c/X_2)$. Equation (5.1.51), therefore, becomes

$$\theta_f = \frac{(X_c - X^*)}{(A/m_S)\,R_c} \ln (X_c/X_2)$$

$$= \frac{(0.5)}{(0.0557)\,(5.0)} \ln (0.5/0.1) = 2.889 \text{ hr}$$

Therefore, $\theta = \theta_c + \theta_f = 4.6125$ hr

Since the dryer can hold 50 kg of dry material per metre of length, the required length of the dryer is

$$= (\text{Feed rate on dry basis})\ \theta/(50.0)$$

$$= \frac{[1000/(1+0.98)](4.6125)}{(50.0)} = \mathbf{46.6\ m}$$

### 5.1.2 Industrial Drying Equipment

Industrial dryers may be broadly classified into two major categories such as *direct dryers* and *indirect dryers.* In the former type, there is direct contact between the drying medium (say, hot air) and the wet material that is being dried. The moisture that is vapourised is carried away by the drying medium. Since the major mode of heat transfer in these dryers is convection, direct dryers might also be termed convection dryers. However, at high temperatures (drying temperatures may range up to 1000° K), radiation also becomes an important heat transfer mechanism. The drying medium is usually hot air or hot flue gases. Since the entire heat for drying is supplied by the drying medium, large amounts of hot gas will have to be circulated through the dryer, particularly when the moisture content of the material is to be reduced to very low values. As a result, the fuel consumption (for preheating the gas before sending to the dryer) per kg of water evaporated is relatively large for direct dryers. While drying very small particles, the exhaust gas shall contain significant amounts of dust, thereby demanding large and expensive dust collection equipment.

In indirect dryers, heat for drying is transferred to the wet material by conduction through a separating solid surface that is usually made of metal. Indirect dryers therefore, are also termed *conduction dryers.* Heating may be done by steam, hot combustion products, hot water or any other heating fluid. Since the vapourised liquid is removed independently of the heating medium (to note that there is no direct contact between the wet solid and the drying medium), drying can be accomplished under reduced pressures or under inert atmospheres to permit recovery of the vapourised liquid and also to prevent formation of explosive mixtures or oxidation of easily decomposed materials. Such dryers can also be conveniently used for handling dusty materials.

Both direct dryers and indirect dryers may be operated continuous or batchwise. Batch dryers are preferred for small capacity installations. Continuous drying is more efficient and more economical when large amounts of wet material are to be dried per unit time. There are special designs such as infra-red dryers and dielectric dryers which are employed for very specific applications only.

## Direct Batch Dryers

*Compartment dryers* are one of the most popular types of direct batch dryers. The so called *tray dryers* fall under this category. The equipment consists of a well-insulated rectangular chamber with provision for air circulation and air preheating. An inert gas or superheated steam may also be used as the heating medium particularly when oxidation of the material is expected. The material to be dried is spread in trays and hot air is passed over the trays, thereby executing a parallel flow over the bed of solids in the tray. In many cases, dryers of this kind have provision for heating the air inside the dryer rather than outside (see Fig. 5.7). Air velocities of 1 to 10 m/s are usually employed to improve the surface heat transfer coefficient and also to eliminate stagnant air pockets. Trays may be square or rectangular, usually with 0.5 to 1.0 $m^2$ per tray and may be fabricated from any material that is compatible with the corrosion and temperature conditions prevailing. Metal trays are preferable to nonmetallic trays, since they conduct heat more readily. Tray dryers may be of the *tray–truck* or the *stationary–tray type.* In the former, the trays are loaded on trucks which are pushed into the dryer (see Fig. 5.8), while in the latter, the trays

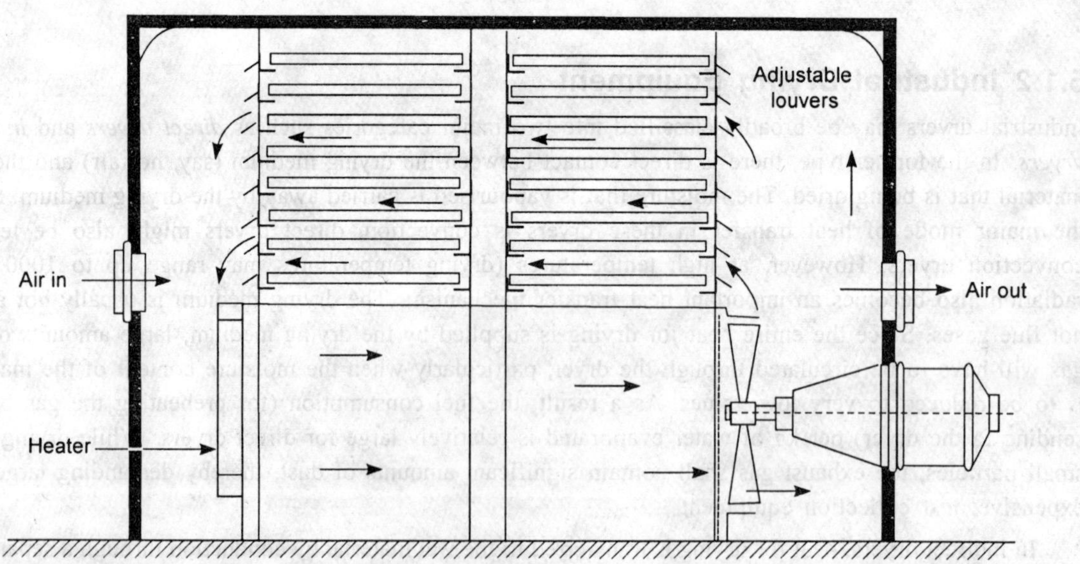


*Figure 5.7: Typical schematic of tray dryer*

are loaded directly into stationary racks within the dryer (Fig. 5.7). When trays are stacked in the truck, a minimum clearance of 4 cm is maintained between the material in one tray and the bottom of the tray immediately above. Trucks usually contain two tiers of trays with 18 to 48 trays per tier and they may be fitted with flanged wheels to run on tracks or with flat swivel wheels.

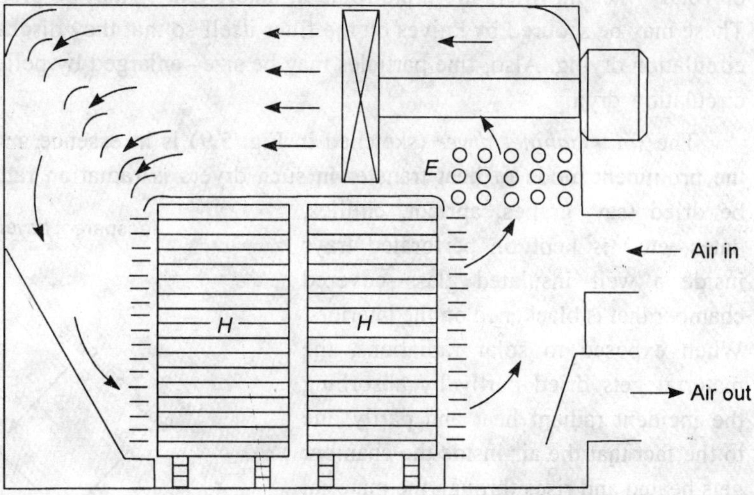


*Figure 5.8: Typical schematic of double truck dryer. E: fin heaters, H: trucks carrying trays*

Compartment dryers are basically expensive mainly due to the high labour requirements usually associated with loading and unloading of the trays/compartments and are usually restricted to small capacity batch operation. *Tray dryers* are used for handling filter cakes, sedimentation sludges, starches, pastes and plastic masses. Granular and crystalline solids can also be dried in trays provided they are larger than 150 microns in size. Examples are salt crystals, ores, sand and fibrous materials like synthetic rubber and potato strips*. Compartment dryers can also be used for drying rayon skeins, lumber, wallboard, leather and foam rubber sheets. In such cases, trays are not used, but alternate methodologies depending on the type of feed stock are employed. For example, rayon skeins are hung from poles, while boardlike or sheet materials are stacked in piles, the layers separated from each other by spacer blocks.

It is also possible to use trays with perforated bottom or removable screen bottom. The heated air passes through the sationary bed of the wet material placed on the trays. Such dryers are called batch *through circulation dryers.* These are very similar to the standard tray dryers described above except that the hot air passes through the wet solid instead of across it. Drying times in these types of dryers are usually much shorter than in parallel–flow tray dryers. However, they are restricted in application to granular materials that permit free through–circulation of air. In other words, the wet material must be in a state of granular or pelleted subdivision so that hot air may be readily blown through it. Many substances such as fibrous, flaky and coarse granular materials (that are more than about 500 microns in size) like silica gel, saw dust, fluorspar, cotton linters, rayon waste, etc. meet this requirement. Others require special pretreatment called *preforming* to render them suitable for through–circulation drying. There are several methods of preforming depending on the physical state of the wet solid. For example, relatively dry materials like cellulose acetate, cryolite, dye intermediates, etc. can be preformed by granulation, while pasty materials like titanium dioxide, white lead, lithopone, zinc stearate, etc. can be

*Fine, free-flowing solids are difficult to be handled on trays since a significant amount of the fines could be carried away in the exhaust air.

preformed by extrusion to form sphaghetti–like rods. Wet pastes may also be preformed (and predried) on a steam–heated finned drum. Some of the filter cakes (such as those of starch, alumina hydrate) formed on rotary vacuum filters are thixotropic in nature and cannot be preformed by any of the above methods. These may be scoured by knives on the filter itself so that they discharge into pieces suitable for through–circulation drying. Also, fine particles may be size–enlarged by pelletising or briquetting prior to through circulation drying.

The *solar cabinet dryer* (sketched in Fig. 5.9) is in essence a *through–circulation batch dryer,* but the prominent mode of heat transfer in such dryers is radiation rather than convection. The material to be dried (say, grapes, apricot, chillies, dates, etc.) is kept on perforated trays inside a well–insulated glass–covered chamber that is blackened on the interiors. When exposed to solar radiations, the material gets dried partly by absorbing the incident radiant heat and partly due to the fact that the air inside the chamber gets heated and rises through the material on the trays. As hot air goes out from the top, fresh air from outside enters from below, thereby maintaining a natural circulation.

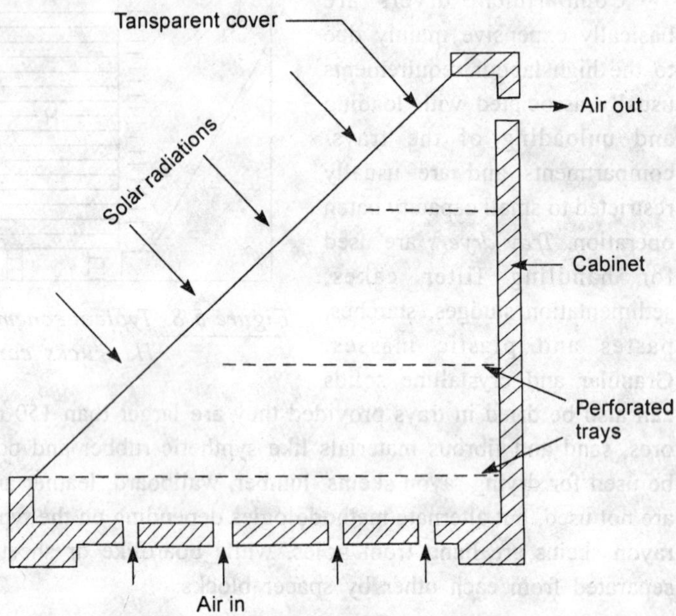

*Figure 5.9: Schematic of solar cabinet dryer*

As stated earlier, uniform air flow (that avoids stagnant pockets) is a detrimental parameter to the operation of tray dryers. As a result, large volumes of heated air must have to be blown over the trays at high velocities (keeping in mind that the solid is not carried off in the high velocity air stream) to ensure a uniform rate of drying and a uniformly dried product. In such cases, it is then common practice to recirculate 80 to 95 percent of the air (see Figs 5.7 and 5.8). In the absence of recirculation, a significant amount of heat gets lost in the discharged air and consequently, the cost of operation of the dryer becomes prohibitive. Due to recirculation, the humidity of air inside the dryer increases which, in turn, tends to lower the rate of drying. This is usually compensated by preheating the air to a fairly high temperature. The dryer must be then adequately insulated so that inside walls are at temperatures much above the dew point of air so as to prevent condensation of moisture on the walls. The two modes of air recirculation employed in the tray dryer of Fig. 5.7 and in the double–truck dryer of Fig. 5.8 are sketched in Figs 5.7 (a) and 5.8 (a) respectively.

If drying takes place in the constant rate period, then the rate of drying in a tray dryer may be estimated from Eq. (5.1.52), provided the *total heat transfer coefficient* $h_t$ can be computed. If we assume that heat transfer takes place essentially by convection, then $h_t = h_c$, where $h_c$ is the convection heat transfer coefficient between the hot gas and the drying surface. Unless at high temperatures, the rate of heat transfer by radiation from the hot gas to the drying surface shall not be significant. The solid may receive radiation from the bottom surface of the tray immediately above it. If this is to be accounted for, then $h_t$ must include a *radiation heat transfer coefficient* $h_r$ also. In other words, the rate of heat transfer by radiation ($q_r$) to the drying surface from the bottom surface of the tray above will be

$$q_r = \sigma_r \in (T_R^4 - T_S^4) \qquad \text{... (5.1.55)}$$

$$= h_r (T_R - T_S) \qquad \text{... (5.1.56)}$$

where

$\sigma_r$ = Stefan-Boltzmann constant

$\in$ = emissivity of drying surface

$T_R$, $T_S$ = absolute temperature (in °K) of radiating surface (that is, bottom surface of tray above) and drying surface respectively.

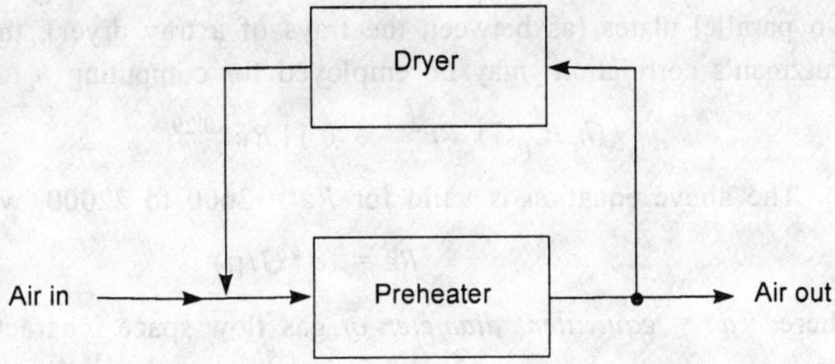


*Figure 5.7 (a): Air circulation in tray dryer of Fig. 5.7*

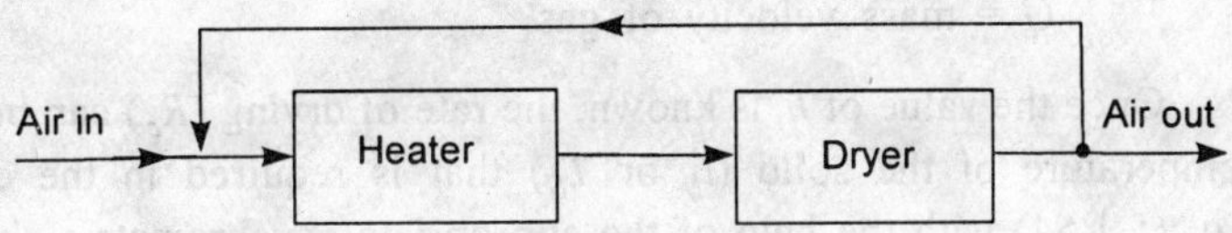


*Figure 5.8 (a): Air circulation in double truck dryer of Fig. 5.8*

It may be further required to account for the heat transfer by conduction through the bottom surface of the tray and then through the solid bed. Since hot air is blown below the tray as well, the bottom surface of the tray receives heat by convection from air (the convection heat transfer coefficient in this case may be assumed to be equal to that for the drying surface, namely $h_c$), heat is then conducted through the bottom surface of the tray and then through the solid bed on the tray. The *combined heat transfer coefficient* ($U_k$) for these can be deduced by the usual way of adding up the resistances. Thus

$$(1/U_k) = (1/h_c) + (z_m/k_m) + (z_S/k_S) \qquad \text{... (5.1.57)}$$

where

$z_m$ = wall thickness of tray bottom

$z_S$ = thickness (or, height) of solid bed on tray

$k_m$, $k_S$ = thermal conductivity of material of construction of tray and that of drying solid respectively.

In the above equation, heat transfer through the sides of the tray has been neglected.

The total rate of heat transfer to the drying surface ($q$) will be, therefore

$$q = q_c + q_k + q_r \qquad \text{... (5.1.58)}$$

where $q = h_t\,(t_G - t_S) = h_t\,(T_G - T_S)$ ... (5.1.59)

$$q_c = h_c\,(T_G - T_S) \qquad \text{... (5.1.60)}$$

$$q_k = U_k\,(T_G - T_S) \qquad \text{... (5.1.61)}$$

$$q_r = h_r\,(T_R - T_S) \qquad \text{(5.1.62)}$$

It is clear from the above four equations that

$$h_t = (h_c + U_k) + h_r\,(T_R - T_S)/(T_G - T_S) \qquad \text{... (5.1.63)}$$

For computing $h_t$ from the above equation, the values of $h_r$ and $U_k$ can be obtained from Eqs (5.1.55), (5.1.56) and (5.1.57). *The convection heat transfer coefficient* $h_c$ is to be computed from any of the experimental correlations reported in literature. For flow of gas parallel to a surface and confined between two parallel plates (as between the trays of a tray dryer), the heat transfer analogue of O'Brien and Stutzman's correlation[3] may be employed for computing $h_c$.

$$(h_c/C_pG)\ Pr^{2/3} = 0.11\,Re^{-0.29} \quad \text{... (5.1.64)}$$

The above equation is valid for $Re$ = 2600 to 22000, where the Reynolds number is defined as

$$Re = (d^*G/\mu) \quad \text{... (5.1.65)}$$

where $d^*$ = *equivalent diameter of* gas flow space (characteristic length)

$$= \frac{(4.0)\,(\text{cross-sectional area for gas flow})}{(\text{Wetted perimeter})} \quad \text{... (5.1.66)}$$

$G$ = mass velocity of gas

Once the value of $h_t$ is known, the rate of drying ($R_c$) can be estimated from Eq. (5.1.52). The surface temperature of the solid ($t_S$ or $T_S$) that is required in the computation can be obtained by solving Eq. (5.1.54) with the help of the appropriate psychrometric chart. It may be noted that if the effects of conduction and radiation are neglected, then $h_t = h_c$ and Eq. (5.1.54) reduces to that for the wet-bulb thermometer, namely Eq. (5.1.26). In other words, if heat transfer by radiation and conduction are negligible, then the drying surface will be at the wet-bulb temperature of air.

For *through–circulation drying,* it is more convenient to express the rate of drying in terms of the *mass transfer coefficient,* $k_y$. For example, if we consider a differential element of height $dz$ of the solid bed, then the rate of drying (considering drying of unbound moisture only) can be expressed using an expression similar to Eq. (5.1.53).

$$dR = k'_Y\,(dS)\,(Y'_{as} - Y') \quad \text{... (5.1.67)}$$

where $k'_Y$ = mass transfer coefficient, kg/(m$^2$·s·$\Delta Y'$)

$dS$ = interfacial area per unit area of bed cross-section

$= a\,dZ \quad \text{... (5.1.68)}$

$a$ = specific interfacial area (that is, interfacial area per unit volume of bed)

In the above expression*, it has been assumed that the humidity of gas at the liquid–gas interface is $Y'_{as}$, which is the *humidity at the adiabatic saturation temperature* ($t_{as}$) of entering gas. Also, if $G_S$ is the superficial mass velocity of gas (on dry basis) then the rate of drying may also be written as

$$dR = G_S\,dY' \quad \text{... (5.1.69)}$$

Combining the above three equations we get

$$G_S dY' = k'_Y\,(a\,dZ)\,(Y'_{as} - Y') \quad \text{... (5.1.70)}$$

---

* The reader must take care to distinguish between the mass transfer coefficients $k_y$ and $k'_Y$. The former ($k_y$) is based on mole fraction driving force, while the latter ($k'_Y$) is based on mass ratio driving force. For values of $Y$ much less than unity, $k'_Y = (M_B k_y)$. See also the discussion given after Eq. (5.1.84).

$$\int_{Y_1'}^{Y_2'} \frac{dY'}{Y_{as}' - Y_2'} = \frac{(k_Y' a)}{G_S} \int_0^Z dZ \qquad \text{... (5.1.71)}$$

$$\ln\left(\frac{Y_{as}' - Y_1'}{Y_{as}' - Y_2'}\right) = (k_Y' a)\, Z/G_S \qquad \text{... (5.1.72)}$$

or

$$Z = \left[\frac{G_S}{(k_Y' a)}\right] \ln\left(\frac{Y_{as}' - Y_1'}{Y_{as}' - Y_2'}\right) \qquad \text{... (5.1.73)}$$

$$= H_{tG} N_{tG} \qquad \text{... (5.1.74)}$$

where $H_{tG}$ = height of a gas phase transfer unit

$N_{tG}$ = number of gas phase transfer units

Incidentally, *the maximum rate of drying* will occur when the gas leaving the bed is saturated at the adiabatic saturation temperature. Thus

$$R_{max} = G_S\,(Y'_{as} - Y'_1) \qquad \text{... (5.1.75)}$$

The rate of drying, in general when the gas leaves the bed at a humidity $Y'_2$ will be

$$R = G_S\,(Y'_2 - Y'_1) \qquad \text{... (5.1.76)}$$

Therefore,

$$(R/R_{max}) = (Y'_2 - Y'_1)/(Y'_{as} - Y'_1) \qquad \text{... (5.1.77)}$$

$$= 1 - \left(\frac{Y_{as}' - Y_2'}{Y_{as}' - Y_1'}\right) \qquad \text{... (5.1.78)}$$

$$= 1 - \exp(-N_{tG}) \qquad \text{... (5.1.79)}$$

*The rate of drying in the constant rate period of a through–circulation drying* process can be estimated from the above equation provided the number of transfer units ($N_{tG}$) could be computed from any of the available experimental correlations. The value of $N_{tG}$ in turn depends on the mass transfer coefficient ($k'_Y$) and the specific interfacial area *a*. Unfortunately, experimental data for the computation of $k'_Y$ and *a* are quite scattered. For the drying of unbound moisture from large particles (of 3.2 to 20 mm in diameter), usually encountered in through–circulation drying, in shallow beds (10 to 64 mm thick), the *mass transfer coefficient* may be computed from the following correlation[4]:

$$k'_Y = j_D G_S / Sc^{2/3} \qquad \text{... (5.1.80)}$$

where $k'_Y$ = mass transfer coefficient, kg $H_2O/(m^2 s \Delta Y')$

$G_S$ = superficial mass velocity of dry gas

$j_D$ = *j-factor for mass transfer*

$= (2.06/\epsilon)\, Re^{-0.575}$, if $Re$ = 90 to 4000 ... (5.1.81)

$= (20.4/\epsilon)\, Re^{-0.815}$, if $Re$ = 5000 to 10300 ... (5.1.82)

$\epsilon$ = void fraction (also called porosity) of the bed

$$Re = (d_p \; G/\mu) \qquad \text{... (5.1.83)}$$

$d_p$ = *volume–surface diameter of each particle** (that is, diameter of a spherical particle having the same specific surface as the particle)

$G$ = Superficial mass velocity of gas through the dryer

The interfacial surface may be taken equal to the total surface of the particles, so that

$$a = 6\,(1 - \epsilon)/d_p \qquad \text{... (5.1.84)}$$

The reader must exercise caution so as not to confuse between $k_y$ and $k'_Y$. The difference between the two, in fact, is not difficult to understand. The mass transfer coefficient $k_y$ is expressed as (kmoles of $H_2O$ evaporated) / ($m^2 \cdot s \cdot$ mole fraction) whereas $k'_Y$ is expressed as (kg of $H_2O$ evaporated) / ($m^2 \cdot s \cdot$ mass ratio). However, if we assume that $Y$ is much smaller than unity, then mole fraction ≈ mole ratio [this is the assumption we used in Eq. (5.1.25)] and $k'_Y \simeq (M_B k_y)$.

Experimental correlations are scarce for the falling rate period and also for the drying of the bound moisture and a generalised treatment is difficult for such cases.

**Example 5.4:** (*a*) A cabinet dryer that is used to dry a wet, granular material (thermal conductivity = 1.73 W/(m K) and dry density = 1385 kg/m$^3$) consists of trays made of galvanised iron that are 1.0 m by 1.0 m in cross-section and 3.8 cm deep (wall thickness = 1.65 mm). A clearance of 5 cm is maintained between material in one tray and the bottom surface of the tray that is immediately above. Air at 95° C (dry-bulb) and humidity 0.05 kg of moisture per kg of dry air is passed parallel to the upper and lower surfaces at 5.25 m/s. Estimate the time required to dry the material from 35 percent moisture to 10 percent moisture (both on dry basis), if the critical moisture content of the material is 8.9 percent (on dry basis). Take thermal conductivity of GI = 45 W/(m K).

(*b*) How would the results be different if the trays were fully insulated?

**Solution:** It is given that $X_1$ = 0.35, $X_2$ = 0.10 and $X_c$ = 0.089. Since $X_2$ is larger than $X_c$, the entire drying process takes place within the constant rate period. The rate of drying can be therefore estimated from Eq. (5.1.52).

$$R_c = h_t\,(t_G - t_S)/\lambda_S = h_t\,(95 - t_S)/\lambda_S \qquad \text{... (i)}$$

The value of $h_t$ is to be obtained from Eq. (5.1.63). Neglecting heat transfer by radiation,

$$h_t = (h_c + U_k)$$

Let us first compute $h_c$ from Eq. (5.1.64).

$$h_c = 0.11\; Re^{-0.29}\,(C_p G)\; Pr^{-2/3}$$

$$= 0.11 \left(\frac{d^* G}{\mu}\right)^{-0.29} C_p G\,\left(C_p\,\mu/k\right)^{-2/3} \qquad \text{... (ii)}$$

---

*If the particles are not of uniform size, an average value of $d_p$ may be used in computations, as defined below:

$$d_p = \sum_i N_i \Big/ \sum_i N_i\, d_i^2$$

where $N_i$ is the number of particles of size $d_i$.

From Eq. (5.1.66),

$$d^* = \frac{(4.0)(1.0\times 0.05)}{(1.0+0.5)\times 2} = 0.0952 \text{ m}$$

The humid volume of air at the inlet is

$$v_H \text{ [from (Eq. 5.1.13)]} = [0.00283 + (0.00456)(0.05)](273 + 95)$$
$$= 1.125 \text{ m}^3/\text{kg of dry air}$$

We have taken $P = 1$ atm. The density of air at the inlet is, therefore,

$$\rho_a = (1.05/1.125) = 0.933 \text{ kg/m}^3$$

Though the density air varies along the dryer (due to change in humidity), this change shall not be significantly large. The mass velocity of air at the inlet is

$$G = (5.25)(0.933) = 4.898 \text{ kg/(m}^2\text{s)}$$

The other property values of air such as viscosity ($\mu$), specific heat ($C_p$) and thermal conductivity ($k$) are taken to be equal to those for dry air at 95° C and 1 atm. Thus

$$\mu = 21.7 \times 10^{-6} \text{ kg/(ms)}$$
$$C_p = 1.009 \text{ kJ/(kg K)}$$
$$k = 0.0317 \text{ W/(m K)}$$

Now,
$$Re = \frac{(0.0952)(4.898)}{\left(21.7\times 10^{-6}\right)} = 21488.0$$

$$Pr = \frac{(1009.0)\left(21.7\times 10^{-6}\right)}{(0.0317)} = 0.69$$

Therefore, from Eq. (ii),

$$h_c = 38.58 \text{ W/(m}^2\text{ K)}$$

Now, from Eq. (5.1.57),

$$(1/U_k) = (1/h_c) + (z_m/k_m) + (z_S/k_S)$$

where $h_c = 38.58$ W/(m$^2$ K)
$z_m = 0.00165$ m
$k_m = 45.0$ W/(m K)
$z_S = 0.038$ m
$k_S = 1.73$ W/(m K)

Therefore,
$$U_k = 20.867 \text{ W/(m}^2\text{ K)}$$
$$h_t = (38.58) + (20.867) = 59.447 \text{ W/(m}^2\text{ K)}$$

The humid heat of inlet air is,

$$C_{S1} = 1.0 + 1.87 Y_1' = 1.0 + (1.87)(0.05) = 1.0935 \text{ kJ/(kg K)}$$

The temperature of the drying surface ($t_S$) can be computed by solving Eq. (5.1.54). By substituting $h_t = (h_c + U_k)$, Eq. (5.1.54) can be rearranged to give

$$\left(1+\frac{U_k}{h_c}\right)(t_G - t_S) = (Y'_S - Y'_1)\ \lambda_S / C_{S1}$$

To note that for air–water system, $(h_c / k_y M_B) = C_S$ and this substitution has been made in the above equation. Inserting the values of $U_k$, $h_c$, $t_G$, $Y'_1$ and $C_{S1}$, we get

$$(1.5408)\ (95 - t_S) = (Y'_S - 0.05)\ \lambda_S/(1.0935)$$

The above equation is to be solved for $t_S$ by trial. Thus, let $t_S$ = 49° C. Now, from the psychrometric chart,

$$Y'_S = 0.082$$

From steam tables, the latent heat of vapourisation at 49° C is, $\lambda_S$ = 2385.12 kJ/kg. Substituting these values in the above equation and solving for $t_S$, we get

$$t_S = 49.7° \text{ C}$$

Since the computed value of $t_S$ does not differ significantly from that assumed at the outset, we can avoid further trial. Now, from Eq. (i),

$$R_c = \frac{(59.447)(95-49)}{(2385.12)(1000)} = 1.1465 \times 10^{-3} \text{ kg}/(\text{m}^2\text{s})$$

The time required to dry the material can be obtained from Eq. (5.1.41).

$$\theta_c = \frac{m_S (X_1 - X_2)}{A R_c} = \frac{m_S (0.35 - 0.10)}{(1.0)\left(1.1465 \times 10^{-3}\right)}$$

The volume of the tray is,

$$= (1.0 \times 1.0)\ (0.038) = 0.038 \text{ m}^3$$

If we assume that the wet solid occupies the entire volume of the tray (this, in fact, is an overestimation and is true only when the void fraction of the bed of particles in the tray is much less than unity), then

$$0.038 = (m_S/\rho_S) + (m_S X_1/\rho_W)$$

where $\rho_S$ = density of dry solid = 1385 kg/m$^3$

$\rho_W$ = density of moisture = 1000 kg/m$^3$

$X_1$ = 0.35.

Therefore, $m_S$ = 35.447 kg

And, $$\theta_c = \frac{(35.447)(0.25)}{\left(1.1465 \times 10^{-3}\right)(3600)} = \mathbf{2.15\ hr}$$

(*b*) If the trays are fully insulated, then the only mode of heat transfer to the drying surface will be convection. As stated earlier, in such cases, the temperature of the drying surface will be equal to the wet-bulb temperature of the inlet air. From the psychrometric chart, corresponding to $t_G$ = 95° C and $Y'_1$ = 0.05, $t_W$ = 46.5° C. Also, at $t_W$ = 46.5° C, the saturation humidity ($Y'_W$) is 0.072. Thus

$$t_S = t_W = 46.5° \text{ C}$$

$$Y'_S = Y'_W = 0.072$$

$$\lambda_S = \lambda_W = 2391.17 \text{ kJ/kg (from steam tables)}$$

$$h_t = h_c = 38.58 \text{ W/(m}^2\text{ K)}$$

Therefore, from Eq. (5.1.52),

$$R_c = \frac{(38.58)(95 - 46.5)}{(2391.17)(1000)} = 7.825 \times 10^{-4} \text{ kg/(m}^2\text{ s)}$$

The time required to dry the material will be,

$$\theta_c = \frac{(35.447)(0.35 - 0.10)}{(1.0)\left(7.825 \times 10^{-4}\right)(3600)} = \mathbf{3.15\ hr}$$

**Example 5.5:** The tray dryer sketched in Fig. 5.7 is being operated with air recirculation. Air entering the dryer is at 60° C (dry-bulb) and has an absolute humidity of 0.06 kg of moisture per kg of dry air. Atmospheric air is supplied at 20° C (dry-bulb) and an absolute humidity of 0.006 kg $H_2O$/kg dry air. If 90 percent of the air is recirculated (that is, 0.90 kg of dry air is recirculated per kg of dry air sent to the heater), compute

(*a*) kg of water evaporated per kg of dry fresh air supplied,
(*b*) temperature of air entering the heater,
(*c*) heat required per kg of water evaporated. Assume adiabatic drying.

**Solution:** The mode of air recirculation in the tray dryer of Fig. 5.7 is sketched in Fig. 5.7 (a). For the sake of convenience, it is resketched here in Fig. 5.5.1 with special reference to the present problem (to note that though the heater has been sketched external to the dryer, it may actually be inside the dryer as shown in Fig. 5.7).

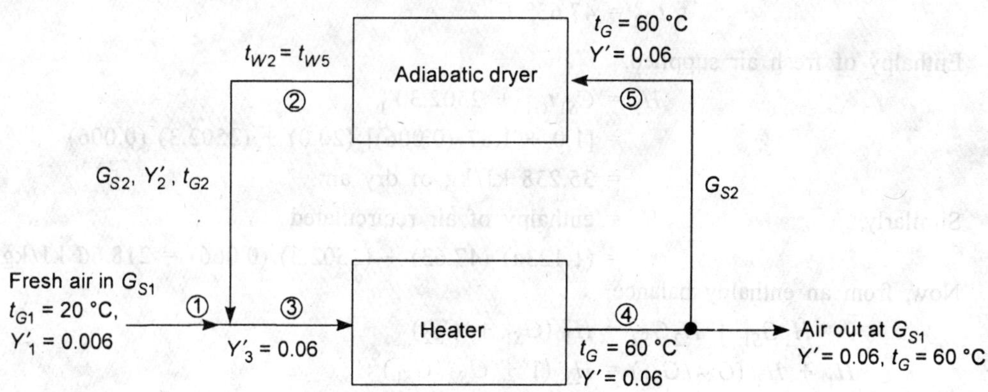


*Figure 5.5.1*

Let the rate at which fresh dry air enters the system be $G_{S1}$ kg/s. An overall water balance therefore gives

$$G_{S1}\ (0.06 - 0.006) = \text{kg of water evaporated}$$

or kg of water evaporated per kg of dry fresh air supplied

$$= \frac{G_{S1}(0.06 - 0.006)}{G_{S1}} = 0.054$$

Since 90 percent of air is recirculated,

$$\frac{G_{S2}}{(G_{S1} + G_{S2})} = 0.90$$

or
$$(G_{S2}/G_{S1}) = 9.0$$

Therefore, kg of water evaporated per kg of dry air sent to dryer

$$= (0.054/9.0) = 0.006$$

The humidity of air leaving the dryer can be now computed as

$$Y'_2 = (0.06 + 0.006) = 0.066$$

Assuming adiabatic drying, the wet-bulb temperature at position (2), namely exit from dryer, shall be the same as that at position (5), namely inlet to the dryer. From the psychrometric chart, wet-bulb temperature corresponding to $t_G$ = 60° C and $Y'$ = 0.06 is 45.5° C.

$$t_{W5} = t_{W2} = 45.5° \text{ C}$$

Also, the saturation humidity at 45.5° C is 0.067 from the chart. Thus, $Y_{W2}$ = 0.067. From steam tables, the latent heat of vapourisation at 45.5° C (that is, $\lambda_{W2}$) is 2393.59 kJ/kg. The humid heat of exit air from dryer is,

$$C_{S2} = 1.0 + 1.87\ (0.066) = 1.1234 \text{ kJ/(kg K)}$$

Now, from Eq. (5.1.26),

$$(t_{G2} - 45.5) = \frac{(0.067 - 0.066)(2393.59)}{(1.1234)}$$

or
$$t_{G2} = 47.63° \text{ C}$$

Enthalpy of fresh air supplied,

$$H_1 = C_{S1} t_{G1} + 2502.3\, Y'_1$$
$$= [1.0 + 1.87\ (0.006)]\ (20.0) + (2502.3)\ (0.006)$$
$$= 35.238 \text{ kJ/kg of dry air}$$

Similarly,
$$H_2 = \text{enthalpy of air recirculated}$$
$$= (1.1234)\ (47.63) + (2502.3)\ (0.066) = 218.66 \text{ kJ/kg of dry air}$$

Now, from an enthalpy balance,

$$H_1 G_{S1} + H_2 G_{S2} = H_3\ (G_{S1} + G_{S2})$$

or
$$H_1 + H_2\ (G_{S2}/G_{S1}) = H_3\ (1 + G_{S2}/G_{S1})$$

Since
$$G_{S2}/G_{S1} = 9.0,$$
$$H_3 = 0.1 H_1 + 0.9 H_2$$
$$= (0.1)\ (35.238) + (0.9)\ (218.66) = 200.32 \text{ kJ/kg dry air}$$

Substituting in Eq. (5.1.18), we get

$$200.32 = [1.0 + 1.87\ (0.06)]\ t_{G3} + (2502.3)\ (0.06)$$

or
$$t_{G3} = 45.12° \text{ C}$$

This is the temperature of air entering the heater.

(*c*) Due to adiabatic operation, the enthalpy of air entering the dryer and that leaving the dryer shall be the same. In other words,

$$H_4 = H_5 = H_2 = 218.66 \text{ kJ/kg dry air}$$

Therefore, the heat supplied by the heater (which is the heat required to evaporate water) will be,

$$= (H_4 - H_3)\ (G_{S1} + G_{S2})$$

Heat required per kg of dry fresh air supplied

$$= (H_4 - H_3)\ (G_{S1} + G_{S2})/G_{S1}$$

$$= (H_4 - H_3)\ [1 + G_{S2}/G_{S1}] = (218.66 - 200.32)\ (10.0) = 183.4 \text{ kJ}$$

Since kg of water evaporated per kg of dry fresh air supplied is 0.054,

Heat required per kg of water evaporated

$$= (183.4)/(0.054) = \mathbf{3396.3\ kJ}$$

**Example 5.6:** In a drying process employing the double–truck batch dryer shown in Fig. 5.8, a part of the discharge air is recycled to control the drying rate and thus to avoid undesirable effects. The air that enters the dryer is at 95° C (dry-bulb) and has an absolute humidity of 0.05 kg of water per kg of dry air, while fresh air supplied from outside is at 25° C (dry-bulb) and humidity 0.01 kg $H_2O$ per kg dry air. The air is circulated through the dryer at an average velocity of 3 m/s and free area for air flow in the dryer may be taken to be 0.7 m$^2$. If the rate at which water is evaporated from the solid is 0.008 kg / s, compute the percentage recirculation of air and the heat duty per kg of water evaporated. Assume that the dryer operates adiabatically.

**Solution:** The mode of air recirculation in the double–truck dryer is sketched in Fig. 5.8 (a), which is slightly different from that employed in the tray dryer of Fig. 5.7. It is resketched in Fig. 5.6.1 to consider the present problem.

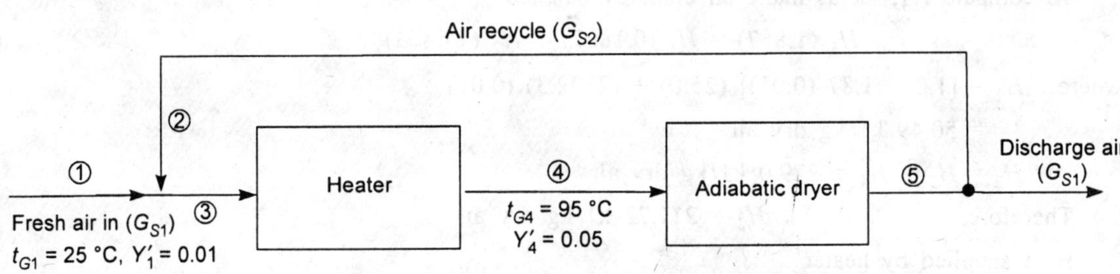


*Figure 5.6.1*

Volume rate of air entering the dryer

$$= (3.0)\ (0.7) = 2.1 \text{ m}^3/\text{s}$$

Since air enters the dryer (see position 4 in figure) at 95° C and a humidity of 0.05, its humid volume will be

$$v_{H4} = [0.00283 + 0.00456\ (0.05)]\ (273 + 95) = 1.125 \text{ m}^3/\text{kg of dry air}$$

Therefore, rate of dry air into the dryer

$$= (2.1/1.125) = 1.867 \text{ kg/s}$$

Since the dryer operates adiabatically, the wet-bulb temperature at position (5) will be the same as that at position (4). From the psychrometric chart, corresponding to $t_G = t_{G4} = 95°$ C and $Y'_4 = 0.05$, the wet-bulb temperature is 46.5° C. Therefore, $t_{W5} = t_{W4} = 46.5$ °C. The saturation humidity at this temperature ($Y'_{W5}$) is 0.072 from the chart and from steam tables, $\lambda_{W5} = 2391.7$ kJ/kg. Since the rate of evaporation of water is 0.008 kg/s,

$$Y'_5 = 0.05 + (0.008/1.867) = 0.05428$$

$$C_{S5} = 1.0 + (1.87)(0.05428) = 1.1015 \text{ kJ/(kg K)}$$

Now, from Eq. (5.1.26), $t_{G5} = t_{W5} + (Y'_{W5} - Y'_5)\,\lambda_{W5}/C_{S5}$

$$= 46.5 + (0.072 - 0.05428)(2391.7)/(1.1015) = 84.98° \text{ C}$$

If dry, fresh air is supplied at $G_{S1}$ kg/s, then an overall water balance gives

$$G_{S1}(0.05428 - 0.01) = 0.008$$

or, $G_{S1} = 0.1807$ kg/s

Rate of recirculation of air

$$= G_{S2} = (1.867 - 0.1807) = 1.6863 \text{ kg/s}$$

To note that $G_{S4} = 1.867 \text{ kg/s} = G_{S3} = (G_{S1} + G_{S2})$.

Percentage recirculation

$$= (1.6863/1.867)(100) = \mathbf{90.32 \text{ percent}}$$

To compute the heat duty of the heater, we are required to know the enthalpy of air entering the heater ($H_3$) and that leaving the heater ($H_4$). $H_4$ can be computed directly from Eq. (5.1.18).

$$H_4 = [1.0 + 1.87(0.05)](95.0) + (2502.3)(0.05) = 229.0 \text{ kJ/kg dry air}$$

To compute $H_3$, let us make an enthalpy balance:

$$H_3(1.867) = H_1(0.1807) + H_2(1.6863)$$

where $H_1 = [1.0 + 1.87(0.01)](25.0) + (2502.3)(0.01)$

$= 50.49$ kJ/kg dry air

$H_2 = H_5 = H_4 = 229.0$ kJ/kg dry air

Therefore, $H_3 = 211.72$ kJ/kg dry air

Heat supplied by heater

$$= (H_4 - H_3)(1.867) = (229.0 - 211.72)(1.867) = 32.26 \text{ kW}$$

Heat duty per kg of water evaporated

$$= (32.26/0.008) = \mathbf{4032.5 \text{ kJ}}$$

**Example 5.7:** A filter cake is being dried by passing hot air through it. The air enters at 120° C (dry-bulb) and with a humidity of 0.1 kg of moisture per kg of dry air and the superficial mass velocity of air through the cake is 3800 kg of dry air/(m$^2$ hr). The cake is 63.5 mm thick and the particles in the cake are nearly 6 mm cubes. The apparent density of the cake is 1040 kg/m$^3$. Density of dry solids = 3300 kg/m$^3$. Estimate the time required for drying the cake from 5.5 percent moisture to 0.1 percent moisture (both on dry basis) if drying is being conducted in the constant rate period. What will be the humidity of the exit air? For air drying of water from solids, $Sc = 0.6$.

**Solution:** Since the particles in the filter cake are within the size range of 3.2 to 20 mm and the cake thickness is also not large, we may use Eq. (5.1.80) for estimating the mass transfer coefficient and therefrom the rate of drying. For this, let us first compute the volume–surface diameter ($d_p$) of the particles. Since the particles are 6 mm cubes,

$$\text{Volume of each particle} = (6 \times 6 \times 6) = 216 \text{ mm}^3$$

$$\text{Surface area of each particle} = 6 \times (6 \times 6) = 216 \text{ mm}^2$$

$$\text{Specific surface of each particle} = (216/216) = 1.0 \text{ mm}^2/\text{mm}^3$$

The volume–surface diameter (also called Sauter diameter) is defined as the diameter of a spherical particle having the same specific surface as the particle. Since the specific surface of a spherical particle of diameter $d_p$ is $(6/d_p)$,

$$(6/d_p) = 1.0 \text{ mm}^2/\text{mm}^3$$

$$\text{or } d_p = 6 \text{ mm} = 0.006 \text{ m}$$

$$\text{Given that, } G_S = (3800/3600) \text{ kg}/(\text{m}^2\text{ s})$$

$$G = G_S (1 + Y'_1)$$

$$= 3800 (1 + 0.1) = 4180 \text{ kg}/(\text{m}^2 \text{ hr})$$

Average viscosity of air may be taken to be equal to that of dry air at 120° C without serious error. Thus

$$\mu = 22.9 \times 10^{-6} \text{ kg}/(\text{ms})$$

$$Re = \frac{(0.006)(4180/3600)}{(22.9 \times 10^{-6})} = 304.22$$

The apparent density of the cake ($\rho_{ap}$) is defined as

$$\rho_{ap} = \frac{\text{Mass of solids}}{\text{Volume of cake}} = 1040.0 \text{ kg}/\text{m}^3$$

Since the density of dry solids is 3300 kg/m$^3$,

$$\rho_s = \frac{\text{Mass of solids}}{\text{Volume of solids}} = 3300.0$$

The volume ratio of solids in the cake is, therefore, $(\rho_{ap}/\rho_S)$ and

$$\in = (\text{void volume})/(\text{total volume of cake})$$

$$= 1 - (\rho_{ap}/\rho_S) = 1 - (1040/3300) = 0.6848$$

Now, from Eq. (5.1.81), $j_D = (2.06/0.6848)(304.22)^{-0.575} = 0.1123$

And from Eq. (5.1.80). $k'_Y = \dfrac{(0.1123)(3800/3600)}{(0.6)^{2/3}} = 0.1666 \text{ kg}/(\text{m}^2\cdot\text{s}\cdot\Delta Y')$

From Eq. (5.1.84), $a = (6.0)(1 - 0.6848)/(0.006) = 315.2 \text{ m}^2/\text{m}^3$

The number of transfer units can be now computed from Eq. (5.1.74).

$$N_{tG} = Z\,(k'_Y a)/G_S = \frac{(0.0635)(0.1666)(315.2)}{(3800/3600)} = 3.159$$

Now, from Eq. (5.1.79),

$$(R/R_{max}) = 1 - \exp\,(-\,3.159) = 0.9575$$

For computing $R_{max}$ from Eq. (5.1.75), we are required to know the value of $Y'_{as}$. For air–water system, $Y'_{as} = Y'_W$. From the psychrometric chart, corresponding to $t_G = 120°$ C and $Y'_1 = 0.1$, $t_W = 57.3°$ C and $Y'_W = 0.1315$. Therefore, from Eq. (5.1.75),

$$R_{max} = (3800/3600)\,(0.1315 - 0.1) = 0.03325 \text{ kg}/(\text{m}^2\,\text{s})$$

Therefore, $R = (0.03325)\,(0.9575) = 0.03184 \text{ kg}/(\text{m}^2\,\text{s})$

Also from Eq. (5.1.79),

$$\left(\frac{Y_W{}' - Y_2{}'}{Y_W{}' - Y_1{}'}\right) = \exp\,(-N_{tG})$$

$$\frac{(0.1315 - Y_2{}')}{(0.1315 - 0.1)} = \exp\,(-3.159) = 0.0425$$

or,

$$Y'_2 = 0.13016 \simeq 0.13$$

Therefore, the humidity of exit air is 0.13 kg of moisture per kg of dry air.

To be precise, the average value of $G$ between the inlet and the outlet is to be used for computing the Reynolds number. Since the value of $Y'_2$ is not known at the outset, a trial and error procedure will have to be adopted. However, since the humidity of air does not vary substantially from the inlet to the outlet as in the present case, a second trial shall be more than sufficient. For example, since $(Y'_1 + Y'_2)/2 = 0.115$, the average value of $G$ is $(3800)\,(1.115) = 4237 \text{ kg}/(\text{m}^2\,\text{hr})$. Then

$$Re = 308.37$$

$$j_D = 0.11145$$

$$k'_Y = 0.1654 \text{ kg}/(\text{m}^2\cdot\text{s}\cdot\Delta Y')$$

$$N_{tG} = 3.136$$

Now, from Eq. (5.1.79),

$$\frac{(0.1315 - Y_2{}')}{(0.1315 - 0.1)} = \exp\,(-3.136) = 0.04345$$

or

$$Y'_2 = 0.13013$$

It can be seen that the above value $Y'_2$ does not differ from that computed in the first trial. Since we know from the definition of apparent density that $\rho_{ap} = (m_S/Az_S)$, $z_S$ being the thickness of the cake,

$$(m_S/A) = (1040.0)\,(0.0635) = 66.04 \text{ kg/m}^2$$

Therefore, from Eq. (5.1.41), the time for drying is

$$\theta_c = (66.04)\,(0.055 - 0.001)/(0.03184) = \mathbf{112.0 \text{ seconds.}}$$

## Indirect Batch Dryers

The *vacuum shelf dryer* is an indirect–heated batch dryer and consists of a vacuumtight cast iron shell, usually rectangular in cross-section, containing a number of hollow shelves that are fastened permanently inside the chamber. During operation, these shelves are filled with steam, hot water or a hot thermic fluid like Dowtherm. The material to be dried is placed in pans or trays on the heated shelves. Vacuum is applied to the chamber and the moisture or solvent that is vapourised is removed through a large pipe that leads to a condenser where the vapour is condensed. The noncondensable exhaust gas goes to the vacuum source which may be a vacuum pump or a steam–jet ejector. Though relatively expensive, vacuum shelf dryers are extensively used for drying pharmaceuticals, temperature–sensitive and easily oxidisable materials and also for drying such materials which are so valuable that labour cost is insignificant. They are particularly suitable for handling materials that are wet with toxic or valuable solvents since the solvent can be conveniently recovered without the danger of passing through an explosive range. In such dryers, dusty materials can be easily dried with negligible dust loss and hygroscopic materials can be completely dried at temperatures much below that required in dryers operating at atmospheric pressure.

The *agitated batch dryer* (also called *vacuum rotary dryer*) consists of a stationary cylindrical shell mounted horizontally, in which a set of agitator blades mounted on a revolving central shaft stir the solids being treated. Heat is supplied by circulating hot water, steam or a thermic fluid like Dowtherm through a jacket surrounding the shell and in larger units, through the hollow central shaft. The agitator is either a single discontinuous spiral or a double continuous spiral. The outer blades are set very close to the shell wall, having a clearance of 0.3 to 0.6 cm. Solids are charged through one or more ports at the top and discharged through discharge nozzles at the bottom. Vacuum is applied by means of a vacuum pump or a steam–jet ejector and a condenser is usually installed between the dryer and the vacuum source to collect the liquid that is vapourised. Modified designs that use a revolving shell or a rotating double–cone chamber are also available. These types of dryers are used for drying large batches of heat–sensitive materials and also in cases where the solvent is valuable and must be recovered.

*Agitated pan dryers* are quite suitable for drying small batches of pastes, slurries, sticky and granular materials. They can be easily cleaned and the solvent that is vapourised can be easily recovered. The material is agitated while being dried. This could be an advantage. The dryer consists of a shallow cast–iron pan with a steam jacket and a central shaft that carries rotating scraper blades. The material to be dried is shoveled into the pan and during the operation it is continuously stirred to expose new material to the heating surface. At the end of drying, the solids are discharged through doors on either side of the pan. These dryers may be kept open and operated at atmospheric pressure or kept covered and operated under vacuum.

*Freeze drying* or *sublimation* drying is a process that is used for very special purposes such as in the drying of foodstuffs. The vapour pressure of water from pure ice is 4.6 mm. Therefore, if the wet solid is exposed to a vacuum of less than this amount, it will freeze and water will sublime from the frozen solid. Vacuum shelf dryers, discussed earlier, can be used to conduct this operation. The process, however, is slow and expensive. It is recommended for pharmaceuticals such as penicillin and blood plasma and other heat–sensitive and readily oxidisable materials that cannot be dried successfully by other means or for those substances which show a markedly improved quality for a rather high average cost (coffee, mushrooms, diced chicken, etc.).

## Direct Continuous Dryers

As stated earlier, continuous dryers are more efficient and more economical for large capacity installations. In a direct continuous dryer, the solids are moved through the dryer while in contact with the moving hot gas stream. If the operation is adiabatic (that is, no heat is supplied in the dryer nor lost to the surroundings), then the gas will cool down and will become more humid as it moves through the dryer since it supplies heat to the moisture to vapourise. Alternately, the dryer may be operated nonadiabatically by installing heating coils within the dryer and the gas may be maintained at constant temperature.

The mode of operation may also be countercurrent or cocurrent. In the former, the gas and the solids move in the opposite directions, while in the latter, they move in the same direction. A direct continuous dryer operating in the countercurrent mode is sketched in Fig. 5.10 and the cocurrent mode in 5.11.

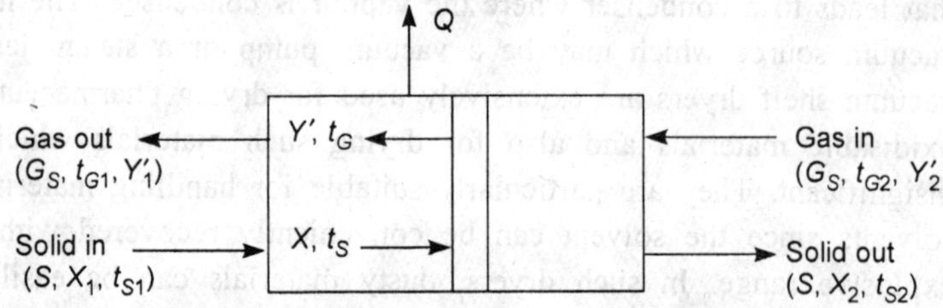


*Figure 5.10: Schematic of direct countercurrent continuous dryer*

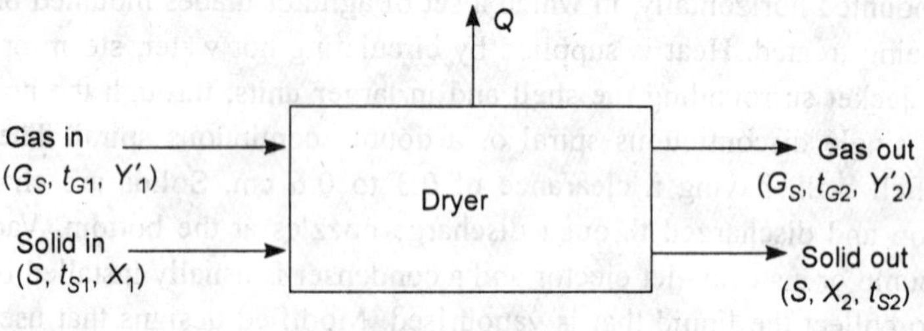


*Figure 5.11: Schematic of direct cocurrent continuous dryer*

It is evident that in an *adiabatic countercurrent operation,* the hottest gas will be in contact with the driest solids (namely, the solids at the discharge). This could often be an advantage, since during the last stages of drying, it is the bound moisture that is usually removed and removal of the last traces of bound moisture often demands high temperatures. However, this mode of operation does have serious disadvantages such as the loss of a considerable amount of sensible heat in the discharged solid (which lowers the overall thermal efficiency of the operation) and the possibility of the solid getting damaged when heated to high temperatures.

If the operation is *adiabatic and cocurrent,* then the wet solid (namely, the feed) will be in contact with the hottest gas. So long as unbound surface moisture is present, the surface temperature of the solid shall seldom exceed the wet-bulb temperature of the entering gas (which will always be less than the boiling point of the liquid at the prevailing pressure) and for this reason, even heat–sensitive solids can frequently be dried by fairly hot gas in cocurrent dryers. In cocurrent flow, the final moisture content of the solid and the rate of drying can be more accurately controlled which helps in avoiding problems like case hardening. Also, there shall be no danger of the discharged solid getting thermally damaged since it is in contact with the cooled gas. However, in cocurrent operation, dusting (entrainment of fines in the outgoing gas) is more serious than in countercurrent operation.

Consider Fig. 5.10. A moisture balance gives

$$SX_1 + G_S Y'_2 = SX_2 + G_S Y'_1$$

$$S\,(X_1 - X_2) = G_S\,(Y_1' - Y_2') \qquad \text{... (5.1.85)}$$

where $S$ = mass velocity of dry solid, kg / (m² s)

$G_S$ = mass velocity of dry gas, kg / (m² s)

$X_1$, $X_2$ = moisture content (on dry basis) of solid at the feed end and discharge end respectively

$Y_2'$, $Y_1'$ = humidity of inlet gas and outlet gas respectively

An enthalpy balance gives,

$$SH_{S1} + G_S H_{G2} = SH_{S2} + G_S H_{G1} + Q \qquad \text{... (5.1.86)}$$

where $Q$ = rate of heat loss from dryer, $W$.

If the operation of the dryer is adiabatic, then $Q = 0.0$. The enthalpy of gas ($H_{G1}$ and $H_{G2}$) can be computed from Eq. (5.1.17). The enthalpy of wet solid ($H_{S1}$ or $H_{S2}$) in kJ/kg of dry solid is given by

$$H_S = C_{pS}\,(t_S - t_{\text{ref}}) + XC_{pL}\,(t_S - t_{\text{ref}}) + \Delta H_A \qquad \text{... (5.1.87)}$$

where

$C_{pS}$, $C_{pL}$ = heat capacity of dry solid and that of moisture respectively, kJ/(kg K)

$\Delta H_A$ = heat of wetting[39] (or of adsorption, hydration or solution) at $t_{\text{ref}}$, kJ/kg dry solid.

Taking $t_{\text{ref}} = 0°$ C, the above equation reduces to

$$H_S = (C_{pS} + XC_{pL})\; t_S + \Delta H_A \qquad \text{... (5.1.88)}$$

Similarly, for the *cocurrent operation* (Fig. 5.11), the material and enthalpy balance equations will be

$$S\,(X_1 - X_2) = G_S\,(Y_2' - Y_1') \qquad \text{... (5.1.89)}$$

$$SH_{S1} + G_S H_{G1} = SH_{S2} + G_S H_{G2} + Q \qquad \text{... (5.1.90)}$$

One of the good examples of direct continuous dryers is the *continuous tunnel dryer,* which in many cases consists of a number of batch truck or tray compartment dryers operated in series. The wet material is placed in trays or on trucks which move progressively through a long tunnel in contact with hot gases or hot air. The trucks may move on tracks or monorails and they may be conveyed mechanically employing chain drives connecting to the bottom of each truck. As one truck moves out from the discharge end, a new truck enters at the inlet end. The operation is thus semicontinuous. Alternately, the solids may be carried on an endless conveyor belt that runs through the tunnel. In such cases, the operation will be fully continuous. The hot air or hot gases may be blown either cocurrently or countercurrently. Frequently, corssflow is also employed in which case the heating air flows back and forth across the trucks in series. In addition, when handling granular, particulate solids that do not offer high resistance to air flow, perforated or screen-type belt conveyors are used with through–circulation of gas to improve heat transfer and mass transfer rates. The operation may be adiabatic or reheat coils may be installed inside the tunnel to maintain constant temperature operation. Air recirculation is also often employed as in batch dryers.

The heat and mass transfer mechanisms in tunnel dryers are exactly similar to those in batch compartment dryers. Practically all forms of particulate solids and large solid objects can be handled in these dryers as in the batch compartment units. Materials like skeins of wet yarn may also be dried in these equipment by suspending them from poles or racks which move continuously through the tunnel.

Textiles may be hung in festoons from moving racks and hides may be stretched on frames which hang from conveyor chains passing through the dryer. Continuous tunnel dryers provide savings in investment and installation and therefore, are more suitable for large quantity production over multiple batch compartments. Labour savings are particularly significant in continuous operation employing belt and screen conveyors.

The *turbo–tray dryer* consists of a vertical, cylindrical or polygonal shell containing a stack of rotating annular trays or shelves. Each tray has one or more slots cut into the tray. The wet material is fed from the top and it falls onto the topmost tray. The material is spread on the tray to a uniform thickness by a levelling scraper. After completing one revolution, the material is wiped by a stationary wiper (or scraper) through the radial slots to fall to the tray below. This action is repeated on each tray, the material being transferred from each tray to the tray below once in every rotation. Every tray is provided with a leveler and a scraper/wiper. From the last tray, the material is discharged through the bottom of the dryer. Hot air or gas is introduced from the bottom through several openings and as it flows upwards, a number of turbo-type fans fitted to a central shaft circulate the air over the trays. Banks of steam–heated finned tubes serve to reheat the air continuously as it circulates. Alternately, external heating arrangement may be provided. Also, the exhaust gas may be sent to a condenser for recovering the vapourised solvent, reheated and recirculated. These types of dryers provide larger drying rates than in tray–equipped tunnel dryers described earlier due to the constant turning over and mixing of solids. Also, by virtue of its vertical construction, the turbo-tray dryer has a stack effect, the resulting draft being frequently sufficient to operate the dryer with natural draft. The turbo-tray dryer can handle materials from thick slurries to fine powders. It has been successfully employed for drying urea crystals, calcium hypochlorite, calcium chloride flakes and sodium chloride crystals. It is not suitable for fibrous materials which mat or for doughy or tacky materials.

The *horizontal conveying screen dryer* is a continuous, through–circulation dryer (see Fig. 5.12). The wet material is conveyed as a layer, 2 to 15 cm thick, on a horizontal mesh screen or perforated apron, while heated air is blown either upward or downward through the bed of the material. This dryer usually consists of a number of individual sections, complete with fan and heating coils, arranged in series to form a housing or tunnel through which the conveying screen travels. As shown in Fig. 5.12, the air circulates through the wet material and is reheated before reentering the bed. It is common to circulate hot air upward at the wet end and downward at the dry end. A portion of the air is exhausted continuously by one or two exhaust fans (not shown in figure). Since each section can be operated independently, extremely flexible operation is possible with high temperature at the wet end and the temperature progressively decreasing towards the discharge end. The rate of drying is much higher in through–circulation dryers than in tray-type tunnel dryers due to the large area of contact and short distance of travel for the internal moisture. As discussed under batch through–circulation dryers, fibrous, flaky and coarse granular materials can be directly handled in through–circulation dryers, while others such as pasty materials and filter cakes are to be preformed before feeding to the dryer.

*Rotary dryers* are the most versatile types of dryers that can be used for drying a large variety of materials from granular, crystalline or fibrous solids to free–flowing powders. They are also suitable for handling pumpable suspensions and colloidal' solutions and emulsions. Pastes, sludges and filter cakes can also be dried in rotary dryers provided they are not too sticky in nature. A rotary dryer consists of a long cylinder (its length being 4 to 10 times its diameter), mounted slightly inclined to the horizontal

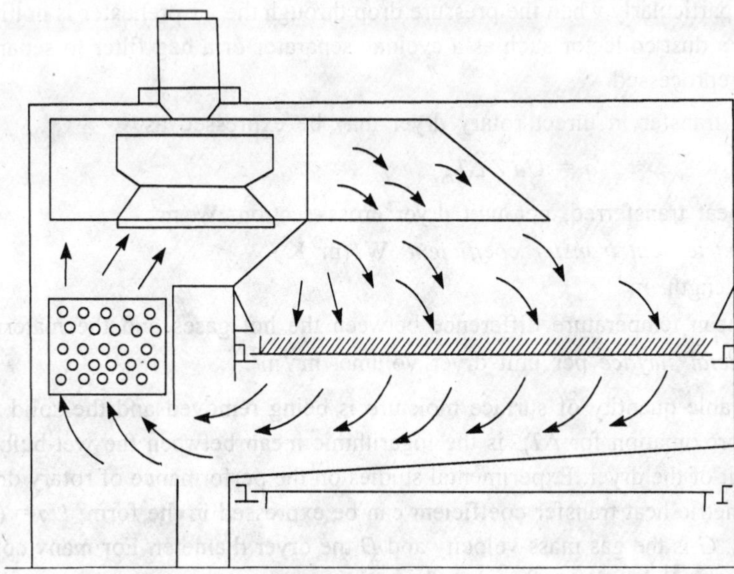

*Figure 5.12: Continuous through circulation dryer*

(the slope varying from zero to 0.08 m/m). It is rotated around its axis at peripheral speeds of 0.25 to 0.5 m/s. The wet solid is fed from one end (tip end) and the dried product is discharged from the other end. In the direct rotary dryer, hot air or gas is passed through the cylinder either countercurrently or cocurrently (Indirect rotary dryers are described subsequently in this section). While flowing cocurrently, the gas aids the movement of the solids, while in the countercurrent operation, it tends to resist the flow of solids. Air mass velocities usually range from 0.5 to 5.0 kg/(m$^2$s). At higher velocities, excessive dusting (entrainment of fine solid particles in the outgoing gas) could occur. Dusting is less serious in countercurrent operation than in cocurrent flow. Countercurrent flow of gas and solids offers greater heat transfer efficiency at a given inlet gas temperature, while cocurrent flow is more suitable for drying heat–sensitive materials at higher inlet gas temperatures because of the rapid cooling of gas during the initial vapourisation of surface moisture.

*Direct rotary dryers* are usually equipped with flights on the interior for lifting and shoveling the solids through the gas stream during passage through the cylinder. The shape of the flights depends on the handling characteristics of the solids. For free-flowing materials, radial flights with 90° lip are employed, while for sticky materials, flat radial flights without any lip are used. When materials change their characteristics during drying, the flight design is also changed along the dryer length. For example, flat flights with no lips are used in the first one-third of dryer length measured from the feed end, flights with 45° lips in the middle one third and flights with 90° lips in the final one-third of dryer length. When cocurrent gas-solid flow is used, flights may be left out to the final one metre length at the exit end to minimise dusting. Also, spiral flights may be provided in the first one metre at the feed end to accelerate forward flow of solids from under the feed chute. The hot gas or air is forced through the cylinder by an exhaust fan or an exhaust fan–blower combination. Though the latter arrangement helps to maintain a precise control of the pressure inside the dryer, an exhaust fan alone is often sufficient to maintain the circulation of air, particularly when the pressure drop through the air preheater is not high. The exhaust gas is usually sent to a dust collector such as a cyclone separator or a bag filter to separate the entrained dust which may be reprocessed.

The rate of heat transfer in direct rotary dryer may be expressed as

$$q = Ua\,Z\,\Delta T_m \quad \text{... (5.1.91)}$$

where $q$ = total heat transferred, per unit dryer cross-section, $W/m^2$

$(Ua)$ = *volumetric heat transfer coefficient,* $W/(m^3\,K)$

$Z$ = dryer length, m

$\Delta T_m$ = true mean temperature difference between the hot gases and the material, K

$a$ = *interfacial surface* per unit dryer volume, $m^2/m^3$.

When a considerable quantity of surface moisture is being removed and the solid temperatures are unknown, a good approximation for $\Delta T_m$ is the logarithmic mean between the wet-bulb temperatures of air at the inlet and exit of the dryer. Experimental studies on the performance of rotary dryers have shown that the overall volumetric heat transfer coefficient can be expressed in the form, $Ua = (K\,G^n/D)$, where $K$ and $n$ are constants, $G$ is the gas mass velocity and $D$ the dryer diameter. For many commercial dryers, the following relationship[31] has been found to be satisfactory:

$$Ua = (237.0)\ G^{0.67}/D \quad \text{... (5.1.92)}$$

Unless material characteristics limit the gas temperature, the inlet temperature is usually fixed by the heating medium employed. The exit gas temperature, which is largely an economic function, may be determined based on the empirical criterion that rotary dryers are most economically operated at $N_{tG}$ = 1.5 to 2.5, where $N_{tG}$ is the *number of gas phase transfer units* for heat transfer. $N_{tG}$ is defined as

$$N_{tG} = \Delta T'_G/\Delta T_m \quad \text{... (5.1.93)}$$

where $\Delta T'_G$ = change in gas temperature owing to heat transfer to solids only (exclusive of heat losses).

The time of passage of solids through the dryer, which is defined as the holdup of solids in the dryer divided by the volumetric feed rate of solids, is an important design parameter. This time of passage or the retention time ($\theta$) should be sufficiently large so that the moisture content of the solids is reduced to the desired limit. However, an accurate estimation of retention time is difficult since the retention time of individual particles could often be widely different from the average retention time. If $\phi$ is the *fractional holdup of solids* in the dryer (that is, the fraction of the dryer volume occupied by the solids), then the average retention time is given by

$$\theta = Z\phi\,\rho_S/S \quad \text{... (5.1.94)}$$

where $Z$ = dryer length

$S$ = mass velocity of dry solids, $kg/(m^2 s)$

$\rho_S$ = density of dry solids, $kg/m^3$

It may be noted that $S/(Z\,\rho_S)$ is the volumetric flow rate of dry solids per unit volume of dryer. *The holdup* $\phi$ may be estimated from the experimental correlation proposed by Friedman and Marshall[33]:

$$\phi = \phi_0 \pm KG \quad \text{... (5.1.95)}$$

where $\phi_0$ = solid holdup when there is no gas flow

$$= \frac{(0.3344)\,S}{\left(\rho_S\, N^{0.9}\, Ds\right)} \qquad \text{... (5.1.96)}$$

$$K = 0.6085/\left(\rho_S \sqrt{d_p}\right) \qquad \text{... (5.1.97)}$$

$s$ = slope of dryer, m/m
$N$ = speed of rotation, rps
$d_p$ = average particle diameter, m
$G$ = mass velocity of gas, kg/(m²s)

Most of the commercial rotary dryers operate with $\phi$ = 0.05 to 0.15. In Eq. (5.1.95), the plus sign is to be used for countercurrent flow and minus sign for cocurrent flow. Substituting Eqs (5.1.96) and (5.1.97) in Eq. (5.1.94), the expression for $\theta$ reduces to

$$\theta = \frac{(0.3344\,Z)}{\left(N^{0.9}\, Ds\right)} \pm \frac{0.6085\,ZG}{S\sqrt{d_p}} \qquad \text{... (5.1.98)}$$

*Fluidised bed dryers* have been extensively used for drying free-flowing fine materials including crystals, granules and short fibres. Fluidised bed units for drying coal, cement, rock and limestone are in general acceptance. Particularly when large amounts of solids are to be handled, such units have been observed to be quite economical.

The operating principle of fluidised beds has already been discussed in Section 2.12 of Chapter 2. Since in a fluidised bed, there is a high degree of turbulence resulting in more intimate fluid–solid contacting, wet solids can be conveniently dried by fluidising them with hot air or hot gas. A typical fluidised bed dryer is shown in Fig. 5.13. Mixing and heat transfer are very rapid in such systems. The rate of drying is several times that in a static bed and the average residence time of solids in the bed is usually between 30 and 60 seconds. The solids must be free-flowing and of a size range of 0.1 to 36 mm. A part of the dried product may be recycled and mixed with the wet feed to improve its handling characteristics and reduce stickiness. Since the mass flow rate of gas required to supply the necessary heat of drying is usually much less than that

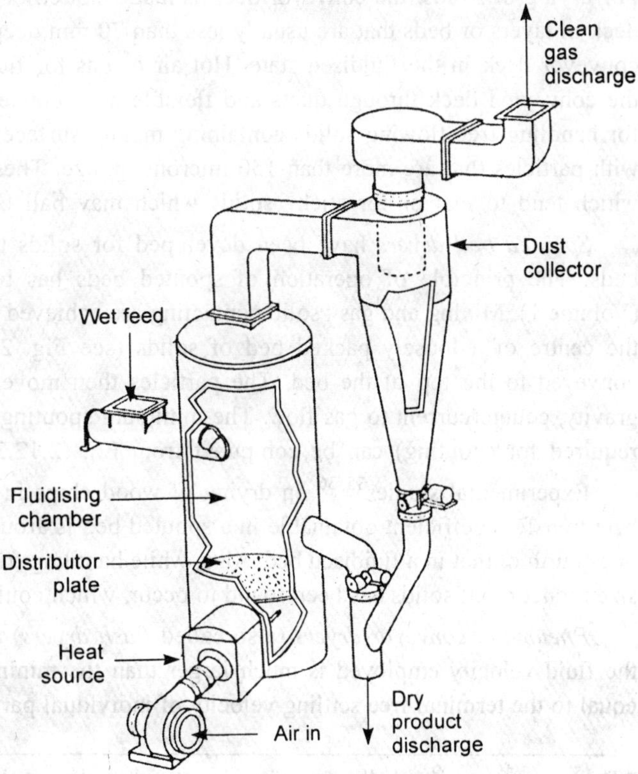


*Figure 5.13: Typical fluidised bed dryer*

required to fluidise the bed, *fluidised bed dryers* can be economically operated at the minimum fluidisation velocity. If large amounts of fines are present, they tend to get carried away by the outgoing gas and a dust collector, such as a cyclone or a bag filter, will have to employed to recover the entrained dust. Gas cyclones are widely used for dust collection. In cyclones, the dust–laden gas is admitted tangentially at a high velocity. As a result, the gas executes a helical or spiral motion inside the separator and consequently, the particles experience very large centrifugal forces and are thrown to the wall, where they settle down and are collected through the bottom outlet. The gas flows up along the axis of the separator and goes out through the top outlet.

When a stationary vessel is employed for fluidisation, it becomes necessary to ensure that all solids being treated are completely fluidised. Otherwise, the nonfluidisable fractions shall fall to the bottom of the bed and may eventually block the gas distributor. The addition of mechanical vibration to the fluidised system, as practised in *vibrating conveyor dryers,* helps in eliminating this disadvantage. Even if nonfluidisable solid fractions are present, directional–throw vibration will cause them to be conveyed to the discharge end of the conveyor. Also, even if the particle size is nonuniform across the bed, due to mechanical vibration, chances of incipient channeling are very much reduced. It has also been observed that in vibrating conveyor dryers, fluidisation can be accomplished at lower pressures and lower gas velocities. The vibrating conveyor consists of a spring-supported horizontal pan or trough which is vibrated either mechanically (by an eccentric) or electrically (by an electromagnet) as it moves forward. For drying of solids, the conveyor deck is made slotted or perforated. Solids are spread on the conveyor deck in layers or beds that are usually less than 70 mm deep and they are carried forward on the vibrating conveyor deck in the fluidised state. Hot air or gas for fluidisation is introduced into a plenum beneath the conveying deck through ducts and flexible hose connections. Vibrating conveyor dryers are suitable for handling free-flowing solids containing mainly surface moisture. They give satisfactory performance with particles that are more than 150 microns in size. These dryers are not suitable for fibrous materials which tend to mat or for sticky solids which may ball or adhere to the deck.

*Spouted bed dryers* have been developed for solids that are too coarse to be handled in fluidised beds. The principle of operation of spouted beds has been discussed in Section 2.12 of Chapter 2 (Volume 1). Mixing and gas–solid contacting are achieved first in the fluid *spout* flowing upward through the centre of a loosely packed bed of solids (see Fig. 2.38). Particles are entrained in the fluid and conveyed to the top of the bed. The particles then move downward in the surrounding annulus under gravity, countercurrent to gas flow. The minimum spouting velocity (minimum superficial velocity of gas required for spouting) can be computed from Eq. (2.12.38) of Chapter 2.

Experimental studies[35, 36] on drying of wood chips in spouted beds have shown that the volumetric heat transfer coefficient obtainable in a spouted bed is around twice that in a direct–heat rotary dryer, but one-fourth of that in a fluidised bed. Also, while handling friable materials like cellulose acetate, significant size reduction of solids has been found to occur, which could indicate a possible limitation of these dryers.

*Pneumatic conveyor dryers* (also called *flash dryers*) are similar to fluidised bed dryers, except that the fluid velocity employed is much larger than the minimum fluidisation velocity and is more than or equal to the terminal free settling velocity of individual particles*. As a result, the solids are carried along

*The gas velocity is usually maintained greater than the terminal free settling velocity of the largest particle in the feedstock.

by the fluidising gas. In other words, the particles are dispersed in a rapidly rising hot gas stream. The gas stream after emerging from the drying column is sent to a cyclone separator or a bag filter to separate the dried solids from the gas. Since the gas velocity is maintained high, the exposure time is very small, usually of the order of a few seconds and as a result, these dryers can be efficiently used only for removing the surface moisture. However, due to the short time of contact, the temperature of the solids rarely rises above 40° C during drying and therefore, heat–sensitive materials can be conveniently handled in these dryers. The solids most be granular and free-flowing when dispersed in the gas stream so that they do not stick on the conveyor walls or agglomerate. As in fluidised bed dryers, slightly sticky materials may also be handled in these dryers if a part of the dried product is recycled and mixed with the wet feed. These dryers are, however, rarely suitable for abrasive solids. A significant size reduction of solids could occur during drying particularly when crystalline or other friable materials are being handled. This may or may not be an advantage. Multistage installations of pneumatic conveyor dryers are often used for drying materials that contain large quantities of moisture or for drying materials that contain internal as well as surface moisture.

*Spray dryers* are extensively used for drying aqueous solutions, pastes and fine suspensions. They are basically expensive equipment and are recommended for those applications in which the feed solution, slurry or paste is one that cannot be dewatered mechanically, or is heat sensitive and cannot be exposed to high temperatures for long periods, or contains ultrafine particles that will agglomerate and fuse if dried in other than a dilute condition. The dryer consists of a large cylindrical and usually vertical chamber into which the material to be dried is sprayed in the form of small droplets and into which is fed a large volume of hot air or gas sufficient to supply the heat required to complete evaporation of the liquid. Heat transfer and mass transfer are accomplished by direct contact of the hot gas with the dispersed droplets. The cooled gas and the dried product are separated partly at the conical bottom of the drying chamber and partly in an external cyclone separator or bag filter (see Fig. 5.14). Most of the commercial spray dryers employ cocurrent flow of gas and solids as shown in the figure (Countercurrent flow dryers are used primarily for drying soaps and detergents; it is also possible to have mixed flow dryers which combine countercurrent and cocurrent operations). Further, as shown in Fig. 5.14, the product may be cooled by ambient air before final discharge. The feed solution may be atomised (that is, dispersed into fine droplets) by pressure nozzles or high-speed centrifugal disks. Two-fluid nozzles, since do not operate efficiently at high capacities, are not popular in commercial spray dryers. Pressure nozzles affect atomisation by forcing the liquid under high pressure (2,700 to 69,000 kPa) and with a high degree of spin through a small orifice (usually 0.25 to 0.4 mm in diameter). Piston pumps furnish the liquids at high pressure. Instead of a single orifice, multiple nozzles may be used. However, this will complicate the drying chamber design and the air flow pattern and may cause collision of particles, leading to nonuniformity of spray and particle size. Pressure nozzles tend to get plugged and eroded while handling suspensions or slurries, while simplicity and low cost are their plus points. Centrifugal disks atomise liquids by extending them in thin sheets which are discharged at high velocities from the periphery of a rapidly rotating, specially designed disk. The speed of rotation of the disk ranges from 4,000 to 20,000 rpm and it may be driven by a high speed electric motor or a steam turbine. Centrifugal disk atomisation is particularly suitable for handling thick pastes and slurries which erode and plug nozzles. Disks are also capable of operating over a wide range of feed rates and disk speeds without significantly affecting the product quality and uniformity.

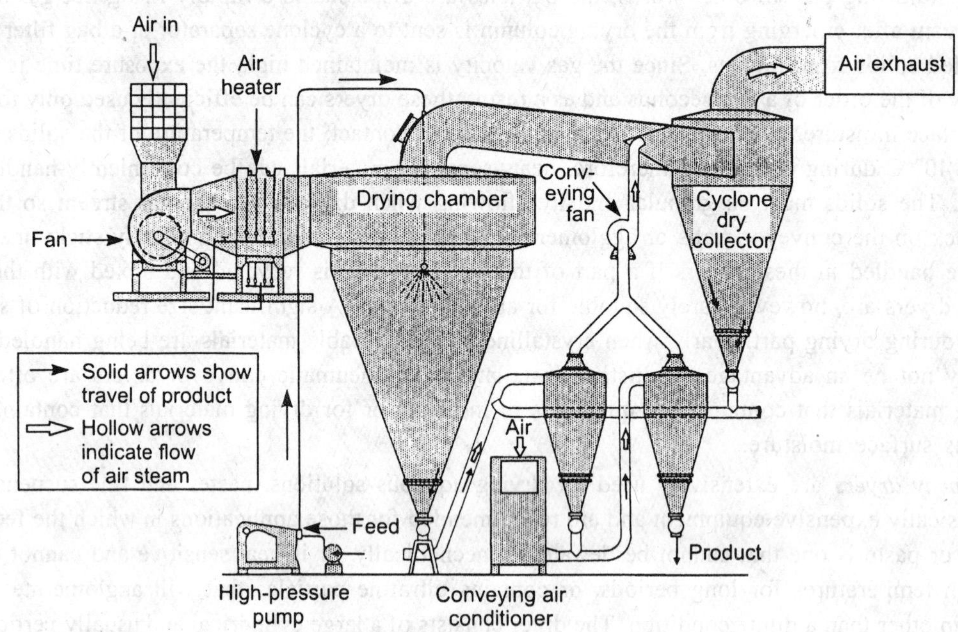


*Figure 5.14: Typical spray dryer (with accessories)*

Due to the large drying surface and small droplet sizes created, the rate of drying in a spray dryer is quite large (the total residence time of solid in the dryer seldom exceeds 30 seconds). Spray drying has also the specific advantage that it produces a product that is in the form of spherical particles. These spherical particles may be solid or hollow. Aqueous solutions of materials like soap, gelatin and water soluble polymers which form tough, casehardened outer skins on drying form hollow spherical particles. The casehardened outer surface prevents liquid from reaching the surface from the particle interior, but due to the high rate of heat transfer to the drop, the liquid in the interior vapourises causing the outer shell to expand and form a hollow sphere. This may also blow a hole through the wall of the spherical shell, leading ultimately to rupture. Spray drying is applicable to heat–sensitive products such as milk powders and other foods and pharmaceuticals due to the short contact time in the dryer hot zone and also due to the fact that though hot gases or hot air at high temperature may be used as drying medium, the material will not be heated much above the wet-bulb temperature of the drying air. Further, the water film on the liquid drop protects the solids from high gas temperatures. The fact that the product will be powdery, spherical and free-flowing is particularly advantageous in the drying of colour pigments. Spray dryers have also been developed for encapsulation processes to convert volatile liquid flavours and perfumes to particulate solid forms.

In spite of the many-sided advantages of *spray drying,* a fully satisfactory analytical method for the design of spray drying equipment has not yet been developed. Most of the design variables are still to be estimated from experimental tests*. The theoretical correlations published by Gluckert[37] have received

*A good discussion on spray dryer design has been given by Gauvin and Katta[41].

a fair degree of acceptance, though these involve a large number of assumptions. For example, it has been assumed that the largest droplet size in a spray population is three times the average drop size and the droplet Nusselt number (defined as $h_c\, d_p / k_f$, where $h_c$ is the convective heat transfer coefficient, $k_f$ the gas-phase thermal conductivity and $d_p$ the average drop size) is equal to 2.0. The drying chamber dimensions have been evaluated based on the largest droplets, since they dry most slowly. It is also assumed that drying conditions are uniform throughout the chamber, or, in other words, the entire chamber is at the gas exit temperature. Further, the temperature difference driving force for drying ($\Delta t$) is assumed to be the difference between the outlet temperature of the drying gas and the gas wet-bulb temperature or adiabatic saturation temperature. Thus, $\Delta t = (t_{G2} - t_{as})$. Based on these assumptions, Gluckert[37] has proposed the following correlations for computing the rate of heat transfer to the spray:

$$Q = 6.38\, k_f V^{2/3}\, (D_0/d_{pm}^2)\, (\Delta t)\, \sqrt{\rho_g / \rho_L}$$

(for single-fluid pressure nozzles) ... (5.1.99)

$$= \left[\frac{4.19\, k_f\, (D - R)\, \Delta t}{4\, d_{pm}^2\, \rho_L}\right] \sqrt{\frac{Q_L\, \rho_g}{R N}}$$

(for centrifugal disk atomisers) ... (5.1.100)

where $Q$ = rate of heat transfer to spray, W

$k_f$ = thermal conductivity of gas film surrounding the droplet (evaluated at the average between the dryer gas and drop temperature), W/(m K)

$V$ = volume of drying chamber, m$^3$

$\Delta t$ = temperature difference driving force, °K

$= (t_{G2} - t_{as})$ ... (5.101)

$D_0$ = diameter of pressure nozzle discharge orifice, m

$D$ = diameter of drying chamber, m

$R$ = radius of disk, m

$N$ = speed of rotation of disk, rps

$Q_L$ = mass flow rate of liquid, kg/s

$\rho_L$, $\rho_g$ = density of liquid and density of drying gas at exit conditions respectively, kg/m$^3$

$d_{pm}$ = maximum drop diameter, m

$= 3\, (d_{vs})$ ... (5.1.102)

$d_{vs}$ = average drop diameter (*volume-surface mean diameter of drops*), m

For single-fluid pressure nozzles,

$$d_{vs} = (0.009516)\ (-\Delta P)^{-1/3} \quad \text{... (5.1.103)}$$

where $(-\Delta P)$ = pressure drop across nozzle, N/m$^2$

For centrifugal disk atomisers[38],

$$(d_{vs}/R) = 0.4 \left(\frac{\Gamma}{\rho_L\, N R^2}\right)^{0.6} (\mu_L/\Gamma)^{0.2} \left(\frac{\sigma \rho_L\, L_W}{\Gamma^2}\right)^{0.1} \quad \text{... (5.1.104)}$$

where $\Gamma$ = spray mass velocity, kg / (s·m of wetted disk periphery)
$\mu_L$ = liquid viscosity, kg / (ms)
$\sigma$ = surface tension, N / m
$L_W$ = wetted disk periphery, m

The rate of evaporation of moisture in kg / s (namely the rate of drying) under constant rate drying conditions will be then given by $(Q/\lambda_{as})$ or $(Q/\lambda_w)$. The constant rate period will end when the drop surface becomes solid. This may occur, as stated earlier, while the interior of the drop is still a relatively dilute solution. If swelling and rupture do not occur, then the rate of drying during the falling rate period will be controlled by diffusion through the shell.

One of the promising new developments in the field of drying is *foam mat drying.* This method offers a wide scope for drying vegetable and fruit juices and sticky, viscous, heat sensitive materials like glue, gelatin, pigments, varnishes etc. The feed solution to a foam mat equipment must not necessarily contain more than 50 percent dissolved solids in order to ensure formation of a stable foam. Recommended values of solid concentration in different feedstocks are 25 percent for tomato paste, 35–45 percent for milk concentrate. 40–60 percent for orange juice, 40 percent for pineapple juice, 60 percent for lemon concentrate and 55 percent for grape juice. The feed solution is first fed to an aerator in presence of a foam inducer and/or foam stabiliser (0.5 to 1.0 percent by weight of the feed) and agitated with air or any other suitable gas. The gas is selected with an eye to foam stability and product characteristics. The foam stabiliser could be a surface active agent, a hydrophilic colloid or a mixture of the two. The foam produced must be capable of being pumped, spread, extruded uniformly and dried in hot air/gas without loss of foam structure. The foam is spread on an endless belt as a 2.0 to 5.0 mm thick layer with the help of a feeder and is sent to a drying chamber in which air at controlled temperature (20 to 65° C) is passed across the material. Vacuum is not generally used. Because of the enormous increase in the gas-liquid interface, drying rates are significantly high in foam mat drying, in spite of the fact that the heat transfer is impeded by the large volume of gas present in the foamed mass. The operating temperatures are relatively low (20 to 65° C) and the finished product is superior to the spray-dried or drum-dried products because of its honeycomb structure and better reconstitution.

**Example 5.8:** Wood chips of bulk density 110 kg/m$^3$ containing 70 percent moisture (on dry basis) are to be dried continuously in a horizontal conveyor dryer that uses 1 m wide belt running at 0.3 m/s. The solids form a 20 mm thick layer on the belt and the space for air flow above the belt is 0.27 m high. Specific heat of solids (when dry) = 2.1 kJ/(kg K). Drying air enters at the rate of 4.7 m$^3$/s at 115° C (dry-bulb) and with an absolute humidity of 0.005 kg of moisture per kg of dry air. It flows across the surface of the solids, countercurrent to the belt. The feedstock enters at 30° C and dried chips are discharged at 47.6° C. The critical moisture content of the material is 0.3 (on dry basis). Assuming that heat and mass transfer take place only to (or, from) the top surface of the solids, compute the length of belt required. The dryer operates adiabatically and heat transfer by conduction and radiation may be neglected.

**Solution:** A typical direct contact continuous dryer operating in the countercurrent mode is sketched in Fig. 5.8.1.

The entire dryer may be assumed to be divided into three zones as shown. Drying of the material takes place essentially in zone II. Zone I is essentially a preheating zone where the feedstock is heated from temperature $t_{S1}$ to the drying temperature $t''_S$. Similarly, zone III acts as a reheating zone where the dried solids are heated to the discharge temperature $t_{S2}$. We shall, therefore, assume that moisture removal takes place only in zone II.

To start the computations, let us make an assumption (which we shall be verifying later) that drying takes place in the constant rate period. If this is so and since heat transfer by conduction and radiation are neglected, the surface temperature of drying solids will be constant and equal to the wet-bulb temperature of air. Since the dryer operates adiabatically, the wet-bulb temperature of air will be the same at the two ends of zone II. Thus, $t'_S = t''_S = t_W$. It is given that

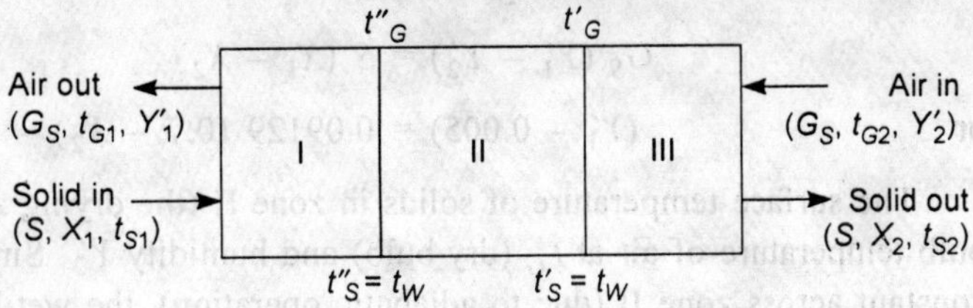

*Figure 5.8.1*

$$Y'_2 = 0.005$$

$$t_{G2} = 115° \text{ C}$$

The enthalpy of inlet air can be, therefore, computed from Eq. (5.1.18):

$$H_{G2} = [1.0 + 1.87\ (0.005)]\ (115.0) + (2502.3)\ (0.005) = 128.587 \text{ kJ/kg}$$

Similarly, enthalpy of the feedstock can be computed from Eq. (5.1.88), since $X_1 = 0.7$, $t_{S1} = 30°$ C and $C_{pS} = 2.1$ kJ/(kg K). Thus, taking $\Delta H_A = 0.0$,

$$H_{S1} = [2.1 + 4.18\ (0.7)]\ (30.0) = 150.78 \text{ kJ/kg}$$

The humid volume of inlet air is (taking $P = 1$ atm),

$$v_{H2} = [0.00283 + 0.00456\ (0.005)]\ (115 + 273)$$

$$= 1.107 \text{ m}^3\text{/kg of dry air}$$

Mass flow rate of air (on dry basis)

$$= (4.7/1.107) = 4.25 \text{ kg/s}$$

Mass flow rate of solids (on dry basis) = $(Bz_S)\ S\rho_b/(1 + X_1)$

where $B$ = belt width
= 1.0 m
$z_S$ = thickness of solid layer on belt
= 0.02 m
$S$ = belt speed
= 0.3 m/s
$\rho_b$ = bulk density of solids
= 110 kg/m$^3$
$X_1$ = 0.7

Therefore, mass flow rate dry solids

$$= (1.0)\ (0.02)\ (0.3)\ (110.0)/(1.0 + 0.7) = 0.388 \text{ kg/s}$$

$$(S/G_S) = (0.388/4.25) = 0.09129$$

To compute the length of the dryer, the outlet conditions of air and those of the solids are to be known. These can be determined by solving the material balance and the heat balance equations. Thus, an overall moisture balance gives

$$G_S\,(Y_1' - Y_2') = S\,(X_1 - X_2)$$

or $$(Y_1' - 0.005) = 0.09129\,(0.7 - X_2) \quad \text{... (i)}$$

The surface temperature of solids in zone II (the drying zone) will be equal to $t_W$, which is the wet-bulb temperature of air at $t_G'$ (dry-bulb) and humidity $Y_2'$. Since the wet-bulb temperature of air remains constant across zone II (due to adiabatic operation), the wet-bulb temperature corresponding to $t_G''$ (dry-bulb) and humidity $Y_1'$ will also be $t_W$. Since $t_W$, $t_G$ and $Y'$ are inter-related by Eq. (5.1.26) or the psychrometric chart, we can write

$$t_W = \text{function of } (t_G', Y_2') \quad \text{... (ii)}$$

Also, $$t_W = \text{function of } (t_G'', Y_1') \quad \text{... (iii)}$$

Now, an overall heat balance yields,

$$(H_{G1} - H_{G2}) = (S/G_S)\,(H_{S1} - H_{S2})$$

or $$(H_{G1} - 128.587) = 0.09129\,(150.78 - H_{S2}) \quad \text{... (iv)}$$

A heat balance for zone III yields,

$$(H_{G2} - H_G') = (S/G_S)\,(H_{S2} - H_S')$$

or $$(128.587 - H_G') = 0.09129\,(H_{S2} - H_S') \quad \text{... (v)}$$

Similarly, a heat balance for zone I gives

$$(H_{G1} - H_G'') = (S/G_S)\,(H_{S1} - H_S'')$$

or $$(H_{G1} - H_G'') = 0.09129\,(150.78 - H_S'') \quad \text{... (vi)}$$

The above six equations are to be solved simultaneously for the six unknowns such as $t_{G1}$, $Y_1'$, $X_2$, $t_G'$, $t_G''$ and $t_W$. However, since two of the equations, such as Eqs (ii) and (iii) are only implicit in $t_W$, a direct solution is difficult and we have to resort to a trial and error procedure. Thus, let $t_W = 35°$ C. Now from the psychrometric chart, at $t_W = 35°$ C and $Y_2' = 0.005$, $t_G' = 110°$ C. Therefore,

$$H_G' = [1.0 + 1.87\,(0.005)]\,(110.0) + 2502.3\,(0.005) = 123.54 \text{ kJ/kg}$$

From Eq. (v),

$$(128.587 - 123.54) = 0.09129\,(H_{S2} - H_S')$$

Since $H_{S2} = (2.1 + 4.18X_2)\,(47.6)$ and $H_S' = (2.1 + 4.18X_2)\,t_W$
$= (2.1 + 4.18X_2)\,(35.0)$, the above equation becomes

$$5.047 = 0.09129\,(2.1 + 4.18X_2)\,(47.6 - 35.0)$$

or $$X_2 = 0.547$$

Once $X_2$ is known, Eq. (i) can be solved for $Y_1'$. Thus

$$Y_1' = 0.019$$

Now, from psychrometric chart, at $t_W = 35°$ C and $Y_1' = 0.019$, $t_G'' = 75.5°$ C.

$$H_{S2} = [2.1 + 4.18\,(0.547)]\,(47.6) = 208.795 \text{ kJ/kg}$$

Now, from Eq. (iv),

$$H_{G1} = 123.291 \text{ kJ/kg} = [1.0 + 1.87\,(0.019)]\, t_{G1} + (2502.3)\,(0.019)$$

Therefore, $t_{G1} = 73.15°$ C

$$H_G'' = [1.0 + 1.87\,(0.019)]\,(75.5) + (2502.3)\,(0.019) = 125.726 \text{ kJ/kg}$$

Now, from Eq. (vi),

$$(123.291 - 125.726) = (0.09129)\,(150.78 - H_S'')$$

or

$$H_S'' = 177.453 \text{ kJ/kg} = [2.1 + 4.18\,(0.7)]\, t_W$$

Therefore, $t_W = 35.3°$ C

Since this value of $t_W$ is very close to that assumed at the outset, no further trials are required. Thus, air leaves the dryer at 73.15° C (dry-bulb) and with 1.9 percent humidity and solids are discharged at 47.6° C with 54.7 percent moisture. Since $X_2$ is greater than $X_c$, our initial assumption that drying takes place in the constant rate period is justified.

To compute the length of the dryer, let us rewrite Eq. (5.1.91) as

$$Q_r = m_{GS}\, C_S\, \Delta T_G' = U\,(ZB)\,\Delta T_m$$

where $Q_r$ = rate of heat transfer, W

$m_{GS}$ = mass flow rate of air (on dry basis)

= 4.25 kg/s

$Z$ = dryer length, m

We have assumed that the heat transfer coefficient is constant and is the same for all zones. The above equation may be rearranged to give

$$Z = \left(\frac{m_{GS} C_S}{U\,B}\right) N_{tG} \qquad \text{... (vii)}$$

Let us, therefore, first compute the number of gas phase transfer units, $N_{tG}$. For zone I, the value of $\Delta T_G'$ is

$$\Delta T_G' = (t_G'' - t_{G1}) = (75.5 - 73.15) = 2.35° \text{ C}$$

The true mean temperature difference ($\Delta T_m$) for zone I is

$$\Delta T_m = \frac{(t_G'' - t_W) - (t_{G1} - t_{S1})}{\ln\left(\dfrac{t_G'' - t_W}{t_{G1} - t_{S1}}\right)} = 41.81° \text{ C}$$

Therefore, $N_{tG}$ (for zone I) = (2.35/41.81) = 0.0562

$$\text{Similarly, } N_{tG} \text{ (for zone III)} = \frac{(t_{G2} - t_G')\ln\left(\dfrac{t_{G2} - t_{S2}}{t_G' - t_W}\right)}{(t_{G2} - t_{S2}) - (t_G' - t_W)} = 0.0703.$$

$$N_{tG} \text{ (for zone II)} = \frac{(t_G' - t_G'') \ln\left(\dfrac{t_G' - t_W}{t_G'' - t_W}\right)}{(t_G' - t_W) - (t_G'' - t_W)} = \ln\left(\frac{t_G' - t_W}{t_G'' - t_W}\right) = 0.6162$$

The total number of gas phase transfer units is, therefore,

$$N_{tG} = (0.0562) + (0.0703) + (0.6162) = 0.7427$$

Since heat transfer by conduction and radiation are neglected, $U = h_c$. The convection heat transfer coefficient $h_c$ may be computed from Eq. (5.1.64):

$$h_c = 0.11\ Re^{-0.29}\ (C_p G)\ Pr^{-2/3} \quad \text{... (viii)}$$

Now $\quad Y'_{av} = (Y'_1 + Y'_2)/2 = 0.012$

$$G_{av} = G_S\,(1 + Y'_{av}) = \frac{(4.25)(1.012)}{(1.0)(0.25)} = 17.204 \text{ kg/(m}^2\text{s)}$$

To note that the area of cross-section for the flow of air is (1.0) (0.27 – 0.02) = 0.25 m$^2$.

$$d^* = \frac{(4.0)(1.0)(0.25)}{(1.0 + 0.25) \times 2} = 0.4 \text{ m.}$$

Since $(t_{G1} + t_{G2})/2 = 94°$ C, the property values such as $C_p$, $\mu$ and $k$ are taken equal to those of dry air at 94° C and 1 atm. Thus

$$C_p = 1.009 \text{ kJ/(kg K)}$$

$$\mu = 0.00002165 \text{ kg/(ms)}$$

$$k = 0.0316 \text{ W/(m K)}$$

Therefore, $\quad Pr = \dfrac{(1009.0)(0.00002165)}{(0.0316)} = 0.69129$

$$Re = (0.4)\ (17.204)/(0.00002165) = 317856.81$$

Therefore, from Eq. (viii),

$$h_c = 61.967 \text{ W/(m}^2\text{ K)}$$

Also, $\quad C_{S,av} = 1.0 + 1.87\,Y'_av = 1.0 + (1.87)\ (0.012) = 1.02244$ kJ/(kg K)

Substituting in Eq. (vii), we get

$$Z = \left[\frac{(4.25)(1022.44)}{(61.967)(1.0)}\right] (0.7427) = \mathbf{52.1 \text{ m}}$$

**Example 5.9:** A crystalline material [average size 0.5 mm, specific heat when dry = 0.836 kJ/(kg K)] is to be dried from 20 percent moisture to 0.3 percent moisture in a direct heat rotary dryer that operates in the countercurrent mode. Air enters the dryer at 155° C (dry-bulb) with 1.0 percent humidity and its maximum superficial mass velocity is restricted at 1.6 kg/(m$^2$s) to avoid excessive dusting. The wet material enters at 21° C and leaves at 120° C. The product delivery rate is to be 500 kg/hr. Neglecting heat losses, specify the required dimensions of the dryer. What should be the speed of rotation of the dryer if the dryer is sloped at 0.1 cm in 10 cm of length? Take the bulk density of the material to be 1250 kg/m$^3$.

**Solution:** Figure 5.8.1 applies to the rotary dryer under consideration as well. As in the previous example, let us assume that drying takes place only in zone II and zones I and III act as preheating and reheating zones respectively. Let us further assume (for want of necessary data) that the critical moisture content of the material is quite low and therefore, drying takes place under constant rate drying conditions. If we go one more step ahead to assume that heat transfer to the solid is by convection only (radiation from dryer wall is neglected), then the surface temperature of the drying solid will be equal to the wet-bulb temperature of air in zone II. Due to adiabatic operation, the wet-bulb temperature of air shall remain constant in zone II and let us denote it as $t_w$. Thus, $t'_S = t''_S = t_w$. The problem is, thus, very similar to that discussed in Example 5.8.

From the given data,

$$X_1 = 0.20/(1 - 0.20) = 0.25$$

$$X_2 = 0.003/(1 - 0.003) \simeq 0.003$$

Since the product delivery rate is 500 kg/hr, the flow rate of dry material is

$$m_S = (SA) = 500\ (1 - X_2) = 498.5 \text{ kg/hr}$$

Now, from Eq. (5.1.85),

$$(SA)\ (X_1 - X_2) = (G_S A)\ (Y'_1 - Y'_2)$$

$$(498.5)\ (0.25 - 0.003) = (G_S A)\ (Y'_1 - 0.01)$$

or $$(G_S A)\ (Y'_1 - 0.01) = 123.1295 \quad \text{... (i)}$$

The outlet temperature of air ($t_{G1}$) is not known. We shall fix it based on the empirical condition that rotary dryers are most economically operated at $N_{tG}$ = 1.5 to 2.5. $N_{tG}$ is approximately given as

$$N_{tG} \approx \frac{(t_{G2} - t_{G1})}{\left[(t_{G1} - t_{S1}) - (t_{G2} - t_{S2})\right] / \ln\left(\dfrac{t_{G1} - t_{S1}}{t_{G2} - t_{S2}}\right)}$$

The above expression is only approximate and should be used only for an initial approximation of $t_{G1}$. Thus, taking $N_{tG}$ = 2.0 and substituting the values of $t_{G2}$, $t_{S1}$ and $t_{S2}$, we get

$$2.0 = \frac{(155 - t_{G1})}{(t_{G1} - 56)} \ln\left(\frac{t_{G1} - 21}{35}\right)$$

or $$t_{G1} = 70° \text{ C}$$

Now, from Eq. (5.1.86), since $Q$ = 0.0 (heat losses from the dryer are negligible),

$$(SA)\ (H_{S2} - H_{S1}) = (G_S A)\ (H_{G2} - H_{G1})$$

From Eq. (5.1.88), taking $\Delta H_A$ = 0.0,

$$H_S = (C_{pS} + XC_{pL})\ t_S = (0.836 + 4.18X)\ t_S$$

Therefore, $$H_{S1} = [0.836 + 4.18\ (0.25)]\ (21.0) = 39.501 \text{ kJ/kg}$$

$$H_{S2} = [0.836 + 4.18\ (0.003)]\ (120.0) = 101.8248 \text{ kJ/kg}$$

From Eq. (5.1.18), $$H_{G2} = [1.0 + 1.87\ (0.01)]\ (155.0) + (2502.3)\ (0.01) = 182.9215 \text{ kJ/kg}$$

$$H_{G1} = (1.0 + 1.87\,Y'_1)\ (70.0) + 2502.3\,Y'_1$$

Substituting in the heat balance equation, we get

$$(498.5)\ (101.8248 - 39.501) = (G_S A)\ [182.9215 - (1.0 + 1.87 Y'_1)\ (70) - 2502.3 Y'_1]$$

or $\quad G_S A\ (112.9215 - 2633.2 Y'_1) = 31068.414 \qquad \text{... (ii)}$

Solving Eqs (i) and (ii) simultaneously, we get

$$Y'_1 = 0.04$$

$$(G_S A) = 4104.3166 \text{ kg / hr}$$

Therefore, $\quad H_{G1} = [1.0 + 1.87\ (0.04)]\ (70.0) + (2502.3)\ (0.04) = 175.328 \text{ kJ/kg}$

$$(S/G_S) = (498.5/4104.3166) = 0.12145$$

Let us now consider zone III, A heat balance for this zone yields

$$(H_{G2} - H'_G) = (S/G_S)\ (H_{S2} - H'_S)$$

Substituting the known values of $H_{G2}$, $H_{S2}$ and $(S/G_S)$ ratio, we get

$$(182.9215 - H'_G) = (0.12145)\ (101.8248 - H'_S) \qquad \text{... (iv)}$$

where $\quad H'_G = C_{S2} t'_G + 2502.3 Y'_2$

$$= [1.0 + 1.87\ (0.01)]\ t'_G + (2502.3)\ (0.01) = 1.0187 t'_G + 25.023$$

$$H'_S = [0.836 + 4.18\ (0.003)]\ t'_S = 0.84854 t_W \text{ (since } t'_S = t_W)$$

Since $t_W$ is the wet-bulb temperature of air corresponding to $t'_G$ (dry-bulb) and humidity $Y'_2$, from Eq. (5.1.26),

$$t'_G = t_W + \lambda_W\ (Y'_W - Y'_2)/C_{S2}$$

$$= t_W + \lambda_W\ (Y'_W - 0.01)/(1.0187) \qquad \text{... (v)}$$

Equations (iv) and (v) are to be now solved for the two unknowns $t'_G$ and $t_W$. However, since Eq. (v) is only implicit in $t_W$, a direct solution is difficult and we have to resort to a trial and error procedure. Thus, let $t'_G = 147.1°$ C. Then

$$H'_G = (1.0187)\ (147.1) + 25.023 = 174.8737 \text{ kJ/kg}$$

Now, from Eq. (iv), $\quad H'_S = 35.561 \text{ kJ/kg}$

Therefore, $\quad t_W = (35.561)/(0.84854) = 41.9085°\text{ C}$

Now, from steam tables, the vapour pressure and latent heat of vapourisation of water at 41.9085° C are

$$P'_A = 8.2272 \text{ kN/m}^2$$

$$\lambda_W = 2402.1576 \text{ kJ/kg}$$

Therefore, $\quad Y'_W = \dfrac{(8.2272)\,(0.622)}{(101.3 - 8.2272)} = 0.05498$

Substituting the above values of $t_W$, $Y'_W$ and $\lambda_W$ in Eq. (v), we get

$$t'_G = 147.9°\text{ C}$$

Since the above-computed value of $t'_G$ does not differ significantly from that assumed at the outset, further trial is unnecessary.

Due to adiabatic operation, the wet-bulb temperature of air will be equal to $t_W$ itself at the other end of zone II also. In other words, $t_W$ is also the wet-bulb temperature corresponding to $t''_G$ (dry-bulb) and humidity $Y'_1$. Therefore, from Eq. (5.1.26),

$$t''_G = t_W + \lambda_W (Y'_W - Y'_1)/C_{S1}$$

where $$C_{S1} = 1.0 + 1.87\ (0.04) = 1.0748$$

Thus, $$t''_G = 41.9085 + (2402.1576)\ (0.05498 - 0.04)/(1.0748) = 75.39°\ \text{C}$$

Now, the heat balance equation for zone I is,

$$(H''_G - H_{G1}) = (S/G_S)\ (H''_S - H_{S1})$$

where $$H''_G = C_{S1} t''_G + 2502.3\, Y'_1$$
$$= (1.0748)\ (75.39) + (2502.3)\ (0.04)$$
$$= 181.1195\ \text{kJ/kg}$$
$$H_{S1} = 39.501\ \text{kJ/kg}$$
$$H''_S = [0.836 + 4.18\ (0.25)]\ (41.9085)$$
$$= 78.33\ \text{kJ/kg}$$

Therefore, $$H_{G1} = 176.343\ \text{kJ/kg}$$

Since $$H_{G1} = C_{S1} t_{G1} + (2502.3)\ Y'_1,$$
$$t_{G1} = [176.343 - 2502.3\ (0.04)]/(1.0748) = 70.94°\ \text{C}$$

It can be seen that the above-computed value of $t_{G1}$ is quite close to the value of $t_{G1}$ assumed at the beginning of the problem. We, therefore, avoid further trial.

Let us thus proceed to compute the dimensions of the dryer. The maximum flow rate of air = $(G_S A)$ $(1 + Y'_1)$ = (4104.3166/3600) (1.04) = 1.1857 kg/s. Since the maximum permissible superficial mass velocity of air is 1.6 kg/(m$^2$·s), the minimum required cross-sectional area of the dryer is

$$A = (1.1857)/1.6 = 0.741\ \text{m}^2$$
$$D = [(4.0)\ (0.741)/\pi]^{0.5} = 0.97\ \text{m}$$

A standard diameter of 1.0 m may be therefore, selected for the dryer. Accordingly, $A = 0.7854$ m$^2$. And

$$G_S = \frac{(4104.3166)}{(3600)(0.7854)} = 1.4516\ \text{kg/(m}^2\cdot\text{s)}$$

$$S = \frac{(498.5/3600)}{(0.7854)} = 0.1763\ \text{kg/(m}^2\cdot\text{s)}$$

To compute the length of the dryer, consider Eq. (5.1.91):

$$q = G_S C_S \Delta T'_G = (Ua)\ Z\, \Delta T_m$$

or, $$Z = \left(\frac{G_s C_s}{Ua}\right)\ (\Delta T'_G/\Delta T_m)$$

But, from Eq. (5.1.93), $(\Delta T'_G/\Delta T_m) = N_{tG}$. Therefore, the above expression becomes

$$Z = \left(\frac{G_s C_s}{Ua}\right) N_{tG} \quad \text{... (vi)}$$

The term within brackets in the above expression is nothing but the height of one gas phase transfer unit, $H_{tG}$. The volumetric heat transfer coefficient, $(Ua)$, may be computed from Eq. (5.1.92):

$$(Ua) = (237.0)\ G^{0.67}/D$$

where

$$G = G_{\text{avg}} = G_S\left[1+\frac{(Y_1' + Y_2')}{2}\right]$$

$$= 1.4516\ (1 + 0.025)$$

$$= 1.488\ \text{kg}/(\text{m}^2\,\text{s})$$

Therefore,
$$(Ua) = (237.0)\ (1.488)^{0.67}/(1.0) = 309.31\ \text{W}/(\text{m}^3\,\text{K})$$

$$C_{S,\,\text{avg}} = (C_{S1} + C_{S2})/2$$

$$= (1.0748 + 1.0187)/2 = 1.04675\ \text{kJ}/(\text{kg K}) = 1046.75\ \text{J}/(\text{kg K})$$

Therefore, from Eq. (vi), $H_{tG} = G_S C_S/(Ua) = (1.4516)\ (1046.75)/(309.31) = 4.9124$ m

The number of transfer units may be now computed for each zone. Thus, for zone I, the true mean temperature difference $(\Delta T_m)$ is

$$\Delta T_m \text{ (for zone I)} = \frac{(t_G'' - t_S'') - (t_{G1} - t_{S1})}{\ln\left(\dfrac{t_G'' - t_S''}{t_{G1} - t_{S1}}\right)}$$

where

$t_S'' = t_W = 41.9085°$ C. Therefore

$$\Delta T_m = \frac{(75.39 - 41.9085) - (70.0 - 21.0)}{\ln\left(\dfrac{75.39 - 41.9085}{70.0 - 21.0}\right)}$$

$$= 40.75°\ \text{C}$$

$$N_{tG} \text{ (for zone I)} = (t_G'' - t_{G1})/\Delta T_m = (75.39 - 70.0)/(40.75) = 0.13227$$

Similarly, for zone II,

$$\Delta T_m \text{ (for zone II)} = \frac{(t_G' - t_S') - (t_G'' - t_S'')}{\ln\left(\dfrac{t_G' - t_S'}{t_G'' - t_S''}\right)}$$

Since $t_S' = t_S'' = t_W$ and $N_{tG} = (t_G' - t_G'')/\Delta T_m$,

$$N_{tG} \text{ (for zone II)} = \ln\left(\frac{t_G' - t_W}{t_G'' - t_W}\right) = \ln\left(\frac{147.1 - 41.9085}{75.39 - 41.9085}\right) = 1.145$$

And for zone III,

$$\Delta T_m \text{ (for zone III)} = \frac{(t_{G2} - t_{S2}) - (t_G{}' - t_S{}')}{\ln\left(\dfrac{t_{G2} - t_{S2}}{t_G{}' - t_S{}'}\right)} = 63.785° \text{ C}$$

$$N_{tG} \text{ (for zone III)} = (t_{G2} - t'_G)/\Delta T_m = (155.0 - 147.1)/(63.785) = 0.12385$$

The total number of gas phase transfer units for the dryer is, therefore,

$$N_{tG} = (0.13227) + (1.145) + (0.12385) = 1.401$$

Thus, from Eq. (vi), the required length of the dryer is

$$Z = H_{tG} N_{tG} = (4.9124)\,(1.401) = 6.88 \text{ m} \simeq 7.0 \text{ m}$$

The length to diameter ratio is, thus, (7.0 / 1.0) = 7.0, which is within the generally prescribed range of 4 to 10. The speed of rotation of the dryer may be estimated from Eq. (5.1.95). Thus, taking $\phi = 0.07$,

$$0.07 = \frac{(0.3344)\,(0.1763)}{(1250)\,N^{0.9}\,(1.0)\,(0.01)} + \frac{(0.6085)\,(1.488)}{(1250)\,\sqrt{0.0005}}$$

Therefore, $N = 0.1$ rps = **6.0 rpm**

**Example 5.10:** A charge stock containing 150 percent of moisture (on dry basis) is to be dried to 10 percent moisture (on dry basis) in a continuous, countercurrent air dryer. The feed rate of the stock is 4.2 kg / min (on dry basis) and air at 60° C (dry-bulb) and humidity 1.55 percent is supplied at 215 kg / min (on dry basis). The air temperature is maintained constant at 60° C throughout the dryer by means of steam coils. Based on laboratory experiments, the rate of drying of this stock has been correlated as given below:

$$-(dX/d\theta) = 0.027\,(X - X^*)\,(Y'_W - Y')$$

where $Y'_W$ = saturation humidity of air at the wet-bulb temperature corresponding to $Y'$

$X^*$ = equilibrium moisture content of the material at 60° C (may be taken equal to 30 percent of the relative humidity of air)

Compute the required length of the dryer if it can hold 46.5 kg of dry material per metre of length.

**Solution:** From the given correlation that describes the rate of drying, the time of drying ($\theta$) can be estimated by integrating it between $X = X_1 = 1.5$ and $X = X_2 = 0.10$. Thus

$$\theta = \int_0^{\theta} d\theta = (1/0.027) \int_{0.10}^{1.5} \frac{dX}{(X - X^*)\left(Y_W{}' - Y'\right)}$$

$$= (1/0.027) \int_{0.10}^{1.5} f(X)\,dX \qquad \text{... (i)}$$

where $f(X) = [1/(X - X^*)\,(Y'_W - Y')]$

To evaluate the above integral, either graphically or numerically, the values of $Y'$, $Y'_W$ and $X^*$ at different values of $X$ are required to be known. From Fig. 5.12 (a), a moisture balance between end 2 (that is, the gas-inlet end) and any section of the dryer gives

$$S\,(X - X_2) = G_S\,(Y' - Y'_2)$$

Since $X_2 = 0.10$, $Y'_2 = 0.0155$ and $(S/G_S) = (4.2/215.0) = 0.01953$, the above equation reduces to

$$Y' = 0.01953\ (X - 0.10) + 0.0155 \qquad \text{... (ii)}$$

For any specified value of $X$, the value of $Y'$ can be computed from the above equation. Now, the wet-bulb temperature of air ($t_W$) at $Y'$ and $t_G$ = 60° C (since the air temperature has been maintained constant at 60° C throughout the dryer) can be obtained from the psychrometric chart Fig. 5.2. Once $t_W$ is known, the value of $Y'_W$ can be computed as

$$Y'_W = (0.622\,P'_{AW})/(P - P'_{AW}) \qquad \text{... (iii)}$$

where $P'_{AW}$ = vapour pressure of water at $t_W$ (to be obtained from steam tables)

To compute the equilibrium moisture content $X^*$, the relative humidity of air is to be known. Now,

$$\text{Relative humidity} = (p_A/P'_A) = (y_A\ P/P'_A)$$

Since $y_A = Y/(1 + Y) = Y'/(0.622 + Y')$ and since it is given that $X^*$ may be taken equal to 30 percent of the relative humidity of air,

$$X^* = \frac{(0.3)\,Y'P}{(0.622 + Y')\,P_A{}'}$$

Now, $P'_A$ (at $t_G$ = 60° C) = 19.94 kN/m$^2$ and taking $P$ = 101.3 kN/m$^2$,

$$X^* = \frac{1.524\,Y'}{(0.622 + Y')} \qquad \text{... (iv)}$$

We shall present here a sample calculation. Let $X$ = 0.7. Then

$Y'$ [from Eq. (ii)] = 0.02722

$t_W$ (at 60° C and $Y'$ = 0.02722, from psychrometric chart) = 35.5° C

$P'_{AW}$ (at 35.5° C, from steam table) = 5.8036 kN/m$^2$

$Y'_W$ [from Eq. (iii)] = 0.038

$X^*$ [from Eq. (iv)] = 0.0639

$f(X)$ = 145.833

Similarly, the values of $f(X)$ at different values of $X$ are computed and the results are tabulated below in Table 5.10.1.

**Table 5.10.1**

| $X$ | $Y'$ | $X^*$ | $Y'_W$ | $f(X)$ |
|---|---|---|---|---|
| 1.5 | 0.4285 | 0.09822 | 0.052 | 77.9654 |
| 1.3 | 0.03894 | 0.08979 | 0.048 | 91.2034 |
| 1.1 | 0.035 | 0.08126 | 0.0446 | 102.25 |
| 0.9 | 0.03113 | 0.07263 | 0.0414 | 117.687 |
| 0.7 | 0.02722 | 0.0639 | 0.038 | 145.833 |
| 0.5 | 0.0233 | 0.055 | 0.0345 | 200.642 |
| 0.3 | 0.0194 | 0.0461 | 0.031 | 339.53 |
| 0.1 | 0.0155 | 0.037 | 0.028 | 1269.84 |

The integral of Eq. (i) can be now evaluated either graphically or numerically. Let us evaluate it numerically using *trapezoidal rule.* This rule is discussed in Example 4.12 of Chapter 4. According to this rule,

$$I = \int_{0.10}^{1.5} f(X)\,dX = (h/2)\,[f_0 + 2\,(f_1 + f_2 + f_3 + f_4 + f_5 + f_6) + f_7]$$

where $h$ = step size = $(X_{i+1} - X_i)$.

In the present example, $h = 0.2$ and $f_0 = 1269.84$, $f_1 = 339.53$, $f_2 = 200.642$ and so on. Finally, $f_7 = 77.9654$. Therefore,

$$I = 334.21$$

$$\theta = (334.21/0.027) = 12378.15 \text{ seconds} = 3.44 \text{ hr.}$$

Since the dryer can hold 46.5 kg of dry material per m of length, the required length of the dryer is

$$Z = \frac{(4.2)(3.44 \times 60)}{(46.5)} = \mathbf{18.64\ m}$$

## Indirect Continuous Dryers

*Drym dryers* are widely used for drying suspensions, sludges and pastes. However, unlike the spray dryers, drum dryers are indirect-heat continuous dryers. The fluid to be dried is spread in a thin film on the outer surface of heated, rotating drums or cylinders. The dryer may be a single-drum (see Fig. 5.15), double drum (Fig. 5.16) or a twin-drum (Fig. 5.17) equipment. The drums are heated usually by steam on the inside. As the drums rotate, the fluid gets dried by the evaporation of the solvent and the product is removed from the drum surface as the drum moves past a knife or scraper. As drying is accomplished in thin films which remain in the heated zone for relatively short periods, this type of dryer is suitable for handling heat-sensitive materials. Due to drying in a thin film, products are removed as flakes. The final moisture content of the product depends on the speed of rotation, drum surface temperature and the thickness of fluid layer on the drum surface. In the double drum dryer, the fluid is fed into the nip between the two rotating drums and in this case, the thickness of the fluid layer held on the drum will be controlled by the interspacing between the drums. Twin-drum dryers are similar to double drum dryers except that they rotate away from each other at the top (see Fig. 5.17). For such dryers, the drum spacing does not influence the thickness of the drying layer.

The simplest mode of feeding the drums is to have the drums dip into the feed solution or sludge and roll through the pan holding the feed solution. This is called *dip feeding.* This method, however, is not satisfactory for suspensions of solids which are apt to settle. In such cases, *top feeding* (as shown in Fig. 5.16) may be employed. During top feeding, the feed may be introduced through a perforated pipe or through a pipe which swings like a pendulum from end to end of the dryer. Top centre feeding permits a thick coat to be formed on the roll surfaces and this could be advantageous when a product of high moisture content and coarse granular structure is desired. For certain materials which do not adhere well to the heated surface, a more forceful method of feeding such as *splash feeding* (shown in Fig. 5.17) is used. Here, the feed trough is kept below the drums and the feed solution is splashed onto the drums by means of rotors.

The drum dryers may be operated in the open at atmospheric pressure or may be totally enclosed for operation under vacuum or for solvent recovery.

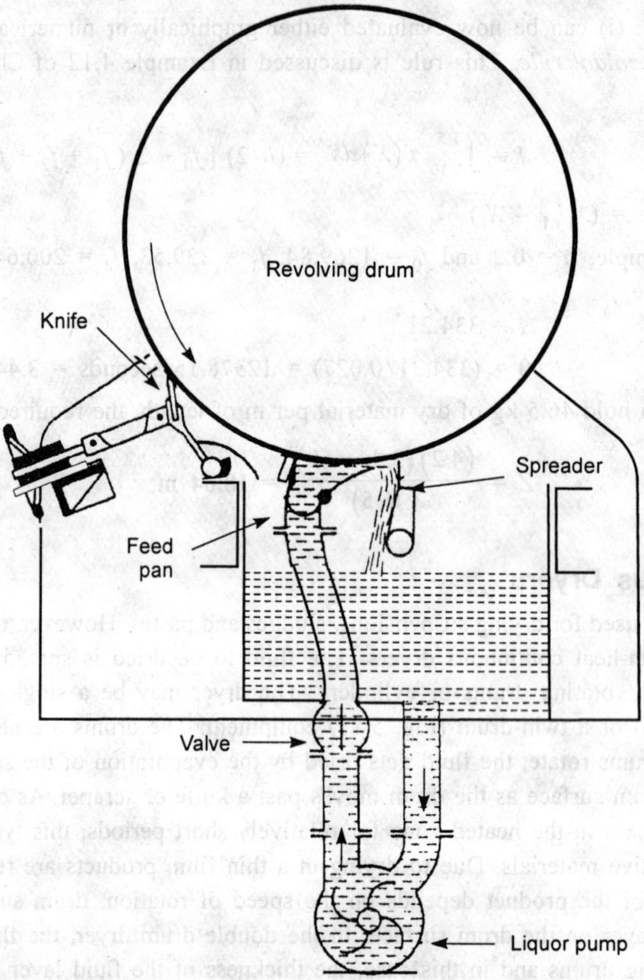


*Figure 5.15: Single drum dryer with pan feed*

*Cylinder dryers* are popularly used for drying continuous sheets of paper and textiles. They consist of a large number of steam–heated rolls, over which the sheet passes continuously. The rolls are usually made of cast iron or welded plate steel and are heated by steam on the inside. In Fig. 5.18, the rolls are arranged in two levels (a three tier arrangement is also used on occasions). The rolls of the bottom row may be geared to one another and driven by motors introduced at various points along the train of rolls. Alternately, the upper row of rolls may be driven from the rolls of the lower row by an idler gear or a chain drive.

The rotary dryers described earlier may be operated also as indirect–heat equipment. *Rotary steam tube dryers* are excellent examples of this category. The dryer consists of a long, rotating cylinder with steam–heated tubes fastened inside in one, two or three concentric rows. These tubes run the full length of the cylinder and rotate with it. Tubes may be simple pipes with condensate draining by gravity into the discharge manifold or the bayonet–type. Steam is admitted to the tubes through a revolving steam

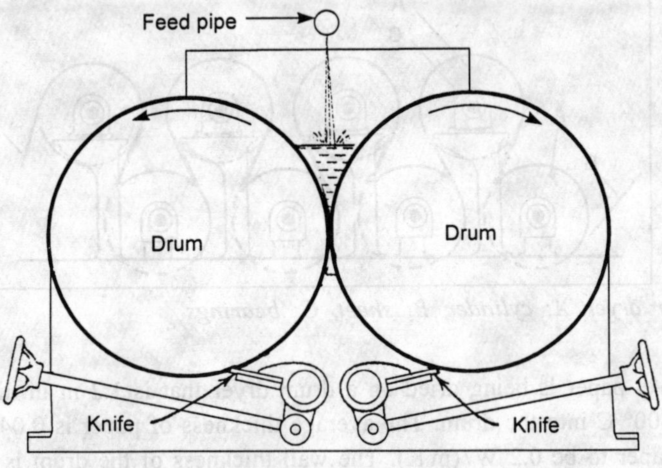


*Figure 5.16: Double drum dryer with top feed*

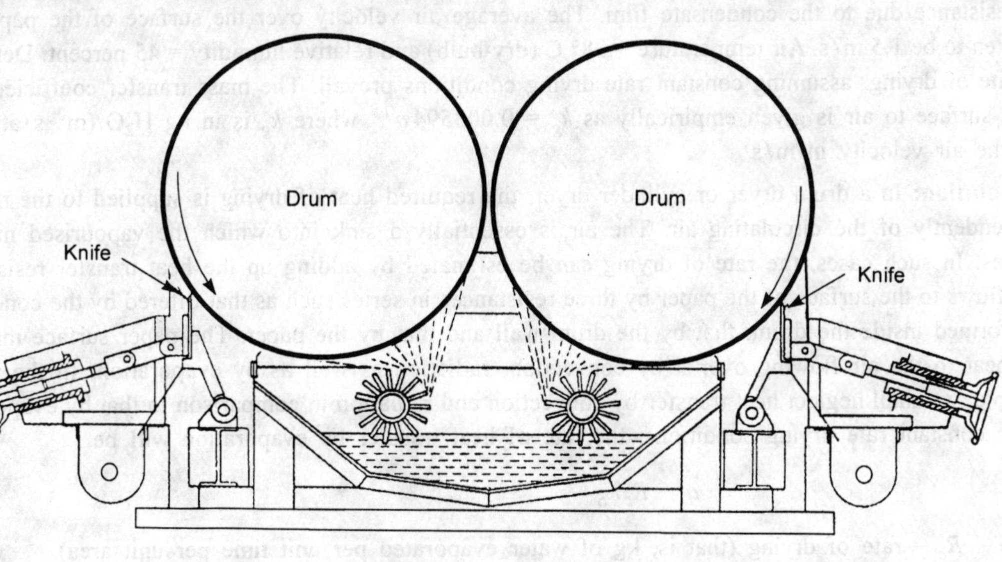


*Figure 5.17: Twin drum dryer with splash feed*

joint. When handling sticky materials, a single row of tubes is preferred. Lifting flights are usually inserted behind the tubes to promote agitation of solids. The wet feed enters the dryer through a chute or screw feeder. The product is discharged through peripheral openings in the shell. These openings also serve to admit purge gas (usually, air) to sweep the solvent vapours from the shell. These vapours are removed at the feed end of the dryer through a natural draft stack and a settling chamber or dust collector.

Rotary steam tube dryers are used for the continuous drying of granular or powdery solids which cannot be exposed to ordinary atmospheric or combustion gases. They are particularly suitable for handling fine, dusty particles which are difficult to be handled in direct–heat rotary dryers. The equipment may be used for drying, solvent recovery as well as for chemical reactions.

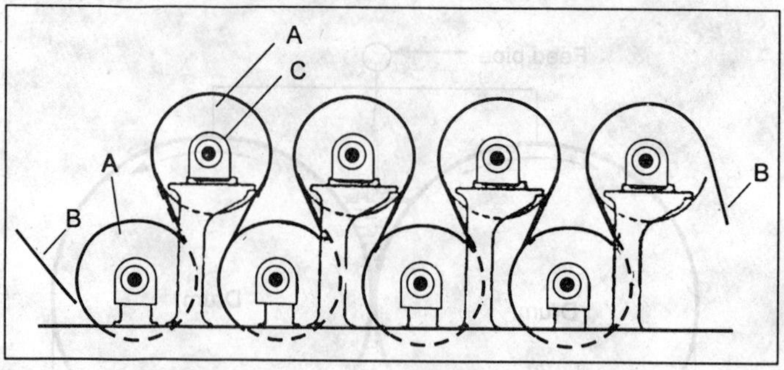


*Figure 5.18: Cylinder dryer, A: cylinder, B: sheet, C: bearings*

**Example 5.11:** Wet paper is being dried on a drum dryer that is 1.2 m in diameter. Dry, saturated steam is supplied at 100° C into the drum. The average thickness of paper is 0.04 mm and take thermal conductivity of wet paper to be 0.2 W/(m K). The wall thickness of the drum is 6.35 mm and thermal conductivity of the metal wall is 45 W/(m K). Assume a coefficient of 4550 W/(m$^2$ K) to account for the resistance due to the condensate film. The average air velocity over the surface of the paper may be taken to be 1.5 m/s. Air temperature = 38° C (dry-bulb) and relative humidity = 45 percent. Determine the rate of drying, assuming constant rate drying conditions prevail. The mass transfer coefficient from paper surface to air is given empirically as $k_g = 0.008594\, v^{0.8}$ where $k_g$ is in kg $H_2O/(m^2 \cdot s \cdot atm)$ and $v$ is the air velocity in m/s.

**Solution:** In a drum dryer or cylinder dryer, the required heat of drying is supplied to the material independently of the circulating air. The air is essentially a sink into which the vapourised moisture escapes. In such cases, the rate of drying can be estimated by adding up the heat transfer resistances. Heat flows to the surface of the paper by three resistances in series such as that offered by the condensate film formed inside the drum, that by the drum wall and that by the paper. The paper surface may also lose heat to the air flowing over it by convection, radiation as well as by evaporation. In the present example, we shall neglect heat transfer by convection and radiation in comparison to that by evaporation. Under constant rate drying conditions, the rate of heat transfer by evaporation will be

$$q = R_c \lambda_S \qquad \text{... (i)}$$

where $R_c$ = rate of drying (that is, kg of water evaporated per unit time per unit area)

$\lambda_S$ = latent heat of vapourisation at temperature $t_S$

$t_S$ = surface temperature of paper

In terms of the mass transfer coefficient, the rate of drying is

$$R_c = k_g\ (P'_{AS} - p_A) \qquad \text{... (ii)}$$

where $k_g$ = mass transfer coefficient

$= (0.008594)\ v^{0.8}$

$= (0.008594)\ (1.5)^{0.8} = 0.012$ kg (m$^2 \cdot$s$\cdot$atm)

$P'_{AS}$ = vapour pressure of water at temperature $t_S$

$p_A$ = partial pressure of water vapour in air

Since the relative humidity of air is 0.45 and $P'_A$ = vapour pressure of water at 38° C = 0.06594 atm, $p_A$ = (0.45) (0.06594) = 0.02967 atm. Equation (i), therefore, reduces to

$$q = 0.012\ (P'_{AS} - 0.02967)\ \lambda_S \quad \text{... (iii)}$$

Now, the rate of heat flow to the surface of the paper by the three resistances in series (which will be equal to $q$ under steady state conditions) is

$$q = U_k\ (t_{St} - t_S) = U_k\ (100 - t_S) \quad \text{... (iv)}$$

where $t_{St}$ = steam temperature = 100° C

$(1/U_k) = (1/h_i) + (x_m/k_m) + (x_p/k_p)$

$h_i$ = coefficient to account for resistance due to condensate film = 4550 W/(m² K)

$x_m$ = dryer wall thickness = 0.00635 m

$k_m$ = thermal conductivity of dryer wall = 45 W/(m K)

$x_p$ = thickness of paper = $0.04 \times 10^{-3}$ m

$k_p$ = thermal conductivity of paper = 0.2 W/(m K)

Therefore, $(1/U_k) = 5.6089 \times 10^{-4}$ (m² K)/W

or $U_K$ = 1782.88 W/(m² K)

Eq. (iv), therefore, becomes

$$q = 1782.88\ (100 - t_S) \quad \text{... (v)}$$

From Eq. (iii) and (v),

$$0.012\ (P'_{AS} - 0.02967)\ \lambda_S = 1782.88\ (100 - t_S)$$

or

$$t_S = 100 - (6.73 \times 10^{-6})\ (P'_{AS} - 0.02967)\ \lambda_S \quad \text{... (vi)}$$

The above equation is only implicit in $t_S$ and therefore, we can solve it only by trial. Thus, let $t_S$ = 89.9° C. Then, from steam table,

$$P'_{AS} = 0.6898 \text{ atm}$$

$$\lambda_S = 2283.456 \text{ kJ/kg}$$

Substituting in Eq. (vi) and solving, we get $t_S$ = 89.85° C, which is very close to that assumed at the outset. The rate of drying can be now obtained from Eq. (ii) as

$$R_c = (0.012)\ (0.6898 - 0.02967)\ (2283.456)\ (1000)$$

$$= 0.00792 \text{ kg/(m}^2\cdot\text{s)} = \mathbf{28.52\ kg/(m^2\cdot hr)}$$

## 5.1.3 Mechanical Drying of Solids

Drying by thermal processes, as described in the earlier subsection, could often be expensive. Solids may be dried by mechanical means, such as by using centrifugals[34], followed by a thermal process. This is usually advantageous since this minimises the energy requirement of the thermal process and thereby enhances the overall process economy.

Centrifugals are nothing but centrifugal filters. They are widely used for drying crystals and granular products in chemical process industries. Drying of sugar crystals, ammonium sulphate crystals, etc. are examples. Batch centrifugals may be overdriven suspended basket type or underdriven link-suspended basket type. The former is more versatile possessing greater stability with unbalanced loads and permits greater speed of operation for a given diameter. It consists of a basket 0.5 to 2.0 m in diameter with a perforated wall. It is driven on a vertical shaft by an electric motor at a speed of 1200 to 2000 rpm. The perforated wall is covered with a suitable filter medium such as canvas cloth or woven metal cloth. The feed is introduced from the top and as the basket rotates, the solids experience large centrifugal forces and are driven to the wall. At the wall, since the filter medium is permeable only to the liquid, the liquid alone passes through the filter cloth and then out through the perforations on the wall and may be collected as the *filtrate* if solvent recovery is required. The solid particles (namely, the dried product) form a layer of cake at the wall. At the end of the batch, these particles are unloaded through the bottom of the basket.

The underdriven batch centrifugal operates in a similar manner except that the basket is driven from the bottom. Such a design is preferred in pharmaceutical industries since the open top permits easy charging and easy cleaning. This type can also be made vapour tight for handling volatile materials and requires less headroom than the suspended overdriven centrifugals.

Batch centrifugals may be operated automatically to save labour and improve efficiency. In automatic batch centrifugals, all the operations (addition of feed, removal of dried product, etc.) are controlled by a timing mechanism and therefore, these machines are run continuously. Fine metal screens are used as the filter media. The feed is introduced through the inlet pipe or chute and after the cake has been built up to the desired thickness, the feed is automatically shut off and the basket is unloaded while still running at full speed by a scraper knife that cuts through the solids and pushes them through the discharge chute. These machines are well suited to free–draining crystalline solids not smaller than 0.1 mm in size. For satisfactory operation, the moisture content of the solids should not exceed 15 to 18 percent. Loading and unloading at full speed subject the crystals to vigorous impacts resulting in breakage of fragile crystals.

In continuous centrifugals, the dried solid particles that form the cake at the filter medium are removed continuously either by using a spiral scraper or by a reciprocating pusher/conveyor. In equipment of this kind, a separate filter medium may not be necessary and the perforated wall of the basket itself could act as the filter medium. In the spiral discharge type, the feed enters the annulus between two concentric horizontal cylinders rotating in the same direction, with the inner cylinder or *hub* rotating at a lower speed than the outer perforate cylinder or bowl. The dried product is retained as the cake on the perforate bowl, while the liquid passes through the perforations. The cake is removed along the inside of the perforate outer drum by a spiral scraper or a series of discharge plows. This type of continuous centrifugal is limited to handling of coarse nonfriable crystals as the scraper tends to break the more friable solids. In the pusher discharge type, the deposited cake is shoved forward along the perforate wall by a reciprocating pusher which is activated by oil pressure acting on the piston in the cylinder. Dry cake reaches the end of the drum and is thrown off by centrifugal force into the collector housing and discharged through the bottom. The pusher then returns and additional cake is deposited before the next stroke. The length and frequency of stroke may be adjusted and is generally set to handle slightly more than the anticipated maximum production of dried solids. The smooth action of the hydraulic pusher is relatively free from any cutting action on the crystals and therefore, this type of centrifugal is suitable for handling more friable crystals.

## 5.2 DISTILLATION

Distillation is a separation process which has perhaps received the widest attention in chemical process industries. In distillation, the feed solution is heated so that a part of it vapourises. The vapours are condensed and collected as what is called the *distillate* ($D$). The liquid that is left behind in the distillation still is called the *residue* ($B$). The feed solution is thus split into two fractions such as the distillate and the residue and the process is then called binary distillation or binary fractionation. It is to be kept in mind that both distillate and residue shall contain all constituents of the feed, though the distillate will be rich in the more volatile constituents and the residue will be rich in less volatile constituents. For example, after the fermentation of molasses, alcohol is recovered from the fermentation broth by distillation. When the alcohol–water mixture is distilled, the distillate shall contain large amounts of alcohol but little water (say, 80 percent ethanol and 20 percent water), while the residue could contain more than 90 percent water and the rest ethanol. Distillation thus differs from evaporation in the sense that in distillation, both the products shall contain necessarily all the constituents though in relatively different proportions, while during evaporation of water from an aqueous solution the vapours evolved shall necessarily be those of water only. Distillation differs from other separation processes such as absorption and stripping in the sense that it does not utilise a third substance (such as the absorbent or the stripping agent) to affect separation.

The feed solution that is being distilled may be binary system (that is, containing only two components) or a multi-component system. The products discharged may be two (as in the case of binary fractionation) or a number of product streams may be discharged at different temperatures from different points on the distillation column. The latter is practised in the distillation of petroleum and petroleum products in plate columns.

The distillation process may be operated with or without reflux. For example, in differential distillation (which is a batch process) or in steam distillation (which may be a batch or continuous process) or in flash distillation (also called equilibrium flash distillation, which is often a continuous process), no reflux is used. But in continuous rectification, the overhead vapours are condensed (either partially or fully) and a part of the condensed liquid is continuously recycled back to the column as reflux. A reboiler is also used at the bottom of the column and a part of the bottom liquid is vapourised and continuously returned to the column. In such cases, there will be continuous counterflow of vapour and liquid across the column and there will be intimate contacting between the two phases.

For the design of any distillation equipment, the *vapour–liquid equilibrium data* are required to be known in advance. This is because for distillation to become successful, the concentrations of the materials to be separated must be widely different between the two phases. When the two phases are at equilibrium, then we obtain the maximum relative difference in concentration between the two phases and consequently, the separation efficiency of the process will also be maximum. Attainment of equilibrium is, therefore, most desirable in distillation and that is why most design methods use equilibrium as one of the boundary conditions for quantitative design calculations.

The *VLE data* are often determined experimentally (by performing the distillation in a standard distillation still such as the Othmer still, Brown still, etc.). They can also be computed analytically based on thermodynamic considerations. If $x$ is the mole fraction of a component $A$ (say, the more volatile component) in the liquid and $y^*$ is its mole fraction in the vapour at equilibrium, then the ratio $(y^*/x)$ is called the *equilibrium ratio or the equilibrium distribution coefficient* and is usually denoted by $K$ or $m$. Thus,

$$K = (y^*/x) \qquad \text{... (5.2.1)}$$

It is to be kept in mind that $K$ is a function of temperature, pressure and concentration. Mathematically,

$$K = f_n\,(T, P, x) \qquad \text{... (5.2.2)}$$

where the notation $f_n$ means *function of.*

Thermodynamically, the *equilibrium ratio* of any component $i$ (let us denote it as $K_i$) can be expressed as the ratio of the fugacity coefficient of component $i$ in the liquid solution to that in the vapour mixture. Thus,

$$K_i = \bar{\phi}_i^{(L)} / \bar{\phi}_i^{(v)} \qquad \text{... (5.2.3)}$$

where $\bar{\phi}_i^{(L)}$ = *fugacity coefficient* of component $i$ in the liquid solution

$\bar{\phi}_i^{(v)}$ = fugacity coefficient of component $i$ in the vapour mixture

The fugacity coefficients are to be determined from any of the thermodynamic equations of state (It may be noted that the above equation assumes that the same equation of state is applicable to both liquid and vapour phases). A brief discussion on the different thermodynamic equations of state that have been proposed in literature is given in Chapter 1 [see, for example, Eq. (1.1.15)]. Here, we shall present how the *Redlich-Kwong-Soave equation* of state can be used for the computation of the fugacity coefficients. This equation is, by far, one of the most reliable equations of state for the computation of $K$-values. According to this equation,

$$\ln \bar{\phi}_i = (b_i/b_m)\,(Z-1) - \ln\,(Z-B) - (A/B)\,\ln\left(1+\frac{B}{Z}\right)\left(\frac{2\,\bar{a}_i}{a_m} - \frac{b_i}{b_m}\right) \qquad \text{... (5.2.4)}$$

where

$$a_i = 0.4278\,(RT_{ci})^2/P_{ci} \qquad \text{... (5.2.5)}$$

$$b_i = 0.0867\,(RT_{ci})/P_{ci} \qquad \text{... (5.2.6)}$$

$$b_m = \sum_{i=1}^{n} z_i b_i \qquad \text{... (5.2.7)}$$

$$a_m = \sum_{i=1}^{n}\sum_{j=1}^{n} z_i z_j\,(1 - C_{ij})\,(\beta_i a_i \beta_j a_j)^{1/2} \qquad \text{... (5.2.8)}$$

$z_i = x_i$ (and $z_j = x_j$), when $\bar{\phi}_i^{(L)}$ is being computed

$z_i = y_i = K_i x_i$ (and $z_j = y_j = K_j x_j$), when $\bar{\phi}_i^{(v)}$ is being computed.

$$\beta_i = \left[1 + S_i\left(1 - \sqrt{T_{ri}}\right)\right]^2 \qquad \text{... (5.2.9)}$$

$$S_i = 0.48508 + 1.55171\,\omega_i - 0.15613\,\omega_i^2 \qquad \text{... (5.2.10)}$$

$\omega_i$ = acentric factor of component $i$ (see Table 5.1)
$T_{ri}$ = reduced temperature = $(T/T_{ci})$

$$\bar{a}_i = \sum_{j=1}^{n} z_j\,(1 - C_{ij})\,(\beta_i a_i \beta_j a_j)^{1/2} \qquad \text{... (5.2.11)}$$

$C_{ij}$ = *binary interaction parameter* (see Table 5.2)

$$A = (a_m\,P)/(RT)^2 \qquad \text{... (5.2.12)}$$

$$B = (b_m\,P)/(RT) \qquad \text{... (5.2.13)}$$

$Z$ = compressibility factor

The value of $Z$ is to be obtained from the solution of the cubic equation

$$Z^3 - Z^2 + Z\,(A - B - B^2) - AB = 0 \qquad \text{... (5.2.14)}$$

For vapour mixture, $Z$ = largest among the real, positive roots of the above equation. For liquid solution, $Z$ = smallest of the real, positive roots of the above equation. Thus, Eq. (5.2.4) can be used for computing both $\bar{\phi}_i^{(L)}$ and $\bar{\phi}_i^{(v)}$.

The *binary interaction parameter* $C_{ij}$ is crucial mainly for polar substances. For hydrocarbon mixtures, $C_{ij} = 0$. Experimental values of $C_{ij}$ are available in literature and a few typical values are given in Table 5.2.

The above method of computation of $K$-values, though has a strong thermodynamic basis, involves computational difficulties. Since $K$ is taken to be a function of $T$, $P$ as well as $x$, an iterative, trial and error, procedure becomes invariable (since $x$ is not known at the outset). Also, for any assumed value of $x$, $K_i$ can be computed from Eq. (5.2.3) only by trial since the term $K_i$ appears on both sides of the equation. The computations, therefore, could often become laborious unless the aid of high speed computers is sought. Computer software packages have been developed[5,7] for the thermodynamic evaluation of $K$-values for multicomponent systems.

When the total pressure is less than about 2 atm and the critical temperature of all components is greater than the system temperature, then $K_i$ can be computed from a simpler relation as given below:

$$K_i = \gamma_i^L P'_i / P \qquad \text{... (5.2.15)}$$

where $\gamma_i^L$ = *liquid phase activity* coefficient of component $i$
$P'_i$ = vapour pressure of pure component $i$
$P$ = total pressure

However, accurate prediction of the *liquid phase activity coefficient* is crucial to the application of the above equation. A reasonably accurate estimate of the activity coefficient is possible from the *regular solution theory of Scatchard and Hildebrand* as given below:

$$\ln(\gamma_i^L) = (\bar{V}_i/RT)\,(\delta_i - \bar{\delta})^2 \qquad \text{... (5.2.16)}$$

where $\bar{V}_i$ = *liquid molar volume* of component $i$, $m^3$/kmole

$\delta_i$ = *solubility parameter* of component $i$ (see Table 5.1), $(J/m^3)^{1/2}$

$$\bar{\delta} = \sum_{i=1}^{n} x_i \bar{V}_i \delta_i \Big/ \sum_{i=1}^{n} x_i \bar{V}_i \qquad \text{... (5.2.17)}$$

Normally, it is sufficient to use the value of $\bar{V}_i$ specified at 25° C in the above equation. Eq. (5.2.16) is independent of pressure and it has the inherent limitation that it does not take into account binary interaction. Therefore, it cannot be applied with confidence to systems containing polar substances. For hydrocarbons, the above equation provides satisfactory estimates of $\gamma_i^L$. The values of solubility parameter ($\delta_i$) and liquid molar volume ($\bar{V}_i$) for some selected substances are listed in Table 5.1. The values of $\delta_i$ listed are those at 25° C. However, $\delta_i$ is not a strong function of temperature.

## Table 5.1

| *Substance* | *Acentric factor*[32] | *Solubility parameter*[8], $(kJ/m^3)^{1/2}$ | *Liquid molar volume*[32] *(at 25° C and 1 atm),* $m^3$*/kmole* |
|---|---|---|---|
| Methane | 0.01083 | 367.404 | 0.0378 |
| Ethane | 0.0989 | 391.332 | 0.05474 |
| Propane | 0.1517 | 413.9766 | 0.07561 |
| *n*-Butane | 0.1931 | 435.322 | 0.09665 |
| Isobutane | 0.177 | 403.627 | 0.0977 |
| *n*-Pentane | 0.2486 | 454.08 | 0.1161 |
| *n*-Hexane | 0.3047 | 474.1315 | 0.13129 |
| *n*-Heptane | 0.3494 | 483.187 | 0.1471 |
| *n*-Decane | 0.4842 | 507.767 | 0.1953 |
| Cyclohexane | 0.2149 | 532.3468 | 0.10886 |
| Ethylene | 0.0852 | 393.277 | 0.04924 |
| Propylene | 0.1424 | 415.90 | 0.0688 |
| 1-Butene | 0.1867 | 457.96 | 0.0896 |
| Benzene | 0.2108 | 595.09 | 0.0895 |
| Toluene | 0.2641 | 578.919 | 0.10656 |
| Ethyl benzene | 0.3036 | 582.80 | 0.12269 |
| *p*-Xylene | 0.3259 | 571.157 | 0.12385 |

For systems containing polar substances, more complex predictive equations for $\gamma_i^L$ that involve binary interaction parameter for each pair of components in the mixture are required for use in Eq. (5.2.15). For binary systems, the value of $\gamma_i^L$ may be predicted from any of the two-parameter expressions such as the Margules equation or the van Laar equation[31, 32]. Other popular expressions are the Wilson equation, UNIQUAC equation and the Wohl equation[31, 9]. As per *Margules equation,*

$$\ln \gamma_1 = [\bar{C}_{12} + 2\,(\bar{C}_{21} - \bar{C}_{12})\,x_1]\,x_2^2 \quad \text{... (5.2.18)}$$

$$\ln \gamma_2 = [\bar{C}_{21} + 2\,(\bar{C}_{12} - \bar{C}_{21})\,x_2]\,x_1^2 \quad \text{... (5.2.19)}$$

where $\gamma_1$ and $\gamma_2$ are the *liquid phase activity coefficients* of component 1 and component 2 respectively (the superscript $L$ has been avoided here). Similarly, as per *Van Laar equation,*

$$\ln \gamma_1 = C_{12}\left(\frac{C_{21}x_2}{C_{12}x_1 + C_{21}x_2}\right)^2 \quad \text{... (5.2.20)}$$

$$\ln \gamma_2 = C_{21}\left(\frac{C_{12}x_1}{C_{12}x_1 + C_{21}x_2}\right)^2 \quad \text{... (5.2.21)}$$

Typical values of the binary interaction parameters $C_{12}$, $C_{21}$, $\bar{C}_{12}$ and $\bar{C}_{21}$ for some selected substances[31] are listed in Table 5.2.

**Table 5.2**
**Binary Interaction Parameters**

| *System* | $C_{12}$ | $C_{21}$ | $\bar{C}_{12}$ | $\bar{C}_{21}$ |
|---|---|---|---|---|
| Actone–water | 2.1041 | 1.5555 | 2.040 | 1.5461 |
| Acetone–methanol | 0.6184 | 0.5797 | 0.6184 | 0.5788 |
| Chloroform–methanol | 0.9356 | 1.8860 | 0.8320 | 1.7365 |
| Ethanol–water | 1.6798 | 0.9227 | 1.6022 | 0.7947 |
| Ethanol–benzene | 1.8570 | 1.4785 | 1.8362 | 1.4717 |
| Hexane–ethanol | 1.9195 | 2.8463 | 1.9398 | 2.7054 |
| Methanol–water | 0.8041 | 0.5619 | 0.7923 | 0.5434 |
| Methanol–benzene | 2.1623 | 1.7925 | 2.1411 | 1.7905 |
| Methyl acetate–methanol | 0.9614 | 1.0126 | 0.9605 | 1.0120 |
| Water–acetic acid | 0.4973 | 1.0623 | 0.4178 | 0.9533 |
| Water–1-butanol | 1.0996 | 4.1760 | 0.8608 | 3.2051 |
| Propanol–water | 2.9095 | 1.1572 | 2.7070 | 0.7172 |
| Carbon tetrachloride–benzene | 0.0951 | 0.0911 | 0.0948 | 0.0922 |

If, in addition, the liquid behaves as an ideal solution, then $\gamma_i^L = 1.0$ and equation 5.2.15 reduces to

$$K_i = (P'_i/P)$$

or

$$y_i^* = (P'_i/P)\,x_i \quad \text{... (5.2.22)}$$

The above equation is nothing but the *Raoult's law* given in Eq. (4.5.11) of Chapter 4. Typical equilibrium plots (in which $y^*$ is plotted against $x$) for some selected systems such as ethanol–water, acetone–chloroform and benzene–toluene are shown in Figs 5.19 and 5.20. From the *VLE data,* we can also compute the bubble point and dew point of the system. It must be kept in mind that unlike pure

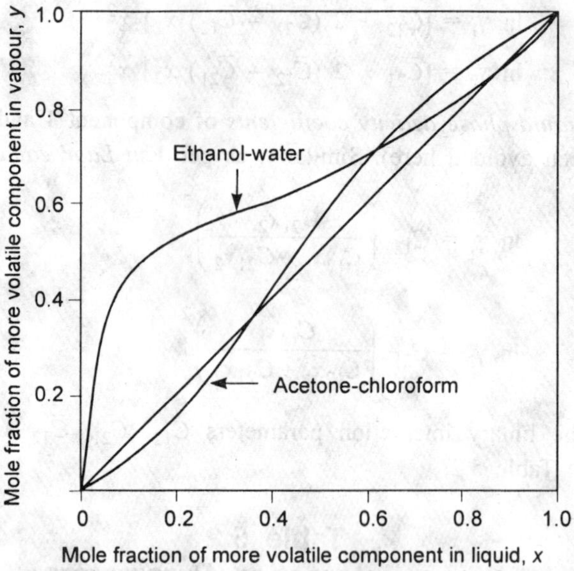


*Figure 5.19: Isobaric equilibrium curves for ethanol-water system and acetone-chloroform system at 1 atm*

liquids, a liquid mixture or liquid solution does not boil at a fixed temperature. When the liquid mixture is heated at a given pressure, the temperature at which the first vapour bubble gets issued is called the *bubble point* of the solution or mixture. In other words, the liquid mixture starts to vapourise at the bubble point. Similarly, when a vapour mixture is cooled, the temperature at which the first drop of condensate is formed is called the *dew point.* For a pure liquid, the bubble point and the dew point are the same and is equal to its boiling point. The difference between the bubble point and the dew point is called the *boiling range* which is a characteristic of every miscible mixture.

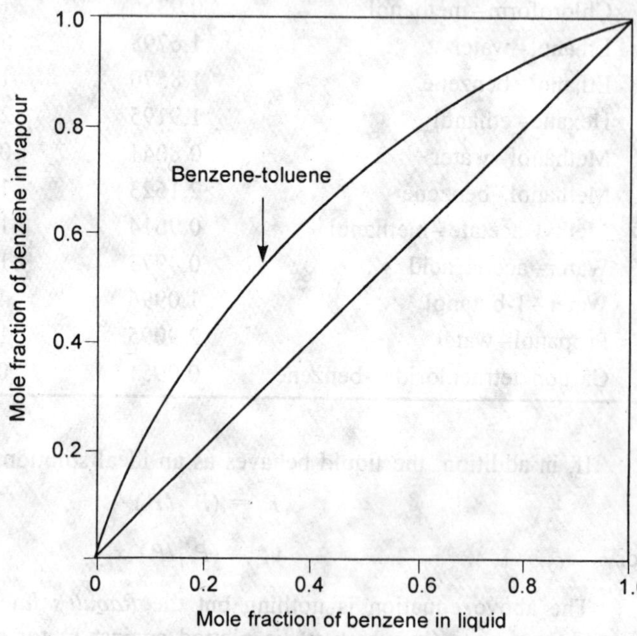


*Figure 5.20: Isobaric equilibrium curve for benzene–toluene system at 1 atm*

When the liquid solution is at its bubble point, it is said to be saturated liquid at that pressure. Similarly, when the vapour mixture is at the dew point, it is said to be a saturated vapour. When the temperature of the vapour mixture is above its dew point, then it becomes superheated.

It is apparent from the above discussion that the bubble point and the dew point depend on the mixture composition. If at any particular composition, the dew point and the bubble point become equal, then the mixture will vapourise at a single temperature, as does a pure liquid. The mixture is then said to form an *azeotrope*. Since at the azeotropic composition, the dew point and bubble point become equal and the mixture vapourises at a single temperature, azeotropes are often called *constant boiling mixtures*. Once the system forms an azeotrope, then it cannot be further separated by conventional distillation since distillation separates materials based on their difference in volatility or boiling temperature. An azeotrope produces an equilibrium vapour that has the same composition as the liquid. In other words, at the azeotropic composition, $y^* = x$ or $K = 1.0$ and consequently, separation becomes impossible. It must be kept in mind that azeotropic composition as well as its boiling point changes with pressure and therefore, it is possible to eliminate azeotropism from the system by changing the pressure. A large number of systems tend to form azeotropes at different pressures and compositions. For example, an ethanol–water mixture containing 89.43 mole percent of ethanol forms an azeotrope at 78.15° C and 1 atm. This azeotrope boils at a temperature that is lower than the boiling point of either pure ethanol or pure water and is, therefore, refered to as a *minimum boiling mixture.* On the other hand, an acetone–chloroform azeotrope is formed at 64.44° C and 1 atm at 34.0 mole percent of acetone. This is called a *maximum boiling mixture* since its boiling point is higher than that of either pure acetone or pure chloroform.

It is, therefore, clear that at $y^* = x$, no separation is possible or in other words, the more is the difference between $y^*$ and $x$, the easier will be the separation. An additional parameter is, therefore, often used to indicate the degree of separability which is called the *relative volatility* (α). For a binary system, if $K_1$ is the equilibrium ratio of the more volatile component and $K_2$ that of the less volatile component, then

$$\alpha = (K_1/K_2) \qquad \text{... (5.2.23)}$$

Since $K_1 = (y_1^*/x_1)$ and $K_2 = (1 - y_1^*)/(1 - x_1)$, we can write

$$\alpha = \frac{(y_1^*/x_1)}{(1-y_1^*)/(1-x_1)} = \frac{y_1^*(1-x_1)}{x_1(1-y_1^*)} \qquad \text{... (5.2.24)}$$

Evidently, when $y_1^* = x_1$, $\alpha = 1.0$ and no separation is possible. The larger the deviation of α from unity, the greater will be the degree of separability. For multicomponent systems, the relative volatility is defined as

$$\alpha_i = (K_i/K_k) \qquad \text{... (5.2.25)}$$

where $K_i$ = equilibrium ratio of component $i$

$K_k$ = equilibrium ratio of the key component (usually, the heavy key)

When the feed is a multicomponent mixture, it is conventional to define one or two *key components.* For example, the *heavy key component is defined in such a way that all* components heavier than it shall not be present in the distillate. In other words, the distillate shall hardly contain components that are heavier than the heavy key and will contain only the heavy key component and components lighter than it. Similarly, we can also define a *light key* such that components lighter than it shall hardly be present in the residue or the bottom product. It must be understood that the terms *lighter* and *heavier* are used to indicate the volatility of the components. For example, when we say component $A$ is lighter than $B$, we mean that $A$ is more volatile than $B$. Similarly, $B$ is heavier than $A$ means $B$ is less volatile than $A$. Let a petroleum fraction containing propane, butane, pentane, hexane and heptane is being distilled. We could choose butane as the light key and hexane as the heavy key. In that case, the distillate shall not contain any heptane and the residue will hardly contain any propane. The key components (such as butane and hexane) and the components in between them (such as pentane) shall be present in both distillate and the residue (in other words, these components shall distribute between the distillate and the residue).

**Example 5.12:** A mixture of *n*-butane and *n*-pentane are at equilibrium at 3 atm and 37.78° C. Compute the liquid and vapour compositions,

(*a*) from Eq. (5.2.3) and if Redlich-Kwong-Soave equation of state is applicable,
(*b*) based on activity coefficients determined from Hildebrand and Scatchard's regular solution theory and from Eq. (5.2.15),
(*c*) from Raoult's law.

**Solution:** (*a*) As stated, the use of Eq. (5.2.3) demands a trial and error procedure. When no information is available regarding the approximate magnitudes of $x$ or $K$-values, then it is recommended that $x$ and $K$-values are first computed assuming that the system behaves ideally (that is, assuming that Raoult's law is being obeyed). The $x$-values, so computed, are now used as the first approximation in the thermodynamic method. For the present system, the $x$-values from Raoult's law have been computed in part (*c*) of this problem. Therefore, let $x_1 = 0.79$ (the suffix 1 denotes *n*-butane).

Even at the above assumed value of $x$, the $K$-values can be computed from Eq. (5.2.3) only by trial, since the term $K_i$ appears on both sides of the equation. Therefore, let

$$K_1 = 1.163$$

Now,

$$x_2 = (1 - x_1) = (1 - 0.79) = 0.21$$
$$y_1 = (K_1 x_1) = (1.163)\ (0.79) = 0.91877$$
$$y_2 = (1 - y_1) = 0.081228$$
$$K_2 = (y_2/x_2) = 0.3868$$

The critical properties of *n*-butane and *n*-pentane are

$$T_{C1} = 425.2°\ K$$
$$P_{C1} = 3797.0\ kN/m^2$$
$$T_{C2} = 469.7°\ K$$
$$P_{C2} = 3369.0\ kN/m^2$$

Therefore, from Eq. (5.2.5),

$$a_1 = 0.4278\ [(8214.4)\ (425.2)]^2/(3797 \times 10^3) = 1.408 \times 10^6$$
$$a_2 = 0.4278\ [(8314.4)\ (469.7)]^2/(3369 \times 10^3) = 1.9366 \times 10^6$$

Similarly, from Eq. (5.2.6),

$$b_1 = (0.0867)\ (8314.4)\ (425.2)/(3797000.0) = 0.080724$$
$$b_2 = (0.0867)\ (8314.4)\ (469.7)/(3369 \times 10^3) = 0.1005$$

The values of acentric factor for *n*-butane and *n*-pentane (from Table 5.1) are,

$$\omega_1 = 0.1931,\ \omega_2 = 0.2486$$

From Eq. (5.2.10), $S_1 = 0.48508 + 1.55171\ (0.1931) - 0.15613\ (0.1931)^2 = 0.77889$

Similarly, $S_2 = 0.8612$

From Eq. (5.2.9) therefore,

$$\beta_1 = \left[1+(0.77889)\left(1-\sqrt{310.78/425.2}\right)\right]^2 = 1.238758$$

$$\beta_2 = 1.347$$

Since the present system is a hydrocarbon mixture, we can take $C_{ij} = 0.0$.

Let us first estimate the liquid-phase fugacity coefficients. Thus,

$$b_m = \Sigma x_i b_i = (x_1 b_1 + x_2 b_2)$$

$$= (0.79)\ (0.080724) + (0.21)\ (0.1005) = 0.08487$$

For a binary system (that is, $n = 2$), Eq. (5.2.8) reduces to (for $C_{ij} = 0$)

$$a_m = x_1^2\ (\beta_1 a_1) + 2x_1x_2\ \sqrt{\beta_1 \beta_2 a_1 a_2}\ + x_2^2\ (\beta_2 a_2)$$

$$= \left(x_1 \sqrt{\beta_1 a_1} + x_2 \sqrt{\beta_2 a_2}\right)^2$$

Substituting the values of $\beta_1$, $\beta_2$, $a_1$, $a_2$ and $x_1$ and $x_2$,

$$a_m = 1.91153 \times 10^6$$

Similarly, for a binary system, Eq. (5.2.11) also reduces to

$$\bar{a}_1 = x_1\ (\beta_1 a_1) + x_2\ (\beta_1 \beta_2 a_1 a_2)^{1/2}$$

$$\bar{a}_2 = x_2\ (\beta_2 a_2) + x_1\ (\beta_1 \beta_2 a_1 a_2)^{1/2}$$

Therefore,

$$\bar{a}_1 = 1.826 \times 10^6$$

$$\bar{a}_2 = 2.233 \times 10^6$$

Now, from Eqs (5.2.12) and (5.2.13),

$$A = \frac{\left(1.91153\times10^6\right)\left(303.975\times10^3\right)}{(8314.4)^2\ (310.78)^2} = 0.087$$

$$B = \frac{(0.08487)\left(303.975\times10^3\right)}{(8314.4)\,(310.78)} = 0.009985$$

To note that $P = 3$ atm $= 303.975$ kN/m$^2$ and $T = 310.78°$ K. Equation (5.2.14) now reduces to

$$Z^3 - Z^2 + Z\ (0.07694) - (8.6894 \times 10^{-4}) = 0$$

The above equation has three real, positive roots and the smallest among them is, $Z = 0.013697$. Now, from Eq. (5.2.4),

$$\bar{\phi}_1^{(L)} = 1.0825$$

$$\bar{\phi}_2^{(L)} = 0.3425$$

The vapour phase fugacity coefficients can be now computed in a similar way. The only modification required is that we have to substitute $y_1$ and $y_2$ in the place of $x_1$ and $x_2$. Thus

$$b_m = y_1 b_1 + y_2 b_2$$

$$= (0.91877)\ (0.080724) + (0.081228)\ (0.1005) = 0.08233$$

Similarly,

$$a_m = 1.8081 \times 10^6$$
$$\bar{a}_1 = 1.7759 \times 10^6$$
$$\bar{a}_2 = 2.1719 \times 10^6$$
$$A = 0.0823$$
$$B = 0.0096853$$

Equation (5.2.14), therefore, reduces to

$$Z^3 - Z^2 + Z\,(0.0725386) - (7.9726 \times 10^{-4}) = 0$$

The above equation also has three real, positive roots and the largest among them is $Z = 0.92023$. Therefore, from Eq. (5.2.4),

$$\bar{\phi}_1^{(v)} = 0.93047$$
$$\bar{\phi}_2^{(v)} = 0.89689$$

The values of equilibrium ratio are therefore,

$$K_1 = \bar{\phi}_1^{(L)} / \bar{\phi}_1^{(v)} = (1.0825)/(0.93047) = 1.1634$$
$$K_2 = (0.3425)/(0.89689) = 0.3818$$

Since the $K$-values computed are very close to those assumed at the outset, we can discontinue further trial.

The $y$-values are now obtained as,

$$y_1 = (K_1 x_1) = (1.1634)\,(0.79) = 0.919086$$
$$y_2 = (K_2 x_2) = (0.3818)\,(0.21) = 0.080178$$
$$\Sigma K_i x_i = (0.919086) + (0.080178) = 0.999264$$

Since $\Sigma K_i x_i$ is very close to unity, the assumed values of $x$ are also accurate and no further iteration is necessary. The liquid and vapour compositions are therefore,

$$x_1 = 0.79 \quad y_1 = 0.919$$
$$x_2 = 0.21 \quad y_2 = 0.08$$

The butane–propane system behaves essentially as an ideal mixture at 3 atm and that is why the compositions computed by the thermodynamic method are essentially the same as those obtained from Raoult's law.

(*b*) To apply the regular solution theory of Scatchard and Hildebrand [Eq. (5.2.16)], let us first retrieve the property values of the components (namely, the liquid molar volume and the solubility parameter) from Table 5.1. Thus

$$\bar{V}_1 = 0.09665 \text{ m}^3/\text{kmole}$$
$$\bar{V}_2 = 0.1161 \text{ m}^3/\text{kmole}$$
$$\delta_1 = 435.322 \text{ (kJ/m}^3)^{1/2}$$
$$\delta_2 = 454.08 \text{ (kJ/m}^3)^{1/2}$$

In this method also, a trial and procedure is invariable. Therefore, let $x_1 = 0.79$. Then, from Eq. (5.2.17),

$$\bar{\delta} = \frac{(x_1 \bar{V}_1 \delta_1) + (x_2 \bar{V}_2 \delta_2)}{(x_1 \bar{V}_1 + x_2 \bar{V}_2)}$$

$$= \frac{(0.79)(0.09665)(435.322) + (0.21)(0.1161)(454.08)}{(0.79)(0.09665) + (0.21)(0.1161)} = 444.99$$

Now, from Eq. (5.2.16), $\ln \gamma_1 = \dfrac{(0.09665)(435.322 - 444.99)^2}{(8.3144)(310.78)}$

or $\gamma_1 = 1.0035$

Similarly, $\ln \gamma_2 = \dfrac{(0.1161)(454.08 - 444.99)^2}{(8.3144)(310.78)}$

or $\gamma_2 = 1.0037189$

The vapour pressure of *n*-butane and *n*-pentane at 310.78 K are

$$P_1' = 2650 \text{ mm Hg}$$

$$P_2' = 830 \text{ mm Hg}$$

It is given that $P$ = 3 atm = 2280 mm Hg. Therefore, from Eq. (5.2.15),

$$K_1 = (1.0035)\ (2650)/(2280) = 1.16635$$

$$K_2 = (1.0037189)\ (830)/(2280) = 0.3654$$

The *y*-values are therefore, $y_1 = (K_1 x_1) = (1.16635)\ (0.79) = 0.9214$

$$y_2 = (0.3654)\ (0.21) = `0.076734$$

$$\Sigma K_i x_i = \Sigma y_i = 0.99815$$

Since the value of $\Sigma K_i x_i$ is not sufficiently close to unity, a second trial is desirable. Therefore, let $x_1$ = 0.795. Then

$$\bar{\delta} = 439.75845$$

$$\gamma_1 = 1.0007364$$

$$\gamma_2 = 1.0092583$$

$$K_1 = 1.163$$

$$K_2 = 0.3674$$

The *y*-values are, $y_1 = (1.163)\ (0.795) = 0.9246$

$$y_2 = (0.3674)\ (0.205) = 0.0753$$

$$\Sigma K_i x_i = (0.9246) + (0.0753) = 0.9999$$

Thus, the liquid and vapour compositions are

$$x_1 = 0.795 \quad y_1 = 0.9246$$

$$x_2 = 0.205 \quad y_2 = 0.0753$$

(*c*) According to Raoult's law [Eq. (5.2.22)],

$$y_1 = (P_1'/P)\ x_1$$
$$y_2 = (P_2'/P)\ x_2$$

If we add the above two equations, we get

$$(y_1 + y_2) = 1.0 = (x_1 P_1' + x_2 P_2')/P = [x_1 P_1' + (1 - x_1)\ P_2']/P$$

Solving for $x_1$, we get $x_1 = (P - P_2')/(P_1' - P_2') = (2280 - 830)/(2650 - 830) = 0.7967$

$$x_2 = (1 - x_1) = 0.2033$$

Now, $y_1 = (P_1'/P)\ x_1 = \left(\dfrac{2650}{2280}\right)(0.7967) = 0.926$

$$y_2 = (1 - y_1) = 0.074$$

The compositions are therefore,

$$x_1 = 0.7967 \quad y_1 = 0.926$$
$$x_2 = 0.2033 \quad y_2 = 0.074$$

The $K$-values are, $K_1 = (y_1/x_1) = 1.1623$

$$K_2 = (y_2/x_2) = 0.364$$

**Example 5.13:** Compute the VLE data for ethanol–water system at 1 atm and 79.8° C from Margules equation. Compare your results with the experimentally reported values given below:

$$x_1 = 0.5079 \quad y_1 = 0.6564$$

where the suffix 1 stands for ethanol. The vapour pressure of ethanol and water may be computed from the Antoine equation:

$$\log P' = A - B/(T + C)$$

where $P'$ is in mm Hg and $T$ is in °C. The constants $A$, $B$ and $C$ are given below:

| | $A$ | $B$ | $C$ |
|---|---|---|---|
| Ethanol | 8.1122 | 1592.864 | 226.184 |
| Water | 8.07131 | 1730.630 | 233.426 |

**Solution:** From Margules equations [Eqs (5.2.18) and (5.2.19)] and with reference to Table 5.2,

$$\ln \gamma_1 = [1.6022 + 2\ (0.7947 - 1.6022)\ x_1]\ x_2^2$$
$$\ln \gamma_2 = [0.7947 + 2\ (1.6022 - 0.7947)\ x_2]\ x_1^2$$

Also, $x_1 + x_2 = 1.0$

The above three equations are to be solved simultaneously with Eq. (5.2.15). However, a direct solution is difficult here since the Margules equations contain exponential terms. We have to, therefore, resort to a trial and error procedure. Therefore, let $x_1 = 0.50$.

$$x_2 = (1 - x_1) = 0.50$$

The margules equations can be now solved for $\gamma_1$ and $\gamma_2$ as

$$\gamma_1 = 1.21978$$
$$\gamma_2 = 1.49264$$

From the Antoine equation, the vapour pressure of ethanol at 79.8° C is obtained as

$$\log P_1' = 8.1122 - 1592.864/(79.8 + 226.184)$$

or
$$P_1' = 806.285 \text{ mm Hg}$$

Similarly, the vapour pressure of water at 79.38° C is,

$$P_2' = 351.66 \text{ mm Hg}$$

From Eq. (5.2.15), therefore,

$$K_1 = (1.21978)\ (806.285)/(760.0) = 1.294$$

$$K_2 = (1.49264)\ (351.66)/(760.0) = 0.69067$$

The $y$-values are,
$$y_1 = (1.294)\ (0.5) = 0.647$$

$$y_2 = (0.69067)\ (0.5) = 0.345335$$

Therefore,
$$\Sigma K_i x_i = \Sigma y_i = (0.647) + (0.345335) = 0.99237$$

Since $\Sigma y_i$ is not sufficiently close to unity, the assumed value of $x_1$ is not accurate and we have to continue the trials. Trials are, therefore, repeated for $x_1$ = 0.51, 0.52 and 0.53. The results are given below:

| | $x_1 = 0.51$ | $x_1 = 0.52$ | $x_1 = 0.53$ |
|---|---|---|---|
| $y_1$ : | 0.65227 | 0.6576 | 0.6630 |
| $y_2$ : | 0.34251 | 0.3396 | 0.3365 |
| $\Sigma y_i$ : | 0.99478 | 0.9972 | 0.9995 |

It can be seen that when $x_1 = 0.53$, $\Sigma y_i$ is very close to unity and we can take this as the final value. The computed VLE data are tabulated below for comparison with the experimentally reported values:

| *Temperature* | $x_1$ | | $y_1$ | |
|---|---|---|---|---|
| | *Computed* | *Reported* | *Computed* | *Reported* |
| 79.8° C | 0.53 | 0.5079 | 0.663 | 0.6564 |

The equilibrium compositions at any other temperature can be similarly computed so as to obtain the complete VLE data at the given pressure.

**Example 5.14:** Compute the bubble point and the dew point at 1720 kN/m$^2$ of a hydrocarbon mixture having the following composition:

*n*-Propane = 12.11 percent, *n*-Butane = 60.0 percent,

*n*-Pentane = 25.0 percent, *n*-Hexane = 2.89 percent

The $K$-values of the above hydrocarbons at 1720 kN/m$^2$ may be computed from the empirical correlation of Amundson and Pontinen[10]:

$$K_i = A_{0i} + A_{1i}T + A_{2i}T^2 + A_{3i}T^3$$

where $T$ is the temperature in °F. The values of the correlations constants are given below:

| *Component* | $A_0$ | $A_1 \times 10^4$ | $A_2 \times 10^6$ | $A_3 \times 10^8$ |
|---|---|---|---|---|
| $C_3$ | 0.84 | −46.6 | 49.4 | −3.033 |
| $C_4$ | −0.177 | 49.5 | −4.15 | 2.220 |
| $C_5$ | −0.0879 | 17.7 | 0.2031 | 1.310 |
| $C_6$ | 0.093 | −15.39 | 10.37 | −0.159 |

**Solution:** Let the bubble point = 232° F = 111.11° C. The *K*-values can be now computed from the given expression and the *y*-values can be obtained as $y_i = K_i x_i$. The results are given below:

| *Component* | $x_i$ | $K_i$ | $y_i = K_i x_i$ |
|---|---|---|---|
| $C_3$ | 0.1211 | 2.039 | 0.24692 |
| $C_4$ | 0.60 | 1.025 | 0.615 |
| $C_5$ | 0.25 | 0.49725 | 0.1243 |
| $C_6$ | 0.0289 | 0.27425 | 0.0079258 |
| | | | $\Sigma y_i = 0.9941458$ |

Since $\Sigma y_i$ is not sufficiently close to unity the assumed value of bubble point is incorrect. Trials are therefore continued by increasing the value of bubble point by 0.1° F. Finally, the results obtained at 233° F are given below:

| *Component* | $x_i$ | $K_i$ | $y_i = K_i x_i$ |
|---|---|---|---|
| $C_3$ | 0.1211 | 2.05244 | 0.24855 |
| $C_4$ | 0.60 | 1.03186 | 0.61912 |
| $C_5$ | 0.25 | 0.50124 | 0.12531 |
| $C_6$ | 0.0289 | 0.2772 | 0.008011 |
| | | | $\Sigma y_i = 1.00099$ |

Since $\Sigma y_i$ is very close to unity, we can discontinue further trials. The bubble point of the mixture at 1720 kN/m$^2$ is, therefore, 233° F or 111.66° C.

The dew point of the mixture can be similarly computed. For example, let the dew point = 250° F. The *K*-values are similarly computed and the *x*-values are obtained as $x_i = (y_i / K_i)$. The results are tabulated below:

| *Component* | $y_i$ | $K_i$ | $x_i = y_i / K_i$ |
|---|---|---|---|
| $C_3$ | 0.1211 | 2.2886 | 0.05291 |
| $C_4$ | 0.60 | 1.148 | 0.5226 |
| $C_5$ | 0.25 | 0.57198 | 0.4370 |
| $C_6$ | 0.0289 | 0.3315 | 0.08718 |
| | | | $\Sigma x_i = 1.09969$ |

Since $\Sigma x_i$ is far from unity, the trials are to be repeated by assuming larger value of dew point. After a good number of trials, the results obtained at 264° F are given below:

| Component | $y_i$ | $K_i$ | $x_i = y_i/K_i$ |
|---|---|---|---|
| $C_3$ | 0.1211 | 2.49467 | 0.04854 |
| $C_4$ | 0.60 | 1.249 | 0.48038 |
| $C_5$ | 0.25 | 0.63457 | 0.39396 |
| $C_6$ | 0.0289 | 0.380 | 0.07605 |
| | | | $\Sigma x_i = 0.99893$ |

The dew point of the mixture is, therefore, 264° F or 128.88° C. It will not be wrong to call the bubble point and the dew point of a mixture its initial boiling point (IBP) and final boiling point (FBP) respectively. The vapourisation commences at the bubble point and gets completed at the dew point when the last drop of liquid also vanishes as vapour. Conversely, the condensation of a vapour mixture starts at its dew point and gets completed at the bubble point when the last trace of vapour also gets converted into liquid.

### 5.2.1 Flash Distillation

Flash distillation, also called *equilibrium distillation,* is a single stage process. It may be conducted either as a batch process or as a continuous process. The feed mixture is introduced to the flash drum or still (see Fig. 5.21) in which the pressure is maintained low. Due to the sudden decrease in pressure, a part of the feed vapourises and the issued vapours are collected and condensed to form the distillate ($D$). It is assumed that these vapours have been in equilibrium with the residual liquid in the still ($B$). Though true equilibrium can never actually be reached, conditions very close to equilibrium can be easily attained in actual practices. The residual liquid may be further flashed in another flash drum and in this way, successive flash vapourisations may be conducted in a series of single-stage operations.

If $x_{Fi}$ is mole fraction of a component $i$ in the feed, $y_{Di}$ that in the vapour and $x_{Bi}$ that in the residual liquid, then a simple material balance yields,

$$F = D + B \quad \text{... (5.2.26)}$$

$$Fx_{Fi} = Dy_{Di} + Bx_{Bi} \quad \text{... (5.2.27)}$$

Since the vapour is in equilibrium with the liquid,

$$(y_{Di}/x_{Bi}) = K_i \quad \text{... (5.2.28)}$$

where $K_i$ is the equilibrium ratio of the component $i$ at the operating pressure and temperature. The values of $K_i$ are to be obtained either from experimental data or from any of the expressions discussed in the earlier subsection such as Eqs (5.2.3), (5.2.15) or (5.2.22).

Equation (5.2.27) can be rewritten as

$$(B + D)\, x_{Fi} = Dy_{Di} + (By_{Di}/K_i)$$

or

$$y_{Di} = \frac{x_{Fi}\,(1 + B/D)}{1 + \left[\dfrac{(B/D)}{K_i}\right]} \qquad \text{... (5.2.29)}$$

Since $\sum_{i=1}^{n} y_{Di} = 1.0$, we can write

$$1.0 = \sum_{i=1}^{n} \frac{x_{Fi}\,(1+q')}{1+(q'/K_i)} \qquad \text{... (5.2.30)}$$

where $q' = (B/D)$. Similarly, by eliminating $y_{Di}$ from Eq. (5.2.27), we get

$$1.0 = \sum_{i=1}^{n} x_{Bi} = \sum_{i=1}^{n} \frac{x_{Fi}\,(1+q')}{(K_i + q')} \qquad \text{... (5.2.31)}$$

Equation (5.2.30) or (5.2.31) can be used for computing the compositions of products from a flash distillation process. For binary systems, the product compositions can also be determined graphically. For example, from Eq. (5.2.27),

$$y_{Di} = (F/D)\, x_{Fi} - (B/D)\, x_{Bi} \qquad \text{... (5.2.32)}$$

Thus, if we plot a straight line of slope $-(B/D)$ and intercept $(F x_{Fi}/D)$, then its point of intersection with the equilibrium curve will give the values of $y_{Di}$ and $x_{Bi}$ (see Fig. 5.22).

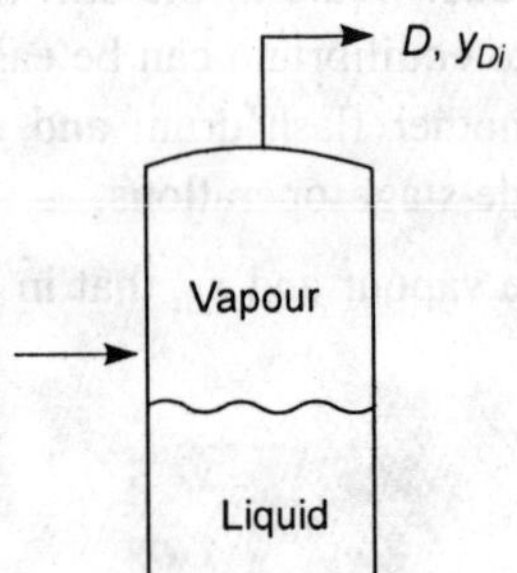


*Figure 5.21: Schematic of flash drum*

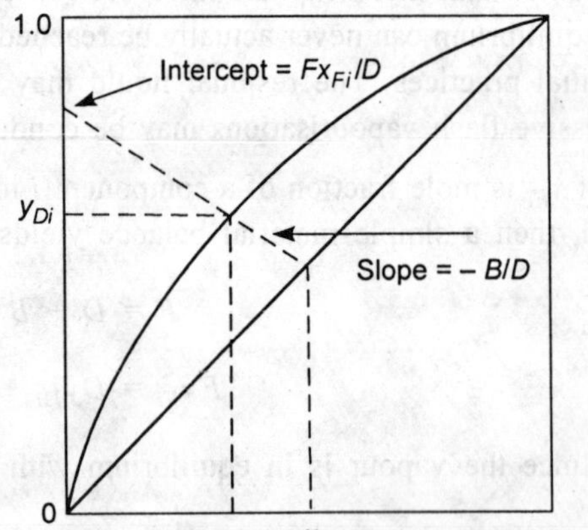


*Figure 5.22: Typical binary flash distillation diagram*

**Example 5.15:** A liquid mixture of methanol and water containing 30 mole percent methanol is flashed at 1 atm pressure so as to vapourise 50 percent of the feed. Determine the compositions of the liquid and vapour products and the temperature of operation. The VLE data of methanol–water system at 1 atm is given below:

| *Temperature, °C* | *Mole fraction of methanol in liquid* | *Mole fraction of methanol in vapour* |
|---|---|---|
| 100.0 | 0.0 | 0.0 |
| 91.2 | 0.06 | 0.304 |
| 89.3 | 0.08 | 0.365 |
| 87.7 | 0.10 | 0.418 |
| 84.4 | 0.15 | 0.517 |
| 81.7 | 0.20 | 0.579 |
| 78.0 | 0.30 | 0.665 |
| 75.3 | 0.40 | 0.729 |
| 73.1 | 0.50 | 0.779 |
| 71.2 | 0.60 | 0.825 |
| 69.3 | 0.70 | 0.870 |
| 67.5 | 0.80 | 0.915 |
| 66.0 | 0.90 | 0.958 |
| 64.5 | 1.00 | 1.00 |

**Solution:** The equilibrium curve for methanol–water system at 1 atm pressure is shown in Fig. 5.15.1. Since 50 percent of feed is vapourised, $(D/F) = 0.50$

$$(B/F) = 1.0 - (D/F) = 0.50$$

$$(B/D) = \frac{(B/F)}{(D/F)} = 1.0$$

We, therefore, draw a straight line of slope $-1.0$ and intercept $= (F\ x_F/D) = (0.3/0.5) = 0.6$, as shown in Fig. 5.15.1. The line intersects the equilibrium curve at the point $P$ and corresponding to this point, the values of $x$ and $y$ are 0.1325 and 0.465 respectively. Therefore,

$y_D$ = mole fraction of methanol in the vapour produced = 0.465

$x_B$ = mole fraction of methanol in the residual liquid = 0.1325

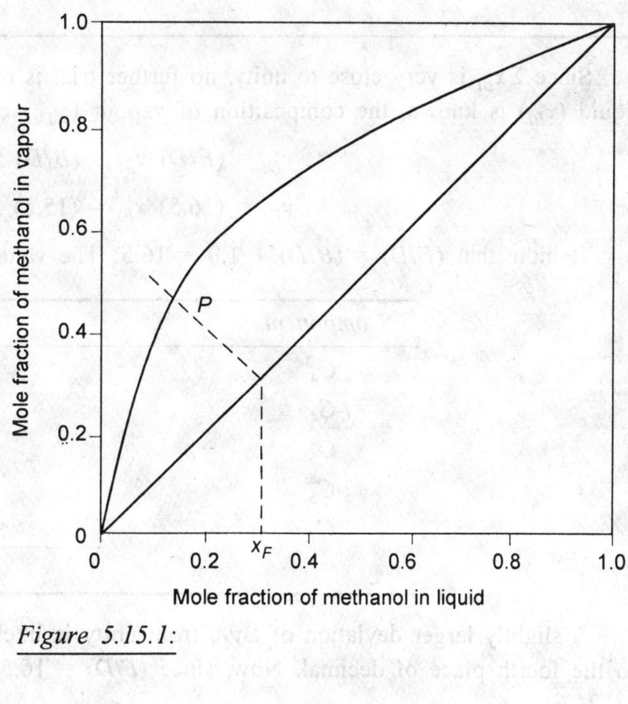

*Figure 5.15.1:*

The vapour thus contains 46.5 mole percent methanol and 53.5 percent water. The residual liquid contains 13.25 mole percent methanol and 86.75 percent water. From the VLE data, the temperature corresponding to $x = 0.1325$ is 85.5° C. The recovery of methanol in the vapour (distillate) is

$$= (Dy_D)/(Fx_F) = (0.5)\ (0.465)/(0.30) = 0.775 = 77.5 \text{ percent}$$

**Example 5.16:** A hydrocarbon mixture containing 16 percent *n*-butane, 5 percent *i*-butane, 10 percent *n*-pentane, 10 percent *i*-pentane and 59 percent *n*-heptane is flashed at 1 atm and 100° F (37.78° C). Determine the composition of the products. What fraction of feed is vapourised? Assume that the substances form ideal solutions.

**Solution:** Since ideal behaviour has been assumed, the *K*-values can be computed from the vapour pressure data available in literature[11, 12]. Thus, from Eq. (5.2.22),

$$K_i = (P'_i/P) = (P'_i/760)$$

where $P'_i$ = vapour pressure of component *i* at 37.78° C, mm Hg.

The $(B/D)$ ratio is now computed from Eq. (5.2.31) by trial. For example, let $(B/D) = q' = 15.5$

| *Component* | $x_{Fi}$ | $P'_i$ | $K_i$ | $x_{Bi} = \dfrac{x_{Fi}(1+q')}{(K_i+q')}$ |
|---|---|---|---|---|
| $C_4$ | 0.16 | 2650 | 3.4868 | 0.139044 |
| $i$-$C_4$ | 0.05 | 3750 | 4.9342 | 0.0403735 |
| $C_5$ | 0.10 | 830 | 1.0921 | 0.099445 |
| $i$-$C_5$ | 0.10 | 1065.0 | 1.40 | 0.097633 |
| $C_7$ | 0.59 | 85.5 | 0.1122 | 0.62355 |
| | | | | $\Sigma x_{Bi}$ = 1.0000455 |

Since $\Sigma x_{Bi}$ is very close to unity, no further trial is required. Once the composition of the residual liquid ($x_{Bi}$) is known, the composition of vapour ($y_{Di}$) can be computed from Eq. (5.2.32):

$$y_{Di} = (F/D)\ x_{Fi} - (B/D)\ x_{Bi}$$

or

$$y_{Di} = (16.5)\ x_{Fi} - (15.5)\ x_{Bi}$$

To note that $(F/D) = (B/D) + 1.0 = 16.5$. The values of $y_{Di}$ are given below:

| *Component* | $y_{Di}$ |
|---|---|
| $C_4$ | 0.4848 |
| $i$-$C_4$ | 0.1992 |
| $C_5$ | 0.1086 |
| $i$-$C_5$ | 0.1367 |
| $C_7$ | 0.070 |
| | $\Sigma y_{Di}$ = 0.9993 |

A slightly larger deviation of $\Sigma y_{Di}$ from unity is likely if the computed $x_{Bi}$-values are rounded off to the fourth place of decimal. Now, since $(F/D) = 16.5$,

$$(D/F) = (1/16.5) = 0.0606$$

Thus, 6.06 percent of the feed has been vapourised.

### 5.2.2 Differential Distillation

Unlike flash distillation, differential distillation (also called *differential weathering*) is a batch process in which the vapour is removed from the still as soon as it is produced. It is, thus, a replica of the conventional laboratory distillation. A specific amount of the feed mixture is charged into the distillation still or pot and either it is heated at a constant pressure or the temperature is maintained constant at the bubble point of the feed and the pressure is gradually reduced. The vapours evolved are continuously removed from the still, condensed in a separate condenser and collected as the distillate. In the case of a multicomponent feed, the most volatile component distills off first and thereafter, the remaining components start distilling off successively in the order of decreasing volatility. The highest boiling (or, the least volatile) component remains behind in the still as the residue at the end of the process. It is, thus, possible to collect a large number of fractions or *cuts,* each being rich in one particular component.

Many standard distillation tests performed in laboratories for determining the boiling ranges and vapourisation characteristics of mixtures such as the petroleum hydrocarbons employ the principle of differential distillation. Examples are the ASTM distillation test, IP distillation test or the Engler distillation test.

Differential distillation of binary mixture is usually conducted at constant pressure. Though the vapours from the differential distillation still are continuously removed as soon as they are produced, the vapour evolved at any instant from the boiling liquid mixture is assumed to be in equilibrium with it. This demands the process be conducted at a very slow rate. The compositions of the liquid as well as the vapour thus change continuously as distillation proceeds, but it is assumed that the composition of the vapour evolved at any instant is equal to the equilibrium composition corresponding to the liquid composition remaining in the still. To analyse the process, we have to, therefore, use a differential approach. Thus, if $L$ is the moles of liquid in the still at any time and if $dL$ is the differential amount that has vapourised, then a material balance for the component $i$ gives

$$Lx_i = (L - dL)(x_i - dx_i) + (y_i^* + dy_i^*)\, dL \quad \text{... (5.2.33)}$$

By neglecting the products of differentials, the above equation can be simplified to

$$L\, dx_i = (y_i^* - x_i)\, dL \quad \text{... (5.2.34)}$$

or

$$\int_B^F \frac{dL}{L} = \ln (F/B) = \int_{x_{Bi}}^{x_{Fi}} \frac{dx_i}{y_i^* - x_i} \quad \text{... (5.2.35)}$$

where $F$ = total moles of feed stock charged into the still

$B$ = total moles of residual liquid left in the still at the end of the process

The above equation is known as the *Rayleigh equation.* Since for binary systems, the more volatile component is chosen as the component $i$, the component notation may be avoided in the above equation.

Equation (5.2.35) can be further simplified if we consider a binary system that exhibits ideal behaviour (thereby obeys Raoult's law) and for which the relative volatility ($\alpha$) is essentially constant. If $\alpha$ is constant, then from Eq. (5.2.24),

$$y^* = \frac{\alpha x}{x(\alpha - 1) + 1} \quad \text{... (5.2.36)}$$

Note that we have avoided the component notation (that is, instead of $x_1$ and $y_1$, $x$ and $y$ are used) since $x$ and $y$ represent mole fractions of the more volatile component in the case of a binary mixture. Substituting Eq. (5.2.36) in Eq. (5.2.35) and integrating, we get

$$\ln (F/B) = \frac{1}{(\alpha - 1)} \left[ \ln (x_F / x_B) + \alpha \ln \left( \frac{1 - x_B}{1 - x_F} \right) \right] \quad \text{... (5.2.37)}$$

or

$$\ln \left( \frac{F x_F}{W x_B} \right) = \alpha \ln \left[ \frac{F (1 - x_F)}{W (1 - x_B)} \right] \quad \text{... (5.2.38)}$$

The above two equations are two useful forms of Rayleigh equation and can be applied to binary systems for which a constant value of relative volatility could be assumed.

In the case of multicomponent feedstocks, the process may be conducted, as stated earlier, either at constant pressure or at constant temperature. In the former case, the feed stock is first heated at constant pressure to its bubble point and the temperature is then gradually increased (keeping the pressure constant) until the composition of the liquid in the still reaches the specified value. A gradual increase in the boiling temperature of the liquid takes place as the lower boiling components are distilled off until the desired quantity of distillate is obtained. In the case of constant temperature operation or *isothermal weathering*, the pressure is gradually decreased while the liquid and vapour remain at a constant temperature throughout the process. An accurate analysis of multicomponent differential distillation is quite complicated. We shall confine our discussion here to some of the simplified situations. For example, let us consider multicomponent systems that form ideal liquid solutions. In such cases, Eq. (5.2.38) can be used after rewriting it as follows:

$$\ln \left( \frac{F x_{Fi}}{B x_{Bi}} \right) = \alpha_i \ln \left( \frac{F x_{Fk}}{B x_{Bk}} \right) \quad \text{... (5.2.39)}$$

where

$x_{Fk}$, $x_{Bk}$ = mole fraction of the key component in the feed mixture and in the final residue respectively

$\alpha_i$ = relative volatility of component $i$ with respect to the key component

$= (K_i/K_k) = (P'_i/P'_k)$ ... (5.2.39a)

$P'_i$, $P'_k$ = vapour pressure of component $i$ and the key component respectively

Also, we know that

$$\sum_{i=1}^{n} x_{Bi} = 1.0 \quad \text{... (5.2.40)}$$

Equation (5.2.39) is to be written separately for all components except the key component and these equations when solved simultaneously with Eq. (5.2.40) give the values of $x_{Bi}$. Once the composition of the residue is known, the distillate composition $y_{Di}$ can be obtained from the material balance equations:

$$F = B + D$$

$$F x_{Fi} = B x_{Bi} + D y_{Di} \quad \text{... (5.2.41)}$$

For a constant temperature process, the initial pressure is the bubble point pressure of the feed (that is, the pressure at which the bubble point of the feed is equal to the operating temperature). The $\alpha$-values are to be specified at the operating temperature from the vapour pressure data and $x_{Bi}$ and $y_{Di}$ are to be computed from Eqs (5.2.39) to (5.2.41) as discussed above. The final pressure is obtained as

$$y_{Di} = K_i x_{Bi} = (P'_i/P)\, x_{Bi}$$

or

$$P\, \Sigma y_{Di} = P = \Sigma x_{Bi} P'_i \qquad \text{... (5.2.42)}$$

If the differential distillation is conducted at constant pressure, then the initial temperature ($T_1$) is the bubble point of the feed at the given pressure. The final temperature ($T_2$) is the bubble point of the residue, which, since unknown at the outset, is to be assumed. The $\alpha$-values are to be specified at $(T_1 + T_2)/2$ and Eqs (5.2.39) and (5.2.40) solved to get the values of $x_{Bi}$ and $y_{Di}$. The bubble point of the residue is now computed which is equal to the final temperature ($T_2$). If this computed value of $T_2$ differs significantly from that assumed at the outset, then the procedure is to be repeated with the newly computed value of $T_2$. However, since $\alpha$-values are not sensitive to moderate changes in temperature, the procedure shall converge rapidly.

It may be noted that an equation similar to (5.2.35) can also be derived for the process of *differential condensation* which is converse to differential distillation in the sense that a given amount of vapour mixture is condensed under equilibrium conditions and the condensate is withdrawn continuously as soon as it is produced. Rayleigh equation for differential condensation will be, therefore,

$$\ln\,(F/D) = \int_{y_{Fi}}^{y_{Di}} \frac{dy_i}{(y_i - x_i^*)} \qquad \text{... (5.2.43)}$$

where $F$ = total moles of vapour feed

$D$ = total moles of vapourous residue (that is, moles of vapour left behind in the condenser at the end of the process)

$y_{Fi}$, $y_{Di}$ = mole fraction of component $i$ in the vapour feed and in the vaporous residue respectively.

**Example 5.17:** If the methanol–water mixture discussed in Example 5.15 is differentially distilled at 1 atm pressure so as to vapourise 50 percent of the feed, determine the compositions of the residue and the distillate.

**Solution:** The product compositions can be computed from Rayleigh equation. Since $(B/F) = 0.5$, $\ln\,(F/B) = \ln\,(2.0) = 0.69314$. Therefore, Eq. (5.2.35) becomes

$$\int_{x_B}^{0.3} \frac{dx}{y^* - x} = 0.69314 \qquad \text{... (i)}$$

The above equation is to be solved for $x_B$ by trial. Thus, let $x_B = 0.06$. Now, the values of $1/(y^* - x)$ for $x = 0.06$ to $x = 0.30$ can be computed from the equilibrium data as given below:

The integral can be now evaluated either graphically or numerically. Let us evaluate it numerically using the trapezoidal rule. This rule has already been discussed in Example 4.12 Chapter 4. Thus,

$$\int_{x_0}^{x_n} f(x)\; dx = (1/2) \sum_{i=1}^{n} h_i\; (f_i + f_{i-1})$$

**Table 5.17.1**

| $x$ | $y^*$ | $f(x) = 1/(y^* - x)$ |
|---|---|---|
| 0.06 | 0.304 | 4.09836 |
| 0.08 | 0.365 | 3.50877 |
| 0.10 | 0.418 | 3.14465 |
| 0.15 | 0.517 | 2.7248 |
| 0.20 | 0.579 | 2.6385 |
| 0.30 | 0.665 | 2.7397 |

where $h_i = (x_i - x_{i-1})$. Here, $x_0 = 0.06$, $x_n = 0.30$, $n = 5$. The step size $h_i$ can be computed as $h_1 = h_2 = 0.02$, $h_3 = h_4 = 0.05$ and $h_5 = 0.1$. The values of $f(x)$ are given in the table above. For example, $f_0 = 4.09836$, $f_1 = 3.50877$ and so on. The value of the integral is, therefore, 0.69234. It can be seen that this value does not differ significantly from that specified in Eq. (i). Therefore, the assumed value of $x_B$ is correct.

The residue thus contains 6.0 percent methanol and 94 percent water. The composition of the distillate can be obtained from a material balance [Eq. (5.2.41)]:

$$x_F = (B/F)\ x_B + (D/F)\ y_D = 0.5\ (x_B + y_D)$$

or

$$y_D = 2x_F - x_B = 2\ (0.30) - 0.06 = 0.54$$

The distillate thus contains 54 mole percent methanol and 46 percent water. The recovery of methanol in the distillate is,

$$= (Dy_D)/(Fx_F) = (0.5)\ (0.54)/(0.3) = 0.90 = \mathbf{90}\ \textbf{percent}$$

It can be seen that the recovery is much higher than that obtained by flash distillation.

**Example 5.18:** 500 kmoles of a propane–pentane mixture containing 20 mole percent propane is subjected to isothermal weathering at 25° C so that the propane content is reduced to 3 percent before the mixture is marketed. Assuming ideal behaviour, specify:

(*a*) the original pressure,

(*b*) the final pressure,

(*c*) the percent recovered as liquid,

(*d*) the loss of pentane in the vapour.

**Solution:** (*a*) The initial pressure of distillation will be the bubble point pressure of the feed (that is, the pressure at which the bubble point of the feed is 25° C). Since ideal behaviour has been assumed, the *K*-values can be computed from Eq. (5.2.22). The vapour pressure of propane and pentane at 25° C are given below[12]:

| | $P'$, *mm Hg* | $K_i$ | $\alpha = (K_1/K_2)$ |
|---|---|---|---|
| Propane | 7139.0 | $(7139/P)$ | 14.025 |
| Pentane | 509.0 | $(509/P)$ | |

Since $\qquad \Sigma y_i = \Sigma K_i x_i = 1.0,$

$$(7139/P)\ x_F + (509/P)\ (1 - x_F) = 1.0$$

where $x_F$ = mole fraction of propane in feed = 0.20

Therefore, $$P = 7139\ (0.2) + 509\ (0.8) = 1835 \text{ mm Hg.}$$

This is the initial pressure of distillation.

(*b*) The final pressure will be the bubble point pressure of the residue. Thus

$$P = 7139 x_B + 509\ (1 - x_B)$$

where $P$ = final pressure

$x_B$ = mole fraction of propane in residue = 0.03

Therefore, $$P = 7139\ (0.03) + 509\ (0.97) = 707.9 \text{ mm Hg}$$

(*c*) Since the system forms ideal solution, we can apply Eq. (5.2.37). Thus,

$$\ln\ (F/B) = \frac{1}{(14.025 - 1.0)}\left[\ln\left(\frac{0.2}{0.03}\right) + 14.025 \ln\left(\frac{0.97}{0.8}\right)\right] = 0.35313$$

or $$(B/F) = 0.7025$$

Therefore, 70.25 percent of the feed is recovered as liquid.

(*d*) The composition of the distillate can be computed from a material balance:

$$x_F = (D/F)\ y_D + (B/F)\ x_B$$

$$0.20 = (1 - 0.7025)\ y_D + (0.7025)\ (0.03)$$

or $$y_D = 0.6014$$

The distillate thus contains 60.14 percent propane and 39.86 percent pentane. The fraction of pentane lost in the vapour (distillate) is,

$$= \frac{D(1 - y_D)}{F(1 - x_F)} = \frac{(0.2975)(0.3986)}{(0.80)} = 0.1482$$

Thus, 14.82 percent of the pentane of the feed is lost in the vapour.

**Example 5.19:** The following high pressure liquid mixture is subjected to isothermal weathering at 25 °C down to a pressure of 73 mm Hg. Compute the resultant liquid yield:

$$C_2 = 0.05,\ C_3 = 0.10,\ i\text{-}C_4 = 0.20 \text{ and } C_7 = 0.65.$$

Assume that the system forms ideal solutions.

**Solution:** Let us choose iso-butane as the key component. Since $K_i = (P'_i/P)$ and $\alpha_i = (K_i/K_k)$,

$$\alpha_i = (P'_i/P'_k) \qquad \text{... (i)}$$

The vapour pressure of the hydrocarbons at 25° C are taken from reference (11) and they are used to compute $\alpha_i$ as given below:

| Component | $x_{Fi}$ | $P'_i$, mm Hg | $\alpha_i$ |
|---|---|---|---|
| $C_2$ | 0.05 | 31510.0 | 12.068 |
| $C_3$ | 0.10 | 7139.0 | 2.7342 |
| $i\text{-}C_4$ | 0.20 | 2611.0 | 1.0 |
| $C_7$ | 0.65 | 47.12 | 0.018 |

Since the $(F/B)$ ratio is unknown, we have to resort to a trial and error procedure. Therefore, let $(F/B) = 1.621$. Equation (5.2.39) is now written separately for each component except for the key component as given below:

$$\ln\left(\frac{1.621\times 0.05}{x_{B1}}\right) = 12.068 \ln\left(\frac{1.621\times 0.2}{x_{B3}}\right) \quad \text{... (ii)}$$

$$\ln\left(\frac{1.621\times 0.1}{x_{B2}}\right) = 2.7342 \ln\left(\frac{1.621\times 0.2}{x_{B3}}\right) \quad \text{... (iii)}$$

$$\ln\left(\frac{1.621\times 0.65}{x_{B4}}\right) = 0.018 \ln\left(\frac{1.621\times 0.2}{x_{B3}}\right) \quad \text{... (iv)}$$

Also, $$x_{B1} + x_{B2} + x_{B3} + x_{B4} = 1.0 \quad \text{... (v)}$$

The above four equations are to be now solved simultaneously for the four unknowns such as $x_{B1}$, $x_{B2}$, $x_{B3}$ and $x_{B4}$. However, a direct solution is difficult since the equations contain exponential terms. Only a trial and error solution is possible. Thus, let $x_{B3} = 0.01$. Now, from Eqs (ii), (iii) and (iv),

$$x_{B1} = 4.745 \times 10^{-20}$$

$$x_{B2} = 1.2 \times 10^{-5}$$

$$x_{B4} = 0.98969$$

Since $\Sigma x_{Bi} = 0.9997$, which is very close to unity, no further trial is necessary. Now, from Eq. (5.2.42),

$$P = \Sigma x_{Bi} P'_i$$

$$= (4.745 \times 10^{-20})\,(31510.0) + (1.2 \times 10^{-5})\,(7139) + (0.01)\,(2611) + (0.98969)\,(47.12)$$

$$= 72.83 \text{ mm Hg}$$

Since this computed value of final pressure is very close to that specified in the problem (namely, 73 mm Hg), the assumed value of $(F/B)$ is correct.

Now, $$(B/F) = (1/1.621) = 0.6169$$

Thus, the liquid yield is 61.69 percent of the feed. The distillate composition also may be computed from the material balance equation,

$$x_{Fi} = (B/F)\,x_{Bi} + (D/F)\,y_{Di} = (0.6169)\,x_{Bi} + (1 - 0.6169)\,y_{Di}$$

The composition is given below:

| Component | $y_{Di}$ |
|---|---|
| $C_2$ | 0.1305 |
| $C_3$ | 0.2610 |
| $i$-$C_4$ | 0.5060 |
| $C_7$ | 0.1030 |
| | $\Sigma y_{Di}$ = 1.0005 |

## 5.2.3 Steam Distillation

If the feedstock is a high boiling mixture, it often becomes difficult to separate it by distillation at ordinary pressures since such high temperatures would be required to vapourise the material that the operation becomes either impractical or too uneconomical. This is also true with substances that are thermally unstable at their normal boiling temperatures. When heated to very high temperatures, these substances not only tend to decompose but also tend to react with other components associated with it. One solution in such cases is to distil under vacuum or at reduced pressure. An alternate method that may be used either alone or in combination with a reduction in pressure is to reduce the partial pressure of the volatile components of the mixture by introducing an inert vapour and thereby reduce the temperature required for vapourisation. The inert vapour must be immiscible with the liquid components being distilled and it should not have any significant effect on the vapour pressure of the liquid components. Steam is widely employed for this purpose since it is readily available at a low cost and the operation is then called *steam distillation.*

Steam or inert distillation is particularly suitable to separate appreciable quantities of high boiling materials or when the material to be subjected to distillation is thermally unstable at the boiling temperature. Steam has the specific advantages that it is immiscible with many high–boiling organic compounds, can be easily removed from most systems by condensation and under proper conditions, it can also provide the necessary heat of vapourisation.

Let us consider a batch steam distillation process. Let the feed stock be a binary mixture containing a volatile component $A$ and a nonvolatile component $B$. The mixture is charged into the distillation still and steam under pressure is supplied from below through a perforated pipe or distributor. Since both $A$ and $B$ are immiscible with water or steam and since $B$ is nonvolatile at the operating temperature and pressure, the outgoing vapours will contain only steam and the vapours of $A$. These vapours are condensed in an overhead condenser and then sent to a separator where the component $A$ and water separate into two different layers and therefore, are separated by gravity.

The total pressure of distillation will be equal to the sum of the partial pressure of $A$ and that of steam. Thus,

$$P = p_A + p_S \quad \text{... (5.2.44)}$$

If apart from being a carrier fluid, steam is also used as the heating fluid (that is, to supply the heat of vapourisation), then part of steam will condense inside the still. Then there will be three phases present in the still such as the vapour phase (containing steam and $A$), the liquid phase (containing $A$ and $B$) and the second liquid phase (containing only water that is formed due to the condensation of steam). In such cases, $p_S$ can be taken equal to the vapour pressure of water at the operating temperature. If we assume that the presence of the nonvolatile component $B$ does not affect the vapourisation of $A$, then $p_A$ will be equal to the vapour pressure of $A$ at the operating temperature. However, in actual practice, since the contact between the carrier fluid (namely, steam) and the material being distilled (namely, $A$) is not perfect, the carrier does not reach equilibrium with the liquid. As a result, the actual partial pressure of $A$ will be equal to its vapour pressure ($P'_A$) multiplied by a factor $E_S$, where $E_S$ is called the *vapourisation efficiency.* In other words, Eq. (5.2.44) can be written as,

$$P = E_S P'_A + p_S \quad \text{... (5.2.45)}$$

The value of $E_S$ varies with the system, the rate of distillation and even with the type of equipment used. However, it usually ranges from 0.6 to 0.95.

Once the operating pressure ($P$) is fixed, then the operating temperature is that value of $T$ which satisfies Eq. (5.2.45). To note that $P'_A$ and $p_S$ are specified at $T$. This temperature will be lower than the boiling point of either pure $A$ or pure water. Thus, distillation of $A$ can be conducted at a temperature much lower than its true boiling point. This improves the economy of the process since the required operating temperature is lowered. This also eliminates any chances of decomposition of $A$ since the operating temperature is below the boiling point of pure $A$.

The molar ratio of carrier fluid to the material being distilled will be equal to the ratio of their partial pressures. Thus, if $N_S$ is the moles of carrier steam and $N_A$ the moles of $A$ distilled, then

$$(N_S/N_A) = (p_S/p_A) \quad \text{... (5.2.46)}$$

From Eq. (5.2.44), $p_S = (P - p_A)$ and from Eq. (5.2.45), $p_A = E_S P'_A$. Therefore, Eq. (5.2.46) becomes

$$(N_S/N_A) = (P - E_S P'_A)/(E_S P'_A)$$

or

$$N_S = \left(\frac{P}{E_S P_A'}\right) N_A \quad \text{... (5.2.47)}$$

The mass of carrier steam will be $(18 N_S)$. The amount of steam used as heating fluid (that is, for heating the charge to the operating temperature and also to supply the latent heat of vapourisation of $N_A$ moles of $A$) can be computed by a heat balance. The *total steam consumption* ($S$) will be then

$$S = (18 N_S) + \dot{m}_S \quad \text{... (5.2.48)}$$

where $\dot{m}_S$ = kg of steam used as heating fluid.

If steam is used only as a carrier fluid and heating of the charge is accomplished by some external means (either by providing heating coils or by jacketing the still and passing a heating fluid through the jacket), then the *total steam consumption* ($S$) will be equal to $(18 N_S)$ and can be computed straighaway from Eq. (5.2.47). However, in such cases, since heating is done externally, the temperature of operation ($T$) can be fixed at any desired value. Usually, when the steam is admitted at a pressure higher than the operating pressure, the operating temperature ($T$) may be fixed at the saturation temperature of inlet steam. The steam gets superheated as it enters the still due to the reduction in pressure and goes out as superheated vapour along with $A$. There will be, therefore, no condensation of steam inside the still. The time required for distillation will be the steam consumption ($S$) divided by the steam flow rate.

It is also possible to supply steam that is considerably superheated to the still and use part or whole of the superheat for heating the charge and vapourising $A$. In such a case also, steam will not condense in the still and will go out as vapour with $A$. However, the steam consumption could be substantially large in such cases, since the superheat is only a small fraction of the total enthalpy of superheated steam.

**Example 5.20:** 18,000 kg of a food extract containing 5 mole percent $n$-hexane is to be steam distilled at 880 mm Hg to reduce its hexane content to 0.1 mole percent. Steam admitted to the batch is saturated steam at 0.2 mN/m$^2$. The initial temperature of the charge is 20° C. Compute the total steam consumption per kg of $n$-hexane vapourised. The vapourisation efficiency may be taken to be 0.6. Approximate molecular weight of the extract is 450 and its specific heat 2.09 kJ/(kg K). Latent heat of vapourisation of $n$-hexane = 325.42 kJ/kg. Neglect radiation losses.

**Solution:** Let us first determine the temperature of the liquid in the still from Eq. (5.2.45). Thus

$$P = E_S P'_A + p_S$$

$$880 = 0.6\,(P'_A) + p_S \qquad \text{... (i)}$$

To start the trial, let $T$ = 76.67° C, Then

Vapour pressure of $n$-hexane at 76.67° C = $P'_A$ = 950 mm Hg

Vapour pressure of water at 76.67° C (from steam table) = 309.8 mm Hg

The right hand side of Eq. (i), therefore, becomes

$$\text{RHS} = 0.6\,(950) + 309.8 = 879.8$$

Since RHS is very close to LHS, we can avoid further trial. The amount of carrier steam required can be now estimated from Eq. (5.2.47):

$$(N_S/N_A) = \left[\frac{(880)}{(0.6)(950)} - 1\right] = 0.54386$$

The total moles of $n$-hexane vapourised ($N_A$) can be obtained from a material balance:

$$F\,(1 - x_F) = B\,(1 - x_B)$$

where $F$ = total moles of original charge
= (18000)/(450) = 40.0 kmoles
$x_F$ = 0.05
$x_B$ = 0.001

Therefore, $B$ = (40.0) (1 – 0.05)/(1 – 0.001) = 38.038 kmoles

Moles of $n$-hexane vapourised,

$$N_A = (Fx_F - Bx_B) = 1.962 \text{ kmoles} = 168.732 \text{ kg}$$

Therefore, $N_S$ = (1.962) (0.54386) = 1.067 kmoles = 19.206 kg

This is the amount of carrier steam required. The amount of heating steam required is to be computed from a heat balance. The total heat that is to be supplied to the charge includes the heat required to raise the temperature of the charge to the operating temperature (that is, 76.67° C) and the latent heat of vapourisation of $n$-hexane. Thus, the total heat supplied ($Q$) is,

$$Q = (2.09)\,(18000)\,(76.67 - 20) + (325.42)\,(168.732)$$
$$= 2186.8077 \times 10^3 \text{ kJ}$$

Since the pressure of inlet steam is higher than the operating pressure, a small amount of heat is supplied from the carrier steam as well. Enthalpy of saturated steam at 0.2 MN/m$^2$ is 2706.7 kJ/kg and that at 880 mm Hg is 2681.76 kJ/kg. Therefore,

Heat supplied from carrier steam

$$= (2706.7 - 2681.76) = 24.94 \text{ kJ/kg} = (24.94)\,(19.206) = 479.0 \text{ kJ}$$

Heat to be supplied from heating steam

$$= (2186807.7 - 479.0) = 2186328.7 \text{ kJ}$$

In the still, steam condenses to water at 76.67° C. Enthalpy of water at 76.67° C is 320.937 kJ/kg. Therefore, amount of heating steam ($m_S$) is,

$$\dot{m}_S = \frac{(2186328.7)}{(2706.7 - 320.937)} = 916.4 \text{ kg}$$

Total steam consumption, $S = (916.4) + (19.206) = 935.6$ kg

Steam consumption per kg of hexane vapourised

$$= (935.6/168.732) = \mathbf{5.545 \text{ kg}}$$

### 5.2.4 Continuous Fractionation—Binary Systems

The processes of differential distillation and flash distillation, discussed in earlier subsections, are unsuitable for large capacity installations. This is due to the fact that by differential distillation or equilibrium flash distillation though the desired concentrations in the distillate and the residue can be reasonably achieved, the yields of these products shall be relatively low in these processes. When not only good product quality but a sizeable product yield is also desired, the process of continuous fractional distillation is more effective and economical. Continuous fractionation may be conducted in a stagewise equipment (such as a sieve–plate column or a bubble–cap plate column) or in a continuous contact equipment (such as in a packed column). Design and analysis of packed bed distillation columns are discussed in one of the subsequent subsections.

It will not be wrong to consider a stagewise continuous fractionator (say, a plate column) as equivalent to a large number of equilibrium distillation stills in series. Each stage receives vapour from the stage below and liquid from the stage above. In turn, the vapour from each stage is conducted to the stage above and the liquid from each stage flows down to the stage below. The vapour and liquid streams thus flow countercurrently along the tower (see Fig. 5.23). At each plate or tray, there will be intimate contacting between the vapour and liquid. It will not be, therefore, irreasonable to assume that the vapour and the liquid leaving any stage are in equilibrium with each other. If this assumption is true, then each plate or tray may be considered to be an equilibrium stage and the temperature at each tray will be the corresponding equilibrium temperature*.

* The extent to which the tray performance deviates from ideality is predicted by the so called *tray efficiency.* We shall be elaborating on this in subsequent pages.

The vapour becomes richer and richer in the low-boiling or more volatile constituents as it moves up from one stage to another. Similarly, the liquid becomes leaner and leaner in low-boiling components but richer and richer in high-boiling or less volatile components as it flows down. As a result, the temperature increases progressively from the topmost plate (which is also the first equilibrium stage) to the bottom–most plate.

As shown in Fig. 5.23, a continuous fractionator is operated with reflux. The feed is introduced at an intermediate plate (more or less centrally between the topmost plate and the bottom–most plate). The section above the feed plate is called the *rectifying* or *enriching* section and the section below the feed plate is called the *stripping* or *exhausting* section. The number of plates in the rectifying and stripping sections may or may not be the same. In the rectifying section, the down-flowing liquid acts as an absorbing agent for the high-boiling components in the vapour, thereby aids the fractionating process by concentrating the low-boiling components in the vapour and the high-boiling components in the liquid. This downflow of liquid is achieved by condensing a part or all of the overhead vapours and returning a part of the condensate liquid to stage 1 (see Fig. 5.23). The liquid that is returned is called the *reflux*. (It is also possible to produces the reflux internally by installing an intercooler inside the column shell). If a total condenser is used, then all of the overhead vapours from stage 1 are condensed into liquid and a part of it is returned as the reflux, while the rest is collected as the *distillate.* Instead, if a partial condenser* is used, then the overhead vapours from stage 1 are only partially condensed, the condensate is returned to the tower as reflux and the remaining vapours are collected as distillate. The ratio of the moles (or mass) of liquid returned as reflux ($L_0$) to the moles (or mass) of distillate ($D$) collected is called the *reflux ratio* or more precisely, the *operating reflux ratio* ($R$). Thus, $R = (L_0/D)$. The $(L/V)$ ratio at any stage/plate, that is the ratio of moles (or mass) of liquid leaving any stage to that of vapour leaving the same stage, is called the *internal reflux ratio.*

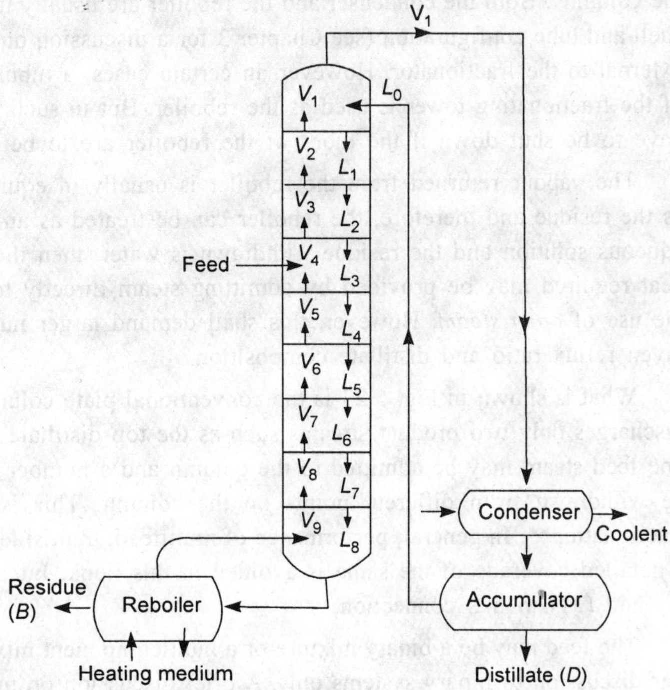

*Figure 5.23: Typical stagewise continuous fractionator*

If no reflux is used, then there would be no liquid returned to stage 1 and as a result, there would be no condensation of $V_2$ to supply liquid leaving stage 1. The vapour leaving stage 1 will be then of the same quantity and composition as that leaving stage 2. This will be true for all stages. As a result,

*The partial condenser is often called the *dephlagmator.*

the vapour leaving every stage will be of the same quantity and composition. The enriching section then loses its meaning since the vapour does not get enriched with the more volatile component as it moves up from plate to plate. Similarly, in the stripping section, the downflowing liquid is stripped off the volatile constituents by the vapour produced in the reboiler by partial vapourisation of the bottom liquid. The rest of the bottom liquid is withdrawn as the *residue* ($B$).

A full-fledged continuous fractionator must be, therefore, equipped with a condenser at the top to supply the reflux and a reboiler at the bottom to produce vapour from the bottom liquid and return to the column*. Both the condenser and the reboiler are usually tubular heat exchangers of the conventional shell and tube configuration (see Chapter 3 for a discussion on heat exchangers). They are normally built external to the fractionator. However, in certain cases, a tubular exchanger that is built into the bottom of the fractionating tower is used as the reboiler. But in such cases, the entire distillation operation will have to be shut down if the tubes of the reboiler are to be cleaned after fouling.

The vapour returned from the reboiler is usually in equilibrium with the liquid that is withdrawn as the residue and therefore, the reboiler can be treated as an ideal, equilibrium stage. If the feed is an aqueous solution and the residue withdrawn is water, then the reboiler may be dispensed with and the heat required may be provided by admitting steam directly to the bottom of the tower. This is called the use of *open steam*. However, this shall demand larger number of plates/trays in the column for a given reflux ratio and distillate composition.

What is shown in Fig. 5.23 is the conventional plate column that receives only one feed stream and discharges only two product streams such as the top distillate and the bottom residue. Often, more than one feed steam may be admitted to the column and a number of intermediate product streams may also be withdrawn from different points on the column. This is what is usually practised in petroleum fractionation**. In general, performance of multifeed, multisidestream columns is complex to analyse and a detailed coverage of the same is avoided in this book. Interested readers may consult references (4), (5) and (31) in this connection.

The feed may be a binary mixture or a multicomponent mixture. In this subsection, we have confined our discussion to binary systems only. A brief discussion on multicomponent distillation is given in one of the subsequent subsections.

The number of ideal stages required for fractionating a binary mixture can be conveniently determined using the method proposed by *McCabe and Thiele*[13]. However, this method is based on the following assumptions:

(*i*) The molar flow rate of liquid and that of vapour is the same for all trays in a particular section (either enriching section or exhausting section). For example, if $L$ and $V$ represent the molar flow rates of liquid and vapour respectively in the enriching section, then for all trays in the enriching section,

$$L_1 = L_2 = L_3 = \ldots\ldots = L \qquad \ldots (5.2.49)$$

$$V_1 = V_2 = V_3 = \ldots\ldots = V \qquad \ldots (5.2.50)$$

* The reboiler thus supplies both heat and vapour to the column.

** A numerical example on rating of a multifeed fractionator is discussed elsewhere in this subsection (see Example 5.24).

In other words, the internal reflux ratio is the same for all trays in the enriching section and is equal to $(L/V)$. Similarly, for every tray in the exhausting section, the internal reflux ratio will be $(L'/V')$. This is what is meant by *constant molar overflow.*

(*ii*) To satisfy the above condition, it is necessary that the molar heats of vapourisation of the two components are equal. It is also necessary that the system operates without any significant heat losses and the heat of mixing or heat of solution is negligibly small.

Fractionators are usually well-insulated and therefore, heat losses are reduced to a minimum. Further the heat of solution is not significantly large for many systems handled in commercial practices. The method of McCabe and Thiele is, therefore, adequate for many purposes in spite of the simplifying assumptions involved. However, for those systems for which these assumptions are far from true, the more rigorous methods such as those of Ponchon and Savarit[14, 15] or Sorel[43] (that are based on enthalpy–concentration data) must be employed. These are discussed subsequently in this section. The method of McCabe and Thiele is outlined below.

Let us apart from the above two inherent assumptions, further assume that the reflux returned to the column is at its bubble point and a total condenser is being used to condense the overhead vapours. Thus, both the reflux $(L_0)$ and the distillate $(D)$ are saturated liquids and $x_D = x_0 = y_1$. Here, $x$ and $y$ denote mole fraction of the more volatile component (say $A$) in the liquid phase and in the vapour phase respectively. A material balance for the envelope $I$, shown in Fig. 5.24 gives

$$V_{n+1} = L_n + D \qquad \text{... (5.2.51)}$$

$$V_{n+1}y_{n+1} = L_n x_n + D x_D$$

or

$$y_{n+1} = (L_n/V_{n+1})\, x_n + (D/V_{n+1})\, x_D \qquad \text{... (5.2.52)}$$

Since all $L$ values are equal and all $V$ values are equal, the subscripts can be dropped and we can write the above equation as

$$y_{n+1} = (L/V)\, x_n + (D/V)\, x_D \qquad \text{... (5.2.53)}$$

Also, from Eq. (5.2.51), putting $n = 0$,

$$V_1 = L_0 + D$$

or

$$(V_1/D) = (V/D) = (L_0/D) + 1 = (R + 1) \qquad \text{... (5.2.54)}$$

where $R = (L_0/D)$ = reflux ratio (defined earlier).

Similarly, $$(V_1/L_0) = (V/L) = 1 + (D/L_0) = (R + 1)/R \qquad \text{... (5.2.55)}$$

Equation (5.2.53) can be, therefore, rewritten as

$$y_{n+1} = \left(\frac{R}{R+1}\right) x_n + \left(\frac{x_D}{R+1}\right) \qquad \text{... (5.2.56)}$$

This is the equation to the operating line for the enriching section. It can be seen that the operating line is a straight line of slope $(R/R + 1)$ and $y$-intercept $(x_D/R + 1)$. Also, when $x_n = x_D$, $y_{n+1} = x_D$. Therefore, the line passes through the point $y = x = x_D$ on the 45° diagonal (see Fig. 5.25). This operating line can be, therefore, easily drawn in the plot of equilibrium curve, as shown in Fig. 5.25.

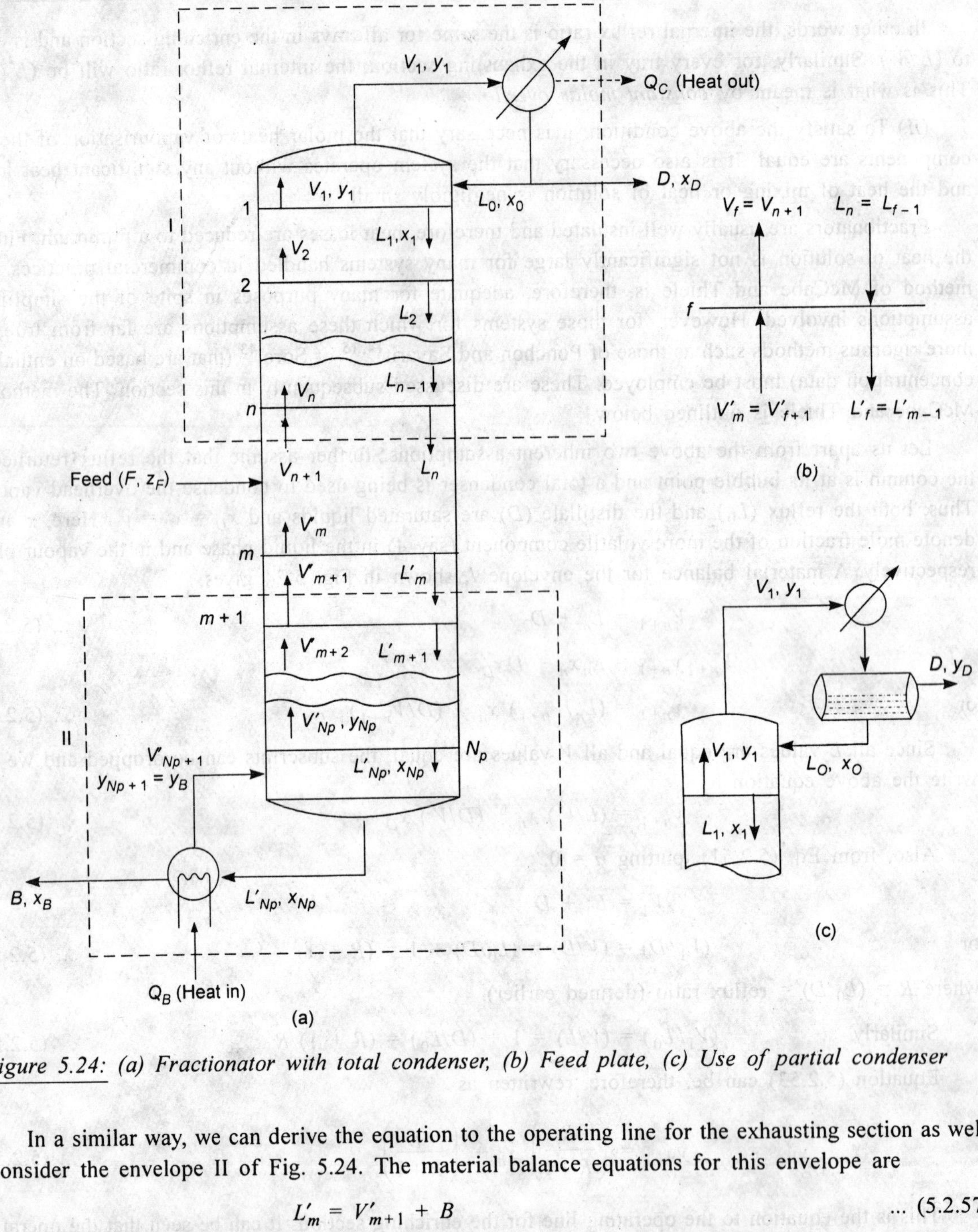


*Figure 5.24: (a) Fractionator with total condenser, (b) Feed plate, (c) Use of partial condenser*

In a similar way, we can derive the equation to the operating line for the exhausting section as well. Consider the envelope II of Fig. 5.24. The material balance equations for this envelope are

$$L'_m = V'_{m+1} + B \qquad \text{... (5.2.57)}$$

$$L'_m x_m = V'_{m+1} y_{m+1} + B x_B$$

or

$$y_{m+1} = (L'_m / V'_{m+1})\, x_m - (B/V'_{m+1})\, x_B$$

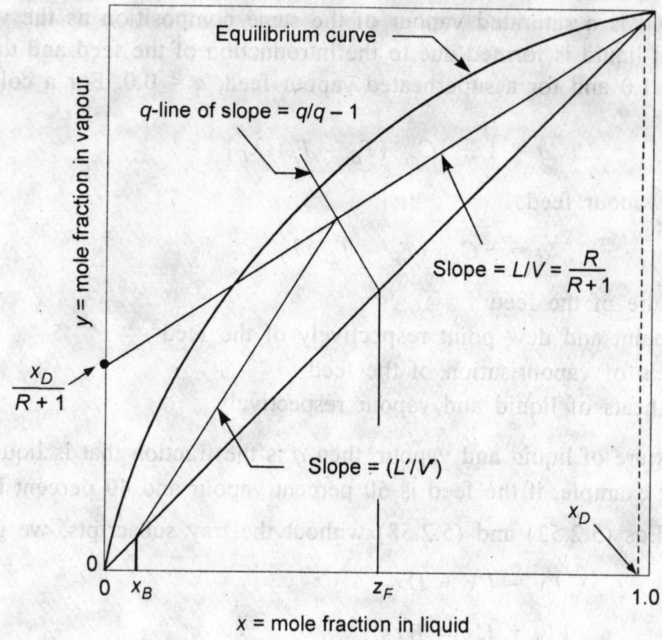


*Figure 5.25: Construction of operating lines*

$$= (L'/V')\, x_m - (B/V')\, x_B \qquad \text{... (5.2.58)}$$

Since all $L'$ values are equal and all $V'$ values are equal, we have dropped the subscripts in Eq. (5.2.58), which is the equation to the operating line for the exhausting section. This line also passes through the point $x = y = x_B$ on the 45° diagonal (see Fig. 5.25).

The intersection of these two operating lines takes place at a point which is a function of feed composition, the thermal condition of the feed and the enthalpy balance relationships around the feed plate. For example, a material balance around the feed plate gives

$$F + V'_{f+1} + L_{f-1} = V_f + L'_f \qquad \text{... (5.2.59)}$$

If we drop the subscripts, the above equation becomes

$$F + V' + L = V + L'$$

or
$$(V' - V) + F = (L' - L) \qquad \text{... (5.2.60)}$$

Dividing throughout by $F$, we get

$$\left(\frac{V' - V}{F}\right) + 1 = \left(\frac{L' - L}{F}\right) = q \qquad \text{... (5.2.61)}$$

or
$$(V' - V) = F\,(q - 1) \qquad \text{... (5.2.62)}$$

where $q$ is the number of moles of saturated liquid formed on the feed plate by the introduction of one mole of feed. If the feed is a saturated liquid of the same composition as the liquid on the feed plate, then $q = 1.0$. If the feed is a saturated vapour of the same composition as the vapour from the feed plate, then no additional liquid is formed due to the introduction of the feed and therefore, $q = 0.0$. For a cold liquid feed, $q > 1.0$ and for a superheated vapour feed, $q < 0.0$. For a cold liquid feed, $q$ may be computed from

$$q = 1 + [C_{pL} (T_B - T_F)/\lambda_F] \quad \text{... (5.2.63)}$$

For a superheated vapour feed,

$$q = -C_{pV} (T_F - T_D)/\lambda_F \quad \text{... (5.2.64)}$$

where $T_F$ = temperature of the feed
$T_B$, $T_D$ = bubble point and dew point respectively of the feed
$\lambda_F$ = latent heat of vapourisation of the feed
$C_{pL}$, $C_{pV}$ = specific heats of liquid and vapour respectively.

If the feed is a mixture of liquid and vapour, then $q$ is the fraction that is liquid. Evidently, in such cases, $0 < q < 1.0$. For example, if the feed is 60 percent vapour and 40 percent liquid, then $q = 0.40$.

Now, if we write Eqs (5.2.53) and (5.2.58) without the tray subscripts, we get

$$Vy = Lx + Dx_D \quad \text{... (5.2.65)}$$

$$V'y = L'x - Bx_B \quad \text{... (5.2.66)}$$

Substracting the first equation from the second,

$$(V' - V)\, y = (L' - L)\, x - (Bx_B + Dx_D) \quad \text{... (5.2.67)}$$

From an overall balance for the entire column,

$$Fz_F = Dx_D + Bx_B \quad \text{... (5.2.67a)}$$

where $z_F$ = mole fraction of the more volatile component in the feed.
= $x_F$, for a liquid feed
= $y_F$, for a vapour feed

Equation (5.2.67) can be, therefore, rewritten as

$$y = \left(\frac{L' - L}{V' - V}\right) x - \left(\frac{Fz_F}{V' - V}\right) \quad \text{... (5.2.68)}$$

But, from Eqs (5.2.61) and (5.2.62), we know that $(L' - L) = qF$ and $(V' - V) = (q - 1) F$. Therefore, the above equation becomes

$$y = \left(\frac{q}{q-1}\right) x - \left(\frac{z_F}{q-1}\right) \quad \text{(5.2.69)}$$

This is the equation to the locus of the point of intersection of the two operating lines. Evidently, it is a straight line (often called the *q-line*) of slope $q/(q - 1)$. It passes through the point $x = y = z_F$ on the 45° diagonal [since from Eq. (5.2.69), when $x = z_F$, $y = z_F$]. The operating lines of the rectifying section and the exhausting section, thus, intersect on the $q$-line. The coordinates of the this point of intersection $(x_a, y_a)$ can be obtained by solving the equation to the $q$-line [Eq. (5.2.69)] and that to the operating line of the enriching section [Eq. (5.2.56)] simultaneously. Thus

$$x_a = [(R + 1)\ z_F + (q - 1)\ z_D]/(R + q) \qquad \text{... (5.2.69a)}$$

$$y_a = [Rz_F + qz_D]/(R + q) \qquad \text{... (5.2.69b)}$$

where $z_D = x_D$ when distillate is liquid (when a total condenser is used) and $z_D = y_D$ when the distillate is collected as vapour (when a partial condenser is used).

Based on the assumption of constant molar overflow, the number of equilibrium stages (ideal stages) required can be now computed by solving the equations to the operating lines and the equilibrium relationship simultaneously. The procedure may be summarised as given below:

1. Specify $F$, $z_F$, $z_D$ and $x_B$ as well as the operating reflux ratio ($R$). Since $D$ and $B$ are not known, compute them from overall mass balance equations as

$$F = B + D$$

$$Fz_F = Bx_B + Dz_D$$

2. Compute $V$ and $L$ as

$$L = L_0 = RD$$

$$V = (L_0 + D) = (R + 1)\ D$$

3. Specify $q$ based on the thermal condition of the feed. Now, compute $V'$ and $L'$ as

$$V' = V + (q - 1)\ F$$

$$L' = V' + B$$

4. Compute $(x_a, y_a)$ from Eqs (5.2.69a) and (5.2.69b).
5. If a total condenser is used,

$$x_D = z_D$$

$$x_0 = y_1 = x_D$$

If a partial condenser is used (use of partial condenser is discussed subsequently in this subsection),

$$y_D = z_D$$

Compute $x_0$ from equilibrium data (corresponding to $y = y_D$) and compute $y_1$ as

$$y_1 = (R/R + 1)\ x_0 + (z_D/R + 1)$$

6. Since $y_1$ is known, compute $x_1$ from equilibrium data.
7. Put $j = 1$.
8. If $x_j > x_a$, compute $y_{j+1}$ from Eq. (5.2.56) as

$$y_{j+1} = \left(\frac{R}{R+1}\right) x_j + \left(\frac{z_D}{R+1}\right) \qquad \text{... (5.2.56a)}$$

If $x_j \le x_a$, compute $y_{j+1}$ from Eq. (5.2.58) as

$$y_{j+1} = (L'/V')\ x_j - (Bx_B/V') \qquad \text{... (5.2.58a)}$$

[If open steam is used, Eq. (5.2.70) must be used to compute $y_{j+1}$ instead of the above equation. Use of open steam is discussed subsequently in this subsection.]

9. Compute $x_{j+1}$ from equilibrium data.
10. If $x_{j+1} > x_B$, put $j = j + 1$ and repeat the computations from step (8). Otherwise, go to step (11).
11. The number of ideal stages required (including reboiler, but not including the partial condenser) $= j + (x_j - x_B)/(x_j - x_{j+1})$. The number of stages required inside the column is, therefore, $j + (x_j - x_B)/(x_j - x_{j+1}) - 1$. This may be rounded off to the nearest larger integer. Often, it is retained as such and the number of actual plates required is computed by dividing it by an appropriately defined plate efficiency. The resultant value is then rounded off to the nearest larger integer.

It must be noted that the above-discussed procedure assumes that the feed is introduced at the intersection of the two operating lines. If is not so, the change over from Eqs (5.2.56a) to (5.2.58a) must be done at the feed plate. In other words, for all stages above the feed plate, $y_{j+1}$ must be computed from Eq. (5.2.56a) and for all stages below the feed plate (including the feed plate), $y_{j+1}$ must be computed from Eq. (5.2.58a). No doubt, in the design of new columns, it shall always be desirable if the point of introduction of feed coincides with the point of intersection of the two operating lines since this shall demand minimum number of stages to affect the given separation.

The number of ideal stages required may also be estimated *graphically* as proposed by McCabe and Thiele. The procedure is as described below:

(*i*) First, plot the equilibrium curve from the available VLE data at the operating pressure of the column.

(*ii*) Draw the operating line for the rectifying section, using the known values of reflux ratio ($R$) and the distillate composition ($x_D$).

(*iii*) Compute or specify $q$ based on the thermal condition of the feed and draw the $q$-line which is of slope $q/(q - 1)$.

(*iv*) Join the point of intersection of the operating line for rectifying section and the $q$-line with the point $x = y = x_B$ on the 45° diagonal. This gives the operating line for the exhausting section.

(*v*) From the point $y = x = x_D$ on the 45° diagonal, draw a horizontal line to intersect the equilibrium curve at the point $(x_1, y_1)$. Draw a vertical line from this point to intersect the operating line for the rectifying section at $(x_1, y_2)$ and continue this *staircase* construction procedure until the liquid composition reached is equal to or slightly less than $x_B$. When a new column is being designed, the optimum feed plate location is at the intersection of the two operating lines. Therefore, once the $q$-line is crossed, the construction of intersections should be done on the operating line for the exhausting section (see Fig. 5.26). However, as stated earlier under the Analytical Method, in the case of an existing column being adapted to a new separation where the feed plate location has already been specified, the staircase construction must be transferred from the enriching-section operating line to the exhausting-section operating line once the feed plate location is reached. For example, if the feed is introduced upon the seventh plate from the top, then beginning with the seventh plate, the staircase construction must be performed on the stripping-section operating line. This is to be done regardless of whether the point of intersection of the two operating lines is before or after the seventh plate.

(*vi*) Count the number of intersections of the equilibrium curve, which gives the number of equilibrium stages required for the separation. If the vapour from the reboiler is in equilibrium with the residue (which, usually, is the case), then the last step of the staircase construction represents the reboiler.

The entire procedure is illustrated in Fig. 5.26 that considers a saturated liquid feed.

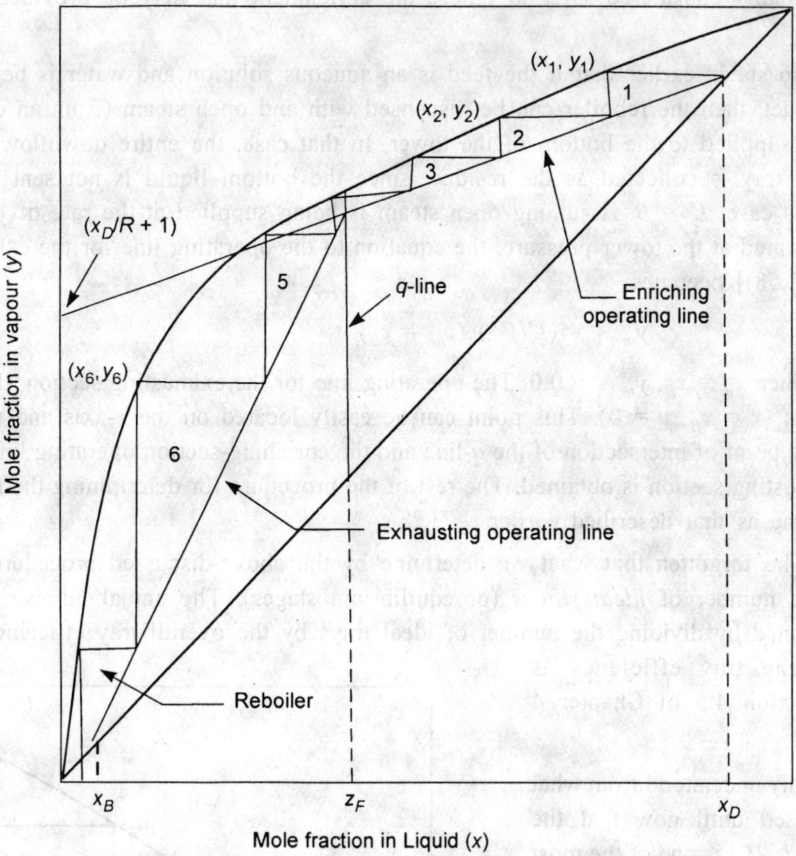


*Figure 5.26: McCabe and Thiele method*

It may be noted that though the graphical method could appear faster for a hand-calculation, the personal error involved is pretty large in procedures like this. Wherever personal computers or programmable calculators are available, the analytical method shall be more preferable.

If, instead of a total condenser, a partial condenser is being used, then, as stated earlier, the overhead vapours are only partially condensed, the condensate is returned to the column as the reflux and the remaining vapours are collected as the distillate. The distillate may be further condensed in a separate condenser. In the partial condenser, either differential condensation or equilibrium condensation can occur. If the condensate is removed as soon as it is formed, then differential condensation occurs. But, if the time of contact between the vapour product and the liquid reflux is sufficiently large, then the two will be in equilibrium with each other and in such a case, the partial condenser acts as one theoretical stage for separation. As a result, in the staircase construction for determining the number of ideal stages, described above, the first step will then represent the partial condenser and the staircase construction is to be started from the point $x = y = y_D$ on the 45° diagonal. However, it is difficult to ensure that equilibrium condensation will occur in the condenser and therefore, in the design of new distillation columns, it is usually considered safer to ignore the enrichment that may be provided by the partial condenser.

We have also stated earlier that if the feed is an aqueous solution and water is being removed as the residue product, then the reboiler can be dispensed with and open steam (from an external source) may be directly supplied to the bottom of the tower. In that case, the entire downflowing liquid from the bottom-most tray is collected as the residue, since the bottom liquid is not sent to the reboiler. Therefore, in this case, $L' = B$. Assuming open steam is being supplied at the rate of $V'$ moles/s and the steam is saturated at the tower pressure, the equation to the operating line for the exhausting section [that is, Eq. (5.2.58)] becomes

$$y_{m+1} = (L'/V')\,(x_m - x_B) \qquad \dots (5.2.70)$$

Evidently, when $x_m = x_B$, $y_{m+1} = 0.0$. The operating line for the exhausting section, therefore, passes through the point ($x = x_B$, $y = 0$). This point can be easily located on the $x$-axis and when this point is joined with the point of intersection of the $q$-line and the enriching-section operating line, the operating line for the exhausting section is obtained. The rest of the procedure for determining the number of ideal stages is the same as that described earlier.

It must not be forgotten that what we determine by the above-discussed procedure (analytical or graphical) is the number of *ideal stages* (or equilibrium stages). The actual number of plates/trays required is obtained by dividing the number of ideal trays by the overall tray efficiency. The method of estimating the tray efficiency is discussed in Section 4.5 of Chapter 4 (Volume 1).

It can be easily understood from what we have discussed until now that the reflux ratio $R$ (= $L_0/D$) is one of the most detrimental parameters in fractional distillation. As the reflux ratio is increased, the number of stages/plates required to affect a given separation decreases. Ultimately, when $R = \infty$, $(L/V) = (L'/V') = 1.0$ and the operating lines for both sections coincide with the 45° diagonal (see Fig. 5.27). The number of plates required is then the minimum ($N_m$). In practice, this happens when all the overhead vapours are condensed and returned to the column as the reflux (this is called *total reflux*), all the residue product is reboiled and when no fresh feed is admitted to the column. Thus, since no distillate is collected and no

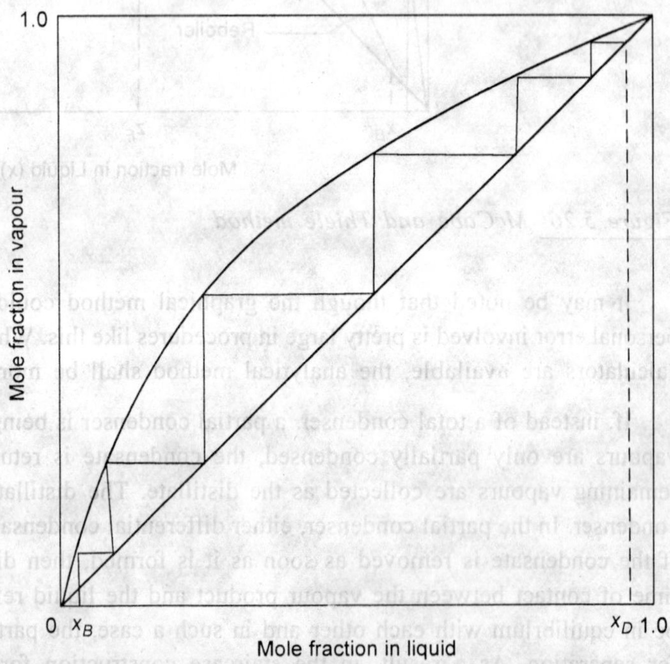


*Figure 5.27: Minimum stages at total reflux (with no feeds or products)*

residue is withdrawn, no fresh feed can be introduced to the column and therefore, the capacity of the column reduces to zero although a definite separation is taking place.

Operation under total reflux is an extreme condition and is difficult to obtain in actual practice. It must also be kept in mind that as the reflux ratio is increased, the cooling load on the condenser and the heat load on the reboiler also increase (thereby demanding larger condenser and reboiler surfaces) and since $L$, $L'$, $V$ and $V'$ all increase, the tower cross-section will have to be increased to accomodate the larger loads. For systems for which the relative volatility ($\alpha$) is essentially constant, the minimum number of ideal stages ($N_m$) required to affect a given separation (at total reflux) may be computed analytically from

$$N_m = \frac{\log\left[\dfrac{x_D(1-x_B)}{x_B(1-x_D)}\right]}{\log(\alpha)} \qquad \text{... (5.2.71)}$$

The above equation is called *Fenske's equation*[16]. In the above equation, $N_m$ includes the reboiler also. For small variations in $\alpha$, an average value of $\alpha$ may be used in the above equation as

$$\alpha_{avg} = \sqrt{\alpha_1 \alpha_B} \qquad \text{... (5.2.72)}$$

where $$\alpha_1 = \frac{y_1(1-x_1)}{x_1(1-y_1)} \qquad \text{... (5.2.73)}$$

$$\alpha_B = \frac{y_B(1-x_B)}{x_B(1-y_B)} \qquad \text{... (5.2.74)}$$

$$y_B = y_{N_p+1} \qquad \text{(5.2.75)}$$

= mole fraction of more volatile component in the reboiler vapour (that is assumed to be in equilibrium with the residue product).

$N_p$ = total number of ideal stages excluding reboiler.

The converse to total reflux is the *minimum reflux.* When the reflux ratio is reduced to the minimum, the number of plates required to affect a given separation will become infinitely large, though the values of $L$, $L'$, $V$ and $V'$ will decrease to the lowest and the condenser and reboiler loads will also reduce to the minimum. Thus, when $R = R_m$, $N_p = \infty$. The minimum reflux ratio also can be determined graphically. For the usual case of the equilibrium curve being concave downward, the value of $R_m$ can be determined by drawing an operating line through the point of intersection of the $q$-line with the equilibrium curve (see Fig. 5.28). The $y$-intercept of this line is equal to $x_D/(R_m + 1)$.

If the system exhibits azeotropic tendencies and the equilibrium curve has large curvature [as shown in Figs 5.29 (a) and (b)], then a tangent to the equilibrium curve in the enriching section will indicate the *minimum reflux ratio* [see Fig. 5.29 (a)]. The $y$-intercept of this tangent will be equal to $x_D/(R_m + 1)$. If the curvature of the equilibrium curve is more pronounced in the bottom portion [as in Fig. 5.29 (b), then a tangent operating line is to be drawn in the exhausting section to intersect the $q$-line at a point, say $P$. The operating line for the enriching section is now drawn through the point $P$ and its $y$-intercept will be equal to $x_D/(R_m + 1)$.

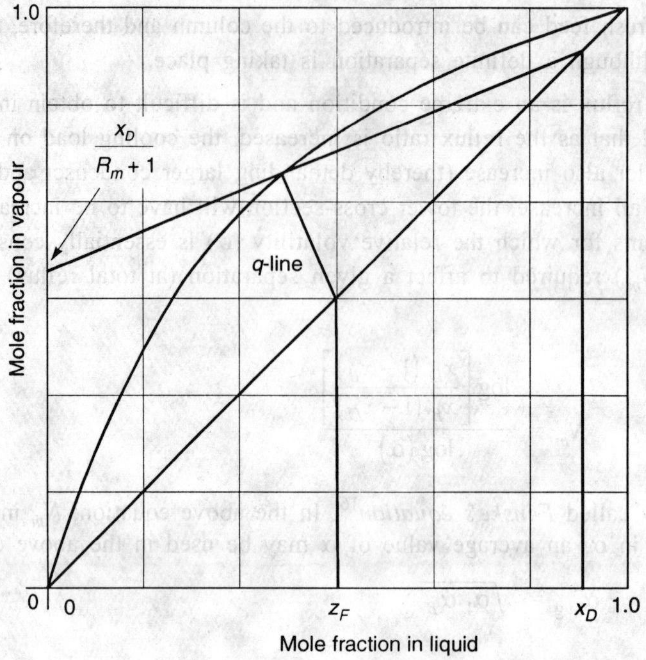


*Figure 5.28: Determination of minimum reflux ratio (general case)*

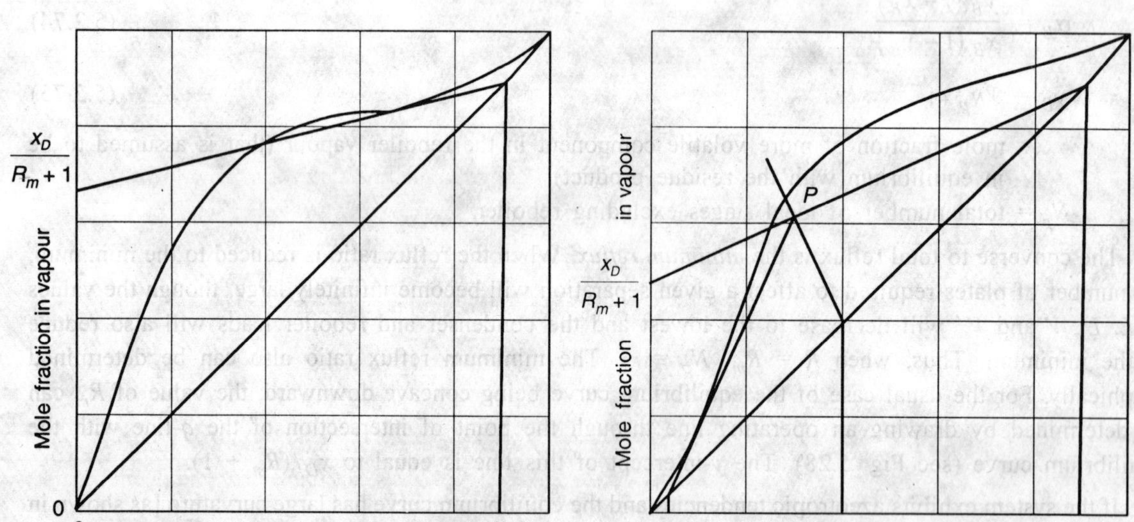


*Figure 5.29 (a) and (b): Determination of minimum reflux ratio for special cases. (a) Equilibrium curve with top curvature, (b) Equilibrium curve with bottom curvature*

The operation of a fractionator should be, therefore, between the total reflux and the minimum reflux. It is common practice to choose an *operating reflux ratio* that is equal to 1.2 to 1.5 times the minimum. That is, usually $R = (1.2R_m)$ to $(1.5R_m)$.

It is possible to propose an *optimum reflux ratio* based on cost considerations. As the reflux ratio is increased, the number of plates required to affect the desired separation decreases. But, the required column diameter increases since the liquid and vapour flow rates increase. The fixed charges for the equipment, therefore, initially decrease with increase in reflux ratio, but after some time, a point is reached where the increase in the required column diameter is more rapid than the decrease in the required number of plates and therefore, the fixed charges tend to increase with further increase in reflux ratio. In other words, the fixed charges first decrease and then increase with increase in reflux ratio. Since the reflux is made by supplying heat at the reboiler and withdrawing it at the condenser, both the steam requirement of the reboiler and the cooling water requirement of the condenser will increase if the reflux ratio is increased. The total cost, which is the sum of the fixed charges on the equipment and the steam and cooling water costs, therefore, first decreases with increase in reflux ratio, passes through a minimum and thereafter increases. The optimum reflux ratio is located at the point where the total cost is a minimum (see Fig. 5.30).

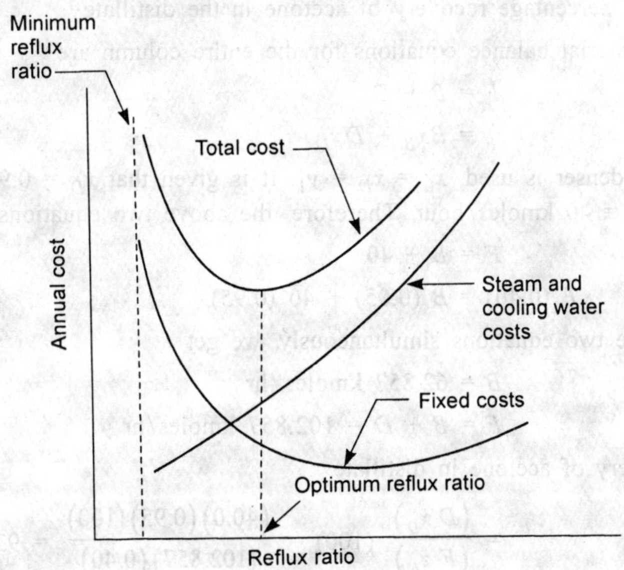


*Figure 5.30: Optimum reflux ratio*

**Example 5.21:** It is desired to produce 960 kmoles of 95 percent acetone per day by continuous fractionation of a feed containing 40 mole percent acetone and 60 mole percent acetic acid at 760 mm Hg. The fractionator is equipped with a total condenser and the reflux is returned at its bubble point. The bottom residue is to contain 5.0 mole percent acetone. Determine the number of theoretical stages required for the following two cases both graphically as well as analytically:

(*a*) the feed is saturated liquid and reflux ratio that is 2.5 times the minimum is used, (*b*) the feed is saturated vapour and a reflux ratio that is 1.5 times the minimum is used.

Neglect heat losses from the column. Equilibrium data for acetone-acetic acid system at 760 mm Hg is given below:

| *Temperature, °C* | *Mole fraction of acetone in liquid* | *Mole fraction of acetone in vapour* |
|---|---|---|
| 112.1 | 0.042 | 0.108 |
| 107.4 | 0.082 | 0.225 |
| 104.6 | 0.158 | 0.356 |
| 94.3 | 0.226 | 0.564 |
| 90.4 | 0.271 | 0.630 |
| 86.3 | 0.307 | 0.709 |
| 78.6 | 0.433 | 0.844 |
| 70.8 | 0.550 | 0.918 |
| 65.6 | 0.668 | 0.966 |
| 60.7 | 0.935 | 0.997 |

What will be the percentage recovery of acetone in the distillate?

**Solution:** The material balance equations for the entire column are

$$F = B + D$$

$$Fz_F = Bx_B + Dx_D$$

Since a total condenser is used, $x_0 = x_D = y_1$. It is given that $x_D = 0.95$, $x_B = 0.05$, $z_F = 0.40$, $D = 960$ kmoles/day $= 40$ kmoles/hour. Therefore, the above two equations become

$$F = B + 40$$

$$F\ (0.40) = B\ (0.05) + 40\ (0.95)$$

Solving the above two equations simultaneously, we get

$$B = 62.857 \text{ kmoles/hr}$$

$$F = B + D = 102.857 \text{ kmoles/hr}$$

Percentage recovery of acetone in distillate

$$= \frac{(D\,x_D)}{(F\,z_F)}\ (100) = \frac{(40.0)(0.95)(100)}{(102.857)(0.40)} = 92.36$$

(*a*) *Graphical method:* The equilibrium curve is plotted as shown in Fig. 5.21.1. The $q$-line will be a vertical line [since the feed is a saturated liquid $q = 1.0$ and the slope of $q$-line $= q/(q-1) = \infty$] as shown. The point $x = y = x_D = 0.95$ is marked on the 45° diagonal and a straight line is drawn through this point and the point of intersection of $q$-line with the equilibrium curve (this line is shown as a dotted line in the figure). The $y$-intercept of this line is read as 0.69. Therefore,

$$\frac{x_D}{R_m + 1} = 0.69$$

or $$R_m = 0.3768$$

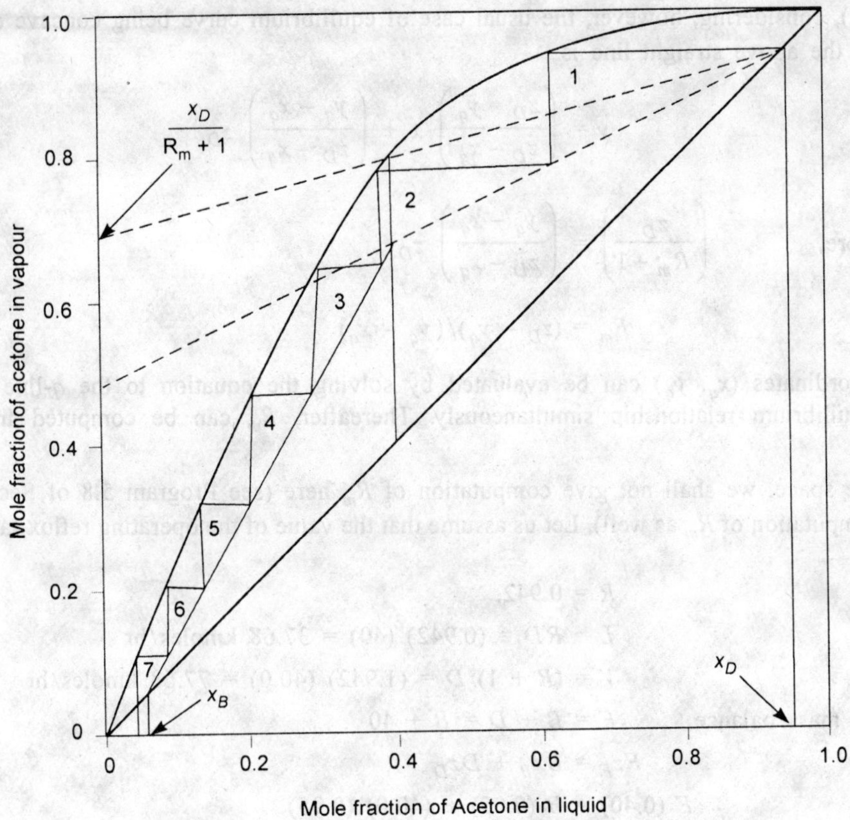


*Figure 5.21.1*

Since the operating reflux ratio is 2.5 times the minimum,

$$R = (2.5)\ R_m = 0.942$$

$$\left(\frac{x_D}{R+1}\right) = \left(\frac{0.95}{1.942}\right) = 0.489$$

The point $y = 0.489$ is marked on the $y$-axis. The straight line joining this point and the point $x = y = 0.95$ (already marked on the 45° diagonal) gives the operating line for the enriching section. This intersects the $q$-line as shown. The point of intersection when joined with the point $x = y = x_B = 0.05$ on the 45° diagonal gives the operating line for the exhausting section. The staircase construction is now performed as shown in the figure. The number of ideal stages required is 6.9, including the reboiler. The feed is to be introduced to the second ideal stage from the top. We are not rounding off the number of ideal stages to the nearest integer because this figure is to subsequently divided by the tray efficiency to obtain the actual number of real trays required.

*Analytical method:* The minimum reflux ratio ($R_m$) also can be computed analytically. Let the point of intersection of the $q$-line and the equilibrium curve is $(x_q, y_q)$. We know that the $y$-intercept of the straight line starting from the point $y = x = z_D$ on the 45° diagonal and passing through $(x_q, y_q)$ is

$z_D/(R_m + 1)$, considering, however, the usual case of equilibrium curve being concave downward. The equation to the above straight line is

$$y = \left(\frac{z_D - y_q}{z_D - x_q}\right) x + \left(\frac{y_q - x_q}{z_D - x_q}\right) z_D \quad \text{... (i)}$$

Therefore, $$\left(\frac{z_D}{R_m + 1}\right) = \left(\frac{y_q - x_q}{z_D - x_q}\right) z_D$$

or $$R_m = (z_D - y_q)/(y_q - x_q) \quad \text{... (ii)}$$

The coordinates $(x_q, y_q)$ can be evaluated by solving the equation to the $q$-line [Eq. (5.2.69)] and the equilibrium relationship simultaneously. Thereafter, $R_m$ can be computed from the above equation.

To save space, we shall not give computation of $R_m$ here (see Program 5.8 of Section 5.3 which includes computation of $R_m$ as well). Let us assume that the value of the operating reflux ratio is available. Thus,

$$R = 0.942.$$

Now, $$L = RD = (0.942)\ (40) = 37.68 \text{ kmoles/hr}$$

$$V = (R + 1)\ D = (1.942)\ (40.0) = 77.68 \text{ kmoles/hr}$$

From a mass balance, $$F = B + D = B + 40$$

$$Fz_F = Bx_B + Dz_D$$

$$F\ (0.40) = B\ (0.05) + (40.0)\ (0.95)$$

Solving simultaneously, $$F = 102.857 \text{ kmoles/hr}$$

$$B = 62.857 \text{ kmoles/hr}$$

Since $$q = 1.0,\ V' = V = 77.68 \text{ kmoles/hr}$$

$$L' = (V' + B) = 140.537 \text{ kmoles/hr}$$

From Eqs (5.2.69a) and (5.2.69b),

$$x_a = 0.4$$

$$y_a = 0.6832$$

Since total condenser is used, $y_1 = x_D = 0.95$. Therefore, from equilibrium data, $x_1 = 0.6287$. Now, putting $j = 1$, from Eq. (5.2.56a),

$$y_2 = 0.794$$

And from equilibrium data, $x_2 = 0.3865$. Now, we put $j = 2$. But, since $x_2 < x_a$, $y_3$ is computed from Eq. (5.2.58a) as

$$y_3 = 0.6587$$

$$x_3 = \text{(from equilibrium data)} = 0.2841$$

The procedure is continued for higher values of $j$ until $x_{j+1} < 0.05$. The results are listed below:

| $j$ | $x_{j+1}$ | $y_{j+1}$ |
|---|---|---|
| 0 | $0.6287 = x_1$ | $0.95 = y_1$ |
| 1 | $0.3865 = x_2$ | $0.794 = y_2$ |
| 2 | $0.2841 = x_3$ | $0.6587 = y_3$ |
| 3 | $0.1964 = x_4$ | $0.4735 = y_4$ |
| 4 | $0.13417 = x_5$ | $0.3149 = y_5$ |
| 5 | $0.07423 = x_6$ | $0.2023 = y_6$ |
| 6 | $0.03649 = x_7$ | $0.09384 = y_7$ |

**If $N_S$ represents the number of ideal stages required (including the reboiler) then**

$$N_S = 6 + (0.07423 - 0.05)/(0.07423 - 0.03649) = \mathbf{6.64}$$

The slight discrepancy between the computed values of $(x_1, y_1)$, $(x_2, y_2)$, etc. and those determined graphically in Fig. 5.21.1 might be attributed to the personal error involved in the graphical procedure. The results show that feed is to be introduced between the first and second equilibrium stages.

(*b*) *Graphical method:* The procedure is same as in (*a*), except that the $q$-line will be a horizontal line [since the feed is saturated vapour, $q = 0.0$ and therefore, the slope of $q$-line $= q/(q - 1) = 0.0$] as shown in Fig. 5.21.2. The operating line for minimum reflux (shown as a broken line in figure)

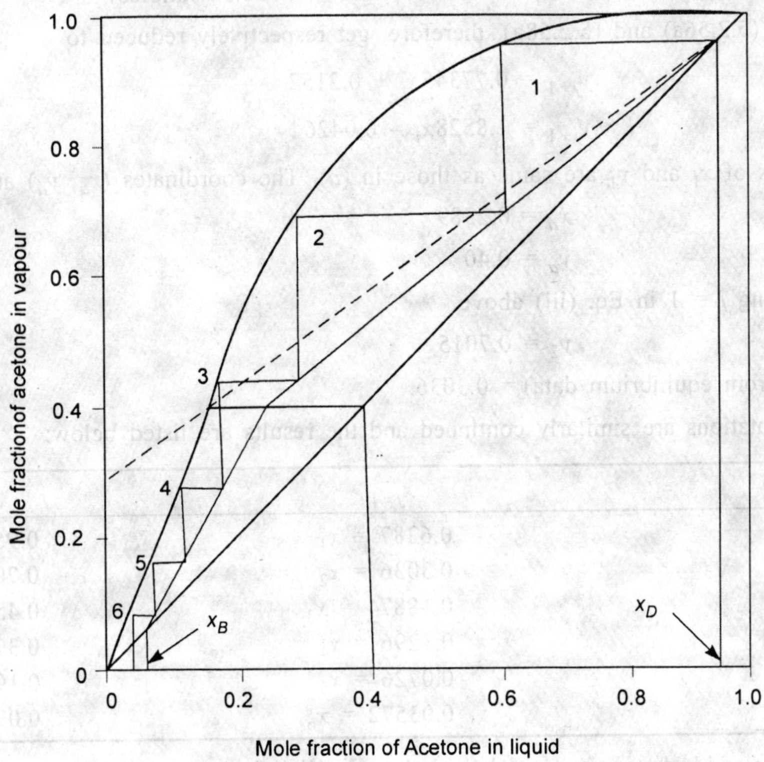


*Figure 5.21.2*

intersects the $y$-axis at $y = 0.29$. Therefore,

$$\left(\frac{x_D}{R_m+1}\right) = 0.29$$

or

$$R_m = 2.276$$

$$R = (1.5)R_m = 3.414$$

$$\left(\frac{x_D}{R+1}\right) = 0.2152$$

The operating lines are now drawn as in (*a*) and the number of ideal stages required is determined, as before, by the staircase construction. It is seen that 5.7 ideal trays, including the reboiler, are required to affect the desired separation (the tower should, therefore, contain 4.7 ideal trays). The feed is to be introduced to third ideal tray from the top.

*Analytical method:* We follow the same procedure as in (*a*). Since $R = 3.414$,

$$V = (4.414)\ (40.0) = 176.56 \text{ kmoles/hr}$$

Since

$$q = 0.0,$$

$$V' = V + (q-1)F = 176.56 - 102.857 = 73.703 \text{ kmoles/hr}$$

$$L' = 73.703 + 62.857 = 136.56 \text{ kmoles/hr}$$

Equations (5.2.56a) and (5.2.58a), therefore, get respectively reduced to

$$y_{j+1} = 0.77345x_j + 0.2152 \quad \text{... (iii)}$$

and

$$y_{j+1} = 1.8528x_j - 0.04264 \quad \text{... (iv)}$$

The values of $x_1$ and $y_1$ are same as those in (*a*). The coordinates $(x_a, y_a)$ are

$$x_a = 0.2389$$

$$y_a = 0.40$$

Now, putting $j = 1$ in Eq. (iii) above,

$$y_2 = 0.7015$$

And, $x_2$ (from equilibrium data) = 0.3036.

The computations are similarly continued and the results are listed below:

| $j$ | $x_{j+1}$ | $y_{j+1}$ |
|---|---|---|
| 0 | $0.6287 = x_1$ | $0.95 = y_1$ |
| 1 | $0.3036 = x_2$ | $0.7015 = y_2$ |
| 2 | $0.18874 = x_3$ | $0.450 = y_3$ |
| 3 | $0.1296 = x_4$ | $0.307 = y_4$ |
| 4 | $0.0726 = x_5$ | $0.1975 = y_5$ |
| 5 | $0.03572 = x_6$ | $0.09186 = y_6$ |

The number of ideal stages required (including reboiler) is $5 + (0.0726 - 0.05)/(0.0726 - 0.03572) = 5.613$ and feed is to be introduced between the second and third ideal stages.

**Example 5.22:** A feedstock containing 22.3 mole percent ethanol and balance water, at 1 atm and an enthalpy of 193.458 kJ/kg, is to be separated by distillation in a continuous, bubble-cap fractionating column equipped with a reboiler and a total condenser. The bottom product is to contain 1.0 mole percent ethanol and the distillate product 75 mole percent ethanol. The feed rate is 43.5 kmoles/hr. The reflux is returned at its bubble point and a reflux ratio that is thrice the minimum is to be employed.

(*a*) How many actual plates (in addition to the reboiler) are required with an overall plate efficiency of 70 percent? Assume the reboiler is equivalent to one ideal plate.

(*b*) What should be the rate of flow of cooling water in the condenser if water enters the condenser at 10° C and leaves at 45° C?

(*c*) What will be the number of plates required if the feed is composed of 40 percent vapour and 60 percent liquid and the reflux ratio employed is same as that in (*a*)?

The equilibrium data and enthalpy data for ethanol-water system at 1 atm are given below:

**Equilibrium Data**

| *Temperature, °C* | *Mole fraction of ethanol in liquid* | *Mole fraction of ethanol in vapour* |
|---|---|---|
| 95.5 | 0.0190 | 0.170 |
| 89.0 | 0.0721 | 0.3891 |
| 86.7 | 0.0966 | 0.4375 |
| 85.3 | 0.1238 | 0.4704 |
| 84.1 | 0.1661 | 0.5089 |
| 82.7 | 0.2337 | 0.5445 |
| 82.3 | 0.2608 | 0.5580 |
| 81.5 | 0.3273 | 0.5826 |
| 80.7 | 0.3965 | 0.6122 |
| 79.8 | 0.5079 | 0.6564 |
| 79.7 | 0.5198 | 0.6599 |
| 79.3 | 0.5732 | 0.6841 |
| 78.74 | 0.6763 | 0.7385 |
| 78.41 | 0.7472 | 0.7815 |
| 78.15 | 0.8943 | 0.8943 |

**Enthalpy Data**

| *Mass fraction of ethanol* | *Saturated liquid enthalpy, kJ/kg* | *Saturated vapour enthalpy, kJ/kg* |
|---|---|---|
| 0.0 | 418.9126 | 2674.9 |
| 0.10 | 371.69 | 2516.732 |
| 0.20 | 335.6418 | 2355.075 |
| 0.30 | 314.01 | 2193.418 |
| 0.40 | 298.1932 | 2030.598 |

(*Contd.*)

**Enthalpy Data** (*Contd.*)

| *Mass fraction of ethanol* | *Saturated liquid enthalpy, kJ/kg* | *Saturated vapour enthalpy, kJ/kg* |
|---|---|---|
| 0.50 | 285.8654 | 1870.104 |
| 0.60 | 273.305 | 1707.284 |
| 0.70 | 258.4186 | 1544.464 |
| 0.80 | 241.4388 | 1386.296 |
| 0.90 | 224.6916 | 1223.476 |
| 1.0 | 207.014 | 1064.145 |

**Solution:** (*a*) it is given that mole fraction of ethanol in the feed is 0.223. Therefore, mass fraction of ethanol in the feed will be,

$$= \frac{(0.223\,M_A)}{(0.223\,M_A + 0.777\,M_B)}$$

where $M_A$ = molecular weight of ethanol = 46.0
$M_B$ = molecular weight of water = 18.0

Therefore, mass fraction of ethanol in feed = 0.423

Corresponding to this mass fraction, the saturated liquid enthalpy is (from the given enthalpy data) 295.3578 kJ/kg. Since it is given that the enthalpy of feed is 193.458 kJ/kg, it is clear that the feed is a cold liquid. Therefore, $x_F = 0.223$. For a cold liquid feed, the value of $q$ is to be computed from Eq. (5.2.63).

$$q = 1 + [C_{pL}\,(T_B - T_F)/\lambda_F] = \frac{\lambda_F + C_{pL}\,(T_B - T_F)}{\lambda_F} \quad \text{... (i)}$$

We know that

$H_L$ = saturated liquid enthalpy = $C_{pL}\,(T_B - T_{ref})$
$H_F$ = feed enthalpy (for cold liquid) = $C_{pL}\,(T_F - T_{ref})$
$T_{ref}$ = reference temperature

Therefore, Eq. (i) becomes

$$q = [\lambda_F + (H_L - H_F)]/\lambda_F$$

Since $\lambda_F = (H_V - H_L)$, where $H_V$ is the enthalpy of saturated vapour,

$$q = (H_V - H_F)/(H_V - H_L) \quad \text{... (ii)}$$

It can be noted that $q$ can be, therefore, defined also as the ratio of the heat required to convert one mole of feed to saturated vapour to the latent heat of vapourisation. From the enthalpy data, corresponding to mass fraction = 0.423, $H_V$ = 1993.684 kJ/kg and $H_L$ = 295.3578 kJ/kg. Therefore,

$$q = \frac{(1993.684 - 193.458)}{(1993.684 - 295.3578)} = 1.06$$

$$q/(q - 1) = 17.67$$

*Graphical method:* The equilibrium curve is shown in Fig. 5.22.1. The $q$-line is drawn with a positive slope of 17.67. The operating line for minimum reflux (not shown in figure) intersects the $y$-axis at $y = 0.45$. Therefore,

$$\left(\frac{x_D}{R_m+1}\right) = 0.45$$

or
$$R_m = (0.75/0.45) - 1 = 0.6667$$

To note that since a total condenser is used, $y_1 = x_0 = x_D = 0.75$. Since the operating reflux ratio is thrice the minimum,

$$R = (3.0)\ (0.6667) = 2.0$$

$$\left(\frac{x_D}{R+1}\right) = (0.75/3.0) = 0.25$$

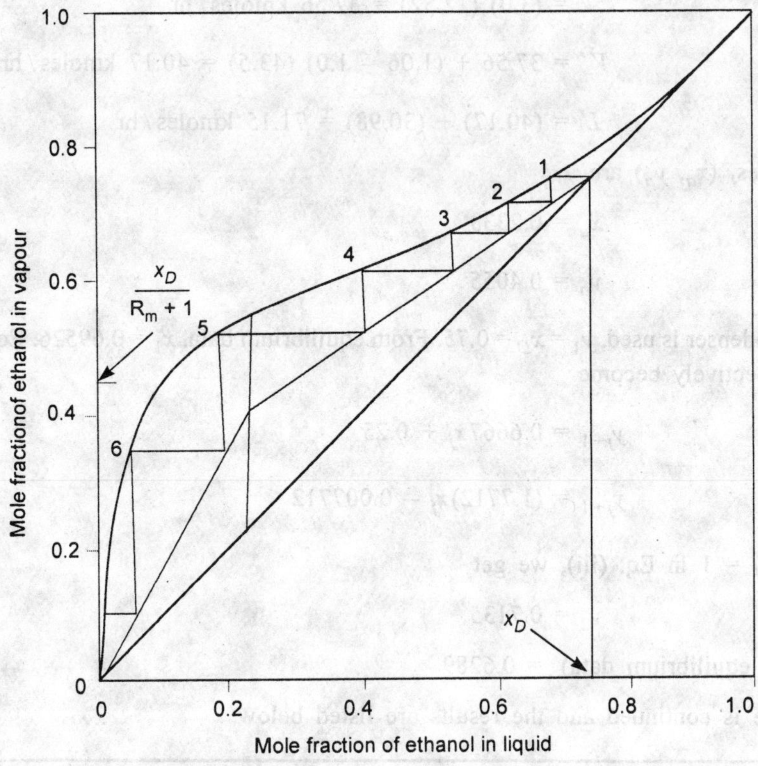


*Figure 5.22.1*

The operating lines are now drawn as shown in Fig. 5.22.1. The number of ideal plates required is determined by performing the staircase construction. It is found that seven ideal plates, including the reboiler, are required to affect the separation. The column should, therefore, contain six ideal plates. Since the overall plate efficiency is 70 percent, the number of actual plates required is (6.0/0.70) = 8.57 ≃ 9.0.

*Analytical method:* As in Example 5.21, the analytical computation of $R_m$ is not given here. We start with the known value of $R$, such as $R = 2.0$. From a material balance for the entire column,

$$F = 43.5 = B + D$$

$$Fz_F = Bx_B + Dz_D$$

$$(43.5)\ (0.223) = B\ (0.01) + D\ (0.75)$$

Solving simultaneously,

$$B = 30.98 \text{ kmoles/hr}$$

$$D = 12.52 \text{ kmoles/hr}$$

Now, $V = (3.0)\ (12.52) = 37.56$ kmoles / hr

$$V' = 37.56 + (1.06 - 1.0)\ (43.5) = 40.17 \text{ kmoles/hr}$$

$$L' = (40.17) + (30.98) = 71.15 \text{ kmoles/hr}$$

The coordinates, $(x_a, y_a)$ are

$$x_a = 0.2333$$

$$y_a = 0.4055$$

Since total condenser is used, $y_1 = x_D = 0.75$. From equilibrium data, $x_1 = 0.69526$. Equations (5.2.56a) and (5.2.58a) respectively become

$$y_{j+1} = 0.6667x_j + 0.25 \qquad \text{... (ii)}$$

and

$$y_{j+1} = (1.7712)x_j - 0.007712 \qquad \text{... (iii)}$$

Now, putting $j = 1$ in Eq. (iii), we get

$$y_2 = 0.7135$$

And $x_2$ (from equilibrium data) $= 0.6289$

The procedure is continued and the results are listed below:

| $j$ | $x_{j+1}$ | $y_{j+1}$ |
|---|---|---|
| 0 | $0.69526 = x_1$ | $0.75 = y_1$ |
| 1 | $0.6289 = x_2$ | $0.7135 = y_2$ |
| 2 | $0.5405 = x_3$ | $0.6693 = y_3$ |
| 3 | $0.3922 = x_4$ | $0.6103 = y_4$ |
| 4 | $0.1709 = x_5$ | $0.5114 = y_5$ |
| 5 | $0.0493 = x_6$ | $0.295 = y_6$ |
| 6 | $0.0089 = x_7$ | $0.0796 = y_7$ |

The number of ideal stages required (including the reboiler) is, therefore, 6 + (0.0493 – 0.01)/(0.0493 – 0.0089) = 6.97. The number ideal stages required inside the column is 5.97 and the number of actual plates required (5.97/0.70) = 8.53 ≃ 9.0. The feed is to be introduced between fourth and fifth ideal stages.

Here also, there is slight discrepancy between the values of ($x_1$, $y_1$), ($x_2$, $y_2$), etc. graphically determined and those analytically computed. This is attributable to the personal error involved in the graphical procedure.

(*b*) For a total condenser and a bubble point reflux, $y_1 = x_D = 0.75$. Now,

$$y_1 \text{ (as mass fraction)} = \frac{(0.75)(46.0)}{(0.75)(46.0)+(0.25)(18.0)} = 0.8846$$

Corresponding to this mass fraction, the enthalpy of saturated liquid (through a linear interpolation of enthalpy data) is $H_L$ = 227.27 kJ/kg and that of saturated vapour ($H_V$) is 1248.55 kJ/kg. The latent heat of condensation is, therefore, ($H_V - H_L$) = 1021.28 kJ/kg. Since the reflux is returned at its bubble point, the heat removed in the condenser is this latent heat. If $\dot{m}$ is the mass flow rate of water in kg/hr, then

$$\dot{m}\, C_p\, (t_2 - t_1) = (1021.28)\, D \qquad \text{... (iv)}$$

where $C_p$ = specific heat of water
= 4.18 kJ/(kg K)
$t_1$ = inlet temperature of water = 10° C
$t_2$ = outlet temperature of water = 45° C
$D$ = amount of distillate, kg/hr

We know, $D$ = 12.52 kmoles/hr
= 12.52 (0.75 × 46 + 0.25 × 18) kg/hr = 488.28 kg/hr

Now, from Eq. (iv),

$$\dot{m} = \frac{(1021.28)(488.28)}{(4.18)(45-10)} = 3408.548 \text{ kg/hr} = \mathbf{0.9468\ kg/s}$$

(*c*) *Graphical method:* The procedure is same as in (*a*), except that since the feed is partially vapourised and contains 60 mole percent liquid, $q$ = 0.6. The slope of the $q$-line is, therefore, (0.6)/(0.6 – 1.0) = −1.5. The $q$-line is, thus, drawn with a negative slope as shown in Fig. 5.22.2. Since the reflux ratio employed is the same, the operating line for the enriching section will be the same as in (*a*). The number of ideal stages required is determined by performing the staircase construction as before. The required number of ideal stages is found to be 7.3, including the reboiler. The actual number of plates required will be (7.3/0.7) = 10.43 ≃ 11.0. This includes the reboiler which has been assumed equivalent to one ideal plate.

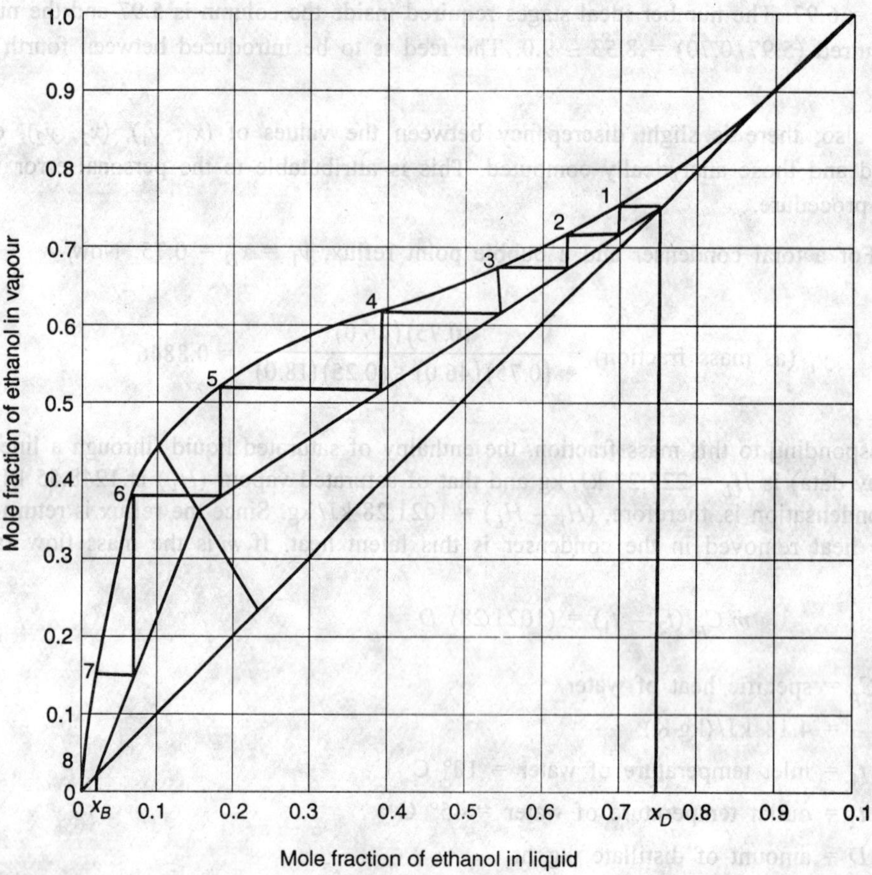


*Figure 5.22.2*

*Analytical method:* Since the value of $R$ is the same, the equation to the operating line for the enriching section shall be the same as in ($a$). Thus

$$y_{j+1} = 0.6667x_j + 0.25 \qquad \text{... (ii)}$$

However, Eq. (5.2.58a) gets modified to

$$y_{j+1} = 2.5367x_j - 0.015367 \qquad \text{... (v)}$$

This is because since $q$ is now 0.6, $V' = 37.56 + (0.6 - 1.0)\ 43.5 = 20.16$ kmoles/hr and $L' = 20.16 + 30.98 = 51.14$ kmoles/hr. The values of $x_a$ and $y_a$ are, $x_a = 0.1419$ and $y_a = 0.3446$. The values of ($x$, $y$) for all values of $j$ up to $j = 4$ shall be the same as those computed in ($a$). From $j = 5$ onwards, the values of ($x$, $y$) change since these are to be obtained from Eq. (v). We shall list the results below:

| $j$ | $x_{j+1}$ | $y_{j+1}$ |
|---|---|---|
| 0 | $0.69526 = x_1$ | $0.75 = y_1$ |
| 1 | $0.6289 = x_2$ | $0.7135 = y_2$ |
| 2 | $0.5405 = x_3$ | $0.6693 = y_3$ |
| 3 | $0.3922 = x_4$ | $0.6103 = y_4$ |
| 4 | $0.1709 = x_5$ | $0.5114 = y_5$ |
| 5 | $0.066 = x_6$ | $0.36395 = y_6$ |
| 6 | $0.017 = x_7$ | $0.152 = y_7$ |
| 7 | $0.0031 = x_8$ | $0.02775 = y_8$ |

The number of ideal stages required (including reboiler) is, therefore, $7 + (0.017 - 0.01)/(0.017 - 0.0031) = 7.5$. The feed is to be introduced between the fifth and the sixth ideal stages.

**Example 5.23:** One hundred kmoles per hour of a solution containing 20 mole percent of acetone and balance water are to be distilled in a plate column with open steam (supplied saturated at 1 atm) and a partial condenser. The column operates at 1 atm and the feed is at its bubble point. The distillate should contain 95.5 mole percent of acetone and no more than 0.1 percent of the acetone content of the feed is to be lost in the residue. Determine the number of ideal stages required (including the partial condenser) if a reflux ratio that is 1.5 times the minimum is to be employed. Compute the number of actual plates required, if the partial condenser may be assumed equivalent to one ideal plate and is separate from the column and the rest of the column has a plate efficiency of 75 percent. The equilibrium data for acetone-water system at 1 atm is given below[2]:

| *Temperature, °C* | *Mole fraction of acetone in liquid* | *Mole fraction of acetone in vapour* |
|---|---|---|
| 74.8 | 0.0500 | 0.6381 |
| 68.53 | 0.10 | 0.7301 |
| 65.26 | 0.15 | 0.7716 |
| 63.59 | 0.20 | 0.7916 |
| 61.87 | 0.30 | 0.8124 |
| 60.75 | 0.40 | 0.8269 |
| 59.95 | 0.50 | 0.8387 |
| 59.12 | 0.60 | 0.8532 |
| 58.29 | 0.70 | 0.8712 |
| 57.49 | 0.80 | 0.8950 |
| 56.68 | 0.90 | 0.9335 |
| 56.30 | 0.95 | 0.9627 |

**Solution:** Let us first write down the material balance equations for the entire column. Since the feed is a saturated liquid, $z_F = x_F$ and since a partial condenser is used, the distillate is vapour with $y_D = 0.955$. Now,

$$F = 100 = B + D$$

$$Fx_F = Bx_B + Dy_D$$

But, it is given that the acetone content of the residue should not be more than 0.1 percent of that of the feed. That is,

$$(Bx_B)/(Fx_F) = 0.001$$

or

$$Bx_B = (0.001)\ (Fx_F) = (0.001)\ (100)\ (0.2) = 0.02 \text{ kmoles/hr}$$

Therefore,

$$Dy_D = (F\ x_F - B\ x_B) = (20.0 - 0.02) = 19.98 \text{ kmoles/hr}$$

$$D = (Dy_D/y_D) = (19.98/0.955) = 20.92 \text{ kmoles/hr}$$

$$B = (F - D) = 79.08 \text{ kmoles/hr}$$

$$x_B = (Bx_B/B) = (0.02/79.08) = 0.00025$$

*Graphical method:* The equilibrium curve is shown in Fig. 5.23.1. It can be seen that the curve has a significant curvature in the upper portion. The operating line for minimum reflux (shown as a broken

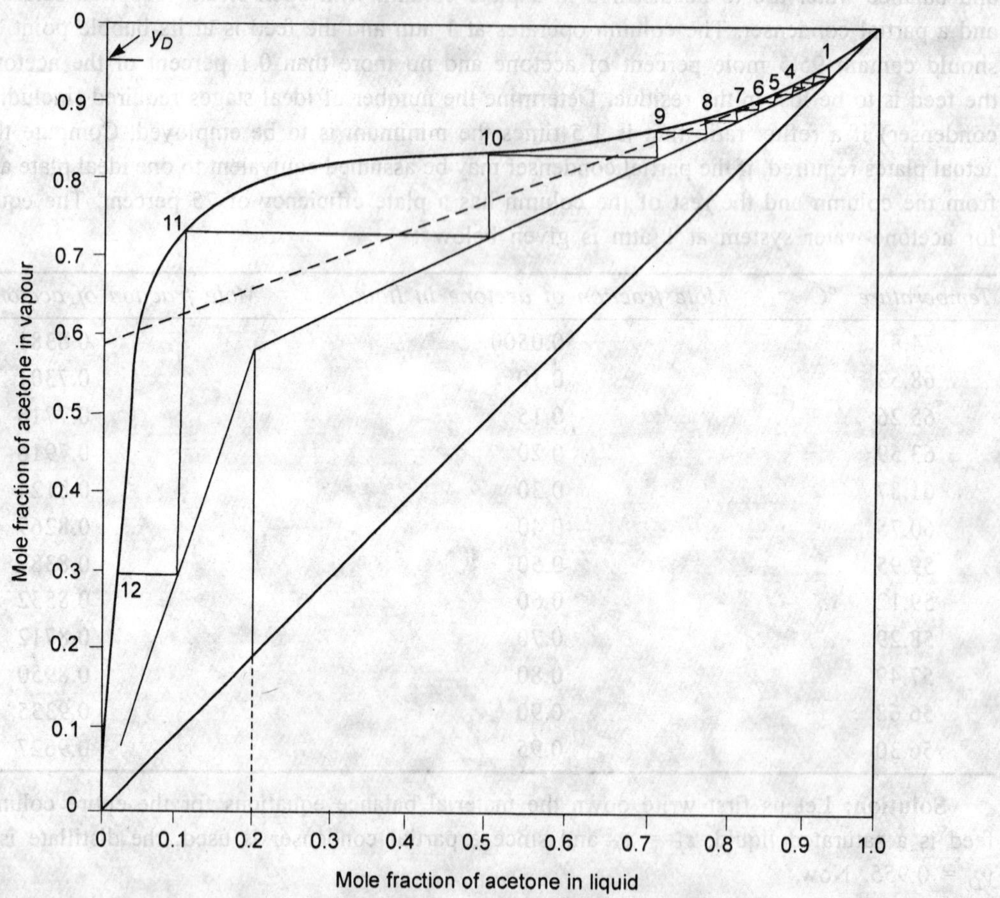


*Figure 5.23.1*

line in the figure) is drawn as tangent to the equilibrium curve and it intersects the $y$-axis at $y = 0.59$. To note that this line is drawn from the point $x = y = y_D$ on the 45° diagonal. Thus,

$$\left(\frac{y_D}{R_m + 1}\right) = 0.59$$

or $$R_m = 0.6186$$

Therefore, $$R = (1.5)\ R_m = 0.9279$$

$$\left(\frac{y_D}{R+1}\right) = 0.49535; \quad \left(\frac{R}{R+1}\right) = 0.4813$$

The operating line for the enriching section is, therefore, drawn with a slope of 0.4813 and a $y$-intercept of 0.49535 and it intersects the $q$-line (which is a vertical line since $q = 1.0$ and the slope $= q/(q-1) = \infty$ for a bubble point feed) as shown. From this point of intersection, a straight line of slope $(L'/V')$ is drawn which becomes the operating line for the exhausting section. The ratio $(L'/V')$ is computed as follows:

$$L' = L + qF = L + F \text{ (since } q = 1.0)$$

$$V' = V + (q - 1)\ F = V$$

Now, $$L = L_0 = RD \text{ and } V = V_1 = (L_0 + D) = (R + 1)\ D.$$

Therefore, $$(L'/V') = (L + F)/V = (RD + F)/(R + 1)\ D$$

Substituting the values of $R$, $D$ and $F$, we get $(L'/V') = 2.97$. The operating line for the exhausting section is, therefore, drawn with a slope of 2.97. Since open steam is being supplied to the bottom of the tower (no reboiler is used), this operating line extends up to the point ($x = x_B$, $y = 0$), which is a point on the $x$-axis. However, due to the graphic difficulty in marking this point accurately (since the value of $x_B$ is very small), we have not drawn the line in full in the figure. For very low values of $x$, we shall use an analytical method for computing the number of stages.

The staircase construction is now made as shown in the figure. It is to be noted that the first stage corresponds to the partial condenser (since we have assumed that equilibrium condensation takes place in the condenser). The first point of intersection with the equilibrium curve corresponds to ($x = x_0$, $y = y_D$), the second point of intersection is ($x_1$, $y_1$), the third ($x_2$, $y_2$) and so on. It can be seen that at the end of the twelveth ideal stage, $x = x_m = 0.0225$. The number of ideal stages below this value of $x$ may be computed analytically as follows:

For low values of $x$, it is not wrong to assume that the equilibrium curve is essentially linear. If this is so, then the number of ideal stages may be computed from Kremser-Brown-Souders[17,18] equation (this equation has already been discussed under Gas Absorption in Section 4.5 of Chapter 4) as given below:

$$(N_p - m + 1) = \frac{\log\left[\frac{(x_m - x_B/m')(1-A)}{(x_B - x_B/m')} + A\right]}{\log(1/A)} \qquad \text{... (5.2.76)}$$

where $N_p$ = total number of ideal stages required (excluding reboiler)

$x_m$ = composition of liquid leaving tray $m$, which is the last ideal tray obtained by graphical work

$m'$ = average slope of equilibrium curve

It is important to keep in mind that the above equation is valid only for those systems for which the relative volatility (α) is essentially constant. If the reboiler is dispensed with and open steam is being used, then the left hand side of the above equation will be $(N_p - m)$ and the term $(x_B/m')$ is to be dropped from the right hand side (since $y_B = y_{N_p+1} = 0.0$). Thus, for the present case, the expression is

$$(N_p - m) = \frac{\log\left[\dfrac{x_m(1-A)}{x_B} + A\right]}{\log(1/A)} \qquad \text{... (5.2.76a)}$$

$$A = L'/(V'm') \qquad (5.2.77)$$

As stated earlier, for the low concentration range the equilibrium curve may be assumed to be a straight line of slope $m'$ with allowable error. From the equilibrium curve, corresponding to $x = x_m = 0.0225$, the value of $y^*$ is 0.29. Therefore,

$$m' = (0.29/0.0225) = 12.88$$

$$A = (L'/V')/m' = (2.97/12.88) = 0.2306$$

Now, $x_B = 0.00025$, $x_m = 0.0225$ and $m = 12$. Therefore,

$$(N_p - m) = 2.89$$

or

$$N_p = 14.89$$

Thus the total number of ideal stages required (including the partial condenser) is 14.89. If we assume that the partial condenser is equivalent to an ideal plate that is separate from the column, then the number of ideal stages required inside the column is 13.89. Since the overall plate efficiency for the column is 0.75, the number of actual plates required is $(13.89/0.75) = 18.52 \simeq 19$.

*Analytical method:* Since the reflux ratio is 0.9279,

$$V = (1.9279)(20.92) = 40.33 \text{ kmoles/hr}$$

$$V' = V + (q-1)F = V \text{ (since } q = 1.0)$$

$$L' = (40.33 + 79.08) = 119.41 \text{ kmoles/hr}$$

Since a partial condenser is used, $y_D = z_D = 0.955$. Now, from equilibrium data (corresponding to $y = y_D$), $x_0 = 0.9368$ and

$$y' = (0.9279/1.9279)(0.9368) + (0.955/1.9279) = 0.9462$$

$$x_1 = \text{(from equilibrium data)} = 0.9218$$

The values of $x_a$ and $y_a$ are, $x_a = 0.20$ and $y_a = 0.5916$. Equation (5.2.56a) reduces to

$$y_{j+1} = 0.4813\, x_j + 0.4953 \qquad \text{... (i)}$$

Since open steam is used, we must use Eq. (5.2.70) to compute $y_{j+1}$ for all values of $x_j \le x_a$. This equation reduces to

$$y_{j+1} = (2.96074)(x_j - 0.00025) \qquad \text{... (ii)}$$

Now, putting $j = 1$, from Eq. (i),

$$y_2 = 0.939$$

and $x_2$ (from equilibrium data) $= 0.9094$

Computations are continued by incrementing $j$ and the results are summarised below:

| $j$ | $x_{j+1}$ | $y_{j+1}$ |
|---|---|---|
| 0 | $0.9218 = x_1$ | $0.9462 = y_1$ |
| 1 | $0.9094 = x_2$ | $0.939 = y_2$ |
| 2 | $0.8989 = x_3$ | $0.9331 = y_3$ |
| 3 | $0.8857 = x_4$ | $0.9280 = y_4$ |
| 4 | $0.8692 = x_5$ | $0.9216 = y_5$ |
| 5 | $0.8486 = x_6$ | $0.9137 = y_6$ |
| 6 | $0.8229 = x_7$ | $0.9038 = y_7$ |
| 7 | $0.7847 = x_8$ | $0.8913 = y_8$ |
| 8 | $0.70776 = x_9$ | $0.87304 = y_9$ |
| 9 | $0.477 = x_{10}$ | $0.8359 = y_{10}$ |
| 10 | $0.0972 = x_{11}$ | $0.725 = y_{11}$ |
| 11 | $0.0225 = x_{12}$ | $0.287 = y_{12}$ |
| 12 | $0.00516 = x_{13}$ | $0.06587 = y_{13}$ |
| 13 | $0.001139 = x_{14}$ | $0.01454 = y_{14}$ |
| 14 | $0.000206 = x_{15}$ | $0.00263 = y_{15}$ |

The number of ideal stages required (excluding the partial condenser) is, therefore, 14 + (0.001139 – 0.00025)/(0.001139 – 0.000206) = 14.953. The feed is to be introduced between the tenth and eleventh ideal stages. If the partial condenser is counted as the first ideal stage, then the feed plate location shall be between the eleventh and twelveth ideal stages.

It can be seen that the number of ideal stages computed analytically differs from that determined graphically (though not substantially). The equilibrium curve has sharp curvature. Also, at high values of $(x, y)$, the staircase construction is quite crowded since the terminal points are close to each other. All these tend to induce a large degree of personal error into the graphical computation. It is to be kept in mind that the accuracy of the graphical procedure depends heavily on the draftsmanship of the person and even on the sharpness of the pencil used.

**Example 5.24:** Sixty eight kmoles per hour of a feed solution containing ethanol and water is divided into two streams in the ratio 40 : 28 and are fed to a continuous laboratory fractionating unit that contains ten plates and is equipped with a total condenser and a reboiler. The larger stream containing 70.5 mole percent ethanol is fed to the fifth ideal stage from the top and the smaller stream containing 39.5 mole percent ethanol to the seventh ideal stage from the top. If all the feed streams are saturated liquids and the column: is operated at 1 atm pressure at a reflux ratio of 2.0 and the distillate is withdrawn at the rate of 50 kmoles/hr, estimate the composition of the product streams. Assume the reboiler is equivalent to one ideal plate and the rest of the column has an overall plate efficiency of 70 percent.

**Solution:** *Graphical method:* The system is that of multiple feed. The solution demands a trial and error procedure since compositions of both product streams (distillate and residue) are unknown. To start with, let us assume that

$$x_D = 0.78$$

Since $R = 2.0$,

$$\left(\frac{x_D}{R+1}\right) = (0.78/3.0) = 0.26$$

The equilibrium curve is plotted as shown in Fig. 5.24.1 from the equilibrium data given in Example 5.22. The first operating line for the top enriching section is drawn in the usual way (the line passes through the point $y = x = x_D = 0.78$ and has a $y$-intercept of 0.26). Since the feed is saturated liquid, the $q$-line is drawn as a vertical line at $x_F = x_{F1} = 0.705$ and it intersects the first operating line as shown. The second operating line (that for the section between the two feed streams) starts from this point of intersection and its slope is computed from Eqs (5.2.61) and (5.2.62) as given below:

$$L' = L + q\,F_1 = L + (1.0)\,(40)$$

Now, $$L = L_0 = (RD) = (2.0)\,(50.0) = 100 \text{ kmoles/hr}$$

Therefore, $$L' = 140 \text{ kmoles/hr}$$

$$V' = V + (q - 1)\,F_1 = V = (L_0 + D) = 150 \text{ kmoles/hr}$$

$$(L'/V') = (140/150) = 0.9333$$

The second operating line is, thus, drawn with a slope of 0.9333 (see Fig. 5.24.1). Another $q$-line is drawn at $x_F = x_{F2} = 0.395$, which is also a vertical line and intersects the second operating line as shown. The third operating line for the bottom stripping section is obtained by joining this point of intersection with the point $x = y = x_B$ on the 45° diagonal. For this, the value of $x_B$ is first computed from a material balance for the entire column:

$$(F_1 x_{F1}) + (F_2 x_{F2}) = B x_B + D x_D$$

where $F_1 = 40$ kmoles/hr

$F_2 = 28$ kmoles/hr

$D = 50$ kmoles/hr

$B = (F_1 + F_2 - D) = 18$ kmoles/hr

$x_{F1} = 0.705$

$x_{F2} = 0.395$

$x_D = 0.78$

Therefore, $$x_B = 0.01444$$

The third operating line is now drawn as shown in the figure. The staircase construction is performed starting from the point $y = x = x_D$ on the 45° diagonal. From the fifth stage onwards, the construction is transferred to the second operating line (since the first feed is introduced to the fifth stage) and from the seventh stage onwards, to the third operating line. To note that the shifting from one operating line to the other is not done at the intersection of the operating lines since the feed streams are not introduced at these intersections but at prespecified stages. It is found that the number of ideal stages required is

7.6. This includes the reboiler. The number of ideal stages required inside the column is, therefore, 6.6. Since the overall plate efficiency is 70 percent, the number of actual plates required is (6.6/0.70) = 9.43 ≃ 10.0. This checks with the specification of the fractionator available. Therefore, the assumed value of $x_D$ is correct and no further trial is necessary. The compositions of the product streams are, thus,

Distillate: 78 percent ethanol, 22 percent water

Residue: 1.444 percent ethanol, 98.557 percent water

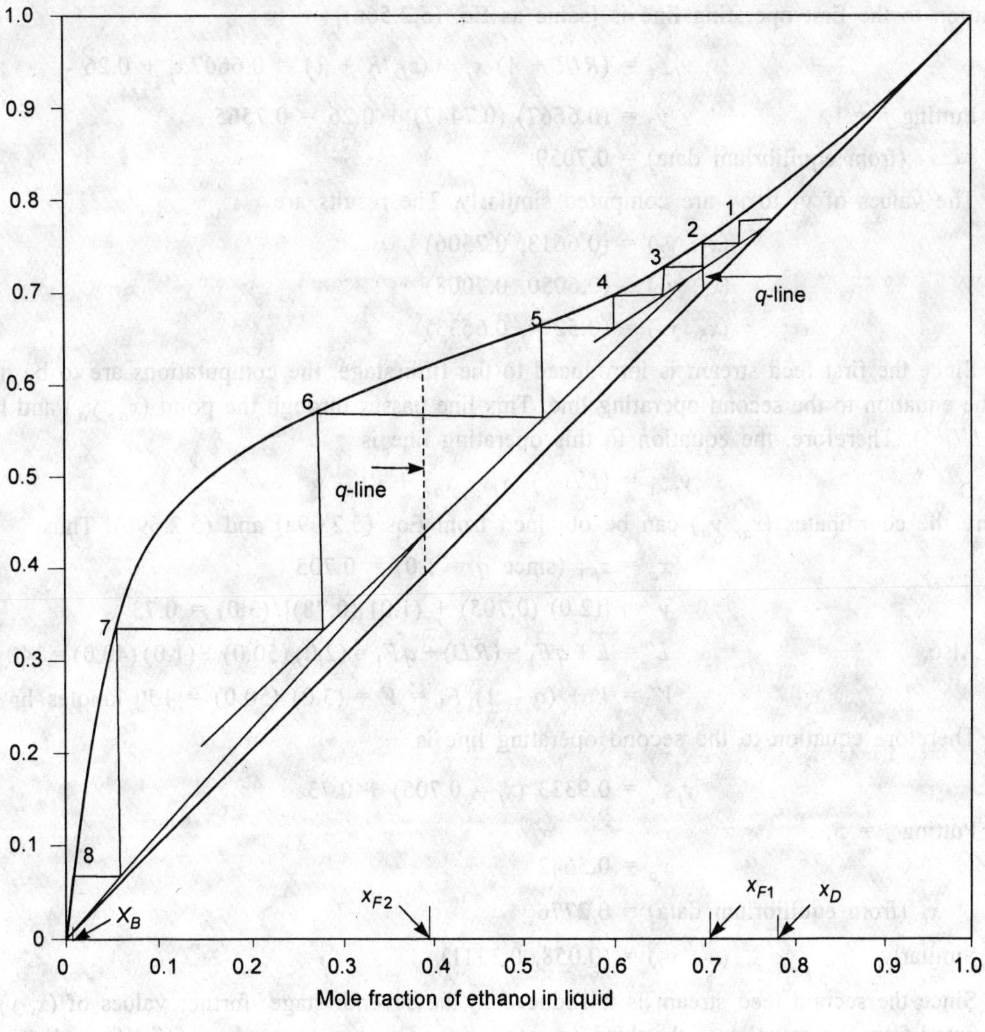


*Figure 5.24.1*

*Analytical method:* The product compositions ($x_D$, $x_B$) are to be computed by trial. Hence, let

$$x_D = z_D = 0.78$$

Now, from material balance,

$$B = F - D = 68 - 50 = 18 \text{ kmoles/hr}$$

$$F_1 z_{F1} + F_2 z_{F2} = Bx_B + Dx_D$$

$$(40.0)\ (0.705) + (28)\ (0.395) = (18)\ x_B + (50)\ (0.78)$$

$$x_B = 0.01444$$

Since total condenser is used, $y_1 = x_D = 0.78$. Therefore, from equilibrium data, $x_1 = 0.7447$. The equation to the first operating line is [same as Eq. (5.2.56a)]

$$y_{j+1} = (R/R + 1)\ x_j + (z_D/R + 1) = 0.6667x_j + 0.26 \qquad \text{... (i)}$$

Putting $j = 1$, $y_2 = (0.6667)\ (0.7447) + 0.26 = 0.7565$

$x_2$ (from equilibrium data) = 0.7059

The values of $y_3$ to $y_5$ are computed similarly. The results are

$$(x_3, y_3) = (0.6613,\ 0.7306)$$

$$(x_4, y_4) = (0.6050,\ 0.7008)$$

$$(x_5, y_5) = (0.5274,\ 0.6633)$$

Since the first feed stream is introduced to the fifth stage, the computations are to be now shifted to the equation to the second operating line. This line passes through the point $(x_a, y_a)$ and has a slope of $(L'/V')$. Therefore, the equation to this operating line is

$$y_{j+1} = (L'/V')\ (x_j - x_a) + y_a$$

where the coordinates $(x_a, y_a)$ can be obtained from Eqs (5.2.69a) and (5.2.69b). Thus

$$x_a = z_{F1} \text{ (since } q = 1.0) = 0.705$$

$$y_a = [(2.0)\ (0.705) + (1.0)\ (0.78)]/(3.0) = 0.73$$

Also, $L' = L + qF_1 = (RD) + qF_1 = (2.0)\ (50.0) + (1.0)\ (40.0) = 140 \text{ kmoles/hr}$

$$V' = V + (q - 1)\ F_1 = V = (3.0)\ (50.0) = 150 \text{ kmoles/hr}$$

Therefore equation to the second operating line is

$$y_{j+1} = 0.9333\ (x_j - 0.705) + 0.73 \qquad \text{... (ii)}$$

Putting $j = 5$,

$$y_6 = 0.5642$$

$x_6$ (from equilibrium data) = 0.2776

Similarly, $(x_7, y_7) = (0.058,\ 0.3311)$.

Since the second feed stream is introduced to the seventh stage, further values of $(x, y)$ are to be computed from the equation to the third operating line. This line has a slope $(L''/V'')$ and passes through the point $x = y = x_B$. Therefore, equation to this operating line is [see Eq. (5.2.58a)]

$$y_{j+1} = (L''/V'')\ x_j + (1 - L''/V'')\ x_B$$

Since $L'' = L' + qF_2 = (140.0) + (1.0)\ (28.0) = 168.0 \text{ kmoles/hr}$

and $V'' = V' = V = 150 \text{ kmoles/hr}$,

$$y_{j+1} = 1.12x_j - 0.0017316 \quad \text{... (iii)}$$

Putting $j = 7$, $y_8 = 0.06328$ and from equilibrium data, $x_8 = 0.007073$. Since $x_8$ is less than $x_B$, the computations are terminated here. The number of ideal stages required (including the reboiler) is, therefore, $7 + (0.058 - 0.01444)/(0.058 - 0.007073) = 7.855$. The number of ideal stages required inside the column is 6.855 and the number of actual plates required is $(6.855/0.70) = 9.793 \simeq 10.0$. This checks with the specification of the fractionator available. The assumed value of $x_D$ is, therefore, correct and we can discontinue further trial. The product compositions are, thus,

$$x_D = 0.78$$

$$x_B = 0.01444$$

In many engineering work, it is not necessary to calculate the number of equilibrium plates with great precision because of the uncertainty of available data on overall plate efficiency and also because additional plates are usually provided as a factor of safety. It is due to this the McCabe-Thiele method is still popular in spite of the simplifying assumptions involved in it. It is a simple and rapid method and is particularly advantageous in fractionation calculations involving extremely dilute solutions where the assumption of constant molal overflow approaches actual conditions very closely. Also, the more rigorous methods such as that of Sorel or Ponchon and Savarit relies tremendously on the accuracy of the enthalpy data available. If accurate and detailed enthalpy data are not available, then much of the additional accuracy of the rigorous methods could get lost in such cases.

The procedure involved in the rigorous method of analysis may be summarised as follows:

Consider Fig. 5.24. For an accurate analysis, we must write down apart from the material balance equations [such as Eqs (5.2.51), (5.2.52) and (5.2.58), the heat balance (or enthalpy balance) equations also. Thus, for envelope I of Fig. 5.24,

$$V_{j+1} = L_j + D \quad \text{... (5.2.51)}$$

$$V_{j+1}\, y_{j+1} = L_j x_j + D z_D \quad \text{... (5.2.52)}$$

$$V_{j+1}\, H_{V,\,j+1} = L_j H_{L,\,j} + D H_D + Q_c + Q_l \quad \text{... (5.2.52a)}$$

where $Q_c$ is the condenser duty (rate of heat transfer or heat removal in the condenser) and $Q_l$ is the rate of heat loss to the surroundings. To note that we have changed the suffix from $n$ to $j$ (for sake of convenience) and $x_D$ has been replaced by $z_D$. If heat losses are assumed negligible, then Eq. (5.2.52a) reduces to

$$V_{j+1}\, H_{V,\,j+1} - L_j H_{L,\,j} = D H_D + Q_c \quad \text{... (5.2.52b)}$$

Substituting $L_j = (V_{j+1} - D)$ and dividing throughout by $V_{j+1}$, we get

$$H_{V,\,j+1} = H_{L,\,j} + (1/V_{j+1})\,(D H_D - D H_{L,\,j} + Q_c) \quad \text{... (5.2.52c)}$$

Similarly, Eq. (5.2.52) reduces to

$$y_{j+1} = x_j + (D/V_{j+1})\,(z_D - x_j) \quad \text{(5.2.52d)}$$

The heat and mass balance equations can be similarly written for the envelope II of Fig. 5.24. Thus

$$L_j' = V_{j+1}' + B \quad \text{... (5.2.57)}$$

$$L'_j x_j = V'_{j+1} y_{j+1} + B x_B \quad \text{... (5.2.58)}$$

$$L'_j H_{L,j} + Q_B = V'_{j+1} H_{V,j+1} + B H_B + Q'_l$$

Neglecting heat losses ($Q'_l = 0$) and rearranging,

$$H_{V,j+1} = H_{L,j} + (1/V'_{j+1})\,(BH_{L,j} - BH_B + Q_B) \quad \text{... (5.2.58b)}$$

where $Q_B$ is the reboiler duty (rate of heat transfer or heat input at reboiler). Similarly, combining Eqs (5.2.57) and (5.2.58), we get

$$y_{j+1} = x_j + (B/V'_{j+1})\,(x_j - x_B) \quad \text{... (5.2.58c)}$$

Now, the number of ideal stages required in the enriching section can be estimated by solving Eqs (5.2.52c) and (5.2.52d) simultaneously and that in the exhausting (stripping) section by solving Eqs (5.2.58b) and (5.2.58c) simultaneously. We could use a graphical procedure similar to that employed in the simplified method of McCabe and Thiele. However, since the $(L/V)$ ratios are not assumed constant and since they vary from plate to plate (or stage to stage), the two operating lines (one for the enriching section and the other for the exhausting section) shall no longer be straight lines but will be curved. As a result, to plot them, we shall require a large number of points. Ponchon[14] and Savarit[15] have therefore developed a separate graphical procedure (that is based on the above performance equations) for the estimation of the number of ideal stages required. We shall be elaborating on this subsequently. The coordinates of the point of intersection of the two operating lines (or curves), namely $(x_a, y_a)$, can be determined as follows by writing down the heat and material balance equations for the entire tower. Thus

$$F = D + B$$

$$F z_F = D z_D + B x_B$$

or, combining the two equations,

$$D/B = (z_F - x_B)/(z_D - z_F) \quad \text{... (5.2.78)}$$

The heat balance equation for the entire tower is,

$$FH_F + Q_B = DH_D + BH_B + Q_c + Q_l + Q'_l \quad \text{... (5.2.78a)}$$

Neglecting heat losses ($Q_l = Q'_l = 0$) and rearranging, we get

$$FH_F = DQ' + BQ'' \quad \text{... (5.2.78b)}$$

where $Q' = (H_D + Q_c/D)$ and $Q'' = (H_B - Q_B/B)$ Eliminating $F$ from the above equation as $F = D + B$, we get

$$D/B = (H_F - Q'')/(Q' - H_F) \quad \text{... (5.2.78c)}$$

Combining Eqs (5.2.78) and (5.2.78c), we get

$$\frac{(z_F - x_B)}{(z_D - z_F)} = \frac{(H_F - Q'')}{(Q' - H_F)} \quad \text{... (5.2.78d)}$$

This is equation to a straight line that passes through the points ($H_F$, $z_F$), ($Q'$, $z_D$) and ($Q''$, $x_B$). This straight line is sketched as $\overline{\Delta' F \Delta}$ in Fig. 5.31. The points $\Delta$ and $\Delta'$, whose coordinates are ($Q'$, $z_D$) and ($Q''$, $x_B$) respectively, are usually called the *difference points*. It must be noted that the point ($H_F$, $z_F$), which is shown as point $F$ in figure, need not always lie below the saturated-liquid enthalpy curve as shown. This is the case when the feed is a cold liquid. If the feed is a saturated liquid at its bubble point, then the point $F$ will lie on the saturated-liquid enthalpy curve ($H_L$ versus $x$ curve) and if it is a saturated vapour, then the point $F$ will lie on the saturated-vapour enthalpy curve ($H_V$ versus $y$ curve). If the feed is partially vapourised, then the point $F$ will be in between the two enthalpy curves and for a superheated feed, $F$ will be above the saturated-vapour enthalpy curve. As shown in Fig. 5.31, the points of intersection of this line with the enthalpy curves provide the values of the coordinates ($x_a$, $y_a$).

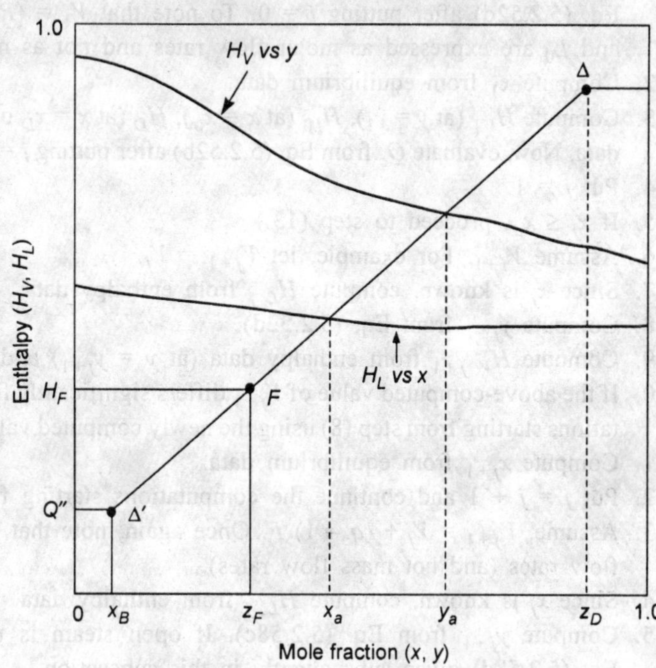


*Figure 5.31: Enthalpy-concentration plots and location of difference points and feed point*

The values of ($x_a$, $y_a$) may also be analytically computed. Considering any two points through which the line $\Delta' F \Delta$ passes such as ($H_F$, $z_F$) and ($Q'$, $z_D$), the equation to the line is

$$H = \left(\frac{Q' - H_F}{z_D - z_F}\right) z + \left(\frac{H_F z_D - Q' z_F}{z_D - z_F}\right) \quad \text{... (5.2.78e)}$$

The coordinates ($x_a$, $y_a$) can be now computed by solving the above equation and the enthalpy data simultaneously. For example, first assume the value of $y_a$ (say, $y_a = z_D - 0.001$). Now, compute the value of $H$ (namely, $H_V$) from the above equation after putting $z = y_a$. Compute the value of $H_V$ from the enthalpy data ($H_V$ versus $y$ data) also at $y = y_a$. If these two values of $H_V$ agree with each other, then the assumed value of $y_a$ is correct. Otherwise, a lower value of $y_a$ is to be assumed and trials repeated. In the same way, the value of $x_a$ can also be determined by solving Eq. (5.2.78e) simultaneously with liquid enthalpy data ($H_L$ versus $x$ data).

Once the coordinates ($x_a$, $y_a$) have been evaluated, the number of ideal stages required can be analytically computed in a similar manner as in McCabe and Thiele's method. The procedure* is summarised below:

---

*The procedure discussed here assumes the feed to be introduced at the intersection of the two operating lines. For a sizing problem (design of a new column), this is most preferable since this shall demand the least number of stages to affect the given separation.

1. If a total condenser is used, then $x_D = z_D$ and $y_1 = x_0 = x_D$. If a partial condenser is used, $y_D = z_D$ and compute $x_0$ from equilibrium data (corresponding to $y = y_D$). Now, compute $y_1$ from Eq. (5.2.52d) after putting $j = 0$. To note that $V_1 = (R + 1) D$ and $L_0 = (RD)$, provided $V_1$, $D$ and $L_0$ are expressed as molar flow rates and not as mass flow rates.
2. Compute $x_1$ from equilibrium data.
3. Compute $H_{V1}$ (at $y = y_1$), $H_{L0}$ (at $x = x_0$), $H_D$ (at $x = x_D$ or $y = y_D$) and $H_B$ (at $x = x_B$) from enthalpy data. Now, evaluate $Q_c$ from Eq. (5.2.52b) after putting $j = 0$ and then compute $Q_B$ from Eq. (5.2.78a).
4. Put $j = 1$.
5. If $x_j \le x_a$, proceed to step (13.)
6. Assume $V_{j+1}$. For example, let $V_{j+1} = V_j$.
7. Since $x_j$ is known, compute $H_{L,j}$ from enthalpy data.
8. Compute $y_{j+1}$ from Eq. (5.2.52d).
9. Compute $H_{V,j+1}$ from enthalpy data (at $y = y_{j+1}$) and then solve Eq. (5.2.52c) for $V_{j+1}$.
10. If the above-computed value of $V_{j+1}$ differs significantly from that assumed earlier, repeat the computations starting from step (8) using the newly computed value of $V_{j+1}$. Otherwise, proceed to step (11).
11. Compute $x_{j+1}$ from equilibrium data.
12. Put $j = j + 1$ and continue the computations starting from step (5).
13. Assume, $V'_{j+1} = V_j + (q - 1) F$. Once again, note that in this relation, $V_j$, $V_{j+1}$ and $F$ are molar flow rates (and not mass flow rates).
14. Since $x_j$ is known, compute $H_{L,j}$ from enthalpy data.
15. Compute $y_{j+1}$ from Eq. (5.2.58c). If open steam is used, then $y_{j+1}$ must be computed from Eq. (5.2.58d) given subsequently in this subsection.
16. Compute $H_{V,j+1}$ from enthalpy data. Now, solve Eq. (5.2.58b) for $V'_{j+1}$. If open steam is used, $V'_{j+1}$ must be obtained by solving Eq. (5.2.58e). This equation is also discussed subsequently in this subsection.
17. If the above-computed value of $V'_{j+1}$ differs significantly from that assumed earlier, repeat the iterations starting from step (15) using the newly computed value of $V'_{j+1}$. Otherwise, proceed to step (18).
18. Compute $x_{j+1}$ from equilibrium data.
19. If $x_{j+1} \le x_B$, proceed to step (20). Otherwise, put $j = j + 1$, assume $V'_{j+1} = V'_j$ and continue the computations starting from step (14).
20. Number of ideal stages required $= j + (x_j - x_B)/(x_j - x_{j+1})$ which includes the reboiler but not the partial condenser.

As we have mentioned earlier, the above-discussed procedure assumes the feed being introduced at the intersection of the two operating lines. For a sizing problem, this is widely recommended. In a rating problem where the position of the feed plate has already been specified, the transfer of computations from the enriching section to the exhausting section (from step 5 to step 13) must be done from that ideal stage at which the feed is introduced (see Example 5.26 and Program 5.10).

If open steam is used and the reboiler is dispensed with, the procedure for computing the number of ideal stages shall be the same as that described above, except that the heat and mass balance equations for the exhausting section get slightly modified. For example, Eq. (5.2.57) gets modified to

$$L'_j + S = V'_{j+1} + B \qquad \text{... (5.2.57a)}$$

where $S$ is the rate of introduction (in kmoles/s) of steam, assumed saturated at the tower pressure. Equation (5.2.58) shall remain the same. As a result, if we combine the above equation with Eq. (5.2.58), we get

$$y_{j+1} = (1 - S/V'_{j+1})\, x_j + (B/V'_{j+1})\, (x_j - x_B) \qquad \text{... (5.2.58d)}$$

Similarly, the heat balance equation for the exhausting section Eq. (5.2.58b) gets modified to (neglecting heat losses)

$$L'_j H_{L,\,j} + S H_S = V'_{j+1} H_{V,\,j+1} + B H_B$$

Combing the above equation with Eq. (5.2.57a), we get

$$H_{V,\,j+1} - H_{L,\,j} = [(B - S)\, H_{L,\,j} + S H_S - B H_B]/V'_{j+1} \qquad \text{... (5.2.58e)}$$

Thus, if open steam is used, Eqs (5.2.58d) and (5.2.58e) are to be used in place of Eqs (5.2.58c) and (5.2.58b) in the above-described procedure for the estimation of the number of ideal stages.

Incidentally, the overall mass balance and heat balance equations also get modified if open steam is being employed. The overall mass balance equations (for the entire tower) become

$$F + S = D + B$$

$$F z_F = D z_D + B x_B$$

Combining the above two equations, we get

$$\left(\frac{D}{B - S}\right) = \frac{z_F - (B/B - S)\, x_B}{(z_D - z_F)} \qquad \text{... (5.2.78f)}$$

This is the modified form of Eq. (5.2.78). The overall heat balance equation [Eq. (5.2.78b)] gets modified to

$$F H_F = (B - S)\, Q''' + Q' D \qquad \text{... (5.2.78g)}$$

where $Q''' = (B H_B - S H_S)/(B - S)$. Eliminating $F$ from the above equation as $F = D + B - S$, we get

$$\left(\frac{D}{B - S}\right) = \frac{(H_F - Q''')}{(Q' - H_F)} \qquad \text{... (5.2.78h)}$$

From Eqs (5.2.78f) and (5.2.78h) therefore,

$$\left[\frac{z_F - (B/B - S)\, x_B}{(z_D - z_F)}\right] = \frac{(H_F - Q''')}{(Q' - H_F)} \qquad \text{... (5.2.78i)}$$

This is equation to a straight line that passes through the points $(H_F, z_F)$, $(Q', z_D)$ and $[Q''', (B/B - S)\, x_B]$. Thus, when open steam is used, the coordinates of the difference point ($\Delta$) and the feed point ($F$) remain unchanged, but the coordinates of the difference point ($\Delta'$) get changed to $[Q''', (B/B - S)\, x_B]$.

In both cases (whether the fractionator is equipped with a reboiler or open steam is used), the two difference points ($\Delta$ and $\Delta'$) and the feed point ($F$) shall lie on a single straight line.

The number of ideal stages may also be computed graphically by the *method devised by Ponchon and Savarit.* (14, 15). This method is outlined below.

First, plot the two enthalpy-concentration curves such as the $H_V$ versus $y$ curve and the $H_L$ versus $x$ curve (see Fig. 5.32). Now, mark the difference points ($\Delta$ and $\Delta'$) and the feed point ($F$). To note that these three lie on a straight line. Draw the line $\overline{\Delta' F \Delta}$. If a total condenser is being used, mark $x_D$ on the liquid enthalpy curve and $y_1$ (which is equal to $x_D$) on the vapour enthalpy curve. Find $x_1$ from the equilibrium data (or equilibrium curve) and mark it on the liquid enthalpy curve. Join this point to $\Delta$ and this line intersects the vapour enthalpy curve at $y_2$. (The line joining $x_1$ and $y_1$, which is shown as a dotted line in

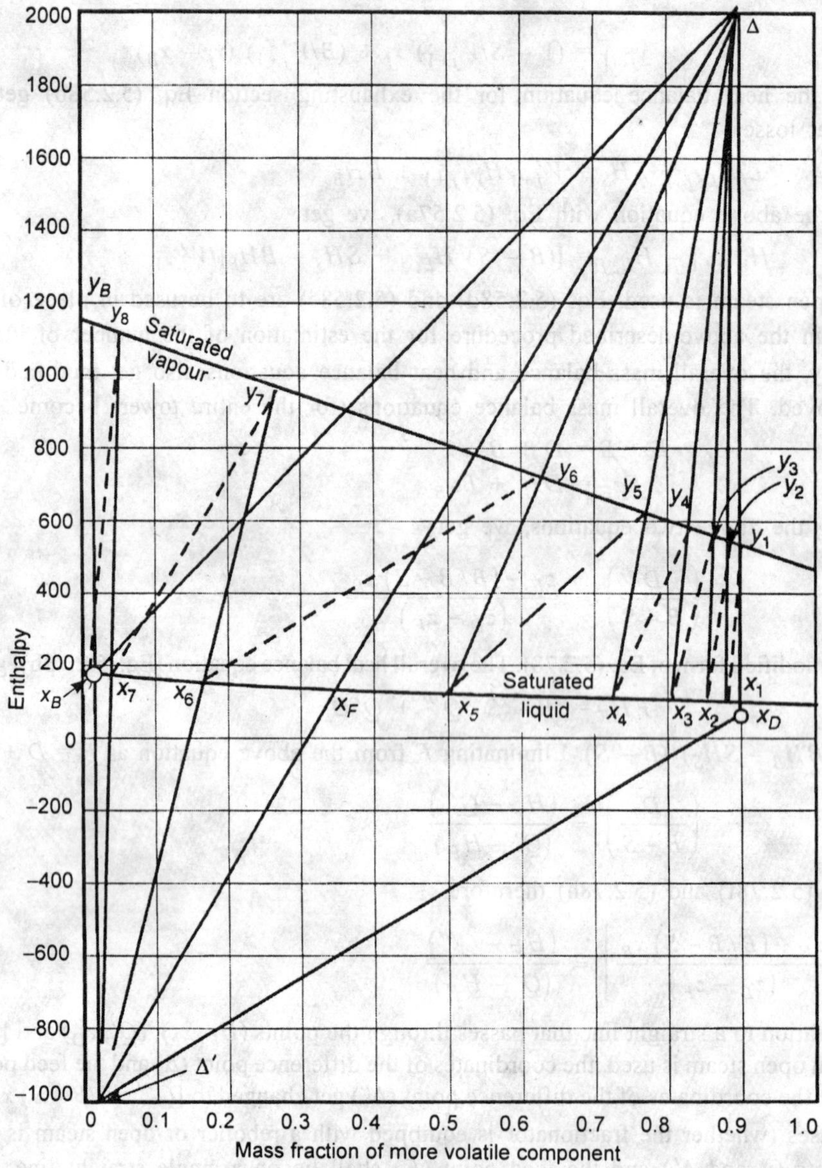


*Figure 5.32: Ponchon-Savarit method (with closed steam)*

Fig. 5.32, is called the *tie line.*) Now, find $x_2$ from equilibrium data and mark it on the liquid enthalpy curve. Join this point to $\Delta$ and this line intersects the vapour enthalpy curve at $y_3$. Continue this procedure until the feed line $\overline{\Delta' F \Delta}$ is crossed. Thereafter, continue the construction using $\Delta'$ as the difference point, until the point marked on the liquid enthalpy curve is equal to or less than $x_B$. The number of ideal stages required (including the reboiler) will be now equal to the number of tie lines drawn.

If a partial condenser is used, then mark $y_D$ on the vapour enthalpy curve. Now, find $x_0$ from the equilibrium data and mark it on the liquid enthalpy curve. Join this point to Δ and this line intersects the vapour enthalpy curve at $y_1$. The rest of the procedure is same as above. In this case, the number of tie lines will be equal to the total number of ideal stages required including the reboiler and the partial condenser.

If open steam is used and the reboiler is dispensed with, then the entire graphical procedure shall be the same as above, except that the coordinates of the bottom difference point (Δ′) will be now $[Q''', (B/B - S)\, x_B]$. The point Δ′ is therefore marked accordingly and the feed line $\overline{\Delta' F \Delta}$ is drawn. Then

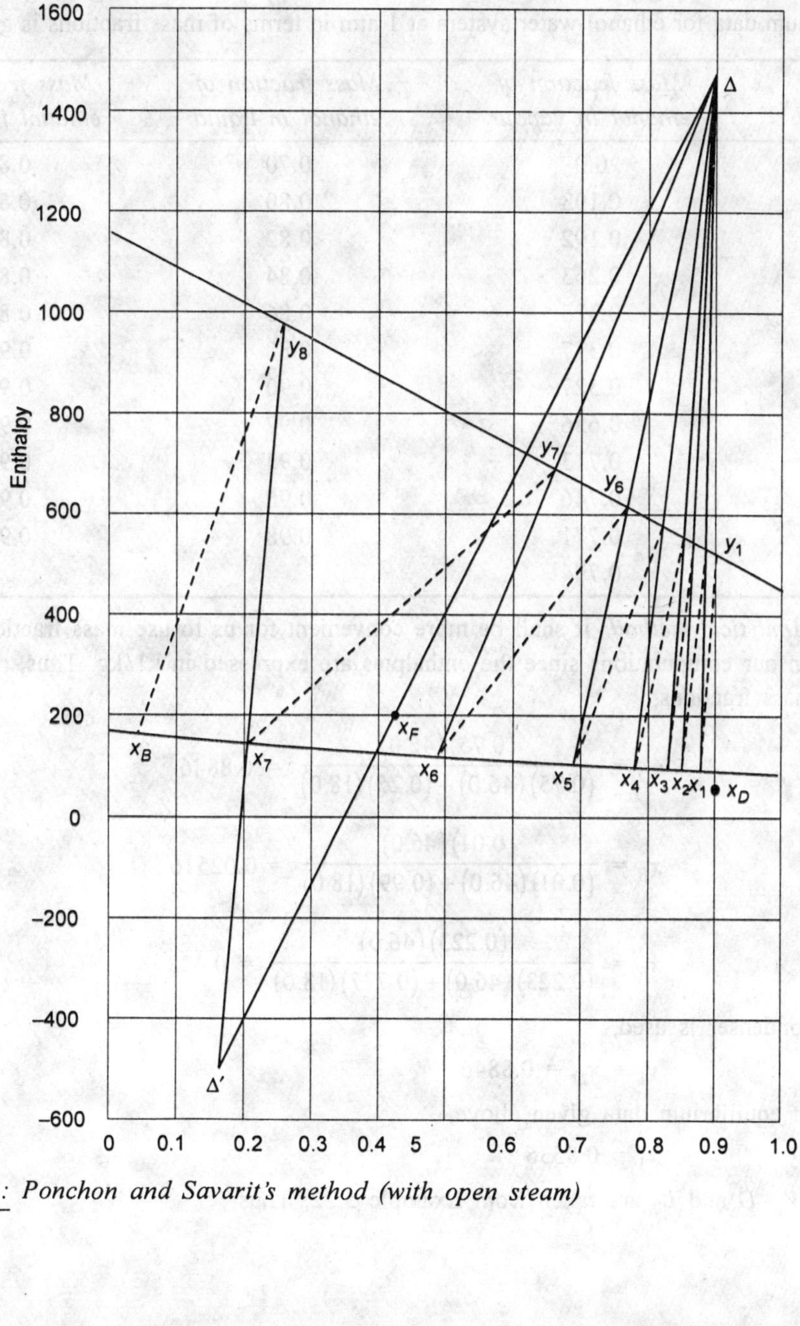


*Figure 5.33: Ponchon and Savarit's method (with open steam)*

the graphical construction is performed as described above (see Fig. 5.33).

If the feed plate location has been specified, then the change over from difference point $\Delta$ to $\Delta'$ must be done at the feed plate. For example, let the feed be introduced at the seventh ideal stage from the top. Then the point $x_7$ marked on the liquid enthalpy curve must be joined to the bottom difference point $\Delta'$ and the rest of the graphical construction performed using $\Delta'$ as the difference point.

**Example 5.25:** Rework Example 5.22 without making the assumption of constant molal overflow. Compute the number of ideal stages required for the feed of enthalpy 193.458 kJ/kg using a reflux ratio of 2.0. The equilibrium data for ethanol-water system at 1 atm in terms of mass fractions is given below:

| *Mass fraction of ethanol in liquid* | *Mass fraction of ethanol in vapour* | *Mass fraction of ethanol in liquid* | *Mass fraction of ethanol in vapour* |
|---|---|---|---|
| 0.0 | 0.0 | 0.70 | 0.822 |
| 0.01 | 0.103 | 0.80 | 0.858 |
| 0.02 | 0.192 | 0.82 | 0.868 |
| 0.03 | 0.263 | 0.84 | 0.877 |
| 0.04 | 0.325 | 0.86 | 0.888 |
| 0.05 | 0.377 | 0.88 | 0.900 |
| 0.10 | 0.527 | 0.90 | 0.912 |
| 0.20 | 0.656 | 0.92 | 0.926 |
| 0.30 | 0.713 | 0.94 | 0.942 |
| 0.40 | 0.746 | 0.96 | 0.959 |
| 0.50 | 0.771 | 0.98 | 0.978 |
| 0.60 | 0.794 | | |

**Solution:** *Analytical method:* It shall be more convenient for us to use mass fractions (instead of mole fractions) in our computations since the enthalpies are expressed in kJ/kg. Thus, reexpressing $x_D$, $x_F$ and $x_B$ as mass fractions,

$$x_D = \frac{(0.75)(46.0)}{(0.75)(46.0) + (0.25)(18.0)} = 0.8846$$

$$x_B = \frac{(0.01)(46.0)}{(0.01)(46.0) + (0.99)(18.0)} = 0.02516$$

$$x_F = \frac{(0.223)(46.0)}{(0.223)(46.0) + (0.777)(18.0)} = 0.423$$

Since a total condenser is used,

$$y_1 = x_D = 0.8846$$

Therefore, from equilibrium data given above,

$$x_1 = 0.8536$$

The values of $V_1$, $D$ and $L_0$ are taken from Example 5.22. Thus

$$V_1 = 37.56 \text{ kmoles/hr}$$

Since molecular weight of vapour = (0.75) (46.0) + (0.25) (18.0) = 39.0,

$$V_1 = (37.56)\ (39.0) = 1464.84 \text{ kg/hr}$$

Since $x_D = x_0 = y_1$, the molecular weight of distillate ($D$) and that of reflux ($L_0$) will also be 39.0. Therefore,

$$D = (12.52)\ (39.0) = 488.28 \text{ kg/hr}$$

$$L_0 = (25.04)\ (39.0) = 976.56 \text{ kg/hr}$$

Molecular weight of bottom residue ($B$) is (0.01) (46.0) + (0.99) (18.0) = 18.28 and therefore

$$B = (30.98)\ (18.28) = 566.3144 \text{ kg/hr}$$

Molecular weight of feed = (0.223) (46.0) + (0.777) (18.0) = 24.244

Therefore, $$F = (43.5)\ (24.244) = 1054.614 \text{ kg/hr}$$

Now, from enthalpy data,

$$H_{V1} \text{ (corresponding to } y = y_1 = 0.8846) = 1248.55 \text{ kJ/kg}$$

$$H_D = 227.27 \text{ kJ/kg}$$

$$H_B = 407.0295 \text{ kJ/kg}$$

$$H_{L0} = H_D$$

Therefore, putting $j = 0$ in Eq. (5.2.52b), we get

$$V_1 H_{V1} - L_0 H_{L0} = D H_D + Q_c$$

$$(1464.84)\ (1248.55) - (976.56)\ (227.27) = (488.28)\ (227.27) + Q_c$$

or $$Q_c = 1496012.205 \text{ kJ/hr}$$

Now, from Eq. (5.2.78a),

$$(1054.614)\ (193.458) + Q_B = (488.28)\ (227.27) + (566.3144)\ (407.0295) + 1496012.205$$

or $$Q_B = 1633466.752 \text{ kJ/hr}$$

Also, $$Q' = H_D + (Q_c/D) = 3291.11 \text{ kJ/kg}$$

$$Q'' = H_B - Q_B/B = -\ 2477.35 \text{ kJ/kg}$$

Before proceeding further, let us determine the coordinates of the point of intersection of the two operating lines (or curves) such as $(x_a, y_a)$. This involves a trial and terror procedure. Thus, let $y_a$ = 0.64. Now, from enthalpy data,

$$H_{Va} = 1642.156 \text{ kJ/kg}$$

From Eq. (5.2.78e), $$H_{Va} = \left(\frac{3291.11 - 193.458}{0.8846 - 0.423}\right)(0.64) + \left[\frac{(193.458)(0.8846) - (3291.11)(0.423)}{(0.8846 - 0.423)}\right]$$

$$= (6710.6845)\ (0.64) - 2645.1616 = 1649.6767 \text{ kJ/kg}$$

Since these two values of $H_{Va}$ do not differ significantly, the assumed value of $y_a$ is correct. Similarly, let $x_a$ = 0.438. Then, from enthalpy data

$$H_{La} = 293.5086 \text{ kJ/kg}$$

From Eq. (5.2.78e),

$$H_{La} = \left(\frac{3291.11 - 193.458}{0.8846 - 0.423}\right)(0.438) + \left[\frac{(193.458)(0.8846) - (3291.11)(0.423)}{(0.8846 - 0.423)}\right]$$

$$= (6710.6845)(0.438) - 2645.1616 = 294.12 \text{ kJ/kg}$$

Since these two values of $H_{La}$ also do not differ significantly, the assumed value of $x_a$ is correct. Thus, the coordinates $(x_a, y_a)$ are (0.438, 0.64).

Let us now start the stage to stage computations. Since $x_1$ = 0.8536, from enthalpy data, $H_{L1}$ = 232.4623 kJ/kg. Putting $j$ = 1 and substituting the values of $H_{L1}$, $H_D$, $Q_c$ and $D$ in Eq. (5.2.52c), we get

$$H_{V2} = 232.4623 + (1493476.908/V_2) \quad \text{... (i)}$$

Similarly, Eq. (5.2.52d) reduces to

$$y_2 = x_1 + (488.28/V_2)(0.8846 - x_1)$$

$$= 0.8536 + 15.13668/V_2 \quad \text{... (ii)}$$

Solving the above two equations simultaneously by trial, we get

$$y_2 = 0.8642$$

$$V_2 = 1428.0 \text{ kg/hr}$$

Now, from equilibrium data, $x_2$ = 0.8124 and correspondingly from enthalpy data, $H_{L2}$ = 239.362 kJ/kg. Putting $j$ = 2, the values of $y_3$ and $V_3$ are now computed in the same way as above. Computations are repeated for higher values of $j$ until $x_{j+1} < x_B$. The results are listed in Table 5.25.1.

**Table 5.25.1**

| $j$ | $x_{j+1}$ | $y_{j+1}$ | $V_{j+1}$ (kg/hr) | $V_{j+1}$ (kmoles/hr) |
|---|---|---|---|---|
| 0 | 0.8536 | 0.8846 | 1464.84 | 37.56 |
| 1 | 0.8124 | 0.8642 | 1428.0 | 33.84 |
| 2 | 0.7444 | 0.8380 | 1377.1 | 33.212 |
| 3 | 0.6107 | 0.797 | 1301.46 | 32.28 |
| 4 | 0.3303 | 0.723 | 1190.9 | 31.14 |
| 5 | 0.09133 | 0.501 | $V'_6$ = 1012.32 | $V'_6$ = 31.607 |
| 6 | 0.0141 | 0.1395 | $V'_7$ = 777.88 | $V'_7$ = 35.51 |

It may be noted that since $x_5 < x_a$, $y_6$, $V'_6$ and $y_7$, $V'_7$ are computed by solving Eqs (5.2.58b) and (5.2.58c) simultaneously. The total number of ideal stages required (including reboiler) is, therefore, 6 + (0.09133 – 0.02516)/(0.09133 – 0.0141) = 6.8567. It can be seen that the above result is very close to that obtained in Example 5.22 based on the assumption of constant molal overflow. Thus, for the present system of ethanol-water, the assumption of constant molal overflow is not too erroneous. This is also clear from the computed values of $V$ and $V'$ (in kmoles/hr, listed above) which show little variation from stage to stage.

*Graphical method:* The above problem can also be solved graphically using Ponchon and Savarit's procedure. The coordinates of the difference points ($\Delta$ and $\Delta'$) are (0.8846, 3291.11) and (0.02516, – 2477.35) respectively and those of the feed point ($F$) are (0.423, 193.458). The line $\overline{\Delta' F \Delta}$ is thus drawn as shown in Fig. 5.25.1. The enthalpy curves ($H_V$ versus $y$ and $H_L$ versus $x$) are plotted and the graphical construction is performed as shown. It is seen that the number of ideal stages required (including reboiler) is 6.8. The feed is to be introduced between the fourth and fifth stages. Readers may notice that there is slight discrepancy between the values of ($x$, $y$) analytically computed and those graphically determined. This must be attributed to the personal error involved in the graphical procedure.

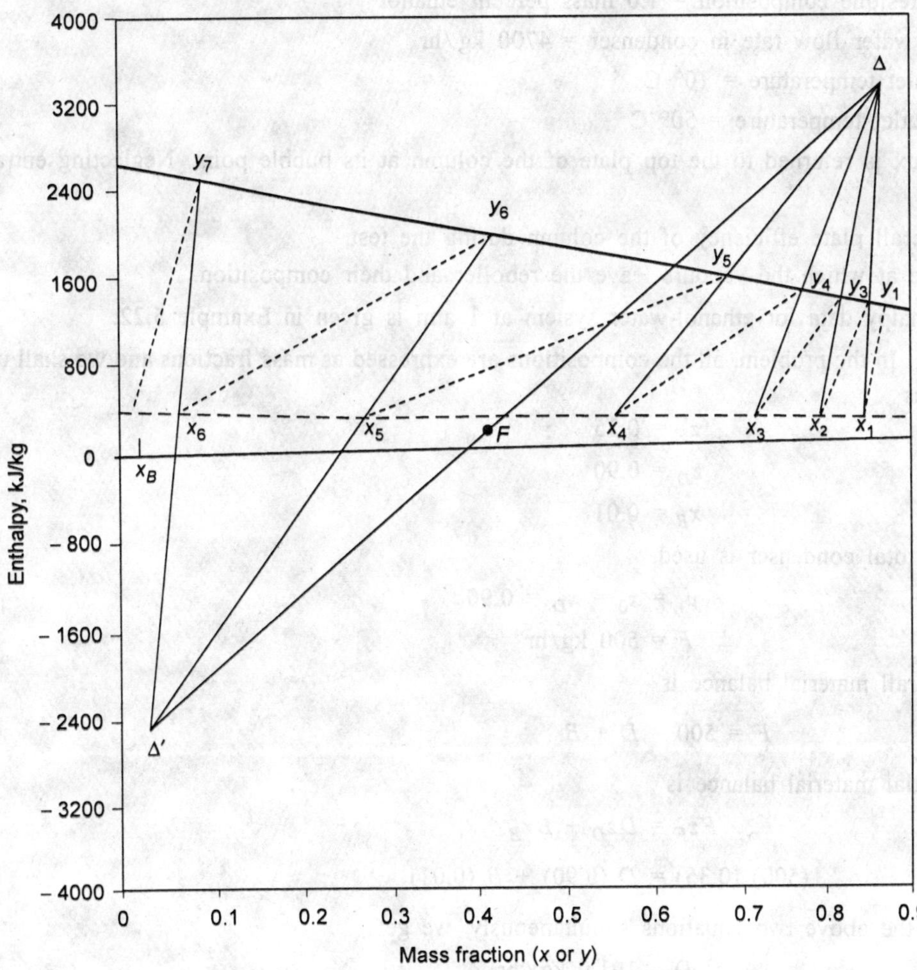


*Figure 5.25.1*

**Example 5.26:** A bubble-cup column operated at atmospheric pressure has been used for the continuous fractionation of an aqueous solution of ethanol. The column was equipped with a total condenser and a reboiler. The following test data were obtained during an 8 hr period during which steady state conditions existed:

Actual number of plates = 10

Feed introduced on 8th plate from top.

Feed rate = 500 kg/hr

Feed composition = 35 mass percent ethanol

Feed enthalpy = 348.9 kJ/kg

Distillate composition = 90 mass percent ethanol

Bottom residue composition = 1.0 mass percent ethanol

Cooling water flow rate in condenser = 4700 kg/hr

Water inlet temperature = 10° C

Water outlet temperature = 50° C

The reflux is returned to the top plate of the column at its bubble point. Neglecting entrainment, compute

(*a*) the overall plate efficiency of the column during the test,

(*b*) the rate at which the vapours leave the reboiler and their composition.

The enthalpy data for ethanol-water system at 1 atm is given in Example 5.22.

**Solution:** In the problem, all the compositions are expressed as mass fractions and we shall use them as such. Thus

$$z_F = 0.35$$

$$z_D = 0.90$$

$$x_B = 0.01$$

Since a total condenser is used,

$$y_1 = x_0 = x_D = 0.90$$

$$F = 500 \text{ kg/hr}$$

The overall material balance is

$$F = 500 = D + B \quad \text{... (i)}$$

The partial material balance is

$$Fz_F = Dz_D + Bx_B$$

$$(500)\ (0.35) = D\ (0.90) + B\ (0.01) \quad \text{... (ii)}$$

Solving the above two equations simultaneously, we get

$$D = 191.0 \text{ kg / hr}$$

$$B = 309.0 \text{ kg / hr}$$

At $z_F = 0.35$, from enthalpy data, $H_L = 306.1$ kJ/kg and $H_V = 2112.008$ kJ/kg. Since the enthalpy of feed ($H_F = 348.9$ kJ/kg) is in between these two values, the feed is partially vapourised. Now, since $y_1 = 0.90$, from equilibrium data, $x_1 = 0.88$. And from enthalpy data (given in Example 5.22),

$$H_{V1} = 1223.476 \text{ kJ/kg}$$
$$H_D = H_{L0} = 224.6916 \text{ kJ/kg}$$
$$H_{L1} = 228.04 \text{ kJ/kg}$$
$$H_B = 414.19 \text{ kJ/kg}$$

The reflux ratio is not specified and therefore, $V_1$ cannot be computed straightaway. However, the condenser duty ($Q_c$) can be computed since the cooling water flow rate and temperatures are known. Thus

$$Q_c = (4700)(4.18)(50 - 10) = 785840.0 \text{ kJ/hr}$$

Putting $j = 0$ and substituting the values of $H_{V1}$, $H_{L0}$, $H_D$, $Q_c$ and $D$ in Eq. (5.2.52c), we get

$$V_1 = 786.8 \text{ kg/hr}$$

And from Eq. (5.2.78a), neglecting heat losses,

$$(500.0)(348.9) + Q_B = (191.0)(224.6916) + (309.0)(414.19) + (785840.0)$$

or $$Q_B = 782290.8056 \text{ kJ/hr}$$

We are required to compute the overall plate efficiency of the column. Since the actual number of plates is 10, it follows that

$$\text{Overall plate efficiency} = (\text{Number of ideal stages required})/10 \qquad \text{... (iii)}$$

To determine the number of ideal stages required, we must know the feed plate location. It is given that the feed is introduced to the eighth plate from the top. However, since plate efficiency is not known, we cannot find out to which ideal stage this plate corresponds to. As a result, a trial and error procedure becomes imminent. At the lack of any other information, it is better to start with a first approximation that the feed is being introduced at the point of intersection of the two operating lines. The coordinates of this point of intersection $(x_a, y_a)$ may be now determined by trial. Thus, let $y_a = 0.549$. Since $Q' = H_D + Q_c/D = (224.6916) + (785840.0)/(191.0) = 4339.037$ kJ/kg, Eq. (5.2.78e) reduces to

$$H = (7254.7948)\, z - (2190.278)$$

Therefore, $$H_{Va} = (7254.7948)(0.549) - 2190.278 = 1792.6043 \text{ kJ/kg}$$

From enthalpy data, corresponding to $y = y_a = 0.549$,

$$H_{Va} = 1790.322 \text{ kJ/kg}$$

Since the above two values of $H_{Va}$ do not differ significantly, the assumed value of $y_a$ is correct. Similarly, let $x_a = 0.344$. From enthalpy data,

$$H_{La} = 307.0506 \text{ kJ/kg}$$

From Eq. (5.2.78e), $$H_{La} = (7254.7948)(0.344) - 2190.278 = 305.37 \text{ kJ/kg}$$

Since the difference between the two values of $H_{La}$ is not large, the assumed value of $x_a$ is correct. The coordinates $(x_a, y_a)$ are, therefore, (0.344, 0.549).

We now start performing the stage to stage computations. Putting $j = 1$ and substituting the values of $H_{L1}$, $Q'$ and $D$ in Eq. (5.2.52c), we get

$$H_{V2} = 228.04 + 785200.4557/V_2$$

And Eq. (5.2.52d) reduces to

$$y_2 = 0.88 + 3.82/V_2$$

Solving the above two equations simultaneously by trial, we get

$$y_2 = 0.885$$

$$V_2 = 769.91 \text{ kg/hr}$$

Computations are similarly repeated for higher values of $j$ until $x_{j+1} < x_B$. The results are listed below in Table 5.26.1.

**Table 5.26.1**

| $j$ | $x_{j+1}$ | $y_{j+1}$ | $V_{j+1}$, kg/hr |
|---|---|---|---|
| 0 | 0.88 | 0.90 | 786.80 |
| 1 | 0.8545 | 0.8850 | 769.91 |
| 2 | 0.8162 | 0.8661 | 749.18 |
| 3 | 0.7455 | 0.8384 | 721.66 |
| 4 | 0.5787 | 0.7891 | 676.82 |
| 5 | 0.2447 | 0.6815 | 597.0 |
| 6 | 0.06016 | 0.4075 | $V'_7 = 445.47$ |
| 7 | 0.00995 | 0.1025 | $V'_8 = 366.18$ |

It may be noted that since $x_6 < x_a$, the values of $y_7$, $V'_7$ and $y_8$, $V'_8$ are computed by solving Eqs (5.2.58b) and (5.2.58c) simultaneously. The total number of ideal stages required is, therefore, $7 + (0.06016 - 0.01)/(0.06016 - 0.00995) = 7.999 \simeq 8.0$. This includes the reboiler. The number of equilibrium plates required inside the column is, thus, 7.0 and the overall plate efficiency is $(7.0/10.0) = 0.70$. The feed is introduced to the 5.7th ideal stage from the top, which corresponds to $(5.7/0.70) = 8.14 \simeq 8$th plate. This agrees with that specified in the problem. Alternately, we may check the feed plate location as follows.

The feed is being introduced between the 5th and 6th ideal stages. The 5th stage corresponds to $(5/0.70) = 7.14$th plate and the 6th stage corresponds to $(6/0.70) = 8.57$th plate. Therefore, if feed is introduced to the 8th plate from the top, it agrees to the above proposition.

Vapours leave the reboiler at the rate of 366.18 kg/hr (namely, $V'_8$) and these contain 10.25 mass percent ethanol and rest water.

**Example 5.27:** A feed of 42.2 mass percent ethanol and balance water at 1 atm and an enthalpy of 465.2 kJ/kg is to be separated by distillation in a plate column with open steam at an enthalpy of 2791.2 kJ/kg and a partial condenser. The bottom product is to contain 10 mass percent ethanol and the distillate (saturated vapour) 90 mass percent ethanol. 0.416 kg of steam are used per kg of feed. Compute the number of equilibrium plates required to affect this separation. Neglect entrainment and the reflux may be assumed to be at its bubble point.

**Solution:** It is given that

$$z_F = 0.422$$

$$H_F = 465.2 \text{ kJ/kg}$$

$$x_B = 0.10$$

Since a partial condenser is used,

$$z_D = y_D = 0.90$$

Also, $S/F = 0.416$ kg/kg, $H_S = 2791.2$ kJ/kg

The operating reflux ratio is not specified. However, from a mass balance, we can compute the ratios $(D/F)$ and $(B/F)$ and also the quantities $Q'$ and $Q'''$. Thus

$$F + S = B + D$$

$$1 + S/F = B/F + D/F$$

$$1.416 = (B/F) + (D/F) \qquad \text{... (i)}$$

Also,
$$Fz_F = Dy_D + Bx_B$$

or
$$z_F = (D/F)\, y_D + (B/F)\, x_B$$

$$0.422 = (0.90)\,(D/F) + (0.10)\,(B/F) \qquad \text{... (ii)}$$

Solving Eqs (i) and (ii) simultaneously, we get

$$B/F = 1.0655$$

$$D/F = 0.3505$$

Now, we know [see Eq. (5.2.78g)]

$$Q''' = (BH_B - SH_S)/(B - S) = \frac{(B/F)\,H_B - (S/F)\,H_S}{(B/F) - (S/F)}$$

From enthalpy data, at $x = x_B = 0.10$, $H_B = 371.69$ kJ/kg. Substituting the values of $B/F$, $S/F$, $H_B$ and $H_S$,

$$Q''' = \frac{(1.0655)\,(371.69) - (0.416)\,(2791.2)}{(1.0655 - 0.416)} = -1178.0 \text{ kJ/kg}$$

Also,
$$D/(B - S) = \frac{(D/F)}{(B/F) - (S/F)} = \frac{(0.3505)}{(1.0655 - 0.416)} = 0.5396$$

Therefore, from Eq. (5.2.78h),

$$0.5396 = \frac{(465.2) - (-1178.0)}{(Q' - 465.2)}$$

or
$$Q' = 3510.1684 \text{ kJ/kg}$$

Since
$$Q' = H_D + Q_c/D,$$

$$Q_c/D = Q' - H_D = (3510.1684) - (1223.476) = 2286.6924 \text{ kJ/kg}$$

To note that from enthalpy data, at $y = y_D = 0.90$, $H_D = 1223.476$ kJ/kg. Let us now solve Eqs (5.2.52c) and (5.2.52d) simultaneously (after putting $j = 0$) for $V_1$ and $y_1$. Thus, from Eq. (5.2.52c),

$$H_{V1} = H_{L0} + (1/V_1)\,(DH_D - DH_{L0} + Q_c) = H_{L0} + (D/V_1)\,(Q' - H_{L0})$$

From equilibrium data, corresponding to $y = y_D = 0.90$, $x = x_0 = 0.88$. And from enthalpy data, $H_{L0} = 228.041$ kJ/kg. Therefore,

$$H_{V1} = (228.041) + (D/V_1)\,(3282.127) \qquad \text{... (iii)}$$

Similarly, Eq. (5.2.52d) reduces to

$$y_1 = x_0 + (D/V_1)\,(y_D - x_0) = 0.88 + (D/V_1)\,(0.02) \qquad \text{... (iv)}$$

Solving the above two Eq. (iii) and (iv) simultaneously (by trial), we get

$$y_1 = 0.8862$$

$$D/V_1 = 0.31$$

Now, from equilibrium data, $x_1 = 0.8567$ and from enthalpy data, $H_{L1} = 231.943$ kJ/kg. Computations are thus continued from stage to stage and the results are listed below in Table 5.27.1. Incidentally, at the lack of any other specific information, it has been assumed that the feed has been introduced at the intersection of the two operating lines. The coordinates $(x_a, y_a)$ of this point of intersection are determined by trial. Thus, after substituting the values of $Q'$, $H_F$, $z_F$ and $z_D$, Eq. (5.2.78e) reduces to

$$H = (6370.227)\,z - 2223.0357 \qquad \text{... (v)}$$

Now, let $y_a = 0.614$. Substituting $z = y_a = 0.614$ in the above equation, we get

$$H_{Va} = (6370.227)\,(0.614) - 2223.0357 = 1688.28 \text{ kJ/kg}$$

From enthalpy data, at $y = y_a = 0.614$, $H_{Va} = 1684.489$ kJ/kg. Since these two values of $H_{Va}$ do not differ substantially, we can discontinue the trial. Similarly, let $x_a = 0.396$. Now, from Eq. (v),

$$H_{La} = (6370.227)\,(0.396) - 2223.0357 = 299.574 \text{ kJ/kg}$$

From enthalpy data, at $x = x_a = 0.396$, $H_{La} = 298.826$ kJ/kg. Once again, these two values of $H_{La}$ check and no further trials are required. Thus, the coordinates $(x_a, y_a)$ are (0.396, 0.614).

**Table 5.27.1**

| $j$ | $y_{j+1}$ | $x_{j+1}$ | $D/V_{j+1}$ |
|---|---|---|---|
| 0 | 0.8862 | 0.8567 | 0.310 |
| 1 | 0.8704 | 0.8253 | 0.3164 |
| 2 | 0.8497 | 0.7769 | 0.3263 |
| 3 | 0.8188 | 0.6886 | 0.3404 |
| 4 | 0.7653 | 0.4772 | 0.3629 |
| 5 | 0.6520 | 0.1969 | 0.4134 |
| 6 | 0.2395 | 0.02669 | $0.70 = (D/V'_7)$ |

It may be noted that since $x_6 < x_a$, further computations are performed using Eqs (5.2.58e) and (5.2.58d). Thus, $y_7$ and $(D/V'_7)$ are computed by solving Eqs (5.2.58e) and (5.2.58d) simultaneously. For example, Eq. (5.2.58d) can be rearranged to

$$y_{j+1} = x_j + (D/V'_{j+1})\left[\frac{(B-S)}{D}\,x_j - (B/D)\,x_B\right]$$

where $$(B - S)/D = \frac{(B/F)-(S/F)}{(D/F)} = (1.0655 - 0.416)/(0.3505) = 1.853$$

$$B/D = (B/F)/(D/F) = 1.0655/0.3505 = 3.04$$

Therefore, $$y_{j+1} = x_j + (1.853x_j - 0.304)\,(D/V'_{j+1}) \quad \text{... (vi)}$$

Similarly, Eq. (5.2.58e) reduces to

$$H_{V,j+1} = H_{Lj} + \left[\frac{(B-S)}{D} H_{Lj} + S H_S / D - B H_B / D\right] (D/V'_{j+1})$$

$$= H_{Lj} + (1.853 H_{Lj} + 2182.87)\,(D/V'_{j+1}) \quad \text{... (vii)}$$

The values of $y_7$ and $(D/V'_7)$ are thus determined by solving the above two Eqs (vi) and (vii) simultaneously (by trial). The number of ideal stages required is

$$= 6 + (0.1969 - 0.1)/(0.1969 - 0.02669) = 6.5693$$

This includes the partial condenser. The feed is introduced between the fifth and sixth stages.

## 5.2.5 Design of Plate Columns

We have already discussed on the design of plate columns (either sieve-plate columns or bubble-cap plate columns) for gas absorption in Chapter 4. The design of plate towers for distillation is exactly analogous.

The diameter of the plate column is to be selected in such a way that the entrainment of liquid in the up-flowing vapour is at a minimum and there is minimum tendency for the plate to flood. The *flooding velocity* ($U_{fn}$) may be, therefore, computed from Eq. (4.5.24) which is the correlation based on Fair and Mathews' data. The only modification required is that the term $\rho_g$ should be replaced by $\rho_V$, where $\rho_V$ is the density of the vapour. The tower cross-section is to be then obtained from Eq. (4.5.31). The tray spacing* ($S$) is to be first assumed and subsequently verified with reference to Table 4.10. The computed tower diameter is to be checked for entrainment by using the *fractional entrainment* plots given in Fig. 4.11. In the case of sieve trays (also called perforated trays), it is also necessary to ensure that the vapour velocity through the perforations is sufficiently above the incipient *weeping velocity* predicted by Eq. (4.5.33).

Once the tower diameter is computed, the *active plate area* ($A_a$) can be finalised from the suitably chosen downcomer area ($A_d$) as discussed in Chapter 4 (see Table 4.11). The hole size and the hole pitch (for sieve plates) and the cap diameter and cap pitch (for bubble-cap plates) are to be selected from the standard sizes available. Bubble caps are usually 5 cm, 7.5 cm, 10 cm or 15.0 cm in diameter. The cap pitch (centre to centre distance between adjacent caps) is normally 1.25 to 1.5 times the cap diameter. The caps are usually laid on equilateral triangular layout. That is, the caps are arranged on the plate at the corners of equilateral triangles with the rows oriented normal to the direction of flow across the plate. The number of caps can be found out only by drawing the caps within the active plate area and finding the maximum number that will best fit into the plate area. The slots may be shaped rectangular, triangular or trapezoidal. In the case of trapezoidal slots, the larger width is usually 50 percent of the smaller width. The

*Note that tray spacing is denoted as $l_t$ in Chapter 4.

slot dimensions (height and width), the number of slots per cap, the area of riser are selected from the standard specifications available in manufacturers' catalogues (see Table 4.8 of Chapter 4).

The hole sizes for sieve plates usually range from 1 to 25 mm, though sizes of 4 to 6 mm are most popular. The pitch to hole diameter ratio normally falls between 2.5 and 4.0. It is also important to maintain the ratio of plate thickness to hole diameter within 0.4 to 0.7. The hole arrangement is usually on a square or triangular layout.

The height of the column depends on the number of plates required which is to be determined using the procedures discussed in the earlier sub-section. Since what is determined by the graphical or analytical methods described in the earlier subsection is the number of ideal stages (or equilibrium stages) required, the number of actual plates required can be computed only if the plate efficiency is known. A reasonable estimate of *murphree plate efficiency* $E_{mG}$ may be obtained from Eqs (4.5.45) to (4.5.69) based on the procedure discussed in Chapter 4. Though the value of $E_{mG}$ may be assumed to be constant for any particular plate, its value does vary significantly from plate to plate and therefore, the specification of an overall plate efficiency ($E_0$) for the entire column is far from realistic for distillation columns. This is because computation of $E_0$ from Eq. (4.5.70) becomes reasonable only if $E_{mG}$ does not vary materially from plate to plate and this happens only if the equilibrium curve is nearly linear. However, in distillation problems, the equilibrium curves are usually strongly curved and are linear only over limited concentration ranges. Nevertheless, computation of $E_0$ is useful for rough estimates in engineering work, if not for final design.

For a reasonably rigorous estimate of the actual number of plates, the value of murphree plate efficiency ($E_{mG}$ or $E_{ma}$) should be determined for each plate (that is, for each value of $x$) and the equilibrium curve redrawn. For example, between zero and 1.0, choose sample values of $x$ from the equilibrium data. For each value of $x$, compute $E_{mG}$ (or $E_{ma}$, if entrainment is expected to be substantial) from Eqs (4.5.45) to (4.5.69). To note that, in these equations, $Q_L = (LM_L)$, where $M_L$ is the molecular weight of liquid which is given by $M_L = xM_A + (1 - x)\ M_B$, $M_A$ and $M_B$ being the molecular weight of more volatile component and less volatile component respectively. Similarly, $Q_G = Q_V = (VM_V)$ and $M_V = yM_A + (1 - y)\ M_B$. Based on the assumption of constant molal overflow, the values of $L$ and $V$ will be the same for all the plates in the enriching section. For all plates in the exhausting section, these values will be $L'$ and $V'$. Also, the value of stripping factor ($\lambda$) required in these equations is to be computed as $\lambda = (m'V/L)$ or $(m'V'/L')$, where $m'$ is the slope of the equilibrium curve and may be taken equal to $\Delta y^*/\Delta x$ at any $x$. Here, $\Delta x$ is the difference between two consecutive $x$-values and $\Delta y^*$ is the difference between the corresponding $y^*$-values. The equilibrium curve is to be now redrawn as shown by a broken curve in Fig. 5.34, such that at any value of $x$ in the enriching section the ratio of the length of the line $BC$ to that of the line $AC$ is equal to $E_{mG}$ (or $E_{ma}$) and at any $x$ in the exhausting section, $(B'C'/A'C') = E_{mG}$ or $E_{ma}$. This broken curve is now used instead of the equilibrium curve to perform the staircase construction and the number of plates so determined will be the number of real trays/plates.

Once the tower diameter and tower height have been essentially finalised, what is left to be determined is the pressure drop across each plate/tray. The term *plate stability*, in fact, refers to the ability of the plate to maintain satisfactory operating characteristics even when the flow rates fluctuate or even when unsteady state conditions exist. For the stable operation of a tray/plate, it is necessary that (*i*) The liquid gradient across the plate is about 12 mm, but not more than 25 mm.

(*ii*) The height of liquid in the downcomer (also called *liquid backup* in downcomer) is less than 50 percent of the plate spacing (to ensure that flooding does not occur).

(*iii*) For bubble-cap plates, the *vapour distribution ratio*, which is the dimensionless ratio of liquid gradient across the plate to the pressure drop caused by the bubble-cap assemblies, is less than 0.4.

(*iv*) The residence time of the downflowing liquid in the downcomer is not less than 5 seconds (this is to ensure the escape of all the vapour that is entrained in the liquid as the liquid enters the downcomer).

(*v*) In the case of perforated plates, the operation is well above the weep point.

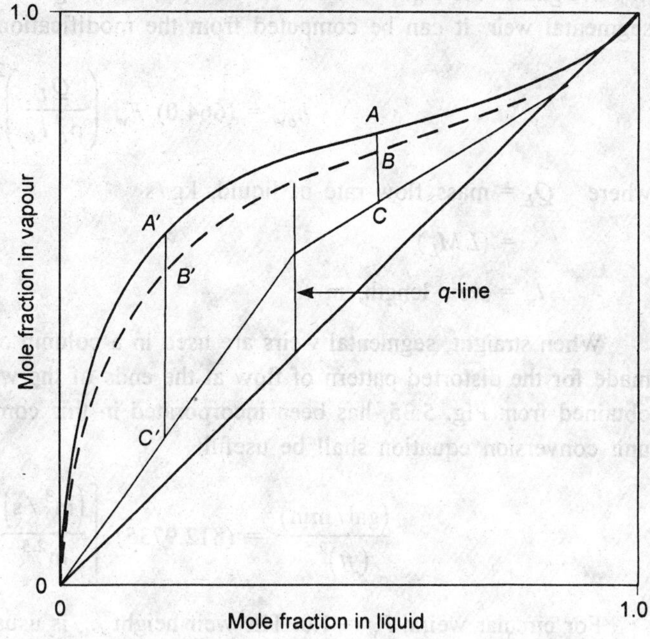


*Figure 5.34*

Broadly speaking, the total pressure drop across the plate is the sum of two terms such as the pressure drop due to vapour flow through the dry cap assembly or the dry perforations ($h_c$ or $h_o$) and the pressure drop for vapour flow through the aerated mass (liquid and froth) which is designated as $h_L$. Thus, the total pressure drop ($\Delta H$) across a bubble cap plate is ($h_c + h_L$) or ($h_{rc} + h_{so} + h_L$) where $h_{rc}$ is the pressure drop through the riser and $h_{so}$ that through the slots and for a sieve plate, ($\Delta H$) is equal to ($h_o + h_L$). The notation $h$ is used to designate pressure drop because it is expressed in mm of clear liquid.

The pertinent equations for computing the total pressure drop across the plate ($\Delta H$, in mm of liquid) and the liquid backup in the downcomer ($H_d$, in mm of liquid) are given below:

For bubble cap plates:

$$\Delta H = (h_{rc} + h_{so}) + h_L \quad \text{... (5.2.79a)}$$

$$= (h_{rc} + h_{so}) + \beta h_{ds} \quad \text{... (5.2.79b)}$$

$$= (h_{rc} + h_{so}) + \beta\,(h_{ss} + h_{ow} + 0.5\,h_g) \quad \text{... (5.2.79c)}$$

$$H_d = (\Delta H + h_w + h_{ow} + h_g + h_d) \quad \text{... (5.2.80)}$$

For sieve plates:

$$\Delta H = h_o + h_L \quad \text{... (5.2.81a)}$$

$$= h_o + \beta h_{ds} + h_{st} \quad \text{... (5.2.81b)}$$

$$= h_o + \beta\,(h_w + h_{ow} + 0.5\,h_g) + h_{st} \quad \text{... (5.2.81c)}$$

$$H_d = [\Delta H + \beta\,(h_w + h_{ow} + 0.5\,h_g) + h_d] \quad \text{... (5.2.82)}$$

All the terms on the right hand side of the above equations are illustrated in Figs 4.6 (b) and 4.9 of Chapter 4. Each term is, in fact, a pressure drop term expressed in mm of clear liquid. For example, $h_{ow}$ designates the liquid head over weir (or, the height of liquid crest over the weir) and for straight, segmental weir, it can be computed from the modification of *Francis weir equation* as given below:

$$h_{ow} = (664.0)\, F_w \left(\frac{Q_L}{\rho_L\, l_w}\right)^{2/3} \quad \text{... (5.2.83)}$$

where $Q_L$ = mass flow rate of liquid, kg/s

$$= (L M_L) \quad \text{... (5.2.83a)}$$

$l_w$ = weir length, m

When straight, segmental weirs are used in a column of circular cross-section, a correction must be made for the distorted pattern of flow at the ends of the weirs. The correction factor $F_w$, which can be obtained from Fig. 5.35, has been incorporated in this connection[19]. For using Fig. 5.35, the following unit conversion equation shall be useful:

$$\frac{(\text{gal/min})}{(ft)^{2.5}} = (812.9735)\left[\frac{(m^3/s)}{m^{2.5}}\right] \quad \text{... (5.2.84)}$$

For circular weirs, $F_w$ = 1.0. The weir height $h_w$ is usually selected as 25 to 50 mm for sieve plates, while for bubble cap plates, it is to be estimated from Eq. (4.5.18) of Chapter 4.

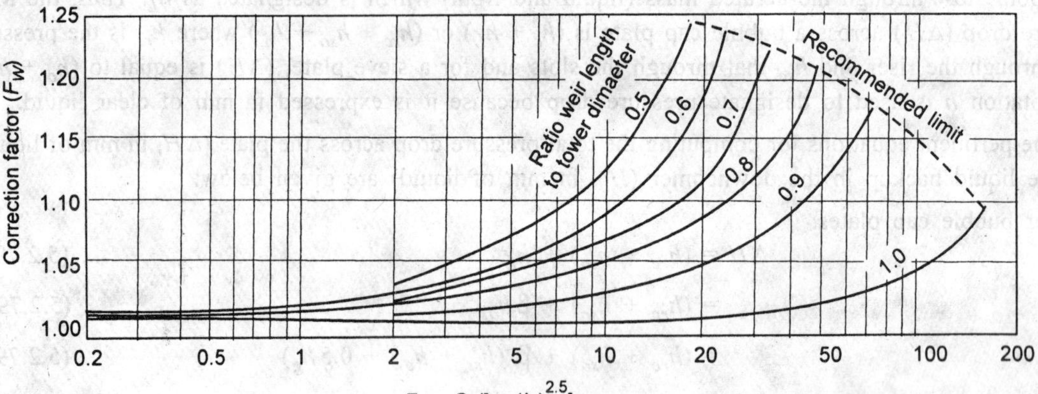


*Figure 5.35: Correction factor for effective weir length. (From Bolles, WL, Pet. Refiner, 25, 613, 1946, by permission)*

The *static submergence* ($h_{ss}$), which is nothing but the liquid head corresponding to the distance between the top of slots and the top of weir (when the static liquid is just ready to flow over the overflow weir), is a design variable and depends on the magnitude of the operating pressure $P$. It is usually selected as zero if the column is operating under vacuum, 12.5 mm if $P$ = 1 atm, 25 mm if $P$ = 7.0 to 8.0 atm and 37.5 mm if $P$ = 20 to 35 atm.

It can be seen from Eqs (5.2.79) and (5.2.81) that $h_L = (\beta h_{ds})$*. Here, β is a dimensionless aeration factor (which takes care of the effect of *aeration* or the bubbling of vapour through the liquid and the consequent foam or froth formation) and $h_{ds}$ is called the *dynamic liquid seal* which represents the liquid head loss (in mm of clear liquid) caused by the resistance to vapour flow through the liquid and the froth above the slots/perforations. It is once again clear from Eqs (5.2.79c) and (5.2.81c) that

$$h_{ds} = h_{ss} + h_{ow} + 0.5 h_g \text{ (for bubble cap plates).} \quad \text{... (5.2.85)}$$

$$h_{ds} = h_w + h_{ow} + 0.5 h_g \text{ (for sieve plates).} \quad \text{... (5.2.86)}$$

The value of β can be obtained from the plots[20] given in Fig. 5.36. In this $V_a$ is the average vapour velocity through the active plate area ($A_a$) or $V_a = Q_V/(\rho_V A_a)$. The parameter φ is called the *relative froth density* (the ratio of froth density to liquid density) and is related to β as[9]

$$\beta = (1 + \phi)/2 \quad \text{... (5.2.87)}$$

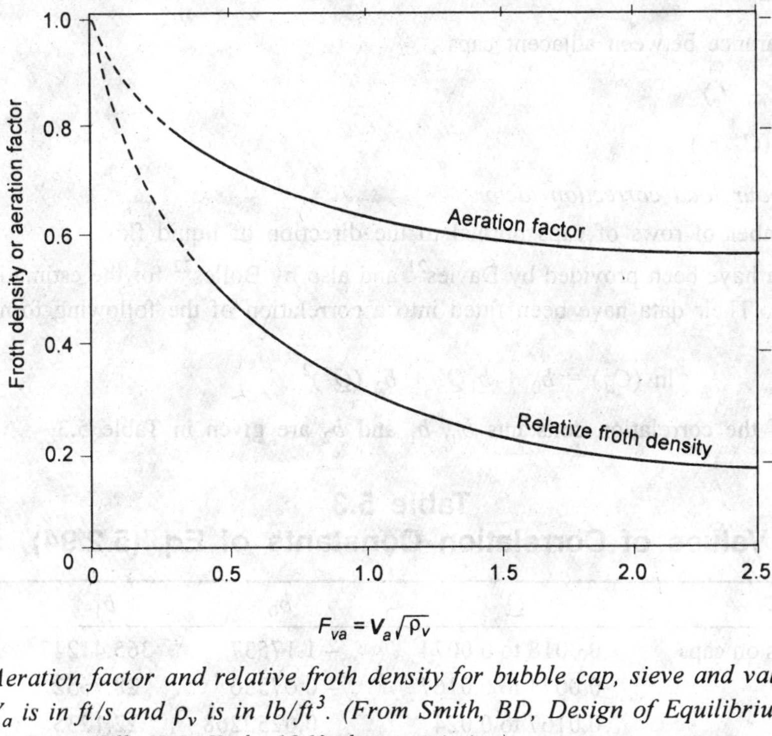


*Figure 5.36: Aeration factor and relative froth density for bubble cap, sieve and value trays. Note that $V_a$ is in ft/s and $\rho_v$ is in lb/ft$^3$. (From Smith, BD, Design of Equilibrium Stage Processes, McGraw-Hill, New York, 1963, by permission)*

* For sieve plates, an additional term $h_{st}$, which is the pressure drop to overcome surface tension effect during bubble formation, is also included in $h_L$.

The term $h_g$ represents the *liquid gradient across the tray.* It is the difference in the height of the liquid crest flowing over the weir and the height of the liquid on the inlet side of the tray (see Fig. 4.9 of Chapter 4). $h_g$ is, in fact, a complex function of the operating variables (liquid flow rate, vapour flow rate), design variables (specifications of caps and accessories, perforations) and the system variables (density of liquid, density of vapour).

For bubble cap plates, a generalised method for computing $h_g$ has been developed by Davies[21], which demands a trial and error solution. A simplified version of the same has been proposed by Bolles[22]. Bolles has, however, assumed that the caps are spaced on equilateral triangular pitch and the riser diameter ($d_r$) is 70 percent of cap diameter. Bolles' correlation is given below:

$$(Q'/C_d) = 0.0417\left(\frac{\gamma}{1+\gamma}\right)\sqrt{h_g'}\;[0.0016\,h_g' + 0.003\,(h_w + h_{ow} + 0.3\,h_{sk}/\gamma)] \quad \text{... (5.2.88)}$$

where $Q' = Q/(\rho_L W_a)$ ... (5.2.89)

$W_a = (D + l_w)/2$ ... (5.2.90)

$C_d$ = *liquid gradient factor*

$\gamma = S_c/d_c$ ... (5.2.91)

$S_c$ = clearance between adjacent caps

$= (p_c - d_c)$ ... (5.2.92)

$h_g' = h_g/(C_{vf} r_c)$ ... (5.2.93)

$C_{Vf}$ = *vapour load correction factor.*

$r_c$ = number of rows of caps normal to the direction of liquid flow.

Graphical data have been provided by Davies[21] and also by Bolles[22] for the estimation of the liquid gradient factor $C_d$. Their data have been fitted into a correlation of the following form by Narayanan and Bhattacharya[5]:

$$\ln(C_d) = b_0 + b_1 Q' + b_2 (Q')^2 \quad \text{... (5.2.94)}$$

The values of the correlation constants $b_0$, $b_1$ and $b_2$ are given in Table 5.3.

## Table 5.3
## Values of Correlation Constants of Eq. (5.2.94)

| | $Q'$ | $b_0$ | $b_1$ | $b_2$ |
|---|---|---|---|---|
| No hold-down bars on caps | 0.0018 to 0.0074 | – 1.17537 | 365.41247 | 29377.214 |
| | 0.0074 to 0.0167 | – 0.07556 | 28.7652 | 0.0 |
| | 0.0167 to 0.024 | 0.0252268 | 22.1235 | 0.0 |
| Hold-down bars on caps | 0.0037 to 0.0074 | – 0.4194485 | 42.2602 | 0.0 |
| | 0.0074 to 0.0167 | – 0.3863218 | 37.80305 | 0.0 |
| | 0.0167 to 0.024 | 0.008015 | 11.578129 | 0.0 |

Bolles[22] has also reported graphical data (a family of plots) for the estimation of the *vapour load correction factor, $C_{Vf}$*. It is possible to fit the reported data into a correlation of the form given below with allowable error:

$$C_{Vf} = \alpha_0 + \alpha_1 Q' \qquad \text{... (5.2.95)}$$

where $Q'$ has already been defined in Eq. (5.2.89). The values of the correlation constants $\alpha_0$ and $\alpha_1$ depend on the magnitude of the factor $F_V$ which is defined as

$$F_V = U_{sup} \sqrt{\rho_V} \qquad \text{... (5.2.96)}$$

where $U_{sup}$ = superficial velocity of vapour based on cross-sectional area of empty column.

The values of $\alpha_0$ and $\alpha_1$ are listed in Table 5.4.

For sieve plates, the liquid gradient (in mm of clear liquid) across the plate may be computed from the correlation proposed by Hughmark and O'Connell[23] as given below:

$$h_g = \left( \frac{1000 \, f \, U_f^2 \, W}{g R_H} \right) \qquad \text{... (5.2.97)}$$

where $W$ = distance between downcomers [see Eq. (4.5.20)], m

$R_H$ = hydraulic radius of the aerated liquid, m

$$= (W_a \, h_f)/(1000 \, W_a + 2 h_f) \qquad \text{... (5.2.98)}$$

$h_f$ = actual height of froth, mm

$$= \phi \, h_L \qquad \text{... (5.2.99)}$$

$U_f$ = velocity of aerated liquid (same as that of clear liquid), m/s

$$= \frac{1000 \, Q_L}{\rho_L \, h_L W_a} \qquad \text{... (5.2.100)}$$

The friction factor $f$ is to be obtained from the friction factor versus Reynolds number ($Re_f$) plots[20] where $Re_f$ is defined as

$$Re_f = R_H U_f \rho_L / \mu_L \qquad \text{... (5.2.101)}$$

However, these plots are essentially linear and may be fitted into a correlation of the form

$$\ln (f) = d_0 + d_1 \ln (Re_f) \qquad \text{... (5.2.102)}$$

For intermediate values of $h_W$, the value of friction factor ($f$) may be computed by linear interpolation with allowable error.

**Table 5.4 Values of Correlation Constants of Eq. (5.2.95)**

| | | | | | $F_V = 0.976$ | | $F_V = 0.61$ | |
|---|---|---|---|---|---|---|---|---|
| | $F_V = 1.708$ | $F_V = 1.464$ | $F_V = 1.342$ | $F_V = 1.22$ | $Q' \geq 827.82 \times 10^{-5}$ | $Q' < 827.82 \times 10^{-5}$ | $Q' \geq 827.82 \times 10^{-5}$ | $Q' < 827.82 \times 10^{-5}$ |
| $\alpha_0$ | 1.2667 | 1.140 | 1.00 | 0.870 | 0.785 | 0.6667 | 0.570 | 0.520 |
| $\alpha_1$ | −16.1065 | −9.664 | 0.0 | 9.664 | 12.08 | 26.844 | 26.575 | 33.8237 |

*Note:* Both $Q'$ and $F_V$ are expressed in SI units.

## Table 5.5
## Correlation Constants of Eq. (5.2.102)

| *Weir height* ($h_W$), *mm* | $d_0$ | $d_1$ |
|---|---|---|
| 10.16 | 7.011 | −1.11345 |
| 17.78 | 7.094 | −1.0704 |
| 25.40 | 7.538 | −1.074 |
| 38.10 | 8.01 | −1.0704 |

Incidentally, for sieve plates, the value of $h_L$ may be more accurately estimated from Eqs (4.5.58) to (4.5.62) given in Chapter 4. The total pressure drop across the plate can be then computed directly from Eq. (5.2.81a). If such a high order of accuracy is not essential, then Eqs (5.2.81b) and (5.2.81c) may be employed as discussed above.

For bubble cap plates, $h_{rc}$ represents the pressure drop (in mm of liquid) due to vapour flow through the riser and the cap and $h_{so}$ is that through the slots. Thus, using a variation of the standard orifice equation,

$$h_{rc} = (273.4)\ K_C\ (\rho_V/\rho_L)\ U_{VC}^2 \quad \text{... (5.2.103)}$$

where $U_{VC}$ = linear velocity of vapour through risers, m/s

$$= Q_V/(\rho_V A_r) \quad \text{... (5.2.104)}$$

$A_r$ = total riser area per plate, $m^2$

The coefficient $K_C$ is to be obtained from Fig. 5.37. However, the graphical data of Fig. 5.37 can be fitted into analytical correlations as given below[5]:

If $A' \le 1.3$,

$$K_C = 2.24569 - 2.4172A' + 0.81967\ (A')^2 \quad \text{... (5.2.105)}$$

If $1.3 < A' \le 1.5$,

$$K_C = 0.844 - 0.278A' \quad \text{... (5.2.106)}$$

where $A' = (A_{an}/A_r)$ ... (5.2.107)

$A_{an}$, $A_r$ = total annular area per plate and total riser area per plate respectively

Similarly, the pressure drop through the slots (in mm of liquid) can be obtained from[20]

$$h_{S0} = (937.0)\left(\frac{\rho_V}{\rho_L - \rho_V}\right)^{0.2} (h_S^2\, U_{VS})^{0.4} \quad \text{...(5.2.108)}$$

where $h_S$ = slot height, m

$U_{VS}$ = linear velocity of vapour through slot, m/s

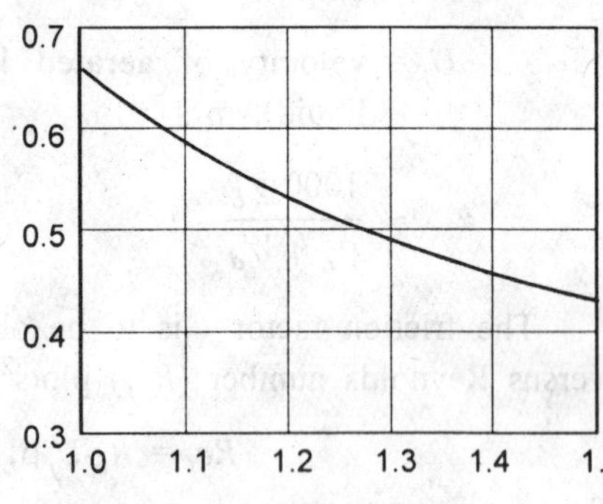


*Figure 5.37: Bubble cap pressure drop constant. (From Bolles, WL, Pet. Process., 11 (2), 64, 1956, by permission)*

$$= Q_V/(\rho_V A_S) \qquad \text{... (5.2.109)}$$

$A_S$ = total slot area per plate, $m^2$

In an analogous way, for sieve plates, $h_0$ represents the pressure drop due to vapour flow through dry perforations. It can also be estimated from a correlation that is similar to Eq. (5.2.103):

$$h_0 = \left(\frac{50.8}{C_V^2}\right) (\rho_V/\rho_L)\, V_h^2 \qquad \text{... (5.2.110)}$$

where $V_h$ = linear velocity of vapour through dry perforations, m/s

$$= Q_V/(\rho_V A_h) \qquad \text{... (5.2.11)}$$

$A_h$ = total hole area per plate [see also Eq. (4.5.28) of Chapter 4], $m^2$

The discharge coefficient $C_V$ can be estimated from the graphical plots of $C_V$ versus $(A_h/A_a)$ at different values of $(t_p/d_h)$ ratio[42]. However, these plots can be conveniently fitted into a correlation of the form

$$C_V = C_0 + C_1\,(A_h/A_a) \qquad \text{.... (5.2.112)}$$

The correlation constants $C_0$ and $C_1$ are listed in Table 5.6.

## Table 5.6
## Correlation Constants of Eq. (5.2.112)

| $(t_p/d_h)$ | $C_0$ | $C_1$ |
|---|---|---|
| 0.1 and less | 0.58825 | 0.705 |
| 0.2 | 0.6441 | 0.706 |
| 0.6 | 0.6703 | 0.74133 |
| 0.8 | 0.7073 | 0.74133 |
| 1.0 | 0.7703 | 0.74133 |
| 1.2 | 0.81206 | 0.7587 |

Finally, the additional term $h_{st}$ in the pressure drop equation for sieve plates represents the pressure drop required to overcome the surface tension effect in forming the froth and forcing the vapour through the aerated mass. It can be reasonably estimated (in mm of liquid) from

$$h_{st} = \frac{(409.0)\,\sigma}{(\rho_L\, d_h)} \qquad \text{... (5.2.113)}$$

where $d_h$ = hole diameter, m

$\sigma$ = surface tension of liquid, N/m

Once the total pressure drop ($\Delta H$) has been computed, it is essential to verify whether the essential conditions for plate stability (given at the beginning of this subsection) are satisfactorily met or not. For example, the liquid gradient ($h_g$) across the plate should not exceed 25 mm. The *vapour distribution ratio*, $(h_g/h_c)$ or $h_g/(h_{rc} + h_{so})$, should not necessarily exceed 0.4. It is also essential to ensure that the *liquid backup in the downcomer* $(H_d)$ is not excessively large to avoid flooding of the column. The value of $H_d$ can be computed from the expressions given earlier. The *pressure drop (in mm of liquid) caused by the liquid flow in the downcomer* $(h_d)$ may be estimated from

$$h_d = 165.2 \left( \frac{Q_L}{\rho_L A_{dm}} \right) \qquad \text{... (5.2.114)}$$

where $A_{dm}$ = minimum area of flow under the downcomer apron, $m^2$
= smaller of downcomer cross-sectional area $(A_d)$ and area under the apron $(A_{ap})$.

The apron clearance $h_{ap}$ (distance between the bottom end of inlet downcomer and the plate) is equal to $(h_w - h_{ls})$, where $h_{ls}$ is called the *downcomer liquid seal* or *downflow baffle seal.* The bottom end of the inlet downcomer thus projects below the upper edge of the outlet weir by $h_{ls}$. Such as arrangement helps to provide a liquid seal for the vapour between two successive plates and thus to prevent vapour from passing up through the downcomer. The value of $h_{ls}$ depends on the weir to baffle distance ($W$). For $W \leq 1.5$ m, $h_{ls}$ may be taken to be 12.5 mm, for $W$ = 1.5 to 3.0 m, $h_{ls}$ should amount to 25 mm and for large towers with $W > 3.0$ m, $h_{ls}$ = 38 mm. The area under the apron, $A_{ap} = (h_{ap}\, l_w) = (h_w - h_{ls})\, l_w$.

To avoid downflow flooding, the magnitude of $H_d$ must be less than 50 percent of the plate spacing.

In fact, the liquid in the downcomer is aerated. The actual backup (which is the height of the aerated liquid in the downcomer) is

$$H'_d = H_d/\phi_{dc} \qquad \text{... (5.2.115)}$$

where $\phi_{dc}$ = relative froth density in the downcomer

For systems with low foamability, $\phi_{dc}$ = 0.5 may be used. For systems of high foamability, values of $\phi_{dc}$ = 0.2 to 0.3 should be used. Design must not permit $H'_d$ to exceed the value of plate spacing, otherwise flooding can be precipitated.

We have also stated earlier that the *residence time of the liquid in the downcomer* $(\theta_d)$ should be sufficiently large so that all the entrained vapour will get sufficient time to escape from the liquid. $\theta_d$ can be estimated as

$$\theta_d = (H_d/1000)\, A_d/(Q_L/\rho_L) \qquad \text{... (5.2.116)}$$

It must be seen that $\theta_d$ is not less than 5 seconds.

In the case of sieve plates, its *weeping tendency* should also be checked with reference to Eqs (4.5.33) and (4.5.36) as stated earlier.

**Example 5.28:** A bubble-cap column operating at 1 atm is used for the continuous fractionation of an aqueous solution of ethanol. The following data are available:

Column diameter = 1.8 m

Actual number of plates = 15

Feed plate = twelfth from top

Feed composition = 0.3 mass fraction of ethanol and balance water

Feed rate = 12,000 kg/hr

Distillate composition = 0.9 mass fraction of ethanol

Bottom residue composition = 0.005 mass fraction of ethanol

Reflux ratio = 2.0

The column is equipped with a reboiler and a total condenser. For the plates in the enriching section, the following layout is suggested:

100 mm caps, arranged in seven rows, mounted flush on 150 mm equilateral triangular centres. Total number of caps = 90. It is also recommended to use a straight, segmental chord weir of length 75 percent of column diameter. Take the active plate area to be 70 percent of column cross-sectional area.

Determine whether the above layout and the chosen column diameter are satisfactory to provide stable plate operation.

**Solution:** For most fractionators, it is not necessary to design each plate of the column separately. It is usually sufficient to size the topmost plate of the enriching section, the feed plate and the bottom-most plate of the stripping section and a common plate layout could be used for all the plates of each section, unless the liquid and vapour compositions and the temperature vary substantially from plate to plate.

In the present case, we shall consider the design of the topmost plate of the enriching section only. Since a total condenser is used, $y_1 = x_0 = x_D = 0.9$ (mass fraction) = 0.7788 (mole fraction). From the equilibrium data (given in earlier examples), corresponding to $y_1 = 0.7788$, $x_1 = 0.7427$ and the temperature is 78.43° C. This is the temperature of the topmost tray. The liquid load on the tray is $L = L_1 = L_0 = (RD)$ and the vapour load is $V = V_1 = (L_0 + D) = (R + 1)\ D$. The distillate rate ($D$) can be obtained by performing a material balance for the entire column :

$$F = 12000 = B + D$$

$$Fz_F = Bx_B + Dx_D$$

or

$$12000\ (0.3) = (12000 - D)\ (0.005) + D\ (0.90)$$

Therefore,

$$D = 3955.3 \text{ kg/hr}$$

Since $x_D = 0.7788$ (mole fraction), the molecular weight of the distillate is

$$= (0.7788)\ (46.0) + (1.0 - 0.7788)\ (18.0) = 39.8064 \text{ kg/kmole}$$

Therefore,

$$D = (3955.3 / 39.8064) = 99.3634 \text{ kmoles/hr}$$

$$L = (RD) = (2.0)\ (99.3634) = 198.7268 \text{ kmoles/hr}$$

Since $x_0 = x_D$, molecular weight of reflux liquid ($L_0$) will be the same as that of distillate. Therefore,

$$Q_L = (198.7268)\ (39.8064) = 7910.6 \text{ kg/hr}$$

Now,

$$V = (R + 1)\ D = (3.0)\ (99.3634) = 298.09 \text{ kmoles/hr}$$

Once again, since $y_1 = x_D$, the molecular weight of overhead vapour will also be equal to 39.8064. Thus,

$$Q_V = (298.09)\ (39.8064) = 11865.9 \text{ kg/hr}$$

At 78.43° C, density of water is 971.8 kg/m$^3$ and density of ethanol is 760.0 kg/m$^3$. Also, $x_1 =$ 0.7427 (mole fraction) = 0.8806 (mass fraction). Therefore, density of liquid ($\rho_L$) is

$$\rho_L = \frac{1}{(0.8806/760.0) + (0.1194/971.8)} = 780.3057 \text{ kg/m}^3$$

Similarly, molar volume of overhead vapours from the topmost tray (assuming ideal gas behaviour) is

$$= (22.4)\ (353)/(273) = 28.964\ \text{m}^3/\text{kmole}$$

Since molecular weight of vapour is 39.8064 (computed earlier),

$$\text{Density of vapour } (\rho_V) = (39.8064/28.964) = 1.3743\ \text{kg/m}^3$$

Let us first verify whether the column diameter chosen is OK or not. From Eq. (4.5.25) of Chapter 4, the flow parameter ($\pi_1$) is

$$\pi_1 = (Q_L/Q_V)\ \sqrt{\rho_V/\rho_L}$$

$$= (7910.6/11865.9)\ \sqrt{1.3743/780.3057} = 0.028$$

Since $\pi_1 < 0.03$, $\pi_1 = 0.03$. Also, since $D = 1.8$ m, let us choose a plate spacing of 0.61 m. Now, from Table 4.9,

$$a_0 = 0.0564\ (0.61) + 0.0207 = 0.0551$$

$$a_1 = 0.0492\ (0.61) + 0.004115 = 0.034127$$

Therefore, from Eq. (4.5.24), the capacity parameter is

$$\pi_2 = 0.10707$$

Now, from Eq. (4.5.26), taking $\sigma = 0.021$ N/m

$$U_{fn} = 2.574\ \text{m/s}$$

Since $(l_w/D) = 0.75$, from Eq. (4.5.20),

$$(W/D) = 0.6614$$

And from Eq. (4.5.23),

$$(A_d/A) = 0.112$$

Therefore,

$$A_d = (0.112)\ [\pi\ (1.8)^2/4] = (0.112)\ (2.5447) = 0.285\ \text{m}^2$$

$$A_n = (A - A_d) = (2.5447 - 0.285) = 2.2597\ \text{m}^2$$

The operating vapour velocity based on $A_n$ is, therefore,

$$\mathbf{V}_{gn} = \frac{(Q_V/\rho_V)}{A_n} = \frac{(11865.9/3600)}{(1.3743)(2.2597)} = 1.0614\ \text{m/s}$$

$$\text{Percent flood} = 100\,\mathbf{V}_{gn}/U_{fn} = (100)\ (1.0614)/(2.574) = 41.235\ \text{percent}$$

This, in fact, is superfluous since a percent flood of 65.0 to 70.0 could very will be used in the present case. Now, from Fig. 4.10 of Chapter 4,

Fractional entrainment ($\psi$) = 0.028

This is well below the permissible limit of 0.1. Thus, the chosen diameter is quite satisfactory for the present system.

Let us now proceed to compute the pressure drop across the plate. From Eq. (5.2.83),

$$h_{ow} = (664.0)\ F_w \left[\frac{7910.6/3600}{(780.3057)(0.75)(1.8)}\right]^{2/3}$$

Since $Q_L/(\rho_L l_w^{2.5}) = 1.3298 \times 10^{-3}\ (\text{m}^2/\text{s})/\text{m}^{2.5} = 1.081\ (\text{gal/min})/(ft)^{2.5}$, from Fig. 5.35,

$$F_w = 1.015$$

Therefore, $h_{ow} = 11.0$ mm

Since the operating pressure is 1 atm, let us choose

$$h_{ss} = 12.5 \text{ mm}$$

Now from standard cap specifications (Table 4.8 of Chapter 4), let us select for $d_c = 100$ mm,

Slot height $(h_s) = 31.75$ mm

Slot type = rectangular

Slot width $(b_s) = 5.0$ mm

Area of slots per cap $(a_s) = 0.00524$ m$^2$

Riser area per cap $(a_r) = 0.0031$ m$^2$

Height of shroud ring (distance between the slot and the lower edge of the cap) = $h_{sr} = 5.0$ mm

Skirt clearance $(h_{sk}) = 25.0$ mm

Ratio of annular area to riser area $(A_{an}/A_r) = 1.25$

From Eq. (5.2.105), $K_c = 0.5049$

Since there are 90 caps per plate, the total riser area per plate $(A_r)$ is

$$A_r = (0.0031)\,(90) = 0.279 \text{ m}^2$$

And, total slot area per plate

$$(A_s) = (0.00524)\,(90) = 0.4716 \text{ m}^2$$

Therefore, from Eq. (5.2.104),

$$U_{VC} = \frac{(11865.9/3600)}{(1.3743)(0.279)} = 8.5963 \text{ m/s}$$

And, from Eq. (5.2.103), $h_{rc} = (273.4)\,(0.5049)\,(1.3743/780.3057)\,(8.5963)^2 = 17.96$ mm

Now, from Eq. (5.2.109), $U_{VS} = \dfrac{(11865.9/3600)}{(1.3743)(0.4716)} = 5.0856 \text{ m/s}$

And, from Eq. (5.2.108), $h_{so} = (937.0)\left[\dfrac{(1.3743)}{(780.3057-1.3743)}\right]^{0.2}\left[(0.03175)^2\,(5.0856)\right]^{0.4}$

$$= 31.988 \text{ mm}$$

This shows that there is a slight inadequacy in the value of slot height $(h_s)$ chosen (since $h_{so}$ slightly exceeds $h_s$), though this is not serious. The weir height $(h_w)$ can be now obtained from Eq. (4.5.18) of Chapter 4:

$$h_w = h_{sk} + h_{sr} + h_s + h_{ss} = (25.0 + 5.0 + 31.75 + 12.5) = 74.25 \text{ mm}$$

Let us now proceed to compute the liquid gradient across the plate $(h_g)$. Thus,

$$W_a = (D + l_w)/2$$

Since $l_w = (0.75)\,D = (0.75)\,(1.8) = 1.35$ m,

$$W_a = (1.8 + 1.35)/2 = 1.575 \text{ m}$$

Liquid load per unit flow width,

$$Q' = Q_L/(\rho_L W_a) = \frac{(7910.6/3600)}{(780.3057)(1.575)} = 0.001788 \text{ (m}^3\text{/s)/m}$$

Therefore, from Table 5.3,

$$b_0 = -1.17537$$

$$b_1 = 365.41247$$

$$b_2 = 29377.214$$

Accordingly, from Eq. (5.2.94),

$$\ln (C_d) = -1.17537 + (365.41247)(0.001788) + (29377.214)(0.001788)^2$$
$$= -0.428095$$

or

$$C_d = 0.65175$$

Now, $\gamma = (S_c/d_c) = (p_c - d_c)/d_c = (150 - 100)/(100) = 0.50$

Equation (5.2.88), therefore, reduces to

$$0.002743 = 0.0139 \sqrt{h_g'}\ (0.0016 h_g' + 0.30075)$$

Solving by trial, we get

$$h_g' = 0.4285 \text{ mm}$$

Now, from Eq. (5.2.96),

$$F_V = \left(\frac{Q_V}{\rho_V A}\right)\sqrt{\rho_V} = Q_V/\left(A\sqrt{\rho_V}\right)$$

$$= \frac{(11865.9/3600)}{(2.5447)(1.3743)^{1/2}} = 1.105 \text{ (m/s)} \sqrt{\text{kg/m}^3}$$

Now, from Table 5.4 and Eq. (5.2.95), the values of $C_{Vf}$ at $F_V = 0.976$ and at $F_V = 1.22$ are

| $F_V$ | $C_{Vf}$ |
|---|---|
| 0.976 | 0.8066 |
| 1.22 | 0.88728 |

Therefore, by linear interpolation, the value of $C_{Vf}$ at $F_V = 1.105$ is

$$C_{Vf} = 0.84925$$

Now, from Eq. (5.2.93), $h_g = h_g' C_{Vf} r_c = (0.4285)(0.84925)(7.0) = 2.55$ mm

The value of $h_g$ is seen to be much below the maximum limit of 25 mm. Now, from Eq. (5.2.85),

$$h_{ds} = (12.5) + (11.0) + (2.55/2) = 24.775 \text{ mm}$$

This value of $h_{ds}$ is slightly outside the usual range of 25 – 65 mm. The deviation however, is not significant. Since $A_a = 0.7A$,

$$F_{Va} = \mathbf{V}_a \sqrt{\rho_V} = \left(\frac{Q_V}{\rho_V A_a}\right) \sqrt{\rho_V} = Q_V / \left(0.7 A \sqrt{\rho_V}\right)$$

$$= \frac{(11865.9/3600)}{(0.7)(2.5447)\sqrt{1.3743}} = 1.5784 \text{ (m/s)} \sqrt{\text{kg/m}^3}$$

From Fig. 5.36, corresponding to $F_{Va} = 1.5784$, $\beta = 0.67$. Therefore,

$$h_L = \beta h_{ds} = (0.67)(24.775) = 16.6 \text{ mm}$$

$$h_c = h_{rc} + h_{so} = (17.96) + (31.988) = 49.948 \text{ mm}$$

The total pressure drop across the tray is, therefore,

$$\Delta H = (h_c + h_L) = 66.548 \text{ mm}$$

$$\text{Vapour distribution ratio} = (h_g / h_c) = (2.55/49.948) = 0.051$$

This is well below the prescribed limit of 0.4. The performance of the plate shall be, therefore, quite stable.

We should also check the magnitude of the liquid backup in the downcomer. Since $(W/D) = 0.6614$ (computed earlier),

$$W = (0.6614)(1.8) = 1.1905 \text{ m}$$

Since $W < 1.5$ m, $h_{ls} = 12.5$ mm. The area under the apron is

$$A_{ap} = (h_w - h_{ls})\, l_w = (74.25 - 12.5)(1.35)/(1000) = 0.083 \text{ m}^2$$

Since $A_{ap}$ is smaller than $A_d$,

$$A_{dm} = 0.083 \text{ m}^2$$

From Eq. (5.2.114),
$$h_d = (165.2)\left[\frac{(7910.6)}{(3600)(780.3057)(0.083)}\right]^2 = 0.2 \text{ mm}$$

From Eq. (5.2.80),
$$H_d = (66.548) + (74.25) + (11.0) + (2.55) + (0.2) = 154.548 \text{ mm}$$

Since $H_d < (S/2)$, the plate is stable against flooding. Taking the relative froth density in downcomer ($\phi_{dc}$) to be 0.5,

$$H'_d = (154.548/0.5) = 309.1 \text{ mm}$$

This is much less than the plate spacing and therefore, the design is adequate to avoid any chances of flooding. The liquid residence time in the downcomer can be now obtained from Eq. (5.2.116):

$$\theta_d = \frac{(0.154548)(0.285)(780.3057)}{(7910.6/3600)} = 15.64 \text{ seconds}$$

Thus, $\theta_d$ is also much above the minimum required value of 5 seconds. Overall, the design specifications of the plate are satisfactory, though the liquid gradient across the plate ($h_g$) is too low.

### 5.2.6 Continuous Fractionation in Packed Columns

As for absorption (discussed in Chapter 4), packed towers are viable substitutes to plate towers for conducting the continuous fractionation of binary/multicomponent mixtures. A broad comparison between packed columns and plate columns and the specifications of commonly used packing materials are already discussed in section 4.5 of Chapter 4.

Packed columns offer relatively lower pressure drop and are, therefore, more suitable for low pressure distillation. An example in this connection is the distillation of heat-sensitive materials which are to be distilled at low pressure and thereby low temperature since at high temperatures, they tend to decompose. Since packing materials made of ceramic substances, porcelain, glass, clay, tile and carbon are chemically inert and are virtually immune to chemical attack, one of the most important applications of packed columns is in the area of handling corrosive substances. This is true in the case of food and pharmaceutical industries as well, where contamination from metals is highly harmful and is to be dispensed with. Packed columns lend themselves well to high vapour-low liquid loading or low vapour-high liquid loading services and also to laboratory and pilot plant applications since they can be effectively built in smaller sizes.

Packed columns are continuous contact equipment unlike plate columns which provide stagewise contact. In packed columns, the transfer of mass takes place throughout the column at the vapour-liquid interface. As the liquid and vapour flow countercurrently through the packed bed, their compositions change continuously. Since the liquid is at its bubble point and the vapour at its dew point, the temperature gradually increases from the top to the bottom of the column. A typical packed-tower fractionator is sketched in Fig. 5.38. As in the case of plate columns, this also requires a reboiler at the bottom and a condenser at the top. The reflux is continuously returned to the top of the column and reboiled vapour to the bottom of the column. Instead of the reboiler, open steam from an external source may be directly admitted to the bottom of the column in case the residue discharged is essentially water. For the introduction of feed, a short, unpacked section may be provided at the feed entry (see figure), but adequate distribution of liquid over the top of the exhausting section must be ensured. Generally, the support plates, which are introduced basically for the purpose of supporting the packing in the column shell, serve also as liquid and vapour distribution plates. These may be perforated plates, interdistributor screens, grid bars, gas injection support plates, orifice plates, weir flow distributors or trough-type distributors (see Fig. 5.39).

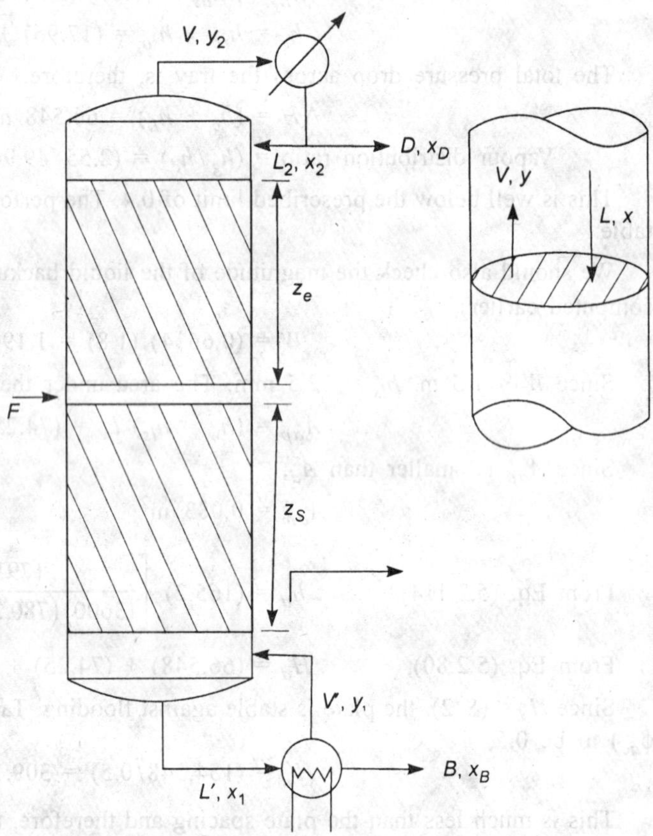


*Figure 5.38: Typical packed column fractionator*

According to Eckert[29], redistribution must be provided at every 2.5 to 3 column diameters for Raschig rings and at every 5 to 8 column diameters for Intalox and Berl saddles. For Pall rings, redistribution is required at each 5 to 10 column diameters. In all cases, redistributors should be installed at every 6 m of column height.

The performance analysis of a packed-bed distillation column is analogous to that of a packed-bed absorber that is discussed in Section 4.5.5 of Chapter 4. We can, therefore, use the HTU-NTU concept here also. We shall, at present confine our discussion to binary feed mixtures. Let us further assume that the assumption of constant molal overflow (used under the McCabe-Thiele method discussed in the earlier subsection) is valid in this case also. Then, the height of a gas phase (or, vapour phase) transfer unit is defined in the same way as in Eq. (4.5.92):

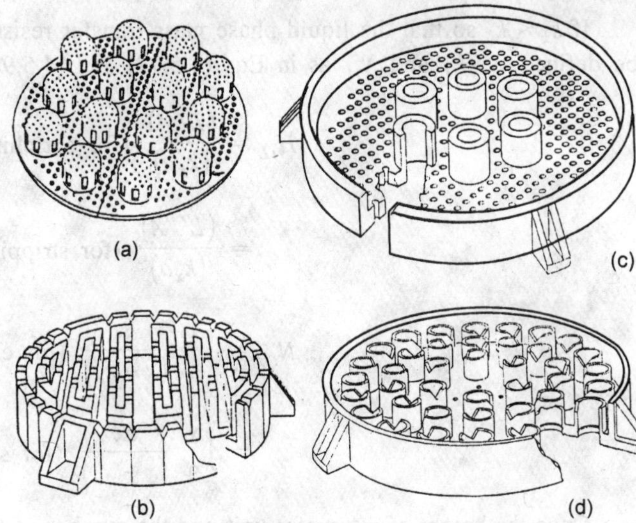

Figure 5.39: Types of distributor plates, (a) Cap type gas injection support plate, (b) Trough type distributor plate, (c) Orifice distributor, (d) Weir flow distributor

$$H_{tG} = \frac{(V/A)}{(k_g a)\, P} \qquad \text{... (5.2.117)}$$

$$= \frac{(V/A)}{(k_y a)} \qquad \text{... (5.2.117a)}$$

where $A$ is the cross-sectional area of the column. For stripping section, $(V/A)$ be replaced by $(V'/A)$. The number of gas phase transfer units is to be computed as

$$N_{tG} = \int_{y_a}^{y_2} \frac{dy}{(y_e - y)} \qquad \text{... (5.2.118)}$$

where $y_2$ = mole fraction of the more volatile component in the overhead vapours leaving the top of the column

$y_e$ = mole fraction of the more volatile component at the vapour-liquid interface

The above equation computes $N_{tG}$ for the enriching section. For the stripping (or, exhausting) section,

$$N_{tG} = \int_{y_1}^{y_a} \frac{dy}{(y_e - y)} \qquad \text{... (5.2.119)}$$

where $y_1$ = mole fraction of more volatile component in the reboiled vapour entering the bottom of the column*.

*It is important to note that the component notation used here are different from those used for plate towers, but are identical with those used for the packed bed absorber discussed in Chapter 4.

Assuming that the feed plate location is at the intersection of the two operating lines (this is most desirable and is so in new column design), $(x_a, y_a)$ are the coordinates of the point of intersection of the two operating lines (through which the $q$-line also passes).

If $k_x < k_y$ so that the liquid phase mass transfer resistance is controlling, then HTU and NTU may be defined as $H_{tL}$ and $N_{tL}$ as in Eqs (4.5.97) and (4.5.98). Thus

$$H_{tL} = \frac{(L/A)}{(k_x a)} \text{ (for enriching section)} \qquad \text{... (5.2.120)}$$

$$= \frac{(L'/A)}{(k_x a)} \text{ (for stripping section)} \qquad \text{... (5.2.121)}$$

$$N_{tL} = \int_{x_a}^{x_2} \frac{dx}{(x - x_e)} \text{ (for enriching section)} \qquad \text{... (5.2.122)}$$

$$= \int_{x_1}^{x_a} \frac{dx}{(x - x_e)} \text{ (for stripping section)} \qquad \text{(5.2.123)}$$

Once the height of a transfer unit and the number of transfer units are computed, the packed height of the distillation column may be obtained from

$$Z_e \text{ (or } Z_s) = (H_{tG} N_{tG}) = (H_{tL} N_{tL}) \qquad \text{... (5.2.124)}$$

where $Z_e$ is the height of the enriching section and $Z_s$ that of the stripping section. The total packed height of the column will be, $Z = (Z_e + Z_s)$.

Incidentally, though the molar rates of flow ($V$, $V'$, $L$, $L'$) may be assumed constant with allowable error, the mass flow rates could vary significantly from point to point (or from section to section) since the average molecular weight of the vapour mixture and that of the liquid solution may vary considerably with change in concentration. As a result, there could be significant variation in the values of the mass transfer coefficient ($k_x$ or $k_y$) and specific interfacial area ($a$), since their values depend on the mass flow rates (and not molar flow rates) of the vapour and the liquid. In such cases, it shall be more accurate to rewrite Eq. (5.2.124) by including the term $(k_y a)$ or $(k_x a)$ also within the integral as

$$Z_e = (V/A) \int_{y_a}^{y_2} \frac{dy}{k_y a (y_e - y)}$$

$$= (L/A) \int_{x_a}^{x_2} \frac{dx}{k_x a (x - x_e)} \qquad \text{... (5.2.124a)}$$

$$Z_s = (V'/A) \int_{y_1}^{y_a} \frac{dy}{k_y a (y_e - y)}$$

$$= (L'/A) \int_{x_1}^{x_a} \frac{dx}{k_x a (x - x_e)} \qquad \text{... (5.2.124b)}$$

The mass transfer coefficients are to be evaluated separately for each section. Incidentally, most of the correlations for mass transfer in packed columns available in literature have been proposed based on experiments on gas absorption that are conducted at relatively low temperatures. Therefore, it cannot be guaranteed that these correlations shall be applicable for distillation as well, where the temperatures are normally relatively high. However, in the absence of more specific data, the same mass transfer correlations as those given in Chapter 4 are to be used in distillation problems as well.

The operating diagram that consists of the equilibrium curve and the two operating lines (such as those for the enriching section and for the exhausting or stripping section) is drawn in exactly the same manner as for plate columns (see Fig. 5.40). Equations for operating lines are same as those for plate columns except that the tray-number subscripts can be omitted. Thus, the equation to the operating line for the enriching section is

$$y = \left(\frac{R}{1+R}\right) x + \left(\frac{x_D}{1+R}\right) \qquad \text{... (5.2.125)}$$

And that to the operating line for the stripping section is

$$y = (L'/V')\, x - (B\, x_B/V') \qquad \text{... (5.2.126)}$$

where $x$ and $y$ are the mole fraction of the more volatile component in the incoming liquid solution and in the outgoing vapour mixture respectively at any particular horizontal section of the tower (see Fig. 5.38).

The slope of the $q$-line depends on the thermal condition of the feed as discussed in the earlier subsection 5.2.4. Since $k_x$ and $k_y$ are mass transfer coefficients and $(x_e, y_e)$ are interfacial concentrations, a simple mass balance yields

$$k_y\,(y - y_e) = k_x\,(x_e - x)$$

or

$$(y - y_e) = -\,(k_x/k_y)\,(x - x_e) \qquad \text{... (5.2.127)}$$

The values of interfacial concentration ($x_e$ or $y_e$) can be thus computed by solving the above equation simultaneously with the equilibrium relationship. For example first a value of $x_e$ is assumed (say, $x_e = x - 0.001$) and then $y_e$ is computed from the above equation. Now, compute $y_e$ from the equilibrium data also (to note that $y_e$ is the equilibrium mole fraction in vapour corresponding to liquid phase mole fraction $x_e$). If the two values of $y_e$ agree with each other, then the assumed value of $x_e$ is correct. Otherwise, an alternate value of $x_e$ is to be assumed and trial repeated.

The values of $(x_e, y_e)$ may also be determined graphically. For example, from any point $(x, y)$ on the operating line, if a straight line of slope $(-k_x/k_y)$ is drawn, then it will intersect the equilibrium curve at $(x_e, y_e)$. This is illustrated in Fig. 5.40.

The column diameter is to be determined with reference to the flooding plots given in Fig. 4.19 of Chapter 4. The same set of plots may be used for estimating the pressure drop across the packed bed during the simultaneous flow of vapour and liquid. The pressure drop should never be more than 90 percent of $(-\Delta P)_{\text{flood}}$ and preferably, not more than 60 percent of $(-\Delta P)_{\text{flood}}$. The maximum permissible pressure drop is usually dictated by the operating pressure. Under atmospheric conditions, a pressure drop of 18 to 50 mm of water per metre of packing is usual, while a pressure drop of not more than 9 mm water per metre of packing is normally specified for operation under vacuum.

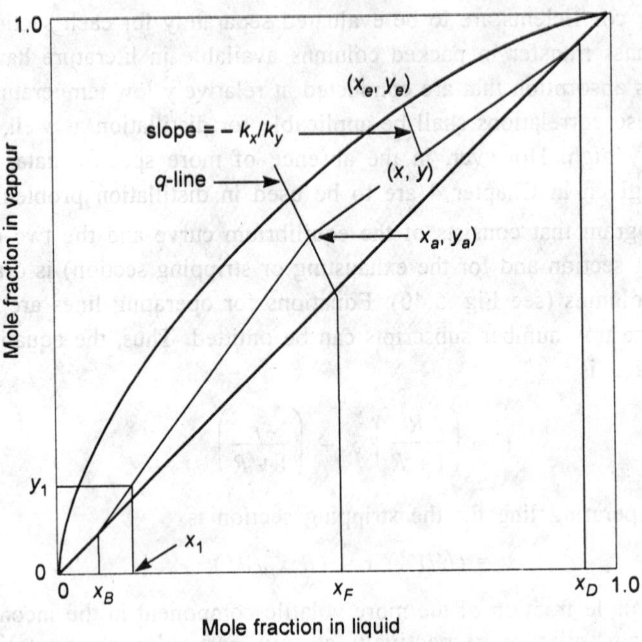


*Figure 5.40: Operating diagram for packed column fractionator*

**Example 5.29:** If the acetone-water system considered in Example 5.23 is distilled in a packed column 1.5 m in diameter and equipped with a total condenser and a reboiler using the same reflux ratio, what will be the required packed height of the column? The packing material is 25 mm ceramic Raschig rings (surface diameter = 35.6 mm).

**Solution:** Since the liquid and vapour compositions vary continuously along the column, for a rigorous design of the packed column, a section to section computation is required. However, we shall solve the present problem based on the following assumptions:

(*i*) The assumption of constant molal overflow holds good. Thus, $L$ and $V$ are the liquid and vapour flow rates (in kmoles/s) in the enriching section and $L'$ and $V'$ are those in the stripping (or exhausting) section.

(*ii*) The change in the value of mass transfer coefficient from one end to the other end of a section (either the enriching section or the stripping section) is neglected.

Based on the above assumptions, the packed bed distillation column can be designed as given below:

*At the top of the column:* At the top of the column, the liquid and vapour compositions are

$$x_2 = x_D = 0.955$$

$$y_2 = x_2 = x_D = 0.955$$

The temperature is taken to be the dew point of the vapour (of composition $y_2 = 0.955$), which from equilibrium data (given in example 5.23) is 56.4° C. Thus

$$T_2 = 56.4° \text{ C}$$

The density and viscosity of acetone are very close to those of water. Therefore, these property values shall not change materially due to the change in composition. They could change due to the change in temperature and this also shall not be substantially large. The assumption of a constant mass transfer coefficient (stated earlier), thus, holds true with allowable error. At 56.4° C,

$\rho_L$ (considering pure water) = 985 kg/m$^3$

$\mu_L$ (considering pure acetone) = 0.000575 kg/(ms)

$\mu_V$ (considering pure acetone vapour) = $7.35 \times 10^{-6}$ kg/(ms)

The molecular weights of the liquid and vapour streams are

$$M_L = x_2\,(58.0) + (1 - x_2)\,(18.0) = (0.955)\,(58.0) + (0.045)\,(18.0) = 56.2$$

$$M_V = y_2\,(58.0) + (1 - y_2)\,(18.0) = 56.2$$

The flow rates of the two steams are,

$$L = L_2 = RD = (0.9279)\,(20.92) = 19.41 \text{ kmoles/hr}$$

$$Q_L = LM_L = (19.41)\,(56.2) = 1090.842 \text{ kg/hr} = 0.303 \text{ kg/s}$$

$$V = V_1 = (L_2 + D) = 19.41 + 20.92 = 40.33 \text{ kmoles/hr}$$

$$Q_V = VM_V = (40.33)\,(56.2) = 2266.546 \text{ kg/hr} = 0.6296 \text{ kg/s}$$

The values of reflux ratio ($R$) and $D$ are taken from Example 5.23. The liquid phase diffusivity ($D_L$) of acetone-water system has been computed in Example 4.9 Chapter 4. After correcting for temperature,

$$D_L = 1.243 \times 10^{-9} \text{ m}^2/\text{s}$$

The gas-phase diffusivity may be computed from Eq. (4.1.2). The critical properties of acetone are $T_c$ = 508° K and $P_c$ = 47 atm and those for water are $T_c$ = 647.15° K and $P_c$ = 218.4 atm. Therefore,

$$D_G = (3.64 \times 10^{-8}) \left[\frac{329.4}{\sqrt{(508)(647.15)}}\right]^{2.334} [(47)\,(218.4)]^{1/3}$$

$$[(508)\,(647.15)]^{5/12} \sqrt{(1/58) + (1/18)} = 1.1645 \times 10^{-5} \text{ m}^{\text{-}} \text{ s}$$

$$Sc_G = \mu_V/(\rho_V\, D_G)$$

where $\rho_V$ = (molecular weight/molar volume)

$$= \frac{(5.6.2)}{\left(\dfrac{22.4}{273}\right)(329.4)} = 2.079 \text{ kg/m}^3$$

Therefore,

$$Sc_G = \frac{\left(7.35 \times 10^{-6}\right)}{(2.079)\left(1.1645 \times 10^{-5}\right)} = 0.3035$$

$$Sc_L = \mu_L/(\rho_L\, D_L) = \frac{(0.000575)}{(985.0)\left(1.243 \times 10^{-9}\right)} = 469.635$$

Let us use Shulman et. al's correlations [given in Eqs (4.5.125) and (4.5.126) for estimating the mass transfer coefficients. First, the total liquid holdup is obtained from Eq. (4.5.128):

$$\phi_t = C_1\ (737.5\,Q_L/A)^{\beta}/d_s^2$$

where $Q_L = 0.303$ kg/s

$A = (\pi/4)\ (1.5)^2 = 1.767\ \text{m}^2$

$d_S = 0.0356$ mm

$\beta = C_2\ d_S^{0.376}$

From Table 4.14, for 25 mm ceramic Raschig rings,

$$C_1 = 2.09 \times 10^{-6}$$

$$C_2 = 1.508$$

Therefore, $\beta = 1.508\ (0.0356)^{0.376} = 0.43$

$$\phi_t = \frac{\left(2.09\times10^{-6}\right)}{(0.0356)^2}\ [(737.5)\ (0.303)/1.767]^{0.43} = 0.0132$$

From Table 4.7, $\in = 0.74$

Substituting all these values in Eq. (4.5.125), we get

$$F_g = 7.1862 \times 10^{-4}\ \text{kmoles}/(\text{m}^2\text{s})$$

And from Eq. (4.5.126), $k_L = 5.4989 \times 10^{-5}$ m/s

Now, $F_L \simeq k_x = k_L\ (\rho_L/M_L)$. Therefore,

$$F_L = (5.4989 \times 10^{-5})\ (985.0/56.2)$$
$$= 9.637 \times 10^{-4}\ \text{kmoles}/(\text{m}^2\text{s})$$

$(F_L/F_g) = 1.341$

Let us also compute the specific interfacial area from Eq. (4.5.130). Since $(Q_L/A) = (0.303/1.767) = 0.17148\ \text{kg}/(\text{m}^2\text{s})$, with reference to Table 4.15, Eq. (4.5.130) becomes

$$a = 34.42\,(Q_L/A)^{0.552} = 13.0\ \text{m}^2/\text{m}^3$$

The equilibrium curve is plotted in Fig. 5.29.1. The operating line for enriching section is drawn, as shown, with an intercept of $x_D/(R+1) = (0.955)/(1.9279) = 0.49535$. It intersects the $q$-line (which is a vertical line since the feed is a saturated liquid) as shown and from the point of intersection, the operating line for the stripping section is drawn with a slope of $(L'/V') = 2.97$. Now, we draw lines of slope $-(F_L/F_g) = -1.341$ from different points on the enriching section

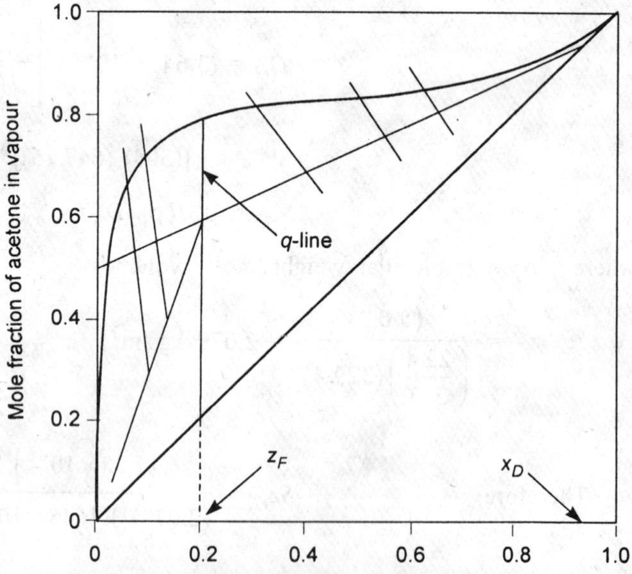


*Figure 5.29.1*

operating line and from the points of intersection of these lines with the equilibrium curve, we get the values of $(x_e, y_e)$. The results are tabulated below in Table 5.29.1. In the figure, we have shown only three such lines and thus the computation of three pairs of values of $(x_e, y_e)$. Other values are obtained in a similar manner.

**Table 5.29.1**

| $x$ | $x_e$ | $1/(x - x_e)$ |
|---|---|---|
| 0.955 | 0.95 | 200.0 |
| 0.80 | 0.7925 | 133.33 |
| 0.75 | 0.735 | 66.667 |
| 0.70 | 0.675 | 40.00 |
| 0.65 | 0.620 | 33.33 |
| 0.60 | 0.555 | 22.22 |
| 0.55 | 0.4925 | 17.3913 |
| 0.50 | 0.425 | 13.33 |
| 0.45 | 0.365 | 11.7647 |
| 0.40 | 0.30 | 10.0 |
| 0.35 | 0.245 | 9.5238 |
| 0.30 | 0.19 | 9.0909 |
| 0.25 | 0.135 | 8.6956 |
| 0.20 | 0.095 | 9.5238 |

Since $F_L$ and $F_g$ are relatively of the same magnitude, it shall not make much difference we use Eq. (5.2.118) or (5.2.122) for computing NTU. Let us, therefore, use Eq. (5.2.122) to compute the number of transfer units ($N_{tL}$). Thus

$$N_{tL} = \int_{x_a = 0.2}^{x_2 = 0.955} \frac{dx}{(x - x_e)}$$

From the data given in the above table, the above integral is evaluated numerically using trapezoidal rule (this rule is discussed in Examples 5.2 and 5.10). Thus

$$N_{tL} = 41.5$$

The height of a liquid phase transfer unit is given by

$$H_{tL} = \frac{(L/A)}{(k_x a)} = \frac{(19.41)}{(3600)(1.767)\left(9.637 \times 10^{-4}\right)(13.0)} = 0.2435 \text{ m}$$

The packed height of the enriching section is now obtained as

$$Z_e = H_{tL} N_{tL} = (0.2435)\ (41.5) = 10.105 \text{ m}$$

*At the bottom of the column:* The liquid and vapour compositions are,

$$y_1 = y_B = 0.00319 \text{ (from equilibrium data corresponding to } x_B = 0.00025)$$

A material balance around the reboiler yields,

$$L'x_1 = V'y_1 + Bx_B$$

where $L' = L + qF = (L + F) = 119.41$ kmoles/hr

$$V' = V + (q - 1)\,F = V = 40.33 \text{ kmoles/hr}$$

$$B = (F - D) = (100 - 20.92) = 79.08 \text{ kmoles/hr}$$

Therefore, $x_1 = 0.001243$

The temperature is taken equal to the bubble point of liquid of composition $x_1$. Thus, from equilibrium data, $T_1 = 99.37°$ C. The density and viscosity of liquid and vapour are (assumed to be equal to those of pure water at temperature $T_1$),

$$\rho_L = 958.8 \text{ kg/m}^3$$

$$\mu_L = 2.5 \times 10^{-4} \text{ kg/(ms)}$$

$$\mu_V = 12.5 \times 10^{-6} \text{ kg/(ms)}$$

The molecular weights of liquid and vapour are,

$$M_L = x_1\,(58.0) + (1 - x_1)\,(18.0) = 18.0497$$

$$M_V = y_1\,(58.0) + (1 - y_1)\,(18.0) = 18.1276$$

$$\rho_V = \frac{(18.1276)}{(22.4/273)(372.37)} = 0.5933 \text{ kg/m}^3$$

The mass flow rates of the two streams are,

$$Q_L = (L'M_L) = 2155.31 \text{ kg/hr} = 0.5987 \text{ kg/s}$$

$$Q_V = (V'M_V) = 731.086 \text{ kg/hr} = 0.203 \text{ kg/s}$$

The liquid-phase diffusivity ($D_L$) is obtained from that determined for the enriching section after correcting it for temperature. Thus,

$$D_L = 1.4055 \times 10^{-9} \text{ m}^2\text{/s}$$

The gas-phase (or vapour-phase) diffusivity is also obtained from that computed for the enriching section by correcting it for temperature. Thus

$$D_G = (1.1645 \times 10^{-5})\left(\frac{372.37}{329.4}\right)^{2.334} = 1.55 \times 10^{-5} \text{ m}^2\text{/s}$$

$$Sc_G = \frac{(12.5 \times 10^{-6})}{(0.5933)(1.55 \times 10^{-5})} = 1.359$$

$$Sc_L = \frac{(2.5 \times 10^{-4})}{(958.8)(1.4055 \times 10^{-9})} = 185.516$$

$$(Q_L/A) = (0.5987/1.767) = 0.3388$$

$$\phi_t = \frac{\left(2.09 \times 10^{-6}\right)}{(0.0356)^2} \left[(737.5)\ (0.3388)\right]^{0.43}$$

The mass transfer coefficients are

$$F_g = 4.841 \times 10^{-4} \text{ kmoles}/(\text{m}^2\text{s})$$

$$k_L = 7.72345 \times 10^{-5} \text{ m/s}$$

$$F_L \simeq k_x = k_L\,(\rho_L/M_L) = 4.1027 \times 10^{-3} \text{ kmoles}/(\text{m}^2\text{s})$$

$$(F_L/F_g) = 8.4749$$

Straight lines are now drawn from different points on the exhausting section operating line and the coordinates of their points of intersection with the equilibrium curve give $x_e$ and $y_e$. The results are tabulated below in Table 5.29.2. Here also, we have shown only two such lines and thereby the determination of two pairs of values of $(x_e, y_e)$ in the figure. Other values of $(x_e, y_e)$ are similarly obtained.

**Table 5.29.2**

| $y$ | $y_e$ | $\left[ln\left(\frac{1-y}{1-y_e}\right)\right]^{-1}$ |
|---|---|---|
| 0.59 | 0.785 | 1.54914 |
| 0.55 | 0.780 | 1.3974 |
| 0.50 | 0.765 | 1.3244 |
| 0.45 | 0.748 | 1.2812 |
| 0.40 | 0.725 | 1.2818 |
| 0.35 | 0.695 | 1.3216 |
| 0.30 | 0.66 | 1.38478 |
| 0.25 | 0.62 | 1.4708 |
| 0.20 | 0.57 | 1.61075 |
| 0.15 | 0.515 | 1.78225 |
| 0.10 | 0.375 | 2.7424 |
| 0.05 | 0.2195 | 5.0883 |
| 0.00319 | 0.0137 | 94.343 |

Since $F_g$ is much smaller then $F_L$, the gas phase resistance is controlling and therefore, Eqs (5.2.117) and (5.2.119) are to be used for computing HTU and NTU respectively. From Eq. (5.2.117), for stripping section,

$$H_{tG} = \frac{(V'/A)}{k_g a P}$$

Since $\qquad (F_g a) = k_g a P\,(1 - y)_{im},$

$$H_{tG} = \frac{(V'/A)(1-y)_{im}}{(F_g a)}$$

Now, from Eq. (5.2.119), $N_{tG} = \int_{0.00319}^{0.59} \frac{dy}{y_e - y}$

Therefore, $$Z_S = \frac{(V'/A)}{(F_g a)} \int_{0.00319}^{0.59} \frac{(1-y)_{im}\, dy}{(y_e - y)}$$

Since $$(1-y)_{im} = \frac{(1-y)-(1-y_e)}{\ln\left(\frac{1-y}{1-y_e}\right)} = \frac{(y_e - y)}{\ln\left(\frac{1-y}{1-y_e}\right)},$$

The above equation reduces to

$$Z_S = \frac{(V'/A)}{(F_g a)} \int_{0.00319}^{0.59} \frac{dy}{\ln\left(\frac{1-y}{1-y_e}\right)}$$

The specific interfacial area can be computed from Eq. (4.5.130) as done for the enriching section. Thus,

$$a = 34.42\ (Q_L/A)^{0.552} = 34.42\ (0.5987/1.767)^{0.552} = 18.94\ \text{m}^2/\text{m}^3$$

The integral in the above equation is once again evaluated by trapezoidal rule and is equal to 3.258. Therefore,

$$Z_S = \frac{(40.33)}{(3600)(1.767)\left(4.841 \times 10^{-4}\right)(18.94)}\ (3.258) = 2.2528\ \text{m}$$

The total packed height required is, therefore,

$$Z = (10.105 + 2.2528) = 12.3578\ \text{m}$$

It may be noted that in the above example, we have used a single value of mass transfer coefficient and a single value of specific interfacial area for the entire enriching section. And similarly for the stripping section. However, as stated earlier, the values of mass transfer coefficient and specific interfacial area depend on the mass flow rates of the vapour and the liquid and these could vary significantly (due to the change in the average molecular weights of the vapour and the liquid with change in concentration) even if the molar flow rates remain constant. A more accurate solution of the problem would be, therefore, that based on Eqs (5.2.124a) and (5.2.124b). The values of $k_y$, $a$ and $y_e$ are computed for different values of $y$ starting from $y_2$ to $y_a$ (in the decreasing order) in the enriching section and then the integral of Eq. (5.2.124a) is evaluated graphically or numerically to get the value of $Z_e$. The same procedure is used for the stripping section to compute $Z_S$ (here, $y$ varies from $y_1$ to $y_a$). Since the computational load is quite large in such a procedure, the aid of personal computers shall be desirable. The computer program for such a solution is given in Section 5.3 (see program 5.12).

## 5.2.7 Azeotropic and Extractive Distillation

We have already discussed at the beginning of this section the phenomenon of azeotrope formation. At the azeotropic composition, the mixture boils at a constant temperature producing vapours that are of the same composition as the liquid. In other words, $y^* = x$ and therefore, the relative volatility $\alpha$ will be equal to 1.0. As a result, separation by simple fractional distillation becomes impossible. For example, from equilibrium data of ethanol-water system at 1 atm, given in Example 5.22, it can be seen that at 78.15° C, $x$ (mole fraction of ethanol in liquid) = $y$ (mole fraction of ethanol in vapour) = 0.8943. The equilibrium curve intersects the 45° diagonal at this composition. Therefore, it will not be possible to concentrate any ethanol-water mixture to beyond 89.43 mole percent ethanol by conventional fractionation at 1 atm.

Azeotropes which exist as one liquid phase in equilibrium with vapour are called *homogeneous azeotropes,* while those exist as two liquid phases in equilibrium with vapour are called *heterogeneous azeotropes.* Ethanol-water and acetone-chloroform systems form homogeneous azeotropes, while *n*-butanol-water, furfural-water and aniline-water systems form heterogeneous azeotropes. Azeotropes can be further either *minimum boiling azeotropes* (as in the case of ethanol-water system) or *maximum boiling azeotropes* (as in the case of acetone-chloroform system). Typical temperature-concentration diagrams for two types of homogeneous azeotrope are given in Figs 5.41 and 5.42.

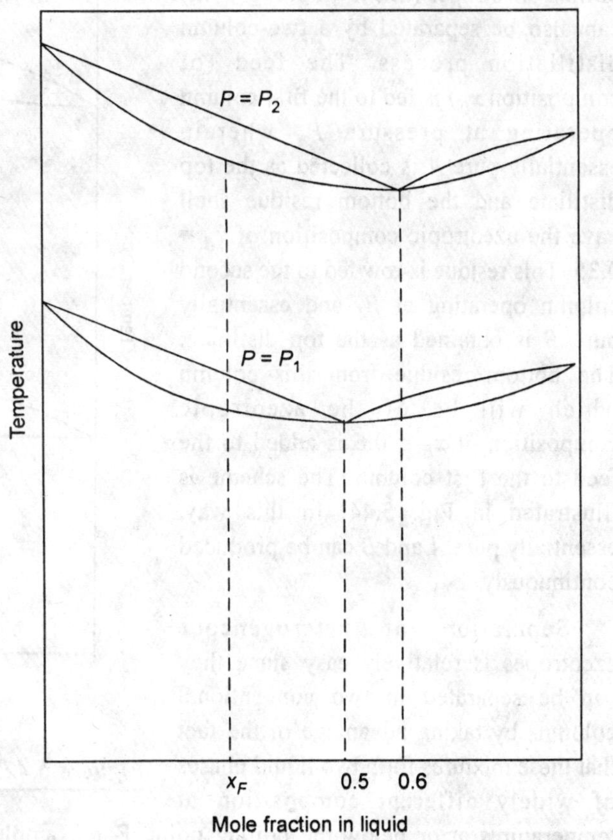

*Figure 5.41: Pressure sensitive minimum boiling azeotrope*

It has been stated earlier that azeotrope formation is very much dependent on the operating pressure. If the azeotrope composition is *pressure sensitive* or varies by at least 4 to 5 percent over a nominal change of total pressure, then separation can be affected by employing a two-column fractionation scheme. For example, consider a binary mixture containing $A$ (the more volatile component) and $B$ (the less volatile component). Now, let the azeotrope composition be 0.5 at pressure $P_1$ and 0.6 at pressure $P_2$ (see Fig. 5.41). In other words, the system forms an azeotrope containing 50 mole percent of $A$ when the operating pressure is $P_1$ and it forms an azeotrope containing 60 mole percent of $A$ when the operating pressure is $P_2$. Now, a feed mixture of composition $x_F$ (with respect to $A$) is fed to a column operating at $P_2$ which will discharge a bottom product containing essentially $B$ and a top distillate that will have the azeotropic composition of $x_D = 0.6$. This distillate which thus contains 60 percent $A$ and rest $B$ is then fed to a second column that operates at a total pressure $P_1$ and essentially pure $A$ is obtained as the bottom product. The top distillate from the second column which will have the azeotropic composition of $x_D = 0.5$, is added to the feed to the first column. The entire scheme is illustrated in Fig. 5.43.

In a similar way, the feed mixture that has a tendency to form maximum boiling azeotropes (as shown in Fig. 5.42) can also be separated by a two-column distillation process. The feed (of composition $x_F$) is fed to the first column operating at pressure $P_2$ wherein essentially pure $A$ is collected as the top distillate and the bottom residue shall have the azeotropic composition of $x_B$ = 0.35. This residue is row fed to the second column operating at $P_1$ and essentially pure $B$ is obtained as the top distillate. The bottom residue from this column which will be of the azeotropic composition of $x_B$ = 0.6 is added to the feed to the first column. The scheme is illustrated in Fig. 5.44. In this way, essentially pure $A$ and $B$ can be produced continuously.

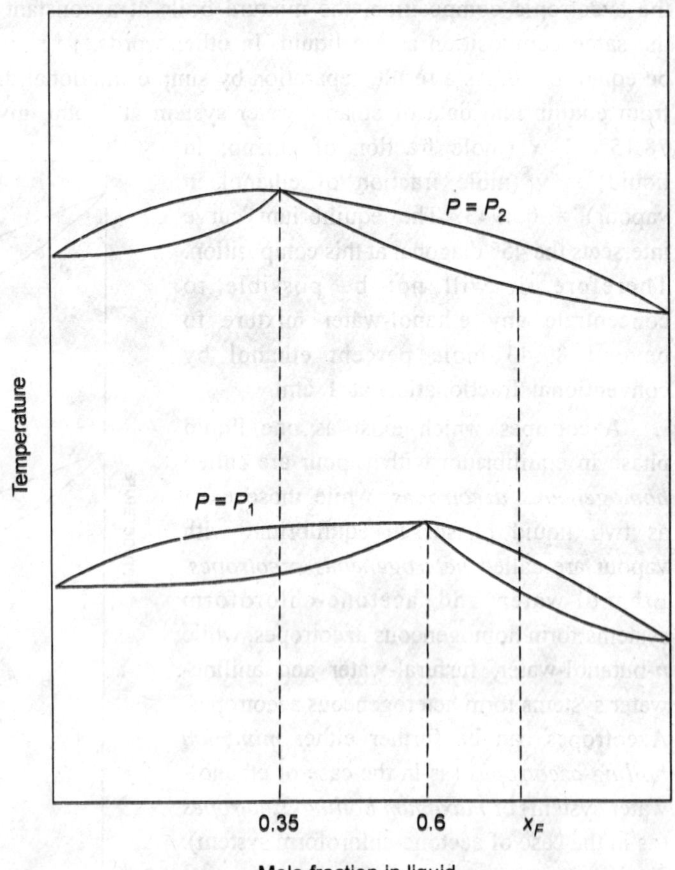


*Figuere 5.42: Pressure sensitive maximum boiling azeotrope*

Separation of heterogeneous azeotropes is relatively easy since they can be separated in two conventional columns by taking advantage of the fact that these mixtures form two liquid phases of widely different composition at temperatures at or below the boiling point. For example, Fig. 5.45 shows the temperature-composition diagram for *n*-butanol-water system at 1 atm pressure which forms a heterogeneous azeotrope at 92.25° C consisting of two liquid phases, one containing 3.0 mole percent butanol and the other containing 40 mole percent butanol, both of which are in equilibrium with vapour containing 25 mole percent butanol. Such a system can be separated by using two fractionators and a liquid-liquid separator (say, a decanter). The feed mixture containing less than 3.0 mole percent butanol is sent to the first fractionator (water column) which will discharge a bottom product containing essentially water and the overhead vapours which are of azeotropic composition are condensed totally (when they form two insoluble liquids of different composition) and sent to the decanter where they separate into two liquid layers. The water-rich layer is sent back to the water column as reflux. The butanol-rich layer is sent to the top plate of the second fractionator (butanol column) which contains only the stripping or exhausting section. The bottom product from the second fractionator will be essentially pure butanol and the overhead vapours from this column

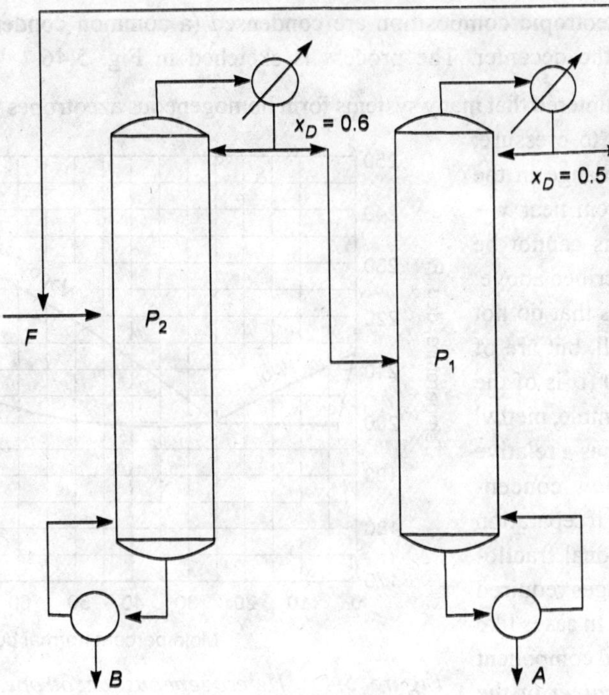


***Figure 5.43:*** *Separation scheme for minimum boiling homogeneous azeotrope*

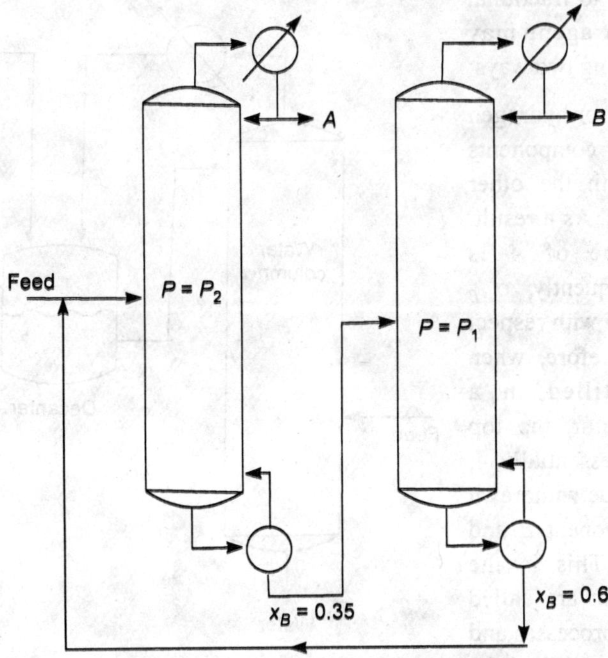


***Figure 5.44:*** *Separation scheme for maximum boiling homogeneous azeotrope*

which will also be of the azeotropic composition are condensed (a common condenser may be used for both columns) and sent to the decanter. The process is sketched in Fig. 5.46.

However, it is often encountered that many systems form homogeneous azeotropes which are not pressure sensitive or even if sensitive to pressure, do not produce significant change in the relative volatility of the system near $x = 1.0$ or $x = 0.0$. Such systems cannot be separated by the methods described above. This is also true with systems that do not form any true azeotrope at all, but are of relative volatility close to 1.0 ($\alpha$ is of the order of 1.01 to 1.05). For example, methyl cyclohexane-toluene system has a relative volatility less than 1.01 at low concentrations of toluene. As a result, if separation is to be affected by conventional fractionation, then the number of stages required will be uneconomically large. In cases like this, to affect separation, a third component which may be called the *entrainer* or the *solvent,* is added to the feed mixture and the system is then subjected to fractional distillation. The addition agent may function in any of the following two ways:

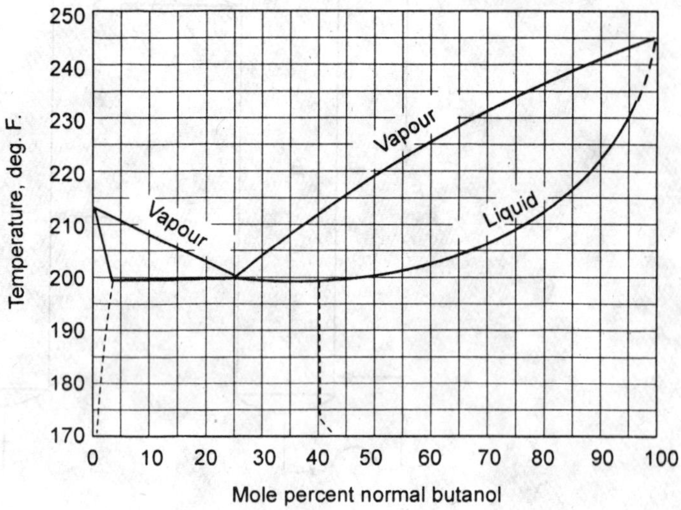


*Figure 5.45: Heterogeneous azeotrope. System: n-butanol-water at 1 atm*

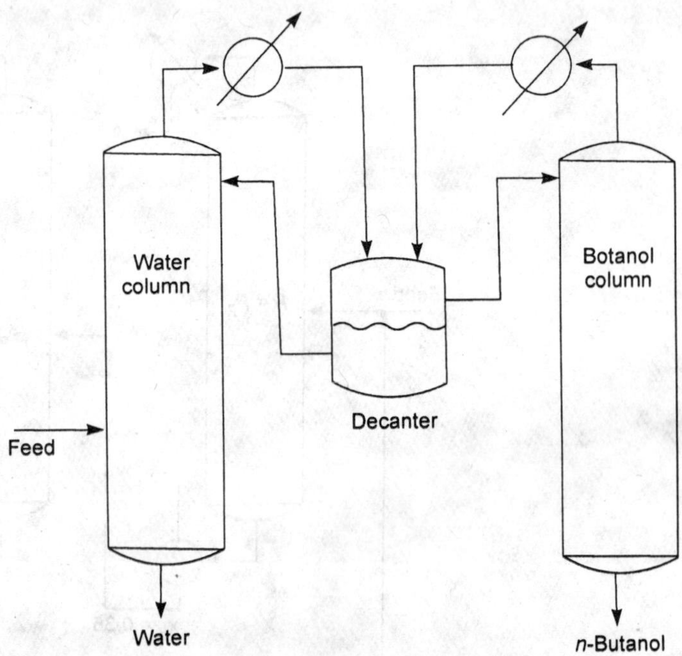


*Figure 5.46: Separation scheme for heterogeneous azeotrope*

(*i*) It may form a complex or hydrogen bond with one of the components (say, $A$), but not with the other component (that is, $B$). As a result, the vapour pressure of $A$ is decreased and consequently, $\alpha_{AB}$ (relative volatility of $A$ with respect to $B$) decreases. Therefore, when the system is distilled in a conventional fractionator, the top distillate will contain essentially $A$, while the bottom residue withdrawn will contain the component $B$ and the addition agent. This is the technique used in what are called *extractive distillation* processes and the addition agent is then called the *solvent.*

(*ii*) The added material may form a minimum boiling azeotrope with one of the components (say, *A*) and as a result, when the system is ultimately distilled in a conventional fractionator, the component *A* and the addition agent will be recovered as the top distillate, while the component *B*, having no change in volatility, will concentrate in the liquid and be removed in the bottom product. Alternately, it is also possible that the addition agent tends to break complexes or hydrogen bonds that have already been formed between *A* and *B* or between molecules of *A*. This, in turn, will increase the effective vapour pressure of *A* and $\alpha_{AB}$ increases. Consequently, when the system is distilled, *A* and the addition agent get collected in the distillate while *B* remains in the residue. This is precisely the basis of *azeotropic distillation* and the addition agent is then called the *entrainer*.

The basic principle behind azeotropic distillation and extractive distillation is, therefore, the same. The principal difference between the two lies in the fact that in extractive distillation, the solvent is recovered in the residue or bottoms, whereas the entrainer is almost completely recovered in the distillate in azeotropic distillation. The selection of an suitable and efficient solvent or entrainer is obviously the most detrimental parameter for these processes. The solvent or entrainer must be noncorrosive, nontoxic, inexpensive and easily available and must be nonreactive with the other components of the feed mixture. It must also be thermally stable and completely miscible with all components of the distilling system at the temperatures and concentrations in the column. Easy recovery and reuse of the solvent/entrainer is also an important factor that has a significant influence on the overall economy of the process. Recovery of solvent in extractive distillation processes is relatively easier since the solvent does not form any azeotrope and could be separated by simple distillation. However, in the case of azeotropic distillation, alternate processes such as solvent extraction may have to be employed for separating the entrainer from the distillate.

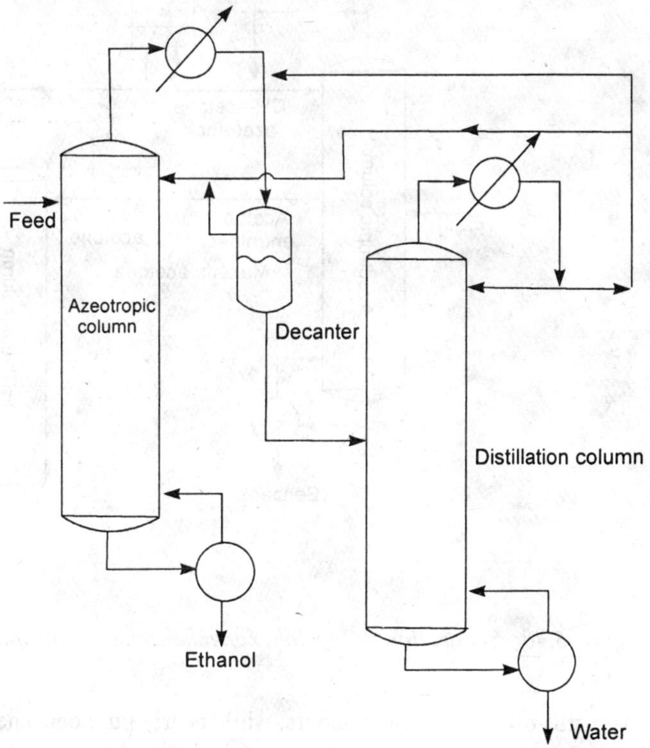


*Figure 5.47: Separation of ethanol-water azeotrope*

A few examples of industrial azeotropic distillation processes are sketched in Figs 5.47 and 5.48. These processes differ from each other based on the type of azeotrope formed and the method employed to recover the entrainer. In Fig. 5.47, dehydration of aqueous alcohol by azeotropic distillation with *n*-pentane as the entrainer is illustrated. Alternately, benzene or diethyl ether may be used as the entrainer. Pentane forms a heterogeneous, ternary, minimum boiling azeotrope with water and small amounts of ethanol and this distills off as overhead vapours from the azeotropic column, while pure ethanol is collected as the bottom product. The vapours when condensed form two insoluble liquid phases (characteristic of heterogeneous azeotropes) and they separate into two layers in the decanter. The top layer which is rich in pentane is returned to the column as reflux, while the bottom layer (containing around 75 mole percent water and 25 mole percent ethanol) is sent to a second fractionator. Almost pure water is collected as bottoms from the second column. The distillate from the second column is combined with the reflux from the decanter and sent to the azeotropic column.

Cyclohexane and benzene have very close boiling points at atmospheric pressure and they form a minimum boiling homogeneous azeotrope at 77.4° C and therefore, cannot be separated by simple distillation. Separation can be affected by *azeotropic distillation* with acetone as the entrainer as shown in Fig. 5.48. Acetone forms a minimum boiling, binary, homogeneous azeotrope with cyclohexane and

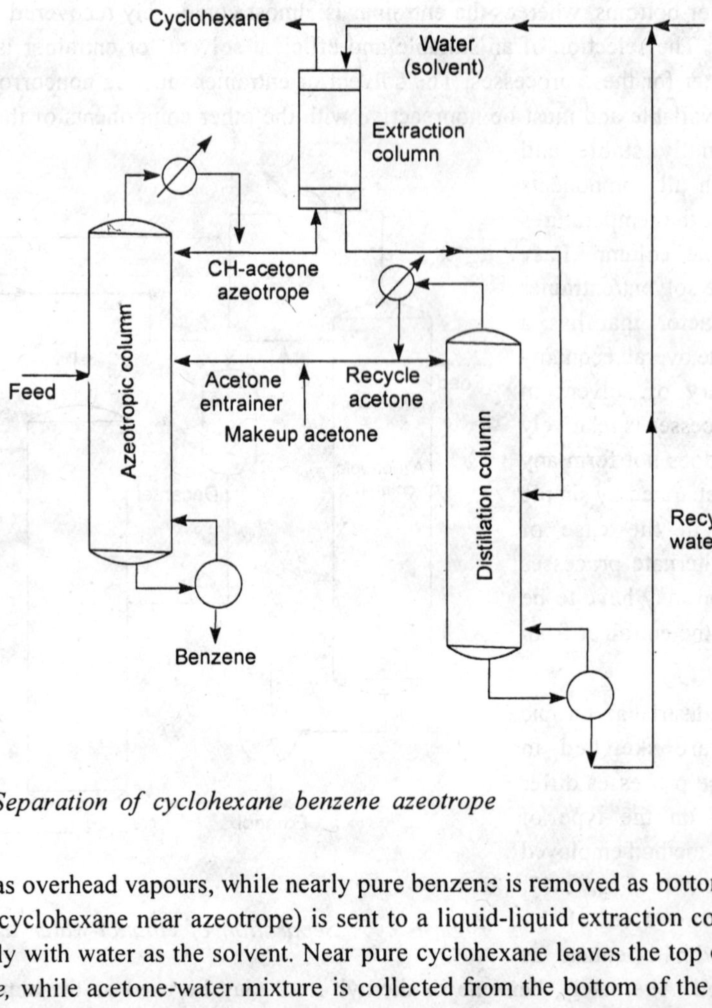


*Figure 5.48: Separation of cyclohexane benzene azeotrope*

this distils off as overhead vapours, while nearly pure benzene is removed as bottoms. The distillate (which is the acetone-cyclohexane near azeotrope) is sent to a liquid-liquid extraction column where it is treated countercurrently with water as the solvent. Near pure cyclohexane leaves the top of the extraction column as the *raffinate,* while acetone-water mixture is collected from the bottom of the extractor as the *extract.* This extract is now separated by simple distillation and the recovered acetone is recycled back to the azeotropic column, while water is recycled back to the top of the extraction column.

A large number of extractive distillation processes are in practice, particularly in the petrochemical industry. Separation of toluene from paraffins using phenol as the solvent is an excellent example. Phenol is a high-boiling solvent and has a boiling point higher than that of toluene or paraffins. From the extractive distillation column, paraffins are collected as the top distillate, while the bottoms product is toluene-phenol mixture. Phenol can be easily separated from this mixture by simple distillation and recycled for reuse (see Fig. 5.49). Another example is the separation of butenes from butanes using either furfural or a mixture containing 85 percent acetone and 15 percent water as the solvent. Many inorganic mixtures have also been successfully processed by extractive distillation. Extractive distillation of nitric acid-water system using sulfuric acid as the solvent is an example. The feed mixture is an azeotrope containing 62 percent nitric acid and 99 percent nitric acid is obtained as the distillate.

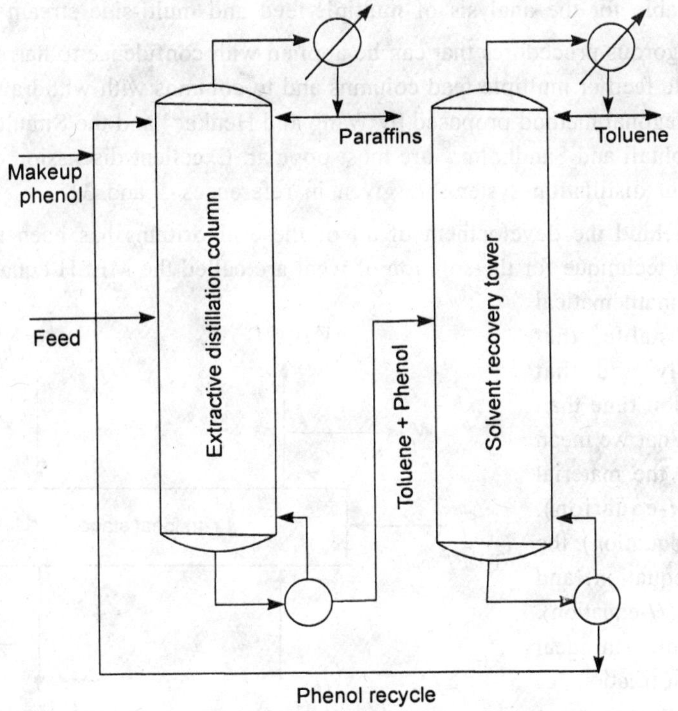


*Figure 5.49: Extractive distillation of toluene from paraffins using phenol as solvent*

## 5.2.8 Multicomponent Distillation

Distillation problems involving multicomponent feed mixtures cannot be solved by direct graphical methods such as those used for binary systems. A stage to stage or plate to plate computation becomes imminent in these cases and therefore, the overall computational load is significantly large. The wide acceptance of digital computers for the solution of multicomponent distillation problems is due to this.

Multicomponent distillation has attracted extensive research and a very large number of methods or algorithms have been proposed in the literature for the analysis of multicomponent distillation processes.

Earliest among them are the approximate methods such as the F-U-G (Fenske-Underwood-Gilliland) method[31,5] and the Smith-Brinkley Group method[25]. Though approximate, these methods are relatively simple and could be performed even on hand calculators. They work well for ideal systems. Improved methods have been later proposed by authors like Thiele and Geddes[26] and Lewis and Matheson[27]. The Thiele-Geddes (*TG*) method is a rating method in the sense that for given column specifications (such as the number of stages, feed plate location, reflux ratio, operating pressure, feed composition and thermal condition of feed), the method predicts the distillate and residue compositions that can be obtained. The Lewis-Matheson algorithm, on the other hand, is a design method in which the number of stages required to obtain specified component distribution in the distillate and the residue is determined. Both of these methods are accurate only for close-boiling mixtures and also to conventional fractionators which receive only one feed stream and discharge only two product streams such as the top distillate and the bottom residue. They are unreliable for the analysis of multiple-feed and multi-side stream columns.

Among the highly rigorous procedures that can be applied with confidence to narrow as well as wide boiling mixtures, to single feed or multiple feed columns and to columns with withdrawal of two or more product streams, the tridiagonal method proposed by Wang and Henke[28] and the Simultaneous Correction method proposed by Naphtali and Sandholm[29] are most popular. Excellent discussion on computer aided design of multicomponent distillation systems is given in references 5 and 31.

The basic attempt behind the development of all of these algorithms has been the formulation of an efficient mathematical technique for the solution of what are called the MESH equations. Each author has tried to propose a mathematical scheme that is more stable, that converges more rapidly and that requires lesser computation time than those proposed earlier. What we mean by MESH equations are the material balance equation (*M*-equation), equilibrium equation (*E*-equation), the summation equation (*S*-equation) and the heat balance equation (*H*-equation). For example, consider the $j$th ideal stage of a multicomponent fractionator as shown in Fig. 5.50. It receives a feed stream at the rate of $F_j$ moles per hour and the composition of the stream is expressed in terms of the mole fraction of any particular component (say, the $i$th component) in it (such as $z_{ij}$). Similarly, the vapour stream entering this stage from the stage below [that is, the $(j + 1)$th stage] is at $V_{j+1}$ moles/hour and the mole fraction of component $i$ in this stream is $y_{i,j+1}$. The liquid stream leaving the stage is at $L_j$ moles/hour and its composition is $x_{ij}$ and so on. $Q_j$ represents the heat removed from the stage in kJ/hr (if heat is being added, then $Q_j$ will be negative). A liquid side stream is withdrawn from this stage at the rate of $U_j$ moles/hour and a vapour sidestream at the rate of $W_j$ moles/hour. We have thus considered the most general case. If no feed is introduced at this stage, then $F_j = 0.0$, if the column

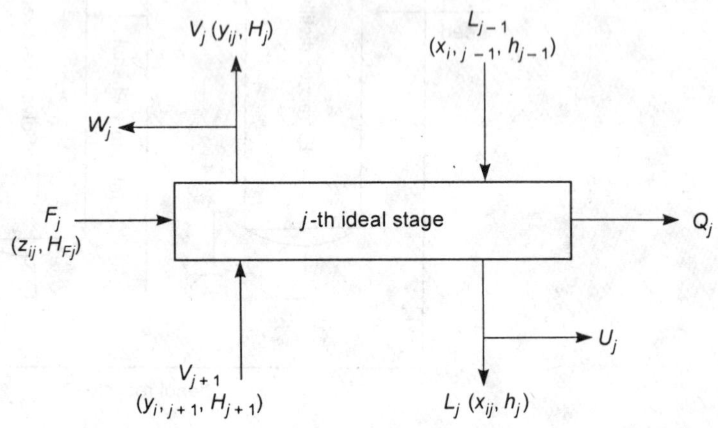


*Figure 5.50: Schematic of jth stage of a multicomponent fractionator*

operates adiabatically (except for the heat added from the reboiler or that removed in the overhead condenser), then $Q_j = 0.0$, if no sidestreams are withdrawn, then either $U_j$ or $W_j$ or both could be equal to zero.

The MESH equations can be now easily written down as follows:

*M*-equation:

$$L_{j-1}\, x_{i,\, j-1} - (L_j + U_j)\, x_{ij} + V_{j+1}\, y_{i,\, j+1} - (V_j + W_j)\, y_{ij} + F_j z_{ij} = 0.0 \quad \text{... (5.2.128)}$$

*E*-equation: $$(y_{ij}/x_{ij}) = K_{ij}$$

or $$y_{ij} - K_{ij} x_{ij} = 0.0 \quad \text{... (5.2.129)}$$

*S*-equation: $$\Sigma y_{ij} = 1.0$$

or $$\Sigma K_{ij} x_{ij} - 1.0 = 0.0 \quad \text{... (5.2.130)}$$

*H*-equation:

$$L_{j-1}\, H_{L,\, j-1} - (L_j + U_j)\, H_{L,\, j} + V_{j+1}\, H_{V,\, j+1} - (V_j + W_j)\, H_{V,\, j} + F_j H_{Fj} - Q_j = 0.0 \quad \text{... (5.2.131)}$$

where $H_{V,\, j}$ and $H_{L,\, j}$ represent enthalpy of vapour mixture and that of liquid solution respectively leaving the $j$th stage. Due to the large number of unknown variables and the complex interdependence of these equations, a direct analytical solution is practically impossible. MESH equations can be, therefore, solved only numerically and an iterative, trail and error procedure becomes invariable.

It may be noted that the *M*-equation and the *E*-equation may be combined so that all the *y*'s of the *M*-equation can be expressed in terms of *x*. Also, all the *L*'s can be expressed as functions of *V*'s by an overall mass balance of all stages from the condenser through the $j$th stage, as given below:

$$L_j = V_{j+1} + \sum_{k=1}^{j} \left(F_k - W_k - U_k\right) - D \quad \text{... (5.2.132)}$$

$$\text{for } 1 \le j \le N_p$$

It can be seen that the above equation is the same as Eq. (5.2.51) except that additional terms such as *F*, *W* and *U* have been included since we are considering a fractionator with multiple feeds and with multiple side streams. The *M*-equation, therefore, reduces to

$$A_j x_{i,\, j-1} + B_j x_{i,\, j} + C_j x_{i,\, j+1} = E_j \quad \text{... (5.2.133)}$$

where $$A_j = L_{j-1} = V_j + \sum_{k=1}^{j-1} \left(F_k - W_k - U_k\right) - D \quad \text{... (5.2.134)}$$

$$B_j = -[L_j + U_j + K_{ij}\,(V_j + W_j)]$$

$$= -[V_{j+1} + \sum_{k=1}^{j} \left(F_k - W_k - U_k\right) - D + U_j + K_{ij}\,(V_j + W_j)] \quad \text{... (5.2.135)}$$

$$C_j = (V_{j+1} K_{i,\, j+1}) \quad \text{... (5.2.136)}$$

$$E_j = -F_j z_{ij} \quad \text{... (5.2.137)}$$

The general procedure for the solution of MESH equations is, therefore, as follows:

(*i*) Assume the values of $V_j$ and $T_j$ ($1 \le j \le N_p$). In other words, the vapour flow rate and the temperature at each plate are assumed.

(*ii*) Compute the *K*-values (that is, values of equilibrium ratio) from Eq. (5.2.3) or (5.2.15) or (5.2.22). If *K*-values are being computed by a thermodynamic method [such as from Eq. (5.2.3)], then the *x*-values are also to be assumed in step (*i*). Often, $K_{ij}$ could be expressed as a function of temperature $T_j$ only, such as

$$K_{ij} = a_{1i} + a_{2i}T_j + a_{3i}T_j^2 + a_{4i}T_j^3 \quad \text{... (5.2.138)}$$

where the coefficients $a_{1i}$ to $a_{4i}$ are constants for a specific component.

(*iii*) Solve the *M*-equation, that is Eq. (5.2.133) to obtain the *x*-values. The *y*-values can be then obtained from the *E*-equation. The compositions of all streams leaving each plate are thus known.

(*iv*) Substitute the computed values of *x* and *K* in the *S*-equation and check whether the equation is satisfied. In other words, check whether the condition $\Sigma y_{ij} = 1.0$ is satisfied at each plate. If so, the assumed values of $V_j$ and $T_j$ are correct and are final. If not, computations are to be repeated with new values of $V_j$ and $T_j$. The usual practice is to solve the *S*-equation for $T_j$. It can be seen that if $K_{ij}$ are expressed as functions of $T_j$ only as in Eq. (5.2.138), then the *S*-equation will be a function of $T_j$ only and it can be solved for each plate using any of the numerical techniques such as the Newton-Raphson iterative method, the Regula Falsi method (also called method of false position), the method of successive bisection or Muller's quadratic interpolation technique[30].

(*v*) Compute the vapour and liquid enthalpies ($H_V$ and $H_L$) at the above-computed values of $T_j$ and compositions. Solve the *H*-equation for $V_j$. Note that all the *L*'s of *H*-equation can also be expressed in terms of *V*'s by using Eq. (5.2.132).

(*vi*) Repeat the computations starting from step (*ii*) with the newly computed values of $V_j$ and $T_j$. Continue the procedure until the values of $T_j$ (and thereby, $V_j$) computed from two successive iterations do not differ significantly.

As stated earlier, different authors have proposed different mathematical techniques for the solution of the *M*-equation in step (*iii*) and the *S*-equation in step (*iv*). It may be noted that the *M*-equation is to be written separately for each component and also for each stage. For example, if the number of ideal stages is ten ($j$ = 10) and the number of components in the feed mixture is six ($i$ = 6), then the *M*-equation will be constituted by a set of 60 equations (since $i \times j = 6 \times 10 = 60$) in 60 unknowns. These equations are to be solved simultaneously by using a suitably devised numerical technique such as Gauss elimination, Kramer's rule[30], etc. The *S*-equation (which is to be solved for each stage), on the other hand, could be highly nonlinear in $T_j$, particularly when a thermodynamic method, like Eq. (5.2.3), is employed for computing the *K*-values. As a result, its solution also demands a numerical technique that is quite stable and fast-converging. The choice of the method/algorithm, therefore, depends on the type of system handled, the computation time available, the degree of accuracy desired and the overall adaptability of the system parameters to the numerical scheme employed in the algorithm.

For a detailed coverage of multicomponent distillation, references (5), (6) and (31) may be consulted. For the interest of the readers, we shall discuss the *F-U-G method* here which, though approximate, predicts the order of magnitude of performance variables. This method can be conveniently executed on a desk *PC* or even on a programmable calculator (see Program 5.13 of Section 5.3).

The F-U-G method utilises Fenske's equation to estimate the number of theoretical plates at total reflux, Underwood's method to compute the minimum reflux ratio and Gilliland's method to estimate the total number of theoretical plates (ideal stages) required. We shall discuss these one by one.

Consider a fractionator that receives only one feed stream and discharges only the two product streams (the top distillate and the bottom residue). Let the feed contain $k$ number of components and let the mole fractions of these components in the feedstock in the order of decreasing volatility be $z_{Fi}$ $(1 \le i \le k)$. Let the feed be introduced at temperature $T_F$ and pressure $P_F$ and for the sake of simplicity, let us assume that the pressure of distillation $(P)$ is equal to $P_F$.

It is usual to specify the splits of any two components in one of the products (distillate or residue). For example, let it be desired that the distillate should contain $p_1$ percent of the $n$th component of the feed and $p_2$ percent of the $m$th component of the feed. In other words, if $z_{Dn}$ and $z_{Dm}$ are the mole fractions of the $n$th and $m$th components in the distillate, then

$$z_{Dn}\, D = (z_{Fn} F)\ (p_1/100) \qquad \text{... (5.2.139)}$$

$$z_{Dm}\, D = (z_{Fm} F)\ (p_2/100) \qquad \text{... (5.2.140)}$$

We have already discussed at the beginning of this section that it is conventional in multicomponent distillation problems to define a light key component and a heavy key component. The choice of these key components is crucial. The following procedure may be adopted for this purpose:

It has been reported by Shiras, Hanson and Gibson[44] that at minimum reflux, the following relationship is *approximately* obeyed:

$$Y_i = (z_{Di} D)/(z_{Fi} F) = \left(\frac{\alpha_i - 1}{\alpha_l - 1}\right)\left(\frac{z_{Dl} D}{z_{Fl} F}\right) + \left(\frac{\alpha_l - \alpha_i}{\alpha_l - 1}\right)\left(\frac{z_{Dh} D}{z_{Fh} F}\right) \qquad \text{... (5.2.141)}$$

where $z_{Di}$ = mole fraction of $i$th component in distillate

$\alpha_i$ = relative volatility of $i$th component with respect to the heavy key at temperature $T_{avg}$

$= (K_i/K_h)$ ... (5.2.142)

$K_i$ = equilibrium ratio of $i$th component at the specified temperature and pressure (here, at temperature $T_{avg}$ and pressure $P$)

$$T_{avg} = (T_D + T_B)/2 \qquad \text{... (5.2.143)}$$

$T_D$, $T_B$ = dew point of top distillate and bubble point of bottom residue respectively

The suffixes $l$ and $h$ stand for the light key and the heavy key respectively. The key components are first assumed. They may or may not be the components $n$ and $m$ whose splits have been specified at the outset. We now write Eq. (5.2.141) separately for $n$th component and $m$th component as given below:

$$Y_n = \left(\frac{\alpha_n - 1}{\alpha_l - 1}\right)\left(\frac{X_l}{z_{Fl} F}\right) + \left(\frac{\alpha_l - \alpha_n}{\alpha_l - 1}\right)\left(\frac{X_h}{z_{Fh} F}\right) \qquad \text{... (5.2.144)}$$

$$Y_m = \left(\frac{\alpha_m - 1}{\alpha_l - 1}\right)\left(\frac{X_l}{z_{Fl}F}\right) + \left(\frac{\alpha_l - \alpha_m}{\alpha_l - 1}\right)\left(\frac{X_h}{z_{Fh}F}\right) \quad \text{... (5.2.145)}$$

where $X_l = (z_{Dl}\,D)$

$X_h = (z_{Dh}\,D)$

$Y_n = (z_{Dn}\,D)/(z_{Fn}\,F) = 0.01\,p_1$ from Eq. (5.2.139)]

$Y_m = (z_{Dm}\,D)/(z_{Fm}\,F) = 0.01\,p_2$ [from Eq. (5.2.140)]

The above two equations, such as Eqs (5.2.144) and (5.2.145) are of the form

$$Y_n = A_1 X_l + B_1 X_h \quad \text{... (5.2.146)}$$

$$Y_m = A_2 X_l + B_2 X_h \quad \text{... (5.2.147)}$$

Solving the above two equations simultaneously for $X_l$ and $X_h$, we get

$$X_h = \frac{(Y_n A_2 - Y_m A_1)}{(A_2 B_1 - A_1 B_2)} \quad \text{... (5.2.148)}$$

$$X_l = (Y_n - B_1 X_h)/A_1 \quad \text{... (5.2.149)}$$

where $A_1 = \dfrac{(\alpha_n - 1)}{(\alpha_l - 1)(z_{Fl}F)}$, $B_1 = \dfrac{(\alpha_l - \alpha_n)}{(\alpha_l - 1)(z_{Fh}F)}$

$A_2 = \dfrac{(\alpha_m - 1)}{(\alpha_l - 1)(z_{Fl}F)}$, $B_2 = \dfrac{(\alpha_l - \alpha_m)}{(\alpha_l - 1)(z_{Fh}F)}$

Once the values of $X_l$ and $X_h$ have been computed, the values of $Y_i$ for all components ($1 \le i \le k$) can be computed from Eq. (5.2.141). If $Y_i$ is less than $-0.01$ or greater than 1.01, then the component $i$ is unlikely to distribute at minimum reflux. Thus, if the keys have been correctly assumed, then for all components lighter than the light key and heavier than the heavy key, $Y_i$ must be less than $-0.01$ or greater than 1.01. If this is not so, then keys are to be rechosen and the trial repeated. For all components between the light key and the heavy key (which will distribute), $Y_i$ will lie between 0.01 and 0.99.

For all undistributed components (components lighter than the light key or heavier than the heavy key), we may safely assume $(z_{Di}\,D) = (z_{Fi}\,F)$ if $Y_i$ is positive and $(z_{Di}\,D) = 0.0$, if $Y_i$ is negative.

It may be noted that the choice of the key components is not necessarily to be started with $l = 1$. The specifications given in the design problem often help to make a probable choice of the light key and computations could be started with that choice. It is also important to keep in mind that the above-discussed procedure based on Eq. (5.2.141) fails to be applicable if $n = h$.

Underwood's method[45] using which the minimum reflux ratio ($R_m$) can be computed approximately, involves solution of the following equation for $j$ values (where $j = h - l$) of the yet undefined factor $\phi$:

$$\sum_{i=1}^{k} \frac{\alpha_i\,(z_{Fi}F)}{(\alpha_i - \phi)} = F\,(1 - q) \quad \text{... (5.2.150)}$$

The parameter $q$ has already been defined earlier under Binary Distillation. If the feed is a saturated liquid, then $q = 1.0$ and if it is a saturated vapour, $q = 0.0$. If the feed is a mixture of vapour and liquid, then $q$ may be computed from the following equations:

$$\sum_{i=1}^{k} \frac{z_{Fi}\left[(L_F/V_F)+1\right]}{1+\left[(L_F/V_F)/K_{Fi}\right]} = 1.0 \qquad \text{... (5.2.151)}$$

$$q = (L_F/F) = \frac{(L_F/V_F)}{(L_F/V_F)+1} \qquad \text{... (5.2.152)}$$

where $K_{Fi}$ = equilibrium ratio of $i$th component ($K_i$) at temperature $T_F$ and pressure $P_F$.

The value of $(L_F/V_F)$ ratio that satisfies Eq. (5.2.151) is to be first computed by trial. It is then substituted in Eq. (5.2.152) to obtain the value of $q$.

Equation (5.2.150) can be solved for $(h - l)$ number of real, positive values of $\phi$ lying between $\alpha_l$ and $\alpha_h$. In other words, we determine the values of $\phi_j$ $(1 \le j \le h - l)$. Now, we can write the following equation separately for each value of $\phi_j$:

$$\sum_{i=1}^{h} \frac{\alpha_i\,(z_{Di}D)}{(\alpha_i - \phi)} = D\,(R_m + 1) \qquad \text{... (5.2.153)}$$

Thus, we get $j$ number of simultaneous equations. These are solved for the $j$ number of unknowns such as the minimum reflux ratio ($R_m$) and the $(z_{Di}D)$ values of components between the keys (that is, $i = l + 1,\ l + 2,\ .....\ h - 1$). It may be noted that for $(l + 1) \le i \le (h - 1)$, there will be $(h - l - 1)$ number of values of $(z_{Di}D)$ that are to be determined and by including $R_m$ also, the total number of unknowns to be evaluated becomes $(h - l)$. Since we have $(h - l)$ number of simultaneous equations, these unknowns can be easily evaluated. If any of the $(z_{Di}D)$ values so computed happens to be negative or exceeds $(z_{Fi}F)$, then it means that the keys have been wrongly chosen. In that case, the keys are to be reselected and the procedure repeated.

The total molar rate of distillate and that of residue can be now computed as follows:

$$D = \sum_{i=1}^{k} (z_{Di}D) \qquad \text{... (5.2.154)}$$

$$(x_{Bi}B) = (z_{Fi}F) - (z_{Di}D) \qquad \text{... (5.2.155)}$$

$$B = \sum_{i=1}^{k} (x_{Bi}B) \qquad \text{... (5.2.156)}$$

The distillate dew point ($T_D$) and the residue bubble point ($T_B$) are initially assumed and ultimately checked by verifying whether the following relationships are satisfied or not:

$$1.0 = \sum_{i=1}^{k} (z_{Di}/K_{Di}) \qquad \text{... (5.2.157)}$$

$$= \sum_{i=1}^{k} K_{Bi}\, x_{Bi} \quad \text{... (5.2.158)}$$

where $K_{Di}$ and $K_{Bi}$ are values of $K_i$ evaluated at $T_D$ and $T_B$ respectively. The average temperature $T_{avg}$ is now computed from Eq. (5.2.143) as

$$T_{avg} = (T_D + T_B)/2 \quad \text{... (5.2.143)}$$

This is checked against the initially assumed value of $T_{avg}$ and if the two are found to differ appreciably, the trial is repeated.

The entire computation procedure discussed above is summarised below:

(*i*) Assume $l$ and $h$ (that is, the light key and the heavy key).

(*ii*) Assume the distillate dew point $T_D$ and the residue bubble point $T_B$ and compute $T_{avg}$ from Eq. (5.2.143).

(*iii*) Compute $K_i$ values for $1 \le i \le k$ at temperature $T_{avg}$ and distillation pressure $P$.

(*iv*) Compute the relative volatilities (with respect to the heavy key) from Eq. (5.2.142).

(*v*) Compute $X_h$ and $X_l$ from Eqs (5.2.148) and (5.2.149).

(*vi*) Compute $Y_i$ $(1 \le i \le k)$ from Eq. (5.2.141).

(*vii*) Check whether $Y_i$ $(h < i \le k)$ and $Y_i$ $(1 \le i < l)$ values are less than $-0.01$ or greater than 1.01. If not, rechoose the keys and repeat the computations starting from step (*iv*). Verify also that $Y_i$ $(l + 1 \le i < h)$ values are between 0.01 and 0.99.

(*viii*) For $i < l$ and $i > h$, if $Y_i$ is negative, then $(z_{Di} D) = 0.0$ and if positive, $(z_{Di} D) = (z_{Fi} F)$. For the key components, the values of $(z_{Di} D)$ are given by $X_l$ and $X_h$ (computed earlier). For those components that distribute (components between the keys), the values of $(z_{Di} D)$ are being computed subsequently in step (*xiv*) using Underwood's method.

(*ix*) Specify the value of $q$ (which depends on the thermal condition of the feed).

(*x*) Solve Eq. (5.2.150) for the $(h - l)$ values of $\phi$ by trial. To start with, assume $\phi = (\alpha_h + 0.00001) = 1.00001$ (say). If $\alpha_l < \alpha_h$, assume $\phi = \alpha_l + 0.00001$.

(*xi*) Compute (LHS) and (RHS) of Eq. (5.2.150) and if they differ by more than 1 percent (this tolerance may be varied if necessary), increase $\phi$ by 0.00001 (say) and repeat the procedure. Continue the trials until a value of $\phi$ that satisfies Eq. (5.2.150) is obtained. Print this value of $\phi$ as $\phi_1$.

(*xii*) Increment $\phi$ and repeat the procedure until $(h - l)$ values of $\phi$ that satisfy Eq. (5.2.150) are obtained.

(*xiii*) Write Eq. (5.2.153) for each value of $\phi$ separately as given below:

$$\sum_{i=1}^{h} \frac{\alpha_i (z_{Di} D)}{(\alpha_i - \phi_1)} = D', \quad \sum_{i=1}^{h} \frac{\alpha_i (z_{Di} D)}{(\alpha_i - \phi_2)} = D', \ \ldots\ldots \ \sum_{i=1}^{h} \frac{\alpha_i (z_{Di} D)}{(\alpha_i - \phi_{h-l})} = D'$$

where $D' = D\,(R_m + 1)$.

(*xiv*) Solve the above equations simultaneously for $D'$ and $(z_{Di} D)$, $i = l + 1, l + 2, \ldots\ldots h - 1$.

(*xv*) If any of the above-computed values of $(z_{Di} D)$ is negative or exceeds $(z_{Fi} F)$, rechoose the keys and repeat the computations starting from step (*iv*).

(*xvi*) Compute $D$ and $B$ from Eqs (5.2.154) to (5.2.156). Also, obtain the values of $z_{Di}$ and $x_{Bi}$ since $z_{Di} = (z_{Di}D)/D$ and $x_{Bi} = (x_{Bi}B)/B$.

(*xvii*) Compute the dew point of the distillate ($T_D$) at pressure $P$ using Eq. (5.2.157). Similarly, compute the bubble point of the residue ($T_B$) at pressure $P$ using Eq. (5.2.158).

(*xviii*) Compute $T_{avg}$ from Eq. (5.2.143) and if this value of $T_{avg}$ differs appreciably from that assumed/computed earlier, repeat the computations with the newly computed value of $T_{avg}$ starting from step (*iii*).

It is to be kept in mind that Underwood's method discussed above assumes constant molal overflow and constant relative volatility at the mean column temperature. Though approximate, this method is still in use since in most of the engineering practices, only a rough estimate of $R_m$ is required.

Fenske's equation for the estimation of the minimum number of theoretical plates at total reflux has already been discussed under Binary-Distillation. For a multicomponent system, Fenske's equation is given by

$$N_m = \frac{\log\left[(z_{Dl}/z_{Dh})(x_{Bh}/x_{Bl})\right]}{\log(\alpha_{al})} \quad \text{... (5.2.159)}$$

where $N_m$ = minimum number of theoretical plates at total reflux (including reboiler and partial condenser)

$$\alpha_{al} = \sqrt{\alpha_{Dl}\,\alpha_{Bl}} \quad \text{... (5.2.160)}$$

$\alpha_{Dl}$, $\alpha_{Bl}$ = relative volatility of light key at $T_D$ and at $T_B$ respectively.

It must not be forgotten that Fenske's equation also assumes constant relative volatility and is basically developed for systems that exhibit nearly ideal behaviour.

Once the minimum reflux ratio ($R_m$) and the number of ideal stages at total reflux ($N_m$) are known, the total number of theoretical plates (ideal stages) required at the operating reflux ratio can be roughly estimated from the graphical correlation of Gilliland[46]. Gilliland has presented a plot of $C_N$ versus $C_R$, where $C_N$ and $C_R$ are respectively defined as

$$C_N = (N - N_m)/(N + 1) \quad \text{... (5.2.161)}$$

$$C_R = (R - R_m)/(R + 1) \quad \text{... (5.2.162)}$$

With allowable error, the plot may be fitted into a correlation of the following form:

$$C_N = a_0 + a_1 C_R + a_2 C_R^2 \quad \text{... (5.2.163)}$$

The values of the correlation constants are given in Table 5.7

For $C_R > 0.50$, the value of $C_N$ may be retrieved from Table (5.8). The data in the Table may be interpolated for intermediate values with permissible error.

It is also possible to estimate the feed plate location approximately using the equation devised by Kirkbride[47]. This equation is given below:

$$\log(m/p) = 0.206 \log\left[(B/D)(z_{Fh}/z_{Fl})(x_{Bl}/z_{Dh})^2\right] \quad \text{... (5.2.164)}$$

**Table 5.7**

| Range of $C_R$ | $a_0$ | $a_1$ | $a_2$ |
|---|---|---|---|
| $\leq 0.03$ | 0.70 | 0.0 | 0.0 |
| $0.03 \leq C_R \leq 0.07$ | 0.775 | $-2.50$ | 0.0 |
| $0.07 \leq C_R \leq 0.50$ | 0.69254 | $-1.377$ | 0.7836 |

**Table 5.8**

| $C_R$ : | 0.50 | 0.60 | 0.70 | 0.80 | 0.90 |
|---|---|---|---|---|---|
| $C_N$ : | 0.20 | 0.16 | 0.11 | 0.07 | 0.04 |

where $m$ = number of ideal stages above the feed plate

$p$ = number of ideal stages below the feed plate

For a computer program that deals with the application of F-U-G method, see Program 5.13 of Section 5.3.

**Example 5.29:** A hydrocarbon mixture containing (by mole) 5 percent *n*-propane, 15 percent *i*-butane, 25 percent *n*-butane, 20 percent *i*-pentane and 35 percent *n*-pentane is to be fractionated at 8.2 atm such that not more than 7.45 percent of *n*-pentane is present in the top distillate and not more than 7.6 percent of *n*-butane is present in the bottom residue. The feed is introduced at its bubble point and at 8.2 atm pressure. The column is equipped with a total condenser and a reboiler and the reflux is returned at its bubble point. The equilibrium ratio of each of these hydrocarbons at 8.2 atm may be assumed to be a function of temperature only as given below:

$$\log (K) = a_0 + a_1 T + a_2 T_2$$

where $T$ is the temperature in degree celsius. The values of $a_0$, $a_1$ and $a_2$ for different hydrocarbons are listed below:

| | $a_0$ | $a_1$ | $a_2$ |
|---|---|---|---|
| *n*-propane | $-0.04$ | 0.0076958 | $-0.00001204$ |
| *i*-butane | $-0.24819$ | 0.0051486 | 0.000006128 |
| *n*-butane | $-0.66733$ | 0.0102395 | $-0.0000122$ |
| *i*-pentane | $-0.94147$ | 0.008875 | 0.0 |
| *n*-pentane | $-2.34556$ | 0.0349 | $-0.0001242$ |

Based on 100 kmoles/hr of the feed and an operating reflux ratio of 2.58, estimate the number of theoretical plates required and the product compositions.

**Solution:** Since our basis is 100 kmoles/hr of the feed, $F$ = 100.0 kmole/hr. It is given that not more than 7.6 percent of the third component (namely, *n*-butane) must be present in the bottom residue or not more than 92.4 percent of the third component must be present in the top distillate. The top distillate is also not to contain more than 7.45 percent of the fifth component (namely, *n*-pentane). Thus, the *n*th and $m$th components (whose splits in the top distillate are specified) are respectively the third and the fifth component. In other words, $n = 3$, $p_1 = 92.4$ percent and $m = 5$, $p_2 = 7.45$ percent.

To begin with, we must first assume the key components. Thus, let $l = 3$ and $h = 5$. Next, we have to assume the value of $T_{avg}$. Let $T_{avg} = 90°$ C. The values of equilibrium ratio ($K_i$) for all components at temperature $T_{avg}$ are now computed from the correlation provided and the values of relative volatility ($\alpha_i$) from Eq. (5.2.142). The results are listed below:

| *Component* | *i* | $K_i$ | $\alpha_i$ |
|---|---|---|---|
| *n*-pentane | 1 | 3.59 | 5.8374 |
| *i*-butane | 2 | 1.84 | 2.992 |
| *n*-butane | 3 | $1.435 = K_l$ | $2.3252 = \alpha_l$ |
| *i*-pentane | 4 | 0.72 | 1.1706 |
| *n*-pentane | 5 | $0.615 = K_h$ | $1.0 = \alpha_h$ |

The values of $A_1$, $A_2$, $B_1$, $B_2$ are, therefore,

$$A_1 = 1.0/25.0 = 0.04$$

$$A_2 = 0.0 \text{ (since } \alpha_m = \alpha_5 = 1.0)$$

$$B_1 = 0.0 \text{ (since } \alpha_l = \alpha_n = \alpha_3)$$

$$B_2 = 1.0/35.0 = 0.02857$$

Therefore, from Eqs (5.2.148) and (5.2.149),

$$X_h = 2.6075$$

$$X_l = 23.1$$

Equation (5.2.141) now reduces to

$$Y_i = \frac{(\alpha_i - 1)}{(1.3252)}\frac{(23.1)}{(25.0)} + \frac{(2.3252 - \alpha_i)}{(1.3252)}\frac{(2.6075)}{(35.0)}$$

$$= 0.69725\ (\alpha_i - 1) + (2.3252 - \alpha_i)\ (0.0562179)$$

The values of $Y_i$ are, therefore,

$$Y_1 = 3.1754$$

$$Y_2 = 1.3514$$

$$Y_3 = Y_n = 0.924$$

$$Y_4 = 0.18386$$

$$Y_5 = Y_h = 0.0745$$

It is seen that the values of $Y_i$ for components below the light key (that is, the values of $Y_1$ and $Y_2$) are above 1.01 and the values of $Y_i$ for components between the keys (that is, the value of $Y_4$) lie between 0.01 and 0.99. Thus, the choice of the keys is correct. Now,

$$(z_{D1} D) = (z_{F1} F) = 5.0$$

$$(z_{D2} D) = (z_{F2} F) = 15.0$$

$$(z_{D3} D) = X_1 = 23.1$$

$$(z_{D5}\,D) = X_h = 2.6075$$

Since the feed is at its bubble point, $q = 1.0$. Eq. (5.2.150) now reduces to

$$\frac{(5.8374)(5.0)}{(5.8374-\phi)} + \frac{(2.992)(15.0)}{(2.992-\phi)} + \frac{(2.3252)(25.0)}{(2.3252-\phi)} + \frac{(1.1706)(20.0)}{(1.1706-\phi)} + \frac{(1.0)(35.0)}{(1-\phi)} = 0$$

or

$$\frac{(29.187)}{(5.8374-\phi)} + \frac{(44.88)}{(2.992-\phi)} + \frac{(58.13)}{(2.3252-\phi)} + \frac{(23.412)}{(1.1706-\phi)} + \frac{(35.0)}{(1-\phi)} = 0$$

The above equation is solved for two values of $\phi$ by trial and these values are $\phi = 1.0925$ and $\phi = 1.581$. For $\phi = 1.0925$, Eq. (5.2.153) reduces to

$$45.162 + 14.988\,(z_{D4}\,D) = D\,(R_m + 1)$$

For $\phi = 1.581$, Eq. (5.2.153) reduces to

$$106.351 - 2.8523\,(z_{D4}\,D) = D\,(R_m + 1)$$

Solving the above two equations simultaneously, we get

$$(z_{D4}\,D) = 3.4297$$

$$D\,(R_m + 1) = 96.5681$$

The above value of $(z_{D4}\,D)$ is neither negative nor exceeds $(z_{F4}\,F)$ and this further confirms that the chosen keys are correct. Since all the values of $(z_{Di}\,D)$ are now known, from Eq. (5.2.154),

$$D = (5.0 + 15.0 + 23.1 + 3.4297 + 2.6075)$$

$$= 49.1372 \text{ kmole/hr}$$

The values of $(x_{Bi}\,B)$ are now obtained from Eq. (5.2.155). The results are listed below:

| *Component* | $(z_{Di}\,D)$ | $z_{Di}$ | $(x_{Bi}\,B)$ | $x_{Bi}$ |
|---|---|---|---|---|
| 1 | 5.0 | 0.101756 | 0.0 | 0.0 |
| 2 | 15.0 | 0.30527 | 0.0 | 0.0 |
| 3 | 23.1 | 0.4701 | 1.90 | 0.037355 |
| 4 | 3.4297 | 0.0698 | 16.5703 | 0.325784 |
| 5 | 2.6075 | 0.053074 | 32.3925 | 0.636861 |
| | $D = 49.1372$ | 1.0 | $B = 50.8628$ | 1.0 |

Also, since $\quad D\,(R_m + 1) = 96.5681$

$$R_m = (96.5681/49.1372) - 1$$

$$= 0.9653$$

Thus, the minimum reflux ratio is very close to 1.0. Let us now verify the assumed value of $T_{avg}$ by computing the dew point of the top distillate ($T_D$) and the bubble point of the bottom residue ($T_B$). These are to be computed by trial. Thus, let $T_D = 72.5°$ C. The $K$-values at 72.5° C are now computed from the given correlation and the results are given below:

| *Component* | $K_{Di}$ | $(z_{Di}/K_{Di})$ |
|---|---|---|
| 1* | 2.84878 | 0.035719 |
| 2 | 1.43646 | 0.212515 |
| 3 | 1.02543 | 0.45844 |
| 4 | 0.50346 | 0.13864 |
| 5 | 0.34030 | 0.155962 |
| | | 1.001276 (checks) |

Since $\Sigma z_{Di}/K_{Di}$ is very close 1.0, the assumed value of $T_D$ is correct. Similarly, let $T_B$ = 107.2° C

| *Component* | $K_{Bi}$ | $(x_{Bi}K_{Bi})$ |
|---|---|---|
| 1 | 4.4323 | 0.0 |
| 2 | 2.3668 | 0.0 |
| 3 | 1.95049 | 0.07286 |
| 4 | 1.023128 | 0.33332 |
| 5 | 0.92989 | 0.59221 |
| | | 0.99839 (checks) |

Thus, the bubble point of bottom residue = 107.2° C. Now

$$T_{avg} = (72.5 + 107.2)/2$$

$$= 89.85° \text{ C}$$

Since this is very close to the initially assumed value of $T_{avg}$, no further trial is necessary.

The minimum number of theoretical plates (ideal stages) required at total reflux can be now estimated from Fenske's equation [Eq. (5.2.159)]. Thus

$$\alpha_{Dl} = (1.02543)/(0.34030) = 3.0133$$

$$\alpha_{Bl} = (1.95049)/(0.92989) = 2.0975$$

Therefore, $$\alpha_{al} = \sqrt{(3.0133)(2.0975)} = 2.514$$

Now, from Eq. (5.2.159), $$N_m = \frac{\log\left[\dfrac{(0.4701)(0.636861)}{(0.053074)(0.037355)}\right]}{\log(2.514)} = 5.44$$

Since the operating reflux ratio ($R$) is 2.58,

$$C_R = (2.58 - 0.9653)/(3.58) = 0.451$$

Now, from Eq. (5.2.163), $C_N = 0.69254 - 1.377\ (0.451) + 0.7836\ (0.451)^2 = 0.230875$

Therefore, $$N = (C_N + N_m)/(1 - C_N)$$

$$= (0.230875 + 5.44)/(1.0 - 0.230875) = 7.373 \simeq 7.5$$

This gives an approximate estimate of the total number of theoretical plates (ideal stages) required. Let us also find out the approximate location of the feed plate from Kirkbride's correlation [Eq. (5.2.164)]. Thus

$$\log (m/p) = 0.206 \log \left[\frac{(50.8628)(0.35)}{(49.1372)(0.25)}\left(\frac{0.037355}{0.053074}\right)^2\right] = -0.029653$$

or $$m/p = 0.934$$

We know, $m + p = 7.5$. Solving simultaneously,

$$m = 3.622$$

$$p = 3.878$$

Thus, the feed be introduced at the 3.6th ideal stage.

## 5.3 COMPUTER PROGRAMS

As in the previous Chapters, we shall present here a good number of sample computer programs for the design and analysis of industrial drying and distillation processes.

Let us first consider the performance of a cross-circulation tray dryer similar to the one discussed in Example 5.4. The computer program is given below:

### PROGRAM 5.1

```
C**    BATCH DRYING IN A CROSS CIRCULATION TRAY DRYER
C**    AIR DRYING OF WATER FROM SOLIDS CONSIDERED.
C**    DRYING TAKES PLACES WITHIN THE CONSTANT RATE PERIOD
       REAL LMD (30)
       REAL L1, KM, KS, KA, LHS, LMDS, MS
       DIMENSION TST (30), PST (30)
       READ (*, *) L1, B1, ZS, ZM, KM, DEL
       READ (*, *) TG1, Y1, P, VA, X1, X2, KS, ROWS
C**    PROPERTIES OF AIR SUCH AS SPECIFIC HEAT,
C**    VISCOSITY AND THERMAL CONDUCTIVITY ARE TAKEN
C**    EQUAL TO THOSE OF DRY AIR AT TEMPERATURE
C**    TG1 AND PRESSURE P.
       READ (*, *) CPA, MEUA, KA
C**    SINCE RATE OF DRYING IS CONSTANT, IT IS
C**    COMPUTED AT DRYER INLET ONLY.
       VH1 = (0.00283 + 0.00456 * Y1) * (TG + 273.0)/P
       ROWA1 = (1.0 + Y1)/VH1
       DC = 4.0 * (B1 * DEL)/[(B1 + DEL) * 2.0]
       G1 = VA * ROWA1
```

```
      RE = DC * G1/MEUA
      PR = CPA * MEUA/KA
C**   COMPUTE CONVECTION HEAT TRANSFER COEFFICIENT
C**   FROM THE HEAT TRANSFER ANALOGUE OF O'BRIEN
C**   AND STUTZMAN'S CORRELATION [EQ. (5.1.64)].
      HC = 0.11 * CPA * G1/[(RE * * 0.29) * PR * * 0.6666]
C**   HEAT TRANSFER THROUGH THE SIDES OF THE TRAY IS NEGLECTED.
      UK = (1.0/HC) + (ZM/KM) + (ZS/KS)
      UK = 1.0/UK
C**   HEAT TRANSFER BY RADIATION IS NEGLECTED.
      HT = HC + UK
      CS1 = 1.0 + 1.87 * Y1
C**   ASSUME THE TEMPERATURE OF DRYING SURFACE.
      TS = TG - 10.0
10    LHS = (1.0 + UK/HC) * (TG - TS)
C**   DETERMINE YS AND LMDS WITH THE HELP OF
C**   STANDARD STEAM TABLES.
      READ (*, *) (TST (I), PST (I), LMD (I), I = 1, 30)
      DO 15 I = 1, 30
      IF [TS. EQ. TST (I)] GO TO 20
      IF [TS. LT. TST (I)] TO TO 25
15    CONTINUE
20    LMDS = LMD (I)
      PA = PST (I)
      GO TO 30
25    LMDS = LMD (I - 1) + [LMD (I) - LMD (I - 1)] * [TS - TST (I - 1)]/
      [TST (I) - TST (I - 1)]
      PA = PST (I - 1) + [PST (I) - PST (I - 1)] * [TS - TST (I - 1)]/[TST
      (I) - TST (I - 1)]
      YS = 0.622 * PA/(P - PA)
      RHS = (YS - Y1) * LMDS/CS1
      CHECK FOR CONVERGENCE.
      DIF = ABS (LHS - RHS)/LHS
      IF (DIF. LT. 0.001) GO TO 35
      TS = TS - 0.1
      GO TO 10
C**   COMPUTE THE RATE OF DRYING.
35    A = L1 * B1
      RC = HT * (TG - TS)/LMDS
```

```
C**    COMPUTE THE MASS OF DRY SOLIDS IN THE TRAY. IT IS ASSUMED THAT THE
       VOID FRACTION OF PARTICLE BED IN THE TRAY IS MUCH LESS THAN UNITY.
       ROWL = 1000.0
       MS = A * ZS/[(1.0/ROWS) + (X1/ROWL)]
C**    COMPUTE TIME FOR DRYING.
       THC = MS * (X1 - X2)/(A * RC)
       WRITE (*, *) THC
       STOP
       END
```

*Note:* It is to be noted that both CS1 and LMDS must expressed in the same units such as in kJ/kg or J/kg. Since in the above program CS1 is computed in kJ/kg, LMDS must also be expressed in kJ/kg.

## PROGRAM 5.2

```
C**    BATCH DRYING IN A THROUGH CIRCULATION DRYER.
C**    AIR DRYING OF WATER FROM SOLIDS IS CONSIDERED.
C**    DRYING TAKES PLACE WITHIN THE CONSTANT RATE PERIOD.
       REAL LMD (30)
       REAL L, JD, LMDW, KY, NTG, MEUA
       DIMENSION TST (30), PST (30)
       READ (*, *) GS, P, TG, Y1, Z, X1, X2, ROWS, ROWAP
C**    IT IS ASSUMED THAT THE PARTICLE SIZE IS
C**    WITHIN 3.2 TO 20 MM AND THE BED DEPTH
C**    IS WITHIN 10.0 TO 64.0 MM.
C**    IF PARTICLE SHAPE IS KNOWN TO BE
C**    CUBICAL, THEN CASE = 1.0. IF PARTICLES ARE
C**    OF IRREGULAR SHAPE BUT OF KNOWN SPHERICITY,
C**    THEN CASE = 2.0.
       READ (*, *) CASE
       IF (CASE = 2.0) GO TO 15
       READ (*, *) L
       VP = L * * 3
       SP = 6.0 * L * L
       SPS = SP/VP
       DVS = 6.0/SPS
       DP = DVS
       GO TO 20
15     READ (*, *) PSI, VP
       PI = 3.14
```

```
      DV = (6.0 * VP/PI) * * 0.3333
      SPS = 6.0/(PSI * DV)
      DVS = 6.0/SPS
      DP = DVS
C**   COMPUTE VOIDAGE OF THE BED.
20    ABSN = 1.0 - ROWAP/ROWS
C**   COMPUTE THE WET BULB TEMPERATURE AND SATURATION
C**   HUMIDITY AT WET BULB TEMPERATURE.
      CS1 = 1.0 + 1.87 * Y1
C**   FIRST ASSUME THE VALUE OF TW.
      TW = TG - 0.1
      TT = TW
C**   DETERMINE LMDW AND YW USING STEAM TABLES.
      READ (*, *) [TST (I), PST (I), LMD (I), I = 1, 30]
25    DO 30 I = 1, 30
      IF [TW. EQ. TST (I)] GO TO 35
      IF [TW. LT. TST (I)] GO TO 40
30    CONTINUE
35    LMDW = LMD (I)
      PAW = PST (I)
      GO TO 45
40    LMDW = LMD (I - 1) + [LMD (I) - LMD (I - 1)] * [TW - TST (I - 1)]/
      [TST (I) - TST (I - 1)]
      PAW = PST (I - 1) + [PST (I) - PST (I - 1)] * [TW - TST (I - 1)]/[TST
      (I) - TST (I - 1)]
45    YW = 0.622 * PAW/(P - PAW)
C**   TO NOTE THAT LMDW MUST BE EXPRESSED IN KJ/KG.
      TW = TG - [LMDW * (YW - Y1)/CS1]
C**   CHECK FOR CONVERGENCE.
      DIF = ABS (TW - TT)/TT
      IF (DIF. LT. 0.001) GO TO 50
      TW = TT - 0.1
      TT = TW
      GO TO 25
C**   COMPUTE MAXIMUM RATE OF DRYING PER UNIT DRYING SURFACE.
50    RMX = GS * (YW - Y1)
C**   COMPUTE THE REYNOLDS NUMBER.
C**   SINCE HUMIDITY OF EXIT AIR IS NOT KNOWN, IT IS FIRST ASSUMED. FOR
C**   EXAMPLE, LET Y2 = 0.0.
      Y2 = 0.0
```

```
      YY = Y2
      YA = (Y1 + Y2)/2.0
      G = GS * (1.0 + YA)
C**   AVERAGE VISCOSITY OF AIR IS TAKEN EQUAL TO
C**   THAT OF DRY AIR AT TEMPERATURE TG AND PRESSURE P.
      READ (*, *) MEUA
55    RE = DP * G/MEUA
C**   COMPUTE MASS TRANSFER COEFFICIENT FROM EQUATION 5.1.80
      IF (RE. LE. 10300.0) JD = (20.4 / ABSN)/(RE * * 0.815)
      IF (RE. LE. 4000.0) JD = (2.06/ABSN)/(RE * * 0.575)
C**   FOR AIR DRYING OF WATER FROM SOLIDS, THE SCHMIDT
C**   NUMBER IS TAKEN EQUAL TO 0.6.
      SC = 0.6
      KY = JD * GS/(SC * * 0.6667)
      SIA = 6.0 * (1.0 - ABSN)/DP
C**   COMPUTE THE NUMBER OF TRANSFER UNITS.
      NTG = Z * KY * SIA/GS
C**   COMPUTE THE HUMIDITY OF EXIT AIR.
      Y2 = YW - [1.0/EXP (NTG)] * (YW - Y1)
      DIF = ABS (Y2 - YY)/Y2
      IF (DIF. LT. 0.001) GO TO 60
      YY = Y2
      YA = (Y1 + Y2)/2.0
      G = GS * (1.0 + YA)
      GO TO 55
C**   COMPUTE ACTUAL RATE OF DRYING PER UNIT DRYING SURFACE.
60    RC = RMX * [1.0 - 1.0/EXP (NTG)]
C**   COMPUTE THE TIME FOR DRYING.
      THC = ROWAP * Z * (X1 - X2)/RC
      WRITE (*, *) THC, RC, Y2
      STOP
      END
```

Readers are advised to take note of all the COMMENT statements given in the above program since they indicate the assumptions involved and thereby the restrictions of the procedure employed.

We shall now consider the design of a direct contact rotary dryer (with special reference to Example 5.9). The computer program is given below. The COMMENT statements in the program explain the stepwise procedure followed as well as the simplifying assumptions (if any) made.

## PROGRAM 5.3

```
C**   DESIGN OF A DIRECT CONTACT ROTARY DRYER
C**   FOR AIR DRYING OF WATER FROM SOLIDS.
C**   DRYER OPERATES COUNTERCURRENTLY AND DRYING
C**   TAKES PLACE WITHIN CONSTANT RATE PERIOD
C**   (CRITICAL MOISTURE CONTENT OF THE MATERIAL
C**   IS QUITE LOW).
C**   OPERATION OF DRYER IS ADIABATIC (HEAT
C**   LOSSES ARE NEGLECTED).
C**   DRYING TAKES PLACE ONLY IN THE MIDDLE ZONE.
C**   THE END ZONES ACT AS PREHEATING AND REHEATING
C**   ZONES.
      REAL K, LHS, M2, MS, MGM, N, NTG, NTG1, NTG2, NTG3, LMDR
      REAL LMD (30)
      DIMENSION TST (30), PST (30)
      READ (*, *) M2, X1, X2, TS1, TS2, DP, CPS, ROWS, SL
      READ (*, *) P, Y2, TG2, GMX
      MS = M2 * (1.0 - X2)
C**   ASSUME OUTLET TEMPERATURE OF AIR. FOR EXAMPLE,
C**   LET TG1 = TG2 - 1.0
      TG1 = TG2 - 1.0
10    TT = TG1
C**   HEAT OF WETTING (DHA) IS ASSUMED NEGLIGIBLE.
      DHA = 0.0
      TREF = 0.0
      LMDR = 2502.3 * 1000.0
      CPL = 4.18 * 1000.0
      HS1 = (CPS + X1 * CPL) * (TS1 - TREF) + DHA
      HS2 = (CPS + X2 * CPL) * (TS2 - TREF) + DHA
      CS2 = (1.0 + 1.87 * Y2) * 1000.0
      HG2 = CS2 * (TG2 - TREF) + LMDR * Y2
C**   COMPUTE HUMIDITY OF EXIT AIR BY SOLVING THE
C**   MASS BALANCE AND HEAT BALANCE EQUATIONS
C**   [EQS (5.1.85) AND (5.1.86)] SIMULTANEOUSLY.
      F = (X1 - X2)/(HS2 - HS1)
      Y1 = [F * (HG2 - TG1) + Y2]/[1.87 * TG1 + (2502.3) * F + 1.0]
      CS1 = (1.0 + 1.87 * Y1) * 1000.0
      HG1 = CS1 * (TG1 - TREF) + LMDR * Y1
      RS = (Y1 - Y2)/(X1 - X2)
```

```
C**    ASSUME TGP1 (THAT IS, TEMPERATURE OF AIR AT
C**    THE INLET TO THE DRYING ZONE, NAMELY ZONE II)
C**    FOR EXAMPLE, LET TGP1 = TG2 - 1.0.
       TGP1 = TG2 - 1.0
30     TTP = TGP1
       HGP1 = CS2 * (TGP1 - TREF) + LMDR * Y2
       HSP1 = HS2 - (HG2 - HGP1)/RS
C**    IT IS ASSUMED THAT HEAT TRANSFER TO THE SOLIDS
C**    IS BY CONVECTION ONLY (RADIATION FROM DRYER
C**    WALL IS NEGLECTED) SO THAT TSP1 = TSP2 =
C**    TWP. ALSO, SINCE THE OPERATION IS ADIABATIC,
C**    THE WET BULB TEMPERATURE OF AIR REMAINS
C**    CONSTANT AT TWP IN ZONE II (DRYING ZONE).
       TSP1 = (HSP1 - DHA)/(CPS + CPL * X2) + TREF
       TWP = TSP1
C**    DETERMINE LMDW AND YW USING STEAM TABLES
       READ (*, *) (TST (I), PST (I), LMD (I), I = 1, 30)
       DO 35 I = 1, 30
       IF [TWP. EQ. TST (I)] GO TO 40
       IF [TWP. LT. TST (I)] GO TO 45
35     CONTINUE
40     LMDW = LMD (I)
       PAW = PST (I)
       GO TO 50
       LMDW = LMD (I - 1) + [LMD (I) - LMD (I - 1)] * [TW - TST (I - 1)]/
       [TST (I) - TST (I - 1)]
       PAW = PST (I - 1) + [PST (I) - PST (I - 1)] * [TW - TST (I - 1)]/[TST
       (I) - TST (I - 1)]
50     YWP = 0.622 * PAW/(P - PAW)
       TGP1 = TWP + LMDW * (YWP - Y2)/CS2
C**    CHECK FOR CONVERGENCE (ON THE VALUE OF TGP1)
       DIF = ABS (TTP - TGP1)
       IF (DIF. LT. 0.01) GO TO 55
       TGP1 = TTP - 0.01
       GO TO 30
C**    COMPUTE TGP2 (THAT IS, TEMPERATURE OF AIR LEAVING ZONE II)
55     TGP2 = TWP + LMDW * (YWP - Y1)/CS1
       HGP2 = CS1 * (TGP2 - TREF) + LMDR * Y1
       TSP2 = TWP
```

```
      HSP2 = (CPS + CPL * X1) * (TSP2 - TREF) + DHA
      HG1 = HGP2 - RS * (HSP2 - HS1)
      TG1 = (HG1 - LMDR * Y1)/CS1 + TREF
C**   CHECK FOR CONVERGENCE (ON THE VALUE OF TG1)
      DIF = ABS (TG1 - TT)
      IF (DIF. LT. 0.01) GO TO 60
      TG1 = TT - 0.01
      GO TO 10
C**   COMPUTE THE REQUIRED CROSS SECTIONAL AREA OF DRYER
60    MGM = (MS/RS) * (1.0 + Y1)
      A = MGM/GMX
      PI = 3.14
      D = SQRT (4 * A/PI)
      S = MS/A
      GS = S/RS
C**   COMPUTE THE NUMBER OF TRANSFER UNITS FOR
C**   ZONE I (PREHEATING ZONE)
      DT1 = TGP2 - TSP2
      DT2 = TG1 - TS1
      DTM1 = (DT1 - DT2)/ALOG (DT1/DT2)
      NTG1 = (TGP2 - TG1)/DTM1
C**   COMPUTE THE NUMBER OF TRANSFER UNITS FOR
C**   ZONE II (DRYING ZONE).
      DT1 = TGP1 - TSP1
      DT2 = TGP2 - TSP2
      DTM2 = (DT1 - DT2)/ALOG (DT1/DT2)
      NTG2 = (TGP1 - TGP2)/DTM2
C**   COMPUTE THE NUMBER OF TRANSFER UNITS FOR
C**   ZONE III (REHEATING ZONE).
      DT1 = TG2 - TS2
      DT2 = TGP1 - TSP1
      DTM3 = (DT1 - DT2)/ALOG (DT1/DT2)
      NTG3 = (TG2 - TGP1)/DTM3
C**   COMPUTE THE TOTAL NUMBER OF GAS PHASE TRANSFER UNITS.
      NTG = NTG1 + NTG2 + NTG3
C**   COMPUTE THE VOLUMETRIC HEAT TRANSFER
C**   COEFFICIENT FROM EQ. (5.1.92).
      G = GS * [1.0 + (Y1 + Y2) * 0.5]
      UV = (237.0 * G * * 0.67)/D
```

```
C**    COMPUTE THE HEIGHT OF A GAS PHASE TRANSFER UNIT.
       CSA = (CS1 + CS2)/2.0
       HTG = GS * CSA / UV
C**    COMPUTE THE REQUIRED LENGTH OF THE DRYER.
       Z = HTG * NTG
C**    TO CHECK THAT Z/D RATIO IS WITHIN 4.0 TO 10.0.
C**    COMPUTE THE DRYER SPEED BASED ON A
C**    CHOSEN VALUE OF SOLID HOLDUP. USUALLY,
C**    PHIL = 0.05 TO 0.15.
       PHIL = 0.075
       K = 0.6085/[ROWS * SQRT (DP)]
       PHILO = PHIL - K * GA
       N = (0.3344 * S)/(PHILO * ROWS * D * SL)
       N = N * * (1.0/0.9)
       WRITE (*, *) A, D, Z, N, SL
       STOP
       END
```

In engineering, one of the earliest applications of computer software is in the analysis of distillation problems (more precisely, multicomponent distillation problems). Distillation is a unit operation the analysis of which often demands a fundamental thermodynamic approach. As a result, the use of computer software has become most attractive in the simulation and analysis of distillation systems. We shall present a good number of examples here. At the outset, let us consider computation of equilibrium ratio (or, equilibrium distribution coefficient) based on a well-established themodynamic equation of state such as the Redlich-Kwong-Soave equation of state. The program (for a binary system) is given below:

## PROGRAM 5.4

```
C**    COMPUTATION OF EQUILIBRIUM RATIO FOR BINARY
C**    SYSTEM FROM REDLICH-KWONG-SOAVE EQUATION
C**    OF STATE.
       REAL K (2)
       DIMENSION TC (2), TR (2), PC (2), PV (2), OM (2), X (2), Y (2), W
       (2), C (2, 2)
       DIMENSION A (2), B (2), S (2), BETA (2), AML (2), AMV (2), AIB (2),
       AK (2), AX (2), AIBL (2), AIBV (2)
       DIMENSION YY (3)
       READ (*, *) P, T
       READ (*, *) [TC (I), PC (I), OM (I), I = 1, 2]
C**    TO NOTE THAT SUFFIX 1 STANDS FOR THE MORE
C**    VOLATILE COMPONENT.
```

```
C**   SPECIFY THE VALUES OF THE BINARY INTERACTION
C**   PARAMETER. FOR HYDROCARBON SYSTEMS, THESE
C**   VALUES REDUCE TO ZERO.
      READ (*, *) [(C (I, J), I = 1, 2), J = 1, 2]
C**   COMPUTE A (I), B (I), S (I), BETA (I) FROM EQS (5.2.5), (5.2.6),
      (5.2.9) AND (5.2.10).
      DO 10 I = 1, 2
      R = 8314.4
      A (I) = {0.4278 * [R * TC (I)] * * 2]/PC (I)
      B (I) = 0.0867 * R * TC (I)/PC (I)
      S (I) = 0.48508 + 1.55171 * OM (I) - 0.15613 * OM (I) * * 2
      TR (I) = T/TC (I)
      BETA (I) = {1.0 + S (I) * [1.0 - SQRT (TR (I))]} * * 2
10    CONTINUE
C**   AS A FIRST APPROXIMATION, ASSUME THAT
C**   THE X-VALUES ARE EQUAL TO THOSE
C**   COMPUTED FROM RAOULT'S LAW.
      READ (*, *) [PV (I), I = 1, 2]
      X (1) = [P - PV (2)]/[PV (1) - PV (2)]
      X (2) = 1.0 - X (1)
C**   THE FIRST APPROXIMATION OF K-VALUES ARE ALSO TAKEN EQUAL TO THOSE
      FROM RAOULT'S LAW.
      K (1) = PV (1)/P
      K (2) = PV (2)/P
20    DO 15 I = 1, 2
      AX (I) = X (I)
      AK (I) = K (I)
      Y (I) = K (I) * X (I)
15    CONTINUE
C**   COMPUTE BM FROM EQ. (5.2.7).
      BML = 0.0
      BMV = 0.0
      DO 25 I = 1, 2
      BML = BML + X (I) * B (I)
25    BMV = BMV + Y (I) * B (I)
C**   COMPUTE AM FROM EQ. (5.2.8). TO NOTE
C**   THAT FOR A BINARY SYSTEM, THIS EQUATION
C**   REDUCES TO THE SIMPLIFIED FORM GIVEN BELOW.
C**   PHASE = 1.0
30    DO 35 I = 1, 2
```

```
      IF [PHASE. EQ. (1.0)] W (I) = X (I)
35    IF [PHASE. EQ. (2.0)] W (I) = Y (I)
      AM = BETA (1) * A (1) * W (1) * * 2 + BETA (2) * A (2) * W (2) * *2
      AM = AM + W (1) * W (2) * [2.0 - C (1, 2) - C (2, 1)]* SQRT [A (1)
      * A (2) * BETA (1) * BETA (2)]
      IF (PHASE. EQ. 1.0) THEN
      AMLIQ = AM
      PHASE = 2.0
      GO TO 30
      ELSE
      AMVAP = AM
      END IF
C**   COMPUTE AIB FROM EQ. (5.2.11).
      DO 40 I = 1, 2
      PHASE = 1.0
45    IF [PHASE. EQ. (1.0)] W (I) = X (I)
      IF [PHASE. EQ. (2.0)] W (I) = Y (1)
      AIB (I) = 0.0
      DO 50 J = 1, 2
      IF [PHASE. EQ. (1.0)] W (J) = X (J)
      IF [PHASE. EQ. (2.0)] W (J) = Y (J)
      AIB (I) = AIB (I) + W (J) * [1.0 - C (I, J)] * SQRT [BETA (I) * A (I)
      * BETA (J) * A (J)]
50    CONTINUE
      IF [PHASE. EQ. (1.0)] THEN
      AIBL (I) = AIB (I)
      PHASE = 2.0
      GO TO 45
      ELSE
      AIBV (I) = AIB (I)
      END IF
40    CONTINUE
C**   COMPUTE AA AND BB FROM EQS (5.2.12) AND (5.2.13).
      AAL = AMLIQ * P/(R * T) * * 2
      AAV = AMVAP * P/(R * T) * * 2
      BBL = BML * P/(R * T)
      BBV = BMV * P/(R * T)
C**   COMPUTE THE COMPRESSIBILITY FACTOR Z BY
C**   SOLVING THE CUBICAL EQ. (5.2.14).
      PHASE = 1.0
```

```
55    IF [PHASE. EQ. (1.0)] THEN
      AA = AAL
      BB = BBL
      ELSE
      AA = AAV
      BB = BBV
      END IF
C**   ALGORITHM FOR THE SOLUTION OF CUBIC
C**   EQUATION OF THE FORM Z^3 + (CB) Z^2 + (CC) Z -
C**   (CD) = 0 WHERE CB, CC, CD ARE NUMERICAL CONSTANTS
      CB = -1.0
      CC = AA - BB - BB * BB
      CD = -AA * BB
      CP = (3.0 * CC - CB * * 2)/3.0
      CQ = (270.0 * CD - 9.0 * CB * CC + 2.0 * CB * * 3)/27.0
      CR = (CP/3.0) * * 3 + (CQ/2.0) * * 2
      C1 = [-0.5 * CQ + SQRT (CR)] * * 0.333
      C2 = [-0.5 * CQ - SQRT (CR)] * * 0.333
      IF (CR) 60, 65, 70
60    THETA = SQRT [-27.0 * CQ * CQ/(4.0 * CP * * 3)]
      THETA = ARCOS (THETA)
      PI = 3.14
      THETA = THETA * 180.0/3.14
      I (CQ. LT. 0.0) F1 = 1.0
      IF (CQ. GT. 0.0) F1 = -1.0
      DO 75 I = 1, 3
      YY (I) = F1 * 2.0 * SQRT (- CP/3.0) * COSD [THETA/3.0 + 120.0 *
      (I - 1)]
      IF [YY (I). LT. 0.0) YY = 0.0
75    CONTINUE
      GO TO 80
65    YY (1) = C1 + C2
      IF [YY (1). LT. 0.0] YY (1) = 0.0
      YY (2) = -(C1 + C2)/2.0
      IF [YY (2). LT. 0.0] YY (2) = 0.0
      YY (3) = YY (2)
      GO TO 80
70    YY (1) = C1 + C2
      YY (2) = YY (1)
      YY (3) = YY (1)
```

```
80    Z1 = YY (1) - CB/3.0
      Z2 = YY (2) - CB/3.0
      Z3 = YY (3) - CB/3.0
      IF [PHASE. EQ. (1.0)] THEN
      ZL = AMIN1 (Z1, Z2, Z3)
      PHASE = 2.0
      GO TO 55
      ELSE
      ZV = AMAX1 (Z1, Z2, Z3)
      END IF
C**   COMPUTE THE FUGACITY COEFFICIENTS (BOTH IN
C**   LIQUID AND VAPOUR PHASES) FROM EQ. (5.2.4)
      DO 85 I = 1, 2
      PHASE = 1.0
90    IF [PHASE. EQ. (1.0)] THEN
      BM = BML
      AM = AMLIQ
      BB = BBL
      AA = AAL
      Z = ZL
      AIB (I) = AIBL (I)
      ELSE
      BM = BMV
      AM = AMVAP
      BB = BBV
      AA = AAV
      Z = ZV
      AIB (I) = AIBV (I)
      END IF
      PHI (I) = [B (I)/BM) * (Z - 1.0) - ALOG (Z - BB)
      PHI (I) = PHI (I) - (AA/BB) * ALOG (1.0 + BB/Z) * [2.0 * AIB (I)/
      AM - B (I)/BM]
      PHI (I) = EXP [PHI (I)]
      IF [PHASE. EQ. (1.0)] THEN
      PHIL (I) = PHI (I)
      PHAE = 2.0
      GO TO 90
      ELSE
      PHIV (I) = PHI (I)
      END IF
```

```
85      CONTINUE
C**     COMPUTE THE EQUILIBRIUM RATIO.
        DO 95 I = 1, 2
        K (I) = PHIL (I)/PHIV (I)
95      CONTINUE
C**     CHECK FOR CONVERGENCE.
        DIF = 0.0
        DO 100 I = 1, 2
        DIF = DIF + {[K (I) - AK (I)]/K (I)} * * 2
        IF (DIF. GT. E - 06) GO TO 20
C**     COMPUTE THE X-VALUES FROM THE SUMMATION EQUATION.
        X (1) = [1.0 - K (2)]/[K (1) - K (2)]
        X (2) = 1.0 - X (1)
C**     CHECK FOR CONVERGENCE.
        DIF = 0.0
        DO 105 I = 1, 2
105     DIF = DIF + {[X (I) - AX (I)]/X (I)} * * 2
        IF (DIF. GT. E - 06) GO TO 20
        WRITE (*, *) [X (I), Y(I), K (I), I = 1, 2]
        STOP
        END
```

The above program can be conveniently extended to a multi component system of known liquid phase composition (*x*-values known). The index *i* (*I* in program) now varies from 1 to *N* (where *N* is the number of components). Since the values of $x_i$ ($i$ = 1 to $N$) are known, the equilibrium temperature* ($T$) and the *K*-values are computed by trial. For the convenience of the readers, the gist of the program is outlined below:

1. The REAL and DIMENSION statements are same as in Program 5.4 except that the subscript = *N*. To note that the subscript of *YY* will remain as 3.
2. The READ statements will be

```
        READ (*, *) P, N
        READ (*, *) [X (I), TC (I), PC (I), OM (I), I = 1, N]
        READ (*, *) {[C (I, J), I = 1, N], J = 1, N}
```

3. Assume the values of *K* (*i*) as in Program 5.4 :

```
        READ (*, *) [PV (I), I = 1, N]
        DO 11 I = 1, N
11      K (I) = PV (I)/P
```

4. Assume *T*. It is better to start with *T* = boiling point of the most volatile component and then increase it subsequently if the required convergence criterion is not satisfied.

---

* *T* is nothing but the bubble point of the mixture.

```
      T = .....
20    DO 15 I = 1, N
      AT = T
      AK (I) = K (I)
      Y (I) = K (I) * X (I)
15    CONTINUE
```

5. Compute *A* (*I*), *B* (*I*), *S* (*I*) and BETA (*I*) as in Program 5.4.
6. Compute BM as in Program 5.4.
7. Compute AM for the multicomponent system from Eq. (5.2.8) as given below:

```
      DO 35 I = 1, N
      PHASE = 1.0
30    IF [PHASE. EQ. (1.0)] W (I) = X (I)
      IF [PHASE. EQ. (2.0)] W (I) = Y (I)
      AM (I) = 0.0
      DO 36 J = 1, N
      IF [PHASE. EQ. (1.0)] W (I) = X (I)
      IF [PHASE. EQ. (2.0)] W (I) = Y (I)
      AM (I) = AM (I) + W (I) * W (J) * [1.0 - C (I, J)] * SQRT [BETA
      (I) * A (I) * BETA (J) * A (J)]
36    CONTINUE
      IF [PHASE. EQ. (1.0)] THEN
      AML (I) = AM (I)
      PHASE = 2.0
      GO TO 30
      ELSE
      AMV (I) = AM (I)
      END IF
35    CONTINUE
      AMLIQ = 0.0
      AMVAP = 0.0
      DO 37 I = 1, N
      AMLIQ = AMLIQ + AML (I)
      AMVAP = AMVAP + AMV (I)
37    CONTINUE
```

8. Compute AIB as in Program 5.4
9. Compute AA, BB as in Program 5.4
10. Compute *Z* by solving the cubic equation as in Program 5.4.
11. Compute the fugacity coefficients PHIL (*I*) and PHIV (*I*) as in Program 5.4.
12. Compute *K* (*I*) and check for convergence as in Program 5.4.

13. Compute $\Sigma y_i$ and check the assumed value of $T$ as given below:

```
        SUM = 0.0
        DO 101 I = 1, N
        Y (I) = K (I) * X (I)
        SUM = SUM + Y (I)
101     CONTINUE
        IF [ABS (SUM - 1.0). GT. 0.001] THEN
        T = T + 0.01
        GO TO 20
        ELSE
        WRITE (*, *) [X (I), Y (I), K (I), I = 1, N]
        STOP
        END
```

As stated earlier, the temperature ($T$) that has been computed in the above program is nothing but the bubble point of the multicomponent mixture.

Let us now discuss a computer program for solving a flash distillation problem (involving a multicomponent mixture, as in Example 5.16). The program is given below:

## PROGRAM 5.5

```
C**   FLASH DISTILLATION OF A MULTICOMPONENT MIXTURE.
C**   FOR WRITING THE TYPE STATEMENTS, WE
C**   ARE REQUIRED TO KNOW THE VALUE OF N
C**   (THAT IS, THE NUMBER OF COMPONENTS). FOR
C**   EXAMPLE, LET N = 10.
      REAL K (10)
      DIMENSION XF (10), XB (10), YD (10), A0 (10), A1 (10), A2 (10),
      A3 (10)
      READ (*, *) P, T, N
      READ (*, *) [XF (I), I = 1, N]
C**   THE K-VALUES ARE COMPUTED FROM THE
C**   EMPIRICAL CORRELATION OF AMUNDSON AND
C**   PONTINEN (GIVEN IN EXAMPLE 5.14).
      READ (*, *) [AO (I), A1 (I), A2 (I), A3 (I), I = 1, N]
C**   ASSUME THE VALUE OF RFB (THAT IS, THE F/B RATIO). IF
C**   THERE IS NO IDEA REGARDING THE ORDER OF
C**   MAGNITUDE OF RFB, THEN ASSUME RFB = 1.001.
C**   THIS IS BECAUSE THE MINIMUM POSSIBLE
C**   VALUE OF RFB IS 1.0.
      RFB = 1.001
```

```
C**     COMPUTE THE VALUE OF QP (THAT IS, THE B/D RATIO)
10      QP = 1.0/(RFB - 1.0)
        SUM = 0.0
        DO 15 I = 1, N
        K (I) = A0 (I) + A1 (I) * T + A2 (I) * T ** 2 + A3 (I) * T ** 3
        XB (I) = XF (I) * (1.0 + QP)/[K (I) + QP]
        SUM = SUM + XB (I)
15      CONTINUE
C**     CHECK FOR CONVERGENCE
        DIF = ABS (SUM - 1.0)
        IF (DIF. LT. 0.00001) GO TO 20
        RFB = RFB + 0.001
        GO TO 10
C**     COMPUTE THE DISTILLATE COMPOSITION.
20      DO 25 I = 1, N
        YD (I) = (1.0 + QP) * XF (I) - QP * XB (I)
25      CONTINUE
C**     COMPUTE RD (THAT IS, THE D/F RATIO).
        RD = 1.0/(1.0 + QP)
        WRITE (*, *) (RD, XB (I). YD (I), I = 1, N)
        STOP
        END
```

We shall now proceed to consider differential distillation of a multicomponent system. Let us assume that the system forms ideal solutions so that Eq. (5.2.39) shall be applicable. The operation may be at constant temperature (isothermal weathering) or at constant pressure. The program for isothermal weathering is given below (Program 5.6). Constant pressure operation is dealt in Program 5.7, given subsequently.

## PROGRAM 5.6

```
C**     ISOTHERMAL WEATHERING OF A MULTICOMPONENT LIQUID MIXTURE.
C**     IT IS ASSUMED THAT THE SYSTEM FORMS IDEAL LIQUID SOLUTIONS.
C**     FOR WRITING TYPE STATEMENTS, WE ARE REQUIRED
C**     TO KNOW THE VALUE OF N (THAT
C**     IS, THE NUMBER OF COMPONENTS). FOR EXAMPLE,
C**     LET N = 10.
        REAL K (10)
        DIMENSION XF (10), XB (10), YD (10), PV (10)
        READ (*, *) T, PP, N
        READ (*, *) [XF (I), PV (I), I = 1, N]
```

```
C**   CHOOSE THE KEY COMPONENT. LET IT BE THE
C**   KTH COMPONENT.
      REAL (*, *) K
C**   ASSUME THE VALUE OF RFB. IF THE ORDER OF
C**   MAGNITUDE OF RFB IS NOT KNOWN, ASSUME
C**   RFB = 1.001. THIS IS BECAUSE THE MINIMUM
C**   POSSIBLE VALUE OF RFB IS 1.0.
      RFB = 1.001
C**   ASSUME THE MOLE FRACTION OF KEY COMPONENT IN THE RESIDUE. LET IT BE
      0.0001.
10    XB (K) = 0.0001
C**   SOLVE EQ. (5.2.39) FOR ALL VALUES OF XB (I).
15    SUM 1 = 0.0
      DO 20 I = 1, N
      ALFA (I) = PV (I)/PV (K)
      XB (I) = RFB * XF (I) * EXP {-ALFA (I) * ALOG [RFB * XF (K)/XB (K)]}
      SUM1 = SUM1 + XB (I)
20    CONTINUE
C**   CHECK FOR THE ASSUMED VALUE OF XB (K).
      DIF = ABS (SUM1 - 1.0)
      IF (DIF. LT. 0.00001) GO TO 25
      XB (K) = XB (K) + 0.0001
      GO TO 15
C**   CHECK FOR THE ASSUMED VALUE OF RFB.
25    SUM2 = 0.0
      DO 30 I = 1, N
      SUM2 = SUM2 + XB (I) * PV (I)
30    CONTINUE
      DIF = ABS (SUM2 - PP)/PP
      IF (DIF. LT. 0.0001) GO TO 35
      RFB = RFB + 0.001
      GO TO 10
C**   COMPUTE THE VALUE OF RB (THAT IS, THE B/F RATIO)
35    RB = 1.0/RFB
C**   COMPUTE THE DISTILLATE COMPOSITION.
      DO 40 I = 1, N
      YD (I) = [XF (I) - RB * XB (I)]/(1.0 - RB)
40    CONTINUE
      WRITE (*, *) (RB, RFB, XB (I), YD (I), I =1, N)
      STOP
      END
```

## PROGRAM 5.7

```
C**   DIFFERENTIAL DISTILLATION OF A MULTI-COMPONENT
C**   MIXTURE AT CONSTANT PRESSURE.
C**   IT IS ASSUMED THAT THE SYSTEM FORMS
C**   IDEAL SOLUTIONS.
C**   FOR WRITING THE TYPE STATEMENTS, WE ARE
C**   REQUIRED TO KNOW THE VALUE OF N (THAT IS,
C**   THE NUMBER OF COMPONENTS). FOR EXAMPLE, LET
C**   N = 10.
      REAL K (10)
      DIMENSION XF (10), XB (10), YD (10), PVA (10), PVB (10), PVF (10),
      PV (10, 100), T (100)
      READ (*, *) P, N
      READ (*, *) [XF (I), I = 1, N]
C**   COMPUTE TBF (THAT IS, THE BUBBLE POINT OF FEED).
C**   FIRST, ASSUME TBF. FOR EXAMPLE, LET TBF = BOILING
C**   POINT OF MOST VOLATILE COMPONENT.
      TBF = .....
C**   THE VAPOUR PRESSURE TABLE IS ASSUMED
C**   AVAILABLE. THIS TABLE LISTS VAPOUR PRESSURE
C**   VALUES OF ALL COMPONENTS AT DIFFERENT
C**   TEMPERATURES RANGING FROM THE BOILING
C**   POINT OF THE MOST VOLATILE COMPONENT TO
C**   THE BOILING POINT OF THE LEAST VOLATILE COMPONENT.
      READ (*, *) {[T (J), PV (I, J), I = 1, N], J = 1,100}
10    SUM = 0.0
      DO 15 I = 1, N
      DO 20 J = 1, 100
      IF [T(J). EQUATION TBF] GO TO 25
      IF [T (J). LT. TBF] GO TO 20
      PVF (I) = PV (I, J - 1) + [PV (I, J) - PV (I, J - 1)] * [TBF - T (J
      - 1)]/[T (J) - T (J - 1)]
      K (I) = PVF (I)/P
      SUM = SUM + K (I) * XF (I)
      GO TO 15
25    PVF (I) = PV (I, J)
      K (I) = PVF (I)/P
      SUM = SUM + K (I) * XF (I)
      GO TO 15
```

```
20    CONTINUE
15    CONTINUE
      DIF = ABS (SUM - 1.0)
      IF (DIF. LT. 0.00001) GO TO 30
      TBF = TBF + 0.01
      GO TO 10
C**   ASSUME TBW (THAT IS, THE BUBBLE POINT OF
C**   THE RESIDUE). FOR EXAMPLE, LET TBW = BOILING
C**   POINT OF THE LEAST VOLATILE COMPONENT.
30    TBW = .....
35    TAV = (TBW + TBF)/2.0
C**   COMPUTE THE VAPOUR PRESSURE VALUES AT TEMPERATURE, TAV.
      DO 40 I = 1, N
      DO 45 J = 1, 100
      IF [T (J). EQUATION TAV] GO TO 50
      IF [T (J). LT. TAV] GO TO 45
      PAV (I) = PV (I, J - 1) + [PV (I, J) - PV (I, J - 1)] * [TAV - T (J
      - 1)]/[T (J) - T (J - 1)]
      GO TO 40
50    PVA (I) = PV (I, J)
      GO TO 40
45    CONTINUE
40    CONTINUE
C**   SELECT THE KEY COMPONENT. LET IT BE
C**   THE M-TH COMPONENT.
      READ (*, *) M
C**   ASSUME THE VALUE OF RFB (THAT IS, THE
C**   F/B RATIO). IF THE ORDER OF MAGNITUDE
C**   OF RFB IS NOT KNOWN, ASSUME RFB = 1.001.
C**   THIS IS BECAUSE THE MINIMUM POSSIBLE
C**   VALUE OF RFB IS 1.0.
      RFB = 1.001
C**   ASSUME THE MOLE FRACTION OF KEY COMPONENT
C**   (M-TH COMPONENT) IN THE RESIDUE.
C**   LET IT BE 0.0001.
55    XB (M) = 0.0001
C**   SOLVE EQ. (5.2.39) FOR ALL VALUES OF XB (I).
60    SUM1 = 0.0
      DO 65 I = 1, N
```

```
      ALFA (I) = PVA (I)/PVA (M)
      XB (I) = RFB * XF (I) * EXP {-ALFA (I) * ALOG [RFB * XF (M)/XB (M)]}
      SUM1 = SUM1 + XB (I)
65    CONTINUE
C**   CHECK FOR THE ASSUMED VALUE OF XB (M).
      DIF = ABS (SUM1 - 1.0)
      IF (DIF. LT. 0.00001) GO TO 70
      XB (M) = XB (M) + 0.0001
      GO TO 60
C**   CHECK FOR THE ASSUMED VALUE OF RFB.
70    SUM2 = 0.0
      DO 75 I = 1, N
      YD (I) = [RFB * XF (I) - XB (I)]/(RFB - 1.0)
      SUM2 = SUM2 + YD (I)
75    CONTINUE
      DIF = ABS (SUM2 - 1.0)
      IF (DIF. LT. 0.00001) GO TO 80
      RFB = RFB + 0.001
      GO TO 55
C**   CHECK FOR THE ASSUMED VALUE OF TBW BY USING THE FACT THAT AT T = TBW,
      ΣK (I) XB (I) = 1.0
80    SUM3 = 0.0
      DO 85 I = 1, N
      DO 90 J = 1, 100
      IF [T (J). EQUATION TBW] GO TO 95
      IF [T (J). LT. TBW] GO TO 90
      PVB (I) = PV (I, J - 1) + [PV (I, J) - PV (I, J - 1)] * [TBW - T
      (J - 1)]/[T (J) - T (J - 1)]
K (I) = PVB (I)/P
      SUM3 = SUM3 + K (I) * XB (I)
      GO TO 85
95    PVB (I) = PV (I, J)
      K (I) = PVB (I)/P
      SUM3 = SUM3 + K (I) * XB (I)
      GO TO 85
90    CONTINUE
85    CONTINUE
      DIF = ABS (SUM3 - 1.0)
      IF (DIF. LT. 0.00001) GO TO 100
      TBW = TBW - 0.01
```

```
      GO TO 35
100   RB = 1.0/RFB
      RD = 1.0 - RB
      WRITE (*, *) [RB, RD, RFB, XB (I), YD (I), I = 1, N]
      STOP
      END
```

Estimation of the required number of stages for the continuous fractionation of a binary mixture can be accomplished by the method of McCabe and Thiele (which nevertheless assumes constant molal overflow) or by the more rigorous method of Sorel or Ponchon and Savarit. In either case, the graphical method could be recommended for hand-calculation. With the help of a personal computer however, the same can be achieved analytically with relative ease. The analytical procedure is more reliable since it is devoid of the personal error involved in the graphical procedure. The computer program for the analytic computation of the number of equilibrium stages based on the method of McCabe and Thiele is given below (Program 5.8). The program includes computation of minimum reflux ratio as well. In the program, ANS is the number of ideal stages which could be a fractional number and NS is its value after rounding off to the nearest larger integer.

## PROGRAM 5.8

```
C**   COMPUTATION OF NUMBER OF EQUILIBRIUM
C**   STAGES REQUIRED FOR THE CONTINUOUS
C**   FRACTIONATION OF A BINARY MIXTURE.
C**   CONSTANT MOLAL OVERFLOW ASSUMED IN
C**   BOTH ENRICHING AND STRIPPING SECTIONS.
C**   IT IS ALSO ASSUMED THAT THE REFLUX IS
C**   RETURNED AT ITS BUBBLE POINT.
      REAL L, LP, LO, LMDF
C**   A SUBSCRIPT OF 100 IS USED IN THE
C**   DIMENSION STATEMENT SINCE THE NUMBER OF
C**   STAGES IS UNLIKELY TO EXCEED 100.
      DIMENSION X (100), Y (100), YE (50), XE (50)
      READ (*, *) F, ZF, ZD, XB
C**   THE EQUILIBRIUM DATA IS ASSUMED AVAILABLE.
      READ (*, *) [YE (K), XE (K), K = 1, 50]
      D = F * (ZF - XB)/(ZD - XB)
      B = F - D
C**   IF TOTAL CONDENSER IS USED, THEN COND =
C**   1.0 AND IF A PARTIAL CONDENSER IS USED,
C**   THEN COND = 2.0. IF REBOILER IS BEING
C**   USED, THEN STEAM = 1.0 AND IF OPEN
C**   STEAM IS BEING EMPLOYED, THEM STEAM = 2.0.
```

```
      READ (*, *) COND, STEAM
C**   SPECIFY THE VALUE OF Q (WHICH DEPENDS
C**   ON THE FEED CONDITION).
C**   IF FEED IS A SATURATED LIQUID, THEN FEED
C**   = 1.0; IF SATURATED VAPOUR, THEN
C**   FEED = 2.0; IF COLD LIQUID, FEED = 3.0;
C**   IF SUPERHEATED VAPOUR, FEED = 4.0 AND
C**   IF VAPOUR-LIQUID MIXTURE, FEED = 5.0.
      READ (*, *) FEED
      IF (FEED. EQ. 1.0) Q = 1.0
      IF (FEED. EQ. 2.0) Q = 0.0
      IF (FEED. LE. 2.0) GO TO 20
      IF (FEED. EQ. 3.0) GO TO 10
      IF (FEED. EQ. 4.0) GO TO 15
      READ (*, *) FL
      Q = FL
      GO TO 20
10    READ (*, *) TF, CPL, LMDF, TBF
      Q = 1.0 + CPL * (TBF - TF)/LMDF
      GO TO 20
15    READ (*, *) TF, CPV, LMDF, TDF
      Q = CPV * (TDF - TF)/LMDF
C**   COMPUTE THE MINIMUM REFLUX RATIO. FOR THIS,
C**   FIRST FIND THE POINT OF INTERSECTION OF
C**   THE Q-LINE AND THE EQUILIBRIUM CURVE, (XQ, YQ).
20    XQ = XB + 0.001
25    YQ = [Q/(Q - 1.0)] * XQ - ZF/(Q - 1.0)
      YY = YQ
      DO 30 K = 1, 50
      UF [XE (K). GE. XQ) GO TO 35
30    CONTINUE
35    YQ = YE (K - 1) + [YE (K) - YE (K - 1)] * [XQ - XE (K - 1)]/[XE (K)
      - XE (K - 1)]
      DIF = ABS (YY - YQ)
      IF (DIF. LT. 0.001) GO TO 40
      XQ = XQ + 0.001
      GO TO 25
40    RMIN = (ZD - YQ)/(YQ - XQ)
C**   SPECIFY THE OPERATING REFLUX RATIO. FOR
```

```
C**   EXAMPLE, LET IT BE 1.5 TIMES THE MINIMUM.
      R = 1.5 * RMIN
      L0 = R * D
      V1 = L0 + D
      L = L0
      V = V1
      LP = L + Q * F
      VP = LP - B
C**   COMPUTE THE POINT OF INTERSECTION OF
C**   THE Q-LINE WITH THE OPERATING LINE OF
C**   THE ENRICHING SECTION, (XA, YA), FROM EQS
C**   (5.2.69A) AND (5.2.69B).
      XA = [(R + 1) * ZF + (Q - 1.0) * ZD]/(R + Q)
      YA = (R * ZF + Q * ZD)/(R + Q)
      IF = (COND. EQ. 1.0) THEN
      X0 = ZD
      Y (1) = X0
      ELSE
      YD = ZD
      D0 50 K = 1, 50
      IF [YE (K). GE. YD] GO TO 55
50    CONTINUE
55    X0 = XE (K - 1) + [XE (K) - XE (K - 1)] * [YD - YE (K - 1)]/[YE (K)
      - YE (K - 1)]
      Y (1) = [R/(R + 1.0)] * X0 + ZD/(R + 1.0)
      END IF
C**   COMPUTE X (1) FROM EQUILIBRIUM DATA.
      DO 60 K = 1, 50
      IF [YE (K). GE. Y (1)] GO TO 65
60    CONTINUE
65    X (1) = XE (K - 1) + [XE (K) - XE (K - 1)] * [Y (1) - YE (K - 1)]/
      [YE (K) - YE (K - 1)]
      J = 1.0
70    IF [X (J). GT. XA] THEN
      Y (J + 1) = [R/(R + 1.0)] * X (J) + ZD/(R + 1.0)
      ELSE
      IF [STEAM. EQ. (1.0)] THEN
      Y (J + 1) = (LP/VP) * X (J) - (B * XB/VP)
      ELSE
      Y (J + 1) = (LP/VP) * [X (J) - XB]
```

```
      END IF
      END IF
C**   COMPUTE X (J + 1) FROM EQUILIBRIUM DATA.
      DO 75 K = 1, 50
      IF [YE (K). GE. Y (J + 1)] GO TO 80
75    CONTINUE
80    X (J + 1) = XE (K - 1) + [XE (K) - XE (K - 1)] * [Y (J + 1) - YE (K
      - 1)]/[YE (K) - YE (K - 1)]
      IF [X (J + 1). LE. XB] GO TO 85
      J = J + 1
      GO TO 70
85    ANS = J + [X (J + 1)/XB]
      NS = J + 1
      WRITE (*, *) [X (J), Y (J), J = 1, NS]
      WRITE (*, *) ANS, NS
      STOP
      END
```

We shall now present a computer program for the rating of a multifeed fractionator (similar to Example 5.24). The product compositions from an available fractionator that receives two feed streams (of known compositions) but discharges as usual only two product streams (the top distillate and the bottom residue) are to be computed. Since both $z_D$ and $x_B$ are unknown, the solution demands a trial and error procedure. The program is given below (Program 5.9).

## PROGRAM 5.9

```
C**   RATING OF A MULTIFEED FRACTIONATOR
C**   (COMPOSITIONS OF PRODUCT STREAMS FROM
C**   AN AVAILABLE FRACTIONATOR ARE TO BE COMPUTED).
C**   CONSTANT MOLAL OVERFLOW ASSUMED.
C**   THE FEED IS A BINARY MIXTURE AND
C**   THE REFLUX IS RETURNED AT ITS BUBBLE POINT.
      REAL L, LP, LPP, LO, LMDF
C**   SUBSCIPT FOR X AND Y IN DIMENSION STATEMENT IS TO
C**   BE (NP * EF + 1). AS AN EXAMPLE, THE
C**   SUBSCRIPT OF 20 IS USED HERE.
      DIMENSION X (20), Y (20), YE (50), XE (50)
      READ (*, *) F1, F2, ZF1, ZF2, D, R, NP, NF1, NF2, EF
      B = F1 + F2 - D
C**   EQUILIBRIUM DATA IS ASSUMED AVAILABLE.
      READ (*, *) [XE (K), YE (K), K = 1, 50]
```

```
C**    IF A TOTAL CONDENSER IS USED, COND = 1.0
C**    AND IF A PARTILA CONDENSER IS USED,
C**    COND = 2.0. IF A REBOILER IS USED, THEN
C**    STEAM = 1.0 AND IF OPEN STEAM IS USED,
C**    THEN STEAM = 2.0.
       READ (*, *) COND, STEAM
C**    SPECIFY THE VALUE OF Q (WHICH DEPENDS
C**    ON THE THERMAL CONDITION OF THE FEED).
C**    THERMAL CONDITION OF BOTH FEED STREAMS
C**    IS ASSUMED TO BE THE SAME.
C**    IF FEED IS A SATURATED LIQUID, THEN
C**    FEED = 1.0; IF SATURATED VAPOUR, FEED =
C**    2.0; IF COLD LIQUID, FEED = 3.0; IF
C**    SUPERHEATED VAPOUR, FEED = 4.0 AND
C**    IF VAPOUR - LIQUID MIXTURE, FEED = 5.0.
       READ (*, *) FEED
       IF (FEED. EQ. 1.0) Q = 1.0
       IF (FEED. EQ. 2.0) Q = 0.0
       IF (FEED. EQ. 3.0) GO TO 10
       IF (FEED. EQ. 4.0) GO TO 15
       IF (FEED. EQ. 5.0) GO TO 20
       GO TO 25
10     READ (*, *) TF, CPL, LMDF, TBF
       Q = 1.0 + CPL * (TBF - TF)/LMDF
       GO TO 25
15     READ (*, *) TF, CPV, LMDF, TDF
       Q = CPV * (TDF - TF)/LMDF
       GO TO 25
20     READ (*, *) FL
       Q = FL
25     L0 = R * D
       L = L0
       LP = L + Q * F1
       LPP = LP + Q * F2
       V1 = (R + 1.0) * D
       V = V1
       VP = V + (Q - 1.0) * F1
       VPP = VP + (Q - 1.0) * F2
C**    ASSUME ZD. FOR EXAMPLE, LET ZD = 0.95.
```

```
      ZD = 0.95
30    XB = (F1 * ZF1 + F2 * ZF2 - D * ZD)/B
      IF (COND. EQ. 1.0) THEN
      X0 = ZD
      Y (1) = X0
      ELSE
      YD = ZD
      DO 35 K = 1, 50
      IF [YE (K). GE. YD] GO TO 40
35    CONTINUE
40    XO = XE (K - 1) + [XE (K) - XE (K - 1)] * [YD - YE (K - 1)]/[YE (K)
      - YE (K - 1)]
      Y (1) = [R/(1.0 + R)] * XO + ZD/(1.0 + R)
      END IF
C**   COMPUTE X (1) FROM EQUILIBRIUM DATA.
      DO 45 K = 1, 50
      IF [YE (K). GE. Y (1)] GO TO 50
45    CONTINUE
50    X (1) = XE (K - 1) + [XE (K) - XE (K - 1)] * [Y1- YE (K - 1)]/[YE
      (K) - YE (K - 1)]
C**   COMPUTE THE COORDINATES (XA, YA) OF THE
C**   POINT OF INTERSECTION OF FIRST OPERATING
C**   LINE WITH THE Q-LINE AT ZF1.
      XA = [(R + 1.0) * ZF1 + (Q - 1.0) * ZD]/(R + Q)
      YA = (R * ZF + Q * ZD)/(R + Q)
      J = 1.0
55    IF (J. GE. NF1) GO TO 60
      Y (J + 1) = [R/(R + 1.0)] * X (J) + ZD/(R + 1.0)
      GO TO 70
60    IF (J. GE. NF2) GO TO 65
      Y (J + 1) = (LP/VP)* [X (J) - XA] + YA
      GO TO 70
65    IF [STEAM. EQ. (1.0) THEN
      Y (J + 1) = (LPP/VPP) * X (J) + XB * (1.0 - LPP/VPP)
      ELSE
      Y (J + 1) = (LPP/VPP) * [X (J) - XB]
      END IF
C**   COMPUTE X (J + 1) FROM EQUILIBRIUM DATA.
70    DO 75 K = 1, 50
```

```
      IF [YE (K). GE. Y (J + 1)] GO TO 80
75    CONTINUE
80    X (J + 1) = XE (K - 1) + [XE (K) - XE (K - 1)] * [Y (J + 1) - YE (K
      - 1)]/[YE (K) - YE (K - 1)]
      IF [X (J + 1). LE. XB] GO TO 85
      J = J + 1
      GO TO 55
85    ANS = J + [X (J + 1)/XB]
      ANPC = (ANS - 1.0)/EF
      IF (ANPC. LE. NP) GO TO 90
      ZD = ZD - 0.001
      GO TO 30
90    IF [(NP - ANPC). GE. 0.4] THEN
      ZD = ZD + 0.001
      GO TO 30
      ELSE
      WRITE (*, *) ZD, XB
      END IF
      STOP
      END
```

Let us now present a computer program that deals with the estimation of the number of ideal stages required using the rigorous method based on enthalpy data. As an example, we shall consider computation of the overall plate efficiency of a binary fractionator that is equipped with a reboiler a total condenser (similar to that discussed in Example 5.26). The program is given below (Program 5.10).

## PROGRAM 5.10

```
C**   COMPUTATION OF OVERALL PLATE EFFICIENCY
C**   OF A FRACTIONATOR THAT DISTILS
C**   A BINARY MIXTURE (USING THE RIGOROUS
C**   METHOD BASED ON ENTHALPY DATA).
C**   THE FRACTIONATOR IS EQUIPED WITH A
C**   TOTAL CONDENSER AND A REBOILER.
C**   ENTRAINMENT IS NEGLECTED.
      REAL MC, NS
      DIMENSION Z (11), SHV (11), SHL (11), XE (50), YE (50)
C**   THE INDEX OF HV, HL, X, Y AND V IS EQUAL
C**   TO THE NUMBER OF STAGES. HERE, IT IS
C**   ARBITRATILY TAKEN TO BE 10.
      DIMENSION HV (10), HL (10), X (10), Y (10), V (10)
```

```
C**    COMPOSITIONS ARE EXPRESSED AS MASS
C**    FRACTIONS AND FLOW RATES AS MASS FLOW
C**    RATES.
       READ (*, *) F, ZF, HF, ZD, XB
C**    FEED IS INTRODUCED TO THE NF-TH
C**    PLATE FROM THE TOP.
       READ (*, *) NP, NF
C**    REFLUX RATIO IS NOT GIVEN. BUT COOLING
C**    WATER FLOW RATE (MC) AND EXIT AND
C**    INLET TEMPERATURES OF COOLING WATER
C**    (T1, T2) ARE SPECIFIED.
       READ (*, *) MC, T1, T2, CPC
       QC = MC * CPC * (T2 - T1)
       D = F * (ZF - XB)/(ZD - XB)
       B = F - D
C**    SINCE TOTAL CONDENSER IS USED, Y (1)
C**    = XO = XD = ZD.
       XO = ZD
       XD = ZD
       Y (1) = ZD
C**    THE ENTHALPY DATA IS ASSUMED AVAILABLE.
       READ (*, *) [Z (K), SHL (K), SHV (K), K = 1, 11]
C**    COMPUTE ENTHALPY OF DISTILLATE (HD).
       DO 10 K = 1, 11
       IF [XD. LE. Z (K)] GO TO 15
10     CONTINUE
15     HD = SHL (K - 1) + [SHL (K) - SHL (K - 1)] * [XD - Z (K - 1)]/[Z (K)
       - Z (K - 1)]
       HLO = HD
C**    COMPUTE ENTHALPY OF BOTTOM RESIDUE.
       DO 20 K = 1, 11
       IF [XB. LE. Z (K)] GO TO 25
20     CONTINUE
25     HB = SHL (K - 1) + [SHL (K) - SHL (K - 1)] * [XB - Z (K - 1)]/[Z (K)
       - Z (K -1)]
C**    COMPUTE ENTHALPY OF OVERHEAD VAPOURS, HV (1).
       DO 30 K = 1, 11
       IF [Y (1). LE. Z (K)] GO TO 35
30     CONTINUE
```

```
35     HV (1) = SHV (K - 1) + [SHV (K) - SHV (K - 1)] * [Y (1) - Z (K - 1)]/
       [Z (K) - Z (K - 1)]
C**    COMPUTE FLOW RATE OF OVERHEAD VAPOURS, V (1).
       V (1) = [D * (HD - HLO) + QC]/[HV (1) - HLO]
C**    COMPUTE REBOILER HEAT LOAD (QB).
       QB = D * HD + B * HB + QC - F * HF
C**    EQUILIBRIUM DATA IS ASSUMED AVAILABLE.
       READ (*, *) [XE (K), YE (K), K = 1, 50]
C**    COMPUTE X (1) FROM EQUILIBRIUM DATA.
       DO 40 K = 1, 50
       IF [Y (1). LE. YE (K)] GO TO 45
40     CONTINUE
45     X (1) = XE (K - 1) + [XE (K) - XE (K - 1)] * [Y (1) - YE (K - 1)]/
       [YE (K) - YE (K - 1)]
C**    IT IS ASSUMED THAT FEED IS INTRODUCED
C**    TO THE NFI-TH IDEAL STAGE. TO START
C**    WITH, LET NFI = 2.
       NFI = 2.0
C**    PERFORM STAGE TO STAGE COMPUTATIONS
C**    TO DETERMINE THE NUMBER OF IDEAL
C**    STAGES REQUIRED.
50     J = 1
C**    ASSUME THE VALUE OF V (J + 1).
55     V (J + 1) = V (J)
       DO 60 K = 1, 11
       IF [X (J). LE. Z (K)] GO TO 65
60     CONTINUE
65     HL (J) = SHL (K - 1) + [SHL (K) - SHL (K - 1)] * [X (J) - Z (K - 1)]/
       [Z (K) - Z (K - 1)]
70     IF (J. LT. NFI) THEN
       Y (J + 1) = X (J) + [D/V (J + 1)] * [ZD - X (J)]
       ELSE
       VP (J + 1) = V (J + 1)
       Y (J + 1) = X (J) + [B/VP (J + 1)] * [X (J) - XB]
       END IF
       DO 75 K = 1, 11
       IF [Y (J + 1). LE. Z (K)] GO TO 80
75     CONTINUE
80     HV (J + 1) = SHV (K - 1) + [SHV (K) - SHV (K - 1)] * [Y (J + 1) - Z
       (K - 1)]/[Z (K) - Z (K - 1)]
```

```
      IF (J. LT. NFI) THEN
      VC (J + 1) = {D * [HD - HL (J)] + QC}/[HV (J + 1) - HL (J)]
      ELSE
      VPC (J + 1) = {B * [HL (J) - HB] + QB}/[HV (J + 1) - HL (J)]
      END IF
C**   CHECK FOR CONVERGENCE.
      IF (J. LT. NFI) THEN
      DIF = ABS [V (J + 1) - VC (J + 1)]/V (J + 1)
      IF (DIF. LE. 0.001) GO TO 85
      V (J + 1) = VC (J + 1)
      GO TO 70
      ELSE
      DIF = ABS [VP (J + 1) - VPC (J + 1)]/VP (J + 1)
      IF (DIF. LE. 0.001) GO TO 85
      V (J + 1) = VPC (J + 1)
      GO TO 70
      END IF
85    DO 90 K = 1, 50
      IF [Y (J + 1). LE. YE (K)] GO TO 95
90    CONTINUE
95    X (J + 1) = XE (K - 1) + [XE (K) - XE (K - 1)] * [Y (J + 1) - YE
      (K - 1)]/[YE (K) - YE (K - 1)]
      IF [X (J + 1). LE. XB] GO TO 100
      J = J + 1
      GO TO 55
100   NS = J + [X (J) - XB]/[X (J) - X (J + 1)]
C**   COMPUTE OVERALL PLATE EFFICIENCY (EF).
      EF = (NS - 1)/NP
C**   COMPUTE FEED PLATE LOCATION. IF NFC
C**   (WHICH IS THE CALCULATED VALUE OF NF)
C**   FALLS BETWEEN NF AND (NF - 1),
C**   THEN TRIALS ARE DISCONTINUED.
      NFC = NFI/EF
      IF [NFC. LE. (NF - 1)] GO TO 105
      IF [NFC. GT. NF] GO TO 105
      GO TO 110
105   NFI = NFI + 1
      GO TO 50
110   WRITE (*, *) EF, NS
      STOP
      END
```

As discussed in the earlier sections, the most popular distillation equipment are plate columns and packed columns. The design of sieve plate column for gas-liquid contacting is discussed in all detail in Chapter 4 (see Programs 4.1 to 4.4). Since design for vapour-liquid contacting is exactly analogous, we shall not discuss the same here. For a difference, let us consider design of a bubble-cap column. Also, we shall program for a rating problem rather than a sizing problem (Example 5.28 discusses rating of a bubble-cap column). In other words, a bubble cap column with a given cap-plate layout is available. We have to determine whether this column shall fit to the given application or not.

In the computer program (Program 5.11) given below, we have used the following table (Table 5.3.1) for computing the aeration factor (β) instead of Fig. 5.36. This table has been prepared from Fig. 5.36 and permits linear interpolation for obtaining the value of β at intermediate values of $F_{Va}$.

**Table 5.3.1**

| $F_{Va} = V_a \sqrt{\rho_V}$ | *Aeration factor (β)* |
|---|---|
| 0.0 | 1.00 |
| 0.61 | 0.79 |
| 1.22 | 0.70 |
| 1.83 | 0.65 |
| 2.44 | 0.62 |
| 2.928 | 0.60 |
| 3.05 | 0.60 |

Also, for checking for entrainment, it is sufficient to ensure that the computed value of the flow parameter ($\pi_1$) is not less than that corresponding to $\psi$ = 0.1, which we denote as $\pi_1$ (min). In other words, if $\pi_1 > \pi_1$ (min), then $\psi$ will be less than 0.1 and this means that the entrainment is reasonably low. The values of $\pi_1$ (min) deduced from Fig. 4.11 of Chapter 4 are listed in Table 5.3.2. This table has been utilised in Program 5.11.

**Table 5.3.2**
**Values of $\pi_1$ (min) for Bubble-cap Trays**

| *Percent flood* | $\pi_1$ *(min)* |
|---|---|
| 40.0 | 0.0135 |
| 50.0 | 0.0160 |
| 60.0 | 0.020 |
| 70.0 | 0.0275 |
| 80.0 | 0.040 |
| 85.0 | 0.055 |

For intermediate values, the data in the above table may also be interpolated with allowable error.

# PROGRAM 5.11

```
C**     RATING OF A BUBBLE CAP COLUMN.
C**     FEED IS A BINARY MIXTURE
C**     THE COLUMN IS EQUIPPED WITH A REBOILER
C**     AND A TOTAL CONDENSER.
        REAL L, LO, LW, M1, M2, MV, ML, KC
        DIMENSION CVF (2), Y (2), X (2), BETA (6), FVA (6), PF (6), PIM (6),
        XE (50), YE (50)
        READ (*,*) P, DT, NP, F, ZF, ZD, XB, R
C**     CROSSFLOW ARRANGEMENT IS CHOSEN. CAPS
C**     ARE ASSUMED LAID ON EQUILATERAL TRIANGULAR CENTRES.
        READ (*, *) DC, NC, RC, AC, SPC, HS, BS, HSK, HSR, RAA, RSR, RSC,
        RANR
        D = F * (ZF - XB)/(ZD - XB)
        B = F - D
10      LO = R * D
C**     DESIGN IS PERFORMED FOR THE TOPMOST PLATE (J = 1.0)
        L = LO
        V = LO + D
C**     SINCE A TOTAL CONDENSER IS USED, Y (1) = XO = ZD
        Y (1) = ZD
C**     FIND X (1) FROM EQUILIBRIUM DATA.
        READ (*, *) [YE (K), XE (K), K = 1, 50]
        DO 15 K = 1, 50
        IF [YE (K). GE. Y (1)] GO TO 20
15      CONTINUE
20      X (1) = XE (K - 1) + [XE (K) - XE (K - 1)] * [Y (1) - YE (K - 1)]/
        [YE (K) - YE (K - 1)]
C**     SPECIFY DENSITY (AT THE PLATE TEMPERATURE
C**     T) AND THE MOLECUL AR WEIGHT OF EACH COMPONENT
        READ (*, *) T, ROWL1, ROWL2, M1, M2
C**     COMPUTE DENSITIES OF LIQUID AND VAPOUR
C**     LEAVING THE TOPMOST PLATE.
        ML = X (1) * M1 + [1.0 - X (1)] * M2
        VL = X (1) * M1/ROWL1 + [1.0 - X (1)] * M2/ROWL2
        ROWL = ML/VL
        MV = Y (1) * M1 + [1.0 - Y (1)] * M2
        RG = 0.082
        ROWV = P * MV/(RG * T)
```

```
C**    CHOOSE THE PLATE SPACING (S).
       IF (DT. LE. 1.2) S = 0.4572
       IF (DT. GT. 1.2) S = 0.610
       IF (DT. GT. 3.0) S = 0.762
       IF (DT. GT. 3.6) S = 0.914
C**    COMPUTE FLOODING VELOCITY FROM FAIR
C**    AND MATHEWS' CORRELATION.
       QL = L * ML
       QV = V * MV
       PI1 = (QL/QV) * SQRT (ROWV/ROWL)
       PIC1 = PI1
       IF (PIC1. LT. 0.03) PIC1 = 0.03
       IF (PIC1. LT. 0.2) THEN
       A0 = 0.0564 * S + 0.0207
       A1 = 0.0492 * S + 0.004115
       ELSE
       A0 = 0.0336 * S + 0.0134
       A1 = 0.0816 * S + 0.014935
       END IF
       PI2 = A0 - A1 * ALOG 10 (PIC1)
C**    THE CORRECTION FOR SURFACE TENSION IS NEGLECTED
       UFN = PI2 * SQRT [(ROWL - ROWV)/ROWV]
C**    IT IS ASSUMED THAT SEGMENTAL DOWNCOMERS ARE EMPLOYED.
C**    CHOOSE RLD (THAT IS, THE LW/DT RATIO).
C**    POPULARLY, LW/DT = 0.70.
       RLD = 0.70
       RWD = SQRT (1.0 - RLD * * 2)
       PI = 3.14
       RADA = ASIN (RLD)/180.0 - (RWD * RLD/PI)
       A = PI * DT * DT/4.0
       AD = RADA * A
       AN = A - AD
       VGN = (QV/ROWV)/AN
       PFC = 100.0 * VGN/UFN
C**    SPECIFY PFM. FOR LOW-FOAMING SYSTEMS,
C**    PFM = 0.85. FOR HIGHLY FOAMING SYSTEMS,
C**    PFM = 0.70.
       READ (*, *) PFM
       IF (PFC. GT. PFM) GO TO 200
C**    CHECK FOR ENTRAINMENT USING TABLE 5.3.2
```

```
      READ (*, *) [PF (I), PIM (I), I = 1, 6]
      IF (PFC. LT. 40.0) GO TO 45
      DO 30 I = 1, 6
      IF [PFC. LE. PF (I)] GO TO 40
30    CONTINUE
40    PIMC = PIM (I - 1) + [PIM (I) - PIM (I - 1)] * [PFC - PF (I - 1)]/
      [PF (I) - PF (I - 1)]
      GO TO 50
45    PIMC = 0.0135
50    IF (PI1. LT. PIMC) GO TO 205
C**   COMPUTE LIQUID HEAD OVER WEIR (HOW). THE
C**   CORRECTION FACTOR (FW) IS ASSUMED EQUAL TO UNITY.
      FW = 1.0
      LW = RLD * DT
      HOW = 664.0 * FW * [QL/(ROWL * LW)] * * 0.6667
C**   SPECIFY STATIC SUBMERGENCE (HSS) BASED
C**   ON THE VALUE OF P. TO NOTE THAT P
C**   IS TO BE EXPRESSED IN ATMOSPHERES.
      IF (P. LT. 1.0) HSS = 0.0
      IF (P. EQ. 1.0) HSS = 12.50
      IF (P. GT. 1.0) HSS = 25.0
      IF (P. GE. 20.0) HSS = 37.5
C**   COMPUTE THE WEIR HEIGHT (HW)
      HW = HS + HSK + HSR + HSS
C**   COMPUTE PRESSURE DROP DUE TO VAPOUR FLOW
C**   THROUGH WET CAP ASSEMBLY (HRC).
      IF (RANR. LE. 1.5) KC = 0.844 - 0.278 * RANR
      IF (RANR. LE. 1.3) KC = 2.24569 - 2.4172 * RANR + 0.81967 * RANR * * 2
      AS = RSC * AC
      AST = AS * NC
      AR = AST/RSR
      UVC = QV/(ROWV * AR)
      HRC = 274.0 * KC * (ROWV/ROWL) * UVC * * 0.2
C**   COMPUTE PRESSURE DROP THROUGH SLOTS (HSO)
      UVS = QV/(ROWL * AST)
      HSO = 937.0 * [(HS * HS * UVS) * * 0.4] * [ROWV/(ROWL - ROUV)] * * 0.2
C**   TO CHECK THAT HSO DOES NOT EXCEED HS.
C**   COMPUTE LIQUID GRADIENT ACROSS THE PLATE (HG). IT
C**   IS ASSUMED THAT (DR/DC) = 0.7 SO THAT BOLLE'S
```

```
C**   CORRELATION IS APPLICABLE.
      SC = SPC - DC
      GAMA = SC/DC
      WA = (DT + LW)/2.0
C**   COMPUTE THE LIQUID GRADIENT FACTOR (CD)
C**   IF HOLD-DOWN BARS ARE USED ON CAPS,
C**   HDB = 1.0. IF NO HOLD-DOWN BARS ARE USED
C**   ON CAPS, HDB = 2.0
      READ (*, *) HDB
      Q1 = QL/(ROWL * WA)
      IF (HDB. EQ. 1.0) GO TO 55
      IF (Q1. LE. 0.024) CD = 0.0252268 + 22.1235 * Q1
      IF (Q1. LE. 0.0167) CD = -0.07556 + 28.7652 * Q1
      IF (Q1. LE. 0.0074) CD = -1.17537 + 365.41247 * Q1 + 29377.214 * Q1
      * * 2
      CD = EXP (CD)
      GO TO 60
55    IF (Q1. LE. 0.024) CD = 0.008015 + 11.578129 * Q1
      IF (Q1. LE. 0.0167) CD = -0.3863218 + 37.80305 * Q1
      IF (Q1. LE. 0.0074) CD = -0.4194485 + 42.2602 * Q1
      CD = EXP (CD)
60    R1 = QL * (1.0 + GAMA)/(ROWL * WA * CD * GAMA)
      R2 = 0.003 * (HW + HOW + 0.3 * HSK/GAMA)
C**   ASSUME THE VALUE OF HGP, FOR EXAMPLE, LET HGP = 1.0 MM
      HGP = 1.0
65    RHS = 0.0417 * (0.0016 * HGP + R2) * SQRT (HGP)
      DIF = ABS (RHS - R1)/R1
      IF (DIF. LT. 0.001) GO TO 70
      HGP = HGP + 0.001
      GO TO 65
C**   COMPUTE THE VAPOUR LOAD CORRECTION FACTOR (CVF).
70    FV = QV/[A * SQRT (ROWV)]
      IF [FV. EQ. (0.61)] GO TO 75
      IF (FV. LE. 0.976) GO TO 80
      IF (FV. LE. 1.22) GO TO 85
      IF (FV. LE. 1.342) GO TO 90
      IF (FV. LE. 1.464) GO TO 95
      CVF (1) = -9.664 * Q1 + 1.14
      CVF (2) = -16.1065 * Q1 + 1.2667
      CVFC = CVF (1) + [CVF (2) - CVF (1)] * (FV - 1.464)/(1.708 - 1.464)
```

```
      GO TO 100
75    CVFC = 26.575 * QP + 0.57
      IF (Q1. LT. 827.82 E - 05) CVFC = 33.8237 * Q1 + 0.52
      GO TO 100
80    IF (Q1. LT. 827.82 E - 05) THEN
      CVF (1) = 33.8237 * Q1 + 0.52
      CVF (2) = 26.844 * Q1 + 0.6667
      ELSE
      CVF (1) = 26.575 * Q1 + 0.57
      CVF (2) = 12.08 * Q1 + 0.785
      END IF
      CVFC = CVF (1) + [CVF (2) - CVF (1)] * (FV - 0.61)/(0.976 - 0.61)
      GO TO 100
85    IF (Q1. LT. 827.82 E - 05) THEN
      CVF (1) = 26.844 * Q1 + 0.6667
      ELSE
      CVF (1) = 12.08 * Q1 + 0.785
      END IF
      CVF (2) = 9.664 * Q1 + 0.87
      CVFC = CVF (1) + [CVF (2) - CVF (1)] * (FV - 0.976)/(1.22 - 0.976)
      GO TO 100
90    CVF (1) = 9.664 * Q1 + 0.87
      CVF (2) = 1.0
      CVFC = CVF (1) + [CVF (2) - CVF (1)] * (FV - 1.22)/(1.342 - 1.22)
      GO TO 100
95    CVF (1) = 1.0
      CVF (2) = -9.664 * Q1 + 1.14
      CVFC = CVF (1) + [CVF (2) - CVF (1)] * (FV - 1.342)/(1.464 - 1.342)
100   HG = HGP * CVFC * RC
      HDS = HSS + HOW + 0.5 * HG
C**   COMPUTE THE AERATION FACTOR (BETA) USING TABLE 5.3.1
      AA = RAA * A
      FVAC = QV/[AA * SQRT (ROWV)]
      READ (*, *) [FVA (I), BETA (I), I = 1, 6]
      IF (FVAC. GE. 2.928) GO TO 108
      DO 105 I = 1, 6
      IF [FVAC. LE. FVA (I)] GO TO 107
105   CONTINUE
107   BETAC = BETA (I) + [BETA (I - 1) - BETA (I)] * [FVA (I) - FVAC]/[FVA
      (I) - FVA (I - 1)]
```

```
      GO TO 109
108   BETAC = 0.60
C**   COMPUTE PRESSURE DROP FOR VAPOUR FLOW THROUGH
C**   AERATED MASS (HL).
109   HL = HDS * BETAC
C**   COMPUTE THE TOTAL PRESSURE DROP ACROSS THE TRAY (DHT)
      DHT = HL + HRC + HSO
      VDR = HG/(HSO + HRC)
C**   COMPUTE LIQUID BACKUP IN DOWNCOMER (HD).
      W = RWD * DT
      IF (W. LE. 1.5) HLS = 12.5
      IF (W. GT. 1.5) HLS = 25.0
      IF (W. GT. 3.0) HLS = 38.0
      HAP = (HW - HLS)/(1000.0)
      AAP = HAP * LW
      ADM = AMIN1 (AAP, AD)
      SHD = (165.20) * [QL/(ROWL * ADM)] * * 2
      HD = DHT + HW + HOW + HG + SHD
C**   THE RELATIVE FROTH DENSITY (PHIDC) IN DOWNCOMER
C**   IS TAKEN TO BE 0.5.
      PHIDC = 0.5
      HDP = HD/PHIDC
C**   COMPUTE LIQUID RESIDENCE TIME IN DOWNCOMER.
      TETAD = (HD/1000.0) * AD * ROWL/QL
C**   CHECK FOR PLATE STABILITY.
      IF (HG. GT. 25.0) GO TO 210
      IF (HD. GT. 0.5 * S) GO TO 210
      IF (HDP. GT. S) GO TO 210
      IF (VDR. GT. 0.4) GO TO 210
      IF (TETAD. LT. 5.0) GO TO 210
      GO TO 215
200   R = R - 0.01
      GO TO 10
205   R = R + 0.01
      GO TO 10
210   WRITE (*, 211)
211   FORMAT (10X, "COLUMN DIAMETER TO BE INCREASED")
      GO TO 220
215   WRITE (*, *) DT, S, LW, W, AA, AD, HW, HLS
220   STOP
      END
```

As stated in the text, packed columns are viable substitutes to plate columns for the continuous fractionation of binary or multicomponent mixtures, though there is dubiousness regarding the applicability of the experimental correlations for the computation of mass transfer coefficients in packed bed distillation columns (since the reported correlations are for gas-liquid absorption and their applicability to distillation is uncertain since distillation temperatures are significantly higher than those employed for gas-liquid absorption). We shall present here the computer program for the design of a packed tower fractionator that is employed for the continuous fractionation of a binary mixture. Since the column diameter and pressure drop are to be computed from the flooding plots or correlation (Fig. 4.19 of Chapter 4) as is done for packed bed absorbers, the same is not included in this program. The program computes the required packed height of the distillation column. In a packed column, point to point (or section to section) computations are necessary. In the program given below, computations are performed at 20 points (or sections) each in the enriching section and in the exhausting (stripping) section. That is, $N = M = 20$.

## PROGRAM 5.12

```
C**   DESIGN OF A PACKED COLUMN FOR THE CONTINUOUS
C**   FRACTIONATION OF A BINARY MIXTURE.
C**   THE COLUMN IS EQUIPPED WITH A TOTAL CONDENSER AND A REBOILER.
C**   CONSTANT MOLAL OVERFLOW ASSUMED.
      REAL L, LO, LP, LMDF, LMY, M1, M2, MEUL, MEUV
      REAL KX (41), KY (41), KL (41), ML (41), MV (41)
      DIMENSION YE (50), XE (50), TE (50), XI (41), YI (41)
      DIMENSION QL (41), QV (41), TL (41), SIA (41), FG (41), FF (41),
      PC (2), TC (2), ROWV (41), ROWL (41)
      READ (*, *) F, ZF, ZD, XB, R, M1, M2
      READ (*, *) P, DT, DP, DS, ABSN
      XD = ZD
      D = F * (ZF - XB)/(XD - XB)
      B = F - D
C**   SPECIFY THE VALUE OF Q (WHICH DEPENDS ON THE FEED CONDITION).
C**   IF FEED IS A SATURATED LIQUID, THEN FEED
C**   = 1.0; IF SATURATED VAPOUR, FEED = 2.0; IF
C**   COLD LIQUID, FEED = 3.0; IF SUPERHEATED
C**   VAPOUR, FEED = 4.0 AND IF VAPOUR-LIQUID
C**   MIXTURE, FEED = 5.0
      READ (*, *) FEED
      IF (FEED. EQ. 1.0) Q = 1.0
      IF (FEED. EQ. 2.0) Q = 0.0
      IF (FEED. EQ. 3.0) GO TO 10
      IF (FEED. EQ. 4.0) GO TO 15
      IF (FEED. EQ. 5.0) GO TO 20
```

```
      GO TO 25
10    READ (*, *) TF, CPL, LMDF, TBF
      Q = 1.0 + CPL * (TBF - TF)/LMDF
      GO TO 25
15    READ (*, *) TF, CPV, LMDF, TDF
      Q = CPV * (TDF - TF)/LMDF
      GO TO 25
20    READ (*, *) FL
      Q = FL
25    LO = R * D
      V1 = LO + D
      L = LO
      V = V1
      LP = L + Q * F
      VP = LP - B
C**   COMPUTE THE COORDINATES (XA, YA) OF THE POINT
C**   OF INTERSECTION OF Q-LINE WITH THE
C**   OPERATING LINE OF ENRICHING SECTION (WHICH
C**   IS ALSO THE POINT OF INTERSECTION OF THE
C**   TWO OPERATING LINES) FROM EQS (5.2.69A)
C**   AND (5.2.69B).
      XA = [(R + 1) * ZF + (Q - 1.0) * ZD]/(R + Q)
      YA = (R * ZF + Q * ZD)/(R + Q)
C**   COMPUTE Y1 (= YB) FROM EQUILIBRIUM DATA.
      READ (*, *) [XE (K), YE (K), TE (K), K = 1, 50]
      DO 30 K = 1, 50
      IF [XE (K). GE. XB] GO TO 35
30    CONTINUE
35    Y1 = YE (K - 1) + [YE (K) - YE (K - 1)] * [XB - XE (K - 1)]/[XE (K)
      - XE (K - 1)]
C**   SINCE TOTAL CONDENSR IS USED, Y2 = XD.
      Y2 = XD
      Y = Y2
      J = 1
      N = 20
      M = 20
      H1 = [Y2 - YA]/N
      H2 = (YA - Y1]/M
C**   COMPUTE X FROM THE EQUATION TO THE
```

```
C**   OPERATING LINE OF ENRICHING SECTION OR
C**   EXHAUSTING (STRIPPING) SECTION.
40    IF [Y. GT. YA] THEN
      X = [Y - XD/(1.0 + R)] * [(1.0 + R)/R]
      ELSE
      X = [Y + B * XB/VP] * (VP/LP)
      END IF
C**   FIND THE BUBBLE POINT OF LIQUID TL (J) FROM
C**   EQUILIBRIUM DATA.
      DO 45 K = 1, 50
      IF [XE (K). GE. X] GO TO 50
45    CONTINUE
50    TL (J) = TE (K - 1) + [TE (K) - TE (K - 1)] * [X - XE (K - 1)]/[XE
      (K) - XE (K - 1)]
C**   FIND THE DEW POINT OF THE VAPOUR TV (J) FROM
C**   EQUILIBRIUM DATA.
      DO 55 K = 1, 50
      IF [YE (K). GE. Y] GO TO 60
55    CONTINUE
60    TV (J) = TE (K - 1) + [TE (K) - TE (K - 1)] * [Y - YE (K - 1)]/[YE
      (K) - YE (K - 1)]
C**   COMPUTE LIQUID PHASE AND GAS PHASE MASS
C**   TRANSFER COEFFICIENTS SUCH AS KX (J) AND KY (J).
C**   VISCOSITY OF LIQUID SOLUTION AND THAT OF
C**   VAPOUR MIXTURE (MEUL, MEUV) ARE ASSUMED
C**   TO BE ESSENTIALLY CONSTANT.
C**   COMPUTE LIQUID PHASE DIFFUSIVITY (DL).
      READ (*, *) PHIL, SV1
      DL = (117.3 E-18) * SQRT (PHIL * M2) * TL (J)/[MEUL * SV1 * * 0.6]
C**   COMPUTE KX (J) FROM EQ. (4.5.126).
C**   SPECIFY LIQUID PHASE DENSITY OF BOTH COMPONENTS
C**   AT TEMPERATURE TL (J).
      READ (*, *) ROWL1, ROWL2
      ROWL (J) = X * ROWL1 + (1.0 - X) * ROWL2
      SCL = MEUL/[ROWL (J) * DL]
      ML (J) = X * M1 + [1.0 - X] * M2
      IF [X. LT. XA] L = LP
      QL (J) = L * ML (J)
      PI = 3.14
```

```
      A = PI * DT * DT/4.0
      REL = DS * [QL (J)/A]/MEUL
      KL (J) = 25.1 * (DL/DS) * (REL * * 0.45) * SQRT (SCL)
      KX (J) = KL (J) * ROWL (J)/ML (J)
C**   COMPUTE GAS PHASE DIFFUSIVITY (DG).
      READ (*, *) [PC (I), TC (I), I = 1, 2]
      D1 = (3.64 E-08) * {TV (J)/SQRT [TC (1) * TC (2)]} * * 2.334
      D2 = [PC (1) * PC (2)] * * 0.333 * [TC (1) * TC (2)] * * 0.4167
      D3 = SQRT (1.0/M1 + 1.0/M2)
      DG = D1 * D2 * D3
C**   COMPUTE FG (J) FROM EQ. (4.5.125).
      MV (J) = Y * M1 + (1.0 - Y) * M2
      RG = 0.082
      ROWV (J) = P * MV (J)/[RG * TV (J)]
      SCG = MEUV/[ROWV (J) * DG]
C**   TABLE 4.14 IS TO BE STORED IN COMPUTER
C**   MEMORY AS A DATABASE. SINCE THE SUBROUTINE
C**   FOR STORING THIS TABLE IS EASY TO
C**   PREPARE, THE SAME IS NOT GIVEN THERE. THE
C**   CONSTANTS C1 AND C2 ARE RETRIEVED
C**   FROM THIS DATABASE.
      READ (*, *) C1, C2
      BETA = C2 * DS * * 0.376
      PHIT = {C1 * [737.5 * QL (J)/A] * * BETA}/(DS * DS)
      ABSO = ABSN - PHIT
      IF (X. LT. XA) V = VP
      QV (J) = V * MV (J)
      FG (J) = (1.195/SCG * * 0.667) * (V/A) * [MEUV * (1.0 - ABSO)/(DS
      * QV (J)/A)] * * 0.36
C**   COMPUTE SPECIFIC INTERFACIAL AREA SIA (J) FROM EQ. (4.5.130)
C**   HERE ALSO, TABLE 4.15 IS TO BE STORED IN
C**   COMPUTER MEMORY AS ANOTHER DATABASE. THE
C**   SUBROUTINE IS NOT GIVEN HERE. VALUES OF CONSTANTS
C**   SM1, SN1, SQ1 ARE TO BE RETRIEVED FROM THIS DATABASE
      READ (*, *) SM1, SN1, SQ1
      SIA (J) = SM1 * {808.0 * [QV (J)/A]/SQRT [ROWV (J)]} * * SN1 * [QL
      (J)/A] * * SQ1
C**   COMPUTE THE INTERFACIAL MOLE FRACTIONS XI (J), YI (J)
C**   FIRST ASSUME XI (J).
```

```
      XI (J) = X - 0.001
65    XX = XI (J)
C**   NOW, COMPUTE YI (J) FROM EQ. (5.2.127).
      YI (J) = Y + [KX (J)/KY (J)] * [X - XI (J)]
C**   COMPUTE XI (J) CORRESPONDING TO THE ABOVE
C**   COMPUTED VALUE OF YI (J) FROM EQUILIBRIUM DATA.
      DO 70 K = 1, 50
      IF [YE (K). GE. YI (J)] GO TO 75
70    CONTINUE
75    XI (J) = XE (K - 1) + [XE (K) - XE (K - 1)] * [YI (J) - YE (K - 1)]/
      [YE (K) - YE (K - 1)]
C**   CHECK FOR CONVERGENCE.
      DFI = ABS [XI (J) - XX]
      IF (DIF. LT. 0.001) GO TO 80
      XI (J) = XI (J) - 0.001
      GO TO 65
C**   IT IS ASSUMED THAT GAS PHASE RESISTANCE IS CONTROLLING IN THE
      ENRICHING SECTION AND LIQUID PHASE RESISTANCE IS CONTROLLING IN THE
      EXHAUSTING SECTION.
80    IF (Y. GT. YA) THEN
      LMY = [YI (J) - Y]/ALOG {[1.0 - Y]/[1.0 - YI (J)]}
      KY (J) = FG (J)/LMY
      FF (J) = 1.0/{KY (J) * SIA (J) * [YI (J) - Y]}
      ELSE
      FF (J) = 1.0/{KX (J) * SIA (J) * [X - XE (J)]}
      END IF
      IF (Y. EQ. Y1) GO TO 85
      IF (Y. GT. YA) Y = Y - H1
      IF (Y. LE. YA) Y = Y - H2
      J = J + 1
      GO TO 40
C**   COMPUTE THE HEIGHT OF THE ENRICHING
C**   SECTION AND THAT OF THE STRIPPING SECTION
C**   FROM EQS (5.2.124A) AND (5.2.124B) BY EVALUATING
C**   THE INTEGRALS USING SIMPSON'S RULE.
85    S1 = FF (1) + FF (N + 1)
      IP = 4
      DO 90 J = 2, N
      S1 = S1 + IP * FF (J)
      IP = 6 - IP
```

```
90      CONTINUE
        S1 = S1 * H/3.0
        ZE = (V/A) * S1
        S2 = FF (N + 1) + FF (N + M + 1)
        IP = 4
        DO 95 J = N + 2, N + M
        S2 = S2 + IP * FF (J)
        IP = 6 - IP
95      CONTINUE
        S2 = S2 * H/3.0
        ZS = (LP/A) * S2
        Z = ZE + ZS
        WRITE (*, *) Z, DT, DP, ABSN
        STOP
        END
```

Let us now program the F-U-G method for the computation of the number of theoretical plates required for the fractionation of a multicomponent mixture. It is assumed that the equilibrium ratio of each component ($K_i$) is a function of temperature ($T$) only such that

$$K_i = A_i + B_i T + C_i T^2 + E_i T^3 \quad \text{... (5.3.1)}$$

where the values of the correlation constants $A_i$, $B_i$, $C_i$ and $E_i$ shall be different for different components.

As discussed in the text, Eq. (5.2.153) has to be written separately for $(h - l)$ values of $\phi$ and these are to be solved simultaneously for the $(h - l)$ unknowns. If the keys are fairly close to each other (that is, $h - l = 1$, 2 or 3), then solution of these equations can be easily accomplished by successive elimination. However, if the number of equations is large, then it shall be more convenient to use methods like Crout's reduction or Gauss-Seidel iteration for the simultaneous solution of these equations. We have discussed on these methods in Chapter 3 under Multiple Effect Evaporation calculations. The former (namely, Crout's reduction), though more reticulate, is a direct method while the latter (namely, Gauss-Seidel method) is more simple but demands an iterative procedure. Solution using Gauss–Seidel iteration is illustrated in the Program given below (Program 5.13). For example, Eq. (5.2.153) can be rearranged to the following form (for $\phi = \phi_1$):

$$\sum_{i=l+1}^{h-1} \frac{\alpha_i DZ_i}{\alpha_i - \phi_1} - D' = \sum_{i=1}^{l} \frac{\alpha_i DZ_i}{\alpha_i - \phi_1} + \sum_{i=h}^{k} \frac{\alpha_i DZ_i}{\alpha_i - \phi_1} \quad \text{... (5.3.2)}$$

where $DZ_i = (z_{Di} D)$ and $D' = D (R_m + 1)$. It can be seen that all terms on the right hand side of the above equation are known and therefore, the entire right hand side can be denoted as $b_1$. If we expand the left hand side of the above equation, we get

$$\left(\frac{\alpha_{l+1}}{\alpha_{l+1} - \phi_1}\right) DZ_{l+1} + \left(\frac{\alpha_{l+2}}{\alpha_{l+2} - \phi_1}\right) DZ_{l+2} + \ldots\ldots +$$

$$\left(\frac{\alpha_{l+n-1}}{\alpha_{l+n+1}-\phi_1}\right) DZ_{l+n-1} - D' = b_1 \qquad \text{... (5.3.3)}$$

where $n = (h - l)$. This is of the form

$$a_{11}x_1 + a_{12}x_2 + \ldots\ldots a_{1,n-1}x_{n-1} + a_{1n}x_n = b_1 \qquad \text{... (5.3.4)}$$

In the same way, for $\phi = \phi_2$,

$$a_{21}x_1 + a_{22}x_2 + \ldots\ldots a_{2,n-1}x_{n-1} + a_{2n}x_n = b_2 \qquad \text{... (5.3.5)}$$

Similarly for all other values of $\phi$. We can, therefore, deduce that the coefficients of the above set of simultaneous equations can be represented as follows:

For $1 \le i \le (h - l)$, $\quad a_{ij} = \alpha_{l+j}/(\alpha_{l+j} - \phi_i)$, for $j = 1, \ldots\ldots (h - l - 1)$ ... (5.3.6)

$= 1.0$ for $j = (h - l)$ ... (5.3.7)

$$b_i = \sum_{j=1}^{l} \frac{\alpha_j DZ_j}{(\alpha_j - \phi_i)} + \sum_{j=h}^{k} \frac{\alpha_j DZ_j}{(\alpha_j - \phi_i)} \qquad \text{... (5.3.8)}$$

Also, $\quad x_j = DZ_{l+j}$ for $j = 1, \ldots\ldots (h - l - 1)$ ... (5.3.9)

$= D'$ for $j = (h - l)$ ... (5.3.10)

The procedure for the solution of the above set of equations by Gauss-Seidel iteration is now as follows:

(*i*) Assume $x_i = (b_i/a_{ii})$ for $1 \le i \le (h - l)$

(*ii*) Put $XX_i = x_i$ $(1 \le i \le h - l)$

(*iii*) Now compute $x_i$ as

$$x_i = (1/a_{ii})\left(b_i - \sum_{j=i+1}^{h-l} a_{ij}\, XX_i - \sum_{j=1}^{i-1} a_{ij} x_j\right) \qquad \text{... (5.3.11)}$$

To not that when $i = 1$, the last term and when $i = h - l$, the second term on the right hand side of the above equation shall be respectively equal to zero.

(*iv*) After computing all values of $x_i$ (for $1 \le i \le h - l$), check whether the following convergence criterion is satisfied:

$$\sum_{i=1}^{h-l} \left[(XX_i - x_i)/x_i\right]^2 \le 10^{-5}\,(h - l) \qquad \text{... (5.3.12)}$$

(*v*) If the above criterion is not satisfied, repeat computations starting from Step (*ii*).

(*vi*) Now,

$$DZ_{l+i} = x_i \text{ for } 1 \le i \le (h - l - 1) \qquad \text{... (5.3.13)}$$

$$D' = x_i \text{ for } i = (h - l) \qquad \text{... (5.3.14)}$$

The computer program given below (Program 5.13) has been prepared in a user-friendly format. The good number of comment statements that have been inserted in the program are intended to help the reader analyse the program more readily and easily.

## PROGRAM 5.13

```
C**    COMPUTATION OF THE NUMBER OF THEORETICAL
C**    PLATES REQUIRED FOR THE FRACTIONATION
C**    OF A MULTICOMPONENT MIXTURE
C**    USING F-U-G METHOD.
C**    IN THE TYPE STATEMENTS, THE NUMBER
C**    OF COMPONENTS IS SPECIFIED AS 10 AS AN
C**    EXAMPLE.
       REAL K (10), KF (10), KD (10), KB (10)
       REAL NM, NS, LHS, NUMR
       DIMENSION ZF (10), A (10), B (10), C 10), E (10), ALFA (10), CY
       (10), PHL (10)
       DIMENSION DZ (10), X (10), SA (10, 10), SB (10), CRT (5), CNT (5)
       READ (*, *) F, TF, PF, R, SP1, SP2, N, M
       READ (*, *) NK
       READ (*, *) [ZF (I), I = 1, NK]
       P = PF
C**    ASSUME THE KEY COMPONENTS.
       NL = 1
       NH = NL + 1
C**    ASSUME THE DISTILLATE DEW POINT (TD) AND
C**    THE RESIDUE BUBBLE POINT (TB)
       READ (*, *) TD, TB
       TAV = (TD + TB)/2.0
10     XT = TAV
C**    THE VALUES OF THE CORRELATION CONSTANTS
C**    FOR THE COMPUTATION OF EQUILIBRIUM RATIO
C**    ARE ASSUMED AVAILABLE.
       READ (*, *) [A I), B (I), C (I), E (I), I = 1, NK]
       DO 15 I = 1, NK
       K (I) = A (I) + B (I) * TAV + C (I) * TAV * * 2 + E (I) * TAV * * 3
15     CONTINUE
16     DO 20 I = 1, NK
20     ALFA (I) = K (I)/K (NH)
       CY (N) = 0.01 * SP1
```

```
      CY (M) = 0.01 * SP2
      A1 = [ALFA (N) - 1.0]/{[ALFA (NL) - 1.0] * F * ZF (NL)}
      A2 = [ALFA (M) - 1.0]/{[ALFA (NL) - 1.0] * ZF (NL) * F}
      B1 = [ALFA (NL) - ALFA (N)]/{[ALFA (NL) - 1.0] * F * ZF (NH)}
      B2 = [ALFA (NL) - ALFA (M)]/{[ALFA (NL) - 1.0] * F * ZF (NH)}
      XH = [CY (N) * A2 - CY (M) * A1]/(B1 * A2 - A1 * B2)
      XL = [CY (N) - B1 * XH]/A1
      DO 25 I = 1. NK
      IF (I. EQ. N. OR. I. EQ. M) GO TO 25
      Y1 = [ALFA (I) - 1.0] * XL/{ZF (NL) * F * [ALFA (NL) - 1.0]}
      Y2 = [ALFA (NL) - ALFA (I)] * XH/{[ALFA (NL) - 1.0] * F * ZF (NH)}
      CY (I) = Y1 + Y2
25    CONTINUE
C**   CHECK THE CHOICE OF KEY COMPONENTS
      DO 30 I = 1, NK
      IF (I. EQ. NL. OR. I. EQ. NH) GO TO 30
      IF (I. LT. NL. OR. I. GT. NH) GO TO 35
      IF [CY (I). LT. 0.01. OR. CY (I). GT. 0.99] THEN
      GO TO 40
      ELSE
      GO TO 30
      END IF
35    IF [CY (I). LT. - 0.01. OR. CY (I). GT. 1.01] GO TO 30
40    IF (NH. LT. NK) THEN
      NH = NH + 1
      GO TO 16
      ELSE
      NL = NL + 1
      NH = NL + 1
      GO TO 16
      END IF
30    CONTINUE
      DO 45 I = 1, NK
      IF (I. LT. NL. OR. I. GT. NH) THEN
      IF [CY (I). LT. 0.0) CY (I) = 0.0
      IF [CY (I). GT. 0.0) CY (I) = 1.0
      ELSE
      GO TO 45
      END IF
```

```
45    CONTINUE
C**   COMPUTE THE VALUES OF DZ (I) FOR COMPONENTS
C**   LIGHTER THAN THE LIGHT KEY
C**   AND HEAVIER THAN THE HEAVY KEY. TO
C**   NOTE THAT DZ (I) = ZD (I) * D.
      DZ (NL) = XL
      DZ (NH) = XH
      DO 46 I = 1, NL - 1
46    DZ (I) = CY (I) * ZF (I) * F
      DO 47 I = NH + 1, NK
47    DZ (I) = CY (I) * ZF (I) * F
C**   SPECIFY THE VALUE OF Q WHICH DEPENDS
C**   ON THE THERMAL CONDITION OF THE FEED.
C**   IF THE FEED IS A SATURATED LIQUID,
C**   THEN FEED = 1.0, IF SATURATED VAPOUR, FEED
C**   = 2.0 AND IF A VAPOUR - LIQUID MIXTURE,
C**   FEE = 3.0
      READ (*, *) FEED
      IF (FEED. LE. 2.0) THEN
      IF (FEED. EQ. 1.0) Q = 1.0
      IF (FEED. EQ. 2.0) Q = 0.0
      GO TO 60
      ELSE
      Q = 1.0 - 0.0001
50    RL = Q/(1.0 - Q)
      SUM = 0.0
      DO 55 I = 1, NK
      KF (I) = A (I) + B (I) * TF + C (I) * TF * * 2 + E (I) * TF * * 3
      SUM = SUM + ZF (I) * (1.0 + RL)/[1.0 + RL/KF (I)]
55    CONTINUE
      LHS = SUM
      DIF = ABS (1.0 - LHS)
      IF (DIF. LE. 0.001) GO TO 60
      Q = Q - 0.0001
      GO TO 50
      END IF
C**   SOLVE EQ. (5.2.150) FOR (NH - NL) VLAUES OF
C**   PHIL. THESE VALUES ARE KEPT STORED IN COMPUTER MEMORY AS PHL (J).
```

```
60      PHIL = 1.00001
        IF [ALFA (NL). LT. ALFA (NH)] PHIL = ALFA (NL) + 0.00001
        J = 1
65      SUM = 0.0
        DO 70 I = 1, NK
70      SUM = SUM + ALFA (I) * ZF (I) * F/[ALFA (I) - PHIL]
        RHS = F * (1.0 - Q)
        IF [RHS. EQ. (0.0)] THEN
        DIF = ABS (LHS - RHS)
        IF (DIF. LE. E-04) GO TO 75
        PHIL = PHIL + 0.00001
        GO TO 65
        ELSE
        DIF = ABS (LHS - RHS)/LHS
        IF (DIF. LE. 0.01) GO TO 75
        PHIL = PHIL + 0.00001
        GO TO 65
        END IF
75      PHL (J) = PHIL
        IF [J. EQ. (NH - NL)] GO TO 80
        PHIL = PHIL + 0.00001
        J = J + 1
        GO TO 65
C**     SOLVE EQ. (5.2.153) WHICH FORMS A SET
C**     OF LINEAR SIMULTANEOUS EQUATIONS BY
C**     GAUSS-SEIDEL METHOD.
80      NI = NH - NL
        NJ = NI - 1
        DO 85 I = 1, NI
        SA (I, NI) = 1.0
        DO 90 J = 1, NJ
90      SA (I, J) = ALFA (NL + J)/[ALFA (NL + J) - PHL (I)]
        SUM1 = 0.0
        DO 95 J = 1, NL
95      SUM1 = SUM1 + ALFA (J) * DZ (J)/[ALFA (J) - PHL (I)]
        SUM2 = 0.0
        DO 100 J = NH, NK
100     SUM2 = SUM2 + ALFA (J) * DZ (J)/[ALFA (J) - PHL (I)]
        SB (I) = SUM1 + SUM2
85      X (I) = SB (I)/SA (I, I)
```

```
105    DO 110 I = 1, NI
110    XX (I) = X (I)
       DO 115 I = 1, NI
       IF (I. EQ. NI) THEN
       SUM3 = 0.0
       ELSE
       SUM3 = 0.0
       DO 120 J = I + 1, NI
120    SUM3 = SUM3 + SA (I, J) * XX (I)
       END IF
       IF [I. EQ. (1.0)] THEN
       SUM4 = 0.0
       ELSE
       SUM4 = 0.0
       DO 125 J = 1, I - 1
125    SUM4 = SUM4 + SA (I, J) * X (J)
       END IF
       X (I) = [SB (I) - SUM3 - SUM4]/SA (I, I)
115    CONTINUE
C**    CHECK FOR CONVERGENCE
       SUM5 = 0.0
       DO 130 I = 1, NI
130    SUM5 = SUM 5 + {[XX (I) - X (I)]/X (I)} * * 2
       IF (SUM5. GT. NI * E-05) GO TO 105
C**    COMPUTE THE VALUES OF DZ (I) FOR
C**    COMPONENTS BETWEEN THE KEYS AND CHECK
C**    THE CHOICE OF KEYS.
       DO 135 J = 1, NJ
       DZ (NL + J) = X (J)
       IF [DZ (NL + J). LT. 0.0. OR. DZ (NL + J). GT. ZF (NL + J) * F] GO TO 140
135    CONTINUE
       GO TO 145
140    IF (NH. LT. NK) THEN
       NH = NH + 1
       GO TO 16
       ELSE
       NL = NL + 1
       NH = NL + 1
       GO TO 16
       END IF
```

```
C**     COMPUTE THE TOTAL MOLAR FLOW RATES AND
C**     COMPOSITIONS OF THE TOP DISTILLATE AND
C**     THE BOTTOM RESIDUE.
145     SUM6 = 0.0
        DO 150 I = 1, NK
150     SUM6 = SUM6 + DZ (I)
        D = SUM6
        B = F - D
        DO 155 I = 1, NK
        ZD (I) = DZ (I)/D
        BX (I) = F * ZF (I) - DZ (I)
        XB (I) = BX (I)/B
155     CONTINUE
C**     COMPUTE THE MINIMUM REFLUX RATIO.
        DP = X (NI)
        RMIN = (DP/D) - 1.0
C**     COMPUTE THE DEW POINT OF TOP DISTILLATE (TD)
160     DO 170 I = 1, NK
170     KD (I) = A (I) + B (I) * TD + C (I) * TD * * 2 + E (I) * TD * * 3
        SUM8 = 0.0
        DO 175 I = 1, NK
175     SUM8 = SUM8 + ZD (I)/KD (I)
        DIF = ABS (1.0 - SUM8)
        IF (DIF. LE. 0.001) GO TO 180
        TD = TD + 0.01
        GO TO 160
C**     COMPUTE THE BUBBLE POINT OF THE BOTTOM
C**     RESIDUE (TB)
180     DO 185 I = 1, NK
185     KB (I) = A (I) + B (I) * TB + C (I) * TB * * 2 + E (I) * TB * * 3
        SUM9 = 0.0
        DO 190 I = 1, NK
190     SUM9 = SUM9 + KB (I) * XB (I)
        DIF = ABS (1.0 - SUM9)
        IF (DIF. LE. 0.001) GO TO 195
        TB = TB + 0.01
        GO TO 180
C**     CHECK THE CHOICE OF TAV.
195     TAV = (TD + TB)/2.0
```

```
      DIF = ABS (XT - TAV)
      IF (DIF. LE. 0.05) GO TO 200
      GO TO 10
C**   COMPUTE THE MINIMUM NUMBER OF THEORETICAL
C**   PLATES REQUIRED AT TOTAL REFLUX
C**   FROM FENSKE'S EQUATION.
200   ALFDL = KD (NL)/KD (NH)
      ALFBL = KB (NL)/KB (NH)
      ALFAL = SQRT (ALFDL * ALFBL)
      NUMR = ZD (NL) * XB (NH)/[ZD (NH) * XB (NL)]
      NM = ALOG (NUMR)/ALOG (ALFAL)
C**   COMPUTE THE TOTAL NUMBER OF THEORETICAL
C**   PLATES REQUIRED FROM GILLILAND'S CORRELATION
      CR = (R - RMIN)/(1.0 + R)
      IF (CR. GT. 0.5) GO TO 205
      CN = 0.69254 - 1.377 * CR + 0.7836 * CR * CR
      IF (CR. LE. 0.07) CN = 0 775- 2.5 * CR
      IF (CR. LE. 0.03) CN = 0.70
      GO TO 220
C**   THE VALUES OF CR AND CN LISTED IN
C**   TABLE 5.8 ARE DESIGNATED AS CRT (I)
C**   AND CNT (I) RESPECTIVELY.
205   READ (*, *) [CRT (I), CNT (I), I = 1, 5]
      DO 210 I = 1, 5
      IF [CR. LE. CRT (I)] GO TO 215
210   CONTINUE
215   CN = CNT (I - 1) + [CNT (I) - CNT (I - 1)] * [CR - CRT (I - 1)]/[CRT (I)
      - CRT (I - 1)]
220   NS = (CS + NM)/(1.0 - CN)
      WRITE (*, *) NS, D, B, NH, NL, RMIN, NM
      WRITE (*, *) [ZD (I), XB (I), I = 1, NK)
      STOP
      END
```

## THINGS TO REMEMBER

1. The term drying principally means removal of small quantities of liquid (usually, water) from a solid or gas. Drying or dehumidification of gases can be accomplished by scrubbing with a suitable liquid (Chapter 4), or by passing over a solid adsorbent (Chapter 6). Removal of moisture from fluids is also possible by membrane-based processes such as ultrafiltration or reverse osmosis (Chapter 7). In this chapter, removal of moisture from solids or nearly solid materials is discussed.

2. Removal of moisture from solids can be achieved either by mechanical means (by using batch or continuous centrifugals) or by thermal vapourisation. Mechanical means are cheaper and they may be employed prior to a thermal process so that the total heat load of the thermal process shall be considerably reduced.
3. Commercial dryers employing thermal vapourisation may be broadly classified into direct dryers (in which there is direct contact between the wet solid and the drying medium) and indirect dryers (in which the wet solid is dried by contacting it with a hot surface that is heated indirectly by steam, hot water or electricity). In the former, heat transfer takes place mainly by convection and to a less extent by radiation, while in the latter, conduction is the chief mode of heat transfer.
4. Any wet solid can be dried only up to its equilibrium moisture content ($X^*$) under specific conditions of purge gas temperature and humidity. The value of $X^*$ thus depends not only on the type of solid but also on the operating conditions such as the temperature and humidity of purge gas (usually, air) used. The difference between the total moisture content of the solid and the equilibrium moisture content under prescribed operating conditions is called its *free moisture* content and it is this free moisture that gets evaporated during the drying process.
5. The total moisture in a solid may be bound moisture or unbound moisture or both. Substances containing bound moisture are called hygroscopic substances. Bound moisture is held in the fine pores or is in physical combination with the solid. Therefore, it exerts a vapour pressure that is less than that of the pure liquid at the given temperature and its removal is relatively more difficult.
6. The temperature of a vapour–gas mixture is usually designated by the *dry-bulb temperature* ($t_G$). This is the temperature a thermometer would record when its bulb is immersed in the vapour–gas mixture. The term *wet-bulb temperature* ($t_W$) is used to designate the steady state temperature (or equilibrium temperature) attained by a tiny mass of liquid when made to evaporate into a very large mass of gas. The term *adiabatic saturation temperature* ($t_{as}$) represents the equilibrium temperature attained by the gas when specific amounts of the gas and the liquid are brought into contact with each other. The value of $t_{as}$ can be estimated from Eq. (5.1.19a) and that of $t_W$ from Eq. (5.1.26) with reasonable accuracy. The plots of $t_G$ versus the absolute humidity ($Y'$) at specific values of $t_W$ are called *psychrometric lines* (or, wet-bulb lines), while plots of $t_G$ versus $Y'$ at specific values of $t_{as}$ are called *adiabatic saturation lines*. For air-water system, these two are identical and are nearly straight lines (since for air-water system, the psychrometric ratio is very close to 1.0 and therefore, $t_W \approx t_{as}$). For other systems, $t_{as}$ is lower than $t_W$ and both psychrometric lines and adiabatic saturation lines are curved concave upward, the former being steeper than the latter. The psychrometric ratio for such systems is to be obtained from experimental correlations such as Eq. (5.1.28).
7. Once the dry-bulb temperature ($t_G$) and humidity ($Y'$) are specified, most of the properties of the vapour–gas mixture such as humid volume ($v_h$), humid heat ($C_S$) and the enthalpy ($H$) can be computed from Eqs (5.1.12) to (5.1.17) or read directly from the *psychrometric charts* (also called humidity charts) such as those shown in Figs 5.2 and 5.3. These charts are developed for a total pressure of 1 atm and if the system pressure is different from this, the values from these charts must be corrected for pressure.
8. The rate at which the moisture content of the solid decreases with time represents the rate of drying ($dX/d\theta$). The plot of ($dX/d\theta$) versus $X$ or $\theta$ is called the *drying curve*. Initially, the rate of drying remains constant (constant rate period) and this period continues until $X = X_c$ where $X_c$ is the *critical*

*moisture content*. The surface temperature of the solid remains constant during this period and the drying rate will depend exclusively on the external operating conditions and will be independent of the mechanism of liquid movement within the solid. To note that the value of $X_c$ depends not only on the nature of the solid but also on the drying conditions.

9. The rate of drying during the constant rate period ($R_c$) can be computed analytically from Eqs (5.1.52) or (5.1.53) and the surface temperature of the solid can be obtained from Eq. (5.1.54).
10. Below the critical moisture content, the rate of drying decreases with increase in time. Since the change of ($dX/d\theta$) with $\theta$ can follow different patterns, there can be more than one (usually, two) falling rate periods. The first falling rate period ends at the second critical point and the last falling rate period ends at $X = X^*$.
11. The rate of drying in the falling rate period is controlled by the mechanism of liquid movement within the solid. In granular or crystalline solids (like sand, minerals, catalysts, etc.) the moisture movement within the solid takes place mainly as a result of gravitational and capillary forces. The critical moisture content of these materials is relatively low and the equilibrium moisture content is very close to zero. These materials are practically insensitive to the drying conditions and the rate of drying decreases essentially linearly with decrease in $X$ in the falling rate period. In another category of materials (like soap, gelatin, glue, etc.) the liquid movement within the solid takes place by diffusion through the solid structure. Since this mode of liquid movement is much slower than that by gravity or capillarity, these materials exhibit very short constant rate periods (the critical moisture content being very close to the initial moisture content). The second falling rate period predominates and the values of $X^*$ are generally high. The material also tends to degrade (shrink or crack) during drying. Equation (5.1.35) gives an order of magnitude of the rate of drying in the falling rate period for materials of this kind.
12. A reliable estimate of drying time or drying rate can be best made by performing experimental drying tests. The loss in the weight of wet solid is recorded at different intervals of time and the drying rate is then computed from Eq. (5.1.37) or (5.1.39) and the drying time from Eq. (5.1.40) or (5.1.42).
13. Among the batch direct dryers, the most popular ones are tray dryers and truck dryers (in which the heating medium, say hot air, is passed across the bed of wet solids) and the through-circulation dryers (in which hot air or gas is passed through the stationary bed of wet solids placed on trays with perforated bottoms). These dryers may be operated with or without air recirculation. The rate of drying during the constant rate period in a tray dryer can be estimated from Eq. (5.1.52), once the combined heat transfer coefficient ($h_t$) is evaluated from Eq. (5.1.63), while that in a through circulation dryer can be estimated from Eqs (5.1.79) to (5.1.84).
14. Examples of continuous direct dryers are the continuous tunnel dryer (which is equivalent to a number of batch truck or tray compartment dryers operated in series), the horizontal conveying screen dryer (which is a continuous through-circulation dryer), rotary dryers (which consist of rotating, slightly inclined cylinders inside which the wet solids and the heating medium are contacted), fluidised bed dryers (in which wet solids are dried by fluidising them with hot air or hot gas), vibrating conveyor dryers (in which wet solids are dried by fluidising them on a vibrating conveyor deck), spouted bed dryers (in which gas-solid contacting is achieved in a fluid spout flowing upward through the centre of a loosely packed bed of solids), pneumatic conveyor dryers

or flash dryers (in which solids are dried while being carried upwards by the fluidising hot gas), spray dryers (in which the material to be dried is sprayed in the form of small droplets into a large volume of hot air or hot gas) and foam mat dryers (in which the feedstock is subjected to dehydration in the form of a mat of foam). Performance equations for most of these dryers are available. These dryers are usually operated adiabatically either in the countercurrent or in the cocurrent mode, though nonadiabatic operation (in which case the temperature of the heating medium is maintained constant by installing heating coils within the dryer) is not unusual.

15. Among the indirect batch dryers, vacuum shelf dryers and vacuum rotary dryers operate under vacuum, while the agitated pan dryer is operated either at atmospheric pressure or under vacuum. Vacuum dryers are excellent for drying heat-sensitive and hygroscopic substances. Dusty materials can also be handled without any significant dust loss. The cylinder dryers (popularly used for drying paper and textiles), drum dryers and the rotary steam tube dryers are continuous indirect dryers. In indirect dryers, the solvent vapourised can be easily recovered.
16. Distillation is a process which separates the components of a liquid mixture by virtue of the difference in their relative volatilities.
17. The vapour-liquid equilibrium (VLE) data, on which the design of any distiliation equipment depends, can be determined experimentally or computed analytically based on thermodynamic consideration. The equilibrium ratio ($K$) is a function of temperature, pressure and composition. It can be evaluated from Eq. (5.2.3) and the Redlich-Kwong-Soave equation of state [Eq. (5.2.4)] with reasonable accuracy. If the total pressure is less than 2.0 atm and the critical teperature of all components is greater than the system temperature, then $K_i$ may be more conveniently computed based on the activity concept from Eqs (5.2.15) and (5.2.16). Instead of Eq. (5.2.16), which is based on Scatchard and Hildebrand's regular solution theory, Margules and van Laar equations may be used to compute the activity coefficient. If the system exhibits ideal behaviour, then the equilibrium relationship is predicted by the Raoult's law Eq. (5.2.22).
18. The separation efficiency of conventional distillation process is predicted by the magnitude of the relative rolatility ($\alpha$) defined in Eqs (5.2.23) and (5.2.25). When $\alpha = 1.0$, no separation is possible and the larger the deviation of $\alpha$ from unity, the larger will be the degree of separability.
19. A liquid solution is said to be saturated when it is at the bubble point and a vapour mixture is said to be saturated when it is at the dew point. Below the bubble point, we get cold liquid and above the dew point, the vapour mixture will be superheated. Between the dew point and the bubble point, the stock exists partly as vapour and partly as liquid or as a vapour-liquid mixture. The difference between the dew point and the bubble point is called the *boiling range,* which is a characteristic of every miscible mixture.
20. In *flash distillation* or *equilibrium distillation* (which is a single stage process), the feed mixture is introduced into a low pressure flash drum or flash still and due to the sudden decrease in pressure, a part of the feed vapourises and the issued vapours ar condensed and collected as the distillate ($D$). The residual liquid ($B$) in the still, which is assumed to be in equilibrium with the issued vapours, may be further flashed successively in a series of flash drums. The compositions of products from a flash distillation process may be computed from Eqs (5.2.30) or (5.2.31). For binary systems, a graphical procedure such as that shown in Fig. 5.22 may be employed.

21. *Differential distillation* or *differential weathering* is a batch process in which the vapours are removed continuously as soon as they are produced. The vapours evolved at any instant from the boiling liquid mixture are assumed to be in equilibrium with it. For binary mixtures, the process is conducted usually at constant pressure and in such cases, the product composition can be computed from the *Rayleigh equation* Eq. (5.2.35). For systems exhibiting ideal behaviour, Eq. (5.2.37) or (5.2.38) may be used. For multicomponent systems, the process may be conducted at constant pressure or at constant temperature (*isothermal weathering*). In the former case, the feed is first heated to its bubble point at the operating pressure and the temperature is then gradually increased (keeping pressure constant) until the composition of the residual liquid in the still reaches the specified value. In isothermal weathering, the feed is first heated to its bubble point at a specified pressure and then the pressure is gradually lowered keeping the temperature constant. The final pressure may be computed from Eq. (5.2.42) provided the system exhibits ideal behaviour. The product compositions are obtained by solving Eqs (5.2.39) and (5.2.40) simultaneously. A trial and error procedure will be required for the solution of these two equations, if the operation is at constant pressure.
22. *Steam or inert distillation* is used to separate high boiling mixtures or when the feed stock is thermally unstable at its normal boiling temperature. For a batch process on a binary system, once the operating pressure ($P$) is fixed, the operating temperature will be that value of $T$ which satisfies Eq. (5.2.45). It is seen that this value of $T$ is less than the boiling point of the component distilled (say, $A$) as well as that of water. Thus, the substance $A$ is distilled at a temperature lower than its true boiling point, which not only improves the economy of the process but also helps in inhibiting thermal decomposition of $A$. The moles of carrier steam required can be computed from Eq. (5.2.47). If steam is also used as the heating fluid, then the total steam consumption ($S$) is obtained from Eq. (5.2.48).
23. For large capacity installations, continuous fractionation is more effective and more economical than either flash distillation or differential distillation. A continuous fractionator (a plate column or a packed column) is equipped with an overhead condenser to supply the reflux (the overhead vapours are either partially or totally condensed and a part of the condensate liquid is returned as reflux to the top of the column or to the topmost plate) and a reboiler at the bottom to produce vapour from the bottom liquid and return to the column (the reboiler, thus, supplies both heat and vapour to the column). As a result of this, the vapour and the liquid will flow countercurrently along the column and there will be intimate vapour-liquid contact at each plate in a plate column or at every point in the packed bed of a packed column. If the vapour and the liquid leaving a plate in the plate column are in equilibrium with each other, then the plate acts as an ideal or equilibrium stage. The deviation of plate performance from ideality is predicted by the *plate efficiency.*
24. The ratio of the moles of liquid returned as reflux ($L_0$) to the moles of distillate collected ($D$) is called the *reflux ratio* ($R$). The ($L/V$) ratio at any stage/plate is called the *internal reflux ratio.* The vapour returned from the reboiler is usually in equilibrium with the liquid that is withdrawn as the residue ($B$) and therefore, the reboiler can be treated as an ideal, equilibrium stage. If the feed is an aqueous solution and the residue withdrawn is water, then the reboiler may be dispensed with and the heat required may be provided by admitting steam (from an external source) directly to the bottom of the tower. This is called the use of *open steam.* However, this shall demand larger number of plates/trays in the column for the same reflux ratio and distillate composition.

25. The number of ideal stages required for fractionating a binary mixture can be determined by the method proposed by McCabe and Thiele or by the more rigorous methods of Sorel or Ponchon and Savarit. McCabe and Thiele's method assumes *constant molal overflow,* that is, the $(L/V)$ ratio is assumed to be the same for all plates in any particular section (enriching/stripping section). The solution can be obtained either graphically or analytically. The methods of Sorel and Ponchon and Savarit utilise simultaneous solution of material balance and enthalpy balance equations. These methods are more accurate through they demand larger computational load. In this case also, the solution can be obtained either graphically or analytically. If the equilibrium curve is linear and $\alpha$ is essentially constant, then the number of ideal stages may be computed from *Kremser-Brown-Souders equation* [Eq. (5.2.76) of Example 5.23].
26. If all the overflow vapours are condensed and returned to column as reflux, then the column is said to be operating under *total reflux.* All the residue product is reboiled and no fresh feed is admitted to the column. The capacity of the column reduces to zero although a definite separation will be taking place. The number of stages required is also reduced to a minimum ($N_m$). The value of $N_m$ can be estimated graphically as shown in Fig. 5.27.

    For systems with constant relative volatility, $N_m$ may be computed from *Fenske's equation* [Eq. (5.2.71)]. Conversely, when the reflux ratio is reduced to a minimum ($R = R_m$), the number of stages required to affect a given separation becomes infinitely large. The value of $R_m$ can be computed analytically (see Program 5.8 of Section 5.3) or estimated graphically as shown in Fig. 5.28 for the general case or as shown in Figs 5.29 (a) and (b) for special cases. The operating reflux ratio is usually chosen to be 1.2 to 1.5 times the minimum. It is also possible to propose an *optimum reflux ratio* at which the total cost (the fixed charges on the equipment as well as the steam and cooling water costs) is a minimum as shown in Fig. 5.30.
27. The number of actual plates required can be computed only when the plate efficiency is known. A reasonable estimate of Murphree plate efficiency ($E_{mg}$ or $E_{ma}$) can be obtained from Eqs (4.5.45) to (4.5.64) of Chapter 4. The number of actual plates required is to be then determined either graphically as shown in Fig. 5.34 or analytically as $N \text{ (actual)} = 1 + \sum_{i=1}^{N_p} (1/E_{mai})$ with allowable error. Here $E_{mai}$ is the murphree stage efficiency of the $i$th plate.
28. The design of plate columns for distillation is exactly analogous to that for gas absorption (discussed in Chapter 4). The column diameter is to be selected based on the flooding and entrainment characteristics. The column height depends on the total number of plates required and the plate spacing. It is also necessary to compute the pressure drop across the plate so as to ensure that the performance of the plate is stable and steady. The pressure drop across a bubble cap plate can be estimated from Eqs (5.2.79) and (5.2.80) and that across sieve plates from Eqs (5.2.81) and (5.2.82).
29. Packed columns are viable substitutes to plate columns particularly when heat-sensitive and corrosive substances are be distilled. For the distillation of binary mixtures and based on the assumption of constant molal overflow, the required packed height of the column may be computed as $(H_{tG} N_{tG})$ or $(H_{tL} N_{tL})$ where the expressions for $H_{tG}$, $H_{tL}$, $N_{tG}$ and $N_{tL}$ are given in Eqs (5.2.117) to (5.2.123). The height should be computed separately for the enriching section and the stripping

section. The interfacial concentrations (for use in these equations) may be determined analytically by trial or graphically as shown in Fig. 5.40. The column diameter is to be computed based on the flooding plots of Fig. 4.19 of Chapter 4. The same plots may be used for estimating the pressure drop across the packed bed.

30. As stated earlier, conventional fractional distillation will be successful only if the relative volatility ($\alpha$) deviates markedly from unity. If $\alpha = 1.0$, then the mixture boils at a constant temperature producing vapours that are of the same composition as the liquid. The mixture is then said to form an *azeotrope* and separation by simple fractional distillation becomes impossible. If the azeotropic temperature is lower than the boiling point of either component; then it is called a *minimum boiling azeotrope.* Similarly, if the boiling point of the azeotrope is more than the boiling point of either component, then it is called a *maximum boiling azeotrope.* Azeotropes which exist as one liquid phase in equilibrium with vapour are called *homogeneous azeotropes,* while those exist as two liquid phases in equilibrium with vapour are called *heterogeneous azeotropes.* The latter can be separated with relative ease in two conventional fractionators since they form two liquid phases of widely different composition at temperatures at or below the boiling point. Homogeneous azeotropes that are highly pressure sensitive can also be separated by employing a two column fractionation scheme as illustrated in Fig. 5.43 or 5.44. To separate homogeneous azeotropes that are not pressure sensitive and also to separate mixtures for which the relative volatility is very close to unity, an additive (called *entrainer* or *solvent*) is added to the feed mixture and then it is fractionated. If the additive is collected in the bottom residue alongwith the less volatile component $B$ (the top distillate containing essentially $A$), then the process is called *extractive distillation* and the additive is called the *solvent.* If the additive distils off alongwith $A$ in the top distillate (the bottom residue containing essentially $B$), then the process is called *azeotropic distillation* and the additive is called the *entrainer.*

31. Analysis of multicomponent distillation problems is fairly complicated. A large of number of numerical algorithms have been proposed in the literature. Each algorithm discusses a mathematical method for the solution of the MESH equations [Eqs (5.2.128) to (5.2.131)]. The choice of the algorithm depends on the type of system handled, the computation time available and the degree of accuracy desired. The F-U-G method, that utilises Fenske's equation to estimate the number of theoretical stages at total reflux, Underwood's method to compute the minimum reflux ratio and Gilliland's method to estimate the total number of theoretical plates (ideal stages) required, is presented through Eqs (5.2.140) to (5.2.164). Though an approximate method, it helps in predicting the order of magnitude of the performance variables. For more rigorous methods such as those of Wang and Henke (tridiagonal method) and Naphtali and Sandholm (simultaneous correction method), references (5) and (31) may be consulted.

## NOMENCLATURE

$A$ Drying area; area of cross-section of dryer/column, m$^2$

$A_a$ active plate area, m$^2$

$A_{an}$ annular area per plate, m$^2$

$A_{ap}$ area under apron, m$^2$

$A_d$ cross-sectional area of each downcomer, $m^2$
$A_{dm}$ minimum area of flow under the downcomer apron, $m^2$
$A_h$ total hole area per plate, $m^2$
$A_n$ net area for flow of vapour, $m^2$
$A_r$ total riser area per plate, $m^2$
$A_S$ total slot area per plate, $m^2$
$a$ specific interfacial area, $m^2/m^3$
$a_{an}$ annular area per cap, $m^2$
$a_c$ cap area (per cap), $m^2$
$a_r$ riser area per cap, $m^2$
$a_S$ slot area per cap, $m^2$
$B$ molar flow rate of residue, kmoles/s
$b_S$ slot width (at the base), m
$C_B$, $C_A$ specific heat at constant pressure of dry gas and that of associated vapour respectively, kJ/(kg K)
$C_d$ liquid gradient factor [see Eq. (5.2.94)]
$C_{ij}$ binary interaction parameter (see Table 5.2)
$C_{pL}$ heat capacity of liquid, kJ/(kg K)
$C_{pS}$ heat capacity of dry solid, kJ/(kg K)
$C_{pV}$ heat capacity of vapour, kJ/(kg K)
$C_S$ humid heat, kJ/(kg K)
$C_V$ discharge coefficient [defined in Eq. (5.2.112)]
$C_{Vf}$ vapour load correction factor, dimensionless
$D$ dryer diameter, m; column diameter, m; molar flow rate of distillate, kmoles/s
$D_G$ gas phase diffusivity, $m^2/s$
$D_L$ liquid phase diffusivity, $m^2/s$
$D_o$ diameter of orifice/nozzle, m
$d$ slab thickness, m
$d^*$ equivalent diameter of flow passage [defined in Eq. (5.1.66)], m
$d_C$ cap diameter, m
$d_h$ hole diameter, m
$d_p$ volume-surface diameter of particle, m; average particle diameter, m
$d_r$ riser diameter, m
$d_S$ surface diameter of packing element, m
$E_{mG}$ murphree plate efficiency, dimensionless
$E_{ma}$ murphree plate efficiency corrected for entrainment, dimensionless
$E_o$ overall column efficiency, dimensionless
$E_S$ vapourisation efficiency, dimensionless
$F$ feed rate, kmoles/s

$F_g$, $F_L$ local gas phase mass transfer coefficient and local liquid phase mass transfer coefficient respectively, kmoles/(m$^2$ s)

$F_j$ feed rate to the $j$th stage, kmoles/s

$F_W$ correction factor for effective weir length [defined in Eq. (5.2.83)]

$f$ friction factor, dimensionless

$G$ superficial mass velocity of gas, kg/(m$^2$ s)

$G_S$ superficial mass velocity of dry gas, kg/(m$^2$ s)

$H$ (or, $H_G$) enthalpy of gas-vapour mixture, kJ/kg dry gas

$H_d$ height of clear liquid (liquid backup) in downcomer, mm

$H'_d$ height of aerated liquid in downcomer, mm

$H_S$ enthalpy of solid, kJ/kg dry solid; enthalpy of steam, kJ/kg or kJ/kmole

$H_{tG}$, $H_{tL}$ height of one gas phase transfer unit and that of one liquid phase transfer unit respectively, m

$H_V$, $H_L$ enthalpy of saturated vapour and that of saturated liquid respectively, kJ/kg

$H_{Vj}$, $H_{Lj}$ enthalpy of vapour mixture leaving the $j$th stage and that of liquid solution leaving the $j$th stage respectively, kJ/kg or kJ/kmole

$h_C$ convective heat transfer coefficient, W/(m$^2$ K); pressure drop due to vapour flow through dry cap assembly, mm clear liquid

$h_d$ pressure drop caused by liquid flow in downcomer, mm clear liquid

$h_{dS}$ dynamic liquid seal, mm

$h_f$ actual height of froth, mm

$h_g$ liquid gradient across the plate, mm

$h'_g$ uncorrected value of $h_g$, mm

$h_{ls}$ downcomer liquid seal (also called downflow baffle seal), mm

$h_L$ pressure drop for vapour flow through the aerated mass, mm liquid

$h_m$ mean liquid depth, mm

$h_o$ pressure drop due to vapour flow through dry perforations, mm clear liquid

$h_{ow}$ liquid head over weir, mm clear liquid

$h_{rc}$ pressure drop due to vapour flow through the riser and the cap, mm clear liquid

$h_S$ slot height, mm

$h_{sk}$ skirt clearance, mm

$h_{so}$ pressure drop due to vapour flow through slots, mm clear liquid

$h_{sr}$ height of shroud ring, mm

$h_{ss}$ static submergence, mm

$h_{st}$ pressure drop to overcome surface tension effect, mm liquid

$h_t$ total heat transfer coefficient [see Eq. (5.1.52)], W/(m$^2$ K)

$h_W$ weir height, mm

$j_D$ $j$-factor for mass transfer, dimensionless

$K$ equilibrium ratio or equilibrium distribution coefficient, dimensionless

$K_i$ equilibrium ratio of component $i$, dimensionless

$K_C$ coefficient defined in Eq. (5.2.103)

$k_g$ gas phase mass transfer coefficient, kmoles/(m$^2$·s·atm)

$k_m$, $k_S$ thermal conductivity of material of construction of tray and that of solids being dried respectively, W/(m K)

$k_x$ liquid phase mass transfer coefficient, kmoles/(m$^2$·s·$\Delta x$)

$k_y$ gas phase mass transfer coefficient, kmoles/(m$^2$·s·$\Delta y$)

$k'_Y$ gas phase mass transfer coefficient, kg/(m$^2$·s·$\Delta Y'$)

$L$ molar flow rate of liquid in the enriching section, kmoles/s

$L'$ molar flow rate of liquid in the exhausting (or, stripping) section, kmoles/s

$L_j$ molar flow rate of liquid leaving the $j$th ideal stage, kmoles/s

$L_n$ molar flow rate of liquid leaving the $n$th stage in the enriching section, kmoles/s

$L'_m$ molar flow rate of liquid leaving the $m$th stage in the exhausting section, kmoles/s

$L_o$ reflux rate, kmoles/s

$Le$ Lewis Number, dimensionless

$l_W$ weir length, m

$M_B$, $M_A$ moleculer weight of dry gas and that of associated vapour respectively, kg/kmole; molecular weight of less volatile component and that of more volatile component respectively, kg/kmole

$m'$ average slope of equilibrium curve

$m_S$ mass of dry solid, kg

$m_W$ mass of wet solid, kg

$N$ speed of rotation of rotary dryer/centrifugal disk, rps

$N_m$ minimum number of ideal stages at total reflux

$N_p$ total number of ideal stages (excluding reboiler)

$N_S$ moles of carrier steam

$N_{tG}$, $N_{tL}$ number of gas phase transfer units and number of liquid phase transfer units, respectively

$p_A$ partial pressure of vapour in gas–vapour mixture, N/m$^2$ or atm

$p_C$ cap pitch, m

$P$ total pressure, N/m$^2$ or atm

$P'_A$ vapour pressure of liquid (see under Drying), N/m$^2$ or atm

$P'_{AW}$ vapour pressure of liquid at temperature $t_W$, N/m$^2$ or atm

$P_{Ci}$ critical pressure of component $i$, N/m$^2$ or atm

$P'_i$ vapour pressure of pure component $i$, N/m$^2$ or atm

$Pr$ Prandtl number, dimensionless

$Q'$ volume flow rate of liquid per unit flow width [see Eq. (5.2.89)], m$^3$/(s·m)

$Q_B$, $Q_C$ reboiler duty and condenser duty respectively, W

$Q_L$, $Q_V$ mass flow rate of liquid and that of vapour respectively, kg/s

$q$ heat flux, W/m$^2$; moles of saturated liquid formed on the feed plate by the introduction of one mole of feed, dimensionless

$R$ rate of drying, kg/(m$^2$·s); gas constant, J/(kmole·K); reflux ratio, dimensionless

$R_C$ rate of drying during constant-rate drying period, kg/(m$^2$·s)

$R_H$ hydraulic radius of aerated liquid [see Eq. (5.2.98)], m
$R_m$ minimum reflux ratio, dimensionless
$Re$ Reynolds number, dimensionless
$Re_f$ modified Reynolds number [see Eq. (5.2.101)], dimensionless
$S$ mass velocity of dry solid, kg/(m$^2$·s); tray spacing m; steam feed rate, kmoles/s or kg/s
$Sc$ Schmidt number, dimensionless
$s$ dryer slope, m/m
$T$ temperature, K
$T_B$ bubble point, K
$T_{Ci}$ critical temperature of component $i$, K
$T_D$ dew point, K
$T_F$ feed temperature, K
$T_j$ temperature of $j$th stage, K
$T_{ri}$ reduced temperature of component $i$, dimensionless
$T_S$ absolute temperature of drying surface, K
$t_{as}$ adiabatic saturation temperature, °C
$t_G$ dry-bulb temperature of gas–vapour mixture, °C
$t_S$ surface temperature of drying solid, °C
$t_W$ wet-bulb temperature, °C
$U$ overall heat transfer coefficient, W/(m$^2$ K)
$(Ua)$ overall volumetric heat transfer coefficient, W/(m$^3$·K)
$U_f$ velocity of aerated liquid [see Eq. (5.2.100)], m/s
$U_{fn}$ flooding vapour velocity based on net area $A_n$, m/s
$U_j$ molar flow rate of liquid side stream withdrawn from the $j$th ideal stage, kmoles/hr
$U_{VC}$ linear velocity of vapour through risers, m/s
$U_{VS}$ linear velocity of vapour through slots, m/s
$V$ molar flow rate of vapour in the enriching section, kmoles/s
$V'$ molar flow rate of vapour in the exhausting section, kmoles/s
$\bar{V}_i$ liquid molar volume of component $i$, m$^3$/kmole
$V_j$ molar flow rate of vapour leaving the $j$th ideal stage, kmoles/s
$V_n$ molar flow rate of vapour leaving the $n$th stage in enriching section, kmoles/s
$V'_m$ molar flow rate of vapour leaving the $m$th stage in the exhausting section, kmoles/s
$\mathbf{V}_a$ velocity of vapour through active plate area, m/s
$\mathbf{V}_{gn}$ operating vapour velocity based on net area $A_n$, m/s
$v_H$ humid volume, m$^3$/kg
$\mathbf{V}_h$ average velocity of vapour through perforations, m/s
$W$ weir to baffle distance (or, distance between downcomers), m
$W_a$ width of flow path of liquid across the plate (measured normal to flow), m
$W_j$ molar flow rate of vapour side stream withdrawn from the $j$th ideal stage, kmoles/hr
$X$ moisture content of solid on dry basis, kg liquid/kg dry solid

$X^*$ equilibrium moisture content, kg liquid / kg dry solid

$\bar{X}$ free moisture content (= $X - X^*$), kg liquid / kg dry solid

$X_C$ critical moisture content, kg liquid / kg dry solid

$x$ mole fraction of more volatile component in liquid, mole / mole

$x_B$, $x_D$ mole fraction of more volatile component in residue and in distillate liquid respectively, mole / mole

$x_{Bi}$, $x_{Di}$ mole fraction of component $i$ in residue and in distillate liquid respectively, mole / mole

$x_F$ mole fraction of more volatile component in liquid feed, mole / mole

$x_{Fi}$ mole fraction of component $i$ in liquid feed, mole / mole

$x_e$, $y_e$ liquid phase and vapour phase mole fraction respectively of more volatile component at the vapour–liquid interface, mole / mole

$x_{ij}$ mole fraction of component $i$ in liquid leaving the $j$th ideal stage, mole / mole

$x_n$ mole fraction of more volatile component in liquid leaving the $n$th stage, mole / mole

$x_o$ mole fraction of more volatile component in reflux, mole / mole

$Y$ molal humidity of gas, mole / mole

$Y'$ absolute humidity of gas, kg / kg

$Y_S$ molal saturation humidity of gas, mole / mole

$Y'_S$ saturation humidity of gas, kg / kg

$Y'_{as}$ adiabatic saturation humidity of gas, kg / kg

$Y'_W$, $Y_W$ saturation humidity of gas and molal saturation humidity of gas respectively at wet-bulb temperature, kg / kg

$y$ mole fraction of more volatile component in vapour, mole / mole

$y^*$ mole fraction of more volatile component in vapour that is in equilibrium with the liquid, mole / mole

$y_{Di}$ mole fraction of component $i$ in distillate (when distillate is vapour), mole / mole

$y_D$ mole fraction of more volatile component in distillate (when distillate is vapour), mole / mole

$y_{ij}$ mole fraction of component $i$ in vapour leaving the $j$th stage mole / mole

$y_n$ mole fraction of more volatile component in vapour leaving the $n$th stage, mole / mole

$Z$ dryer length, m; height of packed bed, m; compressibility factor, dimensionless

$z_F$ mole fraction of more volatile component in feed (= $x_F$, when feed is liquid; = $y_F$, when feed is vapour), mole / mole

$z_{ij}$ mole fraction of component $i$ in the feed stream entering the $j$th stage, mole / mole

$z_m$ wall thickness of tray bottom, m

$z_S$ height of solid bed on tray, m

## GREEK LETTERS

$\alpha$ relative volatility, dimensionless

$\beta$ aeration factor, dimensionless

$\gamma_i^L$ liquid phase activity coefficient of component $i$, dimensionless

$\delta_i$ solubility parameter of component $i$, $(J/m^3)^{1/2}$
$\Delta H$ total pressure drop across the plate, mm clear liquid
$\Delta H_A$ heat of wetting [see Eq. (5.1.87)], kg/kg dry solid
$(-\Delta P)$ pressure drop, $N/m^2$
$\Delta T'_G$ change in gas temperature owing to heat transfer to solids only (not including heat losses), K
$\Delta T_m$ true mean temperature difference between hot gas and the drying solid, K
$\epsilon$ void fraction (also called porosity) of packed bed, dimensionless; emissivity of drying surface, dimensionless
$\theta$ drying time, s
$\theta_C$ drying time during constant-rate drying period, s
$\theta_f$ drying time during falling-rate drying period, s
$\theta_d$ liquid residence time in downcomer, s
$\lambda$ latent heat of vapourisation, kJ/kg; $m^3$ stripping factor, dimensionless;
$\lambda_F$ latent heat of vapourisation of feed, kJ/kg
$\lambda_{as}$ latent heat of vapourisation at the adiabatic saturation temperature, kJ/kg
$\lambda_o$ latent heat of vapourisation at reference temperature, kJ/kg
$\lambda_S$, $\lambda_W$ latent heat of vapourisation at temperature $t_S$ and at temperature $t_W$ respectively, kJ/kg
$\mu_L$, $\mu_V$ viscosity of liquid and that of vapour respectively, kg/(ms)
$\pi_1$ flow parameter, dimensionless
$\pi_2$ capacity parameter, dimensionless
$\rho_L$, $\rho_V$ density of liquid and that of vapour respectively, $kg/m^3$
$\rho_S$ density of dry solid, $kg/m^3$
$\sigma$ surface tension, N/m
$\lambda_r$ Stefan-Boltzmann constant, $W/(m^2 K^4)$
$\phi$ fractional holdup of solids in dryer, $m^3/m^3$; relative froth density, dimensionless
$\phi_{dc}$ relative froth density in downcomer, dimensionless
$\bar{\phi}_i^{(L)}$, $\bar{\phi}_i^{(V)}$ fugacity coefficient of component $i$ in liquid solution and in vapour mixture respectively, dimensionless
$\phi_t$ total liquid holdup in packed tower, $m^3/m^3$
$\omega_i$ acentric factor of component $i$, dimensionless

## COMPUTER NOTATIONS

A = $A$ drying area, area of cross-section of dryer/distillation column, $m^2$
AA = $A$; $A_a$ parameter defined in Eq. (5.2.12); active plate area
AAL, AAV = $A$ value of parameter $A$ for liquid and for vapour respectively
AAP = $A_{ap}$ area under apron, $m^2$
A (I) = $a_i$ parameter in Eq. (5.2.4)
ABSN = $\epsilon$ void fraction of packed bed, dimensionless
AC = $a_C$ cap area (per cap), $m^2$

AD = $A_d$ area of each downcomer, $m^2$
ADM = $A_{dm}$ minimum area of flow under downcomer apron, $m^2$
AIB (I) = $\overline{a}_i$ parameter in Eq. (5.2.11)
AIBL (I), AIBV (I) = $\overline{a}_i$ value of parameter $\overline{a}_i$ for liquid and for vapour respectively
ALFA (I) = $\alpha_i$ relative volatility of component $i$, dimensionless
AM = $a_m$ parameter in Eq. (5.2.8)
AMLIQ, AMVAP = $a_m$ value of parameter am for liquid and for vapour respectively
AN = $A_n$ net area for vapour flow, $m^2$
ANS — number of ideal stages required (without rounding off)
A0 (I) to A3 (I) = $A_{0i}$ to $A_{3i}$ coefficients in the empirical equation of Amundson and Pontinen
A0, A1 = $a_0$, $a_1$ empirical constants of Eq. (4.5.24)
AR = $A_r$ riser area per plate, $m^2$
AS = $a_S$ slot area per cap, $m^2$
AST = $A_S$ slot area per plate, $m^2$
B = $B$ molar flow rate of residue, kmoles/s
B1 = $B_1$ breadth of each tray of tray dryer, m
B (I) = $b_i$ parameter in Eq. (5.2.4)
BB = $B$ parameter in Eq. (5.2.13)
BBL, BBV = $B$ value of parameter $B$ for liquid and for vapour respectively
BETA = $\beta$ parameter in Eq. (4.5.129)
BETA (I) = $\beta_i$, $\beta$ parameter in Eq. (5.2.9), values of aeration factor (dimensionless)
BETAC — calculated value of aeration factor
BM = $b_m$ parameter in Eq. (5.2.7)
BML, BMV = $b_m$ value of parameter $b_m$ for liquid and for vapour respectively
BS = $b_s$ slot width (at the base), m
C (I, J) = $C_{ij}$ binary interaction parameter
CD = $C_d$ liquid gradient factor
CPA = $C_{pa}$ specific heat of dry air, J/(kg·K)
CPC = $C_p$ specific heat of cooling water, J/(kg·K)
CPL = $C_{pL}$ specific heat of liquid/water, J/(kg·K)
CPS = $C_{pS}$ specific heat of dry solids, J/(kg·K)
CPV = $C_{pV}$ specific heat of vapour, J/(kg·K)
CS1 = $C_{S1}$ humid heat of inlet air; that of outlet air for rotary countercurrent dryer, J/(kg·K)
CS2 = $C_{S2}$ humid heat of inlet air to rotary countercurrent dryer, J/(kg·K)
CVF = $C_{Vf}$ vapour load correction factor, dimensionless
CVFC — calculated (by interpolation) value of $C_{Vf}$
D = $D$ diameter of rotary dryer/distillation column, m; molar flow rate of distillate, kmole/s
DC = $d^*$; $d_c$ equivalent diameter; cap diameter, m

DEL — clearance between material in one tray and bottom surface of tray above

DG = $D_G$ gas phase diffusivity, $m^2/s$

DHA = $\Delta H_A$ heat of wetting, J/kg

DHT = $\Delta H$ total pressure drop across the tray, mm

DL = $D_L$ liquid phase diffusivity, $m^2/s$

DP = $d_p$ diameter of particle/packing material, m

DR = $d_r$ riser diameter, m

DS = $d_S$ surface diameter of packing material, m

DT = $D$ tower diameter, m

DTM1, DTM2, DTM3 = $\Delta T_m$ true mean temperature difference between hot gas and solids in zones I, II, III respectively, K

DV = $d_V$ volumetric diameter of particle, m

DVS = $d_{VS}$ volume-surface diameter of particle, m

EF = $E_o$ overall plate (column) efficiency, dimensionless

F = $F$ molar feed rate, kmoles/s

FG (J) = $F_g$ local gas phase mass transfer coefficient at section $j$, kmoles/($m^2 \cdot s$)

FL — fraction of feed that is in liquid state

FV = $F_V$ parameter in Eq. (5.2.96)

FVA = $F_{Va}$ parameter defined in Fig. 5.36

FVA (I) — values of $F_{Va}$ (see Table 5.3.1)

FVAC — calculated value of $F_{Va}$

FW = $F_W$ weir correction factor,

G = $G$ average superficial mass velocity of air, kg/($m^2 \cdot s$)

G1 = $G_1$ superficial mass velocity of inlet air, kg/($m^2 \cdot s$)

GAMA = $\gamma$ ratio of cap spacing to cap diameter

GMX = $G_{max}$ maximum permissible superficial mass velocity of air, kg/($m^2 \cdot s$)

GS = $G_S$ superficial mass velocity of dry air kg/($m^2 \cdot s$)

H = $h$ step size in numerical integration

HAP = $h_{ap}$ apron clearance, m

HB = $H_B$ enthalpy of bottom residue, J/kg

HC = $h_C$ convective heat transfer coefficient, W/($m^2 \cdot K$)

HD = $H_d$; $H_D$ liquid backup in downcomer; enthalpy of distillate, J/kg

HDP = $H'_d$ height of aerated liquid in downcomer

HDS = $h_{ds}$ dynamic liquid seal

HF = $H_F$ enthalpy of feed, J/kg

HG = $h_g$ liquid gradient across the plate

HG1, HG2 = $H_{G1}$, $H_{G2}$ enthalpy of outlet and inlet air respectively (in rotary countercurrent dryer), J/kg

HGP = $h'_g$ uncorrected value of $h_g$

HGP1, HGP2 = $H'_G$, $H''_G$ enthalpy of air entering and leaving zone II respectively, J/kg
HL = $h_L$ pressure drop for vapour flow through aerated mass, mm
HL (J) = $H_{Lj}$ enthalpy of liquid leaving $j$th stage, J/kg or J/kmole
HLO = $H_{LO}$ enthalpy of reflux, J/kg or J/kmole
HLS = $h_{lS}$ downcomer liquid seal
HOW = $h_{ow}$ liquid head over weir
HRC = $h_{rc}$ pressure drop due to vapour flow through riser and cap, mm
HS = $h_S$ slot height, m
HS1, HS2 = $H_{S1}$, $H_{S2}$ enthalpy of solids entering and leaving rotary dryer respectively, J/kg
HSK = $h_{Sk}$ skirt clearance
HSO = $h_{so}$ pressure drop due to vapour flow through slots, mm
HSP1, HSP2 = $H'_S$, $H''_S$ enthalpy of solids leaving and entering zone II respectively, J/kg
HSR = $h_{Sr}$ height of shroud ring
HSS = $h_{SS}$ static submergence, mm
HT = $h_t$ total heat transfer coefficient, W/(m$^2$·K)
HTG = $H_{tG}$ height of gas phase transfer unit, m
HV (J) = $H_{Vj}$ enthalpy of vapour leaving $j$th stage, J/kg or J/kmole
HW = $h_W$ weir height, m
JD = $j_D$ $j$-factor for mass transfer, dimensionless
K = $K$ parameter in Eq. (5.1.97)
K (I) = $K_i$ equilibrium ratio of component $i$, dimensionless
KA = $k_a$ thermal conductivity of air, W/(m·K)
KB (I) — value of $K_i$ for bottom residue (at its bubble point), dimensionless
KC = $K_C$ coefficient defined in Eq. (5.2.103)
KD (I) — values of $K_i$ for top distillate (at its dew point), dimensionless
KF (I) — values of $K_i$ for feed, dimensionless
KM = $k_m$ thermal conductivity of material of construction of trays, W/(m·K)
KL (J) = $k_L$ liquid phase mass transfer coefficient at any $j$, m/s
KS = $k_S$ thermal conductivity of solids being dried W/(m·K)
KY = $k'_Y$ gas phase mass transfer coefficient, kg/(m$^2$·s·$\Delta Y'$)
KX (J) = $k_x$ liquid phase mass transfer coefficient at any $j$, kmoles/(m$^2$·s·$\Delta x$)
KY (J) = $k_y$ gas phase mass transfer coefficient at any $j$, kmoles/(m$^2$·s·$\Delta y$)
L = $L$ molar liquid flow rate in enriching section, kmoles/s
L — linear dimension of cubical particle, m
L1 = $L_1$ length of each tray of tray dryer, m
LMD (I) — latent heat of vapourisation of water at pressure PST (I) and temperature TST (I), J/kg
LMDF = $\lambda_F$ latent heat of vapourisation of feed, J/kg

| | | |
|---|---|---|
| LMDR | = $\lambda_{ref}$ | latent heat of vapourisation of water at reference temperature, J/kg |
| LMDS | = $\lambda_S$ | latent heat of vapourisation of water at temperature $t_S$, J/kg |
| LMDW | = $\lambda_W$ | latent heat of vapourisation of water at temperature, $t_W$, J/kg |
| LMY | = $(1 - y)_{im}$ | logarithmic mean of $(1 - y)$ and $(1 - y_e)$ |
| LO | = $L_o$ | reflux rate, kg/s or kmoles/s |
| LP | = $L'$ | molar liquid flow rate in stripping section kmoles/s |
| LT | = $l_t$ | plate spacing, m |
| LW | = $l_W$ | weir length, m |
| M2 | — | product delivery rate, kg/s |
| M1, M2 | = $M_1, M_2$ | molecular weight of component 1 and that of component 2, kg/kmole |
| MC | — | mass flow rate of cooling water, kg/s |
| MEUA | = $\mu_a$ | viscosity of air, kg/(m·s) |
| MEUL, MEUV | = $\mu_L$, $\mu_V$ | viscosity of liquid solution and that of vapour mixture respectively, kg/(m·s) |
| MGM | — | maximum flow rate of air |
| MS | = $m_S$; $\dot{m}_S$ | mass of dry solid; mass flow rate of dry solids |
| MV, ML | — | molecular weight of vapour mixture and that of liquid solution respectively, kg/kmole |
| N | = $N$; $n$ | speed of rotation of rotary dryer; number of components |
| NC | = $N_C$ | number of bubble caps |
| NS | = $(N_p + 1)$ | total number of equilibrium stages |
| NP | = $N_p$ | total number of plates |
| NTG | = $N_{tG}$ | number of gas phase transfer units |
| NTG1, NTG2, NTG3 | — | value of $N_{tG}$ for zone I, zone II and zone III respectively |
| OM (I) | = $\omega_i$ | acentric factor of component $i$, dimensionless |
| P | = $P$ | operating pressure, N/m$^2$ |
| PA | = $P'_A$ | vapour pressure of water at temperature $t_S$, N/m$^2$ |
| PAW | = $P'_{AW}$ | vapour pressure of water at temperature $t_W$, N/m$^2$ |
| PC (I) | = $P_{Ci}$ | critical pressure of $i$th component |
| PST (I) | = | pressure of saturated steam at temperature TST (I) |
| PSI | = $\psi$ | sphericity of particle; fractional entrainment, dimensionless |
| PV (I) | = $P'_i$ | vapour pressure of $i$th component, N/m$^2$ |
| PV (I, J) | — | vapour pressure of $i$th component at temperature T (J), N/m$^2$ |
| PVA (I) | — | vapour pressure of $i$th component at temperature TAV, N/m$^2$ |
| PVB (I) | — | vapour pressure of $i$th component at temperature TBW, N/m$^2$ |
| PVF (I) | — | vapour pressure of $i$th component at temperature TBF, N/m$^2$ |
| PF (I) | — | values of percentage flood (Table 5.3.2) |
| PFC | — | calculated value of percentage flood |

PFM — maximum permissible value of percentage flood (maximum permissible ratio of operating vapour velocity to flooding velocity)

PHI (I) = $\bar{\phi}_i$ fugacity coefficient of $i$th component, dimensionless

PHIDC = $\phi_{dc}$ relative foam density in downcomer, dimensionless

PHIL = $\phi$ solid holdup in rotary dryer; association parameter for less volatile component

PHIL (I) = $\bar{\phi}_i^{(L)}$ liquid phase fugacity coefficient of $i$th component, dimensionless

PHILO = $\phi_o$ solid hold up in rotary dryer in the absence of gas flow

PHIT = $\phi_t$ total liquid holdup in packed bed, dimensionless

PHIV = $\bar{\phi}_i^{(V)}$ vapour phase fugacity coefficient of $i$th component, dimensionless

PI1 = $\pi_1$ flow parameter, dimensionless

PI2 = $\pi_2$ capacity parameter, dimensionless

PIC1 — calculated value of $\pi_1$

PIM (I) = $\pi_1$ (min) values of $\pi_1$ (min) at different values of percentage flood (Table 5.3.2)

PIMC — calculated value of $\pi_1$ (min).

Q = $q$ $q$-value of feed, dimensionless

Q1 = $Q'$ parameter in Eq. (5.2.89)

QP = $q'$ $B/D$ ratio, dimensionless

QL (J), QV (J) = $Q_L$, $Q_V$ mass flow rate of liquid and that of vapour at any $j$ respectively, kg / s

R = $R$ rate of drying, kg / ($m^2 \cdot s$); reflux ratio, dimensionless

RAA = $(A_a / A)$ ratio of active plate area to tower cross-sectional area

RADA = $(A_d / A)$ ratio of downcomer area to tower cross-sectional area

RANR = $(A_{an} / A_r)$ ratio of annular area to riser area per plate

RB — $(B/F)$ ratio

RC = $R_C$; $r_C$ constant rate of drying, kg / ($m^2 \cdot s$); number of rows of caps

RD — $(D/F)$ ratio

RE = $Re$ Reynolds number, dimensionless

REL = $Re_L$ liquid phase Reynolds number

RFB — $(F/B)$ ratio

RG = $R$ Universal gas constant

RLD = $(l_W / D)$ ratio of weir length to tower diameter

RMIN = $R_m$ minimum reflux ratio, dimensionless

RMX = $R_{max}$ maximum rate of drying, kg / ($m^2 \cdot s$)

ROWA 1 — density of inlet air, kg / $m^3$

ROWAP = $\rho_{ap}$ apparent density of cake / packed bed, kg / $m^3$

ROWL = $\rho_L$ density of liquid / water, kg / $m^3$

ROWL1, ROWL2 — liquid phase density of component 1 and that of component 2, kg / $m^3$

ROWL (J) = $\rho_L$ density of liquid solution at any $j$, kg / $m^3$

ROWS = $\rho_S$ density of dry solids, kg / m$^3$

ROWV (J) = $\rho_V$ density of vapour mixture at any $j$, kg / m$^3$

RS — $(S/G_S)$ ratio

RSC = $(a_S/a_C)$ ratio of slot area to cap area

RSR = $(A_S/A_R)$ ratio of slot area to riser area

RWD = $(W/D)$ ratio of weir-baffle distance to tower diameter

S = $S$ mass velocity of dry solids, kg / (m$^2$·s); plate spacing, m

S (I) = $S_i$ parameter in Eq. (5.2.10)

SC = $Sc$; $S_c$ Schmidt number; cap spacing, m

SCG, SCL = $Sc_G$; $Sc_L$ gas-phase and liquid phase Schmidt number, dimensionless

SHD = $h_d$ pressure drop due to liquid flow in downcomer, mm

SHL (K), SHV (K) — values of saturated liquid enthalpy and saturated vapour enthalpy respectively, J / kg or J / kmole

SIA = $a$ specific interfacial area, m$^2$/m$^3$

SL = $s$ slope of the dryer, m / m

SP = $S_p$ surface area of each particle, m$^2$

SPC = $p_C$ cap pitch, m

SPS — specific surface (surface area per unit volume) of each particle, m$^2$/m$^3$

SV1 = $\bar{V}_A$ molal volume at normal boiling point of more volatile component, m$^3$/ kmole

T = $T$ operating temperature, K

TAV — average of TBF and TBW

TBF, TBW — bubble point of feed and that of residue respectively, K

TC (I) = $T_{Ci}$ critical temperature of $i$th component, K

TDF — dew point of feed, K

TE (K) = $T$ equilibrium temperature, K

TETAD = $\theta_d$ liquid residence time in downcomer, s

TF = $T_F$ feed temperature, K

TG = $t_G$ dry-bulb temperature of inlet air, °C

TG1, TG2 = $t_{G1}$, $t_{G2}$ dry-bulb temperature of air leaving and entering rotary countercurrent dryer respectively, °C

TGP1, TGP2 = $t'_G$, $t''_G$ temperature of air entering and leaving zone II respectively, °C

THC = $\theta_C$ drying time (constant rate), s

TL (J) — bubble point of liquid at any $j$, K

TR (I) = $T_{ri}$ reduced temperature of $i$th component

TREF = $t_{ref}$ reference temperature, °C

TS = $t_S$ surface temperature of drying solids

TS1, TS2 = $t_{S1}$, $t_{S2}$ temperature of solids entering and leaving rotary countercurrent dryer, °C

TSP1, TSP2 = $t'_S$, $t''_S$ surface temperature of solids leaving and entering zone II, °C
TST (I) — temperature of saturated steam at pressure PST (I), °C or K
TV (J) — dew point of vapour at any $j$
TW = $t_W$ wet-bulb temperature of air, °C
TWP = $t_{W2}$ wet-bulb temperature of air at inlet to zone II, °C
UFN = $U_{fn}$ flooding vapour velocity based on $A_n$, m/s
UK = $U_k$ combined heat transfer coefficient Eq. (5.1.57), W/(m$^2$·K)
UV = $(Ua)$ volumetric heat transfer coefficient, W/(m$^3$·K)
UVC, UVS = $U_{VC}$, $U_{VS}$ vapour velocity through riser and through slots respectively, m/s
V = $V$ molar flow rate of vapour in enriching section, kmole/s
V (J) = $V_j$ mass flow rate of vapour leaving $j$th stage, kg/s
VA — average air velocity, m/s
VC (J) — calculated value of $V_j$
VDR — vapour distribution ratio, dimensionless
VGN = $\mathbf{V}_{gn}$ operating vapour velocity based on $A_n$, m/s
VH1 = $\mathbf{V}_{H1}$ humid volume of inlet air, m$^3$/kg
VL — molar volume of liquid leaving the plate
VP = $V'$; $V_p$ molar flow rate of vapour in exhausting section (kmoles/s); volume of particle, m$^3$
VPC (J) — calculated value of $V'_j$
W = $W$ distance between downcomers, m
WA = $W_a$ width of flow path of liquid across the plate, m
X, Y = $x$, $y$ mole fraction of more volatile component in incoming liquid and in outgoing vapour respectively at any section $j$ of packed tower
X (I) = $x_i$ mole fraction of $i$th component in liquid
X (J) = $x_j$ mole fraction of more volatile component in liquid leaving the $j$th stage
X1, X2 = $X_1$, $X_2$ moisture content (on dry basis) of solids entering and leaving the dryer respectively
XA, YA = $(x_a, y_a)$ point of intersection of $q$-line with operating line of enriching section
XB, XD = $x_B$, $x_D$ mole fraction of more volatile component in residue and that in distillate respectively
XB (I) = $x_{Bi}$ mole fraction of $i$th component in residue
XE (K) = $x^*_k$; $x^*$ equilibrium mole fraction of $k$th component in liquid; equilibrium mole fraction of more volatile component in liquid at TE (K)
XF (I) = $x_{Fi}$ mole fraction of $i$th component in liquid feed
XI (J) = $x_e$ interfacial mole fraction of more volatile component at any $j$
XO = $x_o$ mole fraction of more volatile component in reflux

| | |
|---|---|
| (XQ, YQ) = $(x_q, y_q)$ | point of intersection of $q$-line with equilibrium curve |
| Y (I) = $y_i$ | mole fraction of $i$th component in vapour |
| Y (J) = $y_j$ | mole fraction of more volatile component in vapour leaving $j$th stage |
| Y1, Y2 = $Y'_1, Y'_2$ | absolute humidity of inlet air and outlet air respectively; those of exit air and inlet air respectively in case of rotary countercurrent dryer |
| Y1, Y2 = $y_1, y_2$ | mole fraction of more volatile component in reboiled vapour and overhead vapour respectively for a packed tower |
| YA = $(Y'_1 + Y'_2)/2$ | average humidity of air |
| YB = $y_B$ | mole fraction of more volatile component in reboiled vapour |
| YD = $y_D$ | mole fraction of more volatile component in distillate (when it is vapour) |
| YD (I) = $y_{Di}$ | mole fraction of $i$th component in distillate (when distillate is vapour) |
| YE (K) = $y^*_k$; $y^*$ | equilibrium mole fraction of $k$th component in vapour; equilibrium mole fraction of more volatile component in vapour at temperature TE (K) |
| YI (J) = $y_e$ | interfacial mole fraction of more volatile component at any $j$ |
| YW = $Y'_W$ | saturation humidity of air at $t_W$ |
| YWP = $Y'_W$ | saturation humidity of air at TWP entering zone II |
| Z = $Z$ | height of packed bed; length of dryer; compressibility factor |
| ZD = $z_D$ | mole fraction of more volatile component in distillate |
| ZE = $Z_e$ | height of enriching section of packed tower, m |
| ZF = $z_F$ | mole fraction of more volatile component in the feed |
| ZM — | wall thickness of tray (in tray dryer), m |
| ZS = $z_S$; $Z_S$ | height (or depth) of each tray of tray dryer; height of stripping section of packed tower |
| ZV, ZL — | compressibility factor for vapour and for liquid respectively, dimensionless |

## EXERCISES

5.A. Agree with or contradict against the following statements giving reasons. Supplement your answers with examples / sketches wherever possible:

(*a*) For air-water system, the wet-bulb lines and the adiabatic saturation lines are linear and identical,

(*b*) During the constant rate period, the surface temperature of the drying solid remains constant and will be equal to the wet-bulb temperature of the inlet air,

(*c*) Rotary dryers are best operated in the countercurrent mode rather than in the cocurrent mode,

(*d*) In contrast to granular and crystalline solids like sand and minerals, materials like soap and gelatin dry essentially in the constant rate period and therefore, they cannot be dried to very low moisture content in commercial dryers, (*e*) For drying heat-sensitive and hygroscopic substances, direct dryers are more advantageous than the indirect dryers,

(*f*) In distillation, separation is maximum when the relative volatility is equal to unity,

(*g*) Since at total reflux, the number of stages required reduces to a minimum, operation close to total reflux is most desirable,

(*h*) For separating high boiling liquids that are miscible with water, steam distillation is more desirable than flash distillation since the cost of creation of vacuum is much higher than the cost of steam,

(*i*) When the feedstock is an aqueous solution containing a high boiling liquid (for example, aqueous monoethanolamine), the use of open steam is desirable since this eliminates the cost of reboiler.

5.B. Distinguish between:

(*a*) Bound moisture and unbound moisture,

(*b*) Azeotropic distillation and extractive distillation,

(*c*) Equilibrium moisture content and critical moisture content,

(*d*) Fluidised bed dryers and spouted bed dryers,

(*e*) Hygroscopic and nonhygroscopic materials,

(*f*) Wet bulb temperature and adiabatic saturation temperature,

(*g*) Flash distillation and differential distillation,

(*h*) Minimum reflux and total reflux,

(*i*) Heterogeneous azeotropes and homogeneous azeotropes.

5.C. Discuss briefly what do you understand by,

(*a*) Relative volatility,

(*b*) MESH equations,

(*c*) Psychrometric ratio,

(*d*) VLE data,

(*e*) Fenske's equation.

5.D. Recommend a suitable process / equipment for each of the following. Give reasons to your answers. Describe briefly the process / equipment recommended:

(*a*) Manufacture of milk powder from concentrated milk,

(*b*) Production of anhydrous alcohol from an aqueous solution containing 89 mole percent ethanol,

(*c*) Batch drying of fruits like grapes and dates,

(*d*) Drying of large tonnages of paddy,

(*e*) Separation of *n*-heptane from a mixture of *n*-heptane and toluene containing 50 mole percent *n*-heptane.

5.E. In a parallel flow adiabatic humidifier, 4.72 $m^3/s$ of air is admitted at 10° C (dry-bulb) and 4.5° C (wet-bulb) and is brought into contact with liquid water that is admitted at 15.5° C and at the rate of 11.35 $m^3/hr$.

(*a*) If air leaves the humidifier fully saturated and the exit temperature of air is equal to that of water, compute the temperatures of air and water at the exit.

(b) If the operation is not truly adiabatic so that 15 percent of the heat that is theoretically added to air in the adiabatic operation considered above is lost, compute the exit conditions of air and water. Assume that the exit temperature of water is nearly 1.8° C lower than the exit dry-bulb temperature of air.

5.F. The following data were obtained during the drying of a paint pigment in hot air (the sample weighed 3.5 kg when dry):

| *Time, hr* | *Mass of sample, kg* |
|---|---|
| 0.0 | 7.0 |
| 1.0 | 6.65 |
| 2.0 | 6.30 |
| 2.5 | 6.125 |
| 3.0 | 5.95 |
| 3.7 | 5.705 |
| 4.6 | 5.39 |
| 5.7 | 5.00 |
| 6.4 | 4.76 |
| 7.25 | 4.515 |
| 7.80 | 4.375 |
| 9.0 | 4.13 |
| 10.25 | 3.955 |
| 11.35 | 3.85 |
| 13.30 | 3.745 |
| 16.10 | 3.675 |

(a) Plot the drying rate curve and determine the critical moisture content of the material and the rate of drying during the constant rate period.

(b) Compute the time required to dry the same material under the same drying conditions from 95 percent moisture to 8.0 percent moisture (both on dry basis).

5.G. Granular solids (5 mm in diameter, nearly spherical) with dry bulk density 1600 $kg/m^3$ are to be dried from 50 percent moisture to 2.0 percent moisture (both on dry basis) using air that is admitted at 65° C (dry-bulb) and 0.5 percent absolute humidity and at a superficial mass velocity of 6000 $kg/(m^2 hr)$. The critical moisture content of the material is 10 percent (on dry basis) and its equilibrium moisture content is negligible. It may be assumed that the rate of drying falls to zero along a straight line in the falling rate period. The following options are available:

(a) A batch, cross-circulation tray dryer with well-insulated 0.8 × 0.8 m trays, each 25 mm in depth and made of 2 mm carbon steel [thermal conductivity = 36.3 W/(m K)] stacked in such a way that a 45 mm clearance exists between the material in each tray and the bottom surface of the tray above,

(b) the same batch tray dryer as above, except that the trays are not adequately insulated,

(c) a through circulation dryer in which the solids are packed to form a 100 mm thick, 1.2 × 1.2 m bed with an average voidage of 0.55.

Estimate which of the above shall provide the highest rate of drying (or, the lowest time of drying). Take thermal conductivity of the solids to be 2.5 W/(m K) and emissivity = 0.9.

5.H. A clayish sand of bulk density 1765 kg/m$^3$ is to be dried continuously from 150 percent moisture to 20 percent moisture (both on dry basis) by countercurrently contacting with hot air in a direct continuous dryer. Ambient air is available at 27° C (dry-bulb) and 10° C (wet-bulb) and it is mixed with the recycle air (50 percent of the exit air from the dryer is recycled) and preheated to 120° C before entrance to the dryer. The relative humidity of exit air is 0.8 and the wet solids enter the dryer at the adiabatic saturation temperature of the exit air. The critical moisture content of the material is 15 percent (on dry basis). If the dryer operates adiabatically and heat transfer by conduction and radiation may be neglected, compute the length of the dryer required for the following two cases:

(*a*) If a horizontal conveyor dryer with a 1 m wide belt running at 20 m/min is being used (assume that the solids form a layer 25 mm thick on the belt and the space for air flow above the belt is 30 cm high),

(*b*) If a 2 m diameter rotary dryer rotating at 5 rpm and sloped at 0.5 m in 10 m of length is being used [the solid flow rate, on dry basis, remaining the same as in (*a*)].

Take the specific heat of solids (when dry) to be 0.836 kJ/(kg K). Discuss which of the above two dryers you would recommend for this process, giving reasons.

5.I. Simons et. al[40] have reported experimental data on the drying of rayon yarn skeins after centrifuging. It is reported that drying takes place essentially in the falling rate period and their data under constant drying conditions correlate into the following form:

$$-(dX/d\theta) = 0.01378\ G^{1.47}\ (Y'_W - Y')\ (X - X^*)$$

where $Y'_W$ = saturation humidity at the wet-bulb temperature of air.

If a continuous, countercurrent compartment dryer operating at 1 atm is to be used to dry the yarn from 80 percent free moisture (on dry basis) to 1.0 percent free moisture (on dry basis), compute the time required. Air is admitted to the dryer at 65° C (dry-bulb) and with relative humidity = 0.10. The air temperature is maintained constant along the dryer by means of heating coils. The average velocity of air at the inlet is 3.0 m/s. The equilibrium moisture content of the yarn for the above conditions is 3.6 percent (on dry basis) and the feed rate of wet yarn to the dryer is 850 kg/(m$^2$ hr).

5.J. Compute the VLE data for benzene-toluene system at 95° C and 101.3 kPa,

(*a*) from Redlich-Kwong-Soave equation of state,

(*b*) from Hildebrand and Scatchard's regular solution theory,

(*c*) from Raoult's law.

Compare your answers with the experimental values reported in literature. Vapour pressure of benzene at 95° C = 1123 mm Hg and that of toluene at 95° C = 452 mm Hg.

5.K. An ethylene glycol-water mixture containing 75 mole percent glycol is discharged from a dehumidification unit.

(*a*) If this mixture is flashed at 30.4 kPa so as to vapourise 60 percent of the feed, determine the compositions of the products and the operating temperature.

(*b*) If it is differentially weathered at 30.4 kPa so as to vapourise 60 percent of the feed, what will be the compositions of the products?

Which of the above two will provide a higher recovery of glycol in the residue? The VLE data of ethylene glycol-water system at 30.4 kPa is given below:

| *Mole fraction of water in liquid* | *Mole fraction of water vapour* |
|---|---|
| 0.0 | 0.0 |
| 0.03 | 0.53 |
| 0.07 | 0.70 |
| 0.10 | 0.78 |
| 0.15 | 0.87 |
| 0.27 | 0.94 |
| 0.46 | 0.98 |
| 0.60 | 0.99 |
| 0.69 | 0.997 |
| 0.77 | 0.998 |
| 1.0 | 1.0 |

5.L. A sample of natural gasoline with the following composition (percent by mole) is flashed at 345 kPa and 95° C:

Ethane = 0.8, propane = 14.0, *i*-butane = 8.5, *n*-butane = 27.0, *i*-pentane = 6.0, *n*-pentane = 13.25, *n*-hexane = 30.45.

(*a*) What fraction of feed is vapourised?

(*b*) What will be the composition of the products?

(*c*) If the above mixture is isothermally weathered at 95° C to yield the same residue composition, determine the distillate composition, fraction of feed vapourised, the initial and final operating pressure.

Assume that the substances form ideal solutions and base your computations on the experimental vapour pressure data[12] (at 95° C) given below:

| *Hydrocarbon* | *Vapour pressure, atm* |
|---|---|
| Ethane | 120.0 |
| Propane | 38.5 |
| *i*-butane | 17.0 |
| *n*-butane | 13.0 |
| *i*-pentane | 6.0 |
| *n*-pentane | 5.0 |
| *n*-hexane | 1.95 |

5.M. An aqueous solution containing 50 mole percent ethylene glycol is to be fractionated in a sieve plate tower at 30.4 kPa at the rate of 1000 kmoles/hr. The residue should contain 95 mole percent glycol and the distillate 2.0 mole percent glycol. If the feed is introduced as saturated liquid at its bubble point and a total condenser is used to return the reflux at its bubble point, compute (both graphically and analytically)

(*a*) the minimum reflux ratio,

(*b*) the minimum number of ideal stages required,

(*c*) the number of ideal stages required if a reflux ratio that is 1.75 times the minimum is being employed,

(*d*) the number of ideal stages required if the feed is saturated vapour at its dew point and the operating reflux ratio is 1.5 times the minimum,

(*e*) the number of actual plates required with the same reflux ratio as in (*d*), but the feed being 35 percent vapour and 65 percent liquid and if the overall column efficiency is 55 percent.

5.N. It is proposed to separate by fractionation a mixture containing 50 mole percent *n*-heptane and 50 mole percent methyl cyclohexane into a distillate product containing 98 mole percent *n*-heptane and a bottom product containing 2.0 mole percent *n*-heptane. The operating pressure is 1 atm. The feed is admitted as saturated liquid at the rate of 10,000 kg / hr. Determine the number of equilibrium stages required if the reflux ratio employed is three times the minimum. The VLE data for *n*-heptane–methyl cyclohexane system at 1 atm is given below:

| *Mole fraction of n-heptane in liquid* | *Saturation temperature, °C* | *Mole fraction of n-heptane* | |
|---|---|---|---|
| | | *In liquid* | *In vapour* |
| 0.0 | 100.8 | 0.0310 | 0.0350 |
| 0.0787 | 100.55 | 0.0580 | 0.0620 |
| 0.1638 | 100.35 | 0.0950 | 0.1030 |
| 0.2486 | 100.15 | 0.1330 | 0.1430 |
| 0.4126 | 99.70 | 0.1800 | 0.1920 |
| 0.5186 | 99.20 | 0.2160 | 0.2290 |
| 0.6056 | 99.00 | 0.2175 | 0.2890 |
| 0.6993 | 98.85 | 0.3170 | 0.3330 |
| 0.7942 | 98.60 | 0.3630 | 0.3810 |
| 0.9338 | 98.50 | 0.4010 | 0.4200 |
| 1.00 | 98.40 | 0.4560 | 0.4750 |
| – | – | 0.5010 | 0.5210 |
| – | – | 0.5590 | 0.5780 |
| – | – | 0.5990 | 0.6180 |
| – | – | 0.6470 | 0.6660 |
| – | – | 0.7090 | 0.7280 |
| – | – | 0.7560 | 0.7710 |
| – | – | 0.7960 | 0.8100 |
| – | – | 0.8430 | 0.8535 |
| – | – | 0.8790 | 0.8900 |
| – | – | 0.9060 | 0.9130 |
| – | – | 0.9310 | 0.9400 |

5.O. An aqueous ammonia solution containing 30 mole percent ammonia is to be separated into a top product containing 99.5 mole percent ammonia and a bottom product containing 0.05 mole percent ammonia by continuous fractionation at 1240 kPa. The feed solution is available at 38° C and it is fed to the column at the rate of 4550 kg / hr. Open steam saturated at 1310 kPa will be introduced underneath the bottom plate. The column will be equipped with total condenser and the reflux is returned at its bubble point.

(*a*) Compute the number of ideal stages required if a reflux ratio equal to 1.5 times the minimum is used.

(*b*) Determine the heating surface required for a preheater that is to be used to preheat the feed to saturation point before feeding to the column. Assume that the preheater is a countercurrent 1 – 1 exchanger and the overall heat transfer coefficient is 2270 W/(m$^2$ K). The VLE data for ammonia-water system at 1240 kPa is given below:

| *Temperature, °C* | *Mass fraction of ammonia in liquid* | *Mass fraction of ammonia in vapour* |
|---|---|---|
| 189.4 | 0.0 | 0.0 |
| 183.1 | 0.0200 | 0.1245 |
| 177.1 | 0.0400 | 0.2361 |
| 171.3 | 0.06 | 0.3341 |
| 165.6 | 0.08 | 0.4241 |
| 159.9 | 0.1000 | 0.5041 |
| 146.3 | 0.1500 | 0.6643 |
| 134.0 | 0.2000 | 0.7783 |
| 129.2 | 0.2200 | 0.8127 |
| 124.5 | 0.2400 | 0.8417 |
| 119.8 | 0.2600 | 0.8663 |
| 110.9 | 0.3000 | 0.9049 |
| 102.2 | 0.3400 | 0.9319 |
| 93.9 | 0.3800 | 0.9516 |
| 86.0 | 0.4200 | 0.9654 |
| 78.4 | 0.4600 | 0.9760 |
| 71.2 | 0.5000 | 0.9830 |
| 56.9 | 0.6000 | 0.9928 |
| 47.6 | 0.7000 | 0.9966 |
| 40.8 | 0.8000 | 0.9986 |
| 35.8 | 0.9000 | 0.9994 |
| 32.1 | 1.0000 | 1.0000 |

**Enthalpy Data**

| *Mass fraction of ammonia* | *Enthalpy of saturated liquid, kJ/kg* | *Enthalpy of saturated vapour, kJ/kg* |
|---|---|---|
| 0.0 | 804.8 | 2782.6 |
| 0.05 | 702.45 | 2727.7 |
| 0.10 | 608.48 | 2671.87 |
| 0.15 | 519.16 | 2616.75 |
| 0.20 | 436.125 | 2558.60 |
| 0.25 | 355.878 | 2499.3 |
| 0.30 | 288.424 | 2440.0 |
| 0.35 | 222.13 | 2380.20 |
| 0.40 | 162.60 | 2321.35 |
| 0.45 | 112.81 | 2260.87 |
| 0.50 | 72.10 | 2199.46 |
| 0.55 | 42.10 | 2135.96 |
| 0.60 | 26.28 | 2072.23 |
| 0.65 | 20.47 | 2005.94 |
| 0.70 | 24.423 | 1936.40 |
| 0.80 | 51.64 | 1792.88 |
| 0.85 | 71.87 | 1714.26 |
| 0.90 | 93.97 | 1630.06 |
| 0.95 | 120.95 | 1528.18 |
| 1.00 | 152.12 | 1288.84 |

5.P. A fractionating column containing 15 plates and equipped with a reboiler and a total condenser is producing a distillate containing 90 mass per cent ethanol and a bottom product containing 0.5 mass percent ethanol from a feed containing 30 mass percent ethanol and the balance water. The column operates at a pressure of 1 atm absolute. The feed is introduced on the twelfth plate from top at a temperature of 12° C and at the rate of 12000 kg / hr. The reflux is returned to the top plate of the çolumn at its bubble point. Cooling water enters the condenser at 10° C and leaves at 48.9° C and is circulated at the rate of 95.0 $m^3$/ hr.

(*a*) Compute the overall plate efficiency of the column, the reboiler duty and the rate at which vapours are produced from the reboiler.

(*b*) How would the results computed in (*a*) change if constant molal overflow is assumed?

(*c*) What will be the change in composition of the bottom product if the same column is operated with a heat exchanger to recover some of the heat lost in the bottom product by preheating the feed to 54.4° C, with the top product unchanged, the reflux ratio unchanged and assuming plate efficiencies remain unchanged?

5.Q. Compute the number of equilibrium stages required to separate a mixture of 22.2 mole percent ethanol in water into a distillate containing 78 mole percent ethanol and a bottom product containing

3.8 mole percent ethanol by the use of open steam having an enthalpy of 2790 kJ/kg fed under the bottom plate at the rate of 0.56 moles of steam per mole of feed. Feed is supplied with an enthalpy of 11304.4 kJ/kmole and the distillate is to be delivered as saturated vapour by a partial condenser. The reflux pump returns reflux at bubble point.

(*a*) Compute the number of ideal stages required and the condenser duty per mole of distillate product.

(*b*) How many ideal stages will be required if constant molal overflow is assumed?

5.R. An aqueous solution of 0.15 mass fraction ammonia is being fed to the top plate of a stripping column at the rate of 10500 kg/hr. The temperature of the feed is 60° C. Heat is supplied to the reboiler by means of saturated steam at 861.4 kN/m$^2$, the condensate leaving the steam trap as a saturated liquid. If the column is operated at 5.44 atm pressure, determine

(*a*) the number of ideal stages required in the column to obtain a top product containing 0.70 mass fraction ammonia and a bottom product containing 0.005 mass fraction ammonia,

(*b*) the steam requirement (in kg/hour) under these conditions.

5.S Design the sieve plate column employed for the separation of ethylene glycol–water mixture discussed in Problem (13). Size the column based on the performance of the topmost plate. Choose a plate layout of 4.5 mm perforations arranged on 15 mm triangular centres and 2 m diameter column. Check, based on flooding, entrainment and weeping characteristics, whether the chosen column diameter is satisfactory. Verify also whether the operation of the plate shall be stable by estimating the pressure drop across the plate.

5.T If the distillation of aqueous ammonia discussed in Problem (1) is performed in a tower packed with 25 mm ceramic berl saddles (surface diameter = 0.032 m) and equipped with a reboiler, compute the required diameter and packed height of the column. Liquid phase diffusivity of ammonia in water at 298 K = 2.35 × 10$^{-9}$ m$^2$/s. Use Shulman et. al's correlations (given in Chapter 4) for computing the mass transfer coefficients. What will be the pressure drop across the packing?

## REFERENCES


1. Lewis, WK, Trans. AIME, **44,** 325, 1922.
2. Bedingfield, CH and Dew, TB, Ind. Eng. Chem., **42,** 1164, 1950.
3. O'Brien, LJ and Stutzman, LF, Ind. Eng. Chem., **42,** 1181, 1950.
4. Treybal, RE, Mass Transfer Operations, Third edition, McGraw Hill, New York, 1981
5. Narayanan, CM and Bhattacharya, BC, Computer Aided Design of Chemical Process Equipment, New Central Book Agency, Calcutta, 1992.
6. Holland, CD, Multicomponent Distillation, Prentice Hall, NJ, 1963.
7. Narayanan, CM, Pandey, BR and Prabhanjan, PJ, Software Packages for Computer Aided Determination of Equilibrium Distribution Coefficient for Multicomponent Distillation Systems, Proc. Seventh National Convention of Chemical Engineers, New Delhi, 1991.
8. Chao, KC and Seader, JD, AIChE J, **7,** 598, 1961.
9. Winkle, MV, Distillation, McGraw Hill, New York, 1967.
10. Amundson, NR and Pontinen, AJ, Ind. Eng. Chem., **50** (5), 730, 1958.

11. Katz et. al., Handbook of Natural Gas Engineering, McGraw Hill, New York, 1959.
12. Maxwell, JB, Data Book on Hydrocarbons, Van Nostrand Company, New Jersey, 1950.
13. McCabe, WL and Thiele, EW, Ind. Eng. Chem., **17,** 605, 1925.
14. Ponchon, M, Tech. Mod., **13,** 20, 1921.
15. Savarit, R, Arts Metiers, pp. 65, 142, 178, 241, 266, 307, 1922.
16. Fenske, MR, Ind. Eng. Chem., **24,** 482, 1931
17. Kremser, A, Natl. Petrm. News, **22** (21), 42, 1930.
18. Souders, M and Brown, GG, Ind. Eng. Chem., **24,** 519, 1932.
19. Bolles, WL, Petrm. Refiner, **25,** 613, 1946.
20. Smith, B, Design of Equilibrium Stage Processes, McGraw Hill, New York, 1963.
21. Davies, JA, Petrm. Refiner, **29** (9), 121, 1950; Ind. Eng. Chem., **39,** 774, 1947.
22. Bolles, WL, Petrm. Process., **11** (2), 64, 1956.
23. Hughmark, GA and O'Connell, HE, Chem. Eng. Progress, **53** (3), 127, 1957.
24. Eckert, JS, Chem. Eng. Progr., **57,** 64, 1961.
25. Smith, B and Brinklely, W, AIChE J, **6,** 446, 1960.
26. Thiele, EW and Geddes, RL, Ind. Eng. Chem., **25,** 290, 1933.
27. Lewis, WK and Matheson, GL, Ind. Eng. Chem., **24,** 496, 1932.
28. Wang, JC and Henke, GE, Hydrocarbon Processing, **45** (8), 155, 1966.
29. Naphtali, E and Sandholm, G, AIChE J, **17,** 148, 1971.
30. Krishnamurthy, EV and Sen, SK, Computer-based Numerical Algorithms, East-West Press, New Delhi, 1976.
31. Perry, RH, Chemical Engineer's Handbook, Sixth Edition, McGraw Hill, New York, 1984.
32. Daubert, TE, Chemical Engineering Thermodynamics, McGraw Hill, New York, 1985.
33. Friedman, SJ and Marshall, WR, Chem. Eng. Progr., **45,** 482, 1949.
34. Narayanan, CM and Bhattacharya, BC, Mechanical Operations for Chemical Engineers, Khanna Publishers, New Delhi, 1992.
35. Mathur, KB and Epstein, N, Spouted Beds, Academic Press, New York, 1974.
36. Cowan, H, Eng. J, 41 **(5),** 60, 1958.
37. Gluckert, RA, AIChE J, **8** (4), 460, 1962.
38. Friedman, SJ, Gluckert, RA and Marshall, WR, Chem. Eng. Progress, **48,** 181, 1952.
39. International Critical Tables, Volume 5, 142, 1929.
40. Simons, HP, Koffolt, JH and Withrow, JR, Trans. AIChE, **39,** 133, 1943.
41. Gauvin, WH and Katta, S, AIChE J, **22** (4), 713, 1976.
42. Liebson, I, Kelley, RE and Bullington, LA, Petr. Refiner, **36** (2), 127, 1957; **36** (3), 288, 1957.
43. Sorel, M, La Rectification de l'alcool, Paris, 1893.
44. Shiras, RN et. al., Ind. Eng. Chem., **42,** 871, 1950.
45. Underwood, AJV, Chem. Eng. Progress, **44,** 603, 1948; **45,** 609, 1949.
46. Gilliland, ER, Ind. Eng. Chem., **32,** 1101, 1940.
47. Kirkbride, CG, Petrolm. Refiner, **23,** 32, 1944.

# Chapter 6

# Extraction and Adsorption

## 6.1 LIQUID-LIQUID EXTRACTION

Liquid-liquid extraction (the terms liquid-liquid extraction and solvent extraction are often used synonymously, though the term solvent extraction could encompass solid–liquid extraction or leaching as well) is a widely used unit operation particularly in petrochemical and pharmaceutical industries. In principle, it resembles gas absorption (discussed in Chapter 4) except that the feed to an extraction unit is a liquid-liquid mixture or liquid solution, which is contacted with another immiscible liquid (called the solvent). The solvent (let us call it $B$) is so chosen that one of the constituents of the feed (say, $C$) is highly soluble in it while the other component (let us call it $A$) is either insoluble or has very low solubility in $B$. As a result, $B$ selectively or preferentially extracts $C$ (called the solute) forming a solution rich in $C$ which leaves the system as the *extract*. The residual liquor that contains very little $C$ and most of $A$ is called the *raffinate*. If $A$ is totally insoluble in $B$, then we achieve an excellent separation and the extract then will contain only $B$ and $C$. However, in many cases this does not occur. The extract often contains apart from the solvent ($B$), most of $C$ and small amounts of $A$. Similarly, the raffinate will also be a ternary system containing most of $A$ and small amounts of $C$ and $B$. The extract (and often the raffinate also) is further processed to separate the solvent which is then recycled to the extractor. Recovery and reuse of the solvent is an integral part of the extraction plant since it decides the overall economy of the process. The entire scheme is sketched in Fig. 6.1.

Recovery of solvent is most commonly achieved by distillation. The solvent ($B$) and ($C$) usually have widely different volatilities and therefore can be readily separated by distillation. The choice of liquid-liquid extraction as a substitute to distillation is, therefore, attractive in case of those systems for which separation by conventional distillation / fractionation is either too difficult or too expensive. An excellent

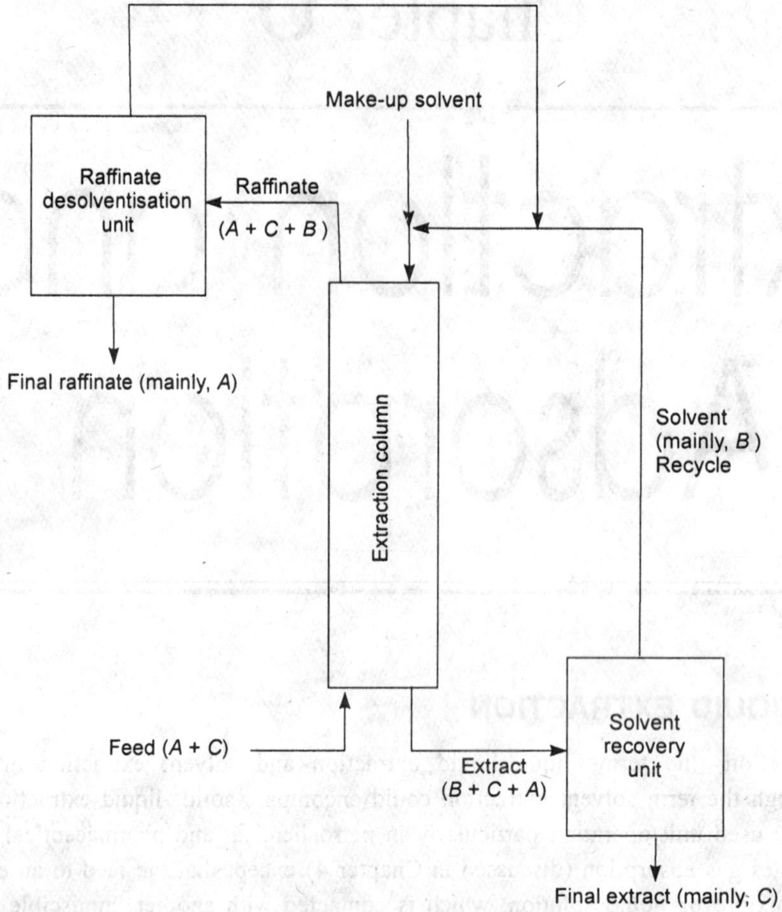


*Figure 6.1: Schematic of a typical liquid-liquid extraction unit*

example is the separation of constant-boiling or closely-boiling mixtures. We have already discussed in Chapter 5 that when the components of the feed solution have essentially the same vapour pressure (or have equal volatility), their separation by distillation is practically impossible*. Liquid-liquid extraction can be conveniently employed in such cases. For example, aromatic hydrocarbons are separated from paraffinic crudes by extraction with solvents like liquefied sulfur dioxide, phenol, furfural and diethylene glycol. Aromatics have essentially the same vapour pressure as paraffinic hydrocarbons and therefore, they can hardly be separated by distillation. Removal of aromatics is necessary since their presence in kerosene and other fuel oils leads to smoke formation and soot deposition during combustion. In lubricating oils, large aromatic content leads to deposition of carbon particles on engine parts and sharp

* Separation can be affected in cases like this by what is called *extractive distillation* (discussed in Chapter 5) which in fact is a combination of fractional distillation and extraction. In extractive distillation, the vapour phase is extracted with a solvent while liquid-liquid extraction involves extraction of a liquid phase.

decrease in oil viscosity with temperature. Solvent extraction is also popularly used for dewaxing diesel oils and lubricating oils. Wax (which is composed of high molecular weight hydrocarbons) is invariably present in most of the high boiling petroleum oils and this reduces the fluidity of the oil. Dewaxing is accomplished by extraction with solvents like liquefied propane or methyl ethyl ketone (MEK). Often, a mixture of MEK and toluene is use as the solvent. The oil has high solubility in MEK while the wax is essentially insoluble. As a result, MEK selectively extracts the oil leaving behind wax in the raffinate.

In the case of dilute aqueous solutions, separation by liquid-liquid extraction is more economical than by distillation (this is in spite of the fact that liquid-liquid extraction involves the additional cost of solvent recovery by distillation) since the latent heat of vapourisation of water is quite large while that of organic solvents is relatively low. A good example is the recovery of acetic acid from a dilute solution with water using isopropyl ether or ethyl acetate as the solvent. Similarly, recovery of phenolics (phenol and its derivatives) from water can be economically accomplished by extraction with butyl acetate or methyl isobutyl ketone (MIK).

Liquid-liquid extraction can also be the most cost-effective method when the components to be separated are heat-sensitive (like antibiotics, food products) or relatively nonvolatile (like mineral salts). Solvent extraction using amyl acetate is employed almost universally for the separation of penicillin from fermentation broth. Removal of fatty acids from vegetable oils requires expensive high-vacuum distillation, but this can be more economically accomplished by extraction with liquefied propane. In metallurgical industries also, liquid-liquid extraction offers a viable alternative to chemical methods since the latter consume chemicals and often lead to sludge disposal problems. It has been reported that recovery of uranium from ore leach liquor by extraction with a suitable solvent (alkyl phosphate in kerosene) is most cost-effective.

Extraction with multiple solvents is not unusual. In the Duosol process, widely used for treating lub oils, two solvents such as liquefied propane and selecto (which is a mixture of phenol and cresylic acid) are employed. The paraffinic hydrocarbons are highly soluble in liquid propane while asphalt and aromatics have high solubility in selecto. As a result, we eventually get two phases, one which is a solution of paraffinic hydrocarbons in liquid propane and the other a solution of asphalt and aromatics in selecto. From the former, liquid propane, being highly volatile, can be easily separated and we are left with a lubricating oil that is rich in paraffinic hydrocarbons and free from asphalt and aromatics. Another example is the separation of a mixture of oxalic acid and succinic acid by extraction with *n*-amyl alcohol and water. Amyl alcohol preferentially extracts succinic and while oxalic acid has a high preferential solubility in water. Thus, the alcohol layer will be rich in succinic acid and the water layer rich in oxalic acid. The solvents are recovered from the respective layers by distillation.

It is now apparent that the choice of a selective solvent is the most detrimental parameter in extraction. The solvent must be of low cost, easily available. It must be nontoxic, noncorrosive and nonflammable. The most important property of the solvent is evidently its selectivity to the solute. *Selectivity* (also called the *separation factor*) may be mathematically defined as

$$\beta = \left(\frac{y^*}{x}\right) \frac{(\text{mass fraction of } A \text{ in raffinate})}{(\text{mass fraction of } A \text{ in extract})} = K_C / K_A \quad \text{... (6.1.1)}$$

where $y^*$, $x$ = mass fraction of solute ($C$) in equilibrium phases (such as in extract and in raffinate respectively)

$K_C$, $K_A$ = equilibrium distribution coefficient of solute ($C$) and that of nonsolute ($A$) respectively.

It can be seen that $\beta$ is analogous to relative volatility ($\alpha$) for dissillation (defined in Chapter 5). To affect separation by extraction, $\beta$ must be larger than 1.0 and the more it exceeds unity, the better will be the separation. If $\beta$ = 1.0, no separation is possible.

Though it is not essential that the equilibrium distribution coefficient $K$ (where $K = y^*/x$) be more than 1.0, larger values of $K$ are always preferable since this reduces the solvent requirement for extraction. The larger the value of $K$, the less will be the amount of solvent required to affect separation.

Another equally important property of the solvent is its recoverability. Since distillation is most commonly used for solvent recovery, the solvent should not form any azeotrope with the extracted solute and the mixtures must exhibit high relative volatility. A solvent with low latent heat of vapourisation is preferred (when the solvent is being vapourised).

A large density difference between the equilibrium phases (namely, the extract and the raffinate phases) and a large interfacial tension between the phases are desirable since this helps in achieving a sharp separation of the phases.

High temperatures normally do not favour liquid-liquid extraction processes. This is because as the temperature is increased, the solubility of nonsolute component ($A$) in the solvent ($B$) increases. Ultimately, at what is called the *critical solution temperature*, $A$ becomes completely soluble in $B$ and hardly any separation becomes possible.

### 6.1.1 Extraction Equipment

As for gas absorption and distillation, the equipment for liquid-liquid extraction can also be broadly classified into stage wise contactors and continuous (or differential) contactors. Mixer-settlers and sieve-tray towers belong to the first category (stage-type extractors). Mixer-settlers consist of a series of mixing tanks (agitated vessels) interspaced by settling tanks as shown in Fig. 6.2. Flat-bladed turbine impellers are most commonly used as agitators in the mixing tanks. The emulsion or dispersion from each mixer is sent to a settling tank

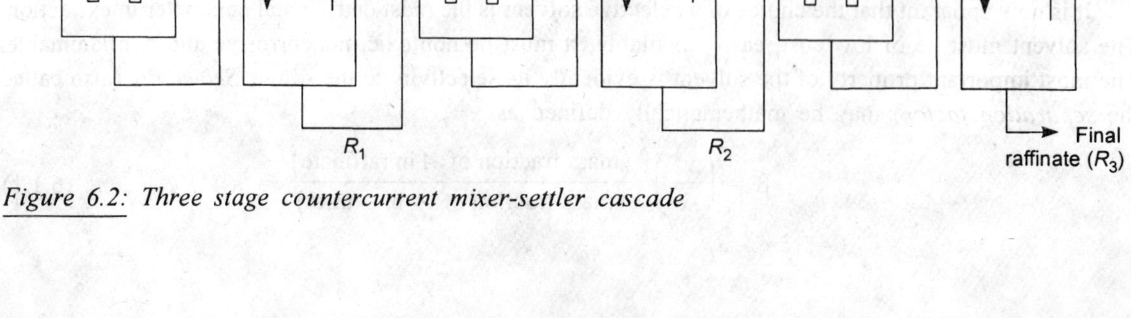


*Figure 6.2: Three stage countercurrent mixer-settler cascade*

where the heavy phase (either extract or raffinate) settles down while the light phase overflows from the top. Each stage thus consists of a mixer and a settler. Liquids are generally pumped from one stage to the next (gravity flow is also possible if sufficient headroom is available). Alternate arrangements to reduce the amount of interstage piping and the corresponding cost have been designed[1, 3]. For example, the stages may be placed one above the other in a vertical stack with mixing impellers mounted on a common shaft, the impellers thus serving as both pumps as well as mixing devices.

The sieve-tray (perforated plate) towers employed for gas absorption and fractional distillation (see Chapters 4 and 5) can be employed for liquid-liquid extraction as well. The design and construction of the tower are essentially the same except that when used as an extraction tower, the overflow weirs can be dispensed with (see Fig. 6.3). What is sketched in Fig. 6.3 is the arrangement used when the light liquid forms the dispersed phase. The operation of the tower is countercurrent. The light liquid is admitted from the bottom and it passes up through the perforations. Since it is the dispersed phase, it forms droplets which rise through the layer of the other liquid that forms the continuous phase (here, the heavy liquid). These liquid droplets coalesce and accumulate beneath each plate as a layer. The heavy liquid forming the continuous phase is admitted from the top, it flows across each plate (cutting across the rising droplets) and then flows down through the downcomer to the plate below. The tower may also be constructed with the plates inverted in which case the downcomers become upspouts through which the light liquid (now forming the continuous phase) flows upward while the heavy liquid (now the dispersed phase) flows down through the perforations. The arrangement thus depends upon which liquid is to be dispersed and is often determined by the interfacial properties of the liquids and the material of construction.

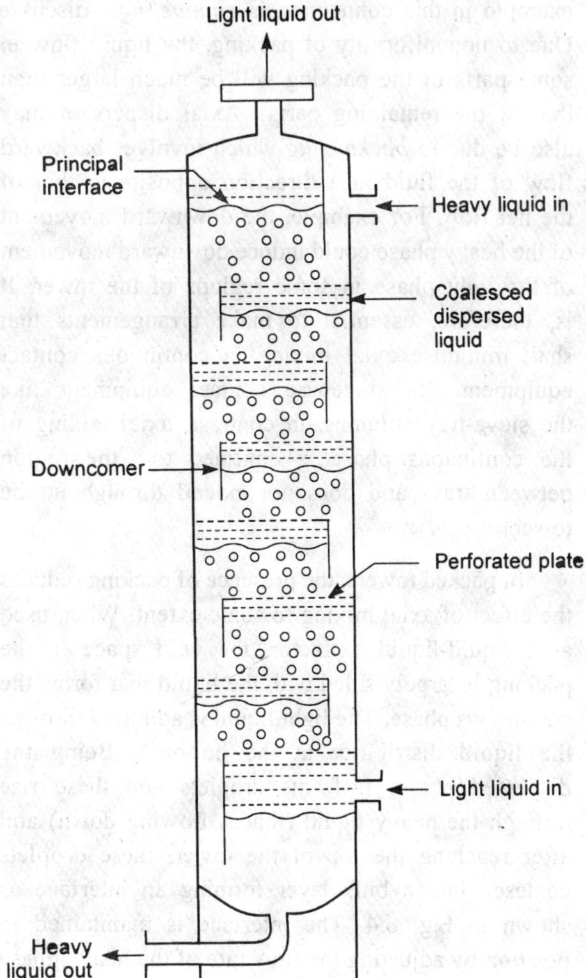


*Figure 6.3: Sieve-tray extraction column (light liquid dispersed)*

Hydraulics of sieve-tray extractor are fairly complex to analyse. A good discussion on this is given by Treybal[1]. Computer simulation of the performance characteristics of a sieve-tray liquid-liquid extractor is discussed by Narayanan[2].

Among the continuous contact extractors mechanically agitated towers and pulsed columns are most popular. Conventional packed towers are also used as liquid-liquid extractors. The problem of *axial mixing* is more acute in continuous contact equipment, as compared to stagewise contact equipment. Due to axial mixing, true plug flow of fluids do not occur in the tower. Axial dispersion reduces the concentration difference between phases and thereby diminishes the rates of mass transfer significantly. In true plug flow, all the fluid elements move with the same velocity at any particular cross-section of the tower. The velocity profile will be thus flat at any cross-section of the tower. If axial dispersion is present, then the fluid elements move with different velocities at a particular cross section. (A good example in this connection is *channeling*—discussed in Chapter 4—that often occurs in packed beds. Due to nonuniformity of packing, the liquid flow in some parts of the packing will be much larger than that in the remaining parts). Axial dispersion may also be due to *backmixing* which involves backward flow of the fluid in a direction opposite to that of the net flow. For example, the downward movement of the heavy phase could induce downward movement of the light phase in some regions of the tower. It is, therefore, essential to make arrangements that shall minimise axial mixing in continuous contact equipment. (In stagewise contact equipment like the sieve-tray columns, in contrast, axial mixing of the continuous phase is confined to the region between trays and does not extend throughout the tower).

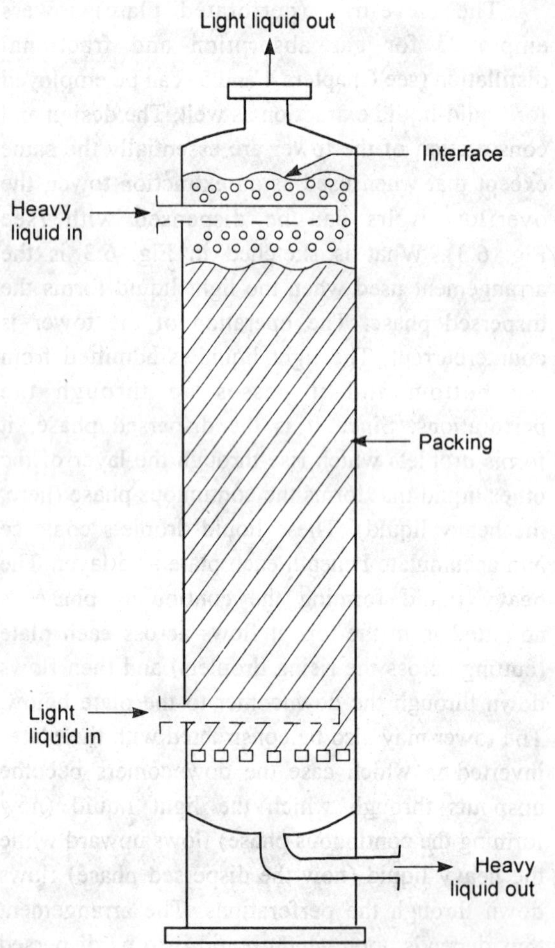


*Figure 6.4: Packed extraction tower (light liquid dispersed)*

In packed towers, the presence of packing reduces the effect of axial mixing to some extent. When used as a liquid-liquid extractor, the void space in the packing is largely filled with the liquid that forms the continuous phase. The light liquid is admitted through the liquid distributor at the bottom*. Being the dispersed phase, it forms droplets and these rise through the heavy liquid (that is flowing down) and after reaching the top of the tower, these droplets coalesce into a bulk layer forming an interface as shown in Fig. 6.4. The interface is maintained in position by adjusting the flow rate of the heavy phase through the bottom outlet by means of a control valve. By adjusting this control valve, the level of the interface may be brought down to below the light liquid distributor. In that case, the light liquid will form the continuous phase and the heavy liquid dispersed. Figure 6.5 shows schematically the method of controlling the level of the interface.

* In most extraction towers operating countercurrently, light liquid enters at the bottom, the heavy liquid at the top.

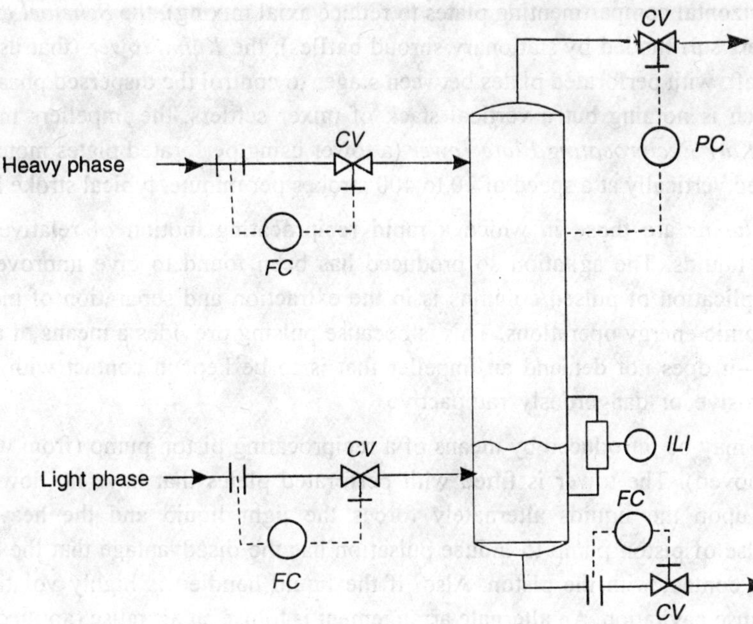


*Figure 6.5: Schematic of control of interface level in packed extraction tower. CV-control valve, FC-flow controller, PC-pressure controller, ILI-interface level indicator*

As in the case of gas absorption, the packing size must be less than one-eighth of tower diameter to minimise wall effects. The packing is to be preferentially wet by the continuous phase. If the dispersed phase droplets wet the packing, then they will coalesce and stream along the packing as rivulets. The dispersed phase distributor is also kept embedded in the packing (as shown in Fig. 6.4). Otherwise, the droplets will tend to coalesce on the packing support and this would lead to premature flooding of the tower.

In spite of the fact that the presence of packing tends to reduce axial mixing, the rates of mass transfer are generally low in packed towers. The choice of packed towers for liquid-liquid extraction is, therefore, not as popular as for gas absorption.

Owing to the small density differences between contacting liquids (unlike in gas absorption where the density difference between the gas and the liquid could be as high as 500 to 1000 $kg/m^3$), the energy available from simple counterflow under force of gravity becomes insufficient to disperse one liquid in the other (and also to produce enough turbulence so as to maintain a high rate of mass transfer) when the systems handled are of high interfacial tension. The conventional sieve-tray and packed bed towers are, therefore, restricted to systems of low interfacial tension (less than 10 mN/m). In systems of high interfacial tension, the desired degree of dispersion can be brought about by mechanically agitating the liquids or by introducing pulsations. An exceedingly large number of designs of mechanically agitated countercurrent extractors have been proposed in the literature (many of them being proprietary devices). A few examples are the *rotary disk contactor* (which consists of a tower divided into compartments by horizontal doughnut–shaped baffles and in each compartment agitation is provided by a rotating, centrally located, horizontal disk), the *Lightnin mixer* (*Oldshue-Rushton*) *tower* (which uses flat-bladed turbine impellers to agitate the liquids and horizontal compartmenting plates to reduce axial mixing), the *Scheibel extractor* (consisting of turbine impellers surrounded by stationary shroud baffles), the *Kühni tower* (that uses shrouded impellers on a central shaft, with perforated plates between stages to control the dispersed phase holdup), the *Treybal extractor* (which is nothing but a vertical stack of mixer settlers, the impellers mounted on a common shaft) and the *Karr Reciprocating Plate Tower* (a tower using perforated plates mounted on a central shaft and reciprocated vertically at a speed of 10 to 400 strokes per minute, typical stroke length being 2.54 cm).

*Pulsed columns* are those in which a rapid reciprocating motion of relatively short amplitude is applied to the liquids. The agitation so produced has been found to give improved rates of extraction. The largest application of pulsed columns is in the extraction and separation of metals from radioactive solutions of atomic energy operations. This is because pulsing provides a means of agitation not requiring moving parts—it does not demand an impeller that is to be kept in contact with the liquids that could be highly corrosive or dangerously radioactive.

Pulsations may be introduced by means of a reciprocating piston pump (from which the check valves have been removed). The tower is fitted with perforated plates that have no downcomers. The pulsing superimposed upon the liquids alternately forces the light liquid and the heavy liquid through the perforations. Use of piston pump to induce pulsation has the disadvantage that the corrosive liquid could come in direct contact with the piston. Also, if the liquid handled is highly volatile, then the too rapid pulsing may cause cavitation. An alternate arrangement is to use an air pulse (applied to a packed column). This keeps corrosive liquids out of contact with the pulsing device and obviates the cavitation problem. However, the power consumption for pulsing shall be higher due to the high compressibility of the gas.

*Centrifugal extractors* provide an excellent substitute to the conventional sieve-tray and packed bed extractors particularly for handling liquids of low density difference and those with tendencies to form emulsions. This is because centrifugal forces are several thousand times larger than the force of gravity and thus provide substantially larger driving forces and equivalent mass transfer rates. The most popular extractor of this category is the *Podbielniak extractor* which consists of a cylindrical drum containing concentric perforated cylinders. The liquids are introduced through the rotating shaft with the help of special mechanical seals; the light liquid is led internally to the drum periphery and the heavy liquid to the axis of the drum. Rapid rotation (up to several thousand revolutions per minute) causes radial counterflow of liquids which are then led out through the shaft. Extractors of this type are characterised by the extremely low holdup of liquid per stage and this has led to their extensive use in the extraction of antibiotics such as penicillin (for which multistage extraction and phase separation have to be done very rapidly to avoid chemical destruction of the product). They are also being used in petroleum processing, for dephenolisation of waste water and for the recovery of uranium from ore leach liquor. Similar devices are the *De Laval extractor* (which consists of a number of perforated cylinders revolving about a vertical shaft, the liquids following a spiral path radially in the countercurrent fashion) and the *Luwesta extractor* (primarily a centrifuge revolving about a vertical axis, used widely for the extraction of acetic acid, pharmaceuticals and similar products).

## 6.1.2 Liquid-Liquid Equilibria

Liquid-liquid extraction involves ternary systems (those composed of three liquid components, *A*, *B*, *C*). The phase relationship is, therefore, essential to be known for analysing the system/process. It is usual to assume while performing design calculations that the extract and the raffinate phases leaving each stage/section are in equilibrium with each other. As in the case of distillation, the value of the equilibrium distribution coefficient $K$ (where $K = y^*/x$) may be determined experimentally or computed analytically using any one of the well-established thermodynamic equations of state. See Treybal[1] for an excellent discussion on the analytical computation of $K$-values.

The *phase diagram* may be represented on triangular coordinates [as shown in Fig. 6.6 (a)] or on rectangular coordinates [as shown in Fig. 6.6 (b)]. In design calculations, the rectangular coordinates are often more convenient since on triangular coordinates the points tend to crowd too much. Any ternary system composed of components $A$, $B$ and $C$ (where $C$ is the solute component of the feed, $A$ the nonsolute component of the feed and $B$ the extraction solvent) may be regarded as composed of three binary systems such as $A$ and $C$, $B$ and $C$, and $A$ and $B$. The most common type of ternary system is that for which two of the binary systems are completely miscible liquids (for example, $A$ and $C$ are completely miscible and so are $B$ and $C$) and the third binary system (namely, $A$ and $B$) is one in which he liquids are only partially miscible. The phase diagram for such a ternary system is what is shown in Figs 6.6 (a) and (b). The curve shown is called the *solubility curve.* Any line RME joining two points $R$ and $E$ on the solubility curve (these two points represent compositions of the two equilibrium phases such as the raffinate and the extract respectively) is called a *tie line.* The point $M$ on the tie line represents the composition of a mixture of two liquid phases which at equilibrium will have the compositions indicated by points $R$ and $E$. It can be noticed that the length of the tie lines decreases and the compositions of the two equilibrium phases approach each other as the mass fraction of $C$ increases. The point corresponding to a tie line of zero length, that is, for which the two equilibrium phases have the same composition is called the *plait point* and is represented by the point $P$.

It is thus clear that the portion of the solubility curve on the left side of the plait point is nothing but the plot of $x$ versus $z_r$ (where $x$ is the mass fraction of $C$ in raffinate and $z_r$ is the mass fraction of $B$ in raffinate) and that portion of the curve on the right side of the plait point is the plot of $y$ versus $z_e$ (where $y$ is the mass fraction of $C$ in extract and $z_e$ that of $B$ in extract). Typical equilibrium plots of $y$ versus $x$ (more correctly $y^*$ versus $x$) are shown in Figs 6.6 (c) and (d). Figure 6.6 (c) illustrates the case when the $y$-values are larger than the corresponding equilibrium values of $x$, while Fig. 6.6 (d) exhibits the case when $y^*$ is less than $x$. The acetone ($C$)-water ($A$)-chlorobenzene ($B$) system belongs to the type shown in Fig. 6.6 (c) while acetic acid ($C$)-water ($A$)-isopropyl ether ($B$) system is an example of the type shown in Fig. 6.6 (d).

The phase diagram and the equilibrium plot may also be prepared on solvent-free coordinates as shown in Figs 6.7 (a) and (b). In many design calculations, such a representation tends to be more convenient. Here, $N$ ($N_R$ or $N_E$) is plotted along the ordinate and $X$ or $Y$ along the abcissa, where $N_R$ and $N_E$ represent mass fraction of $B$ on $B$-free (solvent-free) basis in the raffinate and in the extract respectively and $X$ and $Y$ represent mass fraction of $C$ on $B$-free basis in the raffinate and in the extract respectively. In other words,

$$N = \frac{(\text{Mass of } B)}{(\text{Mass of } A) + (\text{Mass of } C)} \qquad \text{... (6.1.2)}$$

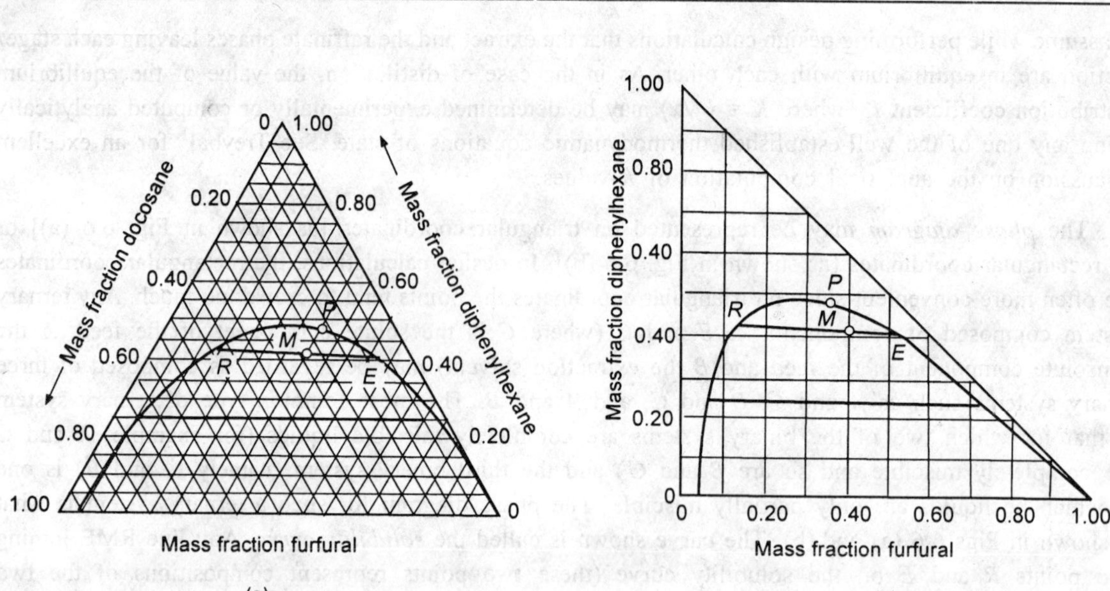


*Figures 6.6 (a) and (b): Phase diagram (solubility curve) for diphenyl hexane (C)-docosane (A)-furfural (B) system at 45° C and 1 atm, (a) On triangular coordinates, (b) On rectangular coordinates*

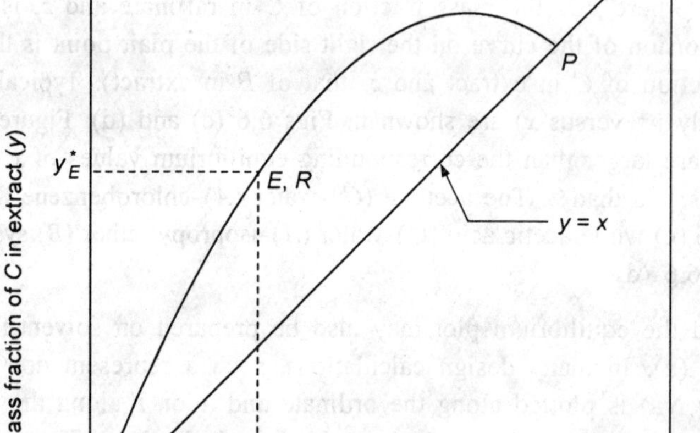


*Figure 6.6 (c): Equilibrium plot for systems with $y^* > x$ or $K > 1.0$*

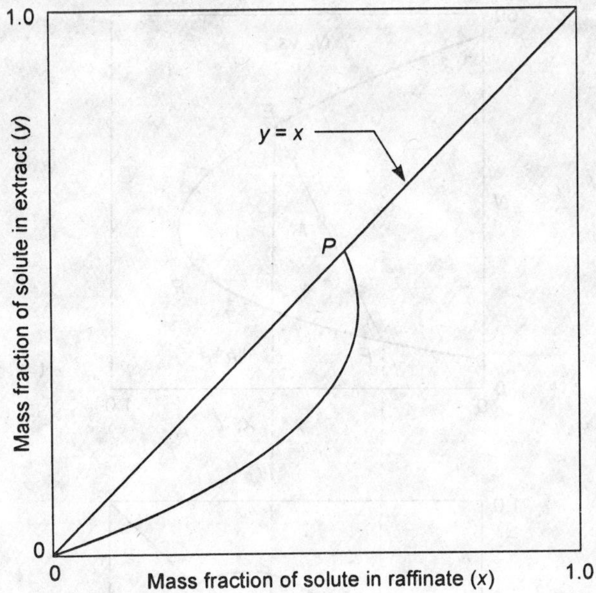


*Figure 6.6 (d): Equilibrium plot for systems with $y^* < x$ or $K < 1.0$*

$$X = \frac{(\text{Mass of } C \text{ in raffinate})}{(\text{Mass of } A \text{ in raffinate}) + (\text{Mass of } C \text{ in raffinate})} \quad \text{... (6.1.3)}$$

$$Y = \frac{(\text{Mass of } C \text{ in extract})}{(\text{Mass of } A \text{ in extract}) + (\text{Mass of } C \text{ in extract})} \quad \text{... (6.1.4)}$$

In Fig. 6.7 (a), the portion of the curve above the plait point $P$ is the plot of $N_E$ versus $Y$ and that portion of the curve below $P$ is the plot of $N_R$ versus $X$. It may be noted that the above defined solvent-free parameters ($N$, $X$, $Y$) are related to $x$, $y$, $z$ as

$$N_R = z_r / (1 - z_r) \quad \text{... (6.1.5)}$$

$$N_E = z_e / (1 - z_e) \quad \text{... (6.1.6)}$$

$$X = x / (1 - z_r) \quad \text{... (6.1.7)}$$

$$Y = y / (1 - z_e) \quad \text{(6.1.8)}$$

It is also not difficult to deduce that at the plait point ($P$), the value of equilibrium distribution coefficient ($K$) is 1.0 and the selectivity ($\beta$) is also equal to 1.0.

The influence of pressure on liquid-liquid equilibrium may usually be neglected unless very high pressures are considered. The equilibrium is, however, sensitive to temperature. Increase in temperature could reduce the area under the solubility curve and could also change the slopes of the tie lines. As

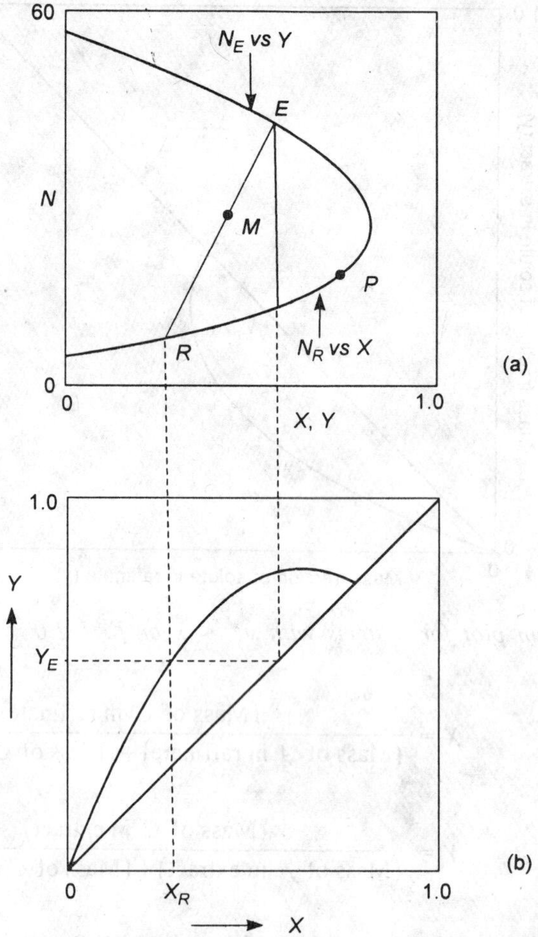


*Figures 6.7 (a) and (b): (a) Phase diagram, (b) Equilibrium plot on solute-free (B-free) coordinates*

we have stated earlier, high temperatures are not generally employed in liquid-liquid extraction since the solubility of the nonsolute ($A$) in $B$ usually increases with increase in temperature.

### 6.1.3 Analysis of Extraction Processes

As in the case of distillation, design computations for liquid-liquid extraction may be performed either analytically or graphically. Once the phase diagram and the equilibrium plot (also called the distribution diagram or distribution isotherm) are available, the graphical procedure becomes more handy for manual computation. However, if the aid of personal computers is available, then the design calculations can be performed analytically with relative ease. We shall be discussing both of these methods in the subsequent paragraphs. The reader shall notice that the graphical procedures outlined below are extremely analogous to those employed for distillation (discussed in Chapter 5). Therefore, once the reader has fully familiarised himself with the graphical methods for distillation, he shall not find any difficulty in assimilating the design procedures discussed below.

## Multistage Crosscurrent Extraction

The scheme is sketched in Fig. 6.8. The feed enters the first stage. The raffinate leaving the first stage is fed to the second stage. In other words, the raffinate leaving any stage forms the feed to the next stage. A stream of fresh solvent is fed to each stage and extract is withdrawn from each stage. All these extract streams may be mixed together to obtain the final (composited) extract. Raffinate from the last stage is collected as the final raffinate. The feed rate of solvent may vary from stage to stage as shown; however, it is usually prefered to feed the same amount of solvent to each stage (that is, $S_1 = S_2 = S_3 = ..... = S_n = S$). The mixer-settler cascades quite often operate in this continuous, crosscurrent mode.

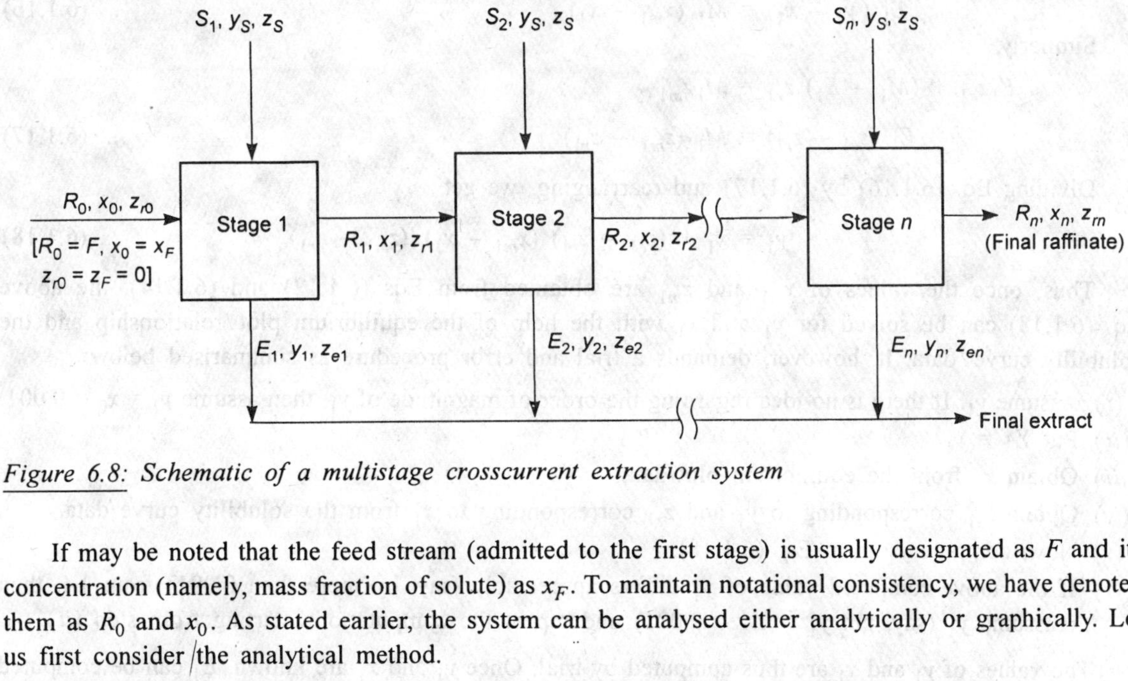


*Figure 6.8: Schematic of a multistage crosscurrent extraction system*

If may be noted that the feed stream (admitted to the first stage) is usually designated as $F$ and its concentration (namely, mass fraction of solute) as $x_F$. To maintain notational consistency, we have denoted them as $R_0$ and $x_0$. As stated earlier, the system can be analysed either analytically or graphically. Let us first consider the analytical method.

## Analytical Method

To understand the method of analysis, let us consider the first stage. The overall material balance for this stage is

$$R_0 + S_1 = E_1 + R_1 \qquad \text{... (6.1.9)}$$

Putting $M_1 = E_1 + R_1$, we can write

$$R_0 + S_1 = M_1 \qquad \text{... (6.1.10)}$$

A solute balance ($C$-balance) gives

$$R_0x_0 + S_1y_s = M_1x_{m1} = (R_0 + S_1)\,x_{m1} \quad \text{... (6.1.11)}$$

or
$$x_{m1} = (R_0x_0 + S_1y_S)/(R_0 + S_1) \quad \text{... (6.1.12)}$$

Also,
$$(S_1/R_0) = (S_1/F) = (x_0 - x_{m1})/(x_{m1} - y_S) \quad \text{... (6.1.13)}$$

Similarly, a solvent balance (*B*-balance) gives

$$R_0z_{r0} + S_1z_S = M_1z_{m1} = (R_0 + S_1)\,z_{m1}$$

or
$$z_{m1} = (R_0z_{r0} + S_1z_S)/(R_0 + S_1) \quad \text{... (6.1.14)}$$

Since the feed to the first stage shall not contain any solvent, $z_{r0} = 0$. Now, since $E_1 + R_1 = M_1$,

$$E_1y_1 + R_1x_1 = M_1x_{m1} \quad \text{... (6.1.15)}$$

Putting $R_1 = (M_1 - E_1)$ and rearranging, we get

$$E_1\,(y_1 - x_1) = M_1\,(x_{m1} - x_1) \quad \text{... (6.1.16)}$$

Similarly,

$$E_1z_{e1} + (M_1 - E_1)\,z_{r1} = M_1z_{m1}$$

or
$$E_1\,(z_{e1} - z_{r1}) = M_1\,(z_{m1} - z_{r1}) \quad \text{... (6.1.17)}$$

Dividing Eq. (6.1.16) by (6.1.17) and rearranging, we get

$$y_1 = x_1 + (z_{e1} - z_{r1})\,(x_{m1} - x_1)/(z_{m1} - z_{r1}) \quad \text{... (6.1.18)}$$

Thus, once the values of $x_{m1}$ and $z_{m1}$ are obtained from Eqs (6.1.12) and (6.1.14), the above Eq. (6.1.18) can be solved for $y_1$ and $x_1$ with the help of the equilibrium plot/relationship and the solubility curve/data. It, however, demands a trial and error procedure as summarised below:

(*i*) Assume $y_1$. If there is no idea regarding the order of magnitude of $y_1$, then assume $y_1 = x_0 = 0.001$.

(*ii*) Put $YA = y_1$.

(*iii*) Obtain $x_1$ from the equilibrium plot/data.

(*iv*) Obtain $z_{e1}$ corresponding to $y_1$ and $z_{r1}$ corresponding to $x_1$ from the solubility curve/data.

(*v*) Now, compute $y_1$ from Eq. (6.1.18).

(*vi*) If the above-computed value of $y_1$ differs appreciably (say, by more than 0.001) from *YA*, then reassume $y_1$ (say, put $y_1 = YA - 0.001$) and repeat the computations starting from step (*ii*).

The values of $y_1$ and $x_1$ are thus computed by trial. Once $y_1$ and $x_1$ are known, $E_1$ can be computed from Eq. (6.1.16) and $R_1 = (M_1 - E_1)$. The flow rates and compositions of both streams leaving the first stage are thus evaluated. This procedure is to be repeated for each subsequent stage. The stagewise computations are continued until a raffinate of the desired composition is obtained.

For the convenience of the readers, we may rewrite Eqs (6.1.12), (6.1.14) and (6.1.18) in the generalised from (for the *i*th stage) as

$$x_{mi} = (R_{i-1}x_{i-1} + S_iy_S)/(R_{i-1} + S_i) \quad \text{... (6.1.12a)}$$

$$z_{mi} = (R_{i-1}z_{r,\,i-1} + S_iz_S)/(R_{i-1} + S_i) \quad \text{... (6.1.14a)}$$

$$y_i = x_i + (z_{ei} - z_{ri})\,(x_{mi} - x_i)/(z_{mi} - z_{ri}) \quad \text{... (6.1.18a)}$$

Now, compute $x_{mi}$ and $z_{mi}$ from the first two equations and then solve the third equation for $x_i$ and $y_i$ with the help of the equilibrium data / plot and the solubility curve/data by a trial and error procedure as described above for stage 1. To summarise again,

(*i*) Assume $y_i$. For example, let $y_i = y_{i-1} - 0.001$.

(*ii*) Put $YA = y_i$.

(*iii*) Obtain $x_i$ from equilibrium data / plot.

(*iv*) Obtain $z_{ei}$ corresponding to $y_i$ and $z_{ri}$ corresponding to $x_i$ from the solubility curve/data.

(*v*) Now, compute $y_i$ from Eq. (6.1.18a) and check whether it deviates appreciably (say, by more than 0.001) from *YA*. If so, reassume $y_i$ (say, put $y_i = YA - 0.001$) and repeat the computations starting from step (*ii*).

(*vi*) Once the values of $y_i$ and $x_i$ have been evaluated by trial, compute $M_i$, $E_i$ as

$$M_i = R_{i-1} + S_i \quad \text{... (6.1.10a)}$$

$$E_i = \frac{M_i\,(x_{mi} - x_i)}{(y_i - x_i)} \quad \text{... (6.1.16a)}$$

Also,
$$R_i = (M_i - E_i).$$

A sample computer program for this analytical method is given at the end of this Chapter (Program 6.1).

It may be noted that all the above-discussed equations may be re-expressed on solvent-free (*B*-free) basis by simply replacing $x$ by $X$, $y$ by $Y$, $z_e$ by $N_E$ and $z_r$ by $N_R$ [see Eqs (6.1.5) to (6.1.8) for the interrelationship between these parameters). The mass flow rates ($R_i$, $E_i$) are also to be replaced by $R_i'$ and $E_i'$ (flow rates on solvent-free basis). To note that

$$R_i' = R_i/(1 + N_{Ri}) = R_i\,(1 - z_{ri}) \quad \text{... (6.1.19)}$$

$$E_i' = E_i/(1 + N_{Ei}) = E_i\,(1 - z_{ei}) \quad \text{... (6.1.20)}$$

## Graphical Method

For the graphical method, we require only Eqs (6.1.12) and (6.1.15). For example, Eq. (6.1.15) can be rewritten as

$$y_1 + (R_1/E_1)\,x_1 = (1 + R_1/E_1)\,x_{m1}$$

or
$$y_1 = -(R_1/E_1)\,x_1 + (1 + R_1/E_1)\,x_{m1} \quad \text{... (6.1.21)}$$

This is equation to a straight line that passes through the points ($x_1$, $y_1$) and ($x_{m1}$, $x_{m1}$). The second point lies on the 45° diagonal. Thus, if we draw a straight line of slope ($-R_1/E_1$) from the point $x = x_{m1}$ on the 45° diagonal, then it will intersect the equilibrium curve at ($x_1$, $y_1$) as shown in Fig. 6.9 (a). Since the value of ($R_1/E_1$) is not known in advance, a trial and error procedure becomes invariable here also:

(*i*) First assume the value of ($R_1/E_1$).

(*ii*) Now, from the point $x = x_{m1}$ on the 45° diagonal [the value of $x_{m1}$ is obtained from Eq. (6.1.12), draw a straight line of slope ($-R_1/E_1$). The point of intersection of this line with the equilibrium curve gives the values of ($x_1$, $y_1$).

(*iii*) Compute the value of $(R_1/E_1)$ from Eq. (6.1.21) as

$$R_1/E_1 = (y_1 - x_{m1})/(x_{m1} - x_1) \quad \text{... (6.1.22)}$$

(*iv*) Check whether the above-computed value of $(R_1/E_1)$ differs appreciably from that assumed at the outset. If so, rechoose $(R_1/E_1)$ and repeat the computations starting from step (*ii*).

Once the values of $x_1$, $y_1$ and the ratio $(R_1/E_1)$ are thus finalised by trial, we can obtain $E_1$ and $R_1$ as $E_1 = M_1/(1 + R_1/E_1)$ and $R_1 = (M_1 - E_1)$. The value of $M_1$ is known from the overall material balance, such as Eq. (6.1.10). By repeating this procedure for each of the subsequent stages, the flow rates and compositions of the raffinate and extract streams leaving each stage can be computed.

This graphical procedure may be performed on solvent-free coordinates as well ($Y$ versus $X$ plot). The procedure shall remain exactly the same except that $R$ and $E$ must be replaced by $R'$ and $E'$. In other words, if from the point $X = X_{mi}$ on the 45° diagonal a straight line of slope $(-R_i'/E_i')$ is drawn, then it will intersect the equilibrium curve at $(X_i, Y_i)$.

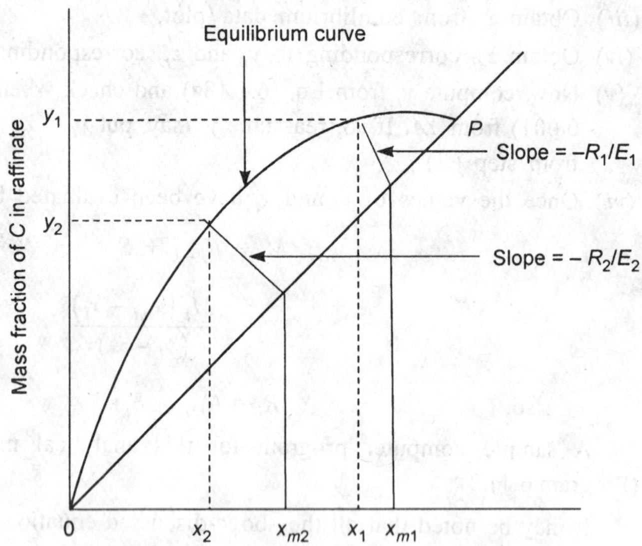


*Figure 6.9 (a): Analysis of multistage crosscurrent extraction using equilibrium plot*

Readers shall note the close analogy of the above-discussed procedure with that employed for flash distillation (see Chapter 5).

It is also possible to perform the graphical construction on the solubility curve as shown in Fig. 6.9 (b). To note that for sake of convenience, mass fraction of $B$ ($z_r$ and $z_e$) has been plotted along the $y$-axis and mass fraction of $C$ ($x$ and $y$) along the abcissa. The points $F$ and $S$ are first marked since their coordiantes, such as $(x_0, 0)$ and $(y_S, z_S)$, are known. Now, mark $M_1$ on the line $\overline{FS}$ corresponding to $x = x_{m1}$ [the value of $x_{m1}$ is obtained from Eq. (6.1.12) or (6.1.12a)]. Then, by trial, the tie line $\overline{x_1 y_1}$ that passes through $M_1$ is located [see Fig. 6.9 (b)]. For example, assume a value of $x_1$ and mark it on the bottom portion (raffinate branch) of the solubility curve. Obtain $y_1$ from the equilibrium plot and mark it on the upper portion (extract branch) of the solubility curve. Join $\overline{x_1 y_1}$ and see whether this line passes through $M_1$ or not. If not, assume an alternate value of $x_1$ and repeat the trial. Once the values of $x_1$ and $y_1$ have been finalised, $E_1$ and $R_1$ can be computed from Eqs (6.1.10a) and (6.1.16a). Now, join $x_1$ to $S$ and mark the point $M_2$ on this line corresponding to $x = x_{m2}$ [the value of $x_{m2}$ once again obtained from Eq. (6.1.12a)] and by trial, locate the tie line $\overline{x_2 y_2}$ that passes through $M_2$. Repeat this procedure for all subsequent stages.

The minimum required and maximum permissible solvent rates may also be determined graphically with the help of the solubility curve. For example, the solvent to feed ratio $(S_1/F)$ or

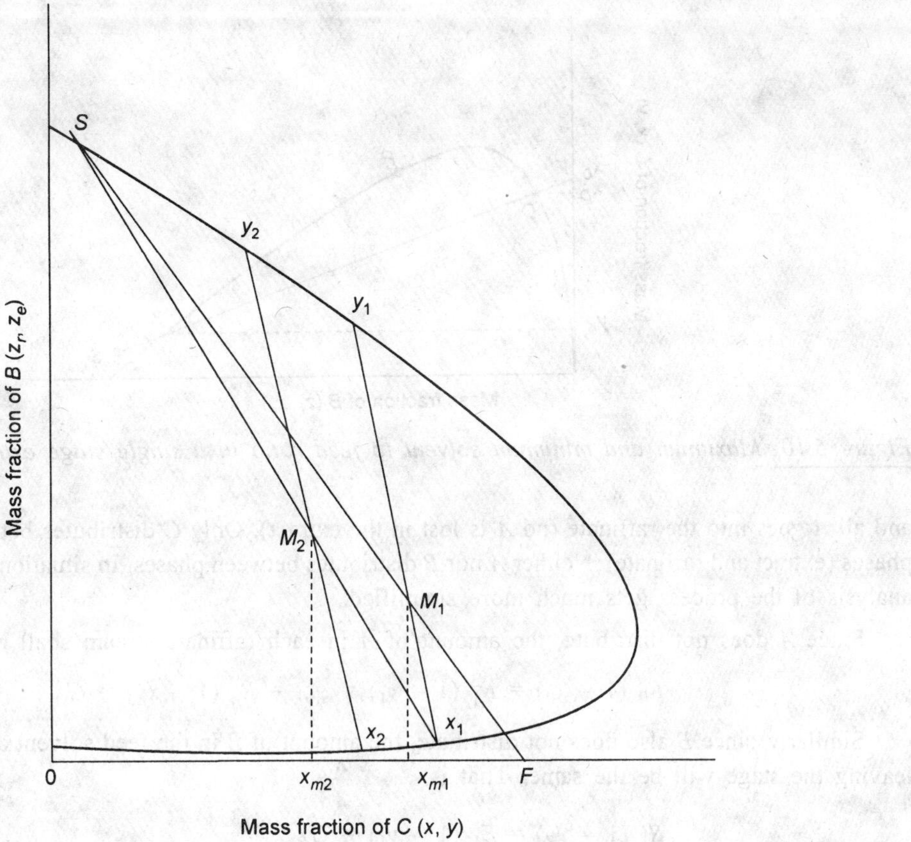


*Figure 6.9 (b): Analysis of multistage crosscurrent extraction using phase diagram*

$(S_1/R_0)$ can be computed from Eq. (6.1.13). The minimum and maximum solvent to feed ratios are then obtained as

$$(S_1/F)\ (\text{min}) = (S_1/R_0)\ (\text{min}) = (x_0 - x_D)/(x_D - y_S) \quad \text{... (6.1.23)}$$

$$(S_1/F)\ (\text{max}) = (S_1/R_0)\ (\text{max}) = (x_0 - y_K)/(y_K - y_S) \quad \text{... (6.1.24)}$$

where $x_D$ and $y_K$ are the values of $x$ and $y$ corresponding to points $D$ and $K$ respectively on the solubility curve (see Fig. 6.10). $D$ and $K$ are nothing but the points of intersection of the line $\overline{FS}$ with the solubility curve. The coordinates of the feed point ($F$) are $(x_F, z_F) = (x_0, z_{r0}) = (x_0, 0)$ and those of the solvent point ($S$) are $(y_S, z_S)$. The construction may be done on solvent-free coordinates as well. It is, however, observed that quite often, there will be the problem of scale when the solubility curve is plotted on solvent-free coordinates. This is due to the wide difference in the values of $N_R$ and $N_E$. For example, the values of $N_R$ usually fall less than 0.1 or 0.5, while the $N_E$ values could go beyond 100.0.

A special case in extraction is that when the nonsolute ($A$) is perfectly insoluble in $B$. In such a case, the extract will be a binary system (consisting of $B$ and $C$ only) and so will be the raffinate (which will contain $A$ and $C$ only). In other words, all the $B$ goes into the extract (no $B$ is lost in the raffinate)

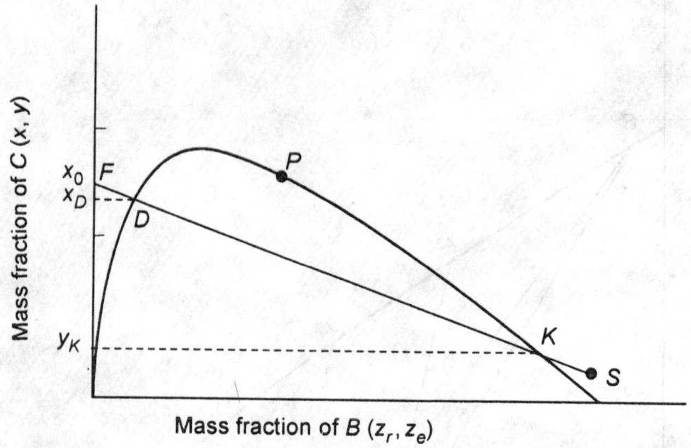


*Figure 6.10: Maximum and minimum solvent to feed ratio in a single stage extraction*

and all $A$ goes into the raffinate (no $A$ is lost in the extract). Only $C$ distributes between the equilibrium phases (extract and raffinate). Neither $A$ nor $B$ distributes between phases. In situations like this, the overall analysis of the process gets much more simplified.

Since $A$ does not distribute, the amount of $A$ in each raffinate stream shall be the same. That is,

$$R_0\,(1 - x_0) = R_1\,(1 - x_1) = \ldots\ldots = R_n\,(1 - x_n) = G_A \qquad \ldots (6.1.25)$$

Similarly, since $B$ also does not distribute, the amount of $B$ in the feed solvent and that in the extract leaving the stage will be the same. That is,

$$S_1\,(1 - y_S) = E_1\,(1 - y_1) = G_{B1} \qquad \ldots (6.1.26a)$$

$$S_2\,(1 - y_S) = E_2\,(1 - y_2) = G_{B2} \qquad \ldots (6.1.26b)$$

$$S_n\,(1 - y_S) = E_n\,(1 - y_n) = G_{Bn} \qquad (6.1.26c)$$

If the same amount of solvent is fed to each stage, then $G_{B1} = G_{B2} = \ldots\ldots = G_{Bn} = G_B$. The solute balance for stage 1 [namely, Eq. (6.1.11)] can be, therefore, rewritten as

$$G_A\left(\frac{x_0}{1-x_0}\right) + G_{B1}\left(\frac{y_S}{1-y_S}\right) = G_A\left(\frac{x_1}{1-x_1}\right) + G_{B1}\left(\frac{y_1}{1-y_1}\right) \qquad \ldots (6.1.27)$$

or

$$G_A x'_0 + G_{B1} y'_S = G_A x'_1 + G_{B1} y'_1 \qquad \ldots (6.1.28)$$

where $x'$ and $y'$ are mass fractions on solute-free basis*. On rearrangement,

$$y'_1 = (-G_A/G_{B1})\,x'_1 + (G_A/G_{B1})\,x'_0 + y'_S \qquad \ldots (6.1.29)$$

---

*In Chapter 4, we have used notations $X$ and $Y$ to denote mass fractions on solute-free basis. However, in this chapter, since $X$ and $Y$ have been used to denote mass fractions on solvent-free basis, mass fractions on solute-free basis are denoted as $x'$ and $y'$.

This is equation to a straight line that passes through the points $(x'_0, y'_S)$ and $(x'_1, y'_1)$. Therefore, if we draw a straight line of slope $(-G_A/G_{B1})$ from the point $(x'_0, y'_S)$, its point of intersection with the equilibrium curve will give the values of $x_1$ and $y_1$ (see Fig. 6.11). Similarly, a straight line of slope $(-G_A/G_{B2})$ drawn from point $(x'_1, y'_S)$ will intersect the equilibrium curve at $(x_2, y_2)$. In this way, the compositions of extract and raffinate streams leaving each stage can be evaluated. Needless to say, the equilibrium curve is to be plotted on solute-free coordinates.

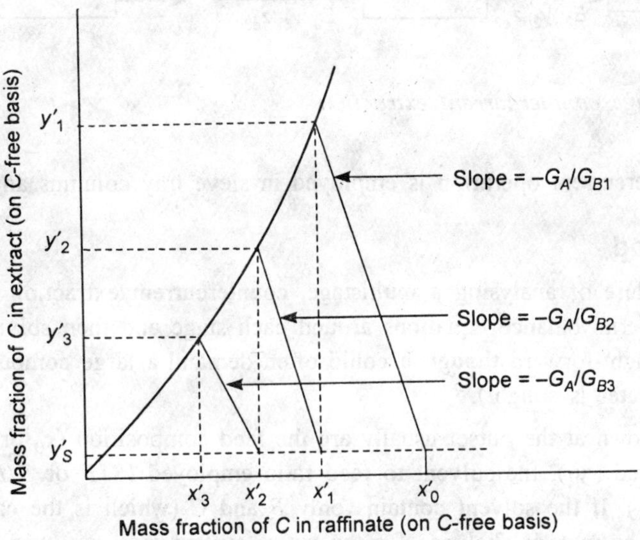


*Figure 6.11: Multistage crosscurrent extraction with insoluble solvent (A and B are totally insoluble)*

Once the values of $(x_i, y_i)$ are known, the flow rates of the phases $(R_i, E_i)$ leaving each stage can be obtained as

$$E_i = G_{Bi}/(1 - y_i) = S_i\,(1 - y_S)/(1 - y_i) \qquad \text{... (6.1.30)}$$

and

$$R_i = G_A/(1 - x_i) = R_0\,(1 - x_0)/(1 - x_i) \qquad \text{... (6.1.30a)}$$

The above graphical procedure shall remain the same even if the mass fractions are replaced by concentrations (expressed in kg/m$^3$ or kg/litre) and the flow rates expressed in volume/time.

## Multistage Countercurrent Extraction

Multistage countercurrent operation is usually more advantageous than crosscurrent operation since the former demands lesser number of stages (for the same amount of solvent) to affect the desired degree of separation. The process of continuous countercurrent extraction is sketched in Fig. 6.12. The extract and raffinate streams move in opposite directions (countercurrent to each other). The feed is admitted to stage 1 and the raffinate from each stage is fed to the next stage. Fresh solvent enters the last stage (namely, the $n$th stage) and the extract from each stage flows to the preceding stage [from $n$th stage to the $(n-1)$th stage, from $(n-1)$th stage to $(n-2)$th stage and so on). The final extract is withdrawn from the first stage and the final raffinate from the $n$th stage. As in the earlier cases, the system can be analysed either analytically or graphically.

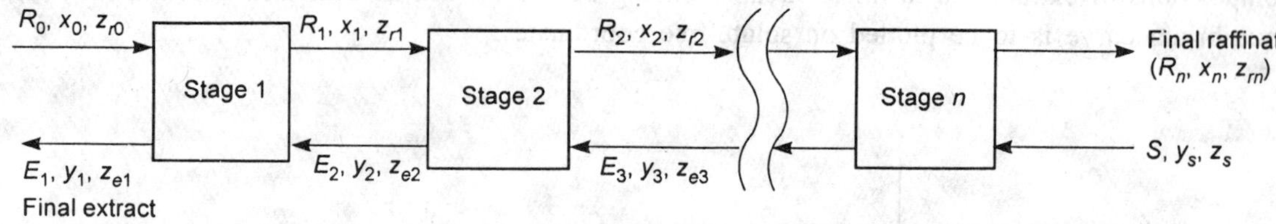


*Figure 6.12: Multistage countercurrent extraction*

Multistage countercurrent operation is employed in sieve tray columns and mixer-settler cascades.

## Analtytical Method

The analytical procedure of analysing a multistage, countercurrent extraction process simply involves writing down the material balance equations around each stage and then solving them simultaneously. The procedure is straight-forward though it could often demand a large computational load (unless the aid of personal computer is sought).

The quantities known at the outset usually are the feed composition ($x_0$ or $x_F$), the composition of the final extract desired ($y_1$), the solvent to feed ratio employed ($S/F$ or $S/R_0$) and the composition of the solvent ($y_S$, $z_S$). If the solvent contains only $B$ and $C$ (which is the case on many occasions), then $z_S = (1 - y_S)$. The material balance for the entire system is

$$R_0 + S = E_1 + R_n$$

or

$$1 + (S/R_0) = (E_1/R_0) + (R_n/R_0) \qquad \text{... (6.1.31)}$$

The solute balance ($C$-balance) for the entire system is

$$R_0 x_0 + S y_S = E_1 y_1 + R_n x_n$$

or

$$x_0 + (S/R_0)\, y_S = (E_1/R_0)\, y_1 + (R_n/R_0)\, x_n \qquad \text{... (6.1.32)}$$

Eliminating $(R_n/R_0)$ between the above two equations and rearranging, we get

$$(E_1/R_0) = [x_0 + (S/R_0)\, y_S - (1 + S/R_0)\, x_n]/(y_1 - x_n) \qquad \text{... (6.1.33)}$$

The solvent balance ($B$-balance) for the entire system is

$$R_0 z_{r0} + S z_S = E_1 z_{e1} + R_n z_{rn} \qquad \text{... (6.1.34)}$$

Since $z_{r0} = 0$ (the feed does not contain any $B$),

$$z_{rn} = (S z_S - E_1 z_{e1})/R_n = \frac{(S/R_0)\, z_S - (E_1/R_0)\, z_{e1}}{1 + (S/R_0) - (E_1/R_0)} \qquad \text{... (6.1.35)}$$

The above two equations, such as Eqs (6.1.33) and (6.1.35), can be now solved simultaneously for $x_n$ and $(E_1/R_0)$ with the help of the solubility data / curve. This, no doubt, demands a trial and error procedure:

(*i*) Assume $x_n$. For example, let $x_n = x_0 - 0.001$

(*ii*) Put $XA = x_n$.

(*iii*) Compute $(E_1/R_0)$ from Eq. (6.1.33).

(*iv*) Compute $z_{rn}$ from Eq. (6.1.35)

(*v*) Obtain the value of $x_n$ from the solubility curve/data corresponding to the above-computed value of $z_{rn}$.

(*vi*) Check whether this value of $x_n$ differs appreciably (say, by more than 0.001) and repeat the computations starting from step (*ii*).

Once the values of $x_n$ and $(E_1/R_0)$ are thus computed, we can proceed to make the stage to stage computations. Consider stage 1. The overall material balance is

$$R_0 + E_2 = E_1 + R_1$$

or

$$1 + (E_2/R_0) = (E_1/R_0) + (R_1/R_0) \quad \text{... (6.1.36)}$$

A solute balance (*C*-balance) yields

$$R_0 x_0 + E_2 y_2 = E_1 y_1 + R_1 x_1$$

or

$$x_0 + (E_2/R_0)\, y_2 = (E_1/R_0)\, y_1 + (R_1/R_0)\, x_1 \quad \text{... (6.1.37)}$$

And a solvent balance (*B*-balance) gives

$$R_0 z_{r0} + E_2 z_{e2} = E_1 z_{e1} + R_1 z_{r1}$$

or

$$z_{r0} + (E_2/R_0)\, z_{e2} = (E_1/R_0)\, z_{e1} + (R_1/R_0)\, z_{r1} \quad \text{... (6.1.38)}$$

The known quantities in the above three equations are $x_0$, $(E_1/R_0)$ and $y_1$. Since $y_1$ is known, $x_1$ can be obtained from the equilibrium plot / data. Also, $z_{r1}$ and $z_{e1}$ can be obtained from the solubility curve /data (corresponding to $x_1$ and $y_1$ respectively). What are to be evaluated are $R_1$, $E_2$ and $y_2$. This can be, therefore, accomplished by solving the above three equations simultaneously with the help of the solubility curve/data. Thus, eliminating $(R_1/R_0)$ from Eq. (6.1.38) by putting $(R_1/R_0) = 1 + (E_2/R_0) - (E_1/R_0)$, we get

$$(E_2/R_0) = [(E_1/R_0)\,(z_{e1} - z_{r1}) + (z_{r1} - z_{r0})]/(z_{e2} - z_{r1}) \quad \text{... (6.1.39)}$$

Similarly, eliminating $R_1/R_0$ from Eq. (6.1.37), we get

$$y_2 = [(E_2/R_0)\, x_1 + (E_1/R_0)\,(y_1 - x_1) + (x_1 - x_0)]/(E_2/R_0) \quad \text{... (6.1.40)}$$

Substituting for $(E_2/R_0)$ from Eq. (6.1.39),

$$y_2 = \frac{(E_1/R_0)(y_1 - x_1) + (x_1 - x_0)}{(E_1/R_0)(z_{e1} - z_{r1}) + (z_{r1} - z_{r0})}\,(z_{e2} - z_{r1}) + x_1 \quad \text{... (6.1.41)}$$

$$= x_1 + G_2 (z_{e2} - z_{r1}) \quad \text{... (6.1.42)}$$

The only unknowns in the above equation are $y_2$ and $z_{e2}$ and these can be, therefore, evaluated by solving the above equation simultaneously with the solubility data. The procedure is as follows:

(*i*) Assume $y_2$. If the order of magnitude of $y_2$ is not known, assume $y_2 = y_1 - 0.001$.

(*ii*) Put $YA = y_2$.

(*iii*) Obtain $z_{e2}$ (corresponding to $y_2$) from the solubility curve / data.

(*iv*) Now, compute $y_2$ from Eq. (6.1.42).

(*v*) Check whether the above-computed value of $y_2$ differs appreciably (say, by more than 0.001) from $YA$. If so, reassume $y_2$ (say, put $y_2 = YA - 0.001$) and repeat the computations starting from step (*ii*).

Once the value of $y_2$ is known, $x_2$ can be obtained from the equilibrium data / plot. Then $(E_2/R_0)$ and $(R_1/R_0)$ can be computed from Eqs (6.1.39) and (6.1.36) respectively. By repeating this procedure for each stage, the flow rates and compositions of streams leaving each stage can be evaluated.

In the generalised form, the above design equations can be expressed as

$$y_{i+1} = x_i + G_{i+1} (z_{e,i+1} - z_{ri}) \quad \text{... (6.1.42a)}$$

where

$$G_{i+1} = \frac{(E_i/R_0)(y_i - x_i) + (R_{i-1}/R_0)(x_i - x_{i-1})}{(E_i/R_0)(z_{ei} - z_{ri}) + (R_{i-1}/R_0)(z_{ri} - z_{r,i-1})} \quad \text{... (6.1.42b)}$$

$$(E_{i+1}/R_0) = [(E_i/R_0)(z_{ei} - z_{ri}) + (R_{i-1}/R_0)(z_{ri} - z_{r,i-1})]/(z_{e,i+1} - z_{ri}) \quad \text{... (6.1.39a)}$$

$$(R_i/R_0) = (R_{i-1}/R_0) + (E_{i+1}/R_0) - (E_1/R_0) \quad \text{... (6.1.36a)}$$

Equation (6.1.42a) is to be first solved for $y_{i+1}$ by trial with the help of the solubility data / curve as described above. $x_{i+1}$ is then obtained from the equilibrium plot / data and the values of $(E_{i+1}/R_0)$ and $(R_i/R_0)$ from Eqs (6.1.39a) and (6.1.36a) respectively. This is to be repeated for all values of $i$ ($i = 1, 2, \ldots n - 1$). In other words, computations are to be continued until $x_{i+1} \leq x_n$. Then, the number of equilibrium stages required is given by, $n = i + (x_i - x_n)/(x_i - x_{i+1})$.

The computer program for this analytical procedure is also given at the end of this Chapter (Program 6.2).

## Graphical Method

The graphical procedure is illustrated in Fig. 6.13 (a). The solubility curve is first plotted as shown (for sake of convenience, the mass fraction of solvent–namely, $z_e$ or $z_r$—is plotted along the ordinate and $x$ or $y$ along the abcissa). As in the analytical method, the quantities known at the outset are the feed composition ($x_0$ or $x_F$), the composition of final extract desired ($y_1$), the solvent to feed ratio employed ($S/F$ or $S/R_0$) and the composition of the solvent ($y_S$, $z_S$). Now, the overall material balance about the entire system [namely, Eq. (6.1.31)] and the solute balance may be written in the modified form as

$$R_0 + S = M \quad \text{... (6.1.43)}$$

where $M = (E_1 + R_n)$.

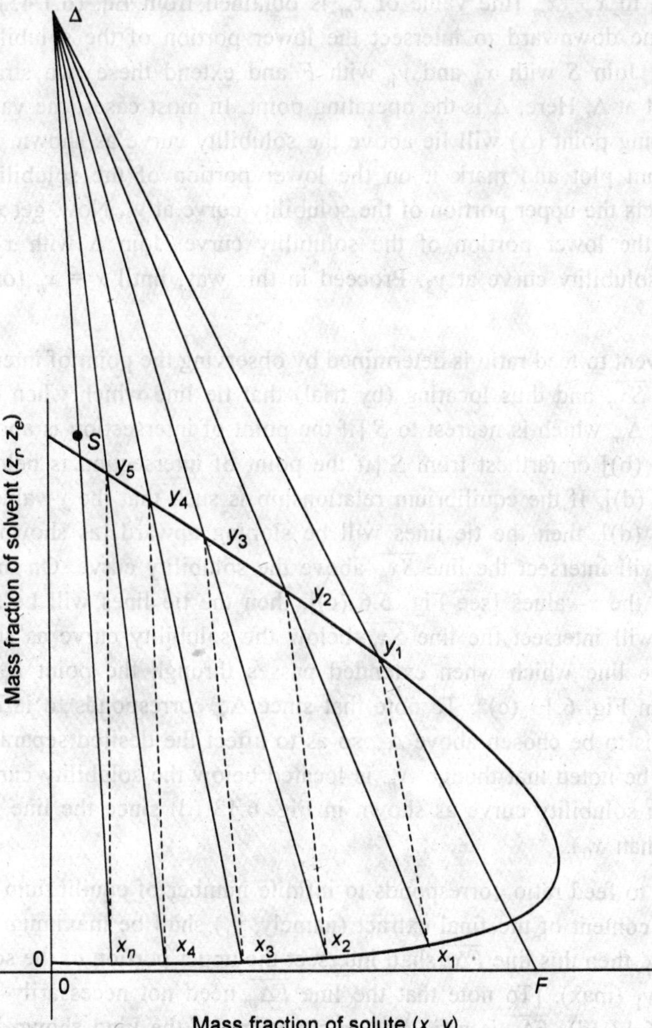


*Figure 6.13 (a): Graphical solution for multistage countercurrent extraction (Tie lines are shown as broken lines)*

Also, $$R_0 x_0 + S y_S = M x_m \quad \text{... (6.1.44)}$$

or $$x_m = (R_0 x_0 + S y_S)/(R_0 + S) = \frac{x_0 + (S/R_0)\,y_S}{1 + (S/R_0)} \quad \text{... (6.1.45)}$$

Thus, once the solvent to feed ratio $(S/R_0)$ is known, $x_m$ can be evaluated from the above equation. Now, the graphical construction is as follows:

Mark the points $F$ and $S$ whose coordinates are $(x_0, 0)$ and $(y_S, z_S)$ respectively. Mark $y_1$ on the upper portion of the solubility curve (namely, the extract curve), as shown. Mark the point $M$ on the line $FS$ corresponding to $x = x_m$ [the value of $x_m$ is obtained from Eq. (6.1.45)]. Join $y_1$ with $M$ and extend this straight line downward to intersect the lower portion of the solubility curve (namely, the raffinate curve) at $x_n$. Join $S$ with $x_n$ and $y_1$ with $F$ and extend these two straight lines upward (or downward) to intersect at $\Delta$. Here, $\Delta$ is the operating point. In most cases, the value of $y_1$ will be lower than $x_0$ and the operating point ($\Delta$) will lie above the solubility curve as shown. Since $y_1$ is known, get $x_1$ from the equilibrium plot and mark it on the lower portion of the solubility curve. Join $\Delta$ with $x_1$ and this line intersects the upper portion of the solubility curve at $y_2$. Now, get $x_2$ from the equilibrium plot and mark it on the lower portion of the solubility curve. Join $\Delta$ with $x_2$ and it intersects the upper portion of the solubility curve at $y_3$. Proceed in this way, until $x = x_n$ (or slightly less than $x_n$) is reached.

The minimum solvent to feed ratio is determined by observing the point of intersection of all extended tie lines with the line $\overline{Sx_n}$ and thus locating (by trial) that tie line which when extended will intersect the line $\overline{Sx_n}$ at a point $\Delta_m$ which is nearest to $S$ [if the point of intersection is above the solubility curve as shown in Fig. 6.13 (b)] or farthest from $S$ [if the point of intersection is below the solubility curve as shown in Fig. 6.13 (d)]. If the equilibrium relationship is such that the $y$-values are smaller than the $x$-values [see Fig. 6.6 (d)], then the tie lines will be sloping upward [as shown in Fig. 6.13 (b)] and when extended, they will intersect the line $\overline{Sx_n}$ above the solubility curve. On the other hand, if the $y$-values are larger than the $x$-values [see Fig. 6.6 (c)], then the tie lines will be sloping downward and when extended, they will intersect the line $\overline{Sx_n}$ below the solubility curve as shown in Fig. 6.13 (d). In many cases, the tie line which when extended passes through the point $F$ will intersect the line $\overline{Sx_n}$ at $\Delta_m$ as shown in Fig. 6.13 (c)*. To note that since $\Delta_m$ corresponds to infinite number of stages, the operating point $\Delta$ is to be chosen above $\Delta_m$ so as to affect the desired separation in a finite number of stages. It may also be noted that though $\Delta_m$ is located below the solubility curve, the operating point $\Delta$ would be above the solubility curve as shown in Fig. 6.13 (d) since the line $\overline{Fy_1}$ slopes upward (as $y_1$ is ordinarily less than $x_0$).

Minimum solvent to feed ratio corresponds to infinite number of equilibrium stages and under these conditions, the solute content of the final extract (namely, $y_1$) shall be maximum. In other words, if the point $F$ is joined to $\Delta_m$, then this line $\overline{F\Delta_m}$ shall intersect the upper portion of the solubility curve (namely, the extract curve) at $y_1$ (max). [To note that the line $\overline{F\Delta_m}$ need not necessarily be a tie line as shown in Figs 6.13 (b) and 6.13 (d). $\overline{F\Delta_m}$ is a tie line in situations of the kind shown in Fig. 6.13 (c)]. Now, joint $y_1$ (max) with $x_n$ and this will intersect the line $\overline{FS}$ at $M_m$. Find $x_m$ (max) corresponding to this point $M_m$. Now, putting $x_m = x_m$ (max) in Eq. (6.1.45), it can be solved for $(S/R_0)$ (min).

The graphical construction may be performed on solvent-free coordinates as well. Equation (6.1.45) still holds except that $x$ is to be replaced by $X$ and $y$ by $Y$ and $S$ and $R_0$ by $S'$ and $R_0'$ respectively. It may be noted that in solvent-free coordinates, $Y_1$ will usually be larger than $X_0$ and the operating point ($\Delta$) will lie below the solubility curve. Also, the tie lines will usually be sloping downwards and therefore, the minimum solvent to feed ratio is determined by locating (by trial) that tie line which when extended will intersect the line $\overline{SX_n}$ at the lowest point ($\Delta_m$). Here also, the tie line which when extended passes

---

* This, however, need not be true always. In Fig. 6.13 (b) for example, it is the tie line $M$ that determines $\Delta_m$ since all other tie lines including tie line $J$ (which when extended passes through $F$) intersect line $\overline{Sx_n}$ at points farther from $S$.

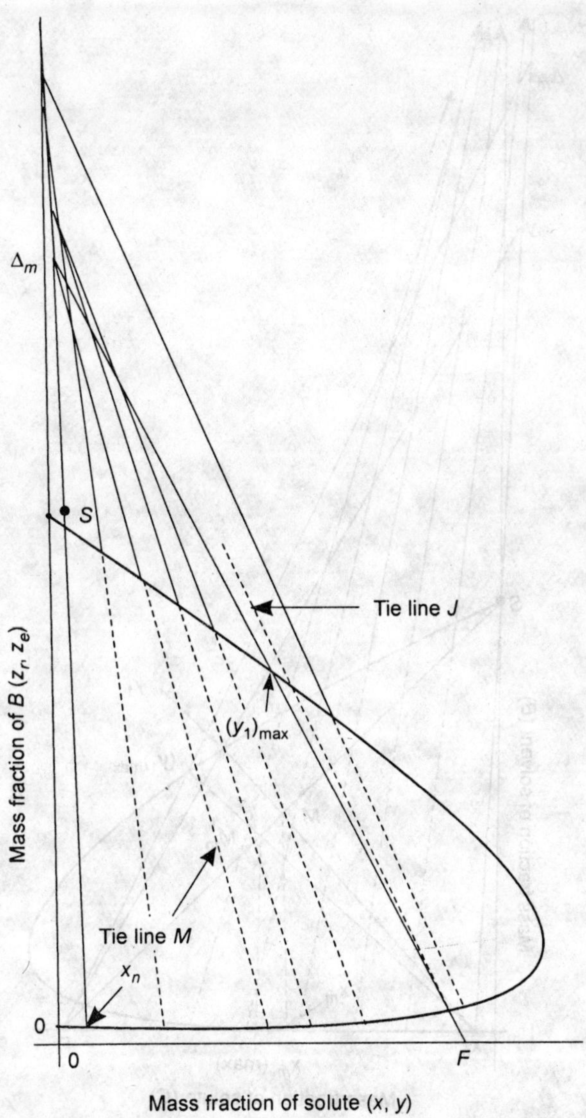


*Figure 6.13 (b): Graphical determination of minimum solvent to feed ratio in multistage countercurrent extraction (general case)*

through $F$ will, in many cases, intersect $\overline{SX_n}$ at $\Delta_m$. Once $\Delta_m$ is located, the minimum value of solvent to feed ratio can be computed as described earlier. The operating point ($\Delta$) will, in this case, lie *below* $\Delta_m$.

As in the case of crosscurrent extraction, a special case of countercurrent extraction is that when $A$ is completely insoluble in $B$. In such a case, both the extract and the raffinate streams will be binary liquid mixtures since the extract will be now composed of only $B$ and $C$ and the raffinate of $A$ and $C$. Since no $A$ is lost in the extract,

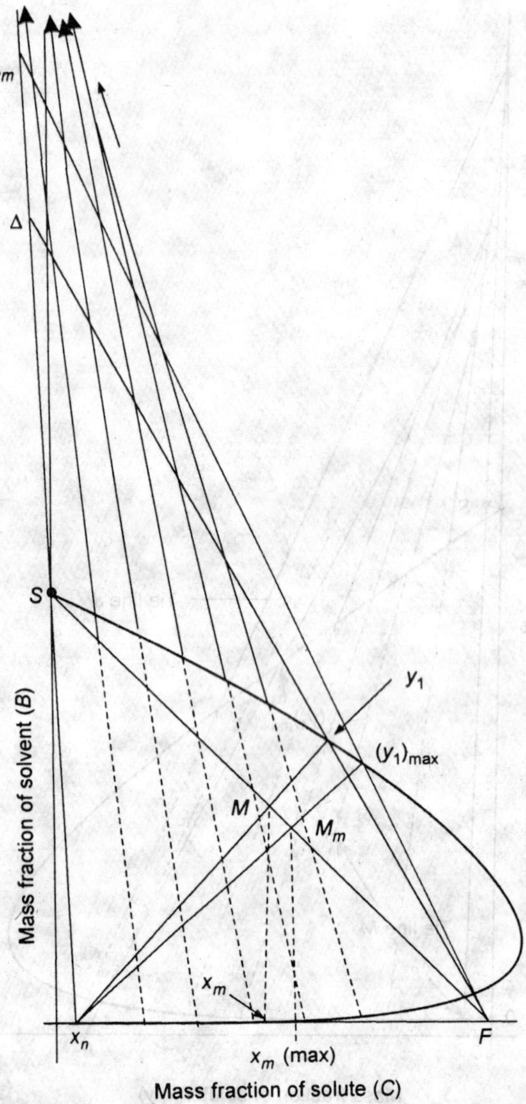


*Figure 6.13 (c): Minimum solvent to feed ratio in multistage countercurrent extraction (Specific case when tie line passing through F decides the minimum ratio).*

$$R_0\ (1 - x_0) = R_1\ (1 - x_1) = \ldots.. = R_n\ (1 - x_n) = G_A \qquad \ldots (6.1.46)$$

Similarly, since no $B$ is lost in the raffinate,

$$S\ (1 - y_S) = E_1\ (1 - y_1) = \ldots.. = E_n\ (1 - y_n) = G_B \qquad \ldots (6.1.47)$$

Therefore, the solute balance for the entire system [(namely, Eq. (6.1.32)] can be re-expressed as

$$G_A\left(\frac{x_0}{1-x_0}\right)+G_B\left(\frac{y_S}{1-y_S}\right)=G_B\left(\frac{y_1}{1-y_1}\right)+G_A\left(\frac{x_n}{1-x_n}\right)$$

or
$$G_A x'_0 + G_B y'_S = G_B y'_1 + G_A x'_n$$

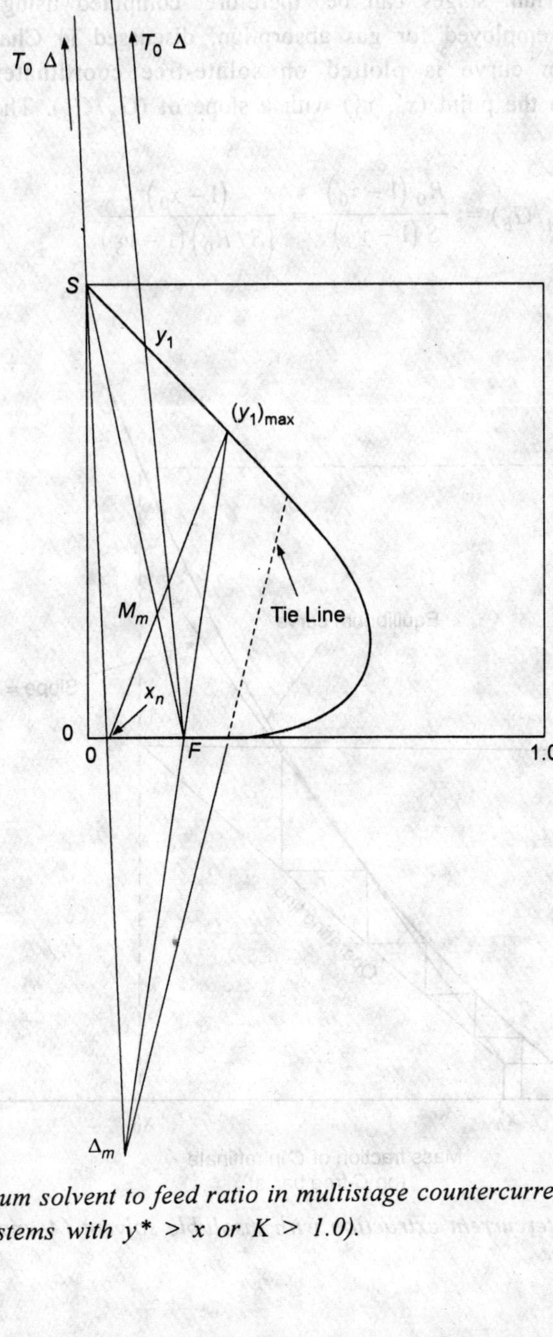


*Figure 6.13 (d): Minimum solvent to feed ratio in multistage countercurrent extraction (alternate case—for systems with $y^* > x$ or $K > 1.0$).*

$$(G_A/G_B) = \frac{(y_1' - y_S')}{(x_0' - x_n')} \quad \text{... (6.1.48)}$$

where $x'$ and $y'$ are mass fractions on solute-free basis. This is equation to a straight line (namely, the operating line) of slope $(G_A/G_B)$ and which passes through the points $(x_0, y_1')$ and $(x_n', y_S')$.

The number of equilibrium stages can be, therefore, computed using a simplified procedure (similar to the one that is employed for gas absorption, discussed in Chapter 4) as illustrated in Fig. 6.14. The equilibrium curve is plotted on solute-free coordinates ($y'$ versus $x'$). The operating line is drawn from the point $(x_0', y_1')$ with a slope of $(G_A/G_B)$. The value of $(G_A/G_B)$ may be computed as

$$(G_A/G_B) = \frac{R_0(1-x_0)}{S(1-y_S)} = \frac{(1-x_0)}{(S/R_0)(1-y_S)} \quad \text{... (6.1.49)}$$

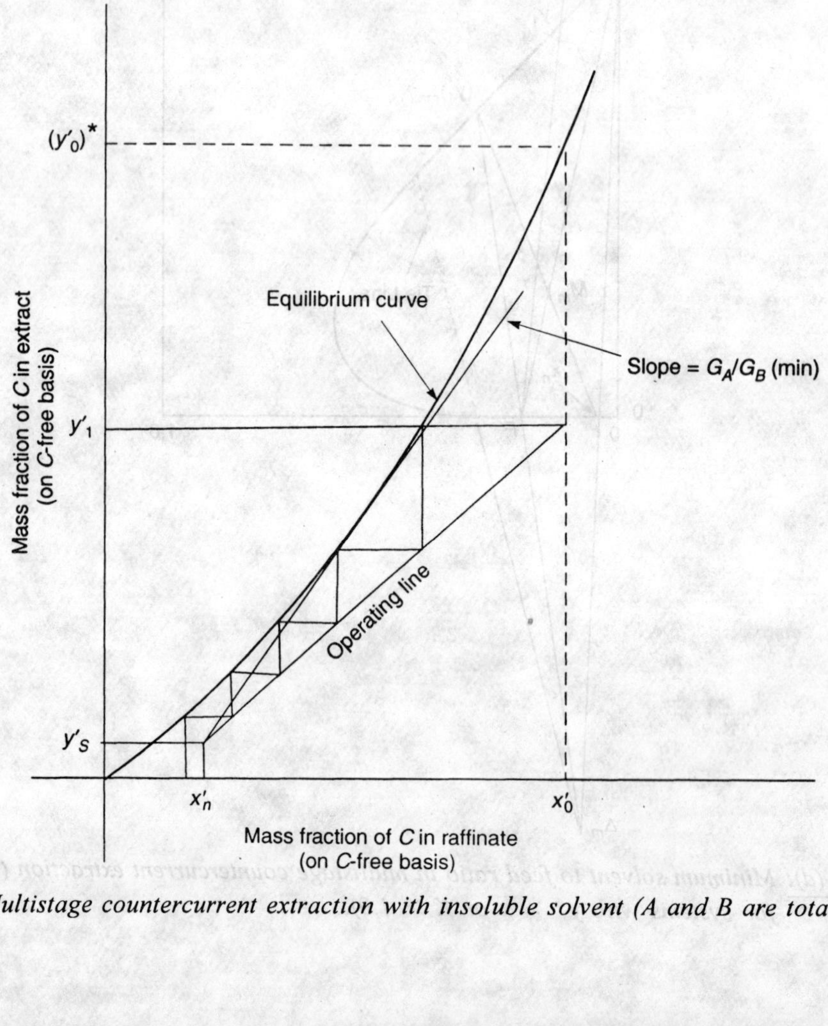


*Figure 6.14: Multistage countercurrent extraction with insoluble solvent (A and B are totally insoluble)*

Now, from the point $(x_0', y_1')$ on the operating line, draw a horizontal line and this will intersect the equilibrium curve at $(x_1', y_1')$. Now, perform the staircase construction as shown until $x' = x_n'$ (or slightly less than $x_n'$) is reached.

As in the case of gas absorption, the minimum solvent to feed ratio (on solute-free coordinates) required, that is $G_B$ (min)$/G_A$, can be determined based on the shape of the equilibrium curve. If the equilibrium curve is concave upward as shown in Fig. 6.14, then $G_A/G_B$ (min) = slope of the tangent to the equilibrium curve from $(x_n', y_S')$. If the equilibrium curve is concave downward or is a straight line, then $G_B$ (min)$/G_A$ can be computed from Eq. (6.1.48) by replacing $y_1'$ with $(y_0')^*$, where $(y_0')^*$ is the equilibrium value of $y'$ corresponding to $x' = x_0'$. In other words, minimum solvent to feed ratio corresponds to infinite number of stages and this would then correspond to the case when the final extract is in equilibrium with the feed.

If, in addition, the equilibrium curve is a straight line of slope $m$, then the number of equilibrium stages may be computed analytically using the analogue of *Kremser-Brown-Souders equation* [originally derived for gas absorption, see Eq. (4.5.38) of Chapter 4]:

$$n = \frac{\log\left[\frac{x_0' - (y_S'/m)}{x_n' - (y_S'/m)}\left(1 - 1/E_f\right) + \left(1/E_f\right)\right]}{\log\left(E_f\right)} \quad \text{... (6.1.50)}$$

where $E_f = (m\,G_B/G_A)$ = extraction factor. It can be seen that this *extraction factor* is analogous to the stripping factor for desorption or stripping (defined in Chapter 4). The above expression is valid only if $E_f$ is not equal to unity. If $E_f = 1.0$, then

$$n = (x_0' - x_n') / [x_n' - (y_S'/m)] \quad \text{... (6.1.51)}$$

To note that the graphical procedure of Fig. 6.14 and the above Eqs (6.1.50) and (6.1.51) shall remain unaltered even if the mass fractions ($x'$ and $y'$) are replaced by concentrations (expressed in kg/m$^3$ or kg/litre) and in that case, $G_A$ and $G_B$ must also be expressed as volumetric flow rates (in m$^3$/s or litres/s).

## Multistage Countercurrent Extraction (with Multiple Feed)

In industrial practices, the plant requirements occasionally demand introduction of more than one feed stream. If apart from the feed stream ($R_0$) which is introduced to stage 1, an additional feed stream ($F_1$) is introduced to any of the intermediate stages as shown in Fig. 6.15, then the graphical procedure of analysis shall remain essentially the same as that described in the earlier section except that in this case, the graphical construction has to be performed with two operating points ($\Delta$ and $\Delta'$) instead of one. To note that the first feed stream ($R_0$) has been assumed to contain larger concentration of $C$ as is usually the case. The operating point $\Delta'$ is the point of intersection of the line $\overline{Jy_1}$ with $\overline{Sx_n}$, while the other operating point $\Delta$ is the point of intersection of the lines $\overline{Fy_1}$ and $F_1\Delta'$. The points $F$ and $S$ can be marked as usual since their coordinates are $(x_0, 0)$ and $(y_S, z_S)$ respectively. The coordinates of $F_1$ are $(x_{F1}, 0)$ where $x_{F1}$ is the mass fraction of solute ($C$) in the intermediate feed stream. The point $J$ lies on the line $\overline{FF_1}$ and it can be marked on that line corresponding to $x = x_J$ (see Fig. 6.16), where the value of $x_J$ may be obtained as

$$R_0 + F_1 = J \quad \text{... (6.1.52)}$$

$$R_0 x_0 + F_1 x_{F1} = J x_J \qquad \text{... (6.1.53)}$$

or
$$x_J = (R_0 x_0 + F_1 x_{F1})/(R_0 + F_1) = \frac{x_0 + (F_1/R_0)\, x_{F1}}{1 + F_1/R_0} \qquad \text{... (6.1.54)}$$

where $F_1$ and $x_{F1}$ are the flow rate (in kg/s) and composition (mass fraction of solute) respectively of the intermediate feed stream. Also, though the value of $y_1$ (the composition of final extract desired) is ordinarily specified at the outset, the value of $x_n$ is not necessarily known. $x_n$ can be located on the solubility curve using a procedure similar to that described in the earlier section. Thus, the overall material balance for the entire system is

$$R_0 + F_1 + S = R_n + E_1$$

or
$$J + S = M \qquad \text{(6.1.55)}$$

where $J = R_0 + F_1$ and $M = R_n + E_1$. The solute balance (C-balance) for the entire system is

$$J x_J + S y_S = M x_m$$

or
$$x_m = (J x_J + S y_S)/(J + S) = \frac{x_J + (S/J)\, y_S}{1 + (S/J)} \qquad \text{... (6.1.56)}$$

where the ratio $S/J$ is given by

$$S/J = S/(R_0 + F_1) = \frac{(S/R_0)}{1 + (F_1/R_0)} \qquad \text{... (6.1.57)}$$

Thus, once the ratios $(S/R_0)$ and $(F_1/R_0)$ are known, $(S/J)$ can be obtained from Eq. (6.1.57) and thereafter, $x_m$ from Eq. (6.1.56). The point $M$ lies on the line $JS$ (and not $FS$) and it can be marked on that line corresponding to $x = x_m$. Now, the line $\overline{y_1 M}$ when extended downward will intersect the solubility curve at $x_n$.

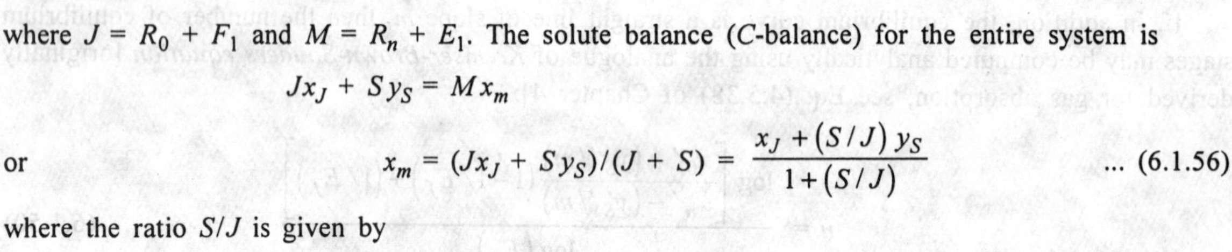


*Figure 6.15: Multistage countercurrent extraction with multiple feed*

Once the operating points (Δ and Δ′) are located as described above, the number of equilibrium stages required and the *correct* stage for the introduction of the intermediate feed (may be called the *optimum* location of the intermediate feed stage) may be determined by following a graphical procedure similar to that described in the earlier section. Thus, first find $x_1$ from the equilibrium plot (since $y_1$ is known) and mark it on the lower portion of the solubility curve (raffinate curve). Join $x_1$ to Δ and this line intersects the solubility curve at $y_2$. Proceed in this way until the value of $x$ falls below $x_{F1}$. For example, let it be found that $x_4$ is less than $x_{F1}$. Now, join $x_4$ to $\Delta'$ (and not $\Delta$) and this line will intersect the solubility curve at $y_5$. Continue this graphical construction using $\Delta'$ as the operating point until $x = x_n$ (or slightly less than $x_n$). The correct stage for the introduction of the intermediate feed is now stage 4. This is called the optimum location because if the intermediate feed is introduced at this stage, then the total number of equilibrium stages required will be a minimum.

The entire graphical procedure is illustrated in Fig. 6.16.

If the stage at which the intermediate feed is introduced has already been specified, then the shifting over from $\Delta$ to $\Delta'$ must be done from that stage onwards. For example, if the intermediate feed is being introduced at the fifth stage, then $x_5$ must be joined to $\Delta'$ (and *not* $\Delta$) and thereafter the graphical construction must be performed using $\Delta'$ as the operating point.

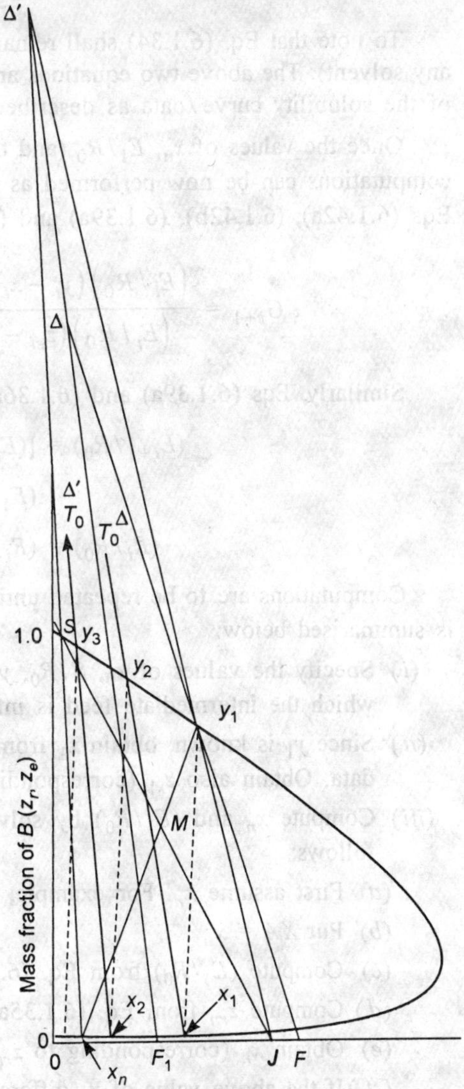


*Figure 6.16: Graphical solution for multistage countercurrent extraction with multiple feed*

## Analytical Method

The analytical procedure of computation of the required number of equilibrium stages in this case is essentially the same as that for the conventional countercurrent extraction (discussed in the earlier section) except that for stage $f$, the additional stream $F_1$ must also be included in the material balance equations. This is evident from the fact that Figs 6.12 and 6.15 are practically alike except for the presence of the additional feed stream ($F_1$) in Fig. 6.15. The material balance equations for the entire system (used for the computation of $x_n$) shall also get slightly modified since the additional term $F_1$ is to be included. Thus, Eqs (6.1.31) and (6.1.32) get modified to

$$1 + (S/R_0) + (F_1/R_0) = (E_1/R_0) + (R_n/R_0) \quad \text{... (6.1.31a)}$$

$$x_0 + (S/R_0)\, y_S + (F_1/R_0)\, x_{F1} = (E_1/R_0)\, y_1 + (R_n/R_0)\, x_n \quad \text{... (6.1.32a)}$$

Accordingly, Eqs (6.1.33) and (6.1.35) also get modified into

$$(E_1/R_0) = [x_0 + (S/R_0)\, y_S + (F_1/R_0)\, x_{F1} - (1 + S/R_0 + F_1/R_0)\, x_n]/(y_1 - x_n) \quad \text{(6.1.33a)}$$

$$z_{rn} = \frac{(S/R_0)\,z_S - (E_1/R_0)\,z_{e1}}{1+(S/R_0)+(F_1/R_0)-(E_1/R_0)} \quad \text{... (6.1.35a)}$$

To note that Eq. (6.1.34) shall remain the same since $z_{F1} = 0$ (as the feed stream does not contain any solvent). The above two equations are to be solved simultaneously for $x_n$ and $(E_1/R_0)$ with the help of the solubility curve/data as described in the earlier section.

Once the values of $x_n$, $E_1/R_0$ (and thereby the value of $R_n/R_0$ also) are known, the stage to stage computations can be now performed as described in the earlier section. For all stages except stage $f$, Eqs (6.1.42a), (6.1.42b), (6.1.39a) and (6.1.36a) apply. For stage $f$, Eq. (6.1.42b) gets modified to

$$G_{i+1} = \frac{(E_i/R_0)(y_i - x_i)+(R_{i-1}/R_0)(x_i - x_{i-1})+(F_1/R_0)(x_i - x_{F1})}{(E_i/R_0)(z_{ei} - z_{ri})+(R_{i-1}/R_0)(z_{ri} - z_{r,i-1})+(F_1/R_0)\,z_{ri}} \quad \text{... (6.1.42c)}$$

Similarly, Eqs (6.1.39a) and (6.1.36a) get modified into

$$(E_{i+1}/R_0) = [(E_i/R_0)(z_{ei} - z_{ri}) + (R_{i-1}/R_0)(z_{ri} - z_{r,i-1}) + (F_1/R_0)\,z_{ri}]/(z_{e,i+1} - z_{ri}) \quad \text{... (6.1.39b)}$$

$$(R_i/R_0) = (R_{i-1}/R_0) + (E_{i+1}/R_0) + (F_1/R_0) - (E_i/R_0) \quad \text{... (6.1.36b)}$$

Computations are to be repeated until $x_{i+1} \le x_n$. For the convenience of the readers, the procedure is summarised below:

(*i*) Specify the values of $x_0$, $S/R_0$, $y_S$, $z_S$, $x_{F1}$, $F_1/R_0$ and $y_1$. Specify also $f$ (that is, the stage to which the intermediate feed is introduced).

(*ii*) Since $y_1$ is known, obtain $x_1$ from the equilibrium data/plot and $z_{e1}$ from the solubility curve/data. Obtain also $z_{r1}$ (corresponding to $x_1$) from the solubility curve/data.

(*iii*) Compute $x_n$ and $(E_1/R_0)$ by solving Eqs (6.1.33a) and (6.1.35a) simultaneously (by trial) as follows:

- (*a*) First assume $x_n$. For example, let $x_n = x_{F1} - 0.001$.
- (*b*) Put $XA = x_n$.
- (*c*) Compute $(E_1/R_0)$ from Eq. (6.1.33a)
- (*d*) Compute $z_{rn}$ from Eq. (6.1.35a).
- (*e*) Obtain $x_n$ (corresponding to $z_{rn}$) from the solubility curve/data.
- (*f*) If the above value of $x_n$ differs significantly (say, by more than 0.001) from $XA$, then put $x_n = XA - 0.001$ and repeat the computations starting from step (*b*).

(*iv*) Put $i = 1$.

(*v*) If $i = f$, compute $G_{i+1}$ from Eq. (6.1.42c), otherwise from Eq. (6.1.42b).

(*vi*) Assume $y_{i+1}$. For example, let $y_{i+1} = y_i - 0.001$.

(*vii*) Put $YA = y_{i+1}$.

(*viii*) Obtain $z_{e,i+1}$ from the solubility curve/data.

(*ix*) Compute $y_{i+1}$ from Eq. (6.1.42a).

(*x*) If the above-computed value of $y_{i+1}$ differs significantly (say, by more than 0.001) from *YA*, then put $y_{i+1} = YA - 0.001$ and repeat the computations starting from step (*vii*).

(*xi*) Obtain $x_{i+1}$ from equilibrium data / plot. Also, obtain $z_{r,i+1}$ from the solubility curve / data.

(*xii*) Compute $(E_{i+1}/R_0)$ and $(R_i/R_0)$ from Eqs (6.1.39a) and (6.1.36a) respectively if $i \neq f$ and from Eqs (6.1.39b) and (6.1.36b) if $i = f$.

(*xiii*) If $x_{i+1} > x_n$, put $i = i + 1$ and go to step (*v*).

(*xiv*) Compute the required number of equilibrium stages (*n*) as

$$n = i + (x_i - x_n)/(x_i - x_{i+1}) \qquad \text{... (6.1.58)}$$

In the above-discussed procedure, it has been assumed that the stage to which the intermediate feed is introduced (namely, stage *f*) has already been specified at the outset. If this is not so and we have to determine the optimum stage at which the intermediate feed is to be introduced (so as to make the total number of equilibrium stages required a minimum), then the procedure is to be slightly modified as given below. We shall use a notation *FS* such that for ail stages up to stage $f$ $(i < f)$, $FS = 0$ and for $i \geq f$, $FS = 1.0$. At $i = f$, *FS* changes its value from zero to 1.0.

(*i*) Specify the values of $x_0$, $(S/R_0)$, $y_S$, $z_S$, $x_{F1}$, $(F_1/R_0)$ and $y_1$. Since $y_1$ is known, obtain $x_1$ from equilibrium data / plot and $z_{e1}$, $z_{r1}$ from solubility data / curve.

(*ii*) Compute $x_n$ and $(E_1/R_0)$ as before.

(*iii*) Put $i = 1$.

(*iv*) Put $FS = 0$.

(*v*) If $FS = 1.0$, go to step (*vii*).

(*vi*) If $x_i < x_{F1}$, then put $FS = 1.0$ and go to step (*viii*).

(*vii*) Compute $G_{i+1}$ from Eq. (6.1.42b) and go to step (*ix*).

(*viii*) Put $f = i$. Compute $G_{i+1}$ from Eq. (6.1.42c).

(*ix*) Assume $y_{i+1}$. For example, let $y_{i+1} = y_i - 0.001$.

(*x*) Put $YA = y_{i+1}$.

(*xi*) Obtain $z_{e,i+1}$ from solubility curve / data.

(*xii*) Compute $y_{i+1}$ from Eq. (6.1.42a).

(*xiii*) If the above-computed value of $y_{i+1}$ differs significantly (say, by more than 0.001) from *YA*, put $y_{i+1} = YA - 0.001$ and repeat the computations starting from step (*x*).

(*xiv*) Obtain $x_{i+1}$ from equilibrium data / plot and $z_{r,i+1}$ from the solubility curve / data.

(*xv*) Compute $(E_{i+1}/R_0)$ and $(R_i/R_0)$ from Eqs (6.1.39a) and (6.1.36a) respectively if $i \neq f$ and from Eqs (6.1.39b) and (6.1.36b) respectively if $i = f$.

(*xvi*) If $x_{i+1} > x_n$, put $i = i + 1$ and go to step (*v*).

(*xvii*) Compute the required number of equilibrium stages (*n*) from Eq. (6.1.58).

(*xviii*) Print *n*, *f*.

**Example 6.1:** One thousand kg / hr of a solution containing 42.5 mass percent diphenyl hexane and balance docosane is to be extracted with furfural at 45° C to reduce the diphenyl hexane content to 2.5 percent. The solvent entering the system contains 0.5 mass percent of diphenyl hexane, the balance being furfural.

(*a*) If a multistage crosscurrent system is employed (each stage being fed with 1100 kg / hr of solvent), compute the number of equilibrium stages required. What will be the composition of the final extract and what will be the total solvent requirement?

(*b*) How much solvent would be required if the same final raffinate concentration were to be obtained in a single crosscurrent stage? Estimate also the minimum solvent rate required to form two phases in this case. What is the maximum solvent rate that will form two phases?

(*c*) If a multistage countercurrent system is being employed with a solvent to feed ratio of 1.1 kg / kg, compute the number of equilibrium stages required. What is the minimum solvent to feed ratio that must be used and what is the maximum concentration of diphenyl hexane that can be achieved in the extract?

(*d*) If separation is being affected with the same number of equilibrium stages and the same amount of solvent as in (*c*) but employing a crosscurrent operation (the total solvent being equally divided among the stages), estimate the percent recovery of solute (diphenyl hexane) that shall be achieved.

The solubility data for diphenyl hexane–docosane–furfural system at 45° C is given below[10]:

| Mass fraction of diphenyl hexane | 0.0 | 0.11 | 0.26 | 0.375 | 0.474 | 0.487 | 0.468 | 0.423 | 0.356 |
|---|---|---|---|---|---|---|---|---|---|
| | 0.274 | 0.185 | 0.09 | 0.0 | | | | | |
| Mass fraction of furfural | 0.04 | 0.05 | 0.07 | 0.10 | 0.20 | 0.30 | 0.40 | 0.50 | 0.60 |
| | 0.70 | 0.80 | 0.90 | 0.993 | | | | | |

Within the range of concentration under consideration, the distribution isotherm for this system is substantially linear and the distribution coefficient = $y^*/x$ = 0.98 = constant.

**Solution:** (*a*) We shall first solve the problem by the analytical method. Putting $i$ = 1, from Eq. (6.1.12a),

$$x_{m1} = (R_0 x_0 + S_1 y_S)/(R_0 + S_1)$$

$$= \frac{x_0 + (S_1/R_0)\, y_S}{1 + (S_1/R_0)} = \frac{0.425 + (1.1)(0.005)}{(1.0 + 1.1)} = 0.205$$

To note that since each stage is fed with the same amount of solvent (that is, 1100 kg / hr), $S_1/R_0$ = $S_2/R_0$ = ......... = $S/R_0$ = 1100 / 1000 = 1.1. Now, from Eq. (6.1.14a),

$$z_{m1} = (R_0 z_{r0} + S z_S)/(R_0 + S) = \frac{z_{r0} + (S/R_0)\, z_S}{1 + S/R_0}$$

Since the feed does not contain any solvent, $z_{r0}$ = 0. Also, the fresh solvent entering the system contains only *B* (furfural) and *C* (diphenyl hexane). Therefore, $z_S = 1 - y_S$ = (1.0 – 0.005) = 0.995. Thus

$$z_{m1} = (1.1)(0.995)/(1.0 + 1.1) = 0.52119$$

Now, Eq. (6.1.18a) becomes

$$y_1 = x_1 + (z_{e1} - z_{r1})\,(0.205 - x_1)/(0.52119 - z_{r1}) \quad \text{... (i)}$$

The above equation is to be solved by trial with the help of the solubility data. The solubility curve is shown in Fig. 6.1.1. Incidentally, once the solubility data has been prepared elaborately, intermediate values can be obtained by interpolation with allowable error. Now, let $y_1 = 0.2035$. From solubility curve / data, $z_{e1} = 0.7792$. Since $y^*/x = 0.98$ = constant, $x_1 = (0.2035/0.98) = 0.20765$ and from the solubility data, $z_{r1} = 0.063$. Substituting these values in Eq. (i),

$$y_1 = 0.20765 + (0.7792 - 0.063)\ (0.205 - 0.20765)/(0.52119 - 0.063)$$
$$= 0.2035077$$

Since the above-computed value of $y_1$ agrees very closely with that assumed at the outset, we can discontinue further trial. Now, from Eq. (6.1.10a),

$$M_1 = R_0 + S$$
$$(M_1/R_0) = 1 + (S/R_0) = 1.0 + 1.1 = 2.1$$

And from Eq. (6.1.16a),

$$E_1/R_0 = (M_1/R_0)\ (x_{m1} - x_1)/(y_1 - x_1) = \frac{(2.1)(0.205 - 0.20765)}{(0.2035 - 0.20765)} = 1.341$$

$$R_1/R_0 = (M_1/R_0) - (E_1/R_0) = 2.1 - 1.341 = 0.759$$

Computations are performed similarly for subsequent stages and the results are listed below:

| $i$ | $y_i$ | $x_i$ | $x_{mi}$ | $E_i/R_0$ | $R_i/R_0$ |
|---|---|---|---|---|---|
| 1 | 0.2035 | 0.20765 | 0.205 | 1.341 | 0.759 |
| 2 | 0.087 | 0.08877 | 0.08774 | 1.0842 | 0.7748 |
| 3 | 0.03925 | 0.04 | 0.03962 | 1.0077 | 0.8671 |
| 4 | 0.0203 | 0.0207 | 0.02045 | 1.2294 | 0.7377 |

It can be seen that $x_4 = 0.0207$ and is less than the desired raffinate concentration $x_n$ (= 0.025). The number of equilibrium stages required is, therefore, 4 (in fact, slightly less than 4.0).

If four equilibrium stages are used, then the composition of the final composited extract will be

$$y_f = (E_1 y_1 + E_2 y_2 + E_3 y_3 + E_4 y_4)/(E_1 + E_2 + E_3 + E_4)$$

$$= \frac{(1.341)(0.2035) + (1.0842)(0.087) + (1.0077)(0.03925) + (1.2294)(0.0203)}{(1.341 + 1.0842 + 1.0077 + 1.2294)}$$

$$= 0.16895 \simeq 0.169$$

Thus, the final extract will contain 16.9 percent diphenyl hexane. The total solvent requirement will be (4.0) (1100) = 4400 kg / hr.

The solution may also be obtained graphically as shown in Fig. 6.1.1. The points $F$ and $S$ are first marked, their coordinates being (0.425, 0) and (0.005, 0.995) respectively. $x_{m1}$ [computed from Eq. (6.1.12a)] is 0.205. The point $M_1$ is now marked on line $\overline{FS}$ corresponding to $x = 0.205$. Then, the tie line $\overline{x_1 y_1}$ that passes through $M_1$ is located by trial as shown in figure. $x_{m2}$ is now obtained from Eq. (6.1.12a) and the point $M_2$ is marked on line $\overline{x_1 S}$ corresponding to $x = x_{m2}$. The tie line $\overline{x_2 y_2}$ that passes through $M_2$ is then located by trial. The procedure is repeated for stage 3 and stage 4 (to note that though the points $M_4$ and $x_4$ are marked in the figure, the tie line $\overline{x_4 y_4}$ is not shown since this crowds the figure too much). It can be seen that since the equilibrium values of $x$ and $y$ are fairly close to each

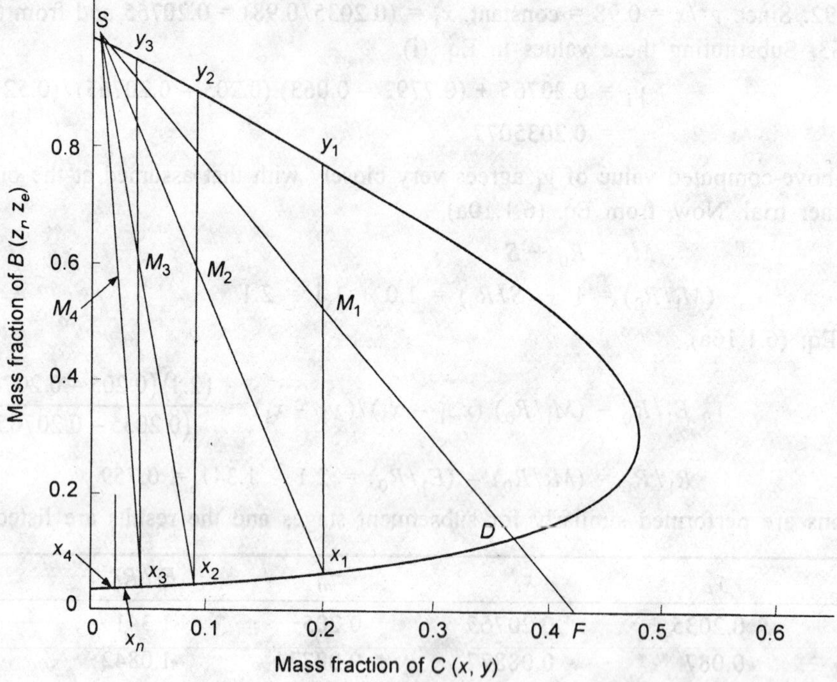


*Figure 6.1.1*

other, the tie lines are close to being vertical. The graphical construction must be, therefore, performed very cautiously using a sharp pencil and as wide a scale as possible.

(*b*) If the same raffinate concentration is required from a single crosscurrent stage, then

$$x_1 = x_n = 0.025$$

$$y_1 = 0.98\ x_1 = 0.0245$$

From the solubility data, $z_{e1}$ (corresponding to $y = y_1 = 0.0245$) = 0.9677 and $z_{r1}$ (corresponding to $x = x_1 = 0.025$) = 0.0423. Equations (6.1.12a), (6.1.14a) and (6.1.18a) respectively get reduced to

$$x_{m1} = \frac{0.425 + (S/R_0)(0.005)}{1 + (S/R_0)}$$

$$z_{m1} = 0.995\ (S/R_0)/(1 + S/R_0)$$

$$0.0245 = 0.025 + (0.9677 - 0.0423)\ (x_{m1} - 0.025)/(z_{m1} - 0.0423)$$

If we substitute for $x_{m1}$ and $z_{m1}$ from the first two equations in the third equation, then it can be solved for $(S/R_0)$. Thus

$$S/R_0 = 20.527$$

The total solvent requirement is, therefore, (20.527) (1000) = 20527.0 kg / hr. The solvent requirement is thus nearly five times that for a four-stage crosscurrent operation.

The ratio $S/R_0$ may be determined graphically as well. First $x_1$ and $y_1$ are marked on the solubility curve (on the raffinate branch and on the extract branch respectively). Now, the line $\overline{x_1 y_1}$ is drawn and it intersects $\overline{FS}$ at $M_1$ (not shown in Fig. 6.1.1) such that $x_{m1} = 0.0245$. The ratio $S/R_0$ may be then computed as

$$S/R_0 = (x_0 - x_{m1})/(x_{m1} - y_S) = (0.425 - 0.0245)/(0.0245 - 0.005) = 20.538$$

This is very close to that computed analytically. The minimum solvent rate required and the maximum permissible solvent rate can be computed from Eqs (6.1.23) and (6.1.24). The line $\overline{FS}$ intersects the solubility curve at $D$ and $K$ (the point $K$ is not shown in Fig. 6.1.1 since it is very close to $S$). From figure,

$$x_D = 0.38$$
$$y_K = 0.015$$

Therefore, $(S/R_0)_{min} = (0.425 - 0.38)/(0.38 - 0.005) = 0.12$

And $(S/R_0)_{max} = (0.425 - 0.015)/(0.015 - 0.005) = 41.0$

The minimum solvent rate required is thus 120 kg / hr and the maximum solvent rate that will form two phases is 41000 kg / hr.

(*c*) For the countercurrent operation also, let us present the analytical solution first. The quantities known are

$$x_0 = 0.425$$
$$x_n = 0.025$$
$$z_{rn} \text{ (from solubility data)} = 0.0377$$
$$y_S = 0.005$$
$$z_S = 0.995$$
$$S/R_0 = 1.1$$

Since the value of $y_1$ is not given, it is to be first computed by solving Eqs (6.1.31), (6.1.32) and (6.1.34) simultaneously. Thus, eliminating $(R_n/R_0)$ among these equations and rearranging, we get

$$y_1 = x_n + G_0 \, (z_{e1} - z_{rn}) \qquad \text{... (i)}$$

where
$$G_0 = \frac{(x_0 - x_n) + (S/R_0)(y_S - x_n)}{(S/R_0)(z_S - z_{rn}) - z_{rn}} \qquad \text{... (ii)}$$

Also, $E_1/R_0 = [(S/R_0)\,(z_S - z_{rn}) - z_{rn}]/(z_{e1} - z_{rn})$ ... (iii)

Substituting the known values of mass fractions and the $S/R_0$ ratio, we get

$$G_0 = \frac{(0.425 - 0.025) + (1.1)(0.005 - 0.025)}{(1.1)(0.995 - 0.0377) - 0.0377} = 0.37229$$

Therefore, $y_1 = 0.025 + 0.37229\,(z_{e1} - 0.0377)$ ... (iv)

This equation has to be now solved by trial. Thus, let $y_1 = 0.271$. Then, from solubility data, $z_{e1} = 0.69634$. Substituting in Eq. (iv), we get

$$y_1 = 0.025 + 0.37229\,(0.69634 - 0.0377) = 0.2702$$

Since this checks closely with the initially assumed value of $y_1$, further trials are not required. Now, $x_1 = (0.271/0.98) = 0.27653$ and from solubility data, $z_{r1} = 0.0743$. Also, from Eq. (iii),

$$E_1/R_0 = \frac{(1.1)(0.995 - 0.0377) - 0.0377}{(0.69634 - 0.0377)} = 1.54155$$

and
$$R_n/R_0 = 1 + S/R_0 - E_1/R_0 = 1.0 + 1.1 - 1.54155 = 0.55845$$

Now, we can proceed with the stage to stage computations. For example, putting $i = 1$ in Eq. (6.1.42b), we get

$$G_2 = \frac{(1.54155)(0.271 - 0.27653) + (1.0)(0.27653 - 0.425)}{(1.54155)(0.69634 - 0.0743) + (1.0)(0.0743 - 0.0)} = -0.157$$

Therefore, Eq. (6.1.42a) gets reduced to

$$y_2 = 0.27653 - 0.157\ (z_{e2} - 0.0743)$$

Solving by trial, $y_2 = 0.1585$ and $z_{e2} = 0.82789$. Now, from Eq. (6.1.39a),

$$E_2/R_0 = \frac{(1.54155)(0.69634 - 0.0743) + (1.0)(0.0743 - 0.0)}{(0.82789 - 0.0743)} = 1.371$$

And from Eq. (6.1.36a),

$$R_1/R_0 = 1.0 + 1.371 - 1.54155 = 0.82945$$

$$x_2 = 0.1585/0.98 = 0.1617$$

$$z_{r2} \text{ (from solubility data)} = 0.05689$$

Computations are similarly performed for subsequent stages and the results are listed below:

| $i$ | $y_i$ | $x_i$ | $E_i/R_0$ | $R_i/R_0$ |
|---|---|---|---|---|
| 1 | 0.271 | 0.27653 | 1.54155 | 0.82945 |
| 2 | 0.1585 | 0.1617 | 1.371 | 0.6801 |
| 3 | 0.08 | 0.08163 | 1.2217 | 0.6073 |
| 4 | 0.0325 | 0.03316 | 1.14886 | 0.5711 |
| 5 | 0.006 | 0.006122 | 1.11268 | |

The number of equilibrium stages required is, therefore, $n = 4 + (0.03316 - 0.025)/(0.03316 - 0.006122) = 4.3$. It can be thus seen that the countercurrent operation affects the desired separation in 4.3 equilibrium stages by using 100 kg of solvent per hour. On the other hand, the crosscurrent operation affects the separation in 4 stages, but demands 4400 kg of solvent per hour. Also, the final composited extract from the 4 stage crosscurrent operation contains only 16.9 percent solute (diphenyl hexane) while that from the countercurrent operation contains 27.1 percent solute.

The graphical solution to this problem is illustrated in Fig. 6.1.2. The points $F$ and $S$ are marked as in (*a*) and $x_n$ is marked on the raffinate branch of the solubility curve. $x_m$ [from Eq. (6.1.45)] is 0.205 and the point $M$ is marked on the line $\overline{FS}$ corresponding to $x = 0.205$. The line $\overline{x_n M}$ when extended intersects the extract branch of the solubility curve at $y_1$. The lines $\overline{Fy_1}$ and $\overline{Sx_n}$ are extended upward to intersect at $\Delta$. The graphical construction is now performed as shown in Fig. 6.1.2 using $\Delta$ as the operating point. To avoid crowding, the tie lines are not shown. The operating line $\overline{x_4 y_5}$ is also omitted in the figure for the same reason. From figure, the compositions of streams leaving each stage are as follow:

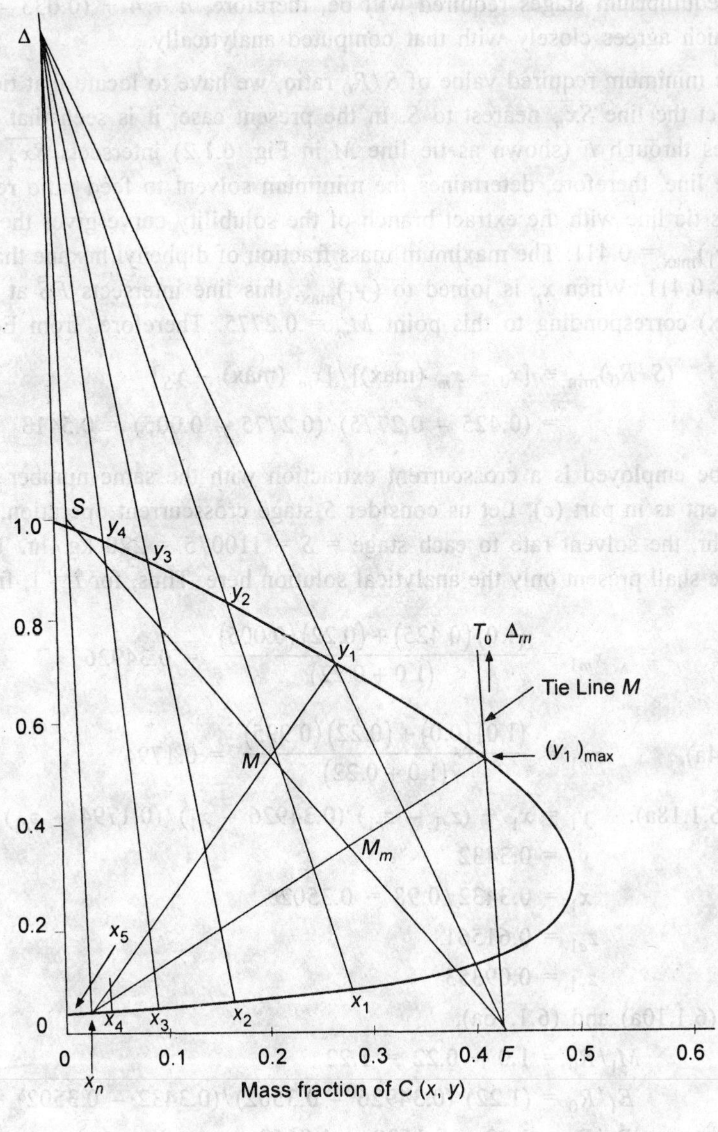


*Figure 6.1.2*

| $i$ | $y_i$ | $x_i$ |
|---|---|---|
| 1 | 0.270 | 0.2755 |
| 2 | 0.160 | 0.16326 |
| 3 | 0.084 | 0.0857 |
| 4 | 0.033 | 0.03367 |
| 5 | 0.0065 | 0.00663 |

The number of equilibrium stages required will be, therefore, $n = 4 + (0.033 - 0.025)/(0.033 - 0.0065) = 4.302$, which agrees closely with that computed analytically.

To determine the minimum required value of $S/R_0$ ratio, we have to locate that tie line which when extended will intersect the line $\overline{Sx_n}$ nearest to $S$. In the present case, it is seen that the tie line which when extended passes through $F$ (shown as tie line $M$ in Fig. 6.1.2) intersects $\overline{Sx_n}$ at a point that is nearest to $S$. This tie line, therefore, determines the minimum solvent to feed ratio required. The point of intersection of this tie line with the extract branch of the solubility curve gives the value of $(y_1)_{max}$. Thus, from figure, $(y_1)_{max} = 0.411$. The maximum mass fraction of diphenyl hexane that can be achieved in the extract is thus 0.411. When $x_n$ is joined to $(y_1)_{max}$, this line intersects $\overline{FS}$ at $M_m$. From figure, the value of $x_m$ (max) corresponding to this point $M_m = 0.2775$. Therefore, from Eq. (6.1.45),

$$(S/R_0)_{min} = [x_0 - x_m\,(max)]/[x_m\,(max) - y_S]$$

$$= (0.425 - 0.2775)/(0.2775 - 0.005) = 0.5413$$

(*d*) What is to be employed is a crosscurrent extraction with the same number of stages and the same amount of solvent as in part (*c*). Let us consider 5 stage crosscurrent operation. The total solvent rate being 1100 kg/hr, the solvent rate to each stage $= S = 1100/5 = 220$ kg/hr. Therefore, $S/R_0 = 220/1000 = 0.22$. We shall present only the analytical solution here. Thus, for $i = 1$, from Eq. (6.1.12a),

$$x_{m1} = \frac{(1.0)(0.425) + (0.22)(0.005)}{(1.0 + 0.22)} = 0.34926$$

From Eq. (6.1.14a),

$$z_{m1} = \frac{(1.0)(0.0) + (0.22)(0.995)}{(1.0 + 0.22)} = 0.1794$$

And from Eq. (6.1.18a), $y_1 = x_1 + (z_{e1} - z_{r1})(0.34926 - x_1)/(0.1794 - z_{r1})$

Solving by trial,

$$y_1 = 0.3432$$
$$x_1 = 0.3432/0.98 = 0.3502$$
$$z_{e1} = 0.61561$$
$$z_{r1} = 0.09353$$

Now, from Eqs (6.1.10a) and (6.1.16a),

$$M_1/R_0 = 1.0 + 0.22 = 1.22$$
$$E_1/R_0 = (1.22)(0.34926 - 0.3502)/(0.3432 - 0.3502) = 0.1638$$

And $R_1/R_0 = 1.22 - 0.1638 = 1.0562$

Computations are performed in the same way for all the five stages. The results are tabulated below:

| $i$ | $y_i$ | $x_i$ | $E_i/R_0$ | $R_i/R_0$ |
|---|---|---|---|---|
| 1 | 0.3432 | 0.3502 | 0.1638 | 1.0562 |
| 2 | 0.2862 | 0.292 | 0.2882 | 0.9879 |
| 3 | 0.236 | 0.2408 | 0.26927 | 0.93868 |
| 4 | 0.1931 | 0.197 | 0.297 | 0.86166 |
| 5 | 0.1555 | 0.15867 | 0.2354 | 0.8462 |

The composition of the final composited extract is

$$y_f = \Sigma E_i y_i / \Sigma E_i = (0.2962)/(1.25367) = 0.2362$$

Thus, a 5 stage crosscurrent extraction with less amount of solvent provides a richer extract than that obtained in a 4 stage crosscurrent extraction discussed in part (*a*), that uses four times larger solvent rate. This is understandable since use of an excessive amount of solvent in each stage will dilute the extract and increase the load on the solvent recovery unit. The solute recovery in the present case is

$$\text{Solute recovery} = \frac{(\text{solute in feed}) - (\text{solute in final raffinate})}{(\text{solute in feed})}$$

$$= \frac{R_0 x_0 - R_n x_n}{R_0 x_0} = 1 - (R_n/R_0)\,(x_n/x_0)$$

$$= 1.0 - (0.8462)\,(0.15867/0.425) = 0.684 \text{ or } 68.4 \text{ percent}$$

To note that in the present case, $x_n = x_5 = 0.15867$ and $R_n/R_0 = R_5/R_0 = 0.8462$. For the countercurrent operation considered in part (c), the solute recovery is

$$= 1.0 - (0.55845)\,(0.025/0.425) = 0.96715 \text{ or } 96.715 \text{ percent}$$

The solute recovery in the 4 stage crosscurrent extraction considered in part (*a*) is

$$= 1.0 - (0.7377)\,(0.0207/0.425) = 0.964 \text{ or } 96.4 \text{ percent}$$

From the results obtained in the above cases, the following observations can be made:

(*i*) The countercurrent operation provides a richer extract ($y_1 = 0.271$) in a limited number of equilibrium stages (4.3 stages) and a higher solute recovery (96.715 percent) using a solvent to feed ratio of 1.1.

(*ii*) The crosscurrent operation using the same amount of solvent ($S$ = 1100 kg / hr; 220 kg / hr per stage) and essentially the same number of equilibrium stages (5 stages) provides much lower solute recovery (68.4 percent) and a relatively leaner extract ($y_f = 0.2362$).

(*iii*) A crosscurrent operation that uses 4 times higher solvent rate ($S$ = 4400 kg / hr; 1100 kg / hr per stage) and 4.0 equilibrium stages provides a much leaner extract ($y_f = 0.169$), but a high solute recovery (96.4 percent).

**Example 6.2:** 100 kg / hr of a feed stream containing 20 mass percent acetic acid and balance water is to be extracted at 25° C with 200 kg / hr of recycle methyl isobutyl ketone (MIBK) that contains 0.1 mass percent acetic acid. The aqueous raffinate is to be extracted down to 1.0 percent acetic acid.

(*a*) Compute the number of equilibrium stages required if a countercurrent multistage operation is employed. What will be the composition of the extract?

(*b*) What is the minimum solvent requirement for the countercurrent operation?

(*c*) If extraction is performed in five crosscurrent stages each stage fed with 50 kg / hr of the solvent, what would be the composition of the final raffinate? Estimate also the composition of the final extract and the fractional solute recovery.

(*d*) If a single stage batch operation is employed to affect the same separation as in (*a*), what will be the solvent requirement?

Within the range of concentration considered, water and MIBK may be considered to be mutually insoluble. The solubility data for the system at 25° C is given below[11]:

| *Water layer* | | *MIBK layer* | |
|---|---|---|---|
| *Mass fraction of acetic acid* | *Mass fraction of MIBK* | *Mass fraction of Acetic acid* | *Mass fraction of MIBK* |
| 0.0 | 0.0155 | 0.0 | 0.9788 |
| 0.0285 | 0.017 | 0.0187 | 0.9533 |
| 0.117 | 0.025 | 0.089 | 0.857 |
| 0.205 | 0.038 | 0.173 | 0.735 |
| 0.262 | 0.060 | 0.246 | 0.609 |
| 0.328 | 0.122 | 0.308 | 0.472 |
| 0.346 | 0.225 | 0.336 | 0.354 |

**Solution:** (*a*) Since water (*A*) and MIBK (*B*) are mutually insoluble, we can use the simplified procedure illustrated in Fig. 6.14. For this, let us first reprepare the equilibrium data in terms of solute-free mass fractions (that is, in terms of $x'$ and $y'$) as given below:

$x'$ : 0.0 0.0293 0.1325 0.2578 0.355 0.488 0.529

$y'$ : 0.0 0.019 0.0977 0.2092 0.3262 0.445 0.506

To note that the water layer is the raffinate layer and the MIBK layer is the extract layer and $x'$ and $y'$ are respectively defined as $x' = x/(1-x)$, $y' = y/(1-y)$. The equilibrium curve ($y'$ versus $x'$) is now plotted as shown in Fig. 6.2.1. From Eq. (6.1.49),

$$G_A/G_B = \frac{(1-0.20)}{(200/100)(1-0.001)}$$

$$= 0.4004$$

Now, $x'_n = (0.01)/(1.0 - 0.01) = 0.0101$ and $y'_S = 0.001/(1.0 - 0.001) \simeq 0.001$. The operating line is thus drawn from the point (0.0101, 0.001) with a slope of 0.4004 as shown (to note that $y'_S$ is very close to zero). Now, starting from $x = x'_0 = 0.2/(1 - 0.2) = 0.25$ on the operating line, the staircase construction is performed until $x' < 0.0101$ is reached. From figure, it can be seen that the number of equilibrium stages required is 4.5. The value of $y'_1$ (as read from figure) is 0.097. Therefore,

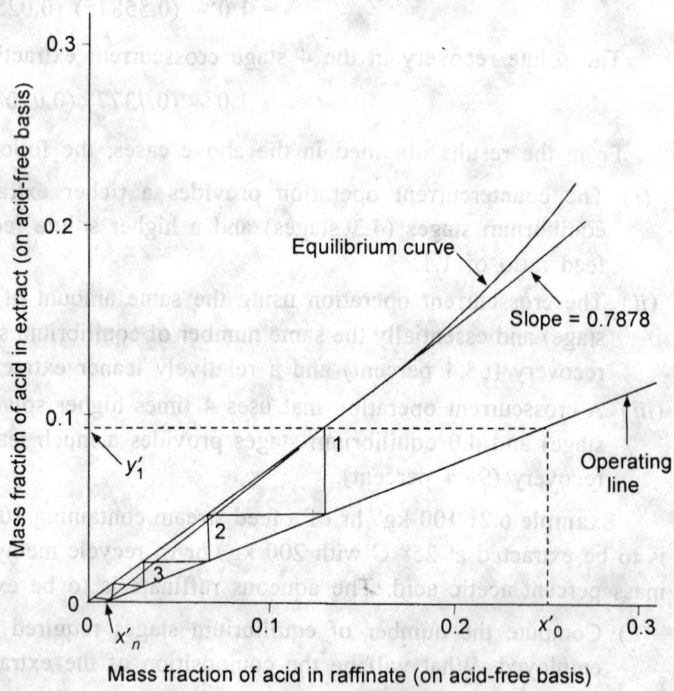


*Figure 6.2.1*

$$y_1 = (0.097)/(1.0 + 0.097) = 0.08842$$

The extract shall thus contain 8.842 percent acetic acid. The solution to this problem by the general graphical method [shown in Fig. 6.13 (a)] has been reported by Perry[3]. The number of equilibrium stages reported is 4.3, which checks closely with that determined by the above procedure.

(*b*) The minimum solvent to feed ratio corresponds to the slope of the tangent to the equilibrium curve from the point $(x'_n, y'_S) = (0.0101, 0.001)$. From Fig. 6.2.1, slope of such a tangent is 0.7878. Therefore

$$G_A/G_B \text{ (min)} = 0.7878$$

Now, $G_A = R_0 (1 - x_0) = 100.0 (1.0 - 0.20) = 80 \text{ kg/hr}$.

Therefore, $G_B \text{ (min)} = (80.0)/(0.7878) = 101.545 \text{ kg/hr}$

$$S \text{ (min)} = G_B \text{ (min)}/(1 - y_S) = 101.545/(1.0 - 0.001) = 101.646 \text{ kg/hr}$$

This is the minimum solvent rate required.

(*c*) In the case of crosscurrent extraction also, since *A* is fully insoluble in *B*, we may employ the simplified procedure illustrated in Fig. 6.11. Since the solvent rate to each stage is the same, the value of $G_B$ will be the same for all stages. Thus

$$G_A/G_B = \frac{R_0 (1 - x_0)}{S (1 - y_S)}$$

$$= \frac{100 (1.0 - 0.20)}{50 (1.0 - 0.001)}$$

$$= 1.6016.$$

The equilibrium plot is drawn as in part (*a*) and is shown in Fig. 6.2.2. Now, starting from $(x'_0, y'_S)$, that is (0.25, 0.001), the graphical construction is performed as shown. To note that the slope of each operating line $= -G_A/G_B = -1.6016$. Incidentally, since the value of $y'_S$ is very small, the horizontal line at $y' = y'_S$ more or less coincides with the abcissa. The graphical construction has been performed accordingly. From the figure, the composition of final raffinate $= x'_5 = 0.0406$. Therefore,

$$x_5 = 0.0406/(1.0 + 0.0406)$$

$$= 0.039$$

Thus, there is larger amount of unrecovered acetic acid in the final raffinate. The compositions and mass flow rates of extract and raffinate streams leaving each stage are listed below:

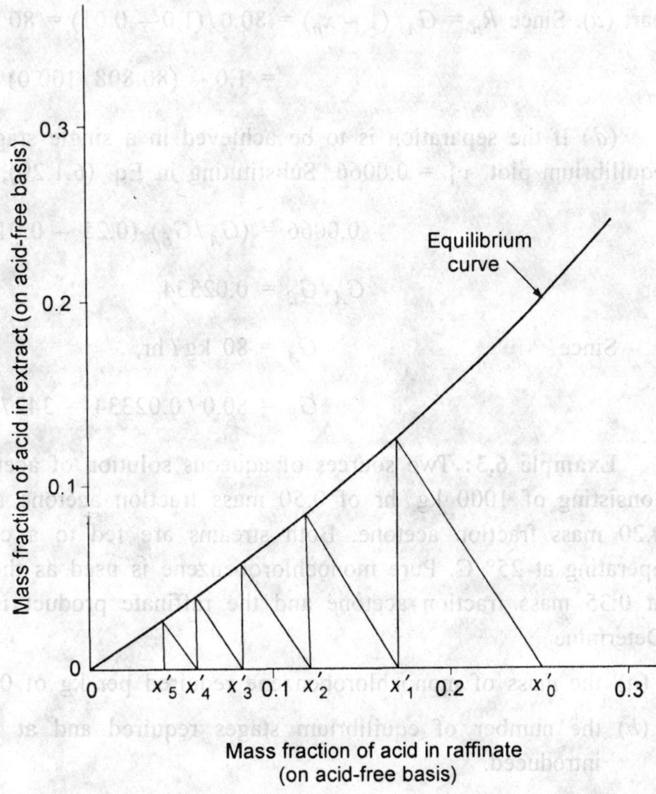


*Figure 6.2.2*

| $i$ | $x'$ | $y'$ | $x$ | $y$ | $E = G_B/(1-y)$ | $R = G_A/(1-x)$ |
|---|---|---|---|---|---|---|
| 1 | 0.17 | 0.1275 | 0.1453 | 0.11308 | 56.32 | 93.60 |
| 2 | 0.1175 | 0.085 | 0.1051 | 0.0783 | 54.20 | 89.4 |
| 3 | 0.081 | 0.0585 | 0.07493 | 0.05527 | 52.872 | 86.48 |
| 4 | 0.057 | 0.0398 | 0.053926 | 0.03827 | 51.938 | 84.56 |
| 5 | 0.0406 | 0.0273 | 0.039 | 0.02657 | 51.3136 | 83.248 |

Mass fraction of acetic acid in the final composited extract is

$$y_f = \Sigma E_i y_i / \Sigma E_i = 16.8858 / 266.6436 = 0.0633$$

This is lower than that attained in countercurrent operation. The fractional solute recovery is

$$= 1 - (R_n/R_0)\ (x_n/x_0)$$

$$= 1.0 - (83.248/100)\ (0.039/0.20) = 0.8376 \text{ or } 83.76 \text{ percent}$$

For sake of comparison, let us also estimate the solute recovery in the countercurrent operation of part (*a*). Since $R_n = G_A/(1 - x_n) = 80.0/(1.0 - 0.01) = 80.808$ kg / hr, the fractional solute recovery is

$$= 1.0 - (80.808/100.0)\ (0.01/0.20) = 0.9596 \text{ or } 95.96 \text{ percent}$$

(*d*) If the separation is to be achieved in a single stage, then $x_1' = x_n' = 0.0101$. Now, from the equilibrium plot, $y_1' = 0.0066$. Substituting in Eq. (6.1.29),

$$0.0066 = (G_A/G_B)\ (0.25 - 0.0101) + 0.001$$

or

$$G_A/G_B = 0.02334$$

Since $G_A = 80$ kg / hr,

$$G_B = 80.0/0.02334 = 3427.59 \text{ kg / hr.}$$

**Example 6.3:** Two sources of aqueous solution of acetone are available in a chemical plant, one consisting of 1000 kg / hr of 0.50 mass fraction acetone and the other consisting of 100 kg / hr of 0.20 mass fraction acetone. Both streams are fed to a countercurrent multistage extraction system operating at 25° C. Pure monochlorobenzene is used as the solvent. The strong extract is discharged at 0.35 mass fraction acetone and the raffinate product is to contain 0.025 mass fraction acetone. Determine

(*a*) the mass of monochlorobenzene required per kg of 0.50 mass fraction acetone solution,

(*b*) the number of equilibrium stages required and at which stage each feed stream should be introduced.

The solubility data for acetone–water–monochlorobenzene system at 25° C and 1 atm pressure is given below[12]:

| *Water layer* | | *Monochloro benzene layer* | |
|---|---|---|---|
| *Mass fraction of acetone* | *Mass fraction of monochloro benzene* | *Mass fraction of acetone* | *Mass fraction of monochloro benzene* |
| 0.0 | 0.0011 | 0.0 | 0.9982 |
| 0.05 | 0.0018 | 0.0521 | 0.9447 |
| 0.10 | 0.0021 | 0.1079 | 0.8872 |
| 0.15 | 0.0024 | 0.1620 | 0.8317 |
| 0.20 | 0.0031 | 0.2223 | 0.7698 |
| 0.25 | 0.0042 | 0.2901 | 0.6982 |
| 0.30 | 0.0058 | 0.3748 | 0.6080 |
| 0.35 | 0.0078 | 0.4328 | 0.5439 |
| 0.40 | 0.0136 | 0.4944 | 0.4751 |
| 0.45 | 0.0224 | 0.5492 | 0.4080 |
| 0.50 | 0.0372 | 0.5919 | 0.3357 |
| 0.55 | 0.0631 | 0.6179 | 0.2438 |
| 0.60 | 0.1259 | 0.6107 | 0.1508 |
| 0.6058 | 0.1376 | 0.6058 | 0.1376 |

**Solution:** The stream containing 0.50 mass fraction acetone, since has larger acetone content, be introduced to stage 1. The stage at which the other stream must be introduced (so that the total number of stages will be the minimum) is to be determined.

(*a*) Let us compute the solvent requirement first by the analytical method. This can be accomplished by solving Eqs (6.1.33a) and (6.1.35a) simultaneously. To note that $F_1/R_0 = (100/1000) = 0.1$ and from solubility data,

$$z_{rn} \text{ (corresponding to } x_n = 0.025) = 0.00145$$

$$z_{e1} \text{ (corresponding to } y_1 = 0.35) = 0.6344$$

Now, substituting the known values in Eq. (6.1.33a), we get

$$E_1/R_0 = [0.50 + (0.1)(0.2) - (1.1 + S/R_0)(0.025)]/(0.35 - 0.025)$$

$$= (0.4925 - 0.025\ S/R_0)/0.325 \quad \text{... (i)}$$

Similarly, Eq. (6.1.35a) gets reduced to

$$0.00145 = \frac{(S/R_0) - (0.6344)(E_1/R_0)}{1.1 + (S/R_0) - (E_1/R_0)} \quad \text{... (ii)}$$

Solving simultaneously, $S/R_0 = 0.9174$

$$E_1/R_0 = 1.4448$$

The amount of monochlorobenzene required is thus 0.9174 kg per kg of 0.50 mass fraction acetone solution.

The $S/R_0$ ratio may also be determined graphically as shown in Fig. 6.3.1. The points $F$, $F_1$ and $S$ are first marked since their coordinates are (0.50, 0), (0.20, 0) and (0, 1.0) respectively. Now, from Eq. (6.1.54),

$$x_J = [0.5 + (0.1)\,(0.20)]/1.1 = 0.4727$$

The point $J$ is therefore marked on the line $\overline{FF_1}$ corresponding to $x = x_J = 0.4727$. $x_n$ (= 0.025) and $y_1$ (= 0.35) are also marked on the raffinate branch and extract branch respectively of the solubility curve. The line $\overline{x_n y_1}$ intersects $\overline{JS}$ at $M$ and $x_m$ (as read from figure) = 0.2577. Therefore, from Eq. (6.1.56),

$$S/J = (0.4727/0.2577) - 1.0 = 0.8343$$

And from Eq. (6.1.57), $S/R_0 = (1.0 + 0.1)\,(0.8343) = 0.9177$

(*b*) Let us also determine the number of equilibrium stages required first by the analytical method. The equilibrium plot is shown in Fig. 6.3.2. From the plot, $x_1$ (corresponding to $y_1 = 0.35$) = 0.29 and from the solubility data, $z_{r1} = 0.00548$. Now, from Eq. (6.1.42b), putting $i = 1$,

$$G_2 = \frac{(1.4448)\,(0.35 - 0.29) + (1.0)\,(0.29 - 0.50)}{(1.4448)\,(0.6344 - 0.00548) + (1.0)\,(0.00548 - 0.0)} = -\,0.13489$$

Equation (6.1.42a) becomes

$$y_2 = 0.29 - 0.13489\,(z_{e2} - 0.00548)$$

Solving by trial with the help of solubility data,

$$y_2 = 0.182$$
$$z_{e2} = 0.8111$$

From equilibrium plot, $x_2 = 0.167$ and from solubility data, $z_{r2} = 0.002638$. Now, from Eq. (6.1.39a),

$$E_2/R_0 = \frac{(1.4448)\,(0.6344 - 0.00548) + (1.0)\,(0.00548)}{(0.8111 - 0.00548)} = 1.1346$$

And from Eq. (6.1.36a),

$$R_1/R_0 = 1.0 + 1.346 - 1.4448 = 0.6898$$

Since $x_2 < x_{F1}$, the *optimum* stage at which the second feed stream (that contains 0.20 mass fraction acetone) is to be introduced is stage 2. Thus, $f = 2$. Therefore, for $i = 2$, $G_{i+1}$ must be computed from Eq. (6.1.42c):

$$G_3 = \frac{(1.1346)\,(0.182 - 0.167) + (0.6898)\,(0.167 - 0.29) + (0.1)\,(0.167 - 0.20)}{(1.1346)\,(0.8111 - 0.002638) + (0.6898)\,(0.002638 - 0.00548) + (0.1)\,(0.002638)}$$

$$= -\,0.077677$$

And from Eq. (6.1.42a), $\quad y_3 = 0.167 - 0.077677\,(z_{e3} - 0.002638)$

Solving by trial, $\quad y_3 = 0.0975$

$$z_{e3} = 0.8979$$

From equilibrium plot, $x_3 = 0.09068$ and from solubility data, $z_{r3} = 0.002044$. Since $i = f$, the values of $E_3/R_0$ and $R_2/R_0$ are also computed from Eqs (6.1.39b) and (6.1.36b) respectively instead of Eqs (6.1.39a) and (6.1.36a). Thus

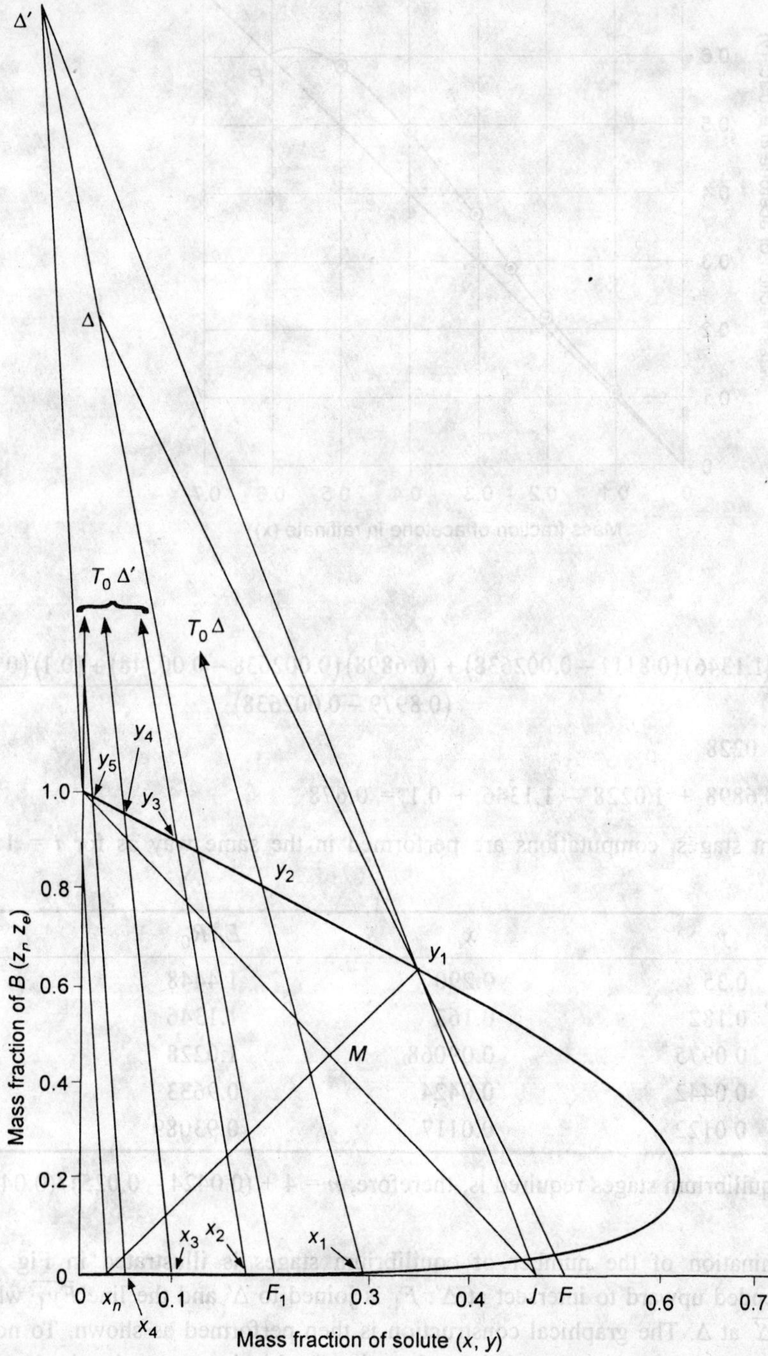
Δ′
Δ
$T_0$ Δ′
$T_0$ Δ
$y_5$
$y_4$
$y_3$
$y_2$
$y_1$
M
$x_3$ $x_2$
$x_1$
$x_n$
$x_4$
$F_1$
J
F
1.0
0.8
0.6
0.4
0.2
0
0.1
0.3
0.4
0.6
0.7
Mass fraction of B ($z_r$, $z_e$)
Mass fraction of solute (x, y)

*Figure 6.3.1*

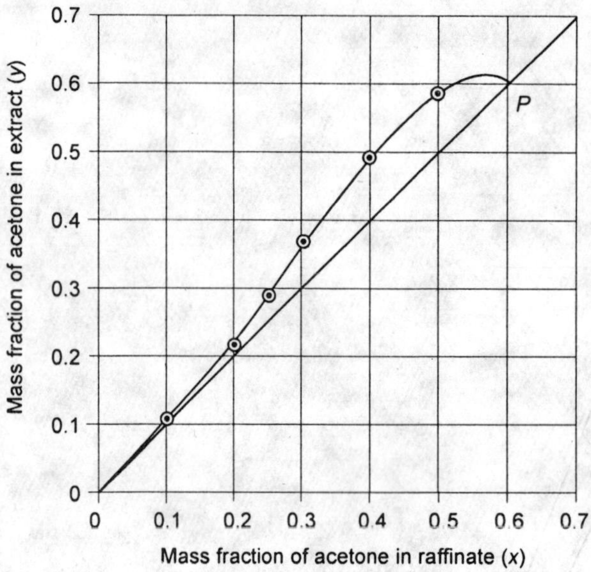


*Figure 6.3.2*

$$E_3/R_0 = \frac{(1.1346)(0.8111-0.002638)+(0.6898)(0.002638-0.00548)+(0.1)(0.002638)}{(0.8979-0.002638)}$$

$$= 1.0228$$

$$R_2/R_0 = 0.6898 + 1.0228 - 1.1346 + 0.1 = 0.678$$

For all subsequent stages, computations are performed in the same way as for $i = 1$. The results are summarised below:

| $i$ | $y_i$ | $x_i$ | $E_i/R_0$ | $R_i/R_0$ |
|---|---|---|---|---|
| 1 | 0.35 | 0.290 | 1.4448 | 0.6898 |
| 2 | 0.182 | 0.167 | 1.1346 | 0.6780 |
| 3 | 0.0975 | 0.09068 | 1.0228 | 0.6185 |
| 4 | 0.0442 | 0.0424 | 0.9633 | 0.58609 |
| 5 | 0.0122 | 0.0117 | 0.93089 | — |

The number of equilibrium stages required is, therefore, $n = 4 + (0.0424 - 0.025)/(0.0424 - 0.0117) = 4.57$.

Graphical determination of the number of equilibrium stages is illustrated in Fig. 6.3.1. Lines $\overline{Jy_1}$ and $\overline{Sx_n}$ are extended upward to intersect at $\Delta'$. $F_1$ is joined to $\Delta'$ and the line $\overline{Fy_1}$ when extended upward intersects $\overline{F_1\Delta'}$ at $\Delta$. The graphical construction is then performed as shown. To note that since $x_2 < x_{F1}$, $x_2$ is joined to $\Delta'$ (and not $\Delta$) and thereafter, the graphical construction is done using $\Delta'$ as the operating point. The results are listed below:

| $i$ | $y_i$ | $x_i$ |
|---|---|---|
| 1 | 0.350 | 0.290 |
| 2 | 0.183 | 0.1675 |
| 3 | 0.10 | 0.0925 |
| 4 | 0.045 | 0.043 |
| 5 | 0.0125 | 0.012 |

The number of equilibrium stages required is, therefore, $n = 4 + (0.043 - 0.025)/(0.043 - 0.012) = 4.58$. The second feed stream (containing 0.20 mass fraction acetone) is to be introduced at the second equilibrium stage.

## Multistage Countercurrent Extraction (with Reflux)

As in distillation, the use of reflux could often be advantageous with countercurrent multistage operation in the case of extraction as well. In the conventional countercurrent operation without reflux, the maximum concentration of solute that can be achieved in the final extract, even if an infinite number of stages is used, is that corresponding to equilibrium with the incoming feed. If the concentration of the solute in the feed is low, it shall be desirable to obtain a higher concentration of solute in the final extract than that corresponds to equilibrium with the feed and this can be accomplished by using reflux at the extract end of the plant (as shown in Fig. 6.17). The feed ($F$) is introduced to one of the intermediate stages (stage $f$). The extract leaving the first stage ($E_1$) is sent to a solvent recovery unit where most of the solvent is removed from it. The resultant liquid is divided into two streams, the reflux stream ($R_0$) which is returned to stage 1 and the stream ($P$) which is collected as the final extract product. The final raffinate is discharged from the last stage ($n$th stage). The recovered solvent ($S_0$) which shall necessarily have the same composition as $S$, is recycled back to the $n$th stage. The use of extract reflux produces an extract product that is richer in solute, but an operation like this demands larger quantity of solvent per unit quantity of feed. It may be noted that in an arrangement like this, the streams $R_0$ and $F$ are not identical since the former denotes the reflux stream and the latter the feed stream.

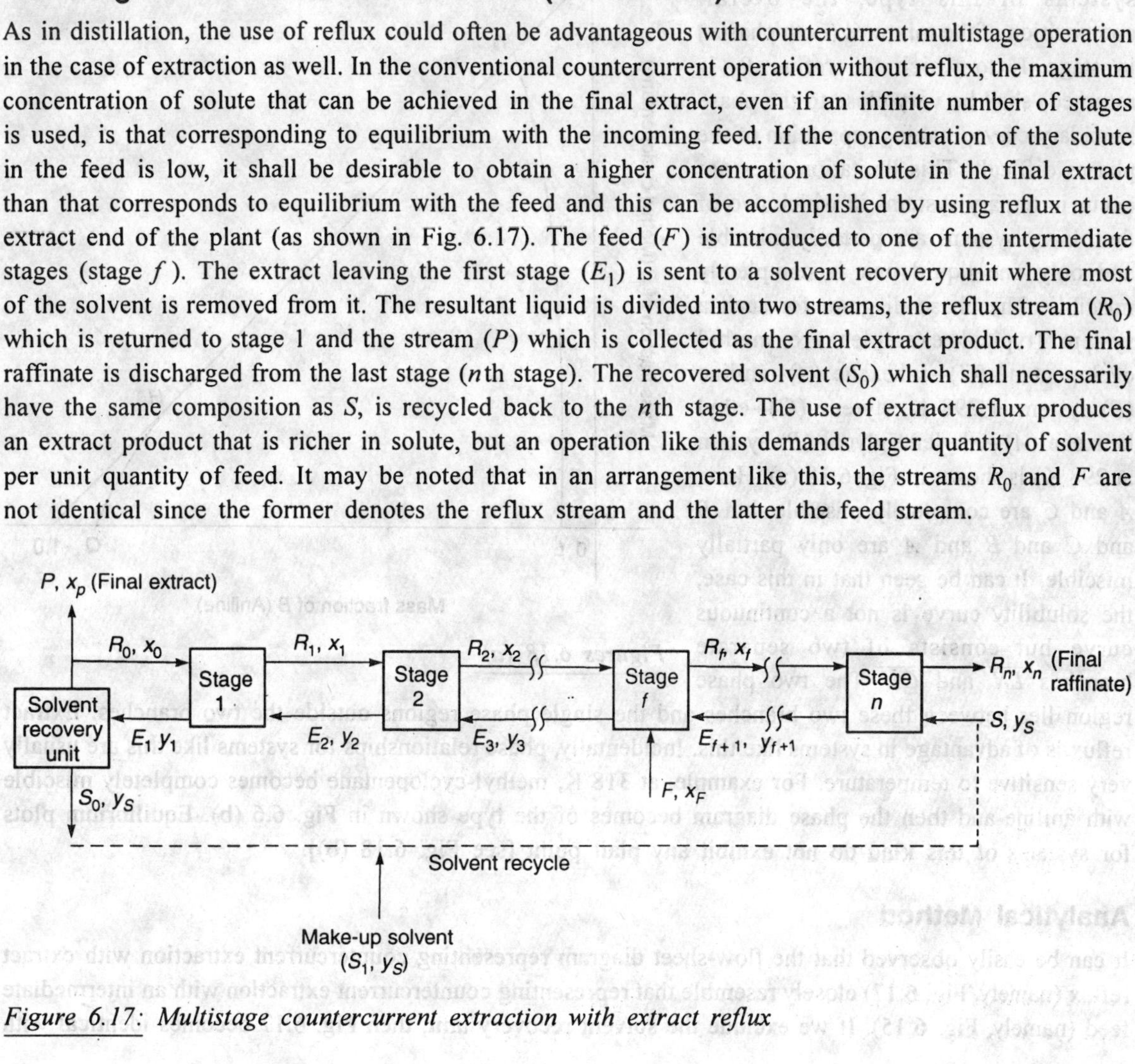


*Figure 6.17: Multistage countercurrent extraction with extract reflux*

All the stages starting from stage 1 to stage *f* (but not including the feed stage) constitute the extract enriching section and all stages beyond stage *f* (up to the *n*th stage) constitute the raffinate striping section. The use of reflux in extraction is thus quite analogous to that for continuous fractionation; there are however a few distinct differences:

(*a*) In extraction, reflux is seldom used at the raffinate end. This is because unlike distillation where heat must be carried in from the reboiler by a vapour reflux, in extraction the solvent (the analog of heat) can enter without a carrier stream.

(*b*) The use of reflux is of little use for those ternary systems in which two of the binary systems are completely miscible liquids and only one pair is partially miscible [see Figs 6.6 (a) and (b)] though these are the most common types of systems in extraction. This is because in systems of this type, the overall composition of the mixtures (both phases) in stage 1 (to which the reflux $R_0$ is returned) shall be very close to the phase boundary curve, making separation of the phases difficult. This limitation does not occur in ternary systems in which two of the binary systems are partially miscible and only one liquid pair is completely miscible. The phase diagram for such a system (typical examples are methyl cyclopentane (*C*)–*n*-hexane (*A*)–aniline (*B*) system at 298 K, styrene (*C*)–ethyl benzene (*A*)–ethylene glycol (*B*) system at 298 K) is shown in Fig. 6.18 (a). Here, *A* and *C* are completely miscible, but *B* and *C* and *B* and *A* are only partially miscible. It can be seen that in this case, the solubility curve is not a continuous curve but consists of two separate branches $\overline{LN}$ and $\overline{QH}$ The two phase region lies between these two branches and the single phase regions outside the two branches. Extract reflux is of advantage in systems like this. Incidentally, phase relationships for systems like this are usually very sensitive to temperature. For example, at 318 K, methyl-cyclopentane becomes completely miscible with aniline and then the phase diagram becomes of the type shown in Fig. 6.6 (b). Equilibrium plots for systems of this kind do not exhibit any plait point [see Fig. 6.18 (b)].

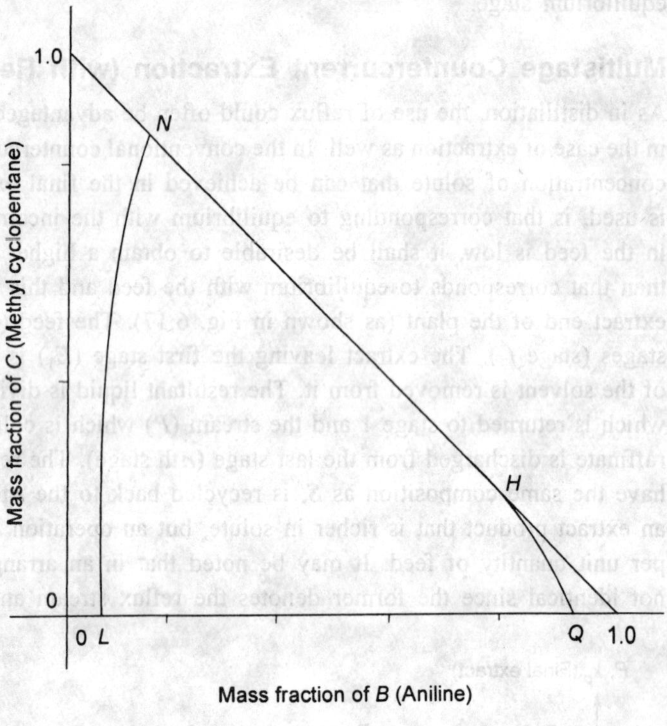


*Figures 6.18 (a)*

## Analytical Method

It can be easily observed that the flow-sheet diagram representing countercurrent extraction with extract reflux (namely, Fig. 6.17) closely resemble that representing countercurrent extraction with an intermediate feed (namely, Fig. 6.15). If we exclude the solvent recovery unit, then Fig. 6.17 becomes identical with

Fig. 6.15 with $F$ replacing $F_1$. As a result, once the end streams ($S$, $R_n$, $R_0$, $E_1$) are quantified, the present system can be analysed using the same material balance equations as those used for countercurrent extraction with intermediate feed (discussed in earlier section). The end streams can be evaluated by performing a material balance for the entire plant and also around the solvent recovery unit. The following quantities are assumed known at the outset:

1. the flow rate ($F$) and composition ($x_F$) of the feed,
2. the composition ($x_P$, $z_P$) of the extract product (to note that $x_0 = x_P$ and $z_{r0} = z_P$),
3. the composition ($x_n$, $z_{rn}$) of the final raffinate,
4. the composition of the solvent ($y_S$, $z_S$),
5. the *reflux ratio*, $RR$ (where $RR = R_0/P$)*.

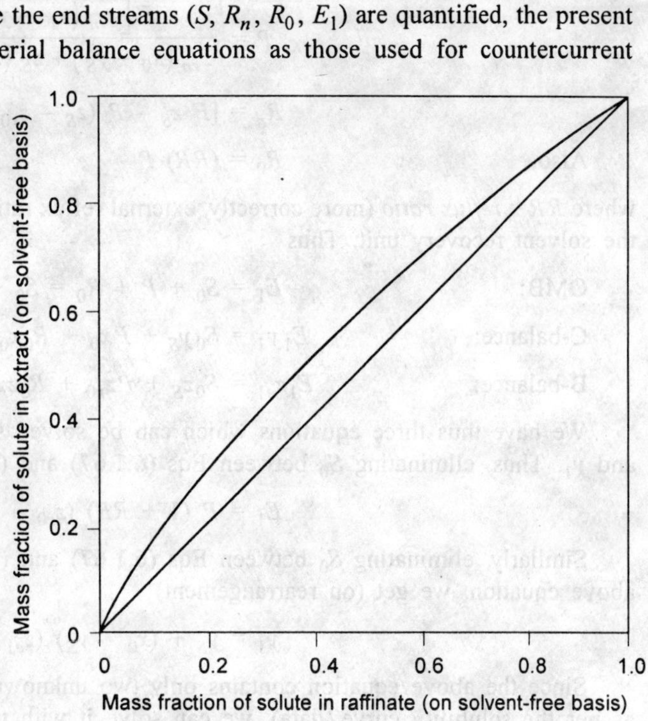


*Figures 6.18 (b)*

Now, the overall material balance for the entire system (not including solvent recycle) is

$$F + S = R_n + P + S_0 \quad \text{... (6.1.59)}$$

The solute balance ($C$-balance) is

$$Fx_F + Sy_S = R_n x_n + Px_0 + S_0 y_S \quad \text{... (6.1.60)}$$

We have substituted $x_0$ for $x_P$ since both $P$ and $R_0$ are of the same composition. It is also usual to assume that $P$ and $R_0$ are saturated phases and therefore, once $x_P$ (= $x_0$) is specified, the value of $z_P$ (= $z_{r0}$) can be obtained from the solubility curve / data. Now, a solvent balance ($B$-balance) for the entire system yields

$$Sz_S = R_n z_{rn} + Pz_{r0} + S_0 z_S \quad \text{... (6.1.61)}$$

We have taken $z_F = 0$, since the feed shall not contain any $B$. Now, multiplying Eq. (6.1.59) by $y_S$ and substracting it from Eq. (6.1.60), we get

$$F(x_F - y_S) = R_n(x_n - y_S) + P(x_0 - y_S) \quad \text{... (6.1.62)}$$

Similarly, multiplying Eq. (6.1.59) by $z_S$ and then substracting it from Eq. (6.1.61), we get

$$Fz_S = R_n(z_S - z_{rn}) + P(z_S - z_{r0}) \quad \text{... (6.1.63)}$$

The above two equations can be now solved simultaneously for $R_n$ and $P$. Thus

* We have used the notation $RR$ to denote the reflux ratio instead of $R$, since the notation $R$ has already been used to designate the raffinate flow rate.

$$P = \frac{F\left[z_{rn}\left(x_F - y_S\right) + z_S\left(x_n - x_F\right)\right]}{z_{rn}\left(x_0 - y_S\right) + z_S\left(x_n - x_0\right) + z_{r0}\left(y_S - x_n\right)} \quad \text{... (6.1.64)}$$

$$R_n = [F\,z_S - P\,(z_S - z_{r0})]\,/\,(z_S - z_{rn}) \quad \text{... (6.1.65)}$$

Also, $$R_0 = (RR)\,P \quad \text{... (6.1.66)}$$

where $RR$ = *reflux ratio* (more correctly, external reflux ratio). Let us now make a material balance around the solvent recovery unit. Thus

OMB: $$E_1 = S_0 + P + R_0 = S_0 + P\,(1 + RR) \quad \text{... (6.1.67)}$$

C-balance: $$E_1 y_1 = S_0 y_S + P x_0 + R_0 x_0 = S_0 y_S + P x_0\,(1 + RR) \quad \text{... (6.1.68)}$$

B-balance: $$E_1 z_{e1} = S_0 z_S + P z_{r0} + R_0 z_{r0} = S_0 z_S + P z_{r0}\,(1 + RR) \quad \text{... (6.1.69)}$$

We have thus three equations which can be solved simultaneously for the three unknowns $E_1$, $S_0$ and $y_1$. Thus, eliminating $S_0$ between Eqs (6.1.67) and (6.1.69), we get

$$E_1 = P\,(1 + RR)\,(z_{r0} - z_S)\,/\,(z_{e1} - z_S) \quad \text{... (6.1.70)}$$

Similarly, eliminating $S_0$ between Eqs (6.1.67) and (6.1.68) and then substituting for $E_1$ from the above equation, we get (on rearrangement)

$$y_1 = y_S + (x_0 - y_S)\,(z_{e1} - z_S)\,/\,(z_{r0} - z_S) \quad \text{... (6.1.71)}$$

Since the above equation contains only two unknowns such as $y_1$ and $z_{e1}$ (which are inter-related as per the solubility curve / data), we can solve it with the help of the solubility data:

(*i*) Assume $y_1$. For example, let $y_1 = x_0 - 0.001$.

(*ii*) Put $YA = y_1$.

(*iii*) Obtain $z_{e1}$ from the solubility curve/data.

(*iv*) Compute $y_1$ from Eq. (6.1.71).

(*v*) If the above-computed value of $y_1$ differs significantly from $YA$ (say, by more than 0.001), then put $y_1 = YA - 0.001$ and repeat the computations starting from step (*ii*).

Once the values of $y_1$ and $z_{e1}$ are known, $E_1$ can be obtained from Eq. (6.1.70) and $S_0$ from (6.1.67). Equation (6.1.59) then gives the value of $S$.

After evaluating $E_1$, $R_0$ and $y_1$, the stage to stage computations are performed in the same way as for countercurrent extraction with intermediate feed (discussed in the earlier section). Thus, for all stages except stage $f$, Eqs (6.1.42a), (6.1.42b), (6.1.39a) and (6.1.36a) are to be used; for stage $f$, Eqs (6.1.42c), (6.1.39b) and (6.1.36b) apply. In the present case however, since the flow rates ($E_1$, $S$, $R_n$, $R_0$) rather than the ratios of the flow rates ($S/R_0$, $E_1/R_0$, $R_n/R_0$) are being known, the expression for $G_{i+1}$ [Eq. (6.1.42b) or (6.1.42c)] may be written in terms of the flow rates ($E_i$, $R_i$, etc.) instead of the mass ratios ($E_i/R_0$, $R_i/R_0$, etc.). Also, $F_1$, $x_{F1}$ are to be replaced by $F$, $x_F$. Computations are to be continued until $x_{i+1} \le x_n$.

## Graphical Method

In the graphical method of analysis described below, it has been assumed that $R_0$ and $P$ are saturated phases and therefore, $x_0$ (or $X_0$) can be marked on the raffinate branch of the solubility curve (see Fig. 6.19)*. It is usually more convenient to use solvent-free coordinates, as shown in Fig. 6.19, since this avoids crowding of points. The points $F$ and $S$ are marked as usual since their coordinates are $(X_F, 0)$ and $(Y_S, N_S)$ respectively. Now, join $X_0$ to $S$ and this line intersects the extract branch of the solubility curve at $Y_1$. The operating point for the extract enriching section ($\Delta$) lies on this line $\overline{X_0 S}$ (in case the solvent is pure $B$, then $\overline{X_0 S}$ will be a vertical line passing through $X_0$ and this will again intersect the extract curve at $Y_1$. In such a case, $Y_1 = X_0$). The point $\Delta$ can be therefore marked on the line $\overline{X_0 S}$ once the value of $N_\Delta$ is known. $N_\Delta$ can be obtained as

$$\Delta = R'_0 - E'_1$$

$$= - S'_0 - P'$$

or

$$\Delta / S'_0 = - (1 + P'/S'_0) \quad \text{... (6.1.72)}$$

Also,

$$\Delta N_\Delta = - (S'_0 N_S + P' N_0)$$

or

$$N_\Delta = \frac{N_S + (P'/S_0')\, N_0}{1 + (P'/S_0')} \quad \text{... (6.1.73)}$$

The ratio $(P'/S'_0)$ can be obtained by solving Eqs (6.1.67) and (6.1.68), after rewriting them on solvent-free basis. Thus

$$E'_1 = S'_0 + P' (1 + RR)$$

or

$$(E'_1 / S'_0) = 1 + (P'/S'_0)(1 + RR) \quad \text{... (6.1.67a)}$$

Similarly,

$$E'_1 Y_1 = S'_0 Y_S + P' X_0 (1 + RR)$$

or

$$(E'_1 / S'_0)\, Y_1 = Y_S + (P'/S'_0)\, X_0 (1 + RR) \quad \text{... (6.1.68a)}$$

Solving simultaneously,

$$(P'/S'_0) = \frac{(Y_1 - Y_S)}{(X_0 - Y_1)(1 + RR)} \quad \text{... (6.1.74)}$$

If the solvent is pure $B$, then $S'_0 = 0$ and $\Delta = -P'$. Also, $E'_1 = P'(1 + RR)$. Now

$$\Delta = R'_0 - E'_1$$

$$\Delta N_\Delta = R'_0 N_0 - E'_1 N_{E1}$$

or

$$N_\Delta = (R'_0 N_0 - E'_1 N_{E1})/\Delta \quad \text{... (6.1.75)}$$

Substituting $\Delta = -P'$ and $E'_1 = P'(1 + RR)$ and rearranging, we get

$$N_\Delta = N_{E1} + RR\,(N_{E1} - N_0) \quad \text{... (6.1.76)}$$

Now, mark $X_n$ on the raffinate curve. Join $X_n$ with $S$ and $\Delta$ with $F$. Extend these two lines $\overline{SX_n}$ and $\overline{\Delta F}$ downward to intersect at $\Delta'$, which is the operating point for the raffinate stripping section. (If the solvent is pure $B$, then the line $\overline{SX_n}$ will be a vertical line through $X_n$). Since $Y_1$ has been located, obtain $X_1$ from the equilibrium plot and mark it on the raffinate branch of the solubility curve. Join $X_1$ to $\Delta$

* In the figure shown however, the raffinate curve practically coincides with the abscissa since $NR$ values are too small (close to zero).

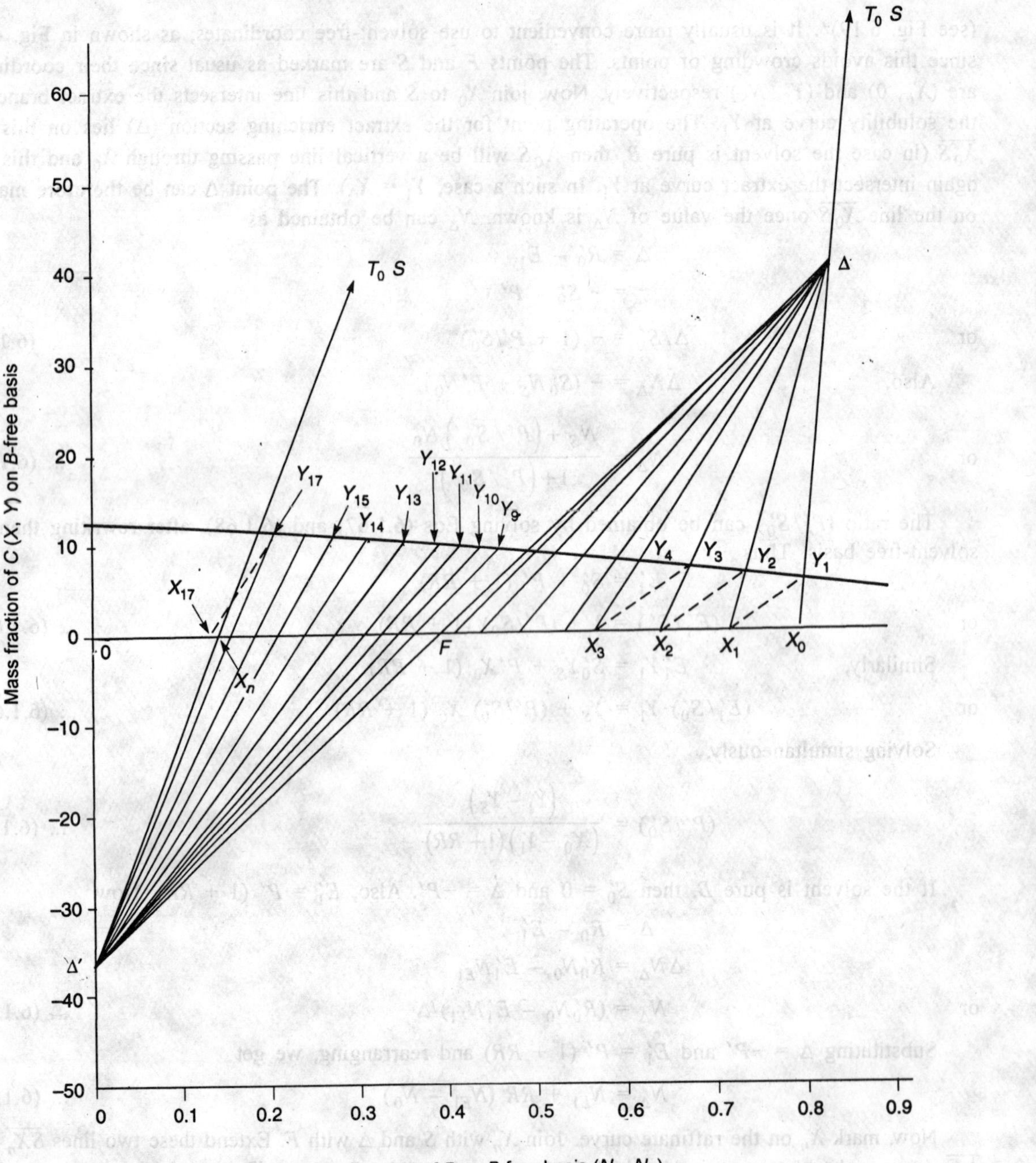


*Figure 6.19*

and this line will intersect the extract branch at $Y_2$. Proceed in this way until the value of $X$ falls below $X_F$. For example, let it be found that $X_6$ is less than $X_F$. Then, join $X_6$ to Δ′ (and not Δ) and this line when extended upward will intersect the extract branch at $Y_7$. Continue this graphical construction using $\Delta'$ as the operating point until $X \le X_n$. The feed stage is now stage 6.

The *minimum reflux ratio* ($RR_m$) can also be estimated graphically by following a procedure similar to that employed for determining the minimum solvent to feed ratio in countercurrent extraction without reflux. A good number of tie lines are drawn and all tie lines to the right of $F$ are extended upward to intersect the line $\overline{X_0\Delta}$ and by trial, the tie line which intersects $\overline{X_0\Delta}$ at a point farthest from the abcissa (farthest from $N = 0$) is located. Let this point of intersection be $L$. The tie lines to the left of $F$ are extended downward to intersect the line $\overline{X_n\Delta'}$ and again by trial, the tie line that intersects $\overline{X_n\Delta'}$ at a point farthest from $N = 0$ is located. Let this point of intersection be $K$. Join $\overline{KF}$ and extend it upward to intersect $\overline{X_0\Delta}$ at $L'$. Then, the point $\Delta_m$ (that is, operating point at minimum reflux) = point $L$ or point $L'$ whichever is farther from $N = 0$. In many cases, $L$ and $L'$ coincide with each other as shown in Fig. 6.20 (a)*. In other words, the tie line which when extended passes through $F$ intersects $\overline{X_0\Delta}$ at $\Delta_m$. This is always true so long as the equilibrium plot ($Y$ versus $X$ plot) is everywhere concave downward. Once $\Delta_m$ is located, the value of $N_{\Delta m}$ can be read from the figure. Now, substituting $N_\Delta = N_{\Delta m}$ in Eq. (6.1.73), it can be solved for $(P'/S_0')$ (max) and then substituting this value of $(P'/S_0')$ (max) in Eq. (6.1.74), it can be solved for $RR_m$. Mathematically,

$$RR_m = \frac{(Y_1 - Y_S)(N_0 - N_{\Delta m})}{(X_0 - Y_1)(N_{\Delta m} - N_S)} \quad (6.1.77)$$

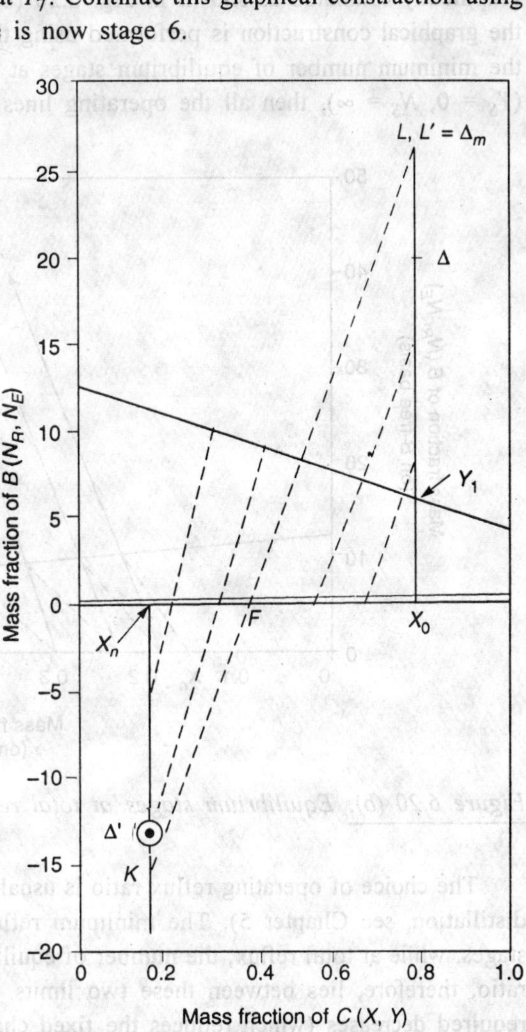


*Figure 6.20 (a): Graphical determination of minimum reflux ratio*

The other extreme to the operation of the extractor is operation at *total reflux*, when the number of equilibrium stages required is the minimum. At total reflux, no products are withdrawn. The entire extract, after recovery of solvent, is recycled back to stage 1 and the entire raffinate is recycled back to stage $n$ along with the solvent. The column accepts no fresh feed. Since $P = 0$, the reflux ratio ($RR$) is infinite. Under these conditions, the operating point $\Delta$ coincides with $S$. As a corollary, $\Delta'$ also coincides with $S$ (since $\Delta'$ is the point of intersection of line $\overline{SX_n}$ with $\overline{\Delta F}$ and here $\overline{\Delta F}$ is nothing but $\overline{SF}$). The number of equilibrium stages at total reflux (namely the minimum number of equilibrium stages, $n_m$) can be

* In Fig. 6.20 (a), it has been assumed that the solvent is pure. Accordingly, the lines $\overline{X_0\Delta}$ and $\overline{X_n\Delta'}$ are vertical lines passing through $X_0$ and $X_n$ respectively.

graphically estimated as shown in Fig. 6.20 (b)*. There is only one operating point ($\Delta = \Delta' = S$) and the graphical construction is performed using this single operating point as shown. In the present case, the minimum number of equilibrium stages at total reflux is 9.0. Incidentally, if the solvent is pure $B$ ($Y_S = 0$, $N_S = \infty$), then all the operating lines will be vertical (see Fig. 6.4.2 of Example 6.4).

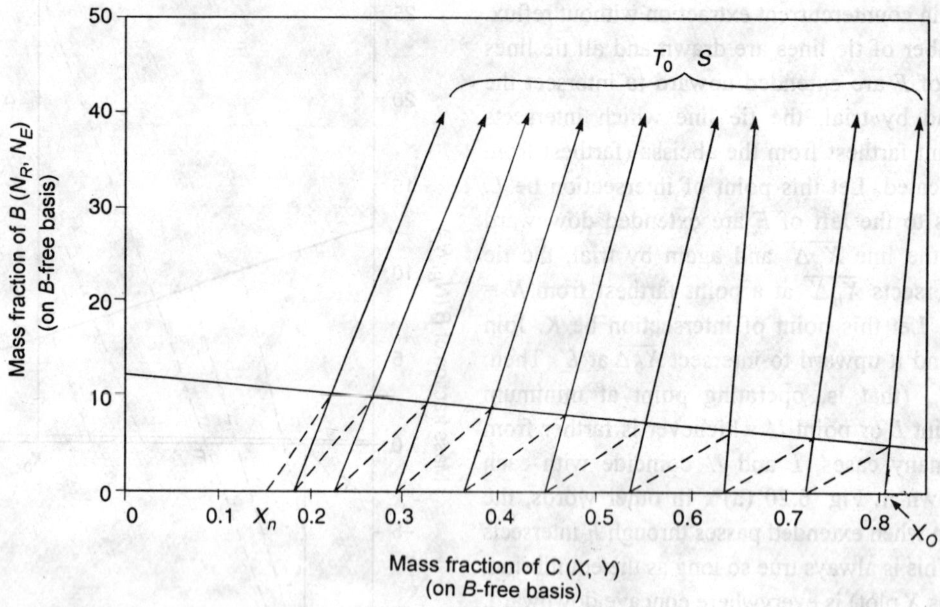


*Figure 6.20 (b): Equilibrium stages at total reflux*

The choice of operating reflux ratio is usually based on economic considerations (as is the case with distillation, see Chapter 5). The minimum reflux ratio corresponds to infinite number of equilibrium stages, while at total reflux, the number of equilibrium stages required is minimum. The operating reflux ratio, therefore, lies between these two limits. As the reflux ratio is increased, the number of stages required decreases (which reduces the fixed charges on the equipment), while the operating cost (cost of solvent recovery and cost of operation of reflux pump) increases. The *optimum reflux ratio* is, therefore, determined from economic considerations so that the total cost (cost of equipment plus operating cost) is a minimum. As stated earlier, in the economic analysis of any extraction process, the cost of solvent recovery (from the extract as well as from the raffinate) plays a crucial role.

## Stage Efficiency

All the computations discussed in the earlier sections are to estimate the number of *equilibrium stages* required in multistage extraction processes. The actual number of real stages (the actual number of plates/trays required in a sieve plate column or the actual number of mixer-settler combinations required)

*In Fig. 6.20 (b) also, the raffinate curve has been shown to be coinciding with the abscissa. This is true in the case of many systems for which the *NR* values are extremely small.

depends on the stage efficiency. In the case of a sieve plate extractor, the Murphree stage efficiency (which varies from stage to stage) for the dispersed phase may be estimated with allowable error from the correlation given by Treybal[4,2]. To note that the Murphree stage efficiency is a complex function of many variables such as the drop diameter, dispersed phase holdup, plate layout, the depth of coalesced dispersed liquid accumulated on each tray, etc. The actual number of plates required is then given by

$$n \text{ (actual)} = \sum_{i=1}^{n} (1/E_{mi}) \quad \text{... (6.1.78)}$$

where $E_{mi}$ is the Murphree plate efficiency of the $i$th stage.

The stage efficiency in extractors is usually much lower than that in gas-liquid absorbers, mainly due to the low operating velocities of phases (which result from the low density difference between phases and the high viscosities of phases).

## Countercurrent Continuous Contact Extraction

If extraction is being carried out in a continuous contact equipment (such as a packed column, spray column, baffled column with / without mechanical agitator), then a precise analysis of the process becomes difficult, mainly due to the unavailability of adequate data on mass transfer coefficients in such equipment. If we consider the specific case of $A$ being completely insoluble in $B$ (so that both the extract and the raffinate streams shall be binary liquid mixtures, the former containing only $B$ and $C$ and the latter containing only $A$ and $C$), then the system can be analysed using the HTU-NTU method that is popularly employed for the analysis of gas-liquid absorption in continuous contact equipment (discussed in detail in Chapter 4). Thus, the height of the column ($Z$) is given by

$$Z = (\text{HTU})_R \, (\text{NTU})_R = (\text{HTU})_E \, (\text{NTU})_E \quad \text{... (6.1.79)}$$

where $(\text{HTU})_R$ = height of a raffinate transfer unit

$$= R / [k_R a \, (1 - x)_{em}] \quad \text{... (6.1.80)}$$

$(\text{NTU})_R$ = number of raffinate transfer units

$$= \int_{x_2}^{x_1} \frac{(1-x)_{em} \, dx}{(1-x)(x-x_e)} \quad \text{... (6.1.81)}$$

$$\simeq \int_{x_2}^{x_1} \frac{dx}{(x-x_e)} + (1/2) \ln \left( \frac{1-x_2}{1-x_1} \right) \quad \text{... (6.1.82)}$$

$$(\text{HTU})_E = E / [k_E a \, (1 - y)_{em}] \quad \text{... (6.1.83)}$$

$$(\text{NTU})_E = \int_{y_2}^{y_1} \frac{(1-y)_{em} \, dy}{(1-y)(y-y_e)} \quad \text{... (6.1.84)}$$

$$\simeq \int_{y_2}^{y_1} \left( \frac{dy}{(y-y_e)} \right) + (1/2) \ln \left( \frac{1-y_2}{1-y_1} \right) \quad \text{... (6.1.85)}$$

$x_1 = x_F$ = *mole fraction* of $C$ (solute) in feed

$x_2$ = mole fraction of solute $C$ in final raffinate

$y_1$ = *mole fraction* of solute $C$ in final extract

$y_2 = y_S$ = mole fraction of solute $C$ in solvent

$x_e$, $y_e$ = interface concentrations (as mole fractions) of $C$

$(1 - x)_{em}$, $(1 - y)_{em}$ = logarithmic mean of $(1 - x)$ and $(1 - x_e)$ and that of $(1 - y)$ and $(1 - y_e)$ respectively

$R$ = *molar* flow rate per unit cross-sectional area of tower (or, molar mass velocity) of raffinate, kmole/(m$^2\cdot$s)

$$\simeq (R_1 + R_2)/2 \qquad (6.1.86)$$

$R_1$ (= $F$), $R_2$ = molar mass velocity of feed and that of final raffinate respectively, kmole/(m$^2\cdot$s)

$E$ = molar mass velocity of extract, kmole/(m$^2\cdot$s)

$$\simeq (E_1 + E_2)/2 \qquad \text{... (6.1.87)}$$

$E_1$, $E_2$ = molar mass velocity of final extract and that of solvent respectively, kmole/(m$^2\cdot$s)

$k_R$, $k_E$ = raffinate phase and extract phase mass transfer coefficients respectively, kmole/(m$^2\cdot$s$\cdot\Delta x$)

$a$ = specific interfacial area for mass transfer, m$^2$/m$^3$

Readers must note that the scheme of notations used in the above equations is the same as that employed for gas absorption (see Chapter 4) and this differs from that used for multistage extractors. Also, the above equations are written in terms of *mole fractions* and not mass fractions.

The interfacial concentration ($x_e$ or $x_e'$) can be determined by using the same graphical procedure as that employed under gas absorption. The equilibrium curve and the operating line are first plotted (in terms of mole fractions). We have already deduced in one of the earlier sections that the operating line / curve in cases like this passes through the point ($x_F'$, $y_1'$) and is of slope $G_A / G_B$ [see Eq. (6.1.48)]. The equation to the operating curve is, therefore, $y' = (G_A / G_B)(x' - x_F') + y_1'$. To be frank, this equation can be easily deduced from Fig. 6.21 by simply making a solute balance ($C$-balance) between the top of the tower and any arbitrary cross-section of the tower. Thus

$$G_A (x' - x_1') = G_B (y' - y_1')$$

or

$$y' = (G_A / G_B)(x' - x_1') + y_1'$$

$$y/(1 - y) = (G_A / G_B)\left(\frac{x}{1-x} - \frac{x_1}{1-x_1}\right) + \left(\frac{y_1}{1-y_1}\right) \qquad \text{... (6.1.88)}$$

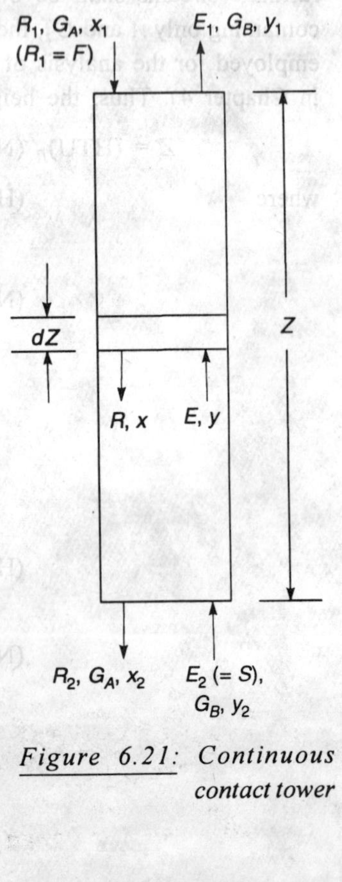


*Figure 6.21: Continuous contact tower*

To note that since $x$ and $y$ are defined as mole fractions (and not mass fractions), $G_A$ and $G_B$ must also be defined as molar flow rates (per unit cross-sectional area of the tower). In other words, $G_A$ and $G_B$ are superficial molar mass velocities, expressed in kmole/(m$^2\cdot$s). We also know that

$$G_A = R_1\ (1 - x_1) = R_2\ (1 - x_2) \quad \text{... (6.1.89)}$$

$$G_B = E_1\ (1 - y_1) = E_2\ (1 - y_2) \quad \text{... (6.1.90)}$$

Once the equilibrium curve and the operating curve are plotted, then if from any point $(x, y)$ on the operating curve a straight line of slope $(-k_R/k_E)$ is drawn, it will intersect the equilibrium curve at $(x_e, y_e)$.

As stated earlier, reported data on individual mass transfer coefficients in extraction are far from adequate. As a result, design calculations are often performed based on overall transfer coefficients ($K_R$ and $K_E$), though such a procedure is strictly valid only if the equilibrium curve is linear and is of constant slope, $m$. In terms of overall coefficients,

$$Z = (\text{HTU})_{OR}\ (\text{NTU})_{OR} = (\text{HTU})_{OE}\ (\text{NTU})_{OE} \quad \text{... (6.1.91)}$$

where

$$(\text{HTU})_{OR} = R/[K_R a\ (1 - x^*)_{\ln}] \quad \text{... (6.1.92)}$$

$$(\text{HTU})_{OE} = E/[K_E a\ (1 - y^*)_{\ln}] \quad \text{... (6.1.93)}$$

$$(\text{NTU})_{OR} = \int_{x_2}^{x_1} \frac{(1-x^*)_{\ln}\ dx}{(1-x)(x-x^*)} \quad \text{... (6.1.94)}$$

$$\simeq \int_{x_2}^{x_1} \frac{dx}{(x-x^*)} + (1/2)\ \ln\left(\frac{1-x_2}{1-x_1}\right) \quad \text{... (6.1.95)}$$

$$(\text{NTU})_{OE} = \int_{y_2}^{y_1} \frac{(1-y^*)_{\ln}\ dy}{(1-y)(y^*-y)} \quad \text{... (6.1.96)}$$

$$\simeq \int_{y_2}^{y_1} \frac{dy}{(y^*-y)} + (1/2)\ \ln\left(\frac{1-y_1}{1-y_2}\right) \quad \text{... (6.1.97)}$$

$(1 - x^*)_{\ln}$ = logarithmic mean of $(1 - x^*)$ and $(1 - x)$

$(1 - y^*)_{\ln}$ = logarithmic mean of $(1 - y)$ and $(1 - y^*)$

$x^*, y^*$ = concentration (mole fraction) of solute ($C$) in equilibrium with $y$ and that in equilibrium with $x$ respectively.

For any value of $x$, the corresponding value of $y$ is obtained from the operating curve [or from Eq. (6.1.88)] and the value of $y^*$ from the equilibrium curve. Conversely, for any value of $y$, the value of $x$ is obtained from the operating curve and the value of $x^*$ from the equilibrium curve. NTU is then estimated by evaluating the integral either graphically or numerically (see Chapter 4 for worked examples on equations of similar type).

The expressions for NTU may be written in terms of mass fractions as well. For example, if $x$ and $y$ are defined as mass fraction of solute ($C$) in the raffinate and in the extract respectively, then

$$(\text{NTU})_{OR} \simeq \int_{x_2}^{x_1} \frac{dx}{(x - x^*)} + (1/2)\ \ln\left(\frac{1-x_2}{1-x_1}\right) + (1/2)\ \ln\left[\frac{x_2(r-1)+1}{x_1(r-1)+1}\right] \quad \text{... (6.1.98)}$$

$$(\text{NTU})_{OE} \simeq \int_{y_2}^{y_1} \frac{dy}{(y^* - y)} + (1/2)\ \ln\left(\frac{1-y_1}{1-y_2}\right) + (1/2)\ \ln\left[\frac{y_1(r-1)+1}{y_2(r-1)+1}\right] \quad \text{... (6.1.99)}$$

where $r = M_A/M_C$ = ratio of molecular weight of nonsolute to that of solute.

Similarly, NTU may be expressed in terms of $x'$ and $y'$ (that is, in terms of mass ratios or mass fractions on solute-free basis) as

$$(\text{NTU})_{OR} \simeq \int_{x_2'}^{x_1'} \frac{dx'}{\left[x' - (x')^*\right]} + (1/2)\ \ln\left(\frac{1+rx_2'}{1+rx_1'}\right) \quad \text{... (6.1.100)}$$

$$(\text{NTU})_{OE} \simeq \int_{y_2'}^{y_1'} \frac{dy'}{\left[(y')^* - y'\right]} + (1/2)\ \ln\left(\frac{1+ry_1'}{1+ry_2'}\right) \quad \text{... (6.1.101)}$$

As in the case of gas absorption, the expressions for NTU and HTU can be further simplified if both the equilibrium curve as well as the operating curve are straight lines (atleast within the concentration range considered). This is true in the case of very dilute solutions and in such cases,

$$Z = (\text{HTU})_{OR}\ (\text{NTU})_{OR} = \left(\frac{R}{K_R a}\right) \frac{(x_1 - x_2)}{(x - x^*)_m} \quad \text{... (6.1.102)}$$

$$= (\text{HTU})_{OE}\ (\text{NTU})_{OE} = \left(\frac{E}{K_E a}\right) \frac{(y_1 - y_2)}{(y^* - y)_m} \quad \text{... (6.1.103)}$$

where

$(x - x^*)_m$ = logarithmic average of $(x_1 - x_1^*)$ and $(x_2 - x_2^*)$

$(y - y^*)_m$ = logarithmic average of $(y_1 - y_1^*)$ and $(y_2 - y_2^*)$.

If, in addition, the equilibrium curve is a straight line passing through the origin ($m = y^*/x = y/x^*$ = constant), then NTU may be computed using an expression similar to Eq. (6.1.50):

$$(\mathrm{NTU})_{OR} = \frac{ln\left[\dfrac{x_1 - y_2/m}{x_2 - y_2/m}\left(1 - 1/E_f\right) + 1/E_f\right]}{\left(1 - 1/E_f\right)} \quad \text{... (6.1.104)}$$

where $E_f = mE/R$ = extraction factor. Similarly,

$$(\mathrm{NTU})OE = \frac{\ln\left[\dfrac{y_2 - mx_1}{y_1 - mx_1}\left(1 - E_f\right) + E_f\right]}{\left(1 - E_f\right)} \quad \text{... (6.1.105)}$$

Equations (6.1.102) to (6.1.105) remain unaltered even if $x$ and $y$ are defined as mass fractions, except that in such cases, $R$ and $E$ must also be expressed in kg / (m$^2$·s). The mass transfer coefficients ($K_R$ and $K_E$) and m must also be defined accordingly. Alternately, $x$ and $y$ may be expressed as molar concentrations (in kmole/m$^3$), and $R$ in m$^3$/(m$^2$·s) and the mass transfer coefficients in kmole (m$^2$·s·kmol/m$^3$) or simply in m/s.

It has been stated at the outset that the above-discussed analysis is restricted to those cases in which both $A$ and $B$ do not distribute between phases. However, if the values of $G_A$ and $G_B$ at the two ends of the column do not differ significantly, then this design procedure may be followed with allowable error using arithmetic average values of $G_A$ and $G_B$.

Also, it may be noted that the scheme sketched in Fig. 6.21 assumes the feed be introduced from the top and the solvent from the bottom of the column. However, even if these directions are interchanged, the design equations given above shall remain unaltered. What is to be kept in mind is that the suffix 1 always represents that end of the tower at which the feed is introduced and suffix 2 always represents that end at which the solvent is introduced.

As for gas absorbers, the diameter of the extraction tower is decided by the flooding characteristics of the tower. Flooding in packed extraction towers has been studied by Crawford and Wilke[5], Hoffing and Lockhart[6], Nemunaitis et. al[7] and a few others. The Crawford-Wilke correlation, that has received fairly large-scale acceptance, is reproduced in Fig. 6.22. Here, the parameters $F_1$ and $F_2$ are respectively defined as (all variables in SI units)

$$F_1 = (35.208 \times 10^4)\ (\sigma/\rho_C)^{0.2}\ (\mu_C/\Delta\rho)\ (a_p/\epsilon)^{1.5} \quad \text{... (6.1.106)}$$

$$F_2 = (V_{CF}^{1/2} + V_{DF}^{1/2})^2\ \rho_C/(a_p\ \mu_C) \quad \text{... (6.1.107)}$$

where

$V_{CF}$, $V_{DF}$ = flooding velocity of continuous phase and that of dispersed phase respectively

$\rho_C$, $\mu_C$ = density and viscosity respectively of continuous phase

$\Delta\rho$ = difference in density between the continuous phase and the dispersed phase

$\sigma$ = interfacial tension

$a_p$, $\epsilon$ = specific packing surface and void fraction respectively of packed bed.

Since the parameter $F_1$ does not contain any velocity terms, it can be evaluated from the available data and then the value of $F_2$ is obtained from Fig. 6.22. If we assume that the operating velocity is taken to be 50 percent of the flooding velocity (which is the most popular choice), then the cross-sectional area ($A_C$) of the tower may be estimated as

$$A_C = \frac{\left[\left(2Q_C/\rho_C\right)^{1/2} + \left(2Q_d/\rho_d\right)^{1/2}\right]^2}{\left(F_2 a_p \mu_C/\rho_C\right)} \quad \text{... (6.1.108)}$$

where

$Q_C$, $Q_d$ = mass flow rate of continuous phase and that of dispersed phase respectively

Usually, the value of $A_C$ is estimated using the values of $Q_C$ and $Q_d$ at the end 1 of the tower (where these flow rates are maximum).

*Figure 6.22: Flooding in packed extraction tower (from Crawford, ME and Wilke, CR; Chem. Eng. Progress, 47, 423, 1951, by permission)*

Flooding phenomena in rotary disc extractors (RDC) are discussed by Strand et al[8].

Some of the popular correlations reported in literature for the estimation of mass transfer coefficients in extraction columns are listed in Table 6.1.

## Table 6.1
## Mass Transfer Coefficients in Packed Extraction Columns

| *Dispersed phase* | *Continuous phase* | *Extracted solute* | *Phase that forms raffinate* | *Phase that wets packing* | *Tower diameter and type of packing* | *Tempe-rature, °C* | $V_C$, *m/hr* | $V_d$, *m/hr* | *Correlation [Ka in hr$^{-1}$, HTU in m]* |
|---|---|---|---|---|---|---|---|---|---|
| Water | Toluene | Benzoic acid | Continuous | Dispersed | 51.56 mm, 12.7 mm saddles | 15–18 | 6.76 | 4.05–13.53 | $K_D a = 6.0 + 0.59\ V_d$ |
| | | | | | | | 13.53 | 4.05–13.53 | $K_D a = 7.5 + 0.59\ V_d$ |
| Water | Benzene | Benzoic acid | Continuous | Dispersed | 48 mm, 12.7 mm saddles | 30 | 2.25–29.26 | 5.85–23.62 | $K_D a = 4.42$ to $12.7$ |
| Toluene | Water | Furfural | Continuous | — | 101.6 mm, 12.7 mm saddles | 25 | 4.48–17.68 | 2.53–8.70 | $(\text{HTU})_{OC} = 1.0\ (V_C/V_d)^{0.95}$ |
| Toluene | Water | Diethyl amine | Continuous | — | 33.3 mm, 4 mm glass beads | 26.8 | 0.405–2.715 | 0.05–1.585 | $(\text{HRU})_{OC} = 0.5273 \left(\dfrac{V_C}{mV_d}\right)$ |
| | | | | | | 38.5 | 0.512–4.303 | 0.024–2.45 | $(\text{HTU})_{OC} = 0.4724 \left(\dfrac{V_C}{mV_d}\right)$ |
| | | | | | | 48.5 | 0.762–5.85 | 0.085–2.83 | $(\text{HTU})_{OC} = 0.3871 \left(\dfrac{V_C}{mV_d}\right)$ |
| | | | | | | 57.5 | 0.74–5.08 | 0.268–1.414 | $(\text{HTU})_{OC} = 0.2408 \left(\dfrac{V_C}{mV_d}\right)$ |
| | | | | | | | | | $m$ = slope of equilibrium curve (molar coordinates) |

(*Contd.*)

**Table 6.1** *(Contd.)*

## Mass Transfer Coefficients in Packed Extraction Columns

| *Dispersed phase* | *Continuous phase* | *Extracted solute* | *Phase that forms raffinate* | *Phase that wets packing* | *Tower diameter and type of packing* | *Temperature, °C* | *$V_C$, m/hr* | *$V_d$, m/hr* | *Correlation [Ka in $hr^{-1}$, HTU in m]* |
|---|---|---|---|---|---|---|---|---|---|
| Methyl isobutyl ketone | Water | Acetic acid | Continuous | Continuous | 90.17 mm, 12.7 mm carbon rings | 23–27 | 3.0–24.0 | 9.0–21.0 | $(HTU)_{OD} = 0.2347 \left(\frac{V_d}{V_C}\right)^{0.648}$ |
| | | | | | Same as above | Same as above | 3.0–18.0 | 3.0 | $K_D a = 7.8 + 0.5545\ V_C$ |
| | | | | | 90.17 mm, 12.7 mm saddles | 23–27 | 3.0–12.0 | 12.0 | $K_D a = 6.988\ V_C$ |
| | | | | | 90.17 mm, 25.4 mm carbon rings | 23–27 | 3.0–12.0 | 12.0 | $K_D a = 15.0 + 2.1\ V_C$ |
| $CaCl_2$ brine | Methyl ethyl ketone | Water | Continuous | Dispersed | 90.17 mm, 12.7 mm saddless | 25–28 | 8.53–20.0 | 5.18–5.27 | $K_C a = 6.26\ V_C^{0.42}$ |
| Methyl ethyl ketone | $CaCl_2$ brine | Water | Dispersed | Continuous | 90.17 mm, 12.7 mm saddles/ ceramic rings | 28–28 | 4.91–10.39 | 7.95–21.18 | $K_D a = 2.4278\ V_d$ |

*Source:* References (1) and (9).

**Example 6.4:** A mixture of methyl cyclopentane (MCP) and *n*-hexane containing 40 mass percent MCP is to be extracted at 25° C with pure aniline. The extract product is to be discharged at 80 mass percent MCP (on solvent-free basis) and the raffinate product is to contain 15 mass percent MCP (on solvent-free basis). A continuous countercurrent multistage extraction system with extract reflux is to be employed for the purpose.

(*a*) Estimate the minimum reflux ratio that may be used.

(*b*) if an operating reflux ratio of 8.25 is being employed, what will be the number of equilibrium stages required? What will be the solvent requirement?

(*c*) What will be the number of equilibrium stages at total reflux?

(*d*) What is the maximum concentration of methyl cyclopentane that can be achieved in the final extract if no reflux is used?

The solubility data for methyl cyclopentane–*n*-hexane–aniline system at 25° C and 1 atm pressure is given below:

| *n-Hexane layer* | | *Aniline layer* | |
|---|---|---|---|
| *Mass fraction of MCP* | *Mass fraction of aniline* | *Mass fraction of MCP* | *Mass fraction of aniline* |
| 0.0 | 0.07206 | 0.0 | 0.92557 |
| 0.0994 | 0.07634 | 0.013 | 0.9162 |
| 0.1988 | 0.0806 | 0.0288 | 0.9068 |
| 0.298 | 0.08595 | 0.04476 | 0.8965 |
| 0.3972 | 0.09235 | 0.06347 | 0.8842 |
| 0.4962 | 0.100 | 0.08598 | 0.8692 |
| 0.5950 | 0.1115 | 0.1132 | 0.8512 |
| 0.6934 | 0.1243 | 0.1472 | 0.8284 |
| 0.7915 | 0.1412 | 0.1936 | 0.7960 |
| 0.8471 | 0.1529 | 0.2279 | 0.7721 |

**Solution:** Since the system consists of only one fully miscible liquid pair, the solubility curve will not be a continuous curve. It will be better to plot the solubility curve on solvent-free coordinates (to avoid crowding of points during graphical construction). Let us, therefore, convert the given solubility data to solvent-free (*B*-free) basis by using the relationships of Eqs (6.1.5) to (6.1.8). The results are tabulated below:

| *n-hexane layer (mass fraction on aniline-free basis)* | | *Aniline layer (mass fraction on aniline-free basis)* | |
|---|---|---|---|
| *MCP* | *Aniline* | *MCP* | *Aniline* |
| 0.0 | 0.07765 | 0.0 | 12.4354 |
| 0.1076 | 0.08265 | 0.155 | 10.933 |
| 0.2162 | 0.08766 | 0.310 | 9.7296 |
| 0.326 | 0.0940 | 0.4324 | 8.6609 |

(*Contd.*)

| n-hexane layer (mass fraction on aniline-free basis) | | Aniline layer (mass fraction on aniline-free basis) | |
|---|---|---|---|
| MCP | Aniline | MCP | Aniline |
| 0.4376 | 0.1017 | 0.5483 | 7.6393 |
| 0.5513 | 0.1111 | 0.6570 | 6.6429 |
| 0.6696 | 0.1255 | 0.761 | 5.720 |
| 0.792 | 0.1419 | 0.8578 | 4.8275 |
| 0.9216 | 0.1644 | 0.9493 | 3.902 |
| 1.00 | 0.1804 | 1.00 | 3.3871 |

(*a*) The minimum reflux ratio is best determined graphically. The solubility curve (on solvent-free coordinates) is plotted as shown in Fig. 6.4.1. Since the $N_R$ values are very small, the raffinate curve

*Figure 6.4.1*

practically coincides with the abscissa. The equilibrium plot for this system at 25° C has already been shown in Fig. 6.18 (b). Since the equilibrium isotherm is everywhere concave downward, the minimum reflux ratio in this case shall be determined by that tie line which when extended passes through $F$. This tie line has been located in Fig. 6.4.1 and it can be seen that it intersects line $\overline{X_0\Delta}$ at $\Delta_m$ such that $N_{\Delta m}$ = 31.0. Also, $N_{E1}$ (read from figure) = 5.3. From solubility data, $N_0$ (corresponding to $X_0$ = 0.80) = 0.143. Therefore, from Eq. (6.1.76),

$$RR_m = (31.0 - 5.3)/(5.3 - 0.143) = 4.9835$$

(*b*) Let us determine the number of equilibrium stages required at $RR$ = 8.25 first by the analytical method. We know

$$X_F = x_F = 0.40$$

$$X_n = 0.15$$

$$X_0 = X_P = 0.80$$

$$Y_1 = X_0 = 0.80 \text{ (since solvent is pure } B)$$

$$N_{E1} = 5.30$$

$$N_0 = 0.143$$

From equilibrium plot, $X_1$ = 0.715 and from the solubility data, $N_{R1}$ = 0.131. Now, if we rewrite Eqs (6.1.59) and (6.1.60) on solvent-free basis, we get

$$F = R_n' + P' \quad \text{... (6.1.59a)}$$

$$FX_F = R_n'X_n + P'X_0 \quad \text{... (6.1.60a)}$$

To note that since the solvent is pure $B$ (does not contain any solute), $S' = S_0' = 0$ and since the feed does not contain any $B$, $F' = F$. Solving the above two equations simultaneously,

$$P'/F = (X_F - X_n)/(X_0 - X_n) = (0.40 - 0.15)/(0.80 - 0.15) = 0.3846$$

$$R_n'/F = 1 - (P'/F) = (1.0 - 0.3846) = 0.6154$$

$$R_0'/F = (RR)\,(P'/F) = (8.25)\,(0.3846) = 3.17295$$

Since $\quad E_1' = P' + R_0',$

$$E_1'/F = 0.3846 + 3.17295 = 3.55755$$

And, $\quad E_1'/R_0' = 3.55755/3.17295 = 1.1212$

Let us now perform the stage to stage computations. Thus, putting $i$ = 1 in Eq. (6.1.42b) (after rewriting it in terms of $B$-free coordinates), we get

$$G_2' = \frac{(E_1'/R_0')(Y_1 - X_1) + (R_0'/R_0')(X_1 - X_0)}{(E_1'/R_0')(N_{E1} - N_{R1}) + (R_0'/R_0')(N_{R1} - N_0)}$$

$$= \frac{(1.1212)(0.8 - 0.715) + (1.0)(0.715 - 0.80)}{(1.1212)(5.3 - 0.131) + (1.0)(0.131 - 0.143)} = 0.0018187$$

Now, from Eq. (6.1.42a), $\quad Y_2 = X_1 + G_2'(N_{E2} - N_{R1}) = 0.715 + 0.0018187\,(N_{E2} - 0.131)$

Solving by trial with the help of solubility data,

$$Y_2 = 0.726$$
$$N_{E2} = 5.387$$

From equilibrium plot, $X_2 = 0.63$ and from solubility data, $N_{R2} = 0.12066$. Now, from Eq. (6.1.39a),

$$E'_2/R'_0 = \frac{(E_1'/R_0')(N_{E1} - N_{R1}) + (N_{R1} - N_0)(R_0'/R_0')}{(N_{E2} - N_{R1})}$$

$$= \frac{(1.1212)(5.3 - 0.131) + (1.0)(0.131 - 0.143)}{(5.387 - 0.131)} = 1.1$$

And from Eq. (6.1.36a),

$$R'_1/R'_0 = 1.0 + 1.1 - 1.1212 = 0.9788$$

Computations are similarly continued for subsequent stages. It is seen that $X_6$ is less than $X_F$ and therefore, $f = 6$. For $i = f = 6$, $G'_{i+1}$ is hence computed from Eq. (6.1.42c) and $(E'_{i+1}/R'_0)$ and $(R'_i/R'_0)$ from Eqs (6.1.39b) and (6.1.36b) respectively. For $i > 6$, computations are performed as for $i = 1$. The results are summarised below:

| $i$ | $Y_i$ | $X_i$ | $E'_i/R'_0$ | $R'_i/R'_0$ |
|---|---|---|---|---|
| 1 | 0.800 | 0.715 | 1.1212 | 0.9788 |
| 2 | 0.726 | 0.630 | 1.10 | 0.7620 |
| 3 | 0.654 | 0.550 | 0.8832 | 0.6860 |
| 4 | 0.5875 | 0.478 | 0.8073 | 0.61195 |
| 5 | 0.530 | 0.420 | 0.733 | 0.5757 |
| 6 | 0.485 | 0.375 | 0.6969 | 0.8581 |
| 7 | 0.440 | 0.333 | 0.6642 | 0.8246 |
| 8 | 0.389 | 0.284 | 0.6307 | 0.7886 |
| 9 | 0.3275 | 0.2325 | 0.5946 | 0.7562 |
| 10 | 0.260 | 0.179 | 0.5623 | 0.7267 |
| 11 | 0.189 | 0.130 | 0.5328 | — |

The number of equilibrium stages required is therefore, $n = 10 + (0.179 - 0.15)/(0.179 - 0.13) = 10.5918$.

The graphical determination of number of equilibrium stages is illustrated in Fig. 6.4.1. The points $F$, $X_0$ and $X_n$ are first marked as shown. Since the solvent is pure $B$, the vertical line through $X_0$ will intersect the extract curve at $Y_1$. Now, from Eq. (6.1.76), taking $N_0 \simeq 0.0$,

$$N_\Delta = (1 + RR)\, N_{E1} = (1.0 + 8.25)\,(5.30) = 49.025$$

The operating point $\Delta$ is thus marked on line $\overline{X_0 Y_1}$ (extended) corresponding to $N_\Delta = 49.025$. The line $\overline{\Delta F}$ when extended downward intersects the vertical line through $X_n$ at $\Delta'$. The graphical construction is now performed as shown using $\Delta$ as the operating point until the point $F$ is crossed and thereafter using $\Delta'$ as the operating point. All the tie lines are shown as broken lines. The tie line $\overline{Y_6 X_6}$ has been however omitted in order to avoid crowding of the figure. The results obtained are listed below:

| $i$ | $Y_i$ | $X_i$ |
|---|---|---|
| 1 | 0.800 | 0.715 |
| 2 | 0.726 | 0.630 |
| 3 | 0.654 | 0.550 |
| 4 | 0.5875 | 0.478 |
| 5 | 0.530 | 0.420 |
| 6 | 0.4875 | 0.378 |
| 7 | 0.441 | 0.335 |
| 8 | 0.390 | 0.287 |
| 9 | 0.3275 | 0.2325 |
| 10 | 0.260 | 0.179 |
| 11 | 0.190 | 0.131 |

The number of equilibrium stages required, is, therefore, $n = 10 + (0.179 - 0.15)/(0.179 - 0.131) = 10.604$.

To determine the solvent requirement, let us first convert the mass flow rates that are on $B$-free basis to total mass flow rates. Thus

$$E_1/F = (E_1'/F)\,(1 + N_{E1}) = (3.55755)\,(1.0 + 5.30) = 22.4125$$

Similarly,

$$P/F = (0.3846)\,(1.0 + 0.143) = 0.4396$$

$$R_n/F = (0.6154)\,(1.0 + 0.0846) = 0.66746$$

To note that from solubility data, $N_{Rn}$ (corresponding to $X_n = 0.15$) $= 0.0846$. Substituting in Eqs (6.1.59) and (6.1.67), we get

$$S/F = 0.66746 + 0.4396 + (S_0/F)$$

and

$$22.4125 = S_0/F + 0.4396\,(1.0 + 8.25)$$

Solving simultaneously,

$$S/F = 19.45$$

$$S_0/F = 18.346$$

Thus, the solvent requirement is 19.45 kg per kg of feed. 18.346 kg of solvent per kg of feed is recycled from the solvent recovery unit and the balance (namely, 1.104 kg per kg of feed) is supplied as make-up solvent. The makeup has become necessary since a small amount aniline is lost in the final extract product ($P$) collected.

(*c*) The number of equilibrium stages at total reflux can be estimated graphically as shown in Fig. 6.4.2. Since the solvent is pure $B$, all the operating lines are vertical. The number of equilibrium stages is seen to be 6.7.

(*d*) In countercurrent operation without reflux, we know that the minimum solvent to feed ratio is decided by that tie line which when extended meets the line $\overline{SX_n}$ (in this case, the vertical line through $X = X_n$) at the lowest point. In the present case, it is seen that the tie line which when extended passes through $F$ intersects $\overline{SX_n}$ at the lowest point. In Fig. 6.4.1, this tie line is nothing but $\overline{F\Delta_m}$ and it intersects

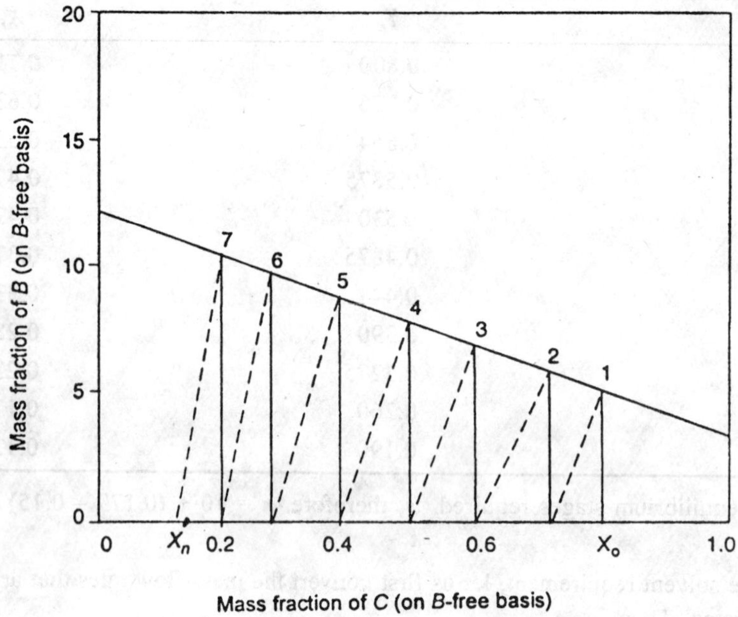


*Figure 6.4.2*

the extract curve at $P$. The value of $Y$ corresponding to this point $P = (Y_1)_{max} = 0.51$. Also, $N_E$ (corresponding to $P$) = 8.0. Therefore,

$$(y_1)_{max} = (0.51)/(1.0 + 8.0) = 0.0566$$

Therefore, the maximum concentration of methyl cyclopentane that can be achieved in the final extract when no reflux is used is 5.66 percent. On solvent-free basis, the maximum extract concentration attainable without reflux is 0.51 (as compared to 0.80 when extract reflux is employed).

**Example 6.5:** The extraction of acetic acid from the aqueous solution of acid by methyl isobutyl ketone (MIBK), considered in Example 6.2, is to be conducted in a packed extraction tower that operates in the countercurrent mode. Under the operating conditions stipulated, MIBK forms the dispersed phase and water the continuous phase and the height of an overall dispersed phase transfer unit may be computed from (see Table 6.1)

$$(\text{HTU})_{OD} = 0.2347\ (V_d / V_C)^{0.648}$$

Estimate the required packed height of the tower.

**Solution:** Since we are considering a continuous contact equipment, let us first redesignate the known process variables as per the notation scheme of Fig. 6.21. Thus

$$R_1 = F = 100 \text{ kg/hr}$$

$$x_1 = x_F \text{ (or } x_0) = 0.20$$

$$E_2 = S = 200 \text{ kg/hr}$$

$$y_2 = y_S = 0.001$$

$$x_2 = x_n = 0.01$$

Since $A$ is totally insoluble in $B$,

$$R_1\ (1 - x_1) = R_2\ (1 - x_2) = G_A = 80\ \text{kg/hr}$$

Therefore,
$$R_2 = 80/(1.0 - 0.01) = 80.808\ \text{kg/hr}$$

$$E_1 = R_1 + E_2 - R_2 = 100 + 200 - 80.808 = 219.192\ \text{kg/hr}$$

Similarly, since $E_1\ (1 - y_1) = E_2\ (1 - y_2) = G_B = 200\ (1 - 0.001) = 199.8$ kg/hr,

$$y_1 = 1 - G_B/E_1 = 1.0 - (199.8/219.192) = 0.08847$$

Equation (6.1.88), namely equation to the operating line, therefore becomes

$$y/(1 - y) = (80.0/199.8)\left[\frac{x}{1-x} - \frac{0.20}{(1.0-0.20)}\right] + \frac{(0.08847)}{(1.0-0.08847)}$$

$$= (0.4004)\left(\frac{x}{1-x} - 0.25\right) + 0.097 \qquad \text{... (i)}$$

To note that in the above equation, $x$ and $y$ are mass fractions (and not mole fractions) and accordingly, $G_A$ and $G_B$ are mass velocities (in kg/m$^2\cdot$s). To estimate the number of transfer units, we have to first evaluate the integral of Eq. (6.1.99). For this, the values of $y^*$ corresponding to different values of $y$ (ranging from $y = y_1 = 0.001$ to $y = y_2 = 0.08847$) are to be computed. For each value of $y$, the value of $x$ is obtained from Eq. (i) and corresponding to this value of $x$, $y^*$ is obtained from the equilibrium plot (the equilibrium curve for this system is shown in Fig. 6.2.1 of Example 6.2). To keep in mind that Fig. 6.2.1 is the plot of $y'$ versus $x'$ where $y' = y/(1 - y)$ and $x' = x/(1 - x)$. The results are listed below:

| $y$ | $x$ [from Eq. (i)] | $y^*$ | $1/(y^* - y)$ |
|---|---|---|---|
| 0.08847 | 0.20 | 0.1683 | 12.5266 |
| 0.07 | 0.1637 | 0.1339 | 15.6494 |
| 0.06 | 0.1432 | 0.1150 | 18.1818 |
| 0.05 | 0.1220 | 0.0959 | 21.7865 |
| 0.04 | 0.1000 | 0.076 | 27.3224 |
| 0.03 | 0.0783 | 0.05815 | 35.5240 |
| 0.02 | 0.0550 | 0.03914 | 52.2466 |
| 0.01 | 0.0319 | 0.02132 | 88.3392 |
| 0.005 | 0.0200 | 0.01321 | 121.8027 |
| 0.002 | 0.01258 | 0.008288 | 159.033 |
| 0.001 | 0.0100 | 0.00658 | 179.2115 |

We can now evaluate the integral of Eq. (6.1.99) either graphically or numerically. Let us evaluate it numerically using the trapezoidal rule. According to this rule,

$$\int_{y_0}^{y_{10}} \frac{dy}{y^* - y} = (1/2) \sum_{i=1}^{10} h_i\ (f_{i-1} + f_i)$$

where $h_i = (y_i - y_{i-1})$. From the table given above, $y_0 = 0.001$, $y_1 = 0.002$, $y_2 = 0.005$ and so on. Finally, $y_{10} = 0.08847$. Similarly, $f_0 = 179.2115$, $f_1 = 159.033$, $f_2 = 121.8027$ and so on and finally $f_{10} = 12.5266$. Thus, the value of the integral is

$$I = 3.4465$$

Therefore, from Eq. (6.1.99),

$$(\text{NTU})_{OE} = (3.4465) + (1/2)\ln\left(\frac{1-0.08847}{1-0.001}\right) + (1/2)\ln\left[\frac{1-(0.08847)(0.7)}{1-(0.001)(0.7)}\right]$$

$$= 3.369$$

To note that $r = M_A / M_C = 18.0/60.0 = 0.3$. Now, from the given correlation

$$(\text{HTU})_{OE} = 0.2347\ (V_d / V_C)^{0.648} = 0.2347\ (E\ \rho_R / R\ \rho_E)^{0.648}$$

where

$$E \simeq (E_1 + E_2)/2 = (200 + 219.192)/2 = 209.6 \text{ kg / hr}$$

$$R \simeq (R_1 + R_2)/2 = (100 + 80.808)/2 = 90.4 \text{ kg / hr}$$

Substituting these values of $E$ and $R$ in the above correlation and taking $(\rho_R/\rho_E)^{0.648} \approx 1.0$, we get

$$(\text{HTU})_{OE} = 0.4064 \text{ m}$$

Therefore, the required packed height of the tower is

$$Z = (0.4064)\ (3.369) = \mathbf{1.369\ m}$$

## 6.2 SOLID–LIQUID EXTRACTION

Solid–liquid extraction (also called *leaching*) is a unit operation that is very much analogous to liquid-liquid extraction. The chief difference between the two operations is that in leaching, the feedstock is usually a mixture of solids (for example, a mixture of ore and the gangue) from which the soluble constituent (namely the solute $C$) is selectively dissolved off or extracted by a suitable solvent ($B$). The solution of $C$ in $B$ so obtained is the extract or the leach solution or the micella (the terminology *micella* is used in the case of extraction of oil from vegetable seeds). It is also called the *overflow* since from the separating vessel (usually a settling tank) this solution overflows from the top (see Fig. 6.23). The leached solids consisting essentially of the nonsolute or the inert solids ($A$) is collected from the bottom of the separating vessel as the *underflow* or the raffinate or the *marc* (this terminology marc is also one that is used in the case of oil extraction from vegetable seeds). The overflow is usually a clear solution of $C$ in $B$, while the underflow usually consists of particles of $A$ suspended in a solution of $C$ in $B$. The schematic of a typical solid-liquid extraction unit is shown in Fig. 6.23.

Alternate terminologies used to denote the process of solid–liquid extraction are *lixiviation, decoction* and *elution.* While decoction refers specifically to the use of solvent at its boiling point, the term elution is usual to designate those leaching operations in which the soluble material is largely on the surface of an insoluble solid and is merely washed off by the solvent.

Conditioning of the solid feed is important in leaching. If the solute is surrounded by a matrix of insoluble matter which is relatively impermeable to the solvent (for example, gold dispersed in rock), then the material must be well-crushed so that all the solute is exposed to the solvent. If the solid has a cellular structure, the extraction rate will be relatively low since the cell walls provide additional

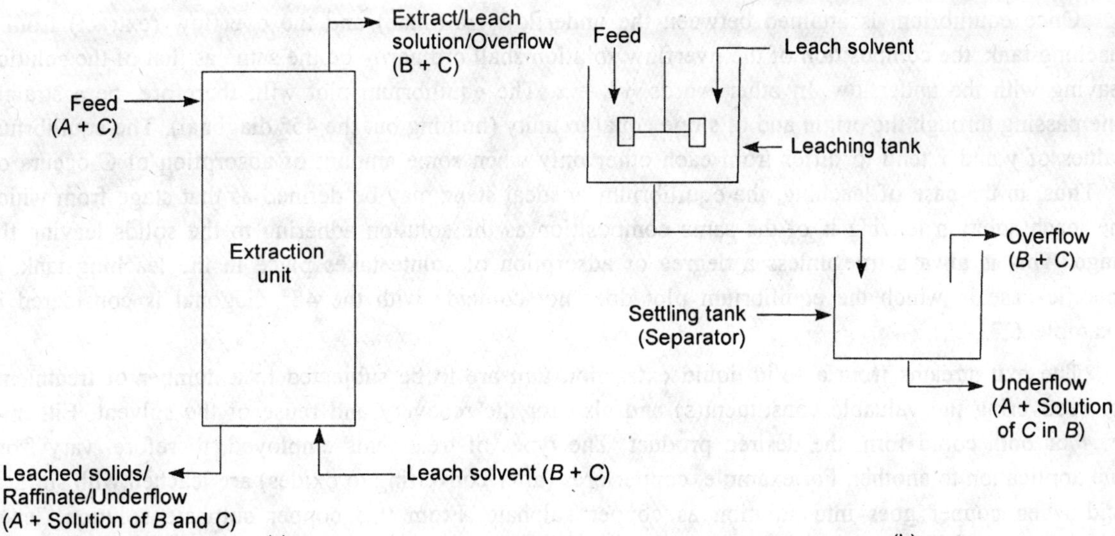


*Figure 6.23: Schematic of solid–liquid extraction system*

resistance. However, complete rupture of the cells is often avoided since the cell walls do perform alternate functions. For example, in the extraction of sugar from beet, the cell walls perform the important function of impeding the passage of undesirable constituents (colloidal and albuminous materials). The beet should therefore be prepared in long strips so that only a small proportion of the cells are ruptured. In the extraction of oil from vegetable seeds or beans, the solute is itself liquid and tends to diffuse towards the solvent. In this case, the seeds or beans are flaked to give particles of size 0.15 to 0.5 mm. The cells are largely ruptured during the flaking process and consequently, the oil is more readily contacted by the solvent.

The choice of operating conditions is obviously influenced by the factors that are responsible for limiting the extraction rate. The dissolution of solute in the solvent is usually very rapid and this has negligible effect on the overall extraction rate. The two factors that largely influence the overall rate of leaching are the rate of diffusion of solute through the solvent in the pores of the solid to outside the particle and the diffusion of solute from the particle surface to the bulk of the solution. If the diffusion of the solute through the porous structure of the residual solids is the controlling factor, the feed solids must be reduced in size so that the distance the solute has to travel is small. On the other hand, if diffusion of solute from the surface of the particles to the bulk of the solution is sufficiently slow to control the process, a high degree of agitation of the fluid is called for.

High operating temperatures are generally advantageous in leaching. The solubility of the solute in the solvent usually increases with increase in temperature. Further, the viscosity of the liquid decreases and diffusivities increase with rise in temperature. As a result, higher temperatures provide higher rates of extraction. In some cases, the upper limit of temperature is determined by secondary considerations, such as the necessity of preventing enzyme action during the extraction of a solute like sugar.

Once equilibrium is attained between the underflow (raffinate) and the overflow (extract) from a leaching tank, the composition of the overflow solution shall ordinarily be the same as that of the solution leaving with the underflow. In other words, $y^* = x$. The equilibrium plot will, therefore, be a straight line passing through the origin and of slope equal to unity (nothing but the 45° diagonal). The equilibrium values of $y$ and $x$ tend to differ from each other only when some amount of adsorption of $C$ occurs on $A$. Thus, in the case of leaching, the equilibrium or ideal stage may be defined as that stage from which the leach solution leaving is of the same composition as the solution adhering to the solids leaving the stage. This is always true unless a degree of adsorption of solute takes place in the leaching tank. A specific case in which the equilibrium plot does not coincide with the 45° diagonal is considered in Example 6.7.

The exit streams from a solid-liquid extraction unit are to be subjected to a number of treatments for recovering the valuable constituent(s) and also for the recovery and reuse of the solvent. Either $C$ or $A$ or both could form the desired product. The types of treatments employed, therefore, vary from one application to another. For example, copper ores (after converting to oxides) are leached with sulfuric acid when copper goes into solution as copper sulphate. From this copper sulphate solution (leach solution), copper is recovered by electrolytic deposition. Sulfuric acid that is regenerated during the electrolytic process may be reused for further leaching of the ore. In the extraction of oil from soyabeans, both the oil as well as the oil-free meal are valuable materials. In addition, the organic solvent (*n*-hexane, benzene) is to be recovered for reuse. The solvent is recovered by evaporation followed by distillation (stripping). The leached seeds are steamed to remove the residual solvent and then air-cooled. After the extraction of sugar from beets, the resultant sugar solution is concentrated by evaporation and sugar crystals are allowed to crystallise out. The mother liquor which is essentially water, may be reused.

The major industrial applications of solid–liquid extraction are listed below:

(*i*) Leaching is extensively used in ore dressing or mineral processing. The soluble ore is leached off by a suitable solvent leaving the insoluble gangue behind. For example, copper is recovered from copper ores as copper sulphate by leaching with sulfuric acid, gold is separated from its ores by leaching with sodium cyanide solution and aluminium recovered from impure bauxite ore as sodium aluminate by leaching with caustic soda.

(*ii*) Leaching is also a popular unit operation in food processing industries. Sugar is leached from beets with hot water, vegetable oil is recovered from seeds and beans (cotton seeds, soybeans, linseeds, castor beans, peanuts) by leaching with organic solvents (*n*-hexane, trichloroethylene, benzene). Oil is extracted from rice bran using *n*-hexane as the solvent and from halibut livers using ethyl ether as the solvent. Domestic preparation of tea or coffee is nothing but a leaching process with hot water.

(*iii*) In many chemical industries, various chemical precipitates are washed of the adhering mother liquor using water as the solvent. Washing of caustic soda from precipitated calcium carbonate in lime-soda process, washing of precipitated barium sulfate in the manufacture of lithopone (pigment), etc. are examples.

## 6.2.1 Leaching Equipment

Leaching equipment can be broadly classified into two categories such as, (*i*) solid bed systems and (*ii*) dispersed contact systems. In the former type of equipment, the solids to be leached form a stationary bed and the leach solvent percolates through it, while in the latter type, the solids are kept dispersed or suspended in the leach solvent.

*Percolation tanks* are large rectangular or cylindrical tanks (they are built in sizes as large as 53 × 20 × 5.5 m) with a false bottom. The solids to be leached are dumped into the tank to a uniform depth. The tank is then filled with the leach solvent which percolates through the solid bed thereby dissolving out the solute. After a specific period of time, strong leach solution is withdrawn through the false bottom. This entire operation constitutes a single stage. The operation may be repeated with fresh batches of solvent until the solute content of the solids is reduced to an economic minimum. Semibatch operation is also possible in which the solvent is continuously admitted from the top and the strong solution is continuously withdrawn through the false bottom. Such an operation is equivalent to many stages. Occasionally, the solvent is sprayed over the top of the solids and allowed to trickle down through the bed without fully immersing the solids (trickle bed operation). Upward flow of solvent is also often used, but this could cause excessive entrainment of fines in the overflow liquid. Closed percolation tanks, often called diffusers, are operated under pressure, the solvent being pumped through the bed of solids (instead of flow under gravity). Closed tanks are particularly useful when the solvent employed is highly volatile or when operating temperatures that are much above the normal boiling point of the solvent are desired to increase the extraction rate. For example, for the recovery of tannin from tree barks, leaching is performed with water at 120° C and 3.45 atm pressure in closed percolation tanks.

Batch or semibatch crosscurrent operation of percolation tanks described above is often insufficient to produce strong leach solutions. The operation may be made close to continuous by using a number of percolation tanks in series and maintaining countercurrent flow of solvent. As stated under liquid-liquid extraction, countercurrent operation produces the strongest leach solution (richest extract). A typical example in this connection is what is called the *Shanks system.* In such a system, a number of percolation tanks are connected in series, but while $(n - 1)$ tanks are in operation, the remaining one tank is being emptied (since it contains completely leached solids) and filled with fresh solids. In this way, though the system operates in the countercurrent mode, movement of solids from tank to tank is avoided. For example, consider five percolation tanks in series as shown in Fig. 6.24. At any instant, tank 5 is disconnected and is being emptied and then filled with fresh solids. The fresh solvent is admitted to tank 1, the leach solution from this tank is pumped (or flows under gravity) to tank 2, that from tank 2 to tank 3 and so on. The final leach solution is collected from tank 4. In the next cycle (the cycle time is prefixed in advance), tank 1 is disconnected for emptying and refilling. Fresh solvent enters tank 2 and the final leach solution is discharged from tank 5. In the third cycle similarly, tank 2 is disconnected while the other four tanks are in operation. This procedure is continued. To note that fresh solids and fresh solvent are continuously added to the system and spent solids and strong leach solution are continuously discharged. Thus, the operation is, in essence, continuous. The only difference is that all the tanks are not in operation at a time.

Continuous percolators, therefore, utilise the principle of the Shanks system. To improve the continuity of the operation, they operate as *moving bed systems.* A good example is the *Bollman extractor* (or *Hansa-Mühle extractor*) that is popularly used for the extraction of oil from vegetable seeds (such as soybeans). It consists of a series of buckets with perforated bottoms suspended on a pair of endless chains that are driven by sprocket wheels. On that side of the equipment where the buckets are moving upward, fresh solvent is sprayed onto the bucket near the top (which contains almost completely leached

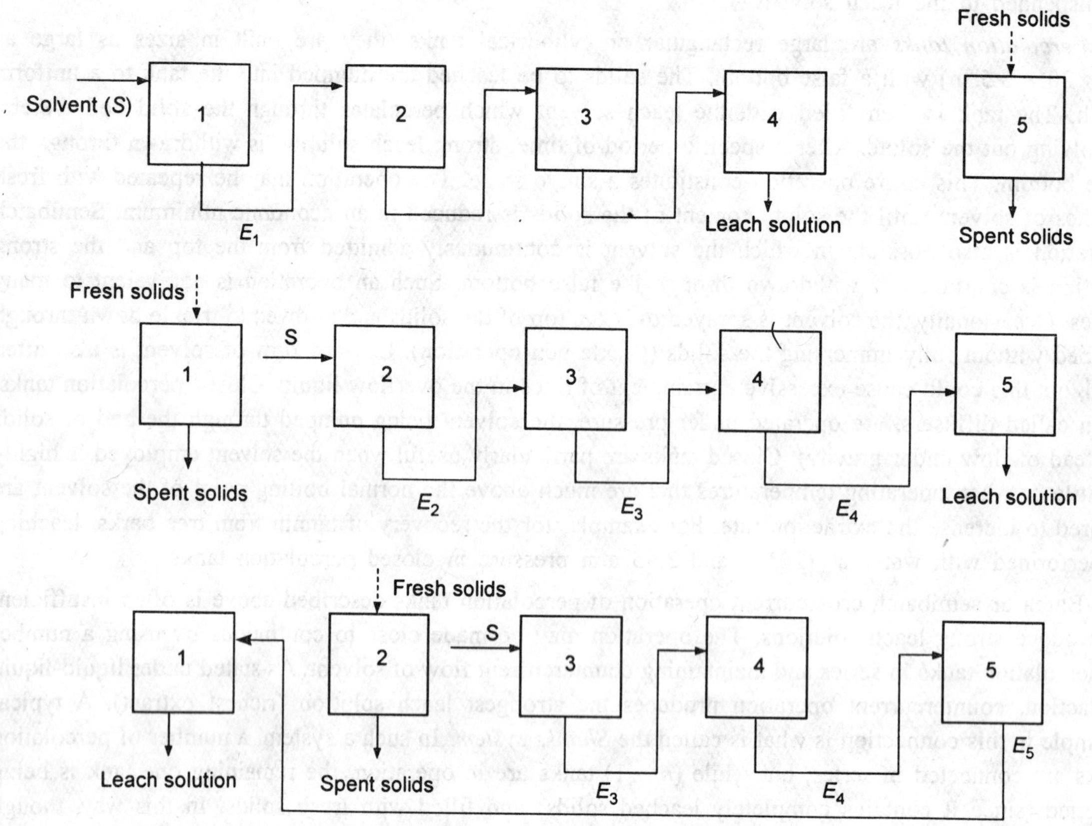


*Figure 6.24: Schematic of Shanks system*

solids). The solvent percolates through the solids, then through the perforated bottom of the bucket and flows down to the bucket below. The leach solution thus flows down from bucket to bucket until it reaches the bottom of the unit where it collects in sump *A* as *half miscella* and is pumped to the intermediate storage tank. A countercurrent multiple-stage operation is thus obtained on this side of the equipment. On the other side of the equipment where the buckets move downward, fresh solids are fed into each descending bucket from the top. The half miscella from the intermediate storage tank is sprayed onto the freshly charged bucket and as the buckets move downward, the leach solution also flows down from bucket to bucket. When it reaches the bottom, it is collected in another sump *B* as the final miscella (final leach solution) from where it is pumped to storage. The operation on this side of the equipment is thus cocurrent, multiple-stage. When a bucket reaches the top of the unit on the ascending side, it is automatically inverted and the leached solid are dumped to the discharge hopper from where they are taken to driers by screw/paddle conveyors. The entire apparatus is enclosed in a vapour-tight housing to prevent loss of solvent vapours. Since the solids are unagitated and the final miscella moves cocurrently, the Bollman extractor permits use of thin flakes while producing extract of good clarity. However, it is a partially countercurrent device and often permits channeling and consequently leads to low stage efficiency.

Another example is the *Rotocel extractor* which also follows the principle of the Shanks system, but the leaching tanks or cells each with a hinged screen bottom are revolved around a central axis using a circular rotor. These cells revolve above a stationary compartmented tank. As the rotor revolves, the cells successively pass a feed point (where the cell is fed with fresh solids by means of a special feeding device), a series of solvent/solution sprays, a drainage section (where the final leach solution is withdrawn) and a discharge section (where the leached solids from the cell are automatically dumped into one of the compartments of the stationary tank below, from where they are continuously conveyed away). The discharge section is circumferentially contiguous to the feed point. Countercurrent operation is achieved by feeding the fresh solvent only to the last cell (containing the most leached solids) just before dumping occurs. This solvent percolates through the solid and then through the screen bottom of the cell and finally flows down to one of the compartments of the tank below, from where it is continuously pumped to the spray to the preceding cell. Thus, each cell (except the last cell) is fed with the leach solution from the succeeding cell, the final leach solution being withdrawn from the freshest solids. The entire machine is enclosed in a vapour-tight housing to prevent escape of solvent vapours. It can be seen that the Rotocel resembles a rotary continuous filter in operation.

The endless belt percolator is similar in principle but the successive feed, solvent spray, drainage and dumping stations are linearly (rather than circularly) disposed. Examples are the *De Smet belt extractor* (uncompartmented) and the *Lurgi frame belt extractor* (compartmented), the latter being a linear equivalent of the Rotocel.

The *Kennedy extractor*, popularly used for oilseed and other chemical leaching operations, is also a stagewise device that operates substantially like a percolator. Here the tanks are moved, but the solids are transferred from one tank to another by a slow-moving impeller. The extractor consists of a number of tubs/chambers in series, each tub fitted with a slow-moving paddle. The impeller lifts the solids above the liquid level and dumps them into the next tub, while the solvent flows by gravity from tub to tub countercurrently to the solids movement. Perforations in the paddles permit drainage of solids between stages.

Finely divided solids that are too fine for treatment by percolation in deep percolation tanks are often filtered and leached in the filter press itself by pumping the leach solvent through the filter cake. This is the common practice followed in washing mother liquor from precipitates. Horizontal table and tilting pan vacuum filters and horizontal belt vacuum filters (that resemble endless belt extractors) are used for the leaching of vegetable seeds and beans.

*Dispersed contact systems* are popularly used for leaching fine solids. Agitated vessels and gravity sedimentation tanks (thickeners) are examples of this category. Both batch as well as continuous operation are possible in the case of agitated vessels. The solids are agitated with the leach solvent by means of a mechanical impeller (turbine, paddle or propeller) until the desired degree of leaching is obtained. The resultant slurry is then transferred to a settling tank where the nonsolute particles settle down to the bottom and get removed as the underflow sludge while the clear leach solution overflows from the top. Instead of a settling tank, a centrifuge or a pressure filter may also be used. Such a mixer-settler combination is equivalent to a single stage. Continuous multistage operation can be, therefore, achieved by connecting a number of such mixer-settler combinations in series. In addition, countercurrent flow of solvent may be employed so as to produce a strong leach solution. The scheme is then same as that sketched in Fig. 6.2 for liquid-liquid extraction.

*Pachuca tanks* that are widely used for the leaching of ores of gold, uranium and other metals employ compressed air for agitation (pneumatic agitation).

Gravity sedimentation tanks (or thickeners) can serve as continuous contacting and separating devices in which fine solids may be leached continuously. Thickeners are large-diameter, shallow tanks, usually fitted with rakes. Rakes are rotating railings with fixed vertical plates and these are positioned slightly above the tank bottom. The rakes not only help in agitating the slurry but also direct the underflow sludge towards the central discharge. A number of thickeners connected in series permit true continuous countercurrent washing of fine solids. Such a system is popularly known as a *CCD unit* (continuous countercurrent decantation unit). The unit can operate with any number of tanks, anywhere from 3 to as high as 16 are in practice. The tanks may be placed at progressively decreasing levels so that liquid can flow from one tank to the other by gravity with a minimum of pumping. It is also possible that solids be washed with different batches of solvent in a single tank. This is then equivalent to a multistage crosscurrent operation.

There are many types of equipment in which the solids are moved countercurrent to the liquid by means of screw conveyors or other mechanical devices. An example is the *rotating plate soybean extractor,* which consists of a vertical cylindrical vessel fitted with a number of horizontal circular plates equally spaced and fixed to a slowly rotating central shaft. Soybean flakes are fed continuously from the top and the solvent pumped from the bottom. The plates are slotted and scraper arms fastened to the shell scrap the surface of each plate and thereby sweep the solids through the slots. The slots are so located that the solids follow a helical path in moving downward through the unit and on reaching the base of the unit, they are discharged by a totally enclosed screw conveyor. The strong leach solution overflows through a screen at the top. (An alternate design is to fasten the scrapers to the central rotating shaft and make the plates stationary).

In general, stationary solid bed extractors (percolators) involve a minimum amount of handling and are usually prefered where large quantities of material are to be treated or where the characteristics of the solid particles are such that continuous movement of the material is undesirable. Solids that tend to form beds of low porosity (either originally or during extraction) are better treated in the dispersed state.

## 6.2.2 Methods of Analysis

Solid–liquid extraction can be analysed by following exactly the same procedure as that followed for liquid-liquid extraction. However, the analysis of solid–liquid extraction is relatively simpler, thanks to the good number of simplifications that are permissible (with allowable error) in the process analysis of the same:

(*i*) The nonsolute ($A$) is completely insoluble in $B$. Consequently, the leach solution (the extractor or the overflow) leaving any stage shall not contain any $A$. In other words, the overflow is a clear solution of $C$ in $B$. This is a fairly reasonable presumption since the overflow shall contain some $A$ only if $A$ is partially soluble in $B$ or if sufficient time has not been provided for all $A$ to settle down into the underflow.

(*ii*) The solute ($C$) is assumed fully dissolved in $B$. The underflow (leached solids or raffinate) leaving each stage contains all $A$ (insoluble solids) and a small amount of solution ($B + C$).

(*iii*) Equilibrium is assumed to have been attained between the overflow (extract or leach solution) and the underflow (raffinate or leached solids) from each stage. In other words, sufficient contact time has been provided in each stage so as to establish equilibrium between the overflow and the underflow. The concentration of $C$ in the solution ($B + C$) leaving with the underflow is equal to the concentration of $C$ in the overflow solution. In other words, $Y = X$ (more precisely, $Y^* = X$). The equilibrium values of $Y$ and $X$ will differ from each other only if a part of $C$ is adsorbed by $A$.

All the above specifications are well-satisfied in an *ideal stage*. The number of actual stages is obtained by dividing with the stage efficiency.

The analytical / graphical method of analysis of solid–liquid extraction is, therefore, exactly analogous to that described for liquid-liquid extraction. The computational load will however be much lower in the case of solid-liquid extraction. Trial and error computations will hardly be required.

Incidentally, in computations pertaining to solid-liquid extraction, it is more convenient to use $A$-free coordinates. Accordingly, the scheme of notations has been slightly modified as given below:

(*i*) $N$ represents mass fraction of $A$ on $A$-free basis (in liquid-liquid extraction, it denoted mass fraction of $B$ on $B$-free basis). That is,

$$N = (\text{mass of } A)/(\text{mass of } B + C) \quad \text{... (6.2.1)}$$

(*ii*) $X$, $Y$ represent mass fractions of $C$ on $A$-free basis (in liquid-liquid extraction, they represented mass fractions of $C$ on $B$-free basis). That is,

$$X = \frac{\left[\text{mass of } C \text{ in solution } (B + C) \text{ leaving with underflow}\right]}{\text{mass of solution } (B + C)} \quad \text{... (6.2.2)}$$

$$Y = \frac{(\text{mass of } C \text{ in overflow solution})}{\text{mass of overflow solution } (B + C)} \quad \text{... (6.2.3)}$$

Since the overflow shall not necessarily contain any $A$, $Y = y$.

(*iii*) $R'$, $E'$ represent mass flow rates on $A$-free basis (in liquid-liquid extraction they represented mass flow rates on $B$-free basis). That is, $R'$, $E'$ = mass flow rates of ($B + C$). When the overflow is free from $A$ (as is usually the case), $E' = E$. Similarly, $S' = S$.

The concentration of $C$ in the solution adhering to the insoluble solids ($A$) in the underflow (namely, $X$) is usually a function of $N_R$ (mass ratio of $A$ to $B + C$ in the underflow). $N_R$ versus $X$ data for any system is experimentally determined and is usually made available. A typical $N_R$ versus $X$ plot is shown in Fig. 6.25. This is thus analogous to the raffinate curve of liquid-liquid system*. Since no $A$ is lost in the overflow, $N_E = 0$ for all $Y$ and the extract curve, therefore, coincides with the abcissa (see Fig. 6.25).

*In solid–liquid extraction terminology, this is called the *underflow curve*.

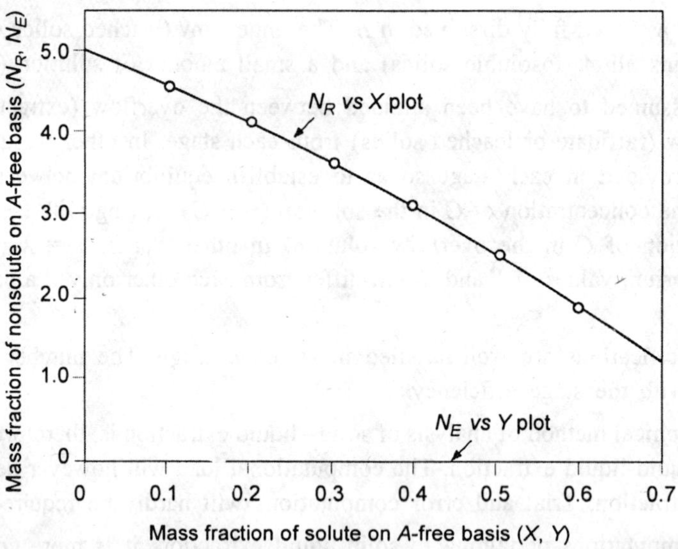


*Figure 6.25: Underflow curve ($N_R$ versus X plot) and overflow curve ($N_E$ versus Y plot) for solid–liquid extraction*

## Multistage Crosscurrent Leaching

The scheme of operation is the same as that sketched in Fig. 6.8. The mass balance equations also remain the same except that the variables now follow the revised nomenclature defined in Eqs (6.2.1) to (6.2.3). Thus, for stage $i$, the overall material balance equation is

$$R'_{i-1} + S'_i = R'_i + E'_i \quad \text{... (6.2.4)}$$

A solute balance (C-balance) gives

$$R'_{i-1}X_{i-1} + S'_iY_S = R'_iX_i + E'_iY_i \quad \text{... (6.2.5)}$$

$$= (R'_i + E'_i)\, X_i \quad \text{... (6.2.5a)}$$

Solving for $X_i$, we get

$$X_i = (R'_{i-1}X_{i-1} + S'_iY_S)/(R'_{i-1} + S'_i) \quad \text{... (6.2.6)}$$

Thus, $X_i$ can be computed from the above equation since all quantities on the right hand side of the above equation are known. To note that $R'_{i-1}$ and $X_{i-1}$ are known from computations performed on the earlier stage.

Once $X_i$ is known, $N_{Ri}$ can be obtained from the underflow curve / data ($N_R$ versus $X$ data / plot). Then

$$R'_i = G_A / N_{Ri} \quad \text{... (6.2.7)}$$

and

$$E'_i = R'_{i-1} + S_i - R'_i \quad \text{... (6.2.8)}$$

To note that since no $A$ is lost in the overflow (leach solution or extract),

$$R'_0 N_{R0} = R'_1 N_{R1} = \ldots\ldots = R'_n N_{Rn} = G_A \quad \text{... (6.2.9)}$$

In this way, the mass flow rates and compositions of all streams leaving the *i*-th stage can be estimated. Since the computations are straightforward (no trial and error solution is demanded), it is not necessary to go for a graphical solution. It may be noted that since the overflow is free from $A$, $E'_i = E$ and similarly, since the solvent also does not contain any $A$, $S'_i = S_i$. We have used the generalised scheme of notation in the above equations for sake of uniformity.

If adsorption of a small amount of $C$ on $A$ is anticipated and thereby the $Y$-values differ from $X$-values, but the equilibrium data ($Y$ versus $X$ data) is available, then computations can still be performed by solving Eqs (6.2.4) and (6.2.5) though a trial and error procedure shall become invariable in such a case. Thus, eliminating $E'_i$ between Eqs (6.2.4) and (6.2.5),

$$X_i = Y_i + (R'_{i-1}/R'_i)\,(X_{i-1} - Y_i) + (S'_i/R'_i)\,(Y_S - Y_i) \quad \text{... (6.2.10)}$$

This can be now solved for $X_i$ by trial as follows:

(*i*) Assume $X_i$. For example, let $X_i = X_{i-1} - 0.001$.

(*ii*) Put $XA = X_i$.

(*iii*) Obtain $Y_i$ from the equilibrium plot/data and $N_{Ri}$ from the underflow curve/data. Now, $R'_i = G_A / N_{Ri}$.

(*iv*) Compute $X_i$ from Eq. (6.2.10).

(*v*) If the above-computed value of $X_i$ differs substantially from $XA$ (say, by more than 0.001), then reassume $X_i$ (put $X_i = X_A - 0.001$) and repeat the computations from step (*ii*).

A graphical procedure may also be employed in cases like this, as shown in Fig. 6.26. The point $F$, that is ($X_0$, $N_{R0}$), and $Y_S$ are first marked as shown. Now, Eqs (6.2.4) and (6.2.5) may be rewritten as

$$R'_{i-1} + S'_i = M'_i \quad (6.2.11)$$

$$R'_{i-1}X_{i-1} + S'_iY_S = M'_iX_{mi} \quad (6.2.12)$$

Solving simultaneously,

$$X_{mi} = (R'_{i-1}X_{i-1} + S'_iY_S)/(R'_{i-1} + S'_i) \quad \text{... (6.2.13)}$$

Now, putting $i = 1$, $X_{m1}$ is obtained from the above equation and the point $M_1$ is marked on line $\overline{FY_S}$ corresponding to $X = X_{m1}$ as shown in figure. Now, by trial, the tie line that passes through $M_1$ is located and its point of intersection with the underflow curve and the abcissa give $X_1$ and $Y_1$. Similarly, the point $M_2$ is marked on line $\overline{X_1 Y_S}$ corresponding to $X = X_{m2}$ [the value of $X_{m2}$ having been obtained from Eq. (6.2.13) by putting $i = 2$] and then by trial, the tie line passing through $M_2$ is located. The procedure is repeated for each subsequent stage.

A special case in leaching is that of *constant underflow.* This means that the mass of insoluble solids ($A$) per kg of solution ($B + C$) in underflow stream discharged from each stage is the same. In other words,

$$N_{R1} = N_{R2} = \ldots = N_{Rn} = NR = \text{constant} \quad \text{... (6.2.14)}$$

The $N_R$ versus $X$ plot will then be a straight line parallel to the abcissa. Since $R'_i = G_A / N_{Ri}$, it follows that

$$R'_1 = R'_2 = \ldots = R'_n = R \quad \text{... (6.2.14a)}$$

Also, from Eq. (6.2.4),

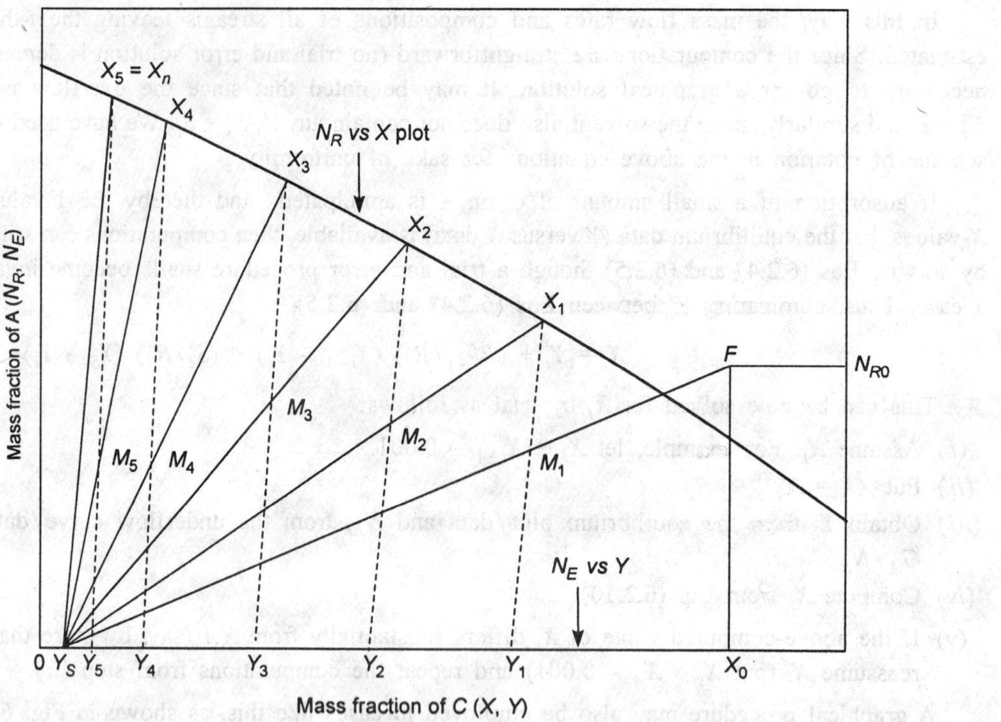


*Figure 6.26: Graphical analysis of multistage crosscurrent leaching. Clear overflow; $Y^* \neq X$ (adsorption of C on A)*

$$E_i' = S_i' \text{ for } i > 1 \quad \text{... (6.2.15)}$$

To note that $R_0'$ is the feed rate to the system and it need not be equal to $R$ and consequently, $E_1'$ also need not be equal to $S_1'$. Equation (6.2.6) therefore reduces to (for $i > 1$)

$$X_i = (RX_{i-1} + S_i'Y_S)/(R + S_i')$$

$$= \frac{(R/S_i')\,X_{i-1} + Y_S}{(1 + R/S_i')} \text{ (for } i > 1) \quad \text{... (6.2.16)}$$

If the solvent rate to each stage is the same and equal to $S'$, then

$$S' = E_2' = E_3' = \ldots.. = E_n = E \quad \text{... (6.2.17)}$$

In other words, for all values of $i$ except $i = 1$, $(R_i'/E_i') = (R/E) =$ constant. Therefore, for $i > 1$,

$$X_i = \frac{(R/E)\,X_{i-1} + Y_S}{(1 + R/E)} \quad \text{... (6.2.18)}$$

For the case of $Y_i \neq X_i$, we may rewrite Eq. (6.2.10) in the simplified form as (for $i > 1$)

$$Y_i = -(R/S_i')\,X_i + Y_S + (R/S_i')\,X_{i-1} \quad \text{... (6.2.19)}$$

$$= -m_i X_i + C_i \quad \text{(6.2.19a)}$$

where $R = G_A/(NR) = R_0'\,N_{R0}/(NR)$. This is equation to a straight line of slope $(-R/S_i')$ and $y$-intercept $C_i$ and this line passes through the point $(X_{i-1}, Y_S)$. Thus, if a straight line of slope $(-R/S_i')$ or $m_i$ is drawn from the point $(X_{i-1}, Y_S)$ then it will intersect the equilibrium curve at $(X_i, Y_i)$ as shown in Fig. 6.27. In this way, the compositions of streams leaving each stage can be computed. To note that for $i = 1$ (for stage 1), Eq. (6.2.10) reduces to

$$Y_1 = -\frac{RX_1}{\left(R_0' + S_1' - R\right)} + \frac{\left(R_0' X_0 + S_1' Y_S\right)}{\left(R_0' + S_1' - R\right)} \quad \text{... (6.2.20)}$$

$$= -m_1 X_1 + C_1 \quad \text{... (6.2.20a)}$$

This line intersects the equilibrium curve at $(X_1, Y_1)$. Once $(X_1, Y_1)$ are known, the higher values of $(X_i, Y_i)$ can be computed using Eq. (6.2.19) as described earlier.

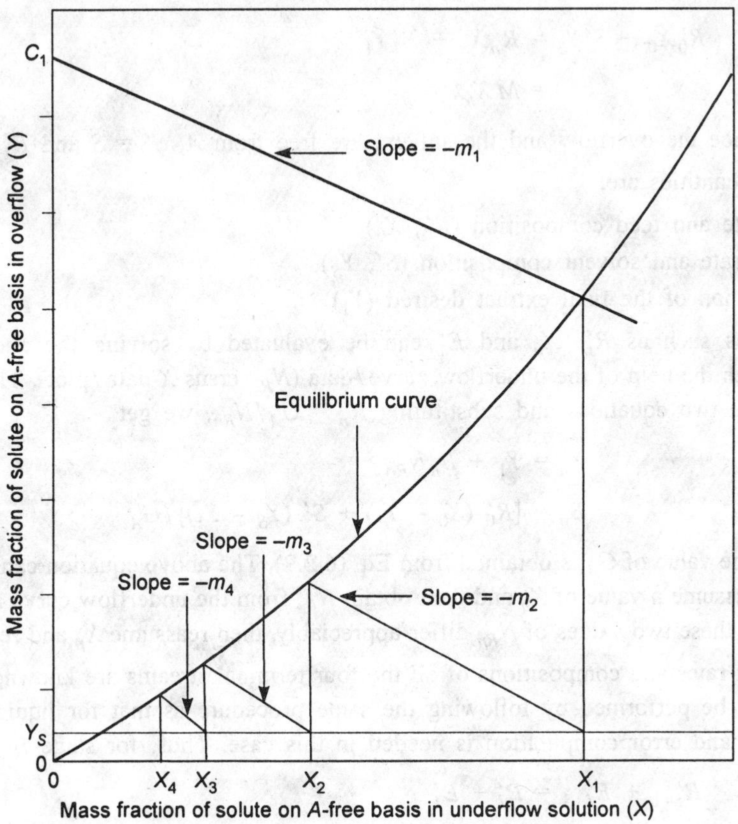


*Figure 6.27: Analysis of multistage crosscurrent leaching. Constant underflow; Clear overflow; $Y^* \neq X$ (adsorption of C on A)*

Equations (6.2.19) and (6.2.20) may also be solved analytically for $X_i$ and $Y_i$ by trial, with the help of the equilibrium plot / data. For example, assume a value of $X_i$ and then obtain $Y_i$ from Eq. (6.2.19) or (6.2.20) as well as from the equilibrium plot /data. If these two values of $Y_i$ differ substantially from each other, reassume $X_i$ and repeat the trial.

Percolation tanks, agitated vessels and washing thickeners are frequently operated in the crosscurrent mode. A single percolation tank or thickener or agitated vessel when fed with different batches of solvent is, in fact, executing a multistage crosscurrent operation, each batch constituting a stage.

## Multistage Countercurrent Leaching

Here also, the scheme of operation is dame as that sketched in Fig. 6.12. The mass balance equations are also the same except for the change in nomenclature. Thus, the overall material balance for the entire system is

$$R'_0 + S' = R'_n + E'_1 \quad \text{... (6.2.21)}$$
$$= M$$

And a solute balance ($C$-balance) for the entire system is

$$R'_0 X_0 + S' Y_S = R'_n X_n + E'_1 Y_1 \quad \text{... (6.2.22)}$$
$$= M X_m \quad \text{... (6.2.22a)}$$

Here also, since the overflow and the solvent are free from $A$, $S' = S$ and $E'_1 = E_1$

The known quantities are,

(*a*) The feed rate and feed composition ($R'_0$, $X_0$)

(*b*) the solvent rate and solvent composition ($S'$, $Y_S$)

(*c*) the composition of the final extract desired ($Y_1$)

The unknowns such as $R'_n$, $X_n$ and $E'_1$ can be evaluated by solving the above two equations simultaneously with the help of the underflow curve / data ($N_R$ versus $X$ data / plot). Thus, eliminating $E'_1$ between the above two equations and substituting $R'_n = G_A / N_{Rn}$, we get

$$X_n = Y_1 + B_n N_{Rn} \quad \text{... (6.2.23)}$$

where $$B_n = [R'_0 (X_0 - Y_1) + S' (Y_S - Y_1)] / G_A \quad \text{... (6.2.24)}$$

To note that the value of $G_A$ is obtained from Eq. (6.2.9). The above equation can be now solved for $X_n$ by trial. First, assume a value of $X_n$ and then obtain $N_{Rn}$ from the underflow curve as well as from the above equation. If these two values of $N_{Rn}$ differ appreciably, then reassume $X_n$ and repeat the trial.

Once the flow rates and compositions of all the four terminal streams are known, the stage to stage computations may be performed by following the same procedure as that for liquid-liquid extraction. However, no trial and error computation is needed in this case. Thus, for stage $i$,

$$R'_{i-1} + E'_{i+1} = R'_i + E'_i \quad \text{... (6.2.25)}$$

$$R'_{i-1} X_{i-1} + E'_{i+1} Y_{i+1} = R'_i X_i + E'_i Y_i \quad \text{... (6.2.26)}$$
$$= (R'_i + E'_i) X_i \quad \text{... (6.2.26a)}$$

Corresponding to $X_i$, obtain $N_{Ri}$ from the underflow curve / data. Then, $R'_i = (G_A / N_{Ri})$ Now, $E'_{i+1}$ is obtained from Eq. (6.2.25) and $Y_{i+1}$ from Eq. (6.2.26a). Computations are to be repeated in this way until $X_{i+1} \le X_n$.

The above procedure is for the usual case when no adsorption of $C$ occurs on $A$ and $Y_i = X_i$. In case adsorption of a small amount of $C$ does occur on $A$ and as a result $Y_i$ differs from $X_i$, then the design procedure shall remain essentially the same as above except that we have to make use of the equilibrium plot / data ($Y$ versus $X$ data / plot) as well. Thus, corresponding to $Y_i$, obtain $X_i$ from the equilibrium plot / data and then $N_{Ri}$ from the underflow curve / data. Now, compute $R'_i$ as $R'_i = G_A / N_{Ri}$. Equation (6.2.25) then gives the value of $E'_{i+1}$ and Eq. (6.2.26) the value of $Y_{i+1}$.

The system may be analysed by the graphical method as well. The graphical method (illustrated in Fig. 6.28) is less reticulate as compared to that for liquid-liquid extraction. The point $F$, whose coordinates

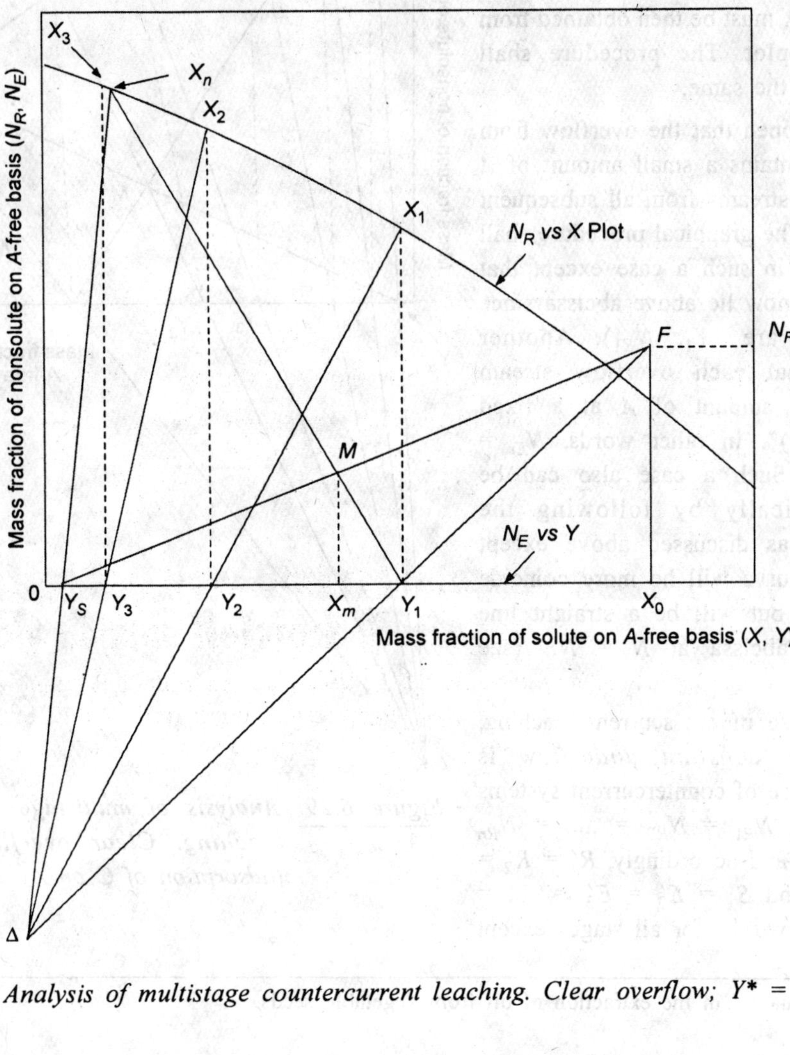


*Figure 6.28: Analysis of multistage countercurrent leaching. Clear overflow;* $Y^* = X$

are $(X_0, N_{R0})$, is first marked. $Y_S$ and $Y_1$ are also marked on the abcissa as shown. Now, from Eq. (6.2.22a),

$$X_m = (R'_0 X_0 + S'Y_S)/(R'_0 + S') \quad \text{... (6.2.27)}$$

The point $M$ is therefore marked on line $\overline{FY_S}$ corresponding to $X = X_m$. The line $\overline{Y_1 M}$ when extended intersects the underflow curve at $X_n$. Join $X_n$ to $Y_S$ and $F$ to $Y_1$. The two lines, $\overline{X_n Y_S}$ and $\overline{FY_1}$, when extended intersect at Δ, which is the operating point. Since $Y_i = X_i$, all tie lines will be vertical. Thus, a vertical through $Y_1$ intersects the underflow curve at $X_1$. Join $X_1$ to Δ and this line intersects the abcissa at $Y_2$. Proceed in this way until $X \leq X_n$.

If $Y_i \neq X_i$ (adsorption of $C$ occurs on $A$), then the tie lines will not be vertical as shown in Fig. 6.29. For each value of $Y$, the corresponding value of $X$ must be then obtained from the equilibrium plot. The procedure shall otherwise remain the same.

It may so happen that the overflow from the first stage contains a small amount of $A$ and the overflow streams from all subsequent stages are clear. The graphical procedure shall remain the same in such a case except that the point $Y_1$ will now lie above abcissa since its coordinates are $(Y_1, N_{E1})$. Another possibility is that each overflow stream contains a small amount of $A$ at a fixed ratio to $(B + C)$*. In other words, $N_{Ei}$ = $NE$ = constant. Such a case also can be analysed graphically by following the same procedure as discussed above except that the extract curve will no more coincide with the abcissa but will be a straight line parallel to the abcissa at $N = NE$ (see Example 6.9).

As in the case of crosscurrent leaching, operation under *constant underflow* is possible in the case of countercurrent systems as well. That is, $N_{R1} = N_{R2} = \ldots = N_{Rn} = NR$ = constant and accordingly, $R'_1 = R'_2 = \ldots = R'_n = R$ and $S' = E'_2 = E'_3 = \ldots = E'_n = E$. In other words, for all stages except

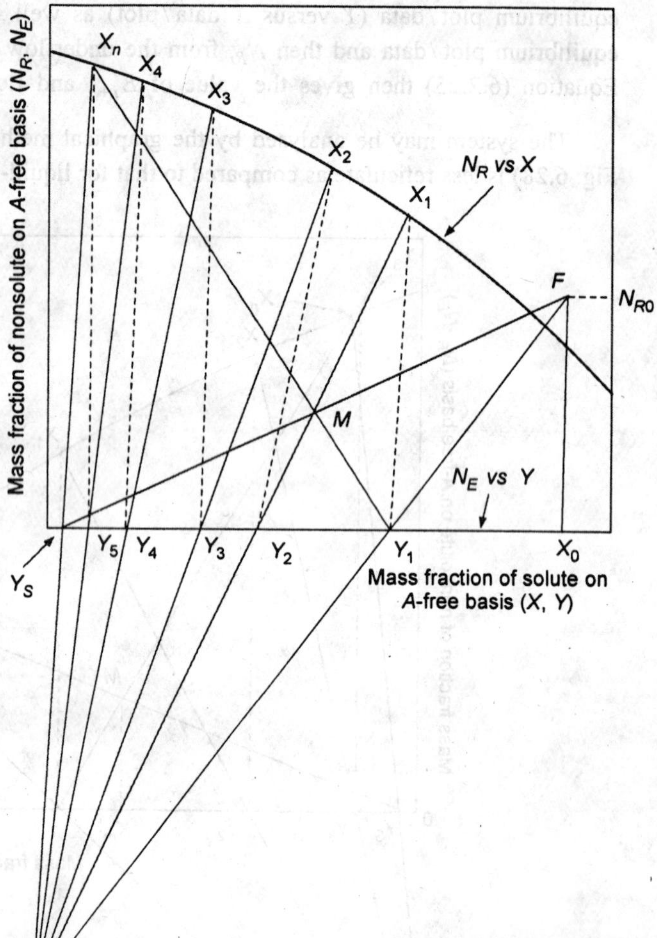


*Figure 6.29: Analysis of multistage countercurrent leaching. Clear overflow; $Y^* \neq X$ (adsorption of C on A).*

* An occasional situation in the extraction of oil from vegetable seeds.

stage 1 (for all values of $i$ greater than 1), $R_i'/E_i' = R/E$ = constant. Therefore, for $i > 1$, Eq. (6.2.26) gets simplified to

$$Y_{i+1} = Y_i + (R/E)\,(X_i - X_{i-1}) \qquad \text{... (6.2.28)}$$

This can be now easily solved with the help of the equilibrium plot. For example, for any known value of $Y_i$, obtain $X_i$ from the equilibrium plot/data; the next higher value of $Y$, namely $Y_{i+1}$, is then obtained from the above equation.

For the usual case of $Y_i = X_i$ (no adsorption of $C$ on $A$), the procedure gets still more simplified and a Kremser-Brown-Souders type correlation [see Eq. (6.1.50)] becomes applicable:

$$\frac{(X_1 - X_n)}{(X_1 - Y_S)} = \frac{(E/R)^n - (E/R)}{(E/R)^n - 1} \qquad \text{... (6.2.29)}$$

or

$$n = \frac{\log\left[1 + \dfrac{(E/R - 1.0)(X_1 - Y_S)}{(X_n - Y_S)}\right]}{\log\,(E/R)} \qquad \text{... (6.2.29a)}$$

Alternately, in terms of only mass fractions,

$$(n-1) = \frac{\log\left[(Y_S - X_n)/(Y_2 - X_1)\right]}{\log\left[(Y_S - Y_2)/(X_n - X_1)\right]} \qquad \text{... (6.2.29b)}$$

If it so happens that the value of $(R_1/E_1)$ is also equal to $(R/E)$, then the above equation can be more generalised by replacing $X_1$ by $X_0$ and $n$ by $(n + 1)$:

$$\frac{(X_0 - X_n)}{(X_0 - Y_S)} = \frac{(E/R)^{n+1} - (E/R)}{(E/R)^{n+1} - 1} \qquad \text{... (6.2.30)}$$

It is thus clear that Eq. (6.2.29) can be applied to that portion of the cascade for which $R/E$ = constant.

Even if $Y_i \neq X_i$ but the equilibrium plot is a straight line of slope $m$, Eq. (6.2.29) remains applicable except that the left hand side of the equation gets modified to $(X_1 - X_n)/(X_1 - m\,Y_S)$. An alternate situation possible is that the mass ratio of nonsolute ($A$) to solvent ($B$) remains the same in all underflow streams. In other words, the underflow streams from all stages contain a fixed ratio of $A$ to $B$. In such a case also, Eq. (6.1.29b) can be applied provided the mass fractions are replaced by the mass ratios of solute ($C$) to solvent ($B$). Thus, the correlation is

$$(n-1) = \frac{\log\left[(Y_S' - X_n')/(Y_2' - X_1')\right]}{\log\left[(Y_S' - Y_2')/(X_n' - X_1')\right]} \qquad \text{... (6.2.29c)}$$

where $\quad Y' = Y/(1 - Y)$ and $X' = X/(1 - X)$

Countercurrent operation produces much stronger leach solution than crosscurrent operation. Mixer-Settler cascades, the CCD unit, the Shanks system and most of the moving bed systems such as the Rotocel, the Kennedy extractor, the rotating plate extractor and the Bollman extractor (partially) operate in the countercurrent mode.

**Example 6.6:** A multistage extraction system is to treat 50 tonnes/hr of wet, sliced sugar beets with fresh water as solvent. The beets have the following composition:

| *Component* | *Mass fraction* |
|---|---|
| Water | 0.48 |
| Pulp | 0.40 |
| Sugar | 0.12 |

It is desired that 97 percent of the sugar in the sliced beets is recovered. It is observed that each tonne of pulp retains 3.0 tonnes of water and the leach solution (overflow) from each stage is clear. Equilibrium is attained at each stage and the compositions of the overflow solution and the solution leaving with the pulp in the underflow are the same.

(*a*) If the operation is to be conducted in the countercurrent mode and the strong solution leaving the system is to contain 0.15 mass fraction sugar, estimate the amount of water required and the number of equilibrium stages.

(*b*) Estimate the number of crosscurrent stages required if each stage is fed with 70 tonnes/hr of pure water. What will be the composition of the final, composited leach solution?

**Solution:** Here, $C$ = sugar, $A$ = pulp and $B$ = water. Given that

$$R'_0 = (50.0)\ (0.48 + 0.12) = 30 \text{ tonnes/hr}$$

$$X_0 = (0.12)/(0.60) = 0.20$$

$$G_A = (50.0)\ (0.40) = 20 \text{ tonnes/hr}$$

Since solvent is pure water, $Y_S = 0$. Also, since it is desired to recover 97 percent of the sugar in the sliced beets,

$$E'_1 Y_1 = (0.97)\ (R'_0 X_0)$$

Since $Y_S = 0$, from Eq. (6.2.22),

$$R'_n X_n = (R'_0 X_0) - (0.97)\ (R'_0 X_0)$$

$$= 0.03\ (R'_0 X_0) = (0.03)\ (30.0)\ (0.20) = 0.18 \text{ tonne/hr}$$

It is given that each tonne of pulp retains 3.0 tonnes of water. In other words, (mass of $A$)/(mass of $B$) = 1/3 = 0.333. Now,

$$N_{Ri} = (\text{mass of } A)/(\text{mass of } B + C)$$

$$= \frac{(\text{mass of } A)}{(\text{mass of } B)} \; \frac{(\text{mass of } B)}{(\text{mass of } B + C)}$$

$$= (0.333)\ (1 - X_i) = -0.333\ X_i + 0.333 \qquad \text{... (i)}$$

Thus, the plot of $N_R$ versus $X$ (the underflow curve) is a straight line. Now, since $R'_n X_n = (G_A / N_{Rn})\ X_n = (20)\ X_n / N_{Rn}$,

$$(20.0)\ X_n / N_{Rn} = 0.18$$

or
$$X_n = (0.18)\ N_{Rn}/(20.0) = \frac{(0.18)(0.333)(1.0 - X_n)}{(20.0)}$$

Solving for $X_n$, we get $X_n = 0.00299$

(*a*) For the countercurrent operation, it is given that $Y_1 = 0.15$. Since $E'_1 Y_1 = (0.97)\ (R'_0 X_0)$,

$$E'_1 = (0.97)\ (30.0)\ (0.20)/(0.15) = 38.8 \text{ tonnes/hr}$$

Also,
$$R'_n = (0.18)/(0.00299) = 60.18 \text{ tonnes/hr}$$

Therefore, from Eq. (6.2.21),

$$S = 60.18 + 38.8 - 30.0 = 68.98 \text{ tonnes/hr}$$

This is the total amount of solvent water required. Since the mass ratio of $A$ to $B$ in the underflow is constant, we can compute the number of equilibrium stages from Eq. (6.2.29c). However, to apply this equation, we must first compute $Y_2$ from the material balance equations around stage 1. That is, from Eqs (6.2.25) and (6.2.26a). Thus,

$$X_1 = Y_1 = 0.15$$

$$N_{R1} = 0.333\ (1.0 - 0.15) = 0.2833$$

$$R'_1 = G_A / N_{R1} = 20.0 / 0.2833 = 70.588 \text{ tonnes / hr}$$

Now, from Eq. (6.2.25), for $i = 1$,

$$E'_2 = (70.588) + (38.8) - (30.0) = 79.388 \text{ tonnes/hr}$$

And from Eq. (6.2.26a),
$$Y_2 = \frac{(70.588 + 38.8)(0.15) - (30.0)(0.20)}{(79.388)} = 0.1311 = X_2$$

Now, $X'_n = (0.00299)/(1.0 - 0.00299) = 0.003$, $X'_1 = (0.15)/(0.85) = 0.17647$, $Y'_2 = 0.1311/(1.0 - 0.1311) = 0.15088$ and $Y'_S = 0.0$. Substituting in Eq. (6.2.29c),

$$n - 1 = \frac{\log\left[0.003/(0.17647 - 0.15088)\right]}{\log\left[0.15088/(0.17647 - 0.003)\right]} = 15.36$$

or
$$n = 16.36$$

Readers may try to resolve this problem by the graphical method. Since the underflow curve is a straight line and all tie lines vertical, the graphical construction shall not be difficult. However, since the total number of stages is large, a wide graph paper has to be used to avoid the problem of scale.

(*b*) For the crosscurrent extraction, since $S'_i = S = 70$ tonnes/hr and $Y_S = 0$, from Eq. (6.2.6),

$$X_i = R'_{i-1} X_{i-1} / (R'_{i-1} + 70)$$

and from Eq. (6.2.8),
$$E'_i = R'_{i-1} - R'_i + 70$$

The stage to stage computations can be now easily performed using the above two equations. For example, for $i = 1$,

$$X_1 = R'_0 X_0 / (R'_0 + 70) = (30.0)\ (0.20)/(30.0 + 70.0) = 0.06$$

$$N_{R1} = (0.333)\ (1.0 - 0.06) = 0.3133$$

$$R'_1 = G_A / N_{R1} = (20.0)/(0.3133) = 63.83 \text{ tonnes/hr}$$
$$E'_1 = (30.0) - (63.83) + (70.0) = 36.17 \text{ tonnes/hr}$$

Similarly for higher values of $i$. The results are given below:

| $i$ | $R'_i$ | $E'_i$ | $X_i$ |
|---|---|---|---|
| 1 | 63.83 | 36.17 | 0.060 |
| 2 | 61.7676 | 72.062 | 0.0286 |
| 3 | 60.8153 | 70.952 | 0.0134 |
| 4 | 60.3763 | 70.439 | 0.00623 |
| 5 | 60.174 | 70.2026 | 0.002886 |

Since $X_5$ is only slightly less than $X_n$, the number of equilibrium stages required can be taken as 5. The composition of the final composited leach solution will be,

$$Y_f = \Sigma E'_i Y_i / \Sigma E'_i = \Sigma E'_i X_i / \Sigma E'_i = (5.8233)/(319.8256) = 0.0182$$

It can be seen that in crosscurrent extraction, the desired separation is affected in five stages, but it demands 350 tonnes/hr of solvent (almost five times that for countercurrent extraction) and the final leach solution is much more dilute.

**Example 6.7:** The causticizing of soda ash follows the reaction

$$Na_2CO_3 + Ca(OH)_2 \rightarrow 2NaOH + CaCO3 \downarrow$$

After the reaction is complete, the product liquor containing 2 kg of $CaCO_3$ per kg of NaOH-free water is fed continuously at the rate of 1000 kg/hr (1 tonne/hr) to a series of thickeners in which it is washed countercurrently with neutral water. It is desired to recover 95 percent of NaOH and the final solution obtained from the system is to contain 0.128 mass fraction NaOH.

(*a*) Compute the amount of neutral water required and the number of thickeners if the thickened sludge from each thickener contains 40 percent solids. The overflow from each thickener is clear (contains no $CaCO_3$) and equilibrium is attained in each thickener such that the overflow solution and the solution in the underflow have the same composition.

(*b*) How many number of thickeners will be required if the amount of solution retained in the underflow is a function of the composition of the solution as given below:

| *kg of NaOH per kg of solution* | *kg of solution per kg of calcium carbonate* |
|---|---|
| 0.0 | 1.50 |
| 0.05 | 1.75 |
| 0.0917 | 2.02 |
| 0.15 | 2.70 |
| 0.20 | 3.60 |

Estimate also the amount of wash water required.

(*c*) The settling characteristics of the slurry as reported by Armstrong and Kammermeyer[14] indicate a degree of adsorption of solute on the solid. As a result, the compositions of the overflow solution and the solution discharged in the underflow differ as tabulated below:

| Mass fraction of NaOH in solution of settled sludge | Mass fraction of NaOH in clear overflow solution |
|---|---|
| 0.0917 | 0.090 |
| 0.0762 | 0.070 |
| 0.0608 | 0.0473 |
| 0.0452 | 0.033 |
| 0.0295 | 0.0208 |
| 0.0204 | 0.01187 |
| 0.01435 | 0.0071 |
| 0.01015 | 0.0045 |

**Compute the number of thickeners required based on the above data.**

**(*d*) Estimate the number of thickeners required and the total solvent requirement if a crosscurrent operation is employed and each thickener is fed with 2500 kg/hr of neutral water. Equilibrium data of part (*c*) hold good and the thickened sludge from each thickener contains 40 percent solids as in part (*a*).**

**Assume that during causticizing, the reactants are used in the stoichiometric proportions and the reaction has gone to 100 percent completion.**

**Solution:** In this example, $C$ = NaOH, $B$ = water, $A$ = $CaCO_3$. Given that

$$Y_S = 0 \text{ (since pure water is used as solvent)}$$

$$Y_1 = 0.128$$

Since the feed contains 2.0 kg of $CaCO_3$ per kg of NaOH-free water,

$$(\text{mass of } A)/(\text{mass of } B) = 2.0$$

Since reactants were used in stoichiometric proportions (none of them in excess) and the reaction is 100 percent complete, the product liquor will contain 2 moles ($2 \times 40 = 80$ kg) of NaOH per 1 mole (100 kg) of $CaCO_3$. Thus

$$\frac{\text{Mass of NaOH}}{\text{Mass of } CaCO_3} = \frac{\text{Mass of } C}{\text{Mass of } A} = (80/100) = 0.8$$

Thus, if $B$ = 1 kg, $A$ = 2 kg and $C$ = 1.6 kg. Therefore,

$$X_0 = 1.6/2.6 = 0.61538$$

$$N_{R0} = 2.0/2.6 = 0.76923$$

$$R_0' = (2.6/4.6)(1000) = 565.217 \text{ kg/hr}$$

$$G_A = (2.0/4.6)(1000) = 434.7826 \text{ kg/hr}$$

Since 95 percent of NaOH is to be recovered,

$$E_1'Y_1 = 0.95\,(R_0'X_0)$$

$$E_1'\,(0.128) = (0.95)\,(565.217)\,(0.61538)$$

or

$$E_1' = 2581.5 \text{ kg/hr}$$

(*a*) Since the underflow sludge from each thickener contains 40 percent solids,

$$N_R = 40/60 = 0.666 = \text{constant.}$$

Also, $R'_1 = R'_2 = ..... = R'_n = R$ and

$$R = G_A/N_R = (434.7826)/(0.666) = 652.1739 \text{ kg/hr}$$

Since $Y_S = 0.0$ and $E'_1 Y_1 = 0.95\ (R'_0 X_0)$, from Eq. (6.2.22),

$$R'_n X_n = R X_n = 0.05\ (R'_0 X_0)$$

Therefore, $X_n = (0.05)\ (565.217)\ (0.61538)/(652.1739) = 0.0266$

Since this is the situation of constant underflow, we may apply Eq. (6.2.29). However, this equation is not applicable to stage 1 ($i = 1$). Thus, for $i = 1$, from Eq. (6.2.25),

$$E'_2 = (652.1739) + (2581.5) - (565.217) = 2668.457 \text{ kg/hr}$$

Since $S = E'_2 = E$,

$$S = 2668.457 \text{ kg/hr}$$

This is the total amount of neutral water required. Now,

$$E/R = (2668.457)/(652.1739) = 4.09$$

Therefore, from Eq. (6.2.29a),

$$n = \frac{\log\left[1.0 + (3.09)(0.128)/0.0266\right]}{\log(4.09)} = 1.96 \simeq 2.0$$

Two thickeners are, therefore, required to affect the desired separation.

(*b*) From the data given, the data for underflow curve are prepared as given below:

| $X$ | $(1/N_R)$ | $N_R$ |
|---|---|---|
| 0.0 | 1.50 | 0.6666 |
| 0.05 | 1.75 | 0.5714 |
| 0.0917 | 2.02 | 0.495 |
| 0.15 | 2.70 | 0.37037 |
| 0.20 | 3.60 | 0.2777 |

The underflow curve is shown in Fig. 6.7.1. Now, since $R'_n X_n = 0.05\ (R'_0 X_0)$ and $R'_n = G_A/N_{Rn}$,

$$X_n = \left(\frac{0.05 R_0{}' X_0}{G_A}\right) N_{Rn}$$

$$= \frac{(0.05)(565.217)(0.61538)}{(434.7826)} N_{Rn} = 0.04 N_{Rn} \qquad \text{... (i)}$$

The value of $X_n$ can be now obtained by trial. Thus, let $X_n = 0.02465$. Then from the underflow curve/data, $N_{Rn} = 0.61625$ and from the above equation

$$X_n = (0.04)\ (0.61625) = 0.02454$$

Since this value of $X_n$ does not differ substantially from that assumed at the outset, we can discontinue further trial.

The value of $X_n$ may also be obtained graphically. For example, Eq. (i) may be rewritten as, $N_{Rn} = (25.0) X_n$. This represents a straight line passing through the origin and of slope 25.0 and this straight line will intersect the underflow curve at $(X_n, N_{Rn})$. This is what is shown in Fig. 6.7.1.

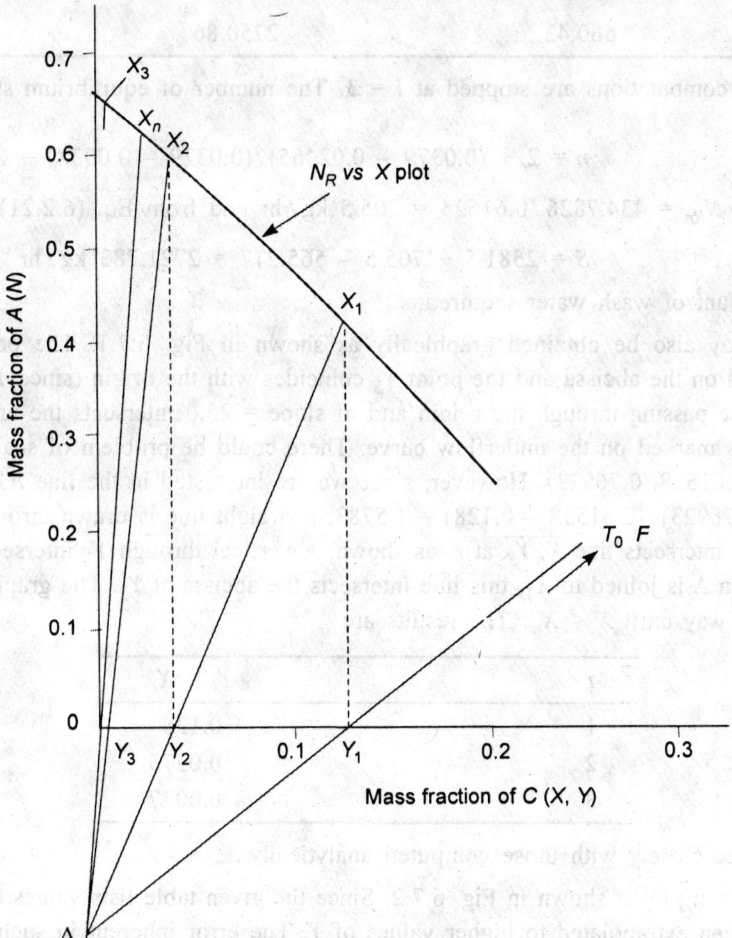


*Figure 6.7.1*

The stage to stage computations can be now performed using Eqs (6.2.25) and (6.2.26a). For example,

$$X_1 = Y_1 = 0.128$$

From underflow curve / data, $N_{R1} = 0.4174$

$$R'_1 = G_A / N_{R1} = 434.7826 / 0.4174 = 1041.638 \text{ kg/hr}$$

From Eq. (6.2.25), $E'_2 = (1041.638) + (2581.5) - (565.217) = 3057.921$ kg/hr

And from Eq. (6.2.26a), $Y_2 = 0.0379 = X_2$

Similarly for higher values of *i*. The results are summarised below:

| $i$ | $R'_i$ | $E'_i$ | $Y_i = X_i$ |
|---|---|---|---|
| 1 | 1041.638 | 2581.5 | 0.128 |
| 2 | 734.56 | 3057.921 | 0.0379 |
| 3 | 660.43 | 2750.86 | 0.0038 |

Since $X_3 < X_n$, computations are stopped at $i = 3$. The number of equilibrium stages required is, therefore,

$$n = 2 + (0.0379 - 0.02465)/(0.0379 - 0.0038) = 2.4$$

Now, $R'_n = G_A/N_{Rn} = 434.7826/0.61625 = 705.5$ kg/hr and from Eq. (6.2.21),

$$S = 2581.5 + 705.5 - 565.217 = 2721.783 \text{ kg/hr}$$

This is the amount of wash water required.

The solution may also be obtained graphically as shown in Fig. 6.7.1. The point $Y_1$, which is (0.128, 0), is marked on the abcissa and the point $Y_S$ coincides with the origin (since $Y_S = 0$). As stated earlier, a straight line passing through the origin and of slope = 25.0 intersects the underflow curve at ($X_n$, $N_{Rn}$). $X_n$ is thus marked on the underflow curve. There could be problem of scale in marking the point $F$, which is (0.61538, 0.76923). However, since we are interested in the line $\overline{FY_1}$ whose slope = $N_{R0}/(X_0 - Y_1) = (0.76923)/(0.61538 - 0.128) = 1.5783$, a straight line is drawn through $Y_1$ with slope 1.5783 and this line intersects line $\overline{X_n Y_S}$ at $\Delta$ as shown. A vertical through $Y_1$ intersects the underflow curve at $X_1$ and when $\Delta$ is joined to $X_1$, this line intersects the abcissa at $Y_2$. The graphical construction is continued in this way until $X < X_n$. The results are

| $i$ | $Y_i = X_i$ |
|---|---|
| 1 | 0.128 |
| 2 | 0.0375 |
| 3 | 0.0037 |

The results agree closely with those computed analytically.

(*c*) The equilibrium plot is shown in Fig. 6.7.2. Since the given table lists values of $Y$ less than 0.1 only, the plot has been extrapolated to higher values of $Y$. The error inherent in such an extrapolation has to be born with (since no extra data are available). From the plot, corresponding to $Y_1 = 0.128$, $X_1 = 0.1205$. Now, from the underflow curve/data,

$$N_{R1} = 0.42446$$

$$R'_1 = G_A/N_{R1} = 434.7826/0.42446 = 1024.3 \text{ kg/hr}$$

From Eq. (6.2.25), $\quad E'_2 = (1024.3) + (2581.5) - (565.217) = 3040.583$ kg/hr

And from Eq. (6.2.26), $\quad Y_2 = 0.03487$

$X_2$ (from equilibrium plot) $= 0.04724$

Computations are repeated for $i = 2$ and $i = 3$. The results are

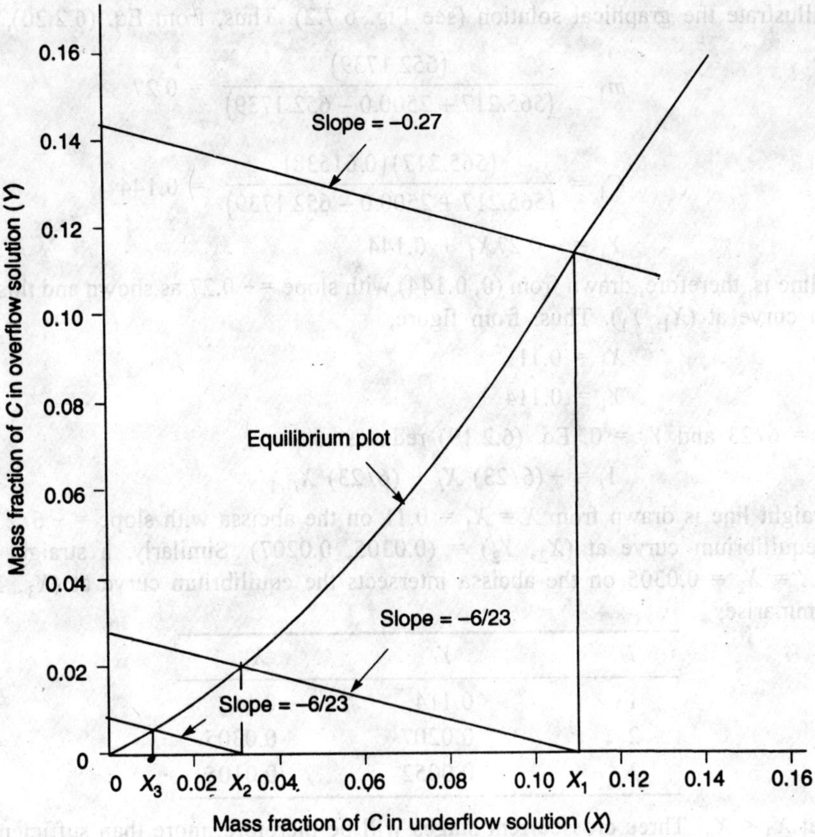


*Figure 6.7.2*

| $i$ | $R'_i$ | $E'_i$ | $Y_i$ | $X_i$ |
|---|---|---|---|---|
| 1 | 1024.3 | 2581.5 | 0.128 | 0.1205 |
| 2 | 768.6 | 3040.583 | 0.03487 | 0.04724 |
| 3 | 682.28 | 2784.9 | 0.006789 | 0.01385 |

The number of equilibrium stages required is, therefore, $n = 2 + (0.04724 - 0.02465)/(0.04724 - 0.01385) = 2.7$.

(*d*) The operation is crosscurrent with constant underflow. Thus

$$N_R = 0.666 = \text{constant.}$$

As determined in part (*a*), $R = 652.1739$ kg/hr

$$X_n = 0.0266$$

Given that, $S_i = S = 2500$ kg/hr

Therefore, $R/S = 652.1739/2500.0 = 0.26087$ or $(6/23)$

We shall illustrate the graphical solution (see Fig. 6.7.2). Thus, from Eq. (6.2.20),

$$m_1 = \frac{(652.1739)}{(565.217 + 2500.0 - 652.1739)} = 0.27$$

$$C_1 = \frac{(565.217)(0.61538)}{(565.217 + 2500.0 - 652.1739)} = 0.144$$

Therefore, $Y_1 = -0.27X_1 + 0.144$

A straight line is, therefore, drawn from (0, 0.144) with slope = – 0.27 as shown and this line intersects the equilibrium curve at $(X_1, Y_1)$. Thus, from figure,

$$X_1 = 0.11$$

$$Y_1 = 0.114$$

Since $R/S = 6/23$ and $Y_S = 0$, Eq. (6.2.19) reduces to

$$Y_i = -(6/23)\, X_i + (6/23)\, X_{i-1}$$

Thus, a straight line is drawn from $X = X_1 = 0.11$ on the abcissa with slope = – 6/23 and this line intersects the equilibrium curve at $(X_2, Y_2)$ = (0.0305, 0.0207). Similarly, a straight line of slope (– 6/23) from $X = X_2 = 0.0305$ on the abcissa intersects the equilibrium curve at $(X_3, Y_3)$ = (0.0105, 0.0052). To summarise,

| $i$ | $Y_i$ | $X_i$ |
|---|---|---|
| 1 | 0.114 | 0.11 |
| 2 | 0.0207 | 0.0305 |
| 3 | 0.0052 | 0.0105 |

To note that $X_3 < X_n$. Three crosscurrent stages will be therefore, more than sufficient to affect the desired separation. If three stages are used, then

$$\text{NaOH in leached solids} = R_3' X_3 = R X_3 = (652.1739)\ (0.0105)$$

Percent of NaOH lost in leached solids

$$= \frac{(652.1739)(0.0105)}{(565.217)(0.61538)} \times 100 = 1.97 \text{ percent}$$

$$\text{Percent of NaOH recovered} = (100.0 - 1.97) = 98.03 \text{ percent}$$

$$\text{Total solvent required} = 3\ (2500.0) = 7500 \text{ kg/hr}$$

This is about 2.8 times that for countercurrent operation.

**Example 6.8:** A leaching battery consists of three thickeners. The solid feed, which contains 50 percent solute and 50 percent inert solids, is divided into two halves and one half is fed to tank 1 and the other half to tank 3. The solvent is pure water and is fed to tank 2 at a rate equal to the total feed rate. The leached solids from tank 3 are fed to tank 2 and that from tank 2 to tank 1. The leach solution (overflow) from tank 2 is fed to tank 1 and that from tank 1 to tank 3. The final leach solution is collected from tank 3 and the final leached solids are discharged from tank 1. Compute the compositions of all streams entering/leaving each tank, if the underflow slurry from each contains 1 kg of solution per kg of suspended solids. The overflow from each tank is clear and no adsorption of solute on solids is reported.

**Solution:** The scheme is sketched in Fig. 6.8.1.

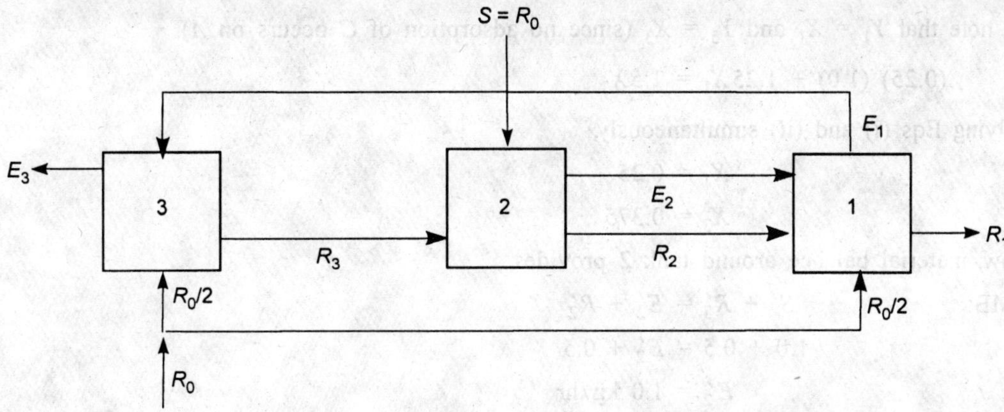


*Figure 6.8.1*

**Basis:** 1.0 kg/hr of total feed ($R_0$).

Since the feed contains 50 percent $C$ and 50 percent $A$,

$$R_0' = 0.5 \text{ kg/hr}$$

$$X_0 = 1.0$$

$$G_A = 0.5 \text{ kg/hr}$$

Also,

$$S = S' = 1.0 \text{ kg/hr}$$

$$Y_S = 0$$

Since underflow from each tank contains 1 kg of solution per kg of $A$,

$$N_R = 1.0 = \text{constant}$$

$$R_1 = R_2 = R_3 = R = G_A / N_R = (0.5/1.0) = 0.5 \text{ kg/hr}$$

An overall material balance around the whole system gives

$$R_0' + S' = R_1' + E_3'$$

$$(0.5) + (1.0) = (0.5) + E_3'$$

or

$$E_3' = 1.0 \text{ kg/hr}$$

A solute balance ($C$-balance) gives

$$R_0' X_0 + S' Y_S = R_1' X_1 + E_3' X_3$$

$$(0.5)\,(1.0) = 0.5X_1 + X_3 \qquad \text{... (i)}$$

Similarly, overall mass balance around tank 3 yields

$$(R_0'/2) + E_1' = E_3' + R_3'$$

$$0.25 + E_1' = 1.0 + 0.5$$

or $$E_1' = 1.25 \text{ kg/hr}$$

A solute balance yields

$$(R_0'/2)\ X_0 + E_1' X_1 = (E_3' + R_3')\ X_3$$

To note that $Y_1 = X_1$ and $Y_3 = X_3$ (since no adsorption of $C$ occurs on $A$).

$$(0.25)\ (1.0) + 1.25X_1 = 1.5X_3 \quad \text{... (ii)}$$

Solving Eqs (i) and (ii) simultaneously,

$$X_1 = 0.25$$

$$X_3 = 0.375$$

Now, material balance around tank 2 provides,

OMB: $$S' + R_3' = E_2' + R_2'$$

$$1.0 + 0.5 = E_2' + 0.5$$

or $$E_2' = 1.0 \text{ kg/hr}$$

$C$-balance: $$S' Y_S + R_3' X_3 = (E_2' + R_2')\ X_2$$

$$(0.5)\ (0.375) = (1.5)\ X_2$$

or $$X_2 = 0.125$$

To summarise, we may compute the total mass fraction of each constituent in the overflow/underflow stream from each tank. For example, consider underflow from tank 1:

$$\text{Solute } (C) = (0.5)\ (0.25) = 0.125 \text{ kg}$$

$$\text{Water } (B) = (0.5)\ (0.75) = 0.375 \text{ kg}$$

$$\text{Insolubles } (A) = 0.5 \text{ kg}$$

$$\text{Total} = 1.0 \text{ kg}$$

Therefore, mass fraction of $C = x_1 = 0.125/1.0 = 0.125$

mass fraction of $B = 0.375/1.0 = 0.375$

mass fraction of $A = 0.5/1.0 = 0.50$

The results are summarised below:

| | *Underflow from tank 1* | *Overflow from tank 1* | *Underflow from tank 2* | *Overflow from tank 2* | *Underflow from tank 3* | *Overflow from tank 3* |
|---|---|---|---|---|---|---|
| Mass fraction of $C$: | 0.125 | 0.25 | 0.0625 | 0.125 | 0.1875 | 0.375 |
| Mass fraction of $B$: | 0.375 | 0.75 | 0.4375 | 0.875 | 0.3125 | 0.625 |
| Mass fraction of $A$: | 0.50 | — | 0.50 | — | 0.50 | — |

**Example 6.9:** A countercurrent leaching unit is fed with 3000 kg/hr of tung meal containing 55 mass percent oil. The solvent that is admitted to the system at the rate of 12000 kg/hr contains 98 mass percent $n$-hexane and 2.0 mass percent tung oil. The leached solids discharged from the system are not to contain more than 2.0 mass percent oil (on inert-free basis).

(*a*) Estimate the number of equilibrium stages required and the percent recovery of oil if the tung meal was so finely divided that some of it goes out suspended in the overflow solution from each stage and this amounts to 0.05 kg of solids (tung meal) per kg of solution. What will be the number of actual stages required if the overall stage efficiency is 60 percent?

(*b*) Estimate the number of equilibrium stages required if the overflow from each stage contains 0.05 kg of oil-free meal per kg of solution.

The solution adhering to the insoluble meal in the underflow was determined experimentally and the results are tabulated below:

| *Mass fraction of oil in solution* | *kg solution/kg inerts* |
|---|---|
| 0.0 | 2.0 |
| 0.2 | 2.5 |
| 0.4 | 3.0 |
| 0.6 | 3.5 |

**Solution:** The given data are

$$R'_0 = (3000)\ (0.55) = 1650 \text{ kg/hr}$$

$$X_0 = 1.0$$

$$N_{RO} = 0.45/0.55 = 0.81818$$

$$Y_S = 0.02$$

$$S' = S = 12000 \text{ kg/hr}$$

$$X_n = 0.02$$

The $N_R$ versus $X$ data are as follows:

| $X$ | $1/N_R$ | $N_R$ |
|---|---|---|
| 0.0 | 2.0 | 0.50 |
| 0.2 | 2.5 | 0.40 |
| 0.4 | 3.0 | 0.333 |
| 0.6 | 3.5 | 0.2857 |

It can be seen that a plot of $1/N_R$ versus $X$ shall be a straight line of equation

$$1/N_R = 2.5X + 2.0 \qquad \text{... (i)}$$

Therefore, $$N_{Rn} = 1.0/[2.5\ (0.02) + 2.0] = 0.4878$$

(*a*) Let us solve this part by the analytical method. Since the overflow is not a clear solution, the material balance equations get modified slightly. Thus, for the entire system,

OMB: $$R'_0 + S = R'_n + E'_1 + E'_1\ (0.05)\ (0.55)$$

$$1650 + 12000 = R'_n + 1.0275 E'_1$$

or, $$R'_n + 1.0275 E'_1 = 13650 \qquad \text{... (ii)}$$

To note that the last term in the above equation accounts for solids in the overflow. Total mass of solids in overflow = (0.05) $E_1'$. On $A$-free basis, mass of solids = (0.05) $E_1'$ (1.0 – 0.45) = $E_1'$ (0.05) (0.55) = $0.0275 E_1'$.

$C$-balance: $(1650)(1.0) + (12000)(0.02) = R_n'(0.02) + E_1'Y_1 + (0.05)E_1'(0.55)$

or, $$0.02R_n' + E_1'(Y_1 + 0.0275) = 1890 \quad \text{... (iii)}$$

$A$-balance: $(1650)(0.81818) = R_n'(0.4878) + (0.05)E_1'(0.45)$

or $$0.4878R_n' + 0.0225E_1' = 1350 \quad \text{... (iv)}$$

Solving Eqs (ii) and (iv) simultaneously,

$$E_1' = 11089.033 \text{ kg/hr}$$

$$R_n' = 2256.018 \text{ kg/hr}$$

Substituting these values in Eq. (iii),

$$Y_1 = 0.13887$$

Now, consider stage $i$. The material balance equations are

OMB: $R_{i-1}' + E_{i+1}' + (0.05)E_{i+1}'(0.55) = R_i' + E_i' + (0.05)E_i'(0.55)$

or $$R_{i-1}' + 1.0275E_{i+1}' = R_i' + 1.0275E_i'$$

Similarly, the $C$-balance gets modified to

$$R_{i-1}'X_{i-1} + E_{i+1}'(Y_{i+1} + 0.0275) = R_i'X_i + E_i'(X_i + 0.0275)$$

And the $A$-balance is

$$R_{i-1}'N_{R,i-1} + 0.0225E_{i+1}' = R_i'N_{Ri} + 0.0225E_i'$$

From the above three equations,

$$R_i' = R_{i-1}'(N_{R,i-1} - 0.0219)/(N_{Ri} - 0.0219)$$

$$E_{i+1}' = (R_i' - R_{i-1}' + 1.0275E_i')/1.0275$$

$$Y_{i+1} = \frac{(R_i' + E_i')X_i - R_{i-1}'X_{i-1} + 0.0275(E_i' - E_{i+1}')}{E_{i+1}'}$$

The stage to stage computations are to be performed using the above equations. Thus, for $i = 1$,

$$X_1 = Y_1 = 0.13887$$

$$N_{R1} = 1.0/[2.5(0.13887) + 2.0] = 0.426$$

$$R_1' = (1650)(0.81818 - 0.0219)/(0.426 - 0.0219) = 3250.973 \text{ kg/hr}$$

$$E_2' = [3250.973 - 1650 + (1.0275)(11089.033)]/(1.0275)$$

$$= 12647.135 \text{ kg/hr}$$

$$Y_2 = \frac{(3250.973 + 11089.033)(0.13887) - (1650)(1.0) + 0.0275(11089.033 - 12647.135)}{(12647.135)} = 0.0236 = X_2$$

Now, for $i = 2$, $N_{R2} = 1.0/[2.5\ (0.0236) + 2.0] = 0.48567$

$$R'_2 = 2833.0 \text{ kg/hr}$$
$$E'_3 = 12240.348 \text{ kg/hr}$$
$$Y_3 = -0.006 = X_3$$

Since $X_3$ is negative, the number of equilibrium stages required is more than 1, but much less than 2. Since the overall stage efficiency is 60 percent, the number of actual stages required is more than $1/0.60 = 1.666$, but much less than $2/0.60 = 3.333$. Two actual stages should, therefore, give the required separation. The percent recovery of oil is

$$= (E'_1 Y_1 / R'_0 X_0)\ (100) = \frac{(11089.033)\,(0.13887)\,(100)}{(1650.0)\,(1.0)} = 93.3 \text{ percent}$$

(*b*) We shall give the graphical solution for this part. In this case also, the overflow is not clear, but it contains only suspended $A$ and

$$N_{Ei} = NE = 0.05 = \text{constant}$$

The graphical solution is shown in Fig. 6.9.1. The underflow curve is first plotted as shown. The overflow curve does not coincide with the abcissa but is a straight line parallel to the abcissa at $N = NE = 0.05$. The point $Y_1$ lies on this line which is to be located. Now, $X_n$ is marked on the underflow curve corresponding to $X = 0.02$ and $Y_S$ in marked on the abcissa (since the solvent does not contain any $A$) corresponding to $Y = 0.02$. There is problem of scale for marking the point $F$. Instead, the slope of line $\overline{FY_S} = N_{RO}/(X_0 - Y_S) = (0.81818)/(1.0 - 0.02) = 0.8348$. A straight line is therefore drawn through $Y_S$ with slope $= 0.8348$ as shown and this forms the line $\overline{FY_S}$. Now, from Eq. (6.2.22a),

$$X_m = \frac{(1650)\,(1.0) + (12000)\,(0.02)}{(1650) + (12000)}$$
$$= 0.1348$$

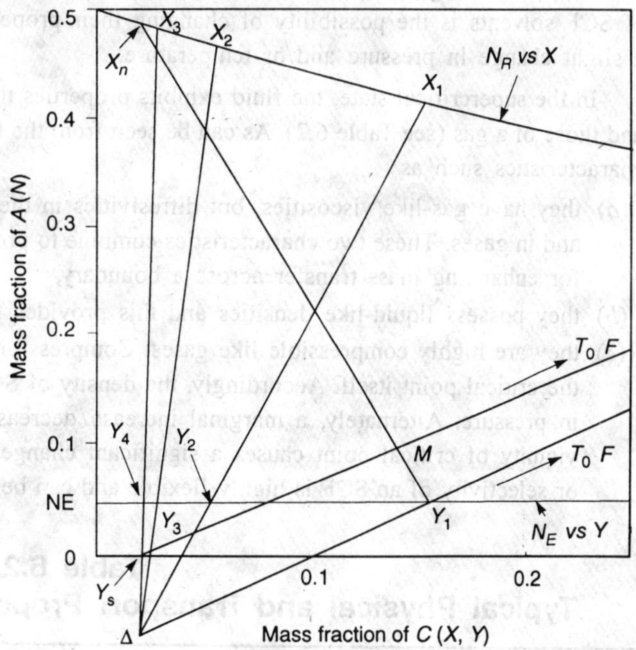


*Figure 6.9.1*

The point $M$ is thus marked on line $\overline{FY_S}$ corresponding to $X = 0.1348$. $X_n$ is joined to $M$ and this line when extended intersects the overflow curve at $Y_1$. As read from figure, $Y_1 = 0.1533$. Now, the slope of line $\overline{FY_1} = (N_{RO} - NE)/(X_0 - Y_1) = (0.81818 - 0.05)/(1.0 - 0.1533) = 0.9073$. The straight line through $Y_1$ with slope $= 0.9073$ forms line $\overline{FY_1}$ and this line when extended intersects line $\overline{X_n Y_S}$ at Δ. (To note that in this case, since $X_n = Y_S = 0.02$, the line $\overline{X_n Y_S}$ is vertical). All tie lines are vertical (assuming no adsorption of $C$ on $A$). Thus, a vertical through $Y_1$ intersects the underflow curve at $X_1$ and the line $\overline{X_1 \Delta}$ intersects the overflow curve at $Y_2$. The graphical construction is continued in this way and the results are summarised below:

| $i$ | $Y_i = X_i$ |
|---|---|
| 1 | 0.1533 |
| 2 | 0.0529 |
| 3 | 0.027 |
| 4 | 0.0216 |
| 5 | 0.0203 |
| 6 | 0.02 |

The last two points, $Y_5$ and $Y_6$, are not shown in the figure since they crowd too much. The number of equilibrium stages required is, therefore, 6.

## 6.3 SUPERCRITICAL EXTRACTION

Supercritical fluids are gases or liquids above their critical temperature ($T_C$) and critical pressure ($P_C$). When a supercritical fluid* is used as the extraction solvent, it is possible to separate the feed mixture on the basis of the difference in component volatilities (as that in distillation) or on the basis of difference in component solubilities in the SCF solvent (as that in solvent extraction). The most important advantage of SCF solvents is the possibility of changing their properties (density, viscosity) over wide limits by a slight change in pressure and/or temperature.

In the supercritical state, the fluid exhibits properties that are intermediate between those of a liquid and those of a gas (see Table 6.2). As can be seen from the table, supercritical fluids possess some unique characteristics such as

(*a*) they have gas-like viscosities, but diffusivities in them are intermediate to those found in liquids and in gases. These two characteristics combine to produce a solvent with ideal transport properties for enhancing mass transfer across a boundary,

(*b*) they possess liquid-like densities and this provides them a high capacity for solutes,

(*c*) they are highly compressible like gases. Compressibility of an SCF, in fact, approaches infinity at the critical point itself. Accordingly, the density of SCF changes substantially with a slight change in pressure. Alternately, a marginal increase/decrease in temperature at constant pressure in the vicinity of critical point causes a significant change in density. As a result, the solvent capacity or selectivity of an SCF is highly flexible and can be easily adjusted to different values as desired.

### Table 6.2
### Typical Physical and Transport Properties of Supercritical Fluids

| *Property* | *Gas* | *SCF* | *Liquid* |
|---|---|---|---|
| Density, $kg/m^3$ | 1.0 | 300.0 | 1000.0 |
| Diffusivity, $cm^2/s$ | 0.1 | 0.001 | $0.5 \times 10^{-5}$ |
| Viscosity, $kg/(m \cdot s)$ | 0.00001 | 0.00001 | 0.001 |

*Values of critical constants for selected substances are listed in the Appendix of Volume 1 (see Table A.6).

Apart from its unique solvent characteristics, the zero surface tension of SCF helps in improving the rate of mass transfer, particularly in mass transfer through micropores.

Supercritical extraction is usually conducted at temperatures that are 1.0 to 10 percent higher than the critical temperature ($T = 1.01$ to $1.10$ times $T_C$) and at pressures that are 1.0 to 50.0 percent higher than the critical pressure ($P = 1.01$ to $1.5$ times $P_C$). Extraction may be carried out in a single stage or in multiple stages, in cocurrent or countercurrent mode with or without reflux. Single stage operation is used for systems with large separation factors and is carried out at a fixed temperature and pressure. In a multistage operation, either the pressure is varied in cascades or the temperature is continuously changed along the column that is operated with reflux.

Recovery of solvent from the extract is easily accomplished by reducing the temperature or pressure. In many cases, the pressure is reduced either to below the critical pressure or to a value between the critical pressure and extraction pressure, while the temperature is adjusted to a value between the extraction temperature and critical temperature. Alternately, separation can be affected by admixing the SCF isobarically with some inert gases and thereby lowering its solvent power.

Carbon dioxide is by far the most widely investigated SCF solvent as it is nontoxic, nonflammable and relatively inexpensive. Also, its low critical temperature (31.04° C) makes it particularly suitable for the extraction of heat sensitive substances such as those in food processing and pharmaceutical industries. A few examples are,

(*a*) SCF carbon dioxide is used for selectively removing caffeine from green coffee beans. During this process, none of those substances which contribute to the aroma formed during roasting are lost. The flavour and aroma of coffee thus remain unaffected.

(*b*) Denicotinisation of tabacco (reduction of nicotine content to the desired level with minimum loss of aroma) is being successfully practised by extraction with SCF $CO_2$.

(*c*) SCF $CO_2$ is used for the extraction of α-acids from hops (hop resins contain α-acids and β-acids and it is the α-acids which after isomerisation during brewing give beer its characteristic bitter taste) and as much as 99.0 percent separation has been reported.

(*d*) Removal of fatty acids and deodorisation of vegetable oils can be carried out simultaneously along with recovery of vitamin *A* and Vitamin *E* concentrates, by extraction with SCF $CO_2$ mixed with propane or a cosolvent such as ethanol.

(*e*) SCF $CO_2$ with an entrainer such as ethanol or methanol has also been successfully used for the extraction of egg yolk lipids from freeze-dried egg yolk and thereby reducing its cholesterol content.

(*f*) For the separation of citric acid from the fermentation broth, compressed carbon dioxide is dissolved into acetone solution of crude citric acid broth. Due to the antisolvent effect of $CO_2$, the residual impurities precipitate out. This process has been reported to be more economical than the conventional calcium salt precipitation and it also produces food grade citric acid of more than 99.5 percent purity.

A few other commercial applications of supercritical extraction are listed below:

1. In the petroleum industry, near-critical (NC) or supercritical propane is being used for refining of lubricating oils (for separating wax, asphalt, colour bodies).
2. In the high pressure process of polyethene manufacture, the polymer (low density polyethene) formed remains dissolved in the SCF ethylene at the operating temperature of 523 K. Downstream

the reactor, this solution is expanded to a lower pressure to precipitate the polymer and then the separated ethylene is recompressed and recycled. In this way, ethylene acts both as reactant as well as solvent for the product.

3. Supercritical water ($T_C$ = 647 K, $P_C$ = 217.6 atm) has been successfully employed in waste water treatment. Virtually, total oxidation of organics is obtained in a short contact time and all the inorganic salts, being insoluble in SCF water at 450–500° C, are almost completely precipitated. The heat generated during the process is used for preheating the feed and the high pressure steam produced for power generation.
4. The use of supercritical water for washing of coal (to reduce its mineral matter content or ash content) is being widely investigated.

Capital and operating costs are relatively high in the case of supercritical extraction processes. However, their application to technically difficult separation problems has been extremely promising since the improved process efficiency and improved product quality usually compensate for the increased operating cost.

Excellent reviews on supercritical extraction are available in references (15) and (16).

## 6.4 ADSORPTION

Adsorption is one of the oldest known unit operations though its large-scale adoption to industrial practices started in the early 1960s. As a separation process, it is analogous to liquid-liquid extraction. The feed solution or feed gas mixture is contacted with a solid adsorbent ($B$) as a result of which the solute component $C$ (also called the *adsorbate*) is selectively adsorbed on $B$ while the other component ($A$) is only sparingly adsorbed or not adsorbed at all. The solids leaving the system shall, therefore, consist of particles of $B$ with most of $C$ and little of $A$ adsorbed on them. This is thus equivalent to the extract stream of liquid-liquid extraction. The exit gas/solution (equivalent to the raffinate stream of liquid-liquid extraction) shall contain most of $A$ and a small amount of unadsorbed $C$. Since solid particles (adsorbent particles) can be easily separated from the fluid, the raffinate fluid will be free from $B$. Adsorption differs from solid–liquid extraction in the sense that adsorption involves transfer of material from fluid phase to the solid phase while in solid–liquid extraction, material is transferred from solid phase to fluid phase*.

It is necessary to distinguish between physical adsorption and chemisorption. Physical adsorption (also called van der Waals adsorption) is due to the intermolecular force of attraction between the solid molecules and the substance adsorbed. The surface of a solid represents a discontinuity of its structure. The forces acting at the surface are unsaturated. Hence, when the solid is exposed to a gas/vapour and the force of attraction between the solid molecules and the gas/vapour molecules is larger than that between gas/vapour molecules themselves, the gas/vapour molecules will form bonds with the solid and become attached. The adsorbed substance may penetrate into the solid through the pores and interstices. The process is exothermic and consequently is accompanied by evolution of large amount of heat. At equilibrium, the partial pressure of adsorbed substance will be equal to that of the contacting gas. This equilibrium is fast established. The adsorbed substance can be desorbed back by increasing the

* Single solute adsorption from liquid solutions, usually performed in mixer-filter combinations, is often termed as *contact filtration*.

temperature (*thermal swing adsorption* or TSA), by lowering the pressure (*pressure swing adsorption* or PSA) or by passing an inert purge gas. Physical adsorption is not confined to gases but is observed with liquids as well.

In *chemisorption* there is chemical interaction between the adsorbed substance and the solid adsorbent and there will be much stronger bonding between the solid and the adsorbate molecules. As a result, it is difficult to reverse the process and even if desorption is accomplished, it releases the reaction products but not the original adsorbate. Heat released during the process is much larger. While practically all gases below critical temperature could be physically adsorbed on solids, chemisorption is restricted to some chemically reactive fluids on some selected solids. Chemisorption is important in catalysis and in the analysis of catalytic reactors (see Chapter 8). In this section, we shall confine our discussion to the process of physical adsorption only.

It is thus apparent that the most important property of an industrial adsorbent is its specific surface (surface area per unit mass or unit volume). Specific surface of an adsorbent is closely related to its porosity and pore size distribution. High porosity usually corresponds to high specific surface. Typical properties and industrial applications of popular commercial adsorbents are listed in Table 6.3. It can be seen that silica gel (porosity = 0.4 to 0.5) has a specific surface of 600 to 800 $m^2/gm$ and activated carbon (porosity = 0.6) has a specific surface of 800 to 1600 $m^2/gm$. Most adsorbents do not have a fixed pore size, but exhibit a pore size distribution. For example, in a representative sample of activated carbon, the pore size varies from 14 angstroms to 60 angstroms (1 angstrom = $10^{-8}$ cm). However, the most advanced types of adsorbents such as the synthetic zeolites (also called *molecular sieves*) are characterised by a specific pore size. For example, 3*A* zeolite has a pore size of 3.0 angstroms and 5*A* zeolite has a pore size of 5.0 angstroms. Molecular sieve carbons have a narrow pore size distribution (3 to 5 angstroms). Other desirable properties of industrial adsorbents are high erosion resistance, abrasion resistance, hardness and compressive strength. Selectivity of adsorbent is defined in the same way as for solvent in liquid-liquid extraction:

$$\beta = \frac{(y^*/x)_C}{(y^*/x)_A} \qquad \text{... (6.4.1)}$$

where $y^*$, $x$ = mass fraction in the solid phase and that in the fluid phase respectively (at equilibrium)

Separation by adsorption can be based on any of the three mechanisms such as steric, kinetic or equilibrium effect. Steric separation is based on the fact that only small and specifically shaped molecules can diffuse into the adsorbent while other molecules are totally excluded. This is thus due to the molecular sieving property of the adsorbent. Steric separation is unique with synthetic zeolites because of their uniform aperture size (pore size). The two largest applications of steric separation are drying of gases and vapours (cracked gas, ethylene, butadiene, ethanol, etc.) by 3 A zeolite (zeolite with pore size = 3.0 angstroms) and separation of normal paraffins from iso-paraffins and aromatics using 5 A zeolite.

Kinetic separation is achieved by virtue of the differences in diffusion rates of different molecules into the adsorbent. Till this date, kinetic separation has been commercially successful essentially with molecular sieve carbon that has a narrow pore size distribution (see Table 6.3). Such a distribution of pores permits different gases to diffuse at different rates into the adsorbent. For example, kinetic separation has been used commercially for the separation of nitrogen from air. The separation is believed

## Table 6.3
## Properties of Adsorbents

| | *Chemical composition* | *Porosity* | *Pore size (angstroms)* | *Surface area ($m^2/gm$)* | *Bulk density, $kg/m^3$* | *Typical Industrial applications* |
|---|---|---|---|---|---|---|
| Alumina | Hydrated alumina heated to expel moisture and activated. | 0.6 | 40–140 | 250 to 360 | 850 | Drying of gases and liquids. |
| Activated bauxite | Bauxite activated by heating to 230–815° C. | 0.35 | 40–50 | ~ 200 | 850 | Drying of gases, decolourising petroleum products. |
| Synthetic zeolites (Molecular sieves) | Metal aluminosilicates specially synthesised with specific, uniform pore size. | | | | | |
| Type 3A | | 0.30 | 3.0 | 700 | 620–680 | Drying cracked gas, ethylene, ethanol, butadiene. |
| Type 4A | | 0.32 | 4.0 | 700 | 610–670 | Drying of natural gas, solvents, liquid paraffinic hydrocarbons. Removal of carbon dioxide from natural gas. |
| Type 5A | | 0.34 | 5.0 | 700 | 600–660 | Recovery of *n*-paraffins from petroleum fractions like kerosene and naphtha. |
| Type 13X | | 0.38 | 10.0 | 600 | 580–640 | Desulfurisation of petroleum fractions and natural gas, dehydration of gases and liquids, simultaneous removal of carbon dioxide and water. |

(*Contd.*)

**Table 6.3** *(Contd.)*
**Properties of Adsorbents**

| *Chemical composition* | | *Porosity* | *Pore size (angstroms)* | *Surface area ($m^2/gm$)* | *Bulk density, $kg/m^3$* | *Typical Industrial applications* |
|---|---|---|---|---|---|---|
| Silica gel | Gel precipitated by acid treatment of sodium silicate solution. | 0.4–0.5 | 20–50 | 600 to 800 | 700–820 | Dehydration of air and other gases, fractionation of hydrocarbons. In gas masks. |
| Clay, acid treated | Bentonite or other clays activated by treating with sulfuric or hydrochloric acid. | 0.33 | — | 180 to 380 | 850 | Removal of colouring matter from petroleum oils, lub oils and food products. |
| Fullers earths | Magnesium aluminium silicates activated by heating and drying. | 0.35 | — | 250 | 800 | Decolourising, drying and neutralising petroleum fractions (kerosene, gasoline, lub oils), vegetable and animal oils. |
| Carbons | Products of destructive distillation (carbonisation) of coconut shells, wood, peat, lignite, coal or coking of petroleum hydrocarbons. Activated by partial oxidation with hot air or steam. | | | | | Decolourising sugar solution, drugs and other industrial chemicals, Water purification, Refining of vegetable and animal oils, Recovery of solvent vapours from gas mixtures, fractionation of hydrocarbon gases. |
| Shell-based | | 0.60 | 14–60 | 800–1600 | 450–550 | |
| Wood-based | | 0.80 | 14–60 | 800–1800 | 250–300 | |
| Peat-based | | 0.55 | 10–40 | 800–1600 | 300–500 | |

*(Contd.)*

**Table 6.3** (*Contd.*)
**Properties of Adsorbents**

| | *Chemical composition* | *Porosity* | *Pore size (angstroms)* | *Surface area ($m^2/gm$)* | *Bulk density, $kg/m^3$* | *Typical Industrial applications* |
|---|---|---|---|---|---|---|
| Lignite-based | | 0.7–0.85 | 10–40 | 400–700 | 400–700 | |
| Bituminous coal-based | | 0.6–0.8 | 20–40 | 900–1200 | 400–600 | |
| Petroleum-based | | 0.80 | 14–60 | 900–1300 | 450–550 | |
| Molecular sieve carbon | Specially synthesised activated carbon with a narrow pore size distribution. | 0.4 | 3 to 5 | 700 | 600–620 | Separation of nitrogen from air, fractionation of organic compounds (alcohols, aldehydes, ketones, acids). |
| Organic polymers | Synthetic polymers/resins. | | | | | |
| Polystyrene | | 0.4–0.5 | 40–90 | 300–700 | 640 | Removal of nonpolar organics (like phenol) from aqueous solutions, recovery of antibiotics. |
| Polyacrylic ester | | 0.5–0.55 | 100–250 | 150–400 | 650–700 | Purification of pulping waste liquors, recovery of antibiotics. |
| Phenolic resin | | 0.45 | — | 80–120 | 420 | Decolourising and deodourising of solutions. |

to be achieved as a result of the slight difference in the kinetic diameters of nitrogen and oxygen which results in a relatively high diffusivity for oxygen. Separation of carbon dioxide from methane or from hydrocarbons is also possible using molecular sieve carbon as a result of the relatively high diffusivity of carbon dioxide.

Large majority of industrial processes involve equilibrium adsorption of the mixture and hence are equilibrium separation processes. The process analysis presented in the subsequent paragraphs is based on adsorption as an equilibrium separation process.

As in extraction, regeneration and reuse of adsorbent is crucial in adsorption. Since adsorption is an exothermic process, it follows that an increase in temperature would diminish the rate of adsorption but enhance the rate of desorption. Consequently, the adsorbate can be desorbed off and the adsorbent regenerated by heating the adsorbent to high temperature. This is what is called *thermal swing adsorption* (TSA). In some industrial applications, the adsorbed organic matter is burnt off from the surfaces of the adsorbent particles. In TSA, each heating–cooling cycle takes a few hours to over a day. Also, high temperatures could damage the products or the adsorbent itself. As a result, TSA is being used almost exclusively for purification purposes in which the amounts of adsorptive gases/vapours being processed are small. The most rapidly growing process is *pressure swing adsorption* (PSA). In this case, the adsorbent is regenerated by lowering the pressure. The operation permits large throughput since rapid cycles, usually in minutes or seconds, are possible. Often, an inert purge gas (like low pressure steam) is passed over the adsorbent which reduces the partial pressure of the adsorbate over the solid and as a result, the adsorbate gets desorbed off into the purge gas and is carried off. Desorption may also be achieved with the help of a solvent in which the adsorbate is more soluble (the process usually termed as *elution*). For example, synthetic polymer adsorbents are frequently regenerated by leaching with low molecular weight alcohols or ketones. Elution is a common method of desorption in chromatographic separations.

*Adsorption chromatography* or *migrational chromatography* is widely used for the separation and analysis of gaseous mixtures and liquid solutions*. The equipment (called the *chromatograph*) consists of a stationary phase (a stationary bed of adsorbent $B$) into which a small amount of the feed (feed solution or feed gas mixture) consisting of $A$ and $C$ is introduced. Both $A$ and $C$ get adsorbed on the top portion of the bed, $C$ being more strongly adsorbed. Now, an elutant is passed through the bed whereupon $A$ gets desorbed into the elutant more readily than $C$. The concentration of $A$ in the elutant therefore increases reaching a maximum value and then starts decreasing when $A$ starts getting adsorbed on the lower part of the bed. Desorption of $C$ starts a little later and the concentration of $C$ in the elutant also first increases to a maximum and then starts decreasing as $C$ also gets adsorbed on the lower part of the bed. Both $A$ and $C$ get desorbed again, only to get readsorbed on still lower portion of the bed. This process continues and it produces concentration waves of $A$ and $C$ (often called adsorption waves or bands) that travel down the bed, that of $A$ moving faster (see Fig. 6.30). The shapes of the waves could change as they pass down the column. Ultimately, on reaching the bottom of the bed, $A$ exits first leaving most of $C$ behind which exits later. If the selectivity of the adsorbent is quite large and the height of each adsorption/desorption zone is small compared to the total height of the bed, a sharp separation is possible.

*Conventionally, all chromatographic methods are fixed bed processes. As a result, the term chromatography is used synonymously with fixed bed adsorption.

This method may also be used for analysing gas mixtures and liquid solutions. Depending on the chemical nature of the solutes, the adsorption bands could appear in different colours. It is owing to this, the process was originally given the name *chromatography* (the term that is still being used).

In *partition chromatography,* the adsorbent is a true liquid, insoluble or nonvolatile with respect to the feed (substrate), contained in the pores of a granular solid (which acts as the supporting material and is relatively inert). The process is that of absorption if the feed is a gas (the solute component *C* being selectively absorbed by the adsorbent liquid *B*, while the nonsolute component *A* remains behind in the feed/substrate) and extraction if the feed is a liquid (the solute component *C* being selectively extracted by the adsorbent liquid *B*). These processes may therefore be called partition absorption and partition extraction chromatography respectively.

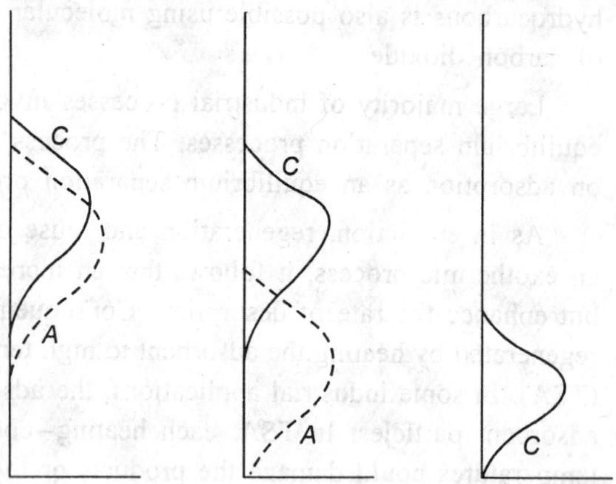


*Figure 6.30: Adsorption waves in migrational chromatography*

*Affinity chromatography* is based on the fact that many molecules such as enzymes, proteins etc. tend to form complexes with other molecules. If these molecules (ligands) are attached to a suitable insoluble matrix, they can adsorb only specific substances from a solution and shall exclude all other components of the solution. This technique is widely used in bioengineering applications.

The process of *ion exchange* is usually termed as a modified form of adsorption process. Precisely speaking, it is not wrong to call so. However, ion exchange involves adsorption of ions (and not molecules). Instead of a single-way adsorption, it involves exchange of ions between the ion exchange resin (the equivalent of adsorbent of conventional adsorption) and the feed solution (substrate). The simplest example of ion exchange process is the softening of water using zeolite. Zeolite exchanges its $Na^+$ ions with the $Ca^{++}$ ions of hard water. Since the $Ca^{++}$ ions thus get immobilised in the zeolite matrix in exchange of $Na^+$ ions of zeolite which go into solution, water gets softened. The process may be represented as

$$Ca^{++} + Na_2R \rightleftharpoons CaR + 2Na^+$$

where *R* represents the residual material of zeolite. The process is reversible and after getting saturated with $Ca^{++}$ ions, zeolite can be regenerated by contacting with sodium chloride solution:

$$CaR + 2NaCl \rightleftharpoons Na_2R + CaCl_2$$

At present, a large number of synthetic ion exchange resins are in commercial use. They are available as porous spherical beads of functional polymers. A functional polymer is one in which a reactive species is chemically bonded to a polymer matrix. This reactive species may be either ionic or ionogenic (which generates ions). Depending on the nature of the functional group, the resins are classified into

(*i*) strong acid resins (cation exchange resins)
(*ii*) strong base resins (anion exchange resins)
(*iii*) weak acid resins
(*iv*) weak base resins

In strong acid and strong base resins, the functional groups are fully ionised, while the other two are practically nonionic (or sparingly ionic) in their free form but can be converted to ionic form by treating with an alkali or an acid. Both strong acid and strong base resins usually contain styrene–divinyl benzene as the base polymer. Strong acid resins are prepared by sulphonating the styrene–DVB network. The $H^+$ ion in the sulfonate group (or sulfonic acid group), $-SO_3H$, is the exchangeable cation, which can be easily exchanged with the cations in the feed solution (substrate). The strong base resin is prepared by chloromethylating the styrene–DVB network followed by amination. The resultant resin has a quaternary ammonium group, $-N(CH_3)_3Cl^-$, as the functional group and it can exchange its anion (the $Cl^-$ ion) with those in the substrate. The exchange capacity of strong acid and strong base resins do not change with the pH of the medium as they remain ionic at all pH.

Currently produced weak acid resins are essentially copolymers of divinyl benzene and acrylic acid or methacrylic acid. When a weak acid resin is treated with alkali, it forms an ionisable salt:

$$R{-}COOH + NaOH \rightarrow R{-}COONa + H_2O$$

The resulting salt can exchange its cation ($Na^+$ ion) with those in the substrate. Weak acid resins are sensitive to the $p_H$ of the medium. Their exchange capacity is practically zero in strong acid solutions ($pH \le 5$) and are most active at $pH > 11$.

A variety of polymer networks are used in weak base resins. For example, type *S* resins have styrene–DVB network, which is chloromethylated and then aminated to a secondary amine (such as dimethylamine). The resulting base is nonionic in its free form, but its salt with an acid has ion pair which can exchange its anion with those in the substrate as shown below:

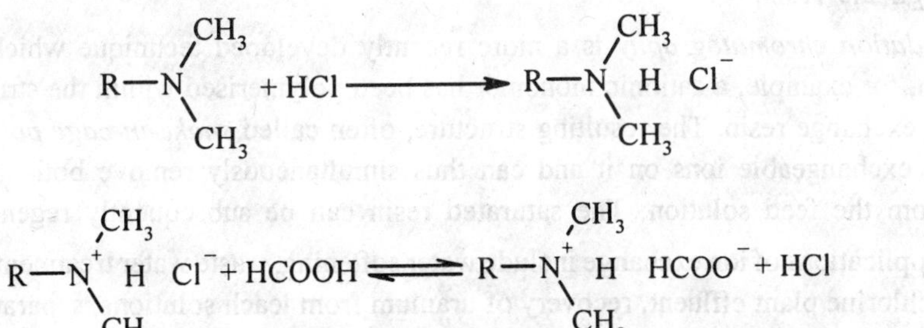

Type-*P* resins are weak base resins that use phenol-formaldehyde network while type-*A* resins contain polyacrylate network.

Apart from those discussed above, a number of ion exchange resins have been synthesised during the recent years that display unusually high selectivity for certain cations. These are called *chelating resins*. An example is the imino diacetate resin whose structural formula is given below:

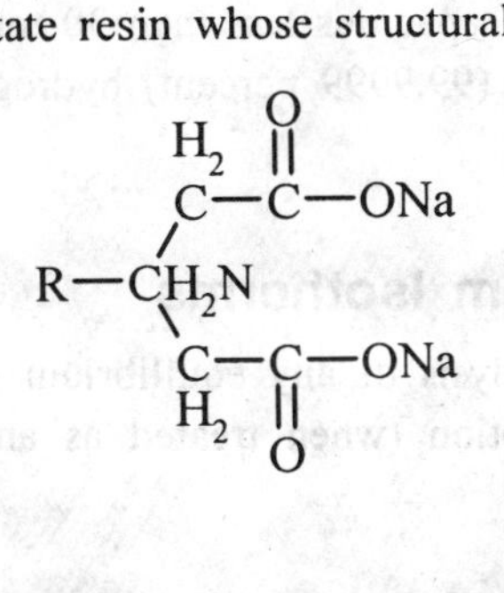

This resin has a high selectivity for divalent and trivalent cations, particularly for those of copper, nickel, cobalt and iron. They are, however, weak acid resins and ionise only at pH above 5 and hence cannot be used in those hydrometallurgical operations where acidic conditions are required. Another example is the poly isothiouronium resins (that contain isothiouronium functional group),

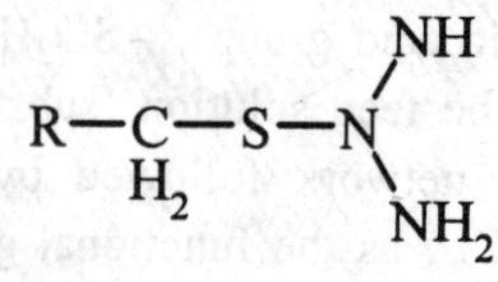

which show very high selectivities for platinum group metals and are widely used for the recovery of these metals from the dilute acid streams from plating operations. They also show good selectivity for metals like gold and mercury. These resins are however difficult to regenerate. Since the metals are expensive, they are often recovered by burning the resin. Sulfhydryl resins, that have –SH group as the functional group, chelate with mercury and are used for the recovery of mercury from industrial waste streams. It is reported that using these resins, the mercury content of effluents can be reduced to as low as 0.5 ppb (parts per billion). Phenol formaldehyde resins in sodium form have high affinity for cesium ions and are used in the nuclear industry.

Ion exchange resins are available either in the gel form or in the macroporous form. *Gel resins* are nearly homogeneous and they tend to swell when soaked in a solvent. This swelling improves the diffusivity of ions through the resin matrix. *Macroporous resins* have biporous structure consisting of both macropores (100 to 1000 angstroms in diameter) and micropores (10 to 100 angstroms in diameter). Most of the active surface of the resin is inside the micropores, while macropores facilitate diffusion of ions through the resin.

*Ion retardation chromatography* is a more recently developed technique which uses a *bifunctional resin* in which, for example, a cationic monomer has been polymerised within the structure of a previously formed anion exchange resin. The resulting structure, often called *snake-in-cage polyelectrolyte*, has both $H^+$ and $OH^-$ exchangeable ions on it and can thus simultaneously remove both the cations as well as the anions from the feed solution. The saturated resin can be subsequently regenerated with water.

Typical applications of ion exchange include water softening, waste water treatment, recovery of mercury from caustic chlorine plant effluent, recovery of uranium from leach solutions, separation and fractionation of proteins from biological solutions and recovery of antibiotics (like streptomycin) from fermentation broth.

The prominent industrial applications of adsorption have been surveyed in Table 6.3. For many difficult separations such as fractionation of air into nitrogen and oxygen, adsorption has been found to be economically superior to distillation. For example, the cost of cryogenic method (liquefaction of air and subsequent fractionation) has been reported[18] to be around 1.3 times that of pressure swing adsorption using 5 A zeolite as the adsorbent (at capacities less than 30 tonnes per day). A similar situations exists for hydrogen production. High purity (99.9999 percent) hydrogen can be produced by PSA separation of steam reformer products.

### 6.4.1 Adsorption Equilibrium Isotherms

The primary requirement for the analysis of any equilibrium separation process is the availability of equilibrium distribution data. Adsorption (when treated as an equilibrium separation process) is no exception. However, equilibrium plots (called *isotherms* since developed at a specific temperature) for adsorption do not follow a specific pattern. As many as five to seven popular varieties of adsorption isotherms are known. Four of them (that are most widely encountered) are sketched in Fig. 6.31. To note

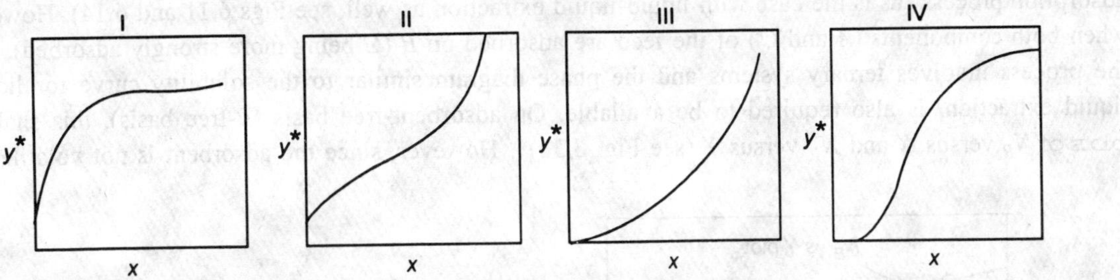


*Figure 6.31: Types of adsorption isotherms. y* = mass fraction of adsorbate (C) in adsorbent, x = mass fraction of C in fluid*

that these isotherms are for single solute adsorption. However, they can be used with confidence for multicomponent mixtures as well if only one component is strongly adsorbed and other components are very poorly adsorbed (in other words, when the adsorption of the principal solute is unaffected by the presence of other components).

A large number of mathematical correlations have been proposed by different authors to describe the above equilibrium relationships. Typical among them are those proposed by Freundlich[19], Langmuir[20] and Brunauer, Emmett and Teller[21] (BET isotherm).

The Langmuir isotherm can be represented as

$$y^* = \frac{(1 + K_L)\, x}{1 + K_L x} \quad \text{... (6.4.2)}$$

where $K_L$ is an empirical constant. This equation fits type I curve (when $K_L > 0$) and type III curve (when $-1 < K_L < 0$) of Fig. 6.31 with reasonable accuracy. The Freundlich equation (which corresponds to an exponential distribution of heats of adsorption) is of the form

$$y^* = kx^n \quad \text{... (6.4.3)}$$

where $k$ and $n$ are empirical constants. For $n < 1$, this correlation fits type I curve favourably and for $n > 1$, it fits type III curve. The BET equation, that fits type I, II and III behaviours, is of the form

$$y^* = \frac{K' x}{\left[1 + (K' + 1)\, x\right](1 - x)} \quad \text{... (6.4.4)}$$

where $K'$ is an empirical constant. Incidentally, all the four types of curves of Fig. 6.31 can be fitted into a trinomial of the form

$$y^* = a_3 x^3 + a_2 x^2 + (1 - a_2 - a_3)\, x \quad \text{... (6.4.5)}$$

such that $-2 < a_3 < (1 - a_2)$, where $a_2$ and $a_3$ are correlation constants. It may be noted that according to Langmuir equation, a plot of $(1/y^*)$ versus $(1/x)$ shall be a straight line, while Freundlich equation predicts that a plot of $y^*$ versus $x$ on log-log coordinates shall be a straight line. As per BET equation, a plot of $1/[y^* (1 - x)]$ versus $(1/x)$ must yield a straight line.

For single solute adsorption (when adsorption of all other components is negligibly small), an equilibrium plot/relationship of the type discussed above shall be sufficient for the analysis of the adsorption process (as is the case with liquid-liquid extraction as well, see Figs 6.11 and 6.14). However, when both components ($A$ and $C$) of the feed are adsorbed on $B$ ($C$ being more strongly adsorbed), then the process involves ternary systems and the phase diagram similar to the solubility curve for liquid-liquid extraction, is also required to be available. On adsorbent-free basis ($B$-free basis), this shall be plots of $N_R$ versus $X$ and $N_E$ versus $Y$ (see Fig. 6.32)*. However, since the adsorbent is not volatile and

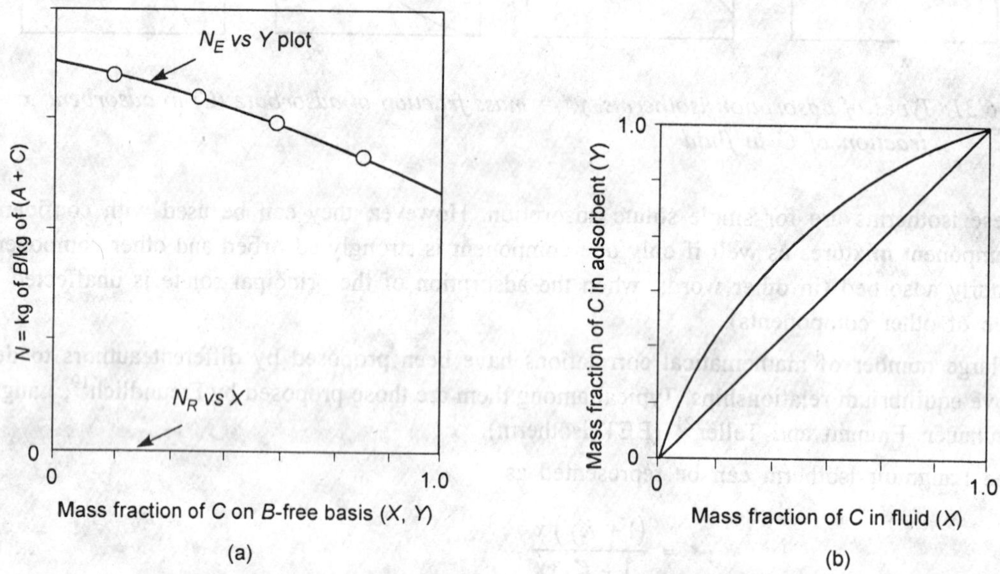


*Figure 6.32: Ternary phase diagram for adsorption (on rectangular coordinates). C = more strongly adsorbed component (solute), A = poorly adsorbed component (nonsolute), B = solid adsorbent*

shall not appear in the fluid phase, the $N_R$ versus $X$ plot shall coincide with the abscissa. Note the similarity between this diagram and that for solid–liquid extraction (leaching). The notations used here are however different, they are same as those for liquid-liquid extraction.

## 6.4.2 Adsorption Equipment

The adsorption equipment can be categorised in the same way as for extraction, such as

1. Stagewise contact equipment:

   (*a*) Mixer-settler or mixer-filter cascades (batch/continuous, countercurrent/crosscurrent operation)

* The notations used are same as those used for liquid-liquid extraction.

(*b*) Sieve-plate column with fluidised bed of adsorbent on each plate (continuous countercurrent operation)

2. Continuous contact equipment:
   (*a*) Moving bed adsorbers (continuous countercurrent operation)
   (*b*) Fixed bed adsorbers (batch/semicontinuous unsteady state operation)

Mixer-settler or mixer-filter cascades are same as those used for extraction, the contacting being done in the mixer or agitated vessel while separation of adsorbent particles from the fluid is accomplished in a settling tank or on a filter.

The sieve-plate column used for adsorption is sketched in Fig. 6.33. The solids (adsorbent) are fed from the top and they move down from one plate to another through downspouts (downcomers) that are fitted with spring-operated valves (see the inset of Fig. 6.33). The feed gas moves up through the perforations of the plate, thereby fluidising the bed of solids on each plate. The fluid velocity is so adjusted that the solids are not carried off to the top plate. The top half of the column is the adsorption section and the bottom half the regeneration section or desorption section. What is shown in Fig. 6.33 is regeneration using an inert purge gas. The regenerated adsorbent particles are transported to the top of the column pneumatically by compressed air through the air lift. The operation is fully continuous and the solids and the fluid move countercurrently.

Continuous contact equipment are either moving bed adsorbers or fixed bed adsorbers. In the former, the solids and the fluid move countercurrently (see Fig. 6.34). The solids are fed from the top and they move down the column under gravity, while the fluid is introduced from the bottom of the column to affect a continuous countercurrent operation. By the time the solids reach the bottom of the tower, they are essentially saturated and are therefore sent to the regenerator. The regenerated adsorbent is transported to the top of the column usually by an air lift. If the feed is liquid, then the solid particles may be conveyed to the top of the column hydraulically using the feed liquid itself as the carrier fluid. An excellent example of moving bed adsorber is the *hypersorber* used widely in the petroleum industry. The major drawback of *moving bed systems* is that the adsorbent tends to get damaged due to mutual collision and attrition between solid particles. The attrition losses are often appreciable with brittle adsorbents like carbon.

It is possible to conduct both adsorption as well as regeneration (desorption) simultaneously in the same column, but in this case the operation has to be made intermittent (and not continuous).

*Fixed bed adsorbers* are increasingly popular in process industries in spite of the fact that a steady state operation is practically impossible in such systems. The popularity of fixed beds is mainly due to the high cost of continuously transporting solid particles to the top of the column in moving bed adsorbers (though they provide steady state operation). Chromatographic separations are exclusively carried out on fixed beds and accordingly, fixed bed adsorption and chromatography are becoming increasingly synonymous.

Batch operation is frequently used in *fixed beds.* If the feed is a liquid solution, it is usually admitted from the top and it percolates down the adsorbent bed (this is to avoid fluidisation of solids). If the feed is a gas mixture, it is admitted from the bottom and it moves up the adsorbent bed. As the process continues, the solid particles get more and more saturated with the adsorbate ($C$ or $A + C$) and as a result, the rate of adsorption decreases. The composition of the outgoing fluid also changes with time. An unsteady state operation, therefore, prevails. Once the bed is fully saturated with the adsorbate*, the

* The term adsorbate is used to designate the substances adsorbed.

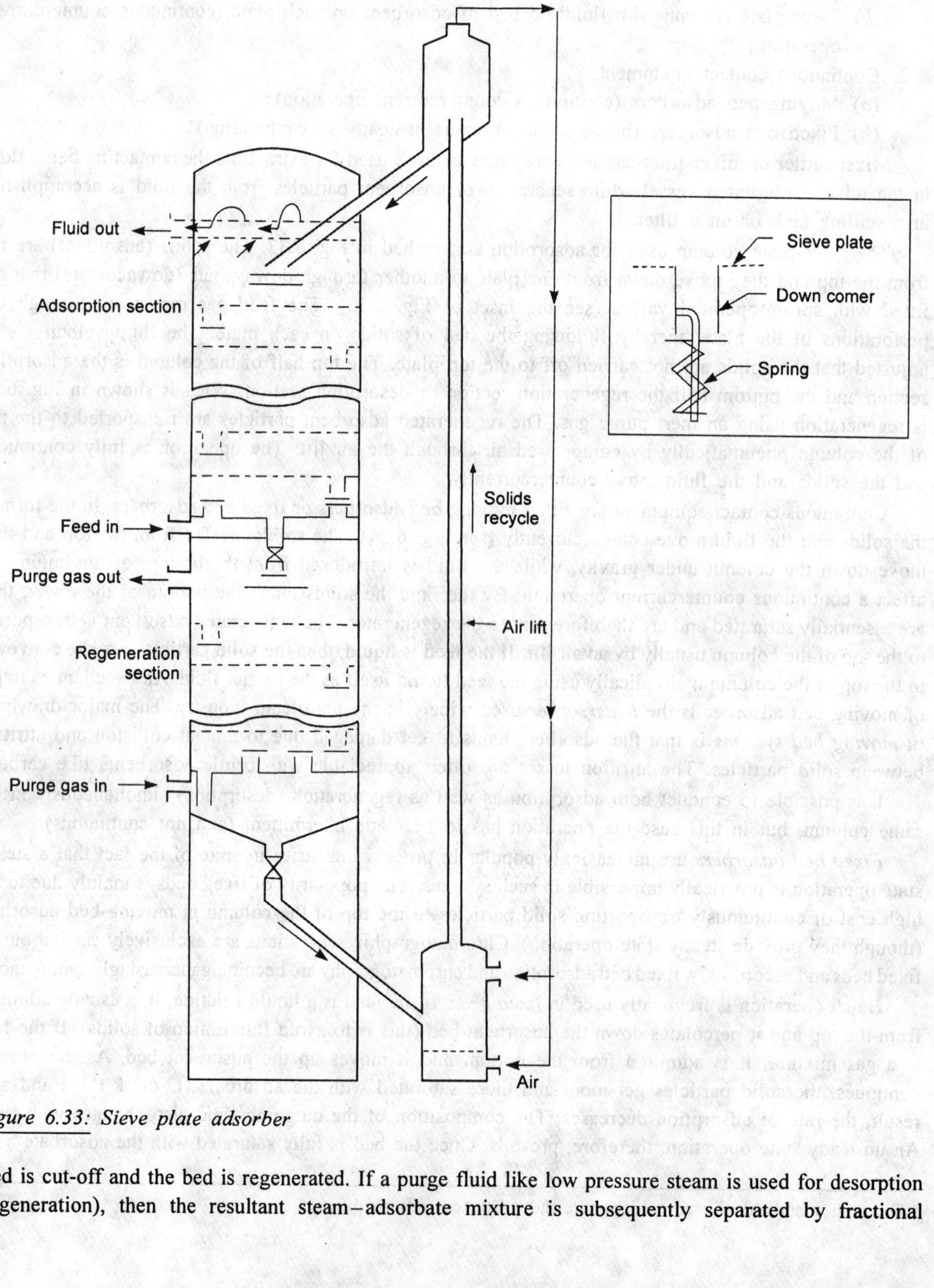


*Figure 6.33: Sieve plate adsorber*

feed is cut-off and the bed is regenerated. If a purge fluid like low pressure steam is used for desorption (regeneration), then the resultant steam–adsorbate mixture is subsequently separated by fractional

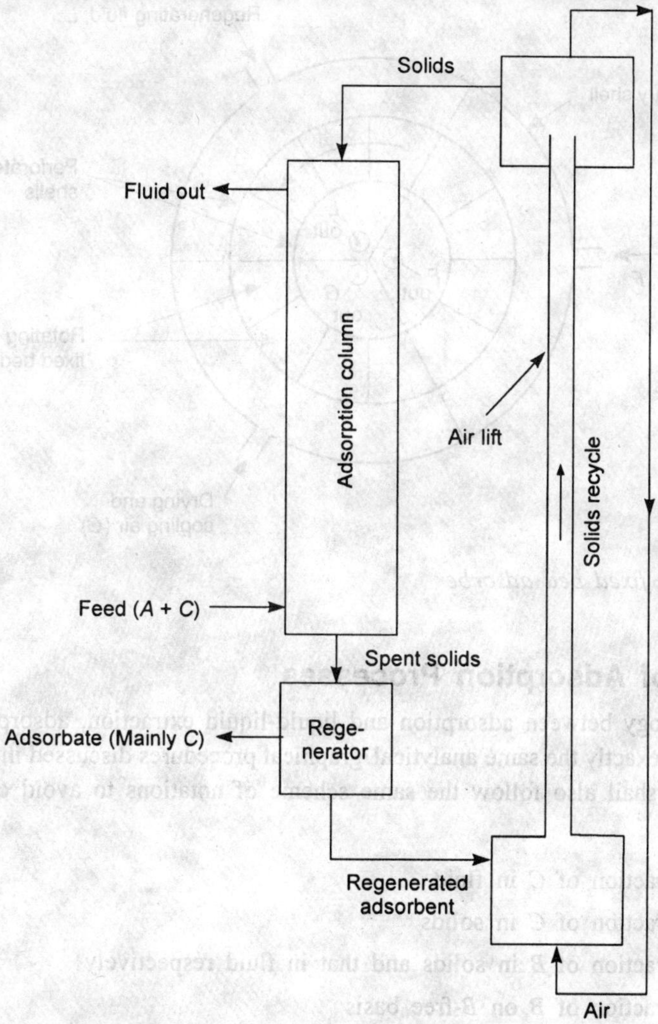


*Figure 6.34: Moving bed adsorber*

distillation. Often, the steam–adsorbate mixture condenses to form two immiscible liquid layers. After regeneration, the bed is often required to be dried and cooled before the next batch of feed is introduced.

To make the operation semicontinuous, a number of fixed beds may be operated in series (the feed fluid moving successively from one bed to another) such that when one bed is under regeneration, the rest are under operation. This is called *cyclic operation* and is similar to the Shanks system employed for solid–liquid extraction.

A fully continuous operation is also possible by using a rotating fixed bed adsorber (sketched in Fig. 6.35) that resembles a rotary continuous filter. As the inner drum rotates, each adsorbent bed is successively exposed to the feed gas/solution, regeneration and drying with cooling.

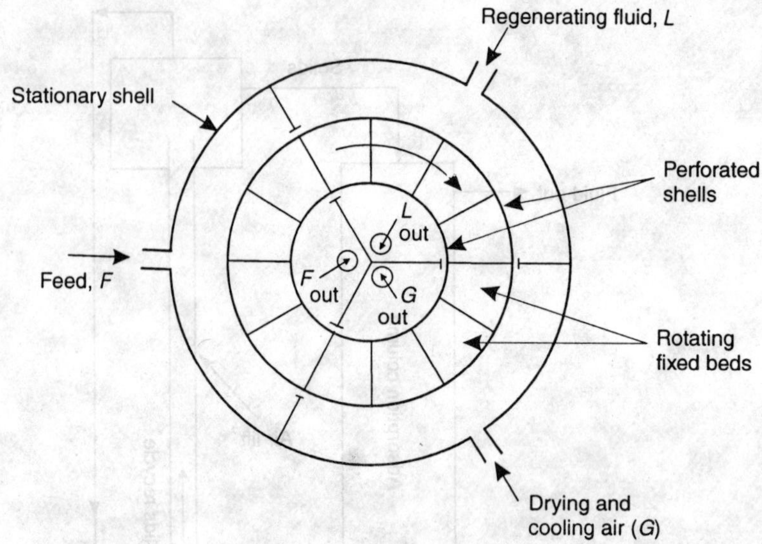


*Figure 6.35: Rotating fixed bed adsorber*

### 6.4.3 Analysis of Adsorption Processes

Due to the close analogy between adsorption and liquid-liquid extraction, adsorption processes can be analysed by following exactly the same analytical/graphical procedures discussed in Section 6.1 for liquid-liquid extraction. We shall also follow the same scheme of notations to avoid confusion.

To recapitulate,

$x$ = mass fraction of $C$ in fluid

$y$ = mass fraction of $C$ in solids

$z_e$, $z_r$ = mass fraction of $B$ in solids and that in fluid respectively

$N$ = mass fraction of $B$ on $B$-free basis

= (kg of $B$/kg of $(A + C)$

$N_E$, $N_R$ = mass fraction of $B$ (on $B$-free basis) in solids and that in fluid respectively

where $C$ is the principal solute adsorbed, $A$ is the unadsorbed or sparingly adsorbed component of the feed and $B$ is the solid adsorbent. Also, $X$, $Y$ denote mass fractions on $B$-free basis and $x'$, $y'$ mass fractions on solute-free ($C$-free) basis. We also know that $x' = x/(1 - x)$, $y' = y/(1 - y)$, $Y = y\ (1 + N_E)$ and $z_e = N_E/(1 + N_E)$. However, since in adsorption $z_r = N_R = 0$ for all values of $x$ (or $X$), owing to the fact that the solid adsorbent is nonvolatile and insoluble in the fluid, the $z_r$ versus $x$ (or $N_R$ versus $X$) plot concides with the abscissa. Consequently, the graphical/analytical procedures get simplified to a large extent. The following nomenclature may also be kept in mind:

$R_i$, $E_i$ = mass flow rate of fluid and that of solids respectively leaving $i$th stage.

$S$ = mass flow rate of fresh or recycle solids being fed to the system.

## 1. Multistage Crosscurrent Adsorption

The scheme is same as that sketched in Fig. 6.8. Instead of solvent, read adsorbent. The raffinate is the treated fluid (containing most of $A$ and little of unadsorbed $C$) and the final extract the spent solids (adsorbent + adsorbate). The mass balance equations also remain the same. However, since $z_{ri} = 0$ for all values of $i$, Eq. (6.1.18a) gets simplified to

$$y_i = x_i + z_{ei}\,(x_{mi} - x_i)/z_{mi} \quad \text{... (6.4.6)}$$

If we substitute the expressions for $x_{mi}$ and $z_{mi}$ from Eqs (6.1.12a) and (6.1.14a) in the above equation, then it gets reduced to a compact from as

$$y_i = x_i + z_{ei}\,[R_{i-1}\,(x_{i-1} - x_i) + S_i\,(y_S - x_i)]/(S_i z_S) \quad \text{... (6.4.7)}$$

Also, $$E_i = S_i z_S / z_{ei} \quad \text{... (6.4.8)}$$

and $$R_i = R_{i-1} + S_i - E_i \quad \text{... (6.4.9)}$$

Equation (6.4.7) is to be thus solved by trial for $y_i$, $x_i$ and $z_{ei}$. First, a value of $y_i$ is assumed and then the value of $z_{ei}$ is obtained from the $z_e$ versus $y$ plot/data and $x_i$ from the equilibrium plot/data. Now, compute $y_i$ from Eq. (6.4.7) and check whether this computed value agrees with that assumed at the outset.

The graphical method remains the same as that illustrated in Fig. 6.9 (b). The only difference is that we have to plot only the $z_e$ versus $y$ curve, since the $z_r$ versus $x$ plot coincides with the abscissa.

If the process involves adsorption of a single solute (only $C$ is adsorbed and $A$ is not adsorbed at all), then Eq. (6.1.29) becomes applicable and the simplified graphical procedure illustrated in Fig. 6.11 can be employed. The solution shall not thus demand any trial and error computations. This is the usual case in many industrial applications such as drying of gases, removal of colouring matter from liquid solutions, etc. To note that unlike that shown in Fig. 6.11, the equilibrium plot (on solute-free coordinates), in many instances, could be concave to the $x$-axis. In a good number of industrial processes, the total number of crosscurrent stages used seldom exceeds two.

## 2. Multistage Countercurrent Adsorption (without Reflux)

The scheme is same as that sketched in Fig. 6.12. The analytical method gets simplified since $z_r = 0$ for all values of $x$. For example, let the quantities known at the outset be $x_0$, $y_1$, $S/R_0$, $y_S$ and $z_S$. Now, Eq. (6.1.34) reduces to

$$S z_S = E_1 z_{e1} = G_B \quad \text{... (6.4.10)}$$

Since the value of $E_1$ or $(E_1/R_0)$ is thus known, we get from Eqs (6.1.31) and (6.1.32),

$$x_n = \frac{x_0 + (S/R_0)\,y_S - (E_1/R_0)\,y_1}{(1 + S/R_0 - E_1/R_0)} \quad \text{... (6.4.11)}$$

The value of $x_n$ can be thus computed straightaway without the aid of any trial and error procedure. Eq. (6.1.42a) reduces to

$$y_{i+1} = x_i + G_{i+1} z_{e,i+1} \quad \text{... (6.4.12)}$$

where $$G_{i+1} = (R_{i-1}/G_B)\,(x_i - x_{i-1}) + (y_i - x_i)/z_{ei} \quad \text{... (6.4.13)}$$

Also, $$E_{i+1}/R_0 = (G_B/R_0)\,/\,z_{e,\,i+1} \quad \text{... (6.4.14)}$$

and $$R_i/R_0 = (R_{i-1}/R_0) + (E_{i+1}/R_0) - (E_i/R_0) \quad \text{... (6.4.15)}$$

The stage to stage computations are to be performed as usual using the above four equations.

The graphical method is the same as that illustrated in Fig. 6.13 (a). Here also $z_r$ versus $x$ plot coincides with the abscissa and we are required to plot only the $z_e$ versus $y$ curve. The *minimum adsorbent to feed ratio*, $(S/R_0)_{min}$, is also to be determined by following the same procedure as that employed for liquid-liquid extraction.

If the process is that of single solute adsorption, then the simplified graphical procedure illustrated in Fig. 6.14 may be employed. If, in addition, the equilibrium plot is a straight line of slope $m$, then the number of ideal stages required can be computed straightaway from Eq. (6.1.50). The minimum adsorbent to feed ratio (on solute-free basis) required, that is $G_B$ (min)/$G_A$, is also determined by following the same methodology as that employed for liquid–liquid extraction. Thus, if the equilibrium curve is concave downward (concave to the abscissa, which is one of the usual cases) or a straight line, then the value of $G_B$ (min)/$G_A$ is obtained from Eq. 6.1.48 once $(y_0')^*$ is substituted for $y_1'$, where $(y_0')^*$ is the equilibrium value of $y'$ corresponding to $x' = x_0'$. If the equilibrium curve is concave upward, then the value of $G_A/G_B$ (min) is given by the slope of the tangent to the equilibrium curve from the point $(x_n', y_S')$.

## 3. Multistage Countercurrent Adsorption (with Reflux)

As in liquid-liquid extraction, reflux is often used in multistage countercurrent adsorption but only at one end (that is, at stage 1). The scheme is same as that sketched in Fig. 6.17. For the convenience of the readers, it is resketched in Fig. 6.36. The spent solids discharged from stage 1 are sent to the regenerator (desorber). The regenerated adsorbent is recycled to the fresh solids inlet. The recovered adsorbate (mostly $C$ + small amounts of $A$) is partly recycled to stage 1 as reflux ($R_0$) and the rest is collected as the final solute product ($P$). The feed ($F$) is introduced at one of the intermediate stages. The stream $P$ is rich in $C$ and contains very little $A$. Similarly, the fluid stream discharged from stage $n$ (namely, $R_n$) shall contain most of $A$ and very little $C$. We have confined our discussion to a binary feed mixture.

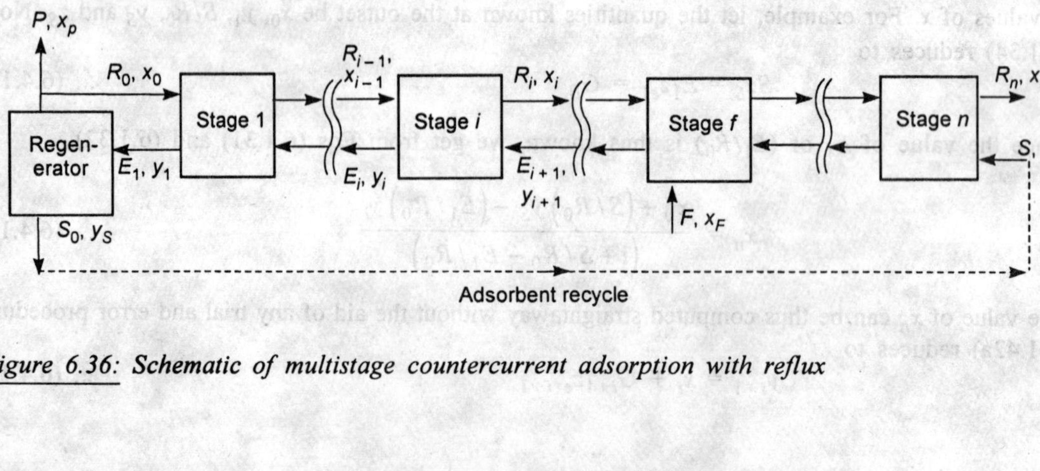


*Figure 6.36: Schematic of multistage countercurrent adsorption with reflux*

The graphical procedure is exactly the same as that illustrated in Fig. 6.19. However, the expression for $N_\Delta$ [Eq. (6.1.73)] gets simplified to

$$N_\Delta = N_S/[1 + (P'/S'_0)] \quad \text{... (6.4.16)}$$

$$= \frac{N_S}{1+(Y_1 - Y_S)/[(X_0 - Y_1)(1+RR)]}$$

If the adsorbent is pure $B$,

$$N_\Delta = N_{E1}\ (1 + RR) \quad \text{... (6.4.18)}$$

As in earlier cases, the plot required is only $N_E$ versus $Y$, since $N_R$ versus $X$ plot coincides with the abscissa. The *minimum reflux ratio* may also be estimated by following the same procedure as that illustrated in Fig. 6.20 (a).

The analytical method also remains the same as that employed for liquid-liquid extraction, except that the design equations get a lot simplified. Thus, the expression for $P'$ [Eq. (6.1.64)] reduces to

$$P' = F'\ (X_n - X_F)/(X_n - X_0) \quad \text{... (6.4.19)}$$

Also,

$$R'_n = F' - P' \quad \text{(6.4.20)}$$

$$R'_0 = (RR)\ P' \quad \text{... (6.4.21)}$$

$$S' = S'_0 \quad \text{... (6.4.22)}$$

And, Eq. (6.1.71) gets simplified to

$$Y_1 = Y_S - (X_0 - Y_S)\ (N_{E1} - N_S)/N_S \quad \text{... (6.4.23)}$$

The above equation can be thus solved by trial (with the help of $N_E$ versus $Y$ plot/data) for $Y_1$ and $N_{E1}$.

The expression for $E'_1$ Eq. (6.1.70) reduces to

$$E'_1 = P'\ (1 + RR)\ N_S/(N_S - N_{E1}) \quad \text{... (6.4.24)}$$

Also, since

$$E'_1 N_{E1} = S'_0 N_S = S' N_S = G_B,$$

$$S'_0 = S' = G_B/N_S \quad \text{... (6.4.25)}$$

If the adsorbent is pure $B$ ($S'_0 = S' = 0$), then

$$E'_1 = P' + R'_0$$

and

$$Y_1 = P'\ (1 + RR) \quad \text{... (6.4.26)}$$

$$= X_0 \quad \text{... (6.4.26a)}$$

The terminal streams and compositions can be thus evaluated from the above equations with relative ease.

The stage to stage computations are to be performed using Eqs (6.4.12) to (6.4.15), after rewriting them on adsorbent-free coordinates. These equations are to be used for all stages except the feed stage (stage $f$). For stage $f$, Eqs (6.4.13) and (6.4.15) get modified to

$$G'_{i+1} = G'_{f+1} = (R'_{i-1}/G_B)\,(X_i - X_{i-1}) + (Y_i - X_i)/N_{Ei} + (F'/G_B)\,(X_i - X_F) \quad \text{... (6.4.13a)}$$

and $$R'_i/R'_0 = R'_f/R'_0 = (R'_{i-1}/R'_0) + (E'_{i+1}/R'_0) - (E'_i/R'_0) + (F'/R'_0) \quad \text{... (6.4.15a)}$$

There will be no change in Eq. (6.4.14).

## 4. Multistage Countercurrent Adsorption (Adiabatic Operation)

Adsorption is an exothermic process and consequently, its rate decreases with increase in temperature. An isothermal operation is, therefore, preferred in many industrial applications. However, if heat effects accompanying the process are significant, then an isothermal operation may turn out to be uneconomical and an adiabatic operation may become more desirable in such cases. If a multistage adsorber is operated adiabatically (no heat is supplied from outside and no heat is permitted to be lost from the system to the surroundings), then the temperature shall change from stage to stage and the process analysis shall involve solution of not only mass balance equation but also heat balance (enthalpy balance) equation at each stage. To simplify the situation, let us assume that

(*i*) the feed is a vapour ($C$) – gas ($A$) mixture,
(*ii*) only a single solute (namely, $C$) is adsorbed,
(*iii*) the operation is countercurrent, but no reflux is used.

The procedure to be employed is exactly the same as that used for the analysis of adiabatic gas-liquid absorption (discussed in Chapter 4). The enthalpy of fluid (vapour–gas mixture) leaving any stage may be computed as

$$H_{Gi} = C_{pA}\,(t_{Gi} - t_{ref}) + x'_i\,[C_{pC}\,(t_{Gi} - t_{ref}) + \lambda_0] \quad \text{... (6.4.27)}$$

where,

$C_{pA}$, $C_{pC}$ = specific heat of gas ($A$) and that of vapour ($C$) respectively, kJ/(kg °C)
$\lambda_0$ = latent heat of vapourisation of $C$ at $t_{\text{ref}}$, kJ/kg
$t_{ref}$ = reference temperature (usually, 0° C)

Enthalpy of solids leaving any stage is given by

$$H_{Si} = C_{pB}\,(t_{Si} - t_{ref}) + y'_i C_{CL}\,(t_{Si} - t_{ref}) + (\Delta H_a) \quad \text{... (6.4.28)}$$

where $C_{pB}$ = specific heat of adsorbent ($B$), kJ/(kg °C)
$C_{CL}$ = specific heat of liquid $C$, kJ/(kg °C)
$\Delta H_a$ = integral heat of adsorption at $t_{\text{ref}}$ and $y'_i$, kJ/kg of $B$

The fluid and the solids leaving each ideal stage are assumed to be in thermal and concentration equilibrium. Accordingly, we may take $t_{Gi} = t_{Si} = t_i$ = temperature at $i$th stage. For many systems, enthalpy-concentration charts are available and the value of enthalpy at any specific concentration and temperature can be read directly from such charts*. A typical enthalpy-concentration chart for silica gel–water system is shown in Fig. 6.37.

* To note that $H_{Gi}$ and $H_{Si}$ are expressed in kJ/kg of $A$ and kJ/kg of $B$ respectively.

Consider the scheme sketched in Fig. 6.12. The feed fluid ($R_0$) enters stage 1 at temperature $t_0$ and treated fluid ($R_n$) leaves stage $n$ at the temperature of stage $n$ (namely, $t_n$). The feed solids ($S$) enter stage $n$ at temperature $t_S$ and the spent solids ($E_1$) leave stage 1 at the temperature of stage 1 (namely, $t_1$). The solute balance for the entire system is given by Eq. (6.1.48):

$$G_A/G_B = (y_1' - y_S')/(x_0' - x_n') \quad \text{... (6.1.48)}$$

The solute balance for stage $i$ is

$$G_A/G_B = (y_i' - y_{i+1}')/(x_{i-1}' - x_i') \quad \text{... (6.4.29)}$$

Similarly, the heat balance for the entire system is

$$G_A\,(H_{G0} - H_{Gn}) = G_B\,(H_{S1} - H_{S0}) \quad \text{... (6.4.30)}$$

and the heat balance for stage $i$,

$$G_A\,(H_{G,i-1} - H_{Gi}) = G_B\,(H_{Si} - H_{S,i+1})$$

or

$$H_{S,i+1} = H_{Si} - (G_A/G_B)(H_{G,i-1} - H_{Gi}) \quad \text{... (6.4.31)}$$

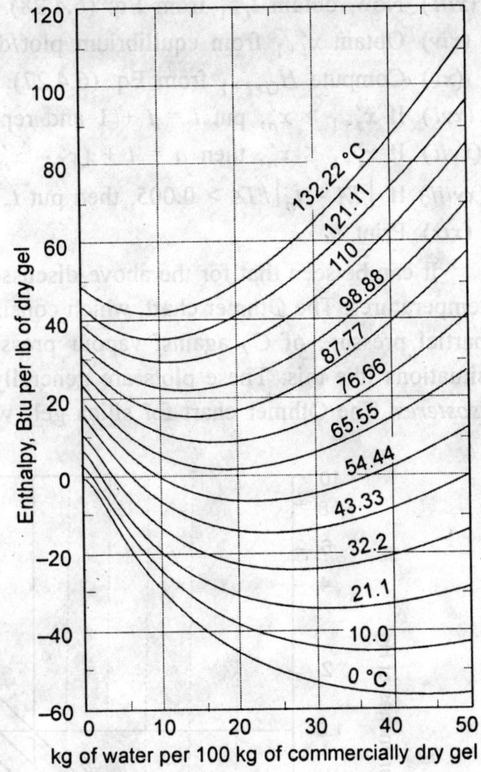


*Figure 6.37: Enthalpy–concentration chart for silica gel–water system (1 Btu/lb = 2.326 kJ/kg). Reference State: Commercially dry gel and liquid water at 273 K. (From Hougen, OA et. al., Chemical Process Principles, Wiley, 1954, by permission).*

The stage to stage computations are to be thus performed based on the above four equations. To note that though $t_0$ and $t_S$ are usually specified in the design problem, the values of $t_n$ and $t_1$ are unknown. As a result, the value of $t_n$ is to be first assumed and checked subsequently at the end of the stage to stage computations. We shall summarise the procedure below:

(*i*) The quantities known at the outset are $R_0$, $x_0$, $t_0$, $x_n$, $S$, $y_S$, $t_S$. Since specific heats are poor functions of temperature, the values of $C_{pA}$, $C_{pB}$, $C_{pC}$ and $C_{CL}$ are assumed constant during the process. The reference temperature ($t_{ref}$) is usually taken equal to 0° C.

(*ii*) Compute ($G_A/G_B$) ratio from Eq. (6.1.49).

(*iii*) Compute $y_1'$ from Eq. (6.1.48)

(*iv*) Compute $H_{G0}$ (corresponding to $x_0'$ and $t_0$) and $H_{S0}$ (corresponding to $t_S$ and $y_S'$) from Eqs (6.4.27) and (6.4.28) respectively. If enthalpy–concentration charts are available, the values of $H_{G0}$ and $H_{S0}$ may be retrieved from them.

(*v*) Assume $t_n$. For example, let $t_n = t_0 + 1.0$

(*vi*) Put $TA = t_n$.

(*vii*) Compute $H_{Gn}$ (corresponding to $t_n$ and $x_n'$) from Eq. (6.4.27). Now, obtain $H_{S1}$ from Eq. (6.4.30).

(*viii*) Obtain $t_1$ from Eq. (6.4.28) since the values of $H_{S1}$ and $y_1'$ are known.

(*ix*) Obtain $x_1'$ from equilibrium data/plot (corresponding to $y_1'$ and $t_1$).

(*x*) Compute $H_{G1}$ from Eq. (6.4.27).

(*xi*) Put $i = 1$.

(*xii*) Compute $y'_{i+1}$ from Eq. (6.4.29). and $H_{S,i+1}$ from Eq. (6.4.31).

(*xiii*) Now, obtain $t_{i+1}$ from Eq. (6.4.28).

(*xiv*) Obtain $x'_{i+1}$ from equilibrium plot/data.

(*xv*) Compute $H_{G,i+1}$ from Eq. (6.4.27).

(*xvi*) If $x'_{i+1} > x'_n$, put $i = i + 1$ and repeat the computations starting from step (*xii*).

(*xvii*) If $x'_{i+1} \le x'_n$, then $n = i + (x'_i - x'_n)/(x'_i - x'_{i+1})$ and $t_n = t_i + (t_{i+1} - t_i)\,(n - i)$.

(*xviii*) If $|TA - t_n|/TA > 0.005$, then put $t_n = t_n + 0.1$ and repeat the procedure starting from step (*vi*).

(*xix*) Print $n$.

It can be seen that for the above-discussed analysis, we require the equilibrium plot/data at different temperatures. The Othmer chart, which consists of plots of vapour pressure of adsorbed $C$ (or equilibrium partial pressure of $C$) against vapour pressure of pure $C$ at different values of $y'$, is quite useful in situations like this. These plots are generally straight lines on logarithmic coordinates. They are called *isosteres*. The Othmer chart for silica gel–water system is shown in Fig. 6.38 (a). This family of plots

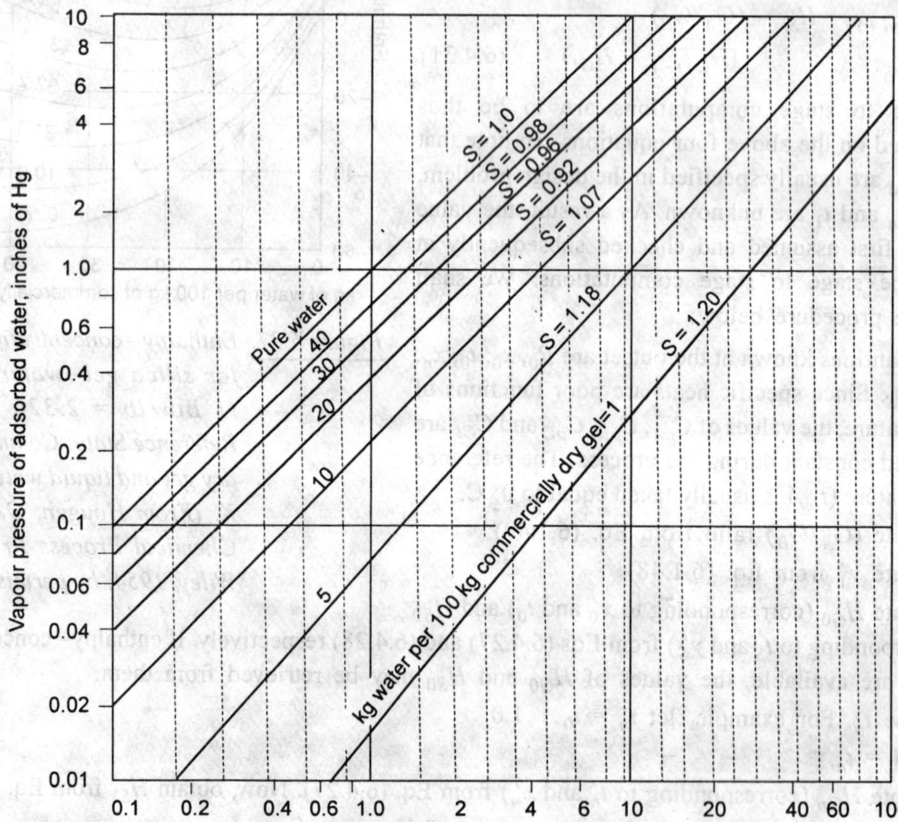


*Figure 6.38 (a): Othmer chart for silica gel–water system. s = Slope of line*

can be condensed into a single curve by plotting $T \ln (\bar{P}/P^*)$ against $y'$ as shown in Fig. 6.38 (b). Here, $\bar{P}$ is the vapour pressure of pure $C$, $P^*$ its equilibrium partial pressure (vapour pressure of adsorbed $C$) and $T$ the absolute temperature (in $K$). Such a plot is accurate within a specific range of temperature and within a specific range of values of $y'$. The plot shown in Fig. 6.38 (b) is for $y' \leq 0.1$ and operating temperatures $\leq 40°$ C. To note that

$$x' = \frac{P^*}{(P - P^*)} (M_C/M_A) \qquad \text{... (6.4.32)}$$

where $P$ is the total pressure (operating pressure of adsorber) and $M_C$, $M_A$ molecular weights of $C$ and $A$ respectively.

The Othmer chart can also be used for estimating the integral heat of adsorption. Incidentally, in the above-discussed computational procedure, we require the value of $\Delta H_a$ at the reference temperature and the specified value of $y'$. Since the reference temperature is fixed (usually at 0° C), $\Delta H_a$ becomes

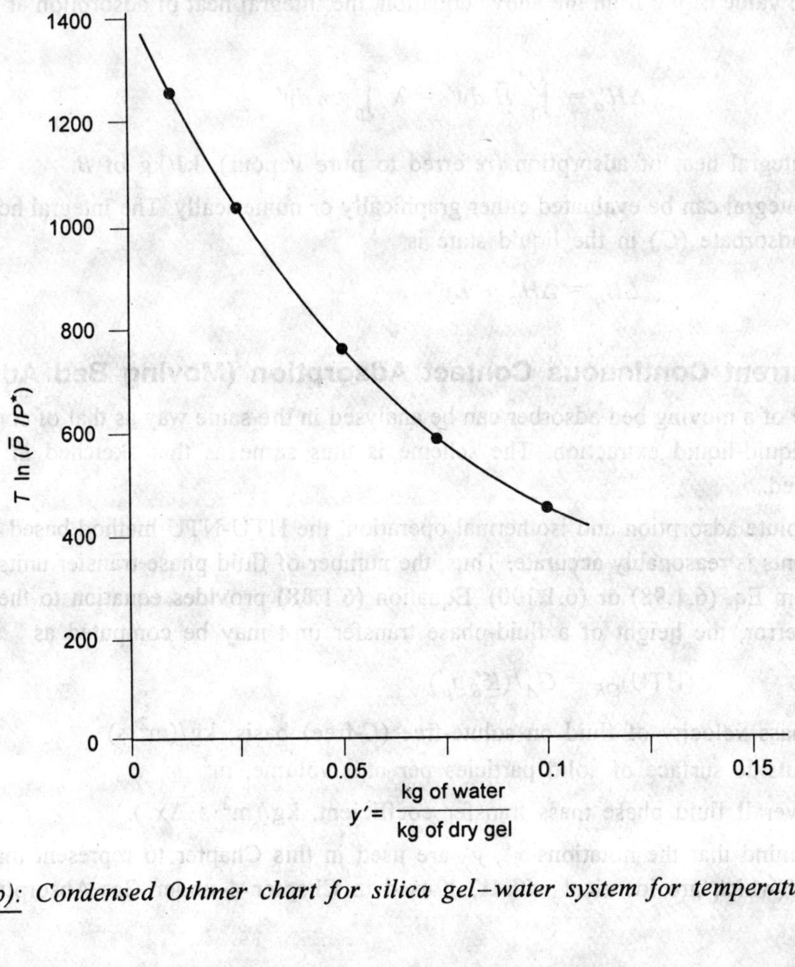


*Figure 6.38 (b): Condensed Othmer chart for silica gel–water system for temperatures ≤ 313 K*

a function of $y'$ only. In other words, we are required to know the integral heat of adsorption at 0° C as a function of adsorbate concentration ($y'$). This can be estimated using the method proposed by Othmer and Sawyer[22]. Thus, the slope of each isostere (each straight line plot) of Fig. 6.38 (a) is

$$m = \frac{d \ln P^*}{d \ln \bar{P}} = \frac{(-\bar{H}) M_C}{\lambda_r M_r} \quad \text{... (6.4.33)}$$

where $\bar{H}$ = differential heat of adsorption (referred to pure vapour) at the specified temperature, kJ/kg of $C$ adsorbed

$\lambda_r$ = latent heat of vapourisation of reference substance at the same temperature, kJ/kg

$M_C$, $M_r$ = molecular weight of $C$ and that of the reference substance respectively.

It is usual to use the substance adsorbed (namely, $C$) itself as the reference substance and in that case, $M_r = M_C$ and

$$(-\bar{H}) = m\lambda \quad \text{... (6.4.34)}$$

where $\lambda$ is the latent heat of vapourisation of $C$ at the specified temperature (in the present case, $\lambda = \lambda_0$). Once the value of $\bar{H}$ at a constant temperature (in this case, at 0° C) is computed for each isostere (for each value of $y'$) from the above equation, the integral heat of adsorption at this temperature is given by

$$\Delta H'_a = \int_0^{y'} \bar{H}\, dy' = \lambda \int_0^{y'} m\, dy' \quad \text{... (6.4.35)}$$

where $\Delta H'_a$ = integral heat of adsorption (referred to pure vapour), kJ/kg of $B$.

The above integral can be evaluated either graphically or numerically. The integral heat of adsorption referred to the adsorbate ($C$) in the liquid state is

$$\Delta H_a = \Delta H'_a + \lambda y' \quad \text{... (6.4.36)}$$

## 5. Countercurrent Continuous Contact Adsorption (Moving Bed Adsorber)

The performance of a moving bed adsorber can be analysed in the same way as that of continuous contact equipment for liquid-liquid extraction. The scheme is thus same as that sketched in Fig. 6.21, with directions reversed.

For single solute adsorption and isothermal operation, the HTU-NTU method based on overall mass transfer coefficients is reasonably accurate. Thus, the number of fluid-phase transfer units, $(\text{NTU})_{OR}$, can be computed from Eq. (6.1.98) or (6.1.100). Equation (6.1.88) provides equation to the operating line. With allowable error, the height of a fluid-phase transfer unit may be computed as

$$(\text{HTU})_{OR} = G_A / (K'_R a_p) \quad \text{... (6.4.37)}$$

where $G_A$ = mass velocity of fluid on solute-free ($C$-free) basis, kg/(m$^2\cdot$s)

$a_p$ = outside surface of solid particles per unit volume, m$^2$/m$^3$

$K'_R$ = overall fluid phase mass transfer coefficient, kg/(m$^2\cdot$s$\cdot\Delta x'$).

To keep in mind that the notations $x'$, $y'$ are used in this Chapter to represent mass fractions on solute-free basis (which are identical with $X$, $Y$ used in Chapter 4 under Gas Absorption).

If both $A$ and $C$ are adsorbed ($C$ being more strongly adsorbed), then the analysis described above shall not be accurate enough. However, due to sparsity of data, the above procedure itself is used for such situations, with a few modifications. Let us assume that the feed is a binary mixture ($A + C$) and the operation is isothermal. It is usual to operate the column with a reflux as shown in Fig. 6.39 (a). The expressions for $(NTU)_{OR}$ and $(HTU)_{OR}$ remain the same as those for single solute adsorption, except that

(*i*) NTU is to be computed separately for the adsorption section (section of column above feed point) and the enriching section (section of column below the feed point). The expression for NTU shall be the same for both sections, only the limits of the integral would change. The limits of the integral shall be $x_2$ and $x_F$ (from $x_2$ to $x_F$) for the adsorption section and $x_F$ and $x_1$ (from $x_F$ to $x_1$) for the enriching section.

(*ii*) The values of $x^*$ are to be obtained using the graphical procedure illustrated in Fig. 6.39 (b). For sake of convenience, adsorbent-free coordinates are used. The operating points $\Delta$ and $\Delta'$ are located in the same way as for multistage equipment. The $N_E$ versus $Y$ plot is first made. The points $X_1$, $X_2$ (which are same as $X_0$ and $X_n$ for multistage equipment) and $X_F$ are marked on the abscissa (to note that $X_1 = x_1$, $X_2 = x_2$, $X_F = x_F$). The point $S$, whose coordinates are $(Y_2, N_{E2})$, is also marked.

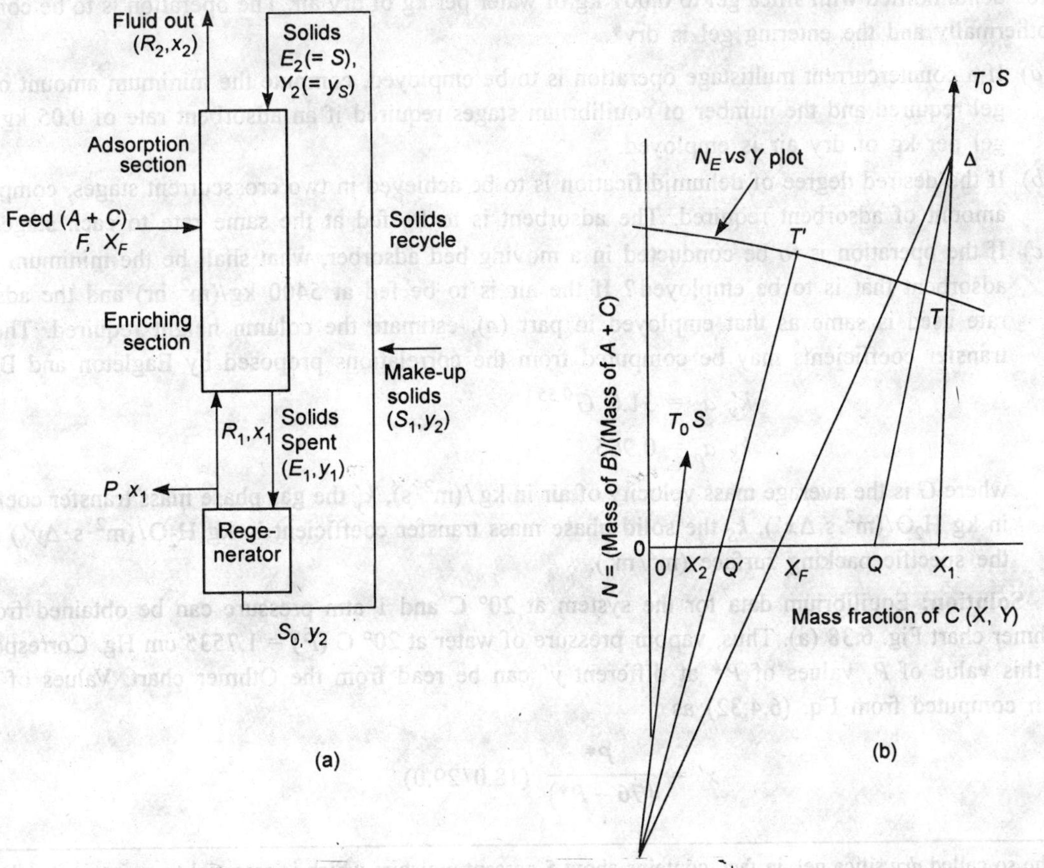


*Figures 6.39 (a) and (b): Analysis of moving bed adsorber with reflux (both components adsorbed)*

The operating point $\Delta$ is now marked on the line $\overline{SX_1}$ corresponding to $N = N_\Delta$, where $N_\Delta$ is obtained from Eq. (6.4.17). To keep in mind that $Y_S$ in Eq. (6.4.17) must be replaced by $Y_2$, $N_S$ by $N_{E2}$ and $X_0$ by $X_1$. The lines $\overline{\Delta X_F}$ and $\overline{SX_2}$ when extended intersect at $\Delta'$. If the adsorbent is pure $B$, then both $\overline{SX_1}$ and $\overline{SX_2}$ shall be vertical lines through $X_1$ and $X_2$ respectively and $N_\Delta$ is obtained from Eq. (6.4.18). Now, take any value of $X$ and mark it on the abscissa (say, point $Q$). Join $Q$ to $\Delta$ and this line intersects the $N_E$ versus $Y$ curve at $T$. Find $Y$ corresponding to $T$ and then obtain $X^*$ corresponding to this value of $Y$ from the equilibrium plot. Thus, for all values of $X$ between $X_1$ and $X_F$, the corresponding values of $X^*$ can be computed. For all values of $X$ between $X_F$ and $X_2$, the corresponding values of $X^*$ are similarly determined, but using the operating point $\Delta'$.

The mass transfer coefficients are to be nevertheless computed from available experimental correlations. Often, enough data are not available for the estimation of these coefficients.

The minimum reflux ratio maybe determined by following the same procedure as that for multistage equipment [see Fig. 6.20 (a)]. As stated under liquid-liquid extraction, in many cases, the tie line which when extended passes through the point $(X_F, 0)$ shall decide the minimum reflux ratio.

**Example 6.10:** Air at 20° C and 1 atm pressure containing 0.01 kg of water per kg of dry air is to be dehumidified with silica gel to 0.001 kg of water per kg of dry air. The operation is to be conducted isothermally and the entering gel is dry*.

(*a*) If a countercurrent multistage operation is to be employed, compute the minimum amount of silica gel required and the number of equilibrium stages required if an adsorbent rate of 0.05 kg of dry gel per kg of dry air is employed.

(*b*) If the desired degree of dehumidification is to be achieved in two crosscurrent stages, compute the amount of adsorbent required. The adsorbent is to be fed at the same rate to each stage.

(*c*) If the operation is to be conducted in a moving bed adsorber, what shall be the minimum rate of adsorbent that is to be employed? If the air is to be fed at 5400 kg/(m$^2\cdot$hr) and the adsorbent rate used is same as that employed in part (*a*), estimate the column height required. The mass transfer coefficients may be computed from the correlations proposed by Eagleton and Bliss[25]:

$$k'_y\, a_p = 31.6\ G^{0.55}$$

$$k_S\, a_p = 0.965$$

where $G$ is the average mass velocity of air in kg /(m$^2\cdot$s), $k'_y$ the gas phase mass transfer coefficient in kg $H_2O$/(m$^2$.s.$\Delta x'$), $k_S$ the solid phase mass transfer coefficient in kg $H_2O$/(m$^2\cdot$s$\cdot\Delta y'$) and $a_p$ the specific packing surface (m$^2$/m$^3$).

**Solution:** Equilibrium data for the system at 20° C and 1 atm pressure can be obtained from the Othmer chart Fig. 6.38 (a). Thus, vapour pressure of water at 20° C ($\bar{P}$) = 1.7535 cm Hg. Corresponding to this value of $\bar{P}$, values of $P^*$ at different $y'$ can be read from the Othmer chart. Values of $x'$ are then computed from Eq. (6.4.32) as

$$x' = \frac{P^*}{(76 - P^*)}\ (18.0/29.0) \qquad \text{... (i)}$$

* The so called *dry* silica gel, in fact, contains about 5 percent moisture which is essential to maintain its adsorptive property. This moisture is not included in the design calculations.

where $P^*$ is in cm Hg. The results are tabulated below:

| $y'$ : | 0.01 | 0.05 | 0.10 | 0.20 | 0.30 | 0.40 |
|---|---|---|---|---|---|---|
| $P^*$ : | 0.0248 | 0.13123 | 0.36435 | 0.6295 | 0.9456 | 1.2695 |
| $x'$ : | 0.0002 | 0.001 | 0.003 | 0.00518 | 0.0078 | 0.0105 |

The equilibrium curve is thus plotted and is shown in Fig. 6.10.1. The plot is essentially linear.

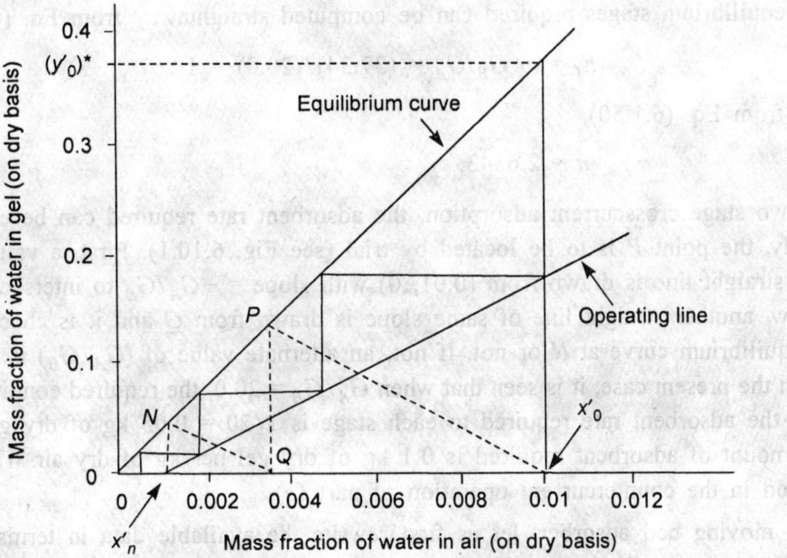


*Figure 6.10.1*

(*a*) The process involves single solute adsorption. It is given that $x'_0 = 0.01$, $x'_n = 0.001$ and $y'_S = 0.0$. Since the equilibrium plot is essentially linear, the minimum adsorbent rate required can be computed from Eq. (6.1.48) by substituting $(y'_0)^*$ for $y_1$. From figure, $(y'_0)^* = 0.382$. Therefore, from Eq. (6.1.48),

$$\frac{G_A}{G_B\,(\min)} = \frac{0.382 - 0}{0.01 - 0.001} = 42.44$$

or $$G_B\,(\min)/G_A = 0.02356 \text{ kg/kg}$$

Thus, the minimum adsorbent rate required is 0.02356 kg of dry gel per kg of dry air.

The adsorbent rate employed is 0.05 kg of dry gel per kg of dry air. That is,

$$G_B/G_A = 0.05$$

or $$G_A/G_B = 20.0$$

The operating line is therefore drawn from the point $(x'_n, y'_S) = (0.001, 0)$ with a slope of 20.0 as shown in Fig. 6.10.1. The staircase construction is then performed as shown and the results are

| $i$ | $y'_i$ | $x'_i$ |
|---|---|---|
| 1 | 0.180 | 0.0046 |
| 2 | 0.072 | 0.00168 |
| 3 | 0.0128 | 0.0003 |

The number of equilibrium stages required is, $n = 2 + (0.00168 - 0.001)/(0.00168 - 0.0003) = 2.49$.

Incidentally, if we consider the equilibrium curve to be a straight line of average slope 37.34, then the number of equilibrium stages required can be computed straightaway from Eq. (6.1.50). Thus,

$$E_f = m\ G_B/G_A = (37.34)/(20.0) = 1.867$$

Therefore, from Eq. (6.1.50),

$$n = 2.6343.$$

(*b*) For a two stage crosscurrent adsorption, the adsorbent rate required can be estimated only by trial. Graphically, the point $P$ is to be located by trial (see Fig. 6.10.1). First, a value of $(G_A/G_B)$ is assumed and a straight line is drawn from (0.01, 0) with slope $= -G_A/G_B$ to intersect the equilibrium curve at $P$. Now, another straight line of same slope is drawn from $Q$ and it is checked whether this intersects the equilibrium curve at $N$ or not. If not, an alternate value of $(G_A/G_B)$ is assumed and the trial repeated. In the present case, it is seen that when $G_A/G_B = 20.0$, the required condition is essentially satisfied. Thus, the adsorbent rate required to each stage is $1/20 = 0.05$ kg of dry gel per kg of dry air. The total amount of adsorbent required is 0.1 kg of dry gel per kg of dry air. This is double the amount employed in the countercurrent operation of part (*a*).

(*c*) For the moving bed adsorber, let us first rewrite the available data in terms of the changed notations (with reference to Fig. 6.21). Thus

$$x'_1 = 0.01$$

$$x'_2 = 0.001$$

$$y'_2 = 0.0$$

The equilibrium curve and the operating line shall be the same as those shown in Fig. 6.10.1. Accordingly, the minimum adsorbent rate required shall also be the same as that determined in part (*a*), that is, 0.02356 kg of dry gel per kg of dry air.

If the $G_A/G_B = 20.0$, then a solute balance gives

$$G_A\ (x'_1 - x'_2) = G_B\ (y'_1 - y'_2)$$

or

$$y'_1 = (G_A/G_B)\ (x'_1 - x'_2) + y'_2 = (20.0)\ (0.01 - 0.001) + 0.0 = 0.18$$

Therefore, the equation to the operating line is [from Eq. (6.1.88)],

$$y' = 0.18 + 20\ (x' - 0.01) \qquad \text{... (ii)}$$

For different values of $x'$ ranging from 0.01, values of $y'$ are obtained from the above equation and the corresponding values of $(x')^*$ from the equilibrium plot Fig. 6.10.1. The results are listed below:

| $x'$ | $y'$ | $(x')^*$ | $1/(x' - x'^*)$ |
|---|---|---|---|
| $x'_1$ = 0.01 | 0.18 | 0.0046 | 185.185 |
| 0.009 | 0.16 | 0.0041 | 204.08 |
| 0.008 | 0.14 | 0.00355 | 224.719 |
| 0.007 | 0.12 | 0.003 | 250.0 |
| 0.006 | 0.10 | 0.0025 | 285.714 |
| 0.005 | 0.08 | 0.00191 | 323.624 |
| 0.004 | 0.06 | 0.00132 | 373.134 |
| 0.003 | 0.04 | 0.00082 | 458.715 |
| 0.002 | 0.02 | 0.0004 | 625.0 |
| $x'_2$ = 0.001 | 0.0 | 0.0 | 1000.0 |

(ii) The integral of Eq. (6.1.100) can be now evaluated either graphically or numerically. Let us evaluate it numerically using the trapezoidal rule. According to this rule,

$$I = \int_{0.001}^{0.01} \frac{1}{(x' - x'^*)}\, dx' = \int_{0.001}^{0.01} f(x)\, dx = (h/2)\,[f_0 + 2\,(f_1 + f_2 + \ldots.. f_{n-1}) + f_n]$$

where $h$ = step size = 0.001, $f_0$ = 1000.0, $f_1$ = 625.0 and so on and finally, $f_n$ = 185.185. Therefore,

$$I = 3.33758$$

Also, $r = M_A/M_C = 29/18 = 1.6111$. Now, from Eq. (6.1.100),

$$(\text{NTU})_{OR} = 3.33758 + (0.5)\ \ln\left[\frac{1.0 + 1.6111(0.001)}{1.0 + 1.6111(0.01)}\right] = 3.83$$

Assuming there is no significant change in the mass velocity of air as it passes through the column,

$$k'_y a_p = 31.6\ (5400/3600)^{0.55} = 39.4945\ \text{kg H}_2\text{O}/(\text{m}^3\cdot\text{s}\cdot\Delta x')$$

$$k_S a_p = 0.965\ \text{kg H}_2\text{O}/(\text{m}^3\cdot\text{s}\cdot\Delta y')$$

With allowable error, the average slope of the equilibrium plot is taken to be 37.34. Thus

$$1/(K'_R\, a_p) = 1/(k'_y a_p) + \frac{(1/m)}{k_S a_p} = (1/39.4945) + (1/37.34)/0.965 = 0.05307$$

$$G_A = (5400/3600)/(1 + 0.01) = 1.485\ \text{kg}/(\text{m}^2\cdot\text{s})$$

Therefore, from Eq. (6.4.37),

$$(\text{HTU})_{OR} = (1.485)\ (0.05307) = 0.0788\ \text{m}$$

The column height required is

$$Z = (0.0788)\ (3.83) = \mathbf{0.3018\ m}$$

Before concluding this example, we would like to state here that the accuracy of any design procedure of this kind weighs heavily on the precision of the reported equilibrium data. The operating conditions during the preparation/activation of the adsorbent determine its final properties and as a result, it is not unusual that the equilibrium data collected on one sample of the adsorbent differs from that on another sample of the same (the empiricity inherent in the experimental data reported).

**Example 6.11:** Derive an expression for the minimum adsorbent rate that is required in a two stage crosscurrent adsorption cascade if the adsorption isotherm

(*a*) is linear

(*b*) is described by Langmuir equation

(*c*) is described by Freundlich equation

Assume single solute adsorption and the adsorbent fed to each stage is free from solute.

**Solution:** (*a*) If the adsorption isotherm is linear (on solute-free coordinates), then $y' = mx'$ (more precisely, $(y')^* = mx'$). We know, from Eq. (6.1.28),

$$G_{B1}/G_A = (x_1' - x_0')/(y_1' - y_S') = (x_1' - x_0')/(mx_1'), \text{ since } y_S' = 0. \quad \text{... (i)}$$

Similarly, $$G_{B2}/G_A = (x_2' - x_1')/(mx_2') \quad \text{... (ii)}$$

Adding the two, we get on rearrangement

$$(G_{B1} + G_{B2})/G_A = (1/\text{m})\ [2 - (x_0'/x_1') - (x_1'/x_2')] \quad \text{... (iii)}$$

To find the minimum total adsorbent, we differentiate the right hand side of above equation with respect to $x_1'$ and set it equal to zero (to note that for a given separation, $x_0'$ and $x_2'$ are constants). This yields

$$x_1' = \sqrt{x_0{}' x_2{}'} \quad \text{... (iv)}$$

If we substitute this in Eq. (i) and (ii), it can be seen that

$$G_{B1}\ (\text{min})/G_A = G_{B2}\ (\text{min})/G_A = (1/\text{m})\left(1 - \sqrt{x_0{}'/x_2{}'}\right) \quad \text{... (v)}$$

Thus, the total adsorbent rate shall be a minimum if the amount of adsorbent fed to each stage is the same.

(*b*) If the equilibrium relationship is described by Langmuir-type equation [Eq. (6.4.2)], then

$$G_{B1}/G_A = (x_1' - x_0')\ (1 + K_L x_1')/(1 + K_L)\ x_1' \quad \text{... (vi)}$$

and $$G_{B2}/G_A = (x_2' - x_1')\ (1 + K_L x_2')/(1 + K_L)\ x_2' \quad \text{... (vii)}$$

Adding up the two and then setting $d\ (G_{B1} + G_{B2})/dx_1' = 0$, we get

$$x_1' = \sqrt{x_0{}' x_2{}'} \quad \text{... (viii)}$$

Substituting this in Eq. (vi) and (vii), we get

$$\frac{G_{B1}(\text{min}) + G_{B2}(\text{min})}{G_A} = \frac{2\left(1 - \sqrt{x_0{}'/x_2{}'}\right) + K_L\left(x_2{}' - x_0{}'\right)}{(1 + K_L)} \quad \text{... (ix)}$$

(*c*) If a Freundlich-type equation [Eq. (6.4.3)] describes the equilibrium relationship, then

$$G_{B1}/G_A = (x_1' - x_0')/k(x_1')^n \quad \text{... (x)}$$

and $$G_{B2}/G_A = (x_2' - x_1')/k\,(x_2')^n \qquad \text{... (xi)}$$

Adding up the two and then setting $d\,(G_{B1} + G_{B2})/dx_1'$ equal to zero, we get

$$(x_1'/x_2')^n - n\,(x_0'/x_1') = (1 - n) \qquad \text{... (xii)}$$

This equation can be solved for $x_1'$ and substituting it in Eqs (x) and (xi), we get the minimum total adsorbent required.

**Example 6.12:** Calculate the number of countercurrent ideal stages and mass of fresh char (adsorbent) required for the isothermal removal of 99 percent of the colouring matter from 100 kg of sugar syrup per hour by agitation with finely divided colour-free activated vegetable char. The feed syrup contains 0.001 mass fraction of colour. The spent char is to contain 0.0004 mass fraction of colour. The equilibrium data for the system, as reported[26], is given below:

**Mass Fractions**

| *Solid phase* | | *Liquid phase* | |
|---|---|---|---|
| *Colour* | *Syrup* | *Colour* | *Syrup* |
| 0.0001 | 0.0005 | 0.00002 | 0.99998 |
| 0.0002 | 0.0005 | 0.00008 | 0.99992 |
| 0.0003 | 0.0004 | 0.0002 | 0.9998 |
| 0.0004 | 0.0004 | 0.0005 | 0.9995 |
| 0.0005 | 0.0003 | 0.001 | 0.9990 |

**Solution:** The process involves adsorption of both $C$ (colouring matter) and $A$ (syrup), $C$ more strongly adsorbed. The data available are

$$R_0 = 100 \text{ kg/hr}$$
$$x_0 = 0.001$$
$$y_S = 0$$
$$y_1 = 0.0004$$

From the equilibrium data provided, we can compute the values of $z_e$. For instance, when the mass fraction of colour in the solid phase (namely, $y$) is 0.0001, that of syrup is 0.0005 and therefore, mass fraction of $B$ (namely, char) is, $z_e = 1.0 - (0.0001 + 0.0005) = 0.9994$. The results are tabulated below:

| $y$ | $z_e$ | $x$ |
|---|---|---|
| 0.0001 | 0.9994 | 0.00002 |
| 0.0002 | 0.9993 | 0.00008 |
| 0.0003 | 0.9993 | 0.0002 |
| 0.0004 | 0.9992 | 0.0005 |
| 0.0005 | 0.9992 | 0.001 |

It can be seen that the value of $z_e$ changes very little and consequently, the $z_e$ versus $y$ plot shall be essentially parallel to the abscissa. The equilibrium plot ($y$ versus $x$) is shown in Fig. 6.12.1. The plot has been prepared on log-log coordinates since this provides a smoother plot.

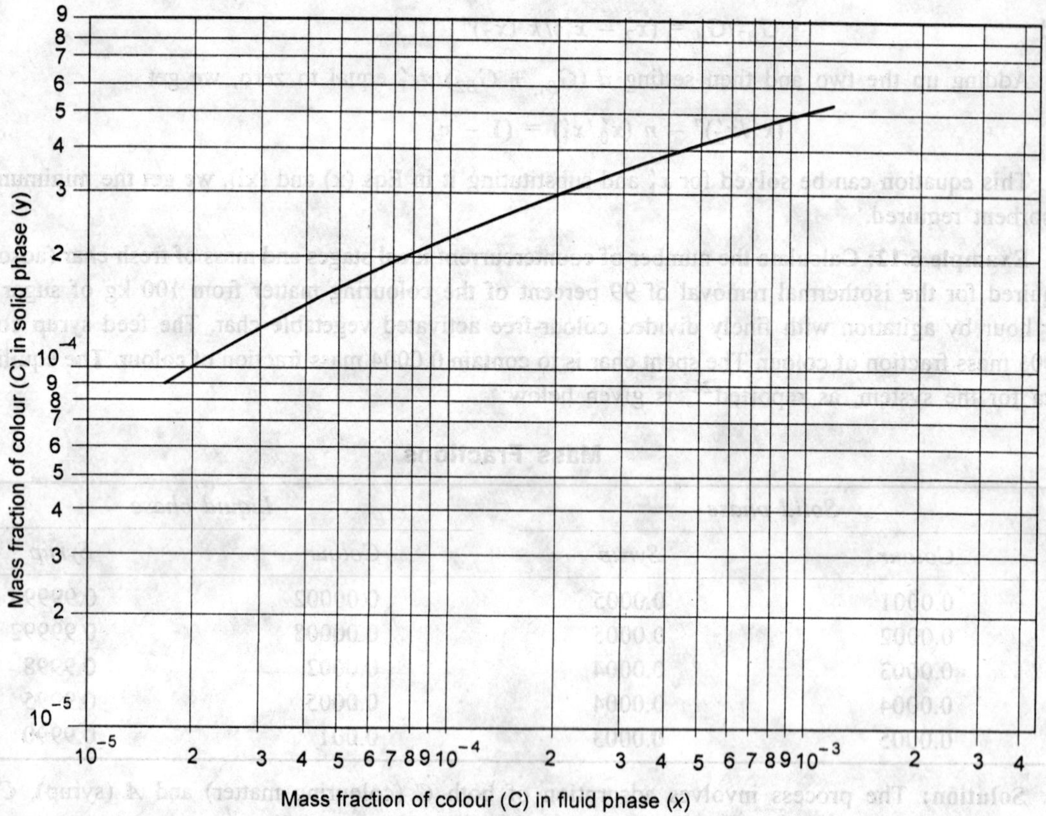


*Figure 6.12.1*

Since 99 percent of the colour is to be removed,

$$R_n x_n = 0.01\ (R_0 x_0)$$

And since $y_S = 0$, from Eq. (6.1.32),

$$E_1 y_1 = 0.99\ (R_0 x_0)$$

or

$$E_1 = (0.99)\ (100)\ (0.001)/(0.0004) = 247.5 \text{ kg/hr}$$

From the table given above, corresponding to

$$y_1 = 0.0004,$$

$$z_{e1} = 0.9992$$

$$x_1 = 0.0005$$

Now,

$$G_B = E_1 z_{e1} = (247.5)\ (0.9992) = 247.302 \text{ kg/hr} = S z_S = S \text{ (since } z_S = 1.0).$$

This is the mass of fresh char required per hour. Now,

$$R_n = 100 + 247.302 - 247.5 = 99.802 \text{ kg/hr}$$

And

$$x_n = (R_n x_n)/R_n = (0.01\ R_0 x_0)/R_n$$

$$= (0.01)\ (100)\ (0.001)/(99.802) = 0.00001$$

Let us compute the number of equilibrium stages by the analytical method. Thus, for $i$ = 1, from Eq. (6.4.13),

$$G_2 = (100/247.302)\ (0.0005 - 0.001) + (0.0004 - 0.0005)/(0.9992)$$

$$= -0.00030226$$

Equation (6.4.12) becomes $y_2 = 0.0005 - (0.000\ 30226)\ z_{e2}$

Solving by trial with the help of the given $z_e$ versus $y$ data,

$$y_2 = 0.000198$$

$$z_{e2} = 0.999302$$

And from the equilibrium plot (Fig. 6.12.1),

$$x_2 = 0.0000788$$

Now, $$E_2 = (247.302)/(0.999302) = 247.4747 \text{ kg/hr}$$

$$R_1 = 100 + 247.4747 - 247.5 = 99.9747 \text{ kg/hr}$$

Similarly, for $i$ = 2, $$G_3 = -0.00005099$$

$$y_3 = 0.0000788 - (0.00005099)\ z_{e3}$$

Solving by trial, $y_3$ = 0.0000278 and $z_{e3}$ = 0.99947. And from equilibrium plot,

$$x_3 = 0.00000556 < x_n$$

Therefore, the number of equilibrium stages required, $n$ = 2 + (0.0000788 – 0.00001) / (0.0000788 – 0.00000556) = 2.939.

Readers may re-solve this problem by the graphical method and check the results

**Example 6.13:** Selective adsorption on silica gel at 25° C and 1 atm pressure is to be used for separating a mixture of hydrocarbon gases that is composed of 53.7 mole percent acetylene and rest ethylene. Silica gel fed to the system is free from both gases. The fractionation is to be conducted in such a way that the effluent gas must not contain more than 2.0 mole percent acetylene and the adsorbate product obtained by desorbing the spent gel should not contain more than 2.0 mole percent ethylene.

(*a*) If a multistage countercurrent operation is employed, estimate the number of equilibrium stages and the gel circulation rate required if the column operates with a reflux ratio of 2.0. What is the minimum reflux ratio?

(*b*) If a moving bed adsorber is used for the purpose with a reflux of 1.2 times the minimum, estimate the number of gas phase transfer units and the adsorbent to feed ratio that is to be employed.

The reported equilibrium data for the system at 25° C and 1 atm pressure is given below[27]:

| *Mole fraction of acetylene in adsorbate* | *Mole fraction of acetylene in gas* | *Gram moles of gas mixture adsorbed per kg of adsorbent* |
|---|---|---|
| 0.9314 | 0.7578 | 1.622 |
| 0.708 | 0.438 | 1.397 |
| 0.542 | 0.286 | 1.298 |
| 0.408 | 0.186 | 1.193 |
| 0.370 | 0.162 | 1.170 |
| 0.136 | 0.068 | 1.078 |

**Solution:** Though the analytical/graphical procedure of analysis shall remain the same even if mole fractions are employed instead of mass fractions, for sake of maintaining uniformity with other Examples, let us first rewrite the equilibrium data provided in terms of mass fractions. For example, for mole fraction of acetylene in adsorbate = 0.9314, the corresponding value of mass fraction is

$$Y = \frac{(0.9314)(26.0)}{(0.9314)(26.0)+(1-0.9314)(28.0)} = 0.9265$$

Similarly, the corresponding equilibrium mass fraction of acetylene in gas is

$$X = \frac{(0.7578)(26.0)}{(0.7578)(26.0)+(1-0.7578)(28.0)} = 0.744$$

The molecular weight of the gas mixture adsorbed (namely, the adsorbate) is (0.9314) (26.0) + (1 – 0.9314) (28.0) = 26.1372. Therefore, mass of adsorbate ($A$ + $C$) per kg of adsorbent is (1.622) (26.1372) = 42.3945 gm. Or

$$N_E = 1000/42.3945 = 23.5879$$

Computations are performed in this way and the results are tabulated below in Table 6.13.1.

## Table 6.13.1

| $X$ | $Y$ | $N_E$ |
|---|---|---|
| 0.744 | 0.9265 | 23.5879 |
| 0.420 | 0.6924 | 26.9264 |
| 0.271 | 0.5236 | 28.6200 |
| 0.175 | 0.3902 | 30.8350 |
| 0.152 | 0.3529 | 31.3536 |
| 0.0635 | 0.1275 | 33.455 |

The $N_E$ versus $Y$ plot is shown in Fig. 6.13.1 and the equilibrium curve ($Y$ versus $X$ plot) in Fig. 6.13.2.

(*a*) It is given that the feed gas mixture contains 53.7 percent acetylene (by mole) and the effluent gas is to contain not more than 2.0 mole percent acetylene. Therefore,

$$x_F = X_F = \frac{(0.537)(26.0)}{(0.537)(26.0)+(0.463)(28.0)}$$

$$= 0.5185$$

$$x_n = X_n = \frac{(0.02)(26.0)}{(0.02)(26.0)+(0.98)(28.0)}$$

$$= 0.0186$$

Also, $Y_S = 0$

and $$x_0 = X_0 = \frac{(0.98)(26.0)}{(0.98)(26.0)+(0.02)(28.0)}$$

$$= 0.9785$$

Since the adsorbent is pure $B$ ($Y_S = 0$),

$$Y_1 = X_0 = 0.9785$$

and $N_{E1}$ (from Fig. 6.13.1) = 22.4

$X_1$ (from Fig. 6.13.2) = 0.88

The minimum reflux ratio in this case is decided by that tie line which when extended passes through the point $F$. Thus, from Fig. 6.13.1

$$N_{\Delta m} = 47.5$$

Therefore,

$$RR_m = (N_{\Delta m}/N_{E1}) - 1$$

$$= (47.5/22.4) - 1.0 = 1.1205$$

Now, from Eq. (6.4.19),

$$P'/F' = P'/F = \frac{(0.0186-0.5185)}{(0.0186-0.9785)}$$

$$= 0.52078$$

And from Eq. (6.4.20),

$$R'_n/F = (1.0 - 0.52078) = 0.4792$$

Since the operating reflux ratio is 2.0,

$$R'_0/F = (2.0)\,(0.52078) = 1.04156$$

$$S' = S'_0 = 0$$

$$E'_1/F = (1.0 + 2.0)\,(0.52078) = 1.56234$$

$$G_B/F = (1.56234)\,(22.4) = 35.0$$

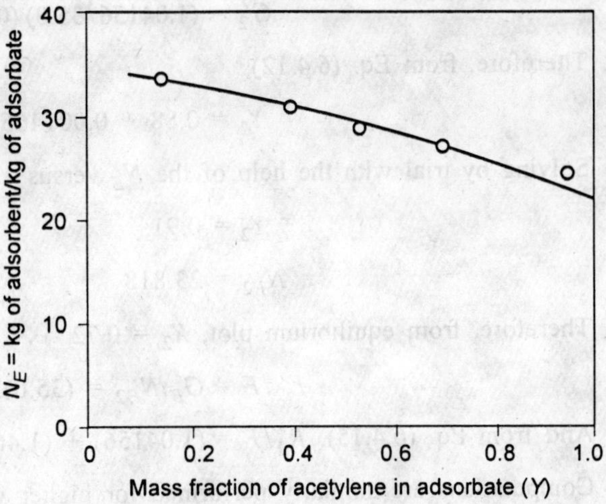


*Figure 6.13.1*

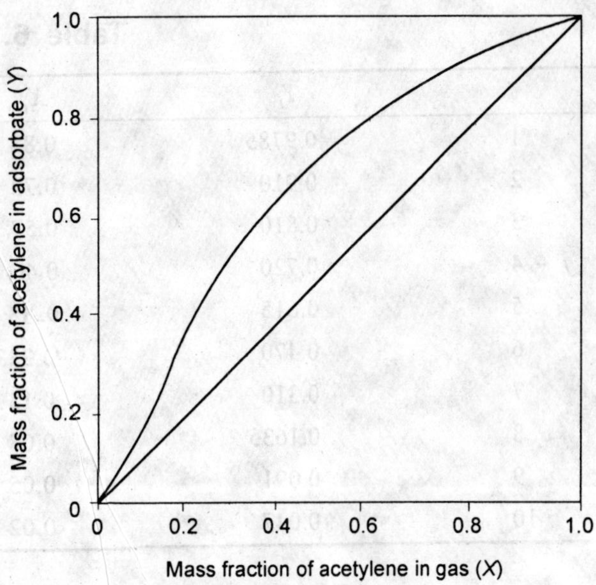


*Figure 6.13.2*

Thus, the amount of gel (namely, $B$) required is 35.0 kg per kg of feed gas mixture.

Let us compute the number of equilibrium stages required first by the analytical method. Thus, for $i = 1$, from Eq. (6.4.13),

$$G'_2 = (1.04156/35.0)\,(0.88 - 0.9785) + (0.9785 - 0.88)/(22.4) = 0.001466$$

Therefore, from Eq. (6.4.12),

$$Y_2 = 0.88 + 0.001466\,N_{E2}$$

Solving by trial with the help of the $N_E$ versus $Y$ plot/data,

$$Y_2 = 0.91$$

$$N_{E2} = 23.818$$

Therefore, from equilibrium plot, $X_2 = 0.72$

$$E'_2/F = G_B/N_{E2} = (35.0)/(23.818) = 1.4695$$

And from Eq. (6.4.15), $R'_1/F = (1.04156) + (1.4695) - (1.56234) = 0.95$

Computations are similarly performed for higher values of $i$. It is found that $X_4$ is less than $X_F$. Therefore, $f = 4$ and for $i = f = 4$, $G'_{i+1}$ is computed from Eq. (6.4.13a) and $R'_i/F$ from Eq. (6.4.15a). The results are listed below in Table 6.13.2.

**Table 6.13.2**

| $i$ | $Y_i$ | $X_i$ | $E'_i/F$ | $R'_i/F$ |
|---|---|---|---|---|
| 1 | 0.9785 | 0.88 | 1.56234 | 0.9500 |
| 2 | 0.910 | 0.72 | 1.4695 | 0.8672 |
| 3 | 0.810 | 0.57 | 1.3867 | 0.7996 |
| $f$ = 4 | 0.720 | 0.46 | 1.3191 | 1.7439 |
| 5 | 0.615 | 0.35 | 1.2634 | 1.6665 |
| 6 | 0.470 | 0.233 | 1.18604 | 1.5827 |
| 7 | 0.310 | 0.13 | 1.1022 | 1.5372 |
| 8 | 0.1635 | 0.08 | 1.0567 | 1.5159 |
| 9 | 0.091 | 0.05 | 1.0355 | 1.500 |
| 10 | 0.047 | 0.02 | 1.0204 | — |

Since $X_{10}$ is more or less equal to $X_n$, the number of equilibrium stages required may be taken as 10. The graphical solution of the problem is also illustrated in Fig. 6.13.1. To note that since $RR = 2.0$,

$$N_\Delta = (3.0)\,(22.4) = 67.2$$

The operating point Δ is thus marked on the vertical line passing through $X_0$. The line $\overline{\Delta F}$ when extended intersects the vertical line through $X_n$ at Δ′. The graphical construction is now performed in the usual manner and the results obtained are shown in Table 6.13.3.

The points $X_{10}$ and $Y_{10}$ are not shown in figure for fear of disturbing the legibility of reproduction.

**Table 6.13.3**

| $i$ | $Y_i$ | $X_i$ |
|---|---|---|
| 1 | 0.9785 | 0.88 |
| 2 | 0.910 | 0.72 |
| 3 | 0.815 | 0.57 |
| 4 | 0.725 | 0.46 |
| 5 | 0.63 | 0.362 |
| 6 | 0.48 | 0.239 |
| 7 | 0.31 | 0.13 |
| 8 | 0.165 | 0.08 |
| 9 | 0.091 | 0.05 |
| 10 | 0.045 | $0.0195 \simeq X_n$ |

(*b*) For the moving bed adsorber, the minimum reflux ratio shall be the same as that determined in part (*a*). Since the operating reflux is 1.2 times the minimum,

$$RR = (1.2)\ (1.1205) = 1.3446$$

Therefore,

$$N_\Delta = (1.0 + 1.3446)\ (22.4) = 52.519$$

The operating point $\Delta$ is thus marked on the vertical line through $X_1$ (this is same as $X_0$ for multistage equipment) as shown in Fig. 6.13.3. The line $\overline{\Delta F}$ is extended to intersect the vertical through $X_2$ (same as $X_n$ for multistage equipment) at $\Delta'$. The graphical procedure described in Fig. 6.39 (b) is now followed. A number of values of $X$ are marked on the abscissa and they are joined with $\Delta$ or $\Delta'$ (depending on whether the value of $X$ is less than $X_F$ or more than $X_F$) and the points of intersection of these lines with the $N_E$ versus $Y$ plot give the values of $Y$ (see figure). The equilibrium values of $X$ (namely, $X^*$) are then obtained from Fig. 6.13.2. The results are tabulated below in Table 6.13.4.

Values below $X = 0.10$ are not sufficiently accurate due to the distinct discontinuity in the equilibrium curve.

The integral of Eq. (6.1.98) can be now evaluated either graphically or numerically. Let us evaluate it numerically by the trapezoidal rule. This rule (in the case of unequal step sizes) can be expressed as

$$\int_{x_0}^{x_n} f(x)\, dx = (1/2) \sum_{i=1}^{n} h_i\, (f_{i-1} + f_i)$$

where $h_i = x_i - x_{i-1}$. In the presents case, $f(x) = f(X) = 1/(X - X^*)$. Thus

$$\int_{X_F}^{X_1} \frac{dX}{X - X^*} = 5.8677$$

and

$$\int_{X_2}^{X_F} \frac{dX}{X - X^*} = 12.9638$$

**Table 6.13.4**

| $X$ | $Y$ | $X^*$ | $1/(X - X^*)$ |
|---|---|---|---|
| $X_1$ = 0.9785 | 0.9785 | 0.88 | 10.1523 |
| 0.90 | 0.93 | 0.76 | 7.1428 |
| 0.85 | 0.905 | 0.71 | 7.1428 |
| 0.80 | 0.88 | 0.67 | 7.6923 |
| 0.75 | 0.855 | 0.635 | 8.6956 |
| 0.70 | 0.83 | 0.60 | 10.0 |
| 0.65 | 0.805 | 0.565 | 11.7647 |
| 0.60 | 0.782 | 0.535 | 15.3846 |
| $X_F$ = 0.5185 | 0.75 | 0.495 | 42.5532 |
| 0.45 | 0.655 | 0.385 | 15.3846 |
| 0.40 | 0.59 | 0.33 | 14.2857 |
| 0.35 | 0.52 | 0.271 | 12.6582 |
| 0.30 | 0.455 | 0.22 | 12.50 |
| 0.25 | 0.380 | 0.17 | 12.50 |
| 0.20 | 0.31 | 0.13 | 14.2857 |
| 0.15 | 0.23 | 0.10 | 20.0 |
| 0.10 | 0.15 | 0.075 | 40.0 |
| 0.05 | 0.07 | 0.035 | 66.6667 |
| $X_2$ = 0.0186 | 0.0186 | 0.0086 | 100.00 |

Now, from Eq. (6.1.98), the value of $(\text{NTU})_{OR}$ for the enriching section is

$$(\text{NTU})_{OR} = (5.8677) + (1/2)\ln\left(\frac{1-0.5185}{1-0.9785}\right) + (1/2)\ln\left[\frac{0.5185\,(0.0769)+1}{0.9785\,(0.0769)+1}\right]$$

$$= 7.405$$

To note that $r = M_A/M_C = 28/26 = 1.0769$. Similarly, the number of overall fluid phase transfer units for the adsorption section is

$$(\text{NTU})_{OR} = 12.9638 + 0.3362 = 13.3$$

The total number of gas phase transfer units required is therefore, 7.405 + 13.3 = 20.705. The value of $(P'/F)$ ratio shall be the same as that computed in part (*a*). [In fact, the expression for $(P'/F)$ can be easily deduced from Fig. 6.39 (a) by a simple material balance]. However, since the operating reflux ratio is 1.3446, the value of $(E_1'/F)$ changes. Thus,

$$E_1'/F = (2.3446)\ (0.52078) = 1.221$$

Therefore,

$$(G_B/F) = (E_1'/F)\ N_{E1} = (1.221)\ (22.4) = 27.35\ \text{kg/kg} = (S/F)$$

This is the adsorbent to feed ratio required.

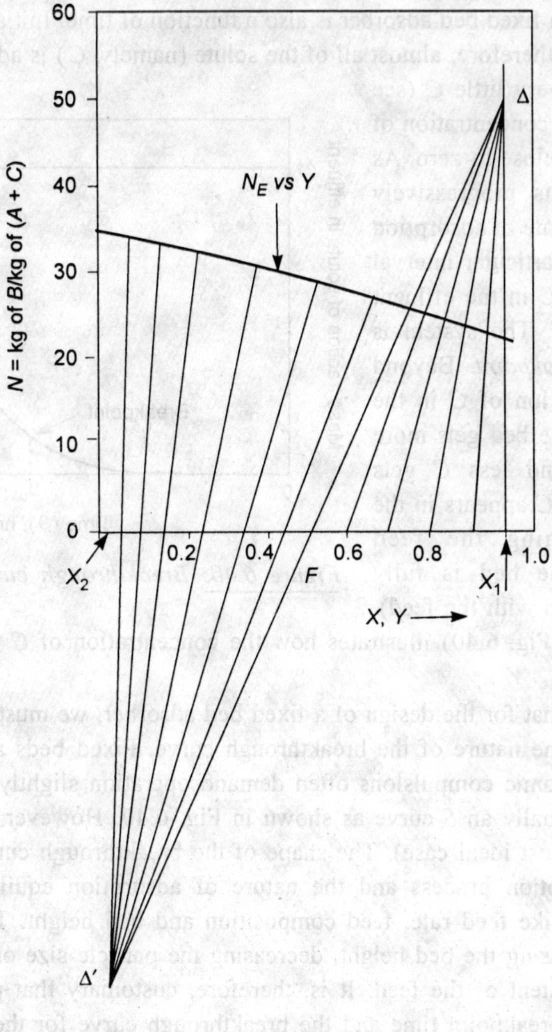


*Figure 6.13.3*

## Fixed Bed Adsorbers

The performance analysis of fixed bed adsorber is relatively more complex. Since the process is under unsteady state conditions, a steady state analysis is not possible and a transient (time-dependent) analysis becomes necessary. As the feed is introduced from below, the adsorption zone shifts and moves up the bed with passage of time. For example, consider a unit volume of feed that has been admitted. All the solute in it are adsorbed on the bottommost portion of the bed. When the next unit volume of feed enters the bed, it will be adsorbed on slightly higher portion of the bed since the bottommost zone is already exhausted. The adsorption zone has thus moved up. This continues and the adsorption zone ultimately reaches the top of the bed and then moves out whence the bed is fully exhausted. The height of the adsorption zone and the rate at which it moves up could vary with time and these are to be determined experimentally prior to design of a new column.

The rate of adsorption in a fixed bed adsorber is also a function of time. Initially, the bed is composed of essentially fresh solids and therefore, almost all of the solute (namely, $C$) is adsorbed and the outgoing fluid (effluent) shall contain very little $C$ (see Fig. 6.40). In other words, the concentration of $C$ in the effluent will be very close to zero. As the time lapses, the bed gets progressively exhausted and as a result, the rate of adsorption starts decreasing and after a particular interval of time, the concentration of $C$ in the effluent rises to a recognizable value. The system is said to have reached the *breakpoint.* Beyond the breakpoint, the concentration of $C$ in the effluent increases steadfast (the bed gets more and more exhausted, less and less $C$ gets adsorbed and more and more $C$ appears in the effluent) ultimately reaching the feed concentration ($x_1$), when the bed is fully exhausted (or is in equilibrium with the feed).

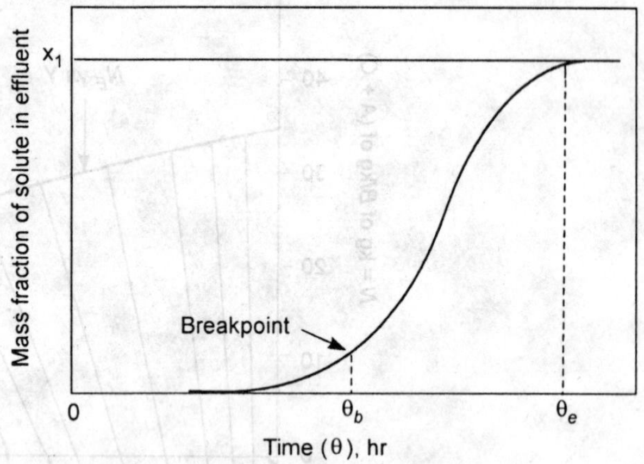

*Figure 6.40: Breakthrough curve*

The *breakthrough curve* (see Fig. 6.40) illustrates how the concentration of $C$ in the effluent increases with time.

It is, therefore, apparent that for the design of a fixed bed adsorber, we must know the time required to reach the breakpoint and the nature of the breakthrough curve. Fixed beds are best operated within the breakpoint, however economic compulsions often demand operation slightly beyond the breakpoint. The breakthrough curve is usually an *S*-curve as shown in Fig. 6.40. However, it could also be linear, distorted and even vertical (most ideal case). The shape of the breakthrough curve depends not only on the mechanism of the adsorption process and the nature of adsorption equilibrium but also on the operating/system parameters like feed rate, feed composition and bed height. The breakpoint time can usually be increased by increasing the bed height, decreasing the particle size or by decreasing the feed flow rate or initial solute content of the feed. It is, therefore, customary that prior to the design of a new fixed bed adsorber, the breakpoint time and the breakthrough curve for the system are determined experimentally using a bench scale or pilot plant scale equipment.

As stated earlier, a rigorous analysis of fixed bed adsorption process is fairly complex. We shall present here a simplified treatment that is due to Michaels[23]. This method is based on the following assumptions:

(*i*) Adsorption is from dilute feed mixtures (liquid or gas) and the operation is isothermal. Single solute adsorption is being considered.

(*ii*) The adsorption isotherm, when plotted on solute-free coordinates ($y'$ versus $x'$), is concave to the $x'$-axis.

(*iii*) The height of adsorption zone ($Z_a$) remains constant as it travels through the column and the height of adsorbent bed is large relative to the height of adsorption zone ($Z_a << Z$).

Incidentally, a good number of industrial systems do satisfy the above requirements. The breakpoint curve is also a bit idealised by assuming that the concentration of $C$ in the effluent is equal to zero below the breakpoint (in actual practices, it is very close to zero). For convenience, the curve is prepared by plotting the mass of solute-free effluent ($W$) rather than $\theta$ along the abscissa and $x'$ (mass fraction of $C$ in fluid on solute-free basis) along the ordinate, as shown in Fig. 6.41.

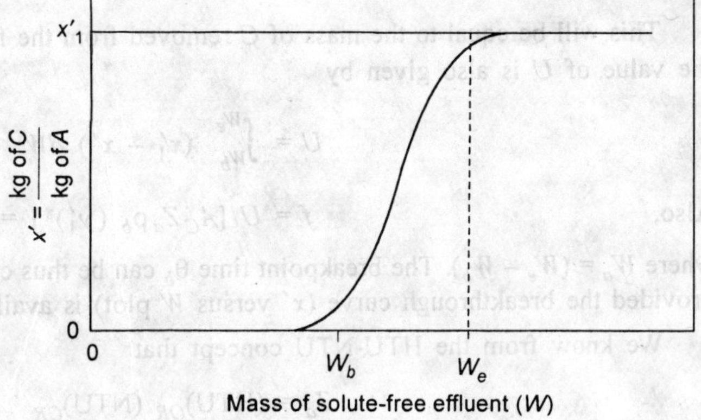


*Figure 6.41: Idealised breakthrough curve (on solute-free coordinates)*

Some low value $x'_b$ (say, 5 percent of $x'_1$) is arbitrarily chosen as the breakpoint concentration and the adsorber is considered as exhausted when the effluent concentration has risen to $x'_e$ (close to $x'_1$, say 95 percent of $x'_1$). The adsorption zone (of constant height $Z_a$) is that part of the bed in which the concentration changes from $x'_b$ to $x'_e$. If $Z$ is the total height of the adsorbent bed, $A_C$ the area of cross-section of the column and $\rho_b$ the bulk density or apparent density of the bed* ($\rho_b$ = mass of adsorbent particles / volume of bed), then the total mass of adsorbent in the bed is ($A_C Z \rho_b$). If at the breakpoint, the entire adsorbent bed has reached equilibrium with the feed, then the total amount of $C$ adsorbed within $\theta_b$ hours will be

$$= A_C Z \rho_b \, (y'_1)^* \quad \text{... (6.4.38)}$$

where $(y'_1)^*$ = equilibrium value of $y'$ corresponding to $x = x'_1$.

Therefore, if $\psi$ is the fractional saturation (degree of saturation) of the bed at the breakpoint, then the total mass of $C$ adsorbed is

$$= A_C Z \rho_b \, (y'_1)^* \, \psi \quad \text{... (6.4.39)}$$

Since the feed rate of $C$ is ($G_A A_C x'_1$), where $G_A$ is the superficial mass velocity of feed fluid on solute-free basis, and the concentration of $C$ in the effluent is zero within the breakpoint time $\theta_b$,

$$\theta_b = A_C Z \rho_b \, (y'_1)^* \, \psi/(G_A A_C x'_1) = Z \rho_b \, (y'_1)^* \, \psi/(G_A x'_1) \quad \text{... (6.4.40)}$$

The degree of saturation of the bed at the breakpoint can be expressed as

$$\psi = (Z - f Z_a)/Z = 1 - f \, (Z_a/Z) \quad \text{... (6.4.41)}$$

where $f$ is the fractional ability of the adsorbent in the adsorption zone still to adsorb solute. Thus

$$f = \frac{\text{Mass of } C \text{ adsorbed in adsorption zone (from breakpoint to exhaustion)}}{\left(\text{Mass of } C \text{ in adsorption zone if it were fully saturated with } C\right)} \quad \text{... (6.4.42)}$$

*See Table 6.3 for typical values of bulk density of popular industrial adsorbents.

If all the adsorbent in the adsorption zone were saturated with the solute, then it would contain $[A_C Z_a \rho_b (y_1')^*]$ mass of solute. Therefore, the mass of $C$ adsorbed in the adsorption zone from the breakpoint to exhaustion (let us call it $U$) is

$$U = [A_C Z_a \rho_b (y_1')^*]\, f \quad \text{... (6.4.43)}$$

$$= A_C Z \rho_b (y_1')^* (1 - \psi) \quad \text{... (6.4.44)}$$

This will be equal to the mass of $C$ removed from the fluid during the same period. In other words, the value of $U$ is also given by

$$U = \int_{W_b}^{W_e} (x_1' - x')\, dW \quad \text{... (6.4.45)}$$

Also,

$$f = U/[A_C Z_a \rho_b (y_1')^*] = U/(W_a x_1') \quad \text{... (6.4.46)}$$

where $W_a = (W_e - W_b)$. The breakpoint time $\theta_b$ can be thus computed from Eqs (6.4.40) through (6.4.45), provided the breakthrough curve ($x'$ versus $W$ plot) is available. This plot can be prepared as follows:

We know from the HTU-NTU concept that

$$Z_a = (\text{HTU})_{OR}\, (\text{NTU})_{OR} \quad \text{... (6.4.47)}$$

where $(\text{NTU})_{OR}$ is the number of fluid phase transfer units *in the adsorption zone.* If we assume that $(\text{HTU})_{OR}$ remains constant with changing concentration, then

$$\frac{Z \text{ at any } x'}{Z_a} = \frac{\displaystyle\int_{x_b'}^{x'} \frac{dx'}{x' - (x')^*}}{\displaystyle\int_{x_b'}^{x_e'} \frac{dx'}{x' - (x')^*}} \quad \text{... (6.4.48)}$$

$$= (W - W_b)/(W_e - W_b) = (W - W_b)/W_a \quad \text{... (6.4.49)}$$

For any value of $x'$, the value of $W$ can be obtained from the above equation by the graphical/numerical evaluation of the integrals. The breakthrough curve ($x'$ versus $W$ plot) can be thus prepared.

However, we must know the equation to the operating line for applying the above equation. Michaels[23] has proposed the following procedure in this connection:

We know that in a fixed bed adsorber, the adsorbent bed is stationary but the adsorption zone moves up the column (when the feed is introduced from the bottom of the column). At any time therefore, there are fresh solids at the top of the bed, while essentially spent solids at the bottom of the bed. Thus, for the moment, let us assume that fresh solids are being fed from the top and spent solids discharged from the bottom of the column [similar to that sketched in Fig. 6.39 (a)]. The most desirable situation is also considered such that the outgoing fluid (effluent) does not contain any solute ($x_2 = 0$), the fresh solids fed from the top are free from solute ($y_2 = 0$) and the outgoing solids are in equilibrium with the feed, that is, $y_1 = y_1^*$ (in actual practice, this shall however demand an infinitely tall column). An overall solute balance therefore gives

$$G_A (x_1' - 0) = G_B [(y_1')^* - 0]$$

or

$$G_A/G_B = (y_1')^*/x_1' \quad \text{... (6.4.50)}$$

A solute balance between any section and the top of the column gives

$$G_A\,(x' - 0) = G_B\,(y' - 0)$$

or

$$y' = (G_A/G_B)\,x' = [(y_1')^*/x_1']\,x' \qquad \text{... (6.4.51)}$$

This is equation to the operating line.

The above-discussed design procedure proposed by Michaels is, no doubt, an approximate method. However, it has been reported to predict reliable results in the case of a good number of industrial systems. Apart from the simplifying assumptions involved, the additional hindrance in applying the above method is the unavailability of reliable data for the estimation of mass transfer coefficients (that are required for computing HTU). Though the fluid phase mass transfer coefficient can be evaluated with reasonably good precision, reliable correlations for the estimation of $k_S$ (mass transfer coefficient in solid phase) are scarce.

Collins[24] has proposed a procedure by which the data on a pilot plant scale or bench-scale fixed bed adsorber can be extended to a large scale production unit. According to him, the adsorbent bed may be assumed to be composed of two sections, one of height $Z_S$ [in which the solute concentration is $(y_1')^*$] and the other of height $Z_{Ub}$ (called height of unused bed). Thus

$$Z = Z_{Ub} + Z_S \qquad \text{... (6.4.52)}$$

Collins reports that the value of $Z_{Ub}$ shall be the same for the pilot plant adsorber and the large scale unit, provided effects of radial mixing and bed channeling are not predominant. Accordingly, $Z_1 = Z_{Ub} + Z_{S1}$ and $Z_2 = Z_{Ub} + Z_{S2}$. Or

$$Z_2 - Z_1 = Z_{S2} - Z_{S1} \qquad \text{... (6.4.53)}$$

where suffixes 1 and 2 refer to the pilot plant adsorber and the large scale unit respectively. The expression for $Z_S$ is obtained by a simple solute balance as

$$G_A\theta_b\,(x_1' - 0) = Z_S\rho_b\,[(y_1')^* - 0]$$

or

$$Z_S = \left[\frac{G_A x_1'}{\rho_b\,(y_1')^*}\right]\theta_b = \beta\theta_b \qquad \text{... (6.4.54)}$$

Equation (6.4.53) can be, therefore, rewritten as

$$Z_2 = Z_1 + \beta\,(\theta_{b2} - \theta_{b1}) \qquad \text{... (6.4.55)}$$

Thus, once the breakpoint time ($\theta_{b1}$) for the bench scale adsorber (of bed height $Z_1$) is known, the breakpoint time for the large scale unit (of bed height $Z_2$) can be estimated from the above equation. To note that the feed mass velocity ($G_A$) and the feed composition ($x_1'$) are to be the same for the pilot plant adsorber and the large scale unit.

**Example 6.14:** An aqueous solution containing 0.12 meq $Na^+$ ions per $cm^3$ is subjected to ion exchange in a fixed bed of sulfonic acid cation exchange resin (bed height = 0.35 m), the solution being passed through the bed at a superficial velocity of 10 m/hr. The operation is isothermal and within the range of concentration considered, the relative adsorptivity of $Na^+$ with respect to $H^+$ may be assumed to be constant and equal to 1.20. At saturation, the resin will contain 2.02 meq $Na^+$ per $cm^3$ resin. The breakpoint concentration is to be taken as 5 percent of initial solution concentration and the bed will be considered exhausted when the effluent concentration is 95 percent of the initial. Estimate the time required to reach the breakpoint and the volume of effluent (per unit bed cross section) at the breakpoint. The overall liquid phase mass transfer coefficient may be computed from the correlation[23]:

$$K_L a_p = 0.86\sqrt{\mathbf{V}_L}$$

where $\mathbf{V}_L$ is the superficial liquid velocity in cm/s and $(K_L a_p)$ is expressed in meq $Na^+/cm^2 \cdot s \cdot (meq/cm^3)$.

**Solution:** The relative adsorptivity (α) of $Na^+$ with respect to $H^+$ is defined as

$$\alpha = \frac{\left(C_S/C_{S0}^*\right)}{1-\left(C_S/C_{S0}^*\right)}\,\frac{1-\left(C_L^*/C_{L0}\right)}{\left(C_L^*/C_{L0}\right)} \quad \text{... (i)}$$

where $C_S$ = concentration of $Na^+$ in solid (in resin) at equilibrium, meq/$cm^3$

$C_L^*$ = concentration of $Na^+$ in solution at equilibrium, meq/$cm^3$

$C_{S0}^*$ = concentration of $Na^+$ in solid at saturation (that is, when all $H^+$ ions were replaced by $Na^+$ ions), meq/$cm^3$

$C_{L0}$ = initial concentration of $(Na^+ + H^+)$ in solution (= total concentration of $Na^+$ and $H^+$ in solution at any time), meq/$cm^3$

It is given that $\alpha = 1.2$, $C_{S0}^* = 2.02$ meq/$cm^3$ and $C_{L0} = 0.12$ meq/$cm^3$. Substituting these values in Eq. (i) and rearranging, we get

$$C_L^* = \frac{0.12}{(2.424/C_S) - 0.2} \quad \text{... (ii)}$$

The equation to the operating line, that is Eq. (6.4.51), may be rewritten in terms of concentration expressed in meq/$cm^3$ as

$$C_S = (C_{S0}^*/C_{L0})\,C_L = (2.02/0.12)\,C_L = 16.8333\,C_L \quad \text{... (iii)}$$

The breakpoint concentration = $C_{Lb} = 0.05\,C_{L0} = (0.05)(0.12) = 0.006$ meq/$cm^3$ and the effluent concentration at exhaustion = $C_{Le} = 0.95\,C_{L0} = (0.95)(0.12) = 0.114$ meq/$cm^3$. Now, for different values of $C_L$ between $C_{Lb}$ and $C_{Le}$, the values of $C_S$ are computed from Eq. (iii) and the corresponding values of $C_L^*$ from Eq. (ii). The integral $\int_{C_{Lb}}^{C_L} dC_L/(C_L - C_L^*)$ is then evaluated, at every value of $C_L$, by numerical integration using trapezoidal rule. The results are tabulated in Table 6.14.1.

We shall show one sample calculation illustrating the evaluation of the integral whose values are listed in column (5) of Table (6.14.1). This integral may be evaluated graphically by plotting the values listed in column (4) against those of column (1), that is by plotting $1/(C_L - C_L^*)$ versus $C_L$, and then finding the area under the curve between $C_L = C_{Lb} = 0.006$ and $C_L$ = the specified value. In this Example, we have evaluated it numerically using trapezoidal rule. For instance, consider evaluation of the integral at $C_L = 0.01$. For different values of $C_L$ between 0.006 and 0.01, the values of $1/(C_L - C_L^*)$ are evaluated from Eqs (ii) and (iii) as listed below:

**Table 6.14.1**

| | $C_L$ | $C_S$ | $C_L^*$ | $1/C_L - C_L^*$ | $\int_{C_{Lb}}^{C_L} \frac{dC_L}{C_L - C_L^*}$ | $(Q_L - Q_{Lb})/Q_{La}$ | $1 - C_L/C_{L0}$ |
|---|---|---|---|---|---|---|---|
| $C_{Lb}$ = | 0.006 | 0.1010 | 0.00504 | 1043.8596 | 0.0 | 0.0 | 0.95 |
| | 0.01 | 0.1683 | 0.00845 | 645.4545 | 3.2457 | 0.1008 | 0.9166 |
| | 0.015 | 0.2525 | 0.01276 | 447.6190 | 5.9118 | 0.1837 | 0.8750 |
| | 0.02 | 0.3366 | 0.01714 | 350.00 | 7.8821 | 0.2449 | 0.8333 |
| | 0.025 | 0.4208 | 0.02158 | 292.6315 | 9.4775 | 0.2945 | 0.7916 |
| | 0.03 | 0.5050 | 0.02609 | 255.5555 | 10.8419 | 0.3369 | 0.7500 |
| | 0.035 | 0.5892 | 0.03066 | 230.2521 | 12.0526 | 0.3745 | 0.7083 |
| | 0.04 | 0.6733 | 0.03529 | 212.5 | 12.9414 | 0.4022 | 0.6666 |
| | 0.045 | 0.7575 | 0.04 | 200.0 | 13.9708 | 0.4342 | 0.6250 |
| | 0.05 | 0.8416 | 0.04477 | 191.4286 | 14.9479 | 0.4645 | 0.5833 |
| | 0.055 | 0.9258 | 0.04962 | 186.014 | 15.8903 | 0.4938 | 0.5416 |
| | 0.06 | 1.01 | 0.05454 | 183.3333 | 16.8126 | 0.5225 | 0.50 |
| | 0.065 | 1.0942 | 0.05954 | 183.2168 | 17.7279 | 0.5509 | 0.4583 |
| | 0.07 | 1.1783 | 0.0646 | 185.7143 | 18.6492 | 0.5795 | 0.4166 |
| | 0.075 | 1.2625 | 0.06976 | 191.1111 | 19.5886 | 0.6087 | 0.3750 |
| | 0.08 | 1.3466 | 0.075 | 200.0 | 20.5647 | 0.6391 | 0.3333 |
| | 0.085 | 1.4308 | 0.08032 | 213.4454 | 21.5962 | 0.6711 | 0.2916 |
| | 0.09 | 1.5150 | 0.08571 | 233.333 | 22.7099 | 0.7057 | 0.25 |
| | 0.095 | 1.5990 | 0.0912 | 263.1579 | 23.9459 | 0.7442 | 0.2083 |
| | 0.10 | 1.6833 | 0.09677 | 310.0 | 25.3695 | 0.7884 | 0.1666 |
| | 0.105 | 1.7675 | 0.10244 | 390.476 | 27.1009 | 0.8422 | 0.1250 |
| | 0.11 | 1.8516 | 0.1082 | 554.5454 | 29.4079 | 0.9139 | 0.0833 |
| $C_{Le}$ = | 0.114 | 1.919 | 0.11287 | 885.9649 | 32.1782 | 1.0 | 0.05 |

| $C_L$ | $C_S$ | $C_L^*$ | $1/(C_L - C_L^*)$ |
|---|---|---|---|
| 0.006 | 0.101 | 0.00504 | 1043.8596 |
| 0.0065 | 0.1094 | 0.005466 | 967.13 |
| 0.007 | 0.1178 | 0.00589 | 901.3906 |
| 0.0075 | 0.1262 | 0.006316 | 844.4444 |
| 0.008 | 0.1346 | 0.00674 | 794.6428 |
| 0.0085 | 0.1431 | 0.007168 | 750.7254 |
| 0.009 | 0.1515 | 0.007595 | 711.7117 |
| 0.0095 | 0.1599 | 0.008022 | 676.8278 |
| 0.01 | 0.1683 | 0.00845 | 645.4545 |

Now, from trapezoidal rule (for equal step sizes),

$$I = \int_{0.006}^{0.01} \frac{dC_L}{C_L - C_L^*} = \int_{x0}^{xn} f(x)\,dx = (h/2)\,[f_0 + 2\,(f_1 + f_2 + \ldots\ldots f_{n-1}) + f_n]$$

Here, $h = 0.0005$, $f_0 = 1043.8596$, $f_1 = 967.13$ and so on and finally, $f_n = 645.4545$. Therefore,

$$I = 3.2457$$

Similarly, all the entries of column (5) are computed. From the table,

$$(\text{NTU})_{OR} \text{ for adsorption zone} = \int_{0.006}^{0.114} \frac{dC_L}{\left(C_L - C_L^*\right)} = 32.1782$$

Therefore, by dividing each entry of column (5) by 32.1782, entries of column (6) can be obtained in accordance with Eq. (6.4.49)

The expression for computing $f$ is given by Eqs (6.4.45) and (6.4.46). These can be combined and rewritten as

$$f = \int_0^{1.0} (1 - C_L / C_{L0})\; d\left(\frac{Q_L - Q_{Lb}}{Q_{La}}\right) \quad \ldots \text{(iv)}$$

The above integral also can be evaluated numerically using trapezoidal rule from the entries of column (6) and column (7) of Table 6.14.1. Thus

$$f = 0.5118$$

From the given correlation,

$$(K_L a_p) = (K_R a_p) = 0.86\,\sqrt{\mathbf{V}_L}$$

Therefore, $$(\text{HTU})_{OR} = \mathbf{V}_L/(K_R a_p) = \mathbf{V}_L/\left(0.86\,\sqrt{\mathbf{V}_L}\right) = \sqrt{\mathbf{V}_L}\,/(0.86)$$

Since $\mathbf{V}_L = 10$ m/hr $= 0.27$ cm/s,

$$(\text{HTU})_{OR} = 0.6128 \text{ cm.}$$

The above expression for HTU can be verified as follows:

$$(\text{HTU})_{OR} = \frac{\left(\text{cm}^3 \text{ of solution}\right)/\left(\text{s}\cdot\text{cm}^2 \text{ bed cross section}\right)}{\left[\dfrac{\left(\text{meq Na}^+\right)}{\left(\text{cm}^2\cdot\text{s}\right)\left(\text{meq Na}^+/\text{cm}^3 \text{ solution}\right)}\right]\dfrac{\left(\text{cm}^2\right)}{\left(\text{cm}^3 \text{ of bed}\right)}}$$

$$= (\text{cm}^3 \text{ of bed})/(\text{cm}^2 \text{ of bed cross section})$$

Therefore, $$Z_a = (0.6128)\,(32.1782) = 19.72 \text{ cm} = 0.1972 \text{ m}$$

And from Eq. (6.4.41), $$\psi = 1.0 - (0.5118)\,(0.1972)/(0.35) = 0.7116$$

We can now compute the breakpoint time from Eq. (6.4.40) after rewriting it in terms of concentrations expressed in meq/cm$^3$. Thus, we know that at saturation, the resin contains $C_{S0}^*$ meq of $Na^+$ per cm$^3$ of resin. Since the volume of resin in the bed is $A_C Z\,(1 - \epsilon)$, where $\epsilon$ is the void fraction of the bed, the amount of $Na^+$ in the resin at the breakpoint is

$$= A_C Z\,(1 - \in)\, C^*_{S0}\, \psi$$

Since the feed rate of $Na^+$ ion is $(A_C \mathbf{V}_L C_{L0})$,

$$\theta_b = \frac{A_C Z\,(1-\in)\, C_{S0}{}^*\,\psi}{A_C\, \mathbf{V}_L\, C_{L0}} = Z\,(1 - \in)\; C^*_{S0}\,\psi/(\mathbf{V}_L C_{L0})$$

Assuming the resin bed to be tightly packed so that $(1 - \in) \simeq 1.0$,

$$\theta_b = \frac{(35.0)\,(2.02)\,(0.7116)}{(0.277)\,(0.12)} = 1509.375 \text{ seconds} = 0.419 \text{ hr}$$

The volume of effluent (per unit bed cross section) at breakpoint is

$$Q_{Lb} = \mathbf{V}_L \theta_b = (10.0)\ (0.419) = 4.19 \text{ m}^3/\text{m}^2 = 419.0 \text{ cm}^3/\text{cm}^2$$

The breakpoint time can be substantially increased by decreasing the feed velocity $(\mathbf{V}_L)$ and increasing the bed height. For example, if $\mathbf{V}_L$ = 1 m/hr (0.0277 cm/s) and $Z$ = 1.0 m, then

$$(\text{HTU})_{OR} = \sqrt{0.0277}\,/0.86 = 0.1938 \text{ cm}$$

$$Z_a = (0.1938)\ (32.1782) = 6.236 \text{ cm}$$

$$y = 1.0 - (0.5118)\ (6.236)/(100.0) = 0.968$$

$$\theta_b = \frac{(100.0)\,(2.02)\,(0.968)}{(0.0277)\,(0.12)} = 58665.87 \text{ seconds} = \mathbf{16.3\ hr}$$

## 6.5 COMPUTER PROGRAMS

Since the three operations dealt in this Chapter such as liquid-liquid extraction, solid–liquid extraction (leaching) and adsorption are quite analogous, sample computer programs are presented here for the analysis of liquid-liquid extraction only. Readers shall not find any difficulty in preparing similar programs for leaching and adsorption operations since they follow quite analogous, in fact relatively simpler, procedures (see Exercises 13, 6.Q). Two sample programs on adsorption have also been included below, one dealing with adiabatic operation of a multistage adsorber and the other batch operation of fixed bed adsorbers.

### PROGRAM 6.1

```
C**     DETERMINATION OF NUMBER OF EQUILIBRIUM STAGES
C**     REQUIRED IN MULTISTAGE CROSSCURRENT EXTRACTION
C**     THE INDEX FOR TYPE STATEMENTS IS ARBITRARILY
C**     TAKEN TO BE 50 SINCE THE VALUE OF N IS NOT
C**     KNOWN AT THE OUTSET.
        REAL MR (50)
        DIMENSION Y (50), X (50), ZE (50), ZR (50), XM (50), ZM (50),
        ERO (50), RRO (50)
        DIMENSION XE (30), YE (30), ZEE (30), ZRR (30)
```

```
C**    THE SOLVENT RATE TO EACH STAGE IS S AND THE
C**    RATIO SRO (= S/R0) IS SPECIFIED.
       READ (*, *) XO, XN, YS, ZS, SRO
       I = 1
C**    ASSUME THE VALUE OF Y (I).
       IF (I. EQ. 1) Y (I) = X0 - 0.001
       IF (I. GT. 1) Y (I) = Y (I - 1) - 0.001
10     YA = Y (I)
C**    THE SOLUBILITY DATA IS ASSUMED AVAILABLE. THE
C**    EQUILIBRIUM VALUES OF X AND Y ARE DESIGNATED
C**    AS XE (J) AND YE (J) RESPECTIVELY. THE EXTRACT
C**    COMPOSITION IS DESIGNATED AS ZEE (J) AND
C**    RAFFINATE COMPOSITION ZRR (J).
       READ (*, *) [XE (J), ZRR (J), YE (J), ZEE (J), J = 1, 30]
       DO 15 J = 1, 30
       IF [Y (I). LE. YE (J)] GO TO 20
15     CONTINUE
20     X (I) = XE (J - 1) + [XE (J) - XE (J - 1)] * [Y (I) - YE (J - 1)]/
       [YE (J) - YE (J - 1)]
       ZE (I) = ZEE (J - 1) + [ZEE (J) - ZEE (J - 1)] * [Y (I) - YE (J - 1)]/
       [YE (J) - YE (J - 1)]
       ZR (I) = ZRR (J - 1) + [ZRR (J) - ZRR (J - 1)] * [X (I) - XE (J - 1)]/
       [XE (J) - XE (J - 1)]
C**    COMPUTE Y (I) FROM EQ. (6.1.18A)
       IF (I. EQ. 1) THEN
       MR (I) = 1.0 + SR0
       XM (I) = [X (I - 1) + SRO * YS]/MR (I)
       ZM (I) = SRO * ZS/MR (I)
       ELSE
       MR (I) = RRO (I - 1) + SRO
       XM (I) = [RRO (I - 1) * X (I - 1) + SRO * YS]/MR (I)
       ZM (I) = [RRO (I - 1) * ZR (I - 1) + SRO * ZS]/MR (I)
       END IF
       Y (I) = X (I) + [ZE (I) - ZR (I)] * [XM (I) - X (I)]/[ZM (I) - ZR (I)]
C**    CHECK FOR CONVERGENCE.
       DIF = ABS [Y (I) - YA]/YA
       IF (DIF. LT. 0.005) GO TO 25
       Y (I) = YA - 0.001
       GO TO 10
25     ERO (I) = MR (I) * [XM (I) - X (I)]/[Y (I) - X (I)]
```

```
      RRO (I) = MR (I) - ERO (I)
      IF [X (I). LE. XN] GO TO 30
      I = I + 1
      Y (I) = Y (I - 1) - 0.001
      GO TO 10
30    N = I
C**   COMPUTE TOTAL SOLVENT REQUIREMENT PER KG. OF FEED
      ST = SRO * N
C**   COMPUTE COMPOSITION OF FINAL COMPOSITED EXTRACT (YF)
      SUM1 = 0.0
      SUM2 = 0.0
      DO 35 I = 1, N
      SUM1 = SUM1 + ERO (I) * Y (I)
      SUM2 = SUM2 + ERO (I)
35    CONTINUE
      YF = SUM1/SUM2
      WRITE (*, *) N, ST, YF
      WRITE (*, *) [X (I), Y (I), RRO (I), ERO (I), I = 1, N]
      STOP
      END
```

## PROGRAM 6.2

```
C**   DETERMINATION OF NUMBER OF EQUILIBRIUM
C**   STAGES IN MULTISTAGE COUNTERCURRENT EXTRACTION
C**   (WITHOUT REFLUX).
C**   THE INDEX FOR TYPE STATEMENTS IS ARBITRARILY
C**   TAKEN TO BE 50 SINCE THE VALUE
C**   OF N IS NOT KNOWN AT THE OUTSET.
      REAL N, NUMR
      DIMENSION X (50), Y (50), ZE (50), ZR (50), ERO (50), RRO (50), G
      (50)
      DIMENSION XE (30), YE (30), ZEE (30), ZRR (30)
      READ (*, *) XO, Y (1), YS, ZS, SRO
C**   THE SOLUBILITY DATA IS ASSUMED AVAILABLE. THE
C**   EQUILIBRIUM VALUES OF X AND Y ARE DESIGNATED
C**   AS XE (J) AND YE (J) RESPECTIVELY. THE EXTRACT
C**   COMPOSITION IS DESIGNATED AS ZEE (J) AND
C**   RAFFINATE COMPOSITION ZRR (J).
      READ (*, *) [XE (J), ZRR (J), YE (J), ZEE (J), J = 1, 30]
```

```
      DO 10 J = 1, 30
      IF [Y (1). LE. YE (J)] GO TO 15
10    CONTINUE
15    X (1) = XE (J - 1) + [XE (J) - XE (J - 1)] * [Y (1) - YE (J - 1)]/
      [YE (J) - YE (J - 1)]
      ZE (1) - ZEE (J - 1) + [ZEE (J) - ZEE (J - 1)] * [Y (1) - YE (J - 1)]/
      [YE (J) - YE (J - 1)]
      ZR (1) = ZRR (J - 1) + [ZRR (J) - ZRR (J - 1)] * [X (1) - XE (J - 1)]/
      [XE (J) - XE (J - 1)]
C**   COMPUTE THE COMPOSITION OF THE FINAL RAFFINATE (XN)
C**   FOR THIS, FIRST ASSUME THE VALUE OF XN.
      XN = X0 - 0.001
20    XA = XN
C**   COMPUTE ERO (1) FROM EQ. (6.1.33).
      ERO (1) = [XO + SRO * YS - (1.0 + SRO) * XN]/[Y (1) - XN]
      RRN = 1.0 + SRO - ERO (1)
C**   COMPUTE ZRN FROM EQ. (6.1.35).
      ZRN = [SRO * ZS - ERO (1) * ZE (1)]/RRN
C**   OBTAIN XN FROM SOLUBILITY DATA.
      DO 25 J = 1, 30
      IF [ZRN. LE. ZRR (J)] GO TO 30
25    CONTINUE
30    XN = XE (J - 1) + [XE (J) - XE (J - 1)] * [ZRN - ZRR (J - 1)]/[ZRR
      (J) - ZRR (J - 1)
C**   CHECK FOR CONVERGENCE.
      DIF = ABS (XA - XN)/XA
      IF (DIF. LT. 0.005) GO TO 40
      XN = XA - 0.001
      GO TO 20
C**   PERFORM STAGE TO STAGE COMPUTATIONS.
40    I = 1
C**   COMPUTE G (I + 1) FROM EQ. (6.1.42B).
45    IF (I. EQ. 1) THEN
      NUMR = ERO (I) * [Y (I) - X (I)] + [X (I) - X (I - 1)]
      DNMR = ERO (I) * [ZE (I) - ZR (I)] + [ZR (I) - ZR (I - 1)]
      ELSE
      NUMR = ERO (I) * [Y (I) - X (I)] + RRO (I - 1) * [X (I) - X (I - 1)]
      DNMR = ERO (I) * [ZE (I) - ZR (I)] + RRO (I - 1) * [ZR (I) - ZR
      (I -1)]
      END IF
```

```
      G (I + 1) = NUMR/DNMR
C**   ASSUME THE VALUE OF Y (I + 1).
      Y (I + 1) = Y (I) - 0.001
50    YA = Y (I + 1)
      DO 55 J = 1, 30
      IF [Y (I + 1). LE. YE (J)] GO TO 60
55    CONTINUE
60    ZE (I + 1) = ZEE (J - 1) + [ZEE (J) - ZEE (J - 1)] * [Y (I + 1) - YE
      (J - 1)]/[YE (J) - YE (J - 1)]
      Y (I + 1) = X (I) + G (I + 1) * [ZE (I + 1) - ZR (I)]
C**   CHECK FOR CONVERGENCE.
      DIF = ABS [Y (I + 1) - YA]/YA
      IF (DIF. LT. 0.005) GO TO 65
      Y (I + 1) = YA - 0.001
      GO TO 50
C**   COMPUTE ERO (I + 1) FROM EQ. (6.1.39A).
65    ERO (I + 1) = DNMR/[ZE (I + 1) - ZR (I)]
      IF (I. EQ. 1) RRO (I) = 1.0 + ERO (I + 1) - ERO (I)
      IF (I. GT. 1) RRO (I) = RRO (I - 1) + ERO (I + 1) - ERO (I)
C**   OBTAIN X (I + 1) FROM EQUILIBRIUM DATA.
      DO 70 J = 1, 30
      IF [Y (I + 1). LE. YE (J)] GO TO 75
70    CONTINUE
75    X (I + 1) = XE (J - 1) + [XE (J) - XE (J - 1)] * [Y (I + 1) - YE (J
      - 1)]/[YE (J) - YE (J - 1)]
      ZR (I + 1) = ZRR (J - 1) + [ZRR (J) - ZRR (J - 1)] * [X (I + 1) - XE
      (J - 1)]/[XE (J) - XE (J - 1)]
      IF [X (I + 1). LE. XN] GO TO 80
      I = I + 1
      GO TO 45
C**   COMPUTE THE NUMBER OF EQUILIBRIUM STAGES REQUIRED.
80    N = I + [X (I) - XN]/[X (I) - X (I + 1)]
      WRITE (*, *) N, RRN
      WRITE (*, *) [X (I), Y (I), ERO (I), I = 1, N]
      WRITE (*, *) [RRO (I), I = 1, N - 1]
      STOP
      END
```

## PROGRAM 6.3

```
C**    DETERMINATION OF NUMBER OF EQUILIBRIUM
C**    STAGES IN MULTISTAGE COUNTERCURRENT
C**    EXTRACTION WITH MULTIPLE FEED.
       REAL N, NUMR
C**    THE INDEX FOR TYPE STATEMENTS IS ARBITRARILY TAKEN TO BE 50 SINCE THE
C**    VALUE OF N IS NOT KNOWN AT THE OUTSET.
       DIMENSION X (50), Y (50), ZE (50), ZR (50), ERO (50), RRO (50),
       G (50)
       DIMENSION XE (30), YE (30), ZEE (30), ZRR (30)
       READ (*, *) XO, Y (1), YS, ZS, SRO, XF1, FRO
C**    THE SOLUBILITY DATA IS ASSUMED AVAILABLE. THE
C**    EQUILIBRIUM VALUES OF X AND Y ARE DESIGNATED
C**    AS XE (J) AND YE (J) RESPECTIVELY. THE EXTRACT
C**    COMPOSITION IS DESIGNATED AS ZEE (J) AND
C**    RAFFINATE COMPOSITION ZRR (J).
       READ (*, *) [XE (J), ZRR (J), YE (J), ZEE (J), J = 1, 30]
       DO 10 J = 1, 30
       IF [Y (1). LE. YE (J)] GO TO 15
10     CONTINUE
15     X (1) = XE (J - 1) + [XE (J) - XE (J - 1)] * [Y (1) - YE (J - 1)]/
       [YE (J) - YE (J - 1)]
       ZE (1) = ZEE (J - 1) + [ZEE (J) - ZEE (J - 1)] * [Y (1) - YE (J - 1)]/
       [YE (J) - YE (J - 1)]
       ZR (1) = ZRR (J - 1) + [ZRR (J) - ZRR (J - 1)] * [X (1) - XE (J - 1)]/
       [XE (J) - XE (J - 1)]
C**    COMPUTE THE COMPOSITION OF THE FINAL RAFFINATE (XN)
C**    FOR THIS, FIRST ASSUME THE VALUE OF XN.
       XN = XF1 - 0.001
20     XA = XN
C**    COMPUTE ERO (1) FROM EQ. (6.1.33A).
       ERO (1) = [XO + SRO * YS + FRO * XF1 - (1.0 + SRO + FRO) * XN]/[Y (1)
       - XN]
C**    COMPUTE ZRN FROM EQ. (6.1.35A).
       RRN = 1.0 + SRO + FRO - ERO (1)
       ZRN = [SRO * ZS - ERO (1) * ZE (1)]/RRN
C**    OBTAIN XN FROM SOLUBILITY DATA.
       DO 25 J = 1, 30
       IF [ZRN. LE. ZRR (J)] GO TO 30
25     CONTINUE
```

```
30     XN - XE (J - 1) + [XE (J) - XE (J - 1)] * [ZRN - ZRR (J - 1)]/[ZRR
       (J) - ZRR (J - 1)]
C**    CHECK FOR CONVERGENCE.
       DIF = ABS (XA - XN)/XA
       IF (DIF. LE. 0.005) GO TO 40
       XN = XA - 0.001
       GO TO 20
C**    PERFORM STAGE TO STAGE COMPUTATIONS.
40     I = 1
       FS = 0.0
45     IF (FS. EQ. 1.0) GO TO 50
       IF [X (I). LT. XF1] GO TO 55
C**    COMPUTE G (I + 1) FROM EQ. (6.1.42B).
50     IF (I. EQ. 1) THEN
       NUMR = ERO (I) * [Y (I) - X (I)] + [X (I) - X (I - 1)]
       DNMR = ERO (I) * [ZE (I) - ZR (I)] + [ZR (I) - ZR (I - 1)]
       ELSE
       NUMR = ERO (I) * [Y (I) - X (I)] + RRO (I - 1) * [X (I) - X (I - 1)]
       DNMR = ERO (I) * [ZE (I) - ZR (I)] + RRO (I - 1) * [ZR (I) - ZR (I
       - 1)]
       END IF
       G (I + 1) = NUMR/DNMR
       GO TO 60
55     FS = 1.0
       NF = I
C**    COMPUTE G (I + 1) FROM EQ. (6.1.42C).
       NUMR = ERO (I) * [Y (I) - X (I)] + RRO (I - 1) * [X (I) - X (I - 1)]
       + FRO * [X (I) - XF1]
       DNMR = ERO (I) * [ZE (I) - ZR (I)] + RRO (I - 1) * [ZR (I) - ZR
       (I - 1)] + FRO * ZR (I)
       G (I + 1) = NUMR/DNMR
C**    ASSUME THE VALUE OF Y (I + 1).
60     Y (I + 1) = Y (I) - 0.001
65     YA = Y (I + 1)
       DO 70 J = 1, 30
       IF [Y (I + 1). LE. YE (J)] GO TO 75
70     CONTINUE
75     ZE (I + 1) = ZEE (J - 1) + [ZEE (J) - ZEE (J - 1)] * [Y (I + 1) - YE
       (J - 1)]/[YE (J) - YE (J - 1)]
C**    COMPUTE Y (I + 1) FROM EQ. (6.1.42A)
```

```
      Y (I + 1) = X (I) + G (I + 1) * [ZE (I + 1) - ZR (I)]
C**   CHECK FOR CONVERGENCE.
      DIF = ABS [Y (I + 1) - YA]/YA
      IF (DIF. LE. 0.005) GO TO 80
      Y (I + 1) = YA - 0.001
      GO TO 65
80    ERO (I + 1) = DNMR/[ZE (I + 1) - ZR (I)]
      IF (I. EQ. 1) RRO (I) = 1.0 + ERO (I + 1) - ERO (I)
      IF (I. GT. 1) RRO (I) = RRO (I - 1) + ERO (I + 1) - ERO (I)
      IF (I. EQ. NF) RRO (I) = RRO (I) + FRO
C**   OBTAIN X (I + 1) FROM SOLUBILITY DATA.
      DO 85 J = 1, 30
      IF [Y (I + 1). LE. YE (J)] GO TO 90
85    CONTINUE
90    X (I + 1) = XE (J - 1) + [XE (J) - XE (J - 1)] * [Y (I + 1) - YE
      (J - 1)]/[YE (J) - YE (J - 1)]
      ZR (I + 1) = ZRR (J - 1) + [ZRR (J) - ZRR (J - 1)] * [X (I + 1) - XE
      (J - 1)]/[XE (J) - XE (J - 1)]
      IF [X (I + 1).LE. XN] GO TO 95
      I = I + 1
      GO TO 45
C**   COMPUTE THE NUMBER OF EQUILIBRIUM STAGES REQUIRED.
95    N = I + [X (I) - XN]/[X (I) - X (I + 1)]
      WRITE (*, *) N, NF, RRN
      WRITE (*, *) [X (I), Y (I), ERO (I), I = 1, N]
      WRITE (*, *) [RRO (I), I = 1, N - 1]
      STOP
      END
```

## PROGRAM 6.4

```
C**   DETERMINATION OF NUMBER OF EQUILIBRIUM STAGES
C**   IN MULTISTAGE COUNTERCURRENT EXTRACTION
C**   WITH REFLUX.
C**   COMPUTATIONS PERFORMED ON SOLVENT-FREE (B-FREE)
C**   COORDINATES.
C**   THE INDEX FOR TYPE STATEMENTS IS ARBITRARILY
C**   TAKEN TO BE 50 SINCE THE VALUE OF N
C**   IS NOT KNOWN AT THE OUTSET.
      REAL N, NUMR, NRN, NRO
      REAL NE (50), NR (50), NEE (30), NRR (30)
```

```
      DIMENSION X (50), Y (50), NE (50), NR (50), EP (50), RP (50), GP (50)
      DIMENSION XE (30), YE (30), NEE (30), NRR (30)
C**   THOUGH THE SAME NOTATIONS X AND Y ARE USED,
C**   HERE THEY REPRESENT MASS FRACTIONS ON B-FREE BASIS.
      READ (*, *) F, XF, XN, XP, YS, NS, RR
      XO = XP
C**   THE SOLUBILITY DATA IN TERMS OF MASS FRACTIONS
C**   ON B-FREE BASIS IS ASSUMED AVAILABLE. THE
C**   EQUILIBRIUM VALUES OF X AND Y ARE DESIGNATED
C**   AS XE (J) AND YE (J) RESPECTIVELY. THE EXTRACT
C**   COMPOSITION IS DESIGNATED AS NEE (J) AND
C**   RAFFINATE COMPOSITION AS NRR (J).
      READ (*, *) [XE (J), YE (J), NEE (J), NRR (J), J = 1, 30]
      DO 10 J = 1, 30
      IF [XN. LE. XE (J)] GO TO 15
10    CONTINUE
15    NRN = NRR (J - 1) + [NRR (J) - NRR (J - 1)] * [XN - XE (J - 1)]/[XE
      (J) - XE (J - 1)]
      DO 20 J = 1, 30
      IF [XO. LE. XE (J)] GO TO 25
20    CONTINUE
25    NRO = NRR (J - 1) + [NRR (J) - NRR (J - 1)] * [XO - XE (J - 1)]/[XE
      (J) - XE (J - 1)]
C**   COMPUTE THE MASS FLOW RATES (ON B-FREE
C**   BASIS) PP, RPN, RPO FROM EQS (6.1.64)
C**   TO (6.1.66)
      FP = F
      NUMR = FP * [NRN * (XF - YS) + NS * (XN - XF)]
      DNMR = NRN * (XO - YS) + NS * (XN - XO) + NRO * (YS - XN)
      PP = NUMR/DNMR
      RPN = [FP * NS - PP * (NS - NRO)]/(NS - NRN)
      RPO = RR * PP
C**   ASSUME THE VALUE OF Y (1).
      Y (1) = XO - 0.001
30    YA = Y (1)
      DO 35 J = 1, 30
      IF [Y (1). LE. YE (J)] GO TO 40
35    CONTINUE
40    NE (1) = NEE (J - 1) + [NEE (J) - NEE (J - 1)] * [Y(1) - YE (J - 1)]/
      [YE (J) - YE (J - 1)]
```

```
      X (1) = XE (J - 1) + [XE (J) - XE (J - 1)] * [Y (1) - YE (J - 1)]/
      [YE (J) - YE (J - 1)]
      NR (1) = NRR (J - 1) + [NRR (J) - NRR (J - 1)] * [X (1) - XE (J - 1)]/
      [XE (J) - XE (J - 1)]
C**   COMPUTE Y (1) FROM EQ. (6.1.71).
      Y (1) = YS + (XO - YS) * [NE (1) - NS]/(NRO - NS)
C**   CHECK FOR CONVERGENCE
      DIF = ABS [Y (1) - YA]/YA
      IF (DIF. LT. 0.005) GO TO 45
      Y (1) = YA - 0.001
      GO TO 30
C**   COMPUTE EP (1) FROM EQ. (6.1.70)
45    EP (1) = PP * (1.0 + RR) * (NRO - NS)/[NE (1) - NS]
      SPO = E (1) - PP * (1.0 + RR)
      SP = RPN + PP + SPO - FP
C**   PERFORM STAGE TO STAGE COMPUTATIONS.
      I = 1
      FS = 0.0
50    IF [FS. EQ. (1.0)] GO TO 55
      IF [X (I). LT. XF) GO TO 60
C**   COMPUTE GP (I + 1) FROM EQ. (6.1.42B)
55    IF (I. EQ. 1) THEN
      NUMR = EP (I) * [Y (I) - X (I)] + [X (I) - X (I - 1)]
      DNMR = EP (I) * [NE (I) - NR (I)] + [NR (I) - NR (I - 1)]
      ELSE
      NUMR = EP (I) * [Y (I) - X (I)] + RP (I - 1) * [X (I) - X (I - 1)]
      DNMR = EP (I) * [NE (I) - NR (I)] + RP (I - 1) * [NR (I) - NR (I -
      1)]
      END IF
      GP (I + 1) = NUMR/DNMR
      GO TO 65
60    FS = 1.0
      NF = I
C**   COMPUTE G (I + 1) FROM EQ. (6.1.42C).
      NUMR = EP (I) * [Y (I) - X (I)] + RP (I - 1) * [X (I) - X (I - 1)]
      + FP * [X (I) - XF]
      DNMR = EP (I) * [NE (I) - NR (I)] + RP (I - 1) * [NR (I) - NR (I -
      1)] + FP * NR (I)
      GP (I + 1) = NUMR/DNMR
C**   ASSUME THE VALUE OF Y (I + 1).
```

```
65    Y (I + 1) = Y (I) - 0.001
70    YA = Y (I + 1)
      DO 75 J = 1, 30
      IF [Y (I + 1). LE. YE (J)] GO TO 80
75    CONTINUE
80    NE (I + 1) = NEE (J - 1) + [NEE (J) - NEE (J - 1)] * [Y (I + 1) - YE
      (J - 1)]/[YE (J) - YE (J - 1)]
C**   COMPUTE Y (I + 1) FROM EQ. (6.1.42A).
      Y (I + 1) = X (I) + GP (I + 1) * [NE (I + 1) - NR (I)]
C**   CHECK FOR CONVERGENCE.
      DIF = ABS [Y (I + 1) - YA]/YA
      IF (DIF. LT. 0.005) GO TO 85
      Y (I + 1) = YA - 0.001
      GO TO 70
85    EP (I + 1) = DNMR/[NE (I + 1) - NR (I)]
      IF (I. EQ. 1) RP (I) = 1.0 + EP (I + 1) - EP (I)
      IF (I. GT. 1) RP (I) = RP (I - 1) + EP (I + 1) - EP (I)
      IF (I. EQ. NF) RP (I) = RP (I) + FP
C**   OBTAIN X (I + 1) FROM SOLUBILITY DATA.
      DO 90 I = 1, 30
      IF [Y (I + 1). LE. YE (J)] GO TO 95
90    CONTINUE
95    X (I + 1) = XE (J - 1) + [XE (J) - XE (J -1)] * [Y (I +1) - YE
      (J - 1)]/[YE (J) - YE (J - 1)]
      NR (I + 1) = NRR (J - 1) + [NRR (J) - NRR (J- 1)] *[X (I+1) - XE
      (J - 1)]/[XE (J) - XE (J - 1)]
      IF [X (I + 1). LE. XN] GO TO 100
      I = I + 1
      GO TO 50
C**   COMPUTE THE NUMBER OF EQUILIBNRIUM STAGES.
100   N = I + [X (I) - XN]/[X (I) - X (I + 1)]
C**   COMPUTE THE MASS FLOW RATES OF EXTRACT AND
C**   RAFFINATE STREAMS.
      DO 105 I = 1, N
105   E (I) = EP (I) * [1.0 + NE (I)]
      DO 110 I = 1, N-1
110   R (I) = RP (I) * [1.0 + NR (I)]
      P = PP * (1.0 + NRO)
      RN = RPN * (1.0 + NRN)
      SO = SPO * (1.0 + NS)
```

```
      S = SP * (1.0 + NS)
      WRITE (*, *) N, NF, RN, P, S, SO
      WRITE (*, *) [X (I), Y (I), E (I), I = 1, N]
      WRITE (*, *) [R (I), I = 1, N - 1]
      STOP
      END
```

We shall now present a sample computer program that deals with the analysis of countercurrent extraction in a continuous contact equipment (such as a packed tower). We shall use the *trapezoidal rule* for the evaluation of the integral of Eq. (6.1.99). Since equal intervals may be used (in Example 6.5, unequal intervals were used), the rule may be written as

$$\int_{y_0}^{y_n} f(y)\ dy = (h/2)\ (f_0 + 2f_1 + 2f_2 + \ldots\ldots + 2f_{n-1} + f_n) \quad \ldots (6.5.1)$$

Since zero cannot be used as index in computer program, we may rewrite the above equation as

$$\int_{y_1}^{y_{n+1}} f(y)\ dy = (h/2)\ (f_1 + 2f_2 + 2f_3 + \ldots\ldots + 2f_n + f_{n+1}) \quad \ldots (6.5.2)$$

where $h = (y_{n+1} - y_1)/n$. Here $h$ is the step size (size iof interval) and $n$ is the number of intervals. Usually, $n = 100$ shall serve the purpose.

## PROGRAM 6.5

```
C**   COUNTERCURRENT EXTRACTION IN CONTINUOUS CONTACT TOWER.
C**   ESTIMATION OF NUMBER OF TRANSFER UNITS.
C**   THE NONSOLUTE (A) IS PERFECTLY INSOLUBLE IN B.
C**   COMPUTATION BASED ON OVERALL TRANSFER COEFFICIENTS.
C**   NOTATIONS WITH REFERENCE TO FIG. 6.21 ARE FOLLOWED.
      REAL NTOE, MA, MC
      DIMENSION Y (101), X (101), XE (30), YE (30), YEC (101), F (101)
      READ (*, *) R1, X1, E2, Y2, X2, MA, MC
      GA = R1 * (1.0 - X1)
      GB = E2 * (1.0 - Y2)
      R2 = GA/(1.0 - X2)
      E1 = R1 + E2 - R2
      Y1 = 1.0 - GB/E1
C**   EVALUATE THE INTEGRAL OF EQ. (6.1.99) BY
C**   TRAPEZOIDAL RULE.
      N = 100
      Y (1) = Y2
      Y (N + 1) = Y1
      H = [Y (N + 1) - Y (1)]/N
      I = 1
```

```
C**     COMPUTE X (I) FROM EQ. (6.1.88)
10      YP = Y (I)/[1.0 - Y (I)]
        YP1 = Y1/(1.0 - Y1)
        XP1 = X1/(1.0 - X1)
        XP = (YP - YP1) * (GB/GA) + XP1
        X (I) = XP/(1.0 + XP)
C**     THE EQUILIBRIUM DATA IS ASSUMED AVAILABLE.
C**     THE EQUILIBRIUM VALUES OF X AND Y ARE
C**     DESIGNATED AS XE (J) AND YE (J).
        READ (*, *) [XE (J), YE (J), J = 1, 30]
        DO 15 J = 1, 30
        IF [X (I). LE. XE (J)] GO TO 20
15      CONTINUE
20      YEC (I) = YE (J - 1) + [YE (J) - YE (J - 1)] * [X (I) - XE (J - 1)]/
        [XE (J) - XE (J - 1)]
        F (I) = 1.0/[YEC (I) - Y (I)]
        IF (I. EQ. N + 1) GO TO 25
        I = I + 1
        Y (I) = Y (I - 1) + H
        GO TO 10
25      SUM = 0.0
        DO 30 I = 2, N
        SUM = SUM + 2.0 * F (I)
30      CONTINUE
        SI = (H/2.0) * [F (1) + SUM + F (N + 1)]
C**     COMPUTE THE NUMBER OF EXTRACT TRANSFER
C**     UNITS FROM EQ. (6.1.99).
        SR = MA/MC
        A1 = 0.5 * ALOG [(1.0 - Y1)/(1.0 - Y2)]
        A2 = 0.5 * ALOG {[Y1 * (SR - 1.0) + 1.0]/[Y2 * (SR - 1.0) + 1.0]}
        NTOE = SI + A1 + A2
        WRITE (*, *) NTOE
        STOP
        END
```

## PROGRAM 6.6

```
C**     ANALYSIS OF ADIABATIC OPERATION OF A
C**     MULTISTAGE COUNTERCURRENT ADSORBER.
C**     THE FEED IS A VAPOUR-GAS MIXTURE.
```

```
C**   SINGLE SOLUTE ADSORPTION IS CONSIDERED. NO
C**   REFLUX IS USED.
C**   SINCE THE NUMBER OF STAGES IS NOT KNOWN,
C**   THE INDEX FOR TYPE STATEMENTS IS ARBITRARILY
C**   TAKEN AS 100.
      REAL LMDO, MA, MC, N
      DIMENSION YCE (10), TLP (10), DHA (10), YC (100), HG (100), HS (100),
      TT (100), PV (100)
      READ (*, *) RO, XO, TO, XN, S, YS, TS, P
      READ (*, *) CPA, CPB, CPC, CCL, LMDO, MA, MC
C**   THE EQUILIBRIUM DATA IS ASSUMED AVAILABLE
C**   IN THE FORM SHOWN IN FIG. 6.38 (B). THAT IS,
C**   VALUES OF TLP (J) AT DIFFERENT VALUES OF
C**   YCE (J) ARE ASSUMED AVAILABLE.
      READ (*, *) YCE (J), TLP (J), J = 1, 10
C**   THE VALUES OF PV (J) AT DIFFERENT VALUES OF
C**   TT (J), NAMELY THE VAPOUR PRESSURE-TEMPERATURE
C**   DATA, ARE ASSUMED AVAILABLE.
      READ (*, *) TT (J), PV (J), J = 1, 100
C**   THE VALUES OF INTEGRAL HEAT OF ADSORPTION AT
C**   REFERENCE TEMPERATURE (0 °C), DHA (J), AT
C**   DIFFERENT VALUES OF YCE (J) ARE ASSUMED AVAILABLE.
      READ (*, *) YCE (J), DHA (J), J = 1, 10
      GR = RO * (1.0 - XO)/[S * (1.0 - YS)]
      XOC = XO/(1.0 - XO)
      XNC = XN/(1.0 - XN)
      YSC = YS/(1.0 - YS)
      YC (1) = YS + GR * (XOC - XNC)
C**   COMPUTE HGO AND HSO FROM EQ. (6.4.27)
C**   AND (6.4.28)
      TREF = 0.0
      HGO = CPA * (TO-TREF) + XOC * [CPC * (TO-TREF) + LMDO]
      IF (YSC. EQ. 0.0) THEN
      DHAC = 0.0
      GO TO 25
      ELSE
      DO 15 J = 1, 10
      IF [YSC. LT. YCE (J)] GO TO 20
15    CONTINUE
```

```
20    DHAC = DHA (J - 1) + [DHA (J) - DHA (J - 1)] * [YSC - YCE (J - 1)]/
      [YCE (J) - YCE (J - 1)]
      END IF
25    HSO = CPB * (TS-TREF) + YSC * CCL * (TS-TREF) + DHAC
C**   ASSUME THE VALUE OF TN. FOR EXAMPLE, TN = TO + 1.0
      TN = TO + 1.0
30    TA = TN
      HGN = CPA * (TN-TREF) + XNC * [CPC * (TN-TREF) + LMDO]
      HS (1) = GR * (HGO - HGN) + HSO
C**   OBTAIN THE VALUE OF DHAC AT YC (I) = YC (1).
      DO 35 J = 1, 10
      IF [YC (1). LT. YCE (J)] GO TO 40
35    CONTINUE
40    DHAC = DHA (J - 1) + [DHA (J) - DHA (J - 1)] * [YC (1) - YCE (J - 1)]/
      [YCE (J) - YCE (J - 1)]
C**   COMPUTE T (1) FROM EQ. (6.4.28).
      T (1) = TREF + [HS (1) - DHAC]/[CPB + YC (1) * CCL]
C**   OBTAIN XC (1) FROM EQUILIBRIUM DATA.
      DO 45 J = 1, 10
      IF [YC (1). LT. YCE (J)] GO TO 50
45    CONTINUE
50    TLPC = TLP (J - 1) + [TLP (J) - TLP (J - 1)] * [YC (1) - YCE (J - 1)]/
      [YCE (J) - YCE (J - 1)]
      DO 55 J = 1, 100
      IF [T (1). LT. TT (J)] GO TO 60
55    CONTINUE
60    PVC = PV (J - 1) + [PV (J) - PV (J - 1)] * [T (1) - TT (J - 1)]/[TT
      (J) - TT (J - 1)]
      PEC = PVC/EXP [TLPC/T (1)]
      XC (1) = [PEC/(P - PEC)] * (MC/MA)
C**   COMPUTE HG (1) FROM EQ. (6.4.27)
      HG (1) = CPA * [T (1) - TREF] + XC (1) * {CPC * [T (1) - TREF] + LMDO}
C**   PERFORM STAGE TO STAGE COMPUTATIONS.
      I = 1
65    IF (I. EQ. 1) THEN
      YC (I + 1) = YC (I) - GR * [XOC - XC (I)]
      HS (I + 1) = HS (I) - GR * [HGO - HG (I)]
      ELSE
      YC (I + 1) = YC (I) - GR * [XC (I - 1) - XC (I)]
      HS (I + 1) = HS (I) - GR * [HG (I - 1) - HG (I)]
```

```
      END IF
      DO 70 J = 1, 10
      IF [YC (I + 1). LT. YCE (J)] GO  TO 75
70    CONTINUE
75    DHAC = DHA (J - 1) + [DHA (J) - DHA (J - 1)] * [YC (I + 1) - YCE
      (J - 1)]/[YCE (J) - YCE (J - 1)]
      T (I + 1) = TREF + [HS (I + 1) - DHAC]/[CPB + YC (I + 1) * CCL]
      DO 80 J = 1, 10
      IF [YC (I + 1). LT. YCE (J)] GO TO 85
80    CONTINUE
85    TLPC = TLP (J - 1) + [TLP (J) - TLP (J - 1)] * [YC (I + 1) - YCE
      (J - 1)]/[YCE (J) - YCE (J - 1)]
      DO 90 J = 1, 100
      IF [T (I + 1). LT. TT (J)] GO TO 95
90    CONTINUE
95    PVC = PV (J - 1) + [PV (J) - PV (J - 1)] * [T (I + 1) - TT (J - 1)]/
      [TT (J) - TT (J - 1)]
      PEC = PVC/EXP [TLPC/T (I + 1)]
      XC (I + 1) = [PEC/(P - PEC)] * (MC/MA)
      HG (I + 1) = CPA * [T (I + 1) - TREF] + XC (I + 1) * {CPC * [T (I + 1) -
      TREF] + LMDO}
      IF [XC (I + 1). GT. XNC] THEN
      I = I + 1
      GO TO 65
      ELSE
      TN = T (I) + [T (I + 1) - T (I)] * [XC (I) - XNC]/[XC (I) - XC (I + 1)]
      END IF
C**   CHECK FOR CONVERGENCE.
      DIF = ABS (TA - TN)/TA
      IF (DIF. GT. 0.005) THEN
      TN = TN + 0.1
      GO TO 30
      ELSE
      N = I + [XC (I) - XNC]/[XC (I) - XC (I + 1)]
      END IF
      WRITE (*, *) N
      STOP
      END
```

## PROGRAM 6.7

```
C**    PERFORMANCE ANALYSIS OF A FIXED BED ADSORBER.
C**    SINGLE SOLUTE ADSORPTION AND ISOTHERMAL
C**    OPERATION ARE ASSUMED.
C**    ADSORPTION ISOTHERM IS CONCAVE TO THE ABSCISSA
C**    AND THE HEIGHT OF ADSORPTION ZONE, ZA, IS CONSTANT.
C**    CONCENTRATION OF SOLUTE IN EFFLUENT IS ZERO BELOW
C**    THE BREAKPOINT.
       REAL KR, NTU (11)
       DIMENSION XCE (50), YCE (50), YCEC (1), XC (101), YC (101)
       DIMENSION XCEC (101), F (101), WR (11), XR (11), H (11)
       READ (*, *) XC1, GA, Z, ROWB
C**    THE EQUILIBRIUM DATA ON SOLUTE-FREE COORDINATES
C**    IS ASSUMED AVAILABLE.
       READ (*, *) XCE (J), YCE (J), J = 1, 50
C**    THE BREAKPOINT CONCENTRATION IS TAKEN EQUAL
C**    TO 5 PERCENT OF FEED CONCENTRATION
C**    AND THE ADSORBER IS CONSIDERED EXHAUSTED
C**    WHEN THE EFFLUENT CONCENTRATION IS 95 PERCENT
C**    OF FEED CONCENTRATION.
       XCB = 0.05 * XC1
       XEE = 0.95 * XC1
       SH = (XCE - XCB)/100.0
C**    COMPUTE THE VALUE YCEC (1) WHICH IS THE VALUE OF YCE (I) CORRESPONDING
       TO XCE (I) = XC1.
       DO 10 J = 1, 50
       IF [XC1. LT. XCE (J)] GO TO 15
10     CONTINUE
15     YCEC (1) = YCE (J - 1) + [YCE (J) - YCE (J - 1)] * [XC1 - XCE (J -
       1)]/[XCE (J) - XCE (J - 1)]
C**    COMPUTE THE VALUES OF NTU (I) AT DIFFERENT
C**    VALUES OF XC (I) BY EVALUATING THE INTEGRAL
C**    OF EQ. (6.4.48) NUMERICALLY USING SIMPSON'S RULE.
       DO 20 K = 1, 101
       IF (K. EQ. 1) XC (K) = XCB
       IF (K. GT. 1) XC (K) = XC (K - 1) + SH
       YC (K) = [YCEC (1)/XC1] * XC (K)
C**    COMPUTE XCEC (K) FROM EQUILIBRIUM DATA.
       DO 25 J = 1, 100
```

```
      IF [YC (K). LT. YCE (J)] GO TO 30
25    CONTINUE
30    XCEC (K) = XCE (J - 1) + [XCE (J) - XCE (J - 1)] * [YC (K) - YCE (J
      - 1)]/[YCE (J) - YCE (J - 1)]
      F (K) = 1/[XC (K) - XCEC (K)]
20    CONTINUE
      NTU (1) = 0.0
      I = 2
      N = 10
35    SUM = 0
      DO 40 K = 1, N - 1
      SUM = SUM + 2.0 * F (K + 1)
40    CONTINUE
      NTU (I) = [F (1) + SUM + F (N + 1)] * (SH/2.0)
      IF (I. LT. 11) THEN
      I = I + 1
      N = N + 10
      GO TO 35
      ELSE
      DO 45 I = 1, 101
      WR (I) = NTU (I)/NTU (11)
45    CONTINUE
      K = 1
      DO 50 I = 1, 11
      XR (I) = 1.0 - XC (K)/XC1
      K = K + 10
50    CONTINUE
      END IF
C**   COMPUTE SF BY EVALUATING THE INTEGRAL USING
C**   TRAPEZOIDAL RULE (FOR UNEQUAL STEP SIZES)
C**   THE EXPRESSION FOR SF IS TO BE FIRST REWRITTEN
C**   INTO THE FORM SHOWN IN EQ. (IV) OF
C**   EXAMPLE 6.14.
      SUM = 0.0
      DO 55 I = 1, 11
      H (I) = WR (I) - WR (I - 1)
      SUM = SUM + H (I) * [XR (I - 1) + XR (I)]
55    CONTINUE
      SF = 0.5 * SUM
```

```
C**   THE MASS TRANSFER COEFFICIENTS ARE TO
C**   BE COMPUTED FROM AVAILABLE EXPERIMENTAL
C**   CORRELATIONS. IN THE PRESENT CASE, THE
C**   VALUES OF KR AND AP ARE ASSUMED AVAILABLE
      READ (*, *) KR, AP
      HTU = GA/(KR * AP)
C**   COMPUTE HEIGHT OF ADSORPTION ZONE.
      ZA = HTU * NTU (11)
C**   COMPUTE FRACTIONAL SATURATION AT BREAKPOINT
C**   FROM EQ. (6.4.41)
      PSI = 1.0 - SF * ZA/Z
C**   COMPUTE THE BREAKPOINT TIME FROM EQ. (6.4.40).
      TETAB = Z * ROWB * YCEC (1) * PSI/(GA * XC1)
C**   COMPUTE THE MASS OF EFFLUENT COLLECTED AT
C**   BREAKPOINT PER UNIT BED CROSS SECTION.
      G = GA * (1.0 + XC1)
      WB = G * TETAB
      U = Z * ROWB * YCEC (1) * (1.0 - PSI)
      WA = U/(SF * XC1)
      WRITE (*, *) TETAB, WB, WA, ZA
      STOP
      END
```

## THINGS TO REMEMBER

1. Liquid-liquid extraction involves separation of a liquid-liquid mixture (*A* and *C*) by contacting it with another immiscible liquid (*B*). The solvent (*B*) selectively extracts the solute *C* forming a solution rich in *C* which leaves the system as the extract. The extract will contain a small amount of *A* also (unless *A* is perfectly insoluble in *B*). The residual liquor that contains most of *A*, very little *C* (that remains unextracted) and a small amount of *B* is discharged from the system as the raffinate. The solvent (*B*) is recovered from the extract (and also from the raffinate) usually by distillation and is recycled back to the system. Recovery and reuse of the solvent is a critical parameter that influences the overall economy of the process.
2. Liquid-liquid extraction offers a viable alternative to distillation for the separation of constant boiling mixtures, dilute aqueous solutions and liquid mixtures which are highly heat-sensitive or relatively nonvolatile. Extraction with multiple solvents is also practised.
3. Selectivity (or separation factor) is defined as the ratio of the equilibrium distribution coefficient of the solute ($K_C$) to that of the nonsolute ($K_A$). Thus, it is analogous to relative volatility for distillation. Separation is possible only if selectivity (β) is larger than 1.0.
4. In liquid-liquid extraction, both the equilibrium phases (the extract phase as well as the raffinate phase) are ternary systems (consisting of three liquid components such as *A*, *B* and *C*). The ternary

systems usually encountered in liquid-liquid extraction are essentially of two types. The type I systems consist of two fully miscible liquid pairs (*A* and *C*, *B* and *C*) and one partially miscible liquid pair (*A* and *B*). Systems of this type are most common. Examples are diphenyl hexane (*C*)-water (*A*)-isopropyl ether (*B*), acetone (*C*)-water (*A*)-monochlorobenzene (*B*), ethanol (*C*)-benzene (*A*)-water (*B*), acetic acid (*C*)-water (*A*)-methyl isobutyl ketone (*B*). For systems like this, the solubility curve is a closed curve as shown in Fig. 6.6 (a) or (b). The equilibrium plot ($y^*$ versus $x$) is as shown in Fig. 6.6 (c) or (d). These systems exhibit a plait point *P* at which the compositions of both equilibrium phases are the same. At *P*, therefore, selectivity ($\beta$) = 1.0 and the degree of separation possible is zero. The portion of solubility curve on the left of *P* [below *P* in Fig. 6.7 (a)] is the raffinate curve ($z_r$ versus $x$) and that on the right of *P* [above *P* in Fig. 6.7 (a)] is the extract curve ($z_e$ versus $y$).

5. The type II systems consist of two partially miscible liquid pairs (*B* and *C*, And *B*) and only one fully miscible liquid pair (*A* and *C*). For such systems, the solubility curve is not a closed curve but consists of two separate branches as shown in Fig. 6.18 (a). The equilibrium plot does not exhibit any plait point and is of the form shown in Fig. 6.18 (b). Examples are methyl cyclopentane (*C*)-*n*-hexane (*A*)-aniline (*B*), styrene (*C*)-ethyl benzene (*A*)-ethylene glycol (*B*), oleic acid (*C*)-cotton seed oil (*A*)-liquid propane (*B*), methyl ethyl ketone (*C*)-chlorobenzene (*A*)-water (*B*). Phase relationships of systems of this kind are highly sensitive to temperature.
6. Liquid-liquid extraction equipment may be stagewise contactors (such as sieve-tray towers, mixer-settler cascades) or continuous contactors (spray towers, packed towers, baffled columns with provision for mechanical agitation). The separation efficiency of continuous contactors is highest when both phases execute plug flow. Any degree of axial or longitudinal dispersion shall diminish the concentration gradient between phases and will thus reduce the rate of transfer of solute. Axial dispersion is maximum in spray towers and consequently, their separation efficiency is relatively low. Presence of packing, baffles and mechanical agitators minimise axial mixing and enhance separation efficiency. Introduction of pulsations also helps in increasing the rates of extraction. Centrifugal extractors, though demand high operating cost, provide substantially larger mass transfer rates.
7. Both batchwise as well as continuous operation are employed in liquid-liquid extraction, though the latter is preferred in large capacity installations. The operation may be in the multistage crosscurrent mode (Fig. 6.8) or in the multistage countercurrent mode (Fig. 6.12). For the same number of stages, the latter demands less amount of solvent to achieve the desired separation and it provides a richer (more concentrated) extract. Countercurrent extraction with multiple feed (Fig. 6.15) is also often employed. All the three processes may be analysed either by the graphical method or by the analytical method. The analytical method involves writing down the mass balance equations around each stage and then solving them simultaneously. This method is more rigorous but demands large computational load. The graphical method is more handy for manual computations. The graphical solutions are illustrated in Figs 6.9 (a) and (b), Fig. 6.13 (a) and Fig. 6.16 respectively for multistage crosscurrent extraction, multistage countercurrent extraction (without reflux) and multistage countercurrent extraction (with multiple feed). In case the nonsolute (*A*) is totally insoluble in *B* (so that no *A* is lost in the extract and no *B* is lost in the raffinate), then a simplified procedure as shown in Fig. 6.11 or Fig. 6.14 may be employed.

8. The minimum solvent to feed ratio for countercurrent extraction may also be determined graphically by locating that tie line which when extended will intersect the line $\overline{Sx_n}$ at a point nearest to the solubility curve [if tie lines are sloping upward—$K < 1.0$ as shown in Fig. 6.13 (b)] or at a point farthest from the solubility curve if $K > 1.0$ and tie lines are sloping down ward [as shown in Fig. 6.13 (b)] or at a point farthest from the solubility curve if $K > 1.0$ and tie lines are sloping down ward [as shown in Fig. 6.13 (d)]. In many cases, but not always, the tie line which when extended passes through the feed point $F$ decides the minimum solvent to feed ratio [as shown in Fig. 6.13 (c)]. If solvent-free coordinates are used, then $\Delta_m$ is located below the solubility curve (the lowest point of intersection of all tie lines with $\overline{SX_n}$). When $S/R_0$ is minimum, the number of equilibrium stages required is infinite. To affect separation in a finite number of stages therefore, the operating point $\Delta$ is to be chosen above $\Delta_m$ (below $\Delta_m$ if solvent-free coordinates are used).
9. Reflux is used in liquid-liquid extraction only at the extract end and that too for the type II systems only. In this case also, the process may be analysed either by the graphical method (as shown in Fig. 6.19) or by the analytical method. Graphical determination of minimum reflux ratio is illustrated in Fig. 6.20 (a) and number of stages at total reflux ($RR = \infty$) can be estimated as shown in Fig. 6.20 (b). The operating reflux ratio, that lies between these two limits, is usually fixed based on economic considerations.
10. Countercurrent extraction in continuous contact towers can be analysed by the conventional HTU-NTU method (same as that used for gas absorption). However, this method is precise only for those systems in which the nonsolute ($A$) is totally insoluble in $B$. The required tower diameter (for packed towers) may be estimated from the flooding plot of Crawford and Wilke (Fig. 6.22).
11. Leaching (solid–liquid extraction) is a process quite analogous to liquid-liquid extraction. The feed to a leaching unit is necessarily a mixture of solids ($A + C$) from which the solute ($C$) is selectively dissolved off or extracted by a suitable solvent ($B$). The solution of $C$ in $B$ so obtained is the extract or leach solution (also called *miscella*). It is also called the *overflow* since from the separator (settling tank, filter, centrifuge), it usually overflows from the top. The leached solids, consisting essentially of $A$, are discharged as the *underflow* or raffinate (also called *marc*). The overflow is usually a clear solution of $C$ in $B$, while the underflow consists of particles of $A$ suspended in a solution of $C$ in $B$.
12. Alternate terminologies such as lixiviation, decoction and elution are also used to denote solid–liquid extraction. In decoction, the solvent is used at its boiling point and elution involves washing off the solute from the surfaces of insoluble solids. Leaching is extensively used in mineral processing industries and food processing industries. It is also widely employed for washing chemical precipitates.
13. The mechanism of leaching involves three major steps such as (*i*) dissolution of solute in solvent, (*ii*) diffusion of solute through the porous structure of the residual solids to the particle surface, (*iii*) diffusion of solute from particle surface to the liquid bulk. Either the second or the third step controls the overall rate of leaching, since the first is usually too rapid. Feed solids are well-crushed to reduce resistance to the second step and the slurry is well-agitated to promote the third step. Vegetable seeds are flaked to rupture the cell walls. In special cases like leaching of sugar from beets, complete rupture of cell walls is avoided since cell walls help in retaining undesirable constituents like colloidal and albuminous substances as the cell walls are impermeable to them

(but permeable to sugar and water). Increase in temperature increases the solubility of solute, decreases viscosity of the liquid and enhances diffusion rates. Rate of leaching is, therefore, high at high temperatures. The exit streams from a leaching unit are subjected to a number of treatments for the recovery of valuable constituents (either $C$ or $A$ or both) and for the recovery and reuse of the solvent. This includes evaporation, distillation (stripping) and often electrolysis (electrolytic deposition).

14. The leaching equipment may be the stationary bed type or the dispersed contact type. In the former, the solids to be leached form a stationary bed and the leach solvent percolates through it, while in the latter, the solids are kept suspended or dispersed in the leach solvent. Percolation tanks or diffusers, filters (filter cake washing) belong to the former category while continuous thickeners or mixer-settler cascades, rotating plate extractors fall under the latter category. A number of percolation tanks may be operated in series such that when $(n - 1)$ tanks are in operation, one tank is being emptied and refilled with fresh solids. Such a scheme (called Shanks system, sketched in Fig. 6.24) permits continuous countercurrent operation without the movement of solids from one tank to another. However, all tanks will not be in operation at a time. The Rotocel extractor and Bollman extractor follow this principle and in addition, the tanks/buckets are continuously moved to maintain continuity of operation. The Kennedy extractor also work on this principle, but the solids are moved from one tank to another by a slow-moving impeller, the tanks being kept stationary.
15. The dispersed contact systems employ mixer-settler combinations (one as contacting device and the other as the separating device). Alternately, both leaching and separation may be affected in the same sedimentor (thickener). To improve capacity, a number of thickeners are operated in series with countercurrent flow of leach solution and the underflow sludge (the CCD unit). Pachuca tanks are pneumatically agitated vessels in which both leaching and separation can be carried out. In equipment like rotating plate extractors, solids are moved countercurrent to liquid by means of screw conveyors or other mechanical devices.
16. As compared to liquid-liquid extraction, the analysis of solid–liquid extraction is relatively simpler. The procedure employed is same as that for liquid-liquid extraction. However, the analysis is more conveniently carried out on $A$-free coordinates. Thus, a modified scheme of notation is employed in which $N$ represents mass fraction of $A$ on $A$-free basis and $X$ and $Y$ represent mass fraction of $C$ on $A$-free basis in the underflow solution and in the overflow solution respectively [see Eqs (6.2.1) to (6.2.3)]. Also $R'$, $E'$, $S'$ denote mass flow rates on $A$-free basis. Since the overflow is usually free from $A$, the plot of $N_E$ versus $Y$ (overflow curve or extract curve) coincides with the abscissa. The $N_R$ versus $X$ plot (underflow curve or raffinate curve) is to be prepared from available experimental data. At any equilibrium stage, the composition of the overflow solution may be assumed equal to that of the solution leaving with the leached solids in the underflow (that is $Y^* = X$) unless adsorption of some amount of $C$ on $A$ is anticipated. Accordingly, the equilibrium plot coincides with the 45° diagonal and all tie lines in the graphical construction shall be vertical.
17. Multistage crosscurrent leaching can be analysed analytically using Eqs (6.2.6) to (6.2.8) with relative case (for the usual case of clear overflow and $Y^* = X$). In case $Y^* \neq X$, then the analysis is to be done by solving Eq. (6.2.10) by trial with the help of the available equilibrium data or graphically as shown in Fig. 6.26. In case of constant underflow [the ratio of $A$ to $(B + C)$ is the same in all underflow streams], the analysis becomes still more simple as Eq. (6.2.6) gets simplified

to Eq. (6.2.16) or (6.2.18). Both of these equations are applicable for $i > 1$ and for the usual case of $Y^* = X$. If $Y^* \neq X$, then Eqs (6.2.19) and (6.2.20) are to be employed (they are to be solved by trial) or solution may be obtained using the graphical procedure illustrated in Fig. 6.27.

18. In case of multistage countercurrent leaching, the number of equilibrium stages required can be computed analytically by solving Eqs (6.2.25) and (6.2.26) or (6.2.26a) simultaneously. Equation (6.2.26a) applies when $Y^* = X$ (usual case) and Eq. 6.2.26 when $Y^* \neq X$. The procedure does not demand any trial and error computations. Alternately, the graphical procedure illustrated in Fig. 6.28 or 6.29 may be employed. In case the overflow stream from each stage contains a specific concentration of $A$ (that is, $N_{Ei} = NE =$ constant), then also the same graphical analysis may be utilised except that in this case, the overflow curve shall not coincide with the abscissa but will be a straight line parallel to the abscissa at $N = NE$ (see Example 6.9). If the operation is under constant underflow, then the number of equilibrium stages can be determined straightaway from a Kremser-Brown-Souders type correlation, such as Eq. (6.2.29a) or (6.2.29b). This is for the usual situation when $Y^* = X$. If $Y^* \neq X$, then also the value of $n$ can be easily determined by solving Eq. (6.2.28) with the help of equilibrium data. If instead of constant underflow, the mass ratio of $A$ to $B$ is constant in all underflow streams, then again the value of $n$ can be directly computed using Eq. (6.2.29c).

19. Extraction using supercritical fluids has been of great interest during the recent years. Supercritical fluids have gas-like viscosities and liquid-like densities. Their surface tension is close to zero and this enhances the mass transfer of solute within the porous structure of the solids. Their gas-like diffusivities help in enhancing mass transfer from particle surface to fluid bulk. Their liquid-like densities give them a high capacity for solutes and since they are highly compressible, their density (and thereby their solvent capacity or selectivity) can be varied over wide ranges by adjusting the pressure or temperature. Also, the solvent can be easily recovered from the extract by reducing the pressure and/or temperature. Supercritical extraction is being widely employed in food processing and pharmaceutical industries, in petroleum and petrochemical industry, in polymer processing and in the detoxification of waste water.

20. Separation of the components of a fluid mixture (gas mixture or liquid solution) using selective adsorption on a solid (called the *adsorbent*) is widely practised in process industries. The substances adsorbed are called the *adsorbate*. As a separation process, adsorption is analogous to liquid-liquid extraction. It differs from solid–liquid extraction in the sense that transfer is being made from fluid phase to solid phase and not vice versa as in solid–liquid extraction. Single solute adsorption from liquid solutions is often termed as *contact filtration*. What is discussed in this Chapter is physical adsorption which is due to intermolecular force of altraction between solid molecules and the substances adsorbed. Chemisorption which involves chemical interaction between the solid and the adsorbate is important in catalysis and in the analysis of catalytic reactors (see Chapter 8). For many difficult separations such as fractionation of air into nitrogen and oxygen, adsorption has been found to be economically superior to distillation.

21. The most important property of an industrial adsorbent is its specific surface (surface area per unit mass or unit volume). This varies from 600 $m^2/gm$ to as large as 1800 $m^2/gm$ (see Table 6.3). The most advanced types of adsorbents are synthetic zeolites (molecular sieves) and molecular sieve carbons. Separation by adsorption can be by steric, kinetic or equilibrium effect.

The majority of industrial processes utilise equilibrium adsorption and hence are equilibrium separation processes.

22. As in extraction, regeneration and reuse of adsorbent is crucial in adsorption as well. Desorption of the adsorbent can be accomplished by heating the spent adsorbent (thermal swing adsorption or TSA) or by lowering the pressure (pressure swing adsorption or PSA). The latter is more popular. Desorption may also be achieved by passing an inert purge gas over the adsorbent or with the help of a solvent in which the adsorbate is more soluble (the process termed as *elution*).
23. Chromatographic separations are fixed bed processes. *Adsorption chromatography* or *migrational chromatography* involves repeated adsorption and desorption of a pulse of feed which produce adsorption waves that travel down the bed, the wave of less strongly adsorbed component moving faster and ahead of that of more strongly adsorbed component. In *partition chromatography,* the adsorbent is a liquid contained in the pores of a granular solid. The process will be, therefore, that of absorption if the feed is a gas mixture and extraction if it is a liquid solution. In *affinity chromatography,* the adsorbent has preferential affinity to some specific substances and therefore can be used for selective removal of these substances from solutions.
24. Ion exchange processes involve exchange of ions (cations or anions) between the ion exchange resin (cation exchange resin or anion exchange resin) and the feed solution (substrate). Synthetic ion exchange resins may be strong acid/strong base resins or weak acid/weak base resins. Resins that display unusually high selectivity for certain cations have been synthesised and these are called *chelating ion exchange resins.* The resins are available in the gel form or in the macroporous form. *Ion retardation chromatography* uses a bifunctional resin which contains exchangeable cations as well as anions and therefore can simultaneously remove both ions from the substrate.
25. Adsorption isotherms follow different patterns and as many as five to eight varieties are popularly known. Four of them are sketched in Fig. 6.31. Among the algebraic correlations that have been proposed to describe the equilibrium relationship, the Freundlich isotherm Eq. (6.4.3), the Langmuir isotherm Eq. (6.4.2), the B-E-T isotherm Eq. (6.4.4) and the trinomial [Eq. (6.4.5)] are most popular. Equilibrium plots are often made available in the form of the Othmer chart, which consists of a family of plots of equilibrium partial pressure of solute ($C$) versus vapour pressure of pure $C$ at different values of $y'$ (mass fraction of $C$ in solids on $C$-free basis). Such a chart Fig. 6.38 (a) could also be used for estimating the integral heat of adsorption through Eqs (6.4.33) to (6.4.35).
26. As in the case of extraction, adsorption equipment may also be either stagewise contact equipment (mixer-settler cascades, mixer-filter cascades, sieve plate columns with fluidised bed of adsorbent on each plate) or continuous contact equipment (moving bed adsorbers, fixed bed adsorbers). Steady state, continuous operation is possible in all except in fixed bed adsorbers which are usually operated batchwise or in the semicontinuous mode (cyclic operation). A fully continuous operation can be made possible by using a rotating fixed bed adsorber (Fig. 6.35).
27. Adsorption processes can be analysed by following exactly the same analytical/graphical procedures that are employed for liquid-liquid extraction. The analysis is relatively simpler since in adsorption, $z_r = N_R = 0$ for all values of $x$ (or $X$) owing to the fact that the adsorbent is nonvolatile and insoluble in the fluid. Accordingly, the $z_r$ versus $x$ (or $N_R$ versus $X$) plot coincides with the abscissa.
28. Though isothermal operation is generally preferred in adsorption, adiabatic operation is also often used especially when the heat affects accompanying the process are significant. The performance

of a multistage adiabatic adsorber can be analysed by the same procedure as that used for the analysis of adiabatic gas-liquid absorption.

29. Performance analysis of fixed bed adsorber is relatively more complex since the operation is under unsteady state conditions and the rate of adsorption is a function of time. The adsorption zone moves up the bed with passage of time. As the adsorption rate falls with lapse of time, the concentration of solute ($C$) in the effluent increases and reaches a recognisable value at the breakpoint. Beyond this point, the effluent concentration further increases steadfast, ultimately reaching the feed concentration when the bed is fully exhausted. The shape of the breakthrough curve depends not only on the mechanism of adsorption process and nature of adsorption equilibrium but also on operating/system parameters like feed rate, feed composition and bed height. A reasonably satisfactory analysis of fixed bed adsorption is possible by the method proposed by Michaels. That is, through Eqs (6.4.40) to (6.4.51). Also, Eq. (6.4.55), which is due to Collins, may be used for scaling up pilot plant data to large scale industrial units.

## NOMENCLATURE

| | |
|---|---|
| $A$ | nonsolute component |
| $A_C$ | area of cross section of tower, $m^2$ |
| $a$ | specific interfacial are, $m^2/m^3$ |
| $a_p$ | specific packing surface, $m^2/m^3$ |
| $B$ | extraction solvent/adsorbent |
| $C$ | solute component |
| $C_{CL}$ | specific heat of component $C$ (as liquid), kJ/(kg·K) |
| $C_{pA}$, $C_{pB}$ | specific heat of component $A$ and that of adsorbent ($B$) respectively, kJ/(kg·K) |
| $C_{pC}$ | specific heat of component $C$ (as vapour), kJ/(kg·K) |
| $E_f$ | extraction factor; adsorption factor, dimensionless |
| $E_i$ | mass flow rate of extract (solids, in case of adsorption) from $i$th stage, kg/s |
| $E_i'$ | same as above but on $B$-free basis, kg/s; mass flow rate of leach solution from $i$th stage on $A$-free basis (in case of leaching), kg/s |
| $F$ | feed rate in countercurrent extraction/adsorption with reflux, kg/s |
| $F_1$ | Mass rate of intermediate feed stream, kg/s |
| $f$ | parameter defined in Eq. (6.4.42), dimensionless |
| $G_A$ | mass flow rate of nonsolute ($A$), kg/s; superficial mass velocity/molar mass velocity of $A$ in case of packed tower/moving bed adsorber, kg/($m^2$·s) or kmole/($m^2$·s) |
| $G_B$ | mass flow rate of pure solvent/adsorbent ($B$), kg/s; superficial mass velocity/molar mass velocity of $B$ in case of packed tower/moving bed adsorber, kg/($m^2$·s) or kmole/($m^2$·s) |
| $G_{Bi}$ | mass flow rate of pure solvent/adsorbent ($B$) to $i$th crosscurrent stage, kg/s |
| $G_{i+1}$ | arbitrarily defined parameter [see Eqs (6.1.42b) and (6.1.42c) |
| $H_{Gi}$ | specific enthalpy of fluid (vapour–gas mixture) leaving $i$th stage, kJ/kg of $A$ |
| $H_{Si}$ | specific enthalpy of solids leaving $i$th stage, kJ/kg of $B$ |
| $H_{GO}$ | specific enthalpy of feed fluid, kJ/kg of $A$ |

$H_{SO}$ specific enthalpy of feed solids, kJ/kg of $A$

$HTU$ height of a transfer unit, m

$K$ equilibrium distribution coefficient, dimensionless

$K_A$, $K_C$ equilibrium distribution coefficient of nonsolute ($A$) and that of solute ($C$) respectively, dimensionless

$K_E$, $K_R$ overall extract phase and raffinate phase mass transfer coefficient respectively, kmole/($m^2 \cdot s \cdot$ mole fraction) or kg/($m^2 \cdot s \cdot$ mass fraction)

$k_E$, $k_R$ extract phase and raffinate phase mass transfer coefficient respectively, kmole/($m^2$.s. mole fraction) or kg/($m^2 \cdot s \cdot$ mass fraction)

$M_A$, $M_C$ molecular weight of $A$ and that of $C$ respectively, kg/kmole.

$n$ number of equilibrium stages.

$N$ mass fraction of $B$ on $B$-free basis; mass fraction of nonsolute ($A$) on $A$-free basis (in case of leaching)

$N_{Ei}$ mass fraction of $B$ on $B$-free basis in extract (solids, in case of adsorption) leaving the $i$th stage; mass fraction of $A$ on $A$-free basis in leach solution (overflow) leaving the $i$th stage

$N_{Ri}$ mass fraction of $B$ on $B$-free basis in raffinate fluid leaving the $i$th stage; mass fraction of $A$ on $A$-free basis in underflow leaving the $i$th stage (in case of leaching)

$N_S$ mass fraction of $B$ on $B$-free basis in solvent (solids, in case of adsorption) fed to the system; mass fraction of $A$ on $A$-free basis in solvent fed to the leaching unit

$NTU$ number of transfer units

$P$ mass flow rate of final extract product/adsorbate product (in countercurrent operation with reflux), kg/s

$P'$ same as above but on $B$-free basis, kg/s

$\bar{P}$ vapour pressure of pure $C$, N/$m^2$

$P^*$ equilibrium partial pressure of $C$, N/$m^2$

$P_C$ critical pressure, N/$m^2$

$Q_L$ volume of effluent collected at any time (see under fixed bed adsorption), $m^3$

$R_i$ mass flow rate of raffinate fluid leaving $i$th stage, kg/s

$R'_i$ same as above but on $B$-free basis, kg/s; mass flow rate of leached solids (underflow) leaving the $i$th stage on $A$-free basis, kg/s

$R_0$ feed rate, kg/s; reflux rate, kg/s

$R'_0$ feed rate/reflux rate on $B$-free basis; feed rate to leaching unit on $A$-free basis, kg/s

$RR$ reflux ratio, dimensionless

$RR_m$ minimum reflux ratio, dimensionless

$S$ solvent rate (feed rate of solids, in case of adsorption), kg/s

$S'$ solvent rate (feed rate of solids, in case of adsorption) on $B$-free basis, kg/s; solvent rate on $A$-free basis (in case of leaching), kg/s

$S_i$ solvent rate (feed rate of solids, in case of adsorption) to $i$th crosscurrent stage, kg/s

$S'_i$ same as above but on $B$-free basis, kg/s; solvent rate to $i$th crosscurrent stage on $A$-free basis (in case of leaching), kg/s

| | |
|---|---|
| $S_0$ | mass flow rate of solvent (solids, in case of adsorption) recycled (see Fig. 6.17), kg/s |
| $S'_0$ | same as above but on *B*-free basis, kg/s |
| $T_C$ | critical temperature, °C |
| $t_i$ | temperature of *i*th equilibrium stage, °C |
| $t_{Gi}$, $t_{Si}$ | temperature of fluid (vapour–liquid mixture) and that of solids respectively leaving the *i*th stage (assumed equal to $t_i$), °C |
| $t_0$, $t_S$ | temperature of feed fluid and that of feed solids respectively, °C |
| $t_{ref}$ | reference temperature, °C |
| $U$ | mass of solute (*C*) adsorbed in the adsorption zone from breakpoint to exhaustion, kg |
| $V_C$, $V_D$ | superficial velocity of continuous phase and that of dispersed phase respectively, m/s |
| $V_{CF}$, $V_{DF}$ | same as above but at flooding, m/s |
| $W$ | mass of solute-free effluent collected at anytime, kg |
| $W_b$, $W_e$ | mass of solute-free effluent collected at breakpoint and at exhaustion respectively, kg |
| $x$ | mass fraction of solute (*C*) in raffinate fluid |
| $x_1$, $x_2$ | mass fraction of solute (*C*) in feed to and in outlet fluid (effluent) from countercurrent continuous contact equipment (same as $x_0$ and $x_n$ for countercurrent multistage equipment) |
| $x'$ | mass fraction of solute (*C*) in raffinate fluid on *C*-free basis |
| $x'_b$, $x'_e$ | mass fraction of *C* (on *C*-free basis) in effluent at breakpoint and at exhaustion respectively. |
| $x_F$ | mass fraction of *C* in feed (in countercurrent extraction/adsorption with reflux) |
| $x_{F1}$ | mass fraction of *C* in intermediate feed stream |
| $x_i$ | mass fraction of *C* in raffinate fluid leaving the *i*th stage. |
| $x'_i$ | same as above but on *C*-free basis. |
| $x_0$ | mass fraction of *C* in feed; mass fraction of *C* in reflux |
| $x'_0$ | same as above but on *C*-free basis |
| $x_P$ | mass fraction of *C* in extract product/adsorbate product (in countercurrent operation with reflux) |
| $X$ | mass fraction of *C* (on *B*-free basis) in raffinate fluid; mass fraction of *C* on *A*-free basis in solution leaving with leached solids (underflow) |
| $X_i$ | mass fraction of *C* on *B*-free basis in raffinate fluid leaving the *i*th stage; mass fraction of *C* on *A*-free basis in solution leaving with leached solids (underflow) from *i*th stage |
| $X_F$ | mass fraction of *C* on *B*-free basis in feed (in countercurrent extraction/adsorption with reflux) |
| $X_0$ | mass fraction of *C* on *B*-free basis in feed; mass fraction of *C* on *B*-free basis in reflux; mass fraction of *C* on *A*-free basis in feed to leaching unit |
| $X_P$ | mass fraction of *C* on *B*-free basis in extract product/adsorbate product (in countercurrent operation with reflux) |
| $y$ | mass fraction of *C* in extract (solids, in case of adsorption) |
| $y'$ | same as above but on *C*-free basis |
| $y_1$, $y_2$ | mass fraction of *C* in final extract (spent solids, in case of adsorption) from and in feed solvent (feed solids, in case of adsorption) to countercurrent continuous contact equipment (same as $y_1$ and $y_S$ for countercurrent multistage equipment) |

$y_i$ mass fraction of $C$ in extract (solids, in case of adsorption) leaving the $i$th stage

$y'_i$ same as above but on $C$-free basis

$(y'_0)^*$ equilibrium value of $y'$ corresponding to $x' = x'_0$

$y_S$ mass fraction of $C$ in solvent (solids, in case of adsorption) fed to the system

$y'_S$ same as above but on $C$-free basis

$Y$ mass fraction of $C$ on $B$ free basis in extract (solids, in case of adsorption); mass fraction of $C$ on $A$-free basis in leach solution (overflow)

$Y_i$ mass fraction of $C$ on $B$-free basis in extract (solids, in case of adsorption) leaving the $i$th stage; mass fraction of $C$ on $A$-free basis in leach solution (overflow) leaving the $i$th stage

$Y_S$ mass fraction of $C$ on $B$-free basis in solvent (solids, in case of adsorption) fed to the system; mass fraction of $C$ on $A$-free basis in solvent fed to leading unit

$z_{ei}$ mass fraction of $B$ in extract (solids, in case of adsorption) leaving the $i$th stage

$z_{ri}$ mass fraction of $B$ in raffinate fluid leaving the $i$th stage

$z_{r0}$ mass fraction of $B$ in the feed; mass fraction of $B$ in the reflux

$z_S$ mass fraction of $B$ in the solvent (solids, in case of adsorption) fed to the system

$Z$ height of continuous contact tower; packed height of packed tower, m

$Z_a$ height of adsorption zone, m

$Z_{Ub}$ height of unused bed [see Eq. (6.4.52)], m

$\beta$ selectivity (separation factor), dimensionless

$\Delta H_a$ integral heat of adsorption, kJ/kg of $B$

$\theta_b$ breakpoint time, s

$\lambda_0$ latent heat of vapourisation of $C$ at reference temperature, kJ/kg

$\in$ void fraction of packed bed, dimensionless

$\mu$ fluid viscosity, kg/(m·s)

$\rho$ fluid density, kg/m$^3$

$\rho_b$ bulk density (also called apparent density) of packed bed, kg/m$^3$

$\sigma$ interfacial tension, N/m

$\psi$ fractional saturation of bed at breakpoint, dimensionless

## COMPUTER NOTATIONS

| | |
|---|---|
| AP = $a_p$ | specific packing surface, m$^2$/m$^3$ |
| CPA = $C_{pA}$ | specific heat of nonsolute component $A$ (gas), kJ/(kg K) |
| CPB = $C_{pB}$ | specific heat of solid adsorbent ($B$), kJ/(kg K) |
| CPC = $C_{pC}$ | specific heat of solute $C$ (as vapour), kJ/(kg K) |
| CCL = $C_{CL}$ | specific heat of $C$ (as liquid), kJ/(kg K) |
| DHA (J) = $\Delta H_a$ | integral heat of adsorption at 0° C at different values of $y'_i$, kJ/kg of $B$ |
| DHAC – | calculated value of $\Delta H_a$ at any specific value of $y'_i$ |

| | | |
|---|---|---|
| E1, E2 | = $E_1$, $E_2$ | inlet and outlet mass flow rates respectively of solids/solvent in a continuous contact equipment (see Fig. 6.21), kg/s |
| E (I) | = $E_i$, | mass flow rate of extract/solids leaving the $i$th stage, kg/s |
| EP (I) | = $E'_i$ | same as above but on $B$-free basis, kg/s |
| ERO (I) | = $(E_i/R_0)$ | ratio, dimensionless |
| F | = $F$ | feed rate in countercurrent operation with reflux, kg/s |
| F (I) | – | function $f(y)$, see Eq. (6.5.2) |
| FP | = $F'$ | same as above but on $B$-free basis |
| FRO | = $(F_1/R_0)$ | ratio, dimensionless |
| G (I + 1) | = $G_{i+1}$ | arbitrarily defined parameter [see Eqs (6.1.42b) and (6.1.42c)] |
| GA, GB | = $G_A$, $G_B$ | mass flow rate/mass velocity of pure $A$ and that of pure $B$ respectively |
| GP (I + 1) | = $G'_{i+1}$ | parameter $G_{i+1}$ on $B$-free basis |
| GR | = $(G_A/G_B)$ | ratio, dimensionless |
| H (I) | = $h_i$ | step size in numerical integration |
| HG (I) | = $H_{Gi}$ | enthalpy of vapour–gas mixture leaving the $i$th stage, kJ/kg of $A$ |
| HGN | = $H_{Gn}$ | enthalpy of vapour–gas mixture leaving the $n$th stage, kJ/kg of $A$ |
| HGO | = $H_{GO}$ | enthalpy of feed fluid, kJ/kg of $A$ |
| HS (I) | = $H_{Si}$ | enthalpy of solids leaving $i$th stage, kJ/kg of $B$ |
| HSO | = $H_{SO}$ | enthalpy of feed solids, kJ/kg of $B$ |
| KR | = $K'_R$ | overall fluid phase mass transfer coefficient, kg/(m$^2$·s·$\Delta x'$) |
| LMDO | = $\lambda_0$ | latent heat of vapourisation of $C$ at 0° C, kJ/kg |
| MA, MC | = $M_A$, $M_C$, | molecular weight of $A$ and that of $C$ respectively, kg/kmole |
| MR (I) | = $(M_i/R_0)$ | ratio, dimensionless |
| $N$ | = $n$ | number of equilibrium stages |
| NE (I) | = $N_{Ei}$ | mass fraction of $B$ on $B$-free basis in extract/solids leaving the $i$th stage |
| NEE (J) | – | values of $N_E$ listed in solubility data |
| NF | = $f$ | the feed stage |
| NR (I) | = $N_{Ri}$, | mass fraction of $B$ on $B$-free basis in the raffinate fluid leaving the $i$th stage |
| NRN, NR0 | = $N_{Rn}$, $N_{R0}$ | mass fraction of $B$ on $B$-free basis in the raffinate fluid leaving the $n$th stage and that in the feed respectively |
| NRR (J) | – | values of $N_R$ listed in solubility data |
| NS | = $N_S$ | mass fraction of $B$ on $B$-free basis in solvent/solids fed to the system |
| NTOE | = $(\text{NTU})_{OE}$ | number of overall extract phase transfer units |
| P | = $P$ | operating pressure, N/m$^2$ |
| PE (I) | = $P_i^*$ | equilibrium partial pressure of $C$ at temperature $t_i$ and concentration $y'_i$, N/m$^2$ |

| | | |
|---|---|---|
| PEC | – | calculated value of $P^*$ at any specified value of $y'_i$ and temperature, $N/m^2$ |
| PP | $= P'$ | mass flow rate on $B$-free basis of extract/adsorbate product in countercurrent operation with reflux, kg/s |
| PSI | $= \psi$ | fractional saturation of bed at breakpoint |
| PV (I) | $= \bar{P}$ | vapour pressure of pure $C$, $N/m^2$ |
| PVC | – | calculated value of $\bar{P}$ at any specified temperature, $N/m^2$ |
| R (I) | $= R_i$ | mass flow rate of raffinate fluid leaving the $i$th stage, kg/s |
| R1, R2 | $= R_1, R_2$ | mass flow rate of feed and that of effluent (final raffinate) in a continuous contact equipment (see Fig. 6.21), kg/s |
| RN | $= R_n,$ | mass flow rate of raffinate fluid leaving the $n$th stage, kg/s |
| R0 | $= R_0$ | feed rate, kg/s |
| RP (I) | $= R'_i$ | mass flow rate of raffinate fluid leaving the $i$th stage on $B$-free basis, kg/s |
| RPN | $= R'_n$ | mass flow rate of raffinate fluid leaving the $n$th stage on $B$-free basis kg/s |
| RPO | $= R'_0$ | feed rate on $B$-free basis, kg/s |
| RR | $= RR,$ | reflux ratio, dimensionless |
| RRN | $= (R_n/R_0)$ | ratio, dimensionless |
| RRO (I) | $= (R_i/R_0)$ | ratio, dimensionless |
| S | $= S$ | feed rate of solvent/solids to the system, kg/s |
| SF | $= f$ | parameter defined in Eq. (6.4.42), dimensionless |
| SH | $= h$ | step size in numerical integration |
| SO | $= S_0$ | mass flow rate of solvent/solids recycled, kg/s |
| SP | $= S'$ | feed rate of solvent/solids to the system on $B$-free basis, kg/s |
| SPO | $= S'_0$ | mass flow rate of solvent/solids recycled on $B$-free basis, kg/s |
| SRO | $= (S/R_0)$ | ratio, dimensionless |
| T (I) | $= t_i$ | temperature of $i$th equilibrium stage, °C |
| TETAB | $= \theta_b$ | breakpoint time, s |
| TLP (J) | $= T_i \ln (\bar{P}_i/P^*_i)$ | see Fig. 6.38 (b) |
| TLPC | – | calculated value of TLP ($J$) at any specific value of $y'_i$, °K or °C |
| TN | $= t_n$ | temperature of $n$th equilibrium stage, °C |
| TO | $= t_0$ | temperature of feed fluid, °C |
| TREF | $= t_{ref}$ | reference temperature, °C |
| TS | $= t_s$ | temperature of solids fed to the system, °C |
| TT (J) | – | values of temperature listed in vapour pressure–temperature data, °C |
| U | $= U$ | mass of $C$ adsorbed in adsorption zone from breakpoint to exhaustion, $kg/m^2$ of bed cross section |
| WB | $= W_b$ | mass of effluent (per unit bed cross section) collected at breakpoint, $kg/m^2$ |

| | |
|---|---|
| WR (I) = $(W_i - W_b)/W_a$ | dimensionless |
| X (I) = $x_i$ | ($X_i$ in Program 6.4), mass fraction (on *B*-free basis in Program 6.4) of *C* in raffinate fluid leaving the *i*th stage |
| X1, X2 = $x_1$, $x_2$ | mass fraction of *C* in the feed and in effluent (final raffinate) respectively in a continuous contact equipment (see Fig. 6.21) |
| XC1 = $x'_1$ | mass fraction of *C* in feed to a continuous contact equipment on *C*-free basis (see Fig. 6.21/6.39) |
| XC (I) = $x'_i$ | mass fraction of *C* on *C*-free basis in raffinate fluid leaving the *i*th stage |
| XCB, XEE = $x'_b$, $x'_e$ | mass fraction of *C* in effluent (on *C*-free basis) at breakpoint and at exhaustion respectively |
| XCE (J), YCE (J) – | values of $x'$ and $y'$ listed in equilibrium data |
| XCEC (J) – | calculated values of $(x'_i)$* |
| XE (J), YE (J) | equilibrium values of *x* and *y* (or, *X* and *Y*) listed in equilibrium data |
| XF = $X_F$ | mass fraction of *C* on *B*-free basis in feed (in countercurrent operation with reflux) |
| XF1 = $x_{F1}$ | mass fraction of *C* in intermediate feed |
| XN = $x_n$ | ($X_n$ in Program 6.4) |
| XNC = $x'_n$ | mass fraction of *C* (on *C*-free basis) in raffinate fluid leaving the *n*th stage |
| XO = $x_0$ | ($X_0$ in Program 6.4) |
| XOC = $x'_0$ | mass fraction of *C* in feed fluid (on *C*-free basis) |
| XP = $X_P$ | mass fraction of *C* on *B*-free basis in extract/adsorbate product (in countercurrent operation with reflux) |
| XR (I) = $(1 - x'_i/x'_1)$ | dimensionless |
| Y (I) = $y_i$ | ($Y_i$ in Program 6.4), mass fraction of *C* (on *B*-free basis in Program 6.4) in extract/solids leaving the *i*th stage |
| Y1, Y2 = $y_1$, $y_2$ | (see Fig. 6.21/6.39) |
| YC (I) = $y'_i$ | mass fraction of *C* (on *C*-free basis) in extract/solids leaving the *i*th stage |
| YCE (J) – | See *XCE (J)* |
| YCEC (J) = | calculated values of $(y'_i)$* |
| YE (J) – | see *XE (J)* |
| YEC (I) = | calculated values of $(y_i)$* |
| YS = $y_S$ | ($Y_S$ in Program 6.4) |
| YSC = $y'_S$ | mass fraction of *C* on *C*-free basis in solvent/solids fed to the system |
| Z = $Z$ | height of packed bed |
| ZA = $Z_a$ | height of adsorption zone |

| | | |
|---|---|---|
| ZE (I) | = $z_{ei}$ | mass fraction of $B$ in extract/solids leaving the $i$th stage |
| ZEE (J), ZRR (J) | – | values of $z_e$ and $z_r$ listed in solubility data |
| ZR (I) | = $z_{ri}$ | mass fraction of $B$ in raffinate fluid leaving the $i$th stage |
| ZRN | = $z_{rn}$ | mass fraction of $B$ in raffinate leaving the $n$th stage |
| ZS | = $z_S$ | mass fraction of $B$ in solvent/solids fed to the system |

## EXERCISES

6.A. Agree with or contradict against the following statements giving reasons. You may supplement your answer with examples/sketches:

(*a*) In liquid-liquid extraction, selectivity of the solvent is highest at the plait point and so is the separation achieved,

(*b*) In multistage countercurrent liquid-liquid extraction, raffinate reflux is always advantageous while extract reflux is of advantage only for those systems that contain two fully miscible liquid pairs,

(*c*) Higher degree of axial dispersion is responsible for the higher separation efficiency of continuous contact extractors as compared to stagewise contact equipment,

(*d*) For the separation of dilute aqueous mixtures, solvent extraction is more economical than distillation,

(*e*) Higher operating temperatures are advantageous in liquid-liquid extraction but not in leaching of solids,

(*f*) In countercurrent multiple-stage extraction, the tie line which when extended passes through the point ($x_F$, 0) determines the minimum amount of solvent required to affect the desired separation,

(*g*) Supercritical fluids exhibit high selectivity due to their high viscosity and high density.

6.B. Briefly outline what do you understand by

(*a*) Supercritical extraction,

(*b*) Shanks system,

(*c*) Axial dispersion in continuous contact towers,

(*d*) Migrational chromatography

(*e*) Chelating ion exchange resins.

6.C. Acetic acid is to be extracted from 2000 kg/hr of a 40 mass percent aqueous solution at 20° C using pure isopropyl ether as the solvent. The exit raffinate is to contain 3.0 mass percent acetic acid.

(*a*) What is the minimum solvent rate required for a multistage countercurrent operation? Estimate the number of equilibrium stages required if a solvent rate of 3000 kg/hr is employed.

(*b*) If a 10 stage crosscurrent operation (each stage fed with 300 kg/hr of pure isopropyl ether) is employed, what would be the composition of the final raffinate? Estimate also the acid content of the final composited extract and the fractional solute recovery and compare them with those obtained in the countercurrent operation of part (*a*).

(*c*) Estimate the solvent requirement for a single batch operation that would give the same raffinate composition (0.03 mass fraction acetic acid).

The solubility data for acetic acid-water-isopropyl ether system at 20° C and 1 atm pressure is given below[12]:

| Water layer | | Isopropyl ether layer | |
|---|---|---|---|
| *Mass fraction of acetic acid* | *Mass fraction of ether* | *Mass fraction of acetic acid* | *Mass fraction of ether* |
| 0.0 | 0.010 | 0.0 | 0.9940 |
| 0.01 | 0.014 | 0.0033 | 0.9895 |
| 0.02 | 0.016 | 0.0063 | 0.9860 |
| 0.05 | 0.019 | 0.014 | 0.9760 |
| 0.10 | 0.0215 | 0.036 | 0.950 |
| 0.20 | 0.028 | 0.070 | 0.908 |
| 0.30 | 0.035 | 0.115 | 0.839 |
| 0.40 | 0.058 | 0.259 | 0.654 |
| 0.44 | 0.096 | 0.301 | 0.591 |
| 0.45 | 0.116 | 0.319 | 0.563 |
| 0.465 | 0.166 | 0.362 | 0.487 |
| 0.448 | 0.313 | 0.448 | 0.313 |

6.D. In the extraction of alcohol from an alcohol-benzene solution by water, ten extraction vessels are used in continuous countercurrent extraction of 5000 kg/hr of feed containing 0.125 mass fraction alcohol. The extract contains 0.40 mass fraction alcohol and the raffinate 0.99 mass fraction benzene. Four of the vessels are put out of service by an accident. What is the concentration of alcohol in the raffinate and what is the increase in loss of alcohol when six vessels are used compared with the original ten? The solubility data for benzene-ethanol-water system at 25° C and 1 atm is given below[13]:

| Benzene layer mass fraction | | Water layer mass fraction | |
|---|---|---|---|
| *Benzene* | *Ethanol* | *Benzene* | *Ethanol* |
| 0.9935 | 0.0060 | 0.0013 | 0.1120 |
| 0.9800 | 0.0185 | 0.0028 | 0.2200 |
| 0.9630 | 0.0350 | 0.0053 | 0.3090 |
| 0.9525 | 0.0445 | 0.0102 | 0.3600 |
| 0.9520 | 0.0450 | 0.0110 | 0.3640 |
| 0.9200 | 0.0740 | 0.0335 | 0.4350 |
| 0.9180 | 0.0755 | 0.0370 | 0.4410 |
| 0.9075 | 0.0845 | 0.0480 | 0.4590 |
| 0.8440 | 0.1385 | 0.1370 | 0.5220 |
| 0.8000 | 0.1745 | 0.2045 | 0.5215 |
| 0.7060 | 0.2475 | 0.3320 | 0.4800 |
| 0.6905 | 0.2585 | 0.3490 | 0.4720 |
| 0.6401 | 0.2930 | 0.4095 | 0.4400 |
| 0.5200 | 0.3740 | 0.5200 | 0.3740 |

6.E. A liquid solution containing 40 mass percent styrene and balance ethyl benzene is to be extracted at 25° C with recycle diethylene glycol solvent that contains 0.005 mass fraction styrene so as to reduce the styrene content of the feed solution to 10.0 mass percent.

(*a*) If continuous countercurrent multistage operation is employed, determine the minimum solvent to feed ratio required.

(*b*) If a solvent to feed ratio that is 1.3 times the minimum is used, what will be the number of equilibrium stages required in the above case?

(*c*) Compute the number of equilibrium stags required if a continuous countercurrent cascade with extract reflux (reflux ratio = 10.0) is employed, the final extract product being discharged at 0.80 mass fraction styrene (on solvent-free basis).

(*d*) Determine the minimum reflux ratio that may be used in the above case. What will be the number of equilibrium stages at total reflux?

The solubility data for styrene-ethyl benzene-diethylene glycol system at 25° C and 1 atm pressure is given below[11]:

| *Hydrocarbon-rich layer (mass fraction on solvent-free basis)* | | *Solvent-rich layer (mass fraction on solvent-free basis)* | |
|---|---|---|---|
| *Styrene* | *Glycol* | *Styrene* | *Glycol* |
| 0.0 | 0.00675 | 0.0 | 8.62 |
| 0.087 | 0.00817 | 0.1429 | 7.71 |
| 0.1883 | 0.00938 | 0.273 | 6.81 |
| 0.288 | 0.01010 | 0.386 | 6.04 |
| 0.384 | 0.01101 | 0.480 | 5.44 |
| 0.458 | 0.01215 | 0.557 | 5.02 |
| 0.464 | 0.01215 | 0.565 | 4.95 |
| 0.561 | 0.01410 | 0.655 | 4.46 |
| 0.573 | 0.01405 | 0.674 | 4.37 |
| 0.781 | 0.01833 | 0.833 | 3.47 |
| 1.00 | 0.0256 | 1.00 | 2.69 |

6.F. A countercurrent multiple-stage extraction system at 20° C is to be used for the recovery of acetic acid from an aqueous solution containing 0.29 mass fraction acetic acid. The acid is to be concentrated by extraction with isopropyl ether, removing the strong extract at 0.08 mass fraction acetic acid, concentrating in a fractionating column and returning the ether at 0.004 mass fraction acetic acid for reuse. Raffinate leaves at 0.03 mass fraction acetic acid. Another aqueous feed stream containing 0.12 mass fraction acetic acid is also to be treated in the extraction system at a rate of 3.0 kg per 14.0 kg of solution containing 0.29 mass fraction acetic acid. How many ideal stages will be required for the separation? At which stage should each stream be introduced? The solubility data is given in Exercise 6. C.

6.G. A 25 percent solution of dioxane in water is to be extracted at the rate of 1000 kg/hr with pure benzene to remove 95 percent of dioxane.

(*a*) Compute the solvent requirement for a single batch operation.

(*b*) If extraction were done with equal amounts of solvent in five crosscurrent stages, how much solvent would be required?

(*c*) If a continuous countercurrent multiple-stage operation is employed, what will be the minimum solvent requirement? How many equilibrium stages will be required if a solvent rate of 900 kg/hr is employed?

(*d*) If extraction is to be performed continuously in a spray tower, determine the number of overall raffinate transfer units.

The equilibrium distribution of dioxane between water and benzene at 25° C is as given below[4]:

| | | | | | |
|---|---|---|---|---|---|
| Mass fraction of dioxane in water | : | 0.0 | 0.051 | 0.189 | 0.252 |
| Mass fraction of dioxane in benzene | : | 0.0 | 0.052 | 0.225 | 0.320 |

At these concentrations, water and dioxane are essentially insoluble.

6.H. A feed solution containing $x_0$ mass fraction of solute ($C$) is to be extracted down to $x_n$ mass fraction solute using a solvent containing $y_S$ mass fraction $C$ and $z_S$ mass fraction $B$ in $N$ equilibrium stages by multistage crosscurrent extraction. Write down a computer program for estimating the solvent rate (per kg of feed) required to each stage, if each stage is being fed with the same amount of solvent.

6.I. A feed solution containing $X_F$ mass fraction solute ($C$) is extracted with a solvent that is pure $B$ in a multistage countercurrent extraction system that uses extract reflux (reflux ratio = $RR$). The extract product is discharged at $X_P$ mass fraction $C$ and the raffinate product at $X_n$ mass fraction $C$ (both on $B$-free basis). Write a computer program for estimating the number of equilibrium stages required and the total solvent requirement.

6.J. It is desired to extract oil from halibut livers by multistage leaching with ethyl ether. The fresh halibut livers contain 25.7 mass percent oil and are fed to the first stage at the rate of 1000 kg/hr. The leach solution from each stage is clear (contains no suspended livers) and equilibrium is attained in each stage such that the concentration of oil in the leach solution and that in the solution withdrawn alongwith the leached solids may be assumed to be the same (no adsorption of solute occurs). The quantity of solution retained by the granulated livers has been determined experimentally as a function of the composition of the solution as given below[17]:

| *kg of oil in 1 kg of solution* | *kg of solution retained by 1 kg of oil-free livers* |
|---|---|
| 0.0 | 0.205 |
| 0.10 | 0.242 |
| 0.20 | 0.286 |
| 0.30 | 0.339 |
| 0.40 | 0.405 |
| 0.50 | 0.489 |
| 0.60 | 0.60 |
| 0.65 | 0.672 |
| 0.70 | 0.765 |
| 0.72 | 0.810 |

(*a*) If leaching is carried out in a six-stage crosscurrent system using 1.2 kg of oil-free ether per kg of fresh liver (this being equally divided among stages), compute what percent of oil in the fresh livers is extracted and the concentration of oil in the final solution.

(*b*) If the operation is to be carried out countercurrently using 0.26 kg of oil-free ether per kg of fresh liver, estimate the number of equilibrium stages required to extract the same percent of oil as in (*a*).

6.K. Titanium dioxide pigment is produced by precipitation and the slurry from the precipitation tank contains $TiO_2$ particles suspended in a 55 percent salt solution, 1.0 kg of solution per kg of suspended solids. This is fed to a Dorr thickener and agitated with 25 kg of fresh water. The batch is allowed to settle, the clear solution withdrawn as overflow and replaced by an equal mass of water and the mixture is agitated again. Estimate how many times this procedure is to be repeated if the final leached product is to be 99.9 percent pure when dried. The plant is to produce 100 kg of $TiO_2$ per day. The reported settling characteristics of the slurry under the existing operating conditions are as follows:

| *kg of solution per kg of $TiO_2$ in settled sludge* | *Mass fraction of dissolved salt in the solution of settled sludge* |
|---|---|
| 0.30 | 0.0 |
| 0.32 | 0.1 |
| 0.34 | 0.2 |
| 0.36 | 0.3 |
| 0.38 | 0.4 |
| 0.40 | 0.5 |

No adsorption of salt on $TiO_2$ is reported.

[**Hint:** Since the final leached product is to be 99.9 percent pure on dry basis, $A/C = 99.9/0.1 = 999.0$ and since $G_A = 100$ kg/day, $C = R'_n X_n = 100/999$. Now find $X_n$ as in Example 6.7.]

6.L. A mineral containing 85.47 percent gangue, 10.26 percent ore and 4.27 percent moisture is to be leached with water countercurrently in a series of extractors. The strong extract solution is to contain 7.0 percent ore and 98 percent of the ore initially loaded to the system is to be recovered. If each kg of inert gangue retains 2 kg of water alongwith whatever ore that is dissolved in that water, compute the required number of equilibrium stages. What will be the total amount of water required per kg of mineral fed? The overflow from each stage is clear solution and no adsorption of ore on the gangue is reported.

6.M. Vegetable seeds containing $M_1$ mass percent oil are to be leached at the rate of $R_0$ kg/hr to recover the oil using a recycle solvent that contains $N_1$ mass percent oil. The final leached solids discharged should not contain more than $X_n$ mass fraction oil (on inert-free basis). It is observed that overflow from each stage contains $m_1$ kg of seeds per kg of solution.

(*a*) Write a computer program for the estimation of the number of equilibrium stages required if the operation is countercurrent and the solvent is fed to the system at the rate of $S$ kg/hr.

(*b*) Write a computer program for determining the solvent rate required per stage if the above separation is to be affected in $N$ cross-current stages.

The $N_R$ versus $X$ data may be assumed available.

6.N. A gas mixture containing 50 mole percent propane and 50 mole percent propylene is to be separated by selective adsorption on silica gel at 25° C and 1 atm pressure. The gel fed to the system is free from both gases.

(*a*) If a rotating plate column that operates countercurrently is employed for the purpose, estimate the minimum gel rate (per kmole of gas) required to restrict the propylene content of effluent gas at 5 mole percent. What would be the number of equilibrium stages required if 300 kg of silica gel are supplied per kmole of feed gas?

(*b*) If the process is çonducted in three crosscurrent stages, each stage being fed with 300 kg of silica gel per kmole of feed gas, what would be the final composition of the effluent gas?

(*c*) If a multistage countercurrent equipment that operates at twice the minimum reflux is used to fractionate the gas mixture into two products, one containing 95 percent propane and the other 5 percent propane, estimate the number of equilibrium stages and the mass of silica gel required per kmole of feed.

(*d*) If a moving bed adsorber that operates at twice the minimum reflux is recommended to affect the fractionation mentioned in part (*c*), estimate the number of gas phase transfer units and the mass of silica gel required per kmole of feed.

The reported equilibrium data[27] for simultaneous adsorption of propane and propylene on silica gel at 25° C and 1 atm pressure is given below:

| *Mole fraction of propane in adsorbate* | *Mole fraction of propane in gas phase* | *kg of silica gel/kmoles of gas mixture adsorbed* |
|---|---|---|
| 0.0 | 0.0 | 416.67 |
| 0.1 | 0.1625 | 450.00 |
| 0.2 | 0.390 | 483.33 |
| 0.3 | 0.560 | 500.00 |
| 0.4 | 0.685 | 516.67 |
| 0.5 | 0.716 | 550.00 |
| 0.6 | 0.842 | 566.67 |
| 0.7 | 0.892 | 585.70 |
| 0.8 | 0.936 | 600.00 |
| 0.9 | 0.970 | 625.00 |
| 1.0 | 1.0 | 640.00 |

6.O. Acetone-air mixture containing 10 mole percent acetone is contacted with activated charcoal at 30° C and 1 atm pressure to reduce its acetone content to 1 mole percent. Charcoal fed to the system is free from acetone and adsorption of air may be neglected.

(*a*) Compute the number of countercurrent equilibrium stages required to affect the desired separation if an adsorbent rate of twice the minimum is to be employed.

(*b*) How much adsorbent shall be required per kmole of feed if the separation is to be affected in two crosscurrent stages?

(*c*) Compute the required number of gas phase transfer units for an adsorbent rate that is twice the minimum.

The equilibrium data for adsorption of acetone on activated charcoal at 30° C and 1 atm is given below[28]:

| | | | | | | |
|---|---|---|---|---|---|---|
| kg acetone adsorbed/kg charcoal | : | 0 | 0.1 | 0.2 | 0.3 | 0.35 |
| Partial pressure of acetone, mm Hg | : | 0 | 2.0 | 12.0 | 42.0 | 92.0 |

6.P. Air with humidity 0.0025 kg water/kg dry air is to be dried using silica gel in a fixed bed adsorber at the rate of 420 kg/hr. Air is to be admitted at 30° C and 1 atm pressure. Experiments on a bench scale fixed bed adsorber (column diameter = 10 cm, bed height = 20 cm, bulk density = 670 kg/m$^3$) showed that the breakpoint (corresponding to a moisture content of 0.0001 kg water/kg dry air in the exit air) occurred at 7 hrs from the start when the air rate (on dry basis) maintained was 0.15 kg/(m$^2$·s). Estimate the required diameter and bed height of the commercial adsorber if a breakpoint time of 20 hrs is desired. Assume isothermal operation. The Othmer chart for silica gel-water system is given in Fig. 6.38 (a).

6.Q. A gas (*A*) contaminated with toxic vapours of *C* is to be purified by selective adsorption of *C* on a solid adsorbent (*B*). The operation is to be conducted adiabatically in a sieve plate column. The feed gas containing $x_0$ mass fraction *C* is admitted to the bottom of the tower at $R_0$ kg/s and at temperature $t_0$ and it is desired that its vapour content is reduced to $x_n$ mass fraction *C*. Pure *B* is fed from the top at *S* kg/s and at temperature $t_S$. Write a computer program for estimating the number of equilibrium stages required. The following data may be assumed available:

(*a*) Values oif enthalpy of vapour-solid system at different temperatures and at different values of vapour concentration,

(*b*) values of vapour pressure of pure *C* at different temperatures,

(*c*) values of correlation constants *m* and *D*, the magnitudes of which depend on the concentration of *C* in solids:

$$\log P^* = m \log \bar{P} + D$$

where $P^*$ is the equilibrium partial pressure of *C* and $\bar{P}$ vapour pressure of pure *C*.

6.R. A hydrocarbon oil containing $x_1$ mass fraction colouring matter is passed through a packed bed of solid adsorbent particles (*B*) at the rate of G kg/(m$^2$·s). The height of the bed is *Zm* and its bulk density is $\rho_b$ kg/m$^3$. The breakpoint corresponds to $x_b$ mass fraction of colouring matter in the effluent and the bed is considered exhausted when the effluent concentration of colouring matter reaches 90 percent of feed concentration. It has been observed that the equilibrium relationship for the system can be expressed by Freundlich equation and the overall fluid phase mass transfer coefficient, $K_R = mG_A^n$ where $G_A$ is the superficial mass velocity of oil (*A*) through the bed and *m* and *n* are constants. Write a computer program for the estimation of the breakpoint time and the total mass of effluent (per unit bed cross section) at break point.

## REFERENCES


1. Treybal, RE, Liquid Extraction, McGraw Hill, New York, 1963.
2. Narayanan, CM, Performance Analysis and Computer Aided Design of Sieve Plate Liquid-Liquid Extractors, Proc. Ninth National Convention of Chemical Engrs. and International Symp. on Biotech., 281–286, Waltair, 1993 (published by Tata McGraw Hill, New Delhi).

3. Perry, RH and Green, DW, Chemical Engineer's Handbook, Sixth Edition, Section 21, McGraw Hill, New York, 1984.
4. Treybal, RE, Mass Transfer Operations, Third Edition, McGraw Hill, New York, 1981.
5. Crawford, ME and Wilke, CR, Chem. Eng. Progr., **47,** 423, 1951.
6. Hoffing, H and Lockhart, M, Chem. Eng. Progr., **50,** 94, 1954.
7. Nemunaitis, RR, Eckert, JS, Foote, EH and Rollison, L, Chem. Eng. Progr., **67** (11), 60, 1971.
8. Strand, CP, Olney, RB and Ackerman, GH, AIChE J, **8,** 252, 1962.
9. Coulson, JM and Richardson, JF, Chemical Engineering, Vol. 2, Pergamon Press, London, 1962.
10. Briggs, SW and Comings, EW, Ind. Eng. Chem., **35,** 411, 1943.
11. Boobar, MG et. al., Ind. Eng. Chem., **43,** 2922, 1951.
12. Othmer, DF, White, RE and Trueger, E, Ind. Eng. Chem., **33,** 1240, 1941.
13. Varteressian, KA and Fenske, MR, Ind. Eng. Chem., **28,** 928, 1936.
14. Armstrong, M and Kammermeyer, A, Ind. Eng. Chem., **34,** 1228, 1942.
15. McHugh, MA and Krukonis, VJ, Supercritical Fluid Extraction—Principles and Practice, Butterworth Publications, Boston, 1985.
16. Paulaitis, ME and Associates (eds.), Chemical Engineering at Supercritical Conditions, Science Publishers, Ann Arbor, 1983.
17. Ravenscroft, EA, Ind. Eng. Chem., **28,** 851, 1936.
18. Yang, RT, Adsorption Separation Processes, Butterworths, London, 1985.
19. Halsey G and Taylor, HS, J Chem. Phys., **15,** 624, 1947.
20. Langmuir, J Am. Chem. Soc., 38, 2267, 1916; **40,** 1361, 1918.
21. Brunauer, S, Emmett, PH and Teller, E, J Am. Chem. Soc., **60,** 309, 1938.
22. Othmer, DF and Sawyer, FG, Ind. Eng. Chem., **35,** 1269, 1943
23. Michaels, AS, Ind. Eng. Chem., **44,** 1922, 1952.
24. Collins, JJ, Chem. Eng. Prog. Symp. Series, **63** (74), 31, 1967.
25. Eagleton, LC and Bliss, H, Chem. Eng. Progr., **49,** 543, 1953.
26. Brown, GG and Associates, Unit Operations, Wiley, New York, 1950.
27. Lewis, WK et. al., J Am. Chem. Soc., **72,** 1157, 1950.
28. Josefewitz and Othmer, DF, Ind. Eng. Chem., **40,** 739, 1948.

# Chapter 7

# Membrane-Based Processes

Membrane-based technology (MBT) is not new to the biological world. Plant tissues have been using the process of osmosis for extracting water and ultra filtration is the process used by human kidneys for the detoxification of blood. However, adaptation of membrane-based technology to industrial processes is a relatively new phenomenon and is yet to catch its full speed. The primary reason behind this is that in spite of the fact that membrane-based processes do exhibit highly promising features with regards to optimisation and economisation of energy consumption in process industries, dubiousness prevails in the minds of many industrialists and practising engineers regarding the overall susceptibility of MBT to the industrial environment and its suitability as a substitute to the conventional operations. No doubt, it is to be admitted that the true mechanisms of many of the membrane-based separation processes are not yet fully understood. However, whatever is known about these processes is useful and what is yet to be known appears full of promise and that makes these processes particularly interesting.

In this chapter, we shall briefly discuss on the basis principles and design features of some of the most popular membrane-based separation processes (such as reverse osmosis, ultrafiltration, electrodialysis, pervaporation) and their technical as well as economical feasibility for application to process industries.

## 7.1 REVERSE OSMOSIS (OR HYPERFILTRATION)

Osmosis has been known for the last 200 years as a natural process involving spontaneous flow of pure water into an aqueous solution (or the flow of any liquid from a less concentrated salt solution to a more concentrated salt solution) across a semipermeable membrane (that is, a membrane that is permeable essentially to water or the liquid only). If we wish to obtain potable water from saline water in a similar process, then the direction of flow is to be reversed. In other words, water must be made to flow from the more concentrated solution (namely, the saline water) to a less concentrated solution. (To note that when we say concentration, we mean concentration of the salt and not that of water). Consequently, the process in which this is made to occur is conveniently termed *reverse osmosis.* However, it will be misleading to describe reverse osmosis as the reverse of osmosis. It must be kept in mind that under isothermal conditions, in either osmosis or reverse osmosis, preferential transport of fluid takes place through the membrane always in the direction of lower chemical potential.

Let us make this more clear. Suppose a salt solution (say, a solution of sodium chloride in water) and pure water are separated by a semipermeable (namely, solvent permeable) membrane. Since the membrane is permeable only to the solvent (namely, water), water will permeate through the membrane from the pure water compartment to the solution compartment, thereby diluting the salt solution further. Thermodynamically, we can say that this happens because the chemical potential of pure water is higher than that of the salt solution. However, the chemical potential (which is a function of temperature, pressure and the concentration of dissolved salts) increases with increase in pressure at a constant temperature. Therefore, if we apply external pressure in the solution compartment, its chemical potential increases. If the applied pressure is gradually increased, then the chemical potential of the solution also increases gradually and ultimately it becomes equal to that of pure water, when the transport of water across the membrane ceases. This pressure which is required to raise the chemical potential of the solvent in any solution to that of the pure solvent at any given temperature is known as the osmotic pressure ($\pi$) of the solution. If the applied pressure is further increased, then the chemical potential of the solution will become higher than that of pure water and as a result, water will start permeating through the membrane from the solution compartment to the pure water compartment. This is what is known as the process of *reverse osmosis* (RO). As a result of RO, more and more water gets separated from the solution and the salt solution gets more and more concentrated.

In the case of an ideal semipermeable membrane (that is, a membrane that is permeable only to water or the solvent), the permeate will be pure water/solvent. However, no commercial membrane available in the market is ideal. As a result, a small amount of solute will also pass through the membrane though the rate of solute flow will be much lower than the solvent flow. If $\pi_1$ is the osmotic pressure of the concentrated solution on the high pressure side of the membrane and $\pi_2$ that of product water (or the permeate), then it is obvious that reverse osmosis will occur only if the transmembrane pressure drop ($\Delta\pi$) exceeds the osmotic pressure difference ($\Delta\pi$), where $\Delta\pi = (\pi_1 - \pi_2)$. If the permeate is pure water/solvent, then $\pi_2 = 0$.

The osmotic pressure data for aqueous sodium chloride solutions at different temperatures are given in Tables 7.4 and 7.5 at the end of this section.

A simple schematic of a reverse osmosis system is shown in Fig. 7.1. The feed solution is pumped into a pressure vessel containing the semipermeable membrane by means of a high pressure pump. The pemeate or product water is collected from the downstream side of the membrane usually at atmospheric pressure. The concentrated solution from the high pressure side (often called the *brine* or the *retentate*) is discharged to atmospheric pressure through a flow regulating value.

Before proceeding further, let us understand the significances of the two important terminologies *solute rejection* and *product recovery.* The overall product recovery (or water recovery) is defined as

$$R = P/F_0 \qquad \text{... (7.1.1)}$$

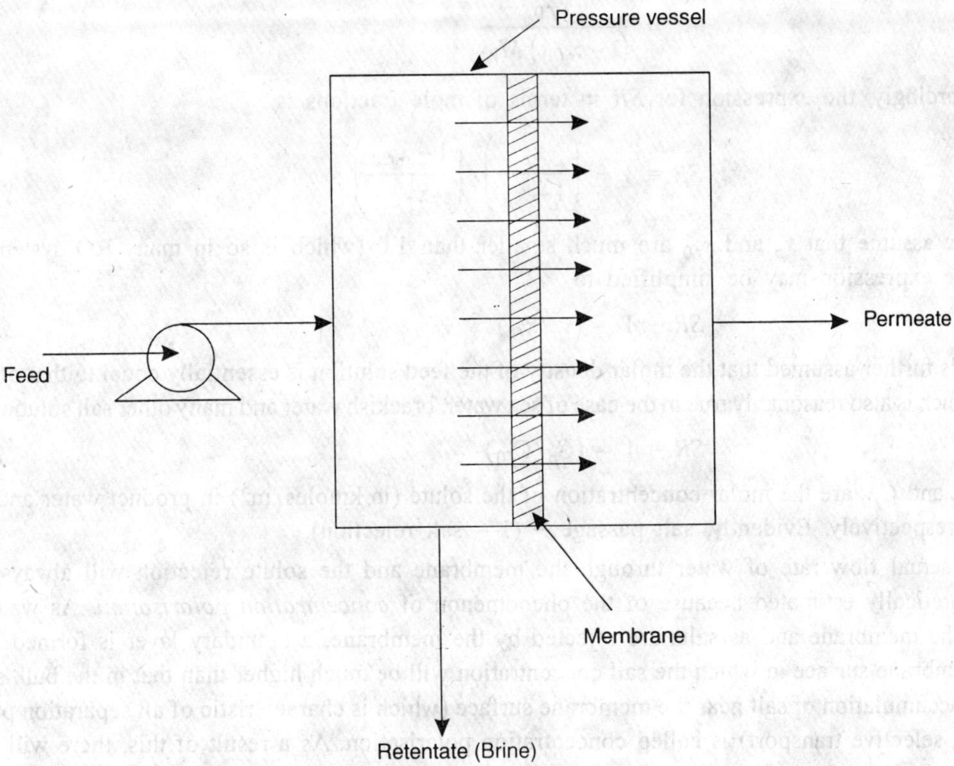


*Figure 7.1: Schematic of a reverse osmosis system*

where $P$ is the total volume of product water collected per unit time (in $m^3/s$) and $F_0$ is the volumetric flow rate of feed solution at the feed inlet. The fractional solute rejection ($SR$) is defined as

$$SR = \frac{(\text{molarity of feed solution}) - (\text{molarity of product water})}{(\text{molarity of feed solution})} \quad \text{... (7.1.2)}$$

Since molarity = moles of solute per litre of water, we can easily express molarity in terms of the mole fraction of solute ($x$) in solution. For example, if $x_{f0}$ is the mole fraction of solute in the feed solution at the feed inlet, then

Moles of solute per mole of water

$$= x_{f0}/(1 - x_{f0})$$

Moles of solute per kg of water

$$= \frac{x_{f0}}{\left(1 - x_{f0}\right) M_W}$$

where $M_W$ is the molecular weight of water in kg/kmole. If we take density of water to be 1.0 gm/cc or 1.0 kg/litre, then the moles of solute per litre of water will be the same as above. Thus, the molarity ($m_{f0}$) of the feed solution is

$$m_{f0} = \frac{x_{f0}}{(1 - x_{f0})\,M_W} \qquad \text{... (7.1.3)}$$

Accordingly, the expression for $SR$ in terms of mole fractions is

$$SR = 1 - \left(\frac{x_p}{1 - x_p}\right)\left(\frac{1 - x_{f0}}{x_{f0}}\right) \qquad \text{... (7.1.4)}$$

If we assume that $x_p$ and $x_{f0}$ are much smaller than 1.0 (which is so in many RO systems), then the above expression may be simplified to

$$SR = 1 - (x_p / x_{f0}) \qquad \text{... (7.1.5)}$$

If it is further assumed that the molar density of the feed solution is essentially equal to that of product water (which is also reasonably true in the case of sea water, brackish water and many other salt solutions), then

$$SR = 1 - (C_p / C_{f0}) \qquad \text{... (7.1.6)}$$

where $C_p$ and $C_{f0}$ are the molar concentration of the solute (in kmoles/$m^3$) in product water and in feed solution respectively. Evidently, salt passage = (1 – salt rejection).

The actual flow rate of water through the membrane and the solute rejection will always be less than theoretically estimated because of the phenomenon of *concentration polarisation.* As water flows through the membrane and as salts are rejected by the membrane, a boundary layer is formed adjacent to the membrane surface in which the salt concentration will be much higher than that in the bulk solution. Such an accumulation of salt near the membrane surface (which is characteristic of all separation processes involving selective transport) is called concentration polarisation. As a result of this, there will be back transport of solute from the concentrated boundary layer to the solution bulk.

Concentration polarisation has many negative effects on the reverse osmosis process. Due to higher salt concentration, osmotic pressure will be much higher at the membrane surface then in the bulk solution. This reduces the net pressure differential across the membrane ($\Delta P - \Delta\pi$) and thereby, the water flow through the membrane decreases. Also, salt flow through the membrane increases due to the higher localised salt concentration and as a result, salt rejection decreases. In addition, the concentration of the salt in the boundary layer often exceeds the solubility of the salt at the operating temperature and consequently, the salt tends to get precipitated on the membrane surface, thereby causing membrane scaling.

## 7.1.1 Performance Analysis of RO Systems

Consider a salt solution flowing between two parallel flat membrane surfaces at a distance $2h$ apart [Fig. 7.2 (a)] or through a tubular membrane of diameter $d$ [Fig. 7.2 (b)]. The former is characteristic of plate and frame type RO modules and the latter represents the tubular and hollow fibre RO systems. As the solution flows along the membrane, its salt concentration gradually increases since more and more water gets separated from it. Thus, both the concentration as well as the flow rate of the feed solution vary continuously from the inlet end to the other end of the membrane. Accordingly, the rate of permeation of water through the membrane also gradually decreases from the inlet end to the other end. It is, therefore, necessary to make a section to selection or segment to segment analysis to study the performance characteristics of an RO system. We shall assume that the membrane system is divided into $n$ number of differential segments each of length $\Delta z$. In other words, the membrane system is assumed to be composed of $n$ differential segments in series such that the concentrated solution leaving each segment becomes the feed to the next segment [see Fig. 7.2 (c)]. For the sake of convenience, the $i$th segment is shown separately in Fig. 7.2 (d). It is assumed that each segment is well-mixed so that the concentration of the solution inside the segment is equal to that of the outlet stream.

The molar flux of water/solvent through the membrane for the $i$th segment is given by

$$N_{Wi} = K_m (\Delta P - \Delta\pi_i) \qquad \text{... (7.1.7)}$$

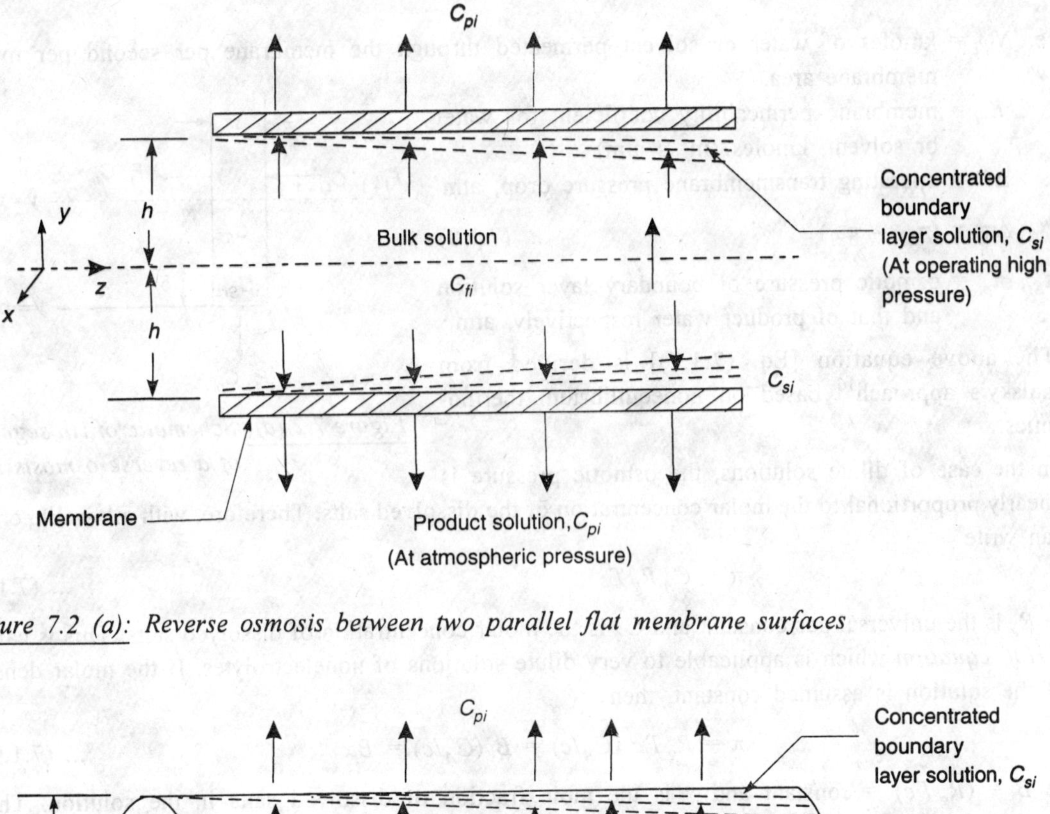


*Figure 7.2 (a): Reverse osmosis between two parallel flat membrane surfaces*

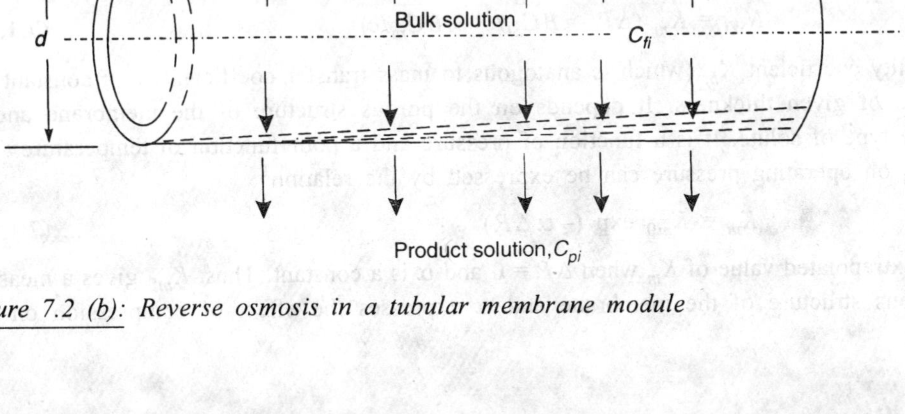


*Figure 7.2 (b): Reverse osmosis in a tubular membrane module*

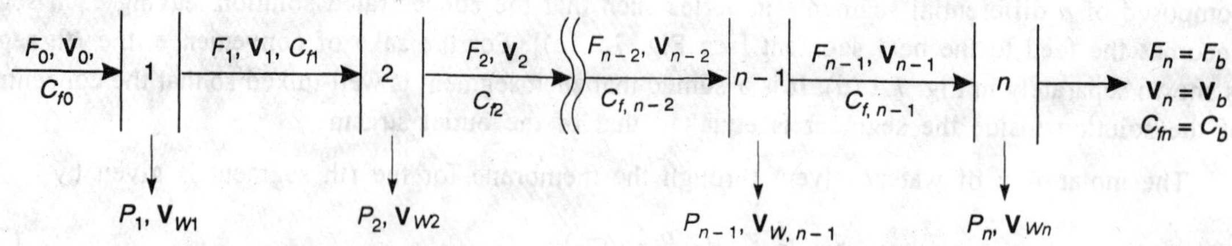


*Figure 7.2 (c): Schematic of n differential segments in series*

where $N_{Wi}$ = kmoles of water or solvent permeated through the membrane per second per $m^2$ of membrane area.

$K_m$ = membrane permeability coefficient for water or solvent, kmoles/($m^2 \cdot s \cdot atm$)

$\Delta P$ = operating transmembrane pressure drop, atm

$$\Delta \pi_i = (\pi_{Si} - \pi_{pi}) \quad \text{... (7.1.8)}$$

$\pi_{Si}, \pi_{pi}$ = osmotic pressure of boundary layer solution and that of product water respectively, atm.

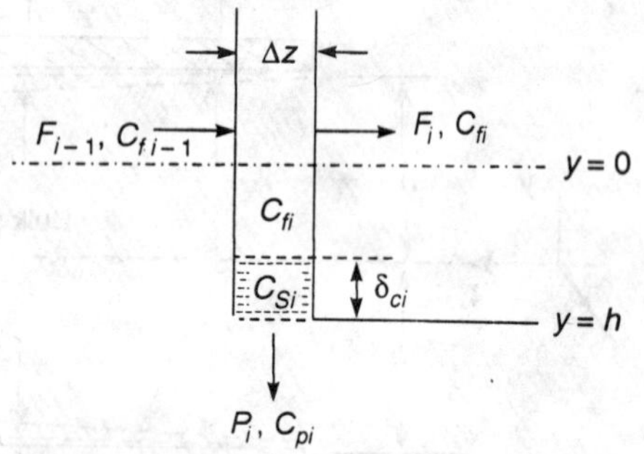


*Figure 7.2 (d): Schematic of ith segment of a reverse osmosis unit*

The above equation [Eq. (7.1.7)] is derived from Katchalsky's approach[16] based on nonequilibrium thermodynamics.

In the case of dilute solutions, the osmotic pressure is very nearly proportional to the molar concentration of the dissolved salts. Therefore, with allowable error, we can write

$$\pi = C_A R_g T \quad \text{... (7.1.9)}$$

where $R_g$ is the universal gas constant and $C_A$ is the molar concentration of dissolved salts. This is called *van't Hoff equation* which is applicable to very dilute solutions of nonelectrolytes. If the molar density ($c$) of the solution is assumed constant, then

$$\pi = R_g T c \, (C_A/c) = B \, (C_A/c) = Bx \quad \text{... (7.1.9a)}$$

where $B = (R_g T c)$ = constant and $x$ is the mole fraction of dissolved salts in the solution. Thus, $\pi_{Si} = (BC_{Si}/c)$ and $\pi_{pi} = (BC_{pi}/c)$. Accordingly, Eq. (7.1.7) may be rewritten as

$$N_{Wi} = K_m \, (\Delta P - BC_{Si}/c + BC_{pi}/c) \quad \text{... (7.1.10)}$$

The permeability coefficient $K_m$ (which is analogous to mass transfer coefficient) is a constant for a given membrane of given thickness. It depends on the porous structure of the membrane and is independent of the type of solute. It is a function of pressure and a poor function of temperature. The dependence of $K_m$ on operating pressure can be expressed by the relation[1]

$$K_m = K_{m0} \exp \, (-\alpha \Delta P) \quad \text{... (7.1.11)}$$

where $K_{m0}$ is the extrapolated value of $K_m$ when $\Delta P = 0$ and $\alpha$ is a constant. Thus, $K_{m0}$ gives a measure of the initial porous structure of the membrane and $\alpha$ expresses the effect of compression on the permeability of the membrane. Though developed for a pressure range of 8 to 100 atm, the above relationship has been found to be applicable at lower pressures as well. The temperature dependency of $K_m$ is indicated by the relation[2]

$$K_m \mu_W = \text{constant} \quad \text{... (7.1.12)}$$

where $\mu_W$ is the viscosity of water. Since the viscosity of water is a poor function of temperature, the variation of $K_m$ with temperature is not significant. The above relationship is based on experiments conducted within a temperature range of 5 to 36° C.

The solute flux ($N_{Ai}$) across the membrane for the $i$th segment is (from the fundamental diffusion equation)

$$N_{Ai} = (D_{AM}/K_{mb}\delta)\ (C_{Si} - C_{pi}) \quad \text{... (7.1.13)}$$

where $D_{AM}$ = diffusivity of solute in the membrane phase, $m^2/s$

$\delta$ = effective thickness of the membrane, m

$K_{mb}$ = distribution coefficient of the solute between the membrane phase and the solution phase, dimensionless

We know that

$$x_{pi} = N_{Ai}/(N_{Ai} + N_{Wi}) \quad \text{... (7.1.14)}$$

However, as stated earlier, the solute flux will be much smaller than the solvent flux ($N_{Ai} << N_{Wi}$) since the membrane is selective to the transport of the solvent. Therefore, we may write with allowable error

$$x_{pi} = N_{Ai}/N_{Wi} \quad \text{... (7.1.15)}$$

or

$$N_{Wi} = N_{Ai}/x_{pi} = N_{Ai}\, c/C_{pi} \quad \text{... (7.1.16)}$$

Substituting for $N_{Ai}$ from Eq. (7.1.13), we get

$$N_{Wi} = (D_{AM}/K_{mb}\delta)\ c\ (C_{Si} - C_{pi})/C_{pi} \quad \text{... (7.1.17)}$$

The parameter $(D_{AM}/K_{mb}\delta)$, which plays the role of a mass transfer coefficient for solute transport through the membrane, is called the *solute transport parameter.* Its value depends on the nature of the solute, the process temperature and pressure.

In the case of cellulose acetate membranes, it has been observed[4] that $(D_{AM}/K_{mb}\delta)$ is essentially independent of feed concentration and feed flow rate at any given operating pressure difference with almost all types of solutes. This has been found to be true for many non-cellulosic membranes as well with a large number of solutes (though not with all solutes). This observation is particularly interesting since it is of great practical utility in reverse osmosis process design.

As reported by Aggarwal and Sourirajan[3], the temperature dependency of solute transport parameter may be expressed as

$$(D_{AM}/K_{mb}\delta) = b_1 \exp\ (b_2 T) \quad \text{... (7.1.18)}$$

where $b_1$ and $b_2$ are constants. The above correlation is based on experimental data collected within a temperature range of 5 to 36° C.

The effect of pressure on solute transport parameter may be expressed by the relation[1]

$$(D_{AM}/K_{mb}\delta)\ (\Delta P)^{\beta} = \text{constant} \qquad \text{... (7.1.19)}$$

where $\beta$ is a constant for a particular solute at a particular temperature. Typical values of $K_m$ and $(D_{AM}/K_{mb}\delta)$ for different types of Loeb-Sourirajan type porous cellulose acetate membranes (CA-NRC-18) are given in Table 7.1.

## Table 7.1
## Specifications of Loeb-Sourirajan Type Cellulose Acetate Membranes (CA-NRC-18)

| *Membrane* | *Solute* | *Operating pressure difference, atm* | $K_m$, *kmoles/(m$^2\cdot$s$\cdot$atm)* | $\left(\frac{D_{AM}}{K_{mb}\,\delta}\right)$, *m/s* |
|---|---|---|---|---|
| Type 1 | Sodium chloride | 102.0 | $0.97 \times 10^{-5}$ | $0.90 \times 10^{-7}$ |
| Type 2 | Sodium chloride | 102.0 | $1.46 \times 10^{-5}$ | $6.0 \times 10^{-7}$ |
| Type 3 | Sodium chloride | 102.0 | $1.87 \times 10^{-5}$ | $20.0 \times 10^{-7}$ |
| Type 4 | Sodium chloride | 102.0 | $2.37 \times 10^{-5}$ | $75.0 \times 10^{-7}$ |
| Type 5 | Sodium chloride | 102.0 | $2.93 \times 10^{-5}$ | $140.0 \times 10^{-7}$ |

*Source.* Sourirajan, S., Reverse Osmosis, Logos Press, London, 1970.

Under steady state conditions, the rate of transport of water from the solution bulk to the membrane will be equal to that across the membrane [see Figs 7.2 (a) and (b)]. Expression for the rate of transport of water from the bulk to the membrane can be easily developed from the basic equations of mass transfer given in Chapter 4. Since $N_{Ai}$ and $N_{Wi}$ are closely inter-related as per Eq. (7.1.14) or (7.1.16), let us proceed to develop the expression for $N_{Ai}$. To note that due to concentration polarisation, there will be back diffusion of solute from the boundary layer to the bulk solution. Now, from Eq. (4.1.24) of Chapter 4, the solute flux is

$$N_{Ai} = (N_{Ai} + N_{Wi})\ x - D_{AS}\,c\,\frac{dx}{dy} \qquad \text{... (7.1.20)}$$

or

$$\frac{dx}{dy} - \frac{(N_{Ai} + N_{Wi})}{cD_{AS}}\,x = -(N_{Ai}/cD_{AS}) = -\frac{(N_{Ai} + N_{Wi})\,x_{pi}}{cD_{AS}} \qquad \text{... (7.1.21)}$$

The above equation can be integrated based on the boundary conditions [with reference to Figs 7.2 (a) and (d)] that

At $\quad y = 0, \quad x = x_{fi}$ ... (7.1.22)

At $\quad y = (h - \delta_{ci}), \quad x = x_{Si}$ ... (7.1.23)

where $\delta_{ci}$ is the thickness of the boundary layer in the $i$th segment. Thus, we get

$$\ln\left(\frac{x_{Si} - x_{pi}}{x_{fi} - x_{pi}}\right) = (N_{Ai} + N_{Wi})/(k_i c) \qquad \text{... (7.1.24)}$$

where $k_i = D_{AS}/(h - \delta_{ci})$ = mass transfer coefficient (in the $i$th segment)

Since $N_{Ai} << N_{Wi}$, $$N_{Wi} = k_i c \ln \left( \frac{x_{Si} - x_{pi}}{x_{fi} - x_{pi}} \right) \qquad \text{... (7.1.25)}$$

Thus, we have three different equations, such as Eqs (7.1.10), (7.1.17) and (7.1.25), to describe the solvent flux across the membrane. Before clubbing them together, it shall be convenient to re-express these three equations in terms of the following variables.

$$U^* = (K_m \, \Delta P / c) \qquad \text{... (7.1.26)}$$

$$\phi = B/(c\Delta P) \qquad \text{... (7.1.27)}$$

$$\theta = (D_{AM}/K_{mb}\delta)/U^* \qquad \text{... (7.1.28)}$$

$$\lambda_i = k_i/(D_{AM}/K_{mb}\delta) = k_i/(U^* \, \theta) \qquad \text{... (7.1.29)}$$

It may be noted that the dimensions of $U^*$ and $\phi$ are (m/s) and (m$^3$/kmoles) respectively, while $\theta$ and $\lambda_i$ are dimensionless parameters. Thus, Eqs (7.1.10), (7.1.17) and (7.1.25) can be respectively rewritten as

$$\mathbf{V}_{Wi} = N_{Wi}/c = U^* \, [1 - \phi \, (C_{Si} - C_{pi})] \qquad \text{... (7.1.30)}$$

$$= U^* \, \theta \, (C_{Si} - C_{pi})/C_{pi} \qquad \text{... (7.1.31)}$$

$$= U^* \, \theta \lambda_i \, \ln \, [(C_{Si} - C_{pi})/(C_{fi} - C_{pi})] \qquad \text{... (7.1.32)}$$

Evidently, $\mathbf{V}_{Wi}$ is the velocity of flow of water though the membrane (in the $i$th segment). Eliminating $\mathbf{V}_{Wi}$ from Eqs (7.1.30) and (7.1.31), we get

$$C_{Si} - C_{pi} = C_{pi}/(C_{pi}\phi + \theta) \qquad \text{... (7.1.33)}$$

Substituting this in Eq. (7.1.31), we get

$$\mathbf{V}_{Wi} = U^* \theta/(C_{pi}\phi + \theta) \qquad \text{... (7.1.34)}$$

Substituting the above two equations in Eq. (7.1.32), we get, on rearrangement,

$$C_{fi} - C_{pi} = \left( \frac{C_{pi}}{C_{pi} \, \phi + \theta} \right) \exp \left( \frac{-1}{\lambda_i \left( C_{pi} \, \phi + \theta \right)} \right) \qquad \text{... (7.1.35)}$$

Let us also write down the material balance equations for the $i$th segment. If we neglect the change in density of the solutions, then an overall material balance yields

$$F_{i-1} = P_i + F_i \qquad \text{... (7.1.36)}$$

A solute balance gives

$$F_{i-1} \, C_{f,\,i-1} = P_i C_{pi} + F_i C_{fi} \qquad \text{... (7.1.37)}$$

Combining the above two equations and noting that $R_i = (P_i/F_{i-1})$, we get

$$C_{f,\,i-1} = R_i C_{pi} + (1 - R_i) \, C_{fi} \qquad \text{... (7.1.38)}$$

If we define a variable $q_i$ such that

$$q_i = C_{fi}/C_{pi} \quad \text{... (7.1.39)}$$

then, Eq. (7.1.38) becomes

$$C_{f,\,i-1} = C_{pi}\;[R_i + (1 - R_i)\; q_i] = C_{pi}\,p_i \quad \text{... (7.1.40)}$$

where $$p_i = R_i + q_i\;(1 - R_i) \quad \text{... (7.1.40a)}$$

Equations (7.1.35) and (7.1.40) form the design equations of an RO system. We may rewrite the expression for product recovery ($R_i$) also in terms of the newly defined variables. Thus

$$R_i = P_i/F_{i-1} = \frac{\mathbf{V}_{Wi}\,(2W\,\Delta z)}{\mathbf{V}_{i-1}\,(W\,2h)} = \mathbf{V}_{Wi}\;\Delta z/(\mathbf{V}_{i-1}h) \quad \text{... (7.1.41)}$$

where $W$ is the width of the membrane and ($2h$) is the inter-spacing between the two parallel surfaces. For the tubular membrane of diameter $d$,

$$R_i = \frac{\mathbf{V}_{Wi}\,(\pi\,d\,\Delta z)}{\mathbf{V}_{i-1}\left(\pi\,d^2/4\right)} = 4\,\mathbf{V}_{Wi}\Delta z/(\mathbf{V}_{i-1}\,d) \quad \text{... (7.1.42)}$$

If we introduce a dimensionless variable $X_i^*$ such that

$$X_i^* = (U^*\Delta z)/(h\mathbf{V}_{i-1}),\ \text{for parallel membranes} \quad \text{... (7.1.43)}$$

$$= (2\,U^*\Delta z)/(d\,\mathbf{V}_{i-1}),\ \text{for tubular membrane} \quad \text{... (7.1.43a)}$$

then the expression for $R_i$ becomes

$$R_i = \mathbf{V}_{Wi}X_i^*/U^*,\ \text{for parallel membranes}$$

$$= 2\,\mathbf{V}_{Wi}X_i^*/U^*,\ \text{for tubular membrane}$$

Substituting for $\mathbf{V}_{Wi}$ from Eq. (7.1.34),

$$R_i = (\theta X_i^*)/(C_{pi}\phi + \theta),\ \text{for parallel membranes} \quad \text{... (7.1.44)}$$

$$= 2\theta\,X_i^*/(C_{pi}\phi + \theta),\ \text{for tubular membrane} \quad \text{... (7.1.44a)}$$

It may also be noted that since $F_i = F_{i-1}\;(1 - R_i)$,

$$\mathbf{V}_i = \mathbf{V}_{i-1}\;(1 - R_i) \quad \text{... (7.1.45)}$$

Therefore, $$X_{i+1}^* = U^*\Delta z/(\mathbf{V}_i h) = \frac{U^*\Delta z}{h\mathbf{V}_{i-1}\left(1 - R_i\right)} = X_i^*/(1 - R_i) \quad \text{... (7.1.46)}$$

As stated earlier, Eqs (7.1.35) and (7.1.40) are the design equations of an RO system.

However, for applying these equations to design problems, we must know the value of the mass transfer coefficient, $k_i$. This can be obtained only from experimental correlations available in the literature. For example, the correlations proposed by Kimura and Sourirajan[5] are reproduced below.

For laminar flow between two parallel flat membrane surfaces at a distance ($2h$) apart,

$$\alpha_0\,Sh = 2.6\,\beta_f^{-0.333} \quad \text{... (7.1.47)}$$

$$\alpha_0\;(Sh)_{av} = 3.9\,\beta_f^{-0.333} \quad \text{... (7.1.48)}$$

where $Sh$ = Sherwood number

$$= (4h)\ k_i/D_{AS}$$

$$= 4k_i/(U^*\alpha_0) \quad \text{... (7.1.49)}$$

$D_{AS}$ = diffusivity of solute in solution

$$\alpha_0 = D_{AS}/(U^* h) \quad \text{... (7.1.50)}$$

$$\beta_f = X_i^*/(3\alpha_0^2) \quad \text{... (7.1.51)}$$

The above two correlations (7.1.47) and (7.1.48) are valid only if $\beta_f < \beta_{as}$, where $\beta_{as}$ is defined as

$$\beta_{as} = [0.65\ \ln\ (1 + 1/Z_1\ \alpha_0^2)]^3 \quad \text{(7.1.52)}$$

For parallel surfaces, $Z_1 = 3.0$. If $\beta_f \geq \beta_{as}$, then the expressions for Sherwood number are to be modified as

$$\alpha_0 Sh = 4.0/\ln\ (1 + 1/Z_1\alpha_0^2) \quad \text{... (7.1.53)}$$

$$\alpha_0\ (Sh)_{av} = \left(\frac{3.9\beta_{as}^{2/3}}{\beta_f}\right) + \left[\frac{4(\beta_f - \beta_{as})/\beta_f}{\ln\left(1 + 1/Z_1\alpha_0^2\right)}\right] \quad \text{... (7.1.54)}$$

For turbulent flow between two parallel flat membrane surfaces, the correlation for the entrance region is

$$\alpha_0\ (Sh)_{av} = 0.48\ Re^{0.25}\ \beta_f^{-0.333} \quad \text{... (7.1.55)}$$

where $Re$ = Reynolds number

$$= (4h)\ \mathbf{V}_{i-1}\rho_f/\mu_f \quad \text{... (7.1.56)}$$

For laminar flow through a tubular membrane of diameter $d$,

$$\alpha_0 Sh = 1.3\,\beta_t^{-0.333} \quad \text{... (7.1.57)}$$

$$\alpha_0\ (Sh)_{av} = 1.95\,\beta_t^{-0.333} \quad \text{... (7.1.58)}$$

where $Sh = k_i d/D_{AS}$ .... (7.1.59)

$$\beta_t = X_i^*/(4\alpha_0^2) \quad \text{... (7.1.60)}$$

$$\alpha_0 = 2\ D_{AS}/(U^* d) \quad \text{... (7.1.61)}$$

As in the case of parallel surfaces, if $\beta_t \geq \beta_{as}$, then Eqs (7.1.53) and (7.1.54) are to be employed, except that in the case of tubular membrane, $Z_1 = 4.0$.

For turbulent flow through tubular membrane, the correlation for the entrance region is

$$\alpha_0 Sh = 0.184\ Re^{0.25}\,\beta_t^{-0.333} \quad \text{... (7.1.62)}$$

$$\alpha_0\ (Sh)_{av} = 0.276\ Re^{0.25}\,\beta_t^{-0.333} \quad \text{... (7.1.63)}$$

where $Re = d\,\mathbf{V}_{i-1}\,\rho_f/\mu_f$ ... (7.1.64)

The design of an RO system means estimation of the required membrane area ($A_m$) for achieving a specified product recovery ($R$). We must also verify the permeate quality, $C_p$. The design procedure, which involves a segment to segment analysis, as stated earlier, is summarised below.

1. Specify the feed concentration ($C_{f0}$), the transmembrane pressure drop and the product recovery desired ($R$). In the case of tubular membrane, specify the diameter ($d$) and in the case of parallel membranes, the inter-spacing ($2h$). Specify also the membrane characteristics such as the permeability coefficient ($K_m$) and the solute transport parameter (assumed constant).
2. Compute $U^*$, $\theta$ and $\phi$ from Eqs (7.1.26) to (7.1.28).
3. Assume $X_1^*$. In the case of tubular membranes and hollow fibers, it is better to assume $L$. For tubular membranes, $L$ = 10 to 20 times $d$, while for hollow fibers, $(L/d)$ = as high as 1000. Now, $\Delta z = (L/n)$ and $\mathbf{V}_0$ can be computed from the specified feed flow rate ($F_0$). The value of $X_1^*$ is then obtained from Eq. (7.1.43a).
4. Put $i = 1$.
5. (*a*) Assume $C_{pi}$.
   (*b*) Compute $R_i$ from Eq. (7.1.44) or (7.1.44a).
   (*c*) Compute the mass transfer coefficient $k_i$ from any of the correlations given in Eqs (7.1.47) to (7.1.63) and compute $\lambda_i$ from Eq. (7.1.29).
   (*d*) Compute $C_{fi}$ from Eq. (7.1.35) and $q_i$ from Eq. (7.1.39).
   (*e*) Compute $p_i$ from Eq. (7.1.40a).
   (*f*) Compute $C_{pi}$ as $C_{pi} = C_{f,i-1}/p_i$
   (*g*) If the above-computed value of $C_{pi}$ differs significantly from that assumed earlier (or computed earlier), then repeat the computations starting from step (*b*).
6. If $i = n$, then go to step (7). Otherwise, put $i = i + 1$ and compute $X_i^*$ as

$$X_i^* = X_{i-1}^*/(1 - R_{i-1})$$

   Now, repeat computations starting from step (5).
7. Compute the total product recovery as

$$R = 1 - \prod_{i=1}^{n} (1 - R_i) \quad \text{... (7.1.65)}$$

8. If the above-computed value of $R$ is equal to or slightly more than that specified in the problem, then proceed to step (9). Otherwise, assume an alternate value of $X_1^*$ (or increase $L$ by 1 cm) and repeat the computations from step (4).
9. Compute $(q/p)$ as

$$q/p = \prod_{i=1}^{n} (q_i / p_i) \quad \text{... (7.1.66)}$$

10. Compute the exit concentration of brine or retentate (also called reject) as

$$C_b = C_{fn} = (q/p)\, C_{f0} \quad \text{... (7.1.67)}$$

11. Since $p = R + q\,(1 - R)$ and since $R$ and $(q/p)$ are known, compute $p$ as

$$p = R/[1 - (q/p)\,(1 - R)] \qquad \text{... (7.1.67a)}$$

12. Compute the permeate concentration as

$$C_p = C_{f0}/p \qquad \text{... (7.1.68)}$$

13. Compute the membrane area per unit permeate rate as

$$A_m/P = (X_1^* n)/(RU^*), \text{ for parallel membranes} \qquad \text{... (7.1.69)}$$

$$= 2\, X_1^* n/(RU^*), \text{ for tubular membrane} \qquad \text{... (7.1.70)}$$

where $A_m = 2\,(LW)$ for parallel membranes and $A_m = \pi dL$ for a tubular membrane.

To note that $P = (F_0 R)$. Also, for tubular membranes and hollow fibers, since $L$ has already been determined, $A_m$ can be directly obtained as $A_m = (\pi dL)$.

Before concluding, we shall summarise here the assumptions involved in the analysis discussed above. While applying the above-given design procedure, these assumptions must be kept in mind.

(*i*) The change in the density of the solution during concentration is neglected while performing the material balance for each segment [see Eq. (7.1.36)]

(*ii*) The osmotic pressure is assumed proportional to the molar concentration of dissolved salts [Eq. (7.1.9)]. A more generalised relationship could be a polynomial fit of the form

$$\pi = A_1 x + A_2 x^2 + A_3 x^3 + \ldots\ldots \qquad \text{... (7.1.71)}$$

(*iii*) The solute transport parameter is assumed independent of feed flow rate and feed concentration. Though very true in the case of cellulose acetate membranes, this need not apply strictly to all non-cellulosic membranes.

Alternate and more rigorous analyses are available in references (6) and (7).

The design of an RO system shall not be complete unless we include the estimation of pressure drop in the system (which decides the power consumption and thereby the operating cost). However, this can be accomplished from the fundamental fluid flow equations. Thus, pressure drop due to friction can be estimated from the Fanning's equation given in Chapter 2. For the $i$th segment of membrane, Fanning's equation can be written as

$$(-\Delta P_f)_i/\rho_f = 2f\,(\Delta z)\,\mathbf{V}_{i-1}^2/d \qquad \text{... (7.1.72)}$$

We also know that for laminar flow through a tubular membrane, the Fanning's friction factor is related to the Reynolds number as

$$f = 16/Re = \frac{16\mu_f}{\left(d\,\mathbf{V}_{i-1}\rho_f\right)} \qquad \text{... (7.1.73)}$$

Equation (7.1.72), therefore, becomes

$$(-\Delta P_f)_i/\rho_f = 32\mu_f\,(\Delta z)\,\mathbf{V}_{i-1}/(\rho_f\, d^2) \qquad \text{... (7.1.74)}$$

The power consumption (in watts) is given by

$$P_i' = [(-\Delta P_f)_i/\rho_f]\; F_{i-1}\rho_f \qquad \text{... (7.1.75)}$$

where $$F_{i-1} = (\pi d^2/4)\ \mathbf{V}_{i-1} \quad \text{... (7.1.76)}$$

Substituting the expression for pressure drop from Eq. (7.1.74) in Eq. (7.1.75), we get

$$P'_i = 8\pi\mu_f\ (\Delta z)\ \mathbf{V}_{i-1}^2 \quad \text{... (7.1.77)}$$

The total power consumption is, therefore,

$$P' = 8\pi\mu_f\ (\Delta z) \sum_{i=1}^{n} \mathbf{V}_{i-1}^2 \quad \text{... (7.1.78)}$$

Now, we know that $\mathbf{V}_i = \mathbf{V}_{i-1}\ (1 - R_i)$.

Therefore,

$$\begin{aligned} \Sigma \mathbf{V}_{i-1}^2 &= \mathbf{V}_0^2 + \mathbf{V}_1^2 + \mathbf{V}_2^2 + \ldots + \mathbf{V}_{n-1}^2 \\ &= \mathbf{V}_0^2\ [1 + (1 - R_1)^2 + (1 - R_1)^2\ (1 - R_2)^2 + \\ &\quad \ldots (1 - R_1)^2\ (1 - R_2)^2 \ldots (1 - R_{n-1})^2] \\ &= \mathbf{V}_0^2 E \end{aligned} \quad \text{... (7.1.79)}$$

where $$E = 1 + \sum_{i=1}^{n-1} \prod_{j=1}^{i} \left(1 - R_j\right)^2 \quad \text{... (7.1.80)}$$

Equation (7.1.78), therefore, becomes

$$P' = 8\pi\mu_f\ (\Delta z)\ \mathbf{V}_0^2 E \quad \text{... (7.1.81)}$$

The power consumption per unit permeate rate will be (note that, $P = F_0\ R = [(\pi d^2/4)\ \mathbf{V}_0 R]$,

$$P'/P = 32\mu_f\ (\Delta z)\ \mathbf{V}_0\, E/(Rd^2) \quad \text{... (7.1.82)}$$

The above equation may also be expressed in terms of the previously defined variables such as $X_1^*$, $\alpha_0$ and the Reynolds number $Re$, where $Re = (d\,\mathbf{V}_0 \rho_f/\mu_f)$, as

$$P'/P = (2\rho_f Sc^3 U^{*2} \alpha_0^3 X_1^* E/R)\ Re^2 \quad \text{... (7.1.83)}$$

where $$Sc = \text{Schmidt number} = \mu_f/(\rho_f\ D_{AS}) \quad \text{... (7.1.84)}$$

The above expression thus predicts the power consumption per unit permeate rate for laminar flow through a tubular membrane. For turbulent flow through the tubular membrane, if we express the *f-Re* relationship as

$$f = 0.04/Re^{0.16} \quad \text{... (7.1.85)}$$

then the expression for the power consumption can be similarly derived as

$$P' = 0.02\pi\ (\Delta z)\ \rho_f\ d\mathbf{V}_0^3 E'/Re^{0.16} \quad \text{... (7.1.86)}$$

where $$E' = \sum_{i=1}^{n-1} \prod_{j=1}^{i} \left(1 - R_j\right)^{2.84} \quad \text{... (7.1.87)}$$

The power consumption per unit permeate rate will be,

$$P'/P = 0.08\ (\Delta z)\ Re^{-0.16}\ (\rho_f \mathbf{V}_0^2/d)\ (E'/R) \quad \text{... (7.1.88)}$$

The expressions for power consumption for flow between parallel membranes can also be similarly developed. For laminar flow between two parallel surfaces that are at a distance $(2h)$ apart, the *f-Re* relationship is given by[8]

$$f = 24/Re \qquad \text{... (7.1.88)}$$

where

$$Re = (4h)\ \mathbf{V}\rho_f/\mu_f \qquad \text{... (7.1.90)}$$

Accordingly, the pressure drop in the $i$th segment may be expressed as

$$(-\Delta P_f)_i/\rho_f = 3\mu_f\,(\Delta z)\ \mathbf{V}_{i-1}/(\rho_f\, h^2) \qquad \text{... (7.1.91)}$$

The power consumption is

$$P_i' = 6\mu_f\,(\Delta z)\ W\ \mathbf{V}_{i-1}^2/h \qquad \text{... (7.1.92)}$$

The total power consumption and the power consumption per unit permeate rate are

$$P' = 6\mu_f\,(\Delta z)\ W\mathbf{V}_0^2 E/h \qquad \text{... (7.1.93)}$$

$$P'/P = 3\mu_f\,(\Delta z)\ \mathbf{V}_0 E/(h^2 R) \qquad \text{... (7.1.94)}$$

In terms of $X_1^*$, $\alpha_0$ and $Re$,

$$P'/P = [(3/16)\ \rho_f\, Sc^3 U^{*2} \alpha_0^3 X_1^*\ E/R]\ Re^2 \qquad \text{... (7.1.95)}$$

where $E$ is defined in Eq. (7.1.80) and $Re = (4h)\ \mathbf{V}_0\ \rho_f/\mu_f$.

For turbulent flow between two parallel surfaces, the *f-Re* relationship is[8]

$$f = 0.079/Re^{0.25} \qquad \text{... (7.1.96)}$$

Accordingly,

$$P' = 0.079\ (\Delta z)\ \mathbf{V}_0^3\, W E^*/Re^{0.25} \qquad \text{... (7.1.97)}$$

$$P'/P = 0.0395\ (\Delta z)\ Re^{-0.25}\ (\mathbf{V}_0^2/h)\ (E^*/R) \qquad \text{... (7.1.98)}$$

where

$$E^* = 1 + \sum_{i=1}^{n-1} \prod_{j=1}^{i} (1 - R_j)^{2.75} \qquad \text{... (7.1.99)}$$

**Example 7.1:** Brackish water containing 3.3 percent (by weight) salt (principally sodium chloride) is to be desalinated in a single stage reverse osmosis unit. The operating temperature is 25° C and the operating transmembrane pressure drop 100 atm. The membrane used is that of cellulose acetate and its specifications (at 25° C) are as given below.

Permeability coefficient for water = $9.7 \times 10^{-6}$ kmoles/(m$^2\cdot$s$\cdot$atm)

Solute transport parameter = $0.9 \times 10^{-7}$ m/s

The module consists of two parallel sheet membranes 0.2 cm apart and the feed solution is admitted at the rate of 54 liters per hour. Width of each membrane = 0.30 m. Compute the membrane area required to obtain at least 40 percent product recovery. Assume that the molar density of the solution remains essentially constant and is equal to 55.35 kmoles/m$^3$ and the osmotic pressure is directly proportional to the mole fraction of the solute (proportionality constant = 2522.35 atm). Diffusivity of salt in solution = $1.475 \times 10^{-9}$ m$^2$/s.

**Solution:** Let us first evaluate the variables $U^*$, $\phi$, $\theta$ and $\alpha_0$. Thus, from Eqs (7.1.26) to (7.1.28),

$$U^* = \frac{(9.7 \times 10^{-6})(100)}{(55.35)} = 17.525 \times 10^{-6} \text{ m/s}$$

$$\phi = \frac{(2522.35)}{(100)(55.35)} = 455.70887 \times 10^{-3} \text{ m}^3/\text{kmole}$$

$$\theta = \frac{(0.9 \times 10^{-7})}{(17.525 \times 10^{-6})} = 0.0051355$$

Now, from Eq. (7.1.50), $\alpha_0 = \dfrac{(1.475 \times 10^{-9})}{(17.525 \times 10^{-6})(0.001)} = 0.084165$

The next step is to assume the value of $X_1^*$. However, when an order of magnitude of $X_1^*$ is not known, it is more convenient to assume the membrane area ($A_m$) or the membrane length ($L$). For example, let

$$A_m = 1.0 \text{ m}^2$$

Since $A_m = 2\,(LW)$ and $W = 0.30$ m,

$$L = \frac{(1.0)}{(2.0)(0.30)} = 1.667 \text{ m}$$

Let us, therefore, choose $L = 1.7$ m. Now, since the feed flow rate ($F_0$) is 54 liters/hr or 0.054 m$^3$/hr,

$$\mathbf{V}_0 = \frac{(0.054/3600)}{(0.3)(0.002)} = 0.025 \text{ m/s}$$

Let us choose $n = 40$. In other words, the membrane module is assumed divided into 40 differential segments (to note that the larger the value of $n$ chosen, the higher will be the accuracy of the computations). Then

$$\Delta z = (1.7/40) \text{ m}$$

And, from Eq. (7.1.43), $X_1^* = \dfrac{(17.525 \times 10^{-6})(1.7/40)}{(0.001)(0.025)} = 0.02979$

For the sake of convenience, let us round off the above value and take $X_1^* = 0.03$.

We have to now proceed to make computations segment by segment. Thus, let us put $i = 1$. From Eq. (7.1.90),

$$Re = \frac{(0.004)(0.025)(1000)}{(0.001)} = 100.0$$

The flow is thus laminar. (We have taken density of solution to be 1000 kg/m$^3$ and viscosity 0.01 poise). Since the feed flow rate decreases from segment to segment, the flow will continue to be in the laminar zone in all the subsequent segments. The check for the flow zone can be, therefore, avoided for all the subsequent segments. Now, from Eq. (7.1.51),

$$\beta_f = \frac{(0.03)}{(3.0)(0.084165)^2} = 1.4116$$

From Eq. (7.1.52), $\beta_{as}$ = 15.9466. Since $\beta_f < \beta_{as}$, Eq. (7.1.48) is applicable. Thus

$$\alpha_0 Sh_{av} = 3.9\ \beta_f^{-1/3}$$

or

$$\lambda_i = (3.9/4\theta)\ \beta_f^{-1/3}$$

Substituting the values of θ and $\beta_f$, we get

$$\lambda_i = 169.2425$$

Now, let $C_{pi} = C_{p1}$ = 0.01 kmoles/m³. Therefore, from Eq. (7.1.35),

$$C_{fi} = 0.5708 \text{ kmoles/m}^3$$

From Eq. (7.1.39), $q_i = (0.5708)/(0.01) = 57.08$

From Eq. (7.1.44), $$R_i = \frac{(0.0051355)(0.03)}{\left[(0.01)\left(455.70887\times10^{-3}\right)+0.0051355\right]} = 0.015895$$

From Eq. (7.1.40a), $p_i = p_1 = 56.1886$

Now, from Eq. (7.1.40), $C_{pi} = C_{p1} = C_{f0}/p_1$

Since the feed contains 3.3 percent salt,

$$C_{f0} = \frac{(3.3/58.5)}{(100/1000)} = 0.5641 \text{ kmoles/m}^3$$

Therefore, $C_{p1} = (0.5641)/(56.1886) = 0.0100394$ kmoles/m³

Though the above value of $C_{p1}$ does not differ significantly from that assumed at the outset, the accuracy can be further improved by performing one more iteration. Thus, assuming $C_{p1}$ = 0.0100394 kmoles/m³, the value of $C_{fi}$ from Eq. (7.1.35) is

$$C_{fi} = C_{f1} = 0.572674 \text{ kmoles/m}^3$$

$$q_i = q_1 = 57.039277$$

$$R_i = R_1 = 0.0158653$$

$$p_i = p_1 = 56.150197$$

Finally, from Eq. (7.1.40), $C_{pi} = C_{p1} = 0.010046269$ kmoles/m³

The percentage error is

$$= \frac{(0.010046269 - 0.0100394)(100)}{(0.0100394)} = 0.0684 \text{ percent}$$

Since the error is quite low, we can avoid further trial. Now, we proceed to the next segment. Thus, $i$ = 2. From Eq. (7.1.46),

$$X_2^* = 0.03/(1.0 - 0.0158653) = 0.03048363$$

$$\beta_f = 1.43443$$
$$\lambda_2 = 168.34282$$

Let $C_{p2} = 0.01027$ kmoles/m$^3$

Then, $C_{f2} = 0.5815175$ kmoles/m$^3$

$$q_2 = 56.6229$$
$$R_2 = 159.489 \times 10^{-4}$$
$$p_2 = 55.7358$$

And, from Eq. (7.1.40), $C_{p2} = (0.572674)/(55.7358) = 0.0102748$

$$\text{Percent error} = (0.0102748 - 0.01027)\,(100)/(0.01027) = 0.04674 \text{ percent}$$

In the same manner, computations are performed for all other segments (until $i = 40$). The overall product recovery is then obtained from Eq. (7.1.65). The result is

$$R = 0.489 = 48.9 \text{ percent}$$

Since this is quite above the minimum specified value, the membrane area chosen is satisfactory for the purpose. Since we have rounded off $X_1^*$ to 0.03, we may recompute $A_m$ [from Eq. (7.1.69)] as

$$A_m = \frac{(X_1^* \, n)}{(RU^*)} \, (F_0 R) = n X_1^* F_0 / U^*$$

$$= (40)\,(0.03)\,(0.054/3600)/(17.525 \times 10^{-6}) = 1.0271 \text{ m}^2$$

$$L = (1.0271)/(2.0)\,(0.30) = \mathbf{1.7118\ m}$$

It may be noted that we have considered only 40 segments ($n = 40$) in the above computation. By employing a larger value of $n$ (for example, $n = 100$), the accuracy of the computations can be further improved. A computer program in this connection (but considering a bundle of tubular membranes instead of parallel membranes) is given in Section 7.7 (Program 7.1).

What has been described in the earlier paragraphs is a single stage RO system. In large capacity installations, a multistage operation is often preferred. A typical schematic of a three-stage RO system is shown in Fig. 7.3. The fresh feed is mixed with the retentate stream from stage 2 and then fed to stage 1. The retentate from stage 1 is collected as the final retentate (or final concentrate). The permeate from stage 1 is mixed with the retentate from stage 3 and then fed to stage 2. Permeate from stage 2 is feed to stage 3 and permeate from stage 3 is collected as the final permeate (product water).

### 7.1.2 Membranes and Membrane Modules

It is obvious that the membrane is the most vital component of an RO system and for that reason, of any membrane-based separation process. A good RO membrane must have the following properties:

(*i*) It must exhibit high solute rejection,

(*ii*) It must provide high solvent flux at reasonably low operating pressures and the flux must remain uniform throughout the useful life of the membrane (often, the flux gets diminished due to compaction or aging of the membrane),

(*iii*) The membrane must have high mechanical strength, must be chemically stable in medium, must be resistant to fouling and must exhibit minimum swelling or shrinkage while in use,

(*iv*) it must be capable of being formed into desirable forms with low wall thickness and at low cost.

Practically, all commercial membranes are polymeric. Ceramic membranes have been found to be quite expensive at the present status.

Cellulose acetate (CA) membranes are the earliest commercial membranes and they are still one of the most popularly used ones. Loeb and Sourirajan were the first to manufacture a cellulose acetate membrane with an asymmetric structure. The membrane had a thin (0.1 to 1.0 micron) dense surface (usually called the *skin*) supported by a porous cellulose acetate substructure. CA membranes provide high permeate fluxes for water and they also provide satisfactorily high salt rejection. Besides, their hydrophilic nature, castability into desired forms at low thicknesses and acceptable mechanical properties even at high pressures have made the CA membranes the most logical choice for many pilot-scale as well as large scale installations.

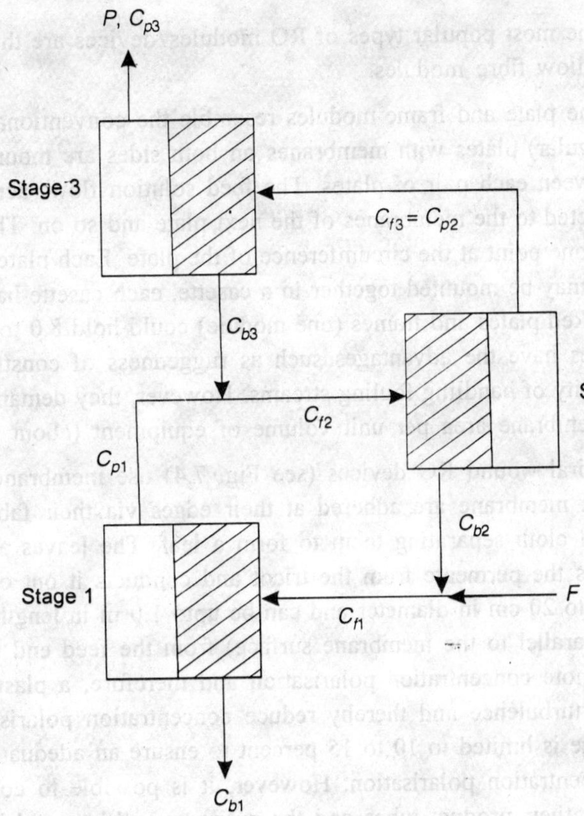

*Figure 7.3: Schematic of a multistage RO system*

Cellulose acetate membranes, however, have two major limitations. First, they are susceptible to degradation from biological attack. The feed solutions, therefore, are to be chlorinated prior to contacting these membranes. Secondly, cellulose acetate hydrolyses back to cellulose under acidic and particularly alkaline conditions. As a result of hydrolysis, salt passage through the membrane increases (in other words, salt rejection decreases). It is, therefore, important that the pH of the medium is maintained between 4.5 and 7.5 while employing CA membranes.

Aromatic polyamide (aramid) membranes were first commercialised by Du Pont, USA. Like CA membranes, these membranes also have an asymmetric structure. Aramid membranes are not susceptible to biological attack and they resist hydrolysis. They can be, therefore, operated at a pH range of 4.0 to 11.0. However, these membranes are degraded by chlorine and therefore, feed solutions containing chlorine must be dechlorinated before contacting them.

Polyamide membranes in thin film composite form have been proved to be of lot of promise. These membranes are formed by an *in situ* interfacial polymerisation technique. A layer of an aqueous solution of a polymeric amine is deposited on a finely porous surface of a polysulfone support membrane and then contacted with reactive difunctional compounds to form the salt-rejecting membrane (polyamide or polyurea thin film composite) with a thickness of 250 to 500 angstroms. If one considers this salt rejecting membrane as a skin and the polysulfone as the supporting substructure, this membrane would have an asymmetric structure. These thin-film composite membranes are not susceptible to biological attack and are resistant to hydrolysis, but they are much more sensitive to chlorine degradation than the aramid membranes.

The most popular types of RO modules/devices are the plate and frame, spiral wound, tubular and the hollow fibre modules.

The plate and frame modules resemble the conventional plate and frame filter presses. Circular (or rectangular) plates with membranes on both sides are mounted horizontally with flow-directing spacers in between each pair of plates. The feed solution flows across the membranes of one plate and then it is directed to the membranes of the next plate and so on. The permeate goes into the plates and is taken out at one point at the circumference of the plate. Each plate may have a permeate outlet or five to seven plates may be mounted together in a casette, each casette having a permeate outlet. One vertical column of stacked plates and frames (one module) could hold 8.0 to 19.0 $m^2$ of membrane area. Plate and frame modules have the advantages such as ruggedness of construction, ability to use rigid membranes and suitability of handling fouling streams. However, they demand high amount of skilled labour and provide low membrane area per unit volume of equipment (about 200 to 400 $m^2/m^3$).

Spiral wound RO devices (see Fig. 7.4) use membrane in the form of a flat film. Two sheets of the flat membrane are adhered at their edges via their fabric support backing with a tricot permeate channel cloth separating them to form a leaf. The leaves are wound spirally about a plastic tube that receives the permeate from the tricot and conducts it out of the device. Conventional spiral cartridges are 10 to 20 cm in diameter and can be upto 1.0 m in length. The feed solution flows in a straight axial path (parallel to the membrane surface) from the feed end to the brine end. Such a flow pattern tends to promote concentration polarisation and therefore, a plastic netting is placed in the feed channel to induce turbulence and thereby reduce concentration polarisation. Usually, the salt rejection per spiral cartridge is limited to 10 to 15 percent to ensure an adequate brine flow rate and to minimise the effect of concentration polarisation. However, it is possible to connect two to six spiral cartridges in series through their product tubes and the entire assembly could be enclosed in a pressure vessel that could be upto 7.0 m long. Such an arrangement can provide a solute rejection as high as 50 percent.

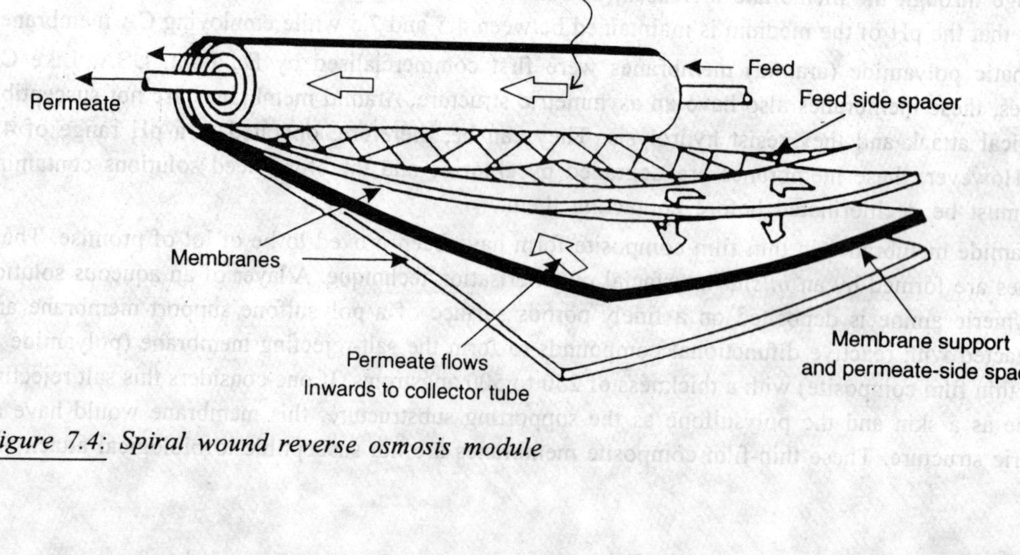


*Figure 7.4: Spiral wound reverse osmosis module*

Spiral wound devices provide large specific membrane areas (300 to 900 $m^2/m^3$) but they are more complex to construct and are not readily amenable to cleaning.

Tubular modules resemble shell and tube heat exchangers and they consist of bundles of porous or perforated tubes (1.3 to 2.5 cm in diameter), the inside walls of which are lined with the membrane (see Fig. 7.5). Pressurised feed flows inside the tubes, the product water permeates through the membrane and the tube and is collected on the outside. The concentrated brine (retentate stream) exits from the other end of the tube bundle. Tubular devices have good resistance to mechanical damage and they are reasonably amenable to cleaning. However, they provide low membrane areas per unit volume (150 to 300 $m^2/m^3$).

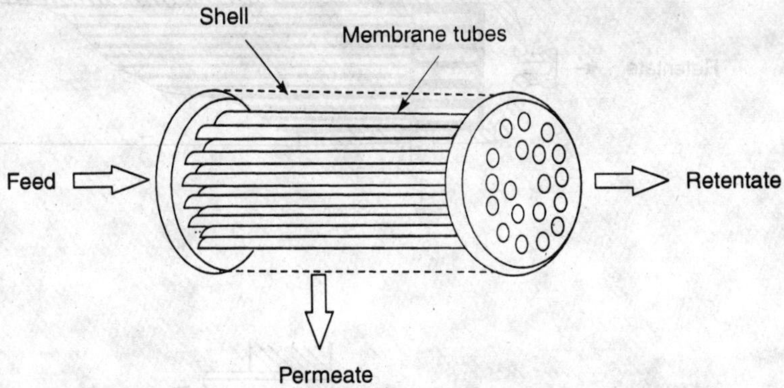

*Figure 7.5: Tubular reverse osmosis module*

Hollow fiber modules consist of fibers with an outside diameter of 25 to 250 microns and a wall thickness of 5 to 50 microns. As many as 4.0 to 4.5 million fibers of this kind are gathered into a bundle. During forming, epoxy adhesive is applied to one end of the bundle which, after curing, becomes a tubesheet. The other end of the fiber bundle is sealed in epoxy to form a nub which prevents short-circuiting of the feed stream to the brine outlet. The bundle is enclosed in a pressure vessel that is upto 1.2 m long and 10 to 25 cm in diameter. Pressurised feed solution enters through a porous distributor tube mounted axially and which extends over the entire length of the unit and then flows radially over the tube bundle. The product water permeates through the fiber wall into the fiber bore. It then flows through the fibers to the tubesheet and leaves the permeator. The concentrated brine (retentate) which flows to the outer perimeter of the fiber bundle exits through the brine port (see Fig. 7.6).

Though more complex in construction, hollow fiber modules are extremely compact and they provide very large membrane areas per unit volume (9,000 to 30,000 $m^2/m^3$). Since solvent flow per unit membrane area is low, concentration polarisation is not significant in hollow fiber devices and they can be operated in the laminar flow regime. However, these devices are more prone to fouling and less amenable to mechanical cleaning.

The feed to any RO system (for that reason, to any membrane-based separation system) requires a good amount of pretreatment. It is to be first filtered using a stainless steel micron cartridge filter or a sand filter to remove suspended particles. If any colloidal matter (that could cause fouling of the membrane) is present, it must be first coagulated (by adding a small amount electrolyte) and then filtered off.

Dissolved salts of calcium, magnesium, etc. tend to precipitate on the membrane surface causing membrane scaling (or precipitation fouling, similar to that occurs in heat exchanger tubes). Acid addition (addition of small amounts of HCl or sulfuric acid) is employed to prevent calcium carbonate scaling. This also helps in reducing the pH of the medium (important for CA membranes to prevent hydrolysis). Scaling due to precipitation of sulfates of calcium, barium and strontium can be minimised by the addition

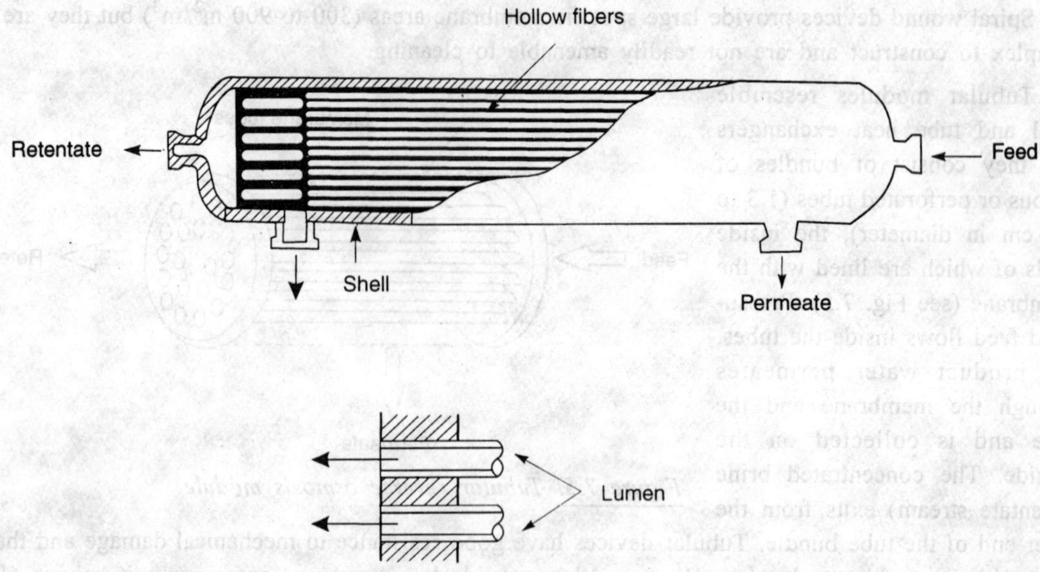


*Figure 7.6: Hollow fibre reverse osmosis module*

of sodium hexameta phosphate or by ion-exchange softening. Silica scaling could be a problem with some brackish waters. This can be minimised by lime/soda ash softening or by increasing the operating temperature to increase the solubility of silica.

If biological fouling is a problem, the feed solution may be chlorinated. This is normally required for brackish and sea waters employing CA membranes, but not admissible for aramid or thin film composite membranes since these are degraded by chlorine. Alternate method of minimising biofouling is addition of chemicals like sodium bisulfite, iodine or hydrogen peroxide. Broadly speaking, the types of unit operations used in the pretreatment system are governed by the feed water composition, the chemical nature of the membrane as well as the configuration of the RO device. Since RO is already being employed on large scale for desalting brackish waters and sea waters, pretreatment requirements for these are accurately defined. However, for special applications (purification of waste streams and industrial effluents), not all the pretreatment requirements are known and pilot testing may be required.

## 7.1.3 Applications of Reverse Osmosis

Reverse osmosis systems are highly energy efficient. They operate at ambient temperatures and they affect separation without any phase change. Thus, there is no consumption of thermal energy and consequently, the energy requirements of RO systems are significantly low. For example, RO modules employed for desalination of sea water consume about 9.0 – 10.0 $kWh/m^3$ of product water, while those employed for desalting brackish waters consume 1.5 – 2.5 $kWh/m^3$ of product water. The energy consumption can be further decreased by employing energy recovery equipment in the retentate stream. The retentate (concentrated brine) that is discharged from the RO devices at a pressure of 50 to 65 atmospheres (5100 to 6550 kPa) may be passed through a hydroturbine or impulse turbine to produce electric power. The energy recovery equipment reduces the plant's low energy requirement by a further 33 percent. In contrast, if desalination of sea water is performed in multiple effect evaporators (employing seven effects), then the steam requirement will be typically 0.09 tonnes per $m^3$ of water vapourised and the electric power consumption 3.0 kWh per $m^3$ of water vapourised. Costwise, one tonne of steam may be assumed equivalent to 312 kWh electricity and therefore, the total energy requirement is 31.08 kWh/$m^3$ of water vapourised, which is more than three times the energy requirement of RO process.

Since RO systems operate at ambient temperatures, heat-sensitive and explosive substances can be handled with relative ease.

RO systems do not discharge any additional waste streams. No noxious gases are evolved nor any solid residue discharged, which otherwise could pollute the environment.

RO units are quite compact and they require little labour and the cost of maintenance is low. Since RO plants are modular in design, scheduled maintenance can be performed without shutting down the entire plant. The modular design also makes expansion an easy option. Frequently, the civil works are designed for a much larger plant than is originally installed. Subsequent expansion requires almost no downtime.

Today, the most extensive application of reverse osmosis is in the production of potable water by desalination of sea water and brackish water. In USA, as much as 500 million $m^3$ of potable water is produced per day from brackish and sea water sources by reverse osmosis. Some of this product water is used for drinking purposes while much goes into industrial and commercial uses. Typical data on the quality of product water obtained from an RO Unit processing brackish water are given in Table 7.2. The operating pressure in brackish water processing units ranges from 27.0 to 40.0 atmospheres while that in sea water desalination units ranges from 55.0 to 70.0 atmospheres.

In power plants that use boilers operating at or close to critical pressure, the feed water has to be *ultrapure.* For example, typical specifications for feed water quality for subcritical boilers (operating at 120.0 to 160.0 atmospheres) and supercritical boilers (operating at more than 215.0 atmospheres) are respectively 0.5 and 0.05 ppm total dissolved solids (TDS). Such stringent water quality is required in many other industries also such as in industries manufacturing electronic components, nuclear power plants and in many pharmaceutical industries. Another example is WFI (Water for Injection) that is indispensable in many hospitals. Production of such ultrapure waters can be efficiently and economically accomplished by the repeated application of reverse osmosis. For example, when a sample of feed water whose hardness = 300 – 800 ppm $CaCO_3$ is subjected to RO process twice, its hardness reduces to 0.125 – 0.225 ppm $CaCO_3$ and this water can be then used as feed for subcritical boilers. An additional RO processing of above water can give water that is suitable as feed to supercritical boilers. The product recovery in each of the above cases is as large as 90 percent and the permeate rates are as high as 350–400 gallons/day/$m^2$ at 65 to 70 atm operating pressure. The overall process is thus highly cost efficient. RO not only desalts water but also removes dissolved silica, colloidal matter and many high molecular weight (molecular weight > 120) organics. It can be, therefore, efficiently used in conjunction with ion exchange in electronics industries to produce ultrapure water for manufacturing electronic components.

At present, in many hospitals, pyrogen-free injection-grade water (WFI) is manufactured by multistage distillation. Accordingly, the cost of WFI is as high as Rs. 1125 per $m^3$. On the other hand, if WFI is produced by RO (the feed water is first deionised by ion-exchange process followed by activated carbon

**Table 7.2**
**Reverse Osmosis for Desalination of Brackish Water. Average Product Water Rate = 150–200 gal/day/m$^2$**

| | *Feed water* | *Product water* | *Concentrated brine* |
|---|---|---|---|
| pH | 7.7 | 7.8 | 8.0 |
| Iron (ppm) | 0.018 | 0.012 | 0.06 |
| Boron (ppm) | 2.75 | 2.25 | 3.50 |
| Bicarbonates (ppm) | 161.7 | 24.4 | 375.2 |
| Chlorides (ppm) | 262.4 | 74.5 | 517.7 |
| Sulfates (ppm) | 1260.5 | 34.2 | 2942.9 |
| Silica (ppm) | 49.2 | 8.4 | 92.8 |
| Calcium (ppm) | 128.7 | 6.9 | 299.4 |
| Magnesium (ppm) | 89.0 | 1.0 | 199.1 |
| Sodium (ppm) | 521.0 | 66.0 | 1168.0 |
| Total dissolved solids (ppm) | 2477.6 | 220.8 | 5592.8 |
| Total hardness (ppm $CaCO_3$) | 692.4 | 25.2 | 1577.7 |
| Total incrustating solids (ppm) | 742.8 | 34.4 | 1673.3 |
| Free $CO_2$ (ppm) | 5.0 | 1.0 | 6.0 |

*Source:* Loeb, S and Johnson, JS, CEP, **63** (1), 90, 1967.

treatment and then fed to the RO unit), its cost will only be Rs. 400/m$^3$ (to note that this includes membrane replacement cost and the cost of feed pretreatment as well).

The application of reverse osmosis for sewage and waste water treatment has been well-established. The conventional sewage treatment plant employing primary and secondary treatment removes only the biochemical oxygen demand (BOD) and the suspended solids. Other contaminants, such as nitrates, phosphates and non-biodegradable surfactants, are not removed. To remove these additional pollutants, tertiary sewage treatment is required. Reverse osmosis can effectively take the place of tertiary treatment and sometimes both secondary and tertiary treatments and offer an effective means of upgrading sewage water to a quality practically suitable for all purposes. Typical results from an RO unit employing cellulose acetate membranes and an operating pressure of 70 atmospheres are presented in Table 7.3. The product recovery reported is 90 percent and the average permeate rate 330 gallons/day/m$^2$.

The waste effluents from metal finishing plants contain numerous constituents such as sulfates of zinc, copper, chromium, nickel and iron, chlorides of nickel, tin and gold and various salts of silver and lead. Many of these salts are toxic to marine life. For example, the toxicity limits to fish life with respect to copper, lead, zinc and chromium are 0.02, 0.1, 0.2 and 1.0 ppm respectively. Since profuse rinsing is a pre-requisite of a sound finish, very large quantities of water are involved in metal finishing operations and as a result, the amount of waste effluent discharged is extremely large. Also, many of the constituents of this waste have good economic value (for example, nickel, chromium, silver, gold, etc.) and therefore, their recovery could fetch attractive financial benefits. Thus, treatment of plating wastes is important from

## Table 7.3
## Reverse Osmosis for Sewage Water Treatment

| *Solute or equivalent* | *Solute concentration in feed, ppm* | *Solute concentration in product water, ppm* |
|---|---|---|
| BOD | 37.0 | 2.0 |
| Nitrates | 0.50 | 0.25 |
| Phosphates | 2.50 | 0.01 |
| Alkyl benzene sulfonate | 1.20 | 0.08 |
| Total dissolved solids | 324.0 | 9.0 |

the points of view of water pollution control, water reuse as well as recovery of valuable chemicals. Successful pilot plant studies have demonstrated that a single stage reverse osmosis unit operating at 70 atmospheres could produce product water (permeate) that contains not more than 10 ppm of dissolved salts from a feed (plating waste) containing 500 ppm of solutes. The reported product recovery is 95 percent and the average permeate rate 200 gallons/day/$m^2$. The permeate is thus suitable for direct reuse. The concentrate (retentate) can be further processed for the recovery of valuable metals.

Reverse osmosis, in conjunction with ultrafiltration (ultrafiltration, which is very similar to reverse osmosis, is discussed in the next section), can be efficiently used for the concentration of black liquor in pulp and paper industry[9]. Black liquor is conventionally concentrated in multiple effect evaporators. Large evaporators having six or seven effects are common in pulp and paper industry and evaporators having as many as 17 effects have been built. Therefore, if black liquor can be partly concentrated by UF (ultra filtration) and/or RO, then a significant saving in energy can be achieved. Black liquor after primary sedimentation and filtration is sent to a UF unit followed by an RO unit, whereby it is concentrated from 10 percent solids to 20–25 percent solids. The permeate from the RO unit will be essentially water, which can be recycled to the digestor for reuse. The concentrate (retentate) is further processed in evaporators. Thus, there will be a drastic reduction in the load on evaporators and consequently a marked saving in energy consumption. To note that paper and pulp industry is one of the largest users of water. An average pulp mill producing 500 tonnes of cellulose pulp consumes as much as 50 million gallons of water per day. Recovery and reuse of water is thus of great economic interest in the case of pulp and paper industries.

Another excellent application of reverse osmosis is in the dairy industry. These industries are typically located in rural areas, away from population centres. Milk is, therefore, required to be transported from dairy farms to regional factories or depots and to city-based milk processors. Reverse osmosis can be used to concentrate raw milk to approximately 50 percent of its original volume and thus could provide approximately 50 percent reduction in transport cost. Studies[10] have demonstrated that RO technology can be introduced without any apparent effect on milk quality. Milk products such as butter and skim milk powder can be more economically manufactured from the RO concentrated milk. Also, yoghurts (or curds) manufactured from RO concentrated milk have been found to possess higher stability and superior physical characteristics.

## Table 7.4
## Osmotic Pressure, Solute Diffusivity and Molar Density of Sodium Chloride-Water System at 25° C

| *Molality* | *Mole fraction* | *Weight percent solute* | *Osmotic pressure, atm* | *Density, $kg/m^3$* | *Molar density, $kmoles/m^3$* | *Solute diffusivity, $m^2/s$* |
|---|---|---|---|---|---|---|
| 0.0 | 0.0 | 0.0 | 0.0 | 997.1 | 55.35 | $1.61 \times 10^{-9}$ |
| 0.1 | 0.001798 | 0.5811 | 4.56 | 1001.1 | 55.35 | $1.483 \times 10^{-9}$ |
| 0.2 | 0.00359 | 1.1555 | 9.05 | 1005.2 | 55.35 | $1.475 \times 10^{-9}$ |
| 0.3 | 0.00537 | 1.7233 | 13.54 | 1009.1 | 55.35 | $1.475 \times 10^{-9}$ |
| 0.4 | 0.00715 | 2.2846 | 17.96 | 1013.0 | 55.34 | $1.475 \times 10^{-9}$ |
| 0.5 | 0.008927 | 2.8395 | 22.52 | 1016.9 | 55.34 | $1.475 \times 10^{-9}$ |
| 0.6 | 0.01069 | 3.3882 | 27.07 | 1020.8 | 55.34 | $1.475 \times 10^{-9}$ |
| 0.7 | 0.01245 | 3.9307 | 31.7 | 1024.8 | 55.34 | $1.475 \times 10^{-9}$ |
| 0.8 | 0.01421 | 4.4671 | 36.33 | 1028.6 | 55.33 | $1.477 \times 10^{-9}$ |
| 0.9 | 0.01595 | 4.9976 | 41.02 | 1032.2 | 55.32 | $1.480 \times 10^{-9}$ |
| 1.0 | 0.01769 | 5.5222 | 45.78 | 1035.7 | 55.30 | $1.483 \times 10^{-9}$ |
| 1.2 | 0.02116 | 6.5543 | 55.37 | 1042.7 | 55.26 | $1.488 \times 10^{-9}$ |
| 1.4 | 0.0246 | 7.5640 | 65.24 | 1050.5 | 55.26 | $1.492 \times 10^{-9}$ |
| 1.6 | 0.02802 | 8.5522 | 75.44 | 1058.1 | 55.26 | $1.497 \times 10^{-9}$ |
| 1.8 | 0.031408 | 9.5194 | 85.85 | 1065.3 | 55.24 | $1.505 \times 10^{-9}$ |
| 2.0 | 0.03477 | 10.4665 | 96.53 | 1072.2 | 55.21 | $1.513 \times 10^{-9}$ |
| 2.2 | 0.03812 | 11.3939 | 107.483 | 1079.0 | 55.17 | $1.521 \times 10^{-9}$ |
| 2.4 | 0.04144 | 12.3022 | 118.710 | 1085.9 | 55.15 | $1.530 \times 10^{-9}$ |
| 2.6 | 0.04474 | 13.1922 | 130.270 | 1092.7 | 55.12 | $1.539 \times 10^{-9}$ |
| 2.8 | 0.04802 | 14.0642 | 142.110 | 1099.1 | 55.07 | $1.548 \times 10^{-9}$ |
| 3.0 | 0.05127 | 14.9190 | 154.420 | 1105.6 | 55.04 | $1.556 \times 10^{-9}$ |
| 3.2 | 0.05451 | 15.7568 | 166.870 | 1112.1 | 55.00 | $1.565 \times 10^{-9}$ |
| 3.4 | 0.05771 | 16.5784 | 180.340 | 1118.5 | 54.97 | $1.570 \times 10^{-9}$ |
| 3.6 | 0.0609 | 17.3840 | 192.790 | 1124.7 | 54.92 | $1.575 \times 10^{-9}$ |
| 3.8 | 0.06407 | 18.1743 | 206.390 | 1130.9 | 54.88 | $1.580 \times 10^{-9}$ |
| 4.0 | 0.06722 | 18.9496 | 220.270 | 1136.9 | 54.84 | $1.585 \times 10^{-9}$ |
| 4.2 | 0.07034 | 19.7103 | 234.420 | 1142.9 | 54.79 | $1.589 \times 10^{-9}$ |
| 4.4 | 0.07344 | 20.4569 | 249.05 | 1149.0 | 54.75 | $1.594 \times 10^{-9}$ |
| 4.6 | 0.07653 | 21.1897 | 263.88 | 1155.0 | 54.72 | $1.593 \times 10^{-9}$ |
| 4.8 | 0.07959 | 21.9092 | 279.18 | 1160.8 | 54.67 | $1.593 \times 10^{-9}$ |
| 5.0 | 0.08263 | 22.6156 | 294.76 | 1166.6 | 54.63 | $1.592 \times 10^{-9}$ |
| 5.2 | 0.08565 | 23.3093 | 310.75 | 1172.3 | 54.58 | $1.592 \times 10^{-9}$ |
| 5.4 | 0.08865 | 23.9908 | 327.00 | 1177.8 | 54.53 | $1.591 \times 10^{-9}$ |

## Table 7.5
## Effect of Temperature on Osmotic Pressure and Molar Density for Sodium Chloride-Water System

| *Molality* | *Osmotic pressure, atm* | | | *Molar density, kmoles/m*$^3$ | | |
|---|---|---|---|---|---|---|
| | *5° C* | *15° C* | *35° C* | *5° C* | *15° C* | *35° C* |
| 0.0 | 0.0 | 0.0 | 0.0 | 55.51 | 55.47 | 55.18 |
| 0.1 | 4.15 | 4.353 | 4.63 | 55.51 | 55.46 | 55.18 |
| 0.3 | 12.110 | 12.585 | 13.40 | 55.51 | 55.45 | 55.20 |
| 0.4 | 16.60 | 17.280 | 18.367 | 55.51 | 55.45 | 55.16 |
| 0.5 | 20.41 | 21.290 | 22.720 | 55.51 | 55.40 | 55.13 |
| 0.6 | 24.97 | 26.05 | 27.823 | 55.54 | 55.46 | 55.13 |
| 0.7 | 28.84 | 30.14 | 32.110 | 55.54 | 55.46 | 55.11 |
| 0.8 | 33.40 | 34.90 | 37.140 | 55.54 | 55.46 | 55.09 |
| 0.9 | 37.28 | 38.98 | 41.50 | 55.57 | 55.47 | 55.09 |
| 1.0 | 41.90 | 43.81 | 46.60 | 55.58 | 55.47 | 55.10 |
| 1.2 | 51.16 | 53.47 | 56.87 | 55.56 | 55.46 | 55.06 |
| 1.4 | 60.41 | 63.47 | 68.03 | 55.58 | 55.45 | 55.05 |
| 1.6 | 69.11 | 72.653 | 77.89 | 55.59 | 55.44 | 55.02 |
| 1.8 | 78.367 | 82.31 | 88.03 | 55.59 | 55.41 | 54.99 |
| 2.0 | 87.483 | 91.43 | 97.75 | 55.56 | 55.38 | 54.94 |
| 2.2 | 97.75 | 102.11 | 109.52 | 55.56 | 55.38 | 54.94 |
| 2.4 | 107.55 | 113.06 | 120.75 | 55.51 | 55.33 | 54.90 |
| 2.6 | 118.16 | 124.50 | 133.13 | 55.49 | 55.31 | 54.86 |
| 2.8 | 128.64 | 135.51 | 144.966 | 55.47 | 55.27 | 54.82 |
| 3.0 | 140.136 | 147.755 | 157.755 | 55.43 | 55.23 | 54.78 |

RO has also found its place in the concentration of fruit juices. Orange juice can be concentrated to 30 percent solids by RO, while apple juice can be concentrated to 20–25° Brix. Several commercial units manufacturing concentrated tomato, apple and orange juices by RO have gone into operation since 1984 and commercial feasibility of using RO for making dietary products such as instant coffee, soups, maple syrup, etc. is being explored.

In metal machining operations, an oil-water emulsion is used to lubricate and cool the tools and the workplace. The emulsion picks up some of the metals. Reverse osmosis (in conjunction with ultrafiltration) can be used to separate this emulsion into a product water (permeate) that can be discharged or reused and an oil concentrate (retentate) that can be easily burnt or further refined to produce reusable oil.

Our discussion on reverse osmosis shall be incomplete unless we shed light on its limitations as well. The major bottleneck with RO is with regards to the magnitude of the operational pressure difference. It has already been pointed out at the beginning of this section that the process of RO can take place only if the transmembrane pressure difference ($\Delta P$) exceeds the osmotic pressure difference ($\Delta\pi$). As the solute concentration increases due to the withdrawal of solvent from the solution, $\Delta\pi$ increases and as a result, the magnitude of $\Delta P$ required to affect separation also increases. When the required value of $\Delta P$ becomes very high, the operation becomes uneconomical. Also, the membrane tends to get degraded when exposed to high pressure differences. Thus, at the present status, RO will not be economically viable at high concentration levels.

The short life span of commercially available membranes is another factor that inducts dubiousness into the commercial acceptability of all membrane-based separation processes. Membranes that are presently available in the market have an average useful life span of 1.5 to 2.0 years. With advances in research on superior membrane materials, the membrane life could get extended upto 10 years in the near future. However, even after accommodating this additional cost of membrane replacement, membrane-based separation processes (reverse osmosis, ultrafiltration) have been found to be economically superior to the energy-intensive thermal processes like evaporation and distillation, at least at low concentration levels.

## 7.2 ULTRAFILTRATION

In principle, ultrafiltration is quite analogous to reverse osmosis since both are pressure-driven processes employing selective transport of one species (usually, the solvent) through a semipermeable membrane. The chief differences between ultrafiltration (UF) and reverse osmosis (RO) are

(*i*) UF is employed for separating solutes of high molecular weight and whose molecular dimensions are ten times or more larger than those of the solvent, while in RO, the molecular size of the solute is only slightly larger than that of the solvent. Accordingly, the pore size of UF men.branes is much larger (20 to 200 angstroms) than that of RO membranes (0.1 to 10 angstroms).

(*ii*) Since the molecular weights of the solutes are large, their osmotic pressures are low and consequently, the operating transmembrane pressure drop for ultrafiltration is also low (3 to 7 atm). For RO, the operating pressure drop is 15 to 70 atm. As a result of this, the energy requirement for pumping and compression and the equipment cost are considerably lower for ultrafiltration than for reverse osmosis.

Only the high molecular weight solutes are retained by the UF membrane. Low molecular weight solutes and the solvent permeate through the membrane.

The membrane materials used for UF are same as those used for RO. Thus, cellulose acetate, aromatic polyamide and thin film composite membranes are most popular. Membranes made of polyacrylonitrile and polyfuran have also been introduced into the market.

Membrane modules for UF are also identical with those for RO. These include plate and frame, tubular, spiral-wound and hollow fiber modules. However, hollow fiber UF modules employ fibers that are 10 to 25 times larger in diameter than those used in RO modules (fibers used for UF are 500 to 1100 microns in diameter). Also, the feed solution usually flows through the fiber bores and the permeate is collected from the outer shell (pressure vessel). An alternate type of UF device is the thin channel tubular module. In this module, the feed solution flows through a narrow channel (0.25 to 0.75 mm) at a high velocity (1.5 to 7.5 m/s). Narrow flow cross-section and high fluid velocity provide high fluid shear rates and this decreases the thickness of the concentrated boundary layer at the membrane surface. As a result, the effect of concentration polarisation is very much reduced.

Mathematical analysis of ultrafiltration can be performed in the same way as that for reverse osmosis. However, since solutions containing high molecular weight solutes are handled in UF, $\Delta\pi$ will be much smaller than $\Delta P$ and with allowable error, we can assume that $(\Delta P - \Delta\pi) \simeq \Delta P$. As a result, Eq. (7.1.7) shall reduce to

$$N_{Wi} = Km\,\Delta P \qquad \text{... (7.2.1)}$$

As in reverse osmosis, concentration polarisation is an undesirable phenomenon that invariably occurs in ultrafiltration as well. Due to the solute build-up at the membrane surface, a concentrated boundary layer (also called *gel layer*) is formed which offers additional resistance to the passage of solvent and low molecular weight solutes. The solute concentration in the gel layer often reaches saturation. As stated earlier, thin channel UF modules have been developed with the main purpose of minimising the thickness of this gel layer. Several studies have been reported in the literature regarding the use of turbulent promoters in the flow field which tend to disturb the gel layer and thereby minimise the effect of concentration polarisation. Pitera and Middleman[11] report that in the case of tubular UF modules, insertion of static mixers enhance mass transfer across the membrane by a factor of 2.7. Narayanan[12] has shown that by employing constricted tubular modules, concentration polarisation can be drastically reduced and significant increase in solute rejection and product recovery can be achieved. It has also been reported by Narayanan that such a construction shall not cause any denaturation of the solute.

## 7.2.1 Applications of Ultrafiltration

Ultrafiltration does have some advantages over reverse osmosis. The energy requirement of UF is significantly lower than that of RO due to lower operating pressure drop. Since the effective driving force for UF is $\Delta P$ and not $(\Delta P - \Delta\pi)$, higher concentration levels can be achieved with UF than with RO. There is also less danger of damaging the product during depressurisation with UF. The higher pressures attendant in RO mean higher shear forces while releasing the concentrated solution (retentate) from the high pressure zone to atmospheric pressure. However, UF membranes retain only the high molecular weight solutes (molecular weight > 300). As a result, it is common practice in many industrial applications to use a UF unit followed by an RO unit. The permeate from the UF unit (which will be free from high molecular weight solutes) is sent to the RO unit which retains the lower molecular weight solutes as well. Consequently, the final permeate from the RO unit will be essentially pure, reusable solvent.

One of the most popular commercial uses of ultrafiltration is in the recovery of electrophoretic paints. UF is used to process the paint by retaining the polymer resins and pigment solids while allowing inorganic salts, water and solvent to permeate through the membrane. The retentate is recycled back to the electropaint tank, while the permeate is used to rinse the freshly painted components as they emerge from the paint and to recover the drag-out excess paint.

The application of UF in conjunction with RO for concentrating the black liquor in the pulp and paper industry and for separating the oil-water emulsion from metal machining operations have already been discussed in the earlier section under reverse osmosis.

Food applications of UF are legion. An excellent example is the concentration of cheese whey. Cheese whey is the fluid portion of the milk obtained after coagulation of casein during the manufacture of cheese or casein. It contains 50 percent of milk solids (most of lactose, 20 percent of protein and most of the vitamins and minerals). Moreover, the yield of whey is also quite large. Typically, 17 kg of milk yield

1 kg of cheese and 1 kg of whey solids. Even though 70 percent of the nutritive value of milk resides in the whey, it is often disposed off as waste. In addition to the loss of potentially valuable food products, this also produces significant pollution problems. Evaporation and spray drying of whey are expensive. Also, the high salt and lactic acid content of dried whey prevents the product from being marketed for human consumption. Ultrafiltration is an efficient and economical process that not only concentrates the whey but also makes it free from unwanted salts, lactic acid and lactose. Except the proteins, all solids pass through the UF membrane. The whey concentrate is thus suitable for human consumption. The permeate from the UF unit may be further processed in an RO unit to recover lactose and other organics and in that case, the product water from RO can be reused within the dairy as fresh water.

Another equally important application of UF is in the concentration of egg albumin (or egg white). The use of egg white in candy and baking industry is predicted on its ability to form foams stable enough to support relatively large quantities of flour and/or sugar. However, the conventional methods (such as spray drying) employed for concentrating egg white impair these functional properties. For example, during spray drying, the tremendous shear forces present in the spray dryer pressure nozzles and centrifugal atomisers denature egg albumin and adversely affect the whipping power (aerating power) of the dried product. Any type of thermal concentration is known to denature the proteins and aggravate what is called the *browning reaction*. This browning phenomenon is due to the presence of glucose which reacts with amino acids (the so called *Maillard reaction*) resulting in the development of caramel–like flavours and odours, evolution of carbon dioxide and eventual separation of an insoluble dark brown material. This also results in a decrease in protein solubility. To alleviate this problem, glucose has to be removed. Fermentation of glucose to gluconic acid by enzyme or microbial fermentation is possible, but this takes as much as 70 hours and adds significantly to the total cost and often also results in off-flavours and odours in the finished product. Ultrafiltration offers an excellent solution to all these problems. UF provides a ready means of removing glucose while concentrating the proteins. Also, the resultant egg white concentrate has excellent functional properties.

Ultrafiltration has direct application to three of the major problems of fermentation such as harvesting the cells, removal of waste products and recovery of the products of fermentation. Centrifugation is the standard method for harvesting the cells. However, since the efficiency of centrifugal separation is proportional to the square root of particle size, smaller bacterial cells (0.5 to 1.0 micron in size) require unduly long centrifugation. Also, when the substrate is a hydrocarbon oil, cell recovery by centrifugation becomes extremely difficult since some cells will be in the hydrocarbon phase and some in the aqueous phase. Thin channel ultrafiltration permits the recovery of these cells without the blocking of the membranes. Unlike with centrifugation, the small sizes of fermentation cells are, if anything, an advantage in UF process. UF is also well-suited for the recovery and concentration of the metabolic products of fermentation. This principle is utilised in membrane fermentors. Since the products of fermentation including toxic metabolites are continuously removed through the UF membrane, fresh substrate can be fed continuously to the fermentor and thus, the overall process becomes fully continuous. All macromolecules produced extracellularly will permeate through the membrane and as a result, quantitative conversion of the feed to product is obtained while maintaining a constant environment for organism survival.

Membrane enzymatic reactor is akin to membrane fermentor with the exception that no micro organisms are present. The reaction mixture containing the enzyme, unreacted substrate, enzyme-substrate complex and the products is pumped continuously to an ultrafiltration unit where the products are separated by permeating through the membrane (see Fig. 7.7). The retentate that contains the enzyme and the unconverted substrate is recycled back to the reactor. This not only permits reuse of enzyme, but also simplifies the problems of purification and enzyme removal from end products. Introduction of membrane enzymatic reactors has given a great boost to the economic prospects of enzyme-catalysed reactions.

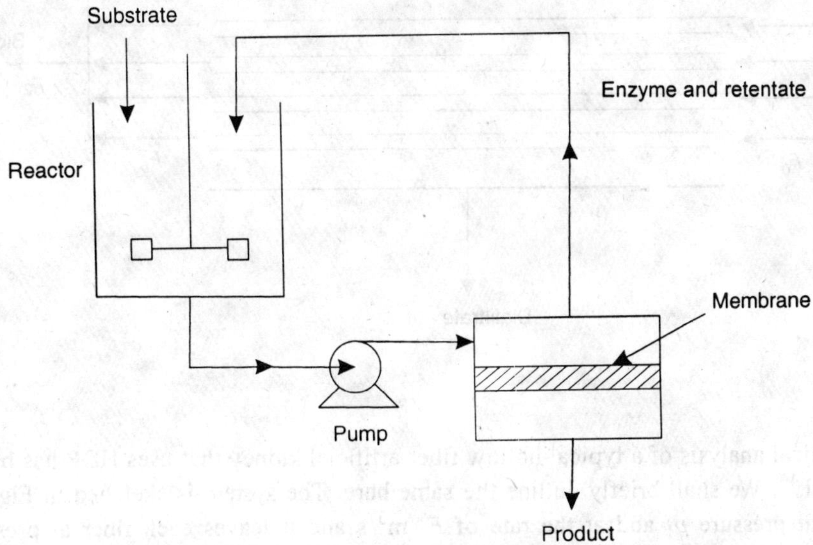


*Figure 7.7: Schematic of membrane enzymatic reactor*

Ultrafiltration does find excellent application in medicine. When UF is used for purifying human blood (as is done in artificial kidney) the process is usually termed as *haemodiafiltration* (HDF).

The purpose of artificial kidney is to replace the function of human kidney in the case of severe or complete renal failure. Its function is to remove certain toxins of low and moderate molecular weight from the blood such as sodium chloride, potassium chloride, urea, creatinine and uric acid. On the other hand, cellular particles and plasmaproteins (macromolecules) should be retained in the blood lumen. Artificial kidney devices employ either haemodiafiltration or haemodialysis for the purification of human blood. Hemodialysers (that employ dialysis for blood purification) are discussed in the next section on Dialysis.

## 7.2.2 Mathematical Analysis of Haemodiafiltration

Haemodiafiltration devices are commonly of hollow fiber configuration, consisting of a bundle of about 10,000 hollow fibers which are perfused with blood. Each fiber has an inside diameter of about 200 to 230 microns and is typically 17 to 20 cm in length. The fiber wall permits water containing small and medium molecular weight toxins to permeate freely while retaining the macro solutes such as cellular particles and plasma proteins. Thus, HDF causes significant reduction in plasma volume, which is restored subsequently in a second stage by a diluting fluid (see Fig. 7.8). The specific advantages of HDF devices are their small size and consequently small priming volume, low pressure drop and low haemolysis (destruction of red blood cells) rates by virtue of the low shear rates.

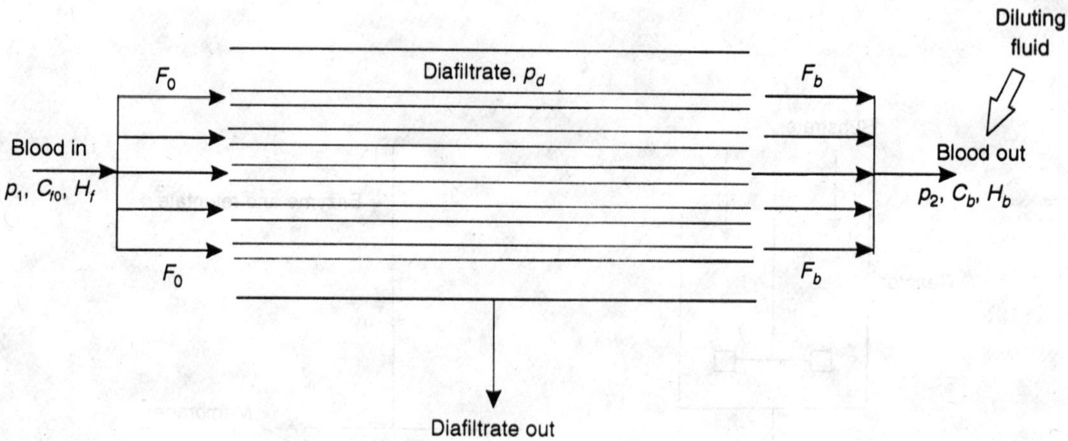


*Figure 7.8:*

The mathematical analysis of a typical hollow fiber artificial kidney that uses HDF has been reported by Papenfuss et. al.[17]. We shall briefly outline the same here. The system is sketched in Fig. 7.8. Blood enters each fiber at pressure $p_1$ and at the rate of $F_0$ m$^3$/s and it leaves each fiber at pressure $p_2$ and at the rate of $F_b$ m$^3$/s. The plasma protein concentration in the inlet blood is $C_{f0}$ and that in outlet blood $C_b$. The haematocrit (volume fraction of red blood cells) in inlet blood and outlet blood are $H_f$ and $H_b$ respectively. The following assumptions are made:

(*i*) The walls of hollow fibers are ideally selective membranes; that is, cellular particles and plasma proteins are fully retained internally, while water and toxins permeate freely.

(*ii*) The membrane permeability coefficient $K_m$ of the fiber walls is constant over the entire fiber length $L$.

(*iii*) The hydraulic pressure $p_d$ of the external diafiltrate (in the shell) is constant.

(*iv*) The two phase character of blood is neglected for the axial fluid transport (Blood, in fact, is a suspension of cellular particles in plasma).

(*v*) Blood executes Newtonian flow through the fibers and the change in blood viscosity along the fibers due to increasing protein concentration and haematocrit is neglected. This assumption has a degree of error in it since blood basically is a non-Newtonian fluid that follows Casson's rheological equation (see Chapter 2).

(*vi*) Radial pressure and concentration gradients in the blood flow are neglected (usually called the *hydraulic approach*).

(*vii*) Diffusive axial transport of proteins in the blood is negligible compared with the convective transport.

(*viii*) Since blood flow through the fibers is at very low Reynolds number ($Re < 10$), the hydraulic entrance effects at the inlet to the fibers may be neglected.

(*ix*) The flow is under steady state conditions.

(*x*) The effect of concentration polarisation is neglected which is fairly true in the case of hollow fiber devices. Accordingly, gel formation at the fiber wall (when the saturation concentration of the proteins is reached) is also assumed absent. It is further assumed that reduction in membrane permeability due to clogging of membrane pores by plasma proteins is not substantial.

The total volume flow rate of blood at any $z$ is the sum of the volume flow rate of plasma and that of the red blood cells (also called erythrocytes). Thus

$$F(z) = F_{pl}(z) + F_e \qquad \text{... (7.2.2)}$$

The solvent flux across the membrane can be expressed by Eq. (7.1.7). Thus

$$J = -(1/\pi d)\frac{dF_{pl}}{dz} = K'_m\left[p(z) - p_d - \pi_{pl}(z) + \pi_d\right] \qquad \text{... (7.2.3)}$$

where $J$ = transmembrane solvent flux per fiber, $m^3/(m^2 \cdot s)$

$d$ = inside diameter of each fiber, m

$K'_m$ = membrane permeability coefficient, $m^3/(m^2 \cdot s \cdot atm)$

$\pi_{pl}$, $\pi_d$ = colloid osmotic pressure in plasma and that in diafiltrate respectively, atm

The difference between $J$ and $N_W$ and that between $K_m$ and $K'_m$ may be noted. $J$ stands for the volumetric solvent flux (in $m^3/m^2 \cdot s$), while $N_W$ is the molar flux (in $kmoles/m^2 \cdot s$). Similarly, $K_m$ is expressed in $kmoles/(m^2 \cdot s \cdot atm)$, while $K'_m$ in $m^3/(m^2 \cdot s \cdot atm)$.

Incidentally, since we have assumed an ideally selective membrane, $\pi_d = 0$. Also, if we neglect the change in the density of blood, then a simple material balance yields

$$F_0 = F_b + P$$

or

$$P = F_0 - F_b \qquad \text{... (7.2.4)}$$

where $P$ is the total diafiltration rate per fiber.

Further, since the red cell volume does not change materially, $F_e$ in Eq. (7.2.2) can be treated as a constant. Accordingly, $dF_{pl}/dz$ in Eq. (7.2.3) may be replaced by $dF/dz$.

Colloid osmotic pressure increases as the protein concentration increases. For small values of protein concentration,

$$\pi_{pl}(z) = a_1 C(z) + a_2[C(z)]^2 + a_3[C(z)]^3 \qquad \text{... (7.2.5)}$$

where $a_0$, $a_1$, $a_2$ are empirical constants and $C(z)$ is the protein concentration in blood plasma at any $z$. As reported by Landis and Pappenheimer[18], the values of $a_0$, $a_1$, $a_2$ at 37° C are

$$a_1 = 0.21 \text{ mm Hg}/(kg/m^3)$$

$$a_2 = 0.0016 \text{ mm Hg}/(kg/m^3)^2$$

$$a_3 = 9.0 \times 10^{-6} \text{ mm Hg}/(kg/m^3)^3$$

Since the fiber wall is impermeable to plasma proteins, mass flow rate of proteins must be constant. Thus

$$F_{pl}(z)\, C(z) = (F_0 - F_e)\, C_{f0} = \text{constant} \quad \text{... (7.2.6)}$$

Also, since $H_f$ is the volume fraction of red cells in blood,

$$H_f = F_e / F_0 \quad \text{... (7.2.7)}$$

From the above two equations, we get

$$C(z) = \frac{C_{f0}\left(1 - H_f\right) F_0}{F(z) - H_f F_0} \quad \text{... (7.2.8)}$$

The local diafiltration rate is, in general, very small compared with the local axial volume flow of blood. Therefore, as a first approximation, we may use the Hagen-Poiseuille equation for expressing the local volume rate of flow of blood in the axial direction.

$$F(z) = \frac{\pi (d/2)^4}{8\,\mu_f} (-dp/dz)$$

or

$$\frac{dp}{dz} = -\frac{8\,\mu_f\, F(z)}{\pi (d/2)^4} \quad \text{... (7.2.9)}$$

Thus, we have to solve the two coupled differential Eqs (7.2.3) and (7.2.9) together with (7.2.5) and (7.2.8). The boundary conditions are

**B.C.1:** At $z = 0,\ p(z) = p_1;\ F(z) = F_0$ ... (7.2.10)

**B.C.2:** At $z = L,\ p(z) = p_2;\ F(z) = F_b$ ... (7.2.11)

It will be more convenient if we express the above differential equations in the dimensionless form. Let us, therefore, define the following dimensionless variables.

$$z^* = z/L$$

$$p^* = [p(z) - p_d]/(p_1 - p_d)$$

$$\pi^* = \pi_{pl}/(p_1 - p_d)$$

$$C^* = C(z)/C_{f0}$$

$$F^* = F(z)/F_{ref} = F(z)\,(\mu_f\, L)/[\pi\, (d/2)^4\, (p_1 - p_d)]$$

$$a_1^* = a_1 C_{f0}/(p_1 - p_d)$$

$$a_2^* = a_2 C_{f0}^2/(p_1 - p_d)$$

$$a_3^* = a_3 C_{f0}^3/(p_1 - p_d) \quad \text{... (7.2.12)}$$

Equations (7.2.3), (7.2.9), (7.2.5) and (7.2.8) thus get respectively transformed into

$$\frac{dF^*}{dz^*} = -2\epsilon\, (p^* - \pi^*) \quad \text{... (7.2.13)}$$

$$\frac{dp^*}{dz^*} = -8F^* \quad \text{... (7.2.14)}$$

$$\pi^* = a_1^* C^* + a_2^* (C^*)^2 + a_3^* (C^*)^3 \qquad \text{... (7.2.15)}$$

$$C^* = \frac{F_0^*(1 - H_f)}{(F^* - H_f F_0^*)} \qquad \text{... (7.2.16)}$$

where
$$\epsilon = \frac{K_m' \mu_f}{(d/2)} (2L/d)^2 \qquad \text{... (7.2.17)}$$

The modified boundary conditions are

**B.C.1:** At $z^* = 0,\ F^* = F_0^*,\ p^* = p_1^* = 1$ ... (7.2.18)

**B.C.2:** At $z^* = 1,\ F^* = F_b^*,\ p^* = p_2^* = (p_2 - p_d)/(p_1 - p_d) = 1 - \sigma$ ... (7.2.19)

Equation (7.2.4) also may be expressed in the dimensionless form as

$$\psi = P/F_{ref} = F_0^* - F_b^* \qquad \text{... (7.2.20)}$$

Since from Eq. (7.2.16), $C_b^* = \dfrac{F_0^*(1 - H_f)}{F_b^* - H_f F_0^*}$ ... (7.2.21)

We may rewrite the expression for $\psi$ as

$$\psi = (1 - H_f)\, F_0^* \,(1 - 1/C_b^*) \qquad \text{... (7.2.22)}$$

The parameter $\epsilon$ defined in Eq. (7.2.17) is often called the *ultrafiltration parameter*. Its significance can be easily understood by rewriting the expression for $\epsilon$ as

$$\epsilon = \frac{1}{16}\ \frac{(8\mu_f L)/\pi (d/2)^4}{1/(\pi dL K_m')} \qquad \text{... (7.2.23)}$$

$$= \frac{\left(\text{Hydrostatic resistance to axial flow in a tube of length } L\right)}{\left(\text{Hydrostatic resistance to transmembrane flow in a tubular membrane of same length}\right)}$$

Equations (7.2.13) to (7.2.18) can be solved numerically by any of the popular techniques such as the fourth order Runga-Kutta method.

It will be better to take the aid of a fast computer in this regard. The computer program is given in Section (7.7) of this Chapter (see Program 7.2).

However, for small values of $\epsilon$ ($0.0001 \le \epsilon \le 0.01$), these equations can be solved analytically[17] in the form of asymptotic expansions:

$$p^* = p^{(1)} + \epsilon\, p^{(2)} + \epsilon^2 p^{(3)} + \ldots\ldots \qquad \text{(7.2.24)}$$

$$F^* = F^{(1)} + \epsilon\, F^{(2)} + \epsilon^2 F^{(3)} + \ldots\ldots \qquad \text{... (7.2.25)}$$

$$\psi = \psi^{(1)} + \epsilon\, \psi^{(2)} + \epsilon^2 \psi^{(3)} + \ldots\ldots \qquad \text{... (7.2.26)}$$

where
$$p^{(1)} = (1 - \sigma z^*) \quad \text{... (7.2.27)}$$

$$p^{(2)} = -(8\sigma/3)\,(z^{*3} - z^*) + 8G\,(z^{*2} - z^*) \quad \text{... (7.2.28)}$$

$$p^{(3)} = 256\,\{(G/24)\,(z^{*4} - 2z^{*3} + z^*) - (\sigma/360)\,(3z^{*5} - 10z^{*3} + 7z^*) - (M/\sigma)\,[(G/6)\,(z^{*3} - z^*) - (\sigma/24)\,(z^{*4} - z^*)]\} \quad \text{... (7.2.29)}$$

$$F^{(1)} = \sigma/8 \quad \text{... (7.2.30)}$$

$$F^{(2)} = (\sigma/3)\,(3z^{*2} - 1) - G\,(2z^* - 1) \quad \text{... (7.2.31)}$$

$$F^{(3)} = -32\,\{(G/24)\,(4z^{*3} - 6z^{*2} + 1) - (\sigma/360)\,(15z^{*4} - 30z^{*2} + 7) - (M/\sigma)\,[(G/6)\,(3z^{*2} - 1) - (\sigma/24)\,(4z^{*3} - 1)]\} \quad \text{... (7.2.32)}$$

$$\psi^{(1)} = 0.0 \quad \text{... (7.2.33)}$$

$$\psi^{(2)} = (2G - \sigma) \quad \text{... (7.2.34)}$$

$$\psi^{(3)} = (8/3)\,[(\sigma/2) + (2M - G) - (6GM/\sigma)] \quad \text{... (7.2.35)}$$

$$G = 1.0 - (a_1^* + a_2^* + a_3^*) \quad \text{... (7.2.36)}$$

$$M = (a_1^*) + 2a_2^* + 3a_3^*)/(1 - H_f) \quad \text{... (7.2.37)}$$

It is important to note that the above solution is not applicable for very small values of $\sigma$ (that is, at $\sigma \to 0$). The parameter $\sigma$ has been defined in Eq. (7.2.19).

**Example 7.2:** A hollow fiber artificial kidney employs ultrafiltration for purifying human blood. The device consists of a fiber bundle containing ten thousand fibers each 220 microns in diameter. Blood from the vein of the renal patient is admitted to the device at the rate of 50 liters per hour. The inlet blood has the following composition:

Plasma proteins = 70 kg/m$^3$

Haematocrit = 0.40

A transmembrane pressure drop of 500 mm Hg is maintained at the fiber inlet.

(*a*) Compute the fiber length required if the protein concentration in the exit blood should be 105 kg/m$^3$.

(*b*) What will be the diafiltration rate?

(*c*) What pressure drop is to be maintained between the inlet and outlet of the fiber bundle?

Take the viscosity of the blood to be 0.01 poise and the membrane permeability coefficient = 3.3275 $\times$ 10$^{-11}$ m$^3$/(m$^2 \cdot$s$\cdot$N/m$^2$). Use Eq. (7.2.5) for computing the colloid osmotic pressure.

**Solution:** (*a*) Let us assume that the ultrafiltration parameter ($\in$) is very small in magnitude (this assumption we shall be verifying subsequently). For such low values of $\in$, the analytical solution given by Papenfuss et. al. could be utilised. Thus, let us first evaluate the dimensionless variables.

$$a_1^* = (0.21)\,(70.0)/(500.0) = 0.0294$$

$$a_2^* = (0.0016)\,(70.0)^2/(500.0) = 0.01568$$

$$a_3^* = (9.0 \times 10^{-6})\ (70.0)^3/(500.0) = 0.006174$$

$$C_b^* = (105.0)/(70.0) = 1.5$$

Since $(p_1 - p_d) = 500$ mm Hg $= 66.66\ \text{kN/m}^2$,

$$F_{ref} = \frac{\pi\left(110\times10^{-6}\right)^4\ (66.66\times1000)}{(0.001)\ L} = (3.0673\ \times\ 10^{-8})/L\ \text{m}^3/\text{s}$$

Since blood is admitted at 50 liters/hr and there are ten thousand fibers in the bundle, the inlet flow rate of blood per fiber is

$$F_0 = (50/10000)\ \text{liters/hr} = 1.39\ \times\ 10^{-9}\ \text{m}^3/\text{s}$$

$$F_0^* = (F_0/F_{ref}) = 0.0453\,L$$

Now, from Eq. (7.2.21), $$1.5 = \frac{(0.0453\,L)(1-0.4)}{F_b{}^* - (0.4)(0.0453\,L)}$$

or

$$F_b^* = 0.03624\,L$$

From Eq. (7.2.17), $$\epsilon = \frac{\left(3.3275\times10^{-11}\right)(0.001)\,L^2}{\left(110\times10^{-6}\right)^3} = 0.025\,L^2$$

The parameters $G$ and $M$ can be evaluated from Eqs (7.2.36) and (7.2.37).

$$G = 1.0 - (0.0294 + 0.01568 + 0.006174) = 0.9487$$

$$M = [0.0294 + (2.0)\ (0.01568) + (3.0)\ (0.006174)]/(1 - 0.4) = 0.132$$

The dimensionless ultrafiltration rate per fiber ($\psi$) is obtained from Eq. (7.2.22).

$$\psi = (1 - 0.4)\ (0.0453\,L)\ (1 - 1/1.5) = 0.00906\,L$$

Substituting the above-computed values in Eq. (7.2.26), we get

$$0.00906\,L = (1.8974 - \sigma)\ \epsilon + (1.33\,\sigma - 1.826 - 2.0/\sigma)\ \epsilon^2 \qquad \text{... (i)}$$

Similarly, Eq. (7.2.25) gets reduced to (after substituting $F^* = F_b^*$ and $z^* = 1$),

$$0.03624\,L = (\sigma/8) + (0.667\,\sigma - 0.9487)\ \epsilon + (0.7369 - 0.711\,\sigma + 1.33575/\sigma)\ \epsilon^2 \qquad \text{... (ii)}$$

where $\epsilon = 0.025\,L^2$. The above two Eqs (i) and (ii) can be now solved simultaneously for the two unknowns, $\sigma$ and $L$. However, since both equations contain higher powers of $L$ and $\sigma$, a direct solution is difficult and we have to resort to a trial and error procedure. Thus, let

$$L = 20\ \text{cm} = 0.20\ \text{m}$$

Then,
$$\epsilon = (0.025)\ (0.2)^2 = 0.001$$

Equation (i) reduces to

$$0.001812 = (1.895574\ \times\ 10^{-3}) - (0.99867\ \times\ 10^{-3}\sigma) - (2.0\ \times\ 10^{-6}/\sigma)$$

or

$$\sigma = 0.06$$

Substituting this value of $\sigma$ in Eq. (ii), we get

$$\text{RHS} = 0.006614$$

$$\text{LHS} = (0.03624)\ (0.2) = 0.007248$$

Since the difference between LHS and RHS of Eq. (ii) is not too substantial, further trial is not essential. Thus, the length of fiber required is

$$L = 0.20 \text{ m}$$

(*b*) The diafiltration rate per fiber is

$$P = \psi F_{ref}$$

where
$$\psi = (0.00906)\ (0.2) = 0.001812$$

$$F_{ref} = (3.0673 \times 10^{-8})/(0.20) = 1.53365 \times 10^{-7} \text{ m}^3/\text{s}$$

Therefore,
$$P = (0.001812)\ (1.53365 \times 10^{-7}) = 2.78 \times 10^{-10} \text{ m}^3/\text{s} = 0.001 \text{ liter/hr}$$

The total diafiltration rate is = (0.001) (10000) = 10 liters/hr.

(*c*) From Eq. (7.2.19),

$$(p_2 - p_d) = (p_1 - p_d)\ (1 - \sigma) = (500.0)\ (1 - 0.06) = 470 \text{ mm Hg}$$

$$(p_1 - p_2) = (p_1 - p_d) - (p_2 - p_d) = (500.0) - (470.0) = 30 \text{ mm Hg}$$

This is the pressure drop that is to be maintained between the two ends of the fiber bundle. To note that since the value of $\in$ is quite low and $\sigma$ is not too small, we are justified in using the analytical solution proposed by Papenfuss et. al. The total membrane area required is

$$A_m = \pi\ (220 \times 10^{-6})\ (0.20)\ (10000) = \mathbf{1.3828\ m^2}$$

## 7.3 DIALYSIS

Like reverse osmosis and ultrafiltration, dialysis is also a membrane-based separation process involving selective transport through a semipermeable membrane. Dialysis is used for the separation of solutes from solutions and this takes place due to the diffusion of solutes from the concentrated solution to the dilute solution through the separating membrane. Separation between solutes can be achieved when the rates of transfer of the solutes through the membrane are substantially different. The rates of solute transfer will evidently depend on solute diffusivites in the membrane and the concentration difference maintained across it. The driving force in dialysis is thus the transmembrane concentration gradient (it is not a pressure-driven process like RO and UF). Since diffusivities of most of the solutes in polymeric membranes are quite low, large concentration differences are to be maintained across the membrane for obtaining acceptably large solute fluxes. This, in turn, implies that practical dialysis is restricted to separation from concentrated solutions. Also, since diffusivites of solutes are weakly size dependent, application of dialysis is limited to the separation of solutes that differ widely in molecular sizes.

Though our principal interest is in the transfer of solute from the concentrated solution to the dilute solution through the separating membrane, there could also be counter diffusion of solvent in the opposite direction (nothing but the process of osmosis). A good dialysis membrane should, therefore, have a low solvent permeability so that this counter diffusion of solvent could be minimised.

As a separation process, dialysis has the distinct advantage that separation occurs under gentle operating conditions (low pressure, ambient temperatures and low shear rates). It is, therefore, ideally suited for handling biological and shear sensitive substances. However, practical applications of conventional concentration driven dialysis are somewhat limited mainly due to the non-availability of suitable membranes. Presently, commercial dialysis membranes are made of

(*i*) regenerated cellulose (often obtained by the hydrolysis of cellulose acetate) which has low water permeability, high compatibility with a variety of feed solutions and permits a wider operating temperature range,

(*ii*) other glassy polymers such as poly acrylonitrile (PAN), polysulfone, polymethyl methacrylate (PMMA) and polyethylene-polyvinyl alcohol copolymer, which have better stability over wide ranges of pH and better shrinkage and swelling properties than cellulosic membranes. However, they have narrower operating temperature range.

Membrane configurations used for dialysis are same as those for RO and UF such as plate and frame, spiral-wound, tubular and hollow fiber modules (see Figs 7.4 to 7.6).

### 7.3.1 Mathematical Analysis

Mathematical simulation of dialysis has been attempted by many authors. The simplest simulation model is one that assumes an ideal membrane—that is, a membrane that permits the passage of solute only. In other words, counter diffusion of solvent is assumed negligible. It is assumed that there are no interactions among solute, solvent and the membrane. In such a case the solute flux across the membrane ($N_A$) can be represented by integrating Fick's first law of diffusion:

$$N_A = K_0 (\Delta C_{Am})_{\ln} \quad \text{... (7.3.1)}$$

$$= K_0 K_{mb} (\Delta C_A)_{\ln} \quad \text{... (7.3.2)}$$

where $N_A$ = solute flux across the membrane, kg/(m$^2$·s)

$K_0$ = overall dialysis coefficient (similar to mass transfer coefficient), m/s

$K_{mb}$ = distribution coefficient of the solute between the membrane phase and the solution phase (= $C_{Am}/C_A$), dimensionless

$C_{Am}$ = concentration of solute (namely, $A$) in the membrane, kg/m$^3$

$C_A$ = concentration of solute in solution, kg/m$^3$

The feed solution is introduced to one side of the membrane and the solution on the other side of the membrane (into which the solute diffuses from the feed solution through the membrane) is called the dialysate. If the feed solution and the dialysate flow countercurrently as shown in Fig. 7.9, then

$$(\Delta C_A)_{\ln} = \frac{(C_{f0} - C_{de}) - (C_{fe} - C_{d0})}{\ln\left(\dfrac{C_{f0} - C_{de}}{C_{fe} - C_{d0}}\right)} \quad \text{... (7.3.3)}$$

If we further assume that no volume changes take place on either side of the membrane, then the solute balance can be written as

$$F\,(C_{f0} - C_{fe}) = D\,(C_{de} - C_{d0}) = N_A A_m \qquad \text{... (7.3.4)}$$

where $A_m$ is the membrane area and $F$ and $D$ are the volume flow rates of feed and dialysate respectively. The performance of a dialyser is commonly expressed in terms of *dialysance* ($\psi$) which is defined as

$$\psi = (N_A A_m)/(C_{f0} - C_{d0}) \qquad \text{... (7.3.5)}$$

Substituting for $N_A$ from Eqs (7.3.2) and (7.3.3) and rearranging, we get

$$\psi = F\left\{\frac{exp\left[K_0 K_{mb} A_m\,(1 - F/D)/F\right] - 1}{exp\left[K_0 K_{mb} A_m\,(1 - F/D)/F\right] - (F/D)}\right\} \qquad \text{... (7.3.6)}$$

The fractional solute extraction ($SE$) is defined as

$$SE = [F\,C_{f0} - (F - P)\,C_{fe}]/(FC_{f0}) \qquad \text{... (7.3.7)}$$

where $P$ is the volume rate of flow of solvent through the membrane from the feed side to the dialysate side (that is, in the same direction as solute transport). For an ideal membrane (which permits the transport of solute only), $P = 0$. Therefore, the expression for $SE$ reduces to

$$SE = (C_{f0} - C_{fe})/C_{f0} \qquad \text{... (7.3.8)}$$

Incidentally, if the inlet concentration of solute in the dialysate is zero (that is, $C_{d0} = 0.0$), then

$$\psi = N_A A_m/C_{f0} = F\,(C_{f0} - C_{fe})/C_{f0} = F\,(SE)$$

or

$$SE = \psi/F \qquad \text{... (7.3.9)}$$

For applying Eq. (7.3.6) to design problems, we must know the values of $K_0$ and $K_{mb}$. The value of the distribution coefficient ($K_{mb}$) is generally taken equal to $\in$, where $\in$ is the volume fraction of pores in the membrane[7] (in other words, the fraction of membrane volume occupied by pores).

The process of mass transfer from the feed compartment to the dialysate compartment involves diffusion through liquid film on each side of the membrane as well as through the membrane itself (see Fig. 7.10). Though not precisely true, the overall dialysis coefficient ($K_0$) may be related to the liquid film coefficients and the membrane permeability coefficient in an analogous manner as the overall heat transfer coefficient is related to the film coefficients. Thus

$$1/K_0 = 1/K_{ms} + 1/k_{L1} + 1/k_{L2} \qquad \text{... (7.3.10)}$$

$$= 1/K_{ms} + 1/k_L \qquad \text{... (7.3.11)}$$

where $K_{ms}$ = membrane permeability coefficient for the solute, m/s

$k_{L1}$, $k_{L2}$ = liquid film mass transfer coefficients on the feed side and the dialysate side of the membrane respectively, m/s

Lane and Riggle[13] propose the following correlation for the estimation of $k_L$ (based on a combined liquid film thickness of 0.6 mm).

$$k_L = (1670.0)\,D_{AS} \qquad \text{... (7.3.12)}$$

where $D_{AS}$ is the diffusivity of solute in solution. The same authors have proposed an expression for the estimation of the membrane permeability coefficient ($K_{ms}$) for the solute.

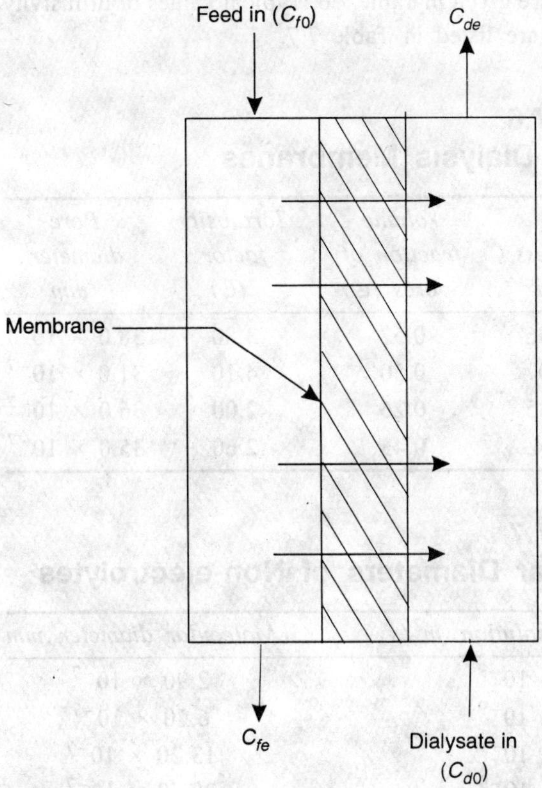


*Figure 7.9*

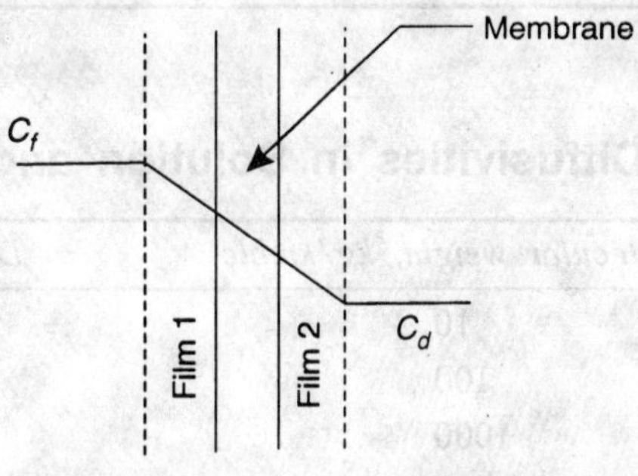


*Figure 7.10*

$$K_{ms} = D_{AS} \, f \in /(\zeta \delta) \qquad \text{... (7.3.13)}$$

where $f$ = drag factor, dimensionless

$\zeta$ = tortuosity factor (ratio of capillary length to membrane wet thickness), dimensionless

$\delta$ = membrane wet thickness, m

The drag factor ($f$) may be estimated from the correlation proposed by Bacon[14].

$$f = 1.0 - 2.104\,(d_p/d_{p0}) + 2.09\,(d_p/d_{p0})^3 - 0.95\,(d_p/d_{p0})^5 \qquad \text{... (7.3.14)}$$

where $d_p$ is the diameter of the diffusing molecule and $d_{p0}$ the average diameter of membrane pores. The diameter of the diffusing molecule can be computed from its molar volume[7]. Thus, for solids,

$$d_p = 1.465 \times 10^{-17}\,(M_A/\rho_S)^{1/3} \qquad \text{... (7.3.15)}$$

For liquids,

$$d_p = 10^{-17}\,(\bar{V})^{1/3} \qquad \text{... (7.3.16)}$$

where $M_A$ = molecular weight of diffusing solute, kg/kmole

$\rho_S$ = density of diffusing solute, kg/m$^3$

$\bar{V}$ = molar volume of the liquid at the boiling point, m$^3$/kmole

Specifications of a few typical dialysis membranes are given in Table 7.6. Typical values of diffusivity ($D_{AS}$) and molecular diameter ($d_p$) of nonelectrolytes are listed in Table 7.7.

**Table 7.6**
**Specifications of Typical Dialysis Membranes**

| *Membrane material* | *Dry thickness, mm* | *Wet thickness, mm* | *Volume fraction of pores ($\in$)* | *Tortuosity factor ($\zeta$)* | *Pore diameter, mm* |
|---|---|---|---|---|---|
| Du Pont cellophane | 0.0191 | 0.0396 | 0.52 | 3.80 | $38.0 \times 10^{-7}$ |
| Avisco cellophane | 0.0254 | 0.0507 | 0.50 | 4.10 | $31.0 \times 10^{-7}$ |
| Parchment paper | 0.0508 | 0.0685 | 0.26 | 2.00 | $36.0 \times 10^{-7}$ |
| Denitrated nitrocellulose (light) | 0.0534 | 0.094 | 0.43 | 2.60 | $35.0 \times 10^{-7}$ |

**Table 7.7**
**Diffusivities in Solution and Molecular Diameters of Non-electrolytes**

| *Molecular weight, kg/kmole* | *Diffusivity in solution, m²/s* | *Molecular diameter, mm* |
|---|---|---|
| 10 | $2.20 \times 10^{-9}$ | $2.90 \times 10^{-7}$ |
| 100 | $0.70 \times 10^{-9}$ | $6.20 \times 10^{-7}$ |
| 1000 | $0.25 \times 10^{-9}$ | $13.20 \times 10^{-7}$ |
| 10000 | $0.11 \times 10^{-9}$ | $28.50 \times 10^{-7}$ |
| 100000 | $0.05 \times 10^{-9}$ | $62.00 \times 10^{-7}$ |
| 1000000 | $0.025 \times 10^{-9}$ | $132.0 \times 10^{-7}$ |

The design of a dialyser based on the above equations can be thus summarised as below:

(*i*) Let us consider a sizing problem in which the values of $F$, $D$, $C_{f0}$, $C_{d0}$ and the fractional solute extraction desired ($SE$) are specified at the outset. We have to compute the membrane area required.

(*ii*) Select the membrane specifications such as pore size ($d_{p0}$), wet thickness ($\delta$), $\in$, tortuosity factor ($\zeta$).

(*iii*) Since $SE$ is known, compute $C_{fe}$ from Eq. (7.3.8) and $C_{de}$ from the solute balance Eq. (7.3.4).

(*iv*) Compute $K_{ms}$ from Eqs (7.3.13) to (7.3.16) and $k_L$ from Eq. (7.3.12). Now, compute $K_0$ from Eq. (7.3.11).

(*v*) Compute the dialysance ($\psi$) from Eqs (7.3.4) and (7.3.5) as

$$\psi = F\,(C_{f0} - C_{fe})/(C_{f0} - C_{d0}) \qquad \text{... (7.3.17)}$$

(*vi*) Substituting the above-computed value of $\psi$ in Eq. (7.3.6), solve it for $A_m$. This is the required membrane area.

Now, consider a rating problem. Here, we have to determine whether the available membrane module is suitable for the given application or not. Thus, the values of $F$, $D$, $C_{f0}$, $C_{d0}$ and $A_m$ are known at the outset and we have to compute the fractional solute extraction (*SE*). The procedure is as follows.

(*i*) Using the known membrane specifications ($d_{p0}$, $\in$, $\zeta$, $\delta$), compute $K_{ms}$, $k_L$ and $K_0$ as in the sizing problem.

(*ii*) Compute the dialysance ($\psi$) from Eq. (7.3.6).

(*iii*) Using the above-computed value of $\psi$, compute $C_{fe}$ from Eq. (7.3.17).

(*iv*) Compute *SE* from Eq. (7.3.8).

(*v*) If the above-computed value of *SE* is equal to or more than the fractional solute extraction desired, then the available membrane module is suitable for the purpose. Otherwise, a module with larger membrane area is to be tried.

### 7.3.2 Applications of Dialysis

Dialysis has found large scale commercial application in the treatment of waste streams and the recovery of acids from metallurgical liquors. For example, dialysis is recommended for the recovery of hydrochloric acid and nitric acid from stainless steel pickling liquor and for the separation of chromic acid from anodising, engraving and etching liquors. Other applications include concentration of fruit juices, purification of high-cost materials in the pharmaceutical industry and the recovery of caustic soda from colloidal hemicellulose in the viscose rayon industry.

At present, the most prominent application of dialysis is in the medical field. Many artificial kidney devices employ dialysis for the purification of human blood. Such devices are, therefore, called *haemodialysers.* The blood from the vein of a renal patient is sent to the haemodialyser where the unwanted solutes such as urea, uric acid, creatinine, phosphates and excess amounts of chlorides are removed from the blood by diffusion through the dialysis membrane. The purified blood is then returned to the body of the patient. For haemodialysis, hollow fiber dialysers arranged in the shell and tube configuration are most commonly used. A broad description of various artificial kidney devices together with an analysis of mass transfer mechanism is given by Cooney[15]

Commercially available artificial kidney systems are the WAK (Wearable Artificial Kidney), FAAK (Filtration Adsorption Artificial Kidney), PAK (Peritoneal Dialysis Artificial Kidney) and the HAK (Haemoperfusion Artificial Kidney). WAK is a haemodialyser that requires very little space (weighs around 3.5 kg) and is portable. However, the wearable module is necessary to be connected daily to a 20 litre dialysate bath for an average of 90 minutes to achieve adequate removal of urea and other impurities. Moreover, since the available dialysis membranes usually have a molecular cut-off of around 5000, they become inefficient in removing uremic substances whose molecular weights are larger than 5000 (the so-called *middle molecules*). The FAAK uses the principle of diafiltration or ultrafiltration and is a closer substitute to the natural kidney (see also the earlier section on Ultrafiltration). It is not dependent on the dialysate and ensures adequate removal of uremic toxins and the middle molecules and can also be made portable. Moreover, it is possible to filter the blood at a low transmembrane pressure difference and after removing the uremic solutes and excess ions from the diafiltrate, it can be reinfused. *Peritoneal dialysis* using PAK is prefered in many cases since in this case, access to uremic substances is through the peritoneal fluid, thereby making it less risky and more safe than other devices which are connected directly to the blood stream. The effectiveness of charcoal haemoperfusion for removing toxins and waste metabolites (as practised in the HAK) has been demonstrated by many authors. However, particulate release and platelet depletion prevent continuous use of such devices.

**Example 7.3:** Compute the membrane area required for a counterflow dialyser for achieving 85 percent solute extraction from the following data.

Volume flow rate of feed solution = 55 liters/hr

Volume flow rate of dialysate = 250 liters/hr

Dextrose (principal solute) content of feed = 5.0 kg/m$^3$

Type of membrane used = Du Pont cellophane

The inlet dialysate does not contain any dextrose and take the diffusivity of dextrose in solution to be $0.65 \times 10^{-9}$ m$^2$/s. Molecular diameter of dextrose = $7.0 \times 10^{-7}$ mm. What will be the concentration of dextrose in the outgoing dialysate?

**Solution:** This is a sizing problem. Since the membrane used is Du Pont cellophane, from Table 7.4,

$$d_{p0} = 38.0 \times 10^{-7} \text{ mm}$$

$$\delta = 0.0396 \times 10^{-3} \text{ m}$$

$$\epsilon = 0.52$$

$$\zeta = 3.80$$

Since the inlet dialysate does not contain any dextrose, $C_{d0}$ = 0.0 and therefore, from Eq. (7.3.9)

$$(\psi/F) = SE = 0.85$$

Since the distribution coefficient ($K_{mb}$) is usually taken equal to $\epsilon$

$$K_{mb} = \epsilon = 0.52$$

From Eq. (7.3.12), $k_L = (1670.0)\,(0.65 \times 10^{-9}) = 10.855 \times 10^{-7}$ m/s

The drag factor is obtained from Eq. (7.3.14).

$$f = 1.0 - 2.104\,(7/38) + 2.09\,(7/38)^3 - 0.95\,(7/38)^5 = 0.62528$$

Now, from Eq. (7.3.13), $K_{ms} = \dfrac{(0.65 \times 10^{-9})(0.62528)(0.52)}{(3.8)(0.0396 \times 10^{-3})} = 14.044 \times 10^{-7}$ m/s

Therefore, from Eq. (7.3.10), $K_0 = 6.1227 \times 10^{-7}$ m/s

Therefore, from Eq. (7.3.10),

$$K_0 = 6.1227 \times 10^{-7} \text{ m/s}$$

Also, it is given that $F = 55$ liters/hr $= 1.5277 \times 10^{-5}$ m$^3$/s

$$F/D = (55/250) = 0.22$$

Substituting the above-computed values of $K$, $K_0$, $(F/D)$ and $F$ in Eq. (7.3.6), we get

$$(\psi/F) = 0.85 = \frac{\exp(0.0162556\,A_m) - 1}{\exp(0.0162556\,A_m) - 0.22}$$

or $A_m = 103.97$ m$^2$

This is the required membrane area. Incidentally, if a hollow fiber module containing fibers that are 200 microns in diameter and 20 cm in length is used, then the number of fibers required will be

$$= \frac{(103.97)}{\pi\left(200\times10^{-6}\right)(0.20)} = 0.827\times10^{6}\ (= 0.827\ \text{million})$$

Now, from Eq. (7.3.8), $0.85 = (5.0 - C_{fe})/5.0$

or $C_{fe} = 0.75\ \text{kg/m}^3$

Therefore, from Eq. (7.3.4),

$$C_{de} = (5.0 - 0.75)\ (55.0)/(250.0) = 0.935\ \text{kg/m}^3$$

This is the concentration of dextrose in the outgoing dialysate.

## 7.4 ELECTRODIALYSIS

Electrodialysis is a unit operation in which an ionic solute is separated from its solution by using ion-selective membranes and in presence of a strong electric field. The electrodialyser consists of a membrane pair (or a number of membrane pairs) and two electrodes placed at the two ends of the membrane pair. Aqueous ionic solution is made to flow between the membranes and a DC field is applied between the two electrodes. Ions produced as a result of dissociation of the solute migrate in the respective directions (cation towards the cathode and anion towards anode). Removal of solute molecules from solution takes place as a result of this migration of cations and anions into the adjoining compartments (see Fig. 7.11).

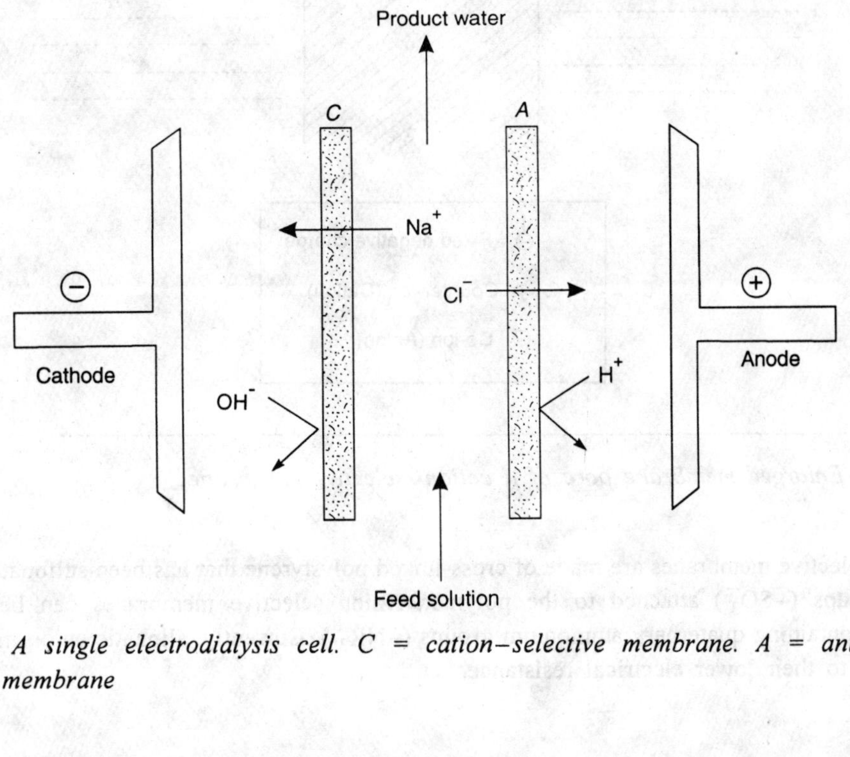


*Figure 7.11: A single electrodialysis cell. C = cation–selective membrane. A = anion–selective membrane*

Thus, membranes used in an electrodialyser are necessarily *ion-selective membranes*. These are membranes that are selectively permeable to one kind of ions with specific charge. For example, cation–selective and anion–selective membranes are permeable to cations and anions respectively. Permeability of an ion–selective membrane for a particular ion is due to the location of fixed electrical charges (opposite to that on permeable ion) inside membrane pores (see Fig. 7.12). These fixed charges exclusively permit flow of counter-ions through its pores and reject co-ions. For example, a cation-selective membrane is permeable to cations only and rejects anions due to the presence of negatively charged fixed functional groups (R—$SO_3^-$ or $RCOO^-$) inside the membrane pores.

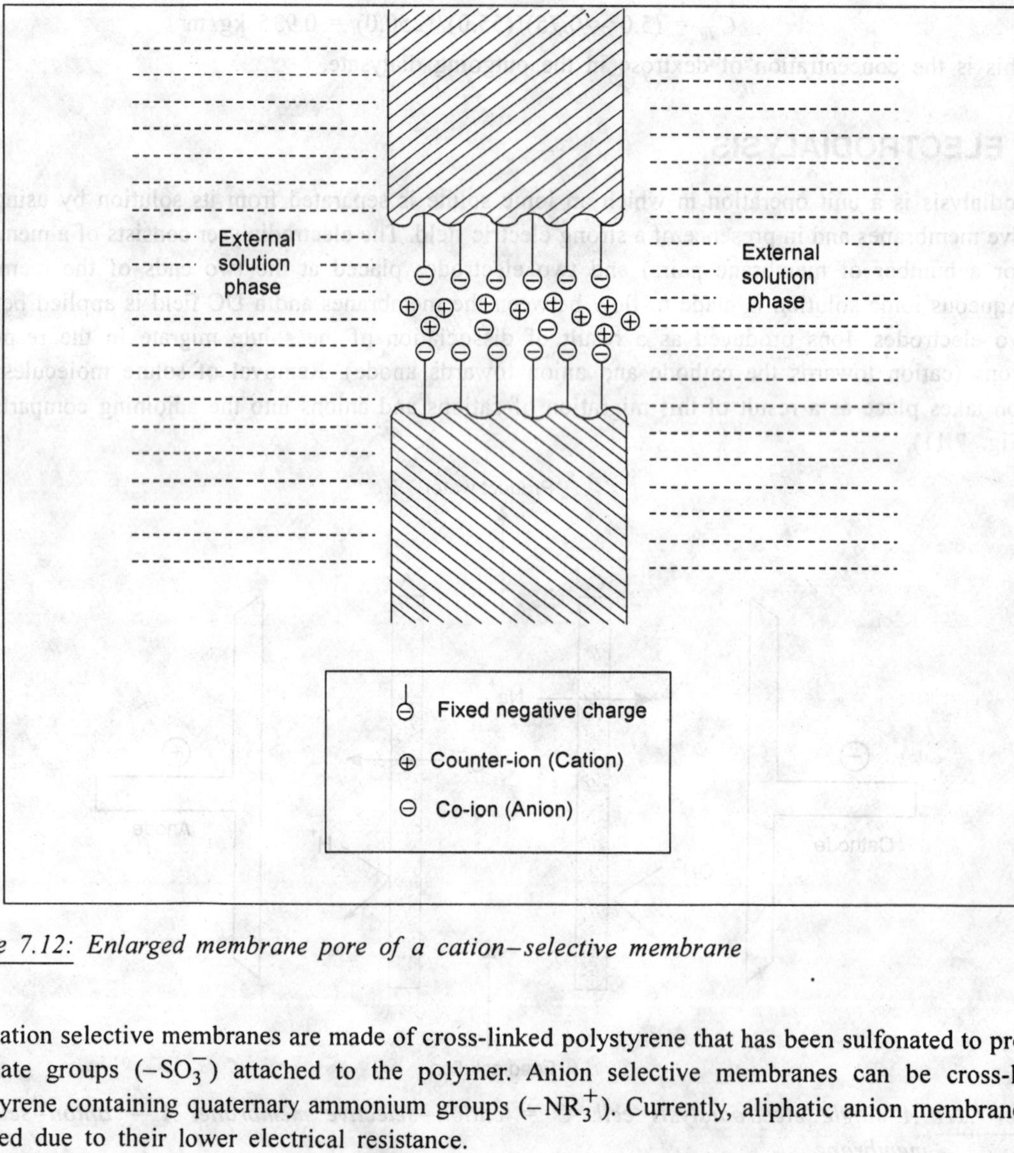


*Figure 7.12: Enlarged membrane pore of a cation–selective membrane*

Cation selective membranes are made of cross-linked polystyrene that has been sulfonated to produce sulfonate groups ($-SO_3^-$) attached to the polymer. Anion selective membranes can be cross-linked polystyrene containing quaternary ammonium groups ($-NR_3^+$). Currently, aliphatic anion membranes are prefered due to their lower electrical resistance.

An ideal ion-selective membrane is, therefore, one that permits only the counter-ions to pass through it and completely rejects the co-ions. In other words, the transport number of counter-ions in the membrane shall be unity, while that of the co-ions zero. Mathematically, for an ideal cation-selective membrane,

$$\bar{t}_+ = 1 \; ; \; \bar{t}_- = 0 \qquad \text{... (7.4.1)}$$

Similarly, for an ideal anion-selective membrane,

$$\bar{t}_- = 1 \; ; \; \bar{t}_+ = 0 \qquad \text{... (7.4.2)}$$

where $\bar{t}_+$ and $\bar{t}_-$ are the transport number of cation and that of anion respectively in the membrane. By transport number, we mean the fraction of the total current carried by the ion. It can be determined by direct transport experiments. It is important to note that the transport number of an ion in the membrane is different from its transport number in the external solution.

The permselectivity ($\psi$) of an ion-selective membrane can be, therefore defined as

$$\psi = (\bar{t}_+ - t_+)/t_-, \text{ for cation-selective membrane} \qquad \text{... (7.4.3)}$$

$$= (\bar{t}_- - t_-)/t_+, \text{ for anion-selective membrane} \qquad \text{... (7.4.4)}$$

where $\bar{t}$ and $t$ stand for transport number in the membrane and that in external solution respectively. Electroneutrality dictates that

$$t_+ + t_- = 1 \qquad \text{... (7.4.5)}$$

It is thus obvious that permselectivity is the most important property of an ion-selective membrane since it denotes the ability of the membrane to pass or discriminate an ion on the basis of its charge. For an ideal ion-selective membrane, $\psi = 1.0$.

It is also important to keep in mind that a membrane behaves as an ideal membrane only when concentration of the ionic soute in the external solution is below a particular value. This is called the *limiting concentration.* If the solute concentration is above this value, then the membrane deviates from ideal behaviour and permits flow of co-ions through it.

The chief requirements of ion-selective membranes that are to be used for electrodialysis can be thus summarised as given below:

(*i*) They must discriminate between counter-ions and co-ions and must be selectively permeable to the counter-ions over a wide range of concentration of the external solution,

(*ii*) They must have high electrical conductivity,

(*iii*) They should have low transference for water or solvent and solute molecules,

(*iv*) They must have good chemical and mechanical stability over a wide range of temperature.

What is shown schematically in Fig. 7.12 (a) is a single cell electrodialyser that uses only one pair of membranes (one being a cation-selective membrane and the other anion-selective). A multicell electrodialyser is sketched in Fig. 7.13. It consists of a number of narrow compartments or cells through which the feed solution is pumped. These compartments are separated by alternating anion-selective and cation-selective membranes. The terminal compartments are bounded by electrodes for passing direct current through the whole stack. The compartments may be stacked horizontally or vertically. The entire assembly resembles a plate and frame filter press with electrodes as pressure plates. The inter-spacing between adjacent membranes is usually held to 1 mm or less in order to reduce ohmic loss and to promote mixing. When the electrodes are connected to a DC source, ion migration begins as shown in Fig. 7.13. Let the feed be an aqueous solution of sodium chloride (say, sea water or brackish water). Under the influence of the electric field, sodium chloride dissociates into $Na^+$ ions and $Cl^-$ ions. It can be seen that every diluate compartment (such as compartments 2, 4, 6 and 8) has a cation–selective membrane on the left and an anion–selective membrane on the right. As a result, $Na^+$ ions migrate towards the left (towards the cathode) through the cation–selective membrane, while $Cl^-$ ions migrate towards the right (towards the anode) through the anion–selective membrane. On the other hand, every concentrate compartment (such as compartments 3, 5 and 7) has anion–selective membrane on the left and cation–selective membrane on the right and consequently, neither the cations ($Na^+$ ions) nor the anions ($Cl^-$ ions) are able to leave the compartment, The $Na^+$ ions tend to move towards left but are stopped by the anion–selective membrane there and similarly, the cation–exchange membrane on the right prevents passage of $Cl^-$ ions. As a result of this, the salt concentration in the diluate compartments goes on decreasing and that in the concentrate compartments goes on increasing. Thus, desalinated water is discharged continuously from the diluate compartments and concentrated brine from the concentrate compartments. Solutions in the electrode compartments (catholyte and anolyte) are contaminated with the products of electrode reactions (typical cathode reaction is $H_2O + e^- \rightarrow 1/2H_2 + OH^-$ and typical anode reaction is $H_2O \rightarrow 2H^+ + 2e^- + 1/2O_2$). A constant flow of catholyte and anolyte is always maintained to minimise accumulation of alkali and acid in these two end compartments.

As with reverse osmosis and ultrafiltration, *concentration polarisation* is a significant factor in electrodialysis as well. Development of boundary layer on either side of the membrane is due to the fact

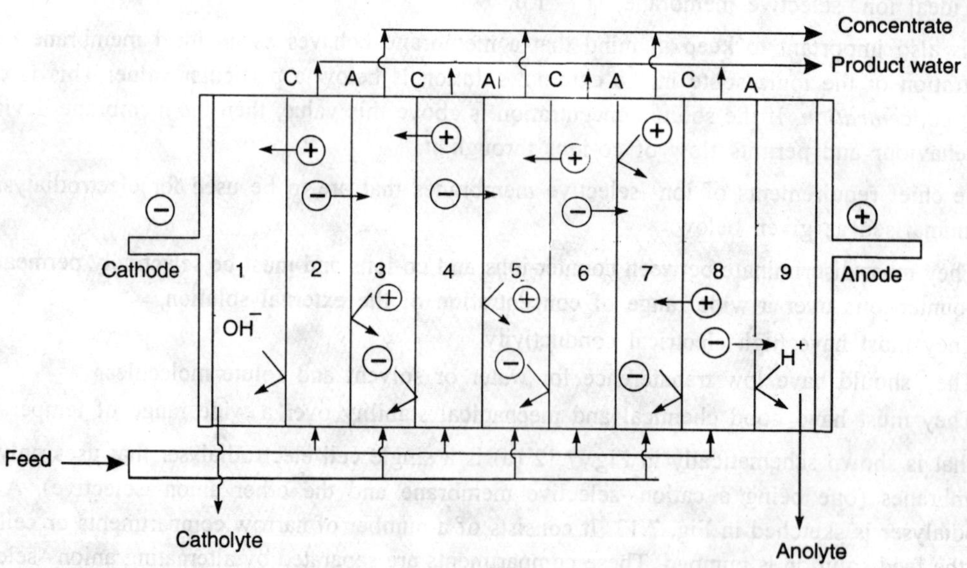


*Figure 7.13: Multicell electrodialyser. 2, 4, 6, 8: diluate compartments. 3, 5, 7: concentrate compartments. 1, 9: cathode and anode compartments respectively. C: cation–selective membrane. A: anion–selective membrane*

that the transport number of the counter ion in the membrane is much higher than that in the external solution. As a result, the surface concentration of counter ion on the diluate side (say, on the right hand side) of the membrane will be much lower than that in the adjoining boundary layer (see Fig. 7.14). Similarly, on the concentrate side of the membrane, since the rate of removal of the ion from membrane surface by diffusion and migration is much slower than the rate of transport through the membrane, the concentration of the ion at the surface will be much higher than that in the bulk, thereby leading to a boundary layer formation.

If we approximate the boundary layer concentration gradient (on the diluate side) as $(C_{dm} - C_d)/\delta_C$ where $\delta_C$ is the thickness of the boundary layer, then the rate of removal of salt through a cation-selective membrane ($J_+$) is given by

$$J_+ = -\frac{D_L\,(C_{dm} - C_d)}{\delta_C\,(\bar{t}_+ - t_+)} \quad \text{... (7.4.6)}$$

where $D_L$ = diffusivity of ionic solute in solution

$C_d$, $C_{dm}$ = molar concentration of diluate in the bulk and that at the membrane surface respectively

Also, if $i$ is the current density at the membrane (in amperes/cm$^2$ of membrane area), then based on Faraday's law,

$$J_+ = i/(N^0 \bar{F}) \quad \text{... (7.4.7)}$$

where $N^0$ = number of electrochemical gm equivalents per gm mole

$\bar{F}$ = Faraday's constant = 96500 coulombs/gm·eq. (Note: 1 ampere = 1 coulomb/s)

Accordingly,
$$\left(\frac{i}{N^0\bar{F}}\right) = -\frac{D_L\,(C_{dm} - C_d)}{\delta_C\,(\bar{t}_+ - t_+)} \quad \text{... (7.4.8)}$$

When $C_{dm}$ approaches zero, the current density approaches a limiting value of

$$i_m = \frac{\bar{F}\,D_L\,N^0\,C_d}{\delta_C\,(\bar{t}_+ - t_+)} \quad \text{... (7.4.9)}$$

To note that if $C_d$ is expressed in gm mole/cm$^3$, then $(N^0 C_d)$ represents the concentration of diluate stream in gm.equivalents/cm$^3$ Or,

$$N^0 C_d = \bar{N}/1000 \quad \text{... (7.4.10)}$$

where $\bar{N}$ is the normality of diluate stream (in gm.eq./liter). Therefore, Eq. (7.4.9) can be re-expressed as

$$(i_m/\bar{N}) = \frac{\bar{F}\,D_L}{\delta_C\,(\bar{t}_+ - t_+)} \quad \text{... (7.4.11)}$$

The above ratio $(i_m/\bar{N})$ is called the *polarisation parameter.* It is the controlling parameter that decides the level of concentration polarisation in an *ED* cell. In actual practice, the operating current density must be maintained lower than the limiting current density. Since the limiting current density is a function of diluate concentration [as evident from Eq. (7.4.11)], it is apparent that with decrease in diluate concentration, the operating current density has to be reduced in proportion in order to prevent it from approaching its limiting value.

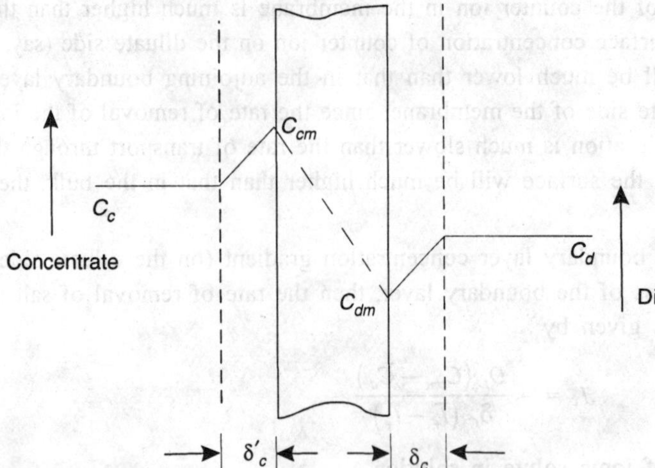


*Figure 7.14: Polarisation of a cation–selective membrane*

If the fluid flow rate through the electrodialyser is maintained high, then the boundary layer thickness will be diminished and this will reduce the effect of concentration polarisation. As a result, the energy requirement of the dialyser will also be reduced. However, increase in fluid flow rate decreases the residence time of the fluid in the dialyser and consequently, the salt removal per pass will also decrease. It is, therefore, recommended that turbulent promoters be placed between membranes so as to generate a high level of turbulence even at low fluid flow rates. Plastic wire meshes or wire screens are commonly employed as turbulent promoters. Eddies are generated due to the back flow created by these wire meshes (also called spacers). If the inter–distance between successive spacers is such that eddies do not die out due to skin effect of the two membranes, then turbulence will be maintained throughout the intermembrane space which will reduce the thickness of the diffusion layer and enhance mass transfer towards the membrane surface. In the absence of such packing, the entire inter–membrane space will be occupied by the two overlapping boundary layers.

When a tortuous path is cut out in the packing sheet to increase the length of fluid flow in the dialyser, the arrangement is known as *tortuous flow arrangement.* This involves installing spacers for the diluate and concentrate compartments in a plane parallel to the membranes. Four holes *A*, *B*, *C* and *D* punched in each spacer match holes punched into the membranes and into the electrodes. When the alternating layers of membranes and spacers are placed one aside the other (or, on top of each other) and compressed tightly, these holes form liquid conduits. Feed enters the diluate compartments only through conduit *A* and leaves only through *C*. Similarly, feed to the concentrate compartments enters only through conduit *B* and leaves only through *D*. Product water and concentrated brine thus leave the respective compartments through separate conduits *C* and *D* respectively.

Salt removal per pass is higher in such a tortuous flow arrangement since the solution has to follow a longer path and consequently its residence time in the dialyser is higher. But, such a flow arrangement is associated with higher pressure drop and hence leads to higher pumping power requirement.

Prevention of scaling and fouling of the membrane surfaces represents a major part of the design task and cost of most electrodialysis plants. The increased salt concentration near the membrane surface in the concentrate compartments may induce scale precipitation if solubility limits are exceeded. Changes in the $p_H$ of the solution promote such scale formation, particularly precipitation of calcium carbonate and magnesium hydroxide. In the electrodialysis of sea water, calcium sulfate scale is common. Deposition of a hard adherent scale on the membrane surfaces causes increased electrical and flow resistance, consequent power loss and often mechanical damage to the membranes. Scrubbing of the membrane surfaces often rejuvenates the membranes, but since the necessary disassembly and reassembly of the stack are time-consuming and expensive, in situ rejuvenation by flushing the compartments with leach solutions (strong alkalies or acids) is usually found more convenient in restoring conductivity and selectivity. Pretreatment of feed solutions by filtration, softening and by adsorption (activated carbon treatment helps in reducing fouling considerably, though the additional cost of these operations shall not be always acceptable.

Current reversal is another technique that is used to impede scaling or fouling of membranes. Whenever the feed contains polyvalent anion or colloidal suspension, due to the nature of negative charge on them, they move towards anion selective membrane and try to pass through. This causes a drop in current density due to blocking of the pores of the membrane. Altering the polarity of the two end electrodes would cause reversal in their movement, thereby minimising this effect. Also, movement of hydrogen ions to places where calcium and magnesium bicarbonates are formed would redissolve these salts, thereby clearing the membrane. To note that whenever the dialyser polarity is reversed, the function of each compartment is also reversed. It is then desirable to have hydraulic interchange of feed and output pipe lines and bifunctional electrodes which can act as either cathode or anode.

The flow arrangement in an electrodialyser may be either cocurrent or countercurrent. Cocurrent flow of diluate and concentrate streams is usually recommended in order to avoid large pressure drops across the membrane. The piping layout is simpler in such an arrangement and whenever the flow is in the upward direction, removal of gases or entrapped air becomes easier. However, a concentration gradient is built up across the membrane as the two streams flow through the two compartments and the gradient is maximum near the outlet. This would polarise the membranes and would reduce the overall current efficiency due to transport of water and the salt across the membrane.

Since electrodialysis is a rate process, rise in temperature enhances its rate due to decrease in the viscosity of the medium and increase in the diffusion rate. Membrane polarisation is reduced with rise in temperature of the solution. However, at elevated temperatures, the rate of transport of salt and water (or solvent) across the membrane will be high, which works against the electrodialysis process.

## 7.4.1 Performance Analysis

*Current efficiency* (η) of an electrodialyser is defined as the ratio of the actual amount of salt (expressed in gm equivalents) transfered to the number of electrical equivalents that have passed through the stack. Thus, based on the well-known Faraday's law, η can be expressed as

$$\eta = F_d \Delta \bar{N} / (I/\bar{F}) \quad \text{... (7.4.12)}$$

The above expression is for a single electrodialysis cell (shown in Fig. 7.11). For a multicell *ED* unit (shown in Fig. 7.13),

$$\eta = (F_t/n_d)\ \Delta\bar{N}/(I/\bar{F}) \qquad \text{... (7.4.13)}$$

where $F_d$ = volumetric flow rate through a single diluate compartment

$F_t$ = total volume flow rate of diluate stream

$n_d$ = number of diluate compartments

$I$ = current, amperes

$\bar{F}$ = Faraday's constant

$\Delta\bar{N}$ = change in normality of diluate stream

By normality, we mean the concentration of the solution expressed in gm. equivalents/liter. For example, consider a sample of brackish water containing 1.5 gm/liter of dissolved salt. Since the equivalent weight of sodium chloride is 58.5, the normality of the solution is

$$= (1.5)/(58.5) = 0.02565 \text{ gm}\cdot\text{eq./liter}$$

In other words, it is a 0.02565 *N* solution. It is usually assumed that the volume change due to salt removal is negligible. Accordingly,

$$F_t = P \qquad \text{... (7.4.14)}$$

where $P$ is the volume flow rate of product solution (product water).

Equations (7.4.12) and (7.4.13) demonstrate that the current efficiency does not depend upon the stack resistance, cell voltage, membrane area and the method of stack operation. The value of η is usually determined empirically on an experimental *ED* unit. Supplementary effects (that work against electrodialysis) such as water (or solvent) transport through the membrane by osmosis and salt transfer by back diffusion are also included in the current efficiency term when it is evaluated empirically using the above definition.

Let us consider a single *ED* cell (Fig. 7.11). In the differential form, the expression for the rate of salt removal is

$$dS = -F_d\, d\bar{N} = \eta\, I/\bar{F} \qquad \text{... (7.4.15)}$$

We know, from Ohm's law,

$$I = (V_p/R_p)\ dA \qquad \text{... (7.4.16)}$$

where $V_p$ = cell voltage, volts

$R_p$ = cell area resistance, ohm.cm$^2$

$dA$ = differential area of membrane

Therefore, Eq. (7.4.15) gets modified to

$$-F_d\, d\bar{N} = \eta\, V_p\, dA/(R_p \bar{F}) \qquad \text{... (7.4.17)}$$

According to Mintz[21], the product of cell area resistance and the average diluate stream normality may be taken as a constant. In other words,

$$R_p \bar{N} = B = \text{constant} \qquad \text{... (7.4.18)}$$

Substituting this in Eq. (7.4.17) and integrating, we get

$$\frac{\eta V_p}{F_d \bar{F}} \int_0^{A_p} dA = -B \int_{\bar{N}_f}^{\bar{N}_p} \frac{d\bar{N}}{\bar{N}}$$

or
$$A_p = \left( \frac{F_d \bar{F} B}{V_p \eta} \right) \ln (\bar{N}_f / \bar{N}_p) \quad \text{... (7.4.19)}$$

where $\bar{N}_f$, $\bar{N}_p$ = normality of feed solution and that of product water respectively

The ratio $q = (\bar{N}_f / \bar{N}_p)$ is generally called the *demineralisation ratio.* The above equation permits us to calculate the required effective cell area (effective area of a membrane pair) for a given demineralisation ratio and fluid flow rate. Conversely, the demineralisation ratio possible at a given flow rate and with a given membrane pair may be determined. In either case, it is necessary to know in advance the current efficiency ($\eta$), the average value of the constant $B$ and to select a value of cell voltage ($V_p$). Since the value of $V_p$ cannot be arbitrarily selected due to concentration polarisation limitations, Eq. (7.4.19) could be rearranged into a more usable form. For this, Mintz[21] showed that the ratio $(V_p / R_p \bar{N})$ or $(V_p / B)$ is related to the *limiting current density* $i_m$ as

$$(i_m / \bar{N}) = (V_p / B) = (V_p / R_p \bar{N}) \quad \text{... (7.4.20)}$$

Equation (7.4.19) could be, therefor, rearranged into the following form.

$$A_p = (F_d \bar{F} / \eta) \; (\bar{N} / i_m) \ln q \quad \text{... (7.4.21)}$$

The ratio $(i_m / \bar{N})$ is nothing but the polarisation parameter. To restrict concentration polarisation to a specified level, this parameter is required to be maintained constant at a specified value. Thus, from an empirically determined value of the polarisation parameter, the membrane area required for obtaining a specified demineralisation ratio can be computed from the above equation.

Equation (7.4.21) has been developed for a single *ED* cell across which a constant voltage is applied. However, it can be extended to a multicell parallel flow membrane stack (such as the one shown in Fig. 7.13), provided a uniform flow distribution and a uniform voltage per cell can be assumed.

The performance equation given above could be further refined if a more accurate relationship between cell area resistance and dilute stream concentration is available. This could be of the form

$$R_p = a \, (\bar{N})^b \quad \text{... (7.4.22)}$$

or
$$(R_p \bar{N}) = a_0 + a_1 \bar{N} + a_2 (\bar{N})^2 + a_3 (\bar{N})^3 \quad \text{... (7.4.23)}$$

where $a$, $b$ and $a_0$ to $a_3$ are all empirical constants. Substitution of any of the above correlations into Eq. (7.4.17) presents no difficulty in integration and the resulting performance equation can be easily converted to the form of Eq. (7.4.19) or (7.4.21). In our analysis that follows, we have assumed that the relationship given in Eq. (7.4.18) is applicable.

The power or energy requirement of an *ED* cell is given by

$$P' = V_p I \quad \text{... (7.4.24)}$$

Substituting for $V_p$ from Eq. (7.4.19) and $I$ from Eq. (7.4.12), we get

$$P' = (F_d \bar{F} / \eta)^2 \; (B / A_p) \; (\bar{N}_f - \bar{N}_p) \ln q \quad \text{... (7.4.25)}$$

The above equation predicts the power requirement of a single *ED* cell. The total power requirement of a multiple cell stack of parallel construction (like the one shown in Fig. 7.13) is simply equal to ($n_d P'$) where $n_d$ is the total number of cells (which is nothing but the total number of diluate compartments or the total number of membrane pairs) and $P$ is the power consumption per cell [computed from Eq. (7.4.25)]. However, this assumes a uniform flow distribution and a uniform voltage per cell. To note that in such a case, the power requirement per unit product flow rate will be the same for a single cell as well as for the multiple-cell stack.

## 7.4.2 Multistage Electrodialysis Systems

Since operating flow rates are necessary to be maintained high in electrodialyser (to impede concentration polarisation), it is not normally possible to achieve a high degree of demineralisation in a single pass through a single membrane stack. In such cases, a multistage system (either externally staged or internally staged) or a continuous process with partial recycle (often called the *feed and bleed* process) is employed so that the desired degree of mineralisation is attained.

In *external staging*, a number of membrane stacks are connected in series in such a way that the product stream from one stack (or stage) becomes the feed stream to the next stack (or stage). If the number of cells (membrane pairs) in each stage is $n_d$ and if the same voltage is applied to each stack and also the product ($R_p \bar{N}$) is essentially constant, then the demineralisation ratio ($q$) will be the same in each stage.

External staging involving three stages is illustrated schematically in Fig. 7.15. The flow is cocurrent in all the stages. To minimise concentration polarisation, feed solution is admitted to the concentrate compartments of the third stage (that is, the last stage). Concentrate streams from the third stage are admitted to the second stage and those from second stage to the first stage. In this way, the concentration gradient across the membrane is maintained low. Since the diluate concentration decreases progressively from first stage to the last, the operating current density also decreases from one stage to the next.

The total power requirement of a multistage system containing $n$ stages will be equal to the sum of the power requirements of individual stages. Let us assume that a constant voltage $V_p$ has been used to give a constant $q$ for all stages. Now the final product is discharged from the nth stage and let us represent its normality as $\bar{N}_{pf}$. Since the demineralisation ratio is $q$, the normality of feed to the $n$th stage is ($q\bar{N}_{pf}$). Therefore, the power requirement of the $n$th stage is [from Eq. (7.4.25)],

$$P'_n = KF_d^2\,(q\bar{N}_{pf} - \bar{N}_{pf}) = KF_d^2\bar{N}_{pf}\,(q-1) \qquad \text{... (7.4.26)}$$

where $$K = n_d\,(\bar{F}/\eta)^2\,(B/A_p)\ln q \qquad \text{... (7.4.27)}$$

Similarly, the power requirement of the $(n-1)$th stage is,

$$P'_{n-1} = KF_d^2\,(q^2\bar{N}_{pf} - q\bar{N}_{pf}) = KF_d^2\bar{N}_{pf}\,q\,(q-1) \qquad \text{... (7.4.28)}$$

and so on. Finally, the power requirement of the first stage is

$$P'_1 = KF_d^2\bar{N}_{pf}\,q^{n-1}\,(q-1) \qquad \text{... (7.4.29)}$$

Therefore, the total power requirement of the externally staged multistage system is

$$P'_{es} = P'_1 + P'_2 + \ldots\ldots + P'_{n-1} + P'_n$$
$$= K F_d^2 \bar{N}_{pf} (q - 1)(1 + q + q^2 + \ldots\ldots q^{n-1}) \quad \ldots (7.4.30)$$

where $K$ is defined in Eq. (7.4.27)

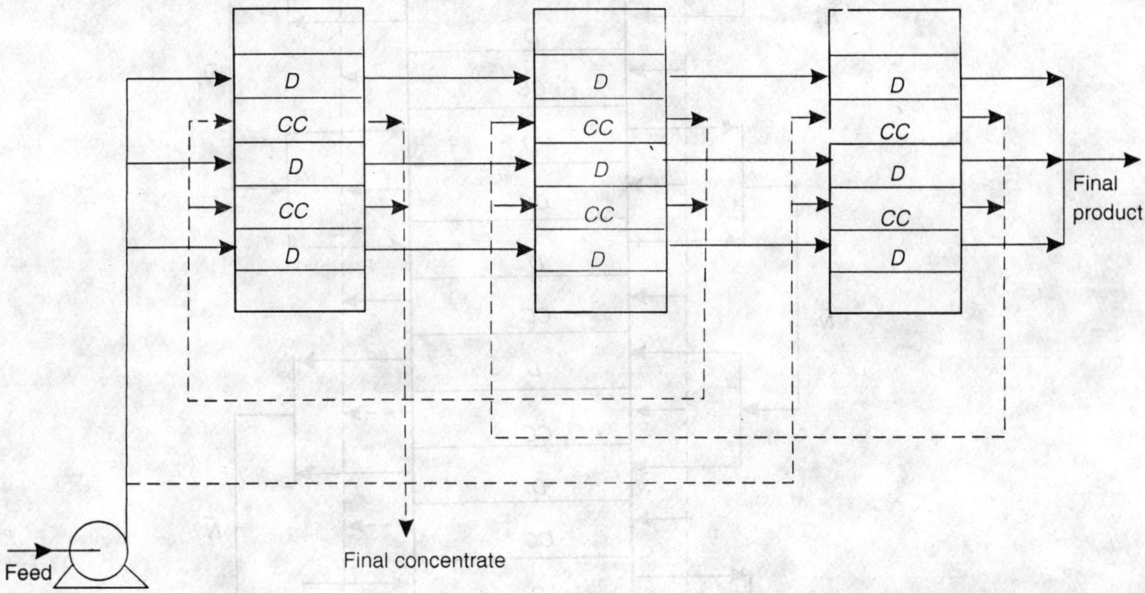


*Figure 7.15: Externally staged electrodialysis unit (showing three stages). D = diluate compartment; CC = concentrate compartment*

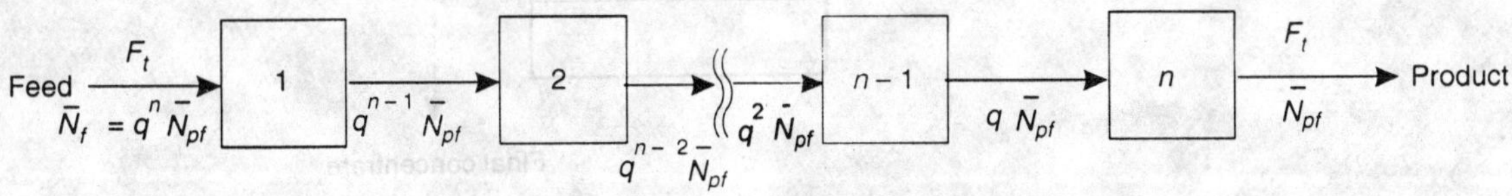


*Figure 7.16: Schematic of externally staged electrodialysis unit (with n stages)*

The chief advantages of external staging are that it provides high capacity and its energy requirements are reasonably low. Also, the process does not involve any recycle or recirculation of the product stream. However, this configuration requires careful balancing between process variables and is sensitive to even a small increase in membrane resistance.

*Internal staging* is represented in Fig. 7.17. In this case also, every stage is a multicell stack (containing 10 to 15 cells or membrane pairs) with a parallel flow arrangement. The stages are connected in series within a single electrodialyser so that the desired degree of demineralisation is achieved when the diluate stream comes out from the last stage. Since the concentrated stream flows in single pass, we get both cocurrent and countercurrent flows in the various stages. However, operating current density is the same

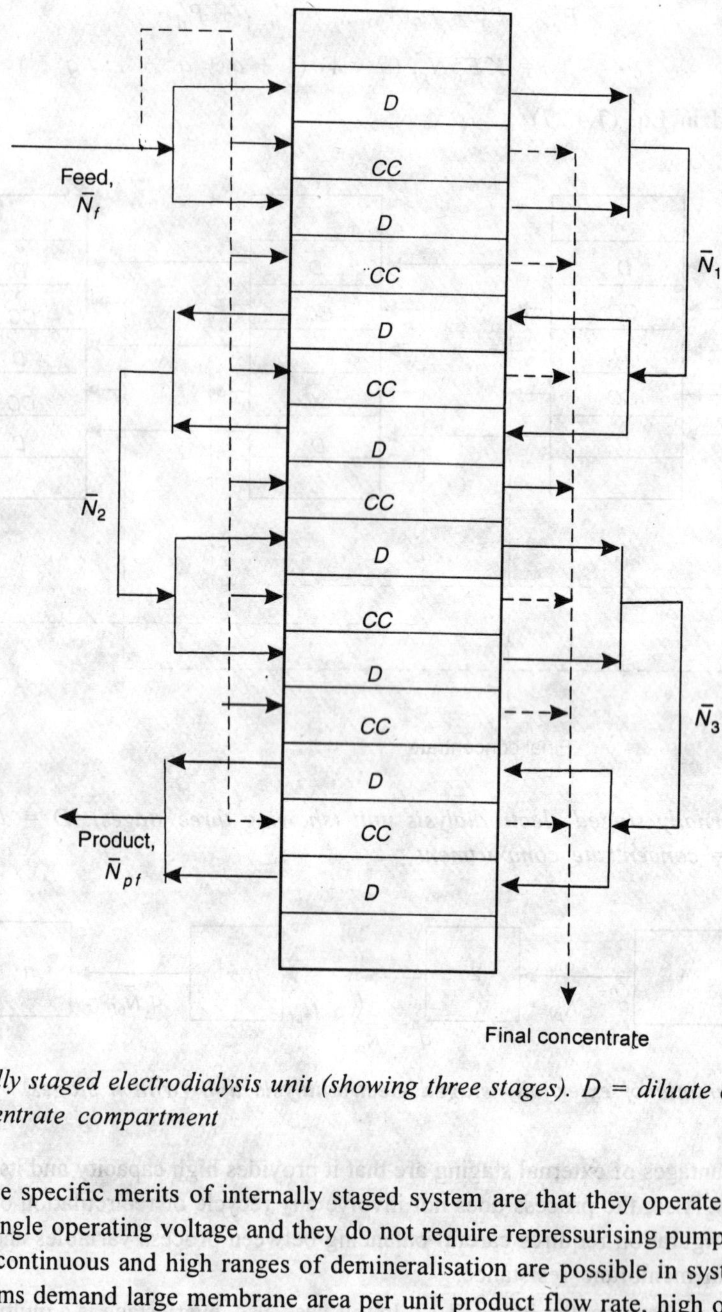


*Figure 7.17: Internally staged electrodialysis unit (showing three stages). D = diluate compartment; CC = concentrate compartment*

for all the stages. The specific merits of internally staged system are that they operate with a single set of electrodes and a single operating voltage and they do not require repressurising pumps between stages. The product flow is continuous and high ranges of demineralisation are possible in systems of this kind. However, these systems demand large membrane area per unit product flow rate, high operating pressure and their performance is quite sensitive to variations in flow rate and membrane resistance.

In an internally staged multistage system, the demineralisation ratio $q$ will not be constant and shall vary from stage to stage. Let $g$ be the number of stages required for reducing the diluate concentration from $\bar{N}_f$ to $\bar{N}_{pf}$. Then, the change in concentration per stage is

$$\Delta \bar{N} = (\bar{N}_f - \bar{N}_{pf})/g \qquad \text{... (7.4.31)}$$

We can, therefore, use Eq. (7.4.25) to compute the power requirement of each stage and by summing them up, we get the total power requirement of the system. Thus,

$$P'_{is} = K_1 \; [\ln (\bar{N}_f / \bar{N}_1) + \ln (\bar{N}_1 / \bar{N}_2) + ..... \ln (\bar{N}_{g-1} / \bar{N}_{pf})] \qquad \text{... (7.4.32)}$$

where

$$K_1 = n_d \, (F_d \bar{F} / \eta)^2 \, (B / A_p) \, \Delta \bar{N} \qquad \text{... (7.4.33)}$$

$\bar{N}_1 = \bar{N}_f - \Delta\bar{N}$; $\bar{N}_2 = \bar{N}_1 - \Delta\bar{N}$ and so on. In general,

$$\bar{N}_1 = \bar{N}_f - \Delta \bar{N} \qquad \text{... (7.4.34)}$$

$$\bar{N}_j = \bar{N}_{j-1} - \Delta \bar{N}, \text{ for } 2 \leq j \leq g - 1 \qquad \text{... (7.4.35)}$$

It has been assumed that the number of cells per stage is constant and is equal to $n_d$.

The number of stages ($g$) required to affect the desired degree of demineralisation can be estimated by noting the fact that the operating current density is the same for all the stages. In other words, the same current ($I$) flows in series through all the stages and therefore, from Eq. (7.4.13),

$$gI = (\bar{F} F_d / \eta) \, (\bar{N}_f - \bar{N}_{pf}) \qquad \text{... (7.4.36)}$$

If $q$ is the demineralisation ratio in the last stage ($g$th stage), then from Eq. (7.4.13), the current required to affect the change in concentration in the last stage is

$$I = (\bar{F} F_d / \eta) \, (q \bar{N}_{pf} - N_{pf}) = (\bar{F} F_d / \eta) \, \bar{N}_{pf} \, (q - 1) \qquad \text{... (7.4.37)}$$

Eliminating $I$ from the above two equations, we get

$$g = \frac{(\bar{N}_f - \bar{N}_{pf})}{\bar{N}_{pf} \, (q-1)} \qquad \text{... (7.4.38)}$$

$$= (q_0 - 1)/(q - 1) \qquad \text{... (7.4.39)}$$

where $q_0$ = overall demineralisation ratio

$$= (\bar{N}_f / \bar{N}_{pf}) \qquad \text{... (7.4.40)}$$

The demineralisation ratio in the last stage (namely, $q$) can be evaluated by applying Eq. (7.4.21) to the last stage. Thus

$$\ln q = (\eta / \bar{F}) \, (i_m / \bar{N}) \, (A_p / F_d) \qquad \text{... (7.4.41)}$$

Whenever there is a variation in the composition of the feed, it is rather difficult to operate the electrodialyser under fixed operating conditions without loss in its performance characteristics. A partial recycle of the product stream with a control on recycle ratio would enable to maintain a constant input composition. Such a process is known as the *feed and bleed* process and is sketched in Fig. 7.18. It may be noted that the final product collected is only that portion of product stream that is bled out of the recirculation loop.

Feed and bleed process provides constant product rate and product quality and it has the unique advantage that it can be adapted to feed water of any salinity. Low current density requirements and the ease of readily observing variations in performance are its additional advantages. However, the operating cost of the process is high and the recirculation system demands sensitive instrumentation.

From Fig. 7.18, a simple material balance yields

$$F_t \bar{N}_f + (m-1)\, F_t \bar{N}_{pf} = m\, F_t \bar{N}_r = m\, F_t\, q\, \bar{N}_{pf} \quad \text{... (7.4.42)}$$

where $q$ = demineralisation ratio

$$= N_r / \bar{N}_{pf} \quad \text{... (7.4.43)}$$

Solving for $m$, we get $$m = \frac{(\bar{N}_f / \bar{N}_{pf}) - 1}{(q-1)} \quad \text{... (7.4.44)}$$

It must be noted that the demineralisation ratio in this case is not equal to $(\bar{N}_f / \bar{N}_{pf})$, but is $(\bar{N}_r / \bar{N}_{pf})$. The total power requirement can be computed using Eq. (7.4.25). Thus

$$P'_{fb} = n_d\, (F_d \bar{F}/\eta)^2\, (B/A_p)\, \bar{N}_{pf}\, (q-1)\, \ln q \quad \text{... (7.4.45)}$$

where $$F_d = (m\, F_t / n_d) \quad \text{... (7.4.46)}$$

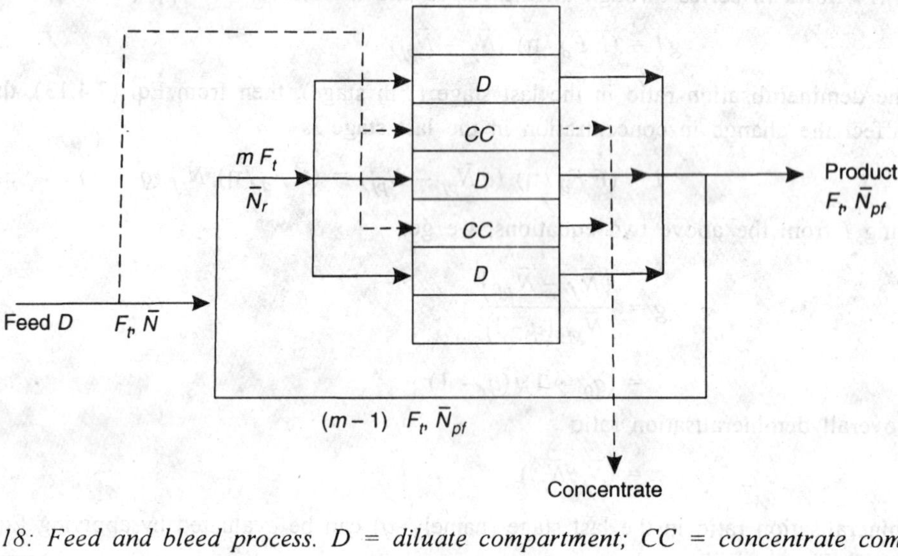


*Figure 7.18: Feed and bleed process. D = diluate compartment; CC = concentrate compartment*

Electrodialysers may also be operated batchwise but with recirculation (see Fig. 7.19). A fixed volume of feed solution is pumped through the dialyser from a holdup tank and the product solution is collected in another holdup tank connected in parallel with the first. The product solution is fully recycled back to the dialyser and the process is continued until the concentration of the product solution has reached the desired value. The product is then discharged and the process stopped. This completes one batch. Though such a batch recirculation process has apparent flexibility, it requires larger amount of instrumentation, piping, values, etc. The total power requirement of this process will be the same as that of a continuous multistage process (with external staging) and is predicted by Eq. (7.4.30). However, unlike the continuous process, the batch process requires continuous recycle of the product solution and the final product is not delivered continuously.

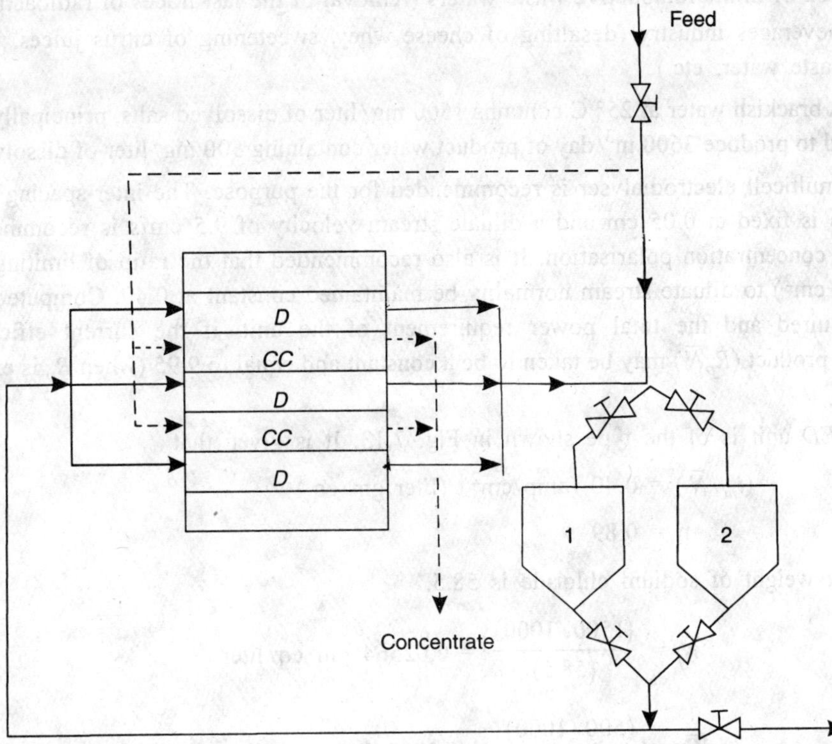


*Figure 7.19: Batch electrodialysis process with recirculation and no mixing (slug flow). D = diluate compartment; CC = concentrate compartment.*

### 7.4.3 Applications of Electrodialysis

Electrodialysis technology has developed rapidly during the last decade and the process has found many diversified applications. A few examples are,

(*i*) Desalination of brackish waters. As a first approximation, the cost of *ED* for desalination may be taken to be directly proportional to the salinity of the feed.

(*ii*) Demineralisation of waste water (removal of mineral salts like chlorides and sulfates of sodium, calcium, magnesium, etc. to produce reusable water).

(*iii*) Demineralisation of sewage effluents. Conventional sewage treatment plants remove organic and biological contaminants but do not reduce salinity. The necessary reduction in salinity can be accomplished by electrodialysis.

(*iv*) Denitrification of agricultural run-off waters.

(*v*) Treatment of cynide containing plating wastes and recovery of precious metals like gold and silver.

(*vi*) Recovery of carboxylic acids (acetic, citric, lactic) from waste streams.

(*vii*) Recovery of sulfuric acid from steel pickling liquors by removing ferrous sulphate.

(*viii*) Recovery of chemicals (such as lignin products) from pulping wastes.

(*ix*) Decontamination of dilute radioactive waste waters (removal of the last traces of radioactive ions).

(*x*) In food and beverages industry (desalting of cheese whey, sweetening of citrus juices, desalting of fishmeal waste water, etc.).

**Example 7.4:** A brackish water at 25° C contains 1500 mg/liter of dissolved salts, principally sodium chloride. It is desired to produce 3600 $m^3$/day of product water containing 500 mg/liter of dissolved salts.

A single stage multicell electrodialyser is recommended for the purpose. The inter-spacing between adjacent membranes is fixed at 0.05 cm and a diluate stream velocity of 9.5 cm/s is recommended in each cell to impede concentration polarisation. It is also recommended that the ratio of limiting current density (in amperes/$cm^2$) to diluate stream normality be maintained constant at 0.40. Compute the total membrane area required and the total power requirement of the unit, if the current efficiency is 89.0 percent and the product ($R_p\bar{N}$) may be taken to be a constant and equal to 9.95 (when $R_p$ is expressed in ohms.$cm^2$).

**Solution:** The *ED* unit is of the type shown in Fig. 7.13. It is given that

$$(i_m/\bar{N}) = 0.40 \text{ (amp/cm}^2\text{) (liter/gm}\cdot\text{eq.)}$$

$$\eta = 0.89$$

Since equivalent weight of sodium chloride is 58.5,

$$\bar{N}_f = \frac{(1500/1000)}{(58.5)} = 0.02564 \text{ gm}\cdot\text{eq/liter}$$

$$\bar{N}_p = \frac{(500/1000)}{(58.5)} = 0.008547 \text{ gm}\cdot\text{eq/liter}$$

$$q = (\bar{N}_f/\bar{N}_p) = 3.0$$

If $L$ is the length (or height) and $W$ is the width of each cell, then

$$F_d = W\,(0.05)\,(9.5) = 0.475\,W \text{ cm}^3\text{/s} = 0.000475\,W \text{ liters/s}$$

$$A_p = (LW)$$

Therefore, from Eq. (7.4.21),

$$LW = \frac{(0.000475\,W)\,(96500)\,ln\,(3.0)}{(0.40)\,(0.89)}$$

or

$$L = 141.45 \text{ cm} \simeq 141.5 \text{ cm}$$

Let us choose a width of 50 cm for each cell. Thus

$$W = 50.0 \text{ cm}$$

$$A_p = (141.5)\,(50.0) = 7075.0 \text{ cm}^2$$

$$F_d = (0.000475)\,(50.0) = 0.02375 \text{ liters/s}$$

Since the total flow rate of product water is 3600 $m^3$/day,

$$P = F_t = (3600)\,(1000)/(24) = 150000.0 \text{ liters/hour} = 41.667 \text{ liters/s}$$

$$\text{Therefore, number of cells} = \frac{(41.667)}{(0.02375)} = 1755$$

Total membrane area required = (7075.0) (1755)/$10^4$ = 1241.66 $m^2$

The power requirement per cell can be computed from Eq. (7.4.25). Thus

$$P' = \frac{(0.02375)^2\,(96500)^2\,(9.95)\,(0.02564 - 0.008547)\,\ln(3.0)}{(0.89)^2\,(7075.0)}$$

$$= 175.13\ W$$

Total power requirement = (175.13) (1755)/(1000) = 307.353 kW

$$\text{Power requirement per unit product rate} = \frac{(307.353)}{(3600/24)} = 2.049\ \text{kWh/m}^3$$

**Example 7.5:** Natural brine is to be desalinated from a concentration of 0.04 to 0.005 N by electrodialysis at the rate of 50 $m^3$/day. Membrane cells 40 cm wide, 100 cm long and 0.1 cm thick are available. Current efficiency = 0.90. The dilute stream velocity is to be maintained at 10 cm/s in each cell to minimise polarisation of membranes. Tests have indicated that in a single *ED* cell of above dimensions, 50 percent demineralisation can be achieved at a solution velocity of 10 cm/s.

(*a*) If an externally staged multistage *ED* unit is to be used for the purpose, compute the number of stages required and the total power requirement. (Assume the demineralisation ratio is the same for all the stages).

(*b*) If an internally staged electrodialyser is being employed, how many stages will be required and what will be the total power requirement?

(*c*) If the desired demineralisation is to be carried out by a feed and bleed process (while maintaining the diluate stream velocity at 10 cm/s itself in each cell), estimate the number of cells that will be required and the total power consumption.

Assume that the cell area resistance (in ohms·$cm^2$) is inversely proportional to the dilute stream normality so that $(R_p \bar{N})$ = constant = 6.0.

**Solution:** It is given that

$$\eta = 0.90$$

$$(R_p \bar{N}) = 6.0$$

It is also given that 50 percent demineralisation is possible at a solution velocity of 10 cm/s in a single *ED* cell. In other words,

$$(\bar{N}_f - \bar{N}_p)/\bar{N}_f = 0.5$$

or
$$1 - (\bar{N}_p/\bar{N}_f) = 0.5$$

or
$$q = \bar{N}_f/\bar{N}_p = 2.0$$

Now,
$$F_d = (40)\,(0.1)\,(10.0) = 40\ \text{cm}^3/\text{s} = 0.04\ \text{liters/s}$$

$$A_p = (100.0)\,(40.0) = 4000.0\ \text{cm}^2$$

Now, applying Eq. (7.4.21) to this *ED* cell, we get

$$4000.0 = (0.04)\ (96500)\ \ln\ (2.0)\ (\overline{N}/i_m)/(0.90)$$

or
$$(i_m/\overline{N}) = 0.7432\ (\text{amp/cm}^2)\ (\text{liter/gm·eq.})$$

(*a*) We know that in an externally staged multistage system (see Fig. 7.16),

$$\overline{N}_f = q^n \overline{N}_{pf}$$

Since $\overline{N}_f = 0.04N$, $\overline{N}_{pf} = 0.005N$ and $q = 2.0$,

$$2^n = 8.0$$

or
$$n = 3$$

The number of stages required is, thus, 3. Since the total product delivery rate is to be 50 m³/day or 0.5787 liters/s and the flow rate per cell is 0.04 liters/s, the number of cells per stage is

$$n_d = (0.5787)/(0.04) = 15$$

Thus three membrane stacks are to be connected in series, each stack containing 15 cells (or membrane pairs). The total power requirement can be now computed from Eqs (7.4.27) and (7.4.30). Thus, from Eq. (7.4.27),

$$K = 15\ (96500/0.9)^2\ (6.0/4000.0)\ \ln\ (2.0) = 1.793 \times 10^8$$

Now, from Eq. (7.4.30), $P'_{es} = (1.793 \times 10^8)\ (0.04)^2\ (0.005)\ (2.0 - 1)\ (1 + 2 + 4) = 10.04$ kW

The power requirement per unit product rate is

$$= (10.04)/(50/24) = 4.82\ \text{kWh/m}^3$$

Energy conversion losses for the production of suitable *DC* power, pumping energy and other plant requirements must be added to the above value when a total economic analysis is required.

(*b*) For an internally staged electrodialyser, the demineralisation ratio in the last stage (namely, $q$) can be computed from Eq. (7.4.41). However, since the values of $A_p$ and $F_d$ are the same as in (*a*), $q = 2.0$. This is the demineralisation ratio in the $g$th stage of the internally staged system. Now,

$$q_0 = (\overline{N}_f/\overline{N}_{pf}) = (0.04/0.005) = 8.0$$

Therefore, from Eq. (7.4.39),

$$g = (8.0 - 1.0)/(2.0 - 1.0) = 7$$

The number of stages required is, thus, seven. The number of cells per stage will be the same as in (*a*) since $F_d$ is maintained the same. That is, $n_d = 15$. The change in concentration per stage is

$$\Delta\overline{N} = (0.04 - 0.005)/7 = 0.005\ \text{gm·eq./liter}$$

Therefore, $\overline{N}_1 = (0.04 - 0.005) = 0.035$ gm·eq/liter

Similarly, $\overline{N}_2 = 0.03$, $\overline{N}_3 = 0.025$, $\overline{N}_4 = 0.02$,

$\overline{N}_5 = 0.015$ and $\overline{N}_6 = 0.01$ gm·eq./liter.

Also, from Eq. (7.4.33), $K_1 = (15)\ [(0.04)\ (96500)/(0.9)]^2\ (6.0/4000.0)\ (0.005) = 2069.4$

Substituting the above values in Eq. (7.4.32), we get the total power requirement as

$$P'_{is} = 4.303\ \text{kW}$$

The power requirement per unit product rate is

$$= (4.303)/(50/24) = 2.065\ \text{kWh/m}^3$$

Though the total power requirement of the internally staged system is less than 50 percent of that of externally staged unit, it must be kept in mind that the number of stages required for the former is larger. The total number of cells in the dialyser will be (7 × 15) = 105, in the internally staged unit.

(*c*) In the feed and bleed process (see Fig. 7.18), since the diluate stream velocity is maintained the same as 10 cm/s in each cell, $F_d$ = 0.04 liter/s. Accordingly, from Eq. (7.4.21), we get $q$ = 2.0. Now, from Eq. (7.4.44),

$$m = \frac{(0.04/0.005) - 1}{(2.0 - 1.0)} = 7.0$$

Since the volume flow rate of the recycle stream is $(m - 1)\ F_t$, the fraction recycled is

$$= (m - 1)\ F_t/(mF_t) = (m - 1)/m = (6/7) \text{ or } 85.7 \text{ percent}$$

Thus, 85.7 percent of the total product stream is recycled and only the remaining 14.3 percent is bled out and collected as the final product. The total fluid flow rate through the diluate compartments is

$$F_r = mF_t = (7.0)\ (50.0) = 350 \text{ m}^3/\text{day} = 4.0509 \text{ liters/s}$$

Therefore, the total number of cells required is

$$n_d = (4.0509)/(0.04) = 102$$

Now, from Eq. (7.4.45), the power requirement is

$$P'_{fb} = 9.615 \text{ kW}$$

The power requirement for the feed and bleed process is practically the same as that for the externally staged system. However, the feed and bleed process demands a larger number of membrane pairs in the dialyser, namely 102.

## 7.5 PERVAPORATION

Pervaporation is a separation process in which a liquid mixture is in direct contact with one side of a membrane and the permeate (also called the *pervaporate*) is removed from the other side in vapour state. Like reverse osmosis and ultrafiltration, pervaporation is also a pressure-driven process. However, it differs from processes like RO and UF in that it involves a phase change.

In Pervaporation, the liquid mixture to be separated is admitted to one side of the membrane and one of the components is allowed to permeate through the membrane. The pressure on the other side of the membrane (that is on the downstream side) is maintained below the saturation pressure of the permeated component at that temperature. This is achieved either by creating vacuum or by sweeping an inert carrier gas. As a result of this, the permeate vapourises from the downstream side of the membrane. The vapours are removed, condensed and collected as the pervaporate (see Fig. 7.20).

The process of pervaporation through a nonporous polymeric membrane involves the following steps:

(*i*) Dissolution or sorption of permeating molecules into the membrane phase (this takes place on the liquid-side or upstream side of the membrane).

(*ii*) Diffusion of these molecules through the membrane.

(*iii*) Evaporation of permeate from the vapour-side or downstream side of the membrane.

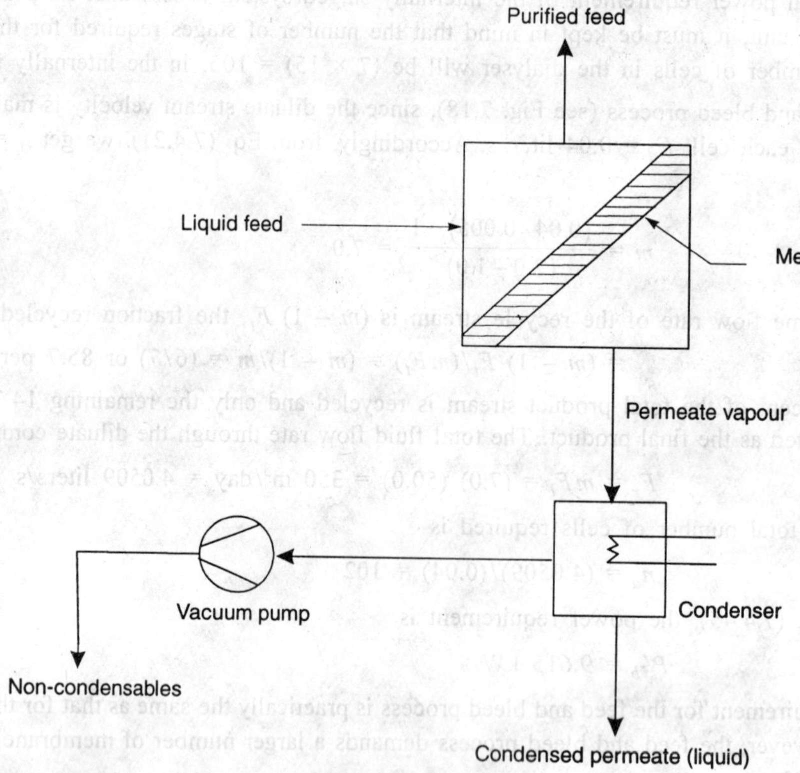


*Figure 7.20: Schematic of pervaporation process*

Thus, the membrane, in fact, acts as a thin solvent layer and sorption or dissolution of the permeating molecules causes swelling of the membrane. This swelling is a unique feature of pervaporation.

For practical use, pervaporation *membranes*[19] have to meet certain criteria such as

(*i*) they should have a thickness commensurate with the performance,

(*ii*) they should not pose a technical resistance to permeate withdrawal,

(*iii*) they should have dimensional stability under swollen conditions.

Eventhough pervaporation membranes are several orders of magnitude thicker than reverse osmosis or ultrafiltration membranes, a composite membrane structure is generally preferred for handling reasons. Also, a necessary condition in pervaporation is that the pressure loss to the permeate vapour be negligible and this condition makes it impossible to use integral–asymmetric membranes in pervaporation.

Commercial pervaporation membranes are either made of glassy polymers (such as cellulosic polymers) or rubbery (elastomeric) polymers (such as polyvinyl pyrrolidone, PVA, PDMS, etc.). Often, the active polymer is alloyed with or grafted to an inert polymer and then used. For example, cellulose acetate may be alloyed with polyphosphonate and the resultant membrane has been successfully used for separating various close-boiling liquid mixtures.

In spite of its high energy efficiency (in comparison to other energy-intensive processes like distillation, cryogenic separation, etc.), flexibility in equipment design and operation (as the systems are modular in nature), minimum space requirement in relation to the throughput and being more effective in processing dilute solutions, commercial applications of pervaporation are somewhat limited mainly due to the fact that the mass transfer phenomena are not still fully understood. Some of the potential applications of pervaporation in the present context are,

(*a*) separation of heat-sensitive substances in pharmaceutical industries and food processing units,
(*b*) separation of liquid mixtures with very low relative volatility which would otherwise require infinite number of plates for separation by distillation (see the section on Distillation in Chapter 5),
(*c*) separation of azeotropic mixtures.

Pervaporation has been proved to be much more cost-efficient than the conventional azeotropic distillation for the separation of liquid mixtures that form azeotropes. An excellent example in this connection is the production of anhydrous alcohol[20]. Ethanol forms an azeotrope with water at a concentration of 89.4 mole percent at 1 atm pressure. To concentrate it beyond this, benzene–azeotropic distillation is to be employed which is an expensive process. Also, the final product will always have some contamination of the entrainer (namely, benzene). By using pervaporation, all these problems can be eliminated and anhydrous alcohol can be conveniently and economically manufactured. Apart from its medicinal values, anhydrous alcohol is also an excellent fuel. It can efficiently substitute gasoline (petrol) in motor vehicles and other gasoline–driven engines. An 80 : 20 blend of gasoline and ethanol (anhydrous) has been widely recommended for use in motor cars. This blend, popularly known as *gasohol,* has high performance efficiency and its use does not demand any significant engine modifications. Therefore, once anhydrous alcohol is manufactured on large scale by using a cost-efficient process like pervaporation, the use of gasohol as a substitute fuel in gasoline engines will be on the increase. This shall minimise the consumption of gasoline and help in conserving petroleum which is the vital need of the day.

## 7.6 LIQUID MEMBRANE TECHNOLOGY

Liquid membrane technology, also called liquid membrane permeation (LMP), is a relatively recent addition to the list of separation processes. However, the technology behind LMP is not complex at all and it has been successfully adapted to a number of industrial processes all over the world.

What is a liquid membrane? In essence, it is nothing but a liquid layer or liquid barrier separating two miscible fluids (one forming the feed phase and the other the receiving phase) and through which selective transport of one of the components of the feed (namely, the solute component) takes place. This concept can be better understood by citing an example, an industrial process in which this technology has been utilised. Let us consider removal (or recovery) of phenolics (phenol and its derivatives) from waste water/industrial effluents. This is probably one of the earliest industrial applications of liquid membrane permeation technology (LMP). Any LMP system shall necessarily consist of three phases such as

(*a*) the feed phase or the continuous phase ($F$),
(*b*) the membrane phase ($M$),
(*c*) the receiving phase ($I$).

In the process of recovery/removal of phenolics from waste water, the feed phase ($F$) is obviously the waste water, the membrane phase ($M$) is a hydrocarbon oil (such as kerosene oil) and the receiving phase ($I$) is aqueous caustic soda solution. First, the membrane phase and the receiving phase (kerosene oil and aqueous caustic soda) are fed to the emulsifying tank where they are strongly agitated by means of a high speed mechanical impeller (see Fig. 7.21) so as to produce a stable, homogeneous emulsion (since these two phases are immiscible with each other, they form an emulsion and not a solution). A small amount of a nonionic surfactant is also added to the tank to reduce the interfacial tension between the two phases and thereby increase the ease of emulsification. This M-I emulsion is now fed to the permeator into which the feed phase (waste water) is also pumped. Once again, since these two are immiscible, the M-I emulsion forms the dispersed phase and gets distributed in waste water (continuous phase) in the form of fine droplets. To note that each droplet is an emulsion, consisting of a droplet of aqueous caustic soda surrounded by a layer of kerosene. The system, therefore, is termed as a *double*

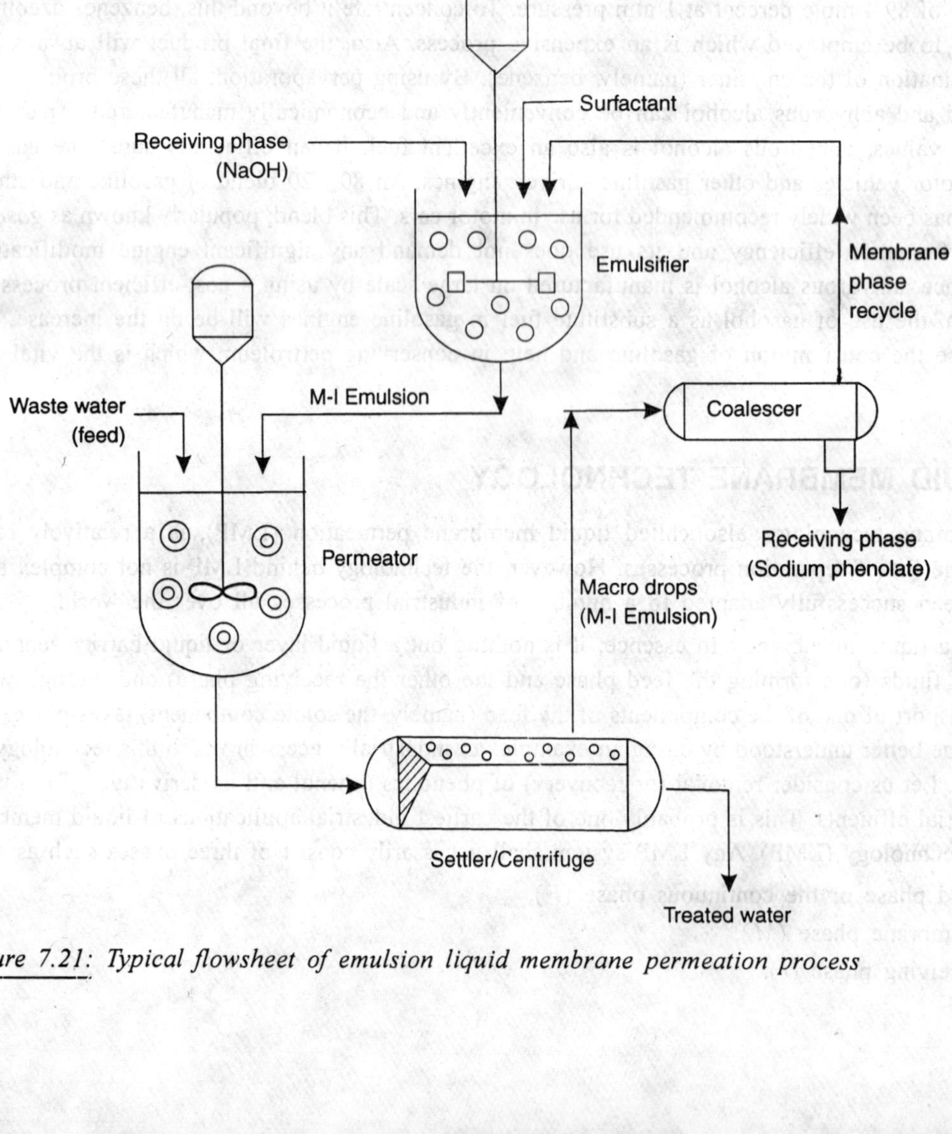


*Figure 7.21: Typical flowsheet of emulsion liquid membrane permeation process*

*emulsion*. Phenolics present in the continuous phase (waste water) diffuse through the kerosene layer into the interior of the droplet, where they react with caustic soda to form sodium phenolate. Since sodium phenolate is insoluble in kerosene oil, it does not permeate back. Thus, once sufficient residence time is provided, practically all the phenolics present in waste water permeate into the multitude of emulsion droplets present, forming sodium phenolate inside each. The entire suspension is now fed to a settling tank or a centrifuge where the droplets settle under gravity (or under the influence of centrifugal forces) and the treated water (that is free from phenolics) overflows out. The M-I droplets now enter an electrostatic coalescer where these droplets are made to coalesce with each other under the influence of a strong electrostatic field. On coalescence, kerosene oil forms a continuous phase in which sodium phenolate droplets remain suspended. This suspension is now separated in a settling tank (sedimentor), from which kerosene oil flows out as the overflow and is recycled back to the emulsifying tank. Sodium phenolate is collected as the underflow (bottom product). The entire scheme is sketched in Fig. 7.21.

Readers can easily notice that in the above process, the kerosene layer acts like a semipermeable membrane that separates the feed phase from the receiving phase and that permits permeation of the solute (namely, phenolics) only. Accordingly, the process is named as *liquid membrane permeation* (LMP).

The chief advantage of LMP process is that separation can be affected economically even if the solute content of the feed (phenol content of waste water) is very low. It also permits selective recovery of solute. To note that in the process described above, phenolics are selectively recovered (as sodium phenolate) and not destroyed. The recovery has been reported to be as high as 95 percent. They are also recovered in the pure form (other constituents of waste water do not permeate through the membrane phase*). The membrane phase is not consumed in the process, but is continuously recovered and recycled back. The mass transfer rates in LMP systems are substantially large since they provide interfacial areas (for mass transport) as high as 3000 $m^2/m^3$. In many applications, the receiving phase is so chosen that the reaction between the solute and the receiving phase is practically instantaneous. As a result, the solute concentration at M-R interface (membrane phase-receiving phase interface) shall be zero (or very close to zero) at any instant. The concentration gradient of the solute is thus maintained maximum across the membrane leading to high rate of mass transfer.

The speed of agitation is one of the crucial parameters that affects the operating efficiency of the permeator described above. It should be sufficiently large to ensure uniform dispersion and formation of uniformly sized macrodrops, but should not be too large to cause rupture of drops and consequent spillage of receiving phase into the feed phase. For example, if caustic soda (receiving phase) leaks out and mixes with feed water, it would form sodium phenolate which is membrane–impermeable and thus will remain behind in the waste water without getting separated.

What has been described above is process involving emulsion liquid membrane (ELM). It is also possible to use a solid-supported liquid membrane (SLM) which is also called immobilised liquid membrane (ILM). In other words, liquid membranes can be broadly classified into two categories:

(*a*) Emulsion liquid membranes (ELM).

(*b*) Solid-supported liquid membranes (SLM) or immobilised liquid membranes (ILM).

* Weak organic acids (such as acetic acid), if present, do permeate through the membrane phase and these from sodium salt (say, sodium acetate) in the receiving phase. These can be thus removed together with phenolics.

In the case of an SLM or ILM, the membrane phase (with or without a carrier) is entrapped within a polymer matrix (solid support) or immobilised within the pores of a polymeric membrane. In many processes employing carrier-mediated transport (discussed subsequently in this section), the carrier is dissolved in an appropriate solvent and the solution is immobilised within the solid support. Popularly used solid supports are polypropylene membrane, CA (cellulose acetate) membranes and PVC (polivinyl chloride) membranes. Silicone rubber membranes and polyether sulfone films are also employed as support materials. The permeator is then constructed as a tubular unit, a hollow fibre unit or as a spriral wound device. These configurations are exactly similar to those employed for reverse osmosis (see Section 7.1.2). For example, a hollow fibre permeator consists of a bundle of hollow fibres made of polypropylene impregnated with the membrane phase and/or the carrier. The bundle is enclosed in a stainless steel shell (the shell may also be made of pyrex glass or ceramic material). The receiving phase forms the tubeside fluid and flows through the fibres, while the feed solution enters the shell and flows over the fibre bundle or vice versa. The solute ($A$) permeates through the fibre wall into the receiving phase.

It is therefore apparent that selection of a suitable membrane phase that has a selective solubility or permeability for the solute is a must in processes like this. In the recovery of phenolics from waste water, we are able to select a hydrocarbon oil (kerosene oil) as the membrane phase which permits only phenolics and weak organic acids to permeate through. However, this need not be practicable always. In such cases, the separation is affected or facilitated by using a *carrier* that may be blended with the membrane phase. The process is then called *carrier-mediated transport* (also called *coupled transport or carrier-facilitated transport*). Most commonly employed carriers are organic compounds such as tertiary amines ($R_3N$), quaternary ammonium halides ($R_4NX$), oxines (which are derivatives of quinoline such as 2, methyl 8, hydroxyquinoline, 5, 7 dichloro 8, hydroxyquinoline, etc.), alkyl phosphoric acids and organophosphorus carriers (such as dihexyl N, N diethyl carbamoyl methyl phosphonate). Once again, the best way to understand this terminology is to discuss an industrial process that involves carrier-mediated or carrier-facilitated transport.

Let us, therefore, consider extraction of hexavalent chromium from acidic waste water discharged from metal processing and electroplating industries. Since chromium content of these effluent streams shall be significantly low, conventional hydrometallurgical operations (chemical/physical methods) shall turn out to be uneconomical for recovery of the same. However, chromium recovery can be efficiently and economically affected by LMP process. Here, the membrane phase employed is a hydrocarbon oil blended with a calculated amount of tertiary amine ($R_3N$) which acts as the mobile carrier. Aqueous caustic soda is used as the receiving phase. The process flowsheet is similar to that for waste water treatment for phenolic removal.

1. First, the M-I emulsion is prepared in a high speed mixing tank (emulsifying tank) in presence of a nonionic surfactant.
2. This emulsion is now dispersed in the feed solution (acidic waste water containing dichromates) in the permeator to produce stable macrodrops of about 1 mm in size. At the surface of these droplets, that is, at the F-M (feed phase-membrane phase) interface, the carrier (tertiary amine) present in the membrane phase reacts with dichromates forming a complex, $(R_3NH)_2\,Cr_2O_7$. The reaction may be represented as follows.

$$Cr_2O_7^{2-} + 2R_3N + 2H^+ \rightleftharpoons (R_3NH)_2\,Cr_2O_7 \quad \text{... (7.6.1)}$$

This complex, so formed, diffuses through the membrane phase to M-I interface (at the interior of each droplet) where it reacts with the receiving phase (caustic soda) as follows (to note that pH at this interface is much larger than that at the F-M interface).

$$(R_3NH)_2\, Cr_2O_7 + 4NaOH \rightleftharpoons 2Na_2CrO_4 + 2R_3N + 3H_2O \qquad \text{... (7.6.2)}$$

Chromium ion is thus rejected into the receiving phase as chromate, while the carrier, $R_3N$, is regenerated into the membrane phase. This regenerated $R_3N$ diffuses back to the F-M interface where it chelates with fresh chromium ions. This process continues and as a result, the chromium content of the feed solution continuously decreases and that of the receiving phase continuously increases. Finally, the receiving phase inside each droplet shall be a rich solution of sodium chromate and the feed solution practically free from dichromates.

3. The contents of the permeater are then fed to a settling tank (sedimentor) where the macrodrops settle and get separated (the feed solution, now free from dichromates, overflows out) and are subsequently de-emulsified by coalescing them in an electrostatic coalescer or a fibre bed coalescer. The separated membrane phase is recycled back, while the rich sodium chromate solution (receiving phase) is sent to storage or to further treatment.

The phenomenon of carrier–mediated transport is sketched schematically in Fig. 7.22. It may be noted that at any instant, the concentration of carrier (let us designate it as $C$) at the F-M interface is zero and that of solute ($A$)-carrier ($C$) complex (let us denote it as $AC$) is practically 100 percent. At the same time, at the M-I interface, the concentration of $AC$ is zero and that of the carrier ($C$) is almost 100 percent. Consequently, the solute-carrier complex diffuses across the membrane from the feed phase (or F-M interface) to the receiving phase, while the carrier diffuses in the reverse direction (from the M-I interface to the feed phase). It will not be, therefore, wrong to state that extraction of hexavalent chromium takes place at the F-M interface. Thus, both extraction and stripping take place in a single step (to note that in a conventional liquid-liquid extraction process discussed in Chapter 6, extraction and stripping (recovery of solvent) are conducted in two separate stages or equipment). Another interesting feature regarding this process is that chromium is being transported against its concentration gradient. As stated earlier, the concentration of chromium in the feed phase decreases with time, while that in

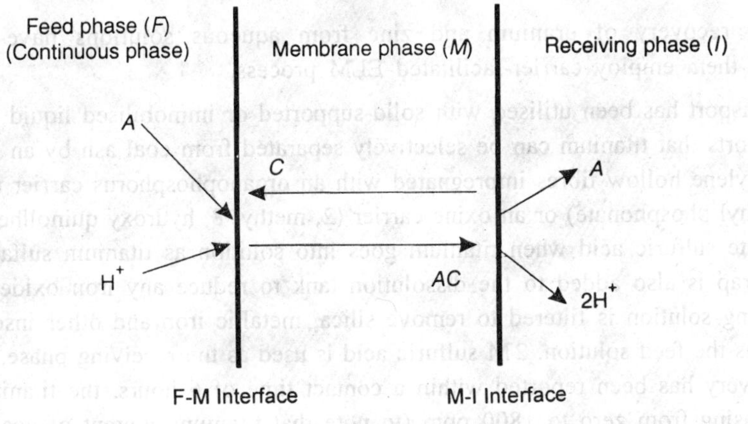


*Figure 7.22: Schematic of carrier-mediated (carrier-facilitated) transport*

the receiving phase increases, yet chromium continues to get transfered from the feed phase to the receiving phase. It is due to this that it is possible to concentrate chromium several fold from ppm level solutions. For the same reason, these LMP systems are often called *pumps*, since the solute is being *pumped* against a steep concentration gradient.

Since the carrier chosen is such that only the solute (chromium ions) chelates with it and forms the complex, the recovery of the solute is quite selective and exclusive. Thus, the sodium chromate solution obtained from the receiving phase shall be practically free from other metal ions originally present in the feed solution.

In a similar way, LMP has been successfully employed for the extraction of copper from dilute buffer solutions of copper sulphate. Here, *a* β-hydroxyoxime is used as the carrier in a hydrocarbon oil membrane (benzoyl acetone has also been reported to be an effective carrier). A strong acid such as 20 mass percent sulfuric acid is used as the receiving phase. The oxime selectively chelates with copper ions at the F-M interface to form a complex that diffuses through the membrane phase. If we denote the oxime carrier as HR, then the reaction may be represented as

$$Cu^{2+} + 2HR \rightleftharpoons CuR_2 + 2H^+ \quad \text{... (7.6.3)}$$

The complex, $CuR_2$, diffuses through the membrane phase and on reaching the M-I interface, it reacts with the acid (receiving phase) releasing $Cu^{2+}$ and regenerating the carrier (HR).

$$CuR_2 + 2H^+ \rightleftharpoons Cu^{2+} + 2HR \quad \text{... (7.6.4)}$$

Here also, transport of copper ions takes place against a steep concentration gradient and we ultimately obtain a strong solution of $Cu^{2+}$ in the receiving phase (the concentration ratio—ratio of $Cu^{2+}$ concentration in receiving phase to that in feed phase—reported is as high as 500 to 1000). This can be easily processed further for the economical recovery of copper. The recovery is also highly selective. With a feed solution containing both $Cu^{2+}$ and $Fe^{2+}$ ions, a separation factor of 1000 has been reported for copper (by separation factor, we mean the ratio of permeation rate of copper ions to that of ferrous ions). The capital cost of the LMP process has been reported to be 40 percent lower than that of solvent extraction process.

Processes involving recovery of uranium and zinc from aqueous solutions have also been commercialised. Both of them employ carrier-facilitated ELM process.

Carrier-mediated transport has been utilised with solid-supported or immobilised liquid membranes as well. Narayanan[25] reports that titanium can be selectively separated from coal ash by an ILM which is composed of polypropylene hollow fibres impregnated with an organophosphorus carrier (dihexyl N, N diethyl carbamoyl methyl phosphonate) or an oxine carrier (2, methyl 8, hydroxy quinoline). Coal ash is first leached with dilute sulfuric acid when titanium goes into solution as titanium sulfate. A small amount of aluminium scrap is also added to the dissolution tank to reduce any iron oxide present to metallic iron. The resulting solution is filtered to remove silica, metallic iron and other insolubles and the clear filtrate is used as the feed solution. 2 M sulfuric acid is used as the receiving phase. More than 91 percent titanium recovery has been reported within a contact time of 6 hours, the titanium content of receiving phase increasing from zero to 1800 ppm (to note that titanium content of coal ash leach solution is less than 250 ppm) and it is practically free from other metal ions. From this rich titanium sulfate solution, titanium can be conveniently recovered by hydrolysing to titanium hydroxide and subsequent calcination to titanium dioxide.

Muscatello and coworkers[26] report recovery of actinides such as americium and plutonium from nitrate–nitric acid waste solutions using an SLM that consists of polypropylene hollow fibres on which bifunctional organophosphorus carriers such as dihexyl N, N diethylcarbamoyl methyl phosphonate and octylphenyl N, N diisobutyl carbamoyl methyl phosphine oxide have been immobilised. Dilute oxalic acid is used as the receiving phase. The feed solution is first pretreated to bring down its nitric acid content to 0.1 M (by neutralisation) and in such a case, 94 percent americium removal has been recorded. It is also recommended that americium is continuously removed from the receiving phase by sorption on an inorganic ion exchanger such as alumina. Once the americium content and plutonium content of the feed solution are reduced to $10^{-7}$ g/L and $10^{-5}$ g/L respectively, it can be disposed as a low level waste.

Separation of gases using SLMs has also been successful. An example is the recovery of ethene and propene from process off gases. Aqueous silver nitrate solution is used as the carrier which is immobilised in the pore structure of commercial reverse osmosis hollow fiber modules. The feed gases are admitted at 4.0 to 8.0 atm and hexane is used as the sweep gas. Permeate streams with propene concentrations in excess of 98 mole percent have been observed in modules of 25.0 to 40.0 $m^2$ membrane area[27]. Similarly, ILMs consisting of a cobalt-based carrier (cobalt-salen complex) in a low volatile organic solvent immobilised in thin microporous films have been used for the separation of oxygen from air. The system is reported to have a separation factor (ratio of oxygen permeability to nitrogen permeability) of 25 and yields a product stream containing 80 percent oxygen[28]. One of the important factors of concern with industrial SLMs or ILMs is the leakage of carrier and/or solvent to the feed phase due to the pressure difference that gets established across the membrane.

### 7.6.1 Mathematical Simulation of LMP Processes

Mathematical analysis of an LMP process is no easy task. However, we have been successful in this context to a very large extent.

Let us start with ELM systems that do not employ a carrier. An excellent example in this connection is the recovery of phenolics from waste water discussed in the earlier paragraphs. Broadly speaking, ELM systems of this kind can be modelled based on two approaches:

(*a*) Advancing reaction front model[29].

(*b*) Reversible reaction model[30].

Let us first discuss the advancing reaction front model. This is very similar to the shrinking core model that is popularly used for describing noncatalysed solid-fluid reactions (see Section 8.6 of Chapter 8). This model assumes that the reaction between the solute ($A$) and the reactive component ($B$) of the receiving phase is irreversible and instantaneous. The solute ($A$) diffuses through the membrane phase and on reaching the M-I interface, it reacts instantaneously with $B$ forming the product $P$. As a result, if $R_0$ is the radius of the macrodrop (that consists of both membrane phase and the receiving phase) and $r_0$ that of the receiving phase droplet, then the value of $r_0$ decreases continuously with time and the reaction front moves towards the centre of the receiving phase (see Fig. 7.23). To simplify the situation, the following assumptions are made:

1. The feed phase or the continuous phase is perfectly mixed. Consequently, the concentration of solute in the continuous phase inside the permeator/reactor (namely, $C_A$) shall be equal to that in the exit stream (namely, $C_{Ae}$). In other words, resistance to transport of solute ($A$) through the continuous phase to the F-M interface is assumed negligible.
2. Each macrodrop (emulsion droplet) is spherical and of radius $R_0$. No coalescence occurs between droplets inside the reactor/permeator.
3. There is no internal circulation of fluid inside each macrodrop.
4. The residence time distribution (RTD) of macrodrops follows an exponential function as shown in Eq. (7.6.30).
5. The diffusional resistance offered by the product ($P$) layer is negligible as compared to that offered by the membrane phase ($M$).

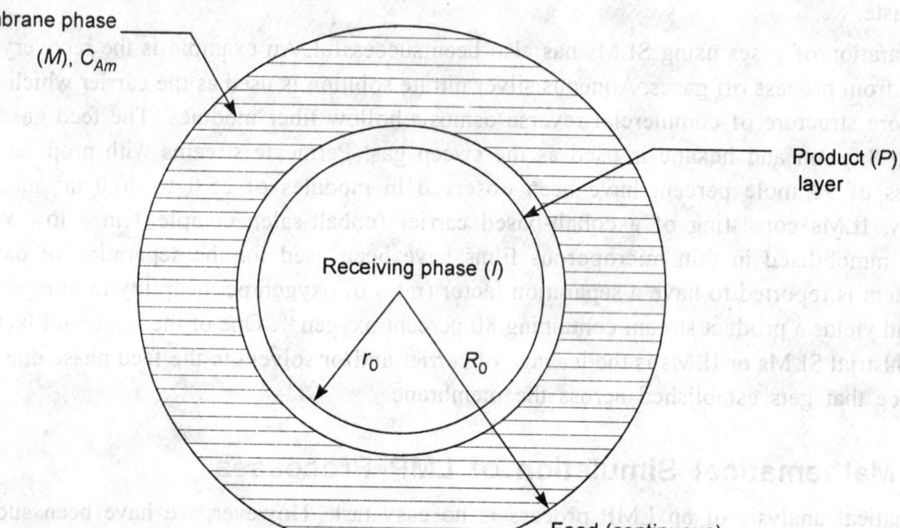


*Figure 7.23: Advancing reaction front model*

Thus, if we consider a single macrodrop, the only resistance to solute transport is the diffusional resistance offered by the membrane phase. The mass transport equation therefore, becomes

$$\frac{d}{dr}\left(r^2 D_e \frac{dC_{Am}}{dr}\right) = 0 \qquad \text{... (7.6.5)}$$

where $C_{Am}$ = concentration of solute ($A$) in the membrane phase, kmoles/m$^3$

$D_e$ = effective diffusivity or permeability of solute ($A$) in the membrane phase, m$^2$/s

If we denote the distribution coefficient (or partition coefficient) of solute between membrane phase and the continuous phase as $K_{mb}$ (which is defined as the ratio of solute concentration in the membrane phase to that in the continuous phase), then the boundary conditions governing the system are

**B.C.1:** At $r = r_0,\ C_{Am} = 0$ ... (7.6.6)

**B.C.2:** At $r = R_0,\ C_{Am} = K_{mb}C_A = K_{mb}C_{Ae}$ ... (7.6.7)

Equation (7.6.5) can be easily integrated twice and the integration constants determined using the above two boundary conditions. The result is

$$C_{Am} = (K_{mb}C_{Ae})\left[\frac{1-(r_0/r)}{1-(r_0/R_0)}\right] \qquad \text{... (7.6.8)}$$

If we define the following dimensionless variables,

$$\phi = \left(\frac{C_{Am}}{K_{mb}C_{Ae}}\right) \qquad \text{... (7.6.9)}$$

$$\eta = (r/R_0) \qquad \text{... (7.6.10)}$$

$$\xi = (r_0/R_0) \qquad \text{... (7.6.11)}$$

then Eq. (7.6.8) becomes

$$\phi = \left(\frac{\eta-\xi}{1-\xi}\right)\left(\frac{1}{\eta}\right) \qquad \text{... (7.6.12)}$$

The boundary conditions become

**B.C.1:** At $\eta = \xi,\ \phi = 0$ ... (7.6.13)

**B.C.2:** At $\eta = 1,\ \phi = 1$ ... (7.6.14)

On differentiation, Eq. (7.6.12) yields (at $\eta = 1$)

$$\left.\frac{\partial\phi}{\partial\eta}\right|_{\eta=1} = \left(\frac{\xi}{1-\xi}\right) \qquad \text{... (7.6.15)}$$

As stated earlier, the radius of inner phase (receiving phase) droplet ($r_0$) decreases continuously with time. It shall be, therefore, of interest to derive an expression for $r_0$ as a function of time. Let the reaction between $A$ and $B$ be represented as.

$$A + B \rightarrow P \qquad \text{... (7.6.16)}$$

As $r_0$ decreases with time, the volume of receiving face [which is equal to $(4/3)\,\pi r_0^3$] also decreases and so is the number of moles of $B$ present in the receiving phase (as $B$ gets consumed by reaction with $A$). Accordingly, we may assume, with allowable error, that the number of kmoles of $B$ per unit volume of receiving phase, namely the molar concentration of $B$ in the receiving phase ($C_B$), remains constant at its initial value, $C_{B0}$. In other words, we assume that

$$C_B = C_{B0},\ \text{for } 0 \le r \le r_0,\ \text{for all } t \qquad \text{... (7.6.17)}$$

Then, if $N_B$ denotes the number of kmoles $B$ present in the receiving phase at any time $t$,

$$\frac{dN_B}{dt} = \frac{d}{dt}\left(\frac{4}{3}\pi r_0^3 C_{B0}\right) = 4\pi\ C_{B0} r_0^2 \frac{dr_0}{dt} \qquad \text{... (7.6.18)}$$

The rate of change of $N_B$ per unit volume of membrane phase is,

$$\left(\frac{1-f}{f}\right)\frac{dN_B}{dt} = 4\pi\, C_{B0}\left(\frac{1-f}{f}\right) r_0^2 \frac{dr_0}{dt} \quad \text{... (7.6.19)}$$

where $f$ = volume fraction of membrane phase in each macrodrop. To be precise, $(1-f)$ = volume fraction of (receiving phase + product layer). However, since volume fraction of product layer is relatively small, it has been assumed that $(1-f) \simeq$ volume fraction of receiving phase.

Since all kmoles of $A$ that reach the reaction front (at $r = r_0$) react instantaneously with $B$,

$$\frac{-dN_{Am}}{dt} = (4\pi r_0^2)\, D_e \left(\frac{dC_{Am}}{dr}\right)_{r=r_0}$$

$$= 4\pi\, D_e K_{mb} C_{Ae}\left(\frac{r_0}{1-(r_0/R_0)}\right) \quad \text{... (7.6.20)}$$

We have obtained the above expression by substituting for $C_{Am}$ from Eq. (7.6.8). Since one mole of $A$ reacts with one mole of $B$ [see Eq. (7.6.16)],

$$\frac{-dN_{Am}}{dt} = -\left(\frac{1-f}{f}\right)\left(\frac{dN_B}{dt}\right) \quad \text{... (7.6.21)}$$

Substituting from Eqs (7.6.19) and (7.6.20) and rearranging, we get

$$\int_0^t dt = \frac{C_{B0}(1-f)/f}{D_e K_{mb} C_{Ae}} \int_{R_0}^{r_0}\left(\frac{r_0^2}{R_0} - r_0\right) dr_0 \quad \text{... (7.6.22)}$$

On integration between the specified limits,

$$t = \frac{C_{B0}(1-f)\, R_0^2}{6 f\, D_e K_{mb} C_{Ae}}\left[1 - 3\,(r_0/R_0)^2 + 2\,(r_0/R_0)^3\right] \quad \text{... (7.6.23)}$$

This can be re-expressed in terms of the following dimensionless variables:

$$C^*_{Ae} = (K_{mb}/K_{mr})\,(C_{Ae}/C_{B0}) \quad \text{... (7.6.24)}$$

$$\alpha_1 = f\,(Q_e/Q_f)\,K_{mb} \quad \text{... (7.6.25)}$$

$$\alpha_2 = (1-f)\,(Q_e/Q_f)\,(K_{mb}/K_{mr}) \quad \text{... (7.6.26)}$$

$$\tau = (tD_e/R_0^2) \quad \text{... (7.6.27)}$$

where $K_{mr}$ = partition coefficient (distribution coefficient) of solute between membrane phase and receiving phase

$$= \frac{\text{(solute concentration in membrane phase)}}{\text{(solute concentration in receiving phase)}}$$

$Q_e$, $Q_f$ = volumetric flow rate of emulsion and that of feed solution respectively

Equation (7.6.23), therefore, becomes

$$\tau = \left(\frac{\alpha_2}{6\alpha_1 C_{Ae}^*}\right)(1 - 3\xi^2 + 2\xi^3) \quad \text{... (7.6.28)}$$

We have so far considered a single macrodrop. The total rate of solute transfer into all macrodrops present in reactor/permeator is

$$\dot{m}_A = Q_f (C_{A0} - C_{Ae}) = \int_0^N \left(4\pi R_0^2\right) D_e f \left.\frac{\partial C_{Am}}{\partial r}\right|_{r=R} dn \quad \text{... (7.6.29)}$$

where $C_{A0}$ = solute concentration in feed solution at the inlet to the reactor/permeator
$N$ = total number of macrodrops in the reactor/permeator
$n$ = number of macrodrops with residence time larger than t.

As stated earlier, we have assumed that the residence time distribution (RTD) of macrodrops in the reactor/permeator follows an exponential function as

$$(n/N) = \exp(-t/\theta_e) \quad \text{... (7.6.30)}$$

where $\theta_e$ = average residence time of emulsion in the permeator/reactor

$$= (V_e/Q_e) \quad \text{... (7.6.31)}$$

$V_e$ = total volume of emulsion (total volume of macrodrops) in the permeator/reactor

After substituting Eq. (7.6.30) in (7.6.29), it can be rewritten in terms of the dimensionless variables defined earlier. Thus

$$\left(\frac{C_{A0}^* - C_{Ae}^*}{C_{Ae}^*}\right) = 3\alpha_1 \int_0^\infty \left.\frac{\partial\phi}{\partial\eta}\right|_{\eta=1} \exp(-\tau/\tau_e)\, d\tau \quad \text{... (7.6.32)}$$

where $C_{A0}^* = (K_{mb}/K_{mr})\,(C_{A0}/C_{B0})$ ... (7.6.33)

$\tau_e = \theta_e D_e / R_0^2$ ... (7.6.34)

If we substitute the expression for $(\partial\phi/\partial\eta)$ and that for $\tau$ from Eqs (7.6.15) and (7.6.28) respectively in Eq. (7.6.32), it gets reduced to

$$(C_{A0}^* - C_{Ae}^*) = 3\alpha_2 \int_0^1 \xi^2 \exp\left[-\frac{\alpha_2\left(1 - 3\xi^2 + 2\xi^3\right)}{6\alpha_1 C_{Ae}^* \tau_e}\right] d\xi \quad \text{... (7.6.35)}$$

The integral on the right hand side of above equation may be evaluated either graphically or numerically. Incidentally, one of the important parameters whose value is required to be known for utilising the above performance equation is the effective diffusivity or permeability of the solute in the membrane phase ($D_e$). $D_e$ is a function of the molecular diffusivity of solute in the membrane phase ($D_{Am}$), its diffusivity in the receiving phase ($D_{Ar}$) and the volume fraction of membrane phase in the emulsion ($f$). Bunge and Noble[30] have developed a correlation that reportedly provides a reliable estimate of $D_e$. Their correlation is given below.

$$D_e = D_{Am}\left(1 - \frac{\pi}{4\,(1+2\,p)^2}\right) + \frac{\pi}{4\,(1+2\,p)}\left(\frac{D_F\,D_{Am}}{D_{Am} + 2\,p\,D_F}\right) \quad \text{... (7.6.36)}$$

where $p = 0.403\ (1 - f)^{-1/3} - 0.5$ ... (7.6.37)

$$D_F = \frac{2\beta\,D_{Am}}{(\beta - 1)}\left(\frac{\beta \ln \beta}{(\beta - 1)} - 1\right) \quad \text{... (7.6.38)}$$

$$\beta = D_{Ar}/(D_{Am}\,K_{mr}) \quad \text{... (7.6.39)}$$

Though Eq. (7.6.35) can be used as the performance equation for the permeator/reactor, it nevertheless demands a trial and error computation. For example, if we have to compute the volume of the reactor/permeator required to affect a specified solute removal rate (we have to compute $\tau_e$ for specified values of $C_{A0}$ and $C_{Ae}$), then the value of $\tau_e$ can be computed only by trial since it appears within the integral on the right hand side of the equation. However, in the case of a rating problem, when $C_{A0}$, $C_{Ae}$ and $\tau_e$ are known and we have to determine whether the given reactor/permeator is suitable for affecting the desired solute removal rate ($\dot{m}_A$), the possible value of $\dot{m}_A$ can be computed straightaway from Eq. (7.6.35) since the integral on the right hand side can be directly evaluated as both $C_{Ae}$ and $\tau_e$ are known. No trial and error computations shall be necessary in the case.

The *most ideal situation* in this case is that when the permeator/reactor acts as a perfectly mixed vessel or as an ideal CSTR (continuous stirred Tank reactor) *with respect to all phases*. In such a case, $C_{A0} = C_{A0}$ (ideal), which represents the maximum feed concentration permissible to obtain a specific solute concentration in the product stream ($C_{Ae}$) when the reactor performs as an ideal CSTR. The value of $C_{A0}$ (ideal) can be straightaway obtained through a macroscopic solute balance. Thus

(Moles of solute in) = (Moles of solute out in continuous phase) + (Moles of solute out in membrane phase) + (Moles of solute out in receiving phase) + (Moles of solute reacted) ... (7.6.40)

$$Q_f\,C_{A0}\ (\text{ideal}) = (Q_f\,C_{Ae}) + C_{Am}\ (Q_e\,f) + C_{Ar}\ (1 - f)\ Q_e + C_{B0}\ (1 - f)\ Q_e \quad \text{... (7.6.41)}$$

where $C_{Ar} = C_{Am}/K_{mr}$ ... (7.6.42)

To note that since the reaction has been assumed to be irreversible, the maximum conversion of $A$ attainable is 100 percent and in such a case, moles of $A$ reacted will be equal to moles of $B$ originally present in the receiving phase. We have used this condition in the above equation. In terms of the dimensionless variables defined earlier, the above equation becomes

$$[C^*_{A0}\ (\text{ideal}) - C^*_{Ae}] = C^*_{Ae}\ (\alpha_1 + \alpha_2) + \alpha_2 \quad \text{... (7.6.43)}$$

The performance efficiency of the permeator/reactor ($E_f$) may be therefore defined as

$$E_f = \frac{(\text{Solute removal rate affected by the reactor})}{\left(\begin{array}{c}\text{Solute removal rate affected if the}\\ \text{permeator were an ideal CSTR}\end{array}\right)}$$

$$= \frac{\dot{m}_A \text{ (actual)}}{\dot{m}_A \text{ (ideal)}} = \frac{\left(C^*_{A0} - C^*_{Ae}\right)}{\left[C^*_{A0} \text{ (ideal)} - C^*_{Ae}\right]} \quad \text{... (7.6.44)}$$

Let us now consider the *reversible reaction model*. The major difference between this model and the advancing reaction front model described above is that this model assumes the reaction between $A$ and $B$ to be reversible and that it attains equilibrium within a short interval of time. The reaction may be thus represented as

$$A + B \underset{K}{\rightleftharpoons} P \quad \text{... (7.6.45)}$$

where K is the equilibrium constant of the reaction. All other assumptions involved in the advancing reaction front model are valid in this case also. Now,

$$K = \frac{(C_{Pr})_{eq}}{(C_{Ar})_{eq}\,(C_B)_{eq}} \quad \text{... (7.6.46)}$$

An overall $B$-balance yields

$$C_{B0} = (C_B)_{eq} + (C_{Pr})_{eq} \quad \text{... (7.6.47)}$$

To note that the second term on the right hand side of the above equation represents kmoles of $B$ reacted (consumed) per unit volume of receiving phase. The suffix *eq* is used to represent equilibrium concentration. Eliminating $(C_B)_{eq}$ between the above two equations, we get

$$(C_{Pr})_{eq} = \frac{KC_{B0}\,(C_{Ar})_{eq}}{1 + K\,(C_{Ar})_{eq}} \quad \text{... (7.6.48)}$$

Since we have assumed that equilibrium has been attained within a short interval of time, we may write, with allowable error,

$$(C_{Pr})_{eq} \simeq C_{Pr} = \frac{KC_{B0}C_{Ar}}{1 + KC_{Ar}} \quad \text{... (7.6.49)}$$

Now, the mass transport equation for a single macrodrop is

$$\frac{\partial C_{Am}}{\partial t} = (\bar{D}_{ef}/r^2)\,\frac{\partial}{\partial r}\left(r^2\,\frac{\partial C_{Am}}{\partial r}\right) - \left(\frac{1-f}{f}\right)(r_A + r_P) \quad \text{... (7.6.50)}$$

where $r_A = \partial C_{Ar}/\partial t$ = [rate of consumption of $A$ (by forward reaction) per unit volume of receiving phase] ... (7.6.51)

$r_P = \partial C_{Pr}/\partial t$ = [rate of formation of $A$ (by backward reaction) per unit volume of receiving phase] ... (7.6.52)

$\bar{D}_{ef}$ = *mean* effective diffusivity for the diffusion of $A$ in the membrane phase and also into the receiving phase

$$= \int_0^1 D_{ef}\, d\phi \quad \text{... (7.6.53)}$$

Bunge and Noble[30] have proposed that $D_{ef}$ may be computed from Eqs (7.6.36) to (7.6.39) given earlier, except that the parameter β has to be redefined as

$$\beta = \frac{f_m D_{Ar}}{D_{Am} K_{mr}} \quad \text{... (7.6.54)}$$

where
$$f_m = 1 + \left(\frac{\alpha_3}{1+\alpha_3 C^*_{Ae}\phi}\right) \quad \text{... (7.6.55)}$$

$$\alpha_3 = K C_{B0} \quad \text{... (7.6.56)}$$

To note that the effective diffusivity, $D_e$ (defined in advancing reaction front model), is independent of kinetics of the chemical reaction between $A$ and $B$ (since the reaction is assumed to be instantaneous), while $D_{ef}$ (defined above) is a function of reaction equilibrium constant, K. After substituting for $C_{Pr}$ from Eq. (7.6.49), Eq. (7.6.50) may be rewritten in terms of the dimensionless variables (defined earlier) as

$$\frac{\partial \phi}{\partial \tau} = (\lambda/\eta^2)\frac{\partial}{\partial \eta}\left(\eta^2 \frac{\partial \phi}{\partial \eta}\right)\left\{\frac{1}{1+(\alpha_2/\alpha_1)\left[1+\alpha_3/\left(1+\alpha_3 C^*_{Ae}\phi\right)^2\right]}\right\} \quad \text{... (7.6.57)}$$

where
$$\lambda = \bar{D}_{ef}/D_e \quad \text{... (7.6.58)}$$

The initial and boundary conditions governing the system are,

**I.C:** At $t = 0,\ C_{Am} = 0,\ \text{for } R_0 > r \geq 0$

At $\tau = 0,\ \phi = 0,\ \text{for } 1 > \eta \geq 0$ ... (7.6.59)

**B.C.1:** At $r = R_0,\ C_{Am} = K_{mb} C_{Ae}\ \text{for } t \geq 0$

At $\eta = 1,\ \phi = 1\ \text{for } \tau \geq 0$ ... (7.6.60)

**B.C.2:** At $r = 0,\ (\partial C_{Am}/\partial r) = 0,\ \text{for all } t$

At $\eta = 0,\ (\partial\phi/\partial\eta) = 0,\ \text{for all } \tau$ ... (7.6.61)

Equation (7.6.57) is a nonlinear partial differential equation in φ and an analytical solution to it is hardly possible. However, its nonlinearity vanishes if we make a simplifying assumption that $(\alpha_3 C^*_{Ae})$ is much smaller than unity so that $(1 + \alpha_3 C^*_{Ae}\phi) \simeq 1.0$. In that case, Eq. (7.6.57) gets reduced to

$$\frac{\partial \phi}{\partial \tau} = \left[\frac{\lambda}{1+(\alpha_2/\alpha_1)(1+\alpha_3)}\right]\frac{1}{\eta^2}\frac{\partial}{\partial \eta}\left(\eta^2\frac{\partial \phi}{\partial \eta}\right) \quad \text{... (7.6.62)}$$

Analytical solution to partial differential equation of the above kind with initial and boundary conditions as specified above is available in most text books on engineering mathematics (we have discussed solutions to partial differential equations of this kind in Section 3.14 of Chapter 3). We, therefore, do not present the procedure of the solution here. The solution is obtained in the form of an infinite series[30]. Once we have got $\phi$ as a function of $\tau$, then it can be differentiated to obtain the expression for solute flux at the surface of the macrodrop. The result is

$$\left.\frac{\partial \phi}{\partial \eta}\right|_{\eta=1} = 2\sum_{n=1}^{\infty} \exp\left[-\frac{n^2\pi^2\alpha_1\lambda\tau}{\alpha_1+\alpha_2(1+\alpha_3)}\right] \quad \text{... (7.6.63)}$$

The total rate of solute transfer ($\dot{m}_A$) can be computed from Eq. (7.6.29) or (7.6.32), provided $D_e$ is replaced by $\bar{D}_{ef}$ and $(3\alpha_1)$ in Eq. (7.6.32) is replaced by $(3\alpha_1\lambda)$. Thus, if we substitute the above Eq. (7.6.63) in Eq. (7.6.32) after replacing $(3\alpha_1)$ by $(3\alpha_1\lambda)$, then it can be integrated to give

$$\left(\frac{C^*_{A0}-C^*_{Ae}}{C^*_{Ae}}\right) = 6\alpha_1\lambda\tau_e \sum_{n=1}^{\infty}\left[1+\frac{n^2\pi^2\alpha_1\lambda\tau_e}{\alpha_1+\alpha_2(1+\alpha_3)}\right] \quad \text{... (7.6.64)}$$

The above equation forms the performance equation for the permeator/reactor when the reaction between $A$ and $B$ is reversible as shown in Eq. (7.6.45).

In this case also, if we assume that the permeator/reactor performs as an ideal CSTR with respect to all phases, then $C_{A0}$ (ideal) can be estimated straightaway from a solute balance [Eq. (7.6.40)]. Thus

$$Q_f\, C_{A0}\,(\text{ideal}) = (Q_f\, C_{Ae}) + C_{Am}\,(Q_e f) + C_{Ar}\,(1-f)\,Q_e + C_{Pr}\,(1-f)\,Q_e \quad \text{... (7.6.65)}$$

After substituting for $C_{Ar}$ from Eq. (7.6.42) and for $C_{Pr}$ from Eq. (7.6.49), the above equation can be rewritten in terms of the dimensionless variables as

$$\left[\frac{C^*_{A0}\,(\text{ideal})-C^*_{Ae}}{C^*_{Ae}}\right] = \alpha_1 + \alpha_2 + \left(\frac{\alpha_2\alpha_3}{1+\alpha_3 C^*_{Ae}}\right) \quad \text{... (7.6.66)}$$

The performance efficiency ($E_f$) of the permeator/reactor may be then estimated from Eq. (7.6.44).

**Example 7.6:** Aqueous effluent from a pharmaceutical unit is to be dephenolised using emulsion liquid membrane permeation process. Kerosene oil is to be used as membrane phase and 0.182 M caustic soda as the receiving phase. The feed solution is fed to the extractor (permeator) at 1.80 L/hr and the kerosene–NaOH emulsion (containing 60 percent by volume of kerosene oil) is fed at the rate of 0.20 L/hr. Determine the volume of reactor required to reduce the phenolic content of feed water from 0.00466 kmoles/m$^3$ to 0.0009 kmole/m$^3$. Assume that the advancing reaction front model is applicable.

[Partition coefficient of solute (phenolics) between membrane phase and continuous phase] = (that between membrane phase and receiving phase) = 1.0

Take average size of macrodrops to be 1.0 mm and diffusivity of phenolics in kerosene oil and that in aqueous caustic soda be $9.298 \times 10^{-10}$ m$^2$/s and $11.158 \times 10^{-10}$ m$^2$/s respectively.

**Solution:** Since advancing reaction front model is applicable, we may make use of Eq. (7.6.35). Now,

$$\frac{\left(C_{A0}^{*}-C_{Ae}^{*}\right)}{(3\alpha_2)} = \frac{(C_{A0}-C_{Ae})}{3C_{B0}(1-f)\left(Q_e/Q_f\right)}$$

$$= \frac{(0.00466-0.0009)}{(3.0)(0.182)(1-0.6)(0.20/1.8)} = 0.1549$$

$$\alpha_1 = f(Q_e/Q_f)\,K_{mb} = (0.6)\,(0.20/1.8)\,(1.0) = 0.06667$$

$$(C_{Ae}^{*}/\alpha_2) = \frac{C_{Ae}}{C_{B0}(1-f)\left(Q_e/Q_f\right)} = \frac{(0.0009)}{(0.182)(1-0.6)(0.20/1.8)} = 0.11126$$

Accordingly, Eq. (7.6.35) reduces to

$$0.1549 = \int_0^1 \xi^2 \exp\left[22.47\left(3\xi^2-2\xi^3-1\right)/\tau_e\right] d\xi \qquad \text{... (i)}$$

This has to be solved for $\tau_e$ by trial. Thus, let $\tau_e = 3.0$. Then the integral on the right hand side of the above equation reduces to

$$I = \int_0^1 \xi^2 \exp\left[7.49\left(3\xi^2-2\xi^3-1\right)\right] d\xi = \int_0^1 F(\xi)\, d\xi$$

Let us evaluate the above integral numerically using Simpson's rule. For this, the values of $F(\xi)$ are computed at different values of $\xi$ and are tabulated in Table 7.6.1 given below.

**Table 7.6.1.**

| $\xi$ | $F(\xi)$ | $\xi$ | $F(\xi)$ |
|---|---|---|---|
| 0.0 | 0.0 | 0.6 | $2578.06 \times 10^{-5}$ |
| 0.1 | $0.689 \times 10^{-5}$ | 0.7 | $9718.00 \times 10^{-5}$ |
| 0.2 | $4.8696 \times 10^{-5}$ | 0.8 | $2936.85 \times 10^{-4}$ |
| 0.3 | $25.35 \times 10^{-5}$ | 0.9 | $6567.57 \times 10^{-4}$ |
| 0.4 | $124.8139 \times 10^{-5}$ | 1.0 | 1.0 |
| 0.5 | $590.890 \times 10^{-5}$ | | |

According to Simpson's rule,

$$I = (h/3)\left[f_0 + f_n + 4\left(f_1 + f_3 + \ldots\ldots f_{n-1}\right) + 2\left(f_2 + f_4 + \ldots\ldots f_{n-2}\right)\right]$$

Here, $f_0 = 0.0$, $f_1 = 0.689 \times 10^{-5}$, $f_2 = 4.8696 \times 10^{-5}$ and so on and finally, $f_n = 1.0$. Also, $h$ = step size = 0.1. Therefore,

$$I = 0.1560$$

This differs little from the left hand side of Eq. (i), namely 0.1549. Therefore, the assumed value of $\tau_e$ is correct and no further trials are required. The value of effective diffusivity ($D_e$) can be now computed from Eqs (7.6.36) to (7.6.39). Thus,

$$\beta = \frac{(11.158 \times 10^{-10})}{(9.298 \times 10^{-10})(1.0)} = 1.2$$

Substituting this in Eq. (7.6.38), we get

$$D_F = 10.48 \times 10^{-10} \text{ m}^2/\text{s}$$

Also, $$p = \frac{(0.403)}{(0.4)^{1/3}} - 0.5 = 0.04695$$

Therefore, from Eq. (7.6.36),

$$D_e = 10.0 \times 10^{-10} \text{ m}^2/\text{s}$$

Now, $$\tau_e = \theta_e D_e / R_0^2$$

Since $$\theta_e = (V_e/Q_e),$$

$$V_e = Q_e R_0^2 \tau_e / D_e = (0.2 \times 10^{-3})(0.001)^2 (3.0)/(10.0 \times 10^{-10}) = 0.6 \text{ m}^3$$

In most continuous stirred tank reactors,

$$(V_e/V_c) = (Q_e/Q_f) = (0.2/1.8) = (1/9)$$

where $V_C$ is the total volume of continuous phase in the extractor/reactor. In other words, $V_e = 0.1 (V_e + V_C) = 0.1$ V. Therefore,

$$V = (0.6/0.1) = \mathbf{6.0 \text{ m}^3}$$

This is the required volume of the reactor/permeator.

Let us now consider carrier-facilitated transport in liquid membranes. Whenever the transport of solute across the membrane (ELM or SLM) is facilitated by a carrier, then the facilitation factor ($F$) becomes the most influencing parameter. This is defined as

$$F = \frac{\begin{pmatrix}\text{Solute flux across the membrane} \\ \text{in presence of carrier}\end{pmatrix}}{\begin{pmatrix}\text{Solute flux across the membrane} \\ \text{in absence of carrier}\end{pmatrix}} \quad \text{... (7.6.67)}$$

Attempts have been made by many researchers for developing analytical expressions for the computation of facilitation factor ($F$). Among these, the correlation developed by Noble et al[31] has received a fair degree of acceptance. Their correlation is given below.

$$F = \frac{(1+\phi)(1+1/Sh)}{1-\phi\left(\dfrac{\tan h\,\lambda}{\lambda}\right)+(1+\phi)(1/Sh)} \quad \text{... (7.6.68)}$$

where $$\phi = \frac{(\alpha K K_{mb} C_A)}{(1 + K K_{mb} C_A)} \quad \text{... (7.6.69)}$$

$$\alpha = \text{mobility ratio} = D_{AC} C_T / (D_{AM} K_{mb} C_A) \quad \text{... (7.6.70)}$$

$$Sh = \text{Sherwood number} = (k\delta / D_{AM}) \quad \text{... (7.6.71)}$$

$$\lambda = (1/2)\left(\frac{1+(\alpha+1)\,K\,K_{mb}C_A}{\in\left(1+K\,K_{mb}C_A\right)}\right) \quad ... (7.6.72)$$

$\in$ = inverse Damkohler number = $D_{AC}/(k_2\delta^2)$ ... (7.6.73)

$D_{AM}$, $D_{AC}$ = diffusivity of solute and that of solute-carrier complex respectively in the membrane phase

$C_T$ = total concentration of carrier in the membrane phase

= (concentration of carrier in the free state) + (that in combined state)

= [C] + [AC] ... (7.6.74)

$\delta$ = thickness of membrane

$k$ = mass transfer coefficient, m/s

To note that,

$$(1/k) = (1/k_C) + (1/k_L) \quad ... (7.6.75)$$

$k_C$, $k_L$ = mass transfer coefficient (for solute transfer) in continuous phase (feed solution) and that in receiving phase respectively, m/s

Often, it is assumed that $D_{AC} = D_{CM}$, where $D_{CM}$ is diffusivity of carrier in the membrane phase. Also, in many cases, when the feed is an aqueous solution, the value of partition coefficient ($K_{mb}$) has been reported to be very close to 1.0. When the feed is a gas mixture (from which the solute gas $A$ is being separated), the solute component ($A$) first dissolves in the solvent (often, water) present in the membrane (SLM) and then reacts with the carrier ($C$). Accordingly, it will not be wrong to assume that ($K_{mb}C_A$) is very close to the saturation concentration of $A$ in the solvent (predicted by the solubility of $A$ in the solvent or by Henry's law). In other words,

$$K_{mb}C_A = y_A/H \quad ... (7.6.76)$$

$$= y_A S \quad ... (7.6.77)$$

where $y_A$ = mole fraction of $A$ in the continuous phase

$H$ = Henry's law constant, m$^3$/kmole

$S$ = solubility of $A$ in the solvent, kmole/m$^3$

The above expression for facilitation factor ($F$) is based on the assumption that the reaction between the solute ($A$) and the carrier ($C$) is reversible and may be represented as

$$A + C \underset{K}{\rightleftharpoons} AC \quad ... (7.6.78)$$

Also, the equilibrium constant (K) for the above reaction is defined as

$$K = (k_1/k_2) \quad ... (7.6.79)$$

where $k_1$, $k_2$ = reaction rate constant for forward reaction [in m$^3$/(kmole·s)] and that for backward reaction (in s$^{-1}$) respectively.

It is important to keep in mind that the above definition of $K$ is valid only if the rate equations for the forward reaction and that for the backward reaction are closely related to the reaction stoichiometry and can be derived directly from the stoichiometric equation. In other words, it has been assumed that the forward reaction is first order in $A$ and first order in $C$ (the overall order being 2.0) and the backward reaction is first order in $AC$. This, in fact, is a possibility and need not be true always. Often, the rate equation for the reaction between $A$ and $C$ (forming the complex $AC$) is highly nonlinear and maintains no direct relation with the reaction stoichiometry.

Many modifications of above correlation are available. For instance, if the reaction between $A$ and $C$ is quite fast and easily attains equilibrium (which is so on many occasions), then $(\tanh \lambda/\lambda)$ shall be very close to zero and the correlation for $F$ reduces to

$$F = \frac{(1+\phi)(1+1/Sh)}{1+(1+\phi)(1+1/Sh)} \qquad \text{... (7.6.80)}$$

If the Sherwood number is quite large (that is, external mass transfer resistance, namely resistance to mass transfer in continuous phase and that in receiving phase, is negligible), then the expression for $F$ gets simplified further to

$$F = (1+\phi) = 1 + \left(\frac{\alpha K K_{mb} C_A}{1 + K K_{mb} C_A}\right) \qquad \text{... (7.6.81)}$$

$$= 1 + \left[\frac{D_{AC} K C_T}{D_{AM}\,(1 + K K_{mb} C_A)}\right] \qquad \text{... (7.6.82)}$$

Once the significance of $F$ is properly understood, we may now proceed to discuss mathematical simulation of carrier-facilitated transport. Let us consider an SLM or ILM (carrier-facilitated transport in an ELM can also be simulated following a similar procedure). As stated earlier, in case of an SLM or ILM, we must specify the membrane configuration employed as well, the most popular configurations being the spiral-wound type and the hollow fiber type. For a simplified (no doubt, approximate) analysis, let us assume that the module is composed of two parallel flat polymeric membranes each of width $W$, set at a distance $2b$ apart and each having been impregnated with the carrier (see Fig. 7.24). The feed solution flows between the two sheets and the receiving phase flows outside the sheets through the annular space between the sheets and the outer shell wall.

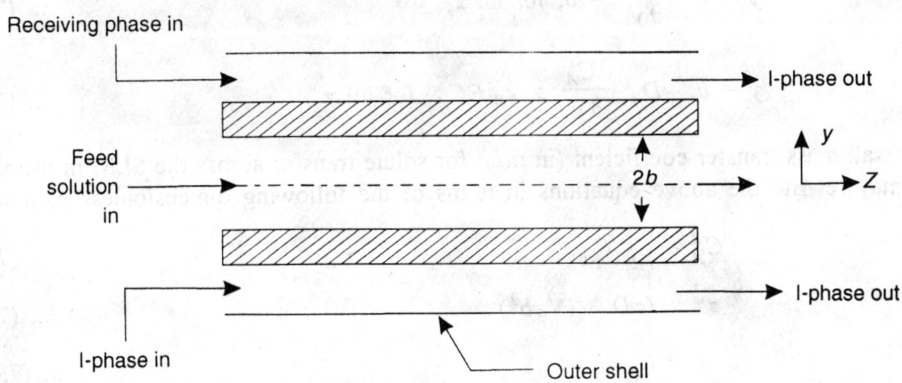


*Figure 7.24: Carrier-facilitated transport across parallel sheet membranes*

Let us further make the following assumptions:

1. The feed solution executes fully developed laminar flow. Therefore, the velocity profile is parabolic and may be presented as

$$\mathbf{V}_z = (3/2)\ \mathbf{V}_0\ [(1 - (y/b)^2] \qquad \text{... (7.6.83)}$$

where $\mathbf{V}_0$ = average velocity of feed solution

$$= Q_f / W\ (2b) \qquad \text{... (7.6.84)}$$

Readers are encouraged to derive the above equation from the equation of continuity (Eq. A of Table 2.6) and equation of motion (Eq. F of Table 2.7) given in Chapter 2 of Volume 1.

2. The flow rate of receiving phase is much larger than that of the feed solution. As a result, whatever solute is rejected into the receiving phase is immediately carried off and we may assume, with allowable error, that the solute concentration in the receiving phase is zero at any instant.
3. The reaction between solute ($A$) and carrier ($C$) is as per Eq. (7.6.78) and is assumed to be sufficiently fast so as to attain equilibrium within a short interval of time. It is further assumed that diffusivity of carrier in membrane phase ($D_{CM}$) and that of solute-carrier complex ($D_{AC}$) are more or less equal. Accordingly, the facilitation factor (equilibrium facilitation factor, to be more precise) is defined as per Eq. (7.6.82), with $D_{AC}$ having been replaced by $D_{CM}$.
4. The reaction between the complex $AC$ and $B$ (reactive component of receiving phase) is practically instantaneous and is not rate-controlling.

The mass transport equation for solute transport can be now written as

$$(3/2)\ \mathbf{V}_0\ [1 - (y/b)^2]\ \frac{\partial C_A}{\partial z} = D_A \frac{\partial^2 C_A}{\partial y^2} \qquad \text{... (7.6.85)}$$

where $D_A$ = diffusivity of solute ($A$) in the continuous phase (feed solution/feed gas)

The boundary conditions governing the system are,

**B.C.1:** At $\quad z = 0,\ C_A = C_{A0}$, for all $y$ ... (7.6.86)

**B.C.2:** At $\quad y = 0,\ \dfrac{\partial C_A}{\partial y} = 0$, for all $z$ ... (7.6.87)

**B.C.3:** At $\quad y = b,\ -D_A \dfrac{\partial C_A}{\partial y} = k_W F C_A$, for all $z$ ... (7.6.88)

where $k_W$ is the wall mass transfer coefficient (in m/s) for solute transfer across the SLM in the absence of carrier. We may rewrite the above equations in terms of the following dimensionless parameters:

$$C_A^* = C_A / C_{A0} \qquad \text{... (7.6.89)}$$

$$z^* = (z D_A)/(\mathbf{V}_0 b^2) \qquad \text{... (7.6.90)}$$

$$\xi = (y/b) \qquad \text{... (7.6.90a)}$$

$$\alpha_4 = (D_{CM} C_T K / D_{AM}) \qquad \text{... (7.6.91)}$$

$$\alpha_5 = (K\,K_{mb}\,C_{A0}) \qquad ...\ (7.6.92)$$

$$Sh_W = (k_W b / D_A) \qquad ...\ (7.6.93)$$

To note that the parameter $\xi$ defined above is different from that defined in Eq. (7.6.11). Also, $\alpha_4$ and $\alpha_5$ are related to the mobility ratio ($\alpha$) as

$$\alpha = (\alpha_4 / \alpha_5) / C_A^* \qquad ...\ (7.6.94)$$

Equations (7.6.85) to (7.6.88), therefore, become

$$(3/2)\,(1 - \xi^2)\,\frac{\partial C_A^*}{\partial z^*} = \frac{\partial^2 C_A^*}{\partial \xi^2} \qquad ...\ (7.6.95)$$

**B.C.1:** At $\quad z^* = 0,\ C_A^* = 1$, for all $\xi$ ... (7.6.96)

**B.C.2:** At $\quad \xi = 0,\ \dfrac{\partial C_A^*}{\partial \xi} = 0$, for all $z^*$ ... (7.6.97)

**B.C.3:** At $\quad \xi = 1,\ -\dfrac{\partial C_A^*}{\partial \xi} = Sh_W F C_A^*$, for all $z^*$ ... (7.6.98)

The facilitation factor ($F$) can also be expressed in terms of the dimensionless variables as

$$F = 1 + \left(\frac{\alpha_4}{1 + \alpha_5 C_A^*}\right) \qquad ...\ (7.6.99)$$

Since the third boundary condition (B.C.3) is nonlinear, it is hardly possible to obtain an analytical solution to the above partial differential equation. We can, however, solve the above equations numerically, by first discretising them and converting to difference equations and then using any of the well-established numerical techniques such as successive overrelaxation (SOR) method, Crank-Nicholson algorithm, Galerkin method, etc. Stroeve and Kim[32] have solved the above equations using finite difference Galerkin Scheme and have reported the results graphically. For the popular case of wall Sherwood number ($Sh_W$) = 1.0, their graphical data can be fitted into a correlation of the following form with allowable error.

$$C_A^*\,(\text{avg}) = a_0 + a_1 z^* + a_2\,(z^*)^2 + a_3\,(z^*)^3 \qquad ...\ (7.6.100)$$

where $C_A^*$ (avg) is the dimensionless average solute concentration of feed solution (namely, the *cup-mixing* concentration) at any $z^*$ and is defined as

$$C_A^*\,(\text{avg}) = \frac{\int_{-1}^{+1} C_A^*\left(1 - \xi^2\right) d\xi}{\int_{-1}^{+1} \left(1 - \xi^2\right) d\xi} \qquad ...\ (7.6.101)$$

The values of correlation constants $a_0$, $a_1$, $a_2$ and $a_3$ against a few typical values of $\alpha_4$ and $\alpha_5$ are listed below in Table 7.8. It is observed that $C_A^*$ is not a strong function of $\alpha_5$, but exhibits a fairly strong dependency on the magnitude of $\alpha_4$.

## Table 7.8
## Values of Correlation Constants of Eq. (7.6.100) Wall Sherwood Number ($Sh_W$) = 1.0

| $\alpha_4$ | $\alpha_5$ | $z^*$ | $a_0$ | $a_1$ | $a_2$ | $a_3$ |
|---|---|---|---|---|---|---|
| 5.0 | 15.0 | $0.1 \le z^* \le 0.39$ | 0.9763 | −0.6993 | −0.02341 | 0.0 |
| | | $0.39 \le z^* \le 1.3$ | 1.0524 | −1.1221 | 0.629 | −0.17666 |
| 10.0 | 15.0 | $0.18 \le z^* \le 0.73$ | 0.954 | −0.8907 | 0.18045 | 0.0 |
| | | $0.73 \le z^* \le 1.62$ | 1.1832 | −1.6547 | 0.952 | −0.212347 |
| 15.0 | 5.0 | $0.065 \le z^* \le 0.283$ | 1.1847 | −5.97173 | 27.6823 | −49.12468 |
| | | $0.283 \le z^* \le 1.39$ | 0.88928 | −1.1604 | 0.49877 | −0.0521 |

Narayanan[25] has reported a more rigorous analysis of carrier-facilitated SLM permeation process, which does not assume zero solute concentration in the receiving phase (I phase) and which considers a hollow fiber module. The receiving phase flows through the fibers and the feed solution forms the shellside fluid. Assuming that the shellside fluid executes true laminar countercurrent flow, the mass transport equation is written as

$$\left(\frac{2V_0}{N-\lambda}\right)[1-(r/R)^2+\lambda \ln (r/R)] = (D_A/r)\frac{\partial}{\partial r}\frac{\partial C_A}{\partial r} \quad ... (7.6.102)$$

where $V_0$ = average velocity of shellside fluid (feed solution)

$$= Q_f/[\pi (R^2 - R_0^2)] \quad ... (7.6.103)$$

$R$ = inside radius of outer shell

$R_0$ = outside radius of hollow fiber

$$\lambda = \frac{(R_0/R)^2 - 1}{\ln (R_0/R)} \quad ... (7.6.104)$$

$$N = 1 + (R_0/R)^2 \quad ... (7.6.105)$$

The boundary conditions governing the system are

**B.C.1:** At $z = 0,\ C_A = C_{A0}$, for all $r$ ... (7.6.106)

**B.C.2:** At $r = R,\ \dfrac{\partial C_A}{\partial r} = 0$ ... (7.6.107)

**B.C.3:** At $r = R_0,\ -D_A \dfrac{\partial C_A}{\partial r} = k_W F (C_A - C_{Ar})$ ... (7.6.108)

The expression for facilitation factor ($F$) given in Eq. (7.6.82) has been retained (this, in fact, is an approximation).

Similarly, the mass transport equation for the receiving phase (assuming true laminar flow) is

$$2U_0\,[1-(r/r_0)^2]\,\frac{\partial C_{Ar}}{\partial z}=(D_{Ar}/r)\,\frac{\partial}{\partial r}\left(r\frac{\partial C_A}{\partial r}\right) \qquad \text{... (7.6.109)}$$

where $U_0$ = average velocity of receiving phase

$$= Q_r/[\pi\,(r_0)^2] \qquad \text{... (7.6.110)}$$

$r_0$ = inside radius of hollow fiber

The boundary conditions are

**B.C.1:** At $z = 0,\ C_{Ar} = 0$, for all $r$ ... (7.6.111)

**B.C.2:** At $r = 0,\ \dfrac{\partial C_{Ar}}{\partial r} = 0$ ... (7.6.112)

**B.C.3 :** At $r = r_0,\ -D_{Ar}\dfrac{\partial C_{Ar}}{\partial r} = k_W F\,(C_A - C_{Ar})$ ... (7.6.113)

To note that the transport equations for shellside flow and those for tubeside flow are coupled with each other. These equations (after rewriting in dimensionless form) have been solved simultaneously using line successive over-relaxation method (the over-relaxation factor, on which the stability and speed of convergence of the scheme critically depend, is finalised by trial and is set at 0.33). The procedure, though iterative, is reportedly fast converging. The results have been reported graphically[25]. The cup-mixing concentrations at any $z$ are obtained by area-averaging the $C_A$-values or $C_{Ar}$-values as shown below.

$$C_A\,(\text{avg})=\frac{\int_{R0}^{R} C_A\left[1-(r/R)^2+\lambda\ln(r/R)\right]r\,dr}{\int_{R0}^{R}\left[1-(r/R)^2+\lambda\ln(r/R)\right]r\,dr} \qquad \text{... (7.6.114)}$$

$$C_{Ar}\,(\text{avg})=\frac{\int_{0}^{r0} C_{Ar}\left[1-(r/r_0)^2\right]r\,dr}{\int_{0}^{r0}\left[1-(r/r_0)^2\right]r\,dr} \qquad \text{... (7.6.115)}$$

Narayanan[25] reports that data computed from the above-discussed mathematical model agree with experimentally determined values with a maximum deviation of 16 percent.

**Example 7.7:** A process gas containing 10 percent (by volume) hydrogen sulfide is to be desulfurised by SLM permeation process. The gas is admitted at 2 atm, 298 K and at the rate of 3.5 m$^3$/hr and its hydrogen sulfide content is to be reduced to 0.001 kmoles/m$^3$. Assume that the membrane module is equivalent to two parallel sheets (each impregnated with the carrier) 1 cm apart and each of width 50 cm. Aqueous solution of ethylene diamine is used as the carrier and helium gas is used as the sweep gas. The total carrier content of the membrane is 2.84 kmoles/m$^3$. Determine the membrane area required. The following data are available.

Solubility of $H_2S$ in water at 298 K = 0.0846 kmole/m$^3$

Equilibrium constant for solute-carrier reaction = 586.3 m$^3$/kmole

Mass transfer coefficient for solute transfer across the membrane (in absence of carrier) = 0.0045 m/s

Take diffusivity of solute in membrane phase as $2.8 \times 10^{-10}$ m$^2$/s and that in feed gas as $2.25 \times 10^{-5}$ m$^2$/s. Also diffusivity of carrier in membrane = that of solute-carrier complex in membrane = $2.52 \times 10^{-12}$ m$^2$/s.

**Solution:** Let us first compute the three influencing parameters $\alpha_4$, $\alpha_5$ and the wall Sherwood number, $Sh_W$. Thus

$$Sh_W = k_W b / D_A = (0.0045)\,(0.005)/(2.25 \times 10^{-5}) = 1.0$$

$$\alpha_4 = D_{CM} C_T K / D_{AM} = (2.52 \times 10^{-12})\,(2.84)\,(586.3)/(2.8 \times 10^{-10})$$
$$= 14.9858 \simeq 15.0$$

$$\alpha_5 = K\,K_{mb}\,C_{A0}$$

Taking $(K_{mb}\,C_{A0})$ = saturation concentration of hydrogen sulfide in water = $(y_{A0}\,S)$,

$$\alpha_5 = K\,y_{A0}\,S = (586.3)\,(0.1)\,(0.0846) = 4.96 \simeq 5.0$$

We can, therefore, use Eq. (7.6.100) that is based on Stroeve and Kim's graphical data[32]. Now, assuming ideal gas behaviour,

$$C_{A0} = (P\,y_{A0})/(RT) = \frac{(0.1)\,(2.0)}{(0.082)\,(298)} = 0.00818 \text{ kmoles/m}^3$$

Therefore, $C^*_{Ae} = (0.001)/(0.00818) = 0.122$

The value of $z^*$ can be now computed only by trial. Thus, let us first assume that the value of $z^*$ falls in the range of 0.283 to 1.39. Then, with reference to Table 7.8,

$$C^*_{Ae} = 0.122 = 0.88928 - 1.1604 z^* + 0.49877\,(z^*)^2 - 0.0521\,(z^*)^3$$

Solving by trial, $z^* = 1.22$

This falls within the range assumed by us at the outset. Hence, no further trial is necessary. Now, from Eq. (7.6.90),

$$z = L = (z^* v_0 b^2)/D_A = \frac{(z^* Q_f b^2)}{W\,(2b)\,D_A} = \frac{(z^* Q_f b)}{(2W\,D_A)}$$

$$= \frac{(1.22)\,(3.5/3600)\,(0.005)}{2\,(0.5)\,(2.25 \times 10^{-5})} = 0.2636 \text{ m}$$

Membrane area required = $2\,(WL) = (2.0)\,(0.5)\,(0.2636)$ = **0.2636 m$^2$**

To note that at the specified flow rate of 3.5 m$^3$/hr, the feed gas would execute laminar flow. Also, we have assumed that the flow rate of sweep gas is quite large that the permeated $H_2S$ is swept off almost instantaneously. For handling 3.5 m$^3$ gas per hour, a single flow channel made of two parallel membranes (SLMs) each of area 0.1318 m$^2$ shall be thus sufficient. For high capacity installations, the permeator/extractor is to consist of a large number of such flow channels in parallel.

**Example 7.8:** In the system considered in Example 7.7, determine the value of facilitation factor ($F$) at the feed inlet. Compare this value with that predicted by Eq. (7.6.68). Comment on the results.

Take membrane thickness = 0.02 cm, $k_C$ = 1.57 m/s, $k_L$ = 1.77 m/s and the forward reaction rate constant $(k_1)$ = 1.3 × $10^{11}$ m³/(kmole.s).

**Solution:** Since at the feed inlet $C_A = C_{A0}$, $C_A^* = 1.0$. Therefore, Eq. (7.6.99) reduces to

$$F = 1 + \left(\frac{\alpha_4}{1+\alpha_5}\right) = 1 + 15/6 = 3.5$$

To compute $F$ from Eq. (7.6.68), we have to first evaluate the parameters $\in$, $\alpha$, $\lambda$. Thus

$$k_2 = (k_1/K) = (1.3 \times 10^{11})/(586.3) = 2.2173 \times 10^8 \text{ s}^{-1}$$

Now, from Eq. (7.6.73), $\quad \in = D_{AC}/(k_2\delta^2) = \dfrac{\left(2.52\times10^{-12}\right)}{\left(2.2173\times10^8\right)(0.0002)^2} = 2.84 \times 10^{-13}$

From Eq. (7.6.94), the mobility ratio $(\alpha)$ is,

$$\alpha = \frac{(15)}{(5)(1.0)} = 3.0$$

Also, $\quad K_{mb}C_A = K_{mb}C_{A0} = y_{A0}S = (0.1)(0.0846) = 0.00846 \text{ kmoles/m}^3$

Therefore,

$$\lambda \text{ [from Eq. (7.6.72)]} = 1.7544 \times 10^6$$

$$(\tanh\lambda)/\lambda = 0.57 \times 10^{-6}$$

It can be seen that the magnitude of $(\tanh\lambda/\lambda)$ is very low and the assumption made in Example 7.7 is fairly accurate. Now,

$$1/k = 1/k_C + 1/k_L$$

Since $\quad k_C = 1.57$ m/s and $k_L = 1.77$ m/s,

$$k = 0.832 \text{ m/s}$$

Therefore, from Eq. (7.6.71),

$$\text{Sherwood number } (Sh) = (0.832)(0.0002)/(2.8 \times 10^{-10}) = 5.9428 \times 10^5$$

The magnitude of Sherwood number is also substantially large and the assumption made in Example 7.7 that the external mass transfer resistance is negligible is also fairly true. Now, from Eq. (7.6.69),

$$\phi = \frac{(3.0)(586.3)(0.00846)}{1+(586.3)(0.00846)} = 2.5$$

Substituting the values of $\phi$, $Sh$ and $\lambda$ in Eq. (7.6.68), we get

$$F = 3.5$$

It can be seen that this value of $F$ is same as that computed from Eq. (7.6.99). This shows that the assumptions that solute-carrier reaction is fairly fast and attains equilibrium within a short interval of time and that external mass transfer resistances are negligible are reasonably accurate in the present case.

## 7.7 MEMBRANE REACTORS

The concept of membrane reactors got developed around two decades ago. The main idea was to prevent the reaction mixture from attaining the equilibrium composition by incorporating a membrane within the reactor, the membrane being used for the continuous and selective removal of products from the reaction zone. We know that the maximum conversion that can be attained in a reversible reaction is the equilibrium conversion. However, if one or all of the products can be continuously removed from the reaction mixture as soon as they are produced, then the backward reaction gets inhibited and the overall reaction can be conducted in the forward direction, theoretically upto 100 percent conversion. This is the principle that is employed in membrane reactors.

Let us consider an example. The dehydrogenation of cyclohexane in presence of platinum or alumina as the catalyst if conducted in a packed bed reactor (at 215° C), the maximum conversion that can be obtained is 33 percent[23]. If hydrogen is continuously removed from the reaction mixture through a glass membrane, then the equilibrium is displaced towards the product side and the conversion has been found to increase to 80 percent. Thus, both reaction and product separation are conducted in the same reaction vessel and as a result, the product yield can be increased to beyond that is permitted by the reaction equilibrium. Since the yield or conversion is increased in this way, it will be possible to employ a lower operating temperature to achieve the desired conversion. Reduced reaction temperature helps in minimising side reactions and in the case of catalytic reactors, this also minimises catalyst fouling. The useful life of the catalyst is thus significantly increased. Higher product selectivity and short contact time are additional advantages of membrane reactors.

Thus, a large number of liquid and gas phase reactions that otherwise involve lower conversions due to unfavourable thermodynamics (that is, equilibrium conversion being too low) or due to undesirable side reactions can be successfully conducted in membrane reactors. Examples are hydrogenation and dehydrogenation reactions, decomposition reactions, hydration and dehydration, oxidation reactions, hydrodealkylation, etc.

The membrane may be made of glass, ceramic materials, metals and metal alloys, metallic oxides (alumina, zirconia, iron oxide) or molecular sieves. The membrane can perform different functions in the reactor. In some applications, the membrane is used as a catalyst support and in some cases, the membrane itself acts as the catalyst. It is called the catalytic membrane.

When the membrane performs only separation and has no catalytic activity, then the most important operating parameters are the relative permeabilities (also called permselectivities) of the reactants and the products and also the separation factor ($\in$). The permselectivity ($S_m$) of any component in the reaction mixture is defined as

$$S_m = \frac{\left(\begin{array}{c}\text{Permeation rate of } i\text{th component}\\ \text{through the membrane}\end{array}\right)}{\left(\begin{array}{c}\text{Permeation rate of fastest permeating}\\ \text{component through the membrane}\end{array}\right)} \quad \text{... (7.7.1)}$$

Evidently, the permselectivity of the fastest permeating component is 1.0.

The separation factor ($\in$) is defined as

$$\in = \frac{\begin{pmatrix}\text{Permeation rate of fastest} \\ \text{permeating component}\end{pmatrix}}{\begin{pmatrix}\text{Reaction rate with respect to the} \\ \text{fastest permeating component}\end{pmatrix}} \quad \text{... (7.7.2)}$$

A good membrane is one that is permeable to the products, but impermeable to the reactants. In the case of dense membranes which are permeable only to one component, the permselectivities of all other components are zero and the extent of equilibrium shift is then determined only by the separation factor (∈). And, if ∈ is very close to unity, then it will be possible to conduct the reaction upto 100 percent completion.

Gas-liquid reactions can be conveniently and more effectively conducted in membrane reactors[24]. We know that one of the major problems connected with gas-liquid catalytic reactions is that the gaseous reactant is often only sparingly soluble in liquid and as a result, the liquid film mass transfer resistance becomes too large and thereby tends to inhibit the overall rate of the process. Interparticle and intraparticle mass transport resistances further inhibit the reaction rate. In such cases, a membrane reactor, which is nothing but a catalytically impregnated ceramic tube, can be conveniently used for conducting the reaction. The gas flows through the tube core and the liquid flows outside the tube in the shell and fills the pores in the tube wall within which the catalytic reaction occurs. The main advantage of this type of design is that the gaseous reactant is supplied directly to the catalyst pores from the gas bulk, thereby avoiding the liquid film mass transfer resistance. Such an arrangement also permits easier and more efficient dissipation of heat and temperature control.

Membrane reactors can also be effectively used for conducting two separate reactions simultaneously. For example, the products that selectively permeate through the membrane of one reactor are made to enter the second reactor situated on the lower pressure side of the first membrane where they are used as reactants to conduct the second reaction. The two reactors are thus coupled to handle two reactions in one unit operation. In cases like this, the membrane performs three-fold function. It serves as a chemical separator, as a partition to prevent the contents of both reactors from intermixing and also as heat transfer medium for the two coupled reactions when one is endothermic and the other is exothermic.

One of the most exciting concepts introduced in the last decade is that of the membrane fermentor. So are the membrane enzymatic reactors which have practically changed the entire economic outlook for enzyme-catalysed reactions. The specific merits of these two have already been outlined in Section 7.2 under Ultrafiltration.

The potential of membrane reactors is thus enormous in the present context. Advanced research on design and simulation of such reactors and development of ultrathin, permselective, high temperature membranes obviously demand attention.

## 7.8 COMPUTER PROGRAMS

The aid of fast computers is particularly advantageous for the process design and simulation of membrane-based processes. Let us first consider a reverse osmosis process employing a tubular membrane module. What is required to be computed is the required membrane area ($A_m$) for a specified product recovery ($R$). We must also evaluate the product quality ($Cp$) and the power consumption. The computer program is given below.

## PROGRAM 7.1

```
C**   SPECIFY THE FEED CONCENTRATION FEED FLOW RATE, TRANS-
C**   MEMBRANE PRESSURE DROP AND PRODUCT
C**   RECOVERY DESIRED.
      REAL L, KM, MEU
      READ (*, *) CFO, FO, DP, RT
      DIMENSION AK (100), ALM (100), CP (100), R (100)
      DIMENSION CF (100), PP (100), Q (100), XD (100), V (100) VW (100)
      N = 100
C**   SPECIFY DIAMETER OF TUBULAR MEMBRANE,
C**   MEMBRANE PERMEABILITY COEFFICIENT,
C**   SOLUTE TRANSPORT PARAMETER AND MOLAR
C**   DENSITY OF SOLUTION (ASSUMED CONSTANT). SPE-
C**   CIFY ALSO THE DIFFUSIVITY OF SOLUTE IN
C**   SOLUTION.
      READ (*, *) D, KM, STP, C, DAS
C**   ASSUMED THAT OSMOTIC PRESSURE IS PROPO-
C**   RTIONAL TO MOLE FRACTION OF SOLUTE.
C**   SPECIFY THE PROPORTIONALITY CONSTANT B.
      READ (*, *) B
      US = KM * DP/C
      TH = STP/US
      PHI = B/(C * DP)
C**   ASSUME THE VALUE OF (L/D) RATIO.
C**   FOR EXAMPLE, LET (L/D) = 10.0
      RATIO = 10.0
      L = D * RATIO
10    DZ = L/N
      PI = 22.0/7.0
      VO = 4.0 * FO/(PI * D ** 2)
      XD (1) = 2.0 * US * DZ/(D * VO)
C**   SPECIFY AVERAGE VALUES OF SOLUTION
C**   DENSITY AND SOLUTION VISCOSITY.
      READ (*, *) ROW, MEU
      DO 15 I = 1, N
      AFO = 2.0 * DAS/(US * D)
      BT = XD (I)/(4.0 * AFO **2)
C**   ASSUME A VALUE OF CP (I). FOR EXAMPLE,
C**   LET CP (I) = 0.00001.
```

```
      CP (I) = 0.00001
20    Y = CP (I)
      R (I) = 2.0 * TH * XD (I)/[TH + CP (I) * PHI]
      IF (I. EQ. 1) V (I) = VO * [1.0 - R (I)]
      IF (I. GT. 1) V (I) = V (I - 1) * [1.0 - R (I)]
      IF (I. EQ. 1) RE = D * VO * ROW/MEU
      IF (I. GT. 1) RE = D * V (I - 1) * ROW/MEU
C**   COMPUTE MASS TRANSFER COEFFICIENT.
      IF (RE. LE. 2100) THEN
      AN = ALOG (1.0 + 0.25/AFO ** 2)
      BAS = (0.65 * AN) ** 3
      BB = (BT - BAS)/BT
      IF (BT. GE. BAS) SH = (3.9 * BAS ** 0.667)/(AFO * BT) + (4.0 * BB)/
      (AFO * AN)
      IF (BT. LT. BAS) SH = 1.95/(AFO * BT ** 0.333)
      ELSE
      SH = (0.276 * RE ** 0.25)/(AFO * BT ** 0.333)
      END IF
      AK (I) = SH * DAS/D
      ALM (I) = AK (I)/(US * TH)
      PTC = CP (I) * PHI + TH
      CF (I) = CP (I) + [CP (I)/PTC]/EXP {1.0/[ALM (I) * PTC]}
      Q (I) = CF (I)/CP (I)
      PP (I) = R (I) + Q (I) * [1.0 - R (I)]
      IF (I. EQ. 1) CP (I) = CFO/PP (I)
      IF (I. GT. 1) CP (I) = CF (I - 1)/PP (I)
      DIF = ABS [Y - CP (I)]/Y
      IF (DIF. GE. 0.001) GO TO 20
      IF (I. EQ. N) GO TO 15
      XD (I + 1) = XD (I)/[1.0 - R (I)]
15    CONTINUE
C**   COMPUTE THE PRODUCT RECOVERY.
      S = 1.0
      DO 25 I = 1, N
      S = S * [1.0 - R (I)]
25    CONTINUE
      RC = 1.0 - S
C**   CHECK THE COMPUTED VALUE OF PRODUCT RECOVERY AGAINST THE SPECIFIED
C**   VALUE.
```

```
      IF (RC. LT. RT) THEN
      L = L + 0.01
      GO TO 10
      ELSE
C**   COMPUTE THE CONCENTRATIONS OF PERMEATE AND RETENTATE.
      S = 1.0
      DO 30 I = 1, N
30    S = S * Q (I)/PP (I)
      CB = S * CFO
      PPT = RC/[1.0 - S * (1.0 - RC)]
      CPT = CFO/PPT
      END IF
C**   COMPUTE MEMBRANE AREA AND PERMEATE RATE.
      AM = PI * D * L
      P = FO * RC
C**   COMPUTE POWER CONSUMPTION.
      IF (RE. LE. 2100) THEN
      ES = 0.0
      DO 35 I = 1, N - 1
      EP = 1.0
      DO 40 J = 1, I
40    EP = EP * [1.0 - R (J)] ** 2
      ES = ES + EP
35    CONTINUE
      E = 1.0 + ES
      PO = 8.0 * PI * MEU * DZ * E * VO * VO
      ELSE
      ES = 0.0
      DO 45 I = 1, N - 1
      EP = 1.0
      DO 50 J = 1, I
50    EP = EP * [1.0 - R (J)] ** 2.84
      ES = ES + EP
45    CONTINUE
      E = 1.0 + ES
      PO = 0.02 * PI * DZ * ROW * D * E * VO ** 3/(RE ** 0.16)
      END IF
      WRITE (*, *) L, AM, RC, CPT, CB, PO
      STOP
      END
```

It may be noted that the two correlations proposed by Kimura and Sourirajan[5] for the estimation of the mass transfer coefficient ($k_i$) are respectively for laminar flow ($Re < 2100$) and well-developed turbulent flow ($Re > 10000$). In the above program however, the correlation for turbulent flow has been used for all values of $Re$ in excess of 2100, which in fact is an approximation. The same approximation is involved in the computation of power consumption as well. Nevertheless, as reported by Kimura and Sourirajan, such an approximation is not expected to cause more than 10 percent error.

Also, confusion could arise in choosing the first approximation of $C_{pi}$ for each segment. Since the product water is expected to contain very little dissolved salts, we have used a first approximation of $C_{pi} = 0.00001$ kmoles/m$^3$ in the above program. This is very close to $C_{pi} = 1$ ppm for aqueous sodium chloride solutions. By assuming $C_{pi}$ more appropriately, the number of iterations can be very well reduced.

Let us now consider the performance analysis of a hollow fiber artificial kidney that employs haemodiafiltration. The performance Eqs are (7.2.13) to (7.2.18) which are to be solved numerically by any of the popular methods available. Let us employ fourth order Runga-Kutta method here. The method is outlined below.

Let $F^* = X$, $p^* = Y$ and $z^* = x$. Then, Eqs (7.2.13) and (7.2.14) get modified into

$$\frac{dX}{dx} = -2\epsilon\ [Y - \pi^*(X)] = f_1(x, X, Y) \qquad \text{... (7.8.1)}$$

$$\frac{dY}{dx} = -8X = f_2(x, X, Y) \qquad \text{... (7.8.2)}$$

To note that $f_1(x, X, Y)$ and $f_2(x, X, Y)$ denote two functions in $x$, $X$ and $Y$. Also $\pi^*$ is a function of $C^*$ and $C^*$ is a function of $F^*$ or $X$. Therefore, $\pi^*$ is a function of $X$ and that is why we have denoted it as $\pi^*(X)$. Now, compute $k_1$ to $k_4$ and $t_1$ to $t_4$ as follows.

Let the step size be 0.01. Thus, $h = 0.01$.

$$k_1 = hf_1(x_i, X_i, Y_i) = -2\pi\ h\ [Y_i - \pi^*(X_i)] \qquad \text{... (7.8.3)}$$

$$t_1 = hf_2(x_i, X_i, Y_i) = -8h\ X_i \qquad \text{... (7.8.4)}$$

$$k_2 = hf_1(x_i + h/2, X_i + k_1/2, Y_i + t_1/2)$$
$$= -2\epsilon\ h\ [Y_i + t_1/2 - \pi^*(X_i + k_1/2)] \qquad \text{... (7.8.5)}$$

$$t_2 = hf_2(x_i + h/2, X_i + k_1/2, Y_i + t_1/2)$$
$$= -8h\ (X_i + k_1/2) \qquad \text{... (7.8.6)}$$

$$k_3 = hf_1(x_i + h/2, X_i + k_2/2, Y_i + t_2/2)$$
$$= -2\epsilon\ h\ [Y_i + t_2/2 - \pi^*(X_i + k_2/2)] \qquad \text{... (7.8.7)}$$

$$t_3 = hf_2(x_i + h/2, X_i + k_2/2, Y_i + t_2/2)$$
$$= -8h\ (X_i + k_2/2) \qquad \text{... (7.8.8)}$$

$$k_4 = h\ f_1(x_i + h, X_i + k_3, Y_i + t_3)$$
$$= -2\epsilon\ h\ [Y_i + t_3 - \pi^*(X_i + k_3)] \qquad \text{... (7.8.9)}$$

$$t_4 = h\, f_2\, (x_i + h,\ X_i + k_3,\ Y_i + t_3)$$

$$= -\,8h\, (X_i + k_3) \qquad \text{... (7.8.10)}$$

$$k = (1/6)\, (k_1 + 2k_2 + 2k_3 + k_4) \qquad \text{... (7.8.11)}$$

$$t = (1/6)\, (t_1 + 2t_2 + 2t_3 + t_4) \qquad \text{... (7.8.12)}$$

Once $k$ and $t$ are computed, the values of $Y_{i+1}$ and $X_{i+1}$ can be obtained as

$$X_{i+1} = X_i + k \qquad \text{... (7.8.13)}$$

$$Y_{i+1} = Y_i + t \qquad \text{... (7.8.14)}$$

As per the initial condition [Eq. (7.2.18)], at $i = 1$, $X = X_1 = F_0^*$, $Y = Y_1 = 1.0$. ... (7.8.15)

Therefore, the values of $k$ and $t$ at $i = 1$ can be computed from which we get the values of $X_2$ and $Y_2$ (since $X_2 = X_1 + k$ and $Y_2 = Y_1 + t$). Now, using the values of $X_2$ and $Y_2$, $X_3$ and $Y_3$ are computed and the procedure is continued until we get the values of $X_{101}$ and $Y_{101}$, where $X_{101} = F_b^*$ and $Y_{101} = p_2^*$. The computer program is given below.

## PROGRAM 7.2

```
C**   PERFORMANCE ANALYSIS OF A HOLLOW
C**   FIBER ARTIFICIAL KIDNEY EMPLOYING
C**   HAEMODIAFILTRATION
      REAL L, KMP, K1, K2, K3, K4, K, N, MEU
      DIMENSION X (101), Y (101), DC (101), DPI (101)
C**   INITIAL VALUES THESE ARE TOTAL INLET
C**   FLOW RATE OF BLOOD (FOT), NUMBER OF
C**   FIBERS (N), DIAMETER AND LENGTH OF
C**   EACH FIBER (D, L) PROTEIN CONCENTRATION
C**   AND HAEMATOCRIT IN INLET BLOOD
C**   (CFO, HF), BLOOD VISCOSITY (MEU),
C**   MEMBRANE PERMEABILITY COEFFICIENT
C**   (KMP), PRESSURE AT BLOOD INLET (P1)
C**   AND HYDRAULIC PRESSURE OF DIAFILTRATE
C**   (PD).
      READ (*, *), FOT, D, L, N, CFO, HF, MEU, KMP, P1, PD
      FO = FOT/N
      PI = 22.0/7.0
      DFP1 = P1 - PD
C**   ASSUMED THAT COLLOID OSMOTIC PRESSURE
C**   VARIES WITH PROTEIN CONCENTRATION
C**   AS PER EQ. (7.2.5). SPECIFY
```

```
C**    THE VALUES OF THE EMPIRICAL CONSTANTS
C**    SUCH AS A1, A2, A3
       READ (*, *) A1, A2, A3
C**    COMPUTE THE DIMENSIONLESS VARIABLES.
       FR = [PI * DFP1 * (D/2.0) ** 4]/(MEU * L)
       DFO = FO/FR
       ABSN = 8.0 * KMP * MEU * L * L/(D ** 3
       DA1 = A1 * CFO/DFP1
       DA2 = A2 * CFO * CFO/DFP1
       DA3 = (A3 * CFO ** 3)/DFP1
C**    SPECIFY STEP SIZE.
       H = 0.01
C**    INITIAL CONDITIONS.
       Y (1) = 1.0
       X (1) = DFO
       DO 10 I = 1, 100
       DC (I) = DFO * (1.0 - HF)/[X (I) - HF * DFO]
       DPI (I) = DA1 * DC (I) + [DA2 * DC (I) ** 2] + [DA3 * DC (I) ** 3]
       K1 = -2.0 * ABSN * H * [Y (I) - DPI (I)]
       T1 = -8.0 * H * X (I)
       DC (I) = DFO * (1.0 - HF)/[X (I) + 0.5 * K1 - HF * DFO]
       DPI (I) = DA1 * DC (I) + [DA2 * DC (I) ** 2] + [DA3 * DC (I) ** 3]
       K2 = -2.0 * ABSN * H * [Y (I) + T1 * 0.5 - DPI (I)]
       T2 = -8.0 * H * [X (I) + 0.5 * K1)
       DC (I) = DFO * (1.0 - HF)/[X (I) + 0.5 * K2 - HF * DFO]
       DPI (I) = DA1 * DC (I) + [DA2 * DC (I) ** 2] + [DA3 * DC (I) ** 3]
       K3 = -2.0 * ABSN * H * [Y (I) + 0.5 * T2 - DPI (I)]
       T3 = -8.0 * H * [X (I) + 0.5 * K2]
       DC (I) = DFO * (1.0 - HF)/[X (I) + K3 - HF * DFO]
       DPI (I) = DA1 * DC (I) + [DA2 * DC (I) ** 2] + [DA3 * DC (I) ** 3]
       K4 = -2.0 * ABSN * H * [Y (I) + T3 - DPI (I)]
       T4 = -8.0 * H * [X (I) + K3]
       K = (K1 + 2.0 * K2 + 2.0 * K3 + K4)/6.0
       T = (T1 + 2.0 * T2 + 2.0 * T3 + T4)/6.0
       X (I + 1) = X (I) + K
       Y (I + 1) = Y (I) + T
10     CONTINUE
       DFB = X (101)
       DP2 = Y (101)
       PSI = DFO - DFB
```

```
DCB = DFO * (1.0 - HF)/(DFB - HF * DFO)
CB = DCB * CFO
P = PSI * FR
FB = DFB * FR
DFP2 = DP2 * DFP1
PT = P * N
FBT = FB * N
WRITE (*, *) CB, FBT, PT, DFP2
STOP
END
```

If the computed values of $C_b$ (protein concentration in outlet blood), $F_b$ (total), $P$ (total) are satisfactory, then the chosen fiber length is final. To note that $F_b$ (total), $P$ (total) and $C_b$ are denoted as FBT, PT and CB respectively in the program. If their values are not satisfactory, then the program may be re-executed using a larger value of $L$. Alternately, the total number of fibers ($N$) may be increased.

## THINGS TO REMEMBER

1. Reverse osmosis involves transfer of solvent through a semipermeable membrane from a more concentrated salt solution to a less concentrated or dilute solution when the transmembrane pressure drop is maintained higher than the osmotic pressure difference. The solution that is collected from the downstream side of the membrane (which will be essentially the solvent) is called the *permeate* and the concentrated solution discharged from the high pressure side of the membane is called the *retentate* or *brine.*
2. Commercially available RO membranes are cellulose acetate, aromatic polyamide (aramid) and thin film composite membranes. CA membranes are quite popular but they cannot be used with highly alkaline or acidic solutions since they tend to get hydrolysed back to cellulose. They also get easily degraded by biological attack and therefore, the feed solution is to be chlorinated prior to contacting these membranes. Aramid membranes and thin-film composites are not susceptible to biological attack and they resist hydrolysis. However, they are degraded by chlorine.
3. RO devices may be of plate and frame, spiral-wound, tubular or hollow fiber configurations. Among these, the hollow fiber modules provide largest membrane area per unit volume, but they are more prone to fouling and least amenable to mechanical cleaning. Plate and frame and tubular modules are more suitable for handling fouling streams since they are reasonably amenable to cleaning, but they provide less membrane area per unit volume.
4. The feed to any RO system must be preconditioned by removing all suspended particulates and colloidal matter. Chemical treatments are also to be given to prevent precipitation fouling (due to dissolved salts of calcium and magnesium) and biological fouling of membranes. The extent of pretreatment depends on the feed water composition, the chemical nature of membrane and the configuration of RO device.
5. To study the performance characteristics of an RO system, a segment to segment (or a section to section) analysis is required since the flow rate as well as the concentration of the salt solution

vary continuously as it flows along the membrane from the inlet end to the outlet and consequently, the rate of permeation of water (solvent) through the membrane also gradually decreases. An additional problem is that of *concentration polarisation.* Due to the accumulation of salt at the membrane surface, a concentrated boundary layer is formed there which decreases the effective pressure differential across the membrane (since osmotic pressure will be much higher in the boundary layer) and thereby decreases the permeate flux. Salt passage across the membrane also increases which diminishes solute rejection. If the concentration of salt in the boundary layer reaches saturation, salt will get precipitated on the membrane surface causing membrane scaling.

6. A reasonably accurate (though simplified) analysis (mainly due to Kimura and Sourirajan) of RO process is described through Eqs (7.1.7) to (7.1.46). This scheme, though fairly satisfactory for design purposes, assumes Van't Hoff's equation to be applicable (that is, osmotic pressure is proportional to the molar concentration), assumes negligible change in the density of the solution and that the solute transport parameter is independent of feed flow rate and feed concentration. The frictional pressure drop in the system may be estimated by using Fanning's equation as shown through Eqs (7.1.72) to (7.1.99).
7. Large capacity RO systems employ multistage operation like the one shown schematically in Fig. 7.3.
8. RO systems are much more energy efficient than many thermal processes like multiple effect evaporation and multistage distillation. They are quite compact requiring little labour and maintenance. They do not discharge any additional waste streams and they are well-suitable for handling heat-sensitive and explosive substances. Some of the major applications of RO are in the desalination of sea water and brackish water, production of ultrapure water for boilers, hospitals and for industries manufacturing electronic components, sewage treatment, treatment of plating wastes, concentration of black liquor in pulp and paper industry (in conjunction with ultrafiltration), concentration of raw milk prior to transportation and in the concentration of fruit juices.
9. The major limitation of RO is that it cannot be used economically at large salt concentrations since the osmotic pressure of solutions increases with increase in salt concentration and therefore, at large concentrations, the transmembrane pressure drop required to affect separation will be very large. Also, at the present status, the RO membranes do not have a useful life span exceeding two years.
10. Ultrafiltration is very similar to reverse osmosis except that UF membranes have larger pore sizes and they retain only the high molecular weight solutes. The solvent and low molecular weight solutes permeate through the membrane. Since high molecular weight solutes have low osmotic pressures, the operating transmembrane pressure drop for UF is much lower than that for RO.
11. UF devices are of the same configuration as RO devices. These are plate and frame, spiral-wound, tubular and hollow fiber modules. An alternate type of UF device is the thin-channel tubular module in which the narrow flow cross-section and high fluid velocity minimise the effect of concentration polarisation. Use of turbulent promoters, insertion of static mixers and the use of constricted tubular modules have been recommended to reduce the thickness of the concentrated boundary layer (gel layer) at the membrane surface and thereby reduce the effect of concentration polarisation and increase the permeate flux.
12. The principle of ultrafiltration is used in artificial kidneys for the purification of blood. The process is then called haemodiafiltration. FAAK (Filtration Adsorption Artificial Kidney) devices are

examples of those employing haemodiafiltration. Mathematical analysis of a hollow fiber haemodiafiltration device is presented through Eqs (7.2.2) to (7.2.37).

13. Ultrafiltration finds extensive applications in the recovery of electrophoretic paints, concentration of black liquor in pulp and paper industry (in conjunction with RO), in food processing industries (concentration of cheese whey, egg white), in fermentation industry (for harvesting cells, recovery of products and removal of waste products), in membrane enzymatic reactors and also in medicine (in artificial kidney devices).
14. Dialysis is used to separate solutes from solutions by permitting the solutes to diffuse from a concentrated solution to a dilute solution through the separating membrane. The driving force in dialysis is thus the transmembrane concentration gradient and it is not a pressure-driven process like RO and UF. The distinct advantage of dialysis lies in its gentle operating conditions (low pressure, ambient temperature, low shear rates).
15. Dialysis membranes may be cellulosic or made of other glassy polymers like PAN, PMMA, polysulfone, etc. Membrane modules follow the same configurations as RO and UF devices such as plate and frame, spiral-wound, tubular and hollow fiber configurations.
16. Assuming countercurrent flow of feed solution and dialysate and negligible volume change on either side of the membrane, the dialysance ($\psi$) may be computed from Eq. (7.3.6) and fractional solute extraction from Eqs (7.3.7) or (7.3.8). Equation (7.3.8) assumes an ideal membrane. Empirical expressions for the estimation of mass transfer coefficient and membrane permeability coefficient are given in Eqs (7.3.12) and (7.3.13).
17. Dialysis has large scale commercial applications in the treatment of waste streams and in the recovery of acids from metallurgical liquors. Its most prominent application is in the medical field. Artificial kidney devices employing dialysis are termed as haemodialysers. WAK (Wearable Artificial Kidney) is the best example in this connection.
18. Electrodialysis involves separation of ionic solute from a solution using ion-selective membranes and in presence of a strong electric field. Ion-selective membranes are those which permit counter ions to pass through, but reject the coions. This is due to the presence of fixed electric charges (opposite to that on the permeable ion) inside the membrane pores (see Fig. 7.12). In an ideal ion-selective membrane therefore, the transport number of counter ion will be unity and that of coion zero. A membrane behaves ideal only when the concentration of the ionic solute in the external solution is below a particular value (the so-called *limiting concentration*). If the solute concentration is above this value, then the membrane deviates from its ideal behaviour and permits flow of coions through it.
19. A multicell electrodialyser consists of a number of narrow compartments or cells separated by alternating anion-selective and cation-selective membranes. Thus, there are alternate diluate and concentrate compartments (see Fig. 7.13). The terminal compartments are electrode compartments from where the catholyte and the anolyte (solutions containing products of electrode reactions) are continuously flushed off. Since diluate compartments are bounded by cation-selective membrane on the left (on the cathode side) and anion-selective membrane on the right (on the anode side), both cations and anions migrate from these compartments to the adjoining concentrate compartments. However, since concentrate compartments are bounded by anion-selective membrane on the left and cation-selective membrane on the right, none of these ions can leave these compartments. Thus,

the solution gets desalinated in diluate compartments from where the product water is withdrawn, while concentrated brine is withdrawn from the concentrate compartments.

20. As with RO and UF, concentration polarisation is a significant factor in ED as well. Development of boundary layer on either side of the membrane is due to the fact that the transport number of counter ion in membrane is much higher than that in the external solution. The ratio ($i_m/\bar{N}$), called the *polarisation parameter*, is the controlling parameter that decides the level of concentration polarisation in an ED cell. The operating current density should always be maintained below $i_m$. Increase in fluid flow rate decreases concentration polarisation, but this also reduces the residence time of fluid in the cell and consequently, salt removal per pass also decreases. Use of spacers (wire meshes) as turbulent promoters is therefore recommended. Often, a tortuous path is cut out in the packing sheet to increase the length of fluid flow and thereby its residence time in the dialyser.
21. Membrane fouling or scaling is a serious problem in electrodialysers. Mechanical scrubbing rejuvenates membranes but is cumbersome. Flushing the compartments with leach solutions and pretreatment of feed are also employed to improve the life of the membranes. Current reversal by interchanging the polarity of the two electrodes is another efficient method to impede scaling or fouling of membranes.
22. Both cocurrent and countercurrent flow are used in electrodialysers. Cocurrent flow is usually recommended in order to avoid large pressure drops across the membranes. But, in cocurrent flow, a concentration gradient is built up across the membrane which will be maximum near the outlet. This would polarise the membranes and would reduce the overall current efficiency of the dialyser.
23. The current efficiency of an electrodialyser can be predicted from Faraday's law as shown in Eq. (7.4.12) or (7.4.13). It does not depend on the stack resistance, cell voltage, membrane area and the method of stack operation.
24. Based on the assumption that the cell area resistance is inversely proportional to the diluate stream normality ($R_p\bar{N}$ = constant), the performance equation of electrodialyser is given by Eq. (7.4.19) or (7.4.21). It could be further refined by using relationships of the kind given in Eqs (7.4.22) and (7.4.23). The power requirement (of a single ED cell) is predicted by Eq. (7.4.25).
25. To obtain high demineralisation ratios, multistage operation is employed. This may be externally staged units [number of ED stacks connected in series as shown in Fig. 7.15 or 7.16] or a single electrodialyser with internal staging (as shown in Fig. 7.17). The total power requirement of the former is predicted by Eq. (7.4.30) and that of the latter by Eq. (7.4.32). Energy requirement per unit permeate rate is generally lower for internally staged systems, but they demand larger number of stages.
26. In the feed and bleed process, part of the product stream is continuously recycled back to the feed inlet. By controlling the recycle ratio, a constant input composition can be thus maintained. The power requirement of such a process is given by Eq. (7.4.45). Batch operation of electrodialysers with recirculation (and no mixing) is also often employed (see Fig. 7.19).
27. Electrodialysis has found many diversified applications. Apart from demineralisation of waste water and sewage and desalination of brackish water, ED has also been successfully employed for the recovery of many useful chemicals from many waste effluents. It has also found its place in food

and beverage industry (for desalting cheese whey, fishmeal waste water and for sweetening citrus juices, etc.).

28. Pervaporation is a separation process in which a liquid mixture is in direct contact with one side of a membrane and the permeate is removed from the other side in vapour state. One of the components of the liquid mixture selectively permeates through the membrane and since the pressure on the downstream side is maintained below the saturation pressure of the permeating component (by creating vacuum or by sweeping an inert gas), the permeate vapourises and these vapours are removed, condensed and collected as the *pervaporate*.
29. Pervaporation membranes may be made of glassy polymers (cellulosic) or rubbery (elastomeric) polymers like PVA, PDMS, etc. Often, the active polymer is alloyed with or grafted to an inert polymer and then used.
30. One of the most potential applications of pervaporation is in the separation of close-boiling liquid mixtures that tend to form azeotropes. Pervaporation is more cost-efficient than the conventional azeotropic distillation and it does not demand the use of any entrainer.
31. In membrane reactors, the products of the reaction are removed continuously by selective permeation through a suitable membrane. As a result, the reaction mixture is prevented from attaining equilibrium composition and the reaction can be led upto as much as 100 percent conversion. Reduced reaction temperature, less catalyst fouling, higher product selectivity and short contact time are other advantages of these reactors.
32. The performance of the membrane is decided by the permselectivities of the components of the reaction mixture and the separation factor [see Eqs (7.6.1) and (7.6.2)]. If the membrane acts also as the catalyst, then it is called the catalytic membrane.
33. Many gas-liquid reactions, fermentation reactions and enzyme-catalysed reactions can be more effectively conducted in membrane reactors. They can also be used for conducting two separate reactions simultaneously in the same reactor.
34. Liquid membrane permeation (LMP) is one of the relatively recent entries to the list of separation processes. A liquid membrane can be an emulsion liquid membrane (ELM), which consists of an immiscible liquid layer (say, a layer of hydrocarbon oil) separating two phases such as the feed solution (continuous phase) and the receiving phase (I phase), or an SLM (solid supported liquid membrane)/ILM (immobilised liquid membrane) in which case the membrane phase is immobilised within the pores of a polymeric membrane (say, a polypropylene membrane). Both ELM and SLM (or ILM) systems can be operated with or without a carrier. If a carrier is blended with the membrane phase, then it will chelate selectively with the solute (present in the feed solution) forming a complex, that diffuses through the membrane phase into the receiving phase (I phase) where it reacts with the reactive constituent ($B$) of I phase and rejects the solute into the I phase, simultaneously regenerating the carrier. The carrier then diffuses back to the M-F interface where it chelates with fresh amount of solute. This is called carrier-mediated or carrier-facilitated transport. To note that the transport of solute occurs *against* a steep concentration gradient.
35. Successful industrial applications of LMP process are removal of phenolics from waste water (ELM process without carrier), recovery of hexavalent chromium and copper from very dilute salt solutions (carrier-facilitated ELM processes), recovery of precious elements such as titanium from coal ash and actinides from nitrate-nitric acid waste solutions (both employ carrier-facilitated SLMs),

separation of gases (separation of propene, ethene from process off gases, oxygen from air employing carrier-impregnated SLMs).

36. ELM permeation processes (without carrier) can be modelled using the Advancing Reaction Front Model [Eq. (7.6.35)] or the reversible Reaction Model [Eq. (7.6.64)]. In the former case, the reaction between the solute and the receiving phase is assumed to be instantaneous, while the latter assumes a reversible reaction with equilibrium constant, K.
37. In case of carrier-facilitated transport, the most influencing parameter is the facilitation factor ($F$). Its value depends on the mobility ratio ($\alpha$), equilibrium constant (K), Sherwood number ($Sh$) and the inverse Damkohler number ($\in$). Analytical expressions such as Eqs (7.6.68), (7.6.80) and (7.6.82) have been proposed in the literature for the computation of $F$. Each has its limitations and specific range of applicability.
38. Analytical simulation models are difficult in the case of carrier-facilitated membrane systems. Simulation models involving numerical solution of transport equations such as those developed by Stroeve and Kim [Eqs (7.6.85) to (7.6.101)] or by Narayanan [Eqs (7.6.102) to (7.6.115)] could however be conveniently employed for the design of such systems.

## NOMENCLATURE

| | |
|---|---|
| $A_m$ | membrane area, $m^2$ |
| $A_p$ | effective cell area (effective area of a membrane pair) |
| $b$ | half of the inter-spacing between two parallel sheet membranes, m |
| $c$ | molar density of solution, $kmoles/m^3$ |
| $C(z)$ | protein concentration in blood plasma at any $z$ |
| $C_A$ | concentration of $A$ (solute) in feed phase/continuous phase, $kmole/m^3$ |
| $C_A^*$ | dimensionless concentration [see Eq. (7.6.6)] |
| $C_{Ae}$ | concentration of solute ($A$) in exit stream, $kmoles/m^3$ |
| $C_{Ae}^*$, $C_{A0}^*$ | dimensionless concentrations defined in Eqs (7.6.24) and (7.6.33) respectively |
| $C_{Am}$ | concentration of solute (namely, $A$) in membrane, $kg/m^3$ |
| $C_{A0}$ | concentration of solute ($A$) in feed solution at the inlet to the reactor/permeator, $kmoles/m^3$ |
| $C_{A0}$ (ideal) | value of $C_{A0}$ when the reactor performs as an ideal CSTR with respect to all phases, $kmoles/m^3$ |
| $C_{Ar}$ | concentration of solute ($A$) in receiving phase, $kmoles/m^3$ |
| $C_B$ | concentration of reactive component ($B$) in receiving phase $kmole/m^3$ |
| $C_{B0}$ | value of $C_B$ at $t = 0$, $kmoles/m^3$ |
| $C_b$ | molar concentration of solute in outgoing concentrate (retentate), $kmoles/m^3$ |
| $C_d$, $C_{dm}$ | molar concentration of ionic solute in diluate bulk and that at membrane-diluate interface, $kmoles/m^3$ |
| $C_{d0}$, $C_{de}$ | concentration of solute in inlet dialysate and that in outgoing dialysate stream respectively, $kg/m^3$ |
| $C_{f,i-1}$ | molar concentration of solute in feed to the $i$th segment, $kmoles/m^3$ |

$C_{fi}$ molar concentration of solute in the concentrate leaving the $i$th segment, kmoles/m$^3$
$C_{f0}$ molar concentration of solute in feed solution, kmoles/m$^3$
$C_p$ molar concentration of solute in the product solution (permeate), kmoles/m$^3$
$C_{pi}$ molar concentration of solute in permeate leaving the $i$th segment, kmoles/m$^3$
$C_{Pr}$ concentration of $P$ (product) in receiving phase, kmole/m$^3$
$C_{Si}$ molar concentration of solute in the boundary layer solution in the $i$th segment, kmoles/m$^3$
$C_T$ total concentration of carrier in membrane, kmoles/m$^3$
$d$ diameter of tubular membrane or hollow fiber, m
$d_p$ diameter of diffusing molecule, m
$d_{p0}$ average diameter of membrane pores, m
$D$ volume flow rate of dialysate, m$^3$/s
$D_A$ diffusivity of solute ($A$) in continuous phase (feed solution/feed gas), m$^2$/s
$D_{AC}$ diffusivity of solute-carrier complex in membrane phase, m$^2$/s
$D_{AM}$ diffusivity of solute in membrane phase, m$^2$/s
$D_{Ar}$ diffusivity of solute ($A$) in receiving phase, m$^2$/s
$D_{AS}$ diffusivity of solute in solution, m$^2$/s
$D_{CM}$ diffusivity of carrier in membrane phase, m$^2$/s
$D_e$ effective diffusivity (permeability) of solute ($A$) in membrane phase, m$^2$/s
$D_{ef}$ value of $D_e$ when reversible reaction exists between $A$ and $B$, m$^2$/s
$\bar{D}_{ef}$ mean value of $D_{ef}$, m$^2$/s
$D_F$ arbitrary parameter defined in Eq. (7.6.38), m$^2$/s
$E_f$ performance efficiency of reactor/permeator [see Eq. (7.6.44)], dimensionless
$D_L$ diffusivity of ionic solute in solution, m$^2$/s
$f$ Fanning's friction factor, dimensionless; drag factor (see under dialysis), dimensionless
$f$ volume fraction of membrane phase in emulsion (in each macrodrop), dimensionless
$f_m$ dimensionless parameter defined in Eq. (7.6.55)
$F$ facilitation factor, dimensionless
$\bar{F}$ Faraday's constant, coulombs/gm.eq.
$F_b$ discharge rate of final concentrate (retentate), m$^3$/s; discharge rate of final concentrate per fiber, m$^3$/s
$F_e$ volume flow rate of red blood cells, m$^3$/s
$F_d$ volumetric flow rate through a single diluate compartment, m$^3$/s
$F_{i-1}$ feed rate to $i$th segment, m$^3$/s
$F_i$ discharge rate of concentrated solution from $i$th segment, m$^3$/s
$F_0$ feed flow rate, m$^3$/s; feed rate per hollow fiber, m$^3$/s
$F_{pl}$ volume flow rate of blood plasma at any $z$, m$^3$/s
$F_{ref}$ reference flow rate, m$^3$/s [see Eq. (7.2.12)]
$F_t$ total volume flow rate of diluate stream, m$^3$/s
$g$ number of stages in an internally staged electrodialysis unit

| | |
|---|---|
| $h$ | half of the inter-spacing between parallel membranes, m |
| $H_b$ | haematocrit (volume fraction of red blood cells) in outlet blood, dimensionless |
| $H_f$ | haematocrit in inlet blood, dimensionless |
| $i$ | current density, amperes/cm$^2$ of membrane area |
| $i_m$ | limiting current density, amperes/cm$^2$ of membrane area |
| $I$ | current, amperes |
| $k_1, k_2$ | rate constant for forward reaction (in m$^3$/kmole.s) and that for backward reaction (in s$^{-1}$) respectively |
| $k_C$ | mass transfer coefficient in continuous phase (feed solution), m/s |
| $k_i$ | mass transfer coefficient in $i$th segment, m/s |
| $k_L$ | mass transfer coefficient in receiving phase; liquid side mass transfer coefficient, m/s |
| $k_W$ | mass transfer coefficient across membrane (SLM) in absence of carrier, m/s |
| $K$ | equilibrium constant for reaction between solute and carrier, m$^3$/kmole |
| $K_m$ | membrane permeability coefficient for water/solvent, kmoles/(m$^2\cdot$s$\cdot$atm) |
| $K'_m$ | membrane permeability coefficient for water/solvent m$^3$/(m$^2\cdot$s$\cdot$atm) |
| $K_{mb}, K_{mr}$ | partition (distribution) coefficient of solute between membrane phase and continuous phase and that between $M$-phase and receiving phase, dimensionless |
| $K_{ms}$ | membrane permeability coefficient for solute, m/s |
| $K_0$ | overall dialysis coefficient, m/s |
| $L$ | length of each hollow fiber, m |
| $\dot{m}_A$ | total solute transfer (removal) rate, kmoles/s |
| $n$ | number of differential segments of RO membrane system; number of stages in an externally staged ED unit |
| $n_d$ | number of cells/membrane-pairs (or number of diluate compartments) |
| $N$ | total number of macrodrops in permeator/extractor |
| $N_{Am}$ | number of kmoles of solute ($A$) in membrane phase at any time t |
| $N_B$ | number of kmoles of $B$ in receiving phase at any time t |
| $\bar{N}$ | normality of diluate stream, gm$\cdot$eq./liter |
| $N_{Ai}$ | transmembrane solute flux in the $i$th segment, kmoles/(m$^2\cdot$s) |
| $\bar{N}_f, \bar{N}_p$ | normality of feed solution and that of product solution respectively, gm.eq./liter |
| $\bar{N}_{pf}$ | normality of final product, gm$\cdot$eq./liter |
| $\bar{N}_r$ | normality of total feed (including the recycled solution) in feed and bleed process, gm$\cdot$eq./liter |
| $N_{Wi}$ | transmembrane solvent flux in the $i$th segment, kmoles/(m$^2\cdot$s) |
| $p_d$ | hydraulic pressure of external diafiltrate, N/m$^2$ |
| $P$ | product flow rate, m$^3$/s; diafiltration rate per hollow fiber, m$^3$/s |
| $P_i$ | product flow rate (permeate rate) from $i$th segment, m$^3$/s |
| $P'$ | power consumption, W |
| $q$ | demineralisation ratio, dimensionless |

| | |
|---|---|
| $Q_e$, $Q_f$, $Q_r$ | volumetric flow rate of emulsion, that of feed solution and that of receiving phase respectively, $m^3/s$ |
| $r_0$ | radius of each receiving phase droplet (see Fig. 7.23), m |
| $R_0$ | radius of each macrodrop (of M-I emulsion), m |
| $R$ | product recovery, dimensionless |
| $R_g$ | gas constant, $(atm \cdot m^3)/(kmoles \cdot K)$ |
| $R_i$ | product recovery in the $i$th segment, dimensionless |
| $R_p$ | cell area resistance, $ohms \cdot cm^2$ |
| $Re$ | Reynolds number, dimensionless |
| $Sc$ | Schmidt number, dimensionless |
| $SE$ | fractional solute extraction, dimensionless |
| $Sh$ | local Sherwood number dimensionless |
| $Sh_W$ | wall Sherwood number (based on $k_W$), dimensionless |
| $(Sh)_{av}$ | average Sherwood number, dimensionless |
| $SR$ | solute rejection, dimensionless |
| $T$ | absolute temperature, K |
| $\overline{t}_+$, $\overline{t}_-$ | transport number of cation and that of anion respectively in membrane, dimensionless |
| $t_+$, $t_-$ | transport numbers of cation and anion in external solution, dimensionless |
| $U^*$ | parameter defined in Eq. (7.1.26), m/s |
| $\mathbf{V}_{i-1}$ | velocity of feed to the $i$th segment, m/s |
| $\mathbf{V}_0$ | inlet velocity of feed solution, m/s |
| $\mathbf{V}_{Wi}$ | velocity of solvent through the membrane (permeate velocity) in $i$th segment, m/s |
| $\overline{V}$ | molar volume of liquid at the boiling point, $m^3/kmole$ |
| $V_p$ | cell voltage volts |
| $V_e$ | total volume of emulsion (macrodrops) in reactor/permeator, $m^3$ |
| $W$ | width of the membrane, m |
| $x_{fi}$ | mole fraction of solute in the concentrated solution leaving the $i$th segment, dimensionless |
| $x_{f0}$ | mole fraction of solute in feed solution, dimensionless |
| $x_p$ | mole fraction of solute in product solution (permeate), dimensionless |
| $x_{Si}$ | mole fraction of solute in the boundary layer solution in the $i$th segment, dimensionless |
| $X_i^*$ | dimensionless parameter defined in Eq. (7.1.43) |
| $z$, $z^*$ | axial coordinate and dimensionless axial coordinate respectively |
| $\alpha_0$, $\beta_f$ | dimensionless parameters defined in Eqs (7.1.50) and (7.1.51) |
| $\alpha$ | mobility ratio, dimensionless |
| $\alpha_1$ to $\alpha_5$ | dimensionless parameters defined in Eqs (7.6.25), (7.6.26), (7.6.56), (7.6.91) and (7.6.92) respectively |
| $\beta$ | dimensionless parameter defined in Eq. (7.6.39) and in Eq. (7.6.54) |
| $\beta_t$ | dimensionless parameter defined in Eq. (7.1.60) |
| $\delta$ | effective thickness of membrane, m |

$\delta_C$ thickness of boundary layer on diluate side, m

$\delta_{Ci}$ thickness of boundary layer in $i$th segment, m

$\Delta \bar{N}$ change in normality of diluate stream, gm·eq./liter

$\Delta P$ transmembrane pressure drop, N/m$^2$ or atm

$(-\Delta P_f)_i$ frictional pressure drop in $i$th segment, N/m$^2$

$\Delta z$ length of each differential segment, m

$\Delta \pi$ osmotic pressure difference, N/m$^2$ or atm

$\in$ ultrafiltration parameter, dimensionless; volume fraction of pores in membrane, dimensionless; inverse Damkohler number, dimensionless

$\zeta$ tortuosity factor, dimensionless

$\eta$ current efficiency of electrodialyser; dimensionless length [Eq. (7.6.10)]

$\theta$, $\lambda_i$ dimensionless parameters defined in Eqs (7.1.28) and (7.1.29)

$\theta_e$ average residence time of emulsion in the reactor/permeator, s

$\lambda$ ratio of $\bar{D}_{ef}$ to $D_e$, dimensionless; parameter defined in Eq. (7.6.72), dimensionless

$\mu_f$ fluid viscosity kg/(m·s)

$\xi$ dimensionless length [see Eqs (7.6.11) and (7.6.90a)]

$\pi$ osmotic pressure, N/m$^2$ or atm

$\pi_{pi}$ osmotic pressure of product solution (permeate) from $i$th segment, N/m$^2$ or atm

$\pi_{pl}$ colloid osmotic pressure in plasma, N/m$^2$ or atm

$\pi_{Si}$ osmotic pressure of boundary layer solution in $i$th segment, N/m$^2$ or atm

$\rho_f$ fluid density, kg/m$^3$

$\sigma$ dimensionless pressure drop = $(p_1 - p_2)/(p_1 - p_d)$

$\tau$ dimensionless time [Eq. (7.6.27)]

$\tau_e$ dimensionless residence time [Eq. (7.6.34)]

$\phi$ parameter defined in Eq. (7.1.27), m$^3$/kmole; dimensionless concentration defined in Eqs (7.6.9) and (7.6.69)

$\psi$ dimensionless diafiltration rate per fiber; dialysance, m$^3$/s; permselectivity of ion-selective membrane, dimensionless

## COMPUTER NOTATIONS

A1, A2, A3 = $a_1$, $a_2$, $a_3$ empirical constants of Eq. (7.2.5)

ABSN = $\in$ ultrafiltration parameter

AF0 = $\alpha_0$ dimensionless parameter defined in Eq. (7.1.50)

AK (I) = $k_i$ mass transfer coefficient in $i$th segment

ALM (I) = $\lambda_i$ dimensionless parameter defined in Eq. (7.1.29)

AM = $A_m$ membrane area

B = $B$ proportionality constant of Eq. (7.1.9a)

BAS = $\beta_{as}$ dimensionless parameter defined in Eq. (7.1.52)

BT = $\beta_t$ dimensionless parameter defined in Eq. (7.1.60)

| | | |
|---|---|---|
| C | = $c$ | molar density of solution |
| CB | = $C_b$ | solute concentration in retentate; protein concentration in outlet blood |
| CF (I) | = $C_{fi}$ | solute concentration in concentrate stream leaving $i$th segment |
| CF0 | = $C_{f0}$ | solute concentration in feed; protein concentration in inlet blood |
| CP (I) | = $C_{pi}$ | solute concentration in permeate from $i$th segment |
| CPT | = $C_p$ | solute concentration in total permeate |
| D | = $d$ | diameter of tubular membrane/hollow fiber |
| DA1, DA2, DA3 | = $a_1^*$, $a_2^*$, $a_3^*$ | dimensionless parameters defined in Eq. (7.2.12) |
| DAS | = $D_{AS}$ | diffusivity of solute in solution. |
| DC (I) | = $C^*$ | dimensionless concentration defined in Eq. (7.2.12) |
| DCB | = $C_b^*$ | dimensionless exit concentration [see Eq. (7.2.21)] |
| DFB | = $F_b^*$ | dimensionless flow rate of exit blood [see Eq. (7.2.12)] |
| DF0 | = $F_0^*$ | dimensionless inlet flow rate of blood [see Eq. (7.2.12)] |
| DFP1 | = $(p_1 - p_d)$ | transmembrane pressure difference at blood inlet |
| DFP2 | = $(p_2 - p_d)$ | transmembrane pressure difference at blood outlet |
| DP | = $\Delta P$ | transmembrane pressure drop |
| DP2 | = $p_2^*$ | dimensionless pressure at blood outlet |
| DPI (I) | = $\pi^*$ | dimensionless osmotic pressure [see Eq. (7.2.12)] |
| DZ | = $\Delta z$ | length of each differential segment |
| FB | = $F_b$ | outlet flow rate of blood from each hollow fiber |
| FBT | = $F_b$ (total) | total exit flow rate of blood from fiber bundle |
| F0 | = $F_0$ | inlet flow rate of feed solution; inlet flow rate of blood to each hollow fiber |
| FOT | = $F_0$ (total) | total inlet rate of blood to fiber bundle |
| FR | = $F_{ref}$ | reference flow rate |
| H | = $h$ | step size in Runga-Kutta method |
| HF | = $H_f$ | haematocrit in inlet blood |
| K | = $k$ | parameter of Runga-Kutta method [see Eq. (7.8.11)] |
| K1, K2, K3, K4 | = $k_1$, $k_2$, $k_3$, $k_4$ | parameters of Runga-Kutta method [see Eqs (7.8.3) to (7.8.9)] |
| KM, KMP | = $K_m$, $K'_m$ | membrane permeability coefficient |
| L | = $L$ | length of tubular module/hollow fiber |
| MEU | = $\mu_f$ | fluid viscosity |
| N | = $n$ | number of differential segments; number of hollow fibers |
| P | = $P$ | permeate rate; diafiltration rate per fiber |
| P1 | = $p_1$ | pressure at blood inlet |
| PD | = $p_d$ | pressure of diafiltrate |
| PHI | = $\phi$ | parameter defined in Eq. (7.1.27) |
| P0 | = $P'$ | power consumption |

| | | |
|---|---|---|
| PP (I) | = $p_i$ | parameter defined in Eq. (7.1.40a) |
| PPT | = $p$ | parameter defined in Eq. (7.1.67a) |
| PSI | = $\psi$ | dimensionless diafiltration rate per fiber |
| PT | = $P$ (total) | total diafiltration rate |
| Q (I) | = $q_i$ | parameter defined in Eq. (7.1.39) |
| R (I) | = $R_i$ | product recovery in $i$th segment |
| RC | = $R$ | computed value of total product recovery |
| RE | = $Re$ | Reynolds number |
| ROW | = $\rho_f$ | fluid density |
| RT | = $R$ | total product recovery desired |
| SH | = $Sh$ | Sherwood number |
| STP | = $(D_{AM}/K_{mb}\ \delta)$ | solute transport parameter |
| T | = $t$ | parameter in Runga-Kutta method [see Eq. (7.8.12)] |
| T1, T2, T3, T4 | = $t_1$, $t_2$, $t_3$, $t_4$ | parameters in Runga-Kutta method [see Eq. (7.8.4) to (7.8.10)] |
| TH | = $\theta$ | dimensionless parameter defined, in Eq. (7.1.28) |
| US | = $U^*$ | parameter defined in Eq. (7.1.26) |
| V (I) | = $\mathbf{V}_i$ | velocity of feed to $(i + 1)$th segment |
| V0 | = $\mathbf{V}_0$ | inlet velocity of feed solution |
| VW (I) | = $\mathbf{V}_{Wi}$ | permeate velocity in $i$th segment |
| X (I) | = $X$ or $F^*$ | dimensionless flow rate of blood at any $z^*$ [see Eq. (7.2.12)] |
| XD (I) | = $X_i^*$ | dimensionless parameter defined in Eq. (7.1.43) |
| Y (I) | = $Y$ or $p^*$ | dimensionless transmembrane pressure difference [see Eq. (7.2.12)] |

## EXERCISES

7.A. Comment on the following statements giving reasons. You may agree with or contradict against any of the statements but do justify your answers. Give examples and/or neat sketches wherever possible to elucidate your answers.

(*a*) Ultrafiltration is not economical at high solute concentrations, but reverse osmosis is.

(*b*) Ideality of an ion-selective membrane depends on the salt concentration of feed solution.

(*c*) For the concentration of egg white, ultrafiltration is superior to spray drying or any type of thermal concentration.

(*d*) Reversible reactions can be conducted upto 100 percent conversion in a membrane reactor even if the equilibrium conversion is below 50 percent.

(*e*) For large scale desalination of sea water, multiple effect evaporation is more energy-efficient than reverse osmosis.

(*f*) An internally staged electrodialysis unit demands less number of stages than an externally staged unit for the same capacity and same demineralisation ratio.

7.B. Explain briefly what do you understand by

(*a*) Concentration polarisation,
(*b*) Feed and bleed process,
(*c*) Ultrafiltration parameter,
(*d*) Solute transport parameter in reverse osmosis,
(*e*) Current efficiency of an electrodialyser,
(*f*) Carrier-facilitated transport.

7.C. Distinguish between
(*a*) Reverse osmosis and ultrafiltration,
(*b*) Dialysis and electrodialysis,
(*c*) Distillation and pervaporation,
(*d*) Haemodialysis and haemodiafiltration,
(*e*) SLM and ELM

7.D. A single stage reverse osmosis unit is to be used for desalinating brackish water that contains 1000 mg/liter of dissolved salts (principally sodium chloride). The solute rejection desired is 50 percent. A tubular membane module containing 50 tubes (each 0.01 m in diameter) is to be employed for the purpose. The tubes are made of 0.05 microns thick polyamide membrane. The feed solution enters the module at the rate of 1.5 $m^3$/hr and the operating transmembrane pressure drop is 60 atm.
(*a*) Compute the length of each tube and total membrane area required.
(*b*) What will be the power consumption per liter of permeate collected?
Diffusivity of salt in membrane = $1.3 \times 10^{-10}$ $cm^2/s$
Distribution coefficient of solute between membrane phase and solution phase = 0.20
Membrane permeability coefficient for water = $2.4 \times 10^{-5}$ kmoles $(s \cdot m^2 \cdot atm)$
The osmotic pressure of solution may be assumed proportional to the mole fraction of solute and the proportionality constant is 2534.9 atm.

7.E. Electrodialysis is being used for desalinating brine from a salt concentration of 1000 mg/liter to 200 mg/liter. Membrane cells 35 cm wide, 90 cm long and 0.05 cm thick are available. Product water is to be collected at the rate of 148 liters/hr. The polarisation parameter is maintained constant at 0.60 $(amp/cm^2)$ (liter/gm·eq.) and the current efficiency = 0.85. The cell area resistance (expressed in ohms·$cm^2$) is found to be inversely proportional to diluate normality and the proportionality constant is 7.5. Determine which of the following arrangements shall be most economical:
(*i*) A number of cells connected in series, the voltage applied to each cell being the same.
(*ii*) A single electrodialyser with the exit stream from each diluate compartment being sent to the next diluate compartment, the product water being collected from the last diluate compartment.
(*iii*) A single electrodialyser with 80 percent of the product solution being continuously recycled back to the feed inlet.

7.F. An anisotropic dialysis membrane has an apparent pore diameter of 30 angstroms ($30 \times 10^{-8}$ cm) and a wet thickness of 50 microns. The tortuosity factor is 2.12 and porosity (volume fraction of pores) = 0.193. Compute the fractional extraction of pepsin (molecular diameter = 46 angstroms) from a feed solution containing 7.5 kg/m$^3$ of pepsin if the membrane module has the following specifications.

Number of hollow fibers = one million

Diameter of each fiber = 220 microns

Length of each fiber = 0.17 m

The feed solution flows through the fibers and its velocity is maintained at 4 cm/s. The dialysate flows countercurrently in the outer shell and the ratio of dialysate flow rate to feed flow rate is fixed at 6.0. The concentration of pepsin in inlet dialysate is 0.05 kg/m$^3$ and diffusivity of pepsin in solution = $0.075 \times 10^{-9}$ m$^2$/s.

7.G. Rework Example 7.6 assuming that the reversible reaction model is applicable. Estimate also the performance efficiency of the extractor/permeator as predicted by the above model. Compare the results obtained with those of Example 7.6.

7.H. Develop a computer program that computes the volume of reactor/permeator ($V$) required to separate a dissolved solute from a feed solution of concentration $C_{A0}$, that is fed at the rate of $Q_f$ m$^3$/hr to the reactor along with the M-I emulsion which is fed at $Q_e$ m$^3$/hr. The concentration of solute in the exit stream should not exceed $C_{Ae}$. Prepare separate programs based on

(*a*) advancing reaction front model

(*b*) reversible reaction model.

## REFERENCES


1. Kimura, S and Sourirajan, S, AIChE J, **13,** 497, 1967.
2. Kopecek, J and Sourirajan, S, J Appl. Polymer Sci., **13,** 637, 1969.
3. Aggarwal, JP and Sourirajan, S, Ind. Eng. Chem. Process Des. Dev., **8,** 439, 1969.
4. Sourirajan, S, Reverse Osmosis, Logos Press Limited, 1970.
5. Kimura, S and Sourirajan, S, Ind. Eng. Chem. process Des. Dev., **7,** 539, 1968.
6. Schweitzer, PA (ed.), Handbook of Separation Techniques for Chemical Engineers, McGraw Hill, New York, 1979.
7. Meares, P (ed.), Membrane Separation Processes, Elsevier, Amsterdam, 1976.
8. Knudsen, JG and Katz, DL, Fluid Dynamics and Heat Transfer, McGraw Hill, New York, 1958.
9. Membrane Technology in Pulp and Paper Industry, Nitto Bullettin, Osaka, Japan, 1987.
10. Dixon, BD, Australian J dairy Tech., 91–95, September, 1985.
11. Pitera, MC and Middleman, M, Ind. Eng. Chem. Proc. Des. Dev., **12,** 52, 1973.
12. Narayanan, CM, Studies on Performance Characteristics of a Constricted Tubular Ultrafiltration Unit, Proc. Thirteenth National Heat Mass Transfer Conf., Surathkal, pp. 895–901, 1995.
13. Lane, A and Riggle, M, Chem. Eng. Progr. Symp. Series, **55,** 127, 1959.
14. Bacon, H, J Franklin Inst., **221,** 251, 1936.
15. Cooney, DO, Biomedical Engineering Fundamentals, Marcel-Dekker, New York, 1976.
16. Katchalsky, A and Curran, PF, Nonequilibrium Thermodynamics in Biophysics, Harvard University Press, Cambridge, Mass., 1965.
17. Papenfuss, HD, Gross, JF and Thorson, ST, AIChE J, **25** (1), 170, 1979.

18. Landis, EM and Pappenheimer, Handbook of Physiology, Section 2, American Physiological Society, Washington, 1963.

19. Bell, CM et. al., J Membrane Sci., **36,** 315, 1988.

20. Phillipe Aptel et. al., J Membrane Sci., **1,** 271, 1976.

21. Mintz, MS, Ind. Eng. Chem, **55** (6), 18, 1963.

22. Wilson, JR, Demineralisation by Electrodialysis, Buterworths, London, 1960.

23. Rao, MB, Catalytic Membrane Reactors, in Ayyanna, C (ed.), Biotechnology in 21st Century, Tata McGraw Hill, New Delhi, 1993.

24. Hsieh, HP, AIChE Symp. Ser., **84** (261), 1–18, 1988; **85** (268), 53–67, 1989.

25. Narayanan, CM, Recovery of Titanium from Coal Ash using Immobilised Liquid Membranes, J Inst. Engrs (I), **85,** 24–30, September, 2004.

26. Muscatello, AC et. al., Actinide Removal from Aqueous Waste Using Solid Supported Liquid Membranes, in Noble, RD and Way, JD (eds), Liquid Membranes–Theory and Applications, ACS Symp. Series, Washington, 1987.

27. Hughes, RD et. al., Recent Developments in Separation Science, Vol. IX, 173–195, CRC Press, 1986.

28. Niederhoffer, EC et. al., Chem. Rev., 84 (2), 137–2003, 1984.

29. Hatton, TA, Lightfoot, EN, Cahn, RP and Li, NN, Ind. Eng. Chem. Fund., **22,** 27, 1983.

30. Bunge, AL and Noble, RD, J Membrane Sci., **21,** 55, 1984.

31. Noble, RD, Way, JD and Powers, LA, Ind. Eng. Chem. Fund., **25,** 450, 1986.

32. Stroeve, P and Kim, J, Seperation in Mass Exchange Devices with Reactive Membranes, in Noble, RD and Way, JD (eds.), Liquid Membranes–Theory and Applications, 39–55, ACS Symp. Series, 1987.

# Chapter 8

# Reaction Kinetics and Reactor Design

Many of the industrial processes involve unit operations as well as unit processes. Unit operations are physical processes that have been discussed in detail in the earlier chapters. Unit processes denote chemical reactions in which the reactants or the raw materials are chemically transformed into products. In many industrial applications, the chemical reactor is a vital component and could occasionally be one of the most expensive equipment used.

The chemical reactions may be classified in a variety of ways. If the reaction takes place in a single phase (in other words, if all the reacting materials are found in a single phase be it gas, liquid or solid), then it is called a *homogeneous reaction.* Most of the gas-phase reactions and liquid-liquid reactions (either non-catalytic or catalysed by a liquid catalyst) may be grouped under this category. If two or more phases are required for conducting the reaction at the specified rate, then it is called a *heterogeneous reaction.* Gas–liquid absorption accompanied by a chemical reaction (discussed under Gas Absorption in Chapter 4) is an excellent example of heterogeneous reaction system. It may be noted that the chemical reaction, in fact, takes place only in the liquid phase and the reaction products also remain in the liquid phase only. (Even then, it is called a heterogeneous reaction system since two phases are required for conducting the process). Thus, in a heterogeneous system, the reaction may take place in one, two or more phases or at the interface. The reactants and the products may be distributed among the phases or may be contained within a single phase. Combustion of coal, roasting of ores, reduction of iron ore to iron (what takes place in the Blast furnace) are other examples of heterogeneous reactions. Most of the catalytic reactions may also be classified under heterogeneous reactions. Examples are catalytic oxidation of sulphur dioxide to sulfur trioxide in the manufacture of sulphuric acid, synthesis of methanol from carbon monoxide and hydrogen, catalytic cracking of heavy petroleum oils, etc. In all the above cases, a solid catalyst is employed and thus, the catalyst is in a phase different from that of the reactants and the products. However, this type of classification is not hard and fast. For example, fermentation reactions (or enzyme–substrate reactions) fall in between these two categories since an enzyme–containing solution is not fully homogeneous, neither truly heterogeneous (since enzymes are of colloidal size, 10 to 100 microns).

## 8.1 FUNDAMENTALS OF REACTION KINETICS

The *rate of a chemical reaction* may be expressed in different ways. It is usually defined as the change in moles of a component (say, $A$) with respect to time, per unit volume of reaction mixture. Thus, the rate will be negative if the component is a reactant and positive if the component is a product. Mathematically,

$$r_A = (1/V)\,\frac{dN_A}{d\theta} \qquad \text{... (8.1.1)}$$

where $r_A$ = rate of chemical reaction, kmoles/$(m^3 \cdot s)$

$V$ = volume of reaction mixture, $m^3$

$N_A$ = number of moles of component $A$ at any time $\theta$

Alternately, $r_A$ may also be defined as

$$r_A = (1/V_r)\,\frac{dN_A}{d\theta} \qquad \text{... (8.1.2)}$$

where $V_r$ = volume of reactor. For homogeneous systems, $V_r$ is often equal to $V$ and in such cases, both of the above equations predict the same value of $r_A$. If the volume of the reaction mixture is constant (such as in a constant volume batch reactor), then

$$r_A = \frac{d\,(N_A/V)}{d\theta} = \frac{dC_A}{d\theta} \qquad \text{... (8.1.3)}$$

where $C_A$ is the molar concentration of $A$ in the reaction mixture. It is also a usual practice to define the reaction rate based on unit interfacial area (for fluid-fluid systems) or based on unit surface of solid (for gas–solid systems). In solid–catalysed fluid-fluid reactions, it is usual to define the rate per unit mass of catalyst used. Thus

$$r_A = (1/W)\,\frac{dN_A}{d\theta} \qquad \text{... (8.1.4)}$$

where $W$ = mass of catalyst used, kg

$r_A$ = rate of reaction, kmoles/(s.kg of catalyst)

As stated earlier, $r_A$ will be negative if $A$ is a reactant and is being consumed as a result of the chemical reaction and $r_A$ will be positive if $A$ is being produced as a result of the reaction.

Let us first consider the kinetics of homogeneous reactions. Kinetics of heterogeneous reactions are discussed in one of the subsequent sections. For homogeneous systems, the rate of reaction will be a function of temperature, pressure and the composition of the reacting substances. However, since these three variables are mutually interdependent (for instance, the pressure is determined once the temperature and the composition of the phase are known), it will not be wrong to consider $r_A$ a function of temperature and composition only. Thus, to develop the rate equation, we have to find out the composition dependency and the temperature dependency of $r_A$.

In the case of the so called *elementary reactions*, the composition dependency of $r_A$ can be readily deduced from the stoichiometric equation representing the reaction. For example, consider a single reaction with stoichiometric equation

$$aA + bB \rightarrow cC + dD \quad \text{... (8.1.5)}$$

Now, suppose the rate equation for the above reaction has been found to be

$$(-r_A) = k\, C_A^a C_B^b \quad \text{... (8.1.6)}$$

where $(-r_A)$ is the rate of disappearance or consumption of $A$ per unit volume of reaction mixture (the negative sign is used to indicate that $A$ is being consumed and not generated) and $k$ is the reaction rate constant. In such a case, the above reaction is called an *elementary reaction*. Thus, for an elementary reaction, the rate equation can be derived directly from the stoichiometric equation since it describes the true reaction mechanism. The *molecularity* of an elementary reaction is the number of moles of reactants involved in the reaction. For example, the molecularity of the above reaction is $(a + b)$. It is important to keep in mind that molecularity is defined only for elementary reactions and its value is 1.0, 2.0 or sometimes 3.0. No elementary reaction with molecularity more than 3.0 has been observed till this date. This inversely means that if the molecularity of a chemical reaction (deduced from the stoichiometric equation) is more than 3.0, then it cannot be an elementary reaction. For example, the reaction $2H_2 + 2NO \rightarrow N_2 + 2H_2O$ is not elementary because had it been an elementary reaction, its molecularity would be more than 3.0.

Only few reactions are elementary in nature. For *nonelementary reactions*, there is hardly any correspondence between the stoichiometric equation and the rate equation. For example, the rate equation for the reaction

$$2N_2O \rightarrow 2N_2 + O_2 \quad \text{... (8.1.7)}$$

has been found to be

$$\left(-r_{N_2O}\right) = \frac{k_1 [N_2O]^2}{1 + k_2 [N_2O]} \quad \text{... (8.1.8)}$$

where $[N_2O]$ stands for the molar concentration of $N_2O$ (that is, $C_{N_2O}$) and $k_1$ and $k_2$ are constants. Similarly, for the reaction between hydrogen and bromine, the stoichiometric equation is

$$H_2 + Br_2 \rightarrow 2HBr \quad \text{... (8.1.9)}$$

while the rate expression is[1]

$$r_{HBr} = \frac{k_1 [H_2][Br_2]^{0.5}}{k_2 + [HBr]/[Br_2]} \quad \text{... (8.1.10)}$$

where $r_{HBr}$ is the rate of formation of hydrogen bromide and the terms in square brackets denote molar concentrations. A still another example is the catalytic synthesis of ammonia (the reputed Haber process) for which the stoichiometric equation is

$$2N_2 + 3H_2 \rightarrow 2NH_3 \quad \text{... (8.1.11)}$$

But, for many catalysts, the rate equation for the above reaction has been found to be

$$(-r_A) = k\,C_A \quad \text{... (8.1.12)}$$

where $A$ stands for nitrogen.

It is possible to assume that a nonelementary reaction is equivalent to a sequence of elementary reactions. Analysis based on this principle, we shall be discussing subsequently.

Let us also understand what is meant by the *order of a chemical reaction.* For example, let the rate equation for the reaction whose stoichiometric equation is Eq. (8.1.5), be given by

$$(-r_A) = k\,C_A^m C_B^n \quad \text{... (8.1.13)}$$

where $m$ and $n$ may or may not be equal to the stoichiometric coefficients $a$ and $b$ respectively. Then, the above reaction is said to be of $m$th order with respect to $A$ and of $n$th order with respect to $B$. The overall order of the reaction is $(m + n)$. If $m = a$ and $n = b$, then the reaction is elementary. If $m = 1$ and $n = 0$, then the reaction is of first order with respect to $A$ and if $m = n = 1$, it is a second order reaction. If $m = n = 0$, the rate of reaction will be independent of composition and will be a function of temperature only. The reaction is then said to be *zeroth order.*

It can be, thus, seen that the order of a reaction is to be determined from the rate equation, which, in turn, is derived empirically from the experimental data. The order of a chemical reaction is, therefore, basically an empirical constant. It need not be an integer, but may also be a fraction. In the case of elementary reactions, the order of the reaction can be deduced from the stoichiometric equation itself since the reaction order agrees with the molecularity. For nonelementary reactions, there is no necessary connection between the order and the reaction stoichiometry. For example, the ammonia synthesis reaction, whose stoichiometry is given in Eq. (8.1.11), is first order in nitrogen and zero order in hydrogen [see Eq. (8.1.12)]. In the case of reactions for which the rate equations are quite complex such as those given in Eqs (8.1.8) and (8.1.10), it is meaningless to speak about the reaction order.

If a reactant is present in an amount that is theoretically required by the stoichiometric equation for conducting the reaction, then it is called the *limiting reactant.* However, in most chemical reactions carried out in industry, the quantities of reactants supplied are usually not in the exact proportions demanded by the stoichiometric equation. At least a few of the reactants are supplied in excess so that the reaction tends to go to completion. If a large excess of one or more of the reactants is used such that the concentration of that reactant hardly changes during the course of the reaction, then the reaction will be zero order with respect to that reactant and the overall order of the reaction gets reduced. For example, if in carrying out a reaction which is normally second order with a rate equation $(-r_A) = k_2 C_A C_B$, a large excess of $B$ is used, then $C_B$ remains essentially constant and equal to the initial value $C_{B0}$. The rate equation may be then written as $(-r_A) = k_1 C_A$, where $k_1 = k_2 C_{B0}$ and the reaction is then said to be *psuedo first order in A*.

The inversion of sucrose in aqueous solution proceeds according to the following bimolecular equation:

$$\underset{\text{Sucrose}}{C_{12}H_{22}O_{11}} + H_2O \rightarrow \underset{\text{Glucose}}{C_6H_{12}O_6} + \underset{\text{Fructose}}{C_6H_{12}O_6}$$

However, since water is used in large excess, the reaction is first order in sucrose. Thus, this is an example of psuedo first order reaction. Synthesis of ammonia ($N_2 + 3H_2 \rightarrow 2NH_3$) is first order in $N_2$, but is zero order in $H_2$. The decomposition of hydrogen peroxide ($H_2O_2$) follows first order kinetics though it has a bimolecular stoichiometry:

$$H_2O_2 + FeCl_3 \rightarrow H_2O_2FeCl_3 \rightarrow H_2O + \frac{1}{2}O_2 + FeCl_3$$

Hydrogen peroxide is fairly stable in the absence of catalysts, but when it combines with a substance such as ferric chloride, the decomposition is rapid. Although the reaction involves two reacting substances, the ferric chloride is released unchanged after the reaction and the overall reaction is first order in $H_2O_2$. Gas phase decomposition of HI into $H_2$ and $I_2$ ($2HI \rightarrow H_2 + I_2$) is a second order reaction[2], where as photochlorination of propane $\left(C_3H_8 + Cl_2 \xrightarrow{\text{Light}} C_3H_7Cl + HCl\right)$ is reported[3] to be second order in chlorine and zero order in propane. Pyrolysis of ethyl nitrate[4] is an example of half order reaction. Also, hydrochlorination of lauryl alcohol with zinc chloride as a homogeneous catalyst has been reported[5] to be half order in alcohol. Decomposition of $NH_4NO_2$ in solution[6] is a third order reaction in $NH_4NO_2$, while the oxidation of nitric oxide to nitrogen dioxide[7] ($2NO + O_2 \rightarrow 2NO_2$) is second order in NO and first order in $O_2$ (the overall order of the reaction being 3.0). Zero order homogeneous reactions are comparatively rare. However, as stated earlier, if one of the reactants is used in large excess, then the reaction will be zero order with respect to that reactant. In the oxidation of NO to $NO_2$ in presence of a large excess of $O_2$, the rate is zero order in $O_2$. In some heterogeneous reactions where the solid phase acts as a catalyst, the rate has been found to be zero order. An example is the decomposition of ammonia on platinum catalyst[8].

The rate constant ($k$), also called *specific reaction rate*, is fundamentally an empirically defined parameter. Its value can be determined only experimentally. If the rate equation conforms to Eq. (8.1.13), then the dimensions of $k$ will be $(s^{-1})$ $(\text{kmoles/m}^3)^{1-q}$, where $q$ is the order of the reaction. Thus, for a first order reaction, the dimensions of $k$ will be $s^{-1}$ and for a second order reaction, $k$ will have the dimensions $m^3/(\text{kmoles.s})$.

It is also important to keep in mind that if the stoichiometric coefficients for two reactants are different, then the rate expressed in terms of one reactant will not be the same as the rate expressed in terms of the other. For example, consider the reaction (8.1.5). Let it be taking place in a constant volume batch reactor. Them, from Eq. (8.1.3), the rate of reaction with respect to $A$ is $(-r_A) = dC_A/d\theta$, that with respect to $B$ $(-r_B) = dC_B/d\theta$, that with respect to $C$ $r_C = dC_C/d\theta$ and so on. But from the stoichiometry,

$$-(1/a)\frac{dC_A}{d\theta} = -(1/b)\frac{dC_B}{d\theta} = (1/c)\frac{dC_C}{d\theta} = (1/d)\frac{dC_D}{d\theta} \quad \text{... (8.1.14)}$$

Therefore, $\boldsymbol{r}_A = (a/b)\,\boldsymbol{r}_B = -(a/c)\,\boldsymbol{r}_C = -(a/d)\,\boldsymbol{r}_D$ ... (8.1.15)

Quite a few of industrial reactions are *reversible* in nature. A typical reversible reaction can be represented as

$$aA + bB \rightleftharpoons cC + dD \quad \text{... (8.1.16)}$$

In the forward reaction, $A$ and $B$ react with each other producing $C$ and $D$, while in the backward reaction, $C$ and $D$ react with each other producing $A$ and $B$. Both of these reactions take place simultaneously. If both of these reactions are elementary, then the net rate of disappearance of $A$ may be described by

$$(-\boldsymbol{r}_A) = k_1 C_A^a C_B^b - k_2 C_C^c C_D^d \quad \text{... (8.1.17)}$$

where $k_1$ is the rate constant of the forward reaction and $k_2$ is that of the backward reaction. In course of time, the rates of both of these reactions become equal and the system will attain equilibrium. Once equilibrium is attained, the composition of the reaction mixture will not change any further. Thermodynamically speaking, the change in free energy will be zero. If the reacting system is an ideal solution, then the *equilibrium constant* K is defined as

$$K = \left(\frac{C_C^c C_D^d}{C_A^a C_B^b}\right) \quad \text{... (8.1.18)}$$

Since the net rate of reaction must be equal to zero at equilibrium, it follows from Eq. (8.1.17) that

$$(k_1/k_2) = \left(\frac{C_C^c C_D^d}{C_A^a C_B^b}\right) = K \quad \text{... (8.1.19)}$$

It is to be noted that a relationship of the above kind can be developed only if the reactions are elementary in nature. In the case of nonelementary reactions, it is difficult to correlate equilibrium, reaction rates and concentrations in a manner as above. For example, though $K$ is defined as per Eq. (8.1.19) or (8.1.20) for both elementary and nonelementary reactions, for nonelementary reactions $K$ will not necessarily be equal to $(k_1/k_2)$.

A more generalised definition of $K$ that can be applied to non-ideal systems as well, is

$$K = \left(\frac{a_C^c a_D^d}{a_A^a a_B^b}\right) \quad \text{... (8.1.20)}$$

where $a_A$, $a_B$, $a_C$ and $a_D$ are the *activities* of components $A$, $B$, $C$ and $D$ respectively at equilibrium. If the components form ideal solutions, then the activities will be proportional to molar concentrations and Eq. (8.1.20) will reduce to (8.1.18). This equilibrium constant is practically unaffected by the pressure of the system, by the presence or absence of inerts and by the kinetics of the reaction, but is affected by the temperature of the system. The rate of change of $K$ with temperature is given by the *Van't Hoff equation:*

$$\frac{d(\ln K)}{dT} = \left(\frac{\Delta H_r}{RT^2}\right) \qquad \text{... (8.1.21)}$$

were $\Delta H_r$ is the heat of reaction (in J/kmole) and $T$ is the absolute temperature (in °K). For exothermic reactions, $\Delta H_r$ is negative and for endothermic reactions, $\Delta H_r$ is positive. In some specific cases, $\Delta H_r$ may be considered to be constant within the temperature interval under consideration. In such cases, Eq. (8.1.21) can be readily integrated to give

$$\ln (K_2/K_1) = (-\Delta H_r/R) \left(\frac{1}{T_2} - \frac{1}{T_1}\right) \qquad \text{... (8.1.22)}$$

It follows that in the case of exothermic reactions, $K$ will decrease with increase in temperature, while for endothermic reactions, $K$ will increase with increase in temperature. Thus, in the case of exothermic reversible reactions (oxidation of $SO_2$ to $SO_3$ is an example), provisions must be made for removing the heat of reaction to avoid a decrease in $K$. Similarly, for endothermic reactions (dehydrogenation of hydrocarbons such as butanes and butenes is an example), heat must be added from outside to maintain the temperature if a decrease in $K$ is to be avoided.

The maximum conversion (that is, fraction of reactants transformed or converted into products) that can be achieved in the case of reversible reactions is the *equilibrium conversion.* It is to be noted that though $K$ is unaffected by pressure or inerts, the equilibrium conversion of reactants can be influenced by these variables. If $K$ is very large (much larger than unity), then practically complete conversion is possible and the reaction can be considered to be essentially irreversible. If $K$ is very small (much smaller than unity), then it means that the reaction will not proceed to any appreciable extent. It is also apparent that the equilibrium conversion will increase with increase in temperature in the case of endothermic reactions, while for exothermic reactions, it will decrease with increase in temperature. For gas-phase reactions that take place with a decrease in total number of moles (for example, $2SO_2 + O_2 \rightleftharpoons 2SO_3$), the equilibrium conversion increases with increase in pressure, while for those that take place with an increase in total number of moles (for example, $2N_2O \rightleftharpoons 2N_2 + O_2$), conversion drops with increase in pressure.

As stated earlier, the rate of homogeneous reactions is a function of composition and temperature. The temperature dependencyof the rate is well-explained, once we determine how the value of the reaction rate constant ($k$) varies with temperature. In the case of elementary reactions, the temperature dependency of $k$ is well represented by *Arrhenius' law*:

$$k = k_0 \exp(-E/RT) \qquad \text{... (8.1.23)}$$

where $k_0$ is called the *frequency factor* and has the same units as $k$. $E$ is the *activation energy* and was considered by Arrhenius as the amount of energy in excess of the average energy level which the reactants must possess in order for the reaction to proceed. Both $k_0$ and $E$ vary little with temperature. However, factors like pressure and presence of a catalyst have been found to influence the value of $E$. Neglecting the temperature dependency of $E$ and $k_0$, Arrhenius' law predicts that a plot of ($\ln k$) versus ($1/T$) will be a straight line of slope equal to ($-E/R$). (Apparently, reactions with high activation energies will be highly temperature–sensitive, while those with low activation energies will be relatively temperature–

insensitive). If the plot of (ln $k$) versus $(1/T)$ deviates significantly from a straight-line plot, then it implies that the reaction is complex.

Though Arrhenius' law is around 100 years old, it predicts the effect of temperature on the rate constant for simple reactions so accurately that it still finds wide application in the analysis of a large number of reaction kinetics problems. More elaborate expressions for improving the Arrhenius' equation have been derived from various theories such as the *collision theory* and the *activated complex theory* (also called *the transition state theory*). According to the *collision theory*, originally postulated by Lewis[9] and Polanyi[10], for a bimolecular (molecularity = 2) gas-phase reaction of the form $A + B \rightarrow C + D$,

$$k = (\sigma_{AB})^2 N_0 \left[\frac{8\pi RT (M_A + M_B)}{(M_A M_B)}\right]^{0.5} \exp(-E/RT) \quad \text{... (8.1.24)}$$

where $M$ is the molecular weight and $\sigma_{AB}$ is the *collision diameter* (that is, the effective molecular diameter of $A$ plus $B$ upon collision). $N_0$ is the Avagadro number. The collision theory predicts results that are in good agreement with experimental data for a number of bimolecular gas-phase reactions (the reaction between $H_2$ and $I_2$ to form $HI$ is an excellent example). It has also been satisfactory for several reactions in solution involving simple ions. However, for many other reactions, this theory predicts substantially larger values of reaction rates. The descrepancy becomes more pronounced when the complexity of the reaction increases. Moreover, unimolecular decompositions are difficult to rationalise by the collision theory.

The *activated complex theory*, originally advanced by Eyring and Coworkers[11], postulates that the reactants (say, $A$ and $B$) first form an activated intermediate complex, which, being in a highly energised state, subsequently decomposes to yield the product (say, $C$). The activated complex is assumed to be in thermodynamic equilibrium with the reactants. The mechanism may be represented as

$$A + B \underset{k_2}{\overset{k_1}{\rightleftharpoons}} AB^* \xrightarrow{k_3} C \quad \text{... (8.1.25)}$$

Since equilibrium is assumed for the first step, the rate controlling step is the decomposition of the activated complex. Therefore the rate of the overall reaction may be expressed as

$$r = k_3 [AB^*] \quad \text{... (8.1.26)}$$

where the term given within the square brackets represents the molar concentration of the activated complex*. It is also shown[12] that the rate constant $k_3$ is given by

$$k_3 = k_B T / h \quad \text{... (8.1.27)}$$

where $k_B$ is *Boltzmann's constant* and $h$ is *Planck's constant* (to note that we have purposefully used the notation $k_B$ to represent Boltzmann's constant, since the notation $k$ has already been used to represent the rate constant). If the equilibrium constant for the formation of $AB^*$ is $K^*$, then in terms of activities,

$$K^* = a_{AB^*}/(a_A a_B) \quad \text{... (8.1.28)}$$

* The concentration of any substance (say, $A$) may be represented either as $C_A$ or as $[A]$. The latter notation is often prefered since it is unsuffixed.

We may rewrite the above expression in terms of *activity coefficients* and the molar concentrations, as

$$K^* = \left(\frac{\gamma_{AB^*}}{\gamma_A \gamma_B}\right) \frac{[AB^*]}{[A][B]} \quad \text{... (8.1.29)}$$

If the substances form ideal solutions, then the activity coefficients will be equal to unity and $K^* = [AB^*]/[A]\ [B]$. Equation (8.1.26), therefore, becomes

$$r = (k_B T/h)\ K^* \left(\frac{\gamma_A \gamma_B}{\gamma_{AB^*}}\right) [A]\ [B] \quad \text{... (8.1.30)}$$

The rate constant for the overall reaction is, therefore, given by

$$k = (k_B T/h)\ K^*\ (\gamma_A \gamma_B / \gamma_{AB^*}) \quad \text{... (8.1.31)}$$

From thermodynamics, the equilibrium constant is related to the free energy change $\Delta F$ (the difference between the free energies of the products and the reactants) as $K = \exp(-\Delta F/RT)$. Therefore

$$K^* = \exp(-\Delta F^*/RT) \quad \text{... (8.1.32)}$$

Since $\Delta F^* = \Delta H^* - T\Delta S^*$, where $\Delta S^*$ is the change in entropy (or, the entropy of activation) and $\Delta H^*$ is the enthalpy change for formation of the activated complex, the above equation reduces to

$$K^* = \exp\left(\frac{\Delta S^*}{R} - \frac{\Delta H^*}{RT}\right) \quad \text{... (8.1.33)}$$

Substituting in Eq. (8.1.31), we get

$$k = (k_B T/h) \left(\frac{\gamma_A \gamma_B}{\gamma_{AB^*}}\right) \exp(\Delta S^*/R)\ \exp(-\Delta H^*/RT) \quad \text{... (8.1.34)}$$

Comparison of the above equation with Arrhenius' equation shows that

$$k_0 = (k_B T/h) \left(\frac{\gamma_A \gamma_B}{\gamma_{AB^*}}\right) \exp(\Delta S^*/R) \quad \text{... (8.1.35)}$$

$$E = \Delta H^* \quad \text{... (8.1.36)}$$

Thus, activated complex theory suggests that $E$ is the enthalpy change for formation of the activated complex. However, to predict this enthalpy, we must know the structure of the activated complex which is very difficult to determine. Even if an experimental value of $E$ is available, we still require to know the structure of activated complex, this time to calculate the entropy of activation ($\Delta S^*$) and thereby the frequency factor $k_0$. This uncertainty about the structure of the activated complex is the serious limitation of this theory. However, it does provide a qualitative interpretation of how molecules react and a reassuring foundation for the empirical rate expressions inferred from experimental data.

The difference in approach between the collision theory and the activated complex theory is not difficult to understand. The collision theory assumes that the reaction rate is governed by the number of energetic collisions between reactants. It assumes that the intermediate complex, if at all formed, breaks down into products so rapidly that it does not influence the rate of the overall reaction. Activated complex theory, on the other hand, views the reaction rate to be governed by the rate of decomposition of the intermediate complex. The rate of formation of the complex is assumed to be so rapid that it is present in equilibrium concentrations at all times. In other words, the collision theory assumes that the first step of Eq. (8.1.25) is slow and rate-controlling, whereas the activated complex theory considers the second step of Eq. (8.1.25) to be rate controlling.

In general, all the above theories predict that

$$k \propto T^m \exp(-E/RT) \qquad \text{... (8.1.37)}$$

where $m = 0$ according to Arrhenius' law, $m = 0.5$ as per collision theory and the activated complex theory postulates that $m = 1.0$. It has been reported that in complex situations, the value of $m$ could be as large as 3.0 to 4.0. However, for most reactions, the effect of the exponential term is so much greater than $T^m$ term that Eq. (8.1.37) essentially reduces to $k \propto \exp(-E/RT)$, which is nothing but the Arrhenius' law.

Reactions with high activation energies are more temperature sensitive than reactions with low activation energies. Also, a high temperature favours the reaction of higher activation energy and a low temperature favours the reaction of lower activation energy. This observation, which is quite important in the case of multiple reactions (two or more reactions taking place simultaneously or consecutively), can be deduced mathematically as well. For example, if $k_1$ is the rate constant of one reaction and $k_2$ that of the other, then from Arrhenius' law

$$(k_1/k_2) = \frac{k_0 \exp\left(-E_1/RT\right)}{k_0{}' \exp\left(-E_2/RT\right)} = (k_0/k_0') \exp[(E_2 - E_1)/RT] \qquad \text{... (8.1.38)}$$

Thus, if $E_2$ is larger than $E_1$, $(k_1/k_2)$ will decrease with increase in temperature. In other words, the rate of the second reaction (with higher activation energy) will be much higher than that of the first at high temperatures. The converse is true if $E_1 > E_2$. This information is very much useful in deciding the operating temperature of multiple reactions like those given below:

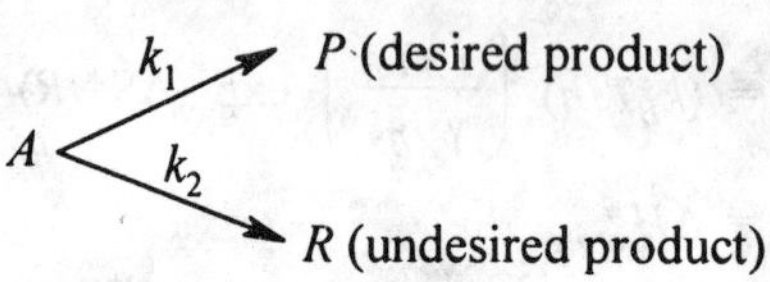


$A \xrightarrow{k_1} P \xrightarrow{k_2} R$ (undesired product). Thus, if $E_1 > E_2$, then the process must be conducted at a high temperature and if $E_1 < E_2$, then a low $T$ must be employed.

We have stated earlier that in the case of elementary reactions, the actual mechanism of the reaction can be predicted directly from the stoichiometric equation. However, in the case of nonelementary or complex reactions, the reaction mechanism cannot be postulated from the stoichiometric equation. Mechanisms of such reactions can be understood only from experimental data. By mechanism, we mean all the individual collisional or elementary processes involving molecules (atoms, free radicals and ions included) that take place simultaneously or consecutively in producing the observed overall rate[12]. The overall reaction may be assumed to be taking place in a number of steps, each step consisting of one or more elementary reactions. This is the basic principle of *chain reactions*. The different steps will be then chain initiation, chain propagation and chain termination. It may so happen that one of the steps or one of the elementary reactions is the slowest and thereby is rate–controlling. The other steps or reactions are much faster and therefore, do not materially influence the overall rate of the process.

The kinetics of the overall reaction will then reflect the kinetics of the rate–controlling step or reaction. Such a reaction will conform to a simple first or second order rate equation, even though its overall mechanism may be quite complex. An example is the pyrolysis (thermal decomposition) of ethane for the manufacture of ethylene. The main overall reaction is

$$C_2H_6 \rightarrow C_2H_4 + H_2 \quad \text{... (8.1.39)}$$

Although there are complications connecting this reaction (for example, apart from ethylene, good amounts of $CH_4$ and $C_2H_2$ are also formed), under most circumstances it is first order[13], the kinetics being largely determined by the first step of a chain mechanism:

$$C_2H_6 \rightarrow 2CH_3\cdot \qquad \text{Initiation}$$

$$\left.\begin{aligned} CH_3\cdot + C_2H_6 &\rightarrow C_2H_5\cdot + CH_4 \\ C_2H_5\cdot &\rightarrow C_2H_4 + H\cdot \\ H\cdot + C_2H_6 &\rightarrow C_2H_5\cdot + H_2 \end{aligned}\right\} \qquad \text{Chain propagation}$$

Eventually, the reaction chains are broken by the termination reactions. It can be seen that *free radicals* such as $CH_3\cdot$, $C_2H_5\cdot$, $H\cdot$ are formed during the process. These free radicals, being highly reactive, further react with reactant molecules and thereby propagate the chain reaction. These free radicals are, therefore, called *chain carriers*. The chain is propagated through a great many cycles until it is terminated by a combination of free radicals. At times, the chain is broken when one of the activated molecules collides with the wall of the reactor or with a foreign material that may be present as an impurity. Free radical reactions are thus greatly affected by the presence of traces of impurities. The existence of free radicals have been adequately evidenced. They are either free atoms or fragments of stable molecules containing one or more unpaired electrons.

The thermal decomposition of propane into methane, ethylene and butane is another example of chain reaction that takes place with the formation of free radicals:

$$C_3H_8 \rightarrow CH_3\cdot + C_2H_5\cdot \qquad \text{Initiation}$$

$$\left.\begin{aligned} CH_3\cdot + C_3H_8 &\rightarrow CH_4 + C_3H_7\cdot \\ C_3H_7\cdot &\rightarrow CH_3\cdot + C_2H_4 \end{aligned}\right\} \qquad \text{Propagation}$$

The $CH_3\cdot$ produced continues to combine with $C_3H_8$ and the chain is propagated. Finally,

$$CH_3\cdot + C_3H_7\cdot \rightarrow C_4H_{10} \qquad \text{Termination}$$

The reaction between $H_2$ and $Br_2$, the rate equation for which is given in Eq. (8.1.10), also follows the free-radical chain reaction mechanism:

$$Br_2 \rightleftharpoons 2Br\cdot \qquad \text{Initiation and Termination}$$

$$\left.\begin{array}{l} Br\cdot + H_2 \rightarrow HBr + H\cdot \\ H\cdot + Br_2 \rightarrow HBr + Br\cdot \\ H\cdot + HBr \rightarrow Br\cdot + H_2 \end{array}\right\} \text{Propagation}$$

Another class of nonelementary reactions do not follow the chain mechanism, but follow an alternate mechanism in which an intermediate product is formed which, being relatively unstable, decomposes readily to yield the products. A good example is the enzyme–catalysed fermentation reaction. The overall reaction may be represented as

$$S \xrightarrow{\text{enzyme }(E)} P \quad \text{... (8.1.40)}$$

where $S$ is the substrate (which is nothing but the reactant) and $P$ is the product. The reaction is viewed to proceed as follows:

$$S + E \rightleftharpoons (ES) \quad \text{... (8.1.41)}$$

$$(ES) \rightarrow E + P \quad \text{... (8.1.42)}$$

The above scheme was first proposed by Michaelis and Menten[16]. The intermediate formed may be molecular (as in the above example) or ionic in nature. Ionic reactions occur mainly in aqueous solutions and they are often catalysed by bases or acids (acid–catalysed oxidation of isopropyl alcohol is an example). Ionic reactions occur in the gas phase only at high temperatures or when the gases are exposed to high voltage electric discharge or X-ray irradiation (photochemical reactions). There is strong experimental evidence to show that such intermediates do get formed in a variety of reactions. However, at any particular instant, the concentration of the intermediate in the reaction mixture will be very small since it rapidly decomposes into the products.

A few other nonelementary reactions tend to follow the mechanism postulated by the activated complex theory, described earlier. In this case, an activated intermediate complex is assumed to be formed from the reactants, which subsequently decomposes to yield the products. Though there is no direct experimental evidence for the existence of such activated complexes, this mechanism does explain observed data in a good number of cases.

In short, determining the mechanism of a reaction is a very difficult task and may require the work of many investigators over a period of many years. It is not necessary to know the mechanism of a reaction in order to design a reactor. What is necessary is a satisfactory rate equation. However, successful procedures for predicting rates of reactions will not be developed until the reaction mechanisms are better understood.

**Example 8.1:** Harkness et al[14] report that for the homogeneous dimerisation of butadiene ($CH_2 = CH\cdot CH = CH_2$), the experimental value of activation energy is 23960 cal/mole.

(*a*) Compute the rate constant at 6000° K from collision theory. Assume that the effective collision diameter is $5.0 \times 10^{-8}$ cm.

(*b*) To what entropy of activation does the above result correspond?

**Solution:** (*a*) According to collision theory, the rate constant is given by Eq. (8.1.24):

$$k = (\sigma_{AB})^2 N_0 \left[\frac{8\pi RT(M_A + M_B)}{M_A M_B}\right]^{1/2} \exp(-E/RT)$$

where $\sigma_{AB} = 5 \times 10^{-8}$ cm

$R = 8.3 \times 10^7$ erg/(mole·K) (or 1.98 cal/mole K)

$T = 600°$ K

$M_A = M_B = 54$

$E = 23960$ cal/mole

$N_0 = 6.02 \times 10^{23}$ molecules/mole

Therefore,

$$k = (5 \times 10^{-8})^2 (6.02 \times 10^{23}) \left[\frac{(8.0)\pi(8.3\times10^7)(600)(108)}{(54.0)(54.0)}\right]^{1/2}$$

$$\exp\left[\frac{-23960.0}{1.98(600)}\right]$$

$$= 3.24 \times 10^{14} \exp(-12101/600)$$

$$= 564398.7 \text{ cm}^3/(\text{mole}\cdot\text{s})$$

$$= 564.3987 \text{ m}^3/(\text{kmole}\cdot\text{s})$$

(*b*) From collision theory, the frequency factor is

$$k_0 = 3.24 \times 10^{14} \text{ cm}^3/(\text{mole}\cdot\text{s})$$

From activated complex theory, the frequency factor [from Eq. (8.1.35)] is

$$k_0 = (k_B T/h) \exp(\Delta S^*/R)$$

Assuming ideal behaviour, we have taken the activity coefficients to be equal to unity. Since $k_B$ = Boltzmann's constant = $1.38 \times 10^{-23}$ J/K and $h$ = Planck's constant = $6.624 \times 10^{-34}$ Js and $R$ = 1.98 cal/(mole·K),

$$3.24 \times 10^{14} = \frac{(1.38\times10^{-23})(600)}{(6.624\times10^{-34})} \exp(\Delta S^*/1.98)$$

or

$$\Delta S^* = \mathbf{6.445\ cal/(mole\cdot K)}$$

**Example 8.2:** Brown and Borkowski[15] have studied the hydrolysis of $(CH_2)_6$ 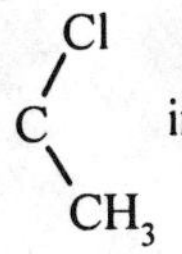

in 80 percent ethanol. The values of first order rate constant reported by them are given below:

| Temperature, °C : | 0 | 25 | 35 | 45 |
|---|---|---|---|---|
| $k$, s$^{-1}$ : | $1.06 \times 10^{-5}$ | $3.19 \times 10^{-4}$ | $9.86 \times 10^{-4}$ | $2.92 \times 10^{-3}$ |

(*a*) Compute the value of activation energy from (*i*) Arrhenius' law, (*ii*) collision theory, (*iii*) activated complex theory.

(*b*) What is the half life of this reaction at 500° C?

**Solution:** (*a*) According to Arrhenius' law,

$$k = k_0 \exp(-E/RT)$$

or

$$\ln k = \ln k_0 - (E/R)(1/T)$$

Thus, a plot of ln *k* versus (1/*T*) must yield a straight line, the slope of which will be equal to (−*E*/*R*). The values of ln *k* and (1/*T*) computed from the given data are listed below:

| | | | | | |
|---|---|---|---|---|---|
| $1/T$, $K^{-1}$ | : | $3.663 \times 10^{-3}$ | $3.356 \times 10^{-3}$ | $3.247 \times 10^{-3}$ | $3.145 \times 10^{-3}$ |
| ln *k* | : | −11.454 | −8.05 | −6.922 | −5.836 |

The plot of ln *k* versus (1/*T*) is shown in Fig. 8.2.1. From the figure,

$$\text{slope} = (-E/R) = -10909.09$$

Since $R = 1.98$ cal/(gmole·K),

$$E = (10909.09)(1.98) = 21600 \text{ cal/gmole}$$

If we express *R* in kJ/(kmole·K) so that $R = 8.313$ kJ/(kgmole K), then

$$E = (10909.09)(8.313) = \mathbf{90687.26\ kJ/kmole}$$

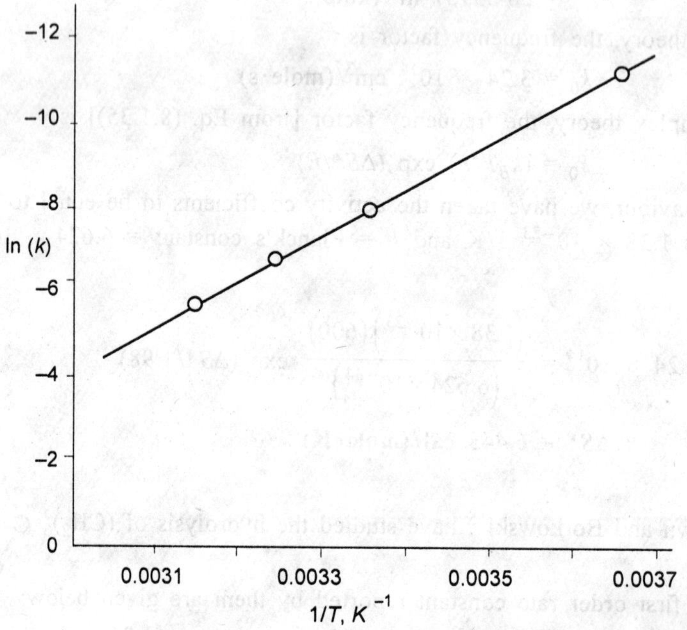


*Figure 8.2.1*

Now, according to collision theory,

$$k \propto T^{1/2} \exp(-E/RT)$$

or $$k = (\text{constant})\ T^{1/2} \exp(-E/RT)$$

Rearranging and taking logarithms,

$$\ln (k/T^{1/2}) = \ln(\text{constant}) - (E/R)(1/T)$$

Thus, a plot $\ln (k/T^{1/2})$ versus $(1/T)$ must yield a straight line of slope equal to $(-E/R)$. The plot is shown in Fig. 8.2.2.

| $1/T$, $K^{-1}$ | : | $3.663 \times 10^{-3}$ | $3.356 \times 10^{-3}$ | $3.247 \times 10^{-3}$ | $3.145 \times 10^{-3}$ |
|---|---|---|---|---|---|
| $\ln (k/T^{1/2})$ | : | $-11.22$ | $-10.9$ | $-9.787$ | $-8.71$ |

From Fig. (8.2.2), the slope of the straight line is $-10900.0$. Therefore,

$$E/R = 10900.0$$

$$E = (10900)(1.98) = 21582.0 \text{ cal/gmole} = \mathbf{90611.7\ kJ/kgmole}$$

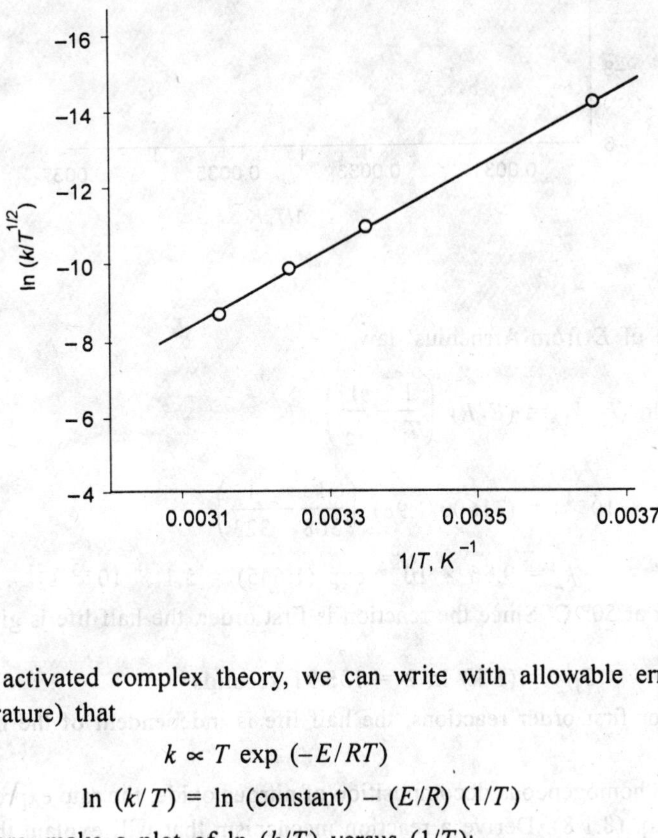


*Figure 8.2.2*

Based on the activated complex theory, we can write with allowable error (since $\Delta S^*$ is a poor function of temperature) that

$$k \propto T \exp(-E/RT)$$

or $$\ln (k/T) = \ln(\text{constant}) - (E/R)(1/T)$$

Let us, therefore, make a plot of $\ln (k/T)$ versus $(1/T)$:

| $1/T$ | : | $3.663 \times 10^{-3}$ | $3.356 \times 10^{-3}$ | $3.247 \times 10^{-3}$ | $3.145 \times 10^{-3}$ |
|---|---|---|---|---|---|
| $\ln (k/T)$ | : | $-17.06$ | $-13.747$ | $-12.65$ | $-11.518$ |

The plot is shown in Fig. 8.2.3. From the figure,

$$\text{slope} = -E/R = -10666.67$$

Therefore, $$E = (10667.67)\ (1.98) = 21120 \text{ cal/mole} = \mathbf{88672.03\ kJ/kmole}$$

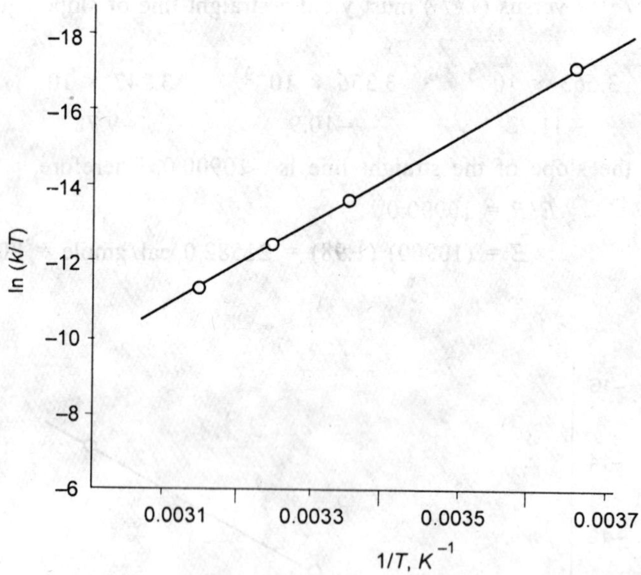


*Figure 8.2.3*

(*b*) Taking the value of $E$ from Arrhenius' law,

$$\ln\ (k_2/k_1) = (E/R)\left(\frac{1}{T_1} - \frac{1}{T_2}\right)$$

$$\ln\ (k_2/9.86 \times 10^{-4}) = (21600/1.98)\left(\frac{1}{308} - \frac{1}{323}\right)$$

or $$k_2 = 9.86 \times 10^{-4} \exp\ (1.645) = 5.1 \times 10^{-3}\ s^{-1}$$

This is the value of $k$ at 50° C. Since the reaction is first order, the half life is given by (from Table 8.1A of Section 8.2)

$$\theta_{1/2} = (1/k) \ln 2 = \mathbf{135.91\ seconds}$$

It can be seen that for first order reactions, the half life is independent of the initial concentration of the reactant.

**Example 8.3:** For the homogeneous decomposition of nitrous oxide, the rate expression derived from kinetic data is given in Eq. (8.1.8). Derive a reaction mechanism that will explain the observed kinetic data.

**Solution:** Let us assume that the reaction mechanism involves formation of an intermediate product $N_2O^*$ which, being unstable, decomposes readily (by further reacting with $N_2O$) to yield the products. Thus, let

$$N_2O \underset{k_2}{\overset{k_1}{\rightleftharpoons}} N_2O^*$$

$$N_2O + N_2O^* \xrightarrow{k_3} 2N_2 + O_2$$

Both of the above steps are elementary reactions. The rate of decomposition of $N_2O$ is, therefore,

$$(-r_A) = k_1\,[N_2O] - k_2\,[N_2O^*] + k_3\,[N_2O]\,[N_2O^*] \qquad \text{... (i)}$$

where $A$ stands for $N_2O$. The concentration of $N_2O^*$ in the reaction mixture at any instant will be very small since it is converted as soon as it is produced and therefore, the rate of change of its concentration with time may be assumed to be equal to zero. This is what is meant by the *steady state hypothesis* (also called *stationary state hypothesis*). Thus,

$$\frac{d[N_2O^*]}{d\theta} = k_1\,[N_2O] - k_2\,[N_2O^*] - k_3\,[N_2O]\,[N_2O^*] = 0$$

or

$$[N_2O^*] = \frac{k_1\,[N_2O]}{k_2 + k_3\,[N_2O]} \qquad \text{... (ii)}$$

Substituting for $[N_2O^*]$ in Eq. (i), we get

$$-r_A = k_1\,[N_2O]\left\{1 + \frac{k_3\,[N_2O] - k_2}{k_3\,[N_2O] + k_2}\right\}$$

$$= \frac{2k_1k_3\,[N_2O]^2}{k_2 + k_3\,[N_2O]} = \frac{(2k_1k_3/k_2)[N_2O]^2}{1 + (k_3/k_2)[N_2O]} = \frac{k\,[N_2O]^2}{1 + k'\,[N_2O]}$$

It can be seen that the above expression is same as Eq. (8.1.8). Thus, the proposed mechanism agrees with the kinetic data.

**Example 8.4:** An enzyme–catalysed fermentation reaction may be represented as follows:

$$S \xrightarrow{\text{Enzyme}} P$$

where $S$ is the substrate (namely, the reactant) and $P$ is the product. Michaelis and Menten[16,17] have developed the following rate equation for the above biochemical reaction:

$$r_P = \frac{k\,[E_0]}{1 + K_m/[S]}$$

where $K_m$ is called the Michaelis-Menten constant. Derive a reaction mechanism that is compatible with the above rate equation.

**Solution:** We shall once again assume that an intermediate product is being formed which further decomposes to yield the product. Thus, let

$$S + E \underset{k_2}{\overset{k_1}{\rightleftharpoons}} ES$$

$$ES \xrightarrow{k_3} P + E$$

The rate of formation of product $P$ is, therefore,

$$r_P = k_3\ [ES] \qquad \text{... (i)}$$

Based on the steady state approximation, the rate of change of $[ES]$ with time is equal to zero. Thus,

$$r_{ES} = \frac{d[ES]}{d\theta} = k_1\ [S]\ [E] - k_2\ [ES] - k_3\ [ES] = 0$$

or

$$[ES] = \frac{k_1 [S][E]}{(k_2 + k_3)} \qquad \text{... (ii)}$$

Since the enzyme will be present either free or complexed,

$$[E_0] = [E] + [ES] \qquad \text{... (iii)}$$

where $[E_0]$ is the total concentration of enzyme in the system. This is given by the concentration of enzyme initially introduced into the mixture. Substituting for $[E]$ in Eq. (ii) and rearranging, we get

$$[ES] = \frac{k_1 [S][E_0]}{k_1 [S] + (k_2 + k_3)} \qquad \text{... (iv)}$$

Therefore, from Eq. (i),

$$r_P = \frac{k_1 k_3 [E_0][S]}{k_1 [S] + (k_2 + k_3)} = \frac{k_3 [E_0]}{1 + K_m / [S]} \qquad \text{... (v)}$$

where $K_m = (k_2 + k_3)/k_1$. It can be seen that the above equation is same as that predicted by Michaelis and Menten. A few interesting features can be observed with respect to this mechanism. The reaction rate is proportional to the total concentration of enzyme, namely $[E_0]$. During the initial stages of reaction when $[S]$ is very large, $r_P \simeq k_3\ [E_0]$ = constant. This means that the reaction is zero order when the concentration of substrate is large. Similarly, at low values of $[S]$, $r_P \simeq (k_3/K_m)\ [E_0]\ [S] = k\ [S]$ and this shows that the reaction is first order in $S$ at low concentrations of $S$.

**Example 8.5:** It has been postulated that the thermal decomposition of diethyl ether occurs by the following chain mechanism:

$$(C_2H_5)_2O \xrightarrow{k_1} CH_3\cdot + \cdot CH_2OC_2H_5 \qquad \text{Initiation}$$

$$\left.\begin{array}{l} CH_3\cdot + (C_2H_5)_2O \xrightarrow{k_2} C_2H_6 + \cdot CH_2OC_2H_5 \\ \cdot CH_2OC_2H_5 \xrightarrow{k_3} CH_3\cdot + CH_3CHO \end{array}\right\} \qquad \text{Propagation}$$

$$CH_3\cdot + \cdot CH_2OC_2H_5 \xrightarrow{k_4} \text{end products} \qquad \text{Termination}$$

Based on the steady state hypothesis, show that the rate of decomposition is first order in ether concentration.

**Solution:** Based on the proposed mechanism, the rate of decomposition of ether (let us denote ether by the symbol $A$) is given by

$$(-r_A) = k_1\,[A] + k_2\,[CH_3\cdot]\,[A] \quad \text{... (i)}$$

According to the steady state hypothesis, $d\,[CH_3\cdot]/d\theta = d\,[\cdot CH_2OC_2H_5]/d\theta = 0$. Thus,

$$\frac{d[\cdot CH_2OC_2H_5]}{d\theta} = k_1\,[A] + k_2\,[CH_3\cdot]\,[A] - k_3\,[\cdot CH_2OC_2H_5] - k_4\,[CH_3\cdot]\,[\cdot CH_2OC_2H_5] = 0$$

or

$$[\cdot CH_2OC_2H_5] = \frac{k_1\,[A] + k_2\,[CH_3\cdot][A]}{k_3 + k_4\,[CH_3\cdot]} \quad \text{... (ii)}$$

Now,

$$d\,[CH_3\cdot]/d\theta = k_1\,[A] - k_2\,[CH_3\cdot]\,[A] + k_3\,[\cdot CH_2OC_2H_5] - k_4\,[CH_3\cdot]\,[\cdot CH_2OC_2H_5] = 0$$

Substituting for $[\cdot CH_2OC_2H_5]$ from Eq. (ii) and rearranging, we get

$$\{k_1 - k_2\,[CH_3\cdot]\}\,\{k_3 + k_4\,[CH_3\cdot]\} + \{k_3 - k_4\,[CH_3\cdot]\}\,\{k_1 + k_2\,[CH_3\cdot]\} = 0$$

or

$$[CH_3\cdot] = \left(\frac{k_1 k_3}{k_2 k_4}\right)^{1/2} = k' \quad \text{... (iii)}$$

Therefore, from Eq. (i),

$$(-r_A) = k_1\,[A] + k_2 k'\,[A] = (k_1 + k_2 k')\,[A] = k\,[A]$$

Thus, the reaction is first order in ether concentration.

**Example 8.6:** Consider the reaction whose stoichiometry is given by $2A_2B \rightarrow 2AB + A_2$. Though no concise rate equation could be developed for this reaction, the kinetic studies show that

(*i*) At high concentrations of $A_2B$ (that is, during the initial stages of the reaction), the reaction is first order in $A_2B$,

(*ii*) At low concentrations of $A_2B$, the reaction is second order in $A_2B$,

(*iii*) Introduction of product $AB$ into the feed does not affect the reaction rate,

(*iv*) Introduction of product $A_2$ into the feed tends to decrease the rate of reaction.

Devise a suitable reaction mechanism that complies with the above observations. Use the steady state hypothesis for your analysis.

**Solution:** Let us first consider the following mechanism:

$$2A_2B \underset{k_2}{\overset{k_1}{\rightleftharpoons}} A_4B_2^*$$

$$A_4B_2^* \xrightarrow{k_3} A_2 + 2AB$$

As per the above mechanism, the rate of decomposition of $A_2B$ (for sake of convenience, let us represent the reactant $A_2B$ as $S$) is

$$(-r_S) = \frac{d[A_2B]}{dt} = k_1\,[A_2B]^2 - 2k_2\,[A_4B_2^*] \quad \text{... (i)}$$

According to the steady state approximation, $d\,[A_4B_2^*]/d\theta$ must be equal to zero. Thus

$$d\,[A_4B_2^*]/d\theta = (1/2)\;k_1\;[A_2B]^2 - k_2\;[A^4B_2^*] - k_3\;[A_4B_2^*] = 0$$

or
$$[A_4B_2^*] = \frac{k_1\left[A_2B\right]^2}{2\left(k_2+k_3\right)} \qquad \text{... (ii)}$$

Substituting in Eq. (i), we get

$$(-r_S) = k_1\;[A_2B]^2\left[1-\frac{k_2}{(k_2+k_3)}\right]$$

$$= \left(\frac{k_1k_3}{k_2+k_3}\right)[A_2B]^2 = k\;[A_2B]^2 \qquad \text{... (iii)}$$

The reaction is thus second order in $A_2B$ for all concentrations and it does not show any shifting order characteristics. The above mechanism, therefore, does not agree with the kinetic observations.

Let us, therefore, try an alternate mechanism:

$$A_2B \underset{k_2}{\overset{k_1}{\rightleftharpoons}} B^* + A_2$$

$$A_2B + B^* \xrightarrow{k_3} 2AB$$

According to this mechanism,

$$(-r_S) = k_1\;[A_2B] - k_2\;[B^*]\;[A_2] + k_3\;[A_2B]\;[B^*] \qquad \text{... (iv)}$$

Now, based on the steady state hypothesis,

$$d\,[B^*]/d\theta = k_1\;[A_2B] - k_2\;[B^*]\;[A_2] - k_3\;[A_2B]\;[B^*] = 0$$

or
$$[B^*] = \frac{k_1\left[A_2B\right]}{k_2\left[A_2\right]+k_3\left[A_2B\right]} \qquad \text{... (v)}$$

Substituting in Eq. (iv) and rearranging, we get

$$(-r_S) = \frac{2k_1k_3\left[A_2B\right]^2}{k_2\left[A_2\right]+k_3\left[A_2B\right]} \qquad \text{... (vi)}$$

At high values of $[A_2B]$,

$$(-r_S) \simeq \frac{2k_1k_3\left[A_2B\right]^2}{k_3\left[A_2B\right]} = 2k_1\;[A_2B] \qquad \text{... (vii)}$$

Thus, the reaction is essentially first order in $A_2B$ at high concentrations of $A_2B$. When $[A_2B]$ is very small,

$$(-r_S) \simeq \frac{2k_1k_3\left[A_2B\right]^2}{k_2\left[A_2\right]} \qquad \text{... (viii)}$$

The reaction is thus second order in $A_2B$ at low concentrations of $A_2B$. From Eq. (vi), it is clear that the rate of decomposition of $A_2B$ is independent of $[AB]$ and therefore, any addition of $AB$ into the feed shall not affect the reaction rate. Introduction of $A_2$ into the feed shall decrease the reaction rate, though the rate of decrease will not be directly proportional to the increase in $[A_2]$. Thus, this second mechanism is in compliance with the kinetic data and therefore, may be accepted as the true mechanism of the reaction.

## 8.2 DETERMINATION OF RATE EQUATIONS FOR HOMOGENEOUS REACTIONS

The first step in the design of reactors is the development of a reliable rate equation. Rate equations for nonelementary or complex reactions are derived from experimental kinetic data, collected by conducting the reaction in a laboratory bench scale reactor. For studying homogeneous reactions, a batch reactor is usually employed due to its simplicity and versatility, whereas heterogeneous reactions are usually studied in a flow reactor.

It is common practice to operate the experimental batch reactor isothermally and at constant volume. Specific amounts of the reactants are charged into the reactor and the progress of the reaction is observed by recording the change in concentration of a particular component with time. In the case of gas-phase reactions, the change in total pressure with time may be recorded and the concentration versus time data computed from Eq. (8.2.60). The data so collected are then fitted into the rate equation either by the *integral method* or by the *differential method.* The former involves comparison of predicted and observed compositions, while the latter compares the predicted and observed rates.

Let us consider the integral method first. For relatively simple reactions with simple rate expressions, this method is, no doubt, the best since it is easy to use and demands lesser amount of experimental data. For more complex situations however, the differential method would be more suitable. The integral method is described below:

For a constant volume batch reactor, the rate expression is that given in Eq. (8.1.3) and we know that the rate is a function of $k$ and the concentrations of materials. In other words,

$$(-r_A) = \frac{-dC_A}{d\theta} = f(k, C) \qquad \text{... (8.2.1)}$$

For many chemical reactions with relatively simple kinetic mechanism, the above equation may be rewritten as

$$(-r_A) = \frac{-dC_A}{d\theta} = kf(C) \qquad \text{... (8.2.2)}$$

If the concentrations of all materials are expressed in terms of $C_A$,

$$\frac{-dC_A}{d\theta} = kf(C_A) \qquad \text{... (8.2.3)}$$

or

$$-\int_{C_{A0}}^{C_A} \frac{dC_A}{f(C_A)} = k\int_0^{\theta} d\theta = k\theta \qquad \text{... (8.2.4)}$$

where $C_{A0}$ is the initial concentration of $A$ in the reaction mixture (at $\theta = 0$). It is clear that if we know the expression for $f(C_A)$, then the above integral can be evaluated in terms of $C_A$ and if this concentration function is plotted against time, a straight line must result. The procedure, therefore, involves assuming a particular rate expression and then evaluating the integral to get the concentration function and checking whether the plot of the concentration function versus time fits into a straight line or not. If so, the assumed rate expression is correct. If not, a new rate expression is to be tried. Let us consider a few examples. Let us assume that the rate expression for the reaction under consideration is

$$(-r_A) = kC_A \qquad \text{... (8.2.5)}$$

This will be true if the reaction is first order in $A$ and zero order in all other reactants. Also, the reaction is to be irreversible. Now, $f(C_A) = C_A$ and Eq. (8.2.4) reduces to

$$-\int_{C_{A0}}^{C_A} \frac{dC_A}{C_A} = k\theta$$

or

$$\ln (C_{A0}/C_A) = k\theta \qquad \text{... (8.2.6)}$$

Thus, if a plot of $\ln (C_{A0}/C_A)$ versus $\theta$ results in a straight line, then the rate expression for the reaction is that given in Eq. (8.2.5). Equation (8.2.6) may also be expressed in terms of the fractional conversion of $A$ (namely, $x_A$), since

$$x_A = (N_{A0} - N_A)/N_{A0} = 1 - (N_A/N_{A0}) \qquad \text{... (8.2.7)}$$

Since the volume of the reaction mixture is constant,

$$x_A = 1 - \left(\frac{N_A/V}{N_{A0}/V}\right) = 1 - (C_A/C_{A0}) \qquad \text{... (8.2.8)}$$

Equation (8.2.6), therefore, becomes

$$-\ln (1 - x_A) = k\theta \qquad \text{... (8.2.9)}$$

Thus, if the assumed rate equation is correct, then a plot of either $\ln (C_{A0}/C_A)$ versus $\theta$ or $-\ln (1 - x_A)$ versus $\theta$ will give a straight line and the slope of this straight line will give the value of the rate constant k.

Now, suppose we assume that the rate equation is

$$(-r_A) = kC_A^{0.5} \qquad \text{(8.2.10)}$$

The above expression is true for an irreversible reaction that is half order in $A$, but zero order in all other reactants. Substituting $f(C_A) = C_A^{1/2}$ in Eq. (8.2.4) and integrating, we get

$$2\,(C_{A0}^{1/2} - C_A^{1/2}) = k\theta \qquad \text{... (8.2.11)}$$

Thus, if the rate expression for the reaction conforms to Eq. (8.2.10), then the measured values of $C_A$ and $\theta$ must fit into the above equation. Similarly, if we suppose

$$(-r_A) = kC_A^2 \qquad \text{... (8.2.12)}$$

which is the rate expression for an irreversible reaction that is second order in $A$, but zero order in all other reactants, then Eq. (8.2.4) can be integrated to give.

$$(1/C_A) - (1/C_{A0}) = k\theta \quad \text{... (8.2.13)}$$

In the same way, if the rate equation is

$$(-r_A) = k\,C_A^3 \quad \text{... (8.2.14)}$$

then Eq. (8.2.4) becomes

$$(1/2)\,[(1/C_A)^2 - (1/C_{A0})^2] = k\theta \quad \text{... (8.2.15)}$$

In general, if the assumed rate expression is of the form

$$(-r_A) = k\,C_A^n \quad \text{... (8.2.16)}$$

which corresponds to an irreversible reaction that is nth order in $A$, but zeroth order in all other reactants, then Eq. (8.2.4) will become (for $n \neq 1$),

$$C_A^{1-n} - C_{A0}^{1-n} = (n-1)\,k\theta \quad \text{... (8.2.17)}$$

To note that the above expression is valid for $n \neq 1$ only.

There is an alternate procedure for fitting the data, which is called the *half-life method*. In this case, instead of concentration versus time data, the time required for one-half of the reactant to disappear (or, the time required for the concentration of a particular reactant to drop to one-half of the original value) is measured starting from different initial concentrations of the reactant. If we denote this half-life as $\theta_{1/2}$, then at $\theta = \theta_{1/2}$, $C_A = (C_{A0}/2)$ and $x_A = 1/2$. If we use this condition in Eq. (8.2.17), we get

$$\theta_{1/2} = \frac{\left(2^{n-1} - 1\right)}{k\,(n-1)}\; C_{A0}^{1-n} \quad \text{... (8.2.18)}$$

Thus, a plot of $\theta_{1/2}$ versus $C_{A0}^{1-n}$ must give a straight line if the assumed rate expression is correct. The reaction is conducted with different initial concentrations of $A$ and in each case, the half-life is recorded. It is then checked whether the data fit into the above equation. This procedure is often more convenient, since, in many cases, it is simpler to measure the half-life than to obtain the concentration–versus–time data.

So far, we have considered cases where the reaction is zeroth order in all reactants except $A$. However, this need not necessarily be so. If the rate equation is expected to contain concentration terms of more than one reactant, then to derive the rate expression by the above method, the reaction stoichiometry must also be known. For example, let us assume that the rate expression is of the form

$$(-r_A) = k\,C_A\,C_B \quad \text{... (8.2.19)}$$

Such a reaction is a second order reaction, but it is first order in $A$ and first order in $B$. It is zero order in all other reactants. Now, to integrate Eq. (8.2.4), $C_B$ must be expressed in terms of $C_A$. This can be done from reaction stoichiometry. Let the stoichiometric equation representing the reaction be

$$aA + bB + \ldots\ldots \rightarrow \text{Products} \quad \text{... (8.2.20)}$$

Since $a$ moles of $A$ react with $b$ moles $B$, $y$ moles of $A$ will react with $(b/a)\,y$ moles of $B$. Thus, concentration of $A$ after a particular interval of time (within which $y$ moles of $A$ have reacted) will be

$$C_A = C_{A0} - (y/V) \quad \text{... (8.2.21)}$$

Accordingly, the concentration of $B$ will be

$$C_B = C_{B0} - (b/a)\, y/V \quad \text{... (8.2.22)}$$

Substituting for $y$ from Eq. (8.2.21) into Eq. (8.2.22), we get

$$C_B = C_{B0} - (b/a)\,(C_{A0} - C_A) \quad \text{... (8.2.23)}$$

Let $M = (C_{B0}/C_{A0})$ and since $x_A = 1 - (C_A/C_{A0})$, the above equation reduces to

$$C_B = C_{A0}\,[M - (b/a)\,x_A] \quad \text{... (8.2.24)}$$

Equation (8.2.4), therefore, becomes

$$-\int_{C_{A0}}^{C_A} \frac{dC_A}{C_A C_{A0}\left[M - (b/a)\,x_A\right]} = k\theta \quad \text{... (8.2.25)}$$

or, in terms of $x_A$,

$$\int_0^{x_A} \frac{dx_A}{(1-x_A)\left[M - \left(\frac{b}{a}\right)x_A\right]} = C_{A0}\,k\theta \quad \text{... (8.2.26)}$$

The above function can be integrated by the *method of partial fractions* (for $M \neq b/a$) to give

$$\left[\frac{1}{M-(b/a)}\right] \ln\left[\frac{M-(b/a)\,x_A}{M\,(1-x_A)}\right] = C_{A0}k\theta \quad \text{... (8.2.27)}$$

In terms of concentrations,

$$\ln\left(\frac{C_{A0}C_B}{C_A C_{B0}}\right) = [C_{B0} - (b/a)\,C_{A0}]\,k\theta \quad \text{... (8.2.28)}$$

By substituting $\theta = \theta_{1/2}$ and $C_A = C_{A0}/2$ (or, $x_A = 1/2$), we get the expression for half-life as

$$\theta_{1/2} = \frac{1}{k\left[C_{B0} - (b/a)\,C_{A0}\right]} \ln\left[\frac{2C_{B0} - (b/a)\,C_{A0}}{C_{B0}}\right] \quad \text{... (8.2.29)}$$

The above three equations are applicable only if $M \neq (b/a)$. If $M = (b/a)$, that is, the reactants are present in stoichiometric proportions, then Eq. (8.2.26) reduces to

$$(a/b)\int_0^{x_A} \frac{dx_A}{(1-x_A)^2} = C_{A0}\,k\theta \quad \text{... (8.2.30)}$$

or

$$x_A/(1 - x_A) = (b/a)\,C_{A0}\,k\theta \quad \text{... (8.2.31)}$$

The expression for half-life is,

$$\theta_{1/2} = \left(\frac{a}{bk}\right)\left(\frac{1}{C_{A0}}\right) \quad \text{... (8.2.32)}$$

Thus, if the observed $C_A$ versus $\theta$ data fit into Eq. (8.2.27) or (8.2.31), or the half-life versus $C_{A0}$ data fit into Eq. (8.2.29) or (8.2.32), then the rate equation will be that given by Eq. (8.2.19).

Let us now consider a few reactions that follow third order kinetics. For example, let

$$(-r_A) = k\,C_A C_B^2 \qquad \text{... (8.2.33)}$$

The above equation corresponds to an irreversible reaction that is first order in $A$ and second order in $B$ (the overall order of the reaction being 3.0). It is zero order in all other reactants. Let the stoichiometry of the reaction be represented by Eq. (8.2.20) itself. Then, Eq. (8.2.4) reduces to

$$-\int_{C_{A0}}^{C_A} \frac{dC_A}{C_A C_{A0}^2\left[M-(b/a)\,x_A\right]^2} = k\theta \qquad \text{... (8.2.34)}$$

In terms of $x_A$ the above equation becomes

$$\int_0^{x_A} \frac{dx_A}{C_{A0}^2\,(1-x_A)\left[M-(b/a)\,x_A\right]^2} = k\theta \qquad \text{... (8.2.35)}$$

Integrating by the method of partial fractions and rearranging, we get

$$\ln\left[\frac{M-(b/a)\,x_A}{M\,(1-x_A)}\right] + \frac{(b/a)\,x_A\left[(b/a)-M\right]}{M\left[M-(b/a)\,x_A\right]} = [M-(b/a)]^2\,C_{A0}^2\,k\theta \qquad \text{... (8.2.36)}$$

A plot of left hand side of above equation (which is a function of $x_A$ only) versus $\theta$ must give a straight line if the assumed rate equation is correct. Equation (8.2.36) may also be expressed in terms of concentrations as

$$\ln\left(\frac{C_{A0}C_B}{C_A C_{B0}}\right) + \frac{\left[(b/a)\,C_{A0}-C_{B0}\right]\left(C_{B0}-C_B\right)}{C_B C_{B0}} = [(b/a)\,C_{A0}-C_{B0}]^2\;k\theta \qquad \text{... (8.2.37)}$$

The expression for half-life is given by

$$\theta_{1/2} = \left(\frac{1}{k\,\alpha_1^2}\right)\ln\,(\alpha_2/C_{B0}) + \left(\frac{(b/a)\,C_{A0}}{C_{B0}\,\alpha_1\alpha_2\,k}\right) \qquad \text{... (8.2.38)}$$

where

$$\alpha_1 = (b/a)\;C_{A0} - C_{B0} \qquad \text{... (8.2.39)}$$

$$\alpha_2 = 2\,C_{B0} - (b/a)\;C_{A0} \qquad \text{... (8.2.40)}$$

The above Eqs (8.2.36) to (8.2.38) are valid only if $M \neq (b/a)$. If $A$ and $B$ are introduced in the stoichiometric ratio so that $M = (b/a)$, then Eq. (8.2.35) reduces to

$$\int_0^{x_A} \frac{dx_A}{(1-x_A)^3} = C_{A0}^2\;(b/a)^2\;k\theta \qquad \text{... (8.2.41)}$$

or

$$(1-x_A)^{-2} - 1 = 2\,C_{A0}^2\;(b/a)^2\;k\theta \qquad \text{... (8.2.42)}$$

In terms of concentrations,

$$\left(\frac{1}{C_A^2}-\frac{1}{C_{A0}^2}\right) = 2\;(b/a)^2\;k\theta \qquad \text{... (8.2.43)}$$

The half-life is given by

$$\theta_{1/2} = \left(\frac{3}{2C_{A0}^2 (b/a)^2 k}\right) \quad \text{... (8.2.44)}$$

It is not unlikely that in the case of reactions that follow third order kinetics, the rate is a function of the concentrations of more than two components. For example, let the rate equation be

$$(-r_A) = k\, C_A\, C_B\, C_D \quad \text{... (8.2.45)}$$

and the stoichiometric equation be

$$aA + bB + dD + \ldots\ldots \rightarrow \text{Products} \quad \text{...(8.2.46)}$$

Such a reaction is first order in $A$, first order in $B$, first order in $D$ and zero order in all other reactants. The overall order of the reaction is 3.0. For this, Eq. (8.2.4) becomes

$$-\int_{C_{A0}}^{C_A} \frac{dC_A}{C_A C_B C_D} = k\theta \quad \text{... (8.2.47)}$$

Now, $C_B$ can be expressed in terms of $C_A$ from Eq. (8.2.24) and similarly, $C_D = C_{A0}\,[M' - (d/a)\,x_A]$, where $M' = (C_{D0}/C_{A0})$. We also know that $C_A = C_{A0}\,(1 - x_A)$. By making these substitutions, Eq. (8.2.47) can be integrated by the method of partial fractions to give

$$\left[\frac{1}{M-(b/a)}\right] \ln\left[\frac{\left(M-\frac{b}{a}x_A\right)}{M(1-x_A)}\right] + \left[\frac{1}{(b/a)M' - M}\right] \ln\left[\frac{M'\left(M-\frac{b}{a}x_A\right)}{M\left(M'-\frac{d}{a}x_A\right)}\right]$$

$$= [M' - (d/a)]\; C_{A0}^2\, k\theta \quad \text{... (8.2.48)}$$

In terms of concentrations,

$$\frac{1}{[C_{B0} - (b/a)\,C_{A0}]} \ln\left(\frac{C_B C_{A0}}{C_{B0} C_A}\right) + \frac{1}{[(b/d)\,C_{D0} - C_{B0}]} \ln\left(\frac{C_B C_{D0}}{C_{B0} C_D}\right)$$

$$= [C_{D0} - (d/a)\;C_{A0}]\;k\theta \quad \text{... (8.2.49)}$$

By putting $x_A = 1/2$ and $\theta = \theta_{1/2}$ in Eq. (8.2.48), we get the expression for half-life which, when expressed in terms of concentrations, will be

$$\theta_{1/2} = \left(\frac{1}{k\alpha_1\alpha_3}\right) \ln\,(\alpha_2/C_{B0}) + \left(\frac{1}{k\alpha_3\alpha_4}\right) \ln\left(\frac{C_{D0}\,\alpha_2}{C_{B0}\,\alpha_2{}'}\right) \quad \text{... (8.2.50)}$$

where $\alpha_1 = C_{B0} - (b/a)\;C_{A0}$

$\alpha_2 = 2C_{B0} - (b/a)\;C_{A0}$

$\alpha_3 = C_{D0} - (d/a)\;C_{A0}$

$\alpha_4 = (b/d)\ C_{D0} - C_{B0}$

$\alpha_2' = 2C_{D0} - (d/a)\ C_{A0}$

For the specific case in which the reactants are fed in stoichiometric proportions such that $M = (b/a)$ and $M' = (d/a)$, Eq. (8.2.47) will get simplified to

$$\int_0^{x_A} \frac{dx_A}{(1-x_A)^3} = C_{A0}^2\ (bd/a^2)\ k\theta \qquad \text{... (8.2.51)}$$

which, as can be noticed, is analogous to Eq. (8.2.41). On integration

$$(1 - x_A)^{-2} - 1 = 2C_{A0}^2\ (bd/a^2)\ k\theta \qquad \text{... (8.2.52)}$$

And the half-life is $\quad \theta_{1/2} = 3/[2C_{A0}^2\ (bd/a^2)\ k]$ ... (8.2.53)

All the cases discussed above are listed in Tables (8.1 A) and (8.1 B).

We have stated at the beginning of this section that in the case of isothermal gas-phase reactions, instead of concentration versus time data, total pressure versus time data may be used since the two are closely inter-related. We shall illustrate this with an example. Consider an isothermal gas-phase reaction that is represented by the stoichiometric equation

$$aA + bB + dD \rightarrow pP + qQ + rR \qquad \text{... (8.2.54)}$$

Let the number of moles of the substances at $\theta = 0$ be $N_{A0}$, $N_{B0}$, $N_{D0}$, $N_{P0}$, $N_{Q0}$ and $N_{R0}$ so that the total number of moles of reaction mixture at $\theta = 0$ is

$$N_0 = N_{A0} + N_{B0} + N_{D0} + N_{P0} + N_{Q0} + N_{R0} \qquad \text{... (8.2.55)}$$

Within a time $\theta$, $y$ moles of $A$ react. Therefore, at the end of time $\theta$, number of moles of $A$ will be $(N_{A0} - y)$, that of $B$ will be $N_{B0} - (b/a)\ y$ and that of $D$, $N_{D0} - (d/a)\ y$. The number of moles of $P$, $Q$ and $R$ will be $N_{P0} + (p/a)\ y$, $N_{Q0} + (q/a)\ y$ and $N_{R0} + (r/a)\ y$ respectively. Thus, the total number of moles of the reaction mixture at time $\theta$ is

$$N = (N_{A0} + N_{B0} + N_{D0} + N_{P0} + N_{Q0} + N_{R0}) - y\left(1 + \frac{b}{a} + \frac{d}{a}\right) + y\left(\frac{p}{a} + \frac{q}{a} + \frac{r}{a}\right)$$

$$= N_0 + (y/a)\ (p + q + r - a - b - d)$$

$$= N_0 + (y/a)\ \Delta n \qquad \text{... (8.2.56)}$$

or

$$y = (N - N_0)\ a/\Delta n \qquad \text{... (8.2.57)}$$

We know that $\quad C_A = C_{A0} - (y/V)$ ... (8.2.58)

Substituting for $y$ from Eq. (8.2.57),

$$C_A = C_{A0} - \left[\frac{(N - N_0)\ a}{V\ \Delta n}\right] \qquad \text{... (8.2.59)}$$

## Table 8.1 A

## Rate Equations for Simple Irreversible Reactions

| *Reaction type* | *Rate equation* | *Integrated rate equation (constant volume)* |
|---|---|---|
| Zero'th order | $(-r_A) = k$ | $k\theta = (C_{A0} - C_A)$<br>$\theta_{1/2} = C_{A0}/2k$ |
| First order in $A$, zero order in all other reactants | $(-r_A) = kC_A$ | $k\theta = \ln (C_{A0}/C_A)$<br>$\theta_{1/2} = (1/k) \ln 2$ |
| Second order in $A$, zero order in all other reactants | $(-r_A) = kC_A^2$ | $k\theta = (1/C_A) - (1/C_{A0})$<br>$\theta_{1/2} = 1/(kC_{A0})$ |
| Half order in $A$, zero order in all other reactants | $(-r_A) = kC_A^{1/2}$ | $k\theta = 2\,(C_{A0}^{1/2} - C_A^{1/2})$<br>$\theta_{1/2} = \dfrac{2\left(1-1/\sqrt{2}\right)}{k}\,(C_{A0}^{1/2})$ |
| Third order in $A$, zero order in all other reactants | $(-r_A) = kC_A^3$ | $k\theta = (1/2)\,[(1/C_A)^2 - (1/C_{A0})^2]$<br>$\theta_{1/2} = \left(\dfrac{3}{2k}\right)(1/C_{A0}^2)$ |
| $n$th order in $A$, zero order in all other reactants | $(-r_A) = kC_A^n$<br>$n \neq 1$ | $k\theta = \left(\dfrac{1}{n-1}\right)(C_A^{1-n} - C_{A0}^{1-n})$<br>$\theta_{1/2} = \left(\dfrac{2^{n-1}-1}{k(n-1)}\right) C_{A0}^{1-n}$ |

Assuming that ideal gas law is valid, $(N/V) = P/RT$ and $(N_0/V) = P_0/RT$, where $P$ and $P_0$ are the values of total pressure at $\theta = 0$ and at time $\theta$ respectively. Therefore, Eq. (8.2.59) becomes

$$C_A = C_{A0} - \left[\frac{(P - P_0)\,a}{RT\,\Delta n}\right] \quad \text{... (8.2.60)}$$

Thus, once $P$ versus $\theta$ data are known, the $C_A$ versus $\theta$ data can be computed from the above equation. Eq. (8.2.60) may also expressed in terms of the partial pressure of $A$ (since $p_A = C_A RT$):

$$p_A = p_{A0} - (a/\Delta n)\,(P - P_0) \quad \text{... (8.2.61)}$$

It is important to keep in mind that to employ the above equation in computations, the exact stoichiometry of the reaction must be known. The procedure becomes inapplicable if more than one

## Table 8.1 B

| *Reaction order and stoichiometry* | *Rate equation* | *Integrated rate equation (constant volume)* |
|---|---|---|
| First order in $A$, first order in $B$, zero order in all other reactants.<br>$aA + bB + \cdots \to$ Products | $(-r_A) = k\,C_A C_B$<br>$(C_{B0}/C_{A0}) \neq (b/a)$ | $k\theta = \left(\frac{1}{C_{A0}}\right)\left[\frac{1}{M-(b/a)}\right]\ln\left[\frac{M-(b/a)\,x_A}{M\,(1-x_A)}\right]$<br>$\theta_{1/2} = \frac{1}{k\left[C_{B0}-(b/a)\,C_{A0}\right]}\ln\left[\frac{2C_{B0}-(b/a)\,C_{A0}}{C_{B0}}\right]$ |
| Same as above. | $(-r_A) = k\,C_A C_B$<br>$(C_{B0}/C_{A0}) = b/a$ | $k\theta = \left(\frac{x_A}{1-x_A}\right)\left[\frac{1}{(b/a)\,C_{A0}}\right]$<br>$\theta_{1/2} = \left(\frac{a}{bk}\right)(1/C_{A0})$ |
| First order in $A$, second order in $B$, zero order in all other reactants.<br>$aA + bB + \cdots \to$ Products | $(-r_A) = k\,C_A C_B^2$<br>$(C_{B0}/C_{A0}) \neq b/a$ | $k\theta = (1/\alpha_1^2)\,\ln\left(\frac{C_{A0}C_B}{C_A C_{B0}}\right) + \frac{(C_{B0}-C_B)}{(\alpha_1 C_B C_{B0})}$<br>$\theta_{1/2} = \left(\frac{1}{k\alpha_1^2}\right)\ln(\alpha_2/C_{B0}) + \left[\frac{(b/a)}{\alpha_1\alpha_2\,kC_{B0}}\right]C_{A0}$<br>$\alpha_1 = (b/a)\,C_{A0} - C_{B0}$<br>$\alpha_2 = 2C_{B0} - (b/a)\,C_{A0}$ |
| Same as above. | $(-r_A) = k\,C_A C_B^2$<br>$(C_{B0}/C_{A0}) = b/a$ | $k\theta = [(1-x_A)^{-2} - 1]/[2\,(C_{A0}b/a)^2]$<br>$\theta_{1/2} = 3/[2k\,(C_{A0}b/a)^2]$ |
| First order in $A$, first order in $B$, first order in $D$, zeroth order in all other reactants.<br>$aA + bB + dD \cdots \to$ Products | $(-r_A) = k\,C_A C_B C_D$<br>$(C_{B0}/C_{A0}) \neq b/a$<br>$(C_{D0}/C_{A0}) \neq d/a$ | $k\theta = \left(\frac{1}{\alpha_1\alpha_3}\right)\ln\left(\frac{C_B C_{A0}}{C_A C_{B0}}\right) + (1/\alpha_3\alpha_4)\ln\left(\frac{C_B C_{D0}}{C_{B0}C_D}\right)$<br>$\alpha_1 = C_{B0} - (b/a)\,C_{A0}$<br>$\alpha_3 = C_{D0} - (d/a)\,C_{A0}$<br>$\alpha_4 = (b/d)\,C_{D0} - C_{B0}$<br>$\theta_{1/2} = \left(\frac{1}{k\,\alpha_1\,\alpha_3}\right)\ln[2-(b/a)\,C_{A0}/C_{B0}] + \left(\frac{1}{k\,\alpha_3\,\alpha_4}\right)\ln\left\{\frac{C_{D0}\left[2C_{B0}-(b/a)\right]}{C_{B0}\left[2C_{D0}-(d/a)\,C_{A0}\right]}\right\}$ |
| Same as above. | $(-r_A) = k\,C_A C_B C_D$<br>$(C_{B0}/C_{A0}) = (b/a)$<br>$(C_{D0}/C_{A0}) = (d/a)$ | $k\theta = [(1-x_A)^{-2} - 1]/(2C_{A0}^2\,bd/a^2)$<br>$\theta_{1/2} = 3/(2C_{A0}^2\;kbd/a^2]$ |

stoichiometric equation is required to represent the reaction. It also becomes invalid when there is no change in number of moles (that is, when $\Delta n = 0$).

For those readers who are familiar with computer programming, we shall present here a computer program that can be conveniently executed for deriving the most probable rate equation for irreversible reactions. All the cases discussed above are included in this program. For each case, based on the assumed rate equation, a straight line fit is tried with the available concentration–versus–time data or the half life–versus–initial concentration data by the *method of least squares.* The average error (also called the truncation error) involved in fitting the straight line, represented as $Q(J)$ in the program, is also computed in each case. The case for which the magnitude of $Q(J)$ is below the prescribed tolerance (say, less than 0.001) provides the rate equation for the reaction under consideration and the slope of the straight line gives the value of the rate constant k [represented as $AK(J)$ in the program]. If it is found that the value of $Q(J)$ is quite large for all of the cases considered, then it means that the reaction does not conform to a simple, irreversible mechanism, it may be a reversible or complex reaction (the analysis of which are discussed subsequently in this section).

The difference between average error and root mean square error should not be confused. For a straight line fit of the form $y_i = mx_i$, the root mean square error is obtained as

$$(\text{Error})^2 = S = \sum_{i=1}^{N} (y_i - mx_i)^2$$

According to the method of least squares, the best fit is that for which $S$ is minimum. Putting $(dS/dm) = 0$ and solving for $m$, we get

$$m = \Sigma x_i y_i / \Sigma (x_i)^2$$

The average error involved in the fit is given by

$$q = (1/N) \sum_{i=1}^{N} (y_i - mx_i)/y_i$$

## PROGRAM 8.1

```
C**    RATE EQUATION FOR IRREVERSIBLE REACTIONS.
C**    METH = 1 USES CONCENTRATION VERSUS TIME DATA.
C**    METH = 2 USES HALF LIFE VERSUS INITIAL CONCENTRATION DATA.
       READ *, METH
       IF (METH, EQ. 2) GO TO 200
       REAL M
       DIMENSION CA (50), THETA (50), X (50), Y (50)
       DIMENSION XA (50), CB (50), Q (8), AK (8)
       READ (*, *) N, CA0, CB0, CD0
       READ *, [CA (I), THETA (I), I = 1, N]
       READ *, A, B, D
```

```
      J = 0
      SUMXY = 0
      SUMXX = 0
100   J = J + 1
      DO 10 I = 1, N
      X (I) = THETA (I)
      IF (J. EQ. 2) GO TO 11
      IF (J. EQ. 3) GO TO 12
      IF (J. EQ. 4) GO TO 13
      IF (J. EQ. 5) GO TO 14
      IF (J. EQ. 6) GO TO 15
      IF (J. EQ. 7) GO TO 16
      IF (J. EQ. 8) GO TO 17
C**   RATE EQUATION OF THE FORM -rA = k.
      Y (I) = CA0 - CA (I)
      GO TO 18
C**   RATE EQUATION OF THE FORM -rA = k CA.
11    Y (I) = ALOG [CA0/CA (I)]
      GO TO 18
C**   RATE EQUATION OF THE FORM -rA = k CA^2
12    Y (I) = -(1.0/CA0) + [1.0/CA (I)]
      GO TO 18
C**   RATE EQUATION OF THE FORM -rA = k CA^1/2
13    Y (I) = 2.0 * {SQRT (CA0) - SQRT [CA (I)]}
      GO TO 18
C**   RATE EQUATION OF THE FORM -rA = k CA^3
14    Y (I) = 0.5 * {[1.0/CA (I)] **2 - (1.0/CA0) **2}
      GO TO 18
C**   RATE EQUATION OF THE FORM -rA = k CA CB
15    AM = CB0/CA0
      XA (I) = 1.0 - CA (I)/CA0
      BA = B/A
      IF (AM. EQ. BA) GO TO 30
      A1 = CA0 * (AM - BA)
      A2 = [AM - BA * XA (I)]/{AM * [1.0 - XA (I)]}
      Y (I) = (1.0/A1) * ALOG (A2)
      GO TO 18
30    Y (I) = A * XA (I)/{B * CA0 * [1.0 - XA (I)]}
      GO TO 18
```

```
C**    RATE EQUATION OF THE FORM -rA = k CA CB^2
16     AM = CB0/CA0
       XA (I) = 1.0 - CA (I)/CA0
       BA = B/A
       IF (AM. EQ. BA) GO TO 31.
       CB (I) = CB0 - BA * [CA0 - CA (I)]
       A1 = BA * CA0 - CB0
       A2 = CA0 * CB (I)/[CA (I) * CB0]
       A3 = [CB0 - CB (I)]/[A1 * CB (I) * CB0]
       Y (I) = (1.0/A1 ** 2) * ALOG (A2) + A3
       GO TO 18
31     A1 = {1.0/[1.0 - XA (I)] ** 2} - 1.0
       Y (I) = (0.5 * A1)/(CA0 * BA) ** 2
       GO TO 18
C**    RATE EQUATION OF THE FORM -rA = k CA CB CD
17     AM = CB0/CA0
       AD = CD0/CA0
       XA (I) = 1.0 - CA (I)/CA0
       BA = B/A
       DA = D/A
       IF. (AM. EQ. BA. AND. AD. EQ. DA) GO TO 32
       CB (I) = CB0 - BA * [CA0 - CA (I)]
       CD (I) = CD0 - DA * [CA0 - CA (I)]
       A1 = CB0 - BA * CA0
       A2 = CA0 * CB (I)/[CA (I) * CB0]
       A3 = CD0 - DA * CA0
       A4 = (BA/DA) * CD0 - CB0
       A5 = CB (I) * CD0/[CD (I) * CB0]
       Y (I) = [1.0/(A1 * A3)] * ALOG (A2) + [1.0/(A3 * A4)] * ALOG (A5)
       GO TO 18
32     A1 = {1.0/[1.0 - XA (I)] ** 2} - 1.0
       Y (I) = A1 * 0.5/(CA0 * CA0 * BA * DA)
18     SUMXY = SUMXY + X (I) * Y (I)
       SUMXX = SUMXX + X (I) * X (I)
10     CONTINUE
       M = SUMXY/SUMXX
       SUMQ = 0.0
       DO 20 I = 1, N
       SUMQ = SUMQ + ABS [Y (I) - M * X (I)]/Y (I)
```

```
20     CONTINUE
       Q (J) = SUMQ/N
       AK (J) = M
       WRITE (*, *) Q (J), AK (J)
       IF (J. LT. 8) GO TO 100
       GO TO 400
C**    RATE EQUATION FROM HALF LIFE VERSUS INITIAL CONCN DATA
200    REAL M
       DIMENSION CA0 (50), HALIFE (50), X (50), Y (50)
       DIMENSION Q (8), AK (8), CB0 (50), CD0 (50), AM (50)
       READ *, N, A, B, D
       READ *, [CA0 (I), HALIFE (I), I = 1, N]
       READ *, [CB0 (I), CD0 (I), I = 1, N]
       J = 0
       SUMXY = 0
       SUMXX = 0
300    J = J + 1
       DO 50 I = 1, N
       X (I) = HALIFE (I)
       IF (J. EQ. 2) GO TO 51
       IF (J. EQ. 3) GO TO 52
       IF (J. EQ. 4) GO TO 53
       IF (J. EQ. 5) GO TO 54
       IF (J. EQ. 6) GO TO 55
       IF (J. EQ. 7) GO TO 56
C**    RATE EQUATION OF THE FORM $-r_A = k$
       Y (I) = 0.5 * CA0 (I)
       GO TO 60
C**    RATE EQUATION OF THE FORM $-r_A = k C_A^2$
51     Y (I) = 1.0/CA0 (I)
       GO TO 60
C**    RATE EQUATION OF THE FORM $-r_A = k C_A^{1/2}$
52     Y (I) = (0.828/1.414) * SQRT [CA0 (I)]
       GO TO 60
C**    RATE EQUATION OF THE FORM $-r_A = k C_A^3$
53     Y (I) = 1.5/[CA0 (I) ** 2]
       GO TO 60
C**    RATE EQUATION OF THE FORM $-r_A = k C_A C_B$
54     BA = B/A
```

```
      DO 65 K = 1, N
      AM (K) = CB0 (K)/CA0 (K)
      IF [AM (K). NE. BA] GO TO 66
65    CONTINUE
      Y (I) = 1.0/[BA * CA0 (I)]
      GO TO 60
66    A1 = CB0 (I) - BA * CA0 (I)
      A2 = 2.0 * CB0 (I) - BA * CA0 (I)
      Y (I) = (1.0/A1) * ALOG [A2/CB0 (I)]
      GO TO 60
C**   RATE EQUATION OF THE FORM -rA = k CA CB^2
55    BA = B/A
      DO 70 K = 1, N
      AM (K) = CB0 (K)/CA0 (K)
      IF [AM (K). NE. BA] GO TO 71
70    CONTINUE
      Y (I) = 1.5/[CA0 (I) * BA] ** 2
      GO TO 60
71    A1 = BA * CA0 (I) - CB0 (I)
      A2 = 2.0 * CB0 (I) - BA * CA0 (I)
      Y (I) = [1.0/(A1 * A1)] * ALOG [A2/CB0 (I)] + BA * CA0 (I) / [A1 * A2 * CB0 (I)]
      GO  TO  60
C**   RATE EQUATION OF THE FORM -rA = k CA CB CD
      BA = B/A
      DA = D/A
      D0 75 I = 1,N
      AM (K) = CB0 (K)/CA0 (K)
      AD (K) = CD0 (K)/CA0 (K)
      IF [AM (K). NE. BA] GO TO 76
      IF [AD (K). NE. DA] GO TO 76
75    CONTINUE
      Y (I) = 1.5/[CA0 (I) * CA0 (I) * BA * DA)
      GO TO 60
76    A1 = CB0 (I) - BA * CA0 (I)
      A2 = 2.0 * CB0 (I) - BA * CA0 (I)
      A3 = CD0 (I) - DA * CA0 (I)
      A4 = (BA/DA) * CD0 (I) - CB0 (I)
      A5 = 2.0 * CD0 (I) - DA * CA0 (I)
      Y1 = [1.0/(A1 * A3)] * ALOG [A2/CB0 (I)]
```

```
      Y2 = [1.0/(A3 * A4)] * ALOG {CD0 (I) * A2/[CB0 (I) * A5]}
      Y (I) = Y1 + Y2
60    SUMXY = SUMXY + X (I) * Y (I)
      SUMXX = SUMXX + X (I) * X (I)
50    CONTINUE
80    M = SUMXY/SUMXX
      SUMQ = 0.0
      DO 25 I = 1, N
      SUMQ = SUMQ + ABS [Y (I) - M * X (I)]/Y (I)
25    CONTINUE
      Q (J) = SUMQ/N
      AK (J) = M
      WRITE (*, *) Q (J), AK (J)
      IF (J. LT. 7) GO TO 300
400   STOP
      END
```

The notations used in the above program are all self-explanatory. For further details, see the nomenclature section at the end of this chapter.

So far, we have considered only the irreversible reactions. A large number of chemical reactions of industrial importance are reversible in nature. As a result, the reaction never goes to completion. The maximum conversion that can be obtained is the equilibrium conversion. Development of rate equation for reversible reactions is basically a complicated process. We shall consider here a few of the relatively simple cases.

For example, let the rate equation be of the form

$$(-r_A) = -\frac{dC_A}{d\theta} = k_1 C_A - k_2 C_P \quad \text{... (8.2.62)}$$

This represents a reversible reaction in which the forward reactions first order in $A$ and the backward reaction is first order in $P$. Let the reaction stoichiometry be

$$A \underset{k_2}{\overset{k_1}{\rightleftharpoons}} P \quad \text{... (8.2.63)}$$

Once the reaction has reached equilibrium, then $dC_A/d\theta$ will be equal to zero and from Eq. (8.2.62),

$$(k_1/k_2) = C_{Pe}/C_{Ae} \quad \text{... (8.2.64)}$$

The suffix $e$ is used to represent concentration at equilibrium. Since the stoichiometric equation shows that one mole of $A$ on complete conversion produces one mole of $P$, if $x_A$ is the fractional conversion of $A$, then

$$C_P = C_{P0} + C_{A0}x_A = C_{A0}(M + x_A) \quad \text{... (8.2.65)}$$

where $M = (C_{P0}/C_{A0})$. Similarly,

$$C_{Pe} = C_{A0}\ (M + x_{Ae}) \quad \text{... (8.2.66)}$$

where $x_{Ae}$ is the equilibrium conversion of $A$. If we substitute Eq. (8.2.64) in Eq. (8.2.62), we get

$$-dC_A/d\theta = k_1\ C_A - (k_1\ C_{Ae}\ C_P/C_{Pe}) \quad \text{... (8.2.67)}$$

Substituting for $C_P$ and $C_{Pe}$ from Eqs (8.2.65) and (8.2.66) and rewriting $C_A$ and $C_{Ae}$ in terms $x_A$ and $x_{Ae}$, we get

$$\frac{dx_A}{d\theta} = \frac{k_1\,(M+1)}{(M+x_{Ae})}\ (x_{Ae} - x_A) \quad \text{... (8.2.68)}$$

On integration,

$$\int_0^{x_A} \frac{dx_A}{(x_{Ae} - x_A)} = \frac{k_1\,(M+1)}{(M+x_{Ae})} \int_0^{\theta} d\theta$$

or,

$$\frac{(M+x_{Ae})}{(M+1)} \ln\left(\frac{x_{Ae}}{x_{Ae} - x_A}\right) = k_1\theta \quad \text{... (8.2.69)}$$

In terms of concentrations,

$$k_1\theta = \frac{(C_{P0} + C_{A0} - C_{Ae})}{(C_{P0} + C_{A0})} \ln\left(\frac{C_{A0} - C_{Ae}}{C_A - C_{Ae}}\right) \quad \text{... (8.2.70)}$$

Now, let it be that the forward reaction is first order, but the backward reaction is of second order so that the rate expression is of the form

$$(-r_A) = -dC_A/d\theta = k_1 C_A - k_2 C_P C_R \quad \text{... (8.2.71)}$$

Let the stoichiometric equation be

$$A \underset{k_2}{\overset{k_1}{\rightleftharpoons}} P + R \quad \text{... (8.2.72)}$$

Let us further assume that $C_{P0} = C_{R0} = 0$. In that case, from stoichiometry, $C_P = C_R = C_{A0}x_A$ and $C_{Pe} = C_{Re} = C_{A0}x_{Ae}$. Also, since $dC_A/d\theta = 0$ at equilibrium,

$$(k_2/k_1) = \left(\frac{C_{Ae}}{C_{Pe}C_{Re}}\right) = \frac{(1 - x_{Ae})}{\left(C_{A0}x_{Ae}^2\right)} \quad \text{... (8.2.73)}$$

Equation (8.2.71) can be, therefore, expressed in terms of $x_A$ and $x_{Ae}$ as

$$\frac{dx_A}{d\theta} = k_1\ [(1 - x_A) - (1 - x_{Ae})\ (x_A^2/x_{Ae}^2)] \quad \text{... (8.2.74)}$$

On rearrangement and integration,

$$x_{Ae}^2 \int_0^{x_A} \frac{dx_A}{(x_{Ae} - x_A)\left[x_{Ae} + x_A\,(1 - x_{Ae})\right]} = k_1\theta \quad \text{... (8.2.75)}$$

The above function can be split into partial fractions and integrated, to give

$$\left[\frac{x_{Ae}}{(2-x_{Ae})}\right] \ln \left[\frac{x_{Ae}(1-x_A)+x_A}{(x_{Ae}-x_A)}\right] = k_1\theta \qquad \text{... (8.2.76)}$$

In terms of concentrations,

$$\left(\frac{C_{A0}-C_{Ae}}{C_{A0}+C_{Ae}}\right) \ln \left[\frac{C_{Ae}^2 - C_A C_{Ae}}{C_{A0}(C_A - C_{Ae})}\right] = k_1\theta \qquad \text{... (8.2.77)}$$

Let us now consider a rate equation of the form

$$(-r_A) = -dC_A/d\theta = k_1 C_A C_B - k_2 C_P C_R \qquad \text{... (8.2.78)}$$

This corresponds to a reversible reaction in which both the forward reaction and the backward reaction are of second order. Let the reaction stoichiometry be

$$A + B \underset{k_2}{\overset{k_1}{\rightleftharpoons}} P + R \qquad \text{... (8.2.79)}$$

We shall further assume that $C_{A0} = C_{B0}$ and $C_{P0} = C_{R0} = 0$. Thus means that $C_A = C_B = C_{A0}$ $(1 - x_A)$ and $C_P = C_R = C_{A0} x_A$.

Also from equilibrium relationship,

$$(k_2/k_1) = \left(\frac{C_{Ae}C_{Be}}{C_{Pe}C_{Re}}\right) = \frac{(1-x_{Ae})^2}{x_{Ae}^2} \qquad \text{... (8.2.80)}$$

Equation (8.2.78) can be now expressed in terms of $x_A$ and $x_{Ae}$ and rearranged to give

$$x_{Ae}^2 \int_0^{x_A} \frac{dx_A}{(x_{Ae}-x_A)\left[x_{Ae}+(1-2x_{Ae})x_A\right]} = k_1 C_{A0}\theta \qquad \text{... (8.2.81)}$$

Splitting into partial fractions and integrating, we get

$$\left[\frac{x_{Ae}}{2C_{A0}(1-x_{Ae})}\right] \ln \left[\frac{x_{Ae}+(1-2x_{Ae})x_A}{(x_{Ae}-x_A)}\right] = k_1\theta \qquad \text{... (8.2.82)}$$

Readers can easily verify that the above relation is valid for the following cases as well (provided $C_{A0} = C_{B0}$ and $C_{P0} = C_{R0} = 0$):

| *Stoichiometric equation* | *Rate equation* |
|---|---|
| $2A \underset{k_2}{\overset{k_1}{\rightleftharpoons}} P + R$ | $(-r_A) = k_1 C_A^2 - k_2 C_P C_R$ |
| $A + B \underset{k_2}{\overset{k_1}{\rightleftharpoons}} 2P$ | $(-r_A) = k_1 C_A C_B - k_2 C_P^2$ |
| $2A \underset{k_2}{\overset{k_1}{\rightleftharpoons}} 2P$ | $(-r_A) = k_1 C_A^2 - k_2 C_P^2$ |

Derivation of rate equation for reversible reactions of higher order or for those which do not conform to any of the cases discussed above is quite difficult by the integral method. For such cases, the search for a suitable rate equation is better performed by the *differential method.* Also, the analysis of reversible reactions can be made simpler by making the experimental measurements during the initial period only. In other words, starting from specific amounts of the reactants (say, $A$ and $B$), the concentration–versus–time data (say, $C_A$ versus $\theta$ data) are recorded and the runs are terminated before appreciable amounts of products (say, $P$ and $R$) are formed. Under these conditions, the rate of backward reaction will be negligibly small and the measured data can be analysed as though the reaction were irreversible to determine the forward rate constant ($k_1$). With this result and the equilibrium constant, the rate constant for the backward reaction ($k_2$) can be obtained. This *initial rate approach* is frequently used to simplify kinetic studies. However, this method does have limitations. In the case of many complex reactions, it often happens that the rate equation or the order of the reaction determined from the $C_A$ versus $\theta$ data at initial conditions is different from that established by data taken at later times. This happens when one of the products act as a catalyst and tends to accelarate the reaction (autocatalytic reactions) or when the products of the reaction tend to inhibit the reaction.

The process of development of rate equation becomes more complex when we consider *multiple reactions.* Yet, such reactions are not uncommon in industrial practices. In processes involving multiple reactions, multiple products are formed, some of which being more desirable and valuable than others. In the case of multiple reactions therefore, we must distinguish between the two parameters such as the *yield* and the *selectivity*. The *yield* of a specific product is defined as the fraction of reactant converted to that product. This is different from *total conversion* which is the total fraction of reactant converted to all products. The *overall selectivity* of a product is the ratio of the amount of that product produced to the amount of another. In other words, it is the ratio of the yield of one product to the yield of anothe. By *point selectivity*, we mean the ratio of the rate of production of one product to the rate of production of another product. With multiple products, there is a separate selectivity based on each pair of products.

We shall illustrate the significance of these terms with an example. Consider two irreversible reactions in parallel (that take place simultaneously) as represented below:

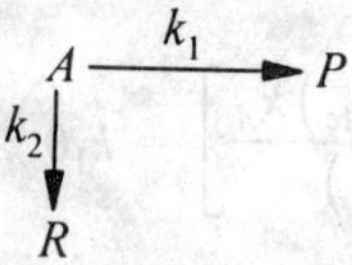


... (8.2.83)

Let $x_A$ be the total conversion of $A$ into $P$ and $R$, while $x_{A1}$ be the fractional conversion of $A$ into $P$ and $x_{A2}$ be that into $R$. The yield of $P$ is, therefore, $x_{A1}$ and that of $R$, $x_{A2}$. The overall selectivity of $P$ is given by

$$S_0 = x_{A1}/x_{A2} \quad \text{... (8.2.84)}$$

The point selectivity of $P$ is

$$S_p = \frac{dC_p/d\theta}{dC_R/d\theta} \quad \text{... (8.2.85)}$$

If both reactions are first order in $A$,

$$S_p = k_1 C_A / k_2 C_A = k_1 / k_2 \quad \text{... (8.2.86)}$$

If $C_{P0} = C_{R0} = 0$, then from the stoichiometric equation, it is clear that $C_P = C_{A0} x_{A1}$ and $C_R = C_{A0} x_{A2}$. Equation (8.2.84) can be then reexpressed as

$$S_0 = C_P / C_R \quad \text{... (8.2.87)}$$

Let the rate expression for the multiple reaction given in Eq. (8.2.83) be

$$(-r_A) = -dC_A/d\theta = k_1 C_A + k_2 C_A = (k_1 + k_2)\, C_A \quad \text{... (8.2.88)}$$

This conforms to our earlier assumption that both reactions are first order in $A$. On integration,

$$\ln\, (C_{A0}/C_A) = (k_1 + k_2)\, \theta \quad \text{... (8.2.89)}$$

Thus, a plot of $\ln\, (C_{A0}/C_A)$ versus $\theta$ must yield a straight line whose slope will be equal to $(k_1 + k_2)$. Now, the rate of production of $P$ is

$$dC_P/d\theta = k_1 C_A \quad \text{... (8.2.90)}$$

Similarly, the rate of production of $R$ is

$$dC_R/d\theta = k_2 C_A \quad \text{... (8.2.91)}$$

Dividing one by the other,

$$dC_P/dC_R = (k_1/k_2) \quad \text{... (8.2.92)}$$

On integration

$$(C_P - C_{P0}) = (k_1/k_2)\, (C_R - C_{R0}) \quad \text{... (8.2.93)}$$

or

$$C_P = (k_1/k_2)\, C_R + [C_{P0} - C_{R0}\, (k_1/k_2)] \quad \text{... (8.2.94)}$$

Thus, a plot of $C_P$ versus $C_R$ will yield a straight line of slope $(k_1/k_2)$. Since $(k_1 + k_2)$ is already known from Eq. (8.2.89), the values of $k_1$ and $k_2$ can be now computed. For deriving rate equation for multiple reactions, we have to thus make at least two plots. The experimental measurements must also record both $C_A$ versus $\theta$ data and $C_P$ versus $\theta$ data. The values of $C_R$ can be then obtained from the stoichiometric relationship:

$$(C_{A0} + C_{P0} + C_{R0}) = (C_A + C_P + C_R) \quad \text{... (8.2.95)}$$

The yield of $P$ can be obtained by dividing Eq. (8.2.90) by (8.2.88) and integrating the resulting expression. Thus

$$-dC_P/dC_A = k_1/(k_1 + k_2) \quad \text{... (8.2.96)}$$

On integration,

$$(C_{P0} - C_P) = \frac{k_1}{(k_1 + k_2)}\, (C_A - C_{A0}) \quad \text{... (8.2.97)}$$

or

$$x_{A1} = \left(\frac{C_P - C_{P0}}{C_{A0}}\right) = \frac{k_1}{(k_1 + k_2)} \left(1 - \frac{C_A}{C_{A0}}\right) = \frac{k_1}{(k_1 + k_2)}\, x_A \quad \text{... (8.2.98)}$$

Similarly, the yield of $R$ is

$$x_{A2} = \left(\frac{C_R - C_{R0}}{C_{A0}}\right) = \left(\frac{k_2}{k_1 + k_2}\right) x_A \quad \text{... (8.2.99)}$$

where $x_A$ is the total conversion of $A$ into $P$ and $R$. The overall selectivity is, therefore,

$$S_0 = x_{A1}/x_{A2} = k_1/k_2 \quad \text{... (8.2.100)}$$

It can be seen that this is also equal to the point selectivity given in Eq. (8.2.86). Although point and overall selectivities are identical for this type of first order simultaneous reactions, the two selectivities differ for most other complex reactions.

Let us consider another type of complex reaction such as two irreversible reactions in series. In other words, two irreversible reactions take place consecutively as given below:

$$A \xrightarrow{k_1} P \xrightarrow{k_2} R \quad (8.2.101)$$

Let the rate equations be

$$(-r_A) = -dC_A/d\theta = k_1 C_A \quad \text{... (8.2.102)}$$

$$r_P = dC_P/d\theta = k_1 C_A - k_2 C_P \quad \text{... (8.2.103)}$$

$$r_R = dC_R/d\theta = k_2 C_P \quad \text{... (8.2.104)}$$

This is equivalent to assuming that the first reaction is first order in $A$ and the second reaction first order in $P$. Let us also assume that $C_{P0} = C_{R0} = 0$. Equation (8.2.102), on integration, gives

$$\ln (C_{A0}/C_A) = k_1\theta$$

or

$$C_A = C_{A0} \exp(-k_1\theta) \quad \text{... (8.2.105)}$$

Equation (8.2.103), therefore, becomes

$$\frac{dC_P}{d\theta} + k_2 C_P = k_1 C_{A0} \exp(-k_1\theta) \quad \text{... (8.2.106)}$$

This is a linear first order differential equation which when solved analytically with the initial condition that at $\theta = 0$, $C_P = C_{P0} = 0$, gives

$$C_P = \frac{C_{A0}k_1}{(k_1 - k_2)} [\exp(-k_2\theta) - \exp(-k_1\theta)] \quad \text{... (8.2.107)}$$

Though the value of $k_1$ is obtained from Eq. (8.2.105), the value of $k_2$ cannot be computed directly from the above equation since it is only implicit in $k_2$. A trial and error solution will have to be resorted to. However, it is possible to compute $k_2$ from the maximum value of $C_P$. Concentration of $P$ initially increases with time (when the rate of formation of $P$ from $A$ predominates over that of conversion of $P$ to $R$), reaches a maximum and thereafter decreases. The value of $C_{Pm}$ (that is, maximum value of $C_P$) can be determined by setting $dC_P/d\theta = 0$. Thus, differentiating Eq. (8.2.107) with respect to $\theta$ and setting it equal to zero and then solving for $\theta$ (max), we get

$$\theta\,(\text{max}) = \frac{\ln(k_1/k_2)}{(k_1 - k_2)} \quad \text{... (8.2.108)}$$

Substituting in Eq. (8.2.107), we get

$$\ln (C_{Pm}/C_{A0}) = \left(\frac{k_2}{k_2 - k_1}\right) \ln (k_1/k_2) \quad \text{... (8.2.109)}$$

By plotting $C_P$ versus θ, we can find out the value of $C_{Pm}$ and then from Eq. (8.2.109), $k_2$ can be computed.

Once the concentrations of $A$ and $P$ are known, the concentration of $R$ at any time θ can be computed from the stoichiometric relationship, $C_{A0} = C_A + C_P + C_R$. Substituting for $C_A$ from Eq. (8.2.105) and for $C_P$ from Eq. (8.2.107), we get

$$C_R = C_{A0} - C_A - C_P$$

$$= C_{A0} \left[1 + \frac{k_2}{k_1 - k_2} \exp(-k_1 \theta) + \frac{k_1}{k_2 - k_1} \exp(-k_2 \theta)\right] \quad \text{... (8.2.110)}$$

The concentration profiles of $A$, $P$ and $R$ for reaction of this kind are sketched in Fig. 8.1

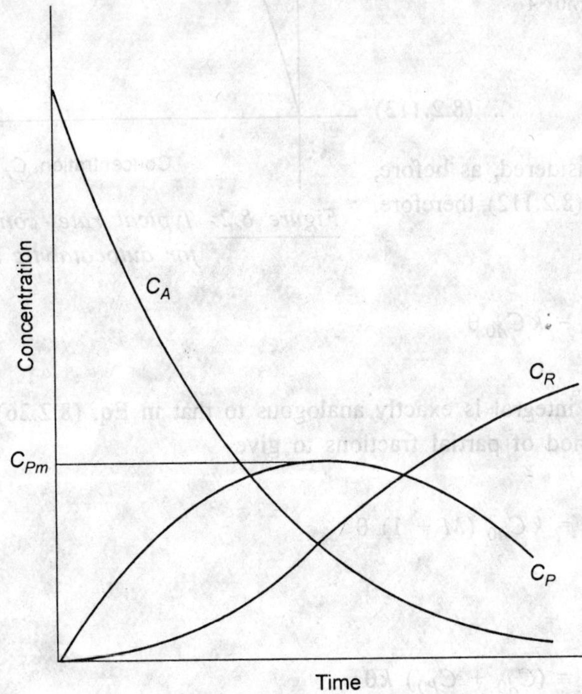


*Figure 8.1: Typical concentration profiles for first order consecutive reactions*

Another category of complex reaction is the autocatalytic reaction. Here, one of the products acts as a catalyst and thereby tends to accelerate the reaction. The simplest form of such a reaction is

$$A + P \rightarrow P + P \qquad \text{... (8.2.111)}$$

The rate of an autocatalytic reaction, therefore, initially increases though the concentration of the reactant (namely, $A$) decreases. The rate increases to a maximum value (when the concentrations of $A$ and $P$ are just equal) and thereafter it starts decreasing (see Fig. 8.2)

Let the rate equation for the autocatalytic reaction represented by Eq. (8.2.111) be

$$(-r_A) = -dC_A/d\theta = k\,C_A C_P \qquad \text{... (8.2.112)}$$

Since from the stoichiometric equation, one mole of $A$ reacts with one mole of $P$ producing two moles of $P$,

$C_P = C_{P0}$ + (moles of $P$ produced per unit volume) – (moles of $P$ reacted per unit volume)

$$= C_{P0} + 2C_{A0}x_A - C_{A0}x_A$$

$$= C_{P0} + C_{A0}x_A$$

$$= C_{A0}\,(M + x_A) \qquad \text{... (8.2.113)}$$

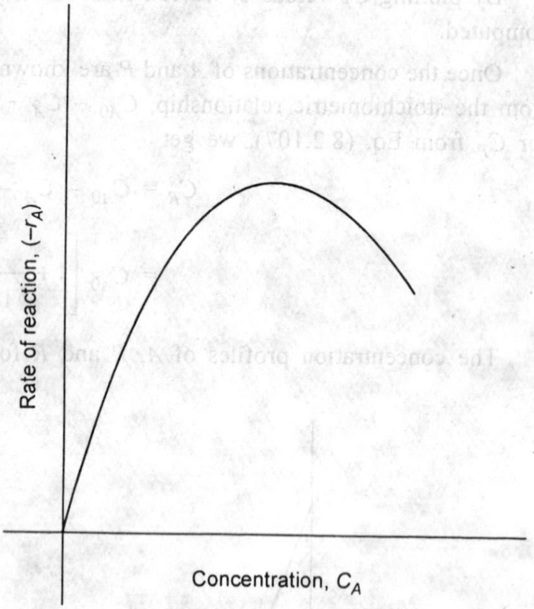


*Figure 8.2: Typical rate–concentration curve for autocatalytic reactions*

where $M = C_{P0}/C_{A0}$. We have considered, as before, a constant volume system. Equation (8.2.112), therefore, becomes (after rearrangement),

$$\int_0^{x_A} \frac{dx_A}{(M + x_A)(1 - x_A)} = k\,C_{A0}\,\theta \qquad \text{... (8.2.114)}$$

It can be seen that the above integral is exactly analogous to that in Eq. (8.2.26). Therefore, this can also be evaluated by the method of partial fractions to give

$$\ln\left[\frac{(M + x_A)}{M(1 - x_A)}\right] = k\,C_{A0}\,(M + 1)\,\theta \qquad \text{... (8.2.115)}$$

In terms of concentrations,

$$\ln\left(\frac{C_P/C_{P0}}{C_A/C_{A0}}\right) = (C_{A0} + C_{P0})\,k\theta \qquad \text{... (8.2.116)}$$

An excellent example of autocatalytic reaction is the *microbial fermentation* in which a microorganism or microbe (yeasts, bacteria, algae, protozoa) catalyses and propagates the reaction. Once a small amount of microorganism is added to the reaction mixture, it will multiply itself and will also get reproduced as a result of the reaction. The overall scheme may be represented as

Substrate + microorganism → Product + (more microorganism)

The reaction is, thus, autocatalytic in nature and its rate first increases, reaches a maximum and then gradually decreases. Also, recycle of a part of product stream to the reactor increases the overall conversion, since the product stream contains some of the microorganisms produced in the reactor. Product recycle is always advantageous in the case of autocatalytic reactions since one of the products is acting as the catalyst.

Microbial fermentation, discussed above, must not be confused with *enzyme–catalysed fermentation* (already discussed in Example 8.4). The chief difference between the two is that unlike microbes, the enzyme does not reproduce itself. Thus, in a continuous enzyme fermenter, fresh enzyme must be added continuously, while in microbial fermenters with sufficient recycle of product fluid, no fresh microorganism need be added to the reactor. Most of the enzyme–catalysed fermentation reactions conform to a rate equation of the form (first proposed by Michaelis and Menten[16])

$$(-r_A) = -dC_A/d\theta = \frac{k_1 C_A}{1 + k_2 C_A} \qquad \text{... (8.2.117)}$$

To integrate the above equation, let us first rearrange it to the following form:

$$-d\theta/dC_A = \frac{1}{k_1 C_A} + \frac{k_2}{k_1} \qquad \text{... (8.2.118)}$$

On integration

$$\ln (C_{A0}/C_A) + k_2 (C_{A0} - C_A) = k_1\theta$$

or

$$\frac{\ln (C_{A0}/C_A)}{(C_{A0} - C_A)} = -k_2 + \left(\frac{k_1 \theta}{C_{A0} - C_A}\right) \qquad \text{... (8.2.119)}$$

Thus, a plot of $[\ln (C_{A0}/C_A)]/(C_{A0} - C_A)$ versus $\theta/(C_{A0} - C_A)$ must yield a straight line (whose slope will be $k_1$ and $y$-intercept $- k_2$) if the rate expression of the reaction is as per Eq. (8.2.117).

An interesting feature can be observed in the above-discussed rate equation. For example, at high concentrations of $A$ ($k_2 C_A >> 1$) when $(1 + k_2 C_A) \approx k_2 C_A$,

$$(-r_A) = (k_1/k_2) = \text{constant} \qquad \text{... (8.2.120)}$$

Thus, the reaction is essentially zero order at high concentrations of $A$. Similarly, at low concentrations ($k_2 C_A << 1$) when $(1 + k_2 C_A) \approx 1.0$,

$$(-r_A) = k_1 C_A \qquad \text{... (8.2.121)}$$

Thus, the reaction is first order in $A$ at low concentrations of $A$. Situations of this kind are encountered with solid–catalysed heterogeneous reactions as well.

All the above-discussed cases of reversible and complex reactions are summarised in Table (8.2).

We have stated at the beginning of this section that an alternate to integral method of analysis is the *differential method* of derivation of rate equation. This method does not require integration of rate equation and might appear more rigorous. However, the method could also be more laborious since it

## Table 8.2
## Rate Equations for Reversible and Complex Reactions

| *Reaction type and stoichiometry* | *Rate equation* | *Integrated rate equation (constant volume)* |
|---|---|---|
| *Reversible reaction:* $A \underset{k_2}{\overset{k_1}{\rightleftharpoons}} P$ | $-r_A = k_1 C_A - k_2 C_P$ | $k_1\theta = \dfrac{(C_{P0} + C_{A0} - C_{Ae})}{(C_{P0} + C_{A0})} \ln \left[\dfrac{C_{A0} - C_{Ae}}{(C_A - C_{Ae})}\right]$ |
| *Reversible reaction:* $A \underset{k_2}{\overset{k_1}{\rightleftharpoons}} P + R$ | $-r_A = k_1 C_A - k_2 C_P C_R$ <br> $C_{P0} = C_{R0} = 0$ | $k_1\theta = \left(\dfrac{C_{A0} - C_{Ae}}{C_{A0} + C_{Ae}}\right) \ln \left[\dfrac{C_{A0}^2 - C_A C_{Ae}}{C_{A0}(C_A - C_{Ae})}\right]$ |
| *Reversible reaction:* <br> $A + B \underset{k_2}{\overset{k_1}{\rightleftharpoons}} P + R$ <br> $A + B \underset{k_2}{\overset{k_1}{\rightleftharpoons}} 2P$ <br> $C_{A0} = C_{B0}$ <br> $C_{P0} = C_{R0} = 0$ | $-r_A = k_1 C_A C_B - k_2 C_P C_R$ <br> $-r_A = k_1 C_A C_B - k_2 C_P^2$ | $k_1\theta = \left[\dfrac{x_{Ae}}{2C_{A0}(1 - x_{Ae})}\right] \ln \left[\dfrac{x_{Ae} + (1 - 2x_{Ae}) x_A}{x_{Ae} - x_A}\right]$ |
| *Reversible reaction:* <br> $2A \underset{k_2}{\overset{k_1}{\rightleftharpoons}} P + R$ <br> $2A \underset{k_2}{\overset{k_1}{\rightleftharpoons}} 2P$ | $-r_A = k_1 C_A^2 - k_2 C_P C_R$ <br> $-r_A = k_1 C_A^2 - k_2 C_P^2$ <br> $C_{P0} = C_{R0} = 0$ | Same as above. |
| *Simultaneous, irreversible reactions:* <br> $A \xrightarrow{k_1} P$, $A \xrightarrow{k_2} R$ | $-r_A = k_1 C_A + k_2 C_A$ | $\ln (C_{A0}/C_A) = (k_1 + k_2)\,\theta$ <br> $C_P = (k_1/k_2)\, C_R + [C_{P0} - C_{R0}\,(k_1/k_2)]$ |
| *Consecutive, irreversible reactions:* <br> $A \xrightarrow{k_1} P \xrightarrow{k_2} R$ <br> $C_{P0} = C_{R0} = 0$ | $-r_A = k_1 C_A$ <br> $r_P = k_1 C_A - k_2 C_P$ <br> $r_R = k_2 C_P$ | $\ln (C_{A0}/C_A) = k_1\theta$ <br> $\ln (C_{Pm}/C_{A0}) = \left(\dfrac{k_2}{k_2 - k_1}\right) \ln (k_1/k_2)$ <br> $C_{Pm}$ = maximum value of $C_P$ |

*(Contd.)*

**Table 8.2**
**Rate Equations for Reversible and Complex Reactions** *(Contd.)*

| *Reaction type and stoichiometry* | *Rate equation* | *Integrated rate equation (constant volume)* |
|---|---|---|
| *Autocatalytic reaction:* | | |
| $A + P \rightarrow P + P$ | $-r_A = kC_AC_P$ | $\ln\left[\frac{(M + x_A)}{M(1 - x_A)}\right] = k\,C_{A0}(M + 1)\,\theta$ <br> $M = C_{P0}/C_{A0}$ |
| *Reaction of shifting order:* | | |
| $A \rightarrow$ Products | $-r_A = \frac{k_1C_A}{1 + k_2C_A}$ | $\frac{\ln(C_{A0}/C_A)}{(C_{A0} - C_A)} = k_1\left(\frac{\theta}{C_{A0} - C_A}\right) - k_2$ |

demands computation of reaction rate from the measured concentration versus time data. The usual practice is, therefore, to try the integral method first and if found unsuitable, resort to the differential method.

In the *differential method of analysis*, the measured values of concentration are plotted against time to get a continuous curve. The values of the reaction rate, namely $-dC_A/d\theta$, at different intervals of time are then determined from this plot. To note that the value of $dC_A/d\theta$ at $\theta = \theta_1$ is equal to the slope of the tangent to the curve at the point $\theta = \theta_1$. Now, it is checked whether a plot of $dC_A/d\theta$ versus $f(C_A)$ yields a straight line or not [see Eq. (8.2.3)]. The expression $f(CA)$ is to be first assumed and by trial, the rate equation is to be finalised. The search for rate equation is similar to that employed in integral method, except that in this case, we do not try a straight line fit with the integrated rate equation, but we try a straight line fit between the reaction rate and the concentration function, $f(C_A)$. Graphical determination $dC_A/d\theta$ values must be accurate since the overall reliability of the method depends very much on this. As in the case of integral method, the procedure can be applied with relative case when the rate expression to be tested conforms to the type shown by Eq. (8.2.3). For example, if the rate equation is of the type of Eq. (8.2.5), then $f(C_A) = C_A$ and if it is of the form of Eq. (8.2.12), $f(C_A) = C_A^2$. Similarly if the assumed rate expression conforms to Eq. (8.2.19), $f(C_A) = (1 - x_A)[M - (b/a)\,x_A]$, for rate equation of the type (8.2.78), $f(C_A) = (1 - x_A)^2 - (1 - x_{Ae})^2\,(x_A/x_{Ae})^2$ and so on. However, if the rate equation to be tested is of the form of Eq. (8.2.1), then the method shall demand either a trial and error adjustment of the constants or a nonlinear least square analysis. Often, special experimental methods that give a partial solution of the problem may have to be used.

We shall illustrate here a specific example for which the differential method is more advantageous than the integral method. Let the rate equation to be tested be

$$(-r_A) = -dC_A/d\theta = k\,C_A^n\,C_B^m \qquad \text{... (8.2.122)}$$

From kinetic experiments, we have the $C_A$ versus $\theta$ data and from the reaction stoichiometry, the values of $C_B$ at different $\theta$ can also be computed. However, determination of *k*, *n* and *m* by integral method shall be difficult since the above equation does not yield easily to integration. Instead, let $C_A$ be plotted against θ and the values of $(dC_A/d\theta)$ at different θ be determined. Now, taking logarithm of Eq. (8.2.122), we get

$$\ln(-r_A) = \ln(k) + n\ln(C_A) + m\ln(C_B) \quad \text{... (8.2.123)}$$

This is of the form

$$y = a_0 + a_1x_1 + a_2x_2 \quad \text{... (8.2.124)}$$

Since the values of $y$ at different values of $x_1$ and $x_2$ are known, we can easily determine the constants $a_0$, $a_1$, and $a_2$ by the method of least squares. For example, based on the method of least squares,

$$a_1 = \frac{A(FC - BH) + C(BG - CD) + N(HD - FG)}{B(2FC - BH) + E\left(HN - C^2\right) - NF^2} \quad \text{... (8.2.125)}$$

$$a_2 = \frac{(AB - DN) - \left(B^2 - EN\right)a_1}{(BC - NF)} \quad \text{... (8.2.126)}$$

$$a_0 = \frac{1}{N}(A - a_1B - a_2C) \quad \text{... (8.2.127)}$$

where $A = \sum_{i=1}^{N} y_i$, $B = \Sigma x_{1i}$, $C = \Sigma x_{2i}$, $D = \Sigma x_{1i}y_i$, $E = \Sigma x_{1i}^2$, $F = \Sigma x_{1i}x_{2i}$, $G = \Sigma x_{2i}y_i$, $H = \Sigma x_{2i}^2$. All the summations are done from $i = 1$ to $i = N$, where $N$ is the number of data points.

**Example 8.7:** Two substances $A$ and $B$ are mixed in a constant volume batch reactor. At the end of one hour, it is observed that 75 percent of $A$ has reacted. How much of $A$ will be left unreacted at the end of two hours if

(*a*) the reaction is first order in $A$ and independent of $B$,

(*b*) the reaction is first order in $A$ and first order in $B$ and the reactants are fed in the stoichiometric proportion,

(*c*) the reaction is zero order in $A$ and independent of $B$.

Assume isothermal operation.

**Solution:** (*a*) If the reaction is first order in $A$ and independent of $B$, then the rate equation will be

$$(-r_A) = -dC_A/d\theta = k_1C_A$$

This is same as Eq. (8.2.5) and therefore, this on integration gives

$$\ln(1 - x_A) = k\theta$$

which is same as Eq. (8.2.9). Now,

$$\frac{\ln(1 - x_{A1})}{\ln(1 - x_{A2})} = (\theta_1/\theta_2)$$

where $\theta_1 = 1$ hr, $x_{A1} = 0.75$, $\theta_2 = 2$ hr. Therefore,

$$\ln(1 - x_{A2}) = 2 \ln(0.25)$$

or

$$x_{A2} = 0.9375 = 93.75 \text{ percent}$$

Therefore, 6.25 percent of $A$ will remain unreacted at the end of two hours.

(*b*) If the reaction is first order in $A$ and first order in $B$, then the rate equation will be

$$(-r_A) = k_2 C_A C_B$$

which is same as Eq. (8.2.19). Since the reactants are fed in the stoichiometric ratio, the integrated form of rate expression will be Eq. (8.2.31):

$$x_A/(1 - x_A) = (b/a)\, C_{A0}\, k_2 \theta$$

Since $x_A = x_{A1}$ at $\theta = \theta_1$ and $x_A = x_{A2}$ at $\theta = \theta_2$,

$$\frac{x_{A1}(1-x_{A2})}{x_{A2}(1-x_{A1})} = (\theta_1/\theta_2)$$

$$\frac{0.75(1-x_{A2})}{0.25\,x_{A2}} = 1/2$$

or

$$x_{A2} = 0.857 = 85.7 \text{ percent}$$

Therefore, 14.3 percent of $A$ will remain unreacted at the end of two hours.

(*c*) If the reaction is zero order in $A$ and independent of $B$, then the overall reaction will be zero order. Thus

$$(-r_A) = -dC_A/d\theta = k$$

On integration,

$$C_{A0} - C_A = k\theta$$

or

$$C_{A0}\, x_A = k\theta$$

$$(x_{A1}/x_{A2}) = (\theta_1/\theta_2)$$

$$(0.75/x_{A2}) = 1/2$$

$$x_{A2} = 1.5 = 150 \text{ percent}$$

This means that conversion of $A$ is complete much before two hours have elapsed. In other words, no $A$ will be remaining unreacted at the end of two hours. We may also compute the time required for the complete conversion of $A$. Thus,

$$(x_{A1}/1.0) = (0.75/1.0) = (1.0/\theta)$$

or

$$\theta = 1.33 \text{ hr}$$

**Example 8.8:** Hydrolysis of ethyl acetate by sodium hydroxide is studied at 250° C in a constant volume batch reactor and the following data are obtained:

| *Time, seconds* | *Concentration of ethyl acetate, kmoles/m*$^3$ |
|---|---|
| 0 | 0.00486 |
| 178 | 0.00398 |
| 273 | 0.00370 |
| 531 | 0.00297 |
| 866 | 0.00230 |
| 1510 | 0.00151 |
| 1918 | 0.00109 |
| 2401 | 0.00080 |

Initial concentration of sodium hydroxide = 0.0098 kmole/m$^3$. Neglecting reversibility of the reaction, determine

(*a*) the rate equation for the reaction,

(*b*) the value of rate constant at 25° C,

(*c*) the time required to saponify 95 percent of the ester at 25° C.

**Solution:** We could start fitting the reported experimental data into different rate equations one by one (starting from the first order equation) and see in which case a close straight line fit is obtained. However, since the reaction stoichiometry is

$$CH_3COOC_2H_5 + NaOH \rightarrow CH_3COONa + C_2H_5OH$$

the intuition tells that the rate equation could be of the form

$$(-r_A) = kC_AC_B$$

where $A$ stands for ethyl acetate and $B$ for sodium hydroxide. Let us, therefore, start our trial with the above equation. The integrated from of the above rate equation is given in Eq. (8.2.27). For the present case,

$$M = C_{B0}/C_{A0} = (0.0098/0.00486) = 2.016$$

$$b/a = 1.0$$

Equation (8.2.27), therefore, reduces to

$$(202.429)\ \ln\left[\frac{2.016 - x_A}{2.016(1 - x_A)}\right] = k\theta$$

or

$$F(x_A) = k\theta \qquad \text{... (i)}$$

To note that since $C_A = C_{A0}(1 - x_A)$, $x_A = 1 - (C_A/C_{A0}) = 1 - (C_A/0.00486) = 1.0 - 205.76\,C_A$. The values of $x_A$ and $F(x_A)$ at different values of $\theta$ are listed below:

| $\theta$, seconds | $C_A$, kmoles/m$^3$ | $x_A$ | $F(x_A)$, m$^3$/kmole |
|---|---|---|---|
| 0 | 0.00486 | 0.0 | 0.0 |
| 178 | 0.00398 | 0.181 | 21.386 |
| 273 | 0.00370 | 0.2387 | 29.6945 |
| 531 | 0.00297 | 0.389 | 56.309 |
| 866 | 0.0023 | 0.5267 | 90.118 |
| 1510 | 0.00151 | 0.6893 | 151.924 |
| 1918 | 0.00109 | 0.7757 | 204.249 |
| 2401 | 0.00080 | 0.8354 | 256.902 |

The plot of $F(x_A)$ versus $\theta$ is shown in Fig. 8.8.1. It is a straight line, showing that the assumed rate equation is correct. From the plot,

$$\text{slope} = k = 6.375 \text{ m}^3/(\text{kmole}\cdot\text{min}) = 0.10625 \text{ m}^3/(\text{kmole}\cdot\text{s})$$

(*a*) The rate equation, therefore, is

$$(-r_A) = 0.10625\, C_A C_B$$

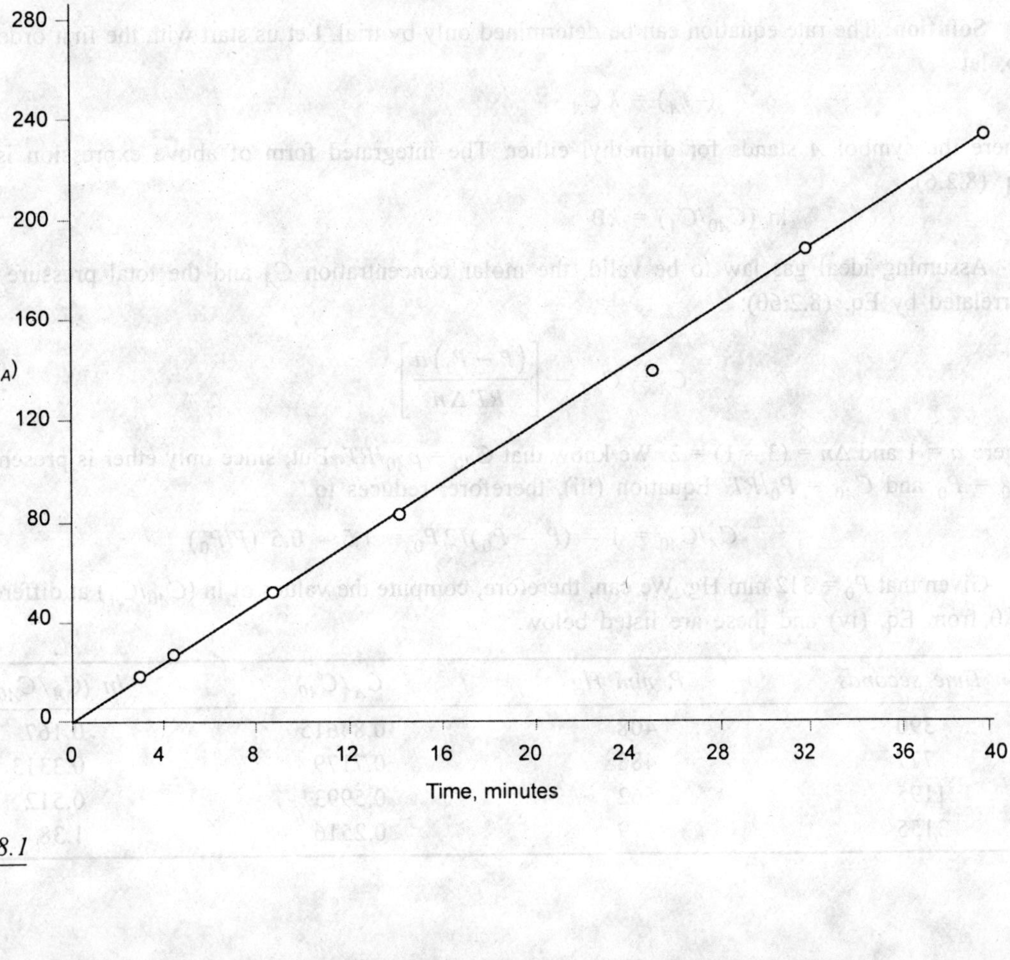


*Figure 8.8.1*

(*b*) The value of $k$ at 25° C is 0.10625 $m^3/(kmole \cdot s)$

(*c*) When $x_A = 0.95$, Eq. (i) becomes

$$(202.429)\ \ln\left[\frac{2.016-0.95}{2.016(0.05)}\right] = 0.10625\,\theta$$

or $$\theta = 4493.5 \text{ seconds} = 74.89 \text{ minutes.}$$

Thus, the time required to saponify 95 percent of the ester is 74.89 minutes.

**Example 8.9:** Hinshelwood and Askey[18] report studies on thermal decomposition of dimethyl ether in the gas phase in a constant volume batch reactor at 504° C. The data reported by them are given below:

| | | | | |
|---|---|---|---|---|
| Time, seconds : | 390 | 777 | 1195 | 3155 |
| Total pressure (mm Hg) : | 408 | 488 | 562 | 779 |

Initial pressure = 312 mm Hg. Assuming that only ether was present initially and that the reaction is

$$(CH_3)_2O \rightarrow CH_4 + H_2 + CO$$

determine a rate equation for the decomposition. What is the numerical value of specific reaction rate at 504° C?

**Solution:** The rate equation can be determined only by trial. Let us start with the first order kinetics. So, let

$$(-r_A) = k\,C_A \qquad \text{... (i)}$$

where the symbol $A$ stands for dimethyl either. The integrated form of above expression is given in Eq. (8.2.6):

$$\ln\,(C_{A0}/C_A) = k\theta \qquad \text{... (ii)}$$

Assuming ideal gas law to be valid, the molar concentration $C_A$ and the total pressure $P$ can be correlated by Eq. (8.2.60):

$$C_A = C_{A0} - \left[\frac{(P-P_0)\,a}{RT\,\Delta n}\right] \qquad \text{... (iii)}$$

where $a = 1$ and $\Delta n = (3 - 1) = 2$. We know that $C_{A0} = p_{A0}/RT$. But, since only ether is present initially, $p_{A0} = P_0$ and $C_{A0} = P_0/RT$. Equation (iii), therefore, reduces to

$$C_A/C_{A0} = 1 - (P - P_0)/2P_0 = 1.5 - 0.5\,(P/P_0) \qquad \text{... (iv)}$$

Given that $P_0 = 312$ mm Hg. We can, therefore, compute the values of ln $(C_{A0}/C_A)$ at different values of $\theta$ from Eq. (iv) and these are listed below:

| *Time seconds* | *P, mm Hg* | $C_A/C_{A0}$ | *ln* $(C_A/C_{A0})$ |
|---|---|---|---|
| 390 | 408 | 0.84615 | 0.167 |
| 777 | 488 | 0.7179 | 0.3313 |
| 1195 | 562 | 0.5993 | 0.512 |
| 3155 | 779 | 0.2516 | 1.38 |

The plot of ln $(C_{A0}/C_A)$ versus θ is shown in Fig. 8.9.1. Since it is a straight line passing through the origin, the assumed rate equation is correct. The slope of the straight line gives the value of $k$. Thus

$$\text{slope} = k = 0.02623 \text{ min}^{-1} = 4.3717 \times 10^{-4} \text{ s}^{-1}$$

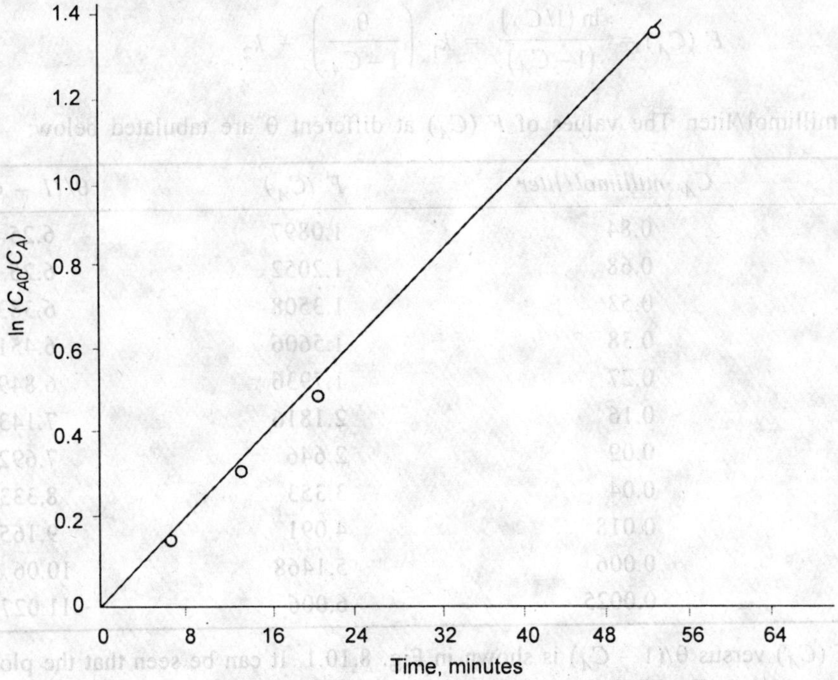


*Figure 8.9.1*

**Example 8.10:** The following data[19] are collected by studying the hydrolysis of sucrose by the catalytic action of the enzyme sucrase in a batch reactor at room temperature:

| Time, hr | : | 1 | 2 | 3 | 4 | 5 | 6 | 7 | 8 | 9 | 10 | 11 |
|---|---|---|---|---|---|---|---|---|---|---|---|---|
| Sucrose concentration, millimol/liter | : | 0.84 | 0.68 | 0.53 | 0.38 | 0.27 | 0.16 | 0.09 | 0.04 | 0.018 | 0.006 | 0.0025 |

Determine the rate equation of the above reaction,

(*a*) by integral method,

(*b*) by differential method.

Initial concentration of sucrose = 1.0 millimol/liter and that of enzyme, $C_{E0}$ = 0.01 millimol/liter.

**Solution:** Since biochemical reactions of the above type tend to follow Michaelis-Menten kinetics, let us first check whether the reported data fit into a rate expression that is of the form of Eq. (8.2.117). Thus, let

$$(-r_A) = \frac{k_1 C_A}{1 + k_2 C_A} \quad \text{... (i)}$$

where the symbol $A$ stands for sucrose. The integrated form of the above expression is Eq. (8.2.119):

$$\frac{\ln(C_{A0}/C_A)}{(C_{A0}-C_A)} = -k_2 + k_1\left(\frac{\theta}{C_{A0}-C_A}\right) \quad \text{... (ii)}$$

or

$$F(C_A) = \frac{\ln(1/C_A)}{(1-C_A)} = k_1\left(\frac{\theta}{1-C_A}\right) - k_2 \quad \text{... (iii)}$$

since $C_{A0}$ = 1.0 millimol/liter. The values of $F(C_A)$ at different $\theta$ are tabulated below:

| $\theta$, hr | $C_A$, millimol/liter | $F(C_A)$ | $\theta/(1 - C_A)$ |
|---|---|---|---|
| 1.0 | 0.84 | 1.0897 | 6.25 |
| 2.0 | 0.68 | 1.2052 | 6.25 |
| 3.0 | 0.53 | 1.3508 | 6.383 |
| 4.0 | 0.38 | 1.5606 | 6.4516 |
| 5.0 | 0.27 | 1.7936 | 6.8493 |
| 6.0 | 0.16 | 2.1816 | 7.143 |
| 7.0 | 0.09 | 2.646 | 7.6923 |
| 8.0 | 0.04 | 3.353 | 8.333 |
| 9.0 | 0.018 | 4.091 | 9.165 |
| 10.0 | 0.006 | 5.1468 | 10.06 |
| 11.0 | 0.0025 | 6.006 | 11.0275 |

The plot of $F(C_A)$ versus $\theta/(1 - C_A)$ is shown in Fig. 8.10.1. It can be seen that the plot is linear, thereby confirming that rate Eq. (i) is being obeyed. From the plot,

$$\text{slope} = k_1 = 1.01\ \text{hr}^{-1}$$

$$y\text{-intercept} = -k_2 = -5.1\ (\text{millimol/L})^{-1}$$

The rate equation, therefore, is

$$(-r_A) = \frac{1.01\,C_A}{1+5.1C_A}$$

Dividing both numerator and the denominator by $(5.1\,C_A)$, we get

$$(-r_A) = \frac{0.19804}{1+(0.19608/C_A)}$$

If we write, $0.19804 = k\,C_{E0} = k\,(0.01)$, then $k$ = 19.804 hr$^{-1}$. Thus, the above equation becomes

$$(-r_A) = \frac{19.804\,C_{E0}}{1+(0.19608/C_A)} \quad \text{... (iv)}$$

The above equation is of the form of Michaelis-Menten equation we derived in Example 8.4 with $K_m$ = Michaelis-Menten constant = 0.19608.

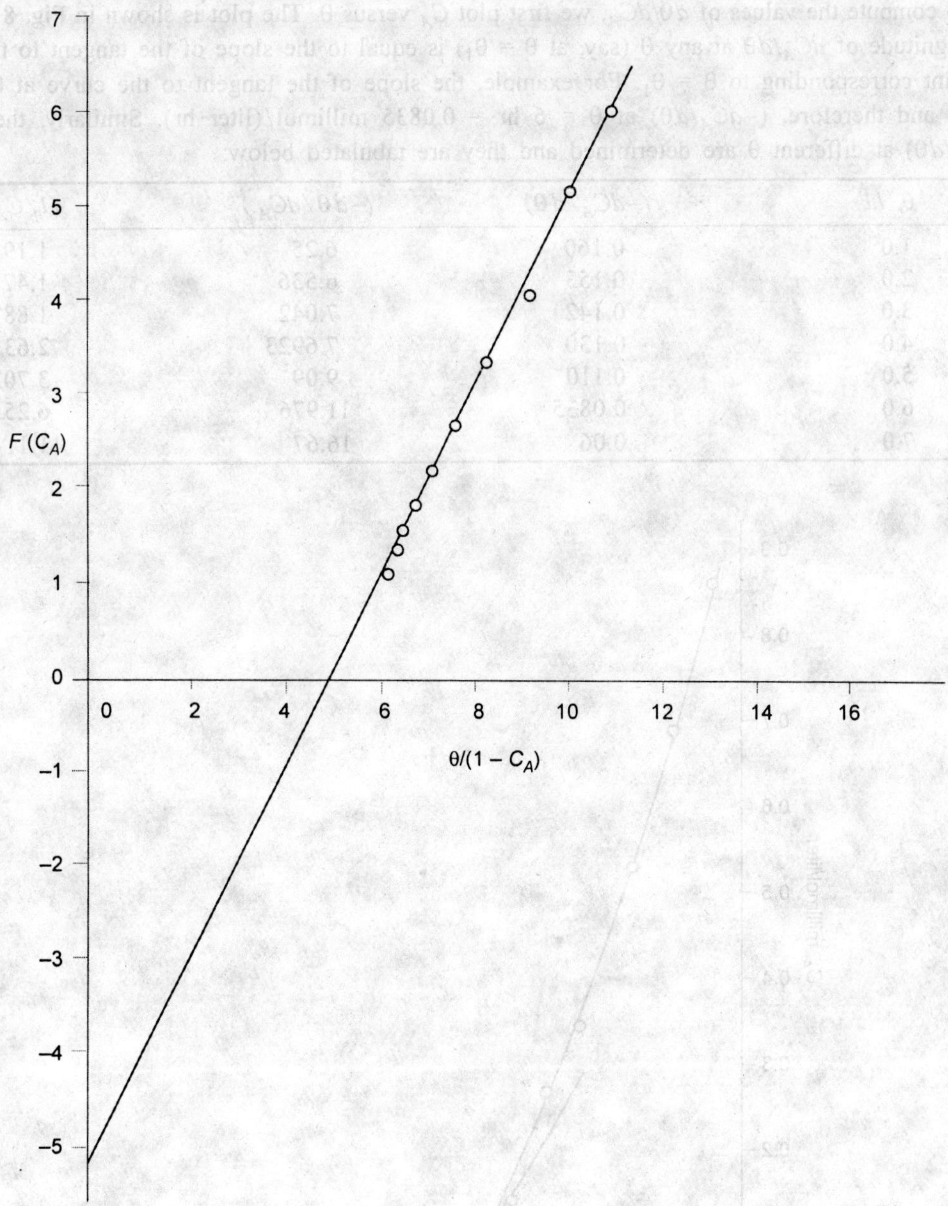


*Figure 8.10.1*

(*b*) To use the *differential method*, let us rewrite Eq. (i) as

$$\frac{1}{(-r_A)} = -\frac{d\theta}{dC_A} = \left(\frac{1}{k_1}\right)(1/C_A) + (k_2/k_1) \qquad \text{... (v)}$$

To compute the values of $d\theta/dC_A$, we first plot $C_A$ versus θ. The plot is shown in Fig. 8.10.2. Now, the magnitude of $dC_A/d\theta$ at any θ (say, at $\theta = \theta_1$) is equal to the slope of the tangent to the curve at the point corresponding to $\theta = \theta_1$. For example, the slope of the tangent to the curve at θ = 6 hr is 0.0835 and therefore, $(-dC_A/d\theta)$ at θ = 6 hr = 0.0835 millimol/(liter·hr). Similarly, the values of $(-dC_A/d\theta)$ at different θ are determined and they are tabulated below:

| θ, hr | $(-dC_A/d\theta)$ | $(-d\theta/dC_A)$ | $(1/C_A)$ |
|---|---|---|---|
| 1.0 | 0.160 | 6.25 | 1.19 |
| 2.0 | 0.153 | 6.536 | 1.47 |
| 3.0 | 0.142 | 7.042 | 1.887 |
| 4.0 | 0.130 | 7.6923 | 2.63 |
| 5.0 | 0.110 | 9.09 | 3.703 |
| 6.0 | 0.0835 | 11.976 | 6.25 |
| 7.0 | 0.06 | 16.67 | 11.11 |

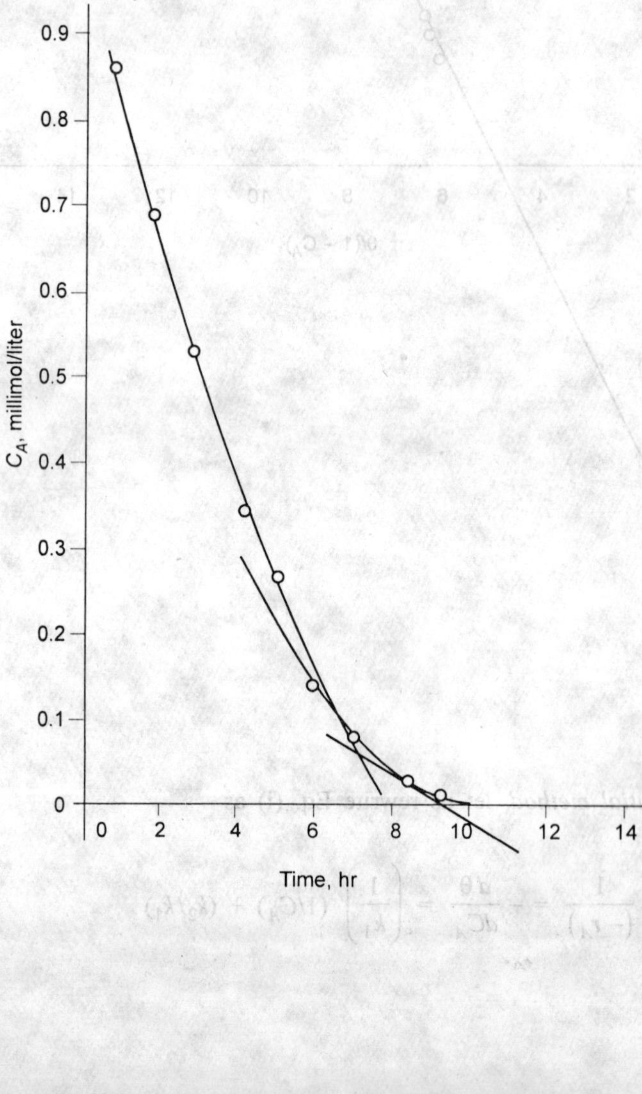


*Figure 8.10.2*

The plot of $(-d\theta/dC_A)$ versus $(1/C_A)$ is shown in Fig. 8.10.3*. The plot is linear with slope equal to 1.032 and $y$-intercept 5.1. Thus

$$(1/k_1) = 1.032$$

$$k_1 = (1/1.032) = 0.97 \text{ hr}^{-1}$$

$$(k_2/k_1) = 5.1$$

or

$$k_2 = (5.1)\,(0.97) = 4.947 \text{ (millimol/liter)}^{-1}$$

The rate equation, therefore, is

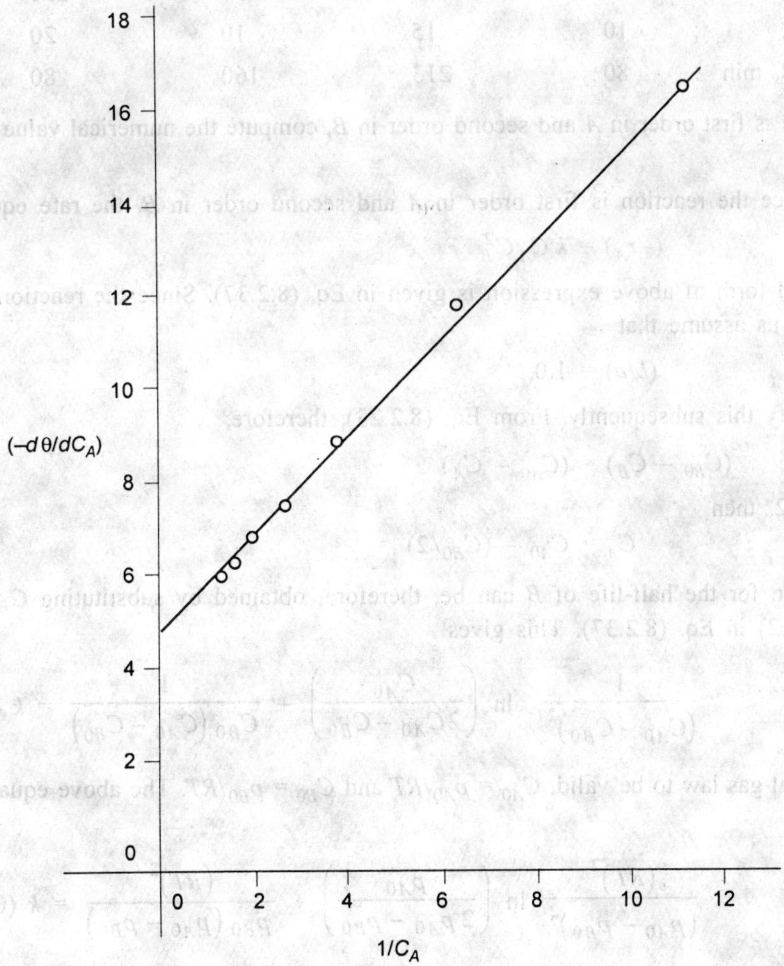


*Figure 8.10.3*

*This type of linear plot is often called the Lineweaver-Burk plot.

$$(-r_A) = \frac{0.97C_A}{1+4.947C_A}$$

This may also be rearranged to the form of Eq. (iv) as

$$(-r_A) = \frac{19.608C_{E0}}{1+(0.202/C_A)} \qquad \text{... (vi)}$$

**Example 8.11:** The reaction between $A$ and $B$ in the gas phase has been studied at 30° C in a batch reactor by measuring the half-life period of $B$ for several initial concentrations of the reactants. The results are given below:

| | | | | | |
|---|---|---|---|---|---|
| $p_{A0}$, mm Hg | : | 500 | 125 | 250 | 250 |
| $p_{B0}$, mm Hg | : | 10 | 15 | 10 | 20 |
| Half-life of $B$, min | : | 80 | 213 | 160 | 80 |

If the reaction is first order in $A$ and second order in $B$, compute the numerical value of the specific reaction rate.

**Solution:** Since the reaction is first order in $A$ and second order in $B$, the rate equation is

$$(-r_A) = k\,C_A\,C_B^2$$

The integrated form of above expression is given in Eq. (8.2.37). Since the reaction stoichiometry is not known, let us assume that

$$(b/a) = 1.0$$

We shall verify this subsequently. From Eq. (8.2.23), therefore,

$$(C_{B0} - C_B) = (C_{A0} - C_A)$$

If $C_B = C_{B0}/2$, then

$$C_A = C_{A0} - (C_{B0}/2)$$

The expression for the half-life of $B$ can be, therefore, obtained by substituting $C_B = C_{B0}/2$ and $C_A = C_{A0} - (C_{B0}/2)$ in Eq. (8.2.37). This gives

$$\frac{1}{(C_{A0}-C_{B0})^2}\ \ln\left(\frac{C_{A0}}{2C_{A0}-C_{B0}}\right) + \frac{1}{C_{B0}(C_{A0}-C_{B0})} = k\ (\theta_{1/2})_B$$

Assuming ideal gas law to be valid, $C_{A0} = p_{A0}/RT$ and $C_{B0} = p_{B0}/RT$. The above equation, therefore reduces to

$$\frac{(RT)^2}{(p_{A0}-p_{B0})^2}\ \ln\left(\frac{p_{A0}}{2p_{A0}-p_{B0}}\right) + \frac{(RT)^2}{p_{B0}(p_{A0}-p_{B0})} = k\ (\theta_{1/2})_B$$

or

$$(3.57 \times 10^8)\ F\ (p_{A0}, p_{B0}) = k\ (\theta_{1/2})_B \qquad \text{... (i)}$$

To note that we have substituted $R = 62.358$ mm Hg·m³/(kmole·K) and $T = 303°$ K so that $(RT)^2 = 3.57 \times 10^8$. The values of $F\,(p_{A0}, p_{B0})$ can be now computed from the given data and are listed below:

| *Half-life of B, min* | *(3.57 × $10^8$) F ($p_{A0}$, $p_{B0}$)* |
|---|---|
| 80 | 71828.4 |
| 213 | 197706.6 |
| 160 | 144570.72 |
| 80 | 73206.42 |

The plot $F'$ ($p_{A0}$, $p_{B0}$) versus half-life of $B$, where $F'$ ($p_{A0}$, $p_{B0}$) = 3.57 × $10^8$ $F$ ($p_{A0}$, $p_{B0}$), is shown in Fig. 8.11.1. The plot is a straight line which confirms our initial assumption that $(b/a) = 1.0$. Now, the average slope of the line is 925.0. Therefore,

$k$ = specific reaction rate = 925.0 $(\text{m}^3/\text{kmole})^2/\text{min}$.

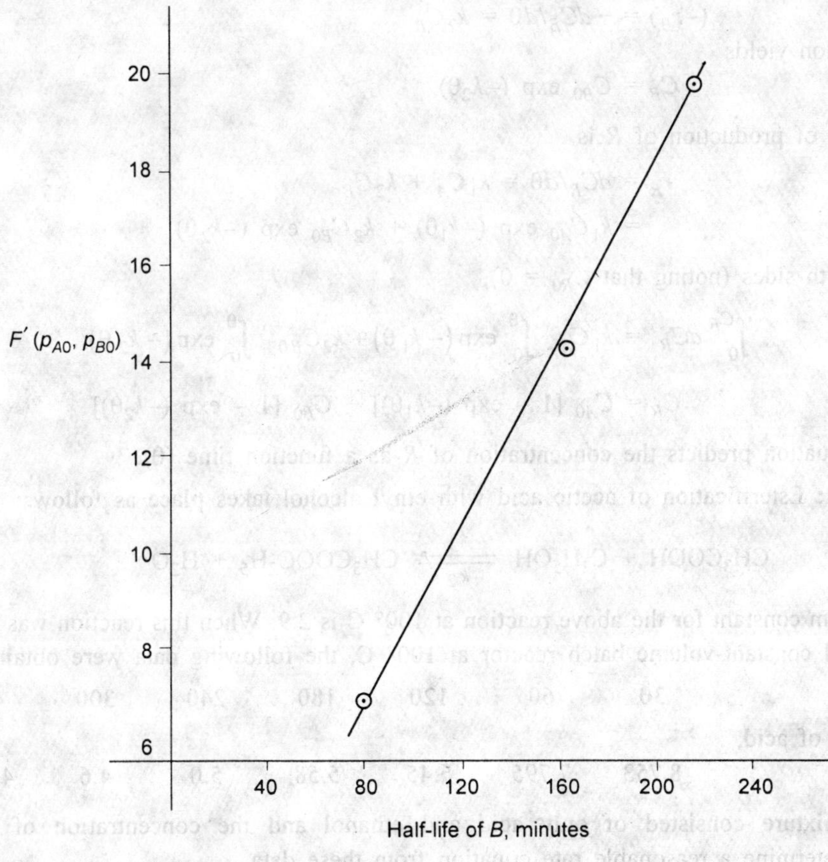


*Figure 8.11.1*

**Example 8.12:** Two irreversible reactions take place in parallel as given below:

$$A \rightarrow R$$

$$B \rightarrow R$$

If the first reaction is first order in $A$ and the second reaction is first order in $B$, derive an expression for the concentration of $R$ as a function of time. Assume that at $\theta = 0$, only $A$ and $B$ were present in the reaction mixture.

**Solution:** Let us assume that the reactions take place in a constant volume, isothermal batch reactor. Then, the rate equation for the first reaction is

$$(-r_A) = -dC_A/d\theta = k_1 C_A$$

On integration,

$$\ln (C_A/C_{A0}) = -k_1\theta$$

or

$$C_A = C_{A0} \exp(-k_1\theta) \quad \text{... (i)}$$

Similarly, for the second reaction, the rate equation is

$$(-r_B) = -dC_B/d\theta = k_2 C_B$$

which on integration yields

$$C_B = C_{B0} \exp(-k_2\theta) \quad \text{... (ii)}$$

Now, the rate of production of $R$ is

$$r_R = dC_R/d\theta = k_1 C_A + k_2 C_B.$$

$$= k_1 C_{A0} \exp(-k_1\theta) + k_2 C_{B0} \exp(-k_2\theta)$$

Integrating both sides (noting that $C_{R0} = 0$),

$$\int_0^{C_R} dC_R = k_1 C_{A0} \int_0^{\theta} \exp(-k_1\theta) + k_2 C_{B0} \int_0^{\theta} \exp(-k_2\theta)$$

$$C_R = C_{A0}[1 - \exp(-k_1\theta)] + C_{B0}[1 - \exp(-k_2\theta)]$$

The above equation predicts the concentration of $R$ as a function time ($\theta$).

**Example 8.13:** Esterification of acetic acid with ethyl alcohol takes place as follows:

$$CH_3COOH + C_2H_5OH \underset{k_2}{\overset{k_1}{\rightleftharpoons}} CH_3COOC_2H_5 + H_2O$$

The equilibrium constant for the above reaction at 100° C is 2.9. When this reaction was conducted in an experimental constant-volume batch reactor at 100° C, the following data were obtained:

| Time, min | 30 | 60 | 120 | 180 | 240 | 300 | 480 |
|---|---|---|---|---|---|---|---|
| Concentration of acid, kmoles/m$^3$ | 8.753 | 7.795 | 6.45 | 5.58 | 5.0 | 4.6 | 4.025 |

The initial mixture consisted of only acid and ethanol and the concentration of each was 10 kmoles/m$^3$. Determine a reasonable rate equation from these data.

**Solution:** Let us start our trial by assuming that the rate expression is

$$(-r_A) = k_1 C_A C_B - k_2 C_P C_R$$

where the symbols $A$, $B$, $P$ and $R$ stand for acetic acid, ethanol, ethyl acetate and water respectively. The reaction stoichiometry will then be

$$A + B \underset{k_2}{\overset{k_1}{\rightleftharpoons}} P + R$$

It is given that $C_{A0} = C_{B0} = 10$ kmoles/m$^3$ and $C_{P0} = C_{R0} = 0$. The integrated form of rate equation under these conditions is that given in Eq. (8.2.82). Also, from Eq. (8.2.80),

$$(k_1/k_2) = K = x_{Ae}^2/(1 - x_{Ae})^2$$

Since $\qquad K = 2.9,$

$$2.9 = x_{Ae}^2/(1 - x_{Ae})^2$$

or $\qquad x_{Ae} = 0.63$

Equation (8.2.82), therefore, becomes

$$(0.085135) \ln\left(\frac{0.63 - 0.26x_A}{0.63 - x_A}\right) = k_1\theta$$

or $\qquad F(x_A) = k_1\theta$

To note that $x_A = 1 - (C_A/C_{A0})$. The values of $x_A$ and $F(x_A)$ at different $\theta$ can be now computed from the given data and they are tabulated below:

| $\theta$, min | $C_A$, kmoles/m$^3$ | $x_A$ | $F(x_A)$, m$^3$/kmole |
|---|---|---|---|
| 30 | 8.753 | 0.1247 | 0.01428 |
| 60 | 7.795 | 0.2205 | 0.02855 |
| 120 | 6.45 | 0.355 | 0.05708 |
| 180 | 5.58 | 0.442 | 0.0858 |
| 240 | 5.0 | 0.50 | 0.1147 |
| 300 | 4.6 | 0.54 | 0.1442 |
| 480 | 4.025 | 0.5975 | 0.2283 |

The plot of $F(x_A)$ versus $\theta$ is shown in Fig. 8.13.1. The plot is linear which confirms that the assumed rate equation is correct. The average slope of the straight line plot is $4.787 \times 10^{-4}$. Therefore,

$$k_1 = 4.787 \times 10^{-4} \text{ m}^3/(\text{kmole}\cdot\text{min})$$

And $\qquad k_2 = k_1/K = (4.787 \times 10^{-4}/2.9) = 1.6507 \times 10^{-4}$ m$^3$/(kmole·min)

The rate equation, therefore, is

$$(-r_A) = (4.787 \times 10^{-4})\, C_A C_B - (1.6507 \times 10^{-4})\, C_P C_R$$

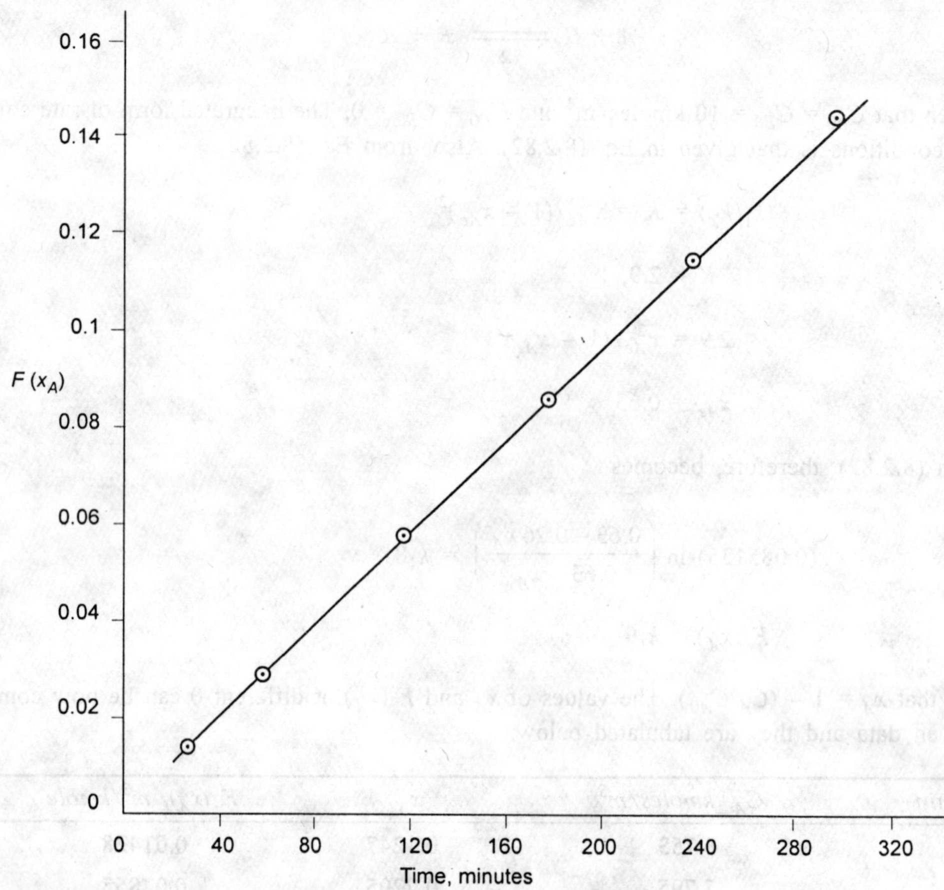


*Figure 8.13.1*

**Example 8.14:** The gas phase reaction, whose stoichiometric equation is given below, is second order in $A$:

$$2A \rightarrow P + 2R$$

When this reaction was conducted in a constant-volume batch reactor, it was observed that the total pressure increases by 40 percent in 2.5 minutes. The initial mixture contained only $A$ and the initial pressure was 2 atm.

(*a*) If this reaction is conducted in a constant pressure batch reactor, estimate the time required to obtain the same conversion. What will be the fractional increase in volume at that time?

(*b*) If the initial mixture contained 70 mole percent $A$ and 30 mole percent inerts, how much time will be required to achieve the same conversion in the constant pressure reactor?

Assume that the reaction is conducted at the same temperature in both reactors and that the presence of inerts does not affect the kinetics of the reaction.

**Solution:** Let us first consider the constant volume batch reactor. The rate equation is

$$(-r_A) = -dC_A/d\theta = kC_A^2$$

On integration,

$$\left(\frac{1}{C_A} - \frac{1}{C_{A0}}\right) = k\theta$$

or
$$\left(\frac{C_{A0} - C_A}{C_A C_{A0}}\right) = k\theta \qquad \text{... (i)}$$

Now, from Eq. (8.2.60),

$$C_A = C_{A0} - \left(\frac{a}{RT\,\Delta n}\right)(P - P_0)$$

where $a = 2$, $\Delta n = 1.0$, $P_0 = 2$ atm. Also, since the initial mixture contained only $A$, $p_{A0} = P_0$ and $C_{A0} = P_0/RT$. Substituting in the above equation, we get

$$C_A = (6.0 - 2P)/RT \qquad \text{... (ii)}$$

Equation (i), therefore, becomes

$$(P - 2)/(6 - 2P) = (k/RT)\,\theta \qquad \text{... (iii)}$$

It is given that at $\theta = 2.5$ min, $P = 1.4\,P_0 = 2.8$ atm. Therefore,

$$(k/RT) = 0.8\ (\text{atm}\cdot\text{min})^{-1}$$

Now, Eq. (i) may also be written as

$$x_A/(1 - x_A) = C_{A0}\,k\theta = (P_0/RT)\,k\theta = (2.0)\,(0.8)\,(2.5) = 4.0$$

or
$$x_A = 0.8$$

(*a*) The analysis of a constant pressure batch reactor or a variable-volume batch reactor is relatively more complex. Let us make a simplifying assumption that the volume of the reaction mixture varies linearly with conversion. In other words,

$$V = V_0\,(1 + \epsilon_A x_A) \qquad \text{... (8.2.128)}$$

where $\epsilon_A$ is defined[19] as the fractional change in the volume of the system between no conversion and complete conversion of reactant $A$. Thus

$$\epsilon_A = \frac{V(\text{at } x_A = 1.0) - V(\text{at } x_A = 0)}{V(\text{at } x_A = 0)} \qquad \text{... (8.2.129)}$$

For example, for the present case, if we start with 2 moles of $A$ (since the initial mixture contains only $A$, the total number of moles at $x_A = 0$ will be 2.0), at complete conversion we will be left with one mole of $P$ and two moles of $R$, the total number of moles being 3.0. Since the volume of a gaseous mixture is proportional to the number of moles,

$$\epsilon_A = (3 - 2)/2 = 0.5$$

For a constant pressure (or, variable volume) batch reactor, the rate equation should be written from Eq. (8.1.1). Thus

$$(-r_A) = -(1/V)\ dN_A/d\theta$$

Since $V = V_0\ (1 + \in_A x_A)$ and $N_A = N_{A0}\ (1 - x_A)$, the above equation becomes

$$(-r_A) = \frac{(N_{A0}/V_0)}{(1+\in_A x_A)}\ (dx_A/d\theta) = \frac{C_{A0}}{(1+\in_A x_A)}\ (dx_A/d\theta) \qquad \text{... (8.2.130)}$$

Now, $$C_A = (N_A/V) = \frac{N_{A0}(1-x_A)}{V_0(1+\in_A x_A)} = \frac{C_{A0}(1-x_A)}{(1+\in_A x_A)} \qquad \text{... (8.2.131)}$$

For the present case, $(-r_A) = k\,C_A^2$ and therefore, Eq. (8.2.130) becomes

$$k\ C_A^2 = \frac{C_{A0}}{(1+\in_A x_A)}\ \frac{dx_A}{d\theta} \qquad \text{... (8.2.132)}$$

Substituting for $C_A$ from Eq. (8.2.131) and rearranging, we get

$$\int_0^{x_A} \frac{(1+\in_A x_A)\,dx_A}{(1-x_A)^2} = k\,C_{A0}\,\theta \qquad \text{... (8.2.133)}$$

Integrating by the method of partial fractions,

$$\frac{(1+\in_A)\,x_A}{(1-x_A)} + \in_A \ln(1 - x_A) = k\,C_{A0}\theta \qquad \text{... (8.2.134)}$$

For the present case,

$$x_A = 0.8$$

$$\in_A = 0.5$$

$$k\,C_{A0} = k\,P_0/RT = (0.8)\ (2.0) = 1.6\ \text{min}^{-1}$$

Therefore, $\theta = 3.247$ minutes.

Fractional increase in volume $= (V - V_0)/V_0 = \in_A x_A = (0.5)\ (0.8) = 0.4$

(*b*) If the initial mixture contains 70 percent *A* and 30 percent inerts, then 10 moles of initial mixture will contain 7 moles of *A* and 3 moles of inerts. 7 moles of *A* on complete conversion gives 3.5 moles of *P* and 7 moles of *R* and since 3 moles of inerts shall remain unconverted, the total number of moles at $x_A = 1.0$ will be 13.5. Therefore,

$$\in_A = (13.5 - 10)/10 = 0.35$$

Since the initial mixture contains 70 mole percent *A*,

$$p_{A0} = (0.70)\ P_0$$

And $$C_{A0} = P_{A0}/RT = 0.7\ P_0/RT$$

Therefore, $$k\,C_{A0} = 0.7P_0\ (k/RT) = (0.7)\ (2.0)\ (0.8) = 1.12\ \text{min}^{-1}$$

Substituting $x_A = 0.8$, $\in_A = 0.35$ and $k\,C_{A0} = 1.12\ \text{min}^{-1}$ in Eq. (8.2.128), we get

$$\theta = 4.32 \text{ minutes}$$

Fractional increase in volume $= \in_A x_A = (0.35)\ (0.8) =$ **0.28**

**Example 8.15:** The following reactions take place in a batch reactor at constant density:

$$A + B \xrightarrow{k_1} P$$

$$P + B \xrightarrow{k_2} R$$

where $P$ is the desired product. The initial mixture contained only $A$ and $B$. If both reactions are second order, derive an expression for the selectivity of $P$ with respect to $R$ in terms of the total conversion of $A$. Also determine the total conversion at which the selectivity will be maximum if $(k_2/k_1) = 1.0$. Will the conversion of $A$ to $P$ (that is, the yield of $P$) also be maximum under these conditions?

**Solution:** Reactions of this kind are called *series–parallel reactions*. Successive chlorination of benzene to produce mono –, di – and trichlorobenzene, photochlorination of various hydrocarbons, reaction between ethylene oxide and ammonia to produce mono –, di – and triethanolamines are examples. Since both reactions are second order, the rate equations are

$$(-r_A) = -dC_A/d\theta = k_1 C_A C_B$$

$$r_P = dC_P/d\theta = k_1 C_A C_B - k_2 C_P C_B$$

Dividing the second equation by the first, we get

$$-\frac{dC_P}{dC_A} = 1 - \left(\frac{k_2 C_P}{k_1 C_A}\right)$$

$$\frac{dC_P}{dC_A} - \left(\frac{k_2}{k_1 C_A}\right) C_P + 1 = 0 \qquad \text{... (i)}$$

This is a linear, first order differential equation [similar to Eq. (8.2.106)] of the form

$$\frac{dy}{dx} + Py + Q = 0$$

where $P$ and $Q$ are functions of $x$ only. The standard solution to such an equation is

$$yI = -\int (QI)\, dx + \text{constant}$$

where $I = \exp\left(\int P dx\right)$. For the present case, $P = -k_2/(k_1 C_A)$ and $Q = 1.0$. Therefore,

$$I = (C_A)^{-k_2/k_1}$$

The solution to the equation with the initial condition that at $\theta = 0$, $C_A = C_{A0}$ and $C_P = C_{P0} = 0$ is therefore,

$$C_P/C_{A0} = \frac{k_1}{(k_1 - k_2)} [(C_A/C_{A0})^{k_2/k_1} - (C_A/C_{A0})] \qquad \text{... (ii)}$$

If $(k_2/k_1) = 1.0$, then Eq. (i) reduces to

$$\frac{dC_P}{dC_A} - (C_P/C_A) + 1 = 0 \qquad \text{... (iii)}$$

For this equation, the integration factor $I = 1/C_A$ and therefore, the solution is

$$C_P = C_A \ln (C_{A0}/C_A) \quad \text{... (iv)}$$

If $x_A$ is the total conversion of $A$, then $x_A = 1 - (C_A/C_{A0})$ and therefore, Eq. (ii) can be rewritten as

$$C_P/C_{A0} = \frac{k_1}{(k_1 - k_2)} [(1 - x_A)^{k_2/k_1} - (1 - x_A)] \quad \text{... (v)}$$

If $(k_2/k_1) = 1.0$, $\quad C_P/C_{A0} = -(1 - x_A) \ln (1 - x_A)$ ... (vi)

From a simple material balance,

$$C_A + C_P + C_R = C_{A0} + C_{P0} + C_{R0} = C_{A0} \text{ (since } C_{P0} = C_{R0} = 0)$$

or

$$C_R/C_{A0} = 1 - (C_A/C_{A0}) - (C_P/C_{A0}) = x_A - (C_P/C_{A0}) \quad \text{... (vii)}$$

The overall selectivity of $P$ with respect to $R$ is, therefore,

$$S_0 = (\text{yield of } P)/(\text{yield of } R)$$

$$= (C_P/C_{A0})/(C_R/C_{A0}) = \frac{(1 - x_A)^\alpha - (1 - x_A)}{x_A (1 - \alpha) - (1 - x_A)^\alpha + (1 - x_A)} \quad \text{... (viii)}$$

where $\alpha = (k_2/k_1)$. When $\alpha = 1.0$, the expression for selectivity is

$$S_0 = \frac{(x_A - 1) \ln (1 - x_A)}{x_A + (1 - x_A) \ln (1 - x_A)} \quad \text{... (ix)}$$

To find the value of $x_A$ at which $S_0$ will be maximum, let us set the derivative of $S_0$ with respect to $x_A$ equal to zero. Thus, $dS_0/dx_A = 0$ (on simplification) gives

$$x_A + \ln (1 - x_A) = 0 \quad \text{... (x)}$$

It can be seen that the exact solution to the above equation is $x_A = 0$. In other words, the selectivity $S_0$ is maximum when the total conversion of $A$ approaches zero. The variation of $S_0$ with $x_A$ is shown in Fig. 8.15.1. It can be seen that $S_0$ decreases sharply with increase in $x_A$.

The yield of $P$ will not be maximum at the conditions of maximum selectivity. To find the value of $x_A$ at which the yield of $P$ will be maximum, let us set the derivative of $(C_P/C_{A0})$ with respect to $x_A$ equal to zero and then solve it for $x_A$. Thus, from Eq. (vi), $d(C_P/C_{A0})/dx_A = 0$ gives

$$1 + \ln (1 - x_A) = 0$$

or

$$x_A = 0.6321$$

If we substitute this in Eq. (vi), we get

$$(C_P/C_{A0})_{max} = 0.3679$$

Thus, the maximum yield of $P$ attainable when $(k_2/k_1) = 1.0$ is 36.79 percent and this is obtained when the total conversion of $A$ is 63.21 percent. The value of selectivity at this conversion is

$$S_0 = \frac{-(0.3679) \ln (0.3679)}{(0.6321) + (0.3679) \ln (0.3679)} = 1.3925$$

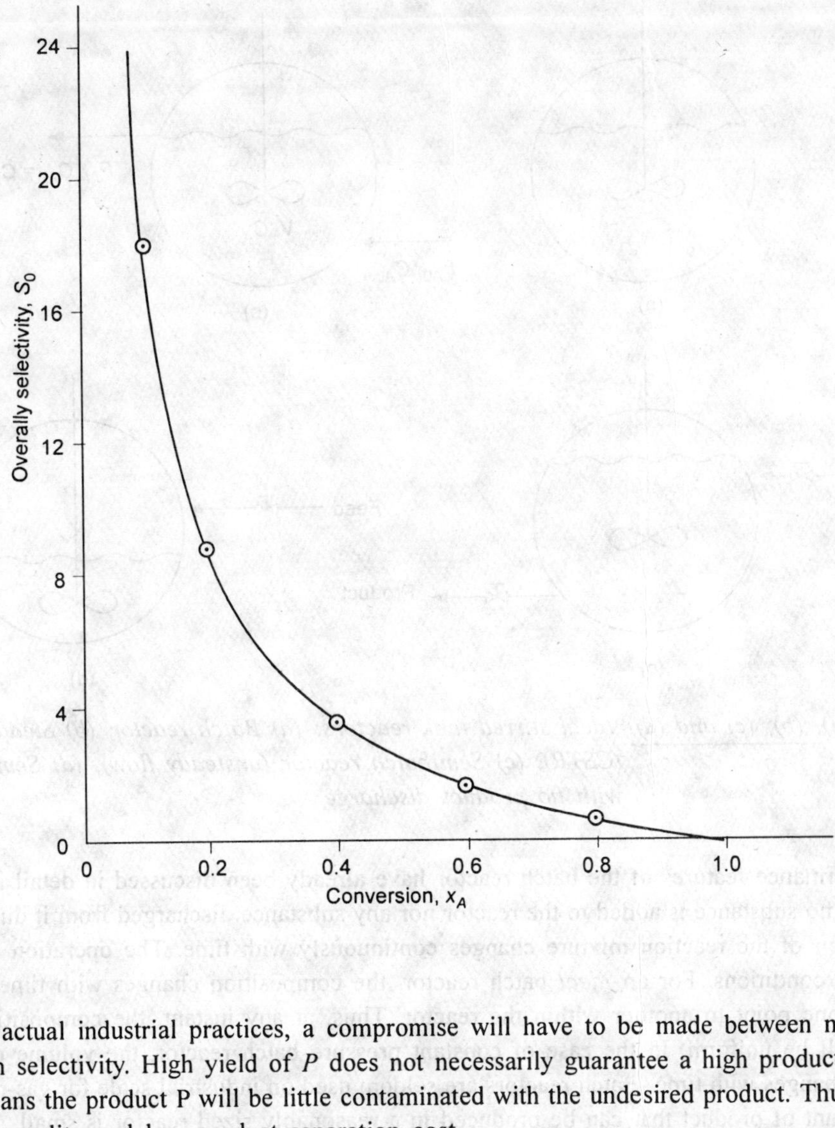


*Figure 8.15.1*

Thus, in actual industrial practices, a compromise will have to be made between maximum yield and maximum selectivity. High yield of *P* does not necessarily guarantee a high product quality. High selectivity means the product P will be little contaminated with the undesired product. Thus, this ensures good product quality and low product separation cost.

## 8.3 DESIGN OF REACTORS FOR HOMOGENEOUS REACTIONS

The most popular types of reactors used for conducting homogeneous reactions are the *tank type* and the *tubular type*. The tank type, commonly called the *stirred tank reactor*, is probably the most common type of reactor in use in the chemical industry. It is equipped with some means of agitation (either stirring, shaking or rocking) and often with provisions for heat transfer (jacket, external or internal heat exchanger). If may be operated as a batch type [Fig. 8.3 (a)], a steady-state flow type [Fig. 8.3 (b)] or as a semibatch [Figs 8.3 (c) and (d)] reactor.

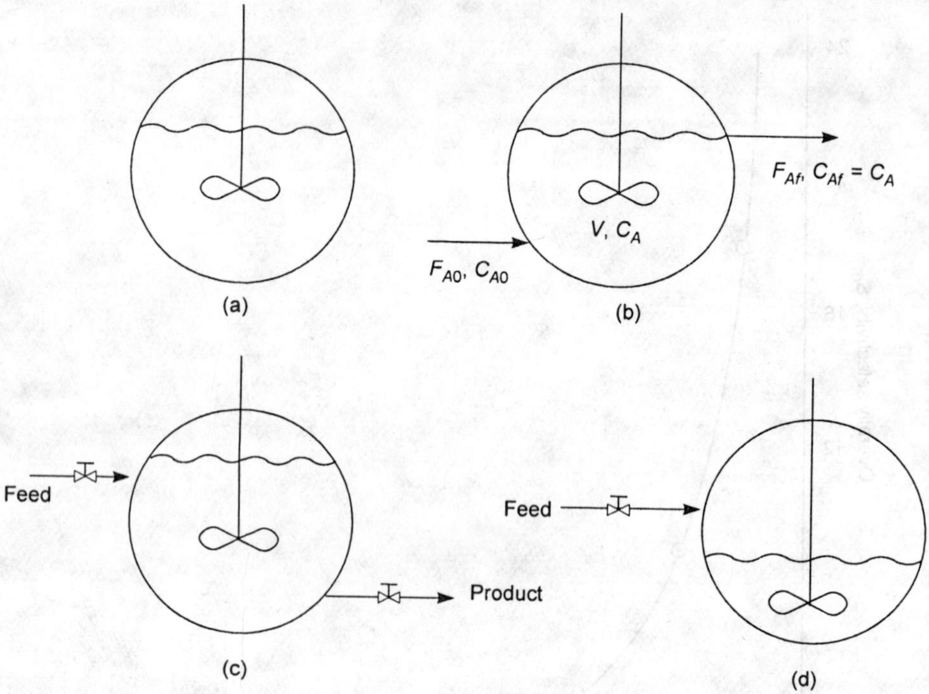


*Figures 8.3 (a), (b), (c) and (d): Ideal stirred tank reactors: (a) Batch reactor, (b) Steady flow reactor (CSTR), (c) Semibatch reactor (unsteady flow), (d) Semibatch reactor with no product discharge*

The performance features of the batch reactor have already been discussed in detail in the previous section. Since no substance is added to the reactor nor any substance discharged from it during operation, the composition of the reaction mixture changes continuously with time. The operation is, thus, under unsteady state conditions. For an *ideal* batch reactor, the composition changes with time, but does not change from one point to another within the reactor. Thus, at any instant, the composition throughout the reactor will be uniform. In the case of constant pressure batch reactor, the volume of the reaction mixture also changes with time. Batch reactors are seldom used on industrial scale for gas-phase reactions since the amount of product that can be produced in a reasonably sized reactor is small. They are often used for liquid-phase reactions in small capacity installations. For the same production rate, operating cost of batch reactors is higher than that of continuous or flow reactors. However, the initial cost of a continuous system is higher due to the instrumentation required. Therefore, for the manufacture of relatively high-priced substances such as pharmaceuticals, where the operating cost is not a predominant factor in the total cost, batch reactors are commonly employed. Also, as discussed in Section (8.2), batch reactors are extensively used as bench-scale reactors for studying the kinetics of industrial reactions. The performance equation of batch reactors is same as Eq. (8.1.1):

$$(-r_A) = -(1/V)\ dN_A/d\theta \qquad \text{... (8.3.1)}$$

In terms of fractional conversion ($x_A$) of $A$,

$$(-r_A) = (N_{A0}/V)\ dx_A/d\theta \quad \text{... (8.3.2)}$$

In the integrated form

$$\theta = N_{A0} \int_0^{x_A} \frac{dx_A}{V(-r_A)} \quad \text{... (8.3.3)}$$

Since $x_A$ is the final conversion desired, it will be more appropriate to denote it as $x_{Af}$ (in the case of flow reactors, such a notation is particularly important). $\theta$ is thus, also the *mean residence time* of the fluid in the reactor. In other words, for a batch reactor, the mean residence time is equal to the average time required to obtain the desired conversion for a given size of batch. If the volume of the system varies linearly with conversion [see Eq. (8.2.128) and Example 8.14] so that $V = V_0\ (1 + \epsilon_A x_A)$,

$$\theta = C_{A0} \int_0^{x_{Af}} \frac{dx_A}{(1+\epsilon_A x_A)(-r_A)} \quad \text{... (8.3.4)}$$

For a constant-volume batch reactor (or, for a reaction mixture of constant density), we can employ Eq. (8.1.3), which on integration gives

$$\theta = -\int_{C_{A0}}^{C_{Af}} \frac{dC_A}{(-r_A)} \quad \text{... (8.3.5)}$$

In the steady-flow stirred tank reactor (also called continuous flow stirred tank reactor, CSTR), there is continuous inflow of feed and continuous outflow of product [see Fig. 8.3 (b)]. The *ideal stirred tank reactor* is one in which there is complete mixing so that the properties of the reaction mixture are uniform in all parts of the vessel and are the same as those in the exit (or product) stream. Thus, if $C_{Af}$ is the concentration of $A$ in the exit stream, then the concentration of $A$ at any point inside the reactor will also be $C_{Af}$. In other words, 100 percent back-mixing is assumed to be taking place. The reactor, therefore, is also called the *back-mix reactor* or the *ideal mixed flow reactor*. In such a reactor, the composition and temperature at which the reaction takes place are the same as the composition and temperature of any exit stream.

The performance equation of an ideal CSTR can be obtained by making a mass balance [see Fig. 8.3 (b)]:

(Rate of input) – (Rate of output) = (Rate of disappearance of reactant by chemical reaction) + (Rate of accumulation) ... (8.3.6)

Since it is a steady flow reactor, rate of accumulation = 0.

Therefore,

Rate of input – Rate of output = (Rate of disappearance by reaction) ... (8.3.7)

$$F_{A0} - F_{Af} = (-r_A)_f\ V \quad \text{... (8.3.8)}$$

where $F_{A0}$ is the molar input of reactant $A$ (in kmoles/s) and $F_{Af}$ is the molar output. Since $F_{Af} = F_{A0}$ $(1 - x_{Af})$, the above equation becomes

$$F_{A0} x_{Af} = (-r_A)_f\ V$$

or

$$V = (F_{A0} x_{Af})/(-r_A)_f \quad \text{... (8.3.9)}$$

The terms *space time* ($\tau$) and *space velocity* ($S$) are widely used to express the performance of flow reactors. These are defined as

$$\tau = \frac{\text{(Reactor volume)}}{\text{(Volumetric flow rate of feed)}} \qquad \text{... (8.3.10)}$$

$$= \frac{V}{(F_0/\rho)} = \frac{V}{(F_{A0}/C_{A0})} = V/Q_0 \qquad \text{... (8.3.11)}$$

where $F_0$ is the total feed rate (in kg/s), $Q_0$ the total volumetric feed rate ($m^3/s$) and $\rho$ is the density of feed mixture ($kg/m^3$).

$$S = 1/\tau \qquad \text{... (8.3.12)}$$

For the present case, from Eq. (8.3.9),

$$\tau = \frac{V\,C_{A0}}{F_{A0}} = C_{A0}x_{Af}/(-r_A)_f \qquad \text{... (8.3.13)}$$

If the density of the reaction mixture remains constant, then $x_A = 1 - (C_A/C_{A0})$ and therefore, Eq. (8.3.9) reduces to

$$V = \frac{F_{A0}\left(C_{A0} - C_{Af}\right)}{C_{A0}(-r_A)_f} \qquad \text{... (8.3.14)}$$

and Eq. (8.3.13) becomes

$$\tau = (C_{A0} - C_{Af})/(-r_A)_f \qquad \text{... (8.3.15)}$$

The mean residence time $\bar{\theta}$ (which is the average of time periods during which individual portions of reaction mixture stay in the reactor) is given by

$$\bar{\theta} = \frac{N_{A0}x_{Af}}{V(-r_A)_f} \qquad \text{... (8.3.16)}$$

For a constant-density system,

$$\bar{\theta} = \frac{C_{A0}x_{Af}}{(-r_A)_f} \qquad \text{... (8.3.17)}$$

Since from Eq. (8.3.9), $x_{Af}/(-r_A)_f = V/F_{A0}$,

$$\bar{\theta} = C_{A0}V/F_{A0} = V/Q_0 = \tau \qquad \text{... (8.3.18)}$$

Thus, for constant density systems, $\bar{\theta} = \tau$. If $V = V_0\,(1 + \epsilon_A x_{Af})$, then

$$\bar{\theta} = \frac{C_{A0}x_{Af}}{\left(1+\epsilon_A x_{Af}\right)(-r_A)_f} \qquad \text{... (8.3.19)}$$

The difference between space time and mean residence time may be noted. The mean residence time is equal to the space time only when the density of the reaction mixture is constant. It is further necessary that the temperature and pressure are constant throughout the reactor and the feed rate is measured at the temperature and pressure in the reactor.

The stirred tank reactor, instead of being operated in steady flow, may also be operated as a *semibatch reactor* as shown in either Fig. 8.3 (c) or (d). In fact, even the steady-flow reactors operate in the semibatch mode at start-up and shut-down, though for short periods. In the semibatch operation shown in Fig. 8.3 (d), all of one of the reactants (say, *B*) is initially charged into the reactor and the other reactant (say, *A*) is then added continuously. No product is withdrawn during the operation. Such a scheme is advantageous when the heat of reaction is large since the heat evolution can be controlled by regulating the rate of addition of *A*. This type of operation also permits a good control of concentration of the reaction mixture and hence the rate of reaction, which is not possible in batch or continuous-flow reactors. In the semibatch operation shown in Fig. 8.3 (c), the reactants are fed continuously and the products are also continuously withdrawn, but the rates of mass flow into and out of the system are unequal. The reactor, thus, operates in non-steady flow.

The analysis of the performance of semibatch reactors is fairly complex since the mass balance equations for semibatch operation could include all the four terms of Eq. (8.3.6). Since both composition and volume of the reaction mixture could change with time, it is often found impossible to solve the mass balance equation analytically and a numerical solution is required. We shall consider here the two cases of semibatch operation shown in Figs 8.3 (c) and (d). We shall also assume ideal reactor performance (complete backmixing).

The mass balance equation is

$$F_{A0} - F_{Af} = (-r_A)\, V + \frac{d(VC_A)}{d\theta} = (-r_A)_f\, V + \frac{d(VC_{Af})}{d\theta} \qquad \text{... (8.3.20)}$$

To note that due to complete backmixing, the concentration of *A* inside the reactor will be the same as that in the exit stream and therefore, $C_A = C_{Af}$. After substituting the appropriate rate expression, the above equation is to be solved analytically or numerically. An important case is that in which the feed and exit flow rates, the feed composition and the density of the reaction mixture are all constant. In such a case, Eq. (8.3.20) reduces to

$$Q_0\, C_{A0} - Q_0\, C_{Af} = (-r_A)_f\, V + V\,(dC_{Af}/d\theta) \qquad \text{... (8.3.21)}$$

or

$$C_{A0} - C_{Af} = (-r_A)_f\, \tau + \tau\,(dC_{Af}/d\theta) \qquad \text{... (8.3.22)}$$

where $\tau$ is the space time. However, since this is a constant density system, $\tau = \bar{\theta}$. If the reaction is first order in *A* and the temperature is constant, then Eq. (8.3.22) becomes

$$\frac{dC_{Af}}{d\theta} + C_{Af}\,(k + 1/\tau) = C_{A0}/\tau \qquad \text{... (8.3.23)}$$

where *k* is the first order rate constant. On integration

$$C_{Af} = \frac{C_{A0}}{(k\tau + 1)} + I \exp\,[-(k + 1/\tau)\,\theta] \qquad \text{... (8.3.24)}$$

where *I* is the integration constant. If initially there is no *A* in the reactor, then at $\theta = 0$, $C_{Af} = 0$. Therefore, $I = -C_{A0}/(k\tau + 1)$ and Eq. (8.3.24) becomes

$$C_{Af} = \frac{C_{A0}}{(k\tau + 1)} \{1 - \exp[-(k + 1/\tau)\,\theta]\} \quad \text{... (8.3.25)}$$

Now, consider Fig. 8.3 (d). Since no product stream is discharged, Eq. (8.3.20) reduces to

$$F_{A0} = (-r_A)\,V + \frac{d(VC_A)}{d\theta} \quad \text{... (8.3.26)}$$

where $C_A$ is the concentration of $A$ inside the reactor at any instant. In this case, $C_A$ may be expressed in terms of the fractional conversion ($x_A$) of $A$:

$$C_A = \frac{(N_{A0} + F_{A0}\,\theta)}{V}(1 - x_A) \quad \text{... (8.3.27)}$$

Substituting for $C_A$ in Eq. (8.3.26) and rearranging, we get

$$F_{A0}\,x_A + (N_{A0} + F_{A0}\theta)\,(dx_A/d\theta) = (-r_A)\,V \quad \text{... (8.3.28)}$$

To note that $N_{A0}$ is the moles of $A$ initially present in the reaction vessel. If only $B$ was present in the reactor initially, then $N_{A0} = 0$. The volume of the reaction mixture changes since $A$ is continuously added to it. Volume also changes due to the change in density of the reaction mixture. However, for many systems, the density of the reaction mixture does not change significantly. For such constant-density systems,

$$V = V_0 + (F_0/\rho)\,\theta = V_0 + Q_0\theta \quad \text{... (8.3.29)}$$

where $V_0$ is the volume of reaction mixture initially present in the reactor. After substituting the appropriate rate expression into Eq. (8.3.28), it can be solved (either analytically or numerically) for $x_A$ as a function of time.

The *tubular reactor* is constructed of either a single continuous coiled tube [Fig. 8.4 (a)] or several tubes in parallel [Fig. 8.4 (b)]. In the latter case, it resembles a shell and tube heat exchanger. The reactants enter at one end of the reactor and the products leave at the other end. The reactor is normally operated at steady state (except at startup or shutdown) so that the properties are constant with respect to time at any point inside the reactor. The long tube and the lack of provision for stirring prevents complete mixing of the fluid in the tube. As a result, the properties of the flowing stream will vary from point to point. In general, variations in properties will occur both in the longitudinal and radial directions. An *ideal tubular flow reactor* is one in which there is no mixing in the axial or longitudinal direction (in the direction of flow), complete mixing in the radial direction and a uniform velocity across the radius (plug flow). Such a reactor is, therefore, called a *plug flow reactor* (PFR). Thus, in a PFR, there is no backmixing at all which is converse to the ideal stirred tank reactor which assumes 100 percent backmixing. In a PFR, the velocity, temperature and concentration profiles are flat over any cross-sectional area perpendicular to the flow, but the composition varies along the flow path.

To derive the performance equation for an ideal tubular-flow reactor (or plug flow reactor), the mass balance must be made for a differential volume element $dV$, as shown in Fig. 8.5. The volume element is chosen extending over the entire cross-section of the tube, since there is no variation in properties or velocity in the radial direction. Since there is no diffusion in the flow direction, the reactant $A$ will enter and leave the volume element only by bulk flow. Thus, for steady-state flow, the mass balance equation for the element is

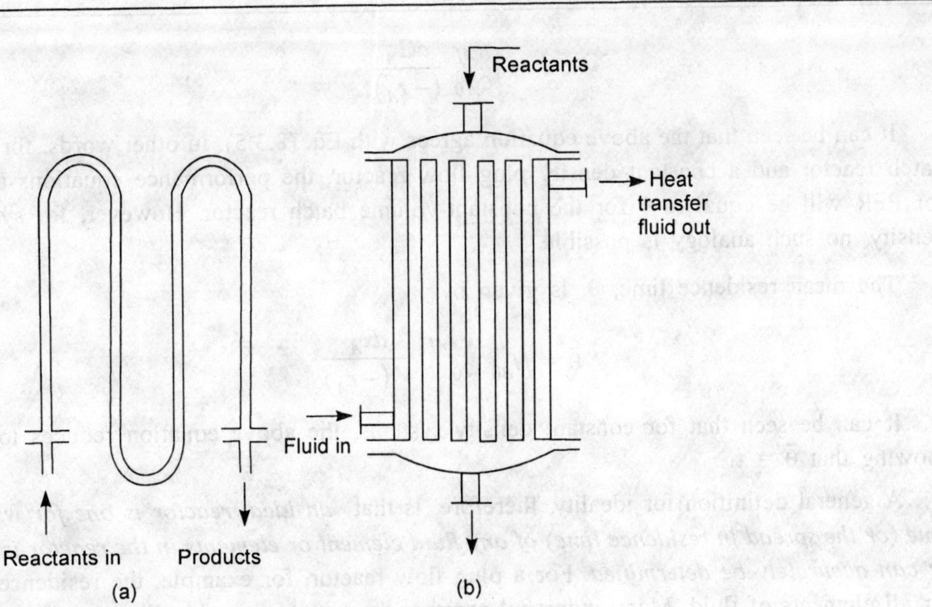


*Figures 8.4: Tubular flow reactors: (a) Single coiled tube, (b) Multiple tube*

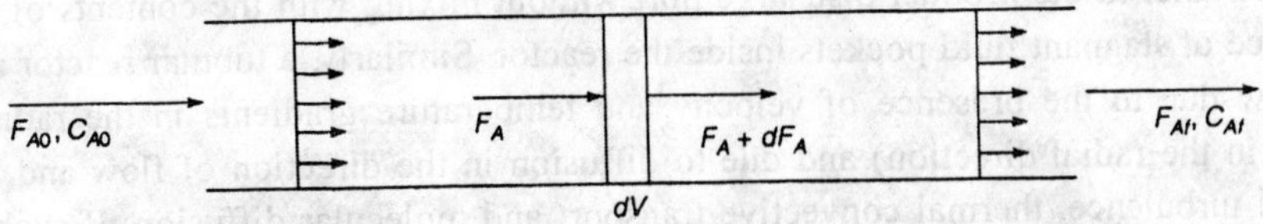


*Figure 8.5*

$$F_A - (F_A + dF_A) = (-r_A)\ dV \qquad \text{... (8.3.30)}$$

or
$$-dF_A = (-r_A)\ dV \qquad \text{... (8.3.31)}$$

Since $dF_A = d\,[F_{A0}\ (1 - x_A)] = -F_{A0}\,dx_A$, the above equation can be rearranged to give

$$\int_0^V dV = V = F_{A0} \int_0^{x_{Af}} \frac{dx_A}{(-r_A)} \qquad \text{... (8.3.32)}$$

$$\tau = VC_{A0}/F_{A0} = C_{A0} \int_0^{x_{Af}} \frac{dx_A}{(-r_A)} \qquad \text{... (8.3.33)}$$

For constant density systems, since $x_A = 1 \quad (C_A/C_{A0})$,

$$V = -(F_{A0}/C_{A0}) \int_{C_{A0}}^{C_{Af}} \frac{dC_A}{(-r_A)} \qquad \text{... (8.3.34)}$$

$$\tau = -\int_{C_{A0}}^{C_{Af}} \frac{dC_A}{(-r_A)} \qquad \text{... (8.3.35)}$$

It can be seen that the above equation agrees with Eq. (8.3.5). In other words, for a constant-volume batch reactor and a constant density plug flow reactor, the performance equations are identical and $\tau$ for PFR will be equal to $\theta$ for the constant-volume batch reactor. However, for systems of changing density, no such analogy is possible.

The mean residence time, $\bar{\theta}$, is given by

$$\bar{\theta} = N_{A0} \int_0^{x_{Af}} \frac{dx_A}{V(-r_A)} \qquad \text{... (8.3.36)}$$

It can be seen that for constant density systems, the above equation reduces to (8.3.33), thereby showing that $\bar{\theta} = \tau$.

A general definition for ideality, therefore, is that *an ideal reactor is one for which the residence time (or the spread in residence time) of any fluid element or elements in the reactor is accurately known or can accurately be determined.* For a plug flow reactor, for example, the residence time is the same for all elements of fluid. Many industrial reactors do deviate from ideality, though this deviation is not large on many occasions. For example, a stirred tank reactor may deviate from ideal behaviour due to short-circuiting and bypassing of the reacting fluid (that is, certain portions of the fluid may proceed directly from the feed inlet to the product discharge port without mixing with the contents of the reactor) or due to the presence of stagnant fluid pockets inside the reactor. Similarly, a tubular reactor may deviate from ideal plug flow due to the presence of velocity and temperature gradients in the radial direction (incomplete mixing in the radial direction) and due to diffusion in the direction of flow and backmixing as the result of fluid turbulence, thermal convective transport and molecular diffusion. If such deviations are significant, then specific experimental studies are to be conducted to determine the residence time distribution (RTD) and mixing pattern in such reactors. This is discussed in detail in Section (8.4).

It will be interesting to compare the performance features of the two most popular ideal reactors such as the stirred tank reactors and the plug flow reactors. Stirred tank reactors are recommended mainly for liquid phase reactions at low or medium pressures. They are difficult to construct for gaseous systems due to the difficulty involved in providing complete mixing. They are also uneconomical to be operated at high pressures since high operating pressures would demand larger wall thickness and complex sealing arrangements for the agitator shaft. Due to the uniformity of temperature, pressure and composition attained as a result of mixing, a stirred tank reactor can be operated under isothermal conditions even when the heat of reaction is high (which is practically an impossibility in long tubular reactors). However, it must be kept in mind that the rate of heat transfer per unit mass of reaction mixture is lower in a tank reactor than in a small-diameter tubular reactor, mainly due to the lower heat transfer surface per unit volume and the lower heat transfer coefficients in tank reactors. Thus, though a stirred tank reactor may be recommended for a highly exothermic reaction, for an endothermic reaction that demands a high operating temperature, a tubular reactor is preferred. In a tubular reactor, high heat transfer coefficients can be easily achieved by forcing the fluid through the tubes at high velocities.

Capacity-wise, a plug flow reactor is to be considered superior to a stirred tank reactor. It is postulated that an infinite number of stirred tank reactors operated in series is equivalent to a single plug flow reactor, provided the total residence time is the same. We can easily illustrate this by an example. For instance, consider three stirred tank reactors operating in series as shown in Fig. 8.6. The exit stream from one

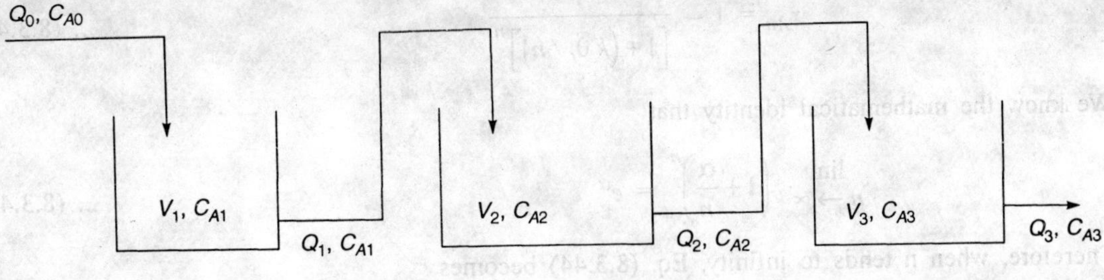


*Figure 8.6: Steady state operation of three ideal stirred tank reactors in series (for constant density systems, $Q_0 = Q_1 = Q_2 = Q_3$)*

serves as the feed to the next. Let us take the specific case of a constant density system and a first order reaction. Thus, for the first reactor, Eq. (8.3.15) becomes

$$\tau_1 = (V_1/Q_0) = (C_{A0} - C_{A1})/(k\,C_{A1}) \qquad \text{... (8.3.37)}$$

To note that since density is constant, the volumetric flow rate ($Q_0$) to each reactor will be the same. Therefore,

$$C_{A1} = C_{A0}/(1 + k\tau_1) \qquad \text{... (8.3.38)}$$

Since for a constant density system, $\tau = \bar{\theta}$,

$$C_{A1} = C_{A0}/(1 + k\bar{\theta}_1) \qquad \text{...(8.3.39)}$$

Similarly, for the second reactor,

$$C_{A2} = C_{A1}/(1 + k\bar{\theta}_2) = \frac{C_{A0}}{(1+k\bar{\theta}_1)(1+k\bar{\theta}_2)} \qquad \text{... (8.3.40)}$$

And, for the third reactor,

$$C_{A3} = \frac{C_{A0}}{(1+k\bar{\theta}_1)(1+k\bar{\theta}_2)(1+k\bar{\theta}_3)} \qquad \text{... (8.3.41)}$$

Thus, if there are $n$ reactors in series, then

$$C_{An} = \frac{C_{A0}}{(1+k\bar{\theta}_1)(1+k\bar{\theta}_2).....(1+k\bar{\theta}_n)} \qquad \text{... (8.3.42)}$$

The total fractional conversion of $A$ in $n$ reactors will be

$$x_{An} = 1 - (C_{An}/C_{A0})$$

$$= 1 - \frac{1}{(1+k\bar{\theta}_1)(1+k\bar{\theta}_2).....(1+k\bar{\theta}_n)} \qquad \text{... (8.3.43)}$$

If the volume of each reactor is the same, then $\bar{\theta}_1 = \bar{\theta}_2 = ..... = \bar{\theta}_n = \bar{\theta}_t/n$, where $\bar{\theta}_t$ is the total residence time in $n$ reactors. Equation (8.3.43) then reduces to

$$x_{An} = 1 - \frac{1}{\left[1+\left(k\bar{\theta}_t/n\right)\right]^n} \quad \text{... (8.3.44)}$$

We know the mathematical identity that

$$\lim_{n\to\infty}\left(1+\frac{\alpha}{n}\right)^n = e^{\alpha} \quad \text{... (8.3.45)}$$

Therefore, when n tends to infinity, Eq. (8.3.44) becomes

$$x_A \text{ (at } n = \infty) = 1 - \exp(-k\bar{\theta}_t) \quad \text{... (8.3.46)}$$

Now, consider a plug flow reactor. From Eq. (8.3.35),

$$t = -\int_{C_{A0}}^{C_{Af}} \frac{dC_A}{kC_A} = (1/k)\ \ln\ (C_{A0}/C_{Af}) \quad \text{... (8.3.47)}$$

or

$$C_{Af} = C_{A0}\ \exp(-k\tau) = C_{A0}\ \exp(-k\bar{\theta}) \quad \text{... (8.3.48)}$$

Now,

$$x_{Af} = 1 - C_{Af}/C_{A0} = 1 - \exp(-k\bar{\theta}) \quad \text{... (8.3.49)}$$

It can be seen that the above equation and Eq. (8.3.46) are identical. Thus, it proves that an infinite number of stirred tank reactors in series is equivalent to a single plug flow reactor, provided the total residence time is the same.

It is also important to note that if, instead of $n$ reactors in series, we use a single stirred tank reactor of the same residence time $\bar{\theta}_t$, the total conversion of $A$ obtained will be

$$x_A = 1 - \frac{1}{\left(1+k\bar{\theta}_t\right)} \quad \text{... (8.3.50)}$$

comparing the above expression with Eq. (8.3.44), it is clear that $n$ reactors in series will provide better performance than a single reactor even if the total residence time is the same in both cases. In other words, even if $V = V_1 + V_2$, two reactors of volumes $V_1$ and $V_2$ operated in series will provide much higher conversion of $A$ than a single large reactor of volume $V$. Increase in the number of reactors will, however, increase the overall fabrication and installation costs.

When the rate equation is a simple one such as that in the case of a first order reaction described above, analytical determination of the performance of a number of stirred tank reactors in series is not difficult. However, when the rate equation is complicated, an analytical evaluation of the performance becomes cumbersome. A graphical method may be employed in such cases. The procedure, for constant density systems, is illustrated in Fig. 8.7. First, the rate of reaction, $(-r_A)$, is plotted against reactant concentration, $C_A$. Now, starting from $C_{A0}$, straight lines are drawn to the curve with slopes equal to $-1/\bar{\theta}_1, -1/\bar{\theta}_2, -1/\bar{\theta}_3$, etc. (see Fig. 8.7). If the volumes of the reactors are known, then the concentration of the final effluent discharged from the last reactor can be estimated by following this procedure. Alternately, the number of reactors required to achieve a given conversion can be determined.

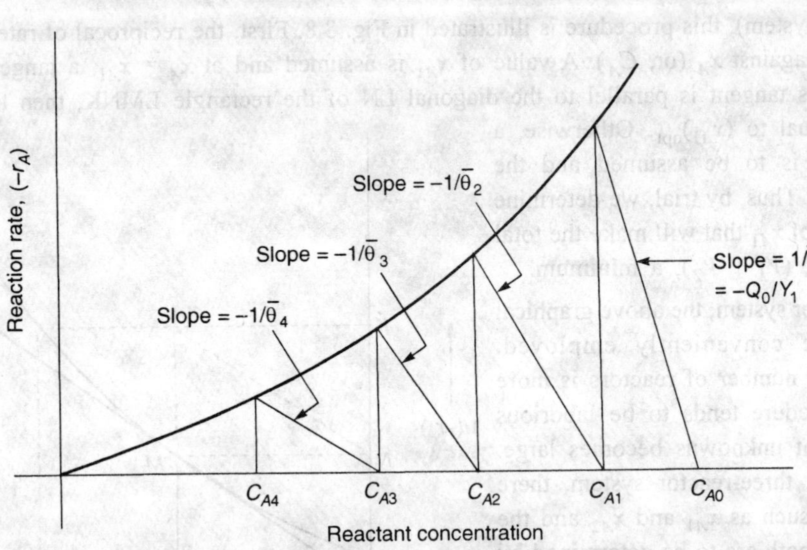


*Figure 8.7: Graphical solution for ideal stirred tank reactors in series*

When the reactors operated in series are of different sizes, then it becomes necessary to estimate the *optimum size ratio* (in fact, the minimum ratio of sizes). For example, consider two stirred tank reactors operated in series (similar to Fig. 8.6). Assuming constant density and first order kinetics,

$$V_1 + V_2 = \frac{Q_0\,(C_{A0} - C_{A1})}{k\,C_{A1}} + \frac{Q_0\,(C_{A1} - C_{A2})}{k\,C_{A2}} \quad \text{... (8.3.51)}$$

$$= (Q_0/k)\left(\frac{C_{A0}}{C_{A1}} + \frac{C_{A1}}{C_{A2}} - 2\right) \quad \text{... (8.3.52)}$$

Since the composition of the fresh feed to the first reactor ($C_{A0}$) and that of the final product discharged from the second reactor ($C_{A2}$) are known, the unknown in the above expression is $C_{A1}$. We have to, therefore, find that value of $C_{A1}$ which will make $(V_1 + V_2)$ a minimum. For the present case, we can determine this by differentiating Eq. (8.3.52) with respected to $C_{A1}$ and putting it equal to zero and then solving it for $(C_{A1})_{\text{opt}}$. Thus, $d\,(V_1 + V_2)/dC_{A1} = 0$ gives

$$(Q_0/k)\left(-\frac{C_{A0}}{C_{A1}^2} + \frac{1}{C_{A2}}\right) = 0$$

or

$$C_{A1} = (C_{A1})_{\text{opt}} = \sqrt{C_{A0}C_{A2}} \quad \text{... (8.3.53)}$$

Once $C_{A1}$ is known, the values of $V_1$ and $V_2$ can be computed from Eq. (8.3.51).

When the rate equation is not as simple as a first order equation and is complicated, the above analytical procedure of finding the optimum will be too cumbersome. A graphical procedure[19] will be more convenient in such cases. For the present case of two reactors in series (with the assumption of a constant density system), this procedure is illustrated in Fig. 8.8. First, the reciprocal of rate of reaction $1/(-r_A)$ is plotted against $x_A$ (or, $C_A$). A value of $x_{A1}$ is assumed and at $x_A = x_{A1}$, a tangent is drawn to the curve. If this tangent is parallel to the diagonal LN of the rectangle LMNK, then the assumed value of $x_{A1}$ is equal to $(x_{A1})_{opt}$. Otherwise, a new value of $x_{A1}$ is to be assumed and the procedure repeated. Thus, by trial, we determine the optimum value of $x_{A1}$ that will make the total volume of reactors, $(V_1 + V_2)$, a minimum.

For a two reactor system, the above graphical procedure can be conveniently employed. However, when the number of reactors is more than two, the procedure tends to be laborious since the number of unknowns becomes large. For example, for a three-reactor system, there are two unknowns such as $x_{A1}$ and $x_{A2}$ and the optimum values of both are to be determined by trial. The overall procedure tends to become too lengthy in such cases. However, it must be kept in mind that the advantage of using unequal but minimum sized reactors in series over equal-size system is usually quite small and therefore, overall economic consideration would mostly recommend the use of equally sized reactors.

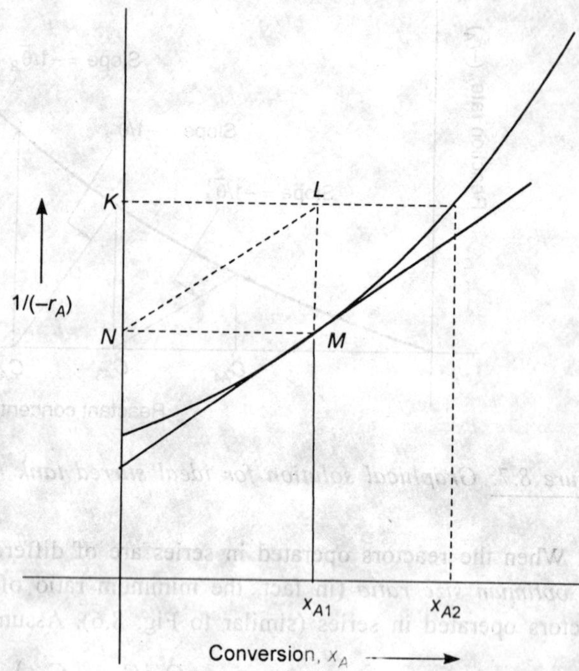


*Figure 8.8: Graphical determination of optimum size ratio of two ideal stirred tanks in series*

Before concluding this section, let us state a few words about *nonisothermal operation* of reactors. What all have been discussed in the previous paragraphs are concerned with isothermal operation of ideal reactors. Isothermal operation has distinct advantages. For example, in the case of endothermic reactions, the rate of reaction decreases with decrease in temperature. Thus, if no arrangement has been made to supply heat to the reactor from outside, the temperature of the reaction mixture will decrease continuously due to the endothermic nature of the reaction and as a result, the rate of reaction also decreases. As the reaction proceeds, the concentration of the reactant decreases and this will further decrease the reaction rate.

Thus, it will be highly advisable to supply heat to the reactor from outside so that the reactor temperature is maintained high enough to obtain a high rate of reaction. In the case of stirred tank reactors (either batch or continuous), the vessel may be jacketted and the heating fluid circulated through the jacket. In the case of tubular reactors that use a multiple of tubes [see Fig. 8.4 (b)], the heating fluid may be passed outside the tubes through the shell. When heat is to be dissipated (as in the case of exothermic reactions), a large amount of diluent may be added to the feed which will absorb the heat of reaction and thus prevent a large temperature drop. If the reaction is reversible, addition of heat provides the further advantage of increasing the maximum attainable conversion (namely, the equilibrium conversion). We have already discussed in Section 8.1 the effect of temperature on equilibrium constant K. The value of $K$ and thereby the equilibrium conversion increases with increase in temperature in the case of endothermic reactions. The reverse is true for exothermic reactions. The magnitude of equilibrium conversion decreases with increase in temperature in the case of exothermic reactions and therefore, it will be highly desirable in such cases to dissipate the heat evolved by a suitable cooling arrangement. This will also prevent the occurrence of undesirable side reactions and the formation of hot spots. However, when the heat of reaction is significantly high, isothermal operation may tend to be uneconomical since the cooling or heating load will become abominantly large. The converse to isothermal operation is the *adiabatic operation*. In this case, no heat is added to or removed from the reactor. In the case of a batch reactor therefore, if the operation is adiabatic, then the temperature of the reaction mixture will change (increase or decrease) continuously with time. Thus, to study the reactor performance, apart from the material balance equation, the energy balance equation will also be required. This is true in the case of a *nonadiabatic operation* as well, in which heat is added or removed from the reactor but the rate of heat addition or dissipation is not sufficient to maintain an isothermal operation*. In the case of an ideal CSTR, the temperature of the reaction mixture will be uniform and constant whether the operation is isothermal or adiabatic due to complete backmixing. But the temperature of the reaction mixture (or, the operating temperature) which will be equal to the temperature of the product stream, will be different from the feed temperature if the operation is adiabatic or nonisothermal. Thus, to find this operating temperature, we have to perform an energy balance. This operating temperature decides the magnitude of the rate constant and thereby the rate of reaction. The dependance of the rate constant on temperature has already been discussed in Section (8.1).

In the case of a PFR (or an ideal tubular flow reactor), if the operation is adiabatic or nonisothermal, there will be no temperature gradient in the radial direction at any section of the reactor but the temperature will vary progressively from the feed end to the discharge end in a similar manner as the concentration changes. As a result, the reactor performance is coupled with the energy balance equation apart from the material balance.

In short, to analyse the nonisothermal operation of reactors, both the material balance equation and the energy balance equation are to be solved simultaneously. The energy balance equation can be written in a similar way as the material balance equation [Eq. (8.3.6)]. Thus

(Rate of heat input in feed stream) – (Rate of heat output in product stream) + (Rate of heat input from surroundings) = (Rate of disappearance of heat by chemical reaction) + (Rate of accumulation of heat) ... (8.3.54)

For a batch reactor, the first two terms on the left hand side are zero. Therefore, the energy balance equation reduces to

$$m C_p \,(dT/d\theta) + (\Delta H_r)\,(-r_A)\, V = Q \qquad \text{... (8.3.55)}$$

where $m$ = mass of reaction mixture, kg

$C_p$ = specific heat of reaction mixture, J/(kg·K)

$\Delta H_r$ = heat of reaction, J/kmole of $A$

$Q$ = rate of heat input to reactor from outside, W

* Thus, a nonadiabatic operation is also a nonisothermal operation. It is intermediate between a fully isothermal operation and a fully adiabatic operation.

To note that $A$ represents the *limiting reactant*. $(\Delta H_r)$ will be a negative quantity for an exothermic reaction and a positive quantity for endothermic reactions. The rate of heat input from surroundings ($Q$) may be expressed as

$$Q = UA_h (T_S - T) \quad \text{... (8.3.56)}$$

where $U$ = overall heat transfer coefficient, W/(m$^2$·K)

$A_h$ = heat transfer area, m$^2$

$T_S$ = temperature of heat source or heat sink, K

$T$ = temperature of reaction mixture, K

If the operation is adiabatic, $Q = 0$. Then, Eq. (8.3.55) becomes

$$mC_p (dT/d\theta) = -(\Delta H_r)(-r_A) V \quad \text{... (8.3.57)}$$

If the density and specific heat of the reaction mixture are essentially constant, then since $m = (V\rho)$ and $(-r_A) = C_{A0} (dx_A/d\theta)$, we get

$$dT/d\theta = -(\Delta H_r C_{A0}/\rho C_p)(dx_A/d\theta) \quad \text{... (8.3.58)}$$

Integrating both sides (with the assumption that at $\theta = 0$, $T = T_0$ and $x_A = 0$), we get

$$T - T_0 = -(\Delta H_r \cdot C_{A0}/\rho C_p) x_A \quad \text{... (8.3.59)}$$

For a plug flow reactor, like the material balance equation, the heat balance equation is also to be written in the differential form:

$$-F_0 C_p dT + Q = (\Delta H_r)(-r_A) dV \quad \text{... (8.3.60)}$$

where $F_0$ = total feed rate, kg/s

$$Q = U (T_S - T) \, dA_h$$

$$= U (T_S - T) \, \pi D \, dz \quad \text{... (8.3.61)}$$

$D$, $z$ = diameter and length respectively of reactor, m

If the operation is adiabatic, $Q = 0$ and therefore Eq. (8.3.60) gets simplified to

$$F_0 C_P dT + (\Delta H_r)(-r_A) \, dV = 0 \quad \text{... (8.3.62)}$$

Now, from Eq. (8.3.32), $dV = F_{A0} \, dx_A/(-r_A)$. Also, $F_0 = Q_0\rho = (F_{A0}/C_{A0}) \, \rho$. Therefore, Eq. (8.3.62) becomes

$$dT = -[(\Delta H_r) C_{A0}/\rho C_p] \, dx_A \quad \text{... (8.3.63)}$$

Assuming $\rho$ and $C_p$ are essentially constant, we can integrate the above equation to give

$$T - T_F = -(\Delta H_r C_{A0}/\rho C_p) \, x_A \quad \text{... (8.3.64)}$$

where $T_F$ is the feed temperature. It can be seen that the above equation is same as Eq. (8.3.59) with $T_0$ replaced by $T_F$.

For an ideal CSTR, as stated earlier, the temperature of the reaction mixture will be uniform and constant throughout. Let us call it $T_e$. The temperature of the exit stream will also be equal to $T_e$. If the operation is isothermal then $T_e = T_F$ = feed temperature. But if the operation is adiabatic or nonisothermal, then $T_e$ will be different from $T_F$. The value of $T_e$ can be then computed from the energy balance equation:

$$F_0 C_p (T_F - T_e) + Q = (\Delta H_r)(-r_A)_f V \quad \text{... (8.3.65)}$$

where
$$Q = UA_h (T_S - T_e) \quad \text{... (8.3.66)}$$

Since $F_0 = Q_0\rho = (F_{A0}/C_{A0})\,\rho$ and from Eq. (8.3.9), $(-r_A)_f = F_{A0}x_{Af}/V$,

$$(T_F - T_e) = (\Delta H_r x_{Af} C_{A0}/\rho C_p) - QC_{A0}/(F_{A0}\rho C_p) \quad \text{... (8.3.67)}$$

If the operation is fully adiabatic, then $Q = 0$ and the above equation will reduce to

$$T_F - T_e = \Delta H_r x_{Af} C_{A0}/(\rho C_p) \quad \text{... (8.3.68)}$$

Equation (8.3.67) or (8.3.68) may be used for computing $T_e$ once the other system variables are known.

**Example 8.16:** *Performance of ideal reactors.* Sucrose is to be hydrolysed by the catalyticaction of the enzyme sucrase at the rate of 0.1 kg/hr. The initial concentration of sucrose ($C_{A0}$) is 1.0 millimole/liter and that of enzyme ($C_{E0}$) is 0.01 millimole/liter:

(*a*) If the process is conducted in a well-mixed, constant volume batch reactor, compute the volume of reactor required to achieve 80 percent conversion.

(*b*) If the same reaction is to be conducted in an ideal continuous stirred tank reactor (CSTR), compute the volume of reactor required to obtain the same fractional conversion of sucrose. Assume that the density of the reaction mixture remains essentially constant.

(*c*) A plug flow reactor (PFR) is recommended for the same purpose and the feed rate of reactant to the reactor is the same as in (*b*). Compute the required reactor volume for obtaining 80 percent conversion, assuming constant density system.

(*d*) What will be the conversion obtained if two, equal-sized ideal stirred tank reactors, whose combined volume is the same as that of the single reactor considered in (*b*), are used in series to conduct the same process?

(*e*) If two ideal stirred tank reactors of unequal sizes are to be used in series for conducting the reaction to 80 percent conversion, determine the minimum size ratio that is to be maintained.

Assume isothermal operation in all the above cases.

**Solution:** The rate equation for this reaction has already been developed in Example 8.10:

$$(-r_A) = \frac{1.01C_A}{1.0 + 5.1C_A}$$

where $r_A$ is expressed in millimol/(liter·hr) and $C_A$ in millimol/liter.

(*a*) For a constant volume batch reactor, from Eq. (8.3.5),

$$\theta = \int_{C_{Af}}^{C_{A0}} \frac{dC_A}{(-r_A)}$$

where $C_{A0}$ = 1.0 millimol/liter and since 80 percent conversion is desired,

$$C_{Af} = C_{A0}\,(1 - x_A) = 1.0\,(1 - 0.8) = 0.2 \text{ millimol/liter}$$

Therefore, $$\theta = \int_{0.2}^{1.0} \frac{(1+5.1C_A)}{1.01C_A}\, dC_A$$

$$= (1/1.01)\ \ln\ (1/0.2) + (5.1/1.01)\ (1.0 - 0.2) = 3.353 \text{ hr}$$

Since the amount of sucrose hydrolysed is 0.1 kg/hr or (0.1/342) = $0.2924 \times 10^{-3}$ kmole/hr (to note that molecular weight of sucrose = 342 kg/kmole)

$$0.2924 \times 10^{-3} = (N_{A0}x_A)/\theta = N_{A0}\ (0.8)/(3.353)$$

or $$N_{A0} = 12.255 \times 10^{-4} \text{ kmole}$$

The reactor must hold this much moles of sucrose and therefore, neglecting the mass of enzyme (which is comparatively very small),

$$V = N_{A0}/C_{A0} = (12.255 \times 10^{-4})/(0.001) = 1.2255 \text{ m}^3 \simeq 1.23 \text{ m}^3$$

It may be noted that $C_{A0}$ = 1.0 millimol/liter = 0.001 kmole/m$^3$.

(*b*) For an ideal stirred tank reactor, from Eq. (8.3.9),

$$V = F_{A0}x_{Af}/(-r_A)_f$$

where $$x_{Af} = 0.8$$

$$(-r_A)_f = \frac{1.01C_{Af}}{1+5.1C_{Af}}$$

$$= \frac{1.01(0.2)}{1+(5.1)(0.2)} = 0.1 \text{ millimol/(liter}\cdot\text{hr)} = 0.0001 \text{ kmoles/(m}^3\cdot\text{hr)}$$

Since sucrose is being hydrolysed at $0.2924 \times 10^{-3}$ kmoles/hr,

$$(0.2924 \times 10^{-3}) = F_{A0}x_{Af}$$

or $$F_{A0} = (0.2924 \times 10^{-3})/(0.8) = 0.3655 \times 10^{-3} \text{ kmoles/hr}$$

Therefore $$V = \frac{(0.3655 \times 10^{-3})(0.8)}{(0.0001)} = 2.924 \text{ m}^3$$

(*c*) For a plug flow reactor, from Eq. (8.3.34),

$$V = (F_{A0}/C_{A0}) \int_{C_{Af}}^{C_{A0}} \frac{dC_A}{(-r_A)}$$

It can be seen that the integral in the above equation is equal to θ of batch reactor and θ = 3.353 hr. Therefore

$$V = (0.3655 \times 10^{-3}/0.001)\ (3.353) = 1.2255 \text{ m}^3 \simeq 1.23 \text{ m}^3$$

This is understandable since for a constant density system, the performance equation for PFR is identical with that of a batch reactor.

(*d*) Since both reactors are of the same size,

$$V_1 = V_2 = (V/2) = (2.924/2) = 1.462 \text{ m}^3$$

For a constant density system, $\bar{\theta} = \tau = (V/Q_0)$ and therefore,

$$\bar{\theta}_1 = \bar{\theta}_2 = (1.462/0.3655) = 4.0 \text{ hr}$$

To note that $Q_0 = (F_{A0}/C_{A0}) = (0.3655 \times 10^{-3}/0.001) = 0.3655 \text{ m}^3/\text{hr}$. Since this is a constant density system, we can employ the graphical procedure illustrated in Fig. 8.7. Thus, the plot of $(-r_A)$ versus $C_A$ is prepared based on the rate equation and it is shown in Fig. 8.16.1.

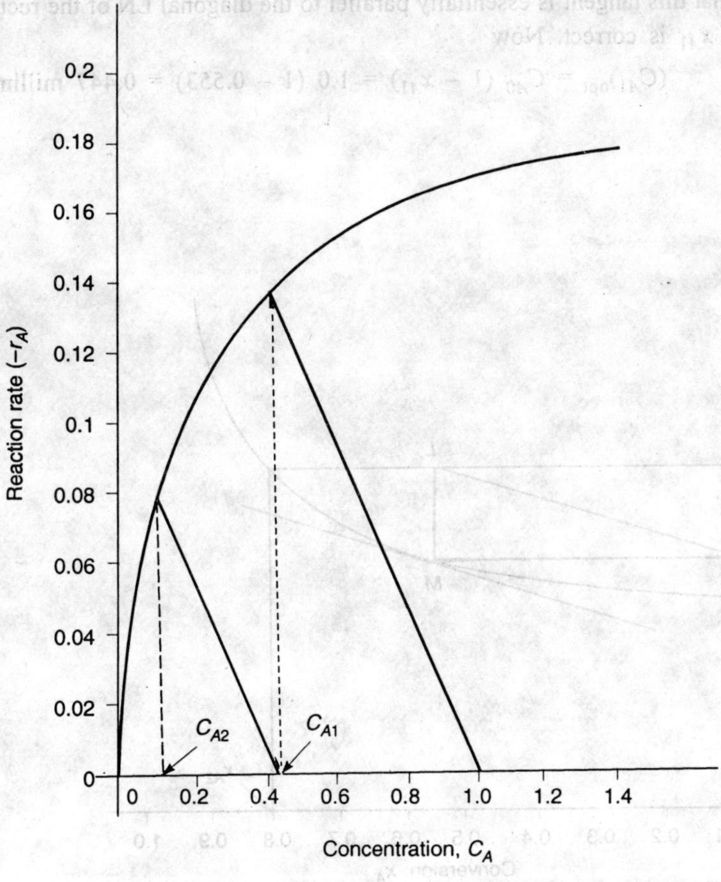


*Figure 8.16.1*

Now, starting from $C_A = C_{A0}$, a straight line of slope – 0.25 is drawn and from its point of intersection with the rate curve, the value of $C_{A1}$ is obtained. From $C_A = C_{A1}$, another straight line of same slope is drawn and its point of intersection with the rate curve gives $C_{A2}$. Thus,

$$C_{A2} = C_{Af} = 0.13 \text{ millimol/liter}$$

Now,
$$x_{Af} = 1 - (C_{Af}/C_{A0}) = 1 - (0.13/1.0) = 0.87$$

Thus, two reactors in series provide higher conversion than a single reactor of same total volume.

(*e*) To determine the minimum size ratio of the two reactors, let us first employ the graphical method illustrated in Fig. 8.8, which is applicable to a constant density system. Thus, the values of $(-r_A)$ at different values of $C_A$ are computed from the rate equation and the values of $x_A$ from $x_A = 1 - (C_A/C_{A0})$. Now, a plot of $1/(-r_A)$ versus $x_A$ is prepared which is shown in Fig. 8.16.2. Since the final conversion desired is 80 percent, $x_{A2} = 0.8$. The optimum value of $x_{A1}$, at which the total volume $(V_1 + V_2)$ of the reactors will be a minimum, is to be determined by trial. Thus, let $x_{A1} = 0.553$. The rectangle LMNK is now drawn as shown in Fig. 8.16.2. At $x_A = x_{A1} = 0.553$, a tangent is drawn to the curve and it is found that this tangent is essentially parallel to the diagonal LN of the rectangle. Therefore, the assumed value of $x_{A1}$ is correct. Now

$$(C_{A1})_{opt} = C_{A0}\ (1 - x_{A1}) = 1.0\ (1 - 0.553) = 0.447 \text{ millimol/liter}$$

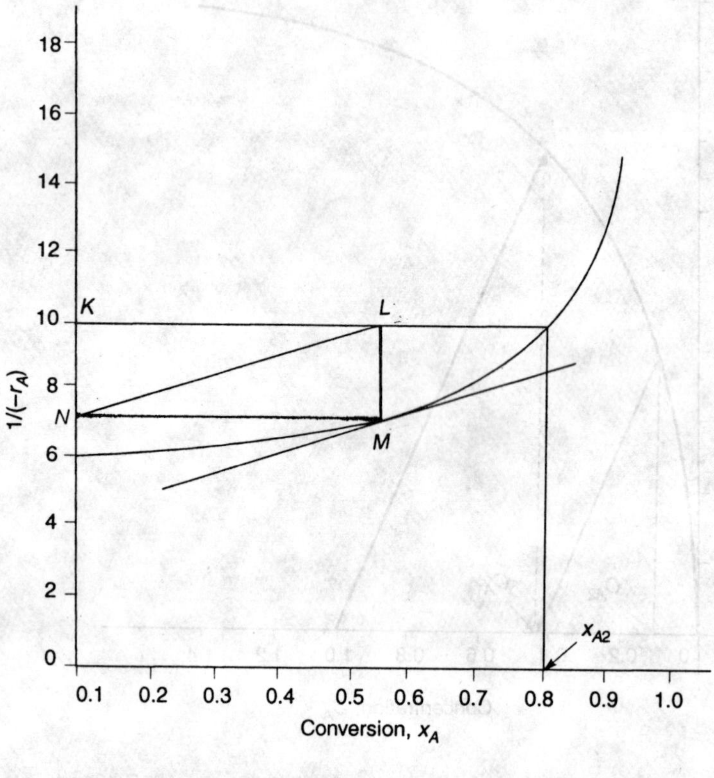


*Figure 8.16.2*

For the present case, since the rate equation is not too complex, the optimum value of $C_{A1}$ can also be analytically determined. Thus

$$V_1 + V_2 = \frac{Q_0\,(C_{A0} - C_{A1})}{(-r_{A1})} + \frac{Q_0\,(C_{A1} - C_{A2})}{(-r_{A2})}$$

$$= \frac{Q_0\,(C_{A0} - C_{A1})(1 + 5.1C_{A1})}{1.01C_{A1}} + \frac{Q_0\,(C_{A1} - C_{A2})(1 + 5.1C_{A2})}{1.01C_{A2}}$$

$d\,(V_1 + V_2)/dC_{A1} = 0$ gives

$$-5.1 - (C_{A0}/C_{A1}^2) + (1 + 5.1\ C_{A2})/C_{A2} = 0$$

or

$$C_{A1} = (C_{A1})_{\text{opt.}} = \sqrt{C_{A0}\,C_{A2}}$$

For the present case,

$$(C_{A1})_{\text{opt}} = \sqrt{(1.0)\,(0.2)} = 0.4472 \text{ millimol/liter}$$

It can be seen that the above value is essentially the same as that determined graphically. The graphical method invariably involves personal error and therefore, it is always preferable to use the analytical method wherever possible. Now,

$$V_1 = \frac{Q_0\,(C_{A0} - C_{A1})(1 + 5.1C_{A1})}{1.01C_{A1}}$$

$$= \frac{(0.3655)\,(1 - 0.4472)\,[1 + 5.1\,(0.4472)]}{(1.01)\,(0.4472)} = 1.4675 \text{ m}^3$$

$$V_2 = \frac{(0.3655)\,(0.4472 - 0.2)\,[1 + 5.1\,(0.2)]}{(1.01)\,(0.2)} = 0.9035 \text{ m}^3$$

$$(V_1/V_2) = 1.6242$$

Thus, it is optimum to use a larger reactor followed by a smaller reactor. The first reactor is to be 62.42 percent larger than the second. By such an arrangement the combined volume of the two reactors is reduced to 2.371 m$^3$.

**Example 8.17:** *Performance of ideal reactors* (*reactor combinations*). The following three reactions are to be conducted on commercial scale:

(*a*) Hydrolysis of acetic anhydride ($A$) at 20° C which in first order is $A$ with rate constant = 0.113 min$^{-1}$, the feed solution containing 0.225 kmole of anhydride per m$^3$ being fed at 20 L/hr.

(*b*) Production of ethyl formate ester at 30° C from ethanol and formic acid in presence of sulfuric acid as the homogeneous catalyst, the feed solution containing the acid and the alcohol in the mole ratio 1 : 5 being fed to the reactor at 7.2 L/hr. It has been reported that the reaction is second order in formic acid ($A$) and zero order in ethanol with specific reaction rate = $2.8 \times 10^{-4}$ m$^3$/(kmole·s). The density of the reaction mixture maybe assumed to remain constant at 820 kg/m$^3$.

(*c*) A reaction with rate equation, $(-r_A) = 1.6\ C_A^{0.56}$ kmoles/(L·min), the feed solution containing 1.2 kmoles of $A$ per liter being charged at 8 L/min. The options available are

(*i*) A plug flow reactor of volume 4.0 L followed by an ideal CSTR of volume 4.0 L,

(*ii*) An ideal CSTR of volume 4.0 L followed by a 4.0 L plug flow reactor.

Determine which of these combinations would provide better performance (larger fractional conversion of $A$) for each of the above reactions. Assume that the density of the reaction mixture remains essentially constant in each case.

**Solution:** (*a*) Here the reaction is psuedo first order in *A*. Let us first consider the PFR – CSTR combination. Since each reactor is of volume 4.0 L and the flow rate of feed/effluent remains constant at 20 L/hr (assuming negligible variation in the density of reaction mixture), the space time ($\tau$) is same for both reactors:

$$\tau = \frac{(4.0)}{(20.0/60)} = 12 \text{ minutes}$$

The concentration of *A* in the effluent from PFR is thus

$$C_{A1} = (0.225) \exp(-0.113 \times 12) = 0.05798 \text{ kmoles/m}^3$$

This solution is now fed to the CSTR. Concentration of *A* in the effluent from CSTR (final effluent) is therefore,

$$C_{Ae} = C_{A1}/(1 + k\tau) = \frac{(0.05798)}{1+(0.113\times 12)} = 0.0246 \text{ kmoles/m}^3$$

The fractional conversion of *A* is then,

$$x_A = 1 - (0.0246/0.225) = 0.8906 = \mathbf{89.06 \text{ percent}}$$

Let us now consider the second combination (CSTR followed by PFR). Thus, concentration of *A* in effluent from CSTR is

$$C_{A1} = C_{A0}/(1 + k\tau) = \frac{(0.225)}{1+(0.113\times 12)} = 0.0955 \text{ kmoles/m}^3$$

This solution is now fed to the PFR. Therefore,

$$C_{Ae} = 0.0955 \exp(-0.113 \times 12) = 0.0246 \text{ kmoles/m}^3$$

And, $x_A = 1 - (0.0246/0.225) = 0.8906 = 89.06$ percent

It can be thus seen that if the reaction follows first order kinetics, both of these combinations would provide the same performance (same fractional conversion of *A*).

(*b*) Here, the reaction is second order in *A* and zero order in the other reactant. Thus,

$$(-r_A) = k\,C_A^2$$

Since molecular weight of formic acid and that of ethanol are the same (= 46.0 kg/kmole), mole ratio of acid to alcohol = mass ratio of acid to alcohol = 1 : 5. Since density of feed solution (and also that of reaction mixture at any instant) is 820 kg/m³, one m³ of feed solution contains (820/6) kg of formic acid or (820/6 × 46) kmoles of formic acid. Thus,

$$C_{A0} = \frac{(820.0)}{(6.0)(46.0)} = 2.971 \text{ kmoles/m}^3$$

Also, $\tau$ (for each reactor) = 4.0/(7.2/3600) = 2000 s. Let us first consider the PFR – CSTR combination. For a PFR and for a second order reaction of above form, we know

$$\frac{1}{C_{A1}} - \frac{1}{C_{A0}} = k\tau$$

or
$$C_{A1} = \frac{C_{A0}}{1 + C_{A0} k\tau}$$

Therefore,
$$C_{A1} = \frac{(2.971)}{1 + \left(2.971 \times 2.8 \times 10^{-4} \times 2000\right)} = 1.11534 \text{ kmoles/m}^3$$

This solution now enters the CSTR. For an ideal CSTR, we know

$$\tau = \frac{(C_{A1} - C_{Ae})}{k C_{Ae}^2}$$

or
$$k\tau C_{Ae}^2 + C_{Ae} - C_{A1} = 0$$

substituting the values of $k$, $\tau$ and $C_{A1}$, we get

$$0.56 C_{Ae}^2 + C_{Ae} - 1.11534 = 0$$

On solving, $C_{Ae} = 0.777$ kmoles/m$^3$

Therefore, $x_A = 1 - (0.777/2.971) = 0.7384 =$ **73.84 percent**

Now, consider the CSTR – PFR combination. For the ideal CSTR, since $C_{A0} = 2.971$ kmoles/m$^3$,

$$0.56 C_{A1}^2 + C_{A1} - 2.971 = 0$$

or
$$C_{A1} = 1.5775 \text{ kmoles/m}^3$$

Now, for the PFR (which receives a feed solution of concentration $C_{A1}$),

$$C_{Ae} = \frac{C_{A1}}{1 + C_{A1} k\tau} = \frac{(1.5775)}{1 + (1.5775 \times 0.56)} = 0.8376 \text{ kmoles/m}^3$$

Therefore, $x_A = 1 - (0.8376/2.971) = 0.718 =$ **71.8 percent**

Thus, for a second order reaction, the first type of reactor combination (PFR followed by CSTR) provides better performance. This, in fact, is true in the case of all reactions with order greater than unity.

(*c*) Here, the reaction is of fractional order (with $n < 1$). Consider the PFR – CSTR combination. For the PFR, since $\tau = (4.0/8.0) = 0.5$ minutes and $C_{A0} = 1.2$ kmoles/L,

$$k\tau = (C_{A0}^{1-n} - C_{A1}^{1-n})/(1 - n)$$

or,
$$C_{A1}^{1-n} = C_{A0}^{1-n} - (1 - n)\, k\tau$$

Since $n = 0.56$,
$$C_{A1}^{0.44} = (1.2)^{0.44} - (0.44 \times 1.6 \times 0.5) = (1.2)^{0.44} - (0.352)$$

or
$$C_{A1} = 0.4914 \text{ kmoles/L}$$

For the ideal CSTR that follows,

$$\tau = 0.5 = \frac{(C_{A1} - C_{Ae})}{1.6 C_{Ae}^{0.56}}$$

or,
$$0.8 C_{Ae}^{0.56} + C_{Ae} - 0.4914 = 0$$

On solving by trial, $C_{Ae} = 0.183$ kmoles/L

Therefore, $x_A = 1 - (0.183/1.2) = 0.8475 =$ **84.75** percent

Now, consider the CSTR – PFR combination. For the first reactor (namely, the CSTR),

$$0.8\,C_{A1}^{0.56} + C_{A1} - 1.2 = 0$$

On solving by trial, $C_{A1} = 0.60$ kmoles/L

Now, for the PFR, which receives the above solution of concentration $C_{A1}$,

$$C_{Ae}^{0.44} = (0.60)^{0.44} - (0.352)$$

or $C_{Ae} = 0.16$ kmoles/L

Therefore, $x_A = 1 - (0.16/1.2) = 0.8667 =$ **86.67** percent

Thus, for the reaction of order less than unity, the second type of reactor combination (an ideal CSTR followed by a PFR) provides higher fractional conversion of $A$.

The conclusions derived may be summarised as follows:

| *Reaction type* | *Reactor combination, recommended* |
|---|---|
| 1. First order reactions | Either |
| 2. Reactions with order greater than unity | PFR – CSTR combination |
| 3. Reactions with order less than unity | CSTR – PFR combination |

**Example 8.18:** *Performance of ideal reactors (reversible reactions).* The gas phase non catalytic dimerisation of $A$ ($2A \rightleftharpoons A_2$) is to be conducted at 640° C and 1 atm pressure. The feed stream containing 0.5 mole of inerts per mole of $A$ is fed to the reactor at the rate of 10 kmoles/hr. Kinetic studies show that the forward reaction is second order in $A$ and the rate constant is given by

$$\ln k_1 = -(12595.0/T) + 18.5657$$

where $k_1$ is expressed in kmoles/(m$^3$·hr·atm$^2$) and $T$ is in degrees Kelvin. The backward reaction is first order and the value of equilibrium constant at 640° C is 95.14 m$^3$/kmole

(*a*) Determine the length of a 0.1 m ID plug flow reactor required to obtain a conversion of 40 percent. What will be the space velocity under these conditions?

(*b*) If an ideal stirred tank reactor of the same volume is used for the purpose, what conversion will be achieved?

**Solution:** The rate equation for the reaction (in terms of partial pressures) is

$$(-r_A) = k_1 p_A^2 - k_2 p_{A_2}$$

Assuming ideal gas law to be valid, $p_A = (A)\,RT$ and $p_{A_2} = (A_2)\,RT$. Therefore

$$(-r_A) = k_1\,[A]^2\,(RT)^2 - k_2\,[A_2]\,RT = k_1'\,[A]^2 - k_2'\,[A_2]$$

where $k_1' = k_1\,(RT)^2$ and $k_2' = k_2\,(RT)$. At equilibrium, $(-r_A) = 0$ and therefore

$$\frac{[A_2]_e}{[A]_e^2} = (k_1'/k_2')$$

But, $[A_2]_e/[A]_e^2 = K =$ equilibrium constant. Therefore

$$(k_1'/k_2') = K = 95.14$$

Since the operating temperature is 913° K,

$$\ln k_1 = -(12595.0/913) + 18.5657$$

or

$$k_1 = 117.98 \text{ kmoles/(hr}\cdot\text{m}^3\cdot\text{atm}^2)$$

$$k_1' = k_1 (RT)^2 = (117.98)(0.08205)^2 (913)^2 = 662067.59 \text{ m}^3/(\text{kmoles}\cdot\text{hr})$$

$$k_2' = (k_1'/K) = (662067.59/95.14) = 6958.877 \text{ hr}^{-1}$$

The rate equation thus becomes

$$(-r_A) = 662067.59\,[A]^2 - 6958.877\,[A]^2 \quad \text{... (i)}$$

Now,

$$[A] = N_A/V = \frac{N_{A0}(1-x_A)}{V_0(1+\epsilon_A x_A)} = \frac{C_{A0}(1-x_A)}{(1+\epsilon_A x_A)}$$

Suppose we start with 2 moles of $A$. Then the number of moles of inerts will be (0.5) (2) = 1.0. Thus, the total number of moles at $x_A = 0$ will be 3.0. Two moles of $A$ on complete conversion give one mole of $A_2$ and one mole of inerts will remain behind. Therefore, the total number of moles at $x_A = 1.0$ will be 2.0. Thus

$$\epsilon_A = (2-3)/3 = -1/3$$

Therefore, the expression for $[A]$ becomes

$$[A] = \frac{C_{A0}(1-x_A)}{(1-0.333\,x_A)} \quad \text{... (ii)}$$

$$[A_2] = \frac{(N_{A0}\,x_A/2)}{V} = \frac{N_{A0}\,x_A}{2V_0(1+\epsilon_A x_A)} = \frac{C_{A0}\,x_A}{2(1-0.333\,x_A)} \quad \text{... (iii)}$$

Also,

$$C_{A0} = p_{A0}/RT = \frac{P y_{A0}}{RT}$$

Since the feed stream contains 0.5 mole of inerts per mole of $A$, the mole fraction of $A$ in the feed stream, $y_{A0} = 1/1.5$ and therefore

$$C_{A0} = \frac{(1.0)}{(1.5)(0.08205)(913)} = 0.0089 \text{ kmoles/m}^3$$

Substituting Eq. (ii) and (iii) and the value of $C_{A0}$ in Eq. (i), we get the rate equation in terms of $x_A$ as

$$(-r_A) = \frac{52.44(1-x_A)^2 - 30.967\,x_A(1-0.333\,x_A)}{(1-0.333\,x_A)^2} \quad \text{... (iv)}$$

(*a*) For a plug flow reactor, from Eq. (8.3.32),

$$V = F_{A0}\int_0^{x_{Af}} \frac{dx_A}{(-r_A)}$$

where

$$F_{A0} = (10)\,y_{A0} = 10\,(1/1.5) = 6.667 \text{ kmoles/hr}$$

$$x_{Af} = 0.40$$

Therefore $$V = 6.667 \int_0^{0.4} F(x_A)\, dx_A$$

where $$F(x_A) = 1/(-r_A) = \frac{(1-0.333x_A)^2}{52.44(1-x_A)^2 - 30.967x_A(1-0.333x_A)}$$

The integral in the above equation can be evaluated either graphically or numerically. Let us evaluate it numerically using Simpson's rule. For this, the values of $F(x_A)$ at different $x_A$ are computed and are given below:

| $x_A$ : | 0.0 | 0.1 | 0.2 | 0.3 | 0.4 |
|---|---|---|---|---|---|
| $F(x_A)$ : | 0.01907 | 0.02367 | 0.031356 | 0.04672 | 0.09223 |

Now, from Simpson's rule, the value of the integral is

$$I = (0.1/3)\ \{F(x_0) + 4\ [F(x_1) + F(x_3)] + 2F(x_2) + F(x_4)\}$$

where $F(x_0) = 0.01907$, $F(x_1) = 0.02367$ and so on. Thus

$$I = 0.015186$$

Therefore $$V = (6.667)\ (0.015186) = 0.10124\ m^3$$

Since $V = [\pi\ (0.1)^2/4]\ L$, where $L$ is the length of the reactor,

$$L = 4\ (0.10124)/\pi\ (0.1)^2 = 12.89\ m$$

The space velocity $S$ is given by

$$S = (1/\tau) = Q_0/V = \frac{F_{A0}}{(V\ C_{A0})} = \frac{(6.667)}{(0.10124)(0.0089)} = 7398.89\ hr^{-1}$$

(*b*) For an ideal stirred tank reactor, from Eq. (8.3.9),

$$V = F_{A0}x_{Af}/(-r_A)_f$$

or $$(-r_A)_f/x_{Af} = (F_{A0}/V) = (6.667/0.10124) = 65.85$$

Substituting the expression for $(-r_A)_f$ from Eq. (iv),

$$\left[\frac{52.44(1-x_{Af})^2}{x_{Af}(1-0.333x_{Af})^2}\right] - \left[\frac{(30.967)}{(1-0.333x_{Af})}\right] = 65.85$$

or $$x_{Af} = 0.31$$

Thus, a stirred tank reactor of same volume gives only 31 percent conversion.

**Example 8.19:** *Rate equation from CSTR data:* An esterification reaction represented by

$$RCOOH + R'OH \rightarrow RCOOR' + H_2O$$

is conducted at constant density in an experimental ideal stirred tank reactor ($V = 0.01\ m^3$) and the following results are obtained:

| *Feed rate of acid kmoles/hour* | *Concentration of acid in the exit stream, kmoles/m³* |
|---|---|
| 0.31 | 0.045 |
| 0.13 | 0.040 |
| 0.07 | 0.035 |
| 0.043 | 0.030 |
| 0.025 | 0.025 |

The concentration of acid in the feed stream was 0.05 kmoles/m³ and that of alcohol 0.1 kmoles/m³.

(*a*) Determine the rate equation for the reaction from the above data.

(*b*) If the above reaction is conducted in a plug flow reactor with a feed rate of 10 m³/hr, compute the volume of reactor required to obtain 75 percent conversion.

**Solution:** (*a*) The rate equation is to be determined by trial. Thus, let the rate equation be

$$(-r_A) = k\, C_A C_B \qquad \text{... (i)}$$

where $A$ stands for acid and $B$ stands for alcohol. In terms of conversion of $A$,

$$C_{Af} = C_{A0}\,(1 - x_{Af}) = 0.05\,(1 - x_{Af})$$

$$C_{Bf} = C_{B0} - C_{A0}x_{Af} = 0.1 - 0.05x_{Af} = 0.05\,(2 - x_{Af})$$

Therefore

$$(-r_A) = k\,(0.05)^2\,(1 - x_{Af})\,(2 - x_{Af}) \qquad \text{... (ii)}$$

Now, from Eq. (8.3.9), for a CSTR

$$V = \frac{F_{A0}x_{Af}}{(-r_A)_f}$$

or

$$\frac{(-r_A)_f V}{x_{Af}} = F_{A0}$$

Substituting the rate equation from Eq. (ii),

$$0.0025\,(1 - x_{Af})\,(2 - x_{Af})\,(0.01)/x_{Af} = (1/k)\,F_{A0}$$

$$F\,(x_{Af}) = (1/k)\,F_{A0}$$

where $F\,(x_{Af}) = (0.000025)\,(1 - x_{Af})\,(2 - x_{Af})/x_{Af}$. Thus, a plot $F\,(x_{Af})$ versus $F_{A0}$ must yield a straight line if the assumed rate equation is correct. To make the plot, the values of $F\,(x_{Af})$ at different values of $x_{Af}$ are computed and are tabulated below. To note that $x_{Af} = 1 - (C_{Af}/C_{A0}) = 1 - (C_{Af}/0.05) = (1 - 20C_{Af})$.

| $F_{A0}$, *kmoles/s* | $C_{Af}$, *kmoles/m³* | $x_{Af}$ | $F\,(x_{Af})$ |
|---|---|---|---|
| $8.61 \times 10^{-5}$ | 0.045 | 0.1 | $4.275 \times 10^{-4}$ |
| $3.61 \times 10^{-5}$ | 0.04 | 0.2 | $1.8 \times 10^{-4}$ |
| $1.944 \times 10^{-5}$ | 0.035 | 0.3 | $9.9167 \times 10^{-5}$ |
| $1.194 \times 10^{-5}$ | 0.03 | 0.4 | $6.0 \times 10^{-5}$ |
| $6.944 \times 10^{-6}$ | 0.025 | 0.5 | $3.75 \times 10^{-5}$ |

The plot of $F(x_{Af})$ versus $F_{A0}$ is shown in Fig. 8.19.1. It is a straight line and therefore, the assumed rate equation is correct. The slope of the straight line is 5.0. Therefore

$$1/k = 5.0$$

or

$$k = 0.2 \text{ m}^3/(\text{kmole}\cdot\text{s})$$

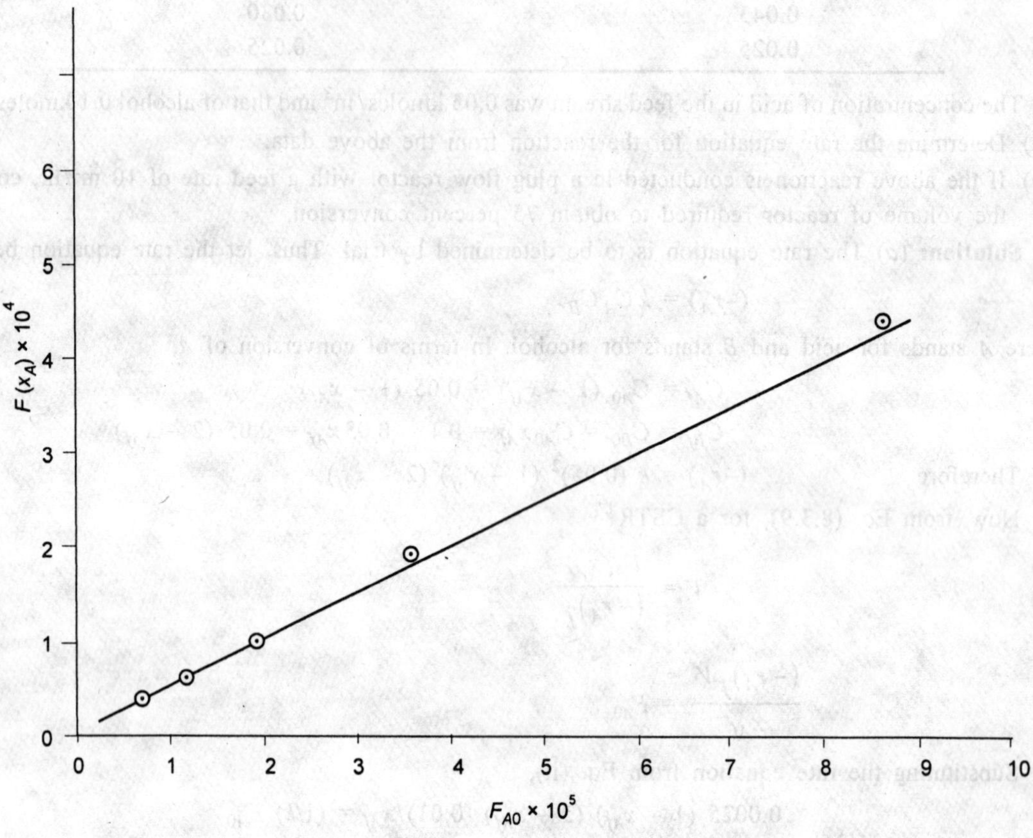


*Figure 8.19.1*

The rate equation is, therefore,

$$(-r_A) = 0.2 C_A C_B \text{ kmoles}/(\text{m}^3\,\text{s})$$

(*b*) For the plug flow reactor, from Eq. (8.3.32),

$$V = F_{A0} \int_0^{x_{Af}} \frac{dx_A}{(-r_A)}$$

where $x_{Af} = 0.75$, $F_{A0} = Q_0 C_{A0} = (10)(0.05)/(3600) = 1.3889 \times 10^{-4}$ kmoles/s. Substituting the rate equation from Eq. (ii) and rearranging,

$$V = 0.2777 \int_0^{0.75} \frac{dx_A}{(1-x_A)(2-x_A)}$$

The integral of the above equation is similar to that of Eq. (8.2.26). Therefore, we can readily write its solution as

$$V = 0.2777 \left\{ \ln\left[\frac{2-x_A}{2(1-x_A)}\right]\right\}_0^{0.75} = (0.2777)\ (0.91629) = \mathbf{0.2544\ m^3}$$

**Example 8.20:** *Rate equation from plug flow reactor data.* The gaseous pyrolysis of acetone was studied at 520° C and 1 atmosphere in an experimental plug flow reactor 0.8 m long and 0.333 m in diameter. The data collected are given below:

| | | | | | |
|---|---|---|---|---|---|
| Feed rate, kg/hr | : | 0.154 | 0.05 | 0.021 | 0.0108 |
| Conversion of acetone | : | 0.05 | 0.13 | 0.24 | 0.35 |

Determine the rate equation for the reaction from the above data. The stoichiometric equation of the reaction is $CH_3COCH_3 \rightarrow CH_2{=}C{=}O + CH_4$

**Solution:** The rate equation can be determined only by trial. As the first trial, let

$$(-r_A) = k\,C_A \quad \text{... (i)}$$

where $A$ stands for acetone. Substituting the above rate expression in Eq. (8.3.32),

$$V = F_{A0} \int_0^{x_{Af}} \frac{dx_A}{kC_A} \quad \text{... (ii)}$$

Now, $$C_A = N_A/V = \frac{N_{A0}(1-x_A)}{V_0(1+\epsilon_A x_A)} = \frac{C_{A0}(1-x_A)}{(1+\epsilon_A x_A)}$$

$$\epsilon_A = (2-1)/(1.0) = 1.0$$

Since pure acetone is fed to the reactor, $p_{A0} = P = 1$ atm and

$$C_{A0} = P/RT = 1.0/(0.08205)\ (793) = 0.01537\ \text{kmoles/m}^3$$

$$V = [\pi\ (0.033)^2/4]\ (8.0) = 6.8424 \times 10^{-3}\ \text{m}^3$$

Equation (ii), therefore, becomes

$$(1.05167 \times 10^{-4}) = (F_{A0}/k) \int_0^{x_{Af}} \frac{(1+\epsilon_A x_A)\,dx_A}{(1-x_A)}$$

On integration and rearrangement,

$$-2\ \ln\ (1-x_{Af}) - x_{Af} = (1.05167 \times 10^{-4} k)\ (1/F_{A0})$$

$$F\ (x_{Af}) = (1.05167 \times 10^{-4} k)\ (1/F_{A0})$$

Therefore, a plot of $F(x_{Af})$ versus $(1/F_{A0})$ must yield a straight line if the assumed rate equation is correct. In other words, $F_{A0}\,F(x_{Af})$ must be a constant at all values of $x_{Af}$ and $F_{A0}$. The values of this product at different $x_{Af}$ are listed below. To note that since the molecular weight of acetone is 58, $F_{A0} = (F_0/58)$.

| $F_0$, kg/hr | $F_{A0}$, kmoles/hr | $x_{Af}$ | $F(x_{Af})$ | $F_{A0}F(x_{Af})$ |
|---|---|---|---|---|
| 0.154 | $2.655 \times 10^{-3}$ | 0.05 | 0.052586 | $1.39617 \times 10^{-4}$ |
| 0.05 | $8.6207 \times 10^{-4}$ | 0.13 | 0.14852 | $1.28038 \times 10^{-4}$ |
| 0.021 | $3.6207 \times 10^{-4}$ | 0.24 | 0.30887 | $1.1183 \times 10^{-4}$ |
| 0.0108 | $1.862 \times 10^{-4}$ | 0.35 | 0.51156 | $0.9525 \times 10^{-4}$ |

It can be seen the values in the last column are far from equal. Therefore, the assumed rate expression is incorrect.

Let us now assume a second order rate equation. Thus

$$(-r_A) = k\,C_A^2$$

Equation (ii), thus, becomes

$$(C_{A0}^2\,kV/F_{A0}) = \int_0^{x_{Af}} \frac{(1+x_A)^2\,dx_A}{(1-x_A)^2}$$

On integration and substituting the values of $C_{A0}$ and $V$, we get

$$(1.6164 \times 10^{-6}k)\,(1/F_{A0}) = 4\ \ln(1 - x_{Af}) + x_{Af} + [4x_{Af}/(1 - x_{Af})] = F(x_{Af})$$

Once again, if the assumed rate equation is correct, then a plot of $F(x_{Af})$ versus $(1/F_{A0})$ must be a straight line, or the product $F_{A0}\,F(x_{Af})$ must be a constant at all values of $x_{Af}$ and $F_{A0}$. The values of this product are given below:

| $F_{A0}$, kmoles/hr | $x_{Af}$ | $F(x_{Af})$ | $F_{A0}F(x_{Af})$ |
|---|---|---|---|
| $2.655 \times 10^{-3}$ | 0.05 | 0.05535 | $1.4695 \times 10^{-4}$ |
| $8.6207 \times 10^{-4}$ | 0.13 | 0.17065 | $1.471 \times 10^{-4}$ |
| $3.6207 \times 10^{-4}$ | 0.24 | 0.4054 | $1.46787 \times 10^{-4}$ |
| $1.862 \times 10^{-4}$ | 0.35 | 0.7807 | $1.4537 \times 10^{-4}$ |

It can be seen that the values in the last column are essentially equal. The plot of $F(x_{Af})$ versus $(1/F_{A0})$ is shown in Fig. 8.20.1, which is a straight line. Thus, the assumed rate expression is correct. From Fig. 8.20.1, the slope of the straight line is $1.4615 \times 10^{-4}$. Therefore

$$(1.6164 \times 10^{-6})\,k = 1.4615 \times 10^{-4}$$

or

$$k = 90.419\ \text{m}^3/(\text{kmole}\cdot\text{hr})$$

The rate equation is $\quad (-r_A) = 90.419\,C_A^2$ kmoles/(m$^3$ hr)

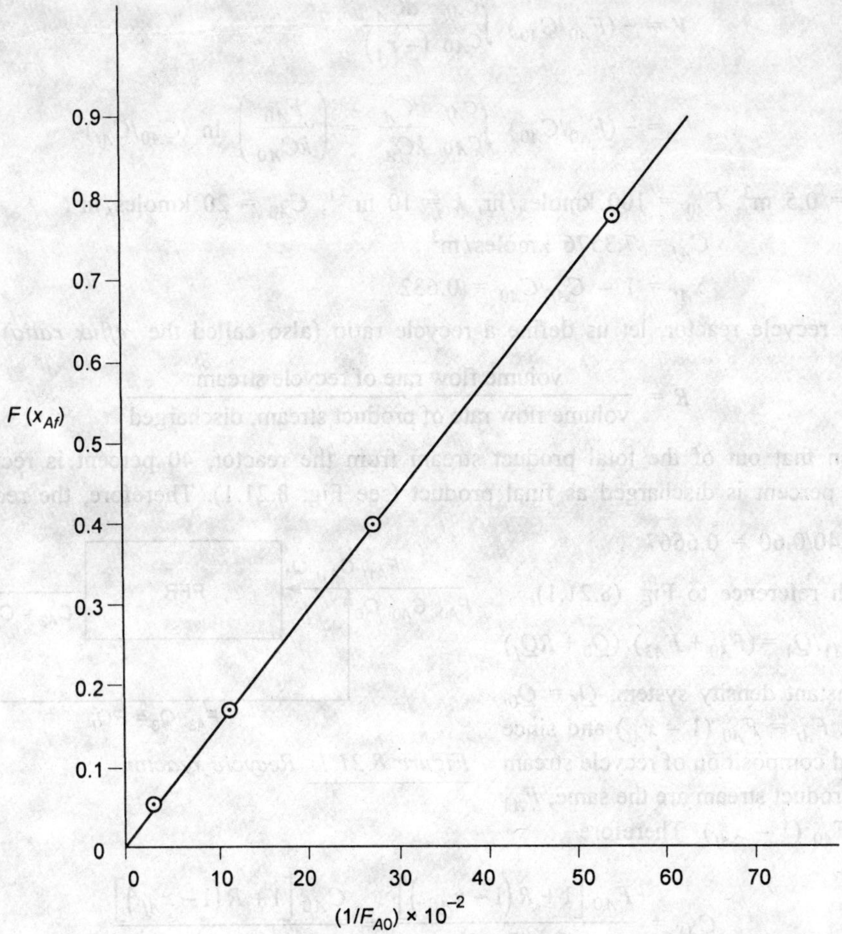


*Figure 8.20.1*

**Example 8.21:** *Plug flow reactor with recycle.* A chemical reaction that is psuedofirst order in $A$ is conducted at constant density in a 0.5 $m^3$ plug flow reactor. The operating temperature is 50° C. Kinetic studies show that the rate constant at 50° C is 10 $hr^{-1}$. The feed (containing only $A$) is fed into the reactor at the rate of 100 kmoles/hr and the concentration of $A$ in the feed is 20 kmoles/$m^3$.

(*a*) Compute the fractional conversion of $A$ obtained.

(*b*) What conversion of $A$ will be obtained if 40 percent of the product stream (by volume) from the reactor is recycled to the feed inlet?

(*c*) If an ideal stirred tank reactor ($V$ = 0.5 $m^3$) is used for the purpose, what conversion of $A$ will be achieved?

**Solution:** (*a*) The conversion obtained in a plug flow reactor may be computed from Eq. (8.3.32). However, since the density of the reaction mixture remains constant, it will be more convenient to use Eq. (8.3.34). Thus

$$V = -(F_{A0}/C_{A0}) \int_{C_{A0}}^{C_{Af}} \frac{dC_A}{(-r_A)}$$

$$= -(F_{A0}/C_{A0}) \int_{C_{A0}}^{C_{Af}} \frac{dC_A}{kC_A} = \left(\frac{F_{A0}}{kC_{A0}}\right) \ln (C_{A0}/C_{Af})$$

Since $V = 0.5\ \text{m}^3$, $F_{A0} = 100$ kmoles/hr, $k = 10\ \text{hr}^{-1}$, $C_{A0} = 20\ \text{kmoles/m}^3$,

$$C_{Af} = 7.3576\ \text{kmoles/m}^3$$

$$x_{Af} = 1 - C_{Af}/C_{A0} = 0.632$$

(*b*) For a recycle reactor, let us define a recycle ratio (also called the *reflux ratio*) $R$ as

$$R = \frac{\text{volume flow rate of recycle stream}}{\text{volume flow rate of product stream, discharged}}$$

It is given that out of the total product stream from the reactor, 40 percent is recycled and the remaining 60 percent is discharged as final product (see Fig. 8.21.1). Therefore, the recycle ratio is

$$R = 0.40/0.60 = 0.6667$$

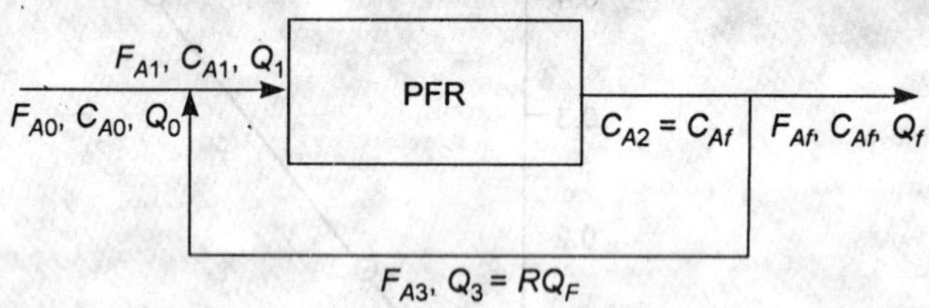


*Figure 8.21.1: Recycle reactor*

Now, with reference to Fig. (8.21.1),

$$C_{A1} = F_{A1}/Q_1 = (F_{A0} + F_{A3})/(Q_0 + RQ_f)$$

For a constant density system, $Q_f = Q_0$. We know that $F_{Af} = F_{A0}\ (1 - x_{Af})$ and since the density and composition of recycle stream and those of product stream are the same, $F_{A3} = R\,F_{Af} = R\,F_{A0}\ (1 - x_{Af})$. Therefore

$$C_{A1} = \frac{F_{A0}\left[1 + R\left(1 - x_{Af}\right)\right]}{Q_0\,(R+1)} = \frac{C_{A0}\left[1 + R\left(1 - x_{Af}\right)\right]}{(R+1)} \quad \text{... (i)}$$

Now, we know that $x_{A1} = 1 - (C_{A1}/C_{A0})$. Substituting the expression for $C_{A1}$, we get

$$x_{A1} = 1 - [1 + R\ (1 - x_{Af})]/(R + 1) = R\,x_{Af}/(R + 1) \quad \text{... (ii)}$$

For a recycle reactor, the plug flow equation [Eq. (8.3.32)] may be used in the modified form as

$$V = F'_{A0} \int_{x_{A1}}^{x_{Af}} \frac{dx_A}{(-r_A)} \quad \text{... (iii)}$$

where $F'_{A0}$ would be the feed rate of $A$ if the stream entering the reactor (fresh feed plus recycle) were unconverted. Since the feed entering the reactor includes both fresh feed and the recycle stream,

$$F'_{A0} = (A \text{ in fresh feed}) + (A \text{ that would enter in an unconverted recycle stream})$$

$$= F_{A0} + R\,F_{A0} = F_{A0} + (R + 1) \quad \text{... (iv)}$$

Equation (iii), therefore, becomes

$$V/F_{A0} = (R+1)\int_{x_{A1}}^{x_{Af}} \frac{dx_A}{(-r_A)} \quad \text{... (v)}$$

where $x_{A1}$ is defined in Eq. (ii). Incidentally, it can be shown[19] that Eqs (ii) and (v) are valid even if the system is not of constant density. Therefore, these two equations may be used as the *general performance equations for recycle reactors.*

For the present case, since $(-r_A) = k\,C_A = k\,C_{A0}\,(1 - x_A)$, the above equation reduces to

$$(k\,C_{A0}\;V/F_{A0}) = (R+1)\int_{x_{A1}}^{x_{Af}} \frac{dx_A}{(1-x_A)}$$

On integration and substituting for $x_{A1}$ from Eq. (ii), we get

$$k\tau = (R+1)\ln\left[\frac{1+R(1-x_{Af})}{(R+1)(1-x_{Af})}\right]$$

It is given that $k = 10\ \text{hr}^{-1}$, $R = 0.6667$, $\tau = C_{A0}V/F_{A0} = (20)\,(0.5)/(100) = 0.1$ hr. Therefore

$$\ln\left[\frac{1+0.6667(1-x_{Af})}{1.6667(1-x_{Af})}\right] = 0.6$$

or

$$x_{Af} = 0.57$$

It can be seen that conversion decreases due to recycle. Recycle of product stream is specifically advantageous for autocatalytic reactions, since in such reactions, one of the products acts as the catalyst. A problem in this connection is discussed in Example (8.22).

(*c*) For an ideal CSTR, from Eq. (8.3.9),

$$V = F_{A0}x_{Af}/(-r_A)_f$$

or

$$(k\,C_{A0}\;V/F_{A0}) = x_{Af}/(1 - x_{Af})$$

Since

$$(k\,C_{A0}\;V/F_{A0}) = (10)\,(20)\,(0.5)/(100) = 1.0,$$

$$x_{Af}/(1 - x_{Af}) = 1.0$$

or

$$x_{Af} = 0.50$$

The results may be summarised as given below:

| | *Conversion of A* |
|---|---|
| Plug flow reactor ($R = 0$) | 0.632 |
| Recycle reactor with $R = 0.6667$ | 0.578 |
| Stirred tank reactor ($R = \infty$) | 0.50 |

The performance of a recycle reactor is, thus, intermediate between $R = 0$ (plug flow) and $R = \infty$ (complete backmixing).

**Example 8.22:** *Optimum recycle (reflux) ratio.* Microbial fermentations, which are basically autocatalytic, follow the following modified Michaelis-Menten equation:

$$(-r_A) = \frac{k_1 C_M C_A}{k_2 + C_A}$$

where $C_M$ is the concentration of the microorganism and $C_A$ that of the substrate (Note the difference between this equation and the Michaelis-Menten equation for enzyme catalysed fermentation discussed in Example 8.10). The reaction may be represented as

$$A + M \rightarrow mM + pP$$

The above reaction is to be conducted in a plug flow reactor with a part of the product stream being recycled to the fed inlet. Derive an expression for the optimum recycle ratio, $R$ (opt), that will minimise the volume of reactor required.

**Solution:** Microbial fermentations are usually conducted under isothermal conditions. We shall also assume that the density of the reaction mixture remains essentially constant. The performance equation for the recycle reactor has already been derived in Example 8.21:

$$V/F_{A0} = (R+1)\int_{x_{A1}}^{x_{Af}} \frac{dx_A}{(-r_A)}$$

where $x_{A1} = R\,x_{Af}/(R+1)$. Substituting the rate equation, we get

$$V/F_{A0} = (R+1)\int_{x_{A1}}^{x_{Af}} \frac{(k_2 + C_A)\,dx_A}{k_1 C_M C_A}$$

$$= \frac{(R+1)\,k_2}{k_1}\int_{x_{A1}}^{x_{Af}} \frac{dx_A}{C_M C_A} + \frac{(R+1)}{k_1}\int_{x_{A1}}^{x_{Af}} \frac{dx_A}{C_M} \quad \text{... (i)}$$

Now,

$$C_A = C_{A0}(1 - x_{Af})$$

$$C_M = C_{M0} - C_{A0}x_{Af} + mC_{A0}x_{Af}$$

$$= C_{M0} + (m-1)\,C_{A0}x_{Af} = C_{A0}(N + ax_{Af})$$

where $N = C_{M0}/C_{A0}$ and $\alpha = (m-1)$. Substituting for $C_A$ and $C_M$ in terms of $x_{Af}$ in $(i)$ and integrating, we get

$$V/F_{A0} = \alpha\,(R+1)\ln\frac{(N + ax_{Af})\left[1 + R(1 - x_{Af})\right]}{\left[R(N + ax_{Af}) + N\right](1 - x_{Af})} +$$

$$\beta\,(R+1)\ln\left[\frac{(N + ax_{Af})(R+1)}{R(N + ax_{Af}) + N}\right] \quad \text{... (ii)}$$

where

$$\alpha = k_2/[k_1 C_{A0}^2\,(N + a)]$$

$$\beta = 1/(k_1 C_{A0}\,a)$$

To find the optimum value of $R$ that minimises $V$, let us differentiate $V$ with respect to $R$ and set $dV/dR$ equal to zero and then solve it for $R_{\text{opt}}$. Thus, $dV/dR = 0$ gives

$$\frac{a(\alpha+\beta)\,x_{Af}}{(MX)} + \frac{(\alpha x_{Af})}{(RX)} = \alpha \ln \frac{(N + a x_{Af})(RX)}{(1 - x_{Af})(MX)} + \beta \ln \left[\frac{(N + a x_{Af})(R_{\text{opt}} + 1)}{(MX)}\right] \quad \text{... (iii)}$$

where $MX = R_{\text{opt}}\,(N + a x_{Af}) + N$

$RX = 1 + R_{\text{opt}}\,(1 - x_{Af})$

The above equation gives implicitly the optimum value of $R$, namely $R$(opt).

To illustrate a numerical example, for $m = 2$, $N = 0.01$, $k_1 = 0.005\ \text{s}^{-1}$, $k_2 = 0.1\ \text{kmole/m}^3$, $C_{A0} = 1.0\ \text{kmole/m}^3$, $x_{Af} = 0.99$, the above equation is solved by trial to get

$$R\ (\text{opt}) = 1.1$$

$$\left(\frac{R}{R+1}\right)(\text{opt}) = 1.1/2.1 = 0.5238$$

It is, thus, optimum to recycle 52.38 percent of the total product from the reactor back to the feed inlet.

For a feed rate ($F_{A0}$) of 1 kmole/hr, the volume of reactor required is

$$V\ (\text{min}) = 0.1265\ \text{m}^3$$

Being an autocatalytic reaction, once the process has started (with recycle), no microorganism is required to be added along with the fresh feed to the reactor. Only at startup, a small amount of microorganism (often called the *inoculum*) is to be added from outside. If we neglect the amount of this inoculum (which is usually very small) in our computations and assume that the fresh feed to the reactor contains no microorganism, even then Eqs (ii) and (iii) can be employed except that in such cases, $N = 0$.

**Example 8.23:** *Reactor performance for parallel reactions.* The following two reactions take place simultaneously at a constant temperature of 50° C

$$A + B \xrightarrow{k_1} P$$

$$A + 2B \xrightarrow{k_2} R$$

where $P$ is the desired product and $R$ is the undesired product. The first reaction is first order in $A$ and first order in $B$, while the second reaction is first order in $A$ and second order in $B$. The values of the rate constants at 50° C are $k_1 = 21.0\ (\text{m}^3/\text{kmole})^2/\text{s}$ and $k_2 = 10.5\ \text{m}^3/(\text{kmole}\cdot\text{s})$. The feed mixture contains only $A$ and $B$ and the concentration of $B$ in the feed is 8.7 kmoles/m³ and that of $A$ 12.5 kmoles/m³. The following two options are available:

(*a*) A plug flow reactor
(*b*) An ideal continuous stirred tank reactor

If 85 percent conversion of $A$ and 85 percent conversion of $B$ are desired, determine which among the above two shall provide higher yield of $P$ and which shall provide higher selectivity of $P$ with respect to $R$. Assume that the density of the reaction mixture remains essentially constant.

**Solution:** (*a*) The analysis of parallel reactions of this kind for a constant-volume batch reactor has already been discussed in Section 8.2. A similar procedure shall be followed here since this is a constant density system and consequently, the performance equations for the batch reactor and the plug flow reactor are identical. Now

$$r_P = dC_P/d\theta = k_1 C_A C_B$$

$$(-r_B) = -dC_B/d\theta = k_1 C_A C_B + k_2 C_A C_B^2$$

Dividing one by the other, we get

$$-dC_P/dC_B = \frac{(k_1 C_A C_B)}{(k_1 C_A C_B) + (k_2 C_A C_B^2)} = \frac{1}{1 + (k_2/k_1) C_B}$$

Rearranging and integrating (to note that since the feed does not contain any $P$ or $R$, $C_{P0} = C_{R0} = 0$), we get

$$C_P = -\int_{C_{B0}}^{C_B} \frac{dC_B}{1 + (k_2/k_1) C_B} = (k_1/k_2) \ln \left[\frac{1 + (k_2/k_1) C_{B0}}{1 + (k_2/k_1) C_B}\right] \quad \text{... (i)}$$

Since $(k_1/k_2) = (21.0/10.5) = 2.0$, $C_{B0} = 8.7$ kmoles/m$^3$, $C_B = C_{B0}(1 - x_B) = 8.7(1 - 0.85) = 1.305$ kmoles/m$^3$,

$$C_P/C_{B0} = 0.27$$

The yield of $P$ is, thus, 27 percent. Now

$$r_R = dC_R/d\theta = k_2 C_A C_B^2$$

$$\frac{(r_R)}{(-r_B)} = -dC_R/dC_B = \frac{1}{1 + (k_1/k_2)(1/C_B)}$$

Integrating both sides,

$$C_R = (C_{B0} - C_B) + (k_1/k_2) \ln \left[\frac{C_B + (k_1/k_2)}{C_{B0} + (k_1/k_2)}\right]$$

Substituting the values of $C_B$, $C_{B0}$ and $(k_1/k_2)$, we get

$$C_R/C_{B0} = 0.58$$

The overall selectivity of $P$ with respect to $R$ is

$$S_0 = 0.27/0.58 = 0.4655$$

(*b*) For a stirred tank reactor, from Eq. (8.3.15),

$$\tau = (C_{B0} - C_{Bf})/(-r_B)_f = \frac{(C_{B0} - C_{Bf})}{(k_1 C_{Af} C_{Bf} + k_2 C_{Af} C_{Bf}^2)} \quad \text{... (ii)}$$

If we write Eq. (8.3.15) for $P$,

$$\tau = (C_{P0} - C_{Pf})/(-r_P)_f = C_{Pf}/(k_1 C_{Af} C_{Bf}) \quad \text{... (iii)}$$

Dividing Eq. (iii) by (ii), we get

$$C_{Pf}/(C_{B0} - C_{Bf}) = \left[\frac{1}{1+(k_2/k_1)\,C_{Bf}}\right]$$

or
$$C_{Pf}/C_{B0} = \frac{(C_{B0} - C_{Bf})/C_{B0}}{1+(k_2/k_1)\,C_{Bf}} = \frac{x_B}{1+(k_2/k_1)\,C_{Bf}}$$

Substituting $x_B = 0.85$, $(k_2/k_1) = 0.5$ and $C_{Bf} = 1.305$ kmoles/m$^3$, we get

$$C_{Pf}/C_{B0} = 0.5144$$

Therefore, the yield of $P$ obtained in a CSTR is 51.44 percent. Now, writing Eq. (8.3.15) for $R$,

$$\tau = (C_{R0} - C_{Rf})/(-r_R)_f = C_{Rf}/(k_2 C_{Af} C_{Bf}^2) \quad \text{... (iv)}$$

Dividing Eq. (iii) by (iv), we get

$$C_{Pf}/C_{Rf} = S_0 = (k_1 C_{Af} C_{Bf})/(k_2 C_{Af} C_{Bf}^2) = (k_1/k_2)/C_{Bf} = (2.0)/(1.305) = 1.5325$$

Thus, the stirred tank reactor provides higher yield of $P$ as well as higher selectivity of $P$ with respect to $R$.

**Example 8.24:** *Reactor performance for consecutive reactions.* The following irreversible first order reactions occur at constant density:

$$A \xrightarrow{k_1} P \xrightarrow{k_2} R$$

The desired product is $P$ and the operating temperature is 65° C. Both reactions are first order and the values of rate constants at 65° C are $k_1 = 0.0025$ s$^{-1}$ and $k_2 = 0.001$ s$^{-1}$. This reaction system is to be conducted in continuous flow reactor with a volumetric feed rate of 12 m$^3$/hr. The following options are available:

(*a*) A single stirred tank reactor of volume $V = 0.4$ m$^3$,
(*b*) A plug flow reactor with a volume of 0.4 m$^3$,
(*c*) Two stirred tank reactors in series each of volume 0.2 m$^3$,
(*d*) Two stirred tank reactors in parallel each of 0.2 m$^3$ in volume and the feed stream being split equally between them,
(*e*) A stirred tank reactor followed by a plug flow reactor each of volume 0.2 m$^3$.

Determine which among the above shall provide the highest yield of $P$. For which among them the selectivity of $P$ with respect to $R$ is the highest?

**Solution:** (*a*) For an ideal stirred tank reactor, from Eq. (8.3.15),

$$\tau = (C_{A0} - C_{Af})/(-r_A)_f = \frac{C_{A0} - C_{Af}}{k_1 C_{Af}}$$

or
$$C_{Af}/C_{A0} = 1/(1 + k_1\tau) \quad \text{... (i)}$$

If we write Eq. (8.3.15) for $P$, we get

$$\tau = (C_{P0} - C_{Pf})/(-r_P)_f$$

Since the feed to the reactor contains only $A$, $C_{P0} = 0$. Also, $(-r_P)_f = k_2 C_{Pf} - k_1 C_{Af}$. Therefore, the above equation becomes

$$-C_{Pf} = (k_2 C_{Pf} - k_1 C_{Af})\ \tau$$

Substituting for $C_{Af}$ from Eq. (i) and rearranging, we get

$$C_{Pf}/C_{A0} = \frac{k_1 \tau}{(1+k_1 \tau)(1+k_2 \tau)} \qquad \text{... (ii)}$$

Since $k_1 = 0.0025\ s^{-1}$, $k_2 = 0.001\ s^{-1}$ and $\tau = (V/Q_0) = (0.4)\ (3600)/(12.0) = 120$ s,

$$C_{Pf}/C_{A0} = 0.206$$

Thus, the yield of $P$ is 20.6 percent. From a simple material balance,

$$C_{A0} + C_{P0} + C_{R0} = C_{Af} + C_{Pf} + C_{Rf}$$

Since $\qquad C_{P0} = C_{R0} = 0,$

$$C_{Rf}/C_{A0} = 1 - (C_{Af}/C_{A0}) - (C_{Pf}/C_{A0})$$

Now, $C_{Af}/C_{A0} = 1/(1 + k_1\tau) = 0.769$. Therefore,

$$C_{Rf}/C_{A0} = 1.0 - (0.769 + 0.206) = 0.25$$

The overall selectivity of $P$ with respect to $R$ is, therefore,

$$S_0 = 0.206/0.025 = 8.24$$

(*b*) For a constant density system like this, the performance equations for a plug flow reactor are identical with those for a batch reactor. This reaction has already been analysed for a constant volume batch reactor in Section (8.2). We can, therefore, use the same equations, such as Eqs (8.2.107) and (8.2.105), after replacing $\theta$ by $\tau$. Thus

$$C_P/C_{A0} = \frac{k_1}{(k_1 - k_2)}\ [\exp\ (-k_2\tau) - \exp\ (-k_1\tau)]$$

$$C_A/C_{A0} = \exp\ (-k_1\tau)$$

Substituting $k_1 = 0.0025\ s^{-1}$, $k_2 = 0.001\ s^{-1}$ and $\tau = 120$ s, we get

$$C_A/C_{A0} = 0.7408$$

$$C_P/C_{A0} = 0.2435$$

The plug flow reactor thus provides 24.35 percent yield of $P$. Now

$$C_R/C_{A0} = 1 - (C_A/C_{A0}) - (C_P/C_{A0}) = 1.0 - (0.7408 + 0.2435) = 0.0157$$

The overall selectivity of $P$ with respect to $R$ is

$$S_0 = (0.2435)/(0.0157) = 15.5095$$

(*c*) Since both reactors are of the same size and the volumetric feed rate ($Q_0$) is the same to each reactor (since the system is of constant density),

$$\tau_1 = \tau_2 = \tau = (0.2)\ (3600)/(12.0) = 60\ s$$

For the first reactor, $C_{A1}/C_{A0} = 1/(1 + k_1\tau)$

For the second reactor,

$$C_{A2}/C_{A1} = 1/(1 + k_1\tau)$$

or

$$C_{A2} = C_{A1}/(1 + k_1\tau) = C_{A0}/(1 + k_1\tau)^2$$

Substituting $k_1 = 0.0025\ s^{-1}$ and $\tau = 60$ s,

$$C_{A2}/C_{A0} = 0.756$$

From what we derived in part (*a*),

$$C_{P1}/C_{A0} = \frac{k_1\,\tau}{(1+k_1\,\tau)(1+k_2\,\tau)}$$

Now, writing the material balance equation of $P$ for the second reactor,

$$F_{P1} - F_{P2} = (-r_P)_2\ V$$

or

$$Q_0\ (C_{P1} - C_{P2}) = (k_2 C_{P2} - k_1 C_{A2})\ V$$

On rearrangement, $$C_{P2}/C_{A0} = \frac{k_1\,\tau}{(1+k_1\,\tau)(1+k_2\,\tau)}\left[\frac{1}{(1+k_1\,\tau)} + \frac{1}{(1+k_2\,\tau)}\right]$$

Substituting the values of $k_1$, $k_2$ and $\tau$, we get

$$C_{P2}/C_{A0} = 0.2231$$

Thus, the yield of $P$ is 22.31 percent which is higher than that obtained in a single stirred tank reactor of volume 0.4 m$^3$. Now

$$C_{R2}/C_{A0} = 1 - (C_{A2}/C_{A0}) - (C_{P2}/C_{A0}) = 1 - (0.756 + 0.2231) = 0.0209$$

The overall selectivity of $P$ with respect to $R$ is,

$$S_0 = 10.6746$$

(*d*) We have two stirred tank reactors operating in parallel each of volume 0.2 m$^3$. The volumetric feed rate to each reactor is $(Q_0/2)$ or 6 m$^3$/hr. Therefore, the value of $\tau$ for each reactor is (0.2) (3600)/(6.0) = 120 s, which is same as that for the single stirred tank reactor considered in part (*a*). Thus, the values of $C_{Af}$, $C_{Pf}$ and $C_{Rf}$ will be the same in this case and the yield of $P$ and the overall selectivity of $P$ with respect to $R$ will also be the same as those obtained in a single reactor in part (*a*). Thus

$$C_{Pf}/C_{A0} = 0.206$$

$$S_0 = 8.24$$

(*e*) Here a stirred tank reactor is followed by a plug flow reactor (see Fig. 8.24.1). For the stirred tank reactor,

$$\tau_m = (0.2)\ (3600)/(12.0) = 60\ s$$

$$C_{A1}/C_{A0} = 1/(1 + k_1\tau_m) = 0.86956$$

$$C_{P1}/C_{A0} = \frac{k_1\,\tau_m}{(1+k_1\,\tau_m)(1+k_2\,\tau_m)} = 0.12305$$

To note that the above expression is the one we have already derived in part (*a*).

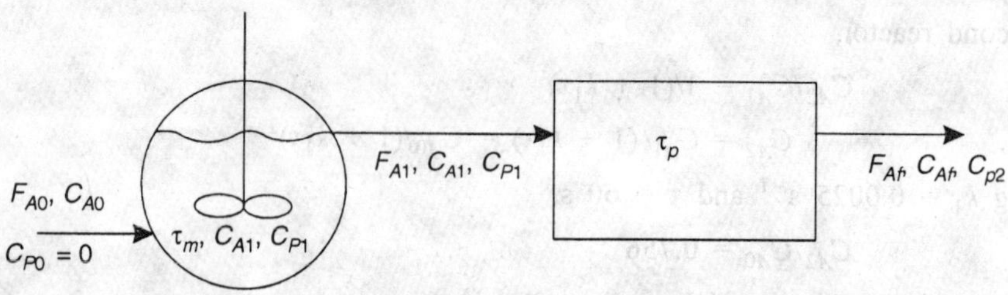


*Figure 8.24.1: A continuous stirred tank reactor (CSTR) followed by a plug flow reactor (PFR). For constant density systems,* $Q_0 = (F_{A0}/C_{A0}) = (F_{A1}/C_{A1}) = (F_{Af}/C_{Af})$

Now, consider the plug flow reactor. For a plug flow reactor that receives a partly converted feed, the performance equation is

$$V = -(F_{A0}/C_{A0}) \int_{C_{A1}}^{C_{Af}} \frac{dC_A}{(-r_A)}$$

or

$$\tau_p = -\int_{C_{A1}}^{C_{Af}} \frac{dC_A}{(-r_A)}$$

Substituting $(-r_A) = k_1 C_A$ and integrating, we get

$$C_{Af} = C_{A2} = C_{A1} \exp(-k_1 \tau_p)$$

Since $k_1 = 0.0025 \text{ s}^{-1}$, $\tau_p = \tau_m = 60$ s, $C_{A1}/C_{A0} = 0.86956$,

$$C_{A2}/C_{A0} = 0.7484$$

If we write the performance equation for $P$,

$$\tau_p = -\int_{C_{P1}}^{C_{P2}} \frac{dC_P}{(-r_P)} = -\int_{C_{P1}}^{C_{P2}} \frac{dC_P}{(k_2 C_P - k_1 C_A)}$$

Differentiating both sides with respect to $C_P$,

$$d\tau_p / dC_P = -1/(k_2 C_P - k_1 C_A)$$

or

$$\frac{dC_P}{d\tau_p} + k_2 C_P = k_1 C_A = k_1 C_{A1} \exp(-k_1 \tau_p)$$

The above differential equation is similar to Eq. (8.2.106) or the one that is solved in Example 8.15. Therefore, its solution with respect to the initial condition that at $\tau_p = 0$, $C_P = C_{P1}$ is

$$C_{P2} = \frac{C_{A1} k_1}{(k_1 - k_2)} [\exp(-k_2 \tau_p) - \exp(-k_1 \tau_p)] + C_{P1} \exp(-k_2 \tau_p)$$

Substituting $k_1 = 0.0025 \text{ s}^{-1}$, $k_2 = 0.001 \text{ s}^{-1}$, $\tau_p = 60$ s, $C_{A1}/C_{A0} = 0.86956$ and $C_{P1}/C_{A0} = 0.12305$, we get

$$C_{P2}/C_{A0} = 0.2333$$

$$C_{R2}/C_{A0} = 1 - (C_{A2}/C_{A0}) - (C_{P2}/C_{A0}) = 1 - (0.7484 + 0.2333) = 0.0183$$

$$S_0 = (0.2333)/(0.0183) = 12.75$$

The results may be summarised as given below:

| | *Yield of P* | *Overall selectivity of P with respect to R* |
|---|---|---|
| Single stirred tank reactor (0.4 $m^3$) | 0.206 | 8.24 |
| Plug flow reactor (0.4 $m^3$) | 0.2435 | 15.5095 |
| Two stirred tank reactors in series each of volume 0.2 $m^3$ | 0.2231 | 10.6746 |
| Two 0.2 $m^3$ stirred tank reactors in parallel with feed split equally between them | 0.206 | 8.24 |
| Stirred tank reactor followed by a plug flow reactor each of volume 0.2 $m^3$ | 0.2333 | 12.75 |

Thus, among all, the highest yield of $P$ and the highest selectivity of $P$ with respect to $R$ are given by the plug flow reactor.

**Example 8.25:** *Reactor volume for maximum yield.* For the consecutive reactions discussed in Example (8.24), determine what should be the volume of an ideal stirred tank reactor so that the maximum possible yield of $P$ shall be obtained. Repeat the same for a plug flow reactor.

**Solution:** For a stirred tank reactor, the expression for the yield of $P$ has already been derived in Example (8.24):

$$C_{Pf}/C_{A0} = \frac{k_1 \tau}{(1+k_1 \tau)(1+k_2 \tau)} \quad \text{... (i)}$$

To find the location of maximum yield, let us set the derivative of $C_{Pf}/C_{A0}$ with respect to $\tau$ equal to zero and then solve it for $\tau$. Thus, $d\,(C_{Pf}/C_{A0})/d\tau = 0$ gives

$$k_1\,(1 + k_1\tau)\,(1 + k_2\tau) - k_1\tau\,[k_2\,(1 + k_1\tau) + (1 + k_2\tau)\,k_1] = 0$$

or

$$\tau = \tau\,(\text{opt}) = 1/\sqrt{k_1 k_2} \quad \text{... (ii)}$$

Substituting $k_1 = 0.0025\ s^{-1}$ and $k_2 = 0.001\ s^{-1}$

$$\tau\,(\text{opt}) = V/Q_0 = 632.455\ s$$

Therefore,

$$V = (12/3600)\,(632.455) = 2.108\ m^3$$

Thus, a fairly large volume reactor is required if the yield of $P$ is to be maximised. The value of the maximum yield so obtained can be computed from Eq. (i) after substituting $\tau = \tau\,(\text{opt}) = 1/\sqrt{k_1 k_2}$ in it. Thus

$$C_P\,(\max)/C_{A0} = 1/\left(1+\sqrt{k_1 k_2}\right)^2 \quad \text{... (iii)}$$

Substituting the values of $k_1$ and $k_2$, we get

$$C_P\ (\max)/C_{A0} = 0.3752$$

For the plug flow reactor, the expressions for $\tau$ (opt) and $C_P$ (max) will be the same as those for a constant-volume batch reactor (since this is a constant density system). For a constant-volume batch reactor, these expressions have already been derived in Section (8.2). Thus, from Eqs (8.2.108) and (8.2.109),

$$\tau\ (\text{opt}) = \frac{\ln (k_1/k_2)}{(k_1 - k_2)} \qquad \text{... (iv)}$$

$$\ln [C_P\ (\max)/C_{A0}] = \frac{k_2}{(k_2 - k_1)} \ln (k_1/k_2) \qquad \text{... (v)}$$

Substituting $k_1 = 0.0025\ \text{s}^{-1}$ and $k_2 = 0.001\ \text{s}^{-1}$,

$$\tau\ (\text{opt}) = 610.86\ \text{s}$$

$$C_P\ (\max)/C_{A0} = 0.543$$

$$V = (12/3600)\ (610.86) = \mathbf{02.036\ m^3}$$

**Example 8.26:** *Reactor performance for series-parallel reactions.* Benzene is chlorinated in the liquid phase in an ideal tubular flow reactor equipped with a recycle system. The HCl produced is separated at the top of the reactor and the liquid stream is recycled back to the reactor. The reactor operates isothermally and the HCl product remains in solution until it reaches the separator at the top of the reactor. At the constant operating temperature, the significant reactions are the two substitution reactions leading to the formation of mono and dichloro benzene. Each reaction is second order and irreversible and the ratio of the rate constants is $(k_1/k_2) = 8.0$. If the total conversion of benzene desired is 80 percent, determine the yield of monochloro benzene and the selectivity of monochloro benzene with respect to dichlorobenzene for the following two cases:

(*a*) zero recycle ratio

(*b*) infinite recycle ratio

The density of the reaction mixture may be assumed constant.

**Solution:** The reactions are

$$C_6H_6 + Cl_2 \xrightarrow{k_1} C_6H_5Cl + HCl$$

$$C_6H_5Cl + Cl_2 \xrightarrow{k_2} C_6H_4Cl_2 + HCl$$

It is given that $(k_1/k_2) = 8.0$.

(*a*) When the recycle ratio (also called, reflux ratio) is zero, the system reduces to a single plug flow reactor. Since the density of the reaction mixture is essentially constant, the performance equations for PFR will be identical with those of a constant-volume batch reactor. Series-parallel reactions of this kind have already been analysed in the case of a batch reactor in Example 8.15 and the expressions for yield and selectivity have been developed. We can, therefore, use the same expressions here. Thus

$$C_P/C_{A0} = \left(\frac{k_1}{k_1 - k_2}\right) \left[(1 - x_A)^{k_2/k_1} - (1 - x_A)\right]$$

$$C_R / C_{A0} = x_A - (C_P / C_{A0})$$

where $A$ stands for benzene, $P$ stands for monochloro benzene and $R$ stands for dichlorobenzene. Now, substituting $(k_2 / k_1) = 1/8 = 0.125$ and $x_A = 0.80$, we get

$$C_P / C_{A0} = 0.706$$

$$C_R / C_{A0} = 0.094$$

The yield of $P$ (namely, monochloro benzene) is thus 70.6 percent and that of $R$ (namely, dichlorobenzene) is 9.4 percent. The overall selectivity of $P$ with respect to $R$ is

$$S_0 = (0.706)/(0.094) = 7.51$$

(*b*) Whenever a plug flow reactor is operated with recycle, certain degree of backmixing is introduced by the recycle stream. When the recycle ratio is infinite, there will be complete backmixing and an ideal stirred tank performance is obtained. Thus, from Eq. (8.3.15),

$$\tau = (C_{A0} - C_{Af})/(-r_A)_f = (C_{A0} - C_{Af})/(k_1 C_{Af} C_{Bf}) \quad \text{... (i)}$$

where $A$ stands for benzene and $B$ stands for chlorine. Now, if we write Eq. (8.3.15) for $P$ (namely, monochloro benzene), we get

$$\tau = (C_{P0} - C_{Pf})/(-r_P)_f = -C_{Pf}/(k_2 C_{Pf} C_{Bf} - k_1 C_{Af} C_{Bf}) \quad \text{... (ii)}$$

Dividing Eq. (ii) by (i), we get

$$-C_{Pf}/(C_{A0} - C_{Af}) = (k_2 / k_1)\ (C_{Pf}/C_{Af}) - 1$$

Rearranging and substituting $C_{Af} = C_{A0}\ (1 - x_A)$, we get

$$C_{Pf}/C_{A0} = \frac{x_A (1 - x_A)}{(1 - x_A) + (k_2 / k_1)\, x_A}$$

Since $x_A = 0.80$ and $(k_2/k_1) = 0.125$,

$$C_{Pf}/C_{A0} = 0.5333$$

$$C_{Rf}/C_{A0} = x_A - (C_{Pf}/C_{A0}) = 0.2667$$

The overall selectivity of $P$ with respect to $R$ is

$$S_0 = (0.5333)/(0.2667) = 2.0$$

It is evident that the yield of monochloro benzene and its selectivity with respect to dichlorobenzene are highest when the recycle ratio is zero.

**Example 8.27:** *Reactor performance for polymerisation reactions:* A homogeneous liquid phase polymerisation is being conducted in an ideal stirred thank reactor which has an average residence time of 35 seconds. The concentration of monomer (namely, $A$) in the feed stream is 1.0 kmole/m$^3$. The polymerisation takes place in two steps such as the initiation reaction in which an active form of the monomer (namely, $P_1$) is produced and the propagation reactions in which the monomer reacts with successive polymers, as shown below:

$$A \xrightarrow{k_1} P_1 \qquad \text{Initiation}$$

$$\left.\begin{array}{l} P_1 + A \rightarrow P_2 \\ P_2 + A \rightarrow P_3 \\ \cdots\cdots\cdots\cdots \\ P_n + A \rightarrow P_{n+1} \end{array}\right\} \text{Propagation}$$

The initiation reaction is first order in $A$ and the rate constant $k_1 = 0.10\ s^{-1}$. The propagation reactions are all second order with the same rate constant $k_2$. It is observed that the mass fraction distribution of total polymer in the exit stream is as given below:

| Polymer : | $P_1$ | $P_2$ | $P_3$ | $P_4$ | $P_7$ | $P_{10}$ | $P_{20}$ |
|---|---|---|---|---|---|---|---|
| Mass fraction : | 0.0180 | 0.0314 | 0.0409 | 0.0470 | 0.0546 | 0.0503 | 0.0250 |

Determine the rate equation for the process from the above data. Compute also the concentrations of the polymer molecules $P_1$ to $P_4$ in the product stream.

**Solution:** Polymerisation reactions are a type of series-parallel reactions. Since the initiation reaction is first order in $A$ and the propagation reactions are all second order with the same rate constant, the rate equation may be written as

$$(-r_A) = k_1 C_A + k_2 C_A \sum_{n=1}^{\infty} [P_n] \quad \text{... (i)}$$

Now, for an ideal stirred tank reactor, from Eq. (8.3.15),

$$\tau = (C_{A0} - C_{Af})/(-r_A)_f \quad \text{... (ii)}$$

where $(-r_A)_f$ is given in Eq. (i). If we write Eq. (8.3.15) for $P_1$, we get

$$\tau = \frac{-[P_1]_f}{-k_1 C_{Af} + k_2 C_{Af} [P_1]_f}$$

or

$$[P_1]_f = \frac{k_1 C_{Af}}{(1/\tau) + k_2 C_{Af}} \quad \text{... (iii)}$$

We shall designate the concentration terms (except for $A$) with square brackets since this notation is more convenient in the case of polymerisation reactions. Now, if we similarly write Eq. (8.3.15) for $P_2$, we get

$$\tau = \frac{-[P_2]_f}{-k_2 C_{Af} [P_1]_f + k_2 C_{Af} [P_2]_f}$$

or

$$[P_2]_f = \frac{k_2 C_{Af} [P_1]_f}{(1/\tau) + k_2 C_{Af}}$$

Substituting for $[P_1]_f$ from Eq. (iii),

$$[P_2]_f = \frac{k_1 k_2 C_{Af}^2}{\left(1/\tau + k_2 C_{Af}\right)^2} \quad \text{... (iv)}$$

In a similar way, the expression for $[P_3]_f$ will be

$$[P_3]_f = \frac{k_1 k_2^2 C_{Af}^3}{\left(1/\tau + k_2 C_{Af}\right)^3} \quad \text{... (v)}$$

Therefore, in general, $$[P_n]_f = \frac{k_1 (k_2)^{n-1} (C_{Af})^n}{\left(1/\tau + k_2 C_{Af}\right)^n} = \frac{(k_1/k_2)}{\left[1 + 1/(\tau k_2 C_{Af})\right]^n} \quad \text{... (vi)}$$

The above equation predicts the concentration of any polymer molecule in the exit stream from the reactor. The mass fraction of $P_n$ in the polymer product is

$$W_n = \frac{(\text{Mass of polymer } P_n)}{(\text{Mass of total polymer product})}$$

The mass of total polymer product is equal to the mass of monomer converted which is $(C_{A0} x_A)$ $Q_0 M_A$, where $M_A$ is the molecular weight of the monomer. Therefore

$$W_n = \frac{[P_n]_f \, Q_0 (n M_A)}{C_{A0} x_A Q_0 M_A} = n \, [P_n]_f /(C_{A0} x_A) \quad \text{... (vii)}$$

To note that if $M_A$ is the molecular weight of the monomer, then the molecular weight of polymer $P_n$ is $(n M_A)$. It may also be noticed that in Eq. (vii), $[P_n]_f / C_{A0}$ is nothing but the yield of $P_n$. Substituting for $[P_n]_f$ from Eq. (vi) in Eq. (vii), we get

$$(W_n/n) = \frac{(k_1/k_2)}{C_{A0} x_A \left[1 + 1/(\tau k_2 C_{Af})\right]^n} \quad \text{... (viii)}$$

Taking logarithm, $$\ln (W_n/n) = \ln \left(\frac{k_1}{k_2 C_{A0} x_A}\right) - n \ln [1 + 1/(\tau k_2 C_{Af})]$$

Thus, a plot of $\ln (W_n/n)$ versus $n$ must yield a straight line. The values of $(W_n/n)$ at different $n$ computed from the given data are listed below:

| $n$ : | 1 | 2 | 3 | 4 | 7 | 10 | 20 |
|---|---|---|---|---|---|---|---|
| $W_n$ : | 0.0180 | 0.0314 | 0.0409 | 0.0470 | 0.0546 | 0.0503 | 0.025 |
| $W_n/n$ : | 0.0180 | 0.0157 | 0.01363 | 0.01175 | 0.0078 | 0.00503 | 0.00125 |

The plot of $\ln (W_n/n)$ versus $n$ is shown in Fig. 8.27.1. It is a straight line of slope = − 0.1379 and $y$-intercept = −3.93. Therefore

$$\ln [1 + 1/(\tau k_2 C_{Af})] = 0.1379$$

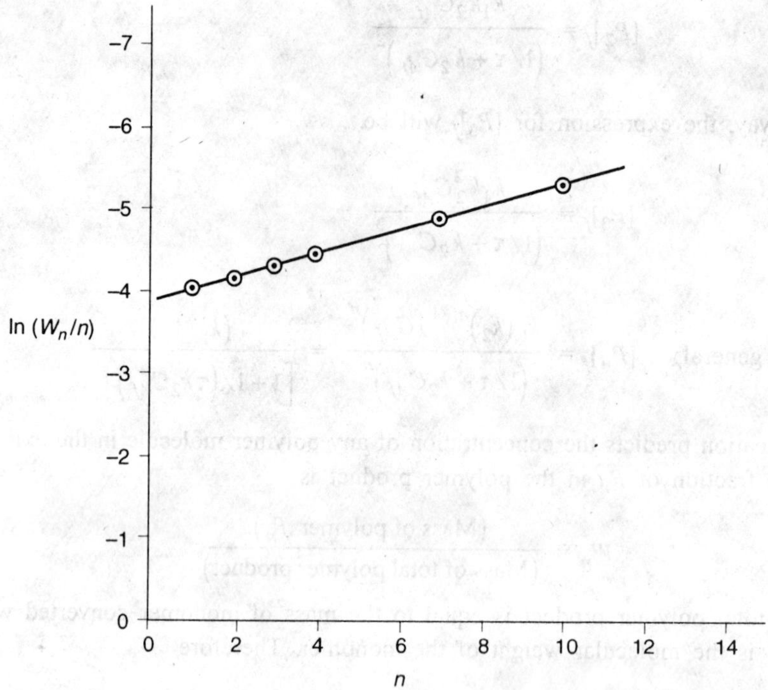


*Figure 8.27.1*

Substituting $\tau = \overline{\theta} = 35$ s, $C_{Af} = C_{A0}\,(1 - x_A) = (1.0)\,(1 - x_A)$, we get

$$k_2\,(1 - x_A) = 0.19323 \qquad \text{... (ix)}$$

$$y\text{-intercept} = \ln\left(\frac{k_1}{k_2\,C_{A0}\,x_A}\right) = -3.93$$

Since $k_1 = 0.1\ \text{s}^{-1}$ and $C_{A0} = 1.0\ \text{kmole/m}^3$,

$$k_2 x_A = 5.0907 \qquad \text{... (x)}$$

Solving (ix) and (x) simultaneously,

$$k_2 = 5.2839\ \text{m}^3/(\text{kmole}\cdot\text{s})$$

$$x_A = 0.9634$$

Thus, the total conversion of monomer is 96.34 percent. The rate equation is

$$(-r_A) = 0.1\,C_A + 5.2839\,C_A \sum_{n=1}^{\infty}[P_n]$$

The concentrations of $P_1$ to $P_4$ in the exit stream can be now easily computed from Eq. (vi). Substituting the values of $\tau$, $k_1$, $k_2$ and $C_{Af}$ in Eq. (vi), we get

$$[P_n]_f = (0.018925)/(1.1533)^n$$

Therefore, $[P_1]_f = 0.0164$ kmoles/m$^3$

$$[P_2]_f = 0.01423 \text{ kmoles/m}^3$$

$$[P_3]_f = 0.01233 \text{ kmoles/m}^3$$

$$[P_4]_f = 0.0107 \text{ kmoles/m}^3$$

Before concluding this example, it may be stated here that it is possible to derive an analytical expression for $x_A$. For example, Eq. (iii) can be rewritten as

$$k_1 C_{Af} = [P_1]_f /\tau + k_2 C_{Af} [P_1]_f$$

Similarly, Eq. (iv) may rearranged to give

$$k_2 C_{Af} [P_1]_f = [P_2]_f /\tau + k_2 C_{Af} [P_2]_f$$

Equation (v), on similar rearrangement, will give

$$k_2 C_{Af} [P_2] = [P_3]_f /\tau + k_2 C_{Af} [P_3]_f$$

and so on. If we sum up all the above equations, we get

$$k_1 C_{Af} = [P_1]_f /\tau + [P_2]_f /\tau + [P_3]_f /\tau + \ldots\ldots = (1/\tau) \sum_{n=1}^{\infty} [P_n]_f \qquad \text{... (xi)}$$

The rate equation, therefore, becomes

$$(-r_A)_f = k_1 C_{Af} (1 + \tau k_2 C_{Af}) \qquad \text{... (xii)}$$

Substituting this in Eq. (ii), we get

$$\tau k_1 (1 + \tau k_2 C_{Af}) = (C_{A0} - C_{Af})/C_{Af}$$

or $\quad \tau k_1 [1 + \tau k_2 C_{A0} (1 - x_A)] = x_A/(1 - x_A) \qquad$ ... (xiii)

The above expression provides the value of $x_A$ once the rate constants, the space time and $C_{A0}$ are known.

**Example 8.28:** *Performance of semibatch reactor*. Butyl Acetate is produced in an ideal stirred tank reactor operating in the semibatch mode. The reaction is conducted at 100° C using sulphuric acid as the catalyst. The rate equation is as given below when an excess of butanol is being used:

$$(-r_A) = kC_A^2$$

where $A$ stands for acetic acid. The value of $k$ at 100° C is 0.0174 m$^3$/(kmole·min). Initially the reactor is charged with 100 gallons of a solution containing butanol, acetic acid and a small amount of sulfuric acid. The concentration of acetic acid in this solution was 1.5 kmole/m$^3$.

The mixture is heated to 100° C and a feed solution containing 3.0 kmoles acetic acid/m$^3$ is added at the rate of 2.0 gallons per minute. The product is withdrawn at the same rate. The density of the reaction mixture maybe assumed constant. Determine the concentration of acid in the reactor effluent at the end of 1 hour. Assume that no water is vapourised in the reactor.

**Solution:** Since the feed rate, feed concentration, product discharge rate and the density of reaction mixture are all constant, we can use Eq. (8.3.22):

$$C_{A0} - C_{Af} = (-r_A)_f \tau + \tau \, dC_{Af}/d\theta = kC_{Af}^2 \tau + \tau \, dC_{Af}/d\theta$$

Separating the variables and integrating, we get

$$k\theta = -\int \frac{dC_{Af}}{\left(C_{Af}^2 + C_{Af}/k\tau - C_{A0}/k\tau\right)}$$

The integral on the right hand side of the above equation may be evaluated either analytically or graphically or numerically. In the present case, we shall evaluate it analytically after substituting the known numerical values of $k$, $\tau$ and $C_{A0}$. Since $\tau = (100/2.0) = 50$ minutes and $k = 0.0174$ m$^3$/(kmole·min), $k\tau = 0.87$ m$^3$/kmole. $C_{A0} = 3.0$ kmoles/m$^3$. Therefore

$$k\theta = -\int \frac{dC_{Af}}{\left(C_{Af}^2 + 1.1494\,C_{Af} - 3.4482\right)} = -\int \frac{dC_{Af}}{\left(C_{Af} + 0.5747\right)^2 - 3.77848}$$

Let us use the formula,

$$\int \frac{dx}{x^2 - a^2} = \frac{1}{2a} \ln\left(\frac{x-a}{x+a}\right) + \text{constant}$$

Therefore,

$$k\theta = (0.2572) \ln\left(\frac{C_{Af} + 2.5185}{C_{Af} - 1.3691}\right) + \text{constant}$$

To evaluate the constant of integration, let us use the initial condition that at $\theta = 0$, $C_{Af} = 1.5$ kmoles/m$^3$. Thus

$$(\text{constant}) = -0.8807$$

or

$$\theta = (886.8965) \ln\left(\frac{C_{Af} + 2.5185}{C_{Af} - 1.3691}\right) - 3036.9$$

where $\theta$ is in seconds. The value of $C_{Af}$ at any $\theta$ can be thus computed from the above equation. When $\theta = 1$ hr $= 3600$ seconds,

$$C_{Af} = 1.3713 \text{ kmoles/m}^3$$

**Example 8.29:** *Performance of semibatch reactor (without continuous product discharge):* Ethyl acetate is to be saponified by adding a solution containing 0.05 kmoles sodium hydroxide per m$^3$ continuously to a vessel containing the ethyl acetate. The reactor is initially charged with 0.4 m$^3$ of an aqueous solution containing 0.0735 kmole/m$^3$ of ethyl acetate and no sodium hydroxide. The sodium hydroxide solution is added at the rate of 0.25 m$^3$/hr until stoichiometric amounts are present. The reaction is second order and irreversible and the specific reaction rate is 6.375 m$^3$/(kmole·min) at 25° C. Assuming that the contents of the vessel are well mixed, determine the concentration of unreacted sodium hydroxide as a function of time. Assume that the density of the reaction mixture does not change significantly during the process.

**Solution:** Here we are concerned with a semibatch reactor of the type shown in Fig. 8.3 (d). No product stream is discharged from the reactor during the operation. From Eq. (8.3.28),

$$F_{A0}\, x_A + (F_{A0}\, \theta)\, (dx_A/d\theta) = (-r_A)\, V \quad \text{... (i)}$$

where $A$ stands for sodium hydroxide. Now,

$$F_{A0} = Q_0 C_{A0} = (0.25/60)\ (0.05) = 2.0833 \times 10^{-4} \text{ kmoles/min}$$

The rate equation is $(-r_A) = kC_A C_B = 6.375\, C_A C_B$ kmoles/(m$^3$·min)

where $B$ stands for ethyl acetate. Now, from Eq. (8.3.27),

$$C_A = (N_{A0} + F_{A0}\theta)\ (1 - x_A)/V$$

Since initially there is no sodium hydroxide in the vessel, $N_{A0} = 0$. Therefore,

$$C_A = (2.0833 \times 10^{-4})\ \theta\ (1 - x_A)/V$$

Now, $$C_B = (N_{B0} - F_{A0}\theta x_A)/V$$

Since initially the vessel contains 0.4 m$^3$ of solution containing 0.0735 kmoles ethyl acetate per m$^3$,

$$N_{B0} = (0.4)\ (0.0735) = 0.0294 \text{ kmoles}$$

Therefore, $$C_B = (0.0294 - 2.0833 \times 10^{-4}\theta x_A)/V$$

The rate expression, therefore, becomes

$$(-r_A)\ V = (1 - x_A)\ \theta\ (10^{-5})\ (3.90467 - 0.027669\,\theta x_A)/V \quad \text{... (ii)}$$

Also, from Eq. (8.3.29), $$V = V_0 + Q_0\theta = 0.4 + (0.25/60)\ \theta \quad \text{... (iii)}$$

To note that $\theta$ is expressed in minutes in all the above equations. We have to now solve Eq. (i) after substituting Eqs (ii) and (iii) in it. However, it is difficult to solve Eq. (i) analytically since it is a nonlinear differential equation. Let us, therefore, solve it numerically. One of the methods of numerical solution is to convert it into a difference equation and then solve it by trial as we solve any algebraic equation. Thus Eq. (i) can be rewritten as

$$F_{A0}\ (x_A d\theta + \theta dx_A) = (-r_A)\ Vd\theta$$

$$d\ (\theta x_A) = (1/F_{A0})\ (-r_A)\ Vd\theta = (4800) - (-r_A)\ Vd\theta$$

Converting into a difference equation, we get

$$\Delta\ (\theta x_A) = 4800\ (-r_A V)_{\text{avg}}\ \Delta\theta \quad \text{... (iv)}$$

The above equation can be now solved for $x_A$ at different increments of time. The smaller the increment chosen, the higher will be the accuracy of computation since this will minimise the approximation involved in using an arithmetic average value of $(-r_A V)$. We shall use a time interval of 2 min here. However, for better accuracy, an increment of 1 min or 0.1 min may be employed. Sample calculations for $\theta = 2$ min and $\theta = 4$ min are given below:

(*i*) $\theta = \theta_0 = 0$

$x_{A0} = 0$

$(-r_A)_0 = r_0 = 0$

(*ii*) $\theta = \theta_1 = 2$ min

$V_1$ [from Eq. (iii)] $= 0.4083$ m$^3$

$\Delta\theta = (\theta_1 - \theta_0) = 2$ min

For this value of $\Delta\theta$, the value of $x_A$ is now computed from Eq. (iv) by trial. Thus

**Trial 1:** Let $x_{A1} = x_{A0} = 0$

From Eq. (ii), $(-r_A)_1 V_1 = (r_1 V_1) = 1.9125 \times 10^{-4}$ kmoles/min

$$(-r_A V)_{avg} = [(r_1 V_1) + (r_0 V_0)]/2 = 9.5625 \times 10^{-5} \text{ kmoles/min}$$

From Eq. (iv), $\Delta(\theta x_A) = 0.918 \text{ min} = (\theta_1 x_{A1} - \theta_0 x_{A0}) = 2x_{A1}$

Therefore, $x_{A1} = 0.459$ (does not check)

**Trial 2:** Let $x_{A1} = (0.459 + 0.0)/2 = 0.2295$

$$r_1 V_1 = 1.46879 \times 10^{-4} \text{ kmoles/min}$$

$$(-r_A V)_{avg} = 7.34395 \times 10^{-5} \text{ kmoles/min}$$

$$\Delta(\theta x_A) = 0.7050 \text{ min}$$

$$x_{A1} = 0.3525 \text{ (does not check)}$$

**Trial 3:** Let $x_{A1} = (0.3525 + 0.2295)/2 = 0.2910$

$$r_1 V_1 = 1.35037 \times 10^{-4} \text{ kmoles/min}$$

$$(-r_A V)_{avg} = 6.75185 \times 10^{-5} \text{ kmoles/min}$$

$$\Delta(\theta x_A) = 0.6482 \text{ min}$$

$$x_{A1} = 0.3241 \text{ (does not check)}$$

**Trial 4:** Let $x_{A1} = (0.291 + 0.3241)/2 = 0.3076$

$$r_1 V_1 = 1.31845 \times 10^{-4} \text{ kmoles/min}$$

$$(-r_A V)_{avg} = 6.5922 \times 10^{-5} \text{ kmoles/min}$$

$$\Delta(\theta x_A) = 0.63285 \text{ min}$$

$$x_{A1} = 0.3164 \text{ (does not check enough)}$$

**Trial 5:** Let $x_{A1} = (0.3076 + 0.3164)/2 = 0.312$

$$r_1 V_1 = 1.31 \times 10^{-4} \text{ kmoles/min}$$

$$(-r_A V)_{avg} = 6.55 \times 10^{-5} \text{ kmoles/min}$$

$$\Delta(\theta x_A) = 0.6288 \text{ min}$$

$$x_{A1} = 0.3144 \text{ (checks)}$$

Therefore, $C_{A1} = (2.0833 \times 10^{-4})(2.0)(0.6856)/(0.4083) = 0.7 \times 10^{-3}$ kmoles/m$^3$

Let us consider the next increment of time. Thus

(*iii*) $\theta_2 = (\theta_1 + \Delta\theta) = (2 + 2) = 4$ min

$V_2 = 0.4167$ m$^3$

**Trial 1:** Let $x_{A2} = x_{A1} = 0.3144$

$$r_2 V_2 = 2.54685 \times 10^{-4} \text{ kmoles/min}$$

$$(-r_A V)_{avg} = (r_2 V_2 + r_1 V_1)/2 = 1.9284 \times 10^{-4} \text{ kmoles/min}$$

$$\Delta(\theta x_A) = 1.8513 \text{ min} = \theta_2 x_{A2} - \theta_1 x_{A1} = 4x_{A2} - 0.6288$$

Therefore, $x_{A2} = 0.62$ (does not check)

**Trial 2:** Let $x_{A2} = (0.62 + 0.3144)/2 = 0.4672$

$$r_2 V_2 = 1.9706 \times 10^{-4} \text{ kmoles/min}$$

$$(-r_A V)_{avg} = 1.6403 \times 10^{-4} \text{ kmoles/min}$$

$$\Delta(\theta x_A) = 1.5747 \text{ min}$$

$x_{A2} = 0.5508$ (does not check)

**Trial 3:** Let $x_{A2} = (0.5508 + 0.4672)/2 = 0.51$

$$r_2 V_2 = 1.81 \times 10^{-4} \text{ kmoles/min}$$

$$(-r_A V)_{avg} = 1.56 \times 10^{-4} \text{ kmoles/min}$$

$$\Delta(\theta x_A) = 1.4976 \text{ min}$$

$x_{A2} = 0.5316$ (does not check)

**Trial 4:** Let $x_{A2} = (0.5316 + 0.51)/2 = 0.521$

$$r_2 V_2 = 1.77 \times 10^{-4} \text{ kmoles/min}$$

$$(-r_A V)_{avg} = 1.54 \times 10^{-4} \text{ kmoles/min}$$

$$\Delta(\theta x_A) = 1.4784 \text{ min}$$

$x_{A2} = 0.5268$ (checks reasonably)

We can take $x_{A2} = (0.5268 + 0.51)/2 = 0.5184$ as the final value. Now,

$$C_{A2} = 0.963 \times 10^{-3} \text{ kmoles/m}^3$$

Similarly the values of $x_A$ and $C_A$ at higher values of $\theta$ such as at $\theta = 6$ min, 8 min, 10 min, etc. can be computed. The results are listed below:

| $\theta$, *min* | $x_A$ | $C_A$, *kmoles/m³* |
|---|---|---|
| 0.0 | 0.0 | 0.0 |
| 2.0 | 0.3144 | $0.7 \times 10^{-3}$ |
| 4.0 | 0.5184 | $0.963 \times 10^{-3}$ |
| 6.0 | 0.6423 | $1.073 \times 10^{-3}$ |
| 8.0 | 0.7137 | $1.101 \times 10^{-3}$ |
| 10.0 | 0.7602 | $1.131 \times 10^{-3}$ |
| 12.0 | 0.794 | $1.144 \times 10^{-3}$ |
| 14.0 | 0.817 | $1.1645 \times 10^{-3}$ |
| 16.0 | 0.835 | $1.178 \times 10^{-3}$ |
| 18.0 | 0.850 | $1.184 \times 10^{-3}$ |

It can be seen that after 10 minutes, the change in the value of $C_A$ is very sluggish. In other words, $C_A$ remains more or less constant after 10 minutes.

The computer program for a problem of above kind (performance analysis of a semibatch reactor) is given below:

## PROGRAM 8.2

```
C**    PERFORMANCE ANALYSIS OF A SEMI BATCH
C**    REACTOR WITH NO CONTINUOUS PRODUCT
C**    DISCHARGE
       DIMENSION THETA (3600), XA (3600), R (3600)
       DIMENSION V (3600), CA (3600), CB (3600)
C**    THE CHEMICAL REACTION IS FIRST ORDER
C**    IN A AND FIRST ORDER IN B
       REAL K, NA0, NB0
       READ (*, *) K, Q0, CA0, V0, CB0
       FA0 = Q0 * CA0
       NA0 = 0.0
       NB0 = V0 * CB0
       THETA (1) = 0.0
       XA (1) = 0.0
       R (1) = 0.0
       CA (1) = CA0
C**    CHOOSE THE TIME INTERVAL. LET IT BE
C**    1.0 SECOND
       DT = 1.0
C**    CONSIDER ONE HOUR OF OPERATION
       N = 3600
       DO 10 I = 2, N
       XA (I) = XA (I - 1)
15     Y = XA (I)
       THETA (I) = THETA (I - 1) + DT
       V (I) = V0 + Q0 * THETA (I)
       CA (I) = [NA0 + FA0 * THETA (I)] * [1.0 - XA (I)]/V (I)
       CB (I) = [NB0 - FA0 * THETA (I) * XA (I)]/V (I)
       R (I) = K * CA (I) * CB (I) * V (I)
       RAVG = [R (I) + R (I - 1)]/2.0
       DTX = (RAVG * DT)/FA0
       XA (I) = [DTX + THETA (I - 1) * XA (I - 1)]/THETA (I)
       DIF = ABS [XA (I) - Y]/Y
       IF (DIF. LT. 0.001) GO TO 10
       XA (I) =[XA (I) + Y]/2.0
       GO TO 15
10     CONTINUE
       WRITE (*, *) [THETA (I), XA (I), CA (I), I = 1, N]
       STOP
       END
```

The nomenclature used in the above program is as follows:

The notations $K$, $Q0$, $CA0$, $CB0$, $V0$, $NA0$, $NB0$, $FA0$, $DT$, THETA ($I$), $XA$ ($I$), $CA$ ($I$), $CB$ ($I$), $V$ ($I$), $R$ ($I$), RAVG and DTX respectively stand for $k$, $Q_0$, $C_{A0}$, $C_{B0}$, $V_0$, $N_{A0}$, $N_{B0}$, $F_{A0}$, $\Delta\theta$, $\theta$, $x_A$, $C_A$, $C_B$, $V$, $(-r_A)$ $V$, $(-r_A V)_{avg}$ and $\Delta$ $(\theta\, x_A)$.

**Example 8.30:** *Adiabatic operation of ideal reactors.* Acetic anhydride is to be hydrolysed in liquid phase. The reaction is irreversible and first order and the specific reaction rate is given by

$$k = (1417 \times 10^4) \exp(-5457.0/T) \text{ min}^{-1}$$

where $T$ is temperature in degrees Kelvin. The heat of reaction is $-209350.0$ kJ/kmole and may be assumed constant.

(*a*) If a batch reactor operating adiabatically is used for the purpose, determine the time required to achieve 80 percent conversion. The reactor is charged with 0.2 $m^3$ of anhydride solution at 15° C and at a concentration of 0.216 kmoles/$m^3$.

(*b*) If a plug flow reactor operating adiabatically is used, compute the volume of reactor required to achieve 80 percent conversion. The feed solution at 15° C containing 0.216 kmoles acetic anhydride per $m^3$ is fed to the reactor at the rate of 0.5 $m^3$/hr.

(*c*) If an ideal stirred tank reactor is used for conducting the reaction, estimate the volume of reactor required to achieve 80 percent conversion. The operation is adiabatic and the feed composition, feed temperature and feed rate are same as those in part (*b*).

(*d*) If instead of a single reactor, two stirred tanks of equal size are used in series, what conversion will be obtained? The volume of each stirred tank is equal to 50 percent of the volume of the single reactor of part (*c*).

Assume that the specific heat and density of the reaction mixture are essentially constant and equal to 3.8 kJ/(kg K) and 1050 kg/$m^3$ respectively.

**Solution:** (*a*) Since the operation is adiabatic and the density and specific heat of the reaction mixture are constant, Eq. (8.3.59) is applicable:

$$(T - T_0) = -(\Delta H_r C_{A0}/\rho C_p)\, x_A$$

Since $\Delta H_r = -209350.0$ kJ/kmole, $C_{A0} = 0.216$ kmoles/$m^3$, $C_p = 3.8$ kJ/(kg·K), $\rho = 1050$ kg/$m^3$, $T_0 = 288°$ K,

$$T = 11.333 x_A + 288 \qquad \text{... (i)}$$

From Eq. (8.3.3), $$\theta = C_{A0} \int_0^{x_A} \frac{dx_A}{(-r_A)}$$

Now, $$(-r_A) = kC_A = (1417 \times 10^4) \exp(-5457/T)\, C_{A0}\, (1 - x_A)$$

Therefore $$\theta = \int_0^{0.8} \frac{dx_A}{(1417 \times 10^4)(1 - x_A) \exp(-5457/T)}$$

$$= \int_0^{0.8} F(x_A)\, dx_A \qquad \text{... (ii)}$$

where $F(x_A) = 1/\{1417 \times 10^4 (1 - x_A) \exp[-5457/(11.333x_A + 288)]\}$

The above integral may be evaluated either graphically or numerically. Let us evaluate it numerically by using Simpson's rule. For this, the values of $F(x_A)$ are computed at different $x_A$ and are listed below:

| $x_A$ | $F(x_A)$ |
|---|---|
| 0.0 | 11.95675 |
| 0.1 | 12.3342 |
| 0.2 | 12.89 |
| 0.3 | 13.693 |
| 0.4 | 14.857 |
| 0.5 | 16.59 |
| 0.6 | 19.308 |
| 0.7 | 23.982 |
| 0.8 | 33.53 |

Then from Simpson's rule,

$$\theta = (0.1/3)\,[11.95675 + 4\,(12.3342 + 13.693 + 16.59 + 23.982) + 2\,(12.89 + 14.857 + 19.308) + 33.53] = \mathbf{13.533\ minutes}$$

(*b*) For the adiabatic operation of PFR with constant $\rho$ and $C_p$, Eq. (8.3.64) is applicable. Since this equation is same as Eq. (8.3.59) used for the batch reactor,

$$T = 11.333x_A + 288$$

Also, we know that for a constant density system like this, Eq. (8.3.35) is applicable which is identical with Eq. (8.3.3) of batch reactor except that $\theta$ is replaced by $\tau$. Since the heat balance equation is also identical with that of batch reactor as given above,

$$\tau = \theta = 13.533 \text{ minutes}$$

Since $\tau = V/Q_0$,

$$V = \tau Q_0 = (13.533)\,(0.5/60) = 0.113 \text{ m}^3$$

(*c*) For the ideal CSTR, we can employ Eq. (8.3.68):

$$(T_F - T_e) = (\Delta H_r\, x_{Af}\, C_{A0})/(\rho\, C_p) = \frac{(-209350.0)(0.8)(0.216)}{(1050)(3.8)} = -9.066$$

or

$$T_e = (288 + 9.066) = 297.066° \text{ K}$$

Now

$$(-r_A)_f = kC_{Af} = kC_{A0}\,(1 - x_{Af})$$
$$= (1417 \times 10^4) \exp(-5457/297.066)\,(0.216)\,(1 - 0.8)$$
$$= 6.442 \times 10^{-3} \text{ kmoles/(m}^3\cdot\text{min)}$$

Therefore, from Eq. (8.3.9),

$$V = F_{A0}x_{Af}/(-r_A)_f = (Q_0 C_{A0})\,x_{Af}/(-r_A)_f$$
$$= (0.5/60)\,(0.216)\,(0.8)/(6.442 \times 10^{-3}) = 0.2235 \text{ m}^3$$

(*d*) When two stirred tank reactors are used in series, we must expect to obtain a higher conversion. Eq. (8.3.68) when written for the first reactor becomes

$$(T_F - T_1) = \Delta H_r x_{A1} C_{A0}/(\rho C_p)$$

Substituting the known values of $T_F$, $\Delta H_r$, $C_{A0}$, $\rho$ and $C_p$, we get

$$T_1 = 11.333 x_{A1} + 288 \quad \text{... (iii)}$$

Equation (8.3.9), for the first reactor, is

$$V_1 = (0.2235/2) = (Q_0 C_{A0})\, x_{A1}/(-r_A)_1$$

$$= \frac{(0.5/60)\, x_{A1}}{\left(1417 \times 10^4\right) \exp\left(-5457/T_1\right)\left(1 - x_{A1}\right)}$$

Substituting for $T_1$ from Eq. (iii) and rearranging, we get

$$1.9 \times 10^8 = \frac{x_{A1}}{\left(1 - x_{A1}\right)} \exp\,[5457/(11.333 x_{A1} + 288)]$$

Solving for $x_{A1}$ by trial, $x_{A1} = 0.641$

$$T_1 = 11.333\ (0.641) + 288 = 295.265 \text{ K}$$

$$C_{A1} = C_{A0}\ (1 - x_{A1}) = (0.216)\ (1 - 0.641) = 0.07754 \text{ kmoles/m}^3$$

Now, writing Eq. (8.3.68) for the second reactor, we get

$$(T_1 - T_2) = (\Delta H_r C_{A1}/\rho C_p)\ x_{A2} = -4.0684 x_{A2}$$

or

$$T_2 = 295.265 + 4.0684 x_{A2} \quad \text{... (iv)}$$

Equation (8.3.9), for the second reactor, is

$$V_2 = (0.2235/2) = Q_0 C_{A1} x_{A2}/(-r_A)_2 = \frac{(0.5/60)\, x_{A2}}{\left(1417 \times 10^4\right) \exp\left(-5457/T_2\right)\left(1 - x_{A2}\right)}$$

Substituting for $T_2$ from Eq. (iv) and rearranging, we get

$$1.9 \times 10^8 = \frac{x_{A2}}{\left(1 - x_{A2}\right)} \exp\,[5457/(4.0684 x_{A2} + 295.265)]$$

Solving for $x_{A2}$, $x_{A2} = 0.68$

$$T_2 = 4.0684\ (0.68) + 295.265 = 298.03 \text{ K}$$

$$C_{A2} = C_{A1}\ (1 - x_{A2}) = 0.07754\ (1 - 0.68) = 0.0248 \text{ kmoles/m}^3$$

The total conversion of acetic anhydride is, therefore,

$$x_A = 1 - (C_{A2}/C_{A0}) = 1 - (0.0248/0.216) = 0.885$$

Thus, we get 88.5 percent conversion by using two reactors in series.

**Example 8.31:** *Nonadiabatic operation of ideal reactors.* (*a*) If the single stirred tank reactor of Example 8.30 is operated in such a way that heat is dissipated at a constant rate of 204 kJ/min (by passing cooling water through the jacket), estimate what conversion shall be obtained.

(*b*) If the batch reactor of Example 8.30 is also jacketed and heat is dissipated at the rate of 204 kJ/min, determine the conversion that will be obtained at the end of 10 minutes.

**Solution:** (*a*) When the operation of the stirred tank reactor is nonisothermal but not adiabatic, we can apply Eq. (8.3.67):

$$(T_F - T_e) = (\Delta H_r C_{A0} x_{Af}/\rho C_p) - [Q/(Q_0 \rho C_p)]$$

where

$$\Delta H_r = -209350.0 \text{ kJ/kmole of } A$$

$$C_{A0} = 0.216 \text{ kmole/m}^3$$

$$\rho = 1050 \text{ kg/m}^3$$

$$C_p = 3.8 \text{ kJ/(kg K)}$$

$$Q = -204 \text{ kJ/min}$$

$$Q_0 = (0.5/60) \text{ m}^3\text{/min}$$

$$T_F = 288° \text{ K}$$

To note that $Q$ represents the rate of input of heat to the reactor. Since in the present case, heat is dissipated from the reactor, $Q$ is a negative quantity. Thus

$$T_e = 11.333 x_{Af} + 281.865 \qquad \text{... (i)}$$

Now, from Eq. (8.3.9), $V = (Q_0 C_{A0})\, x_{Af}/(-r_A)_f$

Since $(-r_A)_f = k C_{Af} = k C_{A0}\,(1 - x_{Af})$,

$$V = \frac{Q_0 x_{Af}}{k\left(1 - x_{Af}\right)}$$

$$0.2235 = \frac{(0.5/60)\, x_{Af}}{\left(1417 \times 10^4\right) \exp\left(-5457/T_e\right)\left(1 - x_{Af}\right)}$$

Substituting for $T_e$ from Eq. (i) and rearranging, we get

$$3.8 \times 10^8 = \frac{x_{Af}}{\left(1 - x_{Af}\right)} \exp\,[5457/(11.333 x_{Af} + 281.865)]$$

Solving for $x_{Af}$ by trial, $x_{Af} = 0.719$

(*b*) For a batch reactor, when the operation is nonisothermal but not adiabatic, Eq. (8.3.58) may be used in the modified form, as

$$dT/d\theta = -(\Delta H_r C_{A0}/\rho C_p)\,(dx_A/d\theta) + [Q/(V\rho C_p)]$$

$$= 11.333\, dx_A/d\theta - 0.2556 \qquad \text{... (ii)}$$

Integrating both sides,

$$(T - T_0) = (T - 288) = 11.333 x_A - 0.2556\theta$$

or

$$T = 11.333 x_A - 0.2556\theta + 288 \qquad \text{...(iii)}$$

The expression for $\theta$ will be the same as that used in Example 8.30.

$$\theta = \int_0^{x_A} \frac{dx_A}{\left(1417 \times 10^4\right)\left(1 - x_A\right) \exp\left(-5457/T\right)} \qquad \text{... (iv)}$$

If we substitute for $T$ from Eq. (iii) in the above equation, then it can be seen that $\theta$ appears on both sides of the equation. Therefore, the above equation can be solved only by trial. Here, we have to find the value of $x_A$ when $\theta = 10$ minutes. We may employ any of the following three methods:

(*i*) Plot $F(x_A)$ versus $x_A$ and by trial, find that value of $x_A$ at which the area under the curve is equal to $\theta$ (namely, 10.0). Those who are at home with plotting graphs, this will be the fastest method.

(*ii*) Assume a value of $x_A$ and then evaluate the integral numerically by using Simpson's rule or trapezoidal rule. Check whether the value of the integral is equal to 10.0. If not, modify the value of $x_A$ and repeat the procedure. When a personal computer or a programmable calculator is available, this method will be the most convenient.

(*iii*) Convert Eqs (ii) and (iv) into difference equations and then solve them algebraically by trial. This is the method we used in Example (8.29). For example, Eqs (ii) and (iv) can be approximated to

$$\Delta T = 11.333\,\Delta x_A - 0.2556\,\Delta\theta \quad \text{... (v)}$$

$$\Delta\theta = [F(x_A)]_{avg}\,\Delta x_A \quad \text{... (vi)}$$

where
$$F(x_A) = \frac{\exp(5457/T)}{(1417\times 10^4)(1-x_A)} \quad \text{... (vii)}$$

Now, Eqs (v) and (vi) can be solved simultaneously at successive increments in $x_A$. For example, consider the first increment. Let $\Delta x_A$ be fixed at 0.1. Then, $x_{A1} = x_{A0} + 0.1$ (in the present example, $x_{A0} = 0$). Now, assume a value of $\Delta\theta$ and compute $\Delta T$ from Eq. (v). Then $T_1 = T_0 + \Delta T$. Once $T_1$ and $x_{A1}$ are known, compute $F(x_{A1})$ from (vii). Now, $[F(x_A)]_{avg} = [F(x_{A0}) + F(x_{A1})]/2$. Compute $\Delta\theta$ from (vi) and check whether it agrees with the assumed value. If not, repeat the procedure with a new value of $\Delta\theta$. Thus, finalise $\Delta\theta$ by trial. Then, $\theta_1 = \theta_0 + \Delta\theta$. Now, proceed to the next increment in $x_A$ such that $x_{A2} = x_{A1} + 0.1$ and once again, finalise $\Delta\theta$ by trial and compute $\theta_2$. Proceed in this way until $\theta = 10$ minutes. On a digital computer, this method can be conveniently employed. It must be noted that transformation of differential equation to a difference equation is an approximation and the smaller the increment used, the lower will be the error involved in this approximation. Therefore, this method may tend to be laborious for hand calculation.

We shall employ the graphical integration method here. After substituting $\theta = 10$ minutes, the expression for $F(x_A)$ becomes

$$F(x_A) = \frac{\exp\left[5457/(11.333x_A + 285.44)\right]}{(1417\times 10^4)(1-x_A)}$$

The plot of $F(x_A)$ versus $x_A$ is shown in Fig. 8.31.1. It can be seen that the area under the curve between $x_A = 0$ and $x_A = 0.6$ is 9.98 which is very close to 10.0. Therefore, the value of $x_A$ at $\theta =$ 10 minutes is 0.6.

A computer program dealing with the analysis of the non-isothermal operation of a batch reactor is given below. The program has been developed with special reference to the above numerical Example 8.31.

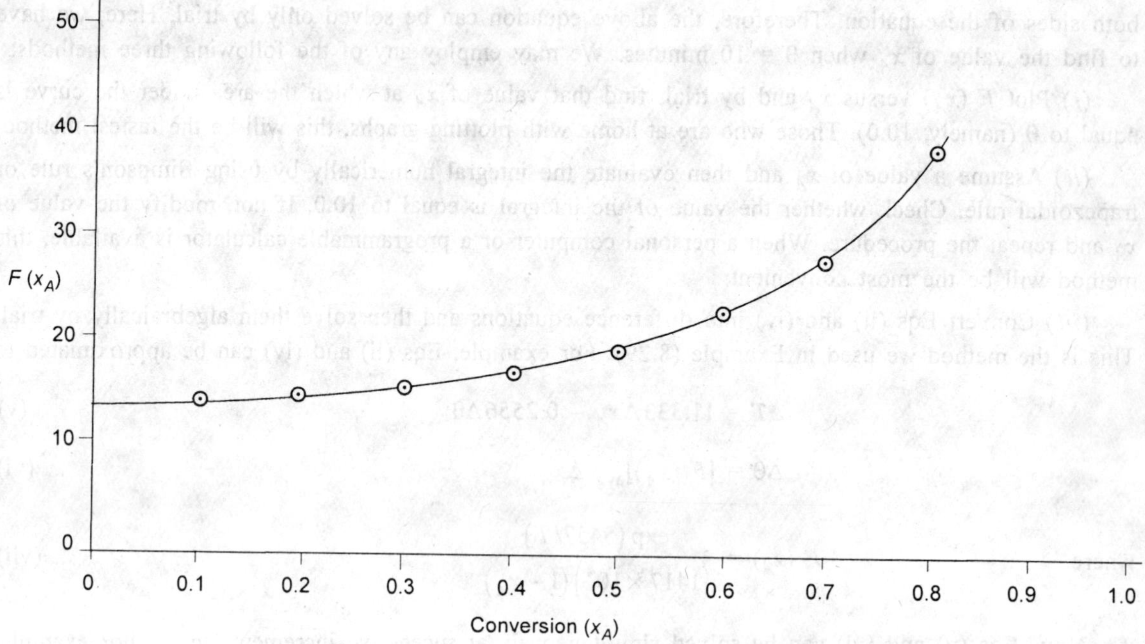


*Figure 8.31.1*

## PROGRAM 8.3

```
C**     ANALYSIS OF NONISOTHERMAL
C**     OPERATION OF A BATCH REACTOR
C**     ASSUME DENSITY AND SPECIFIC HEAT OF
C**     REACTION MIXTURE REMAIN CONSTANT
        DIMENSION T (101), XA (101), F (101)
        REAL K0, K
        READ (*, *) HR, CA0, CP, ROW, Q, V, TO, THETA
C**     REACTION IS FIRST ORDER AND IRREVERSIBLE
        READ (*, *) K0, E
C**     USE TRAPEZOIDAL RULE FOR NUMERICAL
C**     INTEGRATION
        N = 100
        XA0 = 0.0
        XAN = XA0 + 0.001
10      H = (XAN - XA0)/N
        XA (1) = XA0
        XA (101) = XAN
```

```
      DO 15 I = 1, 101
      IF (I.GT.1) XA = XA (I - 1) + H
      B = Q * THETA/(V * ROW * CP)
      T (I) = TO - [HR * CA0/(ROW * CP)] * XA (I) + B
15    F (I) = 1.0/{K0 * [1.0 - XA (I)] * EXP [-E/T (I)]}
      AI = (H/2.0) * [F (I) + F (N + 1)]
      DO 20 I = 2, N
      AI = AI + H * F (I)
20    CONTINUE
      IF (AI. LT. THETA) THEN
      XAN = XAN + 0.001
      GO TO 10
      ELSE
      XAF = XAN
      END IF
      WRITE (*, *) XAF
      STOP
      END
```

The nomenclature used in the above program is as follows:

The notations HR, *CA*0, *CP*, ROW, *Q*, *V*, TO, THETA, *K*, *K*0, *E*, *XA* (*I*) and *T* (*I*) respectively stand for $\Delta H_r$, $C_{A0}$, $C_p$, $\rho$, $Q$, $V$, $T_0$, $\theta$, $k$, $k_0$, $E$, $x_A$ and $T$. To note that the value of $\Delta H_r$ will be a negative quantity for an exothermic reaction and $Q$ will also be a negative quantity when heat is being dissipated from the reactor by means of an external cooling arrangement. It is assumed that the rate constant ($k$) varies with temperature as per the relation, $k = k_0 \exp(-E/RT)$. We are estimating $x_{Af}$ (*XAF* in program) which is the conversion obtained in time $\theta$.

**Example 8.32:** *Stable operating conditions of a stirred tank reactor*. The hydrolysis of acetic anhydride discussed in Example 8.30 is conducted in a 10 liter stirred tank reactor, the feed solution being admitted at 280° K and at the rate of 0.2 liters per second. The feed concentration is 2.573 kmoles acetic anhydride per m$^3$. What are the stable operating conditions (namely, the conversions and temperatures in the product stream) of the reactor? Assume that there is no product in the feed stream and the reactor operates adiabatically. What will be the operating conditions of the reactor if the feed is introduced at 320° K?

**Solution:** The significance of stable operating conditions in a stirred tank reactor is not difficult to understand. We know that for a nonisothermal or adiabatic operation of a CSTR, the heat balance equation [Eq. (8.3.68)] and the mass balance equation [Eq. (8.3.9)] are to be solved simultaneously for $x_{Af}$ and $T_e$ (once $\tau$ is specified in the problem). Often, we get more than one solution when these two equations are solved simultaneously. This can also be graphically illustrated. The plot of $x_{Af}$ versus $T_e$ based on Eq. (8.3.9) is usually a curve (approximately *S*-shaped) since the rate constant varies exponentially with temperature, while the plot of $x_{Af}$ versus $T_e$ from Eq. (8.3.68) will be a straight line once the heat of reaction and the specific heat of the reaction mixture are assumed constants: It is sometimes found that this straight line intersects the curve at more than one point, thereby providing more than one solution to the problem. The stable operating conditions of the reactor correspond to these points of intersection.

For the present example, Eq. (8.3.68) reduces to

$$T_e = -\left(\frac{\Delta H_r C_{A0}}{\rho C_p}\right) x_{Af} + T_F$$

$$= \left[\frac{(209350.0)(2.573)}{(1050.0)(3.8)}\right] x_{Af} + 280 = 135 x_{Af} + 280 \qquad \text{... (i)}$$

From Eq. (8.3.9), $$\tau = \frac{x_{Af}}{k(1 - x_{Af})}$$

Substituting $\tau$ = (10.0)/(200) (60) = 0.833 minutes and $k$ = (1417 × $10^4$) exp (−5457/$T_e$) and rearranging, we get

$$x_{Af} = \frac{(118.083 \times 10^5)}{(118.083 \times 10^5) + \exp(5457/T_e)} \qquad \text{... (ii)}$$

The plot of $x_{Af}$ versus $T_e$ from Eq. (ii) and that from Eq. (i) are shown in Fig. 8.32.1. It can be seen that they intersect at points *A*, *B* and *C*. The points *A* and *C* correspond to the stable operating conditions of the reactor, which are

| | Conversion, $x_{Af}$ | Temperature, $T_e$ |
|---|---|---|
| Point *A* | 0.085 | 292 K |
| Point *C* | 0.95 | 409 K |

For the present case, since the feed temperature is less than 292 K, the reactor will operate at 292 K and the conversion obtained will be 8.5 percent. If the feed enters the reactor partly converted and at a higher temperature (say above 320° K), then the operation of reactor will shift to point *C* whereby a higher conversion of 95 percent will be obtained. The point *B*, though is a point of intersection, represents a metastable point. This is because after small initial displacements from *B*, the system does not return to *B*, whereas disturbances from points *A* and *C* are followed by a return to these stable points. For example, if the feed temperature is below the temperature at point *A*, then the reaction mixture will soon heat up to the operating temperature corresponding to *A*. If the feed temperature is between points *A* and *B*, the reaction mixture will cool to point *A*. Similarly, at initial temperatures between *B* and *C*, transient heating of the reaction mixture will occur until point *C* is reached and at initial temperatures above *C* transient cooling will take place until temperature drops to *C*.

If the feed temperature is 320 K, then Eq. (i) gets modified to

$$T_e = 135 x_{Af} + 320 \qquad \text{.... (iii)}$$

The straight line corresponding to the above equation is also plotted in Fig. 8.32.1 and it can be seen that it intersects the mass balance curve only at one point *D*. The values of $x_{Af}$ and $T_e$ corresponding to point *D* are 0.99 and 454 K respectively. Thus, the reactor will operate at 454 K providing 99 percent conversion in the product stream.

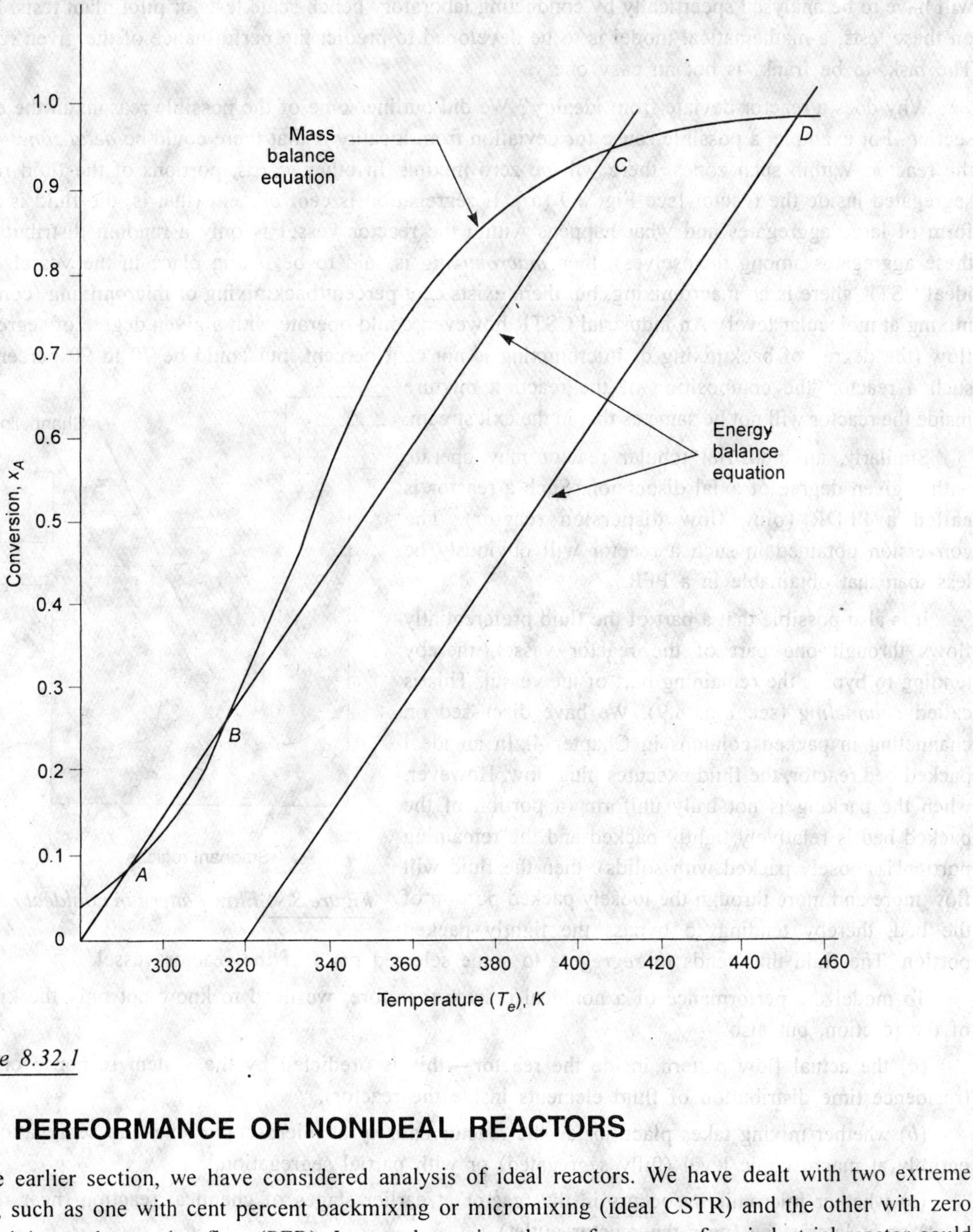


*Figure 8.32.1*

## 8.4 PERFORMANCE OF NONIDEAL REACTORS

In the earlier section, we have considered analysis of ideal reactors. We have dealt with two extreme cases, such as one with cent percent backmixing or micromixing (ideal CSTR) and the other with zero backmixing and true plug flow (PFR). In actual practice, the performance of an industrial reactor could often fall intermediate between those of an ideal CSTR and PFR. When the deviation from ideality is significant (when we cannot approximate the given reactor to an ideal CSTR or PFR), then its performance will have to be analysed specifically by conducting laboratory bench scale tests or pilot plant tests. Based on these tests, a mathematical model is to be developed to predict the performance of the given reactor. The task, to be frank, is not an easy one.

Why does a reactor deviate from ideality? We did outline some of the possible reasons in the earlier section. For example, a possible cause for deviation from ideality is that there could be *dead zones* inside the reactor. Within such zones, there will be zero mixing. In other words, portions of the fluid remain segregated inside the reactor [see Fig. 8.9 (a)]. If segregation is cent percent (that is, the fluid is in the form of large aggregates and what happens within the reactor vessel is only a random distribution of these aggregates among themselves), then *macromixing* is said to be taking place in the vessel. In an ideal CSTR, there is no macromixing, but there exists cent percent backmixing or micromixing (complete mixing at molecular level). An industrial CSTR however, could operate with a given degree of segregated flow (the degree of backmixing or micromixing is not cent percent, but could be 70 to 90 percent). In such a reactor, the composition of the reaction mixture inside the reactor will not be same as that in the exit stream.

Similarly, an industrial tubular reactor may operate with a given degree of axial dispersion. Such a reactor is called a PFDR (plug flow dispersion reactor). The conversion obtained in such a reactor will obviously be less than that obtainable in a PFR.

It is also possible that a part of the fluid preferentially flows through one part of the reactor vessel, thereby tending to bypass the remaining part of the vessel. This is called *channeling* (see Fig. 8.9). We have discussed on channeling in packed columns in Chapter 4. In an ideal packed bed reactor, the fluid executes plug flow. However, when the packing is not truly uniform (a portion of the packed bed is relatively tightly packed and the remaining portion is loosely packed with solids), then the fluid will flow more and more through the loosely packed portion of the bed, thereby tending to bypass the tightly packed portion. The fluid thus tends to *segregate* to some selected parts of the reactor vessel.

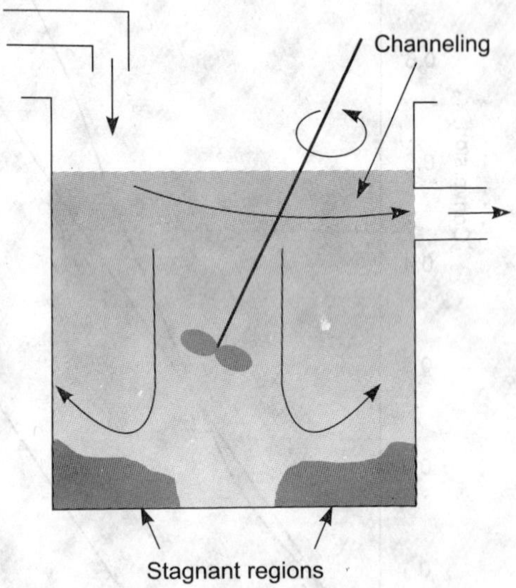


*Figure 8.9: Flow pattern in nonideal reactors*

To model the performance of a nonideal reactor therefore, we need to know not only the kinetics of the reaction, but also

(*a*) the actual flow pattern inside the reactor—this is predicted by the system response or RTD (residence time distribution of fluid elements inside the reactor),

(*b*) whether mixing takes place inside the reactor entirely at microscopic level (no segregation), or entirely at macroscopic level (fully segregated) or with partial segregation,

(*c*) whether micromixing occurs in the reactor at earlier stages of chemical reaction (near reactor inlet) or at later stage (near the reactor outlet).

For obtaining the RTD (residence time distribution) data, we employ what is called the *tracer technique*. A tracer is a substance that does not react with any of the constituents of the reaction mixture, but is easily miscible/soluble in the reaction mixture (since it has similar physical properties as the reaction mixture). It is also important that the substance being used as the tracer does not get adsorbed on the reactor walls or on any of the solid constituents of the reaction mixture. Nonreactive dyes, salts such as sodium chloride, sodium fluoride, etc. are popularly used as tracers. The tracer is injected to the feed stream (in milligrams or milliliters) and the concentration of the tracer in the exit stream from the reactor is recorded at regular intervals of time. This $C(\theta)$ versus $\theta$ data, where $C(\theta)$ is the concentration of tracer in the reactor effluent at any time $\theta$, is called the RTD (residence time distribution) data of the reactor. The RTD data tell us how long each fluid element has stayed inside the reactor (the age or residence time of each fluid element inside the reactor) and this sheds a lot of light on the flow pattern existing in the reactor.

The tracer may be introduced as a *step input* or *pulse input*. These terminologies are not difficult to understand. We may inject a specific amount of tracer into the feed stream instantaneously in one shot. Thereafter, the concentration of tracer in the exit stream is observed at regular intervals of time. The tracer is introduced once (in one shot) and only once. This is called the *impulse input* or impulse stimulus. Alternately, the tracer is added continuously to the feed stream such that the concentration of the tracer in the stream entering the reactor is kept constant at $C_0$ (for all $\theta > 0$). This is called *step input* (see Fig. 8.10). The major hurdle in using a step input is the difficulty in maintaining a constant tracer concentration in the feed stream throughout the experiment. It also demands a large amount of tracer for the entire set of experiments. If the tracer is an expensive species, an impulse input is usually preferred.

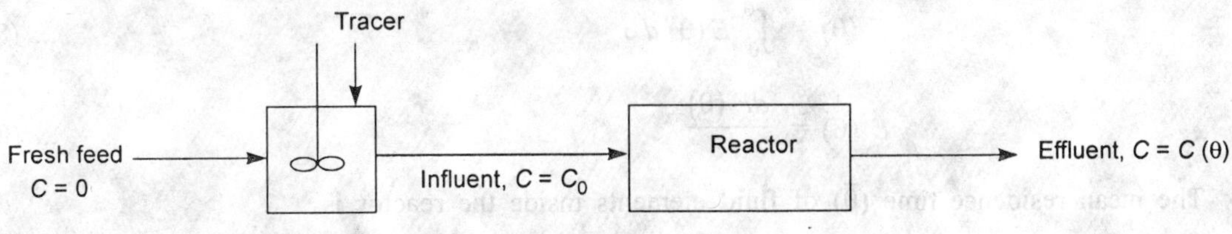


*Figure 8.10: Reactor test using a step input of tracer*

The RTD data may be represented graphically as a $C(\theta)$ versus $\theta$ plot. However, it is usually more convenient to use frequency distribution functions, such as $E(\theta)$ and $F(\theta)$, to represent the data. For example, let an impulse input of tracer is being employed. As stated earlier, in this case, the tracer is introduced once and only once. Let $M$ milligrams of tracer is injected to the feed stream at $\theta = 0$. Now, we record the concentration of the tracer in the outlet stream at regular intervals of time. Let at $\theta = \theta_1$, $m_1$ mg of tracer appeared in the exit stream [corresponding to $C(\theta) = C_1$ mg/L). At $\theta = \theta_2$, $m_2$ mg of tracer is detected in the outlet stream (corresponding to $C(\theta) = C_2$ mg/L) and so on. The tests are continued until the entire $M$ mg of tracer has appeared in the exit stream (at $\theta = \infty$). For a situation like this, the RTD may be represented using the distribution function $E(\theta)$, which is defined as

$$E(\theta) = \frac{C(\theta)}{\int_0^\infty C(\theta)\, d\theta} \quad \text{... (8.4.1)}$$

It can be easily shown that

$$\int_0^{\infty} E(\theta)\ d\theta = 1.0 \quad \text{... (8.4.2)}$$

$E(\theta)$ is often called the *exit age distribution function* and the integral

$$\int_0^{\theta} E(\theta)\ d\theta \quad \text{... (8.4.3)}$$

represents the fraction of the fluid that has resided inside the reactor for a time less than $\theta$. The *mean* residence time $(\bar{\theta})$ of fluid elements in the reactor is

$$\bar{\theta} = \frac{\int_0^{\infty} \theta\, E(\theta)\, d\theta}{\int_0^{\infty} E(\theta)\, d\theta} = \int_0^{\infty} \theta\, E(\theta)\, d\theta \quad \text{... (8.4.4)}$$

Now, consider a step input of tracer. As we know, in this case the concentration of tracer in the feed stream remains constant at $C_0$ and we measure the tracer concentration in the output stream, $C(\theta)$, at regular intervals of time. In such a case, we can define the distribution function as

$$F(\theta) = [C(\theta)/C_0]_{\text{step}} \quad \text{... (8.4.5)}$$

To note that $F(\theta)$ is the fraction of fluid elements that has resided in the reactor for a time less than $\theta$ [or, $dF(\theta)$ is the fraction of fluid elements that has resided in the reactor between $\theta$ and $\theta + d\theta$]. Accordingly, it is related to $E(\theta)$, defined earlier, as

$$F(\theta) = \int_0^{\theta} E(\theta)\, d\theta \quad \text{... (8.4.6)}$$

or

$$E(\theta) = \frac{dF(\theta)}{d\theta} \quad \text{... (8.4.7)}$$

The mean residence time $(\bar{\theta})$ of fluid elements inside the reactor is

$$\bar{\theta} = \int_0^{1} \theta\, dF(\theta) \quad \text{... (8.4.8)}$$

Before proceeding further, let us consider the ideal reactors we are already familiar with (such as the ideal CSTR and the PFR) and examine how these distribution functions would look like in those cases. For example, suppose the reactor under consideration is an ideal CSTR and we use a step input of tracer. If we write a mass balance for the tracer (to note that there will be no chemical reaction term, but the rate of accumulation of tracer inside the reactor is not zero), we get

$$Q_0C_0 - Q_0C = V\frac{dC}{d\theta} \quad \text{... (8.4.9)}$$

or

$$\frac{dC}{d\theta} = (C_0 - C)/\tau \quad \text{... (8.4.10)}$$

where $\tau$ = space time = $(V/Q_0)$. We have assumed a constant density system (that is, the density of the fluid does not change materially due to the addition of tracer). On integrating with the initial condition that at $\theta \le 0$, $C = 0$, we get

$$(C/C_0)_{\text{step}} = F(\theta) = 1 - \exp(-\theta/\tau) \quad \text{... (8.4.11)}$$

Now, let an impulse input of tracer be employed. The distribution function is then $E(\theta)$ and the expression for $E(\theta)$ can be deduced straight away as

$$E(\theta) = \frac{dF}{d\theta} = (1/\tau) \exp(-\theta/\tau) \qquad \text{... (8.4.12)}$$

The mean residence time $(\bar{\theta})$ is [from Eq. (8.4.4)],

$$\bar{\theta} = (1/\tau) \int_0^{\infty} \theta \exp(-\theta/\tau)\, d\theta \qquad \text{... (8.4.13)}$$

$$= \tau \qquad \text{... (8.4.14)}$$

For an ideal CSTR therefore, both functions $E(\theta)$ and $F(\theta)$ are exponential in nature. Also, the mean residence time $(\bar{\theta})$ is equal to the space time $(\tau)$. This observation, however, is not surprising since we have already deduced the same in Section 8.3.

The above expressions for $E(\theta)$ and $F(\theta)$ may also be written in terms of the dimensionless parameter $\theta^*$ such that

$$\theta^* = \theta/\bar{\theta} \qquad \text{... (8.4.15)}$$

Since for an ideal CSTR handling a constant density system, $\bar{\theta} = \tau$, $\theta^* = \theta/\tau$ and

$$E(\theta^*) = \exp(-\theta^*) \qquad \text{... (8.4.16)}$$

$$F(\theta^*) = 1 - \exp(-\theta^*) \qquad \text{... (8.4.17)}$$

Let us now consider a plug flow reactor. In the case of a PFR, its response to a step/pulse input of tracer can be deduced intuitively.

If there is a step input of tracer (the feed stream has a fixed concentration of tracer $C_0$, for all $\theta > 0$), then the tracer concentration in the exit stream will also remain constant at $C_0$ for $\theta \geq \tau$ and there will be no tracer in the exit stream at $\theta < \tau$. In other words,

$$F(\theta) = [C(\theta)/C_0]_{step} = 1.0 \text{ for } \theta \geq \tau \qquad \text{... (8.4.18)}$$

$$= 0 \text{ for } \theta < \tau \qquad \text{... (8.4.19)}$$

On the other hand, if a specific amount of tracer is introduced instantaneously into the feed stream (impulse input), then there will also be a sudden spike in tracer concentration in the exit stream at $\theta = \tau$. And that is all. There will be no tracer in the exit stream at $\theta < \tau$ as well as at $\theta > \tau$.

Accordingly, how much the distribution curves [$E(\theta)$ versus $\theta$ or $F(\theta)$ versus $\theta$] deviate from those for an ideal CSTR or PFR shall tell us how much the performance of the given reactor deviates from ideal behaviour. To estimate this deviation quantitatively, it shall be convenient to employ some of the statistical tools such as the standard deviation $(\sigma)$, the variance $(\sigma^2)$ or the skewness $(\sigma^3)$. These are defined as

$$\sigma^2 = \int_0^{\infty} \left(\theta - \bar{\theta}\right)^2 E(\theta)\, d\theta \qquad \text{... (8.4.20)}$$

$$= \int_0^{\infty} \theta^2 E(\theta)\, d\theta - (\bar{\theta})^2 \qquad \text{... (8.4.21)}$$

In terms of the dimensionless variable $\theta^*$,

$$\sigma^2 (\theta^*) = (\sigma^2/\bar{\theta}^2) = \int_0^\infty (\theta^*)^2 E(\theta^*)\, d\theta^* - 1 \qquad \text{... (8.4.22)}$$

In case a step input of tracer is being employed, then

$$\sigma^2 = \int_0^1 \theta^2\, dF(\theta) - (\bar{\theta})^2 \qquad \text{... (8.4.23)}$$

The parameter skewness ($\sigma^3$) is defined as

$$\sigma^3 = (\sigma^{-1.5}) \int_0^\infty \left(\theta - \bar{\theta}\right)^3 E(\theta)\, d\theta \qquad \text{... (8.4.24)}$$

Application of these statistical tools is illustrated in Example 8.33.

Though RTD tells us regarding the actual flow pattern inside the reactor, it tells little about the level of mixing (or the extent of inter mixing between fluid elements) within the reactor. However, fortunately though, in a number of cases, the reactor performance has been observed to be not too sensitive to the degree of micromixing/macromixing within the reactor. For example, if the reaction is of first order (the rate equation is linear in $C_A$), then the reactor performance shall be independent of the degree of micromixing/macromixing inside the reactor. Even for a second order irreversible reaction, an ideal CSTR with complete micromixing and a CSTR with complete macromixing (complete segregation) give conversions that differ by not more than 10 percent. In such cases, we could predict the reactor performance using mathematical models that are based on RTD studies only. Quite occasionally, the reactor performance is modeled assuming complete micromixing and also by assuming complete macromixing. If the model results do not differ significantly, then it indicates that the degree of micromixing is not a deciding factor.

It is also important to know whether micromixing occurs at the early stages of the chemical reaction or at a later stage. This can be best understood by considering two reactor combinations, such as one in which an ideal CSTR is followed by a PFR and another in which a PFR is followed by an ideal CSTR. In the first combination, in which an ideal CSTR precedes a PFR, micromixing occurs at the initial stages whereas in the second type of reactor combination, micromixing occurs at the final stage. Let a step input of tracer be employed. Then, for the first combination, the tracer concentration at the outlet of the ideal CSTR is

$$C_1(\theta) = C_0 F(\theta) \qquad \text{... (8.4.25)}$$

$$= C_0 [1 - \exp(-\theta/\tau_S)] \qquad \text{... (8.4.26)}$$

where $\tau_S$ is the space time ($= V_S/Q_0$) of the ideal CSTR. When a fluid stream of this concentration is fed to the PFR, the tracer concentration in the outlet stream from PFR will also be $C_1(\theta)$, but at $\theta \geq \tau_p$ only (where $\tau_p$ is the space time of PFR $= V_p/Q_0$). Thus, at the outlet of the reactor combination,

$$C(\theta) = C_0 \{1 - \exp[-(\theta - \tau_p)/\tau_S]\} \text{ for } \theta \geq \tau_p \qquad \text{... (8.4.27)}$$

$$= 0, \text{ for } \theta < \tau_p \qquad \text{... (8.4.28)}$$

In the case of the second reactor combination, a feed stream with tracer concentration $C_0$ enters the PFR. Accordingly, the outlet stream from PFR will also have the same tracer concentration $C_0$ but at

$\theta \geq \tau_p$ only. Now, if this fluid stream enters the ideal CSTR, the tracer concentration in the final outlet stream will be

$$C(\theta) = C_0 F(\theta)$$

$$= C_0 \{1 - \exp[-(\theta - \tau_p)/\tau_S]\} \text{ for } \theta \geq \tau_p \quad \text{... (8.4.29)}$$

$$= 0, \text{ for } \theta < \tau_p \quad \text{... (8.4.30)}$$

It can be seen that the above two equations are identical with Eqs (8.4.27) and (8.4.28). Thus, RTD for both of these reactor combinations is the same. In other words, RTD data is independent of whether micromixing occurs at the initial stages or at a later stage. Accordingly, based on RTD studies, we are not in a position to predict whether the given reactor can be modeled as equivalent to which of these combinations.

Nevertheless, this parameter is also not too trivial on all occasions. For example, if the reaction is first order in $A$, then these two combinations shall provide the same conversion ($x_{Af}$). Thus, if the reaction follows first order kinetics, then the given reactor may be modeled as equivalent to any one these combinations. However, if the reaction order is different from unity, then this information (micromixing occurs at initial stages or at later stage) could make difference. We have already shown in Section 8.3 that for isothermal reactions with order less than 1.0, the first combination (an ideal CSTR followed by a PFR) provides higher conversion and for reactions with order more than 1.0, the second combination provides better performance (see Example 8.17).

It is thus clear that mathematical modelling of a nonideal reactor and thus predicting its performance is no easy task. We have to try a variety of alternatives and attempt to design a model that could predict the reactor performance most realistically. Also, for developing rigorous models, not only the kinetic equation but also the mass transfer/heat transfer equations may have to be written down and these equations (often, partial differential equations) be solved simultaneously (in many cases numerically since analytical solution may not be possible) using appropriate boundary conditions.

We shall discuss here some of the popular mathematical models that are used to describe the performance of nonideal reactors. These are

(*a*) CSTR with a given degree of macromixing or segregated flow
(*b*) PFR with a given degree of axial dispersion (plug flow dispersion reactor, PFDR)
(*c*) Laminar flow tubular reactor (LFTR)
(*d*) Reactor whose performance may be modeled as equivalent to that of a number of ideal CSTRs in series (usually called Tanks-in-Series model)
(*e*) Reactor whose performance may be modeled as equivalent to that of a recycle reactor (a PFR with partial effluent recycle)
(*f*) Reactor combinations.

Let us consider them one by one.

***(a)* CSTR with Macromixing:** This is one of the most straightforward approaches to account for nonideality. We know in an ideal CSTR, cent percent micromixing occurs. The other extreme, therefore, is a CSTR with cent percent macromixing (cent percent segregated flow). An industrial reactor could be then assumed to be operating in between these two extremes. For such a reactor, the fractional conversion that can be achieved may be computed as

$$1 - C_{Af}/C_{A0} = \int_0^{\infty} \left(1 - C_A/C_{A0}\right) E\,(\theta)\, d\theta \qquad \text{... (8.4.31)}$$

or,

$$x_{Af} = \int_0^{\infty} x_A\, E\,(\theta)\, d\theta \qquad \text{... (8.4.32)}$$

RTD studies provide the values of $E$ ($\theta$) at different $\theta$ and based on the kinetic equation, we can express $x_A$ or $(C_A/C_{A0})$ in terms of $\theta$ and thereafter, the above integral can be evaluated numerically. In those cases where $E$ ($\theta$) can be represented as an algebraec expression in $\theta$, the above integral may also be evaluated analytically. For example, consider a CSTR with cent percent macromixing (micromixing is totally absent). For such a reactor, Eq. (8.4.32) becomes

$$x_{Af} = \int_0^{\infty} x_A\ (1/\tau)\ \exp\,(-\theta/\tau)\ d\theta \qquad \text{... (8.4.33)}$$

Now, if the reaction is psuedofirst order in $A$, then

$$x_A = 1 - \exp\,(-k\theta) \qquad \text{... (8.4.34)}$$

Substituting this in Eq. (8.4.33) and integrating, we get

$$x_{Af} = \frac{k_1\tau}{(1 + k_1\tau)} \qquad \text{... (8.4.35)}$$

It can be seen that the conversion achieved is same as that would be achieved in an ideal CSTR [see Eq. (8.3.50) of Section 8.3]. Thus, for reactions following first order kinetics (more precisely, psuedofirst order kinetics), the reactor performance is not affected by the degree of micromixing/macromixing. For a zero order reaction, we know

$$x_A = k\theta/C_{A0} \qquad \text{... (8.4.36)}$$

Accordingly,

$$x_{Af} = (k\tau/C_{A0})\ [1 - \exp\,(-C_{A0}/k\tau)] \qquad \text{... (8.4.37)}$$

If the chemical reaction is psuedo-second order in $A$, then

$$x_A = \frac{C_{A0}kt}{1 + C_{A0}kt} \qquad (8.4.38)$$

Substituting the above expression for $x_A$ in Eq. (8.4.33) and integrating, we get

$$(1 - x_{Af}) = \alpha\, \exp\,(\alpha) \int_{\alpha}^{\infty} \frac{\exp\left[-(\alpha + \theta^*)\right]}{(\alpha + \theta^*)}\, d\,(\alpha + \theta^*) \qquad \text{... (8.4.39)}$$

$$= \alpha\, \exp\,(\alpha)\ ei\,(\alpha) \qquad \text{... (8.4.40)}$$

where $\alpha = 1/(C_{A0}k\tau)$, $\theta^* = \theta/\tau$ and $ei$ ($\alpha$) is the exponential integral (which is a function of $\alpha$ only). The values of exponential integrals are available in the form of standard tables in many books on Engineering Mathematics. The general definition of exponential integral is

$$ei\,(x) = \int_x^{\infty} \frac{\exp\,(-u)}{u}\, du \qquad \text{... (8.4.41)}$$

$$= -0.57721 - \ln x + x - \frac{x^2}{2\angle 2} + \frac{x^3}{3\angle 3} - ..... \quad ...\ (8.4.42)$$

For $x \geq 10$,

$$ei\ (x) = \exp\ (-x)\left(\frac{1}{x} - \frac{1}{x^2} + \frac{\angle 2}{x^3} - \frac{\angle 3}{x^4} + .....\right) \quad ...\ (8.4.43)$$

Another form of exponential integral is,

$$Ei\ (x) = \int_{-\infty}^{x} \frac{\exp(u)}{u}\, du \quad ...\ (8.4.44)$$

$$= 0.57721 + \ln x + x + \frac{x^2}{2\angle 2} + \frac{x^3}{3\angle 3} + ..... \quad ...\ (8.4.45)$$

For $x \geq 10$,

$$Ei\ (x) = \exp\ (x)\left(\frac{1}{x} + \frac{1}{x^2} + \frac{\angle 2}{x^3} + \frac{\angle 3}{x^4} + .....\right) \quad ...\ (8.4.46)$$

Typical values of these integrals are given in Table 8.3.

**Table 8.3**

| $x$ | $Ei\ (x)$ | $ei\ (x)$ | $x$ | $Ei\ (x)$ | $ei\ (x)$ |
|---|---|---|---|---|---|
| 0 | $-\infty$ | $+\infty$ | 1.0 | 1.8951 | 0.2194 |
| 0.01 | −4.0179 | 4.0379 | 1.4 | 3.0072 | 0.1162 |
| 0.02 | −3.3147 | 3.3547 | 2.0 | 4.9542 | 0.0489 |
| 0.05 | −2.3679 | 2.4679 | 2.5 | 7.0738 | 0.0249 |
| 0.10 | −1.6228 | 1.8229 | 3.0 | 9.9338 | 0.01305 |
| 0.20 | −0.8218 | 1.2227 | 5.0 | 40.185 | 0.00115 |
| 0.30 | −0.3027 | 0.9057 | 7.0 | 191.50 | 0.00012 |
| 0.50 | 0.4542 | 0.5598 | | | |

For reactions of fractional order such as an irreversible half order reaction with single reactant $A$,

$$1 - x_A = \left(1 - \frac{k\theta}{2\sqrt{C_{A0}}}\right)^2 \quad ...\ (8.4.47)$$

Accordingly, $$x_{Af} = (k\tau/C_{A0})\ [1 - \exp\ (-2C_{A0}/k\tau)] \quad ...\ (8.4.48)$$

As stated earlier, it is conventional to compute the fractional conversion of $A$ for an ideal CSTR and for a CSTR with complete macromixing (from expressions given above) and a comparison between these two values would tell us how far the reactor performance is influenced by the degree of micromixing/macromixing.

**(b) Plug Flow Dispersion Reactor (PFDR):** Another popular method of accounting for nonideality is to assume that the given reactor is a tubular reactor that operates with a given degree of axial dispersion. In other words, there exists a given degree of backmixing in the axial direction (to note that in a PFR, there is absolutely no backmixing in the axial direction). However, as in a PFR, in this case also, it is assumed that there is cent percent micromixing in the radial direction at any cross-section. Accordingly, neither the fluid velocity nor the concentration varies in the radial direction at any cross-section. Such a reactor is called a plug flow dispersion reactor (PFDR). Many industrial reactors may not perform as a true PFR, but could perform as a PFDR.

The performance equation for a PFDR can be developed by incorporating an axial dispersion term into the mass balance equation. Thus,

$$-\mathbf{V}\frac{dC_A}{dz} + D_d\frac{d^2C_A}{dz^2} = (-r_A) \quad \text{... (8.4.49)}$$

To note that if $D_d = 0$ and if we rewrite $V\,(dC_A/dz)$ as $d\,(\mathbf{V}A\,C_A)/d\,(Az) = dF_A/dV$, then the above equation reduces to that for a PFR [see Eq. (8.3.31) of Section 8.3]. Here, $D_d$ is called the *axial dispersion coefficient* and **V** is the plug flow velocity.

Let us first consider the response of such a reactor to a pulse input of tracer. For a tracer, the chemical reaction term does not exist (the tracer is chemically inert), but since the rate of accumulation term, $\partial C/\partial\theta$, does exist, the mass balance equation becomes

$$-\mathbf{V}\frac{\partial C}{\partial z} + D_d\frac{\partial^2 C}{\partial z^2} = \frac{\partial C}{\partial \theta} \quad \text{... (8.4.50)}$$

In terms of the dimensionless variables $\eta$ and $\theta^*$, where $\eta = (z/L)$ and $\theta^* = \theta/\tau = \theta\mathbf{V}/L$, the above equation becomes

$$-\frac{\partial C}{\partial \eta} + N_d\frac{\partial^2 C}{\partial \eta^2} = \frac{\partial C}{\partial \theta^*} \quad \text{... (8.4.51)}$$

where $\quad N_d = \text{dispersion number} = (D_d/\mathbf{V}L) \quad$ ... (8.4.52)

To note that $N_d$ is nothing but the reciprocal of the Peclet number ($Pe$). $N_d = 0$ for a PFR and $N_d = \infty$ for an ideal CSTR. The above partial differential equation is to be now solved using the appropriate boundary conditions and initial condition. In fact, the boundary conditions can be specified in two ways. Firstly, let us assume that the reactor performs as a *closed* vessel. A closed vessel is one for which plug flow can be assumed to be existing immediately before the entrance to the vessel and immediately after the exit from the vessel. In other words, the fluid enters the vessel in plug flow and leaves the vessel also in plug flow, though true plug flow need not exist inside the vessel. This shall be essentially true if the fluid enters and leaves the vessel (reactor) through a small diameter pipe. In such a case, the boundary conditions pertaining to the system are,

**B.C.1:** $\quad [C - N_d\,(\partial C/\partial\eta)]_{\eta=1} = (C)_{\eta=1}$ for $\theta^* > 0$ ... (8.4.53)

**B.C.2:** $\quad [C - N_d\,(\partial C/\partial\eta)]_{\eta=0} = (C)_{\eta=0}$ for $\theta^* > 0$ ... (8.4.54)

**IC:** At $\quad \theta^* = 0,\ C = 0$ for $0 \le \eta \le 1$ ... (8.4.55)

An analytical solution to Eq. (8.4.51) based on above conditions is hardly possible. However, based on a numerical solution, the following expression has been developed[56] for the variance:

$$\sigma^2\,(\theta^*) = 2N_d - 2N_d^2\,[1 - \exp\,(-\,1/N_d)] \qquad \text{... (8.4.56)}$$

A second possibility is that we may assume that the reactor performs as an *open vessel*. This, in fact, is equivalent to assuming that the reactor is infinitely long and we consider a section of this reactor. Thus, at $z = 0$ or $\eta = 0$ (entrance to the section under consideration) as well as at $z = L$ or $\eta = 1$ (at the exit from the section), the fluid has already undergone a given degree of backmixing. However, since the reactor has been assumed to be infinitely long, the boundary conditions may be specified as

**B.C.1:** At $\eta = -\infty,\ C = 0$ for $\theta \geq 0$ ... (8.4.57a)

**B.C.2:** At $\eta = +\infty,\ C = 0$ for $\theta \geq 0$ ... (8.4.58a)

Equation (8.4.51) can be solved analytically for the above *open* boundary conditions to obtain

$$E\,(\theta^*) = \frac{1}{\sqrt{4\pi\theta^* N_d}}\,\exp\left[-\frac{(1-\theta^*)^2}{4\theta^* N_d}\right] \qquad \text{... (8.4.57)}$$

or,

$$E\,(\theta) = \frac{\mathrm{V}}{\sqrt{4\pi D_d\,\theta}}\,\exp\left[-\frac{(L-\mathrm{V}\theta)^2}{4D_d\,\theta}\right] \qquad \text{... (8.4.58)}$$

Also, the variance is given by

$$\sigma^2\,(\theta^*) = 2N_d + 8\,(N_d)^2 \qquad \text{... (8.4.59)}$$

Data from RTD studies permit us compute the value of the variance ($\sigma^2$). Then, Eq. (8.4.56) or (8.4.59) may be solved for $N_d$. No doubt, the former can be solved for $N_d$ only by trial. The given reactor may be then considered equivalent to a PFDR with the computed value of $N_d$. Further, from the known kinetics of any chemical reaction under consideration, the fractional conversion of the reactant $A$ (namely, $x_A$) obtainable from such a reactor be computed from expressions/charts given subsequently in this Section (see Figs 8.11 and 8.12).

We can also determine the response of a PFDR to a step input of tracer. In this case, considering an *open* vessel, the boundary and initial conditions are

At $\theta^* = 0,\ C = 0$ for $\eta > 0$

$C = C_0$ for $\eta < 0$ ... (8.4.60)

At $\theta^* > 0,\ C = 0$ at $\eta = \infty$

$C = C_0$ at $\eta = -\infty$ ... (8.4.61)

The analytical solution to Eq. (8.4.51) based on the above conditions is

$$F\,(\theta) = (C/C_0)_{\text{step}} = \left(\frac{1}{2}\right)\left\{1 - erf\left[\frac{(1-\theta^*)}{2\sqrt{N_d\,\theta^*}}\right]\right\} \qquad \text{... (8.4.62)}$$

It can be seen that the above expression for $F(\theta)$ is not quite handy and deducing the value of $N_d$ based on the above equation could be cumbersome. What is recommended is that if experimental data based on a step input of tracer are being available, the variance be calculated from Eq. (8.4.23) by numerical integration. Thereafter, the value of $N_d$ could be obtained by solving Eq. (8.4.56) or (8.4.59).

Incidentally, it has been sown[57] that for small deviations from plug flow ($N_d$ anticipated to be less than 0.01), for both open vessel and closed vessel conditions, the variance may be approximated as

$$\sigma^2(\theta^*) = 2N_d \quad \text{... (8.4.63)}$$

Though we could characterise the given reactor as a PFDR with a specific value of dispersion number ($N_d$), to compute the fractional conversion obtainable in the same (for a reaction of known kinetics) is not easy. For this, we have to solve Eq. (8.4.49), after substituting the kinetic equation for $(-r_A)$, using appropriate boundary conditions. For reactions with nonlinear kinetic equations, an analytical solution for Eq. (8.4.49) is practically impossible and we have to resort to numerical solution. If the reaction is psuedo-first order in $A$ such that $(-r_A) = kC_A$, then the following analytical solution[58] has been found to be satisfactory for both closed vessel as well as open vessel conditions:

$$(C_{Af}/C_{A0}) = (1 - x_{Af}) = \frac{4\beta \exp(1/2N_d)}{(1+\beta)^2 \exp(\beta/2N_d) - (1-\beta)^2 \exp(-\beta/2N_d)} \quad \text{... (8.4.64)}$$

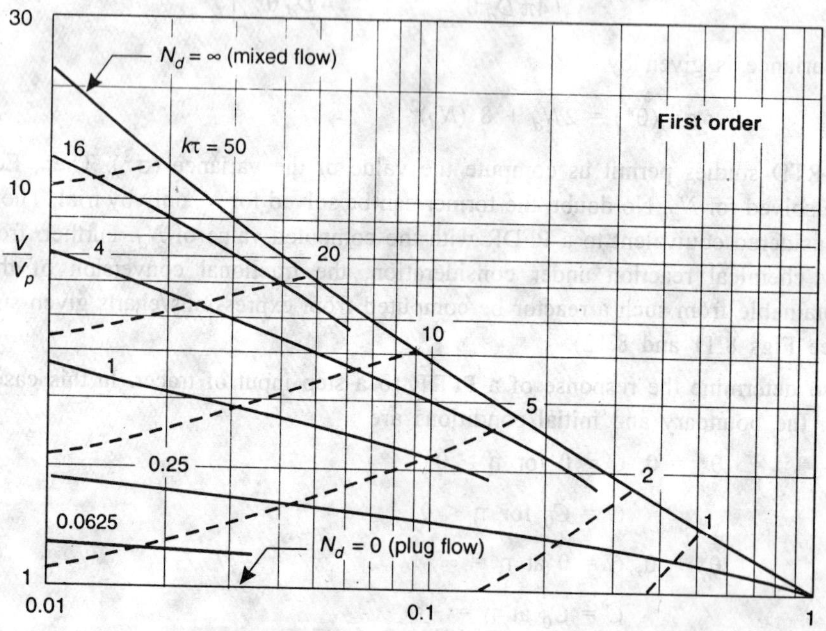


*Figure 8.11: Comparison between PFDR and plug flow reactor for first order reactions of the form $A \rightarrow$ Products, assuming negligible change in density of reaction mixture (from Levenspiel, O, Chemical Reaction Engineering, Wiley, New York, 1999, by permission)*

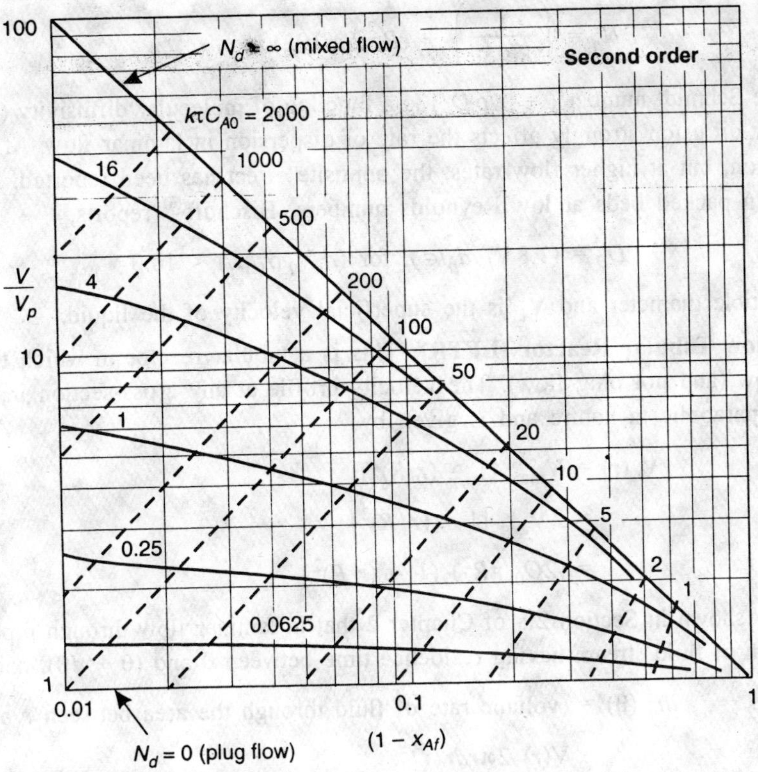


*Figure 8.12: Comparison between PFDR (closed vessel) and plug flow reactor for second order reactions of the form $2A \rightarrow$ Products or $A + B \rightarrow$ Products, with $C_{A0} = C_{B0}$ and assuming negligible change in density of reaction mixture (from Levenspiel, O, Chemical Reaction Engineering Wiley, New York, 1999, by permission)*

where $\beta = \sqrt{1 + 4k\tau N_d}$. The dimensionless parameter, $(k\tau)$, is often called the *Damkohler number for first order reactions*. Levenspiel and Bischoff[59] have prepared two charts (one for first order reactions and the other for second order reactions) using which the fractional conversion of $A$ (namely, $x_{Af}$) in a PFDR (of volume $V$) can be computed. For hand calculations, these charts are quite useful and these are reproduced in Figs 8.11 and 8.12. To note that in these charts, $V_p$ = volume of a PFR that would give the same conversion, $x_{Af}$. The authors recommend that for a reaction of $n$th order ($n$ being different from 1.0 and 2.0), the data from these charts be extrapolated or interpolated with allowable error. Also, to keep in mind that the chart for second order reactions (Fig. 8.12) has been prepared assuming that the reactor performs as closed vessel.

In our discussion on PFDR presented above, one of the most important parameters is the axial dispersion coefficient ($D_d$). Experimental correlations are available in literature[60] for the computation of $D_d$ for flow in pipes, in packed beds, in coiled tubes, etc. For example, for laminar flow of liquids in pipes with $L/D >> 10$, Levenspiel[61] reports the following correlation:

$$N_d = \left(\frac{1}{Re \cdot Sc}\right) + (Re \cdot Sc/192) \quad \text{... (8.4.65)}$$

To note that the Schmidt number ( $= \mu/\rho D_e$) is a function of molecular diffusivity ($D_e$) in the axial direction. Molecular diffusion strongly affects the rate of dispersion in laminar flow. At low flow rates, it promotes dispersion, but at higher flow rates, the opposite effect has been reported. For the flow of Newtonian liquids in packed beds at low Reynolds numbers, Bischoff[60] reports,

$$D_d = (1.8\,\mathbf{V}_L\, d_p/\in), \text{ for } (d_p \mathbf{V}_L \rho_f/\mu_f) < 10.0 \quad \text{... (8.4.66)}$$

where $d_p$ is the particle diameter and $\mathbf{V}_L$ is the superficial velocity of the liquid.

**(c) Laminar Flow Tubular Reactor (LFTR):** This is a tubular reactor in which the fluid stream executes laminar flow (and not plug flow). The velocity profile at any cross-section inside the reactor shall be, therefore, parabolic in nature and is given by

$$\mathbf{V}(r) = \mathbf{V}_{max}\,[1 - (r/R)^2] \quad \text{... (8.4.67)}$$

$$= 2\,\mathbf{V}_{avg}\,[1 - (r/R)^2] \quad \text{... (8.4.68)}$$

$$= (2Q_0/\pi R^2)\,(1 - (r/R)^2] \quad \text{... (8.4.69)}$$

We have already shown in Section 2.8 of Chapter 2 that in laminar flow through pipes/tubes, $\mathbf{V}_{max} = 2\,\mathbf{V}_{avg}$. The fraction of fluid stream having residence time between $\theta$ and $(\theta + d\theta)$, namely $dF(\theta)$, is

$$dF(\theta) = (\text{volume rate of fluid through the area between } r \text{ and } r + dr)/Q_0$$

$$= \mathbf{V}(r)\,2\pi r dr/Q_0 \quad \text{... (8.4.70)}$$

Also, $$\theta = L/\mathbf{V}(r) \quad \text{... (8.4.71)}$$

Substituting for $\mathbf{V}(r)$ from Eq. (8.4.69), we get

$$\theta = \frac{\pi R^2 L}{2Q_0\left[1-(r/R)^2\right]} \quad \text{... (8.4.72)}$$

Since $\tau = \mathbf{V}/Q_0 = (\pi R^2 L/Q_0)$,

$$\theta = \frac{\tau}{2\left[1-(r/R)^2\right]} \quad \text{... (8.4.73)}$$

or $$d\theta = (4\theta^2/R^2\tau)\, r\, dr \quad \text{... (8.4.74)}$$

Equation (8.4.70), therefore, becomes [after substituting for $r\,dr$ from above equation and for $\mathbf{V}(r)$ from Eq. (8.4.69)],

$$dF(\theta) = \left(\frac{\tau^2}{2\theta^3}\right) d\theta \quad \text{... (8.4.75)}$$

or $$\frac{dF(\theta)}{d\theta} = E(\theta) = (\tau^2/2\theta^3) \quad \text{... (8.4.76)}$$

This is the expression for exit age distribution for an LFTR. However, for a reactor of this kind, we cannot integrate the above expression from $\theta = 0$ to $\theta = \theta$ to obtain the expression for $F(\theta)$. This is because the minimum time duration for which a fluid element can reside in this reactor is not zero, but that when its velocity is equal to $\mathbf{V}_{max}$ or $(2\mathbf{V}_{avg})$. Thus, the minimum value of $\theta$ is, $(V/\pi R^2 \mathbf{V}_{max})$ or $(V/2Q_0)$ or $(\tau/2)$. Accordingly,

$$F(\theta) = (C/C_0)_{step} = \int_{\tau/2}^{\theta} \left(\tau^2/2\theta^3\right) d\theta \quad \text{... (8.4.77)}$$

$$= 1 - (\tau/2\theta)^2 \quad \text{... (8.4.78)}$$

To note that for a reactor like this,

$$E(\theta) = 0 \text{ for } \theta < \tau/2 \quad \text{... (8.4.79)}$$

$$= (\tau^2/2\theta^3) \text{ for } \theta \geq \tau/2 \quad \text{... (8.4.80)}$$

We can also estimate the mean residence time using Eq. (8.4.4), but after changing the lower limit of the integral to $(\tau/2)$:

$$\bar{\theta} = \int_{\tau/2}^{\infty} \theta\, E(\theta)\, d\theta \quad \text{... (8.4.81)}$$

$$= \int_{\tau/2}^{\infty} \left(\tau^2/2\theta^2\right) d\theta \quad \text{... (8.4.82)}$$

$$= \tau \quad \text{... (8.4.83)}$$

Thus, the mean residence time $(\bar{\theta})$ is equal to the space time $(\tau)$ for a laminar flow tubular reactor (LFTR) as well.

The values of $E(\theta)$ and $F(\theta)$ obtained from tracer experiment data and those computed from Eqs (8.4.76) and (8.4.78) may be compared with each other by representing them graphically [as plots of $E(\theta)$ or $F(\theta)$ versus $\theta/\tau$] and if they are found to be agreeing closely, then the given reactor may be modelled as an LFTR. It may also be checked that the value of $\bar{\theta}$ computed from tracer experiment data agrees closely with $\tau$ $(= V/Q_0)$.

The fractional conversion $(x_A)$ that can be obtained in such a reactor can be computed from Eq. (8.4.31), but after changing the lower limit of the integral to $(\tau/2)$. Thus

$$x_{Af} = \int_{\tau/2}^{\infty} \left[1 - (C_A/C_{A0})\right] E(\theta)\, d\theta \quad \text{... (8.4.84)}$$

$$= \int_{\tau/2}^{\infty} \left[1 - (C_A/C_{A0})\right] (\tau^2/2\theta^3)\, d\theta \quad \text{... (8.4.85)}$$

By expressing $(C_A/C_{A0})$ as a function of $\theta$ based on the kinetic equation, the above integral may be evaluated numerically or analytically. For example, we know if the reaction is psuedo-first order in $A$, $x_A = 1 - \exp(-k\theta)$ and therefore,

$$x_{Af} = (\tau^2/2) \int_{\tau/2}^{\infty} \left[\frac{1 - \exp(-k\theta)}{\theta^3}\right] d\theta \quad \text{... (8.4.86)}$$

The above integral can be now evaluated by numerical or graphical integration. The fractional conversion so computed may be compared with the results obtained from pilot scale runs on the given reactor so as to ascertain the validity of modelling it as an LFTR.

**(*d*) Tanks-in-Series Model:** This model assumes that the given industrial reactor is equivalent to a number of ideal CSTRs in series. The rationale behind such an approach is not difficult to comprehend. We know that the performance of a series combination of $n$ CSTRs (of total volume $V$) is better than a single ideal CSTR (of volume $V$), but inferior to that of a PFR (of volume $V$). Accordingly, an industrial reactor, whose performance lies in between that of an ideal CSTR and a PFR, may be modelled as equivalent to n ideal CSTRs in series.

The task before us is then to find the value of $n$. This is also not difficult. We know the expression for $E(\theta)$ for an ideal CSTR is that given in Eq. (8.4.12). Then, for $n$ CSTRs in series, it can be easily deduced that

$$E(\theta) = \frac{\exp(-\theta/\tau)}{\tau(n-1)!}(\theta/\tau)^{n-1} \quad \text{... (8.4.87)}$$

where $(n-1)! = \angle(n-1)$ = factorial $(n-1)$. $\tau$ is the space time for *each* reactor (to note that each CSTR is of volume $V/n$) and it has been assumed equal to the mean residence time in each reactor (assuming a constant density system). The variance ($\sigma^2$) is then given by,

$$\sigma^2 = n\tau^2 \quad \text{... (8.4.88)}$$

It shall be more convenient to write the above equation in terms of the space time ($\tau_n$) for the entire system. Since we know that $\tau_n = n\tau = (V/Q_0)$, where $V$ is the total volume of all reactors in series,

$$\sigma^2 = \tau_n^2/n \quad \text{... (8.4.88)}$$

Considering a constant density system (for a tracer experiment, this is a reasonable assumption since addition of tracer does not significantly alter the density of the fluid in most cases), $\tau_n = \bar{\theta}_t$ $(= n\bar{\theta})$ = mean residence time in the entire system and hence,

$$\sigma^2 = (\bar{\theta}_t)^2/n \quad \text{... (8.4.89)}$$

Since the given reactor is assumed to be equivalent to $n$ ideal CSTRs in series, the mean residence time ($\bar{\theta}_t$) in such a reactor and the value of ($\sigma^2$) can be computed from tracer experiment data (RTD data). Then, the value of $n$ can be easily deduced from the above equation.

The fractional conversion ($x_A$) attained in a reactor of this kind is same as that obtained from $n$ ideal CSTRs in series. This therefore, can be computed as discussed in Section 8.3, such as from Eq. (8.3.44).

The expression for $F(\theta)$ for such a reactor shall be

$$F(\theta) = [1 - \exp(-\theta/\tau)]\left[1 + (\theta/\tau) + \frac{(\theta/\tau)^2}{2!} + ..... + \frac{(\theta/\tau)^{n-1}}{(n-1)!}\right] \quad \text{... (8.4.90)}$$

Once again, do note that $\tau$ represents the space time (assumed equal to mean residence time) of each of the reactors in series [that is, $\tau = (V/n)/Q_0$].

**(e) Recycle Reactor Model:** This model assumes that the given industrial reactor is equivalent to a recycle reactor (that is, a PFR with partial effluent recycle). We have already shown in Section 8.3 that the performance of such a recycle reactor is in between that of an ideal CSTR and a PFR. It is therefore apparent that an industrial reactor could be modeled as equivalent to a recycle reactor with a specific recycle ratio, $R$ (see Fig. 8.13).

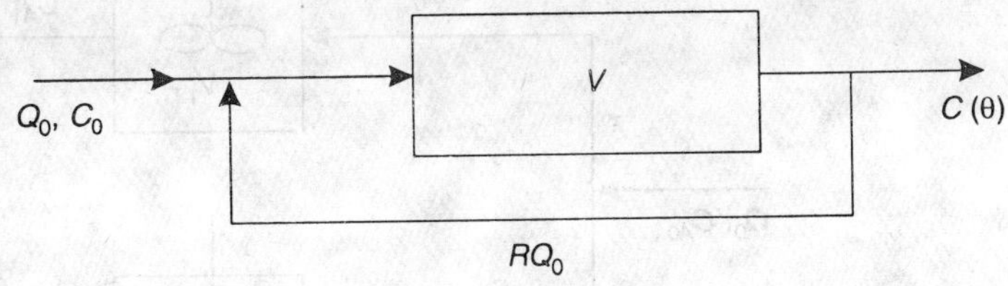


*Figure 8.13: Plug flow reactor with recycle*

The value of $R$ can however be determined only by trial. For instance, the response of such a recycle reactor to a step input of tracer shall be

$$F\ (\theta) = 0.0 \text{ if } \theta \leq \tau = 1 - (R/R + 1) \text{ if } \tau < \theta \leq 2\tau$$

$$= 1 - (R/R + 1)^2 \text{ if } 2\tau < \theta \leq 3\tau$$

$$= 1 - (R/R + 1)^3 \text{ if } 3\tau < \theta \leq 4\tau \quad \text{... (8.4.91)}$$

where the space time ($\tau$) is defined as

$$\tau = \left[\frac{V}{(R+1)\,Q_0}\right] \quad \text{... (8.4.92)}$$

Thus, based on an assumed value of $R$, we have to check whether the values of $F$ ($\theta$) computed from the above equations and those calculated from RTD data (tracer experiment data) tally with each other. If a satisfactory agreement is not observed even after trying a wide range of values of $R$, then it may be concluded that the given reactor cannot be effectively modeled as a recycle reactor.

The fractional conversion ($x_A$) attained in a reactor of this kind shall be same as that obtained in a recycle reactor (with recycle ratio $R$) and this can be therefore computed as discussed in Section 8.3 (see Example 8.21).

**(f) Reactor Combinations:** When neither of the above models could describe the performance of the given reactor satisfactorily, then we have to go for trying different reactor combinations. In other words, the given reactor may be assumed equivalent to a combination of a number of ideal CSTRs and PFRs with allowances for dead zones and recycle streams. Obviously, the number of possible combinations is practically legion. Two of such possible cases are illustrated in Figs 8.14 and 8.15.

Consider the scheme sketched in Fig. 8.14. The given reactor has been assumed to be equivalent to two unequal sized ideal CSTRs operating in parallel. It is further assumed that $V_2$ is much smaller than $V_1$. Readers can easily verify that the response of a reactor of this kind to a pulse input of tracer shall be

$$E\ (\theta) = R\ (1/\tau_1)\ \exp\ (-\theta/\tau_1) + (1 - R)\ (1/\tau_2)\ \exp\ (-\theta/\tau_2) \quad \text{... (8.4.93)}$$

where

$$\tau_1 = V_1/(RQ_0) \quad \text{... (8.4.94)}$$

$$\tau_2 = \frac{V_2}{(1-R)\,Q_0} \quad \text{... (8.4.95)}$$

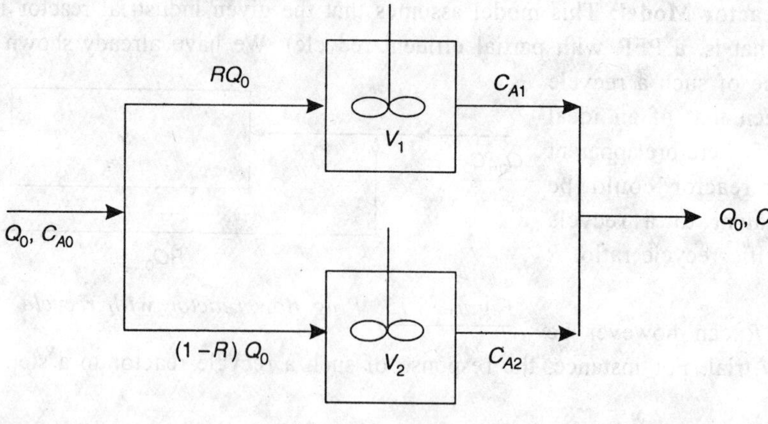


*Figure 8.14: Reactor combination (two CSTRs in parallel)*

A plot of ln $E$ (θ) versus θ will be as shown in Fig. 8.14 (a). As can be seen, at low values of θ, the plot is essentially linear (straight line 1) of slope = $(-1/\tau_1)$ and $y$-intercept = ln $(R/\tau_1)$. Similarly, at large values of θ also, the plot is more or less linear (straight line 2) with slope = $(-1/\tau_2)$. Thus, the model parameters $\tau_1$, $\tau_2$ and $R$ can be determined from the slope and $y$-intercept of these two straight line portions of the curve.

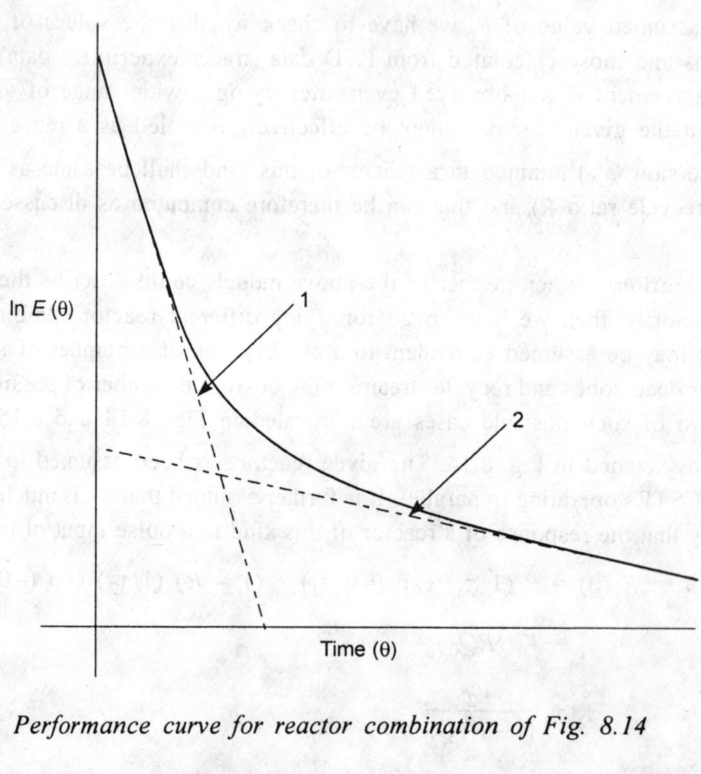


*Figure 8.14 (a): Performance curve for reactor combination of Fig. 8.14*

It is not difficult to estimate the fractional conversion ($x_A$) obtainable in a reactor of this kind as it has been assumed to be composed of two ideal CSTRs, the larger one of volume $V_1$ and the smaller one of volume $V_2$. For example, if the reaction follows first order kinetics, then

$$C_{A1} = \frac{C_{A0}}{(1+k\tau_1)} \qquad \text{... (8.4.96)}$$

Also,

$$C_{A2} = \frac{C_{A0}}{(1+k\tau_2)} \qquad \text{... (8.4.97)}$$

Now, a simple material balance at the exit gives

$$Q_0 C_{Af} = (RQ_0)\ C_{A1} + (1 - R)\ Q_0 C_{A2}$$

or

$$C_{Af} = RC_{A1} + (1 - R)\ C_{A2} \qquad \text{... (8.4.98)}$$

Substituting the expressions for $C_{A1}$ and $C_{A2}$ from Eqs (8.4.96) and (8.4.97), we get

$$(C_{Af}/C_{A0}) = \frac{R}{(1+k\tau_1)} + \frac{(1-R)}{(1+k\tau_2)} \qquad \text{... (8.4.99)}$$

Now, let us consider the scheme sketched in Fig. 8.15. The reactor under consideration is assumed to be composed of a PFR of volume $V_p$ followed by an ideal CSTR of volume $V_m$, with a bypass stream, ($RQ_0$). The reactor is also assumed to have a dead zone of volume $V_d$.

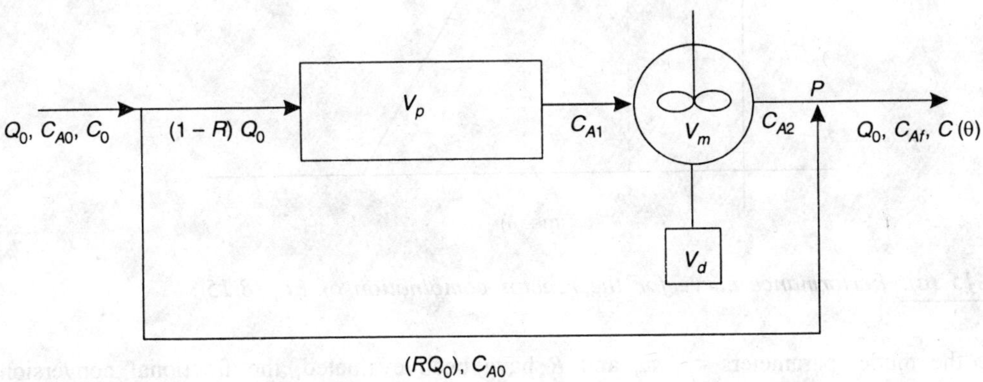


*Figure 8.15: Reactor combination (PFR, CSTR) with dead zone*

The response of such a reactor to a step input of tracer shall be

$$F(\theta) = R \text{ for } \theta \le \tau_p \qquad \text{... (8.4.100)}$$

$$= 1 - (1 - R) \exp\left[(\tau_p/\tau_m) - (\theta/\tau_m)\right] \text{ for } \theta > \tau_p \qquad \text{... (8.4.101)}$$

where

$$\tau_p = \frac{V_p}{(1-R)\,Q_0} \qquad \text{... (8.4.102)}$$

$$\tau_m = \frac{V_m}{(1-R)\,Q_0} \qquad \text{... (8.4.103)}$$

If the given reactor does fit to this model, then the plot of ln $[1 - F(\theta)]$ versus time $(\theta)$ shall be as shown in Fig. 8.15 (a). It can be seen that the value of ln $(1 - R)$ can be read directly from the plot. The equation to the straight line 1 (for $\theta > \tau_p$) is

$$\ln[1 - F(\theta)] = -(1/\tau_m)\,\theta + (\tau_p/\tau_m) + \ln(1 - R) \qquad \text{... (8.4.104)}$$

Thus, the slope of this straight line (straight line 1) is $(-1/\tau_m)$ and its $y$-intercept = ln $(1 - R)$ + $(\tau_p/\tau_m)$. Accordingly, we can obtain the values of $\tau_p$ and $\tau_m$ from its slope and $y$-intercept.

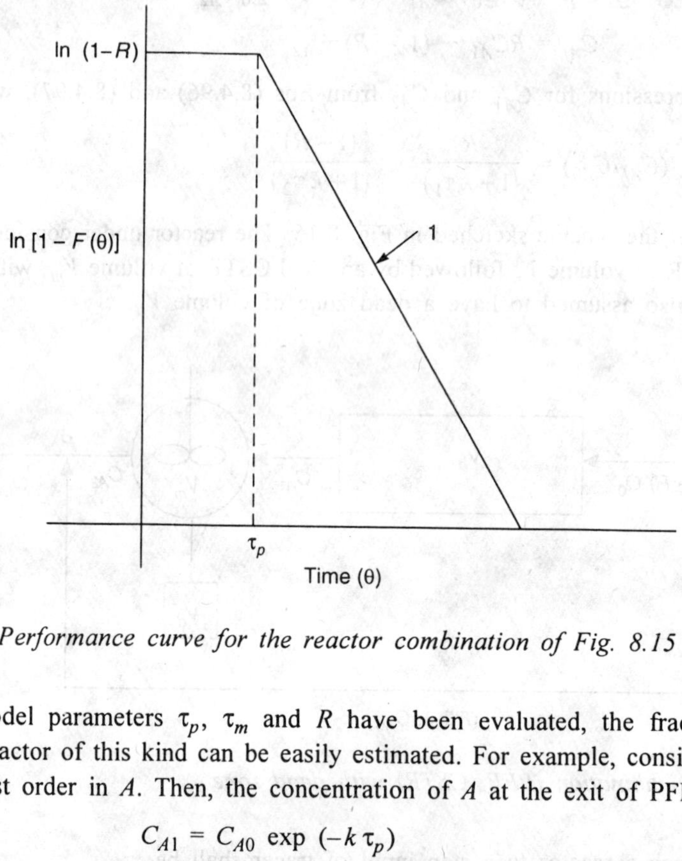


*Figure 8.15 (a): Performance curve for the reactor combination of Fig. 8.15*

Once the model parameters $\tau_p$, $\tau_m$ and $R$ have been evaluated, the fractional conversion $(x_A)$ obtainable in a reactor of this kind can be easily estimated. For example, consider a chemical reaction that is psuedo-first order in $A$. Then, the concentration of $A$ at the exit of PFR is

$$C_{A1} = C_{A0}\exp(-k\,\tau_p) \qquad \text{... (8.4.105)}$$

The concentration of $A$ at the exit of CSTR is

$$C_{A2} = C_{A1}/(1 + k\,\tau_m) = \frac{C_{A0}\exp(-k\,\tau_p)}{(1 + k\,\tau_m)} \qquad \text{... (8.4.106)}$$

Now, a material balance at the exit of the reactor (at the point $P$) yields

$$Q_0 C_{Af} = (1 - R)\,Q_0 C_{A2} + R Q_0 C_{A0} \qquad \text{... (8.4.107)}$$

or $$(C_{Af}/C_{A0}) = (1 - R)\,(C_{A2}/C_{A0}) + R \qquad \text{... (8.4.108)}$$

Substituting for $C_{A2}$ from Eq. (8.4.106) and since $x_{Af} = (1 - C_{Af}/C_{A0})$,

$$x_{Af} = (1 - R)\left[1 - \frac{\exp(-k\tau_p)}{(1 + k\tau_m)}\right] \qquad \text{... (8.4.109)}$$

As stated earlier, the number of possible reactor combinations is practically numerous. By a series of trials, we have to spot that combination which represents the given reactor most satisfactorily.

**Example 8.33:** An industrial reactor of volume 18.5 m$^3$ is being used for conducting first order decomposition of an organic complex ($A$) that is admitted at 222.0 m$^3$/hr. Specific reaction rate = 0.15 min$^{-1}$. When the reaction was conducted in a pilot plant reactor with the same space time, the fractional conversion of $A$ reported was 50 percent. The pilot plant reactor was tested using a pulse of fluoride tracer and the results obtained are given below:

| *Time (min)* | *Tracer concentration in exit stream, mg/L* | *Time (min)* | *Tracer concentration in exit stream, mg/L* |
|---|---|---|---|
| 0.0 | 0.00 | 7.0 | 0.20 |
| 1.0 | 0.05 | 8.0 | 0.10 |
| 2.0 | 0.70 | 9.0 | 0.05 |
| 3.0 | 0.85 | 10.0 | 0.05 |
| 4.0 | 0.75 | 11.0 | 0.025 |
| 5.0 | 0.60 | 12.0 | 0.00 |
| 6.0 | 0.35 | | |

Determine whether the reactor can be modelled as

(*a*) a PFDR

(*b*) a number of CSTRs in series.

Assume negligible change in the density of the reaction mixture.

**Solution:** To determine whether a plug flow dispersion reactor (PFDR) model or a tanks-in-series model fits to the given industrial reactor, we have to first compute the exit age distribution function, $E(\theta)$, at different values of $\theta$ and then the standard deviation ($\sigma$) or the variance ($\sigma^2$), from the tracer experiment data reported. The spreadsheet for the same is given below in Table 8.33.1.

To note that $\bar{C}(\theta)$ is computed as the arithmetic average of two consecutive values of tracer concentration, $C(\theta)$, in the exit stream. For example, at $\theta$ = 3 minutes,

$$\bar{C}(\theta) = (0.85 + 0.70)/2 = 0.775 \text{ mg/L}$$

The exit age distribution function, $E(\theta)$, is to be computed from Eq. (8.4.1). However, if the time interval $\Delta\theta$ chosen is small, then we may approximate this expression as

$$E(\theta) = \frac{\bar{C}(\theta)}{\sum \bar{C}(\theta)\,\Delta\theta} \qquad \text{... (i)}$$

**Table 8.33.1**

| (1) $\theta$, min | (2) C, mg/L | (3) $\bar{C}$, mg/L | (4) E ($\theta$), min$^{-1}$ | (5) $\theta$ E ($\theta$) $\Delta\theta$, min | (6) $\theta^2$ E ($\theta$) $\Delta\theta$, min$^2$ |
|---|---|---|---|---|---|
| 0.0 | 0.00 | – | 0.00 | 0.00 | 0.00 |
| 1.0 | 0.05 | 0.025 | 0.0067 | 0.0067 | 0.0067 |
| 2.0 | 0.70 | 0.375 | 0.10067 | 0.20134 | 0.40268 |
| 3.0 | 0.85 | 0.775 | 0.2080 | 0.6240 | 1.8720 |
| 4.0 | 0.75 | 0.80 | 0.2147 | 0.8588 | 3.4352 |
| 5.0 | 0.60 | 0.675 | 0.1812 | 0.906 | 4.5300 |
| 6.0 | 0.35 | 0.475 | 0.1275 | 0.765 | 4.5900 |
| 7.0 | 0.20 | 0.275 | 0.0738 | 0.5166 | 3.6162 |
| 8.0 | 0.10 | 0.150 | 0.0403 | 0.3224 | 2.5792 |
| 9.0 | 0.050 | 0.075 | 0.0201 | 0.1809 | 1.6281 |
| 10.0 | 0.050 | 0.050 | 0.0134 | 0.1340 | 1.3400 |
| 11.0 | 0.025 | 0.0375 | 0.01 | 0.110 | 1.2100 |
| 12.0 | 0.0 | 0.0125 | 0.0033 | 0.0396 | 0.4752 |
| | | 3.725 | 0.99937 | 4.6653 | 25.6853 |

Since the time interval chosen is constant and equal to 1.0 minute and also, since $\Sigma\ \bar{C}\ (\theta) = 3.725$ mg/L [see column (3) of above table],

$$E\ (\theta) = \frac{\bar{C}\ (\theta)}{(3.725)} \qquad \text{... (ii)}$$

Column (4) of above table has been prepared in this way. Also, from Eq. (8.4.4),

$$\bar{\theta} = \Sigma\theta E\ (\theta)\ \Delta\theta = 4.6653 \text{ min [see column (5) of above table]}$$

To note that the space time ($\tau$) of the reactor is,

$$\tau = (18.5)/(222.0/60) = 5.0 \text{ min}$$

It can be seen that the mean residence time ($\bar{\theta}$) computed is less than $\tau$, indicating that the given reactor does deviate from ideality. We can now compute the variance from Eq. (8.4.21) as

$$\sigma^2 = \Sigma\theta^2 E\ (\theta)\ \Delta\theta - (\bar{\theta})^2 = 25.6853 - (4.6653)^2 = 3.92 \text{ min}^2$$

(*a*) Let us first verify whether the reactor can be modelled as a PFDR. Thus, from Eq. (8.4.22),

$$\sigma^2\ (\theta^*) = (3.92)/(4.6653)^2 = 0.1801$$

Assuming a closed vessel, from Eq. (8.4.56),

$$0.1801 = 2N_d - 2\ (N_d)^2\ [1 - \exp\ (-\ 1/N_d)]$$

Solving by trial, $N_d = 0.1$

Thus, we could assume the reactor to be equivalent to a PFDR with dispersion number = 0.1. The fractional conversion of *A* attainable in such a reactor can be now estimated from Eq. (8.4.64). Thus,

$$\beta = [1 + 4\ (0.15)\ (5.0)\ (0.1)]^{1/2} = 1.14$$

Substituting the values of $\beta$ and $N_d$ in Eq. (8.4.64), we get

$$x_{Af} = 0.506 = 50.6 \text{ percent}$$

This is very close to the reported value of 50 percent.

Let us now assume that the reactor functions as an open vessel. Then from Eq. (8.4.59),

$$0.1801 = 2N_d + 8\ (N_d)^2$$

or

$$N_d = 0.07$$

Now,

$$\beta = [1 + 4\ (0.15)\ (5.0)\ (0.07)]^{1/2} = 1.1$$

Substituting these values of $\beta$ and $N_d$ in Eq. (8.4.64), we get

$$x_{Af} = 0.512 = 51.2 \text{ percent}$$

This is also close to the reported value.

(*b*) Let us now consider the tanks-in-series model. Accordingly, from Eq. (8.4.89),

$$n = (\bar{\theta}_t)^2/\sigma^2 = (4.6653)^2/(3.92) = 5.55$$

As a conservative estimate, we may assume that the given industrial reactor is equivalent to five ideal CSTRs in series. The fractional conversion of $A$ attainable in such a system is [see Eq. (8.3.44)],

$$(1 - x_{Af}) = \frac{1}{\left(1 + k\bar{\theta}\right)^n}$$

Since $k = 0.15\ \text{min}^{-1}$, $\bar{\theta} = (5.0/5) = 1.0$ min,

$$x_{Af} = 0.4972 = 49.72 \text{ percent}$$

This is once again close to the reported value of 50 percent. Thus, the given industrial reactor is best modelled as equivalent to five ideal CSTRs in series, each of volume 3.7 $m^3$. It may also be modelled, with reasonable accuracy, as a PFDR (closed vessel) with dispersion number = 0.1.

**Example 8.34:** An industrial reactor of volume 7.5 $m^3$ is used to conduct hydrolysis (in basis medium) of an organic compound ($A$). The reaction is first order in $A$ with rate constant = 0.5 $\text{min}^{-1}$. The reactor receives $A$ at 36.0 $m^3$/hr. Pilot plant studies showed that 85 percent conversion of $A$ is possible under these conditions. Tracer experiments with a step input of tracer yielded the following results:

| *Time (min)* | *(C/C0)* | *Time (min)* | *(C/C0)* |
|---|---|---|---|
| 0.0 | 0.0 | 12.0 | 0.515 |
| 1.0 | 0.15 | 15.0 | 0.774 |
| 3.0 | 0.15 | 20.0 | 0.937 |
| 5.0 | 0.15 | 25.0 | 0.982 |
| 7.0 | 0.15 | 30.0 | 0.995 |
| 9.0 | 0.15 | 35.0 | 1.00 |
| 10.0 | 0.192 | | |

Determine whether the reactor can be modelled as equivalent to a reactor combination of the type shown in Fig. 8.15. Assume that the density of the reaction mixture remains essentially constant.

**Solution:** In case the reactor performance fits to the scheme shown in Fig. 8.15, then $F(\theta)$ must be equal to $R$ for all values of $\theta$ less than $\tau_p$. From the given table, it can be seen that at $\theta \le 9.0$ minutes, $F(\theta)$ = constant = 0.15. We may, therefore, assume (as the first step) $R = 0.15$. At higher values of $\theta$, Eq. (8.4.104) is required to be satisfied. After substituting the value of $R$, this equation may be rewritten as

$$-\ln[1 - F(\theta)] = (1/\tau_m)\,\theta - (\tau_p/\tau_m) + 0.1625 \quad \text{... (i)}$$

Do note the minus sign on the left hand side of above equation. The values of $-\ln[1 - F(\theta)]$ are now computed at different values of $\theta$ from the data provided and are listed in Table 8.34.1 given below.

**Table 8.34.1**

| $\theta$ (min) | $F(\theta)$ | $-\ln[1 - F(\theta)]$ |
|---|---|---|
| 9.0 | 0.150 | 0.1625 |
| 10.0 | 0.192 | 0.2130 |
| 12.0 | 0.515 | 0.7236 |
| 15.0 | 0.774 | 1.4872 |
| 20.0 | 0.937 | 2.7646 |
| 25.0 | 0.982 | 4.0174 |
| 30.0 | 0.995 | 5.2983 |

The plot of $-\ln[1 - F(\theta)]$ versus time ($\theta$) is shown in Fig. 8.34.1. For $\theta > 10$ minutes, the plot is linear with slope = 0.25 min$^{-1}$ and $y$-intercept = $-2.34$. Therefore, from Eq. (i),

$$1/\tau_m = 0.25$$

or

$$\tau_m = 4.0 \text{ min}$$

And, $$0.1625 - (\tau_p/\tau_m) = -2.34$$

or

$$\tau_p = 10.0 \text{ min}$$

We may, therefore, assume the given reactor to be equivalent to a reactor combination of the form shown in Fig. 8.15 with $\tau_p = 10.0$ min, $\tau_m = 4.0$ min and $R = 0.15$. Now, from Eqs (8.4.102) and (8.4.103),

$$V_p = (10)\,(1 - 0.15)\,(36/60) = 5.1 \text{ m}^3$$

$$V_m = (4.0)\,(1 - 0.15)\,(36/60) = 2.04 \text{ m}^3$$

Therefore, volume of dead zone ($V_d$) = 7.5 − (5.1 + 2.04) = 0.36 m$^3$

The fractional conversion of $A$ attainable in such a reactor can be computed from Eq. (8.4.109). Thus, after substituting $k = 0.5$ min$^{-1}$ and the values of $\tau_p$, $\tau_m$ and $R$ in Eq. (8.4.109), we get

$$x_{Af} = 0.848 = 84.8 \text{ percent}$$

This is very close to the reported value of 85 percent. Thus, we may successfully model the given industrial reactor as equivalent to the system sketched in Fig. 8.15.

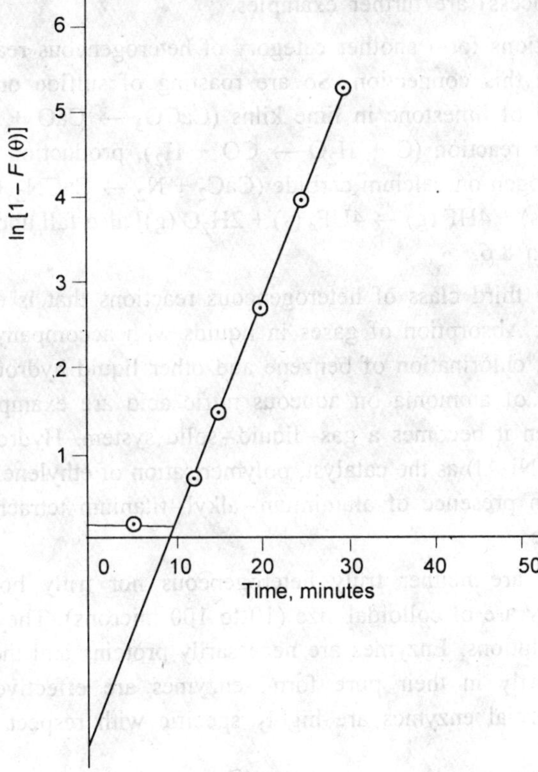


*Figure 8.34.1*

## 8.5 HETEROGENEOUS REACTIONS

Heterogeneous reactions are those in which two or more phases are present. The reaction by itself may be taking place only in one of the phases or at the interface. The reactants and the products may be distributed among the phases or may all be contained in a single phase. The system is still called heterogeneous so long as two or more phases are involved in conducting the reaction.

Solid catalysed fluid-fluid reactions are possibly the most popular types of heterogeneous reactions (see Table 8.4). Most of the reactions conducted in petroleum refineries and petrochemical industries are catalysed by solids and since the reaction temperature is normally very high, the reactants and the products are commonly in the gas (or vapour) phase. Vapour phase cracking of petroleum hydrocarbons in presence of silica–alumina catalyst (for the production of high octane gasoline), reforming of petroleum fractions in presence of platinum (supported on alumina) as the catalyst (also called *platforming*), desulfurisation of petroleum fractions by contacting with hydrogen gas at high temperature and pressure and in presence of cobalt molybdate as the catalyst (the process is called *hydrofining*), isomerisation processes (conversion of *n*-paraffins to *i*-paraffins) where $AlCl_3$ (with HCl as promoter) is used as the catalyst are examples. All these are gas–solid catalytic reactions. Oxidation of $SO_2$ to $SO_3$ in presence of vanadium pentoxide as the catalyst (contact process), synthesis of ammonia from nitrogen and hydrogen in presence of iron oxide as the catalyst (Haber process) are further examples.

Noncatalytic gas–solid reactions form another category of heterogeneous reactions. Combustion of coal is an excellent example in this connection. So are roasting of sulfide ores ($4FeS_2 + 11O_2 \rightarrow 2Fe_2O_3 + 8SO_2$) and calcination of limestone in lime kilns ($CaCO_3 \rightarrow CaO + CO_2$). Manufacture of blue water gas by steam–carbon reaction ($C + H_2O \rightarrow CO + H_2$), production of calcium cyanamide by the action of atmospheric nitrogen on calcium carbide ($CaC_2 + N_2 \rightarrow CaCN_2 + C$), hydrofluorination of uranium dioxide pellets [$UO_2$ (s) + 4HF (g) $\rightarrow$ $4UF_4$ (s) + $2H_2O$ (g)] also fall under this category. These are discussed in detail in Section 8.6.

Gas–liquid reactions form a third class of heterogeneous reactions that is encountered frequently in chemical engineering practice. Absorption of gases in liquids with accompanying chemical reaction (already discussed in Chapter 4), chlorination of benzene and other liquid hydrocarbons, production of ammonium nitrate by the action of ammonia on aqueous nitric acid are examples. If the reaction is catalysed by a solid catalyst, then it becomes a gas–liquid–solid system. Hydrogenation of vegetable oils in presence of Raney-nickel (Ni-Al) as the catalyst, polymerisation of ethylene in a solvent or diluant (usually a liquid hydrocarbon) in presence of aluminium–alkyl–titanium tetrachloride as the catalyst (Ziegler process) are of this type.

*Enzyme catalysed reactions* are neither truly heterogeneous nor truly homogeneous reactions. Enzymes are biocatalysts and they are of colloidal size (10 to 100 microns). They catalyse biochemical reactions normally in aqueous solutions. Enzymes are necessarily proteins and they may be considered as surface catalysts. Though costly in their pure form, enzymes are effective in amazingly small concentrations. However, commercial enzymes are highly specific with respect to the reactions they catalyse.

A large number of enzymes are being used in process industries in their pure form. A few of them are listed below:

| *Enzyme* | *Process application* |
|---|---|
| Papain | Protein hydrolysis |
| Rennin and Pepsin | Cheese manufacture |
| Invertase | Conversion of sugar (sucrose) to dextrose and fructose |
| Amyl glucosidase | Conversion of starch to glucose |

Soluble enzymes in aqueous phase can be used only once since they cannot be economically separated from the product stream for reuse. Since the cost of the enzyme is detrimental to the overall economy of the biochemical process, attempts must have to be made to save enzyme for multiple use. This is best accomplished by attaching the enzyme to an inert solid carrier. This technique is known as *enzyme immobilisation*. Enzymes can be immobilised by adsorption on a solid surface or by the encapsulation techniques. Immobilised enzymes have the distinct advantages such as multiple usage, increased stability, noncontamination of the product by the enzyme and adaptability to continuous processing.

Liquid–solid reactions are quite common in inorganic process industries. Examples are the leaching of uranium ores with sulfuric acid and the manufacture of phosphoric acid by the action of sulfuric acid on phosphate rock. There are also liquid–solid reactions in which the solid acts as the catalyst. Alkylation of petroleum hydrocarbons with solid $AlCl_3$ as the catalyst is an example.

Some of the most popular industrial heterogeneous reactions are listed in Tables 8.4 and 8.4 A.

## Table 8.4
## Catalytic Heterogeneous Reactions

| *Process* | *Principal reaction* | *Catalyst* | *Catalyst poisons* |
|---|---|---|---|
| 1. Ammonia synthesis | $N_2(g) + 3H_2(g) \rightarrow 2NH_3(g)$ | Iron oxide promoted by alumina and $K_2O$ | Compounds of S, P, As and CO, $CO_2$, moisture |
| 2. Synthesis of ethanol | $C_2H_4(g) + H_2O(g) \rightarrow C_2H_5OH(g)$ | Solid phosphoric acid on kieselguhr | S, $O_2$, $NH_3$, organic bases |
| 3. Methanol synthesis | $CO(g) + 2H_2(g) \rightarrow CH_3OH(g)$ | $ZnO + Cr_2O_3$ | Sulphur compounds, Fe, Ni |
| 4. Fischer-Tropsch synthesis | $CO(g) + H_2(g) \rightarrow$ Liquid hydrocarbons | $Fe + Fe_2C + Fe_3O_4$ | Sulphur compounds |
| 5. Synthesis of nitric acid | $4NH_3(g) + 5O_2(g) \rightarrow 4NO(g) + 6H_2O(g)$<br>$2NO(g) + O_2(g) \rightarrow 2NO_2(g)$<br>$3NO_2(g) + H_2O(l) \rightarrow 2HNO_3(aq) + NO(g)$ | Pt on Rh | Compounds of As and $Cl_2$ |
| 6. Manufacture of polyethylene | $nC_2H_4(g) \xrightarrow[\text{Suspended in a diluant}]{\text{Catalyst}} (\text{—}CH_2\text{—}CH_2\text{—}CH_2\text{—})_n$ solid | Al-alkyl-Ti tetrachloride | Moisture, $O_2$, alcohols, $SO_2$, CO, $CO_2$, COS |
| 7. Manufacture of styrene | $C_6H_5C_2H_5(g) \rightarrow C_6H_5CH = CH_2(g) + H_2(g)$ | $Fe_2O_3 + K_2O + Cr_2O_3$ | S compounds, halides, $O_2$, P compounds |
| 8. Oxidation of sulfur dioxide in sulfuric acid manufacture | $2SO_2(g) + O_2(g) \rightarrow 2SO_3(g)$ | $V_2O_5 + K_2O$ on kieselguhr | Arsenic, halides |
| 9. Cracking of petrolium hydrocarbons | High molecular weight hydrocarbons $\rightarrow$ low molecular weight hydrocarbons | Silica-alumina (zeolites) | Organic bases, metals (Ni, V, Cu), carbon deposition |
| 10. Desulfurisation of petroleum fractions (hydrofining) | Interaction of $H_2(g)$ with thiophenes, mercaptanes to form $H_2S$ and S-free hydrocarbons | Cobalt molybdate | Compounds of Na, As, Pb and $H_2S$, CO, $CO_2$ |
| 11. Hydrogenation of vegetable oils | $(C_{17}H_{31}COO)_3C_3H_5(l) + 3H_2(g) \rightarrow (C_{17}H_{33}COO)_3C_3H_5(l)$ | Raney nickel (Ni-Al) | Compounds of sulphur and chlorine |

## Table 8.4A
## Noncatalytic Heterogeneous Reactions

| Process | Principal reaction |
|---|---|
| 1. Combustion of coal | $C\ (s) + O_2\ (g) \rightarrow CO_2\ (g)$ |
| 2. Manufacture of blue water gas | $C\ (s) + H_2O\ (g) \rightarrow CO\ (g) + H_2\ (g)$ |
| 3. Manufacture of calcium cyanamide | $CaC_2\ (s) + N_2\ (g) \rightarrow CaCN_2\ (s) + C\ (s)$ |
| 4. Roasting of sulfide ores | $4FeS_2\ (s) + 11O_2\ (g) \rightarrow 2Fe_2O_3\ (s) + 8SO_2\ (g)$ |
| | $2ZnS\ (s) + 3O_2\ (g) \rightarrow 2ZnO\ (s) + 2SO_2\ (g)$ |
| 5. Calcination of limestone | $CaCO_3\ (s) \rightarrow CaO\ (s) + CO_2\ (g)$ |
| 6. Hydrofluorination of uranium dioxide | $UO_2\ (s) + 4HF\ (g) \rightarrow UF_4\ (s) + 2H_2O\ (g)$ |
| 7. Manufacture of pig iron | $2Fe_2O_3\ (s) + 6C\ (s) + O_2\ (g) \rightarrow 4Fe\ (s) + 2CO_2\ (g) + 4CO\ (g)$ |
| 8. Absorption of carbon dioxide in alkali | $CO_2\ (g) + 2NaOH\ (aq) \rightarrow Na_2CO_3\ (aq) + H_2O\ (l)$ |
| 9. Manufacture of ammonium nitrate | $NH_3\ (g) + HNO_3\ (aq) \rightarrow NH_4NO_3\ (aq)$ |
| 10. Chlorination of benzene | $C_6H_6\ (l) + Cl_2\ (g) \rightarrow C_6H_5Cl\ (l) + HCl\ (g)$ |
| 11. Ion exchange process | $CaCO_3\ (aq) + 2NaR\ (s) \rightarrow Na_2CO_3\ (aq) + CaR_2\ (s)$ |
| 12. Manufacture of phosphoric acid | $Ca_3\ (PO_4)_2\ (s) + 3H_2SO_4\ (l) \rightarrow 3CaSO_4\ (aq) + 2H_3PO_4\ (l)$ |
| 13. Leaching of ores with sulfuric acid | $UO_2\ (s) + 2H_2SO_4\ (aq) \rightarrow U\ (SO_4)_2\ (aq) + 2H_2O\ (l)$ |

### 8.5.1 Solid Catalysed Reactions

As stated in the earlier section, solid catalysed reactions are probably the most important types of heterogeneous reactions. We can broadly define a catalyst as a substance that influences (usually increases) the rate of a chemical reaction but remains chemically unaffected. It is not to be assumed that the catalyst does not take part in the reaction. It does take active part in the reaction, but gets regenerated at the end of the process. It is postulated that a catalyst increases the rate of a chemical reaction because it makes the reaction take place through alternate paths. Each step of this alternate mechanism has a lower activation energy than that for the uncatalysed process. It is popularly assumed that the reactant molecules combine with the active centres on the catalyst surface and thereby form intermediate complexes. These complexes subsequently decompose to yield the final product. For example, if the overall reaction is $A \rightarrow P$ and $X$ is the catalyst, then

$$A + X \rightleftharpoons AX$$

$$AX \rightleftharpoons P + X$$

Each of the above steps has a lower energy of activation than that for the uncatalysed reaction, $A \rightarrow P$. As a result, the overall rate of the reaction gets improved. However, there are alternate theories as well. One of them proposes that the reactant molecules do not attach to the catalyst surface, but move to the close vicinity of the catalyst surface and be under the influence of surface forces. The molecules, thus, remain mobile but will be energised or activated. A still third theory postulates that a free radical is formed at the catalyst surface which moves back into the gas bulk and initiates a chain of reactions. Thus, according to this theory the reaction does not take place at the catalyst surface but takes place in the gas bulk itself, the catalyst surface simply acting as a generator of free radicals.

The actual mechanism of a catalytic reaction is more complex than it appears. It is, therefore, difficult to analyse the process based on a specific theory. Extensive experimentation is warranted in a majority of the cases.

The amount of catalyst required for conducting a given reaction is usually small. In other words, a small amount of catalyst can cause conversion of a large amount of reactant. However, it must be kept in mind that though small, the catalyst concentration is a controlling parameter in most of the catalytic reactions. There are cases in which the reaction rate varies proportionally with catalyst concentration.

In the case of reversible reactions, the equilibrium conversion is not altered by the use of the catalyst. This is because the catalyst accelerates not only the forward reaction but also the backward reaction. As a result, the equilibrium conversion remains unchanged whether a catalyst is used or not.

In the case of multiple reactions, the catalyst can be used for improving the selectivity. In other words, a catalyst can preferentially increase the rate of the desirable reaction and as a result, the rate of production of the desirable product shall be much higher than that of the undesirable products through side reactions. This property of the catalyst is very important in industry, since side reactions can hardly be avoided in most of the industrial processes. For example, consider the multiple reaction

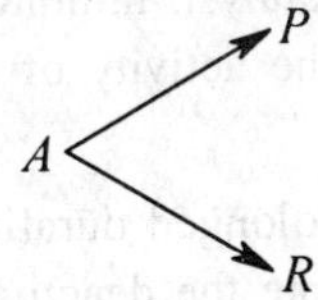

If a catalyst $X_1$ is employed, then conversion of $A$ to $P$ is accelerated and there will be hardly any production of $R$. Alternately, if a different catalyst $X_2$ is used, then $A$ gets converted mainly to $R$ and very little $P$ is produced. Thus, the reaction can be made to proceed in the desirable direction by employing a suitable catalyst.

*Autocatalytic reactions* are those in which one of the products acts as the catalyst. Reactions of this type have already been discussed in Section 8.2. The rate of an autocatalytic reaction increases with increase in conversion. However, after some time when the reactant concentration has fallen to very low value, the rate starts decreasing (see Fig. 8.2). Microbial fermentation (see example 8.22) is an excellent example of autocatalytic reaction.

Strictly speaking, it is necessary to distinguish between *positive catalysts* and *negative catalysts*. The former accelerates a chemical reaction while the latter decreases the rate of the reaction. For example, nitric oxide reduces the rate of decomposition of acetaldehyde[21] and thereby acts as a negative catalyst. So is iodine which is a negative catalyst to the combination of hydrogen and oxygen. However, if the term catalyst is used without any specification, then positive catalysis must be understood.

Though our discussion in this section is confined to solid catalysts, there are also important liquid and gaseous catalysts. For example, $H_2SO_4$ and HF are well-known alkylation and isomerisation catalysts.

An industrial catalyst need not necessarily be a single substance. It may be a composite containing a carrier, a promoter or an inhibitor. Catalysts are often supported or impregnated on a *carrier* since this provides a means of obtaining a large surface area with a small amount of catalytically active material. This is particularly important when expensive substances such as platinum, palladium, ruthenium and silver are used as catalysts. For example, platinum catalyst is often supported on silica or alumina. Vanadium pentoxide is usually supported on kieselguhr, while alumina is the carrier used for nickel catalyst (used in hydrogenation reactions). Impregnation of a catalyst on a carrier is a fairly long process involving a number of steps. For instance, for supporting nickel catalyst on alumina, the evacuated alumina particles are first soaked with nickel nitrate solution. After draining off the excess of solution, these particles are heated in an oven to decompose nickel nitrate to nickel oxide. Finally, the oxide is reduced back to metallic nickel by passing hydrogen. The last step (called *activation*) could very will be carried out in the reactor itself. In that case, it is called *in situ activation*. So long as the solution used to soak the carrier particles does not contain any catalyst poisons, in situ activation is the most desirable method of catalyst preparation.

*Promoters* are substances added during catalyst preparation to improve the activity or selectivity or stability of the catalyst. Promoters tend to increase the number of active centres on the catalyst surface and thereby increase catalyst activity. Some of them prevent sintering of catalyst particles during use and thereby increase the life of the catalyst. Promoters are added in small amounts and by themselves have little activity. *Inhibitors* are converse to promoters. When added in small amounts during catalyst manufacture, they depress the activity of the catalyst. Inhibitors are most useful for improving selectivity since by using a small amount of inhibitor, the activity of a catalyst for an undersirable side reaction can be significantly reduced.

No catalyst can retain its activity for a prolonged duration of time. Once the catalyst is deactivated (the term *spent catalyst* is often used to indicate the deactivated catalyst), the process is required to be shut down and the catalyst is to be replaced or regenerated before restarting the process. The overall cost of shut down and catalyst replacement/regeneration is often prohibitive. It is, therefore, apparent that a good catalyst must have a long, useful life span. The term *catalyst poisoning* is commonly used to describe the deactivation of the catalyst. Catalyst poisoning can occur

(*i*) due to chemisorption of substances present in the reactant stream or produced by the reaction on the catalyst surface,

(*ii*) due to physical deposition of substances on the catalyst surface as a result of which the active centres get blocked (this is also called *fouling*),

(*iii*) due to deterioration of the surface structure of the catalyst (such as sintering) as a result of prolonged exposure to elevated temperature in the reactor.

Catalyst poisons may be classified into four main categories:

(*a*) **Deposited poisons (or diffusion poisons):** Entrained solids in the reactants or solid residues formed as a result of the chemical reaction get deposited on the catalyst surface and cover the active sites and may also partially plug the pores. Blocking the pore mouths prevents the reactants from diffusing into the inner surface. An excellent example is the deposition of coke particles on the silica-alumina catalyst used for the cracking of petroleum hydrocarbons. Coke formation takes place during cracking as a result of the polymerisation reaction. In this case, the catalyst can be regenerated or reactivated by burning off the coke particles from the catalyst surface. This is usually carried out in regenerators by passing hot compressed air through the spent catalyst bed.

(*b*) **Chemisorbed poisons:** Certain substances like compounds of sulfur get chemisorbed on the catalyst surface, thereby covering the active sites and causing a decline in the catalyst activity. However, in course of time, an equilibrium is established between the poison in the reactant stream and that on the catalyst surface. Once this equilibrium is attained, there will be no further decrease in catalyst activity. Sulfur poisoning of platinum and nickel catalysts takes place in this way. If the strength of the adsorption bond is not great, then the catalyst maybe reactivated by removing the poison from the reactant stream.

(*c*) **Selectivity poisons:** There are some substances (usually present in the reactant stream) that get adsorbed on the catalyst surface and then catalyse some of the undesirable side reactions. This, thus, lowers the selectivity. For example, during the cracking of petroleum hydrocarbons if small amounts of iron, nickel or vanadium are present in the feed stock, then these get adsorbed on the catalyst surface and act as dehydrogenation catalysts. As a result, more and more hydrogen and coke will be formed and the yield of the principal product (namely, gasoline) will be poor.

(*d*) **Stability poisons:** Stability poisoning involves structural deterioration of catalyst due to continuous exposure to elevated temperatures. Sintering or localised melting of the solid may occur. It is also possible that some impurities present in the feed stream tend to adversely affect the structure of the catalyst or the catalyst carrier.

Depending on whether the poison is present in the feed stream or is produced during the chemical reaction, the process of catalyst deactivation or poisoning may be grouped into four types:

For example, if the poison is present in the feed stream itself (that is, an impurity present in the feed stream causes catalyst poisoning or deactivation), then it is called *side-by-side deactivation.* If the poison is the product of a parallel side reaction, then it is called *parallel deactivation.* For example,

$$A \rightarrow P \text{ (desired product)}$$

$$A \rightarrow R \text{ (catalyst poison)}$$

Similarly, the poison can be the product of a series reaction as well. For instance, *A* decomposes to produce *P* (which is the desired product) and *P* further decomposes or reacts further with *A* to produce *R* which is a catalyst poison. This is called *series deactivation.* Deactivation of cracking catalyst takes place in this way. If the catalyst is deactivated due to physical deterioration resulting from prolonged exposure to high temperatures, then it is called *independent deactivation.* This is because in this case, the rate of deactivation is independent of reactant or product concentrations. Estimation of the rate of deactivation of catalysts is discussed in Section 8.5.6.

The physical properties of the catalyst such as pre size, pore volume and the surface area are measured experimentally. For example, in the *Helium-Mercury method*, the volume of helium displaced by a sample of catalyst is first measured. The helium is then removed and the volume of mercury displaced by the catalyst is recorded. Since mercury has a significant surface tension, it will not fill the pores of most catalysts at atmospheric pressure and therefore, the difference in volumes measured gives the pore volume of the catalyst sample.

Similarly, the *mercury porosimeter* that is used to measure the pore size of catalysts, makes use of the fact the pressure required to force mercury into the pores depends on the pore size. For example, Ritter and Drake[22] obtained

$$d_{p0} = (4\sigma \cos \theta)/P \qquad \text{... (8.5.1)}$$

where $\sigma$ = surface tension of mercury

$\theta$ = contact angle between mercury and the pore wall (an average value of 140° may be assumed)

$d_{p0}$ = pore diameter

The surface area or the specific surface of the catalyst is measured based on experiments on physical adsorption of a gas on the catalyst surface. For example, the volume of nitrogen adsorbed at equilibrium at the normal boiling point (–195.8° C) is measured over a range of nitrogen pressures below 1 atm. The data are then fitted into any of the popular adsorption isotherms. For example, according to the BET (Brunauer-Emmett-Teller) isotherm,

$$\frac{P_a}{V_0\,(P' - P)} = \frac{1}{(V_m k_a)} + \left[\frac{(k_a - 1)}{k_a V_m}\right](P_a/P') \qquad \text{... (8.5.2)}$$

where $P_a$ = adsorption pressure

$P'$ = vapour pressure of the solute at the operating temperature (for nitrogen at –195.8° C, $P'$ = 1 atm)

$V_0$ = volume of solute adsorbed (usually corrected to 0° C and 1 atm)

$V_m$ = volume of solute adsorbed when the entire catalyst surface (or, all active sites) is covered by adsorbed molecules (that is, when there is a complete monomolecular layer of adsorbed solute)

$k_a$ = a constant for the particular gas-solid system at the particular temperature.

It is clear from the above equation that a plot of $[P_a/V_0\,(P' - P_a)]$ versus $(P_a/P')$ will be a straight line. From the slope and *y*-intercept of such a plot, the values of $k_a$ and $V_m$ can be computed. Since $V_m$ is specified at 0° C and 1 atm, moles of gas adsorbed will be $(V_m/22.4)$ and the number of molecules adsorbed will be $(V_m/22.4)\ N_0$ where $N_0$ = Avogadro number = $6.023 \times 10^{26}$ molecules/kmole. If $\alpha$ is the area covered by one adsorbed molecule, then the total surface area of the catalyst sample will be

$$S_p = (V_m/22.4)\ (6.023 \times 10^{26})\ \alpha \qquad \text{... (8.5.3)}$$

If *m* is the mass of the catalyst sample, then

$$s_p = \text{specific surface} = (S_p/m) \qquad \text{... (8.5.4)}$$

Emmett and Brunauer[23] have proposed that

$$\alpha = (1.09)\left[\frac{(M_A/\rho_A)}{\left(6.023 \times 10^{26}\right)}\right]^{2/3} \qquad \text{... (8.5.5)}$$

where $M_A$ is the molecular weight of solute gas and $\rho_A$ is its density at the operating temperature. For nitrogen at –195.8° C, $\rho_A$ = 808.0 kg/m³ and $M_A$ = 28. Accordingly, $\alpha$ for nitrogen = $1.6245 \times 10^{-19}$ m².

**Example 8.35:** A sample of activated alumina catalyst has the following characteristics:

Density of catalyst particle (determined by mercury displacement) = 1547 kg/m$^3$

True density of solid material in particle = 3675 kg/m$^3$

Surface area of each particle (by adsorption measurement) = 175000 m$^2$/kg

From the above information, compute

(*a*) the porosity of particles,
(*b*) the pore volume per kg,
(*c*) the mean pore radius.

If the bulk density of a bed of above catalyst particles in a 250 cm$^3$ graduate is 810 kg/m$^3$, what fraction of the total volume of the bed is void space between the particles and what fraction is void space within the particles?

**Solution:** The density of a porous solid (namely, mass of particle/total volume of particle) is usually called its *apparent density* ($\rho_p$). This is to distinguish it from the *true density* of the solid material in the particle. In other words,

$$\text{True density } (\rho_S) = \frac{(\text{Mass of particle})}{(\text{Volume of solid})} = \frac{(\text{Mass of particle})}{(\text{Total volume}) - (\text{Pore volume})}$$

It is given that $\rho_p$ = 1547 kg/m$^3$ and $\rho_S$ = 3675 kg/m$^3$. Now,

$$(\rho_p/\rho_S) = \frac{(\text{Volume of solid})}{(\text{Total volume})}$$

or

$$1 - (\rho_p/\rho_S) = \frac{(\text{Total volume}) - (\text{Volume of solid})}{(\text{Total volume})}$$

$$= (\text{Pore volume})/(\text{Total volume}) = \epsilon_p$$

Therefore, $\epsilon_p = 1 - (1547/3675) = 0.579$

Pore volume per kg $(V_{p0}) = (\epsilon_p/\rho_p) = (0.579/1547.0) = 3.7427 \times 10^{-4}$ m$^3$/kg = 374.27 cm$^3$/kg

Assuming cylindrical pores of uniform size (diameter = $d_{p0}$ and length = $L_{p0}$),

$$s_p = \text{specific surface of particle} = (\pi\, d_{p0}\, L_{p0})\, n_{p0}/m_p$$

where $n_{p0}$ is the number of pores and $m_p$ is the mass of the particle. Also,

$$V_{p0} = (\pi\, d_{p0}^2/4)\, L_{p0}\, n_{p0}/m_p$$

Therefore, $(V_{p0}/s_p) = (d_{p0}/4)$

or $d_{p0} = 4\,(V_{p0}/s_p)$

It is given that $s_p$ = 175000 m$^2$/kg. Therefore,

$$d_{p0} = (4.0)\,(3.7427 \times 10^{-4})/(175000.0)$$

$$= 8.555 \times 10^{-9} \text{ m} = 85.55 \text{ angstroms}$$

The mean pore radius is, therefore, (85.55/2) = 42.77 angstroms.

If the bulk density of the packed bed is 810 kg/m$^3$, then it means that

$$\frac{(\text{Mass of particles})}{(\text{Total volume of bed})} = 810.0$$

Since the total volume of the bed is 250 $cm^3$ or $(250 \times 10^{-6})$ $m^3$,

Mass of particles = $(250 \times 10^{-6})$ (810.0) = 0.2025 kg

If $\in$ represents the voidage (or void fraction) of the bed, then

$$\in = \frac{(\text{Void volume between particles})}{(\text{Total volume of bed})}$$

$$= \frac{(\text{Total volume of bed}) - (\text{Volume of particles})}{(\text{Total volume of bed})}$$

$$= \frac{\left(250 \times 10^{-6}\right) - (0.2025/1547)}{\left(250 \times 10^{-6}\right)} = 0.4764$$

Now, void volume within particles = $V_{p0}$ (mass of particles) = $(3.7427 \times 10^{-4})$ (0.2025)

$$\frac{(\text{Void volume within particles})}{(\text{Total volume of bed})} = \frac{\left(3.7427 \times 10^{-4}\right)(0.2025)}{\left(250 \times 10^{-6}\right)} = \mathbf{0.303}$$

**Example 8.36:** A sample of iron-alumina catalyst was studied by low temperature (– 195.8° C) nitrogen adsorption. The weight of the sample was 5.04 gm and the results obtained are given below:

| *Pressure, mm Hg* | *Volume adsorbed (at 0° C and 1 atm), $cm^3$* |
|---|---|
| 8.0 | 103.0 |
| 30.0 | 116.0 |
| 50.0 | 130.0 |
| 102.0 | 148.0 |
| 130.0 | 159.0 |
| 148.0 | 163.0 |
| 233.0 | 188.0 |
| 258.0 | 198.0 |
| 330.0 | 221.0 |

Estimate the specific surface of the catalyst sample.

**Solution:** Let us fit the reported data into the B-E-T isotherm, namely Eq. (8.5.2). Since $P'$ = vapour pressure of nitrogen at – 195.8° C = 1 atm = 760 mm Hg and the values of $P_a$ and $V_0$ are given in the above table, we can compute the values of $(P_a/V_0)/(P' - P_a)$ and $(P_a/P')$ as listed below in Table 8.36.1.

The plot of $(P_a/V_0)/(P' - P_a)$ versus $(P_a/P')$ is shown in Fig. 8.36.1. From the figure,

$$\text{Slope} = \frac{(k_a - 1)}{(k_a V_m)} = 0.0074\ (\text{cm}^3)^{-1}$$

$$\text{Intercept} = 1/(k_a V_m) = 0.000052\ (\text{cm}^3)^{-1}$$

## Table 8.36.1

| $\left[\dfrac{(P_a/V_0)}{(P'-P_a)}\right] \times 10^3,\ (cm^3)^{-1}$ | $(P_a/P')$ |
|---|---|
| 0.10328 | 0.0105 |
| 0.3543 | 0.0395 |
| 0.5417 | 0.0658 |
| 1.0474 | 0.1342 |
| 1.2980 | 0.1710 |
| 1.4836 | 0.1947 |
| 2.3517 | 0.3065 |
| 2.5960 | 0.3395 |
| 3.4725 | 0.4342 |

$$(\text{Slope} + \text{intercept}) = (1/V_m) = 0.007452$$

or
$$V_m = 134.192\ cm^3 = 1.3419 \times 10^{-4}\ m^3$$

Since for nitrogen at $-195.8°$ C, $\alpha = 1.6245 \times 10^{-19}\ m^2$, from Eq. (8.5.3)

$$S_p = (1.3419 \times 10^{-4}/22.4)\,(6.023 \times 10^{26})\,(1.6245 \times 10^{-19}) = 586.14\ m^2$$

Since the mass of the sample is 5.04 gm,

$$s_p = (586.14/5.04) = \mathbf{116.3\ m^2/gm}$$

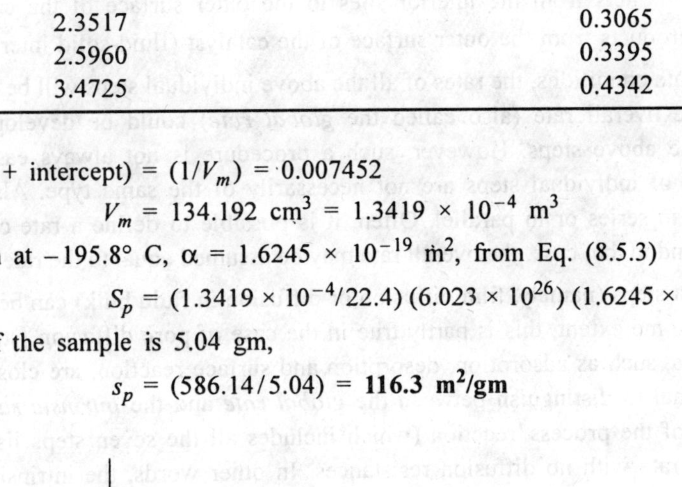


*Figure 8.36.1*

## 8.5.2 Mechanism of Solid–Catalysed Reactions

As stated earlier, the actual mechanism of a solid–catalysed reaction is basically complex. The overall process involves the following steps:

(*i*) Transport of reactants from the fluid bulk to the external surface of the catalyst (fluid–solid interface)

(*ii*) If the catalyst is porous, diffusion of reactants into the catalyst pores (both by molecular diffusion and Knudsen diffusion)

(*iii*) Adsorption of reactants at the interior sites of the catalyst particle

(*iv*) Chemical reaction of adsorbed reactants to adsorbed products (surface reaction)

(*v*) Desorption of adsorbed products

(*vi*) Diffusion of products from the interior sites to the outer surface of the catalyst particle

(*vii*) Transport of products from the outer surface of the catalyst (fluid-solid interface) to the fluid bulk.

Under steady state conditions, the rates of all the above individual steps will be the same. Accordingly, the equation for the overall rate (also called the *global rate*) could be developed by adding up the resistances to all the above steps. However, such a procedure is not always easy since the equations describing the rates of individual steps are not necessarily of the same type. Also, these steps do not necessarily proceed in series or in parallel. Often, it is possible to define a rate controlling step (which is the slowest step) and in that case, the overall rate may be assumed equal to the rate of that individual step.

The first and last steps (namely, film diffusion or diffusion in fluid bulk) can be treated independently during analysis. To some extent, this is partly true in the case of pore diffusion [steps (*ii*) and (*vi*)]. The remaining three steps, such as adsorption, desorption and surface reaction, are closely inter-related. It is, therefore, conventional to distinguish between the *global rate* and the *intrinsic rate*. The former stands for the overall rate of the process/reaction (which includes all the seven steps listed above), while the latter stands for the rate with no diffusion resistances. In other words, the intrinsic rate considers steps (*iii*), (*iv*) and (*v*) only. If the resistances to the first two and last two steps (that is, resistances to internal and external transport) are negligibly small, then the overall rate will be equal to the intrinsic rate.

The usual procedure of finding the overall rate equation is, therefore, as follows:

(*i*) First, assume that the resistance to pore diffusion is insignificant. For such a case, develop an expression for the intrinsic rate. Let us represent it as $(-r_{AS})_{\text{int}}$. To note that, once the pore diffusion is insignificant, then the concentration of any species (say, $A$) in the catalyst pores (namely, $C_A$) will be equal to the concentration of $A$ in the fluid at the catalyst surface (namely, $C_{AS}$).

(*ii*) Once the expression for $(-r_{AS})_{\text{int}}$ is obtained, it is then multiplied by a properly defined effectiveness factor ($\eta$) to get the diffusion-limited rate (that is, the actual rate that includes resistance to pore diffusion as well). Computation of $\eta$ is discussed in Section 8.5.4.

(*iii*) The above rate equation is now clubbed with film diffusion (that is, diffusion in fluid bulk) to obtain the overall rate equation (see Section 8.5.5).

Let us first consider development of expressions for the intrinsic rate, $(-r_{AS})_{\text{int}}$. To be precise, there are two methods of developing the rate equation:

(*a*) The first and the simplest method is to assume a power law type equation similar to that employed for homogeneous reactions, such as $(-r_{AS})_{\text{int}} = k\,C_{AS}^n$ or $k\,C_{AS}^n C_{BS}^m$ and then determine the values of the constants and the exponents ($k$, $n$, $m$) by fitting the equation to experimental data. This method is wholly empirical and does not give any information regarding the actual mechanism of the reaction. It is adopted in industrial reactor design due to its apparent simplicity. It must be admitted that frequently, but not always, such an empirically derived equation could correlate experimental data almost as accurately as (and with fewer adjustable parameters than) more elaborate methods.

(*b*) The second method, which is theoretically more sound, is what is called the *Langmuir-Hinshelwood formulation.* This method is based on the mechanisms of adsorption, surface reaction and desorption. However, since the true mechanisms of these steps are difficult to predict, the rate equation is finalised by trial. We have to first assume a possible mechanism and develop the rate equation based on it. It is then checked whether the available experimental data fit closely into this rate equation. If not, an alternate mechanism is to be assumed and the trial repeated. This method is discussed in detail in the following paragraphs.

### 8.5.3 Rate Equations by Langmuir-Hinshelwood Method

Let us consider an example. Let the reaction be

$$A + B \rightleftharpoons F \qquad \text{... (8.5.6)}$$

It is catalysed by a solid catalyst. Let the mechanism be

$$A + X \rightleftharpoons A{\cdot}X \text{ (adsorption of } A) \qquad \text{... (8.5.7)}$$

$$B + X \rightleftharpoons B{\cdot}X \text{ (adsorption of } B) \qquad \text{... (8.5.8)}$$

Now the question is whether the reaction (namely, surface reaction) is between adsorbed molecules of $A$ and adsorbed molecules of $B$ on adjacent active centers or between adsorbed molecules of $A$ and the gaseous (unadsorbed) molecules of $B$. In former case, the surface reaction can be represented as

$$A{\cdot}X + B{\cdot}X \rightleftharpoons P{\cdot}X + X \qquad \text{... (8.5.9)}$$

And in the latter case,

$$A{\cdot}X + B \rightleftharpoons P{\cdot}X \qquad \text{... (8.5.10)}$$

The next step is the desorption of adsorbed $P$:

$$P{\cdot}X \rightleftharpoons P + X \qquad \text{... (8.5.11)}$$

In all the above equations, $X$ represents a catalyst site or an active center.

Next, we can write down the rate expression for each of the above steps. Thus, the rate expression for the adsorption of $A$ is

$$\begin{aligned} r_a &= k'_a C_{AS} (\overline{C}_m - \overline{C}) - k'_{ar} \overline{C}_A = k'_a C_{AS} \overline{C}_V - k'_{ar} \overline{C}_A \\ &= (k'_a \overline{C}_m)\, C_{AS} \theta_V - (k'_{ar} \overline{C}_m)\, \theta_A = k_a C_{AS} \theta_V - k_{ar} \theta_A \end{aligned} \qquad \text{... (8.5.12)}$$

where $k_a$ is the rate constant for adsorption (the difference between $k_a$ and $k'_a$ may be noticed). $C_{AS}$ is the concentration of $A$ in the fluid at the catalyst surface (kmoles/m$^3$) and $\overline{C}_A$ is the concentration of adsorbed $A$ (kmoles/kg of catalyst). To note that since pore resistance has been assumed to be negligible,

$C_A = C_{AS}$ (where $C_A$ is the concentration of $A$ in catalyst pores). Accordingly, we have used $C_{AS}$ in places of $C_A$. If $\bar{C}_m$ stands for the total concentration of active centres per unit mass of catalyst ($\bar{C}_m$ is nothing but the concentration of adsorbed solute in unit mass of catalyst when all the active centres are covered by the solute), then $(\bar{C}_m - \bar{C}) = \bar{C}_V$ = concentration of vacant sites and $\theta_V = (\bar{C}_V/\bar{C}_m)$ = fraction of total catalyst sites (or active centres) vacant. Similarly, $\theta_A$, $\theta_B$ and $\theta_P$ will represent fractions of total catalyst sites occupied by $A$, $B$ and $P$ respectively. To note that if only $A$ is adsorbed, then $(\bar{C}_m - \bar{C}) = (\bar{C}_m - \bar{C}_A)$. It may also be noted that $r_a$ is expressed in kmoles/(s·kg of catalyst).

If we define an equilibrium constant $K_A$ such that $K_A = (k_a/k_{ar})$, then Eq. (8.5.12) may be rewritten as

$$r_a = k_a\,(C_{AS}\,\theta_V - \theta_A/K_A) \quad \text{... (8.5.13)}$$

Similarly, the rate of adsorption of $B$ is

$$(r_a)_B = (k_a)_B\; C_{BS}\,\theta_V - (k_{ar})_B\;\theta_B \quad \text{... (8.5.14)}$$

$$= (k_a)_B\;[C_{BS}\,\theta_V - (\theta_B/K_B)] \quad \text{... (8.5.15)}$$

If the surface reaction follows Eq. (8.5.9), then its rate expression will be

$$r_S = k'_S\,\bar{C}_A\,\theta_B - k'_{Sr}\,\bar{C}_P\,\theta_V$$

$$= (k'_S\,\bar{C}_m)\;\theta_A\,\theta_B - (k'_{Sr}\,\bar{C}_m)\;\theta_P\,\theta_V$$

$$= k_S\,\theta_A\,\theta_B - k_{Sr}\,\theta_P\,\theta_V \quad \text{... (8.5.16)}$$

$$= k_S\,(\theta_A\,\theta_B - \theta_P\,\theta_V/K_S) \quad \text{... (8.5.17)}$$

where $K_S = (k_S/k_{Sr})$*. $k_S$ is the rate constant for the forward reaction and $k_{Sr}$ that for the backward reaction. Readers may notice that we have written down the above expression for $r_S$ based on the fact the rate of the forward reaction is proportional to the concentration of adsorbed $A$ ($\bar{C}_A$) and the concentration of adsorbed $B$ *on sites (or active centres) adjacent to adsorbed A*. The fraction of adjacent sites occupied by $B$ is proportional to $\theta_B/(1 - \theta_A)$, but for small values of $\theta_A$, it will be proportional to $\theta_B$. Therefore, the rate of forward reaction will be $(k'_S\,\bar{C}_A\,\theta_B)$. Similarly, the rate of backward reaction will be $(k'_{Sr}\,\bar{C}_P\,\theta_V)$ and $r_S$ will be equal to the difference between the two.

If the surface reaction follows Eq. (8.5.10), then

$$r_S = k'_S\,\bar{C}_A\,C_{BS} - k'_{Sr}\,\bar{C}_P = (k'_S\,\bar{C}_m)\;\theta_A\,C_{BS} - (k'_{Sr}\,\bar{C}_m)\;\theta_P$$

$$= k_S\,\theta_A\,C_{BS} - k_{Sr}\,\theta_P \quad \text{... (8.5.18)}$$

$$= k_S\,(\theta_A\,C_{BS} - \theta_P/K_S) \quad \text{... (8.5.19)}$$

If must be noted that unlike in Eq. (8.5.17), $K_S$ is expressed in $(\text{kmoles/m}^3)^{-1}$ in the above equation.

Finally, the rate of desorption of $P$ is

$$r_d = k'_d\,\bar{C}_P - k'_{dr}\,C_{PS}\,\bar{C}_V = (k'_d\,\bar{C}_m)\;\theta_P - (k'_{dr}\,\bar{C}_m)\;\theta_V\,C_{PS}$$

$$= k_d\,\theta_P - k_{dr}\,\theta_V\,C_{PS} \quad \text{... (8.5.20)}$$

*Note that $K_S$ is dimensionless unlike other equilibrium constants (such as $K_A$ and $K_B$) which are expressed in kmoles/m$^3$.

$$= k_d\,(\theta_P - K_P C_{PS}\theta_V) \qquad \text{... (8.5.21)}$$

where $K_P = (k_{dr}/k_d)$. To note that since $K_P$ is the equilibrium constant for desorption, it is defined as $(k_{dr}/k_d)$ and not as $(k_d/k_{dr})$.

Now, we have to assume one of the above steps to be rate-controlling. It is slowest step that is the rate-controlling step. As a result, all other steps are assumed to be quite fast and to be occurring at near-equilibrium conditions. For example, let us assume that adsorption of $A$ is the rate controlling step. Then, the intrinsic rate $(-\boldsymbol{r}_{AS})_{\text{int}}$ will be equal to $\boldsymbol{r}_a$. Thus

$$(-\boldsymbol{r}_{AS})_{\text{int}} = \boldsymbol{r}_a = k_a\,(C_{AS}\theta_V - \theta_A/K_A) \qquad \text{... (8.5.22)}$$

We can eliminate $\theta_A$ and $\theta_V$ from the above equation by making use of the fact that the other steps are at equilibrium. Thus, from Eq. (8.5.21), since $\boldsymbol{r}_d = 0$ at equilibrium,

$$\theta_P = K_P C_{PS}\theta_V \qquad \text{... (8.5.23)}$$

Similarly, from Eq. (8.5.15), at equilibrium,

$$\theta_B = K_B C_{BS}\theta_V \qquad \text{... (8.5.24)}$$

If the surface reaction follows Eq. (8.5.17), then at equilibrium

$$\theta_A = \frac{(\theta_P/\theta_B)\,\theta_V}{K_S} \qquad \text{... (8.5.25)}$$

$$= \left(\frac{K_P C_{PS}}{K_B K_S C_{BS}}\right)\theta_V \qquad \text{... (8.5.26)}$$

Equation (8.5.22), therefore, becomes

$$(-\boldsymbol{r}_{AS})_{\text{int}} = k_a\theta_V\left(C_{AS} - \frac{K_P C_{PS}}{K_A K_B K_S C_{BS}}\right)$$

$$= k_a\theta_V\,(C_{AS} - C_{PS}/K\,C_{BS}) \qquad \text{... (8.5.27)}$$

where $K = (K_A K_B K_S/K_P)$ = equilibrium constant for the overall reaction. It may be noted that in terms of concentrations, $K = (C_P/C_A C_B)_{\text{eq}} = (\theta_P/\theta_A\theta_B)_{\text{eq}}$. We also know that

$$\theta_A + \theta_B + \theta_P + \theta_V = 1.0 \qquad \text{... (8.5.28)}$$

Substituting for $\theta_P$, $\theta_B$ and $\theta_A$ from Eqs (8.5.23), (8.5.24) and (8.5.26), we get

$$\theta_V = 1/[1 + K_B C_{BS} + K_P C_{PS} + (K_A/K)\,(C_{PS}/C_{BS})] \qquad \text{... (8.5.29)}$$

Substituting for $\theta_V$ in Eq. (8.5.27), we get the final expression for the intrinsic rate, as

$$(-\boldsymbol{r}_{AS})_{\text{int}} = \frac{k_a\left[C_{AS} - (1/K)(C_{PS}/C_{BS})\right]}{1 + K_B C_{BS} + K_P C_{PS} + (K_A/K)(C_{PS}/C_{BS})} \qquad \text{... (8.5.30)}$$

Similarly, if adsorption of $A$ is controlling, but the surface reaction follows Eq. (8.5.19), then

$$(-r_{AS})_{\text{int}} = \frac{k_a\left[C_{AS} - (K_P / K_A K_S)(C_{PS}/C_{BS})\right]}{1 + K_B C_{BS} + K_P C_{PS} + (K_P/K_S)(C_{PS}/C_{BS})} \quad \text{... (8.5.31)}$$

Let us now assume that the surface reaction is the rate-controlling step. Then, adsorption of $A$, adsorption of $B$ and desorption of $P$ will occur at equilibrium. Thus, from Eq. (8.5.13), at equilibrium,

$$\theta_A = K_A C_{AS} \theta_V \quad \text{... (8.5.32)}$$

$\theta_P$ and $\theta_B$ will be those given by Eqs (8.5.23) and (8.5.24). Then, from Eq. (8.5.28),

$$\theta_V = 1/(1 + K_A C_{AS} + K_B C_{BS} + K_P C_{PS}) \quad \text{... (8.5.33)}$$

If surface reaction follows Eq. (8.5.17),

$$(-r_{AS})_{\text{int}} = r_S = k_S\,(\theta_A \theta_B - \theta_P \theta_V / K_S)$$

$$= \frac{k_S\left[K_A K_B C_{AS} C_{BS} - (K_P C_{PS}/K_S)\right]}{(1 + K_A C_{AS} + K_B C_{BS} + K_P C_{PS})^2} \quad \text{... (8.5.34)}$$

If surface reaction follows Eq. (8.5.19), then

$$(-r_{AS})_{\text{int}} = \frac{k_S\left[K_A C_{AS} C_{BS} - (K_P C_{PS}/K_S)\right]}{(1 + K_A C_{AS} + K_B C_{BS} + K_P C_{PS})} \quad \text{... (8.5.35)}$$

Proceeding similarly, we can develop expressions for intrinsic rate when desorption of $P$ is the rate controlling step. Rate expressions, so developed for many of the popular cases, are listed in Table 8.5. Once the rate expression has been finalised by trial, then the constants in the expression such as $K_A$, $K_B$, $K_P$ and $k_S$ are to be evaluated by fitting the expression into experimental data (see Examples 8.37 and 8.38).

Though the procedure of developing rate expressions such as those listed in Table 8.5 has a sound theoretical background, there had been disagreement regarding its adaptation to reactor design[19]. The disagreement stems from the argument that many of the solid catalysed reactions can be fitted empirically into simple first order or second order rate equations and therefore, the seemingly cumbersome, trial and error procedure of developing the rate expression that contains as many as five constants which are further to be evaluated from experimental data is unnecessary. It is further argued that the developed expression for intrinsic rate cannot be *easily* clubbed with other resistance steps such as pore and bulk diffusions, which is easily possible in the case of empirically derived easy-to-handle rate equations. No doubt, there is some substance in these arguments. However, the preference to an empirical fit over a deep insight into the process mechanism cannot be justified in full. First of all, with the availability of high speed computers, the trial and error procedure of locating the actual mechanism and the subsequent evaluation of constants is neither too cumbersome nor too lengthy. It must also not be forgotten that once a rate expression has been developed on a firm theoretical basis, it can be conveniently extrapolated to predict the intrinsic rate under alternate operating conditions. Also, a sound knowledge of the mechanism of catalysis is essential for any ambitious research scientist or design engineer who wishes to make developments for the future.

## Table 8.5
## Rate Equations (Langmuir-Hinshelwood Formulations) for Solid Catalysed Reactions

| *S. no.* | *Reaction* | *Mechanism* | *Expression for* $(-r_A)_{int}$, *when the step given against is rate-controlling* | *Remarks* |
|---|---|---|---|---|
| 1. | $A \rightleftharpoons P$ | $A + X \rightleftharpoons A\cdot X$ | $\dfrac{k_a(C_A - C_P/K)}{(1 + K_{PS}C_P)}$ | $K = K_A K_S / K_P$ |
| | | $A\cdot X \rightleftharpoons P\cdot X$ | $\dfrac{k_S K_A(C_A - C_P/K)}{(1 + K_A C_A + K_P C_P)}$ | $K_{PS} = K_P(1 + 1/K_S)$ |
| | | $P\cdot X \rightleftharpoons P + X$ | $\dfrac{k_d K_A K_S(C_A - C_P/K)}{(1 + K_{AS}C_A)}$ | $K_{AS} = K_A(1 + K_S)$ |
| 2. | $A \rightleftharpoons P$ | $2A + X \rightleftharpoons A_2\cdot X$ | $\dfrac{k_a\left(C_A^2 - C_P^2/K^2\right)}{\left(1 + K_{PA}C_P + K_P{}'C_P^2\right)}$ | $R$ stands for $A_2$. |
| | | | | $K = \sqrt{K_A K_{A_2}}\ K_S/K_P$ |
| | | $A_2\cdot X + X \rightleftharpoons 2A\cdot X$ | $\dfrac{(k_a)_{A_2} K_A\left(C_A^2 - C_P^2/K^2\right)}{\left(1 + K_A C_A^2 + K_{PS}C_P\right)^2}$ | $K_{PA} = K_P(1 + 1/K_A)$ |
| | | | | $K'_P = K_P^2/(K_S^2 K_{A_2})$ |
| | | $A\cdot X \rightleftharpoons P\cdot X$ | $\dfrac{k_S\sqrt{K_A K_{A_2}}\,(C_A - C_P/K)}{\left(1 + \sqrt{K_A K_{A_2}}\,C_A + K_A C_A^2 + K_P C_P\right)}$ | $K_{PS} = K_P(1 + 1/K_S)$ |
| | | $P\cdot X \rightleftharpoons P + X$ | $\dfrac{k_d K K_P(C_A - C_P/K)}{\left(1 + K_A C_A^2 + K_S{}'C_A\right)}$ | $K'_S = \sqrt{K_A K_{A_2}}\ (1 + K_S)$ |

*(Contd.)*

**Table 8.5** *(Contd.)*

| S. no. | Reaction | Mechanism | Expression for $(-\mathbf{r}_A)_{int}$, when the step given against is rate-controlling | Remarks |
|---|---|---|---|---|
| 3. | $A \rightleftharpoons P$ | $A + 2X \rightleftharpoons 2A_{1/2}\cdot X$ | $\dfrac{k_a(C_A - K_A C_P/K)}{\left[1 + K_P C_P + \sqrt{(K_P/K_S)\,C_P}\right]^2}$ | $K = K_A K_S/K_P$ |
| | | $2A_{1/2}\cdot X \rightleftharpoons P\cdot X + X$ | $\dfrac{k_S K_A(C_A - C_P/K)}{\left[1 + \sqrt{K_A C_A} + K_P C_P\right]^2}$ | |
| | | $P\cdot X \rightleftharpoons P + X$ | $\dfrac{k_d K_A K_S(C_A - C_P/K)}{\left[1 + \sqrt{K_A C_A} + K_A K_S C_A\right]}$ | |
| 4. | $A \rightleftharpoons P + R$ | $A + X \rightleftharpoons A\cdot X$ | $\dfrac{k_a(C_A - C_P C_R/K)}{(1 + K_P C_P + K_R C_R + K_R{}' C_P C_R)}$ | $K = K_A K_S/(K_P K_R)$ |
| | | $A\cdot X + X \rightleftharpoons P\cdot X + R\cdot X$ | $\dfrac{k_S K_A(C_A - C_P C_R/K)}{(1 + K_A C_A + K_P C_P + K_R C_R)^2}$ | $K'_R = K_P K_R/K_S$ |
| | | $P\cdot X \rightleftharpoons P + X$ | $\dfrac{k_{d1} K K_P(C_A - C_P C_R/K)}{C_R\left[1 + K_A C_A + K_R C_R + K_A{}'(C_A/C_R)\right]}$ | $K'_A = K_A K_S/K_R$ |
| | | $R\cdot X \rightleftharpoons R + X$ | $\dfrac{k_{d2} K K_R(C_A - C_P C_R/K)}{C_P\left[1 + K_A C_A + K_P C_P (K_S K_A/K_P)(C_A/C_P)\right]}$ | |
| 5. | $A \rightleftharpoons P + R$ | $A + X \rightleftharpoons A\cdot X$ | $\dfrac{k_a(C_A - C_P C_R/K)}{\left[1 + K_P C_P + (K_P/K_S)\,C_P C_R\right]}$ | $K = K_A K_S/K_P$ |

*(Contd.)*

**Table 8.5** (*Contd.*)

| S. no. | Reaction | Mechanism | Expression for $(-\mathbf{r}_A)_{int}$, when the step given against is rate-controlling | Remarks |
|---|---|---|---|---|
| | | $A\cdot X \rightleftharpoons P\cdot X+R$ | $\dfrac{k_S K_A (C_A - C_P C_R / K)}{(1+K_A C_A + K_P C_P)}$ | |
| | | $P\cdot X \rightleftharpoons P+X$ | $\dfrac{k_d K_A K_S (C_A - C_P C_R / K)}{C_R\left[1+K_A C_A + K_A K_S (C_A / C_R)\right]}$ | |
| 6. | $A+B \rightleftharpoons P+R$ | $A+X \rightleftharpoons A\cdot X$ | $\dfrac{k_a (C_A C_B - C_P C_R / K)}{C_B\left[1+K_B C_B + K_P C_P + K_R C_R + (K_A / K)(C_P C_R / C_B)\right]}$ | $K=\dfrac{(K_A K_B K_S)}{(K_P K_R)}$ |
| | | $B+X \rightleftharpoons B\cdot X$ | $\dfrac{(k_a)_B (C_A C_B - C_P C_R / K)}{C_A\left[1+K_A C_A + K_P C_P + K_R C_R + (K_B / K)(C_P C_R / C_A)\right]}$ | |
| | | $A\cdot X+B\cdot X \rightleftharpoons P\cdot X+R\cdot X$ | $\dfrac{k_S K_A K_B (C_A C_B - C_P C_R / K)}{(1+K_A C_A + K_B C_B + K_P C_P + K_R C_R)^2}$ | |
| | | $P\cdot X \rightleftharpoons P+X$ | $\dfrac{k_{d1} K K_P (C_A C_B - C_P C_R / K)}{C_R\left[1+K_A C_A + K_B C_B + K_R C_R + K K_P (C_A C_B / C_R)\right]}$ | |
| | | $R\cdot X \rightleftharpoons R+X$ | $\dfrac{k_{d2} K K_R (C_A C_B - C_P C_R / K)}{C_P\left[1+K_A C_A + K_B C_B + K_P C_P + K K_R (C_A C_B / C_P)\right]}$ | |
| 7. | $A+B \rightleftharpoons P+R$ | $A+2X \rightleftharpoons 2A_{1/2}\cdot X$ | $\dfrac{k_a (C_A C_B - C_P C_R / K)}{C_B\left[1+K_B C_B + K_P C_P + K_R C_R + \sqrt{(K_A C_P C_R)/(K C_B)}\right]^2}$ | |

(*Contd.*)

**Table 8.5** *(Contd.)*

| S. no. | Reaction | Mechanism | Expression for $(-\mathbf{r}_A)_{int}$, when the step given against is rate-controlling | Remarks |
|---|---|---|---|---|
| | | $B + X \rightleftharpoons B \cdot X$ | $\frac{(k_a)_B\,(C_A C_B - C_P C_R / K)}{C_A\left[1 + \sqrt{K_A C_A} + K_P C_P + K_R C_R + (K_B C_P C_R)/(K C_A)\right]}$ | $K = \frac{(K_A K_B K_S)}{(K_P K_R)}$ |
| | | $2A_{1/2} \cdot X + B \cdot X \rightleftharpoons P \cdot X + R \cdot X + X$ | $\frac{k_S K_A K_B\,(C_A C_B - C_P C_R / K)}{\left(1 + \sqrt{K_A C_A} + K_B C_B + K_P C_P + K_R C_R\right)^3}$ | |
| | | $P \cdot X \rightleftharpoons P + X$ | $\frac{k_{d1} K\, K_P\,(C_A C_B - C_P C_R / K)}{C_R\left[1 + \sqrt{K_A C_A} + K_B C_B + K_R C_R + (K\, K_P C_A C_B / C_R)\right]}$ | |
| | | $R \cdot X \rightleftharpoons R + X$ | $\frac{k_{d2} K\, K_R\,(C_A C_B - C_P C_R / K)}{C_P\left[1 + \sqrt{K_A C_A} + K_B C_B + K_P C_P + (K\, K_R C_A C_B / C_P)\right]}$ | |
| 8. | $A + B \rightleftharpoons P + R$ | $B + X \rightleftharpoons B \cdot X$ | $\frac{(k_a)_B\,(C_A C_B - C_P C_R / K)}{C_A\left[1 + K_P C_P + (K_P / K_S)(C_P C_R / C_A)\right]}$ | $K = (K_B K_S / K_P)$ |
| | | $A + B \cdot X \rightleftharpoons P \cdot X + R$ | $\frac{k_S K_B\,(C_A C_B - C_P C_R / K)}{(1 + K_B C_B + K_P C_P)}$ | |
| | | $P \cdot X \rightleftharpoons P + X$ | $\frac{k_d K_S K_B\,(C_A C_B - C_P C_R / K)}{C_R\left[1 + K_B C_B + K_B K_S\,(C_A C_B / C_R)\right]}$ | |
| 9. | $A + B \rightleftharpoons P$ | $A + X \rightleftharpoons A \cdot X$ | $\frac{k_a\,(C_A C_B - C_P / K)}{C_B\left[1 + K_B C_B + K_P C_P + (K_A / K)(C_P / C_B)\right]}$ | $K = K_A K_B K_S / K_P$ |

*(Contd.)*

**Table 8.5** (*Contd.*)

| S. no. | Reaction | Mechanism | Expression for $(-\mathbf{r}_A)_{int}$, when the step given against is rate-controlling | Remarks |
|---|---|---|---|---|
| | | $B + X \rightleftharpoons B \cdot X$ | $\dfrac{(k_a)_B\,(C_A C_B - C_P / K)}{C_A\left[1 + K_A C_A + K_P C_P + (K_B / K)(C_P / C_A)\right]}$ | |
| | | $A \cdot X + B \cdot X \rightleftharpoons P \cdot X + X$ | $\dfrac{k_S K_A K_B\,(C_A C_B - C_P / K)}{(1 + K_A C_A + K_B C_B + K_P C_P)^2}$ | |
| | | $P \cdot X \rightleftharpoons P + X$ | $\dfrac{k_d K\, K_P\,(C_A C_B - C_P / K)}{(1 + K_A C_A + K_B C_B + K\,K_P\, C_A C_B)}$ | |
| 10. | $A + B \rightleftharpoons P$ | $A + X \rightleftharpoons A \cdot X$ | $\dfrac{k_a\,(C_A C_B - C_P / K)}{C_B\left[1 + K_P\, C_P + (K_P / K_S)(C_P / C_B)\right]}$ | $K = K_A K_S / K_P$ |
| | | $A \cdot X + B \rightleftharpoons P \cdot X$ | $\dfrac{k_S K_A\,(C_A C_B - C_P / K)}{(1 + K_A C_A + K_P C_P)}$ | |
| | | $P \cdot X \rightleftharpoons P + X$ | $\dfrac{k_d K_A K_S\,(C_A C_B - C_P / K)}{(1 + K_A C_A + K_A K_S C_A C_B)}$ | |
| 11. | $A + (1/2)\,B_2 \rightleftharpoons P$ | $A + X \rightleftharpoons A \cdot X$ | $\dfrac{k_a\left(C_A \sqrt{C_B} - C_P / K\right)}{\sqrt{C_B}\left[1 + K_B \sqrt{C_B} + K_P C_P + (K_A C_P)/\left(K \sqrt{C_B}\right)\right]}$ | $K = K_A K_B K_S / K_P$ |
| | | $1/2\, B_2 + X \rightleftharpoons B \cdot X$ | $\dfrac{(k_a)_B\left(C_A \sqrt{C_B} - C_P / K\right)}{C_A\left[1 + K_A C_A + K_P C_P + (K_B / K)(C_P / C_A)\right]}$ | |

(*Contd.*)

**Table 8.5** (*Contd.*)

| S. no. | Reaction | Mechanism | Expression for $(-\mathbf{r}_A)_{int}$, when the step given against is rate-controlling | Remarks |
|---|---|---|---|---|
| | | $A \cdot X + B \cdot X \rightleftharpoons P \cdot X + X$ | $\dfrac{k_S K_A K_B \left(C_A \sqrt{C_B} - C_P / K\right)}{\left(1 + K_A C_A + K_B \sqrt{C_B} + K_P C_P\right)^2}$ | |
| | | $P \cdot X \rightleftharpoons P + X$ | $\dfrac{k_d K K_P \left(C_A \sqrt{C_B} - C_P / K\right)}{\left(1 + K_A C_A + K_B \sqrt{C_B} + K K_P C_A \sqrt{C_B}\right)}$ | |
| 12. | $A + B_2 \rightleftharpoons P$ | $A + X \rightleftharpoons A \cdot X$ | $\dfrac{k_a \left(C_A C_{B_2} - C_P / K\right)}{C_{B_2}\left[1 + \sqrt{K_B C_{B_2}} + K_P C_P + (K_A / K)\left(C_P / C_{B_2}\right)\right]}$ | $K = K_A K_B K_S / K_P$ |
| | where $P = AB_2$ | $B_2 + 2X \rightleftharpoons 2B \cdot X$ | $\dfrac{(k_a)_{B_2} \left(C_A C_{B_2} - C_P / K\right)}{C_A\left[1 + K_A C_A + K_P C_P + \sqrt{(K_B C_P)/(K C_A)}\right]^2}$ | |
| | | $A \cdot X + 2B \cdot X \rightleftharpoons AB_2 \cdot X + 2X$ | $\dfrac{k_S K_A K_B \left(C_A C_{B_2} - C_P / K\right)}{\left(1 + K_A C_A + \sqrt{K_B C_{B_2}} + K_P C_P\right)^3}$ | |
| | | $AB_2 \cdot X \rightleftharpoons AB_2 + X$ | $\dfrac{k_d K K_P \left(C_A C_{B_2} - C_P / K\right)}{\left(1 + K_A C_A + \sqrt{K_B C_{B_2}} + K K_P C_A C_{B_2}\right)}$ | |

*Note:* If expressions for $(-\mathbf{r}_{AS})_{int}$ are required, then replace the concentrations $C_A$, $C_B$, $C_P$ and $C_R$ by $C_{AS}$, $C_{BS}$, $C_{PS}$ and $C_{RS}$ respectively.

### 8.5.4 Pore Diffusion–Effectiveness Factor

Let us now discuss how the resistance to diffusion can be accounted for. For example, consider a spherical porous catalyst particle of diameter $d_p$ that is submerged in the fluid stream containing the reactant $A$ and the product $P$. Species $A$ diffuses into the pores and gets converted to $P$ on the catalyst sites (or active centres). For the sake of simplicity, let us assume that the intrinsic rate is represented by a first order rate equation such that

$$(-r_A)_{\text{int}} = k\,C_A \quad \text{... (8.5.36)}$$

where $(-r_A)_{\text{int}}$ is in kmoles/(s·kg of catalyst). Alternately, if $(-r'_A)_{\text{int}}$ is the rate in kmoles/($m^3$·s), then

$$(-r'_A)_{\text{int}} = \rho_p\,k\,C_A \quad \text{... (8.5.37)}$$

where $\rho_p$ is the density of catalyst pellet. We can now write down the equation of continuity for mass transfer, assuming steady state transport. Accordingly, equation ($C$) of Table 4.5 of Chapter 4 reduces to

$$\frac{\partial}{\partial r}\,(r^2 N_{Ar}) = -\rho_p\,k\,C_A \quad \text{... (8.5.38)}$$

We have already discussed in Section 4.1.2 of Chapter 4 that diffusion of gases in porous solids could take place both by molecular diffusion as well as by Knudsen diffusion. Accordingly, we have to define a *combined effective diffusivity* ($D_e$) and the molar flux is then expressed as [see Eq. (4.1.60)]

$$N_{Ar} = -D_e\,(dC_A/dr) \quad \text{... (8.5.39)}$$

Strictly speaking, $D_e$ is a function of gas concentration in the pore [see Eq. (4.1.61)]. However, since the variation of $D_e$ with concentration is usually not large, let us assume that $D_e$ is essentially concentration-independent and may be computed from Eq. (4.1.63). In such a case, Eq. (8.5.39) can be inserted into Eq. (8.5.38) and differentiated to give

$$D_e\,\frac{1}{r^2}\,\frac{d}{dr}\left(r^2\,\frac{dC_A}{dr}\right) = \rho_p\,k\,C_A \quad \text{... (8.5.40)}$$

This equation is to be solved with the boundary conditions that at $r = (d_p/2)$, $C_A = C_{AS}$ = concentration of $A$ at the outer surface of the catalyst and at $r = 0$, $C_A$ is finite. If we substitute $(C_A/C_{AS}) = f(r)/r$, where $f(r)$ is any function of $r$, then

$$\frac{d^2 f}{d r^2} = (\rho_p\,k/D_e)\,f \quad \text{... (8.5.41)}$$

Integrating twice, we get

$$C_A/C_{AS} = (C_1/r)\,\cosh\,(\alpha r) + (C_2/r)\,\sinh\,(\alpha r) \quad \text{... (8.5.42)}$$

where $\alpha = \sqrt{\rho_p\,k/D_e}$. Applying the boundary conditions and thereby evaluating the integration constants $C_1$ and $C_2$, we get

$$C_A/C_{AS} = \frac{(d_p/2r)\sinh(\alpha r)}{\sinh(\alpha\,d_p/2)} \quad \text{... (8.5.43)}$$

The rate of reaction $(-r'_A)$ for the whole catalyst pellet is, therefore,

$$(-r'_A) = -(\pi d_p^2) N_{Ar} \big|_{r = dp/2}$$

$$= (\pi d_p^2) D_e (dC_A/dr) \big|_{r = dp/2} \quad \text{... (8.5.44)}$$

Using the expression for $C_A$ from Eq. (8.5.43), we find that

$$(-r'_A) = (2\pi d_p D_e C_{AS}) [\alpha (d_p/2) \coth (\alpha d_p/2) - 1] \quad \text{... (8.5.45)}$$

To note that $(-r'_A)$ is expressed in (kmoles/s). If we express the rate in kmoles/(s·kg of catalyst), then

$$(-r_A) = (12 D_e C_{AS}/\rho_p d_p^2) [(3\phi) \coth (3\phi) - 1] \quad \text{... (8.5.46)}$$

where $$\phi = (\alpha d_p/6) = (d_p/6) \sqrt{\rho_p k/D_e} \quad \text{... (8.5.47)}$$

This dimensionless parameter $\phi$ is called the *Thiele–type modulus*. If the resistance to pore diffusion is negligible, then the rate of reaction will be equal to $(-r_{AS})_{int} = k C_{AS}$. We shall, therefore, define an *effectiveness factor* ($\eta$) such that $\eta$ will be equal to the ratio of the actual rate of reaction to the rate of reaction when the resistance to pore diffusion is insignificant. In other words,

$$\eta = (-r_A)/(-r_{AS})_{int} = (1/3\phi^2) [3\phi \coth (3\phi) - 1]$$

$$= (1/\phi) [\coth (3\phi) - (1/3\phi)] \quad \text{... (8.5.48)}$$

It is thus clear that once the intrinsic rate $(-r_{AS})_{int}$ is known, either through an empirical fit or by the Langmuir-Hinshelwood formulation discussed earlier, the actual rate (that includes the resistance to pore diffusion as well) will be equal to $(-r_{AS})_{int}\ \eta$. If the resistance to pore diffusion is insignificant, then $\eta = 1.0$ (To note that when $\phi < 0.5$, $\eta \simeq 1.0$). If $\eta$ is much less than 1.0 (this happens at large values of $\phi$), then it means that pore resistance is significant and the adsorption and reaction inside the pores are relatively fast. Incidentally, when $\phi > 5.0$, $\eta \simeq (1/\phi)$.

Expressions for $\eta$ for particles of other shapes (other than spherical) have been reported by many investigators[24, 25]. For example, for catalyst particles in the form of thin disks or slabs (the thicknesses of which are so small that diffusion takes place only at the two circular or flat surfaces),

$$\eta = \frac{\tan h\,(\phi)}{\phi} \quad \text{... (8.5.49)}$$

where $$\phi = L \sqrt{\rho_p k/D_e} \quad \text{... (8.5.50)}$$

$L$ = one-half of the thickness of the disk/slab.

If only one surface of the slab/disk is exposed to the reactants and the other surface is sealed, then also Eqs (8.5.49) and (8.5.50) can be used to compute $\eta$ except that $L$ will be then the thickness of the slab/disk.

An interesting observation is that the values of $\eta$ predicted by Eqs (8.5.48) and (8.5.49) are essentially the same with a maximum deviation of 10 percent at $\phi$ = around 1.0. At values of $\phi$ much less than 1.0 or much more than 1.0, both equations predict essentially the same value of $\eta$. This brings us to the conclusion that $\eta$ may be computed from Eq. (8.5.48) or (8.5.49) for particles of all shapes, provided the dimension $L$ is defined as

$$L = V_p / S_p \qquad \text{... (8.5.51)}$$

where $V_p$ is the volume of the catalyst particle and $S_p$ is its external surface area that is exposed to the reactants. For example, for a cylindrical particle (with no axial diffusion),

$$L = \frac{\left(\pi d_p^2/4\right) l}{\left(\pi d_p l\right)} = (d_p/4) \qquad \text{... (8.5.52)}$$

The above-discussed expressions for $\eta$ have been developed for the specific case when the intrinsic rate follows a first order equation which is also the simplest case to analyse. However, for any other type of rate equation as well, $\eta$ may be computed from the same expression, such as Eq. (8.5.48) or (8.5.49), provided the Thiele–type modulus $\phi$ is defined in the most generalised form as

$$\phi = L \rho_p \, (-r_{AS})_{\text{int}} \left[ 2\rho_p \int_0^{C_{AS}} D_e \left(-r_A\right)_{\text{int}} \, dC_A \right]^{-1/2} \qquad \text{... (8.5.53)}$$

For example, if $(-r_A)_{\text{int}} = k\, C_A^n$, then

$$\phi = L \sqrt{(n+1)\, k \rho_p \, C_{AS}^{n-1} / (2 D_e)} \qquad \text{... (8.5.54)}$$

If the intrinsic rate follows an equation which is that of a first order, reversible reaction ($A \rightleftharpoons P$) such that $(-r_A)_{\text{int}} = k_1 C_A - k_2 C_P$, then

$$\phi = L \sqrt{\frac{k_1 (K+1) \rho_p}{K D_e}} \qquad \text{... (8.5.55)}$$

$$= L \sqrt{k_1 \rho_p / (x_{Ae} D_e)} \qquad \text{... (8.5.56)}$$

where $x_{Ae}$ is the equilibrium conversion of $A$ and $K$ is the equilibrium constant ($K = k_1/k_2$). Similarly, if the intrinsic rate follows the Michaelis-Menten scheme (as is the case with enzyme catalysed reactions, already discussed in Section 8.2) such that $(-r_A)_{\text{int}} = k\, C_{E0} / (1 + K_m / C_A)$ where $A$ is the substrate, then the expression for $\eta$ is given by[17]

$$\eta = 1 - \frac{\tan h(\phi)}{\phi} \left[ \frac{\eta_d}{\tan h(\eta_d)} - 1 \right], \text{ for } \eta_d < 1 \qquad \text{... (8.5.57)}$$

$$= \eta_d - \frac{\tan h(\phi)}{\phi} \, [\coth(\eta_d) - 1], \text{ for } \eta_d \geq 1 \qquad \text{... (8.5.58)}$$

where

$$\eta_d = \left(\sqrt{2}/\phi\right) (1 + \beta_m) \sqrt{\beta_m - \ln\left(1 + \beta_m\right) / \beta_m} \qquad \text{... (8.5.59)}$$

$$\beta_m = (C_{AS}/K_m) \qquad \text{... (8.5.60)}$$

Expressions for $\eta$ when the equation for intrinsic rate is of the type of Langmuir-Hinshelwood formulations (listed in Table 8.5) have been reported by Roberts and Satterfield[26]. In all the cases, $\eta$ is computed from Eq. (8.5.48) or (8.5.49) and $\phi$ from Eq. (8.5.53). It can be seen that when the rate equation is of first order, $\phi$ is independent of $C_{AS}$. In all other cases, $\phi$ is a function of surface concentration*.

Weisz and Hicks[27] have extended the computation of $\eta$ to nonisothermal conditions as well. They have shown that when the intrinsic rate follows a first order rate equation, the *nonisothermal effectiveness factor* is a function of the three dimensionless groups such as the Thiele–type modulus evaluated at the catalyst surface temperature ($\phi_S$), the Arrhenius number ($\gamma$) and the heat of reaction parameter ($\beta$), which are defined as given below:

$$\phi_S = L\sqrt{(k_1)_S\,\rho_P/D_e} \quad \text{... (8.5.61)}$$

$$\gamma = (E/RT_S) \quad \text{... (8.5.62)}$$

$$\beta = -(\Delta H_r)\,D_e\,C_{AS}/(k_e\,T_S) \quad \text{... (8.5.63)}$$

$k_e$ = *effective* thermal conductivity of catalyst particle

$(k_1)_S$ = first order rate constant evaluated at temperature $T_S$

$D_e$ = effective diffusivity, evaluated at $T_S$ (the temperature dependence of $D_e$ is often neglected)

$T_S$ = catalyst surface temperature

To note that $(\Delta H_r)$ will be a negative quantity for an exothermic reaction and a positive quantity for an endothermic reaction.

The effective thermal conductivity ($k_e$) of the catalyst pellet is to be experimentally determined and its value usually ranges from 0.15 to 0.7 W/(m K). The more porous the catalyst is, the less will be the value of $k_e$. This is understandable since the void spaces offer a large resistance to the transport of heat. It is due to the same reason that the effective thermal conductivity of a prorous catalyst is significantly lower than that of the bulk solid from which it is prepared.

For example, $k_e$ of alumina (boehmite) catalyst at 60° C is 0.1 W/(m K), but the thermal conductivity of solid alumina is around 1.73 W/(m K). Though temperature does not have a strong influence on the value of $k_e$, the pressure and the nature of the fluid inside the pores do affect the magnitude of $k_e$. For liquids, the effect of pressure is negligible, but for gases at low pressures, $k_e$ increases with increase in pressure. At high pressures however, $k_e$ is essentially independent of pressure.

Figure 8.16 shows $\eta$ as a function of $\phi_S$ and $\beta$ for $\gamma = 20$. Weisz and Hicks have reported similar sets of plots for $\gamma = 10$, 30 and 40. Interestingly, for exothermic reactions (for which $\beta$ is positive), at low values of $\phi_S$ (where pore resistance just begins to intrude), the effectiveness factor can become greater than unity (see Fig. 8.16). This finding is not surprising. For exothermic reactions, heat is released and the particles are hotter than the surrounding fluid and therefore, the nonisothermal rate is always higher

* Strictly speaking, $\phi$ is not a constant even for first order reactions since $D_e$ is a function of concentration as shown in Eq. (4.1.61) of Chapter 4. However, since concentration dependence of $D_e$ is not strong, it is assumed constant in many computations.

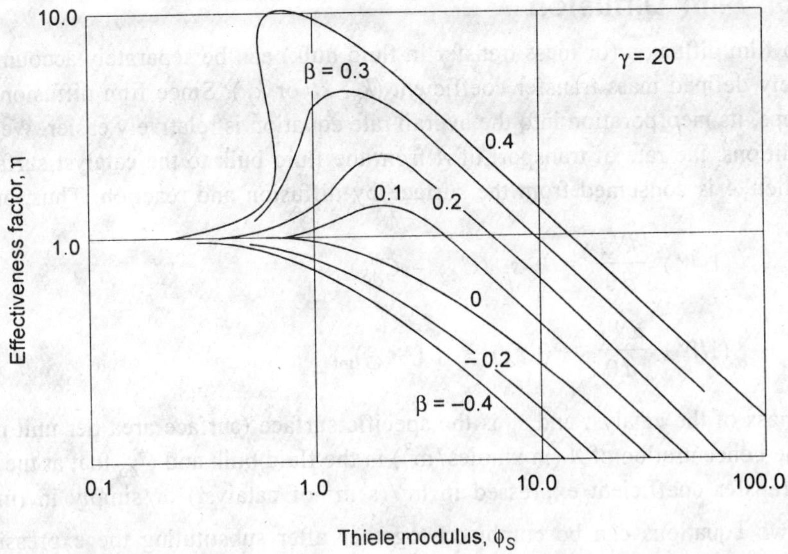


*Figure 8.16: Non-isothermal effectiveness factors for first order reactions in spherical catalyst pellets for the case of Arrhenius number = 20*

than the isothermal rate. The reverse is true in the case of endothermic reactions for which the nonisothermal rate is lower than the isothermal rate since the particles are cooler than the surrounding fluid.

Weisz and Hicks' data are for the specific case when the intrinsic rate follows a first order rate equation. Many other investigators[28, 29] have extended the evaluation of nonisothermal effectiveness factor to alternate cases such as when the intrinsic rate follows an $n$th order rate equation[28] or a first-order reversible rate equation[29], by appropriately defining $\phi_S$ as per Eqs (8.5.54) and (8.5.55). In all cases, the plots are essentially similar. Carberry[29] has also shown that at $\phi_S > 2.5$, $\eta$ may be computed analytically as

$$\eta = (1/\phi_S)\exp(\beta\gamma/5) \qquad \text{... (8.5.64)}$$

where $\phi_S$ is the value of $\phi$ computed from Eq. (8.5.55) at temperature $T_S$ (in other words, $k_1$ and $K$ are to be specified at temperature $T_S$ for computing $\phi_S$. The temperature dependence of $D_e$ is usually neglected).

How far the resistance to pore diffusion influences the reaction rate can be easily identified by noting that at low values of $\phi$, $\eta$ reduces to unity. In other words, at low values of $\phi$, pore resistance is negligible. The commonly accepted criterion is, therefore,

If $\phi^2 < 1$, pore resistance is negligible ... (8.5.65)

If $\phi^2 > 1$, strong pore resistance exists

In the case of first order rate equations, the above criterion can be readily applied since $\phi$ is independent of $C_{AS}$. However, in all other cases, the above criterion can be applied only if $C_{AS}$ is known or if we assume that the resistance to film diffusion is negligible so that $C_{AS} = C_{Ag}$.

### 8.5.5 Bulk or Film Diffusion

The resistance to film diffusion (or mass transfer in fluid bulk) can be separately accounted for by means of an appropriately defined mass transfer coefficient ($k_g$, $k_C$ or $k_L$). Since film diffusion occurs in series with the other steps, its incorporation into the overall rate equation is relatively easier. We know that under steady state conditions, the rate of transport of $A$ from the fluid bulk to the catalyst surface will be equal to the rate at which $A$ is consumed from the surface by diffusion and reaction. Thus, at steady state,

$$(1/W)\frac{dN_A}{d\theta} = k_C s_p\,(C_{Ag} - C_{AS}) \qquad \text{... (8.5.66)}$$

$$(1/W)\frac{dN_A}{d\theta} = (-r_A) = \eta\,(-r_{AS})_{\text{int}} \qquad \text{... (8.5.67)}$$

where $W$ is the mass of the catalyst and $s_p$ is the specific surface (surface area per unit mass) of catalyst particle. $C_{Ag}$ is the concentration of $A$ (in kmoles/m$^3$) in the fluid bulk and $C_{AS}$ that at the catalyst surface. $k_C$ is the mass transfer coefficient expressed in m$^3$/(s·m$^2$ of catalyst) or simply in (m/s).

The above two equations can be combined together after substituting the expression for $(-r_{AS})_{\text{int}}$ and then eliminating the unknown surface concentrations (such as $C_{AS}$). For example, let $(-r_{AS})_{\text{int}} = k\,C_{AS}$. Then, from Eq. (8.5.67),

$$(1/W)\frac{dN_A}{d\theta}\left(\frac{1}{k\eta}\right) = C_{AS} \qquad \text{... (8.5.68)}$$

Similarly, from Eq. (8.5.66),

$$(1/W)\frac{dN_A}{d\theta}\left(\frac{1}{k_C s_p}\right) = C_{Ag} - C_{AS} \qquad \text{... (8.5.69)}$$

If we add the above two equations, we get

$$(1/W)\frac{dN_A}{d\theta}\left(\frac{1}{k_C s_p} + \frac{1}{k\eta}\right) = C_{Ag} \qquad \text{... (8.5.70)}$$

or

$$(1/W)\,(dN_A/d\theta) = \left[\frac{1}{\left(1/k_C s_p\right) + \left(1/k\eta\right)}\right] C_{Ag} \qquad \text{... (8.5.71)}$$

The first term within brackets on the left hand side of Eq. (8.5.70) stands for the resistance to bulk diffusion and the second term represents resistance to pore diffusion, adsorption, surface reaction and desorption. It is thus clear that though the resistance to internal diffusion (namely, pore diffusion) is combined with the rate constant (k) for the chemical step through η, the external resistance (namely, resistance to diffusion in the fluid bulk) can always be treated as a separate, additive resistance.

The mass transfer coefficient in the above equations is to be computed from the available experimental correlations.

**Example 8.37:** Production of ethyl chloride from ethylene and HCl is carried out in presence of inert methane and using a zirconium oxide catalyst (on silica gel). The overall reaction is represented as

$$C_2H_4 + HCl \rightleftharpoons C_2H_5Cl$$

The equilibrium constant for the above reaction at 450° K is 35 $atm^{-1}$.

(*a*) Neglecting external and internal transport resistances, derive an expression for the rate, if the surface reaction between adsorbed ethylene and adsorbed HCl controls the overall kinetics.

(*b*) Evaluate the constants in the rate equation at 450° K from the following data[21, 30]:

| $(-r_A)$, *kmoles/(hr·kg of catalyst)* | *Partial pressures, atm* | | | |
|---|---|---|---|---|
| | $CH_4$ | $C_2H_4$ | $HCl$ | $C_2H_5Cl$ |
| 0.000271 | 7.005 | 0.300 | 0.370 | 0.149 |
| 0.000263 | 7.090 | 0.416 | 0.215 | 0.102 |
| 0.000244 | 7.001 | 0.343 | 0.289 | 0.181 |
| 0.000258 | 9.889 | 0.511 | 0.489 | 0.334 |
| 0.000269 | 10.169 | 0.420 | 0.460 | 0.175 |

**Solution:** Since the reaction is of the form $A + B \rightleftharpoons P$ and the surface reaction between adsorbed $A$ (namely, adsorbed ethylene) and adsorbed $B$ (namely, adsorbed HCl) is controlling, the rate equation will be that given in Eq. (8.5.34). To note that since external and internal transport resistances are negligible, overall rate = $(-r_A) = (-r_{AS})_{int}$. Also, $C_A = C_{AS} = C_{Ag}$ (or $p_A = p_{AS} = p_{Ag}$). This is true for $B$ and $P$ as well. Since $K = K_A K_B K_S / K_P$, Eq. (8.5.34) can be rewritten as

$$(-r_A) = \frac{k_S K_A K_B (C_A C_B - C_P / K)}{(1 + K_A C_A + K_B C_B + K_P C_P + K_M C_M)^2} \quad \text{... (i)}$$

It may be noted that we have added the term $(K_M C_M)$ in the denominator, where $C_M$ is the concentration of inert (namely, methane) and $K_M$ is the adsorption equilibrium constant for methane. We know that $K$ is the equilibrium constant defined in terms of concentrations. In other words,

$$K = (C_P / C_A C_B)_{eq}$$

Using the ideal gas law, $p_i = C_i RT$,

$$K = \left(\frac{p_P}{p_A p_B}\right)_{eq} (RT) = (35.0)\ (RT)$$

Now, rewriting Eq. (i) in terms of partial pressures and rearranging, we get

$$\left[\frac{p_A p_B - (p_P/35)}{(-r_A)}\right]^{1/2} = (RT/a) + (K_A/a)\,p_A + (K_B/a)\,p_B + (K_P/a)\,p_P + (K_M/a)\,p_M \quad \text{... (ii)}$$

where $a = \sqrt{k_S K_A K_B}$. The above equation is of the form

$$y = a_0 + a_1 p_A + a_2 p_B + a_3 p_B + a_3 p_P + a_4 p_M \quad \text{... (iii)}$$

From the given data, the values of $y$ can be computed since the values of $(-r_A)$, $p_A$, $p_B$, $p_P$ and $p_M$ are given. We, thus, get the following five equations in five unknowns:

$$19.8465 = a_0 + 0.3a_1 + 0.37a_2 + 0.149a_3 + 7.005a_4$$

$$18.1382 = a_0 + 0.416a_1 + 0.215a_2 + 0.102a_3 + 7.09a_4$$

$$19.623 = a_0 + 0.343a_1 + 0.289a_2 + 0.181a_3 + 7.001a_4$$

$$30.520 = a_0 + 0.511a_1 + 0.489a_2 + 0.334a_3 + 9.889a_4$$

$$26.450 = a_0 + 0.420a_1 + 0.460a_2 + 0.175a_3 + 10.169a_4$$

The above five equations can be now solved simultaneously for the five unknowns such as $a_0$, $a_1$, $a_2$, $a_3$ and $a_4$. The results are given below:

$$a_0 = 0.43219$$

$$a_1 = 13.161$$

$$a_2 = 16.242$$

$$a_3 = 16.846$$

$$a_4 = 0.99034$$

Therefore, $a = (RT/a_0) = (0.082)\,(450.0)/(0.43219) = 85.379$

$$K_A = (a_1 a) = 1123.6745 \; (\text{kmoles/m}^3)^{-1}$$

$$K_B = (a_2 a) = 1386.7275 \; (\text{kmoles/m}^3)^{-1}$$

$$K_P = (a_3 a) = 1438.2965 \; (\text{kmoles/m}^3)^{-1}$$

$$K_M = (a_4 a) = 84.5548 \; (\text{kmoles/m}^3)^{-1}$$

$$(k_S K_A K_B) = a^2 = 7289.59 \left(\frac{\text{kmoles}}{\text{hr. kg of catalyst}}\right) (\text{kmoles/m}^3)^{-2}$$

Also $K = (35.0)\,(RT) = (35.0)\,(0.082)\,(450.0) = 1291.5 \; (\text{kmoles/m}^3)^{-1}$

Therefore, the rate equation is,

$$(-r_A) = \frac{7289.59\left(C_A C_B - C_P/1291.5\right)}{\left(1 + 1123.6745 C_A + 1386.7275 C_B + 1438.2965 C_P + 84.5548 C_M\right)^2}$$

where $(-r_A)$ is expressed in kmoles/(hr · kg of catalyst) and $C_A$, $C_B$, $C_P$ and $C_M$ in (kmoles/m$^3$). In terms of partial pressures,

$$(-r_A) = \frac{197.55\left(p_A p_B - p_P/35\right)}{\left(36.9 + 1123.6745\, p_A + 1386.7275\, p_B + 1438.2965\, p_P + 84.5548\, p_M\right)^2}$$

where $p_A$, $p_B$, $p_P$ and $p_M$ are in atm.

In the above example, the number of data points is equal to the number of unknowns and therefore, we solved the five equations simultaneously for the five unknowns. Such a procedure has the inherent drawback that it does not attempt to erase the experimental error contained in the data reported. Usually, a large amount of experimental data are collected so that the number of data points will be much larger than the number of unknowns to be evaluated. The data are then fit into the equation and the constants evaluated by the *method of least squares*. The overall procedure may be summarised as given below:

Let the data available be as follows:

| *Dependent variable* | *Independent variables* | | |
|---|---|---|---|
| $y$ | $x_1$ | $x_2$ | $x_3$ ... $x_n$ |
| $y_1$ | $x_{11}$ | $x_{12}$ | $x_{13}$ ... $x_{1n}$ |
| $y_2$ | $x_{21}$ | $x_{22}$ | $x_{23}$ ... $x_{2n}$ |
| $y_3$ | $x_{31}$ | $x_{32}$ | $x_{33}$ ... $x_{3n}$ |
| . | . | . | . . |
| . | . | . | ... |
| . | . | . | . . |
| . | . | . | . . . |
| $y_m$ | $x_{m1}$ | $x_{m2}$ | $x_{m3}$ ... $x_{mn}$ |

We have to fit the above data into an equation of the type

$$y = a_0 + a_1 x_1 + a_2 x_2 + \ldots + a_n x_n \qquad \text{... (8.5.72)}$$

Then, the solution is (in matrix notation)

$$(C)\ (A) = (D) \qquad \text{... (8.5.73)}$$

where $(C) = (X)^t\ (X)$ and $(D) = (X)^t\ (Y)$.

$$(X) = \begin{bmatrix} 1 & x_{11} & x_{12} & x_{13} & \ldots & x_{1n} \\ 1 & x_{21} & x_{22} & x_{23} & \ldots & x_{2n} \\ \vdots & & & & & \\ 1 & x_{m1} & x_{m2} & x_{m3} & \ldots & x_{mn} \end{bmatrix} \qquad \text{... (8.5.74)}$$

$$(Y) = \begin{bmatrix} Y_{11} \\ Y_{21} \\ . \\ . \\ . \\ Y_{m1} \end{bmatrix} = \begin{bmatrix} y_1 \\ y_2 \\ . \\ . \\ . \\ y_m \end{bmatrix} \qquad \text{... (8.5.74a)}$$

$$(A) = \begin{bmatrix} A_{11} \\ A_{21} \\ \cdot \\ \cdot \\ \cdot \\ A_{n'1} \end{bmatrix} = \begin{bmatrix} a_0 \\ a_1 \\ \cdot \\ \cdot \\ \cdot \\ a_n \end{bmatrix} \qquad \text{... (8.5.74b)}$$

where $n' = (n + 1)$. To note that $(X)'$ is the transpose of the matrix $(X)$. In other words, $(X)'$ is the matrix obtained from interchanging the rows and columns of $(X)$. The elements of $(C)$ and $(D)$ are to be obtained as follows:

If $i$ represents the row index and $j$ the column index, then

$$C_{ij} = \sum_{k=1}^{m} \left(XT_{ik} XM_{kj}\right), \text{ for } i = 1 \text{ to } (n + 1), j = 1 \text{ to } (n + 1)$$

$(C)$ is thus a square matrix with $(n + 1)$ rows and $(n + 1)$ columns. $XT_{ij}$ and $XM_{ij}$ are defined below:

$$XM_{ij} = 1, \text{ for } i = 1 \text{ to } m \text{ and } j = 1$$

$$XM_{ij} = x_{i,\, j-1}, \text{ for } i = 1 \text{ to } m \text{ and } j = 2 \text{ to } (n + 1)$$

$$XT_{ij} = 1, \text{ for } i = 1 \text{ and } j = 1 \text{ to } m$$

$$XT_{ij} = x_{j,\, i-1} \text{ for } i = 2 \text{ to } (n + 1) \text{ and } j = 1 \text{ to } m$$

In fact, $XM_{ij}$ are elements of $(X)$ and $XT_{ij}$ are elements of $(X)'$. $(D)$ is a column matrix with only one column and $(n + 1)$ rows. Its elements are

$$D_{i1} = \sum_{k=1}^{m} XT_{ik} Y_{k1}, \text{ for } i = 1 \text{ to } (n + 1)$$

It may be noted that $Y_{k1} = y_k$ $(k = 1$ to $m)$.

Equation (8.5.73), thus, represents a set of $(n + 1)$ equations in $(n + 1)$ unknowns (such as $a_0$, $a_1$ ... $a_n$) and therefore, this can be solved by any of the well-established methods. Successive elimination (which we employed in Example 8.37) is the most straightforward method. However, this method could become too cumbersome if the number of unknowns is large. An iterative method like the Gauss-Seidel Iteration[31] or a direct method like Crout's Reduction[32] is most recommended in such cases. We have discussed these methods in Chapter 3 under multiple effect evaporation and in Chapter 5 under multicomponent distillation. For the convenience of the readers, the Crout's reduction method is illustrated below:

(*i*) Compute $C'_{ij}$ as follows:

$$C'_{ij} = C_{ij}, \text{ if } j = 1$$

$$= C_{ij} - \sum_{k=1}^{j-1} C'_{ik} C'_{kj}, \text{ if } j > 1 \text{ and } i \geq j$$

$$= (C_{ij}/C'_{ii}), \text{ if } i = 1 \text{ and } i < j$$

$$= (1/C'_{ii}) \left( C_{ij} - \sum_{k=1}^{i-1} C_{ik}{}' C_{kj}{}' \right), \text{ if } i > 1 \text{ and } i < j$$

(*ii*) Compute $D'_{i1}$ as follows:

$$D'_{i1} = D_{i1}/C'_{ii}, \text{ if } i = 1 = (1/C'_{ii}) \left( D_{i1} - \sum_{k=1}^{i-1} C_{ik}{}' D_{k1}{}' \right), \text{ if } i > 1$$

(*iii*) Now, compute $A_{i1}$ starting from $i = (n + 1)$ and proceeding down to $i = 1$ as

$$A_{i1} = D'_{i1}, \text{ if } i = (n + 1) = D'_{i1} - \sum_{k=i+1}^{n+1} C'_{ik} A_{k1}, \text{ if } i \le n$$

(*iv*) Finally, find the values of $a_0$, $a_1$, $a_2$ ... $a_n$ as

$$a_{i-1} = A_{i1}, \text{ for } i = 1 \text{ to } (n + 1)$$

The computer program for the above procedure is given below:

## PROGRAM 8.4

```
C**   FITTING EXPERIMENTAL DATA TO AN EQUATION
C**   OF THE FORM y = a0 + a1x1 + a2x2 + ... anxn
C**   DIMENSION X (100, 20), Y (100), XM (100, 20), XT (100, 20)
      DIMENSION C (20, 20), CC (20, 20), D (20, 1), DD (20, 1), A (20, 1)
      READ *, N, M
      READ *, [Y (I), I = 1, M]
      READ *, {[X (I, J), I = 1, M], J = 1, N}
      N1 = N + 1
      DO 10 I = 1, N
      XM (I, 1) = 1.0
      DO 15 J = 2, N1
      XM (I, J) = X (I, J - 1)
15    CONTINUE
10    CONTINUE
      DO 20 J = 1, M
      XT (1, J) = 1.0
      DO 25 I = 2, N1
      XT (I, J) = X (J, I - 1)
25    CONTINUE
20    CONTINUE
      DO 30 I = 1, N1
      DO 35 J = 1, N1
```

```
      SUM = 0.0
      DO 40 K = 1, M
40    SUM = SUM + XT (I, K) * XM (K, J)
      C (I, J) = SUM
35    CONTINUE
30    CONTINUE
      DO 45 I = 1, N1
      SUM = 0.0
      DO 50 K = 1, M
50    SUM = SUM + XT (I, K) * Y (K)
      D (I, 1) = SUM
45    CONTINUE
      DO 55 I = 1, N1
      CC (I, 1) = C (I, 1)
      DO 60 J = 2, N1
      IF (I·EQ·1) GO TO 65
      IF (I·GE·J) GO TO 70
      SUM = 0.0
      DO 75 K = 1, I - 1
75    SUM = SUM + CC (I, K) * CC (K, J)
      CC (I, J) = [C (I, J) - SUM]/CC (I, I)
      GO TO 60
65    CC (I, J) = C (I, J)/CC (I, I)
      GO TO 60
70    SUM = 0.0
      DO 80K = 1, J - 1
80    SUM = SUM + CC (I, K) * CC (K, J)
      CC (I, J) = C (I, J) - SUM
60    CONTINUE
55    CONTINUE
      DD (1, 1) = D (1, 1)/CC (1, 1)
      DO 85 I = 2, N1
      SUM = 0.0
      DO 90 K = 1, I - 1
90    SUM = SUM + CC (I, K) * DD (K, 1)
      DD (I, 1) = [D (I, 1) - SUM]/CC (I, I)
85    CONTINUE
      A (N1, 1) = DD (N1, 1)
      I = N
100   SUM = 0.0
```

```
        DO. 95 K = I + 1, N1
95      SUM = SUM + CC (I, K) * A (K, 1)
        A (I, 1) = DD (I, 1) - SUM
        IF (I·EQ·1) GO TO 105
        I = I - 1
        GO TO 100
105     PRINT *, [A (I, 1), I = 1, N1]
        STOP
        END
```

**Example 8.38:** The reaction between carbon monoxide and chlorine (to produce phosgene or carbonyl chloride) in presence of an activated carbon catalyst is studied at 1 atm and the results show that the surface reaction between adsorbed CO and adsorbed chlorine is the rate-controlling step and that the rate of reaction does not depend on the mass velocity of gases through the reactor. It is also observed that chlorine and phosgene are readily adsorbed on the catalyst, while carbon monoxide is not. The overall reaction is represented by

$$CO + Cl_2 \rightarrow COCl_2$$

The reaction is irreversible. Develop an expression for the intrinsic rate in terms of the partial pressures of the substances in the gas bulk and determine the values of the constants in the rate equation from the following experimental data[21, 33] at 30.6° C and 1 atm (catalyst size = 3.1 mm):

| *Rate, kmoles/(hr. kg of catalyst)* | *Partial pressures, atm* | | |
|---|---|---|---|
| | *CO* | *$Cl_2$* | *$COCl_2$* |
| 0.00414 | 0.406 | 0.352 | 0.226 |
| 0.00440 | 0.396 | 0.363 | 0.231 |
| 0.00241 | 0.310 | 0.320 | 0.356 |
| 0.00245 | 0.287 | 0.333 | 0.376 |
| 0.00157 | 0.253 | 0.218 | 0.522 |
| 0.00200 | 0.179 | 0.608 | 0.206 |

Assume that 3.1 mm catalyst particles are small enough that the pore surface is fully effective.

**Solution:** Since it is observed that the overall rate is not affected by the mass velocity of gases through the reactor, it means that the resistance to external transport (namely, transport in the gas bulk) is not significant. Also, since it is being assumed that for the 3.1 mm catalyst particles the pore surface is fully effective, the effectiveness factor is unity and the resistance to pore diffusion is also negligible. Accordingly,

$$(\text{overall rate}) = (-r_A) = (-r_{AS})_{\text{int}}$$

The reaction is of the type $A + B_2 \rightarrow AB_2$. Since chlorine is a diatomic gas, it may follow dissociative adsorption which is represented by

$$B_2 + 2X \rightleftharpoons 2B{\cdot}X$$

where $B$ stands for chlorine. In other words, the chlorine molecule dissociates upon adsorption with each atom occupying one catalyst site. Since surface reaction between adsorbed $A$ (namely, adsorbed CO) and adsorbed $B$ is controlling, the rate equation will be that given in Eq. (iii) of S. no. 12 of Table 8.5. Thus,

$$(-r_A) = \frac{k_S K_A K_B \left(C_{Ag} C_{Bg} - C_{Pg}/K\right)}{\left(1 + K_B^{1/2} C_{Bg}^{1/2} + K_A C_{Ag} + K_P C_{Pg}\right)^3}$$

where $K = (K_A K_B K_S / K_P)$ and $P$ stands for $AB_2$. We have used bulk concentrations $C_{Ag}$, $C_{Bg}$ and $C_{Pg}$ in places of $C_{AS}$, $C_{BS}$ and $C_{PS}$, since $C_{Ag} = C_{AS} = C_A$ and similarly for $B$ and $P$ as well. Since the reaction is irreversible,

$$(-r_A) = \frac{\left(k_S K_A K_B\right) C_{Ag} C_{Bg}}{\left(1 + K_A C_{Ag} + K_B^{1/2} C_{Bg}^{1/2} + K_P C_{Pg}\right)^3}$$

Since it is aiso given that carbon monoxide is not as readily adsorbed as chlorine and phosgene, the adsorption equilibrium constant of CO (namely, $K_A$) can be considered to be negligible with respect to those of chlorine and phosgene. Accordingly, the above equation simplifies to

$$(-r_A) = \frac{\left(k_S K_A K_B\right) C_{Ag} C_{Bg}}{\left(1 + K_B^{1/2} C_{Bg}^{1/2} + K_P C_{Pg}\right)^3}$$

A further simplification is possible if we neglect 1.0 in the denominator (this is not irreasonable since $K_B$ and $K_P$ are significantly larger than 1.0). Thus

$$(-r_A) = \frac{\left(k_S K_A K_B\right) C_{Ag} C_{Bg}}{\left(K_B^{1/2} C_{Bg}^{1/2} + K_P C_{Pg}\right)^3}$$

In terms of partial pressures (since $p_i = C_i RT$),

$$(-r_A) = \frac{\left(p_{Ag} p_{Bg}\right)}{\left(k_1 p_{Bg}^{1/2} + k_2 p_{Pg}\right)^3} \qquad \text{... (i)}$$

where $k_1 = [K_B^{1/2} (RT)^{1/6}/a]$, $k_2 = [K_P (RT)^{-1/3}/a]$ and $a = (k_S K_A K_B)^{1/3}$. This is the ultimate rate equation. However, we have to verify it by checking whether it fits reasonably to the experimental data provided. Now, on rearrangement, Eq. (i) becomes

$$(1/p_{Pg})\ [p_{Ag} p_{Bg}/(-r_A)]^{1/3} = k_1\ (p_{Bg}^{1/2}/p_{Pg}) + k_2$$

or

$$F\ (p) = k_1\ (p_{Bg}^{1/2}/p_{Pg}) + k_2 \qquad \text{... (ii)}$$

This shows that a plot of $F\ (p)$ versus $(p_{Bg}^{1/2}/p_{Pg})$ must be a straight line. The values of these two parameters can be computed from the available experimental data and are listed below:

| $F(p)$ | $(p_{Bg}^{1/2}/p_{Pg})$ |
|---|---|
| 14.407 | 2.6252 |
| 13.839 | 2.6082 |
| 9.6987 | 1.589 |
| 9.0198 | 1.5347 |
| 6.274 | 0.8944 |
| 18.395 | 3.785 |

The plot of $F(p)$ versus $(p_{Bg}^{1/2}/p_{Pg})$ is shown in Fig. 8.38.1. The plot is essentially rectilinear. This shows that the simplification introduced into the rate equation is not irreasonable. From the plot,

$$\text{Slope} = k_1 = 4.33$$

$$\text{Intercept} = k_2 = 2.75$$

Therefore, the rate equation is

$$(-r_A) = \frac{\left(p_{Ag}\, p_{Bg}\right)}{\left(4.33\, p_{Bg}^{1/2} + 2.75\, p_{Pg}\right)^3}$$

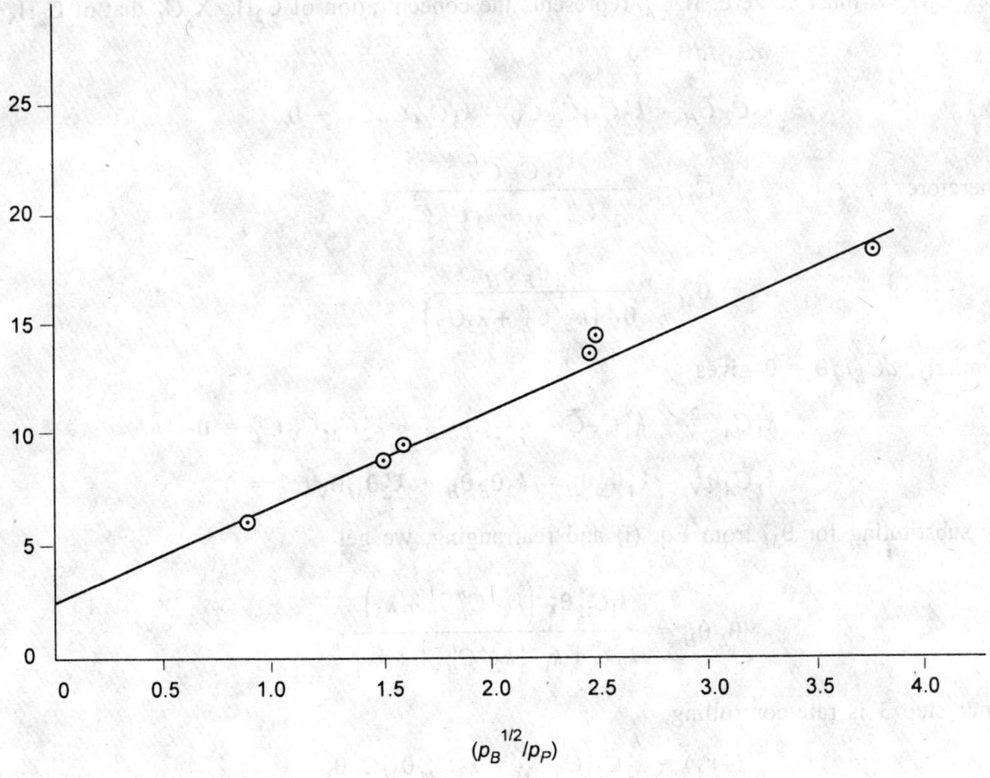


*Figure 8.38.1*

**Example 8.39:** The following mechanism has been proposed for the hydrogenolysis of ethane over a nickel or cobalt catalyst:

$$C_2H_6 + 2X \underset{}{\overset{k_1}{\rightleftharpoons}} C_2H_5{\cdot}X + H{\cdot}X$$

$$C_2H_5{\cdot}X + H{\cdot}X \underset{}{\overset{k_2}{\rightleftharpoons}} C_2H_x{\cdot}X + nH_2 + X$$

$$C_2H_x{\cdot}X + H_2 + X \xrightarrow{k_3} CH_y{\cdot}X + CH{\cdot}X$$

$$CH_y{\cdot}X + CH{\cdot}X + nH_2 \rightarrow 2CH_4$$

(*a*) Derive a rate equation from the above-postulated mechanism based on the steady state hypothesis and by assuming that the rate is controlled by step 3.

(*b*) Experimental data indicate that the rate of reaction may be empirically represented as

$$(-r_A) = k\,C_A^a\,C_B^b$$

where *A* stands for ethane and *B* for hydrogen. How far does the equation derived in (*a*) agree with the above empirical expression?

**Solution:** According to the steady state hypothesis, the net rate of production of the intermediate complex $C_2H_x{\cdot}X$ must be zero. If $\bar{C}_M$ represents the concentration of $C_2H_x{\cdot}X$ $\bar{C}_E$ that of $C_2H_5{\cdot}X$, then

$$d\bar{C}_M/d\theta = 0$$

or

$$k_2\bar{C}_E\bar{C}_H - k_2'\bar{C}_M C_B^n\bar{C}_V - k_3\bar{C}_M C_B\bar{C}_V = 0$$

Therefore

$$\bar{C}_M = \frac{k_2\,\bar{C}_E\,\bar{C}_H}{k_2{'}\,C_B^n\,\bar{C}_V + k_3\,C_B\,\bar{C}_V}$$

or

$$\theta_M = \frac{k_2\,\theta_E\,\theta_H}{\theta_V\left(k_2{'}\,C_B^n + k_3C_B\right)} \qquad \text{... (i)}$$

Similarly, $d\bar{C}_E/d\theta = 0$ gives

$$k_1\,C_A C_V^2 - k_1'\bar{C}_E\bar{C}_H - k_2\bar{C}_E\bar{C}_H + k_2'\bar{C}_M\bar{C}_V C_B^n = 0$$

or

$$k_1\,C_A\theta_V^2 - k_1'\theta_E\theta_H - k_2\theta_E\theta_H + k_2'\theta_M\theta_V C_B^n = 0$$

By substituting for $\theta_M$ from Eq. (i) and rearranging, we get

$$\theta_E\,\theta_H = \frac{k_1 C_A\,\theta_V^2\left(k_2{'}\,C_B^{n-1} + k_3\right)}{k_2k_3 + k_1{'}\left(k_2{'}\,C_B^{n-1} + k_3\right)} \qquad \text{... (ii)}$$

Since step 3 is rate-controlling,

$$(-r_A) = k_3\bar{C}_M C_B C_V = k_3\bar{C}_m^2\,\theta_M C_B\theta_V$$

Substituting for $\theta_M$ from Eq. (i), we get

$$(-r_A) = \frac{k_3 \bar{C}_m^2 k_2 \theta_E \theta_H}{\left(k_2' C_B^{n-1} + k_3\right)}$$

Now, substituting for $(\theta_E \theta_H)$ from Eq. (ii), we get

$$(-r_A) = \frac{k_1 k_2 k_3 \left(\bar{C}_M \theta_V\right)^2 C_A}{k_1' k_2' C_B^{n-1} + k_3 \left(k_1' + k_2\right)} = \frac{\left(k C_V^2\right) C_A}{C_B^{n-1} + k'} \quad \text{... (iii)}$$

where $k = (k_1 k_2 k_3)/(k_1' k_2')$ and $k' = k_3 (k_1' + k_2)/(k_1' k_2')$. At low concentrations of $A$ (namely, ethane) and $B$ (namely, hydrogen), the fraction of catalyst sites covered by the complexes $C_2H_5 \cdot X$ and $C_2H_x \cdot X$ will be small and accordingly, $C_V^2$ will be essentially constant.

(*b*) When $k_1 = k_1'$ and $k_2 = k_2'$, Eq. (iii) reduces to

$$(-r_A) = \frac{\left(k_3 C_V^2\right) C_A}{C_B^{n-1} + k_3 \left[(1/k_1) + (1/k_2)\right]}$$

When $k_1$ and $k_2$ are much larger than $k_3$, the above expression may be further simplified to

$$(-r_A) = (k_3 C_V^2)\, C_A C_B^{1-n} = k C_A C_B^{1-n} \quad \text{... (iv)}$$

The above expression agrees with the empirical equation reported.

**Example 8.40:** A solid–catalysed reaction $A \rightarrow P$, which is irreversible, is studied in an experimental packed bed reactor and the following results are reported:

Feed rate (feed containing only $A$) = 10 kmoles/hr

| kg of catalyst | 1.0 | 2.0 | 3.0 | 4.0 | 5.0 | 6.0 | 7.0 |
|---|---|---|---|---|---|---|---|
| Conversion, $x_{Af}$ | 0.12 | 0.20 | 0.27 | 0.33 | 0.37 | 0.40 | 0.44 |

(*a*) Develop a rate equation for the reaction.

(*b*) If the above reaction is to be conducted in a packed bed reactor with a feed rate (feed containing only $A$) of 400 kmoles/hr, compute the amount of catalyst required to achieve 40 percent conversion. What would be the amount of catalyst required if the same reactor is operated with a very large recycle of the product?

Assume plug flow in the packed bed reactor.

**Solution:** (*a*) The rate equation is to be determined by trial. To start with, let us assume that

(*i*) the resistances to internal and external transport are negligible. Accordingly,

$$(-r_A) = (-r_{AS})_{int}$$

(*ii*) the surface reaction in which the adsorbed $A$ is converted to adsorbed $P$ is the rate-controlling step.

Under these circumstances, the rate equation will be that given in Eq. (ii) of S. no. 1 of Table 8.5. Thus

$$(-r_A) = (-r_{AS})_{int} = \frac{k_S K_A \left(C_A - C_P / K\right)}{\left(1 + K_A C_A + K_P C_P\right)}$$

Since the reaction is irreversible,

$$(-r_A) = \frac{k_S K_A C_A}{(1 + K_A C_A + K_P C_P)}$$

Let us further assume that the reactant $A$ is not as strongly adsorbed as $P$. In other words, the adsorption equilibrium constant for $A$ (namely, $K_A$) is much smaller than $K_P$ and therefore, the term $(K_A C_A)$ can be dropped from the denominator of the above expression. Thus

$$(-r_A) = \frac{(k_S K_A) C_A}{(1 + K_P C_P)} = \frac{k C_A}{(1 + K_P C_P)} \quad \text{... (i)}$$

where $k = (k_S K_A)$. Now, since plug flow is assumed in the packed bed reactor, the performance equation is given by Eq. (8.3.32). However, since $(-r_A)$ is expressed in kmoles/(hr. kg of catalyst), Eq. (8.3.32) is to be modified as

$$(W/F_{A0}) = \int_0^{x_{Af}} \frac{dx_A}{(-r_A)} \quad \text{... (ii)}$$

where $W$ = mass of catalyst used, kg. To apply the above equation, it will be convenient to express $(-r_A)$ in terms of the conversion, $x_A$. Thus,

$$C_A = C_{A0}\ (1 - x_A)/(1 + \epsilon_A x_A)$$

$$C_P = (C_{A0} x_A)/(1 + \epsilon_A x_A)$$

Since there is no change in number of moles during reaction, $\epsilon_A = 0$. Therefore,

$$C_A = C_{A0}\ (1 - x_A)$$

$$C_P = C_{A0} x_A$$

Now, from Eq. (i),

$$1/(-r_A) = \left(\frac{1}{k C_{A0}}\right) \frac{1}{(1 - x_A)} + (K_P/k)\ [x_A/(1 - x_A)]$$

Substituting the above expression in Eq. (ii) and integrating, we get

$$(W/F_{A0}) = (W/10) = -\left(\frac{1}{k C_{A0}}\right) \ln\ (1 - x_{Af}) - (K_P/k)\ [x_{Af} + \ln\ (1 - x_{Af})]$$

or

$$\frac{-(W/10)}{\left[x_{Af} + \ln\left(1 - x_{Af}\right)\right]} = \left(\frac{1}{k C_{A0}}\right) \frac{\ln\left(1 - x_{Af}\right)}{\left[x_{Af} + \ln\left(1 - x_{Af}\right)\right]} + (K_P/k)$$

or

$$F'\ (x_{Af}) = \left(\frac{1}{k C_{A0}}\right) F\ (x_{Af}) + (K_P/k) \quad \text{... (iii)}$$

This shows that a plot of $F'\ (x_{Af})$ versus $F\ (x_{Af})$ must be a straight line. The values of these two functions can be computed from the available data and these are listed below:

| W, kg | $F'(x_{Af})$ | $F(x_{Af})$ |
|---|---|---|
| 1.0 | 12.766 | 16.3192 |
| 2.0 | 8.6417 | 9.6417 |
| 3.0 | 6.7098 | 7.04 |
| 4.0 | 5.6756 | 5.6823 |
| 5.0 | 5.4327 | 5.02 |
| 6.0 | 5.10 | 4.4854 |
| 7.0 | 5.0065 | 4.1469 |

The plot of $F'(x_{Af})$ versus $F(x_{Af})$ is shown in Fig. 8.40.1. The plot is rectilinear, thereby justifying the assumptions made in the development of the rate equation. From the plot,

$$\text{Slope} = 1/(k\,C_{A0}) = 0.666 \text{ (kg·hr)/kmoles}$$

$$y\text{-intercept} = (K_P/k) = 2.2 \text{ (kg·hr)/kmoles}$$

Therefore,

$$k\,C_{A0} = 1/0.66 = 1.5 \text{ kmoles/(kg·hr)}$$

$$(K_P C_{A0}) = (2.2)\ (k\,C_{A0}) = 3.30$$

Therefore, from Eq. (i), the rate equation is

$$(-r_A) = \frac{1.5(1-x_A)}{1+3.3x_A}$$

The following are our conclusions regarding the mechanism of the reaction:

(*i*) The reaction rate is not controlled by the external or internal mass transport resistance,

(*ii*) The surface reaction in which adsorbed $A$ is converted to adsorbed $P$ is the rate-controlling step,

(*iii*) The reactant $A$ is not as strongly adsorbed as the product $P$.

Incidentally, we have followed the *integral method of analysis* in the above computations. Wherever the rate equation can be readily integrated, this method could be the most convenient one. However, when the rate expression is more complicated, the *differential method of analysis* becomes more convenient. As an example, let us re-solve this problem by the differential method of analysis:

From Eq. (8.3.31), $\quad F_{A0}dx_A = (-r_A)\ dW$

or

$$(-r_A) = \frac{dx_A}{dW/F_{A0}} = \frac{dx_A}{d(W/F_{A0})}$$

Thus, a plot of $x_A$ versus $(W/F_{A0})$ is first prepared. This plot is shown in Fig. 8.40.2. The value of $(-r_A)$ at any $x_A$ is the slope of the tangent to this curve. For example, from Fig. 8.40.2, the slope of the tangent to the curve at $x_A = 0.12$ is 0.911. Therefore, the value of $(-r_A)$ at $x_A = 0.12$ is 0.911 kmoles/(hr·kg of catalyst). Similarly, the values of $(-r_A)$ at different values of $x_A$ are determined and are listed below:

| $x_A$ | $(-r_A)$ | $(1 - x_A)/(-r_A)$ |
|---|---|---|
| 0.12 | 0.911 | 0.966 |
| 0.20 | 0.72 | 1.110 |
| 0.30 | 0.52 | 1.346 |
| 0.40 | 0.38 | 1.580 |
| 0.50 | 0.28 | 1.7857 |

Now, from Eq. (i), $1/(-r_A) = 1/(kC_A) + (K_P/k)\,(C_P/C_A)$

Since $C_A = C_{A0}\,(1 - x_A)$ and $C_P = C_{A0}\,x_A$, we get

$$(1 - x_A)/(-r_A) = (K_P/k)\,x_A + (1/k\,C_{A0})$$

*Figure 8.40.1*

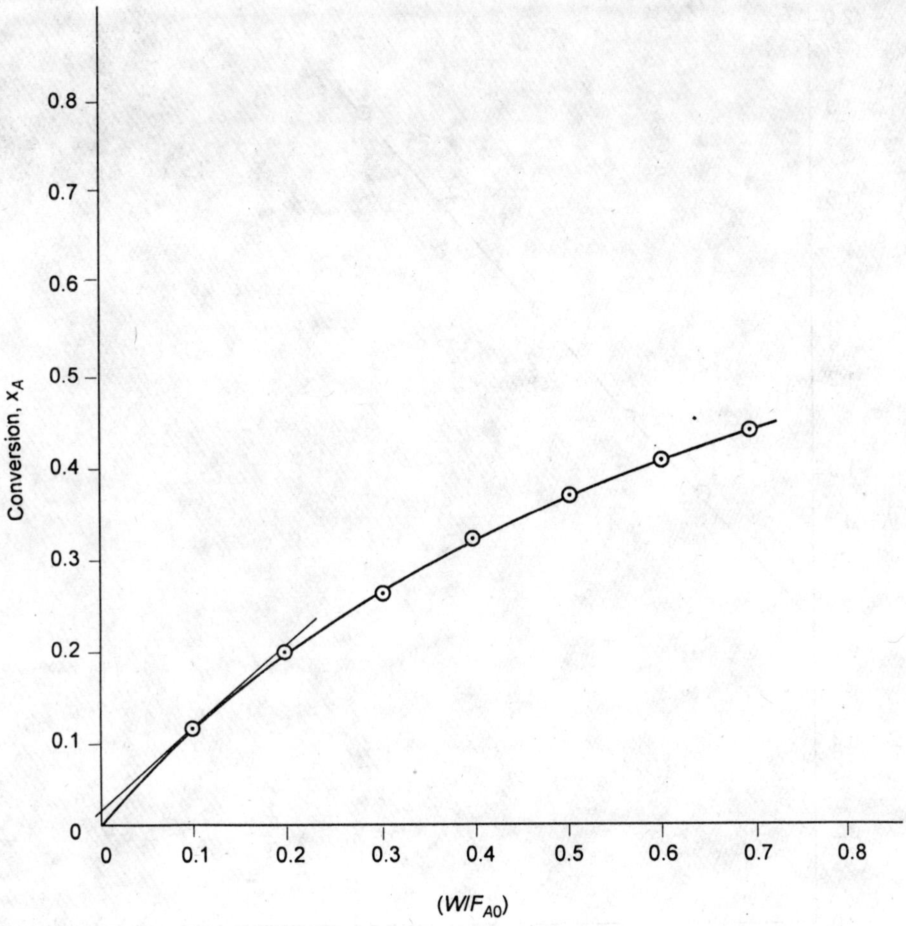


*Figure 8.40.2*

This shows that a plot of $(1 - x_A)/(-r_A)$ versus $x_A$ must be a straight line. This plot is shown in Fig. 8.40.3. Since the plot is rectilinear, the assumptions involved in the development of the rate equation are justified. From the plot,

$$\text{Slope} = (K_P/k) = 2.26 \ (\text{kg} \cdot \text{hr})/\text{kmoles}$$

$$y\text{-intercept} = 1/kC_{A0} = 0.68 \ (\text{kg} \cdot \text{hr})/\text{kmoles}$$

Therefore,

$$kC_{A0} = 1/0.68 = 1.4706 \ \text{kmoles}/(\text{kg} \cdot \text{hr})$$

$$K_P C_{A0} = (2.26)\ (kC_{A0}) = 3.3235$$

The rate equation is, therefore,

$$(-r_A) = \frac{1.4706\,(1 - x_A)}{1 + 3.3235\,x_A}$$

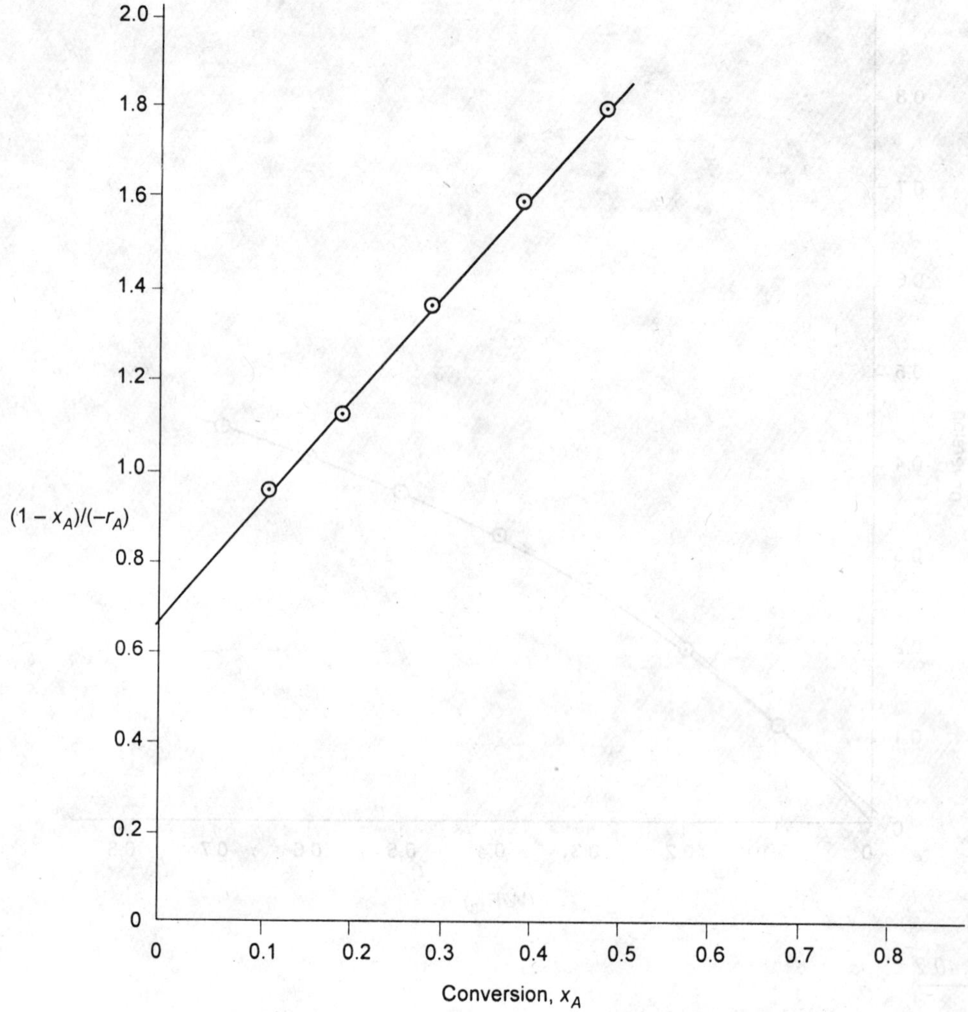


*Figure 8.40.3*

The slight discrepancy in the values of the constants is due to the personal error involved in the graphical determination of the values of $(-r_A)$.

(*b*) We can now rewrite Eq. (iii) in terms of the known values of the constants. Thus

$$(W/F_{A0}) = -(0.666)\ \ln\ (1 - x_{Af}) - (2.2)\ [x_{Af} + \ln\ (1 - x_{Af})]$$

Since $x_{Af} = 0.40$ and $F_{A0} = 400$ kmoles/hr,

$$W = \mathbf{233.75\ kg}$$

This is the amount of catalyst required. In the case of recycle reactor, if the recycle ratio employed is very large, then its performance will approach to that of a back-mix reactor. Accordingly, we can use the performance equation of a back-mix reactor (Eq. 8.3.9) to compute the amount of catalyst required in this case. Thus

$$(-r_A)_f = F_{A0} x_{Af}/W$$

Now, from the developed rate equation,

$$(-r_A)_f = \frac{1.5\left(1 - x_{Af}\right)}{1 + 3.3\,x_{Af}} = \frac{1.5\,(1 - 0.40)}{1 + 3.3\,(0.40)} = 0.3879 \text{ kmoles/(hr}\cdot\text{kg of catalyst)}$$

Therefore, $W = (400.0)\ (0.40)/(0.3879) = \mathbf{412.44\ kg}$

**Example 8.41:** The solid catalysed reaction, $A \rightarrow 3P$ is to be conducted in a tubular, pilot plant reactor that is 50 mm in ID and is packed with a solid mixture containing 25 percent (by weight) catalyst pellets and the rest inert pellets. The reactor operates at 7 atm and 600° C. Both the catalyst pellets and the inert pellets are 3 mm in size and are of the same density (2000 kg/m$^3$). The bulk voidage of the bed is 0.45. The feed containing 50 percent $A$ (by volume) and 50 percent inert gas is introduced at 600° C and 7 atm at the rate of 0.5 kmoles/hr.

(*a*) Assuming plug flow and neglecting the pressure drop across the bed, determine what must be the length of the catalyst bed so that the concentration of $A$ in the exit stream will be one-fourth of that in the inlet stream. When the reaction was initially studied in a basket type backmix reactor at the same temperature and pressure and using 0.005 kg of the same catalyst, the following results were obtained (the feed contained only $A$):

| Feed rate, m$^3$/hr : | 0.0077 | 0.02 | 0.07 | 0.115 | 0.2 | 0.32 | 0.6 |
|---|---|---|---|---|---|---|---|
| $p_{Af}/p_{A0}$ : | 0.1 | 0.2 | 0.4 | 0.5 | 0.6 | 0.7 | 0.8 |

The reaction is exothermic and is presumed to be approximately first order in $A$.

(*b*) Can any conclusions be drawn regarding the influence of (*i*) film diffusion, (*ii*) pore diffusion, (*ii*) temperature gradient in the gas film and (*iii*) temperature gradient in the catalyst pellet on the performance of the pilot plant reactor?

Effective diffusivity in pellet = $9.0 \times 10^{-6}$ m$^2$/hr

Effective thermal conductivity of pellet = 1.163 W/(mK)

Fluid to pellet heat transfer coefficient = 116.3 W/(m$^2$K)

Heat of reaction = 83740 kJ/kmole of $A$

Fluid to pellet mass transfer coefficient = 300 m$^3$/(hr·m$^2$ of catalyst)

**Solution:** (*a*) Since we have an approximate idea that the reaction could be first order in $A$, let us first try to fit the available experimental data into a rate equation of the form, $(-r_A) = k_1 C_A$. Since the reaction was studied in a backmix reactor, let us use this rate equation in the performance equation of the backmix reactor (Eq. 8.3.9):

$$(-r_A)_f = kC_{Af} = F_{A0} x_{Af}/W$$

Since $F_{A0} = Q_0\, C_{A0}$ (where $Q_0$ is the volumetric feed rate in m$^3$/hr),

$$kC_{Af} = Q_0 C_{A0} x_{Af}/W$$

or

$$Q_0 = kW\ (C_{Af}/C_{A0})/x_{Af}$$

Assuming ideal gas law to be valid ($p_i = C_i RT$), $C_{Af}/C_{A0} = (p_{Af}/p_{A0})$. Also, we know that

$$x_{Af} = \frac{1-(C_{Af}/C_{A0})}{1+(\epsilon_A C_{Af}/C_{A0})} = \frac{1-(p_{Af}/p_{A0})}{1+(\epsilon_A p_{Af}/p_{A0})}$$

Since the feed consisted of only $A$, $\epsilon_A = 2.0$. Therefore,

$$Q_0 = (kW)\frac{(p_{Af}/p_{A0})(1+2\,p_{Af}/p_{A0})}{(1-p_{Af}/p_{A0})} = (kW)\ F\ (p)$$

This shows that the plot of $Q_0$ versus $F\ (p)$ must be a straight line. The values of $F\ (p)$ can be computed from the available data and are listed below:

| $Q_0$, m$^3$/hr | : | 0.0077 | 0.02 | 0.07 | 0.115 | 0.2 | 0.32 | 0.6 |
|---|---|---|---|---|---|---|---|---|
| $F\ (p)$ | : | 0.133 | 0.35 | 1.2 | 2.0 | 3.3 | 5.6 | 10.4 |

The plot of $Q_0$ versus $F\ (p)$ is shown in Fig. 8.41.1. It is rectilinear and therefore, the assumed rate equation is correct. From the plot,

$$\text{slope} = (kW) = 0.05769\ \text{m}^3/\text{hr}$$

Since $W = 0.005$ kg,

$$k = (0.05769)/(0.005) = 11.538\ \text{m}^3/(\text{hr}\cdot\text{kg of catalyst})$$

The rate equation is, therefore,

$$(-r_A) = 11.538\, C_A$$

Though this does not shed any light on the mechanism of the reaction, we are successful in obtaining the rate equation for the reaction. In terms of conversion,

$$(-r_A) = 11.538\, C_{A0}\ (1 - x_A)/(1 + \epsilon_A x_A)$$

Now, for the pilot plant reactor, assuming plug flow,

$$W/F_{A0} = \int_0^{x_{Af}} \frac{dx_A}{(-r_A)}$$

or

$$W = \frac{(F_{A0}/C_{A0})}{(11.538)} \int_0^{x_{Af}} \frac{(1+\epsilon_A x_A)\,dx_A}{(1-x_A)}$$

$$= \frac{(F_{A0}/C_{A0})}{(11.538)} [-(1 + \epsilon_A)\ \ln\ (1 - x_{Af}) - \epsilon_A x_{Af}]$$

Consider 2 kmoles of feed (1 kmole of $A$ and 1 kmole of inert gas) at zero conversion. At 100 percent conversion, there will be 3 kmoles of $P$ and 1 kmole of inert gas. Therefore,

$$\epsilon_A = (4 - 2)/2 = 1.0$$

$$x_{Af} = \frac{1-(p_{Af}/p_{A0})}{1+(\epsilon_A p_{Af}/p_{A0})}$$

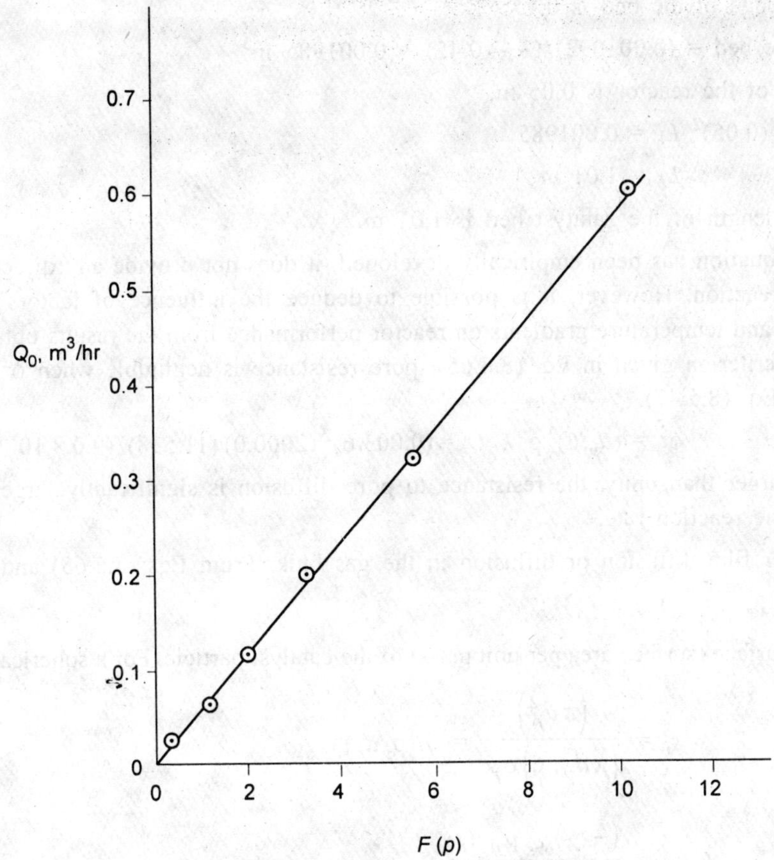


*Figure 8.41.1*

Since $$(C_{Af}/C_{A0}) = (p_{Af}/p_{A0}) = 1/4 = 0.25,$$

$$x_{Af} = (0.75)/(1.25) = 0.6$$

Since the feed contains 50 percent (by volume) $A$,

$$p_{A0} = (0.5)\ (7.0) = 3.5 \text{ atm}$$

$$C_{A0} = p_{A0}/RT = \frac{3.5}{(0.082)(873)} = 0.0489 \text{ kmoles/m}^3$$

$$F_{A0} = (0.5)\ (0.5) = 0.25 \text{ kmoles/hr}$$

Therefore, $$W = 0.546 \text{ kg}$$

Since the reactor is packed with a solid mixture containing 25 percent catalyst pellets and the rest inert pellets (this ensures isothermal operation),

Total mass of packed solids = (0.546)/(0.25) = 2.184 kg

Total volume of packed solids = (2.184/2000) = 0.001092 m$^3$

Since the bulk voidage of the bed is 0.45,

Total volume of the bed = (0.001092)/(1 – 0.45) = 0.001985 $m^3$

Since the diameter of the reactor is 0.05 m,

$$(\pi/4)\ (0.05)^2\ L_b = 0.001985$$

$$L_b = 1.01 \text{ m}$$

Thus, the required length of the catalyst bed is 1.01 m.

(*b*) Since the rate equation has been empirically developed, it does not provide any direct clues on the mechanism of the reaction. However, it is possible to deduce the influence of factors like pore diffusion, film diffusion and temperature gradients on reactor performance from the results obtained. For example, based on the criterion given in Eq. (8.5.65), pore resistance is negligible when $\phi^2 < 1$. For the present case, from Eq. (8.5.47),

$$\phi^2 = (d_p/6)^2\ \rho_p\, k/D_e = (0.003/6)^2\ (2000.0)\ (11.538)/(9.0 \times 10^{-6}) = 641.0$$

Since $\phi^2$ is much larger than unity, the resistance to pore diffusion is significantly large and it is substantially retarding the reaction rate.

Let us now consider film diffusion or diffusion in the gas bulk. From Eqs (8.5.65) and (8.5.67),

$$k_C\, s_p\ (C_{Ag} - C_{AS}) = (-r_A)$$

where $s_p$ is the specific surface (surface area per unit mass) of the catalyst particle. For a spherical particle,

$$s_p = \frac{\left(\pi\, d_p^2\right)}{\left(\pi\, d_p^3/6\right)\rho_p} = 6/(\rho_p d_p)$$

Therefore,
$$(C_{Ag} - C_{AS}) = \frac{(-r_A)_{\text{obs}}\ \rho_p\ d_p}{6\, k_C}$$

where $(-r_A)_{\text{obs}}$ is the observed rate of reaction. For the present case,

$$(-r_A)_{\text{obs}} = k C_{Af} = k\ (0.25)\ C_{A0}$$

$$= (11.538)\ (0.25)\ (0.0489) = 0.141 \text{ kmoles/(hr}\cdot\text{kg of catalyst)}$$

Therefore,
$$C_{Ag} - C_{AS} = \frac{(0.41)(2000)(0.003)}{(6.0)(300.0)} = 4.7 \times 10^{-4} \text{ kmoles/m}^3$$

Since the magnitude of $(C_{Ag} - C_{AS})$ is very small, $C_{Ag} \simeq C_{AS}$ and this apparently means that the resistance to diffusion in the gas bulk is negligible.

Let us now proceed to determine the effect of temperature gradient in the gas film. A simple heat balance between the pellet and the gas bulk gives

$$hS_p\ (T_g - T_S) = (-r_A)_{\text{obs}}\ m_p\ (\Delta H_r)$$

where $m_p$ is the mass of the pellet and $S_p$ is its surface area. Since $s_p = (S_p/m_p)$, we get

$$(T_g - T_S) = \frac{(-r_A)_{\text{obs}}\ (\Delta H_r)}{(h s_p)}$$

Further, since for a spherical particle, $s_p = \rho_p d_p/6$,

$$(T_g - T_S) = (-r_A)_{obs}\ (\Delta H_r)\ \rho_p d_p/(6h)$$

$$= \frac{(0.141)\left(-83740\times 10^3\right)(2000)\,(0.003)}{(6.0)\,(116.3\times 3600)} = -28.2°\ \mathrm{C}$$

Since the reaction is exothermic, $(\Delta H_r)$ is a negative quantity and accordingly, $(T_g - T_S)$ is also negative. This shows that there is a significant temperature gradient in the gas film and the surface temperature of the pellet is much higher than the mean bulk temperature of the gas.

Finally, let us find out the temperature gradient within the catalyst pellet. As shown by Prater[34] (for any particle geometry and any kinetics),

$$k_e\ (T_S - T_C) = D_e\ (C_{AS} - C_{AC})\ (\Delta H_r)$$

where $T_C$ is the temperature at the center of the catalyst pellet and $C_{AC}$ is the value of $C_A$ at the center. $(T_S - T_C)$ becomes maximum when $C_{AC}$ is zero. Thus

$$(T_S - T_C)_{max} = D_e C_{AS}\ (\Delta H_r)/\,k_e$$

Since the resistance to film diffusion is negligible, $C_{AS} = C_{Ag} = (0.25)\ C_{A0} = (0.25)\ (0.0489) = 0.0122$ kmoles/m$^3$. Therefore,

$$(T_S - T_C)_{max} = \frac{\left(9.0\times 10^{-6}\right)(0.0122)\left(-83740\times 10^3\right)}{(1.163\times 3600)} = -0.0022°\ \mathrm{C}$$

Thus, the temperature gradient within the pellet is negligible and the entire pellet is at a uniform temperature. In other words, the temperature gradient within the pellet does not intrude the reaction rate.

**Example 8.42:** The vapour phase dimerisation of ethylene is to be carried out in a fixed bed reactor operating isothermally and in plug flow. The 6 mm nickel–aluminium catalyst pellets are packed in a 0.1 m ID reactor which operates at 38° C and 13.6 atm pressure. The feed rate is to be 10 kmoles/hour of essentially pure ethylene. The bulk density of the packed bed is 1100 kg/m$^3$ and the density of catalyst particle is 1900 kg/m$^3$. Assume that the chief reaction that occurs is only the dimerisation of ethylene to butylene which may be taken as second order and irreversible at the operating conditions. External mass transfer resistance may be neglected, but bench–scale studies indicate that the interparticle diffusion significantly retards the global rate. The bench–scale differential reactor data give the following results at 38° C and 13.6 atm pressure.

Intrinsic, second order rate constant = 0.0195 m$^6$/(s·kmole kg of catalyst)

Effective diffusivity ($D_e$) = $1.2 \times 10^{-7}$ m$^2$/s

(*a*) Estimate the depth of catalyst bed required for a conversion of 50 percent.

(*b*) What catalyst-bed depth would be required for the same conversion if interparticle diffusion had been neglected?

(*c*) Estimate the depth of catalyst bed required for 50 percent conversion if both interparticle diffusion and external mass transfer resistance are to be taken into account. The fluid to particle mass transfer coefficient may be computed from the following correlation[35] (for $Re > 10$):

$$j_D = (0.458/\epsilon)\ Re^{-0.407}$$

where $Re = (d_p V_0 \rho_g/\mu_g)$. Take viscosity of ethylene at the operating conditions to be $1010 \times 10^{-8}$ kg/(m·s).

**Solution:** (*a*) The reaction may be represented as

$$2C_2H_4 \rightarrow C_4H_8$$

This is of the form $2A \rightarrow P$, where $A$ stands for ethylene and $P$ for butylene. Since the intrinsic rate is second order and irreversible,

$$(-r_{AS})_{int} = kC_{AS}^2$$

Since external mass transfer resistance is negligible, $C_{AS} = C_{Ag}$. Therefore,

$$(-r_{AS})_{int} = kC_{Ag}^2 = 0.0195 C_{Ag}^2$$

The global rate, $$(-r_A) = 0.0195 \eta C_{Ag}^2 \quad \text{... (i)}$$

Now, $$C_{Ag} = \frac{C_{A0}(1-x_A)}{(1+\epsilon_A x_A)}$$

Since the feed contains only ethylene, $\epsilon_A = (1-2)/2 = -0.5$. Also assuming ideal gas law, $C_{A0} = (P/RT) = (13.6)/(0.082)(311) = 0.533$ kmoles/m$^3$.

Thus, $$C_{Ag} = \frac{0.533(1-x_A)}{(1-0.5x_A)} \quad \text{... (ii)}$$

The effectiveness factor ($\eta$) may be computed from Eq. (8.5.48), once the thiele–type modulus $\phi$ is known. We can compute $\phi$ from Eq. (8.5.54) after substituting $n = 2$. Thus,

$$\phi = (d_p/6)\,(3kC_{Ag}\rho_p/2D_e)^{1/2}$$

Since $d_p = 6$ mm $= 0.006$ m, $\rho_p = 1900$ kg/m$^3$ and $D_e = 1.2 \times 10^{-7}$ m$^2$/s,

$$\phi = 21.52\sqrt{C_{Ag}} \quad \text{... (iii)}$$

where $C_{Ag}$ is obtained from Eq. (ii). Now, since plug flow is assumed in the reactor, the performance equation is

$$W/F_{A0} = \int_0^{x_{Af}} \frac{dx_A}{(-r_A)} \quad \text{... (iv)}$$

Substituting for $(-r_A)$ from Eq. (i) and since $F_{A0} = 10$ kmoles/hr $= (10/3600)$ kmoles/s,

$$W = (0.14245)\int_0^{0.5} \frac{dx_A}{\eta C_{Ag}^2} = (0.14245)\int_0^{0.5} F(x_A)\,dx_A \quad \text{... (v)}$$

where $F(x_A) = 1/(\eta\, C_{Ag}^2)$. The values of $F(x_A)$ at different values of $x_A$ can be now computed and they are listed below in Table 8.42.1. It may be noted that at large values of $\phi$ ($\phi > 5.0$), $\eta \simeq (1/\phi)$.

The integral of Eq. (v) can be now evaluated either graphically or numerically. Let us evaluate it numerically using trapezoidal rule. According to this rule,

$$\int_0^{0.5} F(x_A)\,dx_A = (h/2)\,[F_0 + 2(F_1 + F_2 + F_3 + F_4) + F_5]$$

where $h = 0.1$, $F_0 = 55.3$, $F_1 = 59.986$, $F_2 = 65.984$ and so on. Finally, $F_5 = 101.62$. Therefore,

**Table 8.42.1**

| $x_A$ | $C_{Ag}$ (kmoles/m³) | $\phi$ | $\eta = 1/\phi$ | $F(x_A)$ |
|---|---|---|---|---|
| 0.0 | 0.533 | 15.71 | 0.0636 | 55.3 |
| 0.1 | 0.5049 | 15.292 | 0.06539 | 59.986 |
| 0.2 | 0.4738 | 14.8125 | 0.0675 | 65.984 |
| 0.3 | 0.4389 | 14.2575 | 0.07 | 74.014 |
| 0.4 | 0.3997 | 13.606 | 0.0735 | 85.144 |
| 0.5 | 0.3553 | 12.828 | 0.0779 | 101.62 |

$$\int_0^{0.5} F(x_A)\, dx_A = 36.36 \text{ m}^6/(\text{kmoles})^2$$

$$W = (0.14245)\,(36.36) = 5.18 \text{ kg}$$

It is given that the bulk density of the packed bed is 1100 kg/m³. It means that

$$(5.18/\text{volume of bed}) = 1100$$

or

$$\text{Volume of bed} = (5.18/1100) = 0.00471 \text{ m}^3$$

Since ID of the reactor is 0.1 m,

$$(\pi/4)\,(0.1)^2\, L_b = 0.00471$$

$$L_b = \mathbf{0.6\ m}$$

This is the required depth of the catalyst bed.

(*b*) If intraparticle diffusion is neglected, then $\eta = 1.0$. The expression for $(-r_A)$ then reduces to

$$(-r_A) = kC_{Ag}^2 = 0.0195\ [0.533\ (1 - x_A)/(1 - 0.5x_A)]^2$$

Equation (iv), therefore, becomes

$$W = \frac{(10/3600)}{(0.0195)\,(0.533)^2} \int_0^{x_{Af}} \frac{(1-0.5x_A)^2}{(1-x_A)^2}\, dx_A$$

$$= 0.5014 \left[\frac{0.25}{1-x_{Af}} - 0.25\left(1-x_{Af}\right) - 0.5 \ln\left(1-x_{Af}\right)\right]$$

Substituting $x_{Af} = 0.5$, we get

$$W = 0.3618 \text{ kg}$$

Volume of the bed $= (0.3618/1100) = 0.0003289$ m³

$$L_b = (0.0003289)\,(4/\pi)/(0.1)^2 = 0.042 \text{ m} = \mathbf{4.2\ cm}$$

It is thus evident that the reaction would have been much faster, had the resistance to interparticle diffusion been negligible.

(*c*) If external mass transfer resistance is also to be accounted for, then from Eqs (8.5.66) and (8.5.67),

$$k_C s_p\,(C_{Ag} - C_{AS}) = (-r_A) = \eta k C_{AS}^2$$

or $$(k_C s_p/k)\,(C_{Ag} - C_{AS}) = \eta C_{AS}^2 \qquad \text{... (vi)}$$

Assuming the particles to be spherical,

$$s_p = \frac{(\pi d_p^2)\,N_p}{(\pi d_p^3/6)\,\rho_p N_p} = 6/(\rho_p d_p) = 6.0/(1900)\,(0.006) = 0.5263\ \text{m}^2/\text{kg}$$

The effectiveness factor $\eta$ is to be computed from Eq. (8.5.48) after obtaining $\phi$ from a modified form of Eq. (iii):

$$\phi = 21.52\,\sqrt{C_{AS}} \qquad \text{... (vii)}$$

Since $(-r_A) = \eta\, k\, C_{AS}^2$, Eq. (iv) becomes

$$W = (F_{A0}/k)\int_0^{x_{Af}} \frac{dx_A}{\eta C_{AS}^2} = (0.14245)\int_0^{0.5} F(x_A)\,dx_A \qquad \text{... (viii)}$$

where $F(x_A) = 1/\eta\, C_{AS}^2$. To apply the above three equations, we must first evaluate the mass transfer coefficient, $k_C$. Now, since the feed consists of only $A$ (namely, ethylene),

$$(\mathbf{V}_0\rho_g) = F_{A0}M_A/(\pi/4)\,(0.1)^2 = \frac{(10/3600)\,(28.0)}{(\pi/4)\,(0.1)^2} = 9.903\ \text{kg}/(\text{m}^2\cdot\text{s})$$

$$Re = (0.006)\,(9.903)/(1010 \times 10^{-8}) = 5882.97$$

Now, since the bulk density of the packed bed is 1100 kg/m$^3$ and the catalyst density ($\rho_p$) is 1900 kg/m$^3$,

$$\frac{(\text{volume of catalyst pellets})}{(\text{volume of bed})} = (1 - \epsilon) = (1100/1900)$$

or $$\epsilon = 1 - (1100/1900) = 0.421$$

Therefore, $$j_D = (0.458/0.421)\,(5882.97)^{-0.407} = 0.03179$$

Now, we know from Eq. (4.3.13) of Chapter 4,

$$k_C = j_D\,\mathbf{V}_0\,Sc^{-2/3}$$

Assuming the fluid density to be that of pure ethylene,

$$\rho_g = C_{A0}M_A = (0.533)\,(28.0) = 14.924\ \text{kg/m}^3$$

Therefore, $$\mathbf{V}_0 = (\mathbf{V}_0\rho_g)/\rho_g = (9.903)/(14.924) = 0.6636\ \text{m/s}$$

For computing the Schmidt number ($Sc$), we must first evaluate the diffusivity of $A$ in the fluid. Assuming that the fluid consists of only ethylene and butylene, we can compute $D_{AB}$ (where $A$ stands for ethylene and $B$ for butylene) from Eqs (4.1.9) and (4.1.10). Thus, since $M_A = 28.0$ and $M_B = 56.0$, from Eq. (4.1.9),

$$B = 10.13 \times 10^{-8}$$

From Table 4.1, $$\sigma_A = 4.232\ \text{angstroms}$$

$$\epsilon_A/k = 205°\ \text{K}$$

For butylene, $T_C$ = 418° K and $P_C$ = 39 atm. Therefore, from Eq. (4.1.5),

$$\sigma_B = 2.44\ (418/39)^{1/3} = 5.3797 \text{ angstroms}$$

From Eq. (4.1.7), $\in_B/k = 0.77\ (418) = 321.86°$ K

Therefore, $\sigma_{AB} = (4.232 + 5.3797)/2 = 4.806$ angstroms

$$\in_{AB}/k = \sqrt{(205)(321.86)} = 256.868° \text{ K}$$

$$kT/\in_{AB} = (418/256.868) = 1.6273$$

From Table 4.2, $I_D = 0.58$

Therefore, from Eq. (4.1.10),

$$D_{AB} = 1.1 \times 10^{-6} \text{ m}^2/\text{s}$$

$$Sc^{-2/3} = \left(\frac{1010 \times 10^{-8}}{14.924 \times 1.1 \times 10^{-6}}\right)^{-2/3} = 1.3824$$

Therefore, $k_C = (0.03179)\ (0.6636)\ (1.3824) = 0.02916$ m/s

Equation (vi), thus, becomes

$$0.787\ (C_{Ag} - C_{AS}) = \eta C_{AS}^2 \qquad \text{... (ix)}$$

Now, at any value of $x_A$, $C_{Ag}$ can be computed from Eq. (ii). Once $C_{Ag}$ is known, the above Eq. (ix) can be solved for $C_{AS}$. However, since Eq. (ix) is only implicit in $C_{AS}$, it is to be solved by trial. First, a value of $C_{AS}$, is to be assumed and $\phi$ computed from Eq. (vii). $\eta$ is then evaluated either from Eq. (8.5.48) or as $\eta = 1/\phi$ if $\phi > 5.0$. Now, it is to be checked whether Eq. (ix) is satisfied or not. If not, a new value of $C_{AS}$ is to be assumed and the trial repeated. Once the value of $C_{AS}$ is obtained, $F(x_A)$ can be computed since $F(x_A) = 1/\eta C_{AS}^2$. In this way, the values of $F(x_A)$ at different $x_A$ can be determined and they are listed below in Table 8.42.2.

**Table 8.42.2**

| $x_A$ | $C_{Ag}$ | $C_{AS}$ | $\phi$ | $\eta = 1/\phi$ | $F(x_A)$ |
|---|---|---|---|---|---|
| 0.0 | 0.533 | 0.511 | 15.3834 | 0.065 | 58.9132 |
| 0.1 | 0.5049 | 0.485 | 14.9869 | 0.0667 | 63.7133 |
| 0.2 | 0.4738 | 0.456 | 14.532 | 0.0688 | 69.887 |
| 0.3 | 0.4389 | 0.4225 | 13.988 | 0.0715 | 78.362 |
| 0.4 | 0.3997 | 0.3855 | 13.36 | 0.0748 | 89.91 |
| 0.5 | 0.3553 | 0.3435 | 12.613 | 0.0793 | 106.894 |

The integral of Eq. (viii) can be now evaluated either graphically or numerically. Let us evaluate it numerically by using the trapezoidal rule. Thus

$$\int_0^{0.5} F(x_A)\ dx_A = (h/2)\ [F_0 + 2\ (F_1 + F_2 + F_3 + F_4) + F_5]$$

where $h = 0.1$, $F_0 = 58.9132$, $F_1 = 63.7133$ and so on. Ultimately, $F_5 = 106.894$. Therefore,

$$\int_0^{0.5} F(x_A)\, dx_A = 38.477 \text{ m}^6/(\text{kmoles})^2$$

Therefore, $$W = (0.14245)(38.477) = 5.48 \text{ kg}$$

Volume of packed bed = $(5.48/1100)$ m$^3$

$$L_b = \frac{(5.48/1100)}{(\pi/4)(0.1)^2} = \mathbf{0.6343\ m}$$

This is the required depth of the catalyst bed. It can be seen that this value does not differ materially from that computed in part (*a*), thereby indicating that the external mass transfer resistance is, in fact, insignificant.

**Example 8.43:** Isomerisation of *n*-butane is to be conducted on 3 mm silica–alumina catalyst in a 0.15 m ID packed bed reactor that operates at 50° C and 5 atm pressure. The intrinsic rate is expected to be first order and reversible. The density of catalyst pellets is 1000 kg/m$^3$. When the reaction was initially studied in a 3 cm ID bench-scale laboratory reactor using the same catalyst and at the same temperature and pressure, the following results were obtained (the feed rate being 0.5 kmoles of pure *n*-butane per hour):

| | | | | | |
|---|---|---|---|---|---|
| Mass of catalyst, kg : | 0.027 | 0.0627 | 0.104 | 0.154 | 0.219 |
| Concentration of *n*-butane in the exit stream, kmoles/m$^3$ : | 0.17 | 0.15 | 0.13 | 0.11 | 0.09 |

It was also observed that the equilibrium conversion at 50° C is 85 percent. The bulk voidage of the catalyst bed was maintained at 0.5 in all the above experimental runs.

(*a*) Develop an expression for the overall (global) rate.

(*b*) Estimate the amount of catalyst required in the 0.15 m ID packed bed reactor to obtain a conversion of 65 percent when the feed rate (feed containing only *n*-butane) is maintained at 15 kmoles/hr. Bulk voidage of the bed = 0.45. Assume plug flow in the reactor. Take the viscosity of *n*-butane and isobutane at the operating conditions to be $800 \times 10^{-8}$ kg/(m·s)

**Solution:** (*a*) The reaction may be represented as

$$A \rightleftharpoons P$$

where $A$ stands for *n*-butane and $P$ for *i*-butane. If the intrinsic rate is first order and reversible, then

$$(-r_A)_{\text{int}} = k_1 (C_A - C_P/K)$$

where $k_1$ is the first order forward rate constant and $K$ is the equilibrium constant. Now, since there is no change in the number of moles during reaction,

$$C_A + C_P = C_t = \text{constant}$$

or $$C_P = C_t - C_A$$

Also, $$K = C_{Pe}/C_{Ae} = (C_t - C_{Ae})/C_{Ae}$$

or $$C_t = (1 + K)\, C_{Ae}$$

Therefore, $$C_P = (1 + K)\, C_{Ae} - C_A$$

Substituting this in the expression for intrinsic rate, we get

$$(-r_A)_{\text{int}} = \frac{k_1(K+1)}{K}(C_A - C_{Ae}) = k(C_A - C_{Ae})$$

where $k = k_1 (K + 1)/K$. Now, the global rate is

$$(-r_A) = \eta\,(-r_{AS})_{\text{int}} = \eta k\,(C_{AS} - C_{Ae})$$

We know that when the intrinsic rate is first order and reversible, $\phi$ is given by Eq. (8.5.55) which is independent of $C_{AS}$. Therefore, $\eta$ is also independent of $C_{AS}$ and can be taken as a constant. Now, if $k_C$ is the fluid to pellet mass transfer coefficient, then under steady state conditions [see Eqs (8.5.66) and (8.5.67)],

$$(-r_A) = k_C s_p\,(C_{Ag} - C_{AS})$$

Combining the above two equations and eliminating $C_{AS}$, we get

$$(-r_A) = \frac{(C_{Ag} - C_{Ae})}{(1/k_C s_p) + (1/\eta k)} = K_0\,(C_{Ag} - C_{Ae}) \qquad \text{... (i)}$$

where $(1/K_0) = \left(\dfrac{1}{k_C\,s_p}\right) + \left(\dfrac{1}{\eta k}\right)$. To note that $k_C$ is a function of fluid flow rate and since the experimental data reported are for a specific flow rate ($F_{A0}$ = 0.5 kmoles/hr), $k_C$ can be taken as a constant. Since plug flow is assumed in the reactor, the performance equation is

$$W/F_{A0} = \int_0^{x_{Af}} \frac{dx_A}{(-r_A)} = (-1/C_{A0}) \int_{C_{A0}}^{C_{Af}} \frac{dC_{Ag}}{K_0\left(C_{Ag} - C_{Ae}\right)}$$

or

$$W = \left(\frac{F_{A0}}{K_0 C_{A0}}\right) \ln\left(\frac{C_{A0} - C_{Ae}}{C_{Af} - C_{Ae}}\right) \qquad \text{... (ii)}$$

This means that a plot of $W$ versus ln $[(C_{A0} - C_{Ae})/(C_{Af} - C_{Ae})]$ must be a straight line. Since the feed contained only $A$,

$$C_{A0} = P/RT = (5.0)/(0.082)\,(323) = 0.188 \text{ kmoles/m}^3$$

$$C_{Ae} = C_{A0}\,(1 - x_{Ae}) = 0.188\,(1 - 0.85) = 0.0283 \text{ kmoles/m}^3$$

**Table 8.43.1**

| $W$, kg | ln $[(C_{A0} - C_{Ae})/(C_{Af} - C_{Ae})]$ |
|---|---|
| 0.027 | 0.11958 |
| 0.0627 | 0.2717 |
| 0.104 | 0.45127 |
| 0.154 | 0.6702 |
| 0.219 | 0.9510 |

The plot of $W$ versus ln $[(C_{A0} - C_{Ae})/(C_{Af} - C_{Ae})]$ is shown in Fig. 8.43.1. The plot is rectilinear and therefore, the assumed rate equation is correct. In other words, the intrinsic rate is truly first order and reversible. From the plot,

$$\text{slope} = (F_{A0}/K_0 C_{A0}) = 0.2225$$

or

$$K_0 = \frac{(0.5/3600)}{(0.2225)(0.188)} = 3.32 \times 10^{-3} \text{ m}^3/(\text{s}\cdot\text{kg of catalyst})$$

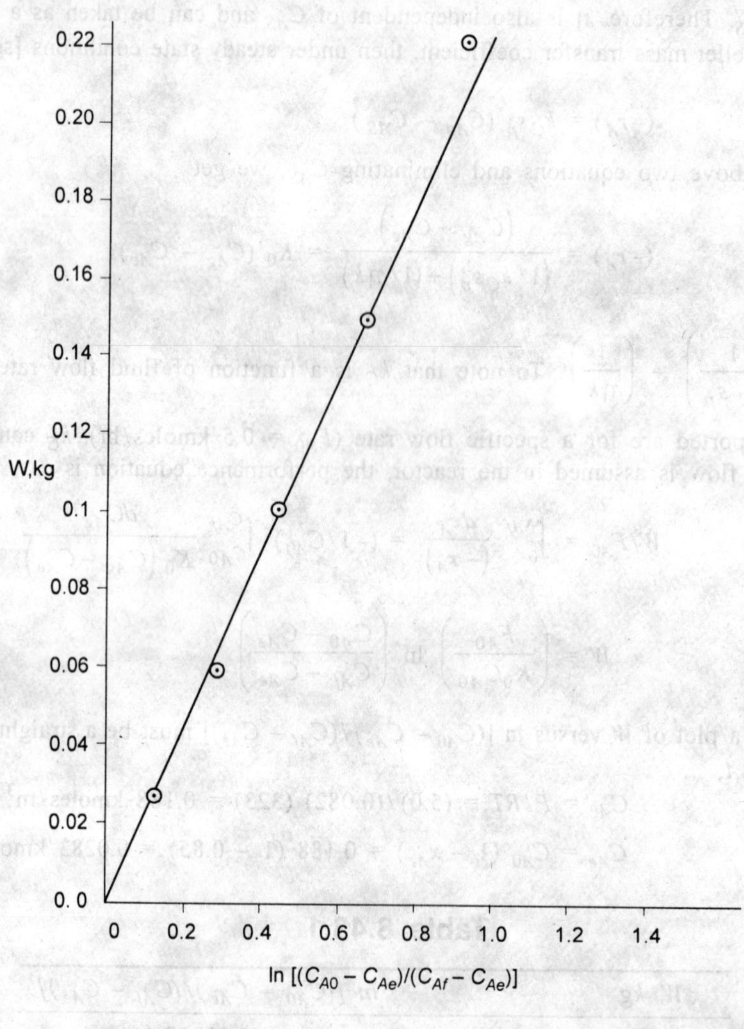


*Figure 8.43.1*

Now, $K_0$ is a function of $\eta$, $k$, $k_C$ and $s_p$. For the given catalyst and at the given temperature and pressure, $\eta$ and $k$ are constants. However, $k_C$ is a process parameter since it depends on the feed rate. It is, therefore, necessary to separate $k_C$ from the rate equation.

To compute $k_C$, let us employ the correlation given in Example 8.42. Thus

$$j_D = (0.458/\in)\ Re^{-0.407}$$

Since $Re = (dp\mathbf{V}_0\rho_g/\mu_g)$ and $k_C = j_D\mathbf{V}_0Sc^{-2/3}$,

above correlation can be re-expressed as

$$k_C = (0.458/\in)\ (d_p\rho_g/\mu_g)^{-0.407}\ \mathbf{V}_0^{0.593}Sc^{-2/3}$$

where $d_p = 0.003$ m, $\rho_g = C_{A0}\ M_A = (0.188)\ (58) = 10.904\ \text{kg/m}^3$, $\mu_g = 800 \times 10^{-8}$ kg (m·s) and the Schmidt number is

$$Sc = \frac{\left(800\times10^{-8}\right)}{(10.904)\ D_{AB}}$$

The diffusivity $D_{AB}$ may be computed from Eqs (4.1.9) and (4.1.10) of Chapter 4. Since $M_A = M_B = 58.0$, from Eq. (4.1.9),

$$B = 10.243 \times 10^{-8}$$

For *n*-butane, $T_C = 425.2°$ K, $P_C = 37.5$ atm and for *i*-butane, $T_C = 408.1°$ K, $P_C = 36.0$ atm. From Eq. (4.1.5),

$$\sigma_A = 2.44\ (425.2/37.5)^{1/3} = 5.4816 \text{ angstroms}$$

$$\sigma_B = 2.44\ (408.1/36.0)^{1/3} = 5.4812 \text{ angstroms}$$

$$\sigma_{AB} = (5.4816 + 5.4812)/2 = 5.4814 \text{ angstroms}$$

Based on Eq. (4.1.7), $\in_{AB}/k = 0.77\ \sqrt{(425.2)(408.1)} = 320.75°$ K

$$kT/\in_{AB} = 323.0/320.75 = 1.007$$

From Table 4.2, $I_D = 0.7197$

Therefore, from Eq. (4.1.10),

$$D_{AB} = 1.02 \times 10^{-6}\ \text{m}^2/\text{s}$$

$$Sc^{-2/3} = \left(\frac{800\times10^{-8}}{10.904\times1.02\times10^{-6}}\right)^{-2/3} = 0.80279$$

Therefore, the expression for $k_C$ reduces to

$$k_C = (0.01246/\in)\ \mathbf{V}_0^{0.593} \quad \text{... (iii)}$$

Since the mass flow rate of feed is $(F_{A0}M_A) = (0.5/3600)\ (58.0) = 8.055 \times 10^{-3}$ kg/s,

$$\mathbf{V}_0 = \frac{\left(8.055\times10^{-3}\right)}{(\pi/4)(0.03)^2(10.904)} = 1.045 \text{ m/s}$$

Therefore, $k_C = (0.01246/0.5)\ (1.045)^{0.593} = 0.0256$ m/s

Now, $(1/K_0) = \dfrac{1}{k_C s_p} + \dfrac{1}{\eta k}$

Since $s_p = 6/\rho_p d_p = 6.0/(1000)\ (0.003) = 2.0\ \text{m}^2/\text{kg},$

$$(1/0.00332) = (19.53125) + 1/(\eta k)$$

or $$1/(\eta k) = 281.69\ (\text{s}\cdot\text{kg of catalyst})/\text{m}^3$$

Thus, from Eq. (i), the rate equation is,

$$(-r_A) = \frac{(C_{Ag} - 0.0283)}{281.69 + (0.5/k_C)}$$

(*b*) For the 0.15 m ID packed bed reactor,

$$F_{A0} = 15\ \text{kmoles/hr}$$

Therefore, $$V_0 = \frac{(15/3600)(58.0)}{(\pi/4)(0.15)^2\ (10.904)} = 1.254\ \text{m/s}$$

Therefore, from Eq. (iii), $k_C = (0.01246/0.45)\ (1.254)^{0.593} = 0.03167\ \text{m/s}$

Now, $$1/K_0 = 281.69 + (0.5/0.03167)$$

or $$K_0 = 0.0033616\ \text{m}^3/(\text{s}\cdot\text{kg of catalyst})$$

Also, $$C_{Af} = C_{A0}\ (1 - x_{Af}) = 0.188\ (1 - 0.65) = 0.0658\ \text{kmoles/m}^3$$

Therefore, from Eq. (ii), $$W = \frac{(15/3600)}{(0.0033616)(0.188)} \ln\left(\frac{0.188 - 0.0283}{0.0658 - 0.0283}\right) = \mathbf{9.553\ kg}$$

This is the amount of catalyst required.

**Example 8.44:** (*a*) Illustrate how the pore diffusion resistance influences the selectivity in the case of the two solid-catalysed simultaneous reactions given below:

$$A \rightarrow P$$
$$B \rightarrow R$$

where $P$ is the desired product. Assume that the intrinsic rates of both reactions are first order and irreversible and the temperature is constant.

(*b*) Wheeler[36] has shown that in the case of two, solid-catalysed successive first order reactions of the form $A \rightarrow P \rightarrow R$ where $P$ is the desired product, the pellet selectivity under conditions of strong pore diffusion resistance and equal effective diffusivities is given by

$$S_p = r_P/(-r_A) = \left(\frac{\alpha}{1+\alpha}\right) - (1/\alpha)\ (C_{Pg}/C_{Ag})$$

where $\alpha = \sqrt{k_1/k_2}$. Based on the above observation, discuss on the effect of pore diffusion on selectivity.

**Solution:** (*a*) Let $k_1$ be the first order rate constant for the first reaction and $k_2$ that for the second reaction. Then

$$(-r_{AS})_{\text{int}} = k_1 C_{AS}$$
$$(-r_{BS})_{\text{int}} = k_2 C_{BS}$$

Therefore, from Eq. (8.5.7),

$$(-\boldsymbol{r}_A) = \boldsymbol{r}_P = \frac{C_{Ag}}{(1/k_C s_p) + (1/\eta_1 k_1)}$$

$$(-\boldsymbol{r}_B) = \boldsymbol{r}_R = \frac{C_{Bg}}{(1/k_C' s_p) + (1/\eta_2 k_2)}$$

Therefore, the selectivity of $P$ with respect to $R$ for a catalyst pellet in the reactor is given by

$$S_P = (\boldsymbol{r}_P/\boldsymbol{r}_R) = \frac{(1/k_C' s_P + 1/\eta_2 k_2)\, C_{Ag}}{(1/k_C s_p + 1/\eta_1 k_1)\, C_{Bg}} \qquad \text{... (i)}$$

If both external and internal diffusion resistances are negligible, then

$$S_P = (k_1 C_{Ag})/(k_2 C_{Bg}) \qquad \text{... (ii)}$$

Since the mass transfer coefficients for $A$ and $B$ (such as $k_C$ and $k_C'$) do not differ appreciably, it is apparent from the above two equations that external diffusion resistance reduces the selectivity. Since $k_1$ is presumably larger than $k_2$ (since $P$ is the desired product), $\phi_1$ is larger than $\phi_2$ and therefore, $\eta_1$ is less than $\eta_2$. Accordingly, it is clear from Eq. (i) that internal diffusion also decreases the selectivity.

To illustrate the effect of internal diffusion more explicitly, let us assume that the external diffusion resistance is negligible. Also, let the intra-pellet resistance be quite strong so that $\phi$ is much larger than 5.0 and $\eta = 1/\phi$. Accordingly, Eq. (i) reduces to

$$S_P = (\phi_2 k_1 C_{Ag})/(\phi_1 k_2 C_{Bg}) = \frac{\sqrt{k_1 (De)_A}\; C_{Ag}}{\sqrt{k_2 (De)_B}\; C_{Bg}}$$

If the effective diffusivities are essentially the same,

$$S_p = (k_1/k_2)^{1/2}\ (C_{Ag}/C_{Bg})$$

Comparison of the above equation with Eq. (ii) shows that strong pore diffusion reduces the selectivity approximately to the square root of its intrinsic value.

(*b*) Since both reactions are intrinsically first order and irreversible,

$$(-\boldsymbol{r}_{AS})_{\text{int}} = k_1 C_{AS}$$

$$(\boldsymbol{r}_{PS})_{\text{int}} = k_1 C_{AS} - k_2 C_{PS}$$

Therefore, in the absence of significant pore diffusion resistance,

$$S_p = (\boldsymbol{r}_{PS})_{\text{int}}/(-\boldsymbol{r}_{AS})_{\text{int}} = 1 - (k_2 C_{PS}/k_1 C_{AS}) \qquad \text{... (iii)}$$

The above equation can be compared with that reported by Wheeler. For a better comparison, consider the case when the feed does not contain any $P$. Then, at the reactor inlet, $C_{Pg} = 0$. Accordingly, Wheeler's expression reduces to

$$S_p = \alpha/(1 + \alpha) \qquad \text{... (iv)}$$

To note that as per Eq. (iii), $S_p$ will be equal to 1.0 under these conditions. In other words, strong pore diffusion resistance reduces $S_p$ from unity to $\alpha/(1 + \alpha)$. This reduction will be particularly severe at small values of $\alpha$.

### 8.5.6 Design of Reactors for Solid–Catalysed Reactions

The most popular types of reactors employed for conducting solid–catalysed reactions are

(*a*) Fixed bed or packed bed reactors,

(*b*) Fluidised bed reactors,

(*c*) Moving bed reactors,

(*d*) Trickle bed reactors,

(*e*) Slurry reactors.

Packed bed or fixed bed reactors are similar to the tubular reactors discussed in Section 8.3 [see Fig. 8.4 (b)] except that the tubes are packed with catalyst particles which may be in the form of granules, pellets, cylinders, spheres, etc. They are usually operated in the vertical position, the feed entering from the bottom and the products leaving from the top [see Fig. 8.17 (a)]. If heat effects are significant, then a coolant or heating fluid may be circulated outside the tubes in the enclosing shell.

In some instances, particularly in the case of metallic catalysts like platinum, wires of metal are made into screens and multiple layers of these screens constitute the catalyst bed. For example, such screen or gauze catalysts are used in the oxidation of ammonia to nitric acid and in the oxidation of acetaldehyde to acetic acid.

If the heat of reaction is large, then small diameter tubes must be employed to facilitate maintenance of uniform temperature within the catalyst bed. If the heat of reaction is not large, then chamber-type

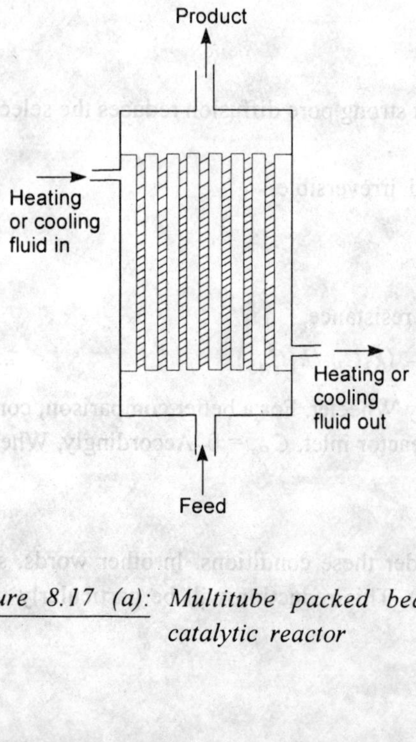


*Figure 8.17 (a): Multitube packed bed catalytic reactor*

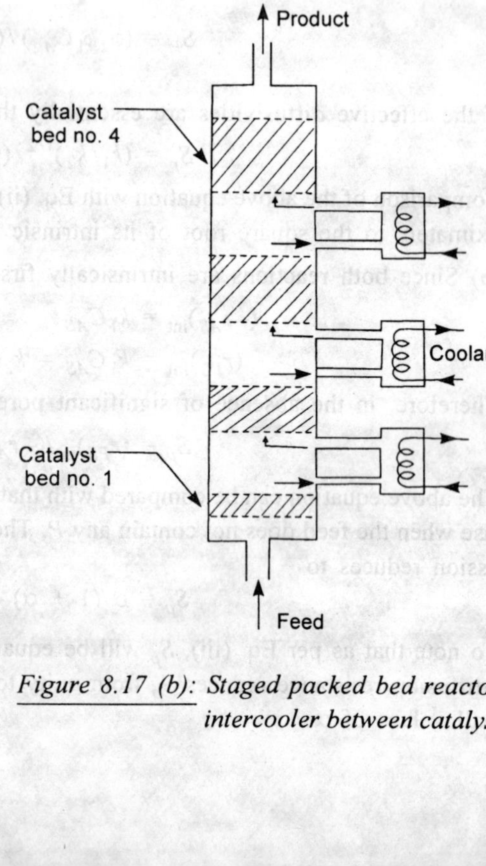


*Figure 8.17 (b): Staged packed bed reactor with intercooler between catalyst beds*

fixed bed reactors that use large diameter tubes may be employed. In this case, the catalyst bed may be divided into a number of parts and an intercooler or inter heater is used between two successive parts [see Fig. 8.17 (b)]. Each part of the catalyst bed operates adiabatically, the heat being supplied or removed through the intercoolers or inter heaters only. Alternately, a number of large-diameter adiabatic catalyst bins may be used in series, with heat exchangers in between. Such types of staged fixed bed reactors are common in sulfur dioxide oxidation, dehydrogenation of butene, etc.

A still alternate method of dissipating the heat of reaction (or supplying heat in the case of endothermic reactions) is to use a large amount of an inert component (usually called the diluant) in the feed stream. For example, in the dehydrogenation of butene (to produce butadiene), steam is used as the diluant. Steam not only supplies heat and thereby helps in maintaining the reaction temperature (the reaction is endothermic) but also helps in minimising the polymerisation of the butadiene product. Also, steam lowers the partial pressure of hydrocarbons and thereby improves the equilibrium conversion.

Packed bed reactors require a minimum of auxiliary equipment and are particularly suitable for small commercial units. However, catalyst regeneration is often a serious problem in this type of reactors. As we have discussed at the beginning of this section, a catalyst can get deactivated or poisoned due to a variety of reasons. Once the activity of the catalyst has decreased to a very low value, the reactor is to be shut down and the catalyst is to be regenerated or reactivated before restart. This is particularly important when the catalyst is too valuable to discard. However, if shut-down is required at frequent intervals, then the overall process may become uneconomical. It is possible to operate two or more reactors in series so that when one is under regeneration, the rest will be under operation. Continuous operation can be thus maintained. However, since this requires two or more reactors, the initial cost of installation will be high. Often, the reactor tube is made longer than required so as to prolong the time gap between regenerations and shutdowns.

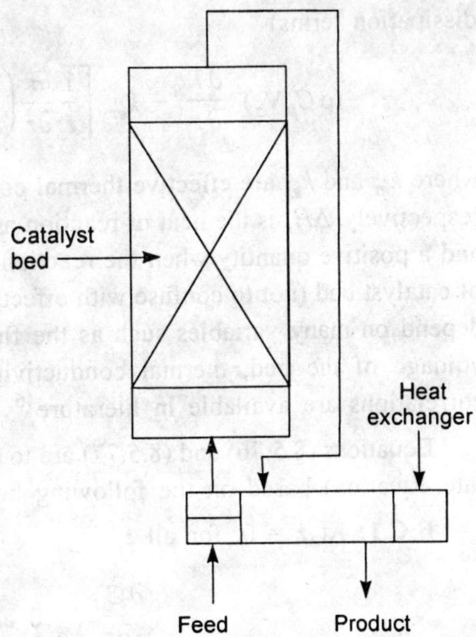

*Figure 8.17 (c): Autothermal operation of packed bed reactor*

When the reaction is highly exothermic, it is always advisable to use the reactor effluent (which will be at a high temperature) to preheat the feed [see Fig. 8.17 (c)]. This is called *autothermal operation.* The heat exchanger may be external to the reactor as shown in Fig. 8.17 (c) or may be built as an integral part of the reactor. Autothermal operation saves a lot of energy and thereby significantly increases the overall energy economy of the process.

For a rigorous analysis of the performance of a fixed bed catalytic reactor, both the mass balance and the heat balance equations (based on a differential volume of the reactor) are to be solved simultaneously. For example, from Eq. (B) of Table 4.6 of Chapter 4, the mass balance equation is

$$\mathbf{V}_z \frac{\partial C_A}{\partial z} - D_{er}\left[\frac{1}{r}\frac{\partial}{\partial r}\left(r\frac{\partial C_A}{\partial r}\right)\right] - D_{ez}\frac{\partial^2 C_A}{\partial z^2} = r_A \rho_p (1-\epsilon) \quad \text{... (8.5.75)}$$

It may be noted that we have defined two average diffusitives such as $D_{er}$ and $D_{ez}$ which are average effective diffusivities in the radial and axial directions respectively. The above equation assumes constant values of diffusivities and it also assumes that the axial velocity $\mathbf{V}_z$ does not vary in the $z$-direction. This is true if the operation is isothermal and for a liquid phase reaction mixture. For a gas-phase reaction mixture, this will be true if the operation is isothermal and if there is no change in the number of moles during the reaction. A more generalised form of the above equation is, therefore,

$$\frac{\partial}{\partial z}\left(\mathbf{V}_z C_A - D_{ez}\frac{\partial C_A}{\partial z}\right) - \frac{1}{r}\frac{\partial}{\partial r}\left(r D_{er}\frac{\partial C_A}{\partial r}\right) = r_A \rho_p (1 - \epsilon) \quad \text{... (8.5.76)}$$

It is important to note that in the above two equations, $r_A$ is the rate of production of $A$ per unit mass of catalyst. For example, if $A$ is the reactant and the global rate is first order in $A$, then $r_A = (-k_1 C_A)$. The rate of production of $A$ per unit volume of catalyst bed is $r_A \rho_p (1 - \epsilon)$, where $\epsilon$ is the bulk voidage of the bed.

Similarly, from Eq. (B) of Table 3.2 of Chapter 3, the energy equation is (after neglecting viscous dissipation terms)

$$(\rho C_p \mathbf{V}_z)\frac{\partial T}{\partial z} - k_{er}\left[\frac{1}{r}\frac{\partial}{\partial r}\left(r\frac{\partial T}{\partial r}\right)\right] - k_{ez}\frac{\partial^2 T}{\partial z^2} = r_A \rho_p (1 - \epsilon)(\Delta H_r) \quad \text{... (8.5.77)}$$

where $k_{er}$ and $k_{ez}$ are effective thermal conductivities of the catalyst bed in the radial and axial directions respectively. $\Delta H_r$ is the heat of reaction and it will be a negative quantity when the reaction is exothermic and a positive quantity when the reaction is endothermic. It must be noted that $k_{er}$ and $k_{ez}$ are properties of catalyst bed (not to confuse with effective thermal conductivity of individual particles) and their values depend on many variables such as the fluid flow rate, temperature of operation, particle diameter, bulk voidage of the bed, thermal conductivity of the fluid and that of the solid particles. Experimental correlations are available in literature[46, 47] for the estimation of $k_{er}$ and $k_{ez}$.

Equations (8.5.76) and (8.5.77) are to be solved simultaneously (after replacing $r_A$ by the corresponding rate equation) based on the following boundary conditions:

**B.C.1:** At $r = 0$, for all $z$,

$$\frac{\partial C_A}{\partial r} = 0 \text{ and } \frac{\partial T}{\partial r} = 0 \quad \text{... (8.5.78)}$$

**B.C.2:** At $r = R$ (where $R$ is the radius of the reactor), for all $z$,

$$\frac{\partial C_A}{\partial r} = 0 \quad \text{... (8.5.79)}$$

$$T = T_W \text{ (for constant wall temperature conditions)} \quad \text{... (8.5.80)}$$

$$-k_{er}\left(\frac{\partial T}{\partial r}\right)_W = q_W = h_W (T_W - T_S),$$

for constant wall heat flux conditions ... (8.5.81)

**B.C.3:** At $z = 0$, for all $r$,

$$(\mathbf{V}_z C_{A0}) = -D_{ez} \frac{\partial C_A}{\partial z}\bigg|_{z>0} + \mathbf{V}_z C_A\bigg|_{z>0} \quad \text{... (8.5.82)}$$

$$T = T_0 \quad \text{... (8.5.83)}$$

**B.C.4:** At $z = L$, for all $r$,

$$\frac{dC_A}{dz} = 0 \text{ and } \frac{dT}{dz} = 0 \quad \text{... (8.5.84)}$$

It can be seen that both the mass balance equation and the heat balance equation are strongly coupled with each other since the reaction rate constants are functions of temperature. They can be, therefore, solved only by numerical methods. An analytical solution is hardly possible. Numerical solutions of these equations with different types of rate equations are reported by many authors. A detailed coverage of the same is beyond the scope of this book. Interested readers may consult references (21) and (37).

A preliminary design of the packed bed reactor can be, however, performed by assuming plug flow in the reactor and neglecting axial diffusion and axial conduction of heat. Once we have assumed plug flow in the reactor, it means that at any cross-section, the velocity does not vary in the radial direction. Accordingly, we can assume that neither the concentration nor the temperature varies in the radial direction at any particular cross-section. Equation (8.5.76), therefore, reduces to

$$\frac{d(\mathbf{V}_z C_A)}{dz} = r_A \rho_p (1 - \epsilon) \quad \text{... (8.5.85)}$$

Since $\mathbf{V}_z$ is the superficial velocity in the $z$-direction, $F_A$ = molal flow rate = $(A_C \mathbf{V}_z C_A)$, where $A_C$ is the cross-sectional area of the reactor. Therefore, $d_{FA} = A_C\, d(\mathbf{V}_z C_A)$. Since $dF_A = d[F_{A0}(1 - x_A)] = -F_{A0}\, dx_A$, $d(\mathbf{V}_z C_A) = -(1/A_C) F_{A0}\, dx_A$. Substituting this in Eq. (8.5.85), rearranging and integrating, we get

$$(-r_A) A_C \rho_p (1 - \epsilon)\, d_z = F_{A0}\, dx_A$$

or

$$(-r_A)\, dW = F_{A0}\, dx_A \quad \text{... (8.5.86)}$$

On integration,

$$W/F_{A0} = \int_0^{x_{Af}} \frac{dx_A}{(-r_A)} \quad \text{... (8.5.86a)}$$

It can be seen that this is nothing but the performance equation of the plug flow reactor [similar to Eq. (8.3.32)]. In a similar way, Eq. (8.5.77) gets simplified to

$$-F_0 C_p dT + h_W (T_W - T)\, dA_h = (-r_A)\, \Delta H_r A_C \rho_p (1 - \epsilon)\, dz$$

$$= (-r_A)\, \Delta H_r\, dW \quad \text{... (8.5.87)}$$

where $F_0 = (A_C \mathbf{V}_z \rho)$ = total feed rate in kg/s and $dA_h = \pi D_i\, dz$ = differential area of heat transfer. The second term in the above equation stands for the rate of heat exchange with the surroundings (with the reactor wall). It can be seen that the above equation is analogous to Eq. (8.5.60). If we eliminate $(-r_A)$ from the above equation by substituting Eq. (8.5.86) in it, we get

$$h_W (T_W - T)\, dA_h = F_{A0} (\Delta H_r)\, dx_A + F_0 C_p dT \quad \text{... (8.5.88)}$$

Correlations for the computation of the wall heat transfer coefficient ($h_W$) in packed beds have been discussed in Section 3.10 of Chapter 3.

Design of packed bed reactors based on Eqs (8.5.86) and (8.5.88), though approximate, does throw a good amount of light on the reactor performance. It provides a good, preliminary estimate of the reactor dimensions and the effects of changes in operating conditions, with minimum computational load. See Example 8.45 for an illustrative example in this connection.

Let us now consider the fluidised bed reactors. The phenomenon of fluidisation has already been discussed in Section 2.12.1 of Chapter 2. Heat transfer in fluidised beds is dealt in Section 3.10.1 of Chapter 3. Fluidised bed reactors do have advantages over packed bed reactors, but they also have their own limitations. When large scale heat effects accompany the reaction (the reaction is highly exothermic or highly endothermic), fluidised bed reactors are more advantageous since rapid mixing of solids in fluidised beds permits easy control of temperature and practically isothermal operation. Accurate control of temperature is particularly important when the reaction mixture is explosive in nature and to avoid formation of hot spots which otherwise could ruin the catalyst. Effective temperature control is difficult in large fixed bed reactors. Fixed bed reactors cannot use very small sizes of catalyst due to plugging and high pressure drop, while fluidised beds can very well use small size particles. In small catalyst particles, the resistance to pore diffusion is usually negligible. Due to the high degree of turbulence in fluidised beds, the resistance to film diffusion is also minimised.

Fluidised bed reactors are particularly preferred when the catalyst is to be frequently regenerated since it deactivates rapidly. The deactivated catalyst is withdrawn continuously from the reactor and after regeneration (or reactivation), it flows back to the reactor (see Fig. 8.18). Excellent examples are fluidised bed catalytic cracking (FCC) and fluidised bed catalytic reforming (FCR) of petroleum hydrocarbons, wherein the catalyst gets deactivated due to the deposition of carbon/coke particles. The spent (or deactivated) catalyst is reactivated by burning off the deposited coke particles. Compressed air is used as the regenerator gas.

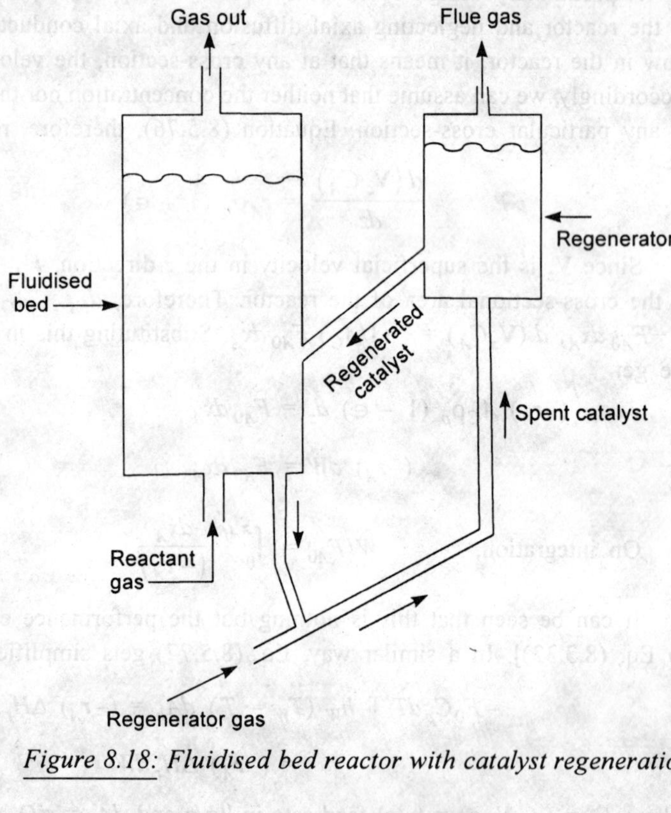


*Figure 8.18: Fluidised bed reactor with catalyst regeneration*

However, though plug flow can be reasonably achieved in fixed bed reactors, in bubbling fluidised beds the flow is complex and is usually far from plug flow and often, there is considerable bypassing of bubble gas. As a result, the contacting is ineffective and larger amounts of catalyst are required to achieve high conversion. This effect is more serious in the case of multiple reactions. For example, in series or consecutive reactions, the rate of production of the intermediate product (which could also be the desired product) could be drastically low if a vigorously bubbling fluidised bed is employed. Another problem with fluidised bed reactors is that due to mutual attrition, the particles decrease in size and a point is reached when they are no longer fluidised, but move off with gas stream. Cyclone separators or electrostatic precipitators are to be employed in the effluent line to remove these catalyst fines. Deterioration of lines and vessels due to the abrasive action of sharp solid particles is another point of concern. This problem has been reported to be serious in fluidised bed catalytic cracking units.

Owing to the complex gas flow and solid movement within the fluidised bed, design procedure of catalytic fluidised bed reactors has not been well established in spite of the fact that a large number of models have been suggested. Many different approaches have been tried by various investigators to analyse and predict the true characteristics of fluidised bed reactors. We shall present here the *three phase model* or the *bubbling bed model* proposed by Kunii and Levenspiel[38]. This model is based on the following assumptions:

1. The fluidised bed is made up of three phases such as the bubble phase, the cloud and the wake and the dense or emulsion phase (see Figs 8.19 and 8.20). It was Davidson and Harrison[39] who first postulated that all the gas that passes through the fluidised bed in the form of bubbles stays with the bubbles, recirculating very much like a smoke ring and penetrating only a small distance into the emulsion. This zone of penetration is called the *cloud* since it envelops the rising bubble.

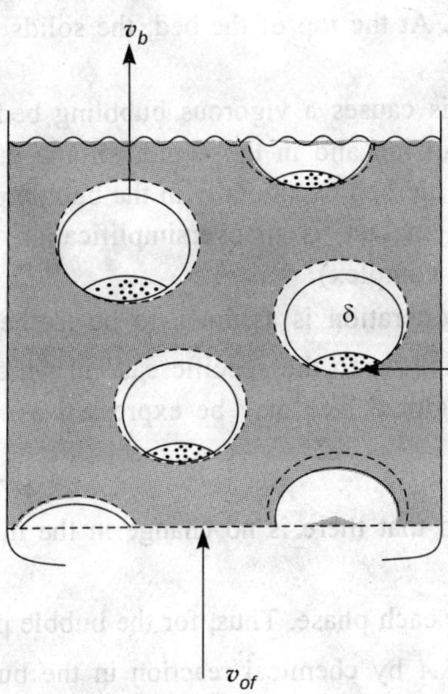


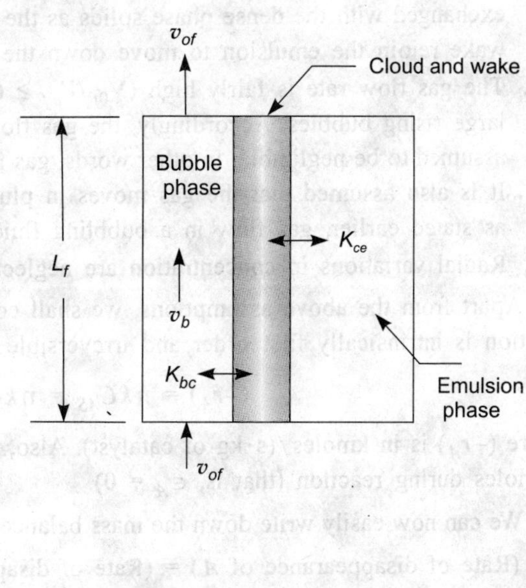


*Figure 8.19 (a) and (b): Flow pattern in fluidised bed according to bubbling bed model*

Subsequently, Rowe and Partridge[40] made the important finding through extensive experimental work that each bubble of gas drags a *wake* of solids up the bed. According to them, the wake occupies approximately 30 percent of the bubble volume.

2. The emulsion phase or dense phase is assumed to be perfectly mixed.
3. Small bubbles form at the distributor (the initial size of the bubbles depends on the type of distributor used). These bubbles coalesce and thereby grow in size and speed up as they rise through the bed. However, most of the growth of the bubbles occurs close to the distributor and accordingly, it is assumed that the bubble size remains constant throughout the bed and is called the effective bubble size ($d_b$).
4. Due to the formation of the wake, there is a circulation of solids in the bed, with upflow behind the bubbles and downflow in the rest of the emulsion. Solids are dragged upwards in the wake by the rising bubbles. These solids are continuously exchanged with the dense phase solids as the bubble rises. At the top of the bed, the solids in the wake rejoin the emulsion to move down the bed.

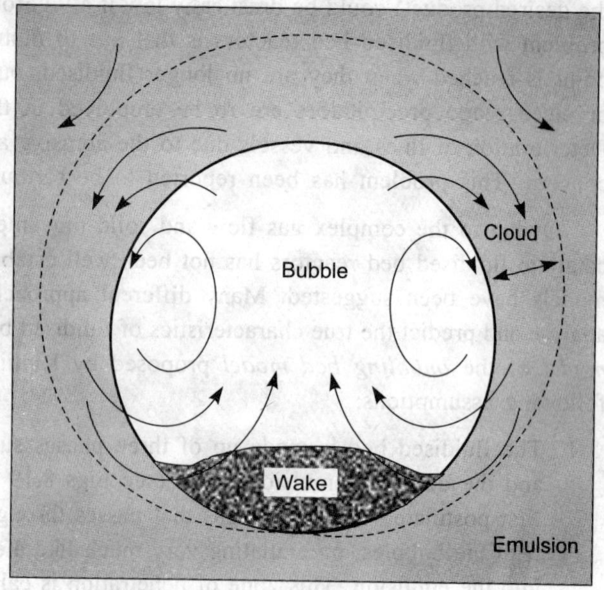


*Figure 8.20: Sketch of an idealised gas bubble in a fluidised bed*

5. The gas flow rate is fairly high ($\mathbf{V}_{0f}/V_{mf} \geq 6$ to 11). This causes a vigorous bubbling bed with large rising bubbles. Accordingly, the gas flow in the emulsion and in the cloud volume may be assumed to be negligible. In other words, gas flow through the bed occurs only in the bubble phase. It is also assumed that the gas moves in plug flow (this, in fact, is an oversimplification since, as stated earlier, gas flow in a bubbling fluidised bed is complex).
6. Radial variations in concentration are neglected and the operation is assumed to be isothermal.

Apart from the above assumptions, we shall confine our analysis to the specific case in which the reaction is intrinsically first order and irreversible so that the global rate may be expressed as

$$(-\boldsymbol{r}_A) = \eta k C_{AS} = \eta k C_{Ag} \qquad \text{... (8.5.89)}$$

where $(-\boldsymbol{r}_A)$ is in kmoles/(s·kg of catalyst). Also, it is assumed that there is no change in the number of moles during reaction (that is, $\epsilon_A = 0$).

We can now easily write down the mass balance equation for each phase. Thus, for the bubble phase,

(Rate of disappearance of $A$) = (Rate of disappearance of $A$ by chemical reaction in the bubble) + (Rate of disappearance of $A$ by transfer to cloud and wake)

$$-\frac{dC_{Ab}}{d\theta} = \gamma_b\,(\eta k \rho_p)\,C_{Ab} + K_{bC}\,(C_{Ab} - C_{AC})$$

or
$$-\mathbf{V}_b \frac{dC_{Ab}}{dz} = \gamma_b k_0 C_{Ab} + K_{bC}(C_{Ab} - C_{AC}) \qquad \text{... (8.5.90)*}$$

where $k_0 = (\eta k \rho_p)$

$C_{Ab}$ = concentration of $A$ in bubbles, kmoles/(m$^3$ of bubbles)

$K_{bc}$ = interchange coefficient between bubble phase and cloud, s$^{-1}$

$\gamma_b$ = volume fraction of solids (catalyst particles) in bubbles

= (volume of solids in bubbles)/(volume of bubbles)

= 0.001 to 0.01 ... (8.5.91)

$\mathbf{V}_b$ is the velocity of gas in the bubble phase (that is, the actual velocity of bubbles or more precisely, the rise velocity of bubbles, clouds and wakes). It is postulated that

$$\mathbf{V}_b = (\mathbf{V}_{0f} - \mathbf{V}_{mf}) + \mathbf{V}_{br} \qquad \text{... (8.5.92)}$$

where $\mathbf{V}_{0f}$ is the operating gas velocity and $\mathbf{V}_{br}$ is the natural rise velocity of any single bubble. It is shown[38] that

$$\mathbf{V}_{br} = 0.711\sqrt{g\, d_b} \qquad \text{... (8.5.93)}$$

The expression for the interchange coefficient $K_{bC}$ is given by Kunii and Levenspiel[38] (based on Higbie's penetration theory) as

$$K_{bC} = 4.5\,(\mathbf{V}_{mf}/d_b) + 5.85\left(\frac{D_A^{1/2} g^{1/4}}{d_b^{5/4}}\right) \qquad \text{... (8.5.94)}$$

where $D_A$ is the diffusivity of $A$ in the gas.

The mass balance equation for the cloud and wake is

(Rate of transfer of $A$ to cloud and wake) = (Rate of disappearance of $A$ by chemical reaction in cloud and wake) + (Rate of disappearance of $A$ by transfer to emulsion)

$$K_{bC}(C_{Ab} - C_{AC}) = \gamma_C k_0 C_{AC} + K_{Ce}(C_{AC} - C_{Ae}) \qquad \text{... (8.5.95)}$$

where $\gamma_C$ = (volume of solids in clouds and wakes)/(volume of bubbles)

$K_{Ce}$ = interchange coefficient between clouds and the emulsion phase, s$^{-1}$

Since there is no flow of gas between clouds and the emulsion phase, diffusion is the only mechanism of transport. The expression for $K_{Ce}$ as given by Kunii and Levenspiel[38] is,

$$K_{Ce} \simeq 6.78\sqrt{\epsilon_{mf} D_A \mathbf{V}_b / d_b^3} \qquad \text{... (8.5.96)}$$

and that for $\gamma_C$ is

$$\gamma_C = (1 - \epsilon_{mf})\left(\frac{3\mathbf{V}_{mf}/\epsilon_{mf}}{\mathbf{V}_{br} - \mathbf{V}_{mf}/\epsilon_{mf}} + \alpha\right) \qquad \text{... (8.5.97)}$$

* In plug flow, the velocity is constant and therefore, $\theta = (z/\mathbf{V}_b)$ or $d\theta = (1/\mathbf{V}_b)\,dz$.

$$\alpha = (\text{volume of wake})/(\text{volume of bubble})$$

$$= 0.25 \text{ to } 1.0 \quad \text{... (8.5.98)}$$

Similarly, the mass balance equation for the emulsion phase is

(Rate of transfer of $A$ to emulsion) = (Rate of disappearance of $A$ by chemical reaction in emulsion)

$$K_{Ce}\,(C_{AC} - C_{Ae}) = \gamma_e k_0 C_{Ae} \quad \text{... (8.5.99)}$$

where $\gamma_e$ = (volume of solids in emulsion)/(volume of bubbles)

$$= \left[\frac{(1-\epsilon_{mf})(1-\delta)}{\delta}\right] - (\gamma_C + \gamma_b) \quad \text{... (8.5.100)}$$

$\delta$ = fraction of fluidised bed consisting of bubbles

$$\simeq (\mathbf{V}_{0f} - \mathbf{V}_{mf})/\mathbf{V}_b \quad \text{... (8.5.101)}$$

If we add Eqs (8.5.90), (8.5.95) and (8.5.99), we get

$$-\mathbf{V}_b \frac{dC_{Ab}}{dz} = k_0\,(\gamma_b C_{Ab} + \gamma_C C_{AC} + \gamma_e C_{Ae}) \quad \text{... (8.5.102)}$$

Now, from Eq. (8.5.99), $C_{Ae} = \dfrac{K_{Ce} C_{AC}}{(K_{Ce} + \gamma_e k_0)}$ ... (8.5.103)

Substituting this in Eq. (8.5.95), we can solve it for $C_{AC}$. Finally, substituting these expressions for $C_{Ae}$ and $C_{AC}$ in Eq. (8.5.102), we get

$$-\mathbf{V}_b \frac{dC_{Ab}}{dz} = K_f C_{Ab} \quad \text{... (8.5.104)}$$

where $$K_f = \left[(\gamma_b k_0) + \cfrac{1}{\cfrac{1}{K_{bC}} + \cfrac{1}{(k_0 \gamma_C + K')}}\right] \quad \text{... (8.5.105)}$$

$$K' = \frac{1}{(1/K_{Ce}) + (1/k_0 \gamma_e)} \quad \text{... (8.5.106)}$$

Since both $\mathbf{V}_b$ and $K_f$ are constants, we can easily integrate Eq. (8.5.104) with the boundary conditions that at $z = 0$, $C_{Ab} = C_{A0}$ and at $z = L_f$, $C_{Ab} = C_{Af}$. Thus

$$-\int_{C_{A0}}^{C_{Af}} \frac{dC_{Ab}}{C_{Ab}} = (K_f/\mathbf{V}_b) \int_0^{L_f} dz$$

or $$\ln\,(C_{A0}/C_{Af}) = (K_f L_f/\mathbf{V}_b) \quad \text{... (8.5.107)}$$

or $$(1 - x_{Af}) = (C_{Af}/C_{A0}) = \exp\,(-K_f L_f/\mathbf{V}_b) \quad \text{... (8.5.108)}$$

The expanded height $L_f$ of the bubbling fluidised bed can be estimated with allowable error from the following relationship[38]:

$$(L_f/\mathbf{V}_b) = \frac{(1-\in)}{(1-\in_{mf})} \; (L_b/\mathbf{V}_{br}) \qquad \text{... (8.5.109)}$$

where $L_b$ and $\in$ are the height and voidage respectively of the packed bed.

Equation (8.5.108) can be conveniently used to predict the performance characteristics of fluidised bed reactors, provided the chemical reaction is first order and irreversible in A. Alternate models for fluidised bed reactors are the *bubble assemblage model* of Kato and Wen[41], the *slugging bed model* of Hovmand and Davidson[42] and the *bubble flow model* of Chen and Doughlas[43]

It is clear from Eq. (8.5.108) that as the bubble diameter ($d_b$) is increased, the conversion decreases (increase in $d_b$ increases $\mathbf{V}_{br}$ and all the factors such as $L_f/\mathbf{V}_b$, $K_{bC}$, $\gamma_C$ and $K_{Ce}$ decrease). At high gas velocities when the bubble size is large, there will be extensive bypassing of bubble gas and consequently, the performance of the bed drops appreciably (even to much below that of a back-mix reactor). However, in industrial reactors, large bed diameters and high gas velocities cannot be avoided since commercial scale operations involve high gas throughput. In such cases, it is desirable to operate the fluidised bed reactor with an appreciable carry over and recycle of solids (see Fig. 8.21). Under these conditions, the bed behaves as a lean emulsion with little gas bubbling and little bypassing. Very high gas velocities can be used here and with proper design, the gas flow can be made to approach plug flow. Alternately, a semi fluidised bed may be employed in which the gas first passes through the fluidised section and then through the packed section (see Section 2.12.2 of Chapter 2 for a good discussion on the characteristics of semi fluidised beds). Insertion of internals into the bed to hinder the bubble growth and to cut down the bubble size is a still another alternative.

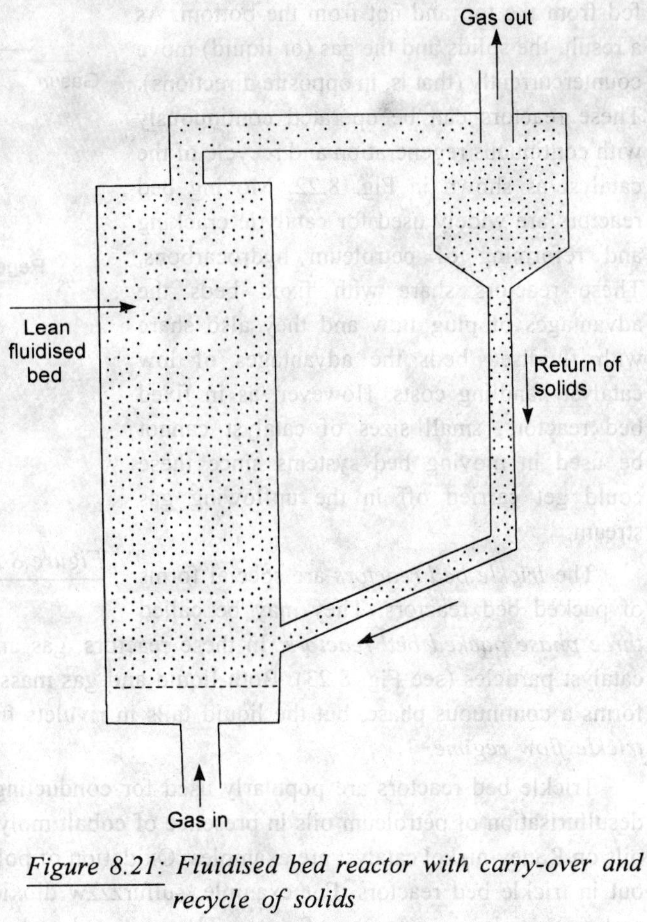


*Figure 8.21: Fluidised bed reactor with carry-over and recycle of solids*

It has been observed that the size of the catalyst particles and their shape play a significant role in deciding the overall performance of fluidised bed reactors. Smooth fluidisation with conventional shape bubbles is observed with catalyst particles having almost round shape and size less than 150 microns. Bigger size and more irregular shaped catalyst particles imply rough fluidisation with bubbles of unconventional and irregular shape. The height of the catalyst bed is also crucial since it decides the occurrence of *slugging* in the fluidised bed (see Section 2.12.1 of Chapter 2). An $(L/D)$ ratio of 1.0 or less is preferred since this eliminates slugging phenomenon. It is to be noted that most of the conversion in a fluidised bed reactor takes place in the lower region of the bed, near the distributor.

The *moving bed reactors* are very similar to the fluidised bed reactors except that in moving bed reactors, the catalyst particles are fed from the top and not from the bottom. As a result, the solids and the gas (or liquid) move countercurrently (that is, in opposite directions). These reactors can be operated continuously with continuous regeneration and recycle of the catalyst as shown in Fig. 8.22. Moving bed reactors are widely used for catalytic cracking and reforming of petroleum hydrocarbons. These reactors share with fixed beds the advantages of plug flow and they also share with fluidised beds the advantages of low catalyst handling costs. However, as in fixed bed reactors, small sizes of catalyst cannot be used in moving bed systems since these could get carried off in the upflowing gas stream.

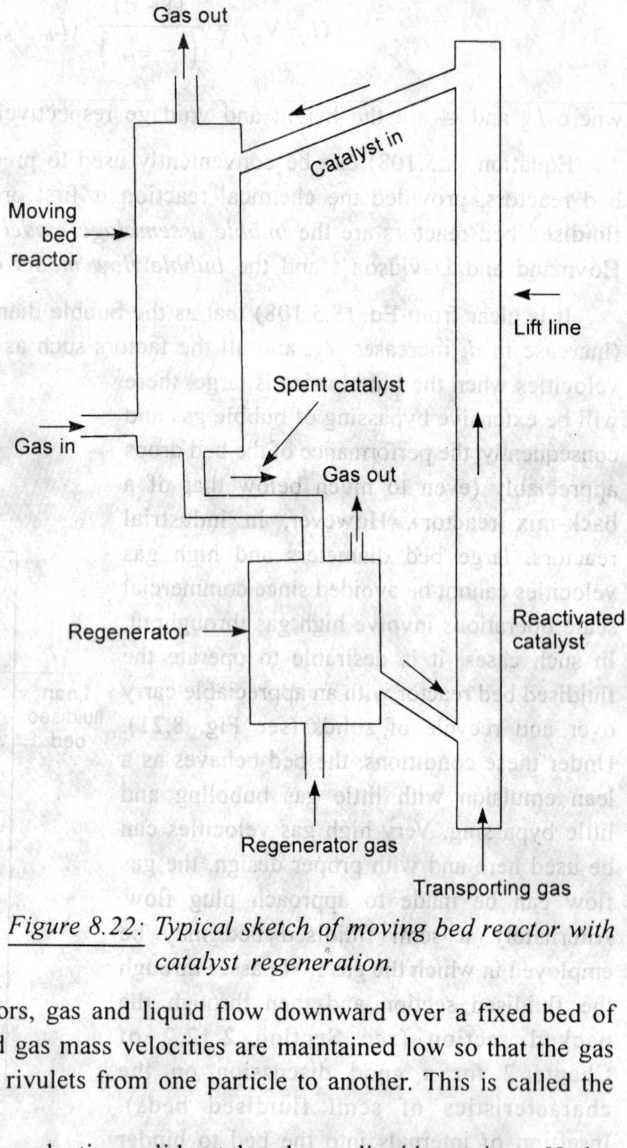


*Figure 8.22: Typical sketch of moving bed reactor with catalyst regeneration.*

The *trickle bed reactors* are special forms of packed bed reactors. They may be called *three phase packed bed reactors*. In these reactors, gas and liquid flow downward over a fixed bed of catalyst particles (see Fig. 8.23). Both liquid and gas mass velocities are maintained low so that the gas forms a continuous phase, but the liquid falls in rivulets from one particle to another. This is called the *trickle flow regime.*

Trickle bed reactors are popularly used for conducting solid catalysed gas–liquid reactions. Hydro desulfurisation of petroleum oils in presence of cobalt molybdate catalyst and hydrogenation of vegetable oils on Raney-nickel catalyst are examples. Oxidation of pollutants dissolved in liquids can also be carried out in trickle bed reactors. For example, sulfurzzzw dioxide is removed from air in a bed of activated carbon particles in presence of water. The activated carbon catalyses the oxidation of sulfur dioxide to sulfur trioxide and the sulfur trioxide dissolves in water to produce sulfuric acid. Thus, sulfur dioxide is not only removed from air but is also recovered as sulfuric acid.

Extensive studies have been conducted to model trickle bed reactors. We shall present here a relatively simplified analysis. The following assumptions are made:

1. The operation is isothermal.
2. Both the gas and the liquid are uniformly distributed across the reactor diameter and therefore, there will be no radial gradients of concentration or velocity. In other words, both gas and liquid execute plug flow through the catalyst bed. In a trickle bed, though gas is distributed uniformly across the diameter, the liquid tends to flow towards the reactor wall. It has been reported that a reactor to particle diameter ratio of 18 or more shall ensure uniform liquid distribution.
3. All the particles are completely covered by the flowing liquid. This is a reasonable assumption for many trickle bed processes.
4. Axial diffusion is neglected in both gas and liquid phases. (For most applications, axial diffusion is negligible in the gas phase, however such dispersion may be important in the liquid on occasions).
5. It is assumed that the reaction is such that a gaseous component ($A$) reacts with a second reactant ($B$) in the liquid phase to form products that may be either gaseous or liquid. The reaction can be, thus, represented as

   $$aA\ (g) + bB\ (l) \rightarrow \text{Products}$$

   It is also assumed that the liquid is nonvolatile.

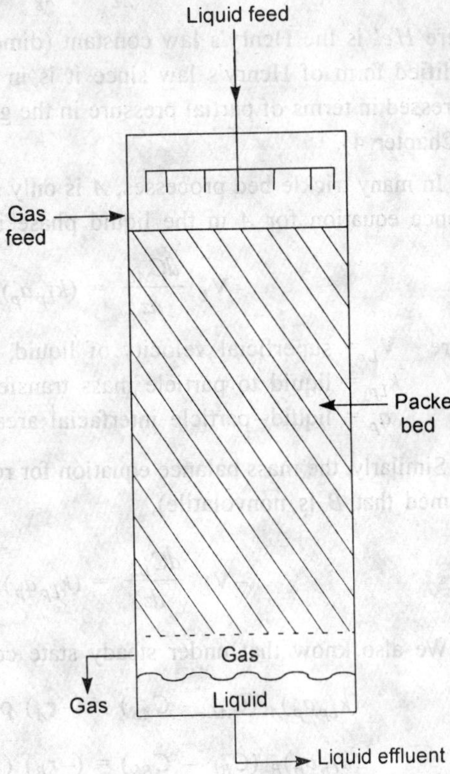


*Figure 8.23: Trickle bed reactor*

6. The reaction is intrinsically first order in $A$ and irreversible so that

$$(-r_A) = \eta k C_{AS} \quad \text{... (8.5.110)}$$

$$(-r_B) = (b/a)\ (-r_A) = (b/a)\ \eta k C_{AS} \quad \text{... (8.5.111)}$$

We can now easily write down the mass balance equations as we did in the case of fluidised bed reactors. The only difference here is that we must write separate equations for $A$ and $B$. Thus, for reactant $A$ in the gas phase

$$-\mathbf{V}_g \frac{dC_{Ag}}{dz} = K_L a_g\ (C^*_{AL} - C_{AL}) \quad \text{... (8.5.112)}$$

where $K_L$ = overall liquid phase mass transfer coefficient, m/s

$a_g$ = gas–liquid interfacial area per unit volume of catalyst bed, $m^2/m^3$

$\mathbf{V}_g$ = superficial velocity of gas, m/s

$C^*_{AL}$ = equilibrium concentration of $A$ in liquid, $kmoles/m^3$

If the system obeys Henry's law, then

$$C_{AL}^{*} = C_{Ag}/He' \qquad \text{... (8.5.113)}$$

where $He'$ is the Henry's law constant (dimensionless). It is to be noted that the above equation is the modified form of Henry's law since it is in terms of concentrations. In the usual form, Henry's law is expressed in terms of partial pressure in the gas and mole fraction in the liquid (see under *gas absorption* in Chapter 4).

In many trickle bed processes, $A$ is only slightly soluble in the liquid so that $K_L \approx k_L$. Now, the mass balance equation for $A$ in the liquid phase is

$$-\mathbf{V}_L \frac{dC_{AL}}{dz} = (k_{Lp} a_p)_A (C_{AL} - C_{AS}) - K_L a_g (C_{AL}^{*} - C_{AL}) \qquad \text{... (8.5.114)}$$

where $\mathbf{V}_L$ = superficial velocity of liquid, m/s
$k_{Lp}$ = liquid to particle mass transfer coefficient, m/s
$a_p$ = liquid–particle interfacial area per unit volume of catalyst bed, $m^2/m^3$

Similarly, the mass balance equation for reactant $B$ in the liquid phase is (to note that we have already assumed that $B$ is nonvolatile),

$$-\mathbf{V}_L \frac{dC_{BL}}{dz} = (k_{Lp} a_p)(C_{BL} - C_{BS}) \qquad \text{... (8.5.115)}$$

We also know that under steady state conditions,

$$(k_{Lp} a_p)_A (C_{AL} - C_{AS}) = (-r_A)\, \rho_p (1 - \epsilon) = \rho_p (1 - \epsilon)\, \eta k C_{AS} = k_0 C_{AS} \qquad \text{... (8.5.116)}$$

$$(k_{Lp} a_p)_B (C_{BL} - C_{BS}) = (-r_B)(1 - \epsilon)\, \rho_p = (b/a)\, k_0 C_{AS} \qquad \text{... (8.5.117)}$$

where $k_0 = [\rho_p (1 - \epsilon)\, \eta k]$.

By expressing $C_{AS}$ in terms of $C_{AL}$ from Eq. (8.5.116) and $C_{AL}$ in terms of $C_{Ag}$ from Eq. (8.5.112), we can solve Eq. (8.5.114) for $C_{Ag}$. For example, from Eq. (8.5.116),

$$C_{AS} = \frac{(k_{Lp} a_p)_A}{k_0 + (k_{Lp} a_p)_A} C_{AL} = K_1 C_{AL} \qquad \text{... (8.5.118)}$$

Similarly, from Eq. (8.5.112),

$$C_{AL} = C_{AL}^{*} + \left(\frac{\mathbf{V}_g}{K_L a_g}\right) \frac{dC_{Ag}}{dz} \qquad \text{... (8.5.119)}$$

Substituting the above two expressions in Eq. (8.5.114) and thereby eliminating $C_{AL}$ and $C_{AS}$, we get an ordinary differential equation in $C_{Ag}$:

$$\frac{d^2 C_{Ag}}{dz^2} + K_2 \frac{dC_{Ag}}{dz} + K_3 C_{Ag} = 0 \qquad \text{... (8.5.120)}$$

where
$$K_2 = \frac{\left(\mathbf{V}_L K_L a_g\right) + \left(\mathbf{V}_g He'\right)\left[\left(k_{Lp}\, a_p\right)_A (1 - K_1) + K_L\, a_g\right]}{\left(\mathbf{V}_g \mathbf{V}_L He'\right)} \quad \text{... (8.5.121)}$$

$$K_3 = (k_{Lp} a_p)_A\ K_L a_g\ (1 - K_1) / (\mathbf{V}_g \mathbf{V}_L He') \quad \text{... (8.5.122)}$$

Equation (8.5.120) can be solved by the method of characteristics if $K_2$ and $K_3$ are constants (not functions of $z$). If we assume that the gas and liquid velocities ($\mathbf{V}_g$ and $\mathbf{V}_L$) do not change significantly as a result of the chemical reaction, then $K_2$ and $K_3$ will be constants and Eq. (8.5.120) can be conveniently solved by the method of characteristics. The result is

$$C_{Ag} = C_1 \exp(\lambda_1 z) + C_2 \exp(\lambda_2 z) \quad \text{... (8.5.123)}$$

where
$$\lambda_1 = \left(-K_2 + \sqrt{K_2^2 - 4K_3}\right)/2 \quad \text{... (8.5.124)}$$

$$\lambda_2 = \left(-K_2 - \sqrt{K_2^2 - 4K_3}\right)/2 \quad \text{... (8.5.125)}$$

The constants $C_1$ and $C_2$ are to be evaluated based on the boundary conditions. Boundary conditions are usually specified by the feed compositions. Thus,

**B.C.1:** At $z = 0,\ C_{Ag} = (C_{Ag})_0$ ... (8.5.126)

**B.C.2:** At $z = 0,\ C_{AL} = (C_{AL})_0$ (8.5.127)

Substituting for $C_{AL}$ from Eq. (8.5.119), the second boundary condition can be rewritten as

$$\left.\frac{dC_{Ag}}{dz}\right|_{z=0} = K_4 = \text{constant} \quad \text{... (8.5.127a)}$$

where
$$K_4 = [(C_{AL})_0 - C^*_{AL}]\ (K_L a_g)/\mathbf{V}_g$$
$$= [(C_{AL})_0 - C_{Ag}/He']\ (K_L a_g)/\mathbf{V}_g \quad \text{... (8.5.128)}$$

Using the above two boundary conditions, we can now evaluate $C_1$ and $C_2$. The results are

$$C_1 = \frac{K_4 - \left(C_{Ag}\right)_0 \lambda_2}{(\lambda_1 - \lambda_2)} \quad \text{... (8.5.129)}$$

$$C_2 = \frac{\left(C_{Ag}\right)_0 \lambda_1 - K_4}{(\lambda_1 - \lambda_2)} \quad \text{... (8.5.130)}$$

Equation (8.5.123) can be thus used to predict the performance characteristics of trickle bed reactors. However, there could be occasions when the assumptions involved in this analysis are not precisely valid. Though the liquid mass velocity and liquid density may not change substantially as a result of the reaction, this assumption may not necessarily hold good for gas. Also, there are cases in which axial diffusion in the liquid phase cannot be neglected. If the effect of axial diffusion is to be included, then Eqs (8.5.114) and (8.5.115) are to be modified as

$$D_{AL} \frac{d^2C_{AL}}{dz^2} - \mathbf{V}_L \frac{dC_{AL}}{dz} = (k_{Lp}a_p)_A (C_{AL} - C_{AS}) - K_L a_g (C^*_{AL} - C_{AL}) \quad \text{... (8.5.114a)}$$

$$D_{BL} \frac{d^2C_{BL}}{dz^2} - \mathbf{V}_L \frac{dC_{BL}}{dz} = (k_{Lp}a_p)_B (C_{BL} - C_{BS}) \quad \text{... (8.5.115a)}$$

Also, the chemical reaction need not necessarily be first order and irreversible. With other forms of rate equations and with axial diffusion terms included, analytical solutions are hardly possible. We have to go for numerical solutions. It is to be noted that if the intrinsic kinetics are not first order, then $\eta$ will not be a constant, but a function of surface concentrations.

*Slurry reactors*, that are widely used for conducting solid–catalysed reactions, are usually three-phase reactors. The catalyst particles are suspended in a liquid (which may or may not take part in the reaction) and the gas is bubbled through this slurry or suspension. The slurry is agitated partly by the bubbling gas and partly by a mechanical impeller (see Figs 8.24 and 8.25). Like fluidised bed reactors, the slurry reactors also employ small size catalyst particles. This minimises intraparticle diffusion resistance. However, unlike the fluidised bed, there is little relative movement between particles and the fluid in a slurry reactor, even though the liquid is agitated mechanically. The particles tend to move with the liquid. The small particle size, low diffusivities in liquids and the low relative velocity all reduce the mass transfer coefficient. As a result, external mass transfer resistance will be significant and can substantially retard the global rate. At the same time, relatively high thermal conductivity of the liquid (as compared to solids in a fixed bed reactor) increases the heat transfer coefficient. This coupled with violent agitation (which helps in uniform distribution of heat) causes uniform temperature throughout the slurry and there will be little temperature difference between particles and the liquid. Thus, external temperature differences can be normally neglected in slurry reactors. The major operational disadvantage of slurry reactors is the difficulty involved in retaining the small catalyst particles in the reactor without getting lost in the product discharged. Screens are often used in the effluent lines but they tend to get clogged or otherwise be unreliable.

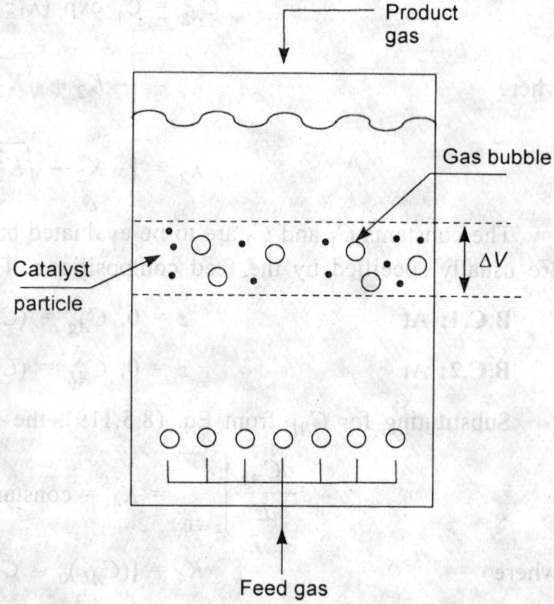


*Figure 8.24: Slurry reactor (batch liquid, continuous flow of gas)*

For a three phase slurry reactor, there are a large number of possible operating conditions. For example, the reactants may all be in the gas phase and the liquid be inert. The liquid simply serves as a medium to suspend the catalyst particles. Examples are hydrogenation of hydrocarbon gases (such as hydrogenation of ethylene in a slurry of Raney nickel catalyst particles suspended in toluene) and the oxidation of gases such as sulfur dioxide and hydrogen sulfide in aqueous slurries of activated carbon particles. In both of these cases, the liquid (toluene or water) acts as a medium to suspend the catalyst and does not take part in the reaction. Similarly, in the manufacture of high-density polyethylene (HDPE), ethylene is polymerised in a slurry of Ziegler-Natta catalyst in cyclohexane. Here also, cyclohexane acts as the diluant and does not take part in the chemical reaction. On the other hand, in the hydrogenation of vegetable oils, the slurry liquid (namely, the oil) reacts with the gas (namely, hydrogen). Thus, depending on the operating conditions, mass balance equations are to be written either for the gas phase or for the liquid phase or for both phases and solved to obtain the performance characteristics of slurry reactors.

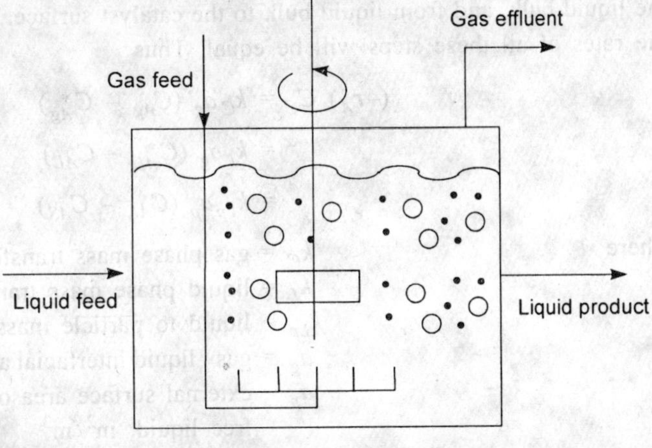


*Figure 8.25: Slurry reactor (continuous flow of gas and liquid)*

Let us discuss here one of these cases. For example, let the reactants be gaseous $A$ and gaseous $B$ and let the product $P$ be also in the gas phase. The liquid is thus inert and acts only as a medium to suspend the catalyst. As stated earlier, this is the case with hydrogenation of hydrocarbon gases and the removal of pollutants (sulfur dioxide, hydrogen sulfide) from air by oxidation. The reaction can be, therefore, represented as

$$A\ (g) + B\ (g) \rightarrow P\ (g)$$

To simplify the situation further, let us assume that the intrinsic rate is first order (in $A$) and irreversible so that $(-r_A) = \eta k C_{AS}$. To note that, $(-r_A)$ is expressed in kmoles/(s·kg of catalyst). The global rate in kmoles/(s·m$^3$ of gas-free liquid) will be, therefore, $(-r_A)\ C_C$ where $C_C$ is the catalyst concentration in the slurry (that is, kg of catalyst per m$^3$ of gas-free liquid). We shall also make the following additional assumptions:

(*i*) The slurry is well mixed (as in a stirred tank reactor) so that the reactant concentration as well as the catalyst concentration remain uniform throughout the slurry,

(*ii*) the gas bubbles execute plug flow through the slurry,

(*iii*) the operation is isothermal.

The first and third assumptions stated above are fairly true in most of the slurry reactors, while the second one involves some approximation. Once the slurry is perfectly mixed and the gas moves in plug flow through it, the performance equation for the reactor will be the same as that for an ideal plug flow reactor (see Eq. 8.3.31). Thus

$$-dF_A = (-r_A)\ C_C\, dV$$

or

$$F_{A0}\, dx_A = (-r_A)\ C_C\, dV \qquad \text{... (8.5.131)}$$

On integration, we get

$$V/F_{A0} = \frac{1}{C_C} \int_0^{x_{Af}} \frac{dx_A}{(-r_A)} \qquad \text{... (8.5.132)}$$

where $V$ is the volume of gas-free liquid in the reactor. To integrate the above equation, we have to first express $(-r_A)$ in terms of bulk concentrations. This can be easily accomplished by summing up the resistances to transfer of $A$ from gas bulk to the gas–liquid interface, from the gas–liquid interface to the liquid bulk, and from liquid bulk to the catalyst surface. We know that under steady state conditions, the rates of all these steps will be equal. Thus

$$(-r_A)\ C_C = k_C a_g\ (C_{Ag} - C^*_{Ag}) \quad \text{... (8.5.133)}$$

$$= k_L a_g\ (C^*_{AL} - C_{AL}) \quad \text{... (8.5.134)}$$

$$= k_{Lp} a_p\ (C_{AL} - C_{AS}) \quad \text{... (8.5.135)}$$

where
$k_C$ = gas phase mass transfer coefficient, m/s
$k_L$ = liquid phase mass transfer coefficient, m/s
$k_{Lp}$ = liquid to particle mass transfer coefficient, m/s
$a_g$ = gas–liquid interfacial area per unit volume of gas-free liquid, $m^2/m^3$
$a_p$ = external surface area of catalyst particles per unit volume of gas-free liquid, $m^2/m^3$
$C^*_{Ag}$, $C^*_{AL}$ = equilibrium concentration of $A$ in gas and that in liquid respectively, kmoles/$m^3$

We have assumed that equilibrium exists at the gas-liquid interface. Further, if Henry's law is valid, then

$$C^*_{Ag} = He' C^*_{AL} \quad \text{... (8.5.136)}$$

Also, in the case of spherical catalyst particles, the total external surface area per unit mass of particle is

$$= \frac{\left(\pi d_p^2 N_p\right)}{\left(\pi d_p^3/6\right)\rho_p N_p} = 6/(\rho_p d_p) \quad \text{... (8.5.135a)}$$

where $N_p$ is the number of particles. Since $C_C$ is the catalyst concentration in the slurry,

$$a_p = \left(\frac{6}{\rho_p d_p}\right) C_C \quad \text{... (8.5.135b)}$$

Similarly, for spherical gas bubbles, the surface area per unit volume of bubbles is $(6/d_b)$ and if $\in_g$ is the gas holdup (volume of gas/volume of gas-free liquid) in the slurry, then

$$a_g = (6/d_b) \in_g \quad \text{... (8.5.135c)}$$

All the above four equations such as Eqs (8.5.133) to (8.5.136) and the rate equation [such as $(-r_A) = \eta k C_{AS}$] can be combined to eliminate all the intermediate concentrations such as $C^*_{Ag}$, $C^*_{AL}$, $C_{AL}$ and $C_{AS}$. The result is

$$(-r_A)\ C_C = k_0 C_{Ag} \quad \text{... (8.5.137)}$$

where

$$(1/k_0) = \left(\frac{1}{k_C a_g}\right) + He' \left(\frac{1}{k_L a_g} + \frac{1}{k_{Lp} a_p} + \frac{1}{\eta k C_C}\right) \quad \text{... (8.5.138)}$$

It is to be noted that in summing up the resistances as shown above, we have assumed that the presence of solid particles does not interfere with gas-liquid mass transfer.

### 8.5.7 Rate Equations with Deactivating Catalysts

So far, we have discussed development of rate equations and design of catalytic reactors assuming that the activity of the catalyst remains constant throughout the process. In actual practice, this is not the case. The catalyst could get deactivated due to a variety of reasons and as a result, its activity decreases with time. We have already discussed on the different mechanisms of catalyst deactivation or poisoning in the early paragraphs of this section (see also Table 8.4). The activity ($\alpha_a$) of a catalyst at any time $\theta$ can be, therefore, defined as

$$\alpha_a = \frac{(\text{rate of reaction at time } \theta)}{(\text{rate of reaction with unpoisoned catalyst})} \quad \text{... (8.5.139)}$$

If $\alpha_a$ remains equal to unity for a prolonged duration of time, then it means that the catalyst is quite stable and has a long useful life. The rate of deactivation of the catalyst, namely $d\alpha_a/d\theta$, is a function of time, temperature and concentration. Thus

$$\frac{d\alpha_a}{d\theta} = f_n\,(T,\ C_i,\ \alpha_a) \quad \text{... (8.5.140)}$$

In the case of independent deactivation, $d\alpha_a/d\theta$ will be independent of $C_i$. For side-by-side deactivation, $C_i$ will be the concentration of poison/impurity in the feed stream. For parallel deactivation, $C_i$ will be the concentration of the reactant (namely, $C_A$), while for series deactivation, $C_i$ will be $C_P$ (where $P$ is the desired product) or $(C_A + C_P)$.

Like the rate equation for the reaction, the expression for $(d\alpha_a/d\theta)$ is also to be developed by trial based on the experimental data available. An expression is to be first assumed and then checked whether it fits the experimental data. If not, a different expression is to be tried and so on. Quite often, an expression of the following form is satisfactory:

$$-\frac{d\alpha_a}{d\theta} = k_d C_i^m \alpha_a^d \quad \text{... (8.5.141)}$$

$$= [k_{d0}\ \exp\ (-E_d/RT)]\ C_i^m \alpha_a^d \quad \text{... (8.5.142)}$$

where the constants $k_{d0}$, $E_d$, $m$ and $d$ are to be evaluated from experimental data. Incidentally, for independent deactivation at constant temperature,

$$-\frac{d\alpha_a}{d\theta} = k_d \alpha_a^d \quad \text{... (8.5.143)}$$

On integration with the initial condition that at $\theta = 0$, $\alpha_a = 1.0$, we get

$$\alpha_a = \exp\ (-k_d\theta),\ \text{if } d = 1.0 \quad \text{... (8.5.144)}$$

$$= [1 + k_d\,(d - 1)\,\theta]^{1/(1-d)},\ \text{if } d \neq 1.0 \quad \text{... (8.5.145)}$$

The above types of exponential (when $d = 1.0$) and hyperbolic (when $d \neq 1.0$) decay functions have been reported by Weekman[44], Wojciechowski[45] and several other authors.

The actual rate of reaction with a deactivating catalyst is thus,

$$(\text{Actual rate of reaction}) = (\text{Rate of reaction without catalyst deactivation})\,(\alpha_a) \quad \text{... (8.5.146)}$$

It is thus evident that the rate equation for the chemical reaction and that for catalyst deactivation are strongly coupled. When they follow first order kinetics, the expression for the actual rate of reaction (global rate) will be simpler and design of reactors based on it will be easier. However, with nonlinear and more complicated rate equations, the analysis becomes difficult and often demands numerical methods for solution.

**Example 8.45:** *Non-isothermal operation of fixed bed catalytic reactor:* Phthalic anhydride is manufactured by the catalytic oxidation of naphthalene by air over vanadium pentoxide (on silica gel) catalyst. Analysis of available data indicates that the global rate of reaction (kmoles of naphthalene converted to phthalic anhydride per hour per kg of catalyst) can be represented empirically by the expression

$$(-r_A) = 305 \times 10^5 p_A^{0.38} \exp(-14090/T)$$

where $p_A$ is the partial pressure of naphthalene in atmospheres and $T$ is in degrees Kelvin. It is also satisfactory to assume that the overall reaction is

$$C_{10}H_8 + 4.5O_2 \rightarrow C_8H_4O_3 + 2H_2O + 2CO_2$$

The above reaction is to be conducted in a multitube type fixed bed reactor that uses 50 mm ID tubes with a heat transfer salt circulated through the jacket. The feed to the reactor consists of 0.10 mole percent naphthalene vapour and 99.9 percent air and it is preheated to 340° C. The circulating heat transfer salt will maintain the inside of the reactor tube walls at 340° C. The catalyst consists of 5 mm by 5 mm cylinders and the bulk density of the packed bed is 803 kg/m$^3$. The superficial mass velocity of the gases through each tube is 1950 kg/(m$^2\cdot$hr) and the average pressure in reactor tubes is 1 atm. The reactor is to be designed to operate at a conversion of 80 percent and have a production rate of 3000 kg/day of phthalic anhydride. The temperature in the catalyst bed is not to exceed 400° C under any circumstance.

(*a*) Estimate the depth of the catalyst bed and the total amount of catalyst required. The reaction is exothermic and the heat of reaction is $21.75 \times 10^5$ kJ/kmole of naphthalene. The properties of the reaction mixture may be taken as equivalent to those for air.

(*b*) What will be the depth of catalyst bed required if reactor is operated adiabatically?

**Solution:** (*a*) Let us solve the problem based on the simplified analysis, that is, based on Eqs (8.5.86) and (8.5.88). Thus, Eq. (8.5.86) can be rewritten as

$$[A_C \rho_p (1 - \epsilon)]\, dz = F_{A0}\, dx_A/(-r_A)$$

or

$$dz/dx_A = \frac{F_{A0}}{A_C \rho_p (1-\epsilon)(-r_A)} \quad \text{... (i)}$$

It is given that the production rate of phthalic anhydride is 3000 kg/day or (3000/148) kmoles/day. Therefore,

$$F_{A0}x_{Af} = \frac{(3000/148)}{(24)} \text{ kmoles/hr}$$

Since $x_{Af} = 0.80$,

$$F_{A0} = \frac{(3000)}{(148)(24)(0.8)} = 1.0557 \text{ kmoles/hr}$$

Also, $F_0 = (F_{A0}/y_{A0})$ (molecular weight of feed)

Since $y_{A0}$ = mole fraction of $A$ in feed = 0.001,

$$F_0 = \frac{(1.0557)}{(0.001)} (0.001 \times 128 + 0.999 \times 29) = 30719.814 \text{ kg/hr}$$

Since mass velocity through each tube is 1950 kg/(m²·hr),

$$F_0 \text{ (per tube)} = (195.0)\ [\pi\ (0.05)^2/4] = 3.829^{*} \text{ kg/hr}$$

Therefore, number of tubes $(N_t) = (30719.814/3.829) = 8023$

$$A_C = (\pi/4)\ (0.05)^2\ (8023) = 15.753 \text{ m}^2$$

Now, since bulk density of the bed is 803 kg/m³,

$$\rho_p\ (1 - \in) = \frac{(\text{Mass of solids})}{(\text{Volume of solids})} \frac{(\text{Volume of solids})}{(\text{Total volume of bed})}$$

$$= (\text{Mass of solids})/(\text{Total volume of bed}) = 803 \text{ kg/m}^3$$

The rate equation is given in terms of $p_A$. We can reexpress it in terms of $x_A$. Thus, let the initial number of moles of feed mixture (at $x_A = 0$) be 1000 (that is, 999 kmoles of air and 1 kmole of naphthalene vapour). At any conversion $x_A$, the number of moles of each component will be

Naphthalene $= 1 - x_A$

Oxygen $= (999 \times 0.21) - 4.5x_A$

Nitrogen $= (999 \times 0.79)$

Phthalic anhydride $= x_A$

Carbon dioxide $= 2x_A$

Water vapour $= 2x_A$

Total number of moles $= (1000 - 0.5x_A)$

Since the total pressure is 1 atm,

$$p_A = (1 - x_A)/(1000 - 0.5x_A)$$

Substituting the values of $F_{A0}$, $A_C$, $\rho_p\ (1 - \in)$ and the expression for $(-r_A)$ in Eq. (i), we get

$$\frac{dz}{dx_A} = (2.73626 \times 10^{-12}) \left(\frac{1000 - 0.5x_A}{1 - x_A}\right)^{0.38} \exp\ (14090/T) \quad \text{... (ii)}$$

Now, the heat balance equation from Eq. (8.5.88) is

$$h_W (613 - T)\, \pi\, (0.05)\, (8023)\, dz = -(1.0557)\, (21.75 \times 10^8)\, dx_A + (30719.814)\, C_p\, dT \quad \text{... (iii)}$$

Since the properties of air (assumed to be equal to those of the reaction mixture) do not vary significantly within the temperature range of 340 to 400° C, we can use average values in our computations. Thus

$$C_p = 1050 \text{ J/(kg·K)}$$

$$\mu_g = 31.0 \times 10^{-6} \text{ kg/(m·s)}$$

$$k_g = 0.049 \text{ W/(m·K)}$$

The wall heat transfer coefficient ($h_W$) can be obtained from Eq. (3.10.11) of Chapter 3. Thus

$$Re_{bm} = D_{VS} \mathbf{V}_0 \rho_g / \mu_g$$

where

$$(\mathbf{V}_0 \rho_g) = 1950 \text{ kg/(m}^2\text{·hr)} = 0.54167 \text{ kg/(m}^2\text{·s)}$$

$$D_{VS} = d_p = 5 \text{ mm} = 0.005 \text{ m}$$

To note that for a cylindrical particle of length equal to diameter, the volume-surface diameter ($D_{VS}$) is equal to its diameter. Thus

$$Re_{bm} = (0.005)\, (0.54167)/(31.0 \times 10^{-6}) = 87.366$$

$$Pr = (1050)\, (31.0 \times 10^{-6})/(0.049) = 0.6643$$

Therefore, from Eq. (3.10.11),

$$h_W = 125.84 \text{ W/(m}^2\text{·K)} = 453024.0 \text{ J/(hr·m}^2\text{·K)}$$

Equation (iii), therefore, becomes

$$\frac{dT}{dx_A} = (17.7)\, (613 - T)\, (dz/dx_A) + 71.18 \quad \text{... (iv)}$$

Equations (ii) and (iv) are to be solved simultaneously. But, since both of them are nonlinear equations, we can solve them only by any of the numerical methods. Let us, therefore, solve them by Euler's method (the one we used in Example 8.29). Thus, rewriting Eq. (iv) as a difference equation, we get

$$\Delta T/\Delta x_A = (17.7)\, (613 - T_{avg})\, (dz/dx_A)_{avg} + 71.18 \quad \text{... (v)}$$

Let us choose an increment of 0.1 in $x_A$. In other words $\Delta x_A = 0.1$ (the smaller this increment, the better will be the accuracy of Euler's method). Now, $(x_A)_0 = 0$, $T_0 = (340 + 273) = 613°$ K.

From Eq. (ii), $(dz/dx_A)_0 = 0.3627$ m

$$(x_A)_1 = (x_A)_0 + \Delta x_A = 0.1$$

The value of $T_1$ is to be obtained by trial. Thus, let $T_1 = 619°$ K. Now, from Eq. (ii),

$$(dz/dx_A)_1 = 0.3021 \text{ m}$$

$$(dz/dx_A)_{avg} = (0.3627 + 0.3021)/2 = 0.3324 \text{ m}$$

$$T_{avg} = (613 + 619)/2 = 616° \text{ K}$$

Therefore, from Eq. (v), $\Delta T = 5.353°$ K

Now, $T_1 = (T_0 + \Delta T) = (613 + 5.353) = 618.353°$ K

Since this value of $T_1$ is very close to that assumed, no further trials are required. Now,

$$\Delta z = (dz/dx_A)_{avg}\ (\Delta x_A) = (0.3324)\ (0.1) = 0.03324 \text{ m}$$

$$z_1 = z_0 + \Delta z = (0.0) + (0.03324) = 0.03324 \text{ m}$$

Following the above procedure, $x_A$ is increased by 0.1 in a stepwise manner and the values of $z$ and $T$ are computed at each $x_A$ until $x_A = 0.8$. The results are listed below in Table 8.45.1.

**Table 8.45.1**

| $x_A$ | $T$, °K | $z$, m |
|---|---|---|
| 0.1 | 618.353 | 0.03324 |
| 0.2 | 621.95 | 0.063 |
| 0.3 | 624.03 | 0.09153 |
| 0.4 | 625.30 | 0.119 |
| 0.5 | 626.08 | 0.1473 |
| 0.6 | 626.14 | 0.1779 |
| 0.7 | 625.535 | 0.212 |
| 0.8 | 624.24 | 0.2524 |

Thus, the required depth of catalyst bed is 0.2524 m.

$$\text{Mass of catalyst required} = (0.2524)\ A_C\ (1 - \in)\ \rho_p = (0.2524)\ (15.753)\ (803) = 3192.77 \text{ kg}$$

It can be seen from Table 8.45.1 that as we go up the bed, the temperature first increases, reaches maximum value (at $x_A = 0.6$, $z = 0.1779$ m) and then starts decreasing. Such a temperature profile is characteristic of exothermic reactions. At low conversions (in the bottom portion of the bed), the temperature increases due to the exothermic nature of the reaction. The rate of reaction also increases as temperature rises. But at high conversion, the concentration of A has fallen to very low value and as a result, the rate decreases and temperature starts falling. The maximum temperature attained (namely the *hot spot*) in the present case is 626.14° K or 353.14° C which is well below the maximum permissible value of 400° C.

(*b*) If the operation is adiabatic, then Eq. (iv) will reduce to

$$dT/dx_A = 71.18$$

On integration, we get

$$T - T_0 = 71.18 x_A$$

or

$$T = T_0 + 71.18 x_A = 613 + 71.18 x_A \qquad \text{... (vi)}$$

Substituting the above expression for $T$ in Eq. (ii) and integrating, we get

$$\int_0^{L_b} dz = L_b = \int_0^{0.8} F(x_A)\ dx_A \qquad \text{... (vii)}$$

where $F(x_A) = 2.73626 \times 10^{-12} \left(\dfrac{1000 - 0.5\,x_A}{1 - x_A}\right)^{0.38} \exp\left(\dfrac{14090.0}{613 + 71.18\,x_A}\right)$

The integral of Eq. (vii) may be evaluated either graphically or numerically. Let us evaluate it numerically by using the trapezoidal rule. For this, the values of $F(x_A)$ at different values of $x_A$ are first computed and they are listed below in Table 8.45.2.

**Table 8.45.2**

| $x_A$ | $F(x_A)$ |
|---|---|
| 0.0 | 0.3627 |
| 0.1 | 0.290 |
| 0.2 | 0.2343 |
| 0.3 | 0.19159 |
| 0.4 | 0.15877 |
| 0.5 | 0.1337 |
| 0.6 | 0.11496 |
| 0.7 | 0.1018 |
| 0.8 | 0.09476 |

Now, as per the trapezoidal rule,

$$\int_0^{0.8} F(x_A)\, dx_A = (h/2)\,[F_0 + 2\,(F_1 + F_2 + \ldots\ldots F_7) + F_8]$$

where $h = 0.1$, $F_0 = 0.3627$, $F_1 = 0.290$, $F_2 = 0.2343$ and so on. Finally, $F_8 = 0.09476$. Therefore,

$$L_b = \int_0^{0.8} F(x_A)\, dx_A = 0.1454 \text{ m}$$

This is the required depth of catalyst bed.

**Example 8.46:** *Rate equation for a deactivating catalyst.* The solid-catalysed reaction, $A \rightarrow P$ is studied at 700° C in a basket-type back-mix reactor containing 1 kg of catalyst pellets. The feed containing pure $A$ is admitted into the reactor and the feed rate is continuously lowered so that the concentration of $A$ in the reactor remains constant. The results obtained are given below:

| $C_{A0} = 1.0$ kmole/m$^3$ | | $C_{A0} = 3.0$ kmoles/m$^3$ | |
|---|---|---|---|
| $C_{Af} = 0.5$ kmoles/m$^3$ | | $C_{Af} = 1.0$ kmoles/m$^3$ | |
| *Time, hr* | *Feed rate, kmoles/hr* | *Time, hr* | *Feed rate, kmoles/hr* |
| 0.0 | 0.5 | 0.0 | 0.75 |
| 1.0 | 0.035267 | 1.0 | 0.03745 |
| 2.0 | 0.025 | 2.0 | 0.0265 |
| 3.0 | 0.0204 | 3.0 | 0.02164 |

Develop expressions for the rate of reaction and the rate of deactivation of the catalyst.

**Solution:** From the available data, it is difficult to predict the actual mechanism of the reaction. However, we can attempt to develop a rate equation by trying to fit the available data empirically into a power-law type rate expression. For example, let

$$(-r_A) = k_n C_{Ag}^n \alpha_a$$

Since $C_{Ag}$ is maintained constant in the present experiment, we can write

$$(-r_A) = k_0 \alpha_a \qquad \text{... (i)}$$

where $k_0 = (k_n C_{Ag}^n)$. Since the feed is pure $A$ (no catalyst poison is present in the feed), side-by-side deactivation must be absent. Let us, therefore, test for parallel or independent deactivation. Thus, let

$$-d\alpha_a/d\theta = k_d C_{Ag}^m \alpha_a^d = k_d' \alpha_a^d \qquad \text{... (ii)}$$

If independent deactivation exists, then $m = 0$. Now, for a back-mix reactor, the performance equation is [similar to Eq. (8.3.9)]

$$W/F_{A0} = x_{Af}/(-r_A)_f = x_{Af}/k_0\alpha_a$$

or

$$WC_{A0}/F_{A0} = (C_{A0} - C_{Af})/(k_0\alpha_a) \qquad \text{... (iii)}$$

Equation (ii) is similar to Eq. (8.5.143). Therefore, on integration, it gives

$$\alpha_a = \exp(-k_d'\theta), \text{ if } d = 1.0 = [1 + k_d'(d-1)\theta]^{1/1-d} \text{ if } d \neq 1.0$$

Equation (iii), therefore, becomes (for $d = 1.0$)

$$\exp(-k_d'\theta) = \frac{(C_{A0} - C_{Af})}{k_0(WC_{A0}/F_{A0})}$$

or

$$\ln(WC_{A0}/F_{A0}) = \ln\left(\frac{C_{A0} - C_{Af}}{k_0}\right) + k_d'\theta \qquad \text{... (iv)}$$

Similarly, for $d \neq 1.0$, Eq. (iii) reduces to

$$(WC_{A0}/F_{A0})^{d-1} = C_1\theta + C_2 \qquad \text{... (v)}$$

where

$$C_1 = k_d'(d-1)[(C_{A0} - C_{Af})/k_0]^{d-1}$$

$$C_2 = [(C_{A0} - C_{Af})/k_0]^{d-1}$$

From the available data, the values of $(WC_{A0}/F_{A0})$ at different $\theta$ can be computed and they are listed below in Table 8.46.1.

It is seen that neither the plot of $\ln(WC_{A0}/F_{A0})$ versus $\theta$ nor that of $(WC_{A0}/F_{A0})$ versus $\theta$ is satisfactorily linear. Therefore, from Eqs (iv) and (v), $d$ is neither equal to 1.0 nor equal to 2.0. However, the plot of $(WC_{A0}/F_{A0})^2$ versus $\theta$ does yield a straight line as shown in Fig. 8.46.1. Thus, $d = 3.0$. Now, from the plot at $C_{A0} = 3.0$ kmoles/m$^3$,

$$\text{Slope} = C_1 = 6401 \text{ kg}^2\cdot\text{hr/m}^6$$

$$y\text{-intercept} = C_2 = 16.0 \text{ (kg}\cdot\text{hr/m}^3)^2$$

## Table 8.46.1

| $\theta$, hr | $(WC_{A0}/F_{A0})$, kg hr/m³ | |
|---|---|---|
| | $C_{A0}$ = 1.0 kmole/m³ | $C_{A0}$ = 3.0 kmoles/m³ |
| 0.0 | 2.0 | 4.0 |
| 1.0 | 28.355 | 80.107 |
| 2.0 | 40.0 | 113.2075 |
| 3.0 | 49.0 | 138.632 |

From Eq. (v), $C_1/C_2 = k'_d\,(d-1)$

or $(6401/16) = k'_d\,(2.0)$

$$k'_d = 200.03 \text{ hr}^{-1}$$

Since $k'_d = k_d C_{Ag}^m$ and $C_{Ag} = C_{Af} = 1.0$ kmole/m³,

$$k_d = 200.03\left(\frac{1}{\text{hr}}\right)(\text{m}^3/\text{kmole})^m$$

Now, $C_2 = 16.0 = [(C_{A0} - C_{Af})/k_0]^2$

Since $C_{A0} = 3.0$ kmoles/m³ and $C_{Af} = 1.0$ kmole/m³,

$$k_0 = 0.5 \text{ kmoles/(hr}\cdot\text{kg of catalyst)} = k_n C_{Ag}^n = k_n\,(1.0)^n$$

or

$$k_n = 0.5\left(\frac{\text{kmoles}}{\text{hr. kg of catalyst}}\right)(\text{m}^3/\text{kmole})^n$$

Now, from the plot at $C_{A0} = 1.0$ kmole/m³,

$$\text{Slope} = C_1 = 800 \text{ kg}^2\cdot\text{hr/m}^6$$

$$y\text{-intercept} = C_2 = 4.0\ (\text{kg}\cdot\text{hr/m}^3)^2$$

$$k'_d\,(d-1) = C_1/C_2 = 200 \text{ hr}^{-1}$$

or $k'_d = 100 \text{ hr}^{-1}$

$$k_d\,(0.5)^m = 100$$

Since $k_d = 200.03$,

$$(0.5)^m = 100/200.03 = 0.5$$

or $m = 1.0$

Now, $C_2 = 4.0 = [(1.0 - 0.5)/k_0]^2$

or $k_0 = 0.25$ kmoles/(hr·kg of catalyst)

Therefore $k_n\,(0.5)^n = 0.25$

Since $k_n = 0.5$, $n = 1.0$. Thus, the rate equations are

$$(-r_A) = 0.5\,C_{Ag}\alpha_a, \text{ kmoles/(hr}\cdot\text{kg of catalyst)}$$

$$-d\alpha_a/d\theta = 200.03\,C_{Ag}\,\alpha_a^3, \text{ hr}^{-1}$$

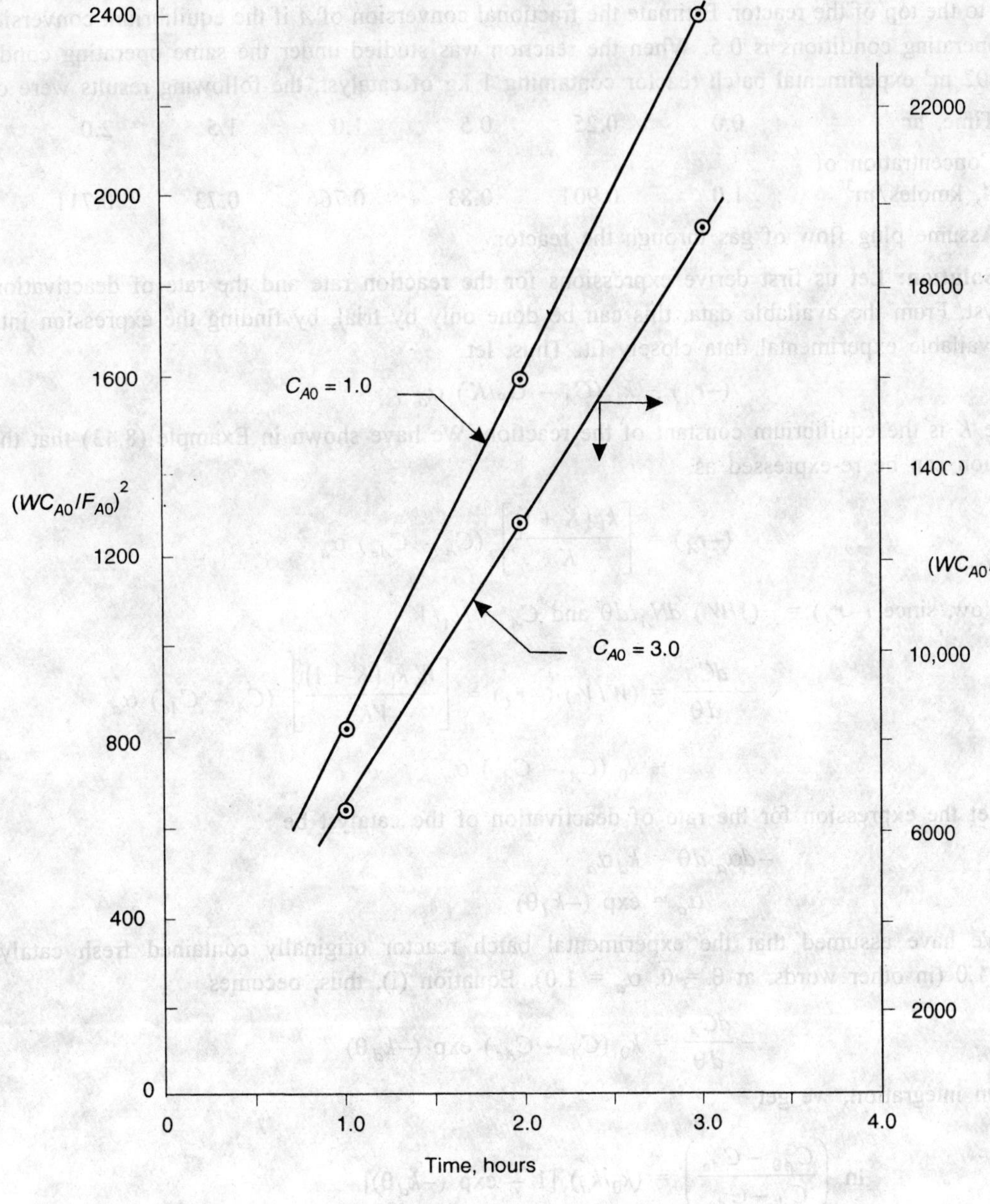


*Figure 8.46.1*

**Example 8.47:** *Performance of moving bed reactor with a deactivating catalyst.* The solid-catalysed, reversible reaction $A \rightleftharpoons P$ is to be conducted in a moving bed reactor. The catalyst is fed from the top at the rate of 2500 kg/hr and the feed (consisting of only $A$) is introduced from below at the rate of 2.0 m$^3$/hr. The spent catalyst whose activity has reduced to 0.3 is discharged from the bottom of the reactor and is sent to the regenerator. The regenerated catalyst which has an activity of 0.95 is recycled back to the top of the reactor. Estimate the fractional conversion of $A$ if the equilibrium conversion under the operating conditions is 0.5. When the reaction was studied under the same operating conditions in a 0.002 $m^3$ experimental batch reactor containing 1 kg of catalyst, the following results were obtained:

| Time, hr | : | 0.0 | 0.25 | 0.5 | 1.0 | 1.5 | 2.0 | ∞ |
|---|---|---|---|---|---|---|---|---|
| Concentration of $A$, kmoles/$m^3$ | : | 1.0 | 0.901 | 0.83 | 0.766 | 0.73 | 0.711 | 0.684 |

Assume plug flow of gas through the reactor.

**Solution:** Let us first derive expressions for the reaction rate and the rate of deactivation of the catalyst. From the available data, this can be done only by trial, by finding the expression into which the available experimental data closely fit. Thus, let

$$(-r_A) = k_1 (C_A - C_P/K)\, \alpha_a$$

where $K$ is the equilibrium constant of the reaction. We have shown in Example (8.43) that the above equation can be re-expressed as

$$(-r_A) = \left[\frac{k_1 (K+1)}{K}\right] (C_A - C_{Ae})\, \alpha_a$$

Now, since $(-r_A) = -(1/W)\, dN_A/d\theta$ and $C_A = N_A/V$,

$$-\frac{dC_A}{d\theta} = (W/V)\,(-r_A) = \left[\frac{W k_1 (K+1)}{VK}\right] (C_A - C_{Ae})\, \alpha_a$$

$$= k_0 (C_A - C_{Ae})\, \alpha_a \qquad \text{... (i)}$$

Let the expression for the rate of deactivation of the catalyst be

$$-d\alpha_a/d\theta = k_d \alpha_a$$

or

$$\alpha_a = \exp(-k_d\theta)$$

We have assumed that the experimental batch reactor originally contained fresh catalyst with $\alpha_a = 1.0$ (in other words, at $\theta = 0$, $\alpha_a = 1.0$). Equation (i), thus, becomes

$$-\frac{dC_A}{d\theta} = k_0 (C_A - C_{Ae}) \exp(-k_d\theta)$$

On integration, we get

$$\ln\left(\frac{C_{A0} - C_{Ae}}{C_A - C_{Ae}}\right) = (k_0/k_d)\,[1 - \exp(-k_d\theta)]$$

When $\theta = \infty$, $C_A = C_{A\infty}$. Therefore,

$$\ln\left(\frac{C_{A0} - C_{Ae}}{C_{A\infty} - C_{Ae}}\right) = k_0/k_d$$

Combining the above two equations (by substracting one from the other), we get

$$\ln\left(\frac{C_A - C_{Ae}}{C_{A\infty} - C_{Ae}}\right) = (k_0/k_d)\ \exp\ (-k_d\theta)$$

Taking logarithm of both sides,

$$\ln\ [\ln\ (C_A - C_{Ae})/(C_{A\infty} - C_{Ae})] = \ln\ (k_0/k_d) - k_d\theta$$

or

$$F\ (C_A) = \ln\ (k_0/k_d) - k_d\theta \qquad \text{... (ii)}$$

Thus, a plot of $F\ (C_A)$ versus $\theta$ must yield a straight line if the assumed rate expressions are correct. The values of $F\ (C_A)$ can be computed from the available experimental batch reactor data (see Table 8.47.1 given below). To note that since $C_{A0} = 1.0$ kmole/m$^3$, $C_{A\infty} = 0.684$ kmole/m$^3$ and $C_{Ae} = C_{A0}\ (1 - x_{Ae}) = 1.0\ (1 - 0.5) = 0.5$ kmole/m$^3$,

$$F\ (C_A) = \ln\ [\ln\ (C_A - 0.5)/0.184]$$

**Table 8.47.1**

| *Time* ($\theta$), *hr* | $F\ (C_A)$ |
|---|---|
| 0.0 | – 0.0003277 |
| 0.25 | – 0.2497 |
| 0.50 | – 0.5376 |
| 1.00 | – 0.998 |
| 1.50 | – 1.50 |
| 2.00 | – 1.988 |

The plot of $F\ (C_A)$ versus $\theta$ is shown in Fig. 8.47.1. The plot is rectilinear, showing that the assumed rate expressions are correct. From the plot,

$$\text{Slope} = -k_d = -0.999$$

or

$$k_d = 0.999\ \text{hr}^{-1}$$

The $y$-intercept can be deduced directly from Table 8.47.1. Thus,

$$y\text{-intercept} = \ln\ (k_0/k_d) = -0.0003277$$

or

$$k_0/k_d = 0.99967$$

$$k_0 = (0.99967)\ (0.999) = 0.99867\ \text{hr}^{-1}$$

Since for the experimental batch reactor, $W = 1.0$ kg and $V = 0.002$ m$^3$,

$$k_1\ (K + 1)/K = k_0\ (V/W) = 0.99867\ (0.002) = 1.9973 \times 10^{-3}\ \text{m}^3/(\text{kg}\cdot\text{hr})$$

The rate expressions are, therefore,

$$(-\boldsymbol{r}_A) = (1.9973 \times 0^{-3})\ (C_A - C_{Ae})\ \alpha_a$$

$$-d\alpha_a/d\theta = 0.999\,\alpha_a$$

The moving bed reactor is sketched in Fig. 8.47.2. Let us assume that the solids are completely back mixed in the reactor. Such an assumption usually causes only allowable error. For such a mixed flow, from Eq. (8.3.15),

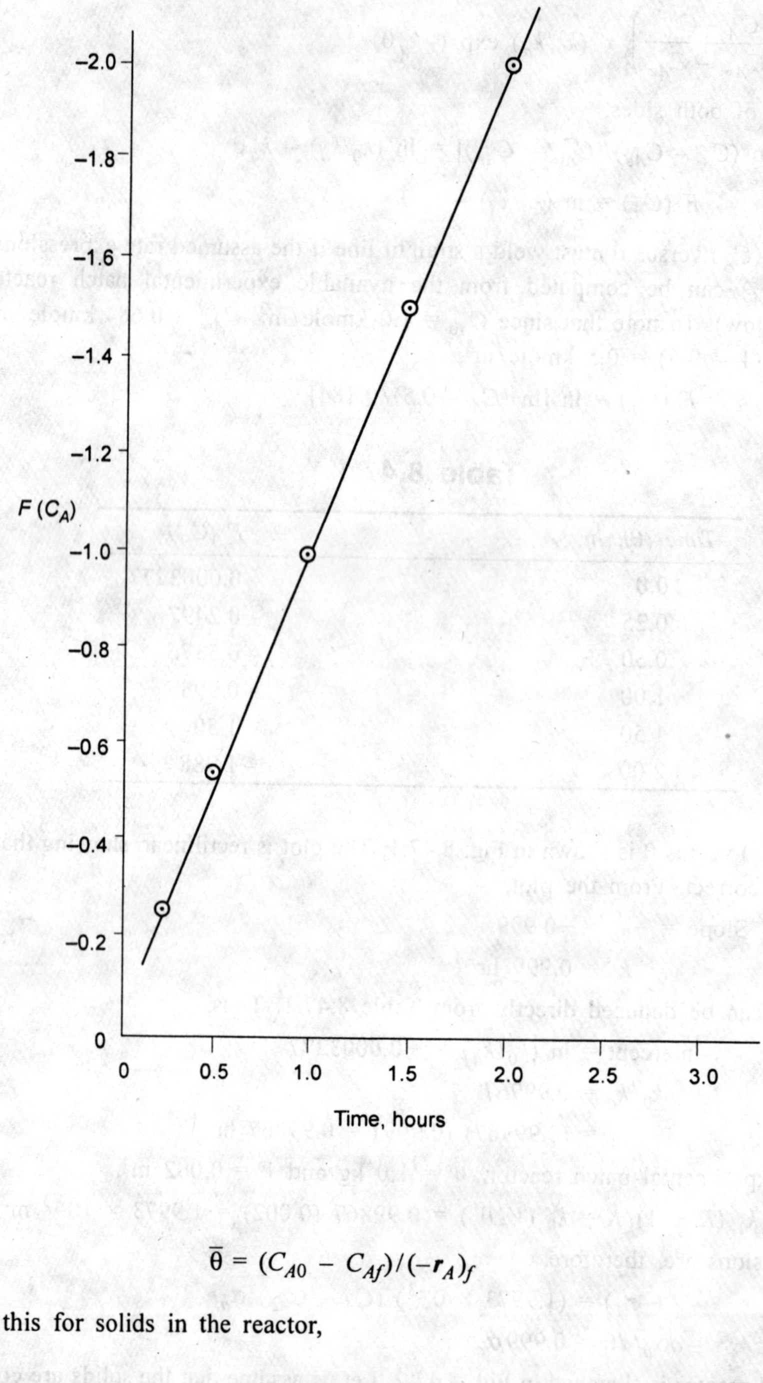


*Figure 8.47.1*

$$\bar{\theta} = (C_{A0} - C_{Af})/(-r_A)_f$$

Applying this for solids in the reactor,

$$\bar{\theta}_1 = (W_1/F_S) = \frac{(\alpha_a)_0 - (\alpha_a)_f}{k_d\,(\alpha_a)_f} \qquad \text{... (iii)}$$

$$= \frac{(0.95 - 0.3)}{(0.999)(0.3)} = 2.17 \text{ hr}$$

Since $F_S$ = 2500 kg/hr, $W_1$ = 5425 kg

Since plug flow of gas is assumed in the reactor, the performance equation is

$$WC_{A0}/F_{A0} = (W/Q_0) = -\int_{C_{A0}}^{C_{Af}} \frac{dC_A}{(-r_A)}$$

Substituting the expression for $(-r_A)$, we get

$$(5425/2.0) = 2712.5 = -\frac{1}{(0.3)\left(1.9973 \times 10^{-3}\right)} \int_{C_{A0}}^{C_{Af}} \frac{dC_A}{C_A - C_{Ae}}$$

To note that due to complete backmixing of solids in the reactor, the averagè activity of catalyst particles in the reactor $(\alpha_a)$ = that in the outlet stream [namely, $(\alpha_a)_f$]. We have, therefore, substituted $\alpha_a = (\alpha_a)_f = 0.3$ and taken it outside the integral. On integration and rearrangement, we get

$$\frac{\left(C_{Af} - C_{Ae}\right)}{\left(C_{A0} - C_{Ae}\right)} = \exp(-1.6253)$$

or $$(x_{Ae} - x_A)/x_{Ae} = 0.197$$

Therefore $$x_A = \mathbf{0.4015}$$

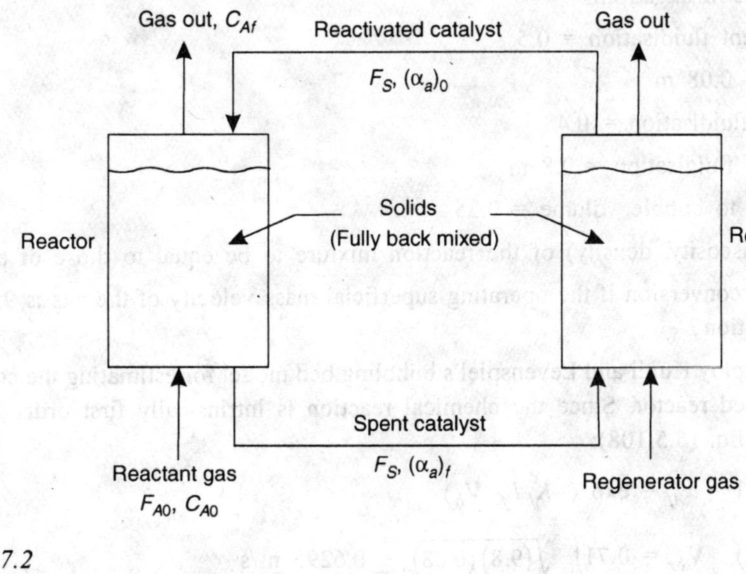


*Figure 8.47.2*

It is important to note that in the above example, in applying Eq. (8.3.15) (namely, the performance equation of backmix reactor) to solids, we have treated the solids as a microfluid. Such an assumption is permissible since deactivation is of first order. For alternate situations, $(\alpha_a)_f$ is to be computed as

$$(\alpha_a)_f = \int_0^{\infty} \alpha_a \left(1/\bar{\theta}\right) \exp\left(-\theta/\bar{\theta}\right) d\theta \qquad \text{... (8.5.147)}$$

where $\bar{\theta} = (W/F_S)$ = average residence time of solids in the reactor. It can be seen that if $-(d\alpha_a/d\theta) = k_d \alpha_a$ so that $\alpha_a = (\alpha_a)_0 \exp(-k_d\theta)$, then the above equation on integration gives

$$(\alpha_a)_f = (\alpha_a)_0/(1 + k_d\bar{\theta})$$

or

$$\bar{\theta} = \frac{(\alpha_a)_0 - (\alpha_a)_f}{k_d(\alpha_a)_f} \qquad \text{... (8.5.148)}$$

which is same as Eq. (iii) of above example.

**Example 8.48:** *Performance of fluidised bed reactor*. An irreversible reaction, $A \rightarrow P$ which is intrinsically first order in $A$ is to be conducted in a fluidised bed reactor that is composed of catalyst particles 200 microns in size. The feed gas (consisting of only $A$) is admitted at 200° C and 1 atm and the operation is essentially isothermal. The operating superficial mass velocity of the gas is six times that at incipient fluidisation.

(*a*) Estimate the fractional conversion of $A$ from the following data:

Diffusivity of $A$ in gas phase = $0.2 \times 10^{-4}$ m²/s

First order rate constant = 2.0 $s^{-1}$

Density of catalyst pellet = 1200 kg/m³

Porosity of catalyst pellet = 0.40

Average pore size = 4500 angstroms

Bed voidage at incipient fluidisation = 0.5

Average bubble size = 0.08 m

Bed voidage prior to fluidisation = 0.45

Packed height prior to fluidisation = 0.8 m

Ratio of wake volume to bubble volume = 0.35

Take the properties (viscosity, density) of the reaction mixture to be equal to those of air.

(*b*) What would be the conversion if the operating superficial mass velocity of the gas is 95 percent of that at incipient fluidisation?

**Solution:** (*a*) Let us employ Kunii and Levenspiel's bubbling bed model for estimating the conversion achieved in the fluidised bed reactor. Since the chemical reaction is intrinsically first order in $A$ and irreversible, we can apply Eq. (8.5.108):

$$1 - x_{Af} = \exp\left(-K_f L_f / \mathbf{V}_b\right) \qquad \text{... (i)}$$

Now, from Eq. (8.5.93), $\mathbf{V}_{br} = 0.711\sqrt{(9.8)(0.08)} = 0.6295$ m/s

From Eq. (8.5.92), $$V_b = V_{0f} - V_{mf} + V_{br}$$
$$= 6\,V_{mf} - V_{mf} + 0.6295 = 5\,V_{mf} - 0.6295 \qquad \text{... (ii)}$$

We can compute the minimum fluidisation velocity from Eq. (2.12.30) of Chapter 2. At 200° C and 1 atm, the properties of air (which are taken equal to those of the reaction mixture) are

$$\mu_g = 26.0 \times 10^{-6}\ \text{kg/(m·s)}$$
$$\rho_g = 0.746\ \text{kg/m}^3$$

Therefore, $$Ar_m = \frac{\left(200\times10^{-6}\right)^3 (0.746)(1200-0.746)(9.8)}{\left(26.0\times10^{-6}\right)^2} = 103.757$$

Hence, from Eq. (2.12.30),

$$Re'_{mf} = [(33.67)^2 + 0.0408\ (103.757)]^{0.5} - 33.67 = 0.0628$$

or $$V_{mf} = \frac{(0.0628)\left(26.0\times10^{-6}\right)}{\left(200\times10^{-6}\right)(0.746)} = 0.011\ \text{m/s}$$

Therefore, from Eq. (ii), $V_b = 5.0\ (0.011) + 0.6295 = 0.6843$ m/s

Now, from Eq. (8.5.96), $K_{Ce} = 6.78\ [(0.5)\ (0.2 \times 10^{-4})\ (0.6843)/(0.08)^3]^{0.5} = 0.7838\ \text{s}^{-1}$

It is given that, $k\rho_p = 2.0\ \text{s}^{-1}$

The effectiveness factor ($\eta$) is very likely to be close to unity. Let us, however, verify this. Since it is given that $D_A = 0.2 \times 10^{-4}$ m$^2$/s, let us compute the Knudsen diffusivity. Thus, from Eq. (4.1.59) of Chapter 4,

$$\bar{u}_A = [(8.0)\ (8314.0)\ (473)/(29\pi)]^{0.5} = 587.63\ \text{m/s}$$

From Eq. (4.1.58), $D_{A,K} = (4500 \times 10^{-10})\ (587.63)/(3.0) = 8.8 \times 10^{-5}\ \text{m}^2/\text{s}$

Therefore, from Eq. (4.1.63),

$$D = \frac{1}{\left(1/0.2\times10^{-4}\right)+\left(1/8.8\times10^{-5}\right)} = 1.63 \times 10^{-5}\ \text{m}^2/\text{s}$$

And from Eq. (4.1.65), $D_e = (0.4)^2\ (1.63 \times 10^{-5}) = 2.608 \times 10^{-6}\ \text{m}^2/\text{s}$

Therefore, $\phi = (200 \times 10^{-6}/6)\ (2.0/2.608 \times 10^{-6})^{1/2} = 0.0146$

Since $\phi$ is much less that 0.5, $\eta \simeq 1.0$. Thus,

$$k_0 = (\eta k\rho_p) = (1.0)\ (2.0) = 2.0\ \text{s}^{-1}$$

Since the value of $\gamma_b$ varies between 0.001 and 0.01, let us choose $\gamma_b = 0.01$. It is given that $\alpha = 0.35$. Thus, from Eq. (8.5.97),

$$\gamma_C = (1 - 0.5)\left[\frac{(3.0)(0.011/0.5)}{0.6295-(0.011/0.5)} + 0.35\right] = 0.2293$$

From Eq. (8.5.101), $\delta = (6\,V_{mf} - V_{mf})/V_b = (5.0)\ (0.011)/(0.6843) = 0.08$

From Eq. (8.5.100), $\gamma = \dfrac{(1-0.5)(1-0.08)}{(0.08)} - (0.2293 + 0.01) = 5.508$

Thus, from Eq. (8.5.106), $K' = \dfrac{1}{(1/0.7838) + [1/(2.0)(5.508)]} = 0.73174 \text{ s}^{-1}$

Now, from Eq. (8.5.94), $K_{bC} = \dfrac{(4.5)(0.011)}{(0.08)} + \dfrac{(5.85)\left(0.2\times10^{-4}\right)^{1/2}(9.8)^{1/4}}{(0.08)^{5/4}} = 1.7067 \text{ s}^{-1}$

Therefore, from Eq. (8.5.105),

$$K_f = (0.01)(2.0) + \frac{1}{(1/1.7067) + \left[\dfrac{1}{2(0.2293) + 0.73174}\right]} = 0.72125 \text{ s}^{-1}$$

From Eq. (8.5.109), $(L_f/\mathbf{V}_b) = \dfrac{(1-0.45)(0.8)}{(1-0.5)(0.6295)} = 1.3979 \text{ s}$

Therefore, from Eq. (i), $x_{Af} = 0.635 = 63.5$ percent

(*b*) If the operating gas velocity is less than the minimum fluidisation velocity, then the catalyst bed will not be fluidised and the reactor will operate as a fixed bed reactor. Assuming plug flow of gas through the reactor, the performance equation will be, therefore,

$$W/F_{A0} = \int_0^{x_{Af}} \frac{dx_A}{(-r_A)} \qquad \text{... (iii)}$$

Now, $(-r_A) = \eta k C_{Ag} = k C_{Ag} = k C_{A0}(1 - x_A)$

Substitution of the above expression in Eq. (iii), integration and rearrangement yields

$$1 - x_{Af} = \exp(-kWC_{A0}/F_{A0}) = \exp(-kW/Q_0)$$

Now, $W = A_C L_b (1 - \epsilon)\rho_p = A_C (0.8)(1 - 0.45)(1200)$ kg

$Q_0 = A_C \mathbf{V}_0 = A_C (0.95)(0.011) \text{ m}^3/\text{s}$

$k = (k_0/\eta\rho_p) = 2.0/(1.0)(1200) \text{ m}^3/(\text{kg}\cdot\text{s})$

Therefore, $kW/Q_0 = 84.21$

$(1 - x_{Af}) = \exp(-84.21)$

or $x_{Af} \simeq 1.0$

Thus, the fixed bed reactor provides close to 100 percent conversion, while the conversion achieved in the bubbling fluidised bed is only 63.5 percent.

**Example 8.49:** *Performance of slurry reactors and trickle bed reactors.* Sulfur dioxide is to be removed from air by catalytically oxidising it to sulfur trioxide on 0.542 mm activated carbon particles at 25° C and 1 atm pressure. The sulfur trioxide formed is to be dissolved in water to produce sulfuric acid solution. The catalyst density ($\rho_p$) is 800 kg/m$^3$. The feed gas contains 2.5 percent (by volume) sulfur dioxide and 97.5 percent air and it is to be introduced at 108 $m^3/hr$. The following two options are available:

(*i*) A slurry reactor that holds a suspension of carbon particles in water, the catalyst concentration being 30 kg of catalyst per $m^3$ of water. For the type of distributor that is being employed, the average bubble size is 3 mm and the gas holdup (bubble volume per unit volume of liquid) is 0.08.

(*ii*) A trickle bed reactor into which the feed gas and pure water are fed from the top such that the superficial mass velocity of gas is 0.06 kg /($m^2 \cdot s$) and that of the liquid 50 kg/($m^2 \cdot s$). Bulk density of catalyst bed = 500 kg/$m^3$.

If 70 percent of the sulfur dioxide is to be removed by converting it to sulfuric acid, determine which of the above two reactors is more suitable for the application. The intrinsic rate at the carbon site is first order in oxygen and zero order in sulfur dioxide so that $(-r_B) = \eta k C_{BS}$, where $k = 4.566 \times 10^{-4}$ $m^3/(kg \cdot s)$. The effectiveness factor for 0.542 mm particles is 0.098.

Henry's law constant for oxygen in water at 25° C = 35.4 (kmole/$m^3$ of gas) /(kmole/$m^3$ of liquid)

Diffusivity of oxygen in liquid water at 25° C = $2.0 \times 10^{-9}$ $m^2/s$

Take the liquid to particle mass transfer coefficient ($k_{Lp}$) for oxygen in the slurry reactor to be $1.9 \times 10^{-4}$ m/s and compute $k_L$ from the following correlation*:

$$(k_L d_b / D_L) = 2.0 + 0.31\ Ra^{1/3}$$

where $Ra$ = Rayleigh number = $d_b^3 (\rho_L - \rho_g)\, g/(D_L \mu_L)$. For the trickle bed reactor, use the following two correlations[48,49] for computing the mass transfer coefficients:

$$\left| \begin{array}{l} k_L a_g / D_L = 11094.253\,(G_L/\mu_L)^{0.4}\ Sc_L^{0.5} \\ (k_{Lp} a_p)/(V_L a_t) = 1.64\,(Re_L)^{-0.331}\ Sc_L^{-2/3} \end{array} \right.$$

for $0.2 < Re_L < 2400$, where $Re_L = (d_p G_L / \mu_L)$.

**Solution:** (*a*) Let us first consider the slurry reactor. We shall take *B* (namely, oxygen) as the limiting reactant, since the rate of reaction depends on the concentration of *B* and is independent of $C_A$ (where *A* stands for sulfur dioxide). Assuming that the slurry is well mixed and the gas bubbles move individually in plug flow, we can apply Eqs (8.5.132) and (8.5.137) after rewriting them for *B*. Thus

$$V/F_{B0} = \int_0^{x_{Bf}} \frac{dx_B}{k_0 C_{Bg}} \qquad \text{... (i)}$$

Now, assuming that $k_C$ is much larger than $k_L$, Eq. (8.5.138) reduces to

$$1/k_0 = He' \left( \frac{1}{k_L a_g} + \frac{1}{k_{Lp} a_p} + \frac{1}{\eta k C_C} \right) \qquad \text{... (ii)}$$

where $H_e' = 35.4$

*Note that this correlation does not contain impeller power or impeller speed. Correlations of this kind assume taht since ($\rho_L - \rho_g$) is quite large, bubble motion due to gravitational force rather than due to impeller motion determines the rate of mass transfer.

$$\eta = 0.098$$
$$k = 4.566 \times 10^{-4}\ \text{m}^3/(\text{kg}\cdot\text{s})$$
$$C_C = 30\ \text{kg/m}^3$$

From Eq. (8.5.135b), $a_p = \left(\dfrac{6}{\rho_p d_p}\right) C_C = \dfrac{(6.0)(30.0)}{(800.0)(0.000542)} = 415.13\ \text{m}^2/\text{m}^3$ of liquid

Since the gas holdup ($\in_g$) is 0.08, from Eq. (8.5.135c),

$$a_g = (6/d_b) \in_g = (6.0)\ (0.08)/(0.003) = 160\ \text{m}^2/\text{m}^3 \text{ of liquid}$$

The value of $k_L$ can be obtained from the available correlation. Thus

$$Ra = \frac{(0.003)^3\ (1000 - 1.2)\ (9.8)}{\left(2.0 \times 10^{-9}\right)(0.001)} = 1.3214 \times 10^8$$

To note that we have substituted $\rho_g = (PM/RT) = (1.0)\ (29.0)/(0.082)\ (298) = 1.2\ \text{kg/m}^3$ in the above expression. Now, from the given correlation,

$$k_L = (2.0 \times 10^{-9}/0.003)\ [2.0 + 0.31\ (1.3214 \times 10^8)1/3]$$
$$= 1.066 \times 10^{-4}\ \text{m/s}$$

Therefore, from Eq. (ii), $k_O = 3.4608 \times 10^{-5}\ \text{s}^{-1}$

Since $y_{BO}$ = mole fraction of $B$ in feed gas = (0.975) (0.21) = 0.20475,

$$C_{BO} = p_{BO}/RT = Py_{BO}/RT = \frac{(1.0)\ (0.20475)}{(0.082)\ (298)} = 8.379 \times 10^{-3}\ \text{kmoles/m}^3$$

Therefore, $F_{BO} = Q_0 C_{BO} = (108/3600)\ (8.379 \times 10^{-3}) = 2.513 \times 10^{-4}\ \text{kmoles/s}$

Equation (i) can be more conveniently integrated if $C_{Bg}$ and $x_B$ are expressed in terms of $x_A$. Thus, since one mole of $A$ (namely, sulfur dioxide) reacts with half mole of $B$,

$$x_B = (1/2)\ N_{AO} x_A / N_{BO} \quad \text{... (iii)}$$
$$= (1/2)\ (2.5/20.475)\ x_A = 0.06105 x_A$$

Similarly, $$C_{Bg} = \frac{C_{BO} - (1/2)\ C_{AO} x_A}{(1 + \in_A x_A)} \quad \text{... (iv)}$$

where $C_{BO} = 8.379 \times 10^{-3}\ \text{kmoles/m}^3$

$$C_{AO} = \frac{(1.0)\ (0.025)}{(0.082)\ (298)} = 1.023 \times 10^{-3}\ \text{kmoles/m}^3$$

To find $\in_A$, consider 100 kmoles of feed gas at $x_A = 0$. Thus, at $x_A = 0$, $SO_2 = 2.5$, $O_2 = 20.475$ and $N_2 = 77.025$ kmoles. At complete conversion ($x_A = 1.0$), $SO_3 = 2.5$, $N_2 = 77.025$ and $O_2 = 20.475 - (1/2)\ (2.5) = 19.225$ kmoles. Thus, total number of moles at $x_A = 1.0$ is (2.5 + 77.025 + 19.225) = 98.75. Therefore

$$\epsilon_A = (100 - 98.75)/100 = 0.0125$$

Equation (i) can be thus re-expressed in terms of $x_A$ as

$$V/F_{B0} = \frac{0.06105}{k_0} \int_0^{x_{Af}} \frac{1+\epsilon_A x_A}{C_{B0} - (C_{A0}/2)\,x_A}\, dx_A$$

On integration and rearrangement, we get

$$\frac{k_0 V C_{A0}^2}{0.06105 F_{B0}} = (4\epsilon_A C_{B0} + 2C_{A0}) \ln\left(\frac{2C_{B0}}{2C_{B0} - C_{A0}x_{Af}}\right) - (2\epsilon_A C_{A0} x_{Af})$$

Substituting $x_{Af} = 0.7$ and the values of $k_0$, $F_{B0}$, $\epsilon_A$, $C_{A0}$ and $C_{B0}$, we get

$$V = 34.77 \text{ m}^3$$

This is the required volume of gas-free liquid in the reactor. Since the catalyst concentration is 30 kg/m$^3$, the amount of catalyst required ($W$) is

$$W = (34.77)\,(30) = 1043.1 \text{ kg}$$

Evidently, the amount of slurry liquid required is fairly large. This is mainly due to the large particle size which decreases the liquid to particle interfacial area ($a_p$) and increases the resistance to pore diffusion (note that $\eta$ is as low as 0.098). It may also be noted that we have based our computations on the steady state period before the sulfuric acid concentration has increased enough to significantly retard the reaction rate.

(*b*) Let us now consider the trickle bed reactor. The reactions are

$$SO_2\,(g) + (1/2)\,O_2\,(g) \rightarrow SO_3\,(g)$$

$$SO_3\,(g) + H_2O\,(l) \rightarrow H_2SO_4\,(l)$$

However, the mass balance Eqs (8.5.112), (8.5.114) and (8.5.116) hold good (except that we have to rewrite them in terms of $B$). Accordingly, Eq. (8.5.123) is also applicable since it is derived from the above mass balance equations. We shall, nevertheless, assume that no oxygen is lost in the outgoing liquid effluent. In other words, all the unreacted oxygen is present in the exit gas stream. To apply Eq. (8.5.123), we must know the values of the constants $K_1$ to $K_4$. Let us, therefore, evaluate them one by one. At the outset, let us compute the mass transfer coefficients from the available correlations. Thus

$$Sc_L = \mu_L/\rho_L D_L = \frac{(0.001)}{(1000)\left(2.0\times10^{-9}\right)} = 500.0$$

$$G_L = 50 \text{ kg}/(\text{m}^2\cdot\text{s})$$

Therefore, from the given correlation,

$$(k_L a_g) = (2.0 \times 10^{-9})\,(11094.253)\,(50.0/0.001)^{0.4}\,(500)^{0.5} = 0.0376 \text{ s}^{-1}$$

Since oxygen is only slightly soluble in water, $(K_L a_g) \simeq (k_L a_g) = 0.0376$ s$^{-1}$. Now,

$$Re_L = (0.000542)\,(50)/(0.001) = 27.1$$

We can, therefore, apply the given correlation for $k_{Lp}$ since the value of $Re_L$ lies within the specified range of 0.2 to 2400.0. In the given correlation, $a_t$ is the total external surface area of particles per unit volume of catalyst bed. Assuming the catalyst particles to be spherical, the total external surface area per unit volume of particles is

$$= \frac{\left(\pi d_p^2\right) N_p}{\left(\pi d_p^3/6\right) N_p} = 6/d_p$$

where $N_p$ is the number of particles. Since $(1 - \in)$ = (volume of particles)/(volume of bed),

$$a_t = (6/d_p)\ (1 - \in)$$

It is given that $\rho_p\ (1 - \in)$ = bulk density of catalyst bed = 500 kg/m$^3$. Also, $\rho_p$ = 800 kg/m$^3$. Therefore,

$$a_t = (6/0.000542)\ (500/800) = 6918.82\ \text{m}^2/\text{m}^3$$

Now, from the given correlation,

$$(k_{Lp} a_p) = (50/1000)\ (6918.82)\ (1.64)\ (27.1)^{-0.331}\ (500)^{-2/3} = 3.0215\ \text{s}^{-1}$$

From Eq. (8.5.117), $k_0 = \rho_p\ (1 - \in)\ \eta k = (500.0)\ (0.098)\ (4.56 \times 10^{-4}) = 0.02237\ \text{s}^{-1}$

From Eq. (8.5.118), $K_1 = (3.0215)/(0.02237 + 3.0215) = 0.99265$

Since $\mathbf{V}_L = (50/1000) = 0.05$ m/s and $\mathbf{V}_g = (0.06/1.2) = 0.05$ m/s, from Eq. (8.5.121),

$$K_2 = \frac{(0.05)(0.0376) + (0.05)(35.4)\left[3.0215(1 - 0.99265) + 0.0376\right]}{(0.05)(0.05)(35.4)}$$

$$= 1.21735\ \text{m}^{-1}$$

From Eq. (8.5.122), $$K_3 = \frac{(3.0215)(0.0376)(1 - 0.99265)}{(0.05)(0.05)(35.4)} = 9.4342 \times 10^{-3}\ \text{m}^{-2}$$

From Eq. (8.5.128), $$K_4 = \frac{0.0376\left[0.0 - \left(8.379 \times 10^{-3}/35.4\right)\right]}{(0.05)} = -1.78 \times 10^{-4}\ \text{kmoles/m}^4$$

To note that we have taken $(C_{BL})_0 = 0.0$ [an alternate assumption possible is to take $(C_{BL})_0 = (C_{Bg})_0/He'$, which assumes that the feed water is saturated with oxygen].

Now, from Eqs (8.5.124) and (8.5.125),

$$\lambda_1 = -\ 0.0078\ \text{m}^{-1}$$

$$\lambda_2 = -1.12095\ \text{m}^{-1}$$

And, from Eqs (8.5.129) and (8.5.130),

$$C_1 = 8.2853 \times 10^{-3}\ \text{kmoles/m}^3$$

$$C_2 = 9.3733 \times 10^{-5}\ \text{kmoles/m}^3$$

Since $x_{Af} = 0.70$, from Eq. (iv) of part (a),

$$C_{Bg} = \frac{\left(8.379\times10^{-3}\right)-(0.5)\left(1.023\times10^{-3}\right)(0.7)}{1.0+(0.0125)(0.70)} = 7.95\times10^{-3}\,\text{kmoles/m}^3$$

Therefore, from Eq. (8.5.123),

$$(7.95\times10^{-3}) = (8.2853\times10^{-3})\exp(-0.0078\,z)+(9.3733\times10^{-5})\exp(-1.12095\,z)$$

or

$$z = 5.3\ \text{m}$$

This is the required depth of catalyst bed. Since the volumetric flow rate of gas is 108 $m^3$/hr, the area of cross-section of the reactor is

$$A_C = (108/3600)\,(1.2)/(0.06) = 0.6\ \text{m}^2$$

Volume of catalyst bed = (0.6) (5.3) = 3.18 $m^3$

Since the bulk density of the bed is 500 kg/$m^3$,

$$\text{Mass of catalyst } (W) = (3.18)\,(500) = 1590\ \text{kg}$$

Thus, the trickle bed reactor demands larger amount of catalyst than the slurry reactor to achieve the same conversion.

The operational difficulties of both reactors should also not be overlooked while selecting between the two. The chief difficulty with slurry reactors, as stated earlier, is in retaining the catalyst particles in the reactor without getting lost in the outgoing effluent. Also, the operating cost of the slurry reactor could be fairly high since the power consumption of the mechanical impeller and that for bubbling the gas up through the large volume of liquid would necessarily be substantially large. In contrast, in a trickle bed reactor, the liquid trickles down by gravity and there is hardly any danger of catalyst particles getting lost in the outgoing effluent. However, the difficulty involved in ensuring uniform liquid distribution is an important operational limitation of trickle beds. Also, isothermal operation is less readily achieved in trickle beds than in slurry reactors.

## 8.6 NONCATALYTIC FLUID–SOLID REACTIONS

In the previous section, we have discussed on solid catalysed reactions. In such reactions, the solid is not a reactant, it only catalyses the reaction. The noncatalytic fluid-solid reactions are those in which both the fluid and the solid are reactants. The products may be either solid or fluid or both. A large number of examples are listed in Table 8.4 A of Section 8.5.

Reactor design problems for noncatalytic fluid-solid reactions are similar to those for solid–catalysed reactions, except that in the case of noncatalytic fluid–solid reactions, the rate of reaction is a function of time. If the solids are in continuous flow as in a moving bed reactor, steady state operation of the reactor is possible. But, if the particles remain in the reactor as in a fixed bed or batch fluidised bed reactor, then steady state operation will not be possible.

Most of the noncatalytic fluid–solid reactions fall under the following two types*:

$$A\ (g) + bB\ (s) \rightarrow pP\ (s) + rR\ (g)$$

*In these two equations and in the analysis that follows, the reactant $A$ and the product $R$ are assumed to be in the gaseous state. However, the analysis is equally valid even if they are liquids.

$$A\ (g) + bB\ (s) \rightarrow pP\ (g) + rR\ (g)$$

Roasting of ores, reduction of iron oxide to metallic iron (in blast furnace), etc. are examples of the first kind, while combustion of coal, gasification of coal to form blue water gas, regeneration of deactivated cracking catalyst by burning off the deposited carbon particles, etc. are of the latter type.

Let us first consider the reactions of the first kind in which one of the products is a solid. This solid product (namely, $P$) forms a layer on the surface of the solid reactant $B$. The thickness of this layer (often called the *ash*) increases as the reaction progresses. The gaseous reactant $A$, therefore, has to diffuse through this ash layer to reach the surface of the unreacted core of $B$, where the reaction takes place. Among all the models that have been proposed to explain processes of this kind, the *shrinking core model* (also called the *unreacted core model*) is quite satisfactory since it fits reasonably to many of the real situations. This model assumes that the radius of the unreacted core of $B$ (namely, $R_C$) decreases with time, while the total radius $R_S$, which is the combined radius of the unreacted core of $B$ and the *ash* layer surrounding it, remains constant (see Fig. 8.26). The overall process is to consist of the following steps:

(*i*) Transfer of $A$ from the gas bulk to the outer surface of the *ash* layer,

(*ii*) Diffusion of $A$ through the *ash* layer to the surface of the unreacted core of $B$,

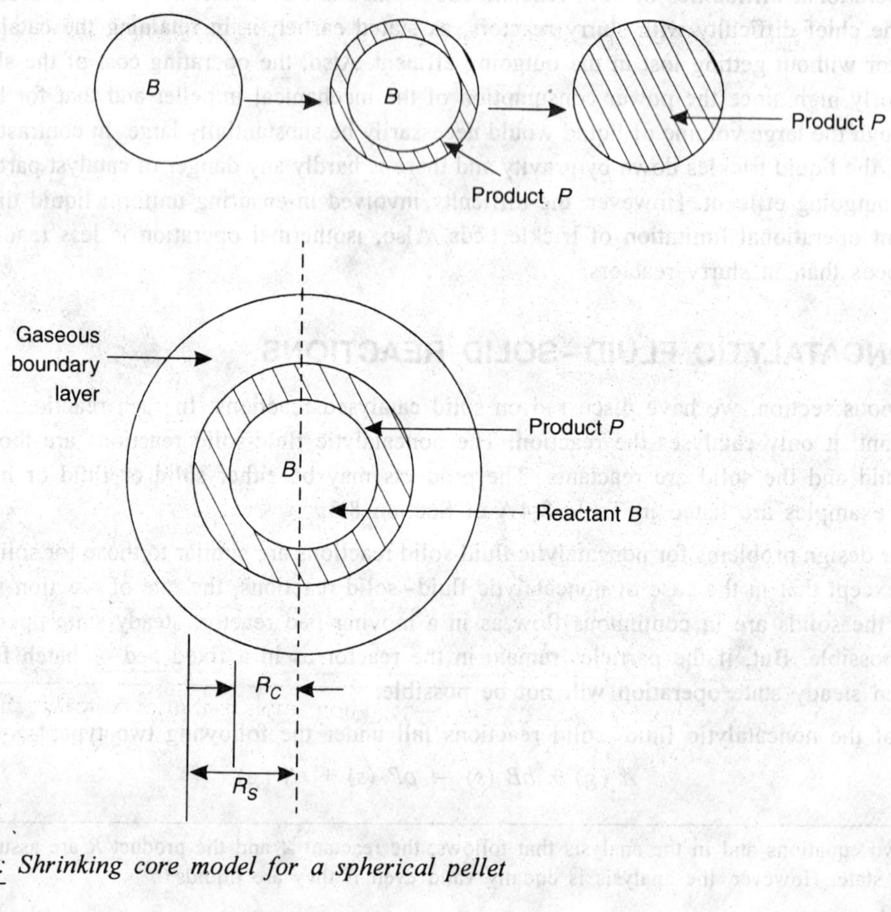


*Figure 8.26: Shrinking core model for a spherical pellet*

(*iii*) Chemical reaction at the surface to form *P* and *R*,

(*iv*) Diffusion of product *R* through the *ash* layer and subsequently, transfer to the gas bulk.

If no gaseous product is formed, then the last step shall be absent. Let us further assume, for the sake of simplicity, that the chemical reaction is intrinsically first order in *A* and is irreversible. The expression for the rate of each step can be now written down as follows (considering a single spherical solid particle):

$$-\frac{dN_A}{d\theta} = (4\pi R_S^2)\, k_C\, (C_{Ag} - C_{AS}) \quad \text{... (8.6.1)}$$

$$-\frac{dN_A}{d\theta} = (4\pi R_C^2)\, D_e\, (dC_A/dr)_{r = R_C} \quad \text{... (8.6.2)}$$

$$-\frac{dN_A}{d\theta} = (4\pi R_C^2)\, kC_{AC} \quad \text{... (8.6.3)}$$

where $k_C$ is the gas phase mass transfer coefficient for *A*, $D_e$ is the effective diffusivity of *A* through the *ash* layer (incidentally, the ash layer has been assumed to be porous) and *k* is the first order rate constant in $m^3/(m^2 \cdot s)$. $C_{Ag}$, $C_{AS}$ and $C_{AC}$ are the concentrations of *A* in the gas film, at the surface of the *ash* layer and at the surface of the unreacted core of *B* respectively. It may be noted that the rate of chemical reaction has been assumed to be proportional to the outer surface area of the unreacted core of *B*. Strictly speaking, the actual area for reaction would be much larger even for a nonporous particle due to small-scale indentations in the surface. Also, the above three equations are based on the steady state concept. That is, under steady conditions, the rates of all the three steps shall be equal. However, in actual practice, the rate at which the reaction interface at $R_C$ recedes (that is, $dR_C/d\theta$) is often smaller than the velocity of diffusion of *A* through the *ash* layer. But, if the density of the gas in the pores of the *ash* layer is small with respect to the density of the solid reactant *B* (which is usually the case), a psuedo-steady state may be assumed with allowable error.

Equations (8.6.1) and (8.6.3) can be directly applied to design problems once the values of the mass transfer coefficient ($k_C$) and the first order rate constant (*k*) are known. However, to apply Eq. (8.6.2), we must know the expression for the concentration gradient ($dC_A/dr$) at $r = R_C$. This can also be easily derived from the equation of mass transfer given in Chapter 4. Thus, for the steady state radial diffusion of *A*, Eq. (C) of Table 4.6 reduces to

$$\frac{d}{dr}\left(r^2 \frac{dC_A}{dr}\right) = 0 \quad \text{... (8.6.4)}$$

We can easily integrate the above equation twice with the boundary conditions that at $r = R_S$, $C_A = C_{AS}$ and at $r = R_C$, $C_A = C_{AC}$ and the result is

$$C_A - C_{AC} = \frac{(C_{AS} - C_{AC})(1 - R_C/r)}{(1 - R_C/R_S)} \quad \text{... (8.6.5)}$$

Therefore,
$$\left.\frac{dC_A}{dr}\right|_{r = R_C} = \frac{(C_{AS} - C_{AC})}{R_C\,(1 - R_C/R_S)} \quad \text{... (8.6.6)}$$

Equation (8.6.2), thus, becomes

$$\frac{dN_A}{d\theta} = \frac{4\pi R_C D_e (C_{AS} - C_{AC})}{(1 - R_C / R_S)} \quad \text{... (8.6.7)}$$

or

$$-\frac{dN_A}{d\theta} \frac{(1 - R_C / R_S)}{(4\pi R_C D_e)} = C_{AS} - C_{AC} \quad \text{... (8.6.7a)}$$

Equations (8.6.1) and (8.6.3) may also be re-expressed in the above form as

$$-\frac{dN_A}{d\theta} \frac{1}{4\pi R_S^2 k_C} = C_{Ag} - C_{AS} \quad \text{... (8.6.1a)}$$

$$-\frac{dN_A}{d\theta} \frac{1}{4\pi R_C^2 k} = C_{AC} \quad \text{... (8.6.3a)}$$

If we add the above three equations, we get

$$-\frac{dN_A}{d\theta} = \frac{4\pi R_C^2 C_{Ag}}{(1/k) + \left(R_C^2 / R_S^2\right)(1/k_C) + \left(R_C / D_e\right)\left(1 - R_C / R_S\right)} \quad \text{... (8.6.8)}$$

It must be kept in mind that though $R_S$ is a constant, $R_C$ is a function of time. If $m_B$ is the mass of the solid particle at any time θ, then

$$N_B = m_B/M_B = (4/3)\ \pi R_C^3 \rho_B/M_B \quad \text{... (8.6.9)}$$

or

$$\frac{dN_B}{d\theta} = (4\pi R_C^2 \rho_B/M_B)\ \frac{dR_C}{d\theta} \quad \text{... (8.6.10)}$$

where $\rho_B$ is the density and $M_B$ is the molecular weight of the solid reactant $B$. Since from reaction stoichiometry, $dN_A/d\theta = (1/b)\ dN_B/d\theta$, where $b$ is the stoichiometric coefficient (that is, the number of moles of $B$ reacting with one mole of $A$), Eq. (8.6.8) can be rewritten as

$$-\frac{dR_C}{d\theta} = \frac{(bM_B/\rho_B)\, C_{Ag}}{(1/k) + \left(R_C^2 / R_S^2\right)(1/k_C) + \left(R_C / D_e\right)\left(1 - R_C / R_S\right)} \quad \text{... (8.6.11)}$$

The above equation can be integrated to give an expression for $R_C$ in terms of $C_{Ag}$ and θ. Substituting this in Eq. (8.6.8), we get the desired expression for the global rate in terms of $C_{Ag}$ and θ. We shall consider here some specific cases.

For example, let the concentration of $A$ in the gas bulk, namely $C_{Ag}$, be maintained constant throughout the reactor. This is true in the case of a well-mixed stirred tank reactor (complete backmixing). This is also true in the case of a fixed bed or fluidised bed reactor, when there is a large excess of the reactant $A$ in the feed. In such a case, $C_{Ag}$ will be so large that its variation with time or position can be neglected. Once $C_{Ag}$ is a constant, then Eq. (8.6.11) can be easily integrated since $R_C$ will be now a function of θ only. Thus, integrating Eq. (8.6.11) with the initial condition that at θ = 0, $R_C = R_S$, we get

$$\theta^* = \left(\frac{D_e}{kR_S}\right)(1 - R_C/R_S) + \frac{1}{2}(1 - R_C^2/R_S^2)$$

$$- \frac{1}{3}\left(1 - \frac{D_e}{k_C\ R_S}\right)(1 - R_C^3/R_S^3) \quad \text{... (8.6.12)}$$

where $\theta^*$ = dimensionless time

$$= (bD_e M_B C_{Ag}\theta)/(R_S^2 \rho_B) \quad \text{... (8.6.13)}$$

Though Eq. (8.6.12) expresses the inter-relationship between $R_C$ and $\theta$, it is difficult to substitute this expression in Eq. (8.6.8) to get the global rate since Eq. (8.6.12) is only implicit in $R_C$. However, if our interest is to find the fractional conversion of $A$ or $B$, this can be done without resorting to Eq. (8.6.8). For instance,

$$x_B = \frac{(\text{initial mass of } B - \text{mass of } B \text{ at time } \theta)}{(\text{initial mass of } B)}$$

$$= \frac{(4/3)\pi R_S^3 \rho_B - (4/3)\pi R_C^3 \rho_B}{(4/3)\pi R_S^3 \rho_B} = 1 - (R_C^3/R_S^3) \quad \text{... (8.6.14)}$$

Thus, from Eqs (8.6.12) and (8.6.14), the value of $x_B$ at any time $\theta$ can be computed.

If is further possible to specify a rate-controlling step as we have done in the case of solid–catalysed reactions. As stated in the previous section, the slowest step is the rate-controlling step. If one of the steps is much slower than the other two, then it can be treated as the rate-controlling step and the overall rate (namely, the global rate) will be essentially equal to the rate of this step. For instance, if the gas phase velocity relative to that of the solid particle is high (as in a fixed bed reactor), then the external mass transfer resistance will be negligible. In addition, for a highly porous ash layer and for low conversions (when the ash layer will be very thin), the resistance to diffusion through the ash layer will also be negligibly small. Under such conditions, the chemical reaction at $R_C$ will be rate-controlling and Eq. (8.6.11) will reduce to

$$-\frac{dR_C}{d\theta} = k\,bM_B C_{Ag}/\rho_B \quad \text{... (8.6.15)}$$

On integration, $$\theta = \rho_B\,(R_S - R_C)/(bkM_B C_{Ag}) \quad \text{... (8.6.16)}$$

Since $x_B = 1 - (R_C/R_S)^3$, $$\theta = \frac{\rho_B R_S}{b\,k\,M_B\,C_{Ag}}\,[1 - (1 - x_B)^{1/3}] \quad \text{... (8.6.17)}$$

The above result can also be directly deduced from Eq. (8.6.12). Since the chemical reaction is rate-controlling,

$$\left(\frac{D_e}{k\ R_S}\right) = \frac{(1/k)}{(R_S/D_e)} = \frac{(\text{chemical reaction resistance})}{(\text{resistance to diffusion through the ash layer})}$$

will be very large. As a result, the first term on the right hand side of Eq. (8.6.12) will be very large and therefore, the remaining terms can be neglected with respect to this. Thus, Eq. (8.6.12) reduces to

$$\theta^* = (D_e/kR_S)\ (1 - R_C/R_S)$$

or

$$\theta = \frac{\rho_B R_S (1 - R_C/R_S)}{b k M_B C_{Ag}}$$

which is same as Eq. (8.6.16). It may be noted that the time required for complete conversion ($\theta_t$) can be estimated from Eq. (8.6.17) by putting $x_B = 1.0$. Thus

$$\theta_t = \rho_B R_S/(b k M_B C_{Ag}) \quad \text{... (8.6.18)}$$

Equation (8.6.16) may be, therefore, re-expressed as

$$\theta/\theta_t = 1 - (R_C/R_S) \quad \text{... (8.6.19)}$$

Now, suppose the chemical reaction is very rapid and the gas phase velocity is also very high. But the diffusivity $D_e$ is quite small. In such a situation, diffusion through the ash layer will be the rate-controlling step even at low conversions. Equation (8.6.11), therefore, reduces to

$$-\frac{dR_C}{d\theta} = \frac{(b M_B/\rho_B)\, C_{Ag}}{(R_C/D_e)(1 - R_C/R_S)} \quad \text{... (8.6.20)}$$

On integration,

$$\theta = \frac{\rho_B R_S^2}{6 b M_B D_e C_{Ag}} [1 - 3\ (R_C/R_S)^2 + 2\ (R_C/R_S)^3] \quad \text{... (8.6.21)}$$

Readers can notice that the above expression also can be directly deduced from Eq. (8.6.12). For instance, since $D_e$ is very small and $k$ is very large, $(D_e/kR_S)$ is negligibly small and therefore, the first term on the right hand side of Eq. (8.6.12) can be dropped. Also, in the last term, $(1 - D_e/k_C R_S)$, will be very close to 1.0. Accordingly, Eq. (8.6.12) gets reduced to

$$\theta^* = (1/2)\ (1 - R_C^2/R_S^2) - (1/3)\ (1 - R_C^3/R_S^3)$$

$$= (1/6) - (1/2)\ (R_C/R_S)^2 + (1/3)\ (R_C/R_S)^3$$

or

$$\theta = \frac{\rho_B R_S^2}{6 b M_B D_e C_{Ag}} [1 - 3\ (R_C/R_S)^2 + 2\ (R_C/R_S)^3]$$

which is same as Eq. (8.6.21). At complete conversion, $x_B = 1.0$ or $R_C = 0$. Therefore, the time required for complete conversion is

$$\theta_t = \rho_B R_S^2/(6 b M_B D_e C_{Ag}) \quad \text{... (8.6.22)}$$

Accordingly, Eq. (8.6.21) may be reoriented as

$$\theta/\theta_t = 1 - 3\ (R_C/R_S)^2 + 2\ (R_C/R_S)^3 \quad \text{... (8.6.23)}$$

A still another alternate situation is when the mass transfer through the gas film is rate-controlling. For a highly porous ash layer and for low conversions, the resistance to diffusion through the ash layer will be negligibly small. If, in addition, the chemical reaction is very rapid, then external mass transfer through the gas film will determine the rate. In such a case, Eq. (8.6.11) reduces to

$$-\frac{dR_C}{d\theta} = (k_C b M_B C_{Ag}/\rho_B)\ (R_S^2/R_C^2) \qquad \text{... (8.6.24)}$$

On integration,

$$\theta = \frac{\rho_B R_S}{3 k_C b M_B C_{Ag}} (1 - R_C^3/R_S^3) \qquad \text{... (8.6.25)}$$

Since at $\theta = \theta_t$, $x_B = 1.0$ and $R_C = 0$,

$$\theta/\theta_t = 1 - (R_C/R_S)^3 \qquad \text{... (8.6.26)}$$

Let us now consider the second class of fluid-particle reactions in which no solid product is formed. Thus, the reaction is of the form

$$A\ (g) + bB\ (s) \rightarrow pP\ (g) + rR\ (g)$$

Since no solid product is formed, there will be no ash layer on the surface of the solid. The gas film surrounds the unreacted core of $B$ and the reactant $A$ diffuses through the gas film directly to the surface of the unreacted core. Equation (8.6.11), therefore, gets modified to

$$-\frac{dR_C}{d\theta} = \frac{(b M_B/\rho_B) C_{Ag}}{(1/k) + (1/k_C)} \qquad \text{... (8.6.27)}$$

It is important to note that the mass transfer coefficient $k_C$ is a function of particle size (namely, $R_C$) and since $R_C$ decreases continuously with time, $k_C$ is also a function of time. In our earlier analysis, we have taken $k_C$ to be a constant since the reactant $A$ is transferred to the surface of the ash layer and the total radius ($R_S$) of this layer had been assumed constant. In the present case, therefore, to integrate Eq. (8.6.27), we have to first express $k_C$ in terms of $R_C$. This can be done with the help of any of the reported experimental correlations. For example, according to the correlation proposed by Ranz and Marshall[50],

$$k_C d_p y_A/D_A = 2.0 + 0.6 Re_p^{1/2} Sc^{1/3} \qquad \text{... (8.6.28)}$$

where $y_A$ is the mole fraction of $A$ in the gas phase, $Sc$ is the Schmidt number and $Re_p$ = particle Reynolds number = $(d_p V_t \rho_g/\mu_g)$, $V_t$ being the relative velocity between the particle and the gas. At low Reynolds number (at low gas velocities and with small particles),

$$k_C d_p y_A/D_A \simeq 2.0$$

or

$$k_C = 2D_A/y_A d_p = D_A/(R_C y_A) \qquad \text{... (8.6.29)}$$

To note that since $C_{Ag}$ is assumed constant, $y_A$ is also a constant.

If the chemical reaction is rate-controlling, then Eq. (8.6.27) will reduce to Eq. (8.6.15), which on integration gives Eq. (8.6.16).

If external mass transfer through the gas film is the rate-controlling step, then Eq. (8.6.27) reduces to

$$-\frac{dR_C}{d\theta} = k_C b M_B C_{Ag}/\rho_B \qquad \text{... (8.6.30)}$$

If we substitute the expression for $k_C$ from Eq. (8.6.29) and integrate, we get

$$\theta = \frac{\rho_B \, y_A R_0^2}{2 b M_B D_A C_{Ag}} \, (1 - R_C^2/R_0^2) \quad \text{... (8.6.31)}$$

where $R_0$ is the initial radius of the solid particle. In other words, at $\theta = 0$, $R_C = R_0$. We know that when $\theta = \theta_t$, $R_C = 0$ and therefore,

$$\theta/\theta_t = 1 - (R_C/R_0)^2 \quad \text{... (8.6.32)}$$

$$= 1 - (1 - x_B)^{2/3} \quad \text{... (8.6.33)}$$

where $x_B = 1 - (R_C/R_0)^3$.

All the $\theta$ versus $x_B$ relationships developed above are for single spherical particles, isothermal operation and for the case when the chemical reaction is intrinsically first order in $A$ and irreversible. It is also assumed that $C_{Ag}$ remains constant throughout the reactor. In the case of a fixed bed reactor with a large excess of reactant $A$ in the feed, these expressions may be used straightaway. If there is a continuous flow of particles through the reactor but the particles execute plug flow so that all have the same residence time in the reactor, then also these expressions can be applied as such. These conditions often occur in a moving bed reactor (in which the fluid and the particles move counter currently) or in a transport reactor (with cocurrent flow of both fluid and particles), when the particles are of the same size and the reactant $A$ is in large excess. If the solid particles are of different sizes but they do move in plug flow, then the average conversion $\bar{x}_B$ may be computed from

$$\bar{x}_B = \sum_{i=1}^{n} x_{Bi} \, W_i \quad \text{... (8.6.34)}$$

where $x_{Bi}$ the conversion for particles of size $d_{pi}$ and $W_i$ is the mass fraction of particles of size $d_{pi}$. If the particle density changes significantly due to reaction, then it will be more accurate to use volume fraction in place of mass fraction in the above equation.

If the solids have a residence time distribution $E(\theta)$, then the average conversion is to be computed from

$$1 - \bar{x}_B = \int_0^{\infty} [1 - x_B(\theta)] \, E(\theta) \, d\theta \quad \text{... (8.6.35)}$$

If the particles are well-mixed as in a continuous fluidised bed reactor, then

$$E(\theta) = (1/\bar{\theta}) \exp(-\theta/\bar{\theta}) \quad \text{... (8.6.36)}$$

where $\bar{\theta}$ is the mean residence time of the particles in the reactor (that is, volume of solids in the reactor divided by the volumetric feed rate of solids).

Incidentally, if a particle remains in the reactor for a time longer than that required for complete conversion (that is, $\theta > \theta_t$), then the calculated conversion will be greater than 1.0. Since this is physically inconceivable, $x_B$ must remain equal to 1.0 for particle residence times greater than $\theta_t$. To ensure that such particles do not contribute to the fraction unconverted, we must modify Eq. (8.6.35) to read

$$1 - \bar{x}_B = \int_0^{\theta_t} [1 - x_B(\theta)] \, E(\theta) \, d\theta \quad \text{... (8.6.37)}$$

For mixed flow of solids, the expression for $E(\theta)$ may be substituted from Eq. (8.6.36). If particles are of different sizes, then

$$\bar{x}_B = \sum_{i=1}^{n} \bar{x}_{Bi}\, w_i \qquad \text{... (8.6.38)}$$

where $\bar{x}_{Bi}$ is to be computed for each particle size $d_{pi}$ from Eq. (8.6.37).

In actual practice, we do come across situations where the assumption that $C_{Ag}$ remains constant throughout the reactor is no longer valid. However, if $C_{Ag}$ does vary from point to point or from section to section inside the reactor, then the analysis becomes more complex. We shall illustrate here a relatively simple case of a *moving bed reactor*. In such a reactor, the fluid and the solids move countercurrently (see Fig. 8.27). To simplify the situation, we shall assume that both the fluid and the solids are in plug flow.

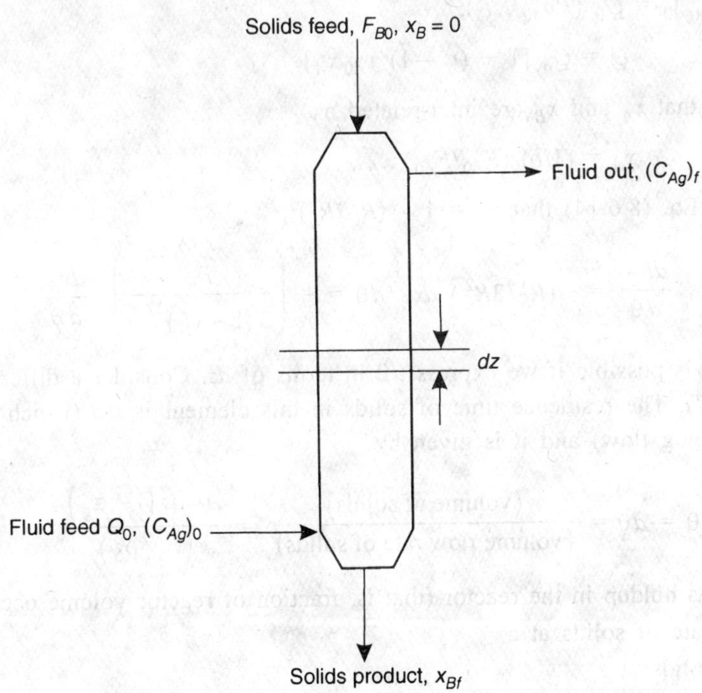


*Figure 8.27: Moving bed reactor for noncatalytic fluid–solid reaction*

However, $C_{Ag}$ does vary from the fluid inlet to the outlet. We can express $C_{Ag}$ in terms of $x_B$ through a simple material balance between the bottom of the reactor and any section at $z$. Thus, with reference to Fig. 8.27,

$$\text{Moles of } A \text{ reacted} = Q_0\,(C_{Ag})_0 - Q\,C_{Ag} \qquad \text{... (8.6.39)}$$

where $Q_0$ is the volumetric flow rate of feed gas and $Q$ is the volume flow rate of gas at $z$. As per stoichiometry, this must be equal to $(1/b)$ (number of moles of $B$ reacted). Since moles of $B$ reacted = $F_{B0}\,(x_{Bf} - x_B)$, where $F_{B0}$ is the molar feed rate of $B$,

$$Q_0\ (C_{Ag})_0 - Q\ C_{Ag} = (F_{B0}/b)\ (x_{Bf} - x_B) \qquad \text{... (8.6.40)}$$

or

$$C_{Ag} = (Q_0/Q)\ (C_{Ag})_0 - \left(\frac{F_{B0}}{bQ}\right)(x_{Bf} - x_B) \qquad \text{... (8.6.41)}$$

If there is no change in the number of moles of gas (that is, if one mole of gaseous product $R$ is produced for each mole of $A$ reacted) and if the operation is isothermal then $Q = Q_0$ (neglecting the effect of change in pressure). If there is a change in number of moles (but the operating temperature and pressure are essentially constant), then

$$Q = [F_{A0}\ (1 - x_A) + \boldsymbol{r} F_{A0} x_A + F_{in}]\ (RT/P) \qquad \text{... (8.6.42)}$$

where $\boldsymbol{r}$ is the number of moles of $R$ produced per each mole of $A$ reacted and $F_{in}$ is the number of moles of inerts in the feed gas. If the feed gas consists of only $A$, then $F_{in} = 0$. Since $Q_0 = (F_{A0} + F_{in})(RT/P)$ and $F_{A0} = Q_0\ (C_{Ag})_0 = Q_0\ (Py_{A0}/RT)$,

$$Q = Q_0\ [1 + (\boldsymbol{r} - 1)\ y_{A0} x_A] \qquad \text{... (8.6.43)}$$

It may also be noted that $x_A$ and $x_B$ are inter-related as

$$x_A = (1/b)\ (F_{B0}/F_{A0})\ x_B \qquad \text{... (8.6.44)}$$

Since we know from Eq. (8.6.14) that $x_B = 1 - (R_C/R_S)^3$,

$$\frac{dR_C}{d\theta} = -(R_S^3/3R_C^2)\ dx_B/d\theta = -\left(\frac{R_S}{3(1-x_B)^{2/3}}\right)\frac{dx_B}{d\theta} \qquad \text{... (8.6.45)}$$

A further modification is possible if we express $d\theta$ in terms of $dz$. Consider a differential element of length $dz$ (see Fig. 8.27). The residence time of solids in this element is $d\bar{\theta}$ (which will be equal to $d\theta$ since solids are in plug flow) and it is given by

$$d\theta = d\bar{\theta} = \frac{\text{(volume of solids)}}{\text{(volume flow rate of solids)}} = \frac{A_C\ dz\,(1-\epsilon_g)}{(F_S/\rho_S)} \qquad \text{... (8.6.46)}$$

where $\epsilon_g$ = fractional gas holdup in the reactor (that is, fraction of reactor volume occupied by gas)
$F_S$ = mass flow rate of solids at $z$
$\rho_S$ = density of solids at $z$

If the change in density of solids due to chemical reaction is negligible (that is, density of $P \simeq$ density of $B$), then $\rho_S = \rho_B$. Also,

$$F_S = F_{B0}\ (1 - x_B)\ M_B + (p/b)\ F_{B0} x_B M_P \qquad \text{... (8.6.47)}$$

If the molecular weights of the solid product $P$ and that of the solid reactant $B$ are equal ($M_P = M_B$) and if one mole of $P$ is produced per each mole of $B$ reacted ($p/b = 1.0$), then $F_S = (F_{B0} M_B)$. If we substitute Eq. (8.6.46) in Eq. (8.6.45), we get, on rearrangement,

$$\frac{dx_B}{dz} = \frac{3A_C\,(1-\epsilon_g)(1-x_B)^{2/3}}{R_S\,(F_S/\rho_S)}\left(\frac{-dR_C}{d\theta}\right) \qquad \text{... (8.6.48)}$$

Now, the expression for $(dR_C/d\theta)$ can be substituted from Eq. (8.6.11), after expressing $R_C$ in terms of $x_B$ from Eq. (8.6.14) and $C_{Ag}$ in terms of $x_B$ as per Eq. (8.6.41). Thereafter, Eq. (8.6.48) can be integrated from $x_B = 0$ to $x_B = x_{Bf}$ to determine the required length of the reactor. See Example 8.52 for a numerical illustration of this procedure.

The above analysis can be applied to a *transport reactor* as well with little modification. The basic difference between a transport reactor and a moving bed reactor is that the fluid and the solids move cocurrently in the former, while in the latter they move countercurrently. In the case of fixed bed reactors with varying fluid composition, the analysis is much more complex and demands numerical methods for solution. See Smith[21] for a good discussion on this.

**Example 8.50:** A gas-solid noncatalytic reaction of the following stoichiometry is to be conducted in a rotary kiln:

$$A\ (g) + B\ (s) \rightarrow P\ (s) + R\ (g)$$

The gas $A$ in contact with the solids is of uniform composition and the whole process is isothermal. The solids move in plug flow from one end of the kiln to the other at a velocity of 9.0 m/hr. The chemical reaction is intrinsically first order in $A$ and is irreversible. The substance $B$ is nonporous, but the solid product $P$ forms a porous layer around the unreacted core of $B$ as the reaction proceeds. When the reaction was studied in a laboratory batch reactor operated at the same temperature and gas composition, the following data were obtained:

| *Particle size, mm* | *Conversion, $x_B$* | *Reaction time, $\theta$* |
|---|---|---|
| 3.0 | 0.875 | 60 minutes |
| 6.0 | 0.657 | 84 minutes |

The study also indicated that diffusion resistance between particle surface and the gas is negligible.

(*a*) Compute the length required for the kiln to obtain 90 percent conversion of $B$, if the solid feed consists of 6 mm spherical particles of pure $B$.

(*b*) If the feed of $B$ consists of 20 percent (by weight) 3 mm particles, 50 percent (by weight) 6 mm particles and 30 percent (by weight) 9.5 mm particles, compute the average conversion in the product using the reactor designed in part (*a*). Assume that the density of solid product $P$ is approximately equal to that of the solid reactant $B$.

**Solution:** Let us assume that the shrinking core model is applicable. Since the gas phase mass transfer resistance is negligible, Eq. (8.6.12) reduces to

$$\theta^* = \left(\frac{D_e}{kR_S}\right)(1 - R_C/R_S) + (1/2)\ (1 - R_C^2/R_S^2) - (1/3)\ (1 - R_C^3/R_S^3)$$

Rewriting in terms of conversion $x_B$, we get

$$K\ (\theta/R_S^2) = \left(\frac{D_e}{kR_S}\right)[1 - (1 - x_B)^{1/3}] + (1/2)$$

$$[1 - (1 - x_B)^{2/3}] - (x_B/3) \quad \text{... (i)}$$

where $K = (bD_e M_B C_{Ag}/\rho_B)$. From the given experimental data, we have to solve the above equation for the two unknowns $K$ and $(D_e/k)$. Thus, since $x_B = 0.875$, $\theta = 3600$ s and $R_S = (3.0/2) = 1.5$ mm for the 3 mm particles, Eq. (i) becomes

$$(1600 \times 10^6)\ K = 333.33\ (D_e/k) + 0.0833 \qquad \text{... (ii)}$$

Similarly, for the 6 mm particles, Eq. (i) reduces to

$$(560 \times 10^6)\ K = 100\ (D_e/k) + 0.036 \qquad \text{... (iii)}$$

Solving the above two equations simultaneously, we get

$$K = 1.375 \times 10^{-10}\ \text{m}^2/\text{s}$$

$$(D_e/k) = 4.1 \times 10^{-4}\ \text{m}$$

(*a*) To compute the required length of the reactor, we must first determine the residence time of solids in the reactor. Thus, substituting $x_B = 0.90$ and $R_S = (6.0/2) = 3$ mm $= 0.003$ m in Eq. (i), we get

$$\theta = 10800\ \text{s} = 3\ \text{hr}$$

Since the solids move in plug flow through the kiln at a velocity of 9 m/hr, the length of kiln required is

$$L = (3.0)\ (9.0) = 27.0\ \text{m}$$

(*b*) If there is a mixture of particles of different sizes, then the conversion $x_B$ is to be computed separately for each particle size. Since the particles move in plug flow, the residence time of all particles in the kiln will be the same and will be equal to 3.0 hr. We know that for this residence time, the conversion of 6 mm particles is 90 percent. For the 3 mm particles, the time for complete conversion $(\theta_t)$ can be computed from Eq. (i) by substituting $x_B = 1.0$ and $R_S = (3.0/2) = 1.5$ mm $= 0.0015$ m. The result is

$$\theta_t\ (\text{for 3 mm particles}) = 2.0\ \text{hr}$$

Since the residence time of particles in the reactor is more than $\theta_t$, $x_B = 1.0$ for 3 mm particles. For 9.5 mm particles, substituting $R_S = (9.5/2) = 4.75$ mm $= 0.00475$ m and $\theta = (3 \times 3600) = 10800$ s Eq. (i), we get

$$x_B = 0.675$$

The results are listed below:

| *Particle size, mm* | *Mass fraction, $w_i$* | *Conversion, $x_{Bi}$* | *$(x_{Bi} w_i)$* |
|---|---|---|---|
| 3.0 | 0.20 | 1.0 | 0.20 |
| 6.0 | 0.50 | 0.90 | 0.45 |
| 9.5 | 0.30 | 0.675 | 0.2025 |
| | | | $\Sigma x_{Bi} w_i = 0.8525$ |

Therefore, the average conversion is

$$\bar{x}_B = 0.8525 = \textbf{85.25 percent}$$

**Example 8.51:** A noncatalytic reaction between solids 6 mm in size and a gas $A$ is to be conducted in a steady flow fluidised bed reactor 1 m in diameter. The reaction stoichiometry is as follows:

$$A\ (g) + B\ (s) \rightarrow P\ (s) + R\ (g)$$

When the reaction was studied in a bench scale fluidised bed reactor with a large excess of $A$, 76.0 percent conversion of $B$ was obtained when the mean residence time of solids in the reactor was 100 s. It was also observed that the chemical reaction at the surface of the unreacted core of $B$ is the rate-controlling step. Compute the height of the fluidised bed in the commercial reactor if 98 percent conversion of $B$ is desired and the feed rate of $B$ is maintained at 4000 kg/hr. The operating voidage of the fluidised bed is 0.5. Assume that the concentration of $A$ in the gas phase remains essentially constant and the solids are of unchanging size and unchanging density (3000 kg/m$^3$).

**Solution:** In a fluidised bed reactor, it is reasonable to assume mixed flow of solids. Accordingly, the average conversion is given by Eqs (8.6.37) and (8.6.36):

$$1 - \bar{x}_B = \int_0^{\theta_t} (1 - x_B)\ (1/\bar{\theta})\ \exp\ (-\theta/\bar{\theta})\ d\theta \qquad \text{... (i)}$$

Based on the shrinking core model, Eq. (8.6.19) is applicable since the chemical reaction is rate-controlling. Thus

$$\theta/\theta_t = (1 - R_C/R_S) = 1 - (1 - x_B)^{1/3}$$

or

$$1 - x_B = (1 - \theta/\theta_t)^3$$

Substituting the above expression in Eq. (i), we get

$$1 - \bar{x}_B = (1/\bar{\theta}) \int_0^{\theta_t} (1 - \theta/\theta_t)^3\ \exp\ (-\theta/\bar{\theta})\ d\theta$$

Integrating by parts and rearranging, we get

$$\bar{x}_B = 3\ (\bar{\theta}/\theta_t) - 6\ (\bar{\theta}/\theta_t)^2 + 6\ (\bar{\theta}/\theta_t)^3\ [1 - \exp\ (-\theta_t/\bar{\theta})] \qquad \text{... (ii)}$$

In the bench scale reactor, $\bar{x}_B = 0.76$ and $\bar{\theta} = 100$ s. Substituting these values in the above equation and solving for $\theta_t$ by trial, we get

$$\theta_t = 120\ \text{s}$$

In the case of the commercial reactor, the conversion desired is 98 percent. Thus, substituting $\bar{x}_B = 0.98$ in Eq. (ii), it is solved for $(\bar{\theta}/\theta_t)$ to get

$$\bar{\theta}/\theta_t = 13.0$$

or

$$\bar{\theta} = (13.0)\ (120) = 1560\ \text{s} = 26\ \text{minutes}$$

We know that $\bar{\theta}$ = mean residence time = (volume of solids in reactor)/(volumetric feed rate of solids). But, since the density of solids remains unchanged, $\bar{\theta}$ = (mass of solids) /(mass feed rate of solids). Therefore,

Mass of solids in the reactor = $\bar{\theta}$ (4000/60) = (26.0) (4000/60) = 1733.33 kg

Volume of solids in the reactor = (1733.33/3000) = 0.577 m$^3$

Since the operating voidage of the bed is 0.5,

Volume of the bed = (0.577)/(1 − 0.5) = 1.155 m$^3$

Since the area of cross-section of the reactor is $(\pi/4)\ (1.0)^2$ = 0.7854 m$^2$,

Height of fluidised bed = (1.155/0.7854) = **1.47 m**

**Example 8.52:** The following noncatalytic gas-solid reaction is to be conducted in a moving bed reactor with upflow of pure hydrogen and downflow of solids:

$$FeS_2\,(s) + H_2\,(g) \rightarrow FeS\,(s) + H_2S\,(g)$$

The reactor is 1.0 m in diameter and operates at 495° C and 1 atm. $FeS_2$ pellets are fed from the top at the rate of 3000 kg/hr. Tests have shown that gas phase diffusion resistance is negligible. The sizes and RTDs for the particles in the reactor are as follows:

| | | | | | |
|---|---|---|---|---|---|
| Particle radius, mm | : | 0.05 | 0.10 | 0.15 | 0.20 |
| Mass fraction | : | 0.10 | 0.30 | 0.40 | 0.20 |
| $\theta/\bar{\theta}$ | : | 1.40 | 1.10 | 0.95 | 0.5 |

The chemical reaction is intrinsically first order in hydrogen and irreversible. Within the above particle size range, the first order rate constant and the diffusivity of hydrogen through the *FeS* product layer are given by

$$k = 3800 \exp\,(-15150/T)\ \mathrm{m^3/(m^2\cdot s)}$$

$$D_e = 3.6 \times 10^{-10}\ \mathrm{m^2/s}$$

where $T$ is in degrees Kelvin. Assume that the density of FeS is essentially equal to that of $FeS_2$ and is 5000 kg/m³.

(*a*) Compute the average conversion of $FeS_2$ to FeS if the mole fraction of hydrogen in the gas is essentially constant and equal to unity and the mean residence time of solids in the reactor is 60 minutes.

(*b*) If specially prepared 10 mm $FeS_2$ pellets are fed to the top of the reactor at the same rate and pure hydrogen is injected from below at 2.6 m³/s, compute the reactor height required to achieve 80 percent conversion of $FeS_2$ to FeS. Assume that both the gas and the solids are in plug flow. The fractional gas holdup in the reactor is 0.5. For these large size $FeS_2$ pellets, the first order rate constant is 100 times that given in part (*a*) and the effective diffusivity of hydrogen through the FeS product layer is 1000 times that given in part (*a*).

**Solution:** (*a*) Let us assume that the shrinking core model is applicable. Since the hydrogen concentration is constant throughout the reactor and gas phase mass transfer resistance is negligible, Eq. (8.6.12) is applicable in the following modified form:

$$\theta^* = \left(\frac{D_e}{kR_S}\right)[1-(1-x_B)^{1/3}] + (1/2)$$

$$[1-(1-x_B)^{2/3}] - (x_B/3) \qquad \text{... (i)}$$

Now,

$$k = 3800 \exp\,(-15150/768) = 1.0295 \times 10^{-5}\ \mathrm{m^3/(m^2\cdot s)}$$

$$D_e/k = (3.6\times10^{-10})/(1.0295\times10^{-5}) = 3.5\times10^{-5}\ \mathrm{m}$$

$$\theta^* = \frac{\left(b\,D_e M_b C_{Ag}\right)\theta}{\rho_B R_S^2}$$

where

$$b = 1.0$$

$$M_B = 120.0\ \mathrm{kg/kmole}$$

$$\rho_B = 5000 \text{ kg/m}^3$$

$$C_{Ag} = Py_A/RT = \frac{(1.0)(1.0)}{(0.082)(768)} = 0.01588 \text{ kmoles/m}^3$$

Therefore, $$\theta^* = (1.3719 \times 10^{-13})\ (\theta/R_S^2)$$

Equation (i), therefore, becomes

$$(1.3719 \times 10^{-13})\ \theta/R_S^2 = (3.5 \times 10^{-5}/R_S)\ [1 - (1 - x_B)^{1/3}] + (1/2)\ [1 - (1 - x_B)^{2/3}] - (x_B/3) \quad \text{... (ii)}$$

We can now compute the conversion of $B$ for each particle size from the above equation. For example, for $R_S = 0.05$ mm $= 5.0 \times 10^{-5}$ m, $\theta = 1.4\bar{\theta} = (1.4)(3600) = 5040$ s and therefore, from Eq. (ii), by trial,

$$x_B = 0.7$$

Similarly, the value of $x_B$ for each particle size can be computed and the results are listed below:

| *Particle radius, mm* | *Mass fraction, $w_i$* | *Conversion, $x_{Bi}$* | *$(x_{Bi}\,w_i)$* |
|---|---|---|---|
| 0.05 | 0.10 | 0.7 | 0.07 |
| 0.10 | 0.30 | 0.345 | 0.1035 |
| 0.15 | 0.40 | 0.215 | 0.086 |
| 0.20 | 0.20 | 0.135 | 0.027 |
| | | | $\Sigma x_{Bi} w_i = 0.2865$ |

The average conversion of $FeS_2$ to FeS is, therefore,

$$\bar{x}_B = 0.2865 = 28.65 \text{ percent}$$

(*b*) Since hydrogen concentration in the gas phase varies from the bottom to the top of the reactor, we must first express $C_{Ag}$ in terms of $x_B$. Since the operation is isothermal and there is no change in the total number of moles of gas (since one mole of $H_2S$ is produced per each mole of hydrogen reacted), $Q = Q_0$. Since $b = 1.0$, Eq. (8.6.41) reduces to

$$C_{Ag} = (C_{Ag})_0 - (F_{B0}/Q_0)\ (0.8 - x_B)$$

where $$Q_0 = 2.6 \text{ m}^3/\text{s}$$

$$F_{B0} = \frac{(3000.0)}{(3600)(120)} = 6.944 \times 10^{-3} \text{ kmoles/s}$$

Since pure hydrogen is injected from below,

$$(C_{Ag})_0 = P/RT = \frac{(1.0)}{(0.082)(768)} = 0.01588 \text{ kmoles/m}^3$$

Therefore, $$C_{Ag} = 0.01374 + (2.6709 \times 10^{-3})\ x_B \quad \text{... (iii)}$$

Since particles are of unchanging size (as per shrinking core model) and of unchanging density (since density of FeS is essentially equal to that of $FeS_2$), we can take

$$F_S = (3000/3600) = 0.8333 \text{ kg/s}$$

Since $A_C = (\pi/4)(1.0)^2 = 0.7854$ m², $\in_g = 0.5$, $R_S = 0.01/2 = 0.005$ m and $\rho_S = \rho_B = 5000$ kg/m³, Eq. (8.6.48) reduces to

$$\frac{dx_B}{dz} = (14.137 \times 10^5)(1 - x_B)^{2/3}(-dR_C/d\theta) \qquad \text{... (iv)}$$

Since the gas-phase mass transfer resistance is negligible, Eq. (8.6.11) reduces to

$$-\frac{dR_C}{d\theta} = \frac{[(1.0)(120.0)/(5000.0)]\,C_{Ag}}{(1/k) + (R_C/D_e)(1 - R_C/R_S)}$$

where

$$k = (100)(1.0295 \times 10^{-5}) = 1.0295 \times 10^{-3} \text{ m}^3/(\text{m}^2\cdot\text{s})$$

$$D_e = (1000)(3.6 \times 10^{-10}) = 3.6 \times 10^{-7} \text{ m}^2/\text{s}$$

$$R_C/R_S = (1 - x_B)^{1/3}$$

Also, substituting the expression for $C_{Ag}$ from Eq. (iii), we get

$$-\frac{dR_C}{d\theta} = \frac{(3.2976 \times 10^{-4}) + (6.41 \times 10^{-5})\,x_B}{(971.3453) + (13.889 \times 10^3)\left[1 - (1 - x_B)^{1/3}\right](1 - x_B)^{1/3}}$$

Substituting the above expression in Eq. (iv), we get on rearrangement,

$$\frac{dz}{dx_B} = \frac{\left\{1 + 14.3\left[1 - (1 - x_B)^{1/3}\right](1 - x_B)^{1/3}\right\}}{(0.0933\,x_B + 0.48)(1 - x_B)^{2/3}} = F(x_B) \qquad \text{... (v)}$$

On integration, $\int_0^L dz = L = \int_0^{0.8} F(x_B)\,dx_B$

For evaluating the above integral numerically, let us compute the values of $F(x_B)$ at different $x_B$ and they are listed below:

| $x_B$ | $F(x_B)$ |
|---|---|
| 0.0 | 2.08326 |
| 0.1 | 2.6945 |
| 0.2 | 4.5413 |
| 0.3 | 6.0504 |
| 0.4 | 7.84786 |
| 0.5 | 10.0708 |
| 0.6 | 12.9657 |
| 0.7 | 17.0395 |
| 0.8 | 23.5745 |

Therefore, from Simpson's rule,

$$\int_0^{0.8} F(x_B)\,dx_B = 7.326$$

or

$$L = \mathbf{7.326\ m}$$

## 8.7 GAS–LIQUID REACTIONS

Gas–liquid reactions (either catalysed on noncatalysed) are important in industrial practices either to remove unwanted constituents from a gas stream or for the production of industrially important chemicals. Typical examples are listed in Tables 8.4 and 8.4A given at the beginning of Section 8.5.

Gas–liquid reactions (noncatalysed) can be conveniently conducted in fixed bed reactors. Here, the packing is an inert material and the packed bed helps in providing intimate contacting between the two fluid streams. Gas absorption with chemical reaction in packed bed towers is discussed in detail in Section 4.5.6 of Chapter 4.

Solid–catalysed gas–liquid reactions may be conducted in trickle bed reactors or slurry reactors. Performance characteristics of these two types of reactors have already been outlined in Section 8.5. Fluidised bed reactors may also be employed for this purpose. However, analysis of the performance of three phase fluidised beds is quite complex (Hydrodynamics in three phase fluidised beds is discussed in Chapter 2. A brief survey of mass transfer in three phase fluidised beds is given by Muroyama and Fan[52]).

The most recent advent into the family of gas-liquid reactors are the membrane reactors. We have briefly discussed their performance characteristics in Section 7.6 of Chapter 7.

A still another type of gas–liquid reactor is the foam bed reactor[53]. Typically, a foam bed reactor consists of a shallow pool of liquid over a gas distributor. Above the liquid pool is a height of foam consisting of a large number of gas bubbles separated by thin films of liquid. The bubbles in the foam bed normally have a uniform dodecahedral structure. Thus, each bubble is surrounded by twelve flat liquid films of regular pentagonal shape[53]. As the foam rises, the gas diffuses into the liquid film from both sides and reacts with it. The unreacted gas leaves from the top of the bed whereas the liquid drains down to the pool through the plateau borders (ten plateau borders for one dodecahedron). It has been observed that drainage of liquid takes place through plateau borders only and hardly any liquid drains down through the liquid films. From one end of the liquid pool, a stream of fresh liquid is continuously fed in and from the other end an equal volume of liquid is taken out. As a result, the volume of liquid in the pool (usually called the storage section) remains constant.

It shall be, therefore, not wrong to assume that a foam bed reactor is typically composed of two sections, a storage section and a foam section. The volume of liquid in the storage section, as stated earlier, is maintained constant. A part of the liquid is converted into foam by the gas. However, the entire amount of the liquid that is carried into the foam section by the gas does not move to the top of the bed. There is continuous drainage of liquid as the foam moves up the bed. To analyse a situation of this kind, the foam section may be assumed to be divided into $n$ number of subsections, each having a constant but different value of liquid holdup ($\in$). This is illustrated schematically in Fig. 8.28.

The shallow pool of liquid at the bottom thus acts as a storage section and it will not be wrong to assume that no gas absorption takes place in this section. This is substantiated by the fact that the contact time of gas in the pool is much less than (less than 5 percent of) the total contact time in the reactor. We can further assume without serious error that in the foam section, absorption with chemical reaction takes place only in the liquid films and little absorption is taking place in the plateau borders. This is reasonably true since the total surface area of the plateau borders is much less than that of the liquid film. Drainage of liquid, however, takes place only through the plateau borders.

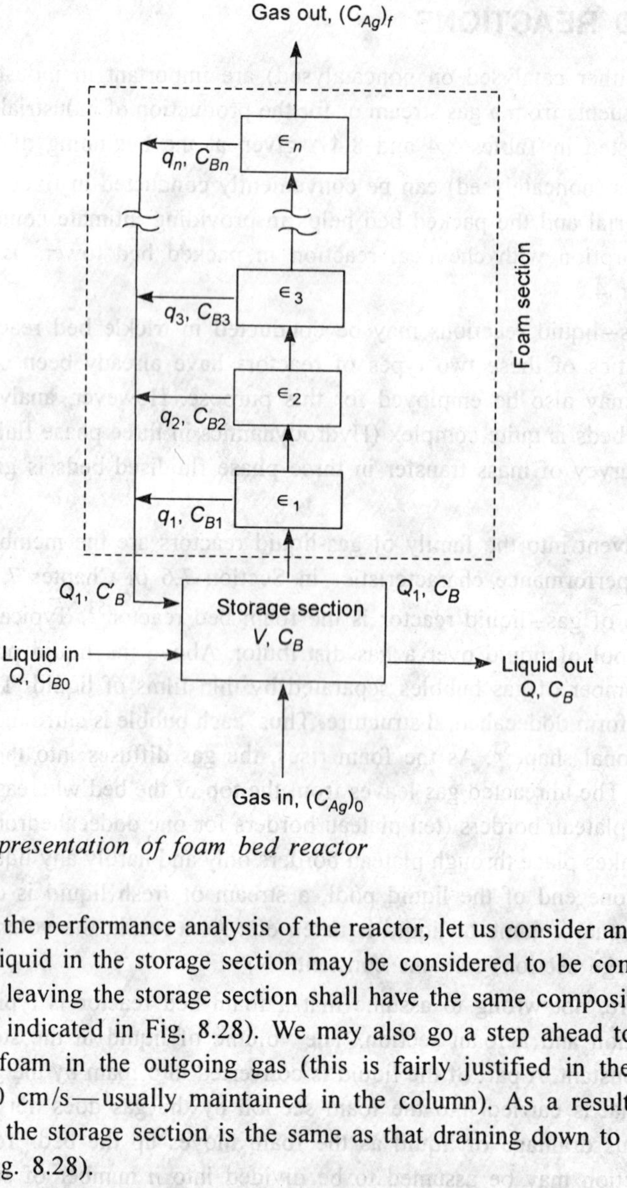


*Figure 8.28: Model representation of foam bed reactor*

To further simplify the performance analysis of the reactor, let us consider an ideal foam bed reactor. In such a reactor, the liquid in the storage section may be considered to be completely back-mixed. In other words, the liquid leaving the storage section shall have the same composition ($C_B$) as that in the section (this is what is indicated in Fig. 8.28). We may also go a step ahead to assume that no liquid is entrained from the foam in the outgoing gas (this is fairly justified in the wake of the low gas velocities—8.0 to 20.0 cm/s—usually maintained in the column). As a result, the amount of liquid entering the foam from the storage section is the same as that draining down to the storage through the plateau borders (see Fig. 8.28).

The material balance for the component $B$ (which is the reactive component of the liquid) in the storage section can be now written as

$$-V\frac{dC_B}{d\theta} = Q_1 C_B + QC_B - QC_{B0} - Q_1 C'_B$$

$$= (Q_1 + Q)\, C_B - QC_{B0} - Q_1 C'_B \qquad \text{... (8.7.1)}$$

It can be seen from Fig. 8.28,

$$Q_1 = q_1 + q_2 + q_3 + \ldots\ldots q_n = \sum_{i=1}^{n} q_i \qquad \text{... (8.7.2)}$$

$$Q_1 C'_B = q_1 C_{B1} + q_2 C_{B2} + q_3 C_{B3} + \ldots\ldots q_n C_{Bn} = \sum_{i=1}^{n} q_i\, C_{Bi} \qquad \text{... (8.7.3)}$$

Equation (8.7.1), therefore, reduces to

$$-V\,\frac{dC_B}{d\theta} = (Q_1 + Q)\, C_B - Q C_{B0} - \sum_{i=1}^{n} q_i\, C_{Bi} \qquad \text{... (8.7.4)}$$

To solve the above equation, the concentrations $C_{B1}$, $C_{B2}$, ..... $C_{Bn}$ are to be expressed in terms of $C_B$ and other system parameters. For this, let us analyse the chemical reaction in the liquid film. The thickness of the liquid film is ($2a$). In other words, the film occupies the space $-a \le x \le a$.

It must be however noted that the liquid film thickness varies from one foam subsection to another. The value of $\alpha$ for any particular subsection may be computed from Eq. (8.7.29) given subsequently.

The component $A$ of the gas diffuses into the liquid film and reacts with the component $B$ of the liquid and let the reaction be irreversible and psuedo first order in $A$. Without significant error, we may also assume that the gas-phase resistance is negligible. The equation of mass transfer for the liquid film is then given by

$$\frac{\partial C_{AL}}{\partial \theta_C} = D_A\,\frac{\partial^2 C_{AL}}{\partial x^2} - k_1 C_{AL} \qquad \text{... (8.7.5)}$$

where $C_{AL}$ = concentration of $A$ in liquid

$\theta_C$ = contact time (which varies from one subsection to another)

$D_A$ = diffusivity of $A$ in liquid

$k_1$ = first order reaction rate constant

The boundary conditions governing the system are

**B.C.1:** At $\theta_C = 0$, for $-a \le x \le a$, $C_A = 0$ ... (8.7.6)

**B.C.2:** At $\theta_C > 0$, for $x = \pm a$,

$$V_g\,\frac{\partial C^*_{Ag}}{\partial \theta_C} = \pm D_A S\,\frac{\partial C_{AL}}{\partial x} \qquad \text{... (8.7.7)}$$

where $S$ is the surface area of the liquid film. The second boundary condition predicts that the rate at which $A$ leaves the gas phase is equal to the rate at which it enters the liquid film. It has been shown[53] that the gas pocket on each liquid film is a five-sided pyramid with the liquid film as the base and its volume is equal to one-twelveth of the volume of a foam bubble. In other words, $V_g = (V_b/12)$. Also, if $K_d$ is the equilibrium distribution coefficient, then $C^*_{Ag} = C_{AL}/K_d$. Equation (8.7.7) can be, therefore, rewritten as

$$\frac{\partial C_{AL}}{\partial \theta_C} = \alpha\,\frac{\partial C_A}{\partial x} \qquad \text{... (8.7.8)}$$

where $\alpha = (12\, K_d S D_A / V_b)$

Solution to differential equations of the form (8.7.5) subject to boundary conditions of the form (8.7.6) and (8.7.8) is given by Crank[54]:

$$(M/M_\infty) = 1 - \sum_{j=1}^{\infty} \frac{\exp\left(p_j t_C\right)}{1+F\left(p_j\right)} \qquad \text{... (8.7.9)}$$

where $$F\,(p_j) = (\alpha a / 2 D_A) + p_j/(2 D_A \lambda_j^2) + p_j^2 a/(2\alpha D_A \lambda_j^2) \qquad \text{... (8.7.10)}$$

The $p_j$'s are the nonzero roots of

$$p_j = \alpha \lambda_j \tan\,(\lambda_j a) \qquad \text{... (8.7.11)}$$

and $$\lambda_j^2 = -\,(p_j + k_1)/D_A \qquad \text{... (8.7.12)}$$

Equation (8.7.9) is the general expression for $M$ (that is, moles of $A$ absorbed in half the film in contact time $t_C$). To note that $M_\infty$ is the value of $M$ at infinite time.

The concentrations of the unreacted liquid reactant, $C_{B1}$, $C_{B2}$, ..... $C_{Bn}$, at the ends of different foam subsections corresponding to the contact times $\theta_{C1}$, $(\theta_{C1} + \theta_{C2})$, $(\theta_{C1} + \theta_{C2} + \theta_{C3})$, ..... $(\theta_{C1} + \theta_{C2} + \ldots\ \theta_{Cn})$ respectively can be now expressed as[53]

$$C_{B1} = C_B - (2 z M_1 / V_1) \qquad \text{... (8.7.13)}$$

For $n > 1$ $$C_{Bn} = C_B - (2 z M_1/V_1) - \sum_{i=2}^{n} 2z\left(M_i - M_{i-1}\right)/V_i \qquad \text{... (8.7.14)}$$

where $V_i$ = volume of liquid film in the $i$th subsection.
$z$ = stoichiometric coefficient (moles of $B$ reacting with one mole of $A$)

Equation (8.7.4) can be, therefore, written in the modified form as

$$\frac{dC_B}{d\theta} + P_1 C_B = P_2 \qquad \text{... (8.7.15)}$$

where $P_1 = Q/V$ ... (8.7.16)

$$P_2 = (1/V)\left[Q\,C_{B0} - \left(2zQ_1 M_1/V_1\right) - 2z \sum_{i=2}^{n}\left(Q_1 - \sum_{j=1}^{i-1} q_j\right)\frac{\left(M_i - M_{i-1}\right)}{V_i}\right] \qquad \text{... (8.7.17)}$$

Solving the above linear first order differential equation with the initial condition

$$\text{at } \theta = 0,\ C_B = C_{B0} \qquad \text{... (8.7.18)}$$

we get the expression for $C_B$ as

$$C_B = C_{B0} \exp\,(-P_1\theta) + (P_2/P_1) - (P_2/P_1) \exp\,(-P_1\theta) \qquad \text{... (8.7.19)}$$

The above expression can be simplified by expanding the last term $\exp\,(-P_1\theta)$ and then neglecting the higher powers. Thus

$$C_B = C_{B0} \exp(-P_1\theta) + P_2\theta - (P_1P_2\theta^2/2) - (P_1^2P_2\theta^3/6) \quad \text{... (8.7.20)}$$

The above equation is the general expression for predicting the concentration of liquid phase reactant $B$ at any time $\theta$ in a foam bed reactor. It could also be used to compute the fractional conversion of $B$. If the gas phase concentration is desired, it could be easily obtained from

$$(C^*_{Ag})_i = (C^*_{Ag})_{i-1} - 12\,(M_i - M_{i-1})/V_b \quad \text{... (8.7.21)}$$

where $i = 1, 2, \ldots\ n$.

The final outlet concentration of gas at any time $\theta$ is given by

$$(C_{Ag})_f = (C_{Ag})_0 - 12M_n/V_b \quad \text{... (8.7.22)}$$

A special case of foam bed reactor operation is the semibatch mode in which there will be no inflow of liquid into the storage section. A definite volume of liquid is initially fed into the reactor and the operation is commenced. In such a case,

$$Q = 0,\ P_1 = 0 \quad \text{... (8.7.23)}$$

Accordingly, Eq. (8.7.20) reduces to

$$C_B = C_{B0} + P_2\theta \quad \text{... (8.7.24)}$$

For utilising Eq. (8.7.20) or (8.7.24) for computing the instantaneous value of $C_B$ or the fractional conversion of $B$, we must know the value of $\alpha$ as well as the values of $q_i$, $V_i$, $a_i$ and $t_{Ci}$ for each subsection (that is, for $i = 1, 2 \ldots\ n$). Biswas and Kumar[53] report that these may be computed as

$$(D_A/\alpha) = \frac{4V_b/\left(K_d\, d_b^2\right)}{(10.05)\tan(54°)} \quad \text{... (8.7.25)}$$

$$q_i = \frac{\epsilon_i Q_g}{(1-\epsilon_i)} - \frac{\epsilon_{i+1} Q_g}{1-\epsilon_{i+1}} \quad \text{... (8.7.26)}$$

$$V_i = V_b\epsilon_i/6\,(1-\epsilon_i) \quad \text{... (8.7.27)}$$

$$\theta_{Ci} = A_C h\,(1-\epsilon_i)/Q_g \quad \text{... (8.7.28)}$$

$$a_i = (\pi d_b/2)\,\epsilon_i/[(1-\epsilon_i)\,(7.537)\tan 54°] \quad \text{... (8.7.29)}$$

where $A_C$ = cross sectional area of reactor column = $(\pi D^2/4)$
$h$ = height of each foam subsection
$d_b$ = average diameter of foam bubble
$V_b$ = average volume of each foam bubble = $(\pi d_b^3/6)$
$Q_g$ = volume flow rate of gas through the reactor
$\epsilon_i$ = liquid holdup (volume fraction of liquid) in the $i$th subsection

Biswas and Kumar have however assumed that the average bubble size ($d_b$) and the experimental values of $\epsilon_i$ are available. Once values of $d_b$ and $\epsilon_i$ are available, parameters $\alpha$ to $a_i$ (note that the liquid film thickness varies from one subsection to another) can be evaluated from Eqs (8.7.25) through

(8.7.29). Incidentally, the value of $d_b$ may be estimated from any of the bubble formation models reported in literature[55].

Biswas and Kumar report that the results from the above discussed mathematical model agree with experimental data with a maximum deviation of 22 percent. This deviation may be considered tolerable in view of the simplifying assumptions made during the modelling.

The specific characteristics of foam bed reactors may be summarised as follows:

(*i*) they offer large gas-liquid interface per unit volume of the bed,

(*ii*) they provide large contact time,

(*iii*) the pressure drop in these reactors is not substantial but is far greater than that in a spray column or a packed bed.

Foam bed reactors have been found to be particularly attractive when large quantities of lean gases have to be treated as in the case of pollutant removal from air streams.

## 8.8 COMPUTER PROGRAMS

The following computer programs are already given in the text of this chapter:

1. Program 8.1 (Development of rate equation for irreversible reactions) in Section 8.2.
2. Program 8.2 (Performance analysis of semibatch reactor without continuous product discharge) in Section 8.3.
3. Program 8.3 (Analysis of nonisothermal operation of a batch reactor) in Section 8.3
4. Program 8.4 (Derivation of rate equation by Langmuir-Hinshelwood method—Evaluation of $n$ constants of a polynomial by fitting experimental data) in Section 8.5.

We shall now present a computer program for the design of a packed bed reactor for a solid-catalysed reaction that is intrinsically second order (assuming plug flow and isothermal operation). A problem of this kind is dealt in Example (8.42). The program is, therefore, prepared with special reference to this Example. A few generalisations are however incorporated. The reaction is represented as

$$aA \rightarrow rR$$

In Example 8.42, it has been assumed that the feed to the reactor consists of only $A$ (namely, ethylene). Here, we shall assume that $y_{A0}$ is the mole fraction of $A$ in the feed gas (the rest being inerts). Also, the mass transfer coefficient ($k_C$) is computed based on the feed inlet conditions in Example 8.42. This shall necessarily cause only allowable error unless the gas phase mass transfer resistance is most controlling. In the program given below, $k_C$ is computed at every section (that is, at every increment in $x_A$).

### PROGRAM 8.5

```
C**   DESIGN OF A PACKED BED REACTOR FOR
C**   A SOLID CATALYSED REACTION THAT IS
C**   INTRINSICALLY SECON ORDER (PLUG FLOW
C**   AND ISOTHERMAL OPERATION ASSUMED)
      REAL MG (101), JD (101), KC (101)
```

```
      REAL K, MA, MR, MIN, LB, LHS, MEUG
      DIMENSION XA (101), CAG (101), CAS (101), FXA (101)
      DIMENSION FF (101), G (101), RE (101), ROWG (101)
      DIMENSION YA (101), YR (101), YIN (101), VO (101), SC (101)
      READ (*, *) Q0, YA0, T, PT, D, DP, DE, ABSN
      READ (*, *) K, MEUG, ROWP, DA, MA, MR, MIN
C**   SPECIFY CONVERSION DESIRED
      READ (*, *) XAF
C**   SPECIFY THE STOICHIOMETRIC COEFFICIENTS
      READ (*, *) A1, R1
      ABSA = (R1/A1 - 1.0) * YA0
      R = 0.082
      CA0 = PT * YA0/(R * T)
      SP = 6.0/(ROWP * DP)
C**   SPECIFY THE STEP SIZE
      N = 100
      XA0 = 0.0
      DX = (XAF - XA0)/N
      N1 = N + 1
      DO 10 I = 1, N1
      XA (1) = XA0
      IF (I·GT·1) XA (I) = XA (I - 1) + DX
      CAG (I) = CA0 * [1.0 - XA (I)]/[1.0 + XA (I) * ABSA]
C**   ASSUME A VALUE OF CAS (I)
      CAS (I) = CAG (I) * 0.01
15    Y = CAS (I)
C**   COMPUTE THE EFFECTIVENESS FACTOR
      PHIL = (DP/6.0) * SQRT [1.5 * K * ROWP * CAS (I)/DE]
      PHI = 3.0 * PHIL
      ETA = (1.0/PHIL) * [1.0/TANH (PHI) - 1.0/PHI]
C**   COMPUTE THE MASS TRANSFER COEFFICIENT FROM
C**   AN AVAILABLE EXPERIMENTAL CORRELATION. HERE WE PRESENT DWIVEDI AND
      UPADHYAY'S CORRELATION AS AN EXAMPLE.
      FF0 = Q0 * PT/(R * T)
      FA0 = FF0 * YA0
      FF (I) = FF0 + FA0 * XA (I) * (R1/A1 - 1.0)
      YA (I) = FA0 * [1.0 - XA (I)]/FF (I)
      YR (I) = FA0 * XA (I) * (R1/A1)/FF (I)
      YIN (I) = 1.0 - YA (I) - YR (I)
```

```
      MG (I) = YA (I) * MA + YR (I) * MR + YIN (I) * MIN
      PI = 22.0/7.0
      G (I) = FF (I) * MG (I)/(PI * D * D/4.0)
C**   THE CHANGE IN VISCOSITY OF GAS IS NEGLECTED.
      RE (I) = DP * G (I)/MEUG
      JD (I) = (0.458/ABSN)/RE (I) ** 0.407
      ROWG (I) = PT * MG (I)/(R * T)
      VO (I) = G (I)/ROWG (I)
      SC (I) = MEUG/[ROWG (I) * DA]
      KC (I) = JD (I) * VO (I)/SC (I) ** 0.6667
      LHS = [KC (I) * SP/K) * [CAG (I) - CAS (I)]
      RHS = ETA * CAS (I) ** 2
      DIF = ABS (LHS - RHS)/LHS
C**   CHECK FOR CONVERGENCE
      IF (DIF·GT·0.001) THEN
      CAS (I) = CAS (I) + 0.001 * CAG (I)
      GO TO 15
      ELSE
      FXA (I) = 1.0/[ETA * CAS (I) ** 2]
      END IF
10    CONTINUE
C**   EVALUATE INTEGRAL BY TRAPEZOIDAL RULE.
      H = DX
      AI = (H/2.0) * [FXA (1) + FXA (N + 1)]
      DO 20 I = 2, N
20    AI = AI + H * FXA (I)
      W = (FA0/K) * AI
      ROWB = ROWP * (1.0 - ABSN)
      LB = (W/ROWB)/(PI * D * D/4.0)
      WRITE (*, *) LB, W
      STOP
      END
```

Let us now consider nonisothermal operation of a multitube packed bed reactor. Example 8.45 deals with a problem of this kind. It is assumed that the chemical reaction is of the form

$$aA + bB \rightarrow cC + dD$$

and the rate equation is

$$(-r_A) = k_0 \exp(-E_0/T)\, C_A^m = k_0 \exp(-E_0/T)\, (p_A/RT)^m \quad \text{... (8.8.1)}$$

where $k_0$, $E_0$ and $m$ are constants. From the reaction stoichiometry, it can be deduced (as illustrated in Example 8.45) that

$$p_A = \frac{y_{A0}(1-x_A)P_t}{1-(\Delta n)\,y_{A0}x_A} \quad ...\ (8.8.2)$$

where
$$\Delta n = 1 + (b/a) - (c/a) - (d/a) \quad ...\ (8.8.3)$$

$$P_t = \text{total pressure}$$

The material balance equation reduces to

$$\frac{dz}{dx_A} = K_m \exp\,(E_0/T)\,(T/p_A)^m \quad ...\ (8.8.4)$$

where $K_m = (F_{A0}R^m)/[A_C\rho_p\,(1-\epsilon)\,k_0]$. The heat balance equation reduces to

$$\frac{dT}{dx_A} = [h_W\,(T_W - T)\,\pi D N_t/F_0 C_p]\,(dz/dx_A) - (F_{A0}\Delta H_r/F_0 C_p) \quad ...\ (8.8.5)$$

The above two equations are now solved by Euler's method to obtain the depth of the catalyst bed required for attaining the specified conversion. The computer program is given below:

## PROGRAM 8.6

```
C**   NON-ISOTHERMAL OPERATION OF A MULTITUBE
C**   PACKED BED REACTOR
      REAL K0, KM, LB, M
      DIMENSION XA (101), T (101), PA (101), Z (101), DZX (101)
      READ (*, *) F0, YA0, GT, CP, PT, D, ROWP, TF, TW, ABSN
      READ (*, *) K0, E0, DHR, M, A1, B1, C1, D1
C**   THE WALL HEAT TRANSFER COEFFICIENT MAY
C**   BE COMPUTED FROM EXPERIMENTAL CORRELATION.
C**   HERE ITS VALUE IS ASSUMED AVAILABLE.
      READ (*, *) HW
C**   SPECIFY TOTAL CONVERSION DESIRED
      READ (*, *) XAF
      FA0 = F0 * YA0
      PI = 22.0/7.0
      FOT = GT * PI * D * D/4.0
      NT = F0/F0T
      AC = (PI * D * D/4.0) * NT
      R = 0.082
C**   SINCE THE GAS CONSTANT (R) IS TAKEN TO
C**   BE 0.082, THE TOTAL PRESSURE (PT) MUST BE
C**   EXPRESSED IN ATMOSPHERES
```

```
      XA0 = 0.0
C**   CHOOSE INCREMENT IN XA
      N = 100
      DXA = (XAF - XA0)/N
      N1 = N + 1
      Z (1) = 0.0
      DO 10 I = 1, N1
      IF (I·EQ·1) XA (I) = XA0
      IF (I·GT·1) XA (I) = XA (I - 1) + DXA
C**   ASSUME THE VALUE OF T (I).
      IF (I·EQ·1) T (I) = TF
      IF (I·GT·1) T (I) = T (I - 1) + 0.1
15    Y = T (I)
      DN = 1.0 + (B1 - C1 - D1)/A1
      PA (I) = YA0 * PT * [1.0 - XA (I)]/[1.0 - DN * XA (I) * YA0]
      KM = (FA0 * R ** M)/[AC * ROWP * (1.0 - ABSN) * K0]
      DZX (I) = {KM * [T (I)/PA (I)] ** M} * EXP (E0/T (I)]
      IF (I·EQ·1) GO TO 10
      DZXA = [DZX (I) + DZX (I - 1)]/2.0
      TA = [T (I) + T (I - 1)]/2.0
      F1 = (HW * PI * D * NT)/(F0 * CP)
      F2 = - (FA0 * DHR)/(F0 * CP)
      DT = F1 * (TW - TA) * DZXA * DXA + (F2 * DXA)
      T (I) = T (I - 1) + DT
C**   CHECK FOR CONVERGENCE
      DIF = ABS [T (I) - Y]
      IF (DIF·GT·0.1) THEN
      T (I) = [T (I) + Y]/2.0
      GO TO 15
      ELSE
      DZ = DZXA * DXA
      Z (I) = Z (I - 1) + DZ
      END IF
10    CONTINUE
      LB = Z (N + 1)
      W = LB * AC * (1.0 - ABSN) * ROWP
      WRITE (*, *) LB, W
      STOP
      END
```

We shall now consider the performance analysis of a fluidised bed reactor based on Kunii and Levenspiel's bubbling bed model (see Example 8.48). The reaction is assumed to be intrinsically first order in *A* and irreversible and what is required to be estimated is the fractional conversion in *A* that can be achieved in the reactor. The computer program is given below:

## PROGRAM 8.7

```
C**   PERFORMANCE OF A FLUIDISED BED REACTOR
C**   FROM KUNII AND LEVENSPIEL'S BUBBLING
C**   BED MODEL
C**   REACTION IS INTRINSICALLY FIRST ORDER IN
C**   A AND IRREVERSIBLE
      REAL K, KCE, K0, KP, KBC, KF, LF, LB, MA, MEUG
      READ (*, *) K, T, PT, MA, DA
      READ (*, *) DP, ROWP, DP0, ABSP
      READ (*, *) ABSN, LB, DB, ABSMF, ALFA
C**   SPECIFY PROPERTIES OF FLUIDISING GAS AT T AND PT
      READ (*, *) MEUG, ROWG
C**   COMPUTE MINIMUM FLUIDISATION VELOCITY.
      G = 9.8
      ARM = (DP ** 3) * ROWG * (ROWP - ROWG) * G/MEUG ** 2
      REMF = SQRT (33.67 ** 2 + 0.0408 * ARM) - 33.67
      VMF = REMF * MEUG/(ROWG * DP)
C**   OPERATING GAS VELOCITY IS TAKEN TO BE
C**   SIX TIMES THE MINIMUM FLUIDISATION VELOCITY
      VOF = 6.0 * VMF
C**   COMPUTE NATURAL RISE VELOCITY OF BUBBLE
      VBR = 0.711 * SQRT (G * DB)
C**   COMPUTE VELOCITY OF GAS IN BUBBLE PHASE
      VB = VOF - VMF + VBR
C**   COMPUTE INTERCHANGE COEFFICIENT BETWEEN
C**   CLOUDS AND EMULSION PHASE
      KCE = 6.78 * SQRT (ABSMF * DA * VB/DB ** 3)
C**   COMPUTE EFFECTIVE DIFFUSIVITY
      R = 8314.0
      PI = 22.0/7.0
      UA = SQRT [8.0 * R * T/(PI * MA)]
      DAK = DP0 * UA/3.0
      DEE = (1.0/DA) + (1.0/DAK)
      DE = (ABSP ** 2)/DEE
```

```
C**    COMPUTE EFFECTIVENESS FACTOR
       PHIL = (DP/6.0) * SQRT (K * ROWP/DE)
       PHI = 3.0 * PHIL
       ETA = (1.0/PHIL) * [1.0/TANH (PHI) - 1.0/PHI]
C**    SINCE VOLUME FRACTION OF SOLIDS IN
C**    BUBBLES (GAMB) VARIES FROM 0.001 TO 0.01,
C**    CHOOSE GAMB = 0.01
       GAMB = 0.01
       VM = (VMF/ABSMF)
       GAMC = (1.0 - ABSMF) * [3.0 * VM/(VBR - VM) + ALFA]
       DELT = (VOF - VMF)/VB
       GAME = (1.0 - ABSMF) * (1.0 - DELT)/DELT - (GAMC + GAMB)
       K0 = ETA * K * ROWP
       KP = (1.0/KCE) + 1.0/(K0 * GAME)
       KP = 1/KP
       KBC = (4.5 * VMF/DB) + 5.85 * SQRT (DA * G ** 0.5/DB ** 2.5)
       KF = (1.0/KBC) + 1.0/(K0 * GAMC + KP)
       KF = (GAMB * K0) + (1.0/KF)
       LF = (1.0 - ABSN) * (LB/VBR) * VB/(1.0 - ABSMF)
       XAF = 1.0 - 1.0/EXP (KF * LF/VB)
       WRITE (*, *) XAF, LF
       STOP
       END
```

Let us now take up the performance analysis of a trickle bed reactor in which a gas-liquid reaction is conducted. We shall follow the analysis presented through Eqs (8.5.110) to (8.5.130). See also Example 8.49, The computer program is given below:

## PROGRAM 8.8

```
C**    GAS-LIQUID REACTION IN A TRICKLE BED REACTOR
C**    REACTION IS INTRINSICALLY FIRST ORDER
C**    IN A AND IRREVERSIBLE
C**    BOTH GAS AND LIQUID EXECUTE PLUG FLOW
C**    AXIAL DIFFUSION IS NEGLECTED BOTH IN
C**    LIQUID AND GASEOUS PHASES
       REAL KLA, KOLA, KLPA, K, KO, MG, LB, K1, K2, K3, K4, MEUL
       READ (*, *) T, PT, D, GG, MG, GL, CALO, CAGO
       READ (*, *) K
       READ (*, *) DP, ABSN, ROWP, DE, MEUL, ROWL, DL, HE
C**    SPECIFY CONVERSION DESIRED
```

```
      READ (*, *) XAF
C**   COMPUTE LIQUID PHASE MASS TRANSFER COEFFICIENT
C**   FROM AVAILABLE EXPERIMENTAL
C**   CORRELATION. HERE WE PRESENT GOTO AND
C**   SMITH'S CORRELATION AS AN EXAMPLE
      SCL = MEUL/(ROWL * DL)
      KLA = (11094.253) * DL * (GL/MEUL) ** 0.4 * SQRT (SCL)
C**   ASSUMED THAT THE GASEOUS REACTANT A IS
C**   ONLY SLIGHTLY SOLUBLE IN THE LIQUID.
      KOLA = KLA
C**   COMPUTE THE LIQUID-TO-PARTICLE MASS TRANSFER
C**   COEFFICIENT FROM AVAILABLE CORRELATION. HERE
C**   WE PRESENT DHARWADKAR AND SYLVESTER'S
C**   CORRELATION AS AN EXAMPLE. THIS IS VALID
C**   FOR REL = 0.2 TO 2400.
      REL = DP * GL/MEUL
      AT = (6.0/DP) * (1.0 - ABSN)
      VL = GL/ROWL
      KLPA = 1.64 * VL * AT/(REL ** 0.331 * SCL ** 0.6667)
C**   COMPUTE EFFECTIVENESS FACTOR.
      PHIL = (DP/6.0) * (K * ROWP/DE) ** 0.5
      PHI = 3.0 * PHIL
      ETA = (1.0/PHIL) * [1.0/TANH (PHI) - 1.0/PHI]
      K0 = K * ROWP * (1.0 - ABSN) * ETA
      K1 = (KLPA)/(K0 + KLPA)
      R = 0.082
C**   EXPRESS PT IN ATMOSPHERES
      ROWG = PT * MG/(R * T)
      VG = GG/ROWG
      V1 = VL * KOLA
      V2 = VG * HE
      V3 = VG * VL * HE
      K2 = {V1 + V2 * [KLPA * (1.0 - K1) + KOLA]}/V3
      K3 = KLPA * KOLA * (1.0 - K1)/V3
      K4 = (KOLA/VG) * (CALO - CAGO/HE)
      R1 = [-K2 + SQRT (K2 ** 2 - 4.0 * K3)] * 0.5
      R2 = [-K2 - SQRT (K2 ** 2 - 4.0 * K3)] * 0.5
      C1 = (K4 - CAGO * R2)/(R1 - R2)
      C2 = (CAGO * R1 - K4)/(R1 - R2)
```

```
C**    ASSUME A VALUE OF Z (DEPTH OF CATALYST BED).
       Z = 0.1
10     CAG = C1 * EXP (R1 * Z) + C2 * EXP (R2 * Z)
       XAFC = 1.0 - CAG/CAG0
       IF (XAFC·LT·XAF) THEN
       Z = Z + 0.01
       GO TO 10
       ELSE
       LB = Z
       END IF
       PI = 22.0/7.0
       AC = PI * D* D/4.0
       W = AC * LB * (1.0 - ABSN) * ROWP
       WRITE (*, *) LB, W
       STOP
       END
```

So far, we have considered reactors dealing with solid-catalysed fluid-fluid reactions. We shall now take up performance analysis of a moving bed reactor in wich a non-catalytic gas-solid reaction is being conducted (similar to the problem dealt in Example 8.52). The reaction stoichiometry is

$$A\ (g) + bB\ (s) \rightarrow pP\ (s) + rR\ (g)$$

The reaction is intrinsically first order in $A$ and irreversible and the rate constant varies with temperature as per the relation, $k = k_0 \exp(-E_0/T)$, where $k_0$ and $E_0$ are constants.

## PROGRAM 8.9

```
C**    GAS-SOLID (NON-CATALYTIC) REACTION IN A
C**    MOVING BED REACTOR.
C**    REACTION IS INTRINSICALLY FIRST ORDER IN
C**    A (GASEOUS REACTANT) AND IRREVERSIBLE.
C**    BOTH GAS AND SOLIDS EXECUTE PLUG FLOW.
       REAL K, K0, MB, MP, L
       DIMENSION XB (101), XA (101), FS (101), Q (101), CAG (101), VS
       (101), ROWS (101)
       DIMENSION FXB (101), DRCT (101), RC (101), DNMR (101)
C**    SPECIFY THE STOICHIOMETRIC COEFFICIENTS
       READ (*, *) B1, P1, R1
       READ (*, *) T, PT, D, Q0, FB0, YA0
       READ (*, *) DP, ROWB, ROWP, MB MP, DE, ABSG
       READ (*, *) K0, E0
       K = K0 * EXP (-E0/T)
```

```
C**      SPECIFY CONVERSION DESIRED IN B.
         READ (*, *) XBF
         PI = 22.0/7.0
         AC = PI * D * D/4.0
         RS = DP/2.0
         R = 0.082
         CAGO = PT * YA0/(R * T)
C**      TO NOTE THAT PT MUST BE EXPRESSED IN ATM.
         FA0 = Q0 * CAGO
         XB0 = 0.0
C**      SPECIFY THE STEP SIZE
         N = 100
         H = (XBF - XB0)/N
         N1 = N + 1
         DO 10 I = 1, N1
         IF (I·EQ·1) XB (I) = XB0
         IF (I·GT·1) XB (I) = XB (I - 1) + H
         FS (I) = FB0 * MB * [1.0 - XB (I)] + (P1/B1) * FB0 * XB (I) * MP
         XA (I) = (1.0/B1) * (FB0/FA0) * XB (I)
         Q (I) = Q0 * [1.0 + (R1 - 1.0) * YA0 * XA (I)]
         CAG (I) = [Q0/Q (I)] * CAG0 - (FB0/B1) * [XBF - XB (I)]/Q (I)
         VS (I) = FB0 * [1.0 - XB (I)] * MB/ROWB + (P1/B1) * FB0 * XB (I) *
         MP/ROWP
         ROWS (I) = FS (I)/VS (I)
         RC (I) = RS * [1.0 - XB (I)] ** 0.3333
C**      COMPUTE DRC/DTHETA FROM EQ. (8.6.11).
C**      GAS PHASE MASS TRANSFER RESISTANCE ASSUMED
C**      NEGLIGIBLE.
         DNMR (I) = (1.0/K) + [RC (I)/DE] * [1.0 - RC (I)/RS]
         DRCT (I) = (B1 * MB/ROWB) * CAG (I)/DNMR (I)
         F1 = RS * (FS/ROWS)/[3.0 * AC * (1.0 - ABSG)]
         FXB (I) = F1/{[1.0 - XB (I)] ** 0.6667 * DRCT (I)}
10       CONTINUE
C**      EVALUATE INTEGRAL BY TRAPEZOIDAL RULE.
         AI = (H/2.0) * [FXB (1) + FXB (N + 1)]
         DO 15 I = 2, N
15       AI = AI + H * FXB (I)
         L = AI
         WRITE (*, *) L
         STOP
         END
```

## THINGS TO REMEMBER

1. Chemical reactions may be roughly classified into heterogeneous and homogeneous reactions. If all the reaction molecules are in a single phase, then it is called a homogeneous reaction. If two or more phases are required to conduct the reaction at a specified rate, then it is called a heterogeneous reaction. In a heterogeneous system, the reaction may take place in one, two or more phases or at the interface. The reactants and products may be distributed among the phases or may be contained in a single phase.
2. The reaction rate ($r_A$) may be defined in alternate ways as shown in Eqs (8.1.1), (8.1.2) and (8.1.4). It will be negative if $A$ is being consumed during the reaction and positive if $A$ is being produced.
3. For homogeneous reactions, $r_A$ is a function of temperature, pressure and the composition of the reacting substances. If the stoichiometric equation representing the reaction describes the true mechanism of the reaction so that the rate equation can be directly deduced from it, then the reaction is said to be an elementary reaction. The molecularity of an elementary reaction is the number of moles of reactants involved in the reaction and it can be one, two or three. If the molecularity of any reaction is more than 3.0, then it cannot be an elementary reaction. In the case of non-elementary reaction, there is hardly any correspondence between stoichiometric equation and the rate equation. Examples are given in Eqs (8.1.8), (8.1.10) and (8.1.12).
4. The order of a chemical reaction is an empirical constant (it may be an integer or a fraction). It is to be determined from the rate equation, which in turn is derived empirically from experimental data. For elementary reactions, the reaction order agrees with the molecularity. If any one of the reactants is in large excess, then the reaction will be zero order with respect to that reactant and the overall order of the reaction gets diminished.
5. The rate constant or the specific reaction rate ($k$) is also an empirically defined parameter. Its value can be determined only experimentally. If the rate equation conforms to Eq. (8.1.13). then the dimensions of $k$ will be $(s^{-1})$ $(kmoles/m^3)^{1-q}$, where $q$ is the order of the reaction.
6. In the case of reversible reactions, the maximum conversion that can be attained is the equilibrium conversion. The equilibrium constant K is defined by Eq. (8.1.20) or (8.1.18). For elementary reactions, $K = (k_1/k_2)$. $K$ is a function of temperature [as per Eq. (8.1.21)] but is practically unaffected by pressure and the presence of inerts. The equilibrium conversion, however, is influenced by temperature, pressure and the presence of inerts. It increases with increase in temperature for endothermic reactions and vice versa for exothermic reactions. For gas-phase reactions that take place with a decrease in the total number of moles, the equilibrium conversion increases with increase in pressure, while for those that take place with an increase in the total number of moles, it decreases with increase in pressure.
7. The temperature dependency of the rate constant ($k$) is expressed by Arhrenius' law Eq. (8.1.23), by collision theory Eq. (8.1.24) or by activated complex theory Eq. (8.1.34).
8. Reactions with high activation energies are more temperature-sensitive than those with low values of $E$. Also, higher temperatures favour reactions of higher activation energy and vice-versa. This observation is particularly important in the case of multiple reactions.
9. The mechanisms of nonelementary or complex chemical reactions can be deduced only from experimental data. Popularly, these fall into the following three categories:

(*i*) In the first category, the overall reaction is assumed to be taking place in three steps such as chain initiation, chain propagation and chain termination each step consisting of one or more elementary reactions. It is assumed that freeradicals are formed during the process wich propagate the chain reaction. It may so happen that one of the steps or one of the elementary reactions is the slowest and is thereby rate-controlling. The kinetics of the overall reaction will then reflect the kinetics of this rate-controlling step.

(*ii*) In the second category, an intermediate product is assumed to have been formed which, being unstable, decomposes readily to yield the products. Enzyme–catalysed reactions (that follow Michaelis-Menten scheme) and many of the ionic reactions are examples.

(*iii*) In the third category of non-elementary reactions, an activated intermediate complex is assumed to have been formed from the reactants which subsequently decomposes to yield the products (see Examples 8.1 to 8.6 for the various modes of deducing reaction mechanism).

10. The first step in the design of reactors is the development of a reliable rate equation. Rate equations for non-elementary or complex reactions are derived from experimental kinetic data collected by conducting the reaction in a laboratory bench scale reactor. For studying homogeneous reactions, a batch reactor is usually employed while heterogeneous reactions are usually studied in a flow reactor. The data so collected are fitted into the rate equation either by the integral method or by the differential method.

11. In the integral method (for homogeneous reactions), the data (concentration versus time) from a constant volume batch reactor is fitted into Eq. (8.2.4) by trial. A rate equation is first assumed, the integral of Eq. (8.2.4) is then evaluated and the resultant concentration expression is plotted against time. If a straight line results, then the assumed rate equation is true. Alternately, the expression for half life ($\theta_{1/2}$) is derived from the assumed rate equation which will be of the form $\theta_{1/2} = f(C_{A0})$. It is then checked whether a plot of $\theta_{1/2}$ versus $f(C_{A0})$ yields a straight line or not. Integrated forms of Eq. (8.2.4) and expressions for half life for different reactions are listed in Tables 8.1 and 8.2. Table 8.1 considers relatively simple, irreversible reactions while more complex reactions such as reversible reactions, multiple reactions (simultaneous reactions and consecutive reactions), autocatalytic reactions and reactions of shifting order (enzyme catalysed reactions, for example) are considered in Table 8.2.

12. Instead of a constant-volume batch reactor, data from a constant-pressure batch reactor may also be used to develop the rate equation. In the integral method, this involves fitting into Eq. (8.2.130) by trial [Eq. (8.2.130) is derived in Example 8.14]. Equation (8.2.130), however, assumes that the volume of the reaction mixture varies linearly with conversion.

13. In the differential method of analysis, measured values of concentration are plotted against time to get a continuous curve and the values of $dC_A/d\theta$ at different $\theta$ are obtained from this curve. It is then checked whether a plot of $dC_A/d\theta$ versus $f(C_A)$ yields a straight line or not [see Eq. (8.2.3)]. As in the integral method, the expression $f(C_A)$ is first assumed and the rate equation finalised by trial. For rate expressions that cannot be readily integrated, the differential method is more convenient.

14. In the case of multiple reactions, it is necessary to distinguish between yield and selectivity. The yield of a specific product is the fraction of the reactant converted to that product. This is different from total conversion which is the fraction of the reactant converted to all products. The overall

selectivity ($S_0$) of a product is the ratio of the amount of that product produced to the amount of another (or, the ratio of the yield of that product to the yield of another). By point selectivity ($S_p$), we mean the ratio of the rate of production of one product to that of another. With multiple products, we can define a separate selectivity based on each pair of products.

15. In autocatalytic reactions, one of the products acts as the catalyst and tends to accelerate the reaction. The rate of an autocatalytic reaction, therefore, increases first, reaches a maximum and thereafter decreases (see Fig. 8.2). Microbial fermentation is an excellent example in this connection.
16. The most popular types of reactors used for conducting homogeneous reactions are the stirred tank reactors and the tubular reactors. The former can be operated as a batch reactor, steady flow reactor or a semibatch reactor. For an *ideal* batch reactor (in which concentration changes with time but not from point to point within the reactor), the performance equation is given by Eq. (8.3.3) or (8.3.4). For an *ideal* continuous flow stirred tank reactor (CSTR), the performance equation is given by Eq. (8.3.9) or (8.3.13). For constant density systems, Eq. (8.3.14) or (8.3.15) may be used. To note that in an *ideal* CSTR, the reaction mixture is well-mixed so that its properties are uniform throughout the vessel and are the same as those in the exit stream (complete backmixing). For semibatch reactors (with complete backmixing), the performance equation is Eq. (8.3.20) or (8.3.28) depending on the mode of operation. For the specific case of constant density systems with a first order reaction, Eq. (8.3.20) may be solved to get Eq. (8.3.25). For an *ideal* tubular reactor (also called the plug flow reactor), there is no variation in properties or velocity in the radial direction and the performance equation is given by Eq. (8.3.32) or (8.3.33). For constant density systems, the performance equation of the plug flow reactor becomes identical with that of a constant-volume batch reactor.
17. The performance equation of a recycle reactor (PFR with recycle) is given by Eq. (v) of Example 8.21 and the estimation of optimum recycle ratio is discussed in Example 8.22. The performance of a recycle reactor is in between those of a PFR and a back-mix reactor (ideal CSTR). At low values of the recycle ratio (reflux ratio), its performance is close to that of a PFR while at high values of the recycle ratio, its performance approaches that of a backmix reactor.
18. In the performance equations mentioned above, $\tau$ is the space time which is defined in Eq. (8.3.10). It is different from the mean residence time $\bar{\theta}$ (which is the average of the time periods during which individual portions of the reaction mixture stay in the reactor and is defined by Eq. (8.3.16) or (8.3.19). $\bar{\theta}$ will be equal to $\tau$ only when the density of the reaction mixture is constant. It is also necessary that the temperature and pressure are constant throughout the reactor and the feed rate is measured at the temperature and pressure of the reactor.
19. The general definition of ideality is that an ideal reactor is one for which the residence time or the spread in residence time is accurately known or can be accurately determined. A CSTR is ideal when there is complete backmixing and a tubular reactor is ideal when there is cent percent plug flow. If there is significant deviation from ideality, then specific experimental studies are to be conducted to determine the RTD and mixing pattern in such reactors.
20. Capacity-wise, a PFR is superior to a CSTR. In fact, an infinite number of stirred tank reactors in series is equivalent to a single PFR if the total residence time is the same. This is illustrated by an example through Eqs (8.3.46) and (8.3.49).

21. To obtain high conversions, it is better to operate stirred tank reactors in series. It is seen that $n$ reactors in series provide much better performance than a single reactor even if the total residence time is the same in both cases. If $n = \infty$, then the performance becomes equal to that of a PFR. When $n$ is large, the performance of the reactor-system may be better determined graphically. For constant density systems, this procedure is illustrated in Fig. 8.7. Also, when the reactors are of different sizes, the optimum size ratio (minimum size ratio) can also be estimated graphically as shown in Fig. 8.8. To note that when $n > 2$, it is usually more economical to use equally sized reactors.
22. Though isothermal operation of reactors has distinct advantages, the same becomes difficult and uneconomical when the heat of reaction is significantly large. Adiabatic or nonisothermal operation is then preferred. In such cases, both material balance and energy balance equations are required to study reactor performance and since these two are strongly coupled, they are to be solved simultaneously. Equations (8.3.55) and (8.3.59) are energy balance equations for ideal batch reactors and Eqs (8.3.60) and (8.3.64) those for PFRs. It can be seen that for constant density systems and for adiabatic operation, the energy balance equations for PFR and for the batch reactor [such as Eqs (8.3.59) and (8.3.64)] are identical. When an ideal CSTR is operated non-isothermally, the temperature will be uniform and constant inside the reactor and will be equal to that in the exit stream (due to complete backmixing) but will not be equal to the feed temperature. When the energy balance equation [Eq. (8.3.67) or (8.3.68)] and material balance equation of a CSTR are solved simultaneously, we often get more tan one solution, thereby indicating that there are more than one stable operating conditions for the reactor. Example 8.32 illustrates this point.
23. The performance analysis of ideal reactors is numerically illustrated in Examples 8.16, 8.17 (for irreversible reactions), 8.18 (for reversible reactions), 8.21 (for recycle reactors), 8.23 (for parallel or simultaneous reactions), 8.24 (for consecutive reactions or reactions in series), 8.26 (for series-parallel reactions), 8.27 (for polymerisation reactions), 8.28 and 8.29 (for semibatch reactors), 8.30 and 8.31 (for adiabatic and nonisothermal operations). Determination of rate equation from CSTR data and PFR data are illustrated in Examples 8.19 and 8.20 respectively.
24. Often, the performance of an industrial reactor deviates from ideality and falls in between that of a PFR and an ideal CSTR. This can be due to presence of dead zones, channeling or other inconsistencies. To model such a reactor, we must know the RTD (residence time distribution) of fluid elements in the reactor (this is determined by tracer experiments), the mixing pattern (whether mixing takes place at microscopic/macroscopic level or with partial segregation) and whether micromixing occurs near the reactor inlet or near the outlet. Tracer experiments may be conducted with a step input of tracer or a pulse input of tracer. From the experimental data [$C(\theta)$ versus $\theta$ data], the distribution functions such as $E(\theta)$ and $F(\theta)$ are to be computed [from Eqs (8.4.1) and (8.4.5)]. How far these functions deviate from those for an ideal CSTR [Eqs (8.4.11) and (8.4.12)] or for a PFR [Eq. (8.4.19)] would tell us regarding the actual performance of the reactor. The deviation may be expressed in terms of standard deviation ($\sigma$), variance ($\sigma^2$) or skewness ($\sigma^3$). These are defined in Eqs (8.4.20) to (8.4.24).
25. Though RTD tells us regarding the actual flow pattern inside the reactor, it tells little about the level of mixing or the extent of inter-mixing between fluid elements within the reactor. However, in a number of cases (though not always), the reactor performance is not too sensitive to the degree

of micromixing/macromixing. This is particularly true in the case of first order reactions. Also, the RTD data is independent of whether micromixing occurs at the early stages of the chemical reaction or at a later stage. For example, RTD for the two types of reactor combinations, such as an ideal CSTR followed by a PFR (here micromixing occurs at the initial stages) and a PFR followed by an ideal CSTR (here micromixing occurs at the final stage), has been found to be the same. Accordingly, based on RTD studies alone, we cannot predict whether the given reactor can be modelled as equivalent to which of these combinations. For first order reactions, however, this parameter is not important since both of these combinations provide the same performance in case of first order reactions (see Example 8.17).

26. Mathematical modelling of a nonideal reactor is thus no easy task since we have to try a multitude of alternatives to arrive at the model that would predict the performance of the given reactor most satisfactorily. A few of the popular models are CSTR with a given degree of macromixing Eqs (8.4.31) to (8.4.48), plug flow dispersion reactor or PFDR Eqs (8.4.49) to (8.4.64), laminar flow tubular reactor or LFTR Eqs (8.4.69) to (8.4.86), tanks-in-series model Eqs (8.4.87) to (8.4.90), recycle reactor or PFR with partial product recycle Eq. (8.4.91) and reactor combinations (two examples are shown in Figs 8.14 and 8.15.
27. The various types of heterogeneous reactions are solid–catalysed fluid-fluid reactions, gas–solid reactions (noncatalytic), gas–liquid reactions, enzyme-catalysed reactions and liquid-solid reactions. Excellent examples are listed in Tables 8.4 and 8.4A.
28. Solid-catalysed reactions are one of the most popular heterogeneous reactions. A catalyst necessarily increases the rate of the reaction (though there are negative catalysts too). In multiple reactions, the catalyst can be used to improve the selectivity. The equilibrium conversion of a reversible reaction is not altered by the use of a catalyst. Catalysts are often supported or impregnated on a carrier. Promoters are often added during catalyst preparation to improve the activity, selectivity or stability of the catalyst, while inhibitors are used to depress or control the activity of the catalyst.
29. Catalyst deactivation or catalyst poisoning can occur due to chemisorption or physical deposition of substances that are either present in the reactant stream (side-by-side deactivation) or produced during the reaction/side reaction (parallel deactivation or series deactivation) or due to deterioration of the surface structure of the catalyst as a result of prolonged exposure to elevated temperatures (independent deactivation).
30. The physical properties of the catalyst (pore size, pore volume, surface area) are experimentally determined by the helium-mercury method, by mercury porosimeter or by low temperature nitrogen adsorption (see Examples 8.35, 8.36).
31. Any solid–catalysed reaction involves the seven steps such as transport of reactants from fluid bulk to catalyst surface (external mass transport), diffusion into catalyst pores, adsorption at the active sites, surface reaction, desorption of products, diffusion of products to catalyst surface and transport of products to fluid bulk (external mass transport). Often, it will be possible to identify a rate-controlling step (which is the slowest step) and in that case, the overall rate (the global rate) will be equal to the rate of that step. The expression for the intrinsic rate (which considers the three steps such as adsorption, surface reaction and desorption only) is first determined (either empirically or by Langmuir-Hinshelwood formulation) and it is then multiplied by a properly defined effectiveness factor ($\eta$) to include the effect of pore diffusion as well. The resultant rate equation is then clubbed

with film diffusion (resistance to film diffusion or external mass transport can be treated as a separate additive resistance and can be separately accounted for by means of an appropriately defined mass transfer coefficient) to obtain the overall rate (global rate) equation as shown in Eq. (8.5.71). Expressions for intrinsic rate developed by Langmuir-Hinshelwood formulation for different types of reactions are listed in Table 8.5. This formulation has a sound theoretical basis though it demands evaluation of as many as five constants from experimental data. The expression for $\eta$ for a spherical catalyst when the intrinsic rate is first order in $A$ is given in Eq. (8.5.48). This expression may be used for particles of all shapes by defining the dimension $L$ as per Eq. (8.5.51) and the Thiele-type modulus ($\phi$) as per Eq. (8.5.50). It can be further extended to reactions following alternate kinetics by defining $\phi$ in the most generalised form shown in Eq. (8.5.53). Prediction of non-isothermal effectiveness factor is possible from Weisz and Hicks' graphical data (see Fig. 8.16). To note that for first order reactions, $\phi$ is independent of $C_{AS}$ (neglecting concentration dependence of $D_e$) while in all other cases, $\phi$ is a function of surface concentration.

32. Examples 8.37 to 8.40 illustrate derivation of rate equation for different types of solid–catalysed reactions based on the above principles. Examples 8.41 and 8.42 illustrate how the resistances to different steps individually influence the reaction rate or the reactor performance. The influence of pore diffusion resistance on the selectivity of solid-catalysed multiple reactions is illustrated in Example 8.44.

33. The most popular types of reactors employed for conducting solid-catalysed reactions are the fixed bed (packed bed), fluidised bed, moving bed, trickle bed and slurry reactors. Design equations of packed bed reactors (assuming plug flow and neglecting axial diffusion and axial conduction of heat) are given by Eqs (8.5.86) and (8.5.88). For more rigorous analysis, Eqs (8.5.76) and (8.5.77) are to be solved simultaneously (by numerical method) using the boundary conditions given in Eqs (8.5.78) to (8.5.84). Packed bed reactors require minimum auxiliary equipment and are particularly suitable for small commercial units. They may be tubular-type [Fig. 8.17 (a)] or chamber-type [8.17 (b)]. When heat affects are large, heat may be dissipated or supplied by external cooling/heating, by using inter-stage coolers or heaters or by using a large amount of diluant in the feed stream. In the case of highly exothermic reactions, it is advisable to use the reactor effluent to preheat the feed (autothermal operation) so as to economise the energy consumption of the plant. Continuous regeneration and reuse of deactivated catalyst is practically not possible in packed bed reactors.

34. Fluidised bed reactors are continuous in operation, permits continuous regeneration and reuse of catalyst, permits use of small sizes of catalyst (this minimises resistance to pore diffusion), permits easy control of temperature and provides nearly isothermal operation. However, the performance of bubbling fluidised beds is usually inferior to that of packed bed reactors due to considerable bypassing of bubble gas. Unlike in packed beds, plug flow can hardly be achieved in fluidised beds. Since high gas velocities cannot be avoided in commercial installations, it is recommended to use the fluidised bed reactor with an appreciable carry-over and recycle of solids as shown in Fig. 8.21. Under these conditions, the bed behaves as a lean emulsion with little gas bubbling and little bypassing and the gas flow can be made to approach plug flow (even at very high gas velocities) with proper design.

35. Analysis of a bubbling fluidised bed reactor is enormously complex. The bubbling bed model of Kunii and Levenspiel (for a first order, irreversible reaction) is presented through Eqs (8.5.90) to (8.5.108).
36. In moving bed reactors, the catalyst and the fluid move countercurrently. They are also continuous in operation, permits continuous regeneration and reuse of catalyst and they share with fixed beds the advantage of plug flow. However, small sizes of catalyst cannot be used in these reactors since they tend to be carried off in the upgoing gas stream.
37. In trickle bed reactors, gas and liquid flow downward over a fixed bed of catalyst particles. Both liquid and gas velocities are maintained low so that the gas forms a continuous phase but the liquid falls in rivulets from one particle to another. A relatively simplified analysis of trickle bed reactor is presented through Eqs (8.5.112) to (8.5.123).
38. Slurry reactors are also three-phase reactors. Catalyst particles are suspended in a liquid (which may or may not take part in the reaction) and the gas is bubbled through the slurry or suspension. The slurry is agitated partly by the bubbling gas and partly by a mechanical impeller. Since small particle sizes are employed, the resistance to pore diffusion is minimised but external mass transfer resistance could be significant in slurry reactors. Due to the high thermal conductivity of the liquid and the violent agitation provided, there will be hardly any temperature difference between particles and the liquid. However, difficulty arises in retaining the small catalyst particles in the reactor without getting lost in the product discharged. The performance equation of a slurry reactor for a first order gas-phase reaction is represented by Eqs (8.5.132) and (8.5.137).
39. The actual rate of reaction with a deactivating catalyst is obtained by multiplying the rate of reaction with an unpoisoned catalyst by the catalyst activity ($\alpha_a$). Like the rate equation for the reaction, the expression for the rate of deactivation ($d\alpha_a/d\theta$) is also to be determined by trial from experimental data. Often, an expression of the form of Eq. (8.5.142) shall be satisfactory. See Examples 8.46 and 8.47.
40. Noncatalytic fluid-solid reactions are similar to solid-catalysed reactions except that in this case, the rate of reaction is a function of time (since the solid is also a reactant). As a result, steady state operation is possible only when solids are in continuous flow (as in a moving bed reactor) but not in a fixed bed reactor. The shrinking core model (that explains the mechanism of these reactions reasonably well) assumes that though the radius of the unreacted core of the solid reactant ($B$) decreases with time, the total radius $R_S$ (the combined radius of the unreacted core of $B$ and the ash layer surrounding it) remains constant. Based on this model, the rate equation (for a chemical reaction that is intrinsically first order in $A$ and irreversible) can be then obtained from Eqs (8.6.11) and (8.6.8). For specific cases in which $C_{Ag}$ (concentration of $A$ in the gas bulk) is constant (this happens in back-mix reactors or when there is a large excess of $A$), Eq. (8.6.11) can be integrated to give Eq. (8.6.12). It further reduces to Eq. (8.6.17) when the chemical reaction is the rate-controlling step, to Eq. (8.6.21) if diffusion through the ash layer is rate-controlling and to Eq. (8.6.25) if mass transfer through the gas film is rate-controlling.
41. If there is no ash layer (no solid product is formed), Eq. (8.6.11) gets modified to Eq. (8.6.27). This reduces further to Eq. (8.6.16) if the chemical reaction is rate-controlling and to Eq. (8.6.31) if external mass transfer through the gas film is rate-controlling. To note that in this case, the mass transfer coefficient $k_C$ has been expressed in terms of $R_C$ from Eq. (8.6.29).

42. The above Eqs (8.6.8) to (8.6.31) are for a single spherical particle. If there is a mixture of particles of different sizes, then the average conversion can be computed from Eqs (8.6.37) and (8.6.38).
43. The performance equation for a moving bed reactor in which $C_{Ag}$ does vary from fluid inlet to the outlet but both the gas and solids are in plug flow is given by Eq. (8.6.48). This can be integrated after substituting Eqs (8.6.11) and (8.6.41) in it to obtain the length of the reactor required ($L$) for achieving a specified conversion. See Example 8.52 for a numerical illustration.
44. Gas–liquid reactions can be conducted in packed beds, fluidised beds, trickle bed reactors, slurry reactors and also in foam bed and membrane reactors. Gas absorption with chemical reaction in packed beds is discussed in detail in Section 4.5.6 of Chapter 4. Principles of membrane reactors are outlined in Section 7.6 of Chapter 7. Performance characteristics of foam bed reactors can be predicted from Eqs (8.7.1) to (8.7.29) which are based on the model proposed by Kumar and Biswas[53].

## NOMENCLATURE

$a_A$, $a_B$ activities of $A$ and $B$ respectively

$a_g$ gas-liquid interfacial area per unit volume of catalyst bed or per unit volume of gas-free liquid, $m^2/m^3$

$a_i$ half the thickness of liquid film in $i$th foam subsection, m

$a_p$ liquid-particle interfacial area per unit volume of catalyst bed; external surface area of particles per unit volume of gas-free liquid, $m^2/m^3$

$[A]$ molar concentration of $A$ (same as $C_A$), kmoles/$m^3$

$A_C$ tower (reactor) cross-sectional area, $m^2$

$A_h$ heat transfer area, $m^2$

$C(\theta)$ concentration of tracer in exit stream from reactor, kg/$m^3$

$C_A$, $C_B$ molar concentrations of $A$ and $B$ respectively, kmoles/$m^3$

$\bar{C}_A$ concentration of adsorbed $A$, kmoles/kg of catalyst

$C_{Ab}$ concentration of $A$ in bubbles, kmoles/$m^3$ of bubbles

$C_{AC}$ concentration of $A$ at the surface of ash layer; concentration of $A$ in cloud and wake, kmoles/$m^3$

$C_{Ae}$ concentration of $A$ at equilibrium (for reversible reactions); concentration of $A$ in the emulsion phase, kmoles/$m^3$

$C_{Ag}$ concentration of $A$ in fluid bulk (gas bulk), kmoles/$m^3$

$C^*_{Ag}$ equilibrium concentration of $A$ in gas, kmoles/$m^3$

$(C_{Ag})_0$ concentration of $A$ in inlet gas to trickle bed/foam bed/moving bed reactor, kmoles/$m^3$

$C_{AL}$ concentration of $A$ in bulk liquid, kmoles/$m^3$

$C^*_{AL}$ equilibrium concentration of $A$ in liquid, kmoles/$m^3$

$(C_{AL})_0$ concentration of $A$ in inlet liquid to trickle bed reactor/moving bed reactor, kmoles/$m^3$

$C_{A0}$ concentration of $A$ at $\theta = 0$ in a batch reactor, kmoles/$m^3$

$C_{A0}$, $C_{Af}$ concentration of $A$ in inlet stream (feed stream) and outlet stream (product stream) respectively for a flow reactor, kmoles/$m^3$

$C_{AS}$ concentration of $A$ in fluid at the catalyst surface, kmoles/m$^3$

$C_{Bi}$, $C'_B$ molar concentration of $B$ at the end of $i$th subsection and in feed back liquid respectively

$C_C$ concentration of catalyst in slurry, kmoles/m$^3$ of gas-free liquid.

$\overline{C}_m$ concentration of active centres per kg of catalyst

$C_p$ specific heat of reaction mixture, J/(kg·K)

$\overline{C}_V$ concentration of vacant sites per unit mass of catalyst

$d_b$ bubble diameter, m

$d_p$ diameter of catalyst particle, m

$d_{p0}$ pore diameter, m

$D$ ID of reactor, m

$D_A$ diffusivity of $A$ in fluid, m$^2$/s

$D_d$ axial dispersion coefficient, m$^2$/s

$D_e$ effective diffusivity, m$^2$/s

$D_{er}$, $D_{ez}$ average effective diffusivities in radial and axial directions in packed bed, m$^2$/s

$E$ activation energy, J/kmole

$E(\theta)$ exit age distribution function at any $\theta$, s$^{-1}$

$F(\theta)$ fraction of tracer that has resided in reactor for time less than $\theta$, dimensionless

$F_{A0}$, $F_{Af}$ molar input and output respectively of $A$, kmoles/s

$F_{in}$ number of moles of inerts in feed gas [see Eq. (8.6.42)]

$F_0$ total feed rate to reactor, kg/s

$F_S$ feed rate of solids (catalyst); mass flow rate of solids at any $z$ (see Eq. 8.6.46), kg/s

$G$ mass velocity, kg/(m$^2$·s)

$h$ planck's constant, J·s; height of each foam section, m

$h_w$ wall heat transfer coefficient in packed beds, W/(m$^2$·K)

$He'$ Henry's law constant, (kmoles/m$^3$ of gas) / (kmoles/m$^3$ of liquid)

$k$ rate constant (specific reaction rate), $\dfrac{\text{kmoles}/(\text{s}\cdot\text{m}^3)}{(\text{kmoles}/\text{m}^3)^n}$

$k_1$, $k_2$ rate constants for adsorption of reactant $A$ and desorption of reactant $A$ respectively (see under Langmuir-Hinshelwood mechanism in Section 8.5.3)

$k_B$ Boltzmann's constant, J/K

$k_C$ fluid to particle mass transfer coefficient, m$^3$/(s·m$^2$ of catalyst); gas phase mass transfer coefficient, m/s

$k_d$ rate constant for catalyst deactivation [see Eq. (8.5.141)], s$^{-1}$ (m$^3$/kmoles)$^m$

$k_d$, $k_{dr}$ rate constants for desorption and adsorption of product $P$ (see Section 8.5.3 under Langmuir-Hinshelwood mechanism)

$k_{er}$, $k_{ez}$ effective thermal conductivity of catalyst bed in the radial direction and axial direction respectively, W/(m·K)

$k_L$ liquid phase mass transfer coefficient, m/s

$k_{Lp}$ liquid to particle mass transfer coefficient, m/s

$k_S$, $k_{Sr}$ rate constants for forward and backward surface reactions respectively (see Section 8.5.3) under Langmuir-Hinshelwood mechanism)

$K$ equilibrium constant for reversible reaction

$K_A$, $K_B$, $K_P$ equilibrium constants for adsorption of $A$, adsorption of $B$ and desorption of $P$ respectively (see Section 8.5.3 under Langmuir-Hinshelwood mechanism)

$K_{bC}$ interchange coefficient between bubble phase and cloud, $s^{-1}$

$K_{Ce}$ interchange coefficient between cloud and emulsion, $s^{-1}$

$K_d$ equilibrium distribution coefficient, dimensionless.

$K_L$ overall liquid phase mass transfer coefficient, m/s

$K_m$ Michaelis-Menten constant, kmoles/$m^3$

$K_S$ equilibrium constant for surface reaction (see Section 8.5.3 under Langmuir-Hinshelwood mechanism)

$L_b$ length (or height) of catalyst bed, m

$L_f$ expanded height of fluidised bed, m

$M$ total moles of $A$ adsorbed in half the liquid film of surface area $S$, kmole

$M_A$, $M_B$ molecular weight of $A$ and that of $B$ respectively, kg/kmole

$M_i$ value of $M$ at the end of $i$-th foam subsection, kmole

$M_\infty$ value of $M$ at infinite time, kmole

$n$ order of reaction; total number of foam subsections

$N_A$ number of moles of $A$

$N_{Ar}$ molar flux of $A$ in the radial direction, kmoles/($m^2 \cdot s$)

$N_d$ dispersion number, dimensionless

$p_A$ partial pressure of $A$, N/$m^2$

$P$ total pressure, N/$m^2$

$P_0$ total pressure at $\theta = 0$ (for batch reactors), N/$m^2$

$q_i$ volume flow rate of liquid draining from the $i$th subsection, $m^3$/s

$q_W$ wall heat flux, W/$m^2$

$Q$ rate of heat input to reactor from outside, $W$; volume flow rate of gas at any $z$ [see Eq. (8.6.39)], $m^3$/s volume flow rate of inlet liquid to foam bed reactor, $m^3$/s

$Q_0$ total volumetric flow rate of feed (gas), $m^3$/s

$Q_1$ total flow rate of liquid draining back to the storage section of foam bed

$Q_g$ volume flow rate of gas through the foam bed reactor, $m^3$/s

$r_a$ rate of adsorption of $A$, kmoles/(s·kg of catalyst)

$(r_a)_B$ rate of adsorption of $B$, kmoles/(s·kg of catalyst)

$r_A$ rate of production (or generation) of $A$, kmoles/($m^3 \cdot s$) or kmoles/(s·kg of catalyst)

$(-r_A)$ rate of disappearance (consumption) of $A$, kmoles/($m^3 \cdot s$) or kmoles/(s·kg of catalyst)

$(-r_{AS})_{int}$ intrinsic rate of consumption of $A$ based on surface concentrations, kmoles/(s·kg of catalyst)

$r_d$ rate of desorption of products, kmoles/(s·kg of catalyst)

$r_S$ rate of surface reaction, kmoles/(s·kg of catalyst)

$R$ gas constant, atm·m³/(kmole·K) or J/(kmole·K); recycle ratio (or reflux ratio) in recycle reactor (see Example 8.21)

$Ra$ Rayleigh number, dimensionless

$Re$ Reynolds number, dimensionless

$R_C$ radius of unreacted core of solid reactant (see Fig. 8.26), m

$R_S$ total radius of unreacted solid and ash (see Fig. 8.26), m

$S_p$ specific surface of particle, $m^2$/kg or $m^2/m^3$

$S$ space velocity [see Eq. (8.3.12)], m/s; surface area of a liquid film, $m^2$

$Sc$ Schmidt number, dimensionless

$S_0$ overall selectivity, dimensionless

$S_p$ point selectivity, dimensionless; total surface area of particles, $m^2$

$T$ absolute temperature, K

$T_e$ temperature of reaction mixture inside CSTR, K

$T_F$ feed temperature, K

$T_0$ fluid inlet temperature, K; temperature of reaction mixture inside a batch reactor at $\theta$ = 0, K

$T_S$ temperature of heat source/heat sink, K; surface temperature of catalyst, K

$T_W$ wall temperature, K

$U$ overall heat transfer coefficient, W/($m^2$·K)

$\mathbf{V}_b$ rise velocity of bubbles, clouds and wakes, m/s

$\mathbf{V}_{br}$ natural rise velocity of a single bubble, m/s

$\mathbf{V}_{mf}$, $\mathbf{V}_{0f}$ minimum fluidisation velocity and operating fluidisation velocity respectively of gas, m/s

$\mathbf{V}_g$ superficial velocity of gas in trickle bed reactor, m/s

$\mathbf{V}_L$ superficial velocity of liquid in trickle bed reactor, m/s

$\mathbf{V}_z$ superficial velocity of fluid in $z$-direction, m/s

$V$ volume of reaction mixture, $m^3$

$V_b$ volume of foam bubble, $m^3$

$V_i$ volume of liquid film in $i$th subsection, $m^3$

$V_0$ value of $V$ at zero conversion (at $x_A = 0$), $m^3$

$V_p$ volume of catalyst particle, $m^3$

$w_i$ mass fraction of particles of size $d_{pi}$

$W$ mass of catalyst, kg

$x_A$ fractional conversion of $A$ [see Eq. (8.2.7)]

$x_{Ae}$ equilibrium conversion of $A$

$x_{Af}$ final conversion of $A$; conversion of $A$ in reactor effluent

$\bar{x}_B$ average conversion of $B$

$y_A$ mole fraction of $A$ in gas

$z$ reactor length, m; stoichiometric coefficient, dimensionless

## GREEK LETTERS

$\alpha$ ratio of volume of wake to volume of bubble, dimensionless

$\alpha_a$ activity of catalyst at any time $\theta$, dimensionless

$\beta$ heat of reaction parameter [see Eq. (8.5.63)], dimensionless

$\gamma$ Arrhenius number, dimensionless [see Eq. (8.5.62)]

$\gamma_A$, $\gamma_B$ activity coefficients, dimensionless

$\gamma_b$ volume fraction of solids (catalyst particles) in bubbles, dimensionless

$\gamma_C$ ratio of volume of solids in clouds and wakes to volume of bubbles, dimensionless

$\gamma_e$ ratio of volume of solids in emulsion to volume of bubbles, dimensionless

$\delta$ fraction of fluidised bed consisting of bubbles, dimensionless

$\Delta H^*$ enthalpy change for formation of activated complex, J/kgmole

$\Delta H_r$ heat of reaction, J/kgmole

$\Delta S^*$ entropy of activation, J/(kmole·K)

$\in$ void fraction of packed bed, dimensionless

$\in_i$ liquid holdup (volume fraction of liquid) in $i$th subsection, dimensionless

$\in_A$ fractional change in volume of system between no conversion and complete conversion [see Eq. (8.2.130) of Example 8.14], dimensionless

$\in_g$ fractional gas holdup, dimensionless

$\in_{mf}$ void fraction of bed at minimum (incipient) fluidisation, dimensionless

$\in_p$ porosity of catalyst particle, dimensionless

$\theta$ time, s

$\bar{\theta}$ mean residence time, s

$\theta^*$ dimensionless time

$\theta_{1/2}$ half life (time required for the concentration of a particular reactant to drop to one-half of the original value), s

$\theta_A$ fraction of catalyst sites (active centres) occupied by $A$, dimensionless

$\theta_B$, $\theta_P$ fraction of catalyst sites (active centres) occupied by $B$ and that by $P$ respectively, dimensionless

$\theta_{Ci}$ contact time in $i$th foam subsection, s

$\theta_V$ fraction of total catalyst sites (active centres) vacant, dimensionless

$\eta$ effectiveness factor, dimensionless

$\rho$ density of feed to reactor, kg/m$^3$

$\rho_B$ density of solid reactant, kg/m$^3$

$\rho_p$, $\rho_S$ apparent density and true density respectively of porous solid (catalyst), kg/m$^3$

$\sigma$, $\sigma^2$ standard deviation ($s$) and variance ($s^2$) respectively

$\sigma_{AB}$ collision diameter, cm

$\tau$ space time [see Eq. (8.3.10)], s

$\phi$ Thiele-type modulus, dimensionless

$\phi_S$ value of $\phi$ at temperature $T_S$ (surface temperature of catalyst), dimensionless

# COMPUTER NOTATIONS

| | | |
|---|---|---|
| ABSA | = $\in_A$ | fractional change in volume of system between no conversion and complete conversion |
| ABSG | = $\in_g$ | fractional gas holdup |
| ABSMF | = $\in_{mf}$ | void fraction of bed at incipient fluidisation |
| ABSN | = $\in$ | void fraction of packed bed |
| ABSP | = $\in_p$ | porosity of particle |
| AC | = $A_C$ | tower (reactor) cross-sectional area |
| ALFA | = $\alpha$ | (volume of wake)/(volume of bubble) |
| ARM | = $Ar_m$ | Archimedes number |
| AT | = $a_t$ | total external surface area of catalyst particles per unit volume of catalyst bed |
| A1, B1, C1, D1 | = $a$, $b$, $c$, $d$ | stoichiometric coefficients (Program 8.6) |
| B1, P1, R1 | = $b$, $p$, $\boldsymbol{r}$ | stoichiometric coefficients (Program 8.9) |
| A1, R1 | = $a$, $\boldsymbol{r}$ | stoichiometric coefficients (Program 8.5) |
| C1, C2 | = $C_1$, $C_2$ | Parameters defined in Eqs (8.5.129) and (8.5.130) |
| CAG | = $C_{Ag}$ | concentration of $A$ in gas |
| CAG0 | = $(C_{Ag})_0$ | concentration of $A$ in inlet gas |
| CAL0 | = $(C_{AL})_0$ | concentration of $A$ in inlet liquid |
| CA0 | = $C_{A0}$ | concentration of $A$ in feed to packed bed reactor (Program 8.5) |
| CAS | = $C_{AS}$ | concentration of $A$ at catalyst surface |
| CP | = $C_p$ | specific heat of reaction mixture (assumed constant) |
| D | = $D$ | ID of reactor |
| DA | = $D_A$ | diffusivity of $A$ in gas |
| DAK | = $D_{A,K}$ | Knudsen diffusivity of $A$ |
| DB | = $d_b$ | bubble diameter |
| DE | = $D_e$ | effective diffusivity |
| DELT | = $\delta$ | fraction of fluidised bed consisting of bubbles |
| DHR | = $\Delta H_r$ | heat of reaction |
| DL | = $D_L$ | liquid phase diffusivity |
| DN | = $\Delta n$ | defined in Eq. (8.8.3) |
| DP | = $d_p$ | particle diameter |
| DP0 | = $d_{p0}$ | pore diameter |
| DRCT | = $-(dR_C/d\theta)$ | see Eq. (8.6.11) |
| DT | = $\Delta T$ | increment in temperature. |
| DZ | = $\Delta z$ | increment in $z$ |
| DZX | = $\Delta z/\Delta x_A$ | difference form of $dz/dx_A$ |
| DZXA | = $(\Delta z/\Delta x_A)_{\text{avg}}$ | average value of $\Delta z/\Delta x_A$ |
| ETA | = $\eta$ | effectiveness factor |

E0, K0 = $E_0$, $k_0$ empirical constants (Programs 8.6 and 8.9)
FA0 = $F_{A0}$ molar feed rate of $A$
FB0 = $F_{B0}$ molar feed rate of $B$
FF = $F_f$ molar flow rate of gas at any $z$
FF0 = $F_{f0}$ molar feed rate of gas
F0 = $F_0$ total feed rate (mass flow rate) to reactor
F0T = $F_0$ (per tube) feed rate to each tube of multitube packed bed reactor
FS = $F_S$ mass flow rate of solids at any $z$
FXA, FXB = $F(x_A)$, $F(x_B)$ function of $x_A$ and that of $x_B$.
G = $G$; $g$ mass velocity of $A$ (Program 8.5); acceleration due to gravity (Program 8.7)
GAMB = $\gamma_b$ volume fraction of solids in bubbles
GAMC = $\gamma_C$ (volume of solids in clouds and wakes)/(volume of bubbles)
GAME = $\gamma_e$ (volume of solids in emulsion)/(volume of bubbles)
GG = $G_g$ mass velocity of gas
GL = $G_L$ mass velocity of liquid
GT = $G_t$ mass velocity in each tube of multitube packed bed reactor
H = $h$ step size
HE = $He'$ Henry's law constant (dimensionless)
HW = $h_W$ wall heat transfer coefficient
JD = $j_D$ $j$-factor for mass transfer
K = $k$ rate constant (for units, refer to the corresponding numerical Example based on which the program is prepared).
K1, K2, K3, K4 = $K_1$, $K_2$, $K_3$, $K_4$ parameters defined in Eqs (8.5.118), (8.5.121), (8.5.122) and (8.5.128)
KBC = $K_{bC}$ interchange coefficient between bubble and cloud
KC = $k_C$ fluid to particle mass transfer coefficient
KCE = $K_{Ce}$ interchange coefficient between cloud and emulsion
KF = $K_f$ parameter defined in Eq. (8.5.105)
KLA = $(k_L a_g)$ product of liquid phase mass transfer coefficient and gas-liquid interfacial area per unit volume of catalyst bed
KLPA = $(k_{Lp} a_p)$ product of liquid to particle mass transfer coefficient and liquid-particle interfacial area per unit volume of catalyst bed
KM = $K_m$ parameter defined in Eq. (8.8.4)
K0 = $k_0$ empirical constant (Programs 8.6 and 8.9); = $(\eta\, k\, \rho_p)$ in Program (8.7); = $\eta k \rho_p\,(1-\epsilon)$ in Program (8.8).
KOLA = $K_L a_g$ product of overall liquid phase mass transfer coefficient and gas-liquid interfacial area per unit volume of catalyst bed
KP = $K'$ parameter defined in Eq. (8.5.106)
L = $L$ reactor height

LB = $L_b$ length (or height) of catalyst bed
LF = $L_f$ expanded height of fluidised bed
M = $m$ exponent in Eq. (8.8.1)
MA, MB, MR = $M_A$, $M_B$, $M_R$ molecular weight of $A$, $B$ and $R$
MEUG, MEUL = $\mu_g$, $\mu_L$ viscosity of gas and that of liquid
MG = $M_g$ molecular weight of gas
MIN = $M_{in}$ molecular weight of inerts
MP = $M_P$ molecular weight of solid product $P$
NT = $N_t$ number of tubes
PA = $p_A$ partial pressure of $A$
PHIL = $\phi$ Thiele-type modulus
PT = $P$ total pressure
Q = $Q$ volume flow rate of gas at any $z$
Q0 = $Q_0$ total volumetric flow rate of feed gas
R = $R$ gas constant
R1, R2 = $\lambda_1$, $\lambda_2$ parameters defined in Eqs (8.5.124) and (8.5.125)
RC = $R_C$ radius of unreacted core of solid reactant
RE = $Re$ Reynolds number
REL = $Re_L$ Reynolds number in Dharwadkar and Sylvester's correlation
REMF = $Re_{mf}$ Reynolds number at incipient fluidisation
ROWB = – bulk density of packed bed (in Program 8.5)
ROWB = $\rho_B$ density of solid reactant $B$ (Program 8.9)
ROWG = $\rho_g$ density of gas
ROWL = $\rho_L$ density of liquid
ROWP = $\rho_p$; $\rho_P$ apparent density of catalyst; density of solid product $P$ (in Program 8.9 only)
ROWS = $\rho_S$ density of solids at any $z$
RS = $R_S$ combined radius of unreacted solid and surrounding ash layer
SC = $SC$ Schmidt number $SCL = SC_L$ liquid phase Schmidt number
$SP$ = $s_p$ specific surface of particles (surface area per unit mass)
T = $T$ temperature
TA = $T_{avg}$ average value of $T$
TF = $T_F$ feed temperature
TW = $T_W$ wall temperature
UA = $\bar{u}_A$ average molecular velocity of $A$
VB = $\mathbf{V}_b$ rise velocity of bubbles, clouds and wakes
VBR = $\mathbf{V}_{br}$ natural rise velocity of a single bubble
VG = $\mathbf{V}_g$ superficial velocity of gas
VL = $\mathbf{V}_L$ superficial velocity of liquid

VMF = $V_{mf}$ minimum fluidisation velocity
V0 = $V_0$ superficial velocity of gas in packed bed (Program 8.5)
VOF = $V_{0f}$ operating gas velocity in fluidised bed
VS = $V_S$ volume flow rate of solids at any $z$
W = $W$ mass of catalyst
XA = $x_A$ fractional conversion of $A$
XAF = $x_{Af}$ final conversion of $A$ desired
XAFC = $x_{Af}$ (cal) calculated value of $x_{Af}$
XA0 = $x_{A0}$ initial conversion of $A$
XB = $x_B$ fractional conversion of $B$
XBF = $x_{Bf}$ final conversion of $B$ desired
XB0 = $x_{B0}$ initial conversion of $B$
YA0 = $y_{A0}$ mole fraction of $A$ in feed gas
YA, YR, YIN = $y_A$, $y_R$, $y_{in}$ mole fraction of $A$, mole fraction of $R$ and mole fraction of inerts respectively

## EXERCISES

8.A. Agree with or contradict against the following statements giving reasons. You may use sketches or diagrams to elucidate your answers:

(*a*) The order of an elementary reaction is equal to its molecularity,

(*b*) The rate of an autocatalytic reaction increases with time, though the reaction concentration decreases,

(*c*) For a reversible reaction, the equilibrium conversion is increased by the use of a catalyst,

(*d*) For a psuedo-first order reaction in $A$, the half life of $A$ is directly proportional to the initial concentration of $A$, while for a zero-order reaction, the half life is independent of the initial concentration of the reactant,

(*e*) According to activated complex theory, the frequency factor is proportional to the square root of absolute temperature, while as per Arrhenius law and collision theory, the frequency factor is independent of temperature,

(*f*) An infinite number of stirred tank reactors operating in series is equivalent to a single plug flow reactor,

(*g*) For a first order reaction, if a part of the product stream is recycled back to the feed inlet, the overall conversion increases in the case of a plug flow reactor, but decreases in the case of a stirred tank reactor.

(*h*) For a series reaction like $A \rightarrow P \rightarrow R$, the yield of $P$ will be maximum when the selectivity of $P$ with respect to $R$ is a minimum,

(*i*) The effectiveness factor of a porous catalyst is independent of reactant concentration if the intrinsic rate is first order and reversible,

(*j*) In solid-catalysed simultaneous reactions, strong pore diffusion reduces the selectivity of the desired product to the square root of its intrinsic value,

(*k*) Fluidised bed reactors are more advantageous than fixed bed reactors since they provide higher conversion for the same feed rate at the same temperature and pressure,

(*l*) In fluidised bed reactors, high gas flow rates are always preferred since this increases turbulence in the bed and thereby permits isothermal operation at high conversions,

(*m*) In a noncatalytic fluid-particle reaction, the time required to achieve a given conversion is proportional to the square of the particle size, when the chemical reaction at the particle surface is rate controlling,

(*n*) The rate of side-by-side deactivation of a solid catalyst is independent of both feed and product compositions, while the rate of parallel deactivation is strongly dependent on the product concentration in the reaction mixture,

(*o*) For a reactor combination consisting of a PFR and an ideal CSTR, the exit age distribution function shall be the same even if PFR precedes the CSTR or vice versa.

8.B. Outline briefly what do you understand by

(*a*) Lineweaver-Burk plot,

(*b*) Catalyst poisons,

(*c*) Semibatch reactors,

(*d*) Bubbling bed model,

(*e*) Shrinking core model,

(*f*) Foam bed reactors,

(*g*) Tanks-in-series model.

8.C. Distinguish between

(*a*) Yield and selectivity

(*b*) Side-by-side deactivation and independent deactivation,

(*c*) Series reactions and parallel reactions,

(*d*) Integral method of analysis and differential method of analysis,

(*e*) Packed bed reactors and trickle bed reactors,

(*f*) Collision theory and activated complex theory,

(*g*) Space time and mean residence time,

(*h*) Autocatalytic reactions and enzyme catalysed reactions,

(*i*) Selectivity poisons and stability poisons

(*j*) An LFTR and a PFDR

8.D. The following data are reported by Wiig[20] for the first order decomposition of acetone dicarboxylic acid in aqueous solution:

| Temperature, °C | : | 0 | 20 | 40 | 60 |
|---|---|---|---|---|---|
| Rate constant, $s^{-1}$ | : | $2.46 \times 10^{-5}$ | $47.5 \times 10^{-5}$ | $576 \times 10^{-5}$ | 0.0548 |

(*a*) Compute the value of activation energy from (*i*) Arrhenius' law, (*ii*) collision theory and (*iii*) activated complex theory

(*b*) What is the half-life of acetone carboxylic acid at 80° C?

8.E. The following chain reaction mechanism has been proposed for the thermal decomposition of acetaldehyde:

$$CH_3CHO \xrightarrow{k_1} CH_3\cdot + CHO\cdot \qquad \text{Initiation}$$

$$\left.\begin{array}{l} CH_3\cdot + CH_3CHO \xrightarrow{k_2} CH_3CO\cdot + CH_4 \\ CH_3CO\cdot \xrightarrow{k_3} CH_3\cdot + CO \end{array}\right\} \text{Propagation}$$

$$CH_3\cdot + CH_3\cdot \xrightarrow{k_4} C_2H_6 \qquad \text{Termination}$$

The overall reaction is

$$CH_3CHO \rightarrow CH_4 + CO$$

Based on the steady state hypothesis, derive a rate expression for the above reaction. Does the order of the reaction agree with molecularity?

8.F. The reaction between hydrogen and bromine $\left(H_2 + Br_2 \rightleftharpoons 2HBr\right)$ follows a free radical chain mechanism which is presented in Section 8.1 of this chapter. Show that this mechanism leads to a rate expression of the form of Eq. (8.1.10). Use the steady state approximation for your analysis.

8.G. The rate equation for a homogeneous, irreversible reaction $A + B + D \rightarrow R + S$ has been found to be

$$(-r_A) = \frac{k_1 C_A C_B C_D}{1 + k_2 C_R}$$

Devise a reaction mechanism that explains the above kinetic equation.

8.H. The rate of hydrolysis of 17 percent sucrose in 0.1 N hydrochloric acid solution is studied at 35° C in a constant volume batch reactor and the results obtained[20] are given below:

| Time, min : | 9.82 | 59.6 | 93.18 | 142.9 | 294.8 |
|---|---|---|---|---|---|
| Sucrose remaining, percent : | 96.5 | 80.3 | 71.0 | 59.1 | 32.8 |

Determine the order of the reaction and the numerical value of specific reaction rate at 35° C.

8.I. The oxidation of sodium succinate to form sodium fumarate by dissolved oxygen in presence of the enzyme succinoxidase when studied in a constant volume batch reactor, the following data were obtained:

| *Concentration of sodium succinate, kmoles/m³* | *Rate of oxidation, kmoles/(m³·s)* |
|---|---|
| 0.01 | $1.17 \times 10^{-6}$ |
| 0.002 | $0.99 \times 10^{-6}$ |
| 0.001 | $0.79 \times 10^{-6}$ |
| 0.0005 | $0.62 \times 10^{-6}$ |
| 0.00033 | $0.50 \times 10^{-6}$ |

Determine the rate equation of the reaction.

8.J. The gas phase reaction between $A$ and $B$ ($A + 2B \rightarrow P + R$) is studied by measuring the change in pressure in a constant-volume reaction vessel. At 25° C and an initial pressure of 100 mm Hg, the following data are obtained:

| Time, min | : | 1.54 | 2.5 | 3.9 | 6.2 | 10.3 | 19.1 | 48.0 |
|---|---|---|---|---|---|---|---|---|
| Total pressure, mm Hg | : | 92.5 | 90.0 | 87.5 | 85.0 | 82.5 | 80.0 | 77.5 |

If the initial mixture contained equimolar amounts of $A$ and $B$, determine the rate equation of the reaction. What is the magnitude of the specific reaction rate at 25° C?

8.K. The following two reactions take place simultaneously as shown:

$$A \xrightarrow{k_1} D + E$$

$$A + B \xrightarrow{k_2} D + F$$

The first reaction is first order in $A$, while the second reaction is first order in $A$ and first order in $B$. Show that for a constant-density reaction mixture, the rate equations can be integrated to give

$$k_2\theta = \frac{1}{(k_1/k_2) + (C_{B0} - C_{A0})} \ln\left[\frac{C_{A0}\,(k_1/k_2 + C_B)}{C_A\,(k_1/k_2 + C_{B0})}\right]$$

8.L. The homogeneous gas phase reaction $A \rightarrow 3R$ is first order in $A$. This reaction was first studied in a constant-pressure batch reactor. The initial mixture was composed of 75 mole percent $A$ and 25 mole percent inerts and the operating pressure was 1 atm. It was observed that the volume increased by 70 percent in 10 minutes. If the same reaction is carried out in a constant-volume batch reactor and the initial pressure is 1 atm, compute the time required for the pressure to reach 2.2 atm. Assume isothermal operation.

8.M. The following reaction takes place at constant density and at 25° C:

$$A + B \rightarrow P$$

The reaction is first order in $A$ and first order in $B$ (the overall order of the reaction being 2.0) and the value of the specific reaction rate at 25° C is 10 $m^3/(kmole \cdot S)$. The following options are available:

(*a*) A single stirred tank reactor of volume 0.1 $m^3$,

(*b*) A single plug flow reactor of volume 0.1 $m^3$,

(*c*) Two stirred tank reactors in series each of volume 0.05 $m^3$,

(*d*) Two stirred tank reactors in parallel each of 0.05 $m^3$ volume and with the feed stream split equally between them,

(*e*) A stirred tank reactor followed by a plug flow reactor each of volume 0.05 $m^3$.

If feed containing equimolar quantities of $A$ and $B$ is introduced at 0.003 $m^3/hr$ and the concentration of $A$ in the feed is 0.02 kmoles/$m^3$, determine which among the above would provide the highest fractional conversion of $A$.

8.N. The following autocatalytic reaction is to be conducted at 30° C in a plug flow reactor with part of the product being recycled back to the reactor inlet:

$$A + P \rightarrow P + P$$

The feed mixture contains 1.1 kmoles of $A$ per $m^3$ and 0.01 kmoles of $P$ per $m^3$ and it is charged at the rate of 0.06 $m^3/hr$.

(*a*) If the fractional conversion of $A$ desired is 90 percent, compute the optimum recycle ratio that would minimise the volume of reactor needed. What will be the volume of the reactor under these optimum conditions?

(*b*) If a plug flow reactor without recycle is employed, what reactor volume will be required to obtain 90 percent conversion of $A$?

(*c*) If a reactor combination, in which a stirred tank reactor is followed by a plug flow reactor each of volume $(V/2)$ m$^3$ where $V$ is the reactor volume computed in part (*b*), is used without any recycle, what conversion of $A$ will be obtained?

(*d*) If the reaction is to be conducted in a single stirred tank reactor, what reactor volume will be required to obtain 90 percent conversion of $A$?

The value of specific reaction rate at 30° C is 75 m$^3$/(kmole·hr) and the rate equation is $(-r_A) = kC_AC_P$. Assume that the density of the reaction mixture remains essentially constant.

8.O. Acetic an hydride is hydrolysed in three ideal stirred tank reactors operated in series. The volumes of the reactors are 0.001 m$^3$, 0.002 m$^3$ and 0.0015 m$^3$ respectively and the operating temperatures of these reactors are 15° C, 40° C and 25° C respectively. The reaction is first order and irreversible and the values of the rate constant are as given below:

| *Temperature, °C* | *Specific reaction rate, min$^{-1}$* |
|---|---|
| 15.0 | 0.0567 |
| 40.0 | 0.380 |
| 25.0 | 0.158 |

The feed is introduced to the first reactor at the rate of 0.024 m$^3$/hr and the density of the reaction mixture may be assumed to be essentially constant. Compute the fraction hydrolysed in the effluent from the third reactor. What will be the fraction hydrolysed if all the reactors were operated at 25° C?

8.P. The following two reactions occur simultaneously at constant density:

$$A \xrightarrow{k_1} B$$

$$A \xrightarrow{k_1} C$$

Both reactions are first order in $A$. For this reaction system, derive an expression for the selectivity of $B$ with respect to $C$ in the case of

(*a*) an ideal stirred tank reactor,

(*b*) a plug flow reactor.

Assume that the operation is isothermal and the feed to the reactor contains only $A$. (It will be seen that the selectivity obtained in the stirred tank reactor is the same as that for the plug flow reactor. Can you explain why?)

8.Q. Rework the above problem (17.) if the reactions are second order and are represented as

$$A + B \xrightarrow{k_2} C$$

$$A + B \xrightarrow{k_2} D$$

Derive an expression for the selectivity of $C$ with respect to $D$ that would be obtained in an ideal stirred tank reactor and show that this is equal to that obtained in a plug flow reactor of same volume.

8.R. A hydrocarbon ($RH_3$) is to be photochlorinated in the liquid phase. Both chlorine and the hydrocarbon are dissolved in an inert solvent and then charged into the reactor which operates isothermally. The HCl produced also remains in solution. The reaction proceeds based on a chain mechanism as given below[21]:

Initiation:

$$Cl_2 \xrightarrow[k_1]{\text{Light}} 2Cl^*$$

Propagation:

$$Cl^* + RH_3 \xrightarrow{k_2} RH_2^* + HCl$$

$$RH_2^* + Cl_2 \xrightarrow{k_3} RH_2Cl + Cl^*$$

$$Cl^* + RH_2Cl \xrightarrow{k_2} RHCl^* + HCl$$

$$RHCl^* + Cl_2 \xrightarrow{k_3} RHCl_2 + Cl^*$$

Termination:

$$RH_2^* \xrightarrow{k_4} \text{end product}$$

$$RHCl^* \xrightarrow{k_4} \text{end product}$$

The overall reactions are

$$RH_3 + Cl_2 \rightarrow RH_2Cl + HCl$$

$$RH_2Cl + Cl_2 \rightarrow RHCl_2 + HCl$$

(*a*) Deduce the rate equations based on the steady state hypothesis. It may be assumed that the rate constant for the termination steps is much less than that for the propagation steps.

(*b*) Derive expressions for the yield of $RH_2Cl$ with respect to $RHCl_2$ in terms of the total conversion of $RH_3$ for the following two cases: (*i*) when a plug flow reactor is being used, (*ii*) when an ideal stirred tank reactor is being used.

The concentration of $RH_3$ in the feed = $C_{A0}$

8.S. Consider the multiple reactions discussed in Example (8.15):

$$A + B \xrightarrow{k_1} P$$

$$P + B \xrightarrow{k_2} R$$

The rate constants vary with temperature as given below:

$$k_1 = 3.5 \times 10^{14} \exp(-9000/T) \ \text{m}^3/(\text{kmole}\cdot\text{hr})$$

$$k_2 = 1.0 \times 10^7 \exp(-4000/T) \ \text{m}^3/(\text{kmole}\cdot\text{hr})$$

where $T$ is in degrees kelvin. Derive expressions for the maximum yield $P$ (to note that $k_2/k_1 \neq 1.0$) and for the overall selectivity of $P$ with respect to $R$ for

(*a*) a plug flow reactor

(*b*) an ideal stirred tank reactor

Show how the maximum yield of $P$ varies with temperature and determine the temperature at which the value of maximum yield of $P$ will be the highest. Consider a temperature range of 10 to 60° C. Neglect change in density of the reaction mixture with temperature.

8.T. An industrial reactor of volume 37.5 $m^3$ is to be used for dephenolising an industrial effluent. The reactor is tested using a step input of fluoride tracer and the results obtained are given below:

| *Time (hr)* | $(C/C_0)$ | *Time (hr)* | $(C/C_0)$ |
|---|---|---|---|
| 0.0 | 0.0 | 3.00 | 0.551 |
| 0.5 | 0.0 | 3.33 | 0.700 |
| 1.0 | 0.0 | 3.67 | 0.800 |
| 1.33 | 0.017 | 4.00 | 0.876 |
| 1.67 | 0.106 | 4.33 | 0.915 |
| 2.00 | 0.220 | 4.67 | 0.957 |
| 2.33 | 0.313 | 5.00 | 0.987 |
| 2.67 | 0.431 | 5.50 | 1.00 |

The feed rate to the reactor during tracer test was 285 $m^3$/d.

(*a*) Plot the *F*-curve and the *E*-curve.

(*b*) If the reactor may be assumed equivalent to a plug flow dispersion reactor (PFDR), find the dispersion number.

(*c*) If the reactor is equivalent to $n$ ideal CSTRs in series, find the value of $n$.

8.U. A laminar flow tubular reactor (LFTR) is to be used for conducting a second order reaction,

$$A + B \rightarrow \text{Products}$$

for which the second order rate constant = 0.1 $m^3/(\text{kmole}\cdot\text{s})$. The feed solution enters the reactor at 2880 $m^3$/hr and it contains equal concentrations of both $A$ and $B$, such as 1.0 kmole/$m^3$. If the reactor volume is 8.0 $m^3$, find the fractional conversion of $A$ attained. Assume a constant density system.

8.V. A 10 $m^3$ industrial reactor that is used to conduct a dimerisation reaction is tested using a pulse input of tracer and the results obtained are given below:

| *Time (min)* | $E(\theta)$, $min^{-1}$ | *Time (min)* | $E(\theta)$, $min^{-1}$ |
|---|---|---|---|
| 0.0 | 0.314 | 3.0 | 0.0280 |
| 0.2 | 0.2594 | 4.0 | 0.01842 |
| 0.3 | 0.2362 | 5.0 | 0.01470 |
| 0.4 | 0.2150 | 6.0 | 0.01316 |
| 0.5 | 0.1958 | 7.0 | 0.01245 |
| 0.6 | 0.1785 | 8.0 | 0.01200 |
| 0.8 | 0.1486 | 9.0 | 0.01173 |
| 1.0 | 0.1241 | 10.0 | 0.01147 |
| 1.5 | 0.0805 | 15.0 | 0.01037 |
| 2.0 | 0.0540 | 20.0 | 0.00938 |
| 2.5 | 0.0379 | | |

The reaction is first order in the monomer ($A$) with a rate constant = 0.3 $min^{-1}$. Pilot plant studies show that this reactor provides 72 percent conversion of $A$ when operated at a space time of 36.0 minutes. Determine whether this reactor could be modelled as equivalent to the reactor combination sketched in Fig. 8.14.

8.W. A biochemical reaction whose stoichiometry may be represented as

$$A \xrightarrow[k_1]{} B \xrightarrow[k_2]{} C$$

is conducted in a laboratory batch reactor starting with 0.5 mM of $A$ and the data recorded are listed below:

| *Time (min)* | *[C], mM* | *Time (min)* | *[C], mM* |
|---|---|---|---|
| 0.0 | 0.0 | 7000 | 0.16 |
| 1000 | 0.01 | 8000 | 0.19 |
| 2000 | 0.02 | 9000 | 0.21 |
| 3000 | 0.05 | 10000 | 0.24 |
| 4000 | 0.07 | 12000 | 0.28 |
| 5000 | 0.10 | 14000 | 0.32 |
| 6000 | 0.13 | 18000 | 0.38 |

If conversion of $A$ to $B$ is first order in $A$ with rate constant = $10^{-4}$ $min^{-1}$, determine the rate constant for the second step. Discuss on which step is rate-controlling. State and explain your assumptions, if any.

8.X. The following data are reported by Rao and Smith[51] for the first order, reversible reaction, $o-H_2 \rightleftharpoons p-H_2$ at −196° C and 1 atm pressure using a NiO–on–Vycor catalyst:

| *Catalyst size* | *Reaction rate, kmole/(s·kg catalyst)* |
|---|---|
| Small particles of average size = 58 microns | $(-r_A) = (5.29 \times 10^{-5})\,(y_{AS} - y_{Ae})$ |
| Pellets 13 mm in diameter and 3 mm long | $(-r_A) = (2.18 \times 10^{-5})\,(y_{AS} - y_{Ae})$ |

where $y_A$ is the mole fraction of $o-H_2$. The density of pellet = 1460 kg/m³. The 58 micron particles are so small in size that for them, $\eta$ may be assumed to be equal to 1.0. As for the pellets, only one face was exposed to the reactants, while the other face and the cylindrical surface were sealed.

(*a*) Evaluate the effectiveness factor for the pellets from the above experimental data.

(*b*) Using the random pore model to estimate $D_e$, compute the effectiveness factor of the pellets and compare it with the answer obtained in (*a*). Only micropores (pore radius = 45 angstroms) exist in Vycor and the porosity of the pellet = 0.304.

8.Y. Ethyl alcohol is manufactured by the vapour phase catalytic hydration of ethylene at 150° C and 2.8 atm pressure in a fixed bed tubular reactor 0.15 m in diameter and packed with 6 mm phosphoric acid–on–kieselguhr pellets. Feed containing 15 moles of steam per mole of ethylene is admitted to the reactor at the rate of 1200 m³/hour. The reaction is intrinsically first order in ethylene and first order in steam and the value of intrinsic rate constant at 150° C is 0.213 $m^6/(s \cdot kmole \cdot kg$ of catalyst)

Compute the conversion of ethylene in the reactor effluent if the length of the catalyst bed is 1.7 m.

Density of catalyst pellets = 1500 kg/m$^3$

Bulk density of packed bed = 1200 kg/(m$^3$ of reactor volume)

Effective diffusivity of ethylene in the catalyst pellets at 150° C and 2.8 atm pressure = 1.05 × 10$^{-4}$ m$^2$/s

Fluid to pellet mass transfer coefficient = 0.01 m/s

Heat of reaction ($\Delta H_r$) = −92100.0 kJ/kmole

Assume that the reactor operates in plug flow and under isothermal conditions. Neglect the pressure drop across the catalyst bed and the reversibility of the reaction.

8.Z. Polyethene is manufactured by the catalytic polymerisation of ethylene using Ziegler-Natta catalyst in a slurry reactor. The slurry of the catalyst particles in cyclohexane (diluant) is fed to the reactor at the rate of 0.06 m$^3$/hr and the volume of liquid in the vessel is 0.01 m$^3$. Pure ethylene gas is admitted from the bottom at the rate of 6.0 m$^3$/hr. The catalyst particles are 0.10 mm in diameter (particle density = 1000 kg/m$^3$) and for them the effectiveness factor may be assumed to be equal to unity. Though the kinetics are complex, the rate of disappearance of ethylene may be assumed to be intrinsically first order in ethylene and may be expressed as

$$(-r_A) = \eta \ (k\, a_p)\ C_{AS}$$

where $k$ (under the present operating conditions) = 0.0001 m/s. Compute the rate of production of polyethene (in terms of moles of ethylene reacted per second) from the following data:

Liquid to particle mass transfer coefficient ($k_{Lp}$) = 0.0003 m/s

Liquid phase mass transfer coefficient ($k_L$) = 0.0007 m/s

Average bubble size = 3.0 mm

Gas holdup = 0.09

Catalyst concentration in slurry = 100.0 kg/m$^3$

Henry's law constant for ethylene = 5.0 (kmole/m$^3$ of gas)/(kmole/m$^3$ of liquid)

Assume that the feed slurry contains no dissolved ethylene or polyethene.

8.Z$_1$. Hydrodesulfurisation of a sample of gas oil is to be conducted in a trickle bed reactor operating at 200° C and 40 atm pressure. The catalyst used is cobalt–molybdenum supported on alumina. The oil containing 1000 ppm of thiophene and no dissolved hydrogen and the feed gas containing 50 percent $H_2$ and 50 percent $N_2$ (by volume) are fed to the top of the catalyst bed. The superficial velocities of gas and liquid are maintained at 0.2 m/s and 0.05 m/s respectively. The principle reaction is that between thiophene in the oil and hydrogen in the feed gas ($C_4H_4S + 4H_2 \rightarrow C_4H_{10} + H_2S$) and the intrinsic rate may be assumed to be psuedo first order in hydrogen, the value of rate constant at 200° C and 40 atm being 0.00011 m$^3$/(kg·s). Compute the depth of the catalyst bed required to remove 75 percent of thiophene. Take the volumetric mass transfer coefficient from liquid to particle $(k_{Lp}a_p)_{H_2}$ to be 0.5 s$^{-1}$ and that from gas to liquid $(k_L a_g)_{H_2}$ to be 0.03 s$^{-1}$. The bulk density of catalyst bed = 960 kg/m$^3$. Henry's law constant for hydrogen at 200° C = 50 (kmole/m$^3$ gas)/(kmole/m$^3$ liquid). Assume that there is no vapourisation of thiophene from the liquid and neglect the solubility of nitrogen in the liquid. Also, the intrinsic rate is slow enough that the effectiveness factor is unity.

8.$Z_2$. A solid-catalysed reaction, $A \rightarrow P$ is studied in a basket-type back-mix reactor in which the concentration of $A$ was maintained unchanged by continuously decreasing the feed rate of $A$ with time (the feed rate is required to be lowered since the catalyst is getting deactivated). The reactor contained 1.5 kg of catalyst. The results obtained are given below:

$C_{A0}$ = 1.0 kmole/m$^3$ $\qquad$ $C_{A0}$ = 3.0 kmoles/m$^3$

$C_{Af}$ = 0.5 kmole/m$^3$ $\qquad$ $C_{Af}$ = 1.0 kmole/m$^3$

| *Time, hr* | *Feed rate, kmoles/hr* | *Time, hr* | *Feed rate, kmoles/hr* |
|---|---|---|---|
| 0.0 | 90.0 | 0.0 | 135.0 |
| 1.0 | 33.1 | 0.5 | 49.66 |
| 2.0 | 12.18 | 1.0 | 18.27 |
| 3.0 | 4.48 | 1.5 | 6.72 |

No catalyst poisons were present in the feed and the catalyst is not affected by the operating temperature. Develop expressions for the rate of reaction and the rate of deactivation of the catalyst.

8.$Z_3$. A gas–solid noncatalytic reaction of the type, $A\ (g) + bB\ (s) \rightarrow pP\ (s) + rR\ (g)$, is investigated by measuring the time required for complete conversion of $B$ as a function of particle diameter. The results are given below:

| Particle diameter, mm | : | 0.063 | 0.125 | 0.250 | 0.380 |
|---|---|---|---|---|---|
| Time for complete conversion, minutes | : | 5.0 | 10.0 | 20.0 | 30.0 |

(*a*) If the diffusion resistance in the gas phase around the particle is negligible, determine what mechanism controls the rate of reaction.

(*b*) The above reaction is to be carried out by passing the reactant gas crosswise over a moving grate carrying the solid particles. The solid particles. The velocity of the grate is such that the particles are exposed to the gas stream for 9.0 minutes. If the size distribution of particles is as given below, determine the average conversion of $B$ leaving the reactor:

| Particle diameter, mm | : | 0.063 | 0.125 | 0.25 | 0.50 |
|---|---|---|---|---|---|
| Mass fraction | : | 0.25 | 0.35 | 0.35 | 0.05 |

Assume that the gas composition does not change significantly during flow across the plate.

## REFERENCES


1. Bodenstein, M and Lind, SC, Z Physik. Chem. (Leipzig), **57**, 168, 1906.
2. Kistiakowsky, GB, J Am. Chem. Soc., 50, 2315, 1928.
3. Cassano, AE and Smith, JM, AIChE J, **12**, 1124, 1966.
4. Houser and Lee, J Phys. Chem., **71** (11), 3422, 1967.
5. Kingsley, HA and Bliss, H, Ind. Eng. Chem., **44**, 2479, 1952.
6. Kolarov, Popyankov and Angelov, Monatsh. Chem., **96** (3), 949, 1965.
7. Treacy, H and Daniels, F, J Am. Chem. Soc., **77**, 1307, 1955.
8. Hinshelwood, CN and Burk, RE, J Chem. Soc. **127**, 1051, 1114, 1925.

9. Lewis, WCM, J Chem. Soc. (London), **113,** 471, 1918.
10. Polanyi, M, Z Elektrochem, **26,** 48, 1920.
11. Glasstone, S, Laidler, KJ and Eyring, H, The Theory of Rate Processes, McGraw Hill, New York, 1941.
12. Basolo, F and Pearson, RG, Mechanisms of Inorganic Reactions, Wiley, New York, 1958.
13. Laidler, KJ, Chemical Kinetics, McGraw Hill, New York 1965
14. Harkness, JB, Kistiakowsky, GB and Mears, WH, J Chem. Phys., **5,** 682, 1937.
15. Brown, GG and Borkowski, H, J Am. Chem. Soc., **74,** 1896, 1952.
16. Michaelis, L and Menten, ML, Biochem. Z, **49,** 333, 1913.
17. Bailey, JE and Ollis, DF, Biochemical Engineering Fundamentals, McGraw Hill, New York, 1986.
18. Hinshelwood, CN and Askey, PJ, Proc. Roy. Soc. (London), **A 115,** 215, 1927.
19. Levenspiel, O, Chemical Reaction Engineering, Wiley, New York, 1999.
20. Daniels, F and Alberty, RA, Physical Chemistry, Wiley, New York, 1955
21. Smith, JM, Chemical Engineering Kinetics, Third Ed., McGraw Hill, New York, 1981.
22. Ritter, HL and Drake, LC, Ind. Eng. Chem. **17,** 787, 1945.
23. Emmett, PH and Brunauer, S, J Am. Chem. Soc., **59,** 1553, 1937.
24. Aris, R, Chem. Eng. Sci., **6,** 262, 1957.
25. Wheeler, A in WG Frankenburg and Associates (eds), Advances in Catalysis, Vol. III, Academic Press, New York, 1951.
26. Roberts, GW and Satterfield, CN, Ind. Eng. Chem. Fund. Quart., **4,** 288, 1965; **5,** 317, 325, 1966.
27. Weisz, PB and Hicks, JS, Chem. Eng. Sci., **17,** 265, 1962.
28. Bischoff, KB, Chem. Eng. Sci., **22,** 525, 1967.
29. Carberry, JJ, AIChE J, **7,** 350, 1961.
30. Thodos, G and Stutzman, LF, Ind. Eng. Chem., **50,** 413, 1958.
31. Krishnamurthy, EV and Sen, SK, Computer-based Numerical Algorithms, East-West press, New Delhi, 1976.
32. Hildebrand, FB, Introduction to Numerical Analysis, McGraw Hill, New York, 1956.
33. Potter, C and Baron, S, Chem. Eng. Progr. **47,** 473, 1951.
34. Prater, CD, Chem. Eng. Sci., **8,** 284, 1958.
35. Dwivedi, PN and Upadhyay, SN, Ind. Eng. Chem. Proc. Des. Dev., **16,** 157, 1977.
36. Wheeler, A in WG Frankenburg et. al. (eds.), Advances in Catalysis, Vol. III, Academic Press, New York, 1951.
37. Lapidus, L and Amundson, NR, Chemical Reactor Theory, Prentice-Hall, NJ, 1977.
38. Kunii, D and Levenspiel, O, Fluidisation Engineering, Wiley, New York, 1970.
39. Davidson, JF and Harrison, D, Fludised Particles, Cambridge University Press, New York, 1963.
40. Rowe, PN and Partridge, BA, Trans. Inst. Chem. Engrs., **43,** 157, 1965.
41. Kato and Wen, Chemical Eng. Sci., **24,** 1351, 1969.
42. Hovmand, S and Davidson, JF, Trans. Inst. Chem. Engrs., **46,** 190, 1968.
43. Chen, BH and Doughlas, WJM, J Chem. Eng., **47,** 113, 1969.
44. Weekan, VW, Ind. Eng. Chem. Process Des. Dev., **7,** 90, 1968
45. Wojciechowski, Ind. Eng. Chem. Process Des. Dev., **12,** 254, 1973.
46. Votruba, J, Hlavacek, V and Marek, M, Chem. Eng. Sci., **27,** 1845, 1972

47. Beek, J, Design of Packed Catalytic Reactors, in Advances in Chemical Engineering, Vol. 3, pp. 229, Academic Press, New York, 1962.

48. Goto, S and Smith, JM, AIChE J, **21,** 706, 1975.

49. Satterfield, CN, van Eeek, MW and Bliss, GS, AIChE J, **24,** 709, 1978.

50. Ranz, WE and Marshall, WR, Chem. Eng. Progr., **48,** 173, 1952.

51. Rao, MR and Smith, JM, AIChE J, **10,** 293, 1964.

52. Narayanan, CM, Proc. Adv. Workshop on CAD of Ind. Process Equipment, Durgapur, February, 1995.

53. Biswas, J and Kumar, R, Chem. Eng. Sci., **36** (9), 1547, 1981.

54. Crank, J, The Mathematics of Diffusion, Clarendon Press, Oxford, 1956.

55. Kumar, R, Chem. Eng. Sci., **26,** 177, 1971.

56. Van der Laan, ET, Chem. Eng. Sci., **7,** 187, 1958.

57. Aris, R, Chem. Eng. Sci., **9,** 266, 1959.

58. Wehner, JF and Wilhelm, RH, Chem. Eng. Sci., **6,** 89, 1956.

59. Levenspiel, O and Bischoff, KB, Ind. Eng. Chem., **51,** 1431, 1959; **53,** 313, 1961.

60. Levenspiel, O, The Chemical Reactor Omnibook, OSU Bookstores, Corvallis, OR 97339, 1996.

61. Levenspiel, O, Ind. Eng. Chem., **50,** 343, 1958.

# Index